International Association of Fire Chiefs

NFPA

Association



# Facility Fire Brigade

## Principles and Practice

SECOND EDITION

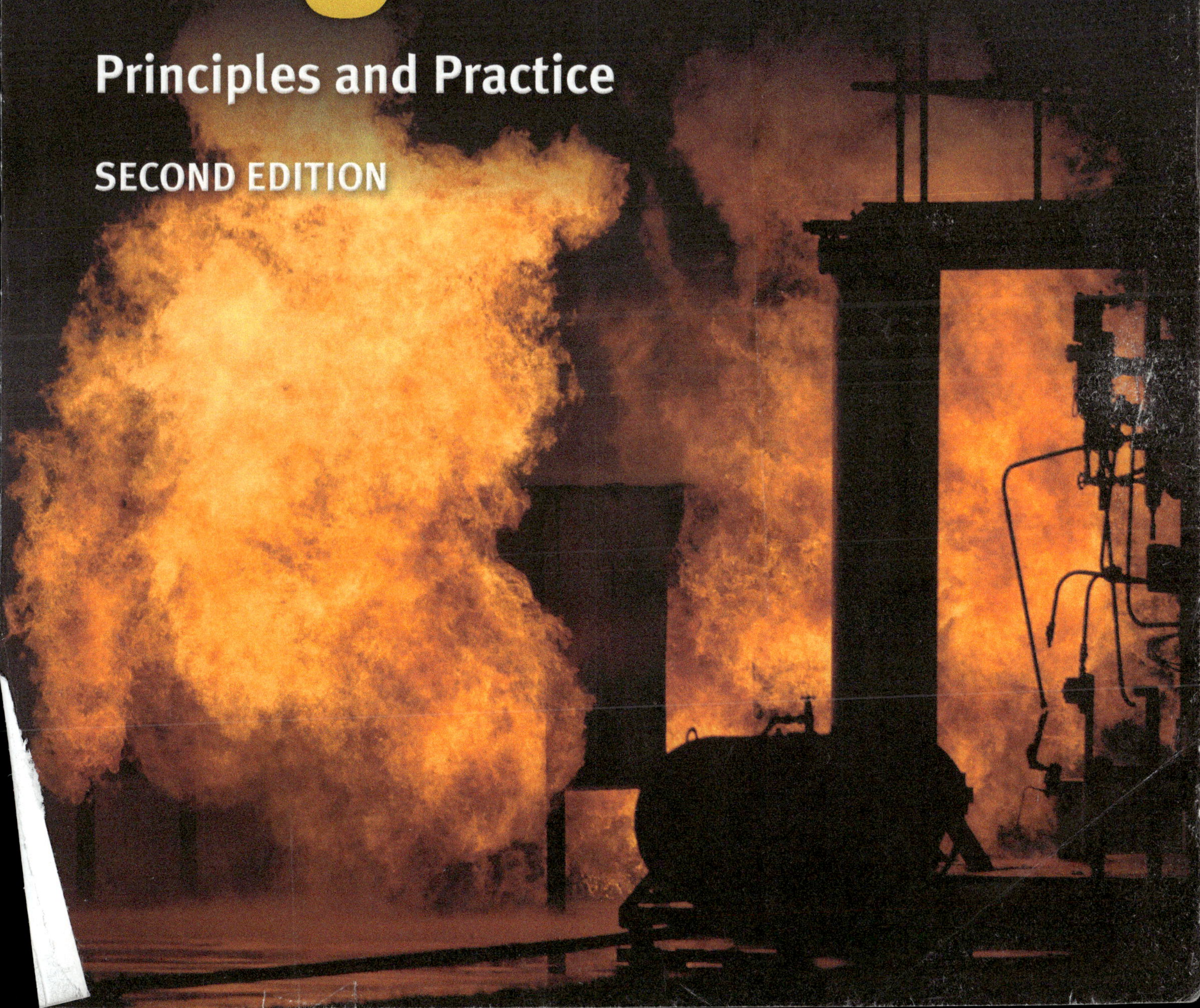

International Association of Fire Chiefs

National Fire Protection Association

# Facility Fire Brigade

## Principles and Practice

SECOND EDITION

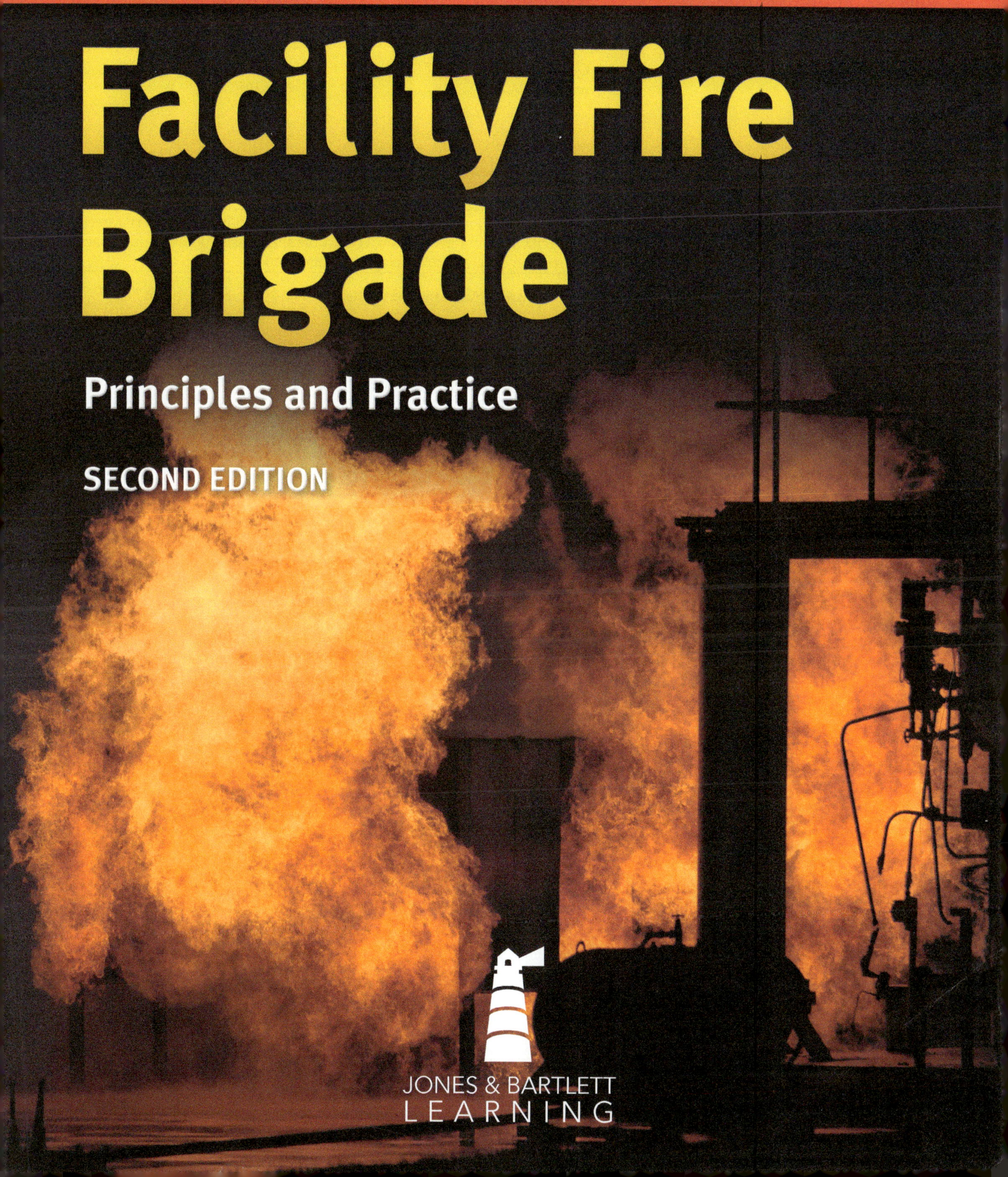

JONES & BARTLETT LEARNING

**Jones & Bartlett Learning**
*World Headquarters*
25 Mall Road
Burlington, MA 01803
978-443-5000
info@jblearning.com
www.jblearning.com
www.psglearning.com

**National Fire Protection Association**
1 Batterymarch Park
Quincy, MA 02169-7471
www.NFPA.org

**International Association of Fire Chiefs**
4025 Fair Ridge Drive
Fairfax, VA 22033
www.IAFC.org

Jones & Bartlett Learning books and products are available through most bookstores and online booksellers. To contact the Jones & Bartlett Learning Public Safety Group directly, call 800-832-0034, fax 978-443-8000, or visit our website, www.psglearning.com.

Substantial discounts on bulk quantities of Jones & Bartlett Learning publications are available to corporations, professional associations, and other qualified organizations. For details and specific discount information, contact the special sales department at Jones & Bartlett Learning via the above contact information or send an email to specialsales@jblearning.com.



20313-4

## Production Credits

General Manager and Executive Publisher: Kimberly Brophy
VP, Product Development: Christine Emerton
Senior Managing Editor: Donna Gridley
Executive Editor: William Larkin
Editorial Assistant: Alexander Belloli
Project Specialist: Meghan McDonagh
Digital Products Manager: Jordan McKenzie
Director of Marketing Operations: Brian Rooney
Production Services Manager: Colleen Lamy
Manufacturing and Inventory Control Supervisor: Amy Bacus
Composition: S4Carlisle Publishing Services
Project Management: S4Carlisle Publishing Services
Cover Design: Scott Moden
Text Design: Scott Moden
Rights Specialist: John Rusk
Cover Image (Title Page, Chapter Opener): © Jones & Bartlett Learning. Photographed by Glen E. Ellman
Printing and Binding: Lakeside Book Company
Cover Printing: Lakeside Book Company

**Library of Congress Cataloging-in-Publication Data**
Library of Congress Cataloging-in-Publication Data unavailable at time of printing.
LCCN: 2019951184

6048

Printed in the United States of America
28 27 26 25 24 10 9 8 7 6 5 4 3 2

# Brief Contents

# Table of Contents

## Chapter 1



## Chapter 2



## Chapter 3

## Chapter 4



## Chapter 5

## Chapter 12



## Chapter 13



## Chapter 14

# Chapter 15



# Chapter 16

## Chapter 17



## Chapter 18



## Chapter 19



## Chapter 20

## Chapter 21



## Chapter 22



## Chapter 23

## Chapter 24



## Chapter 25



## Chapter 26

## Chapter 27



## Appendix A



## Appendix B

# Skill Drills

## Chapter 13



## Chapter 14



## Chapter 15



## Chapter 16

# Skill Drills

### Chapter 16 (continued)



### Chapter 17



### Chapter 19



### Chapter 21

# Acknowledgments

## Editorial Board


**Joe A. Cocciardi**
Cocciardi and Associates, Inc.
Mechanicsburg, Pennsylvania

**Scott Dornan**
ConocoPhillips
Anchorage, Alaska
Kuparuk Fire Department
Kuparuk, Alaska

**Carl Peterson**
National Fire Protection Association
Quincy, Massachusetts

**Shawn P. Stokes**
International Association of Fire Chiefs
Fairfax, Virginia


## Contributing Authors and Editors


**Tyler Bones**
Flint Hills Resources Alaska
North Pole, Alaska

**Joe A. Cocciardi**
Cocciardi and Associates, Inc.
Mechanicsburg, Pennsylvania

**Anthony R. Cole**
Saudi Aramco
Dhahran, Saudi Arabia

**Mike Dalton**
Knox County Fire Investigation Unit
Knoxville, Tennessee

**Scott Dornan**
ConocoPhillips
Anchorage, Alaska
Kuparuk Fire Department
Kuparuk, Alaska

**Michael J. Fagel**
University of Chicago—Master of Threat Risk Program
Chicago, Illinois

**Rick Haase**
ConocoPhillips Wood River Refinery
Roxana, Illinois

**Attila Hertelendy**
University of Nevada, Reno—Fire Science Academy
Carlin, Nevada

**Greg Jakubowski**
Lingohocken Fire Company
Bucks County, Pennsylvania

**Ronald E. Kanterman**
Merck & Co., Inc.
Rahway, New Jersey

**William C. Kelly**
Northeast Refining, Sunoco, Inc.
Marcus Hook, Pennsylvania

**Frederick J. Knipper**
Duke University—Duke Health System
Durham, North Carolina

**Richard W. Kolomay**
Schaumburg Illinois Fire Department
Schaumburg, Illinois

**Randy J. Krause**
Boeing Fire Department
Seattle, Washington

**Bob La Plante**
Navajo Generating Station
Chandler, Arizona

**Marc J. Lawrence**
Northrop Grumman
Fire Department, Former Chevron USA Fire Department
El Segundo, California

**John David Lowe**
Refinery Terminal Fire Company
Corpus Christi, Texas

**Raymond Lussier**
Auburn Fire & Rescue
Auburn, Maine

**Louis N. Molino, Sr.**
LNM Emergency Services Consulting Services (LNMECS)
Bryan, Texas

**Jim Philp**
Beaumont Emergency Services Training (BEST) Complex
Beaumont, Texas

**Roy Robichaux, Jr.**
ConocoPhillips—Alliance Refinery
Belle Chasse, Louisiana

**Craig H. Shelley**
Advanced Fire Training Center
Saudi Aramco
Dhahran, Saudi Arabia

**John Shrader**
Cooper Nuclear Station
Brownville, Nebraska

**Ronald K. Sommers**
Chevron Phillips Chemical Company
Pasadena Plastics Complex Fire & Rescue
Pasadena, Texas

### John Welling, III
Bristol-Myers Squibb Company
Princeton, New Jersey

### David White
Industrial Fire World
College Station, Texas

### Mike Wisby
Texas Engineering Extension Service (TEEX)
Texas A&M University
College Station, Texas

## Reviewers

### Richard R. Anderson
Anderson Risk Consultants
Lambertville, New Jersey

### John H. Austin
Sunoco Inc., Eagle Point Facility
Westville, New Jersey

### Oren Bersagel-Briese
Castle Rock Fire Department
Castle Rock, Colorado

### Tyler Bones
Flint Hills Resources Alaska
North Pole, Alaska

### John Cannon
Johns Manville Corp.
Waterville, Ohio

### Harry J. Cusick
Philadelphia Fire Department (Retired)
Royersford, Pennsylvania

### Robert J. Davidson
Davidson Code Concepts, LLC
Tinton Falls, New Jersey

### Stanford E. Davis
PPL Susquehanna LLC
Sweet Valley Volunteer Fire Company
Sweet Valley, Pennsylvania

### Blair Elliot
Aspen Fire Department
Aspen, Colorado

### Douglas D. Frey
Greater Prudhoe Bay Fire Department, BPX Alaska
Anchorage, Alaska

### Thomas L. "Tommy" Harper
Jefferson College of Health Sciences
Roanoke, Virginia

### David A. Herr
Refinery Terminal Fire Company Training (RTFC) – Training Academy
Corpus Christi, Texas

### Lonnie D. Inzer
Emergency Management Degree Programs
Pikes Peak Community College
Colorado Springs, Colorado

### Kurt P. Keiser
Colorado Mountain College
Edwards, Colorado

### William C. Kelly
Northeast Refining, Sunoco, Inc.
Marcus Hook, Pennsylvania

### David J. Kirk
Tucson Electric Power Company, Springerville Generating Station
Springerville, Arizona

### Gordon L. Lohmeyer
Texas Engineering Extension Service (TEEX)
Texas A&M University
College Station, Texas

### John David Lowe
Refinery Terminal Fire Company
Corpus Christi, Texas

### Wilburn F. "Will" Lyons
Lamar Institute of Technology
Beaumont, Texas

### Ken McMahon
State of Delaware Fire Prevention Commission
Dover, Delaware

### John McPhee
Iowa Fire Service Training Bureau
Ames, Iowa

### William R. Montrie
Springfield Township Fire
Holland, Ohio
Owens Community College
Toledo, Ohio

### Robert Moore
Texas Engineering Extension Service (TEEX)
Texas A&M University
College Station, Texas

### Charles Moye
Ciba Specialty Chemicals
McIntosh, Alabama

### Perry Otero
Inter Canyon Fire Rescue
Morrison, Colorado

### George H. Quick
University of Nevada, Reno—Fire Science Academy
Carlin, Nevada

### Patrick J. Robinson
Valero Fire Department
Paulsboro, New Jersey

### Michael R. Schick
Louisville Fire Department
Louisville, Colorado

### Lawrence G. Schwarz
Training Division, Colorado Springs Fire Department
Colorado Springs, Colorado

### Ronald K. Sommers
Chevron Phillips Chemical Company
Pasadena Plastics Complex Fire & Rescue
Pasadena, Texas

**Roger D. Sylvestre**
Mashantucket Pequot Tribal Nation
Department of Fire & Emergency Services
Mashantucket, Connecticut

**Dennis L. Teegardin, Sr.**
Day-Glo Color Corporation
Emergency Response Team
Cleveland, Ohio

**Ed Waterson**
Schering-Plough Pharmaceutical Corporation
Summit, New Jersey

**Ward Watson**
Wheat Ridge Fire Protection District
Wheat Ridge, Colorado

**John A. Welling, III**
Bristol-Myers Squibb Company
Princeton, New Jersey

**Kevin Whelan**
Rifle Fire Protection District
Rifle, Colorado

**David White**
Industrial Fire World
College Station, Texas

**Kevin M. Wilson**
Poudre Fire Authority
Fort Collins, Colorado

**Jeff Windham**
Lyondell, Houston Refinery
Houston, Texas

**Carl Wren**
Austin Fire Department
Austin, Texas

# Brigade Member Qualifications and Safety

Chapter 1

# Brigade Member Qualifications and Safety

## NFPA 1081 Standard

**Incipient Facility Fire Brigade Member**

**4.1.1** *Qualification or Certification.* For qualification or certification at the incipient facility fire brigade level, the facility fire brigade member shall meet the entrance requirements in Chapter 1 and Sections 4.1 and 4.2, the site-specific requirements in Section 4.3 as defined by the management of the facility fire brigade, and the requirements defined in Chapter 4 of NFPA 1072. Standard for Hazardous Materials/Weapons of Mass Destruction Emergency Response Personnel Professional Qualifications.

**4.3*** *Site-Specific Requirements.* The management of the Facility fire brigade shall determine the site-specific requirements that are applicable to the incipient Facility fire brigade members operating on their site. The process used to determine the site-specific requirements shall be documented, and these additional JPRs added to those identified in Sections 4.1 and 4.2.

**Advanced Exterior Facility Fire Brigade Member**

**5.1*** *General.*

**5.1.1** *Qualification or Certification.* For qualification or certification at the advanced exterior facility fire brigade member level, the facility fire brigade member shall meet the entrance requirements in Chapter 1 and Sections 4.1, 4.2, 5.1, and 5.2, the site-specific requirements in Sections 4.3 and 5.3 as defined by the management of the facility fire brigade, and the requirements defined in Chapter 5 or NFPA 1072, Standard for Hazardous Materials/Weapons of Mass Destruction Emergency Response Personnel Professional Qualifications.

**5.1.2** *Basic Advanced Exterior Facility Fire Brigade Member JPRs.*

**5.2.1*** Use thermal protective clothing during exterior firefighting operations, given thermal protective clothing, so that the clothing is correctly donned within 2 minutes (120 seconds), worn, and doffed.

**(A)** *Requisite Knowledge.* Conditions that require personal protection, uses and limitations of thermal protective clothing, components of thermal protective clothing ensemble, and donning and doffing procedures.

**(B)** *Requisite Skills.* The ability to correctly don and doff thermal protective clothing and to perform assignments while wearing thermal protective clothing.

**5.2.2** Use self-contained breathing apparatus (SCBA) during emergency operations, given SCBA and other personal protective equipment (PPE), so that the SCBA is correctly donned, the SCBA is correctly worn, controlled breathing techniques are used, emergency procedures are enacted if the SCBA fails, all low-air warnings are recognized, respiratory protection is not intentionally compromised, and hazardous areas are exited prior to air depletion, and the SCBA is correctly doffed.

**(A)** *Requisite Knowledge.* Conditions that require respiratory protection, uses and limitations of SCBA, components of SCBA, donning procedures, breathing techniques, indications for and emergency procedures used with SCBA, and physical requirements of the SCBA wearer.

**(B)** *Requisite Skills.* The ability to control breathing, replace SCBA air cylinders, use SCBA to exit through restricted passages, initiate and complete emergency procedures in the event of SCBA failure or air depletion, and complete donning procedures.

**5.2.3 (A)** *Requisite Knowledge.* Principles of fire streams; types, design, operation, nozzle pressure effects, and flow capabilities of nozzles; precautions to be followed when advancing handlines to a fire; observable results that a fire stream has been correctly applied; dangerous conditions created by fire; principles of exposure protection; potential long-term consequences of exposure to products of combustion; physical states of matter in which fuels are found; the application of each size and type of attack line; the role of the backup team in fire attack situations; attack and control techniques; and exposing hidden fires.

**5.3*** *Site-Specific Requirements.* The JPRs in 5.3.1 through 5.3.11 shall be considered as site-specific functions of the advanced exterior facility fire brigade member. The management of the facility fire brigade shall determine the site-specific requirements that are applicable to the advanced exterior facility fire brigade member operating on their site. The process used to determine the site-specific requirements shall be documented, and these additional JPRs added to those identified in Sections 5.1 and 5.2. Based on the assessment of the site-specific hazards of the facility and the duties that facility fire brigade members are expected to perform, the management of the facility fire brigade shall determine the specific requirements of Chapters 5 or 6 of NFPA 472 or the corresponding requirements in OSHA 29 CFR 1910.120(q) that apply.

**Interior Structural Facility Fire Brigade Member**

**6.1** *General.*

**6.1.1** *Qualification or Certification.* For qualification or certification at the interior structural facility fire brigade member level, the member shall meet the entrance requirements in Chapter 1 and Sections 4.1, 4.2, 6.1, and 6.2, the site-specific requirements in Sections 4.3 and 6.3 as defined by the management of the facility fire brigade, and requirements defined by Chapter 5 of NFPA 1072.

**6.1.2*** *Basic Interior Structural Fire Brigade Member JPRs.*

**6.1.2.1** Use thermal protective clothing during structural firefighting operations, given thermal protective clothing, so that the clothing is correctly donned within 2 minutes (120 seconds), worn, and doffed.

**(A)** *Requisite Knowledge.* Conditions that require personal protection, uses and limitations of thermal protective clothing, components of thermal protective clothing ensemble, and donning and doffing procedures.

**(B)** *Requisite Skills.* The ability to correctly don and doff thermal protective clothing and perform assignments while wearing thermal protective clothing.

**6.1.2.2*** Use self-contained breathing apparatus (SCBA) during emergency operations, given SCBA and other personal protective equipment (PPE), so that the SCBA is correctly donned, the SCBA is correctly worn, controlled breathing techniques are used, emergency procedures are enacted if the SCBA fails, all low-air warnings are recognized, respiratory protection is not intentionally compromised, and hazardous areas are exited prior to air depletion and correctly doffed.

**(A)** *Requisite Knowledge.* Conditions that require respiratory protection, uses and limitations of SCBA, components of SCBA, donning procedures, breathing techniques, indications for and emergency procedures used with SCBA, and physical requirements of the SCBA wearer.

**(B)** *Requisite Skills.* The ability to control breathing, replace SCBA air cylinders, use SCBA to exit through restricted passages, initiate and complete emergency procedures in the event of SCBA failure or air depletion, and complete donning and doffing procedures.

**6.1.2.4** Given a structural fire-fighting ensemble including respiratory protection and decontamination equipment on the fireground or at the fire station, the interior structural fire brigade member shall describe and perform field reduction of contaminants of PPE so that the PPE is handled in the proper manner and maintained in a safe working condition.

**(A)** *Requisite Knowledge.* Working knowledge of field reduction of contaminants, policies and procedures for the reduction of carcinogenic or radiological particulates, etiological and chemical hazards of contaminated PPE on the fireground or at the fire station, manufacturer's specifications, and industry best practices for transporting, cleaning, inspecting, and identifying needed repairs.

**(B)** *Requisite Skills.* The ability to perform PPE field reduction of contaminants at the fireground or at the fire station.

**6.2.1 (A)** *Requisite Knowledge.* Principles of conducting initial fire size-up; principles of fire streams; types, design, operation, nozzle pressure effects, and flow capabilities of nozzles; precautions to be followed when advancing hose lines to a fire; observable results that a fire stream has been correctly applied; dangerous building conditions created by fire; principles of exposure protection; potential long-term consequences of exposure to products of combustion; physical states of matter in which fuels are found; common types of accidents or injuries and their causes; and the application of each size and type of handlines, the role of the backup team in fire attack situations, attack and control techniques, and exposing hidden fires.

**6.3*** *Site-Specific Requirements.* The management of the facility fire brigade shall determine the site-specific requirements that are applicable to the interior structural facility fire brigade member operating on their site. The process used to determine the site-specific requirements shall be documented, and these additional JPRs added to those identified in Sections 6.1 and 6.2. Based on the assessment of the site-specific hazards of the facility and the duties that facility fire brigade members are expected to perform, the management of the facility fire brigade shall determine the specific requirements of Chapters 5 and 6 of NFPA472, *Standard for Competence of Responders to Hazardous Materials/ Weapons of Mass Destruction Incidents*, or the corresponding requirements in OSHA29 CFR 1910.120(q) that apply.

## Knowledge Objectives

After completing this chapter, you will be able to:

- Discuss the educational, medical, physical fitness, and emergency medical care requirements for becoming a brigade member.
- Describe how standards and procedures, personnel, training, and equipment are related to the prevention of brigade member injuries and deaths.
- List safety precautions you need to take during training, during emergency responses, at emergency incidents, at the station, and outside your workplace.
- Describe the protection provided by personal protective equipment (PPE).
- Explain the importance of standards for PPE.
- Describe the limitations of PPE.
- Describe how to properly maintain PPE.
- Describe the hazards of smoke and other toxic environments.
- Explain why respiratory protection is needed in the fire service.
- Describe the differences between open-circuit breathing apparatus and closed-circuit breathing apparatus.
- Describe the limitations associated with self-contained breathing apparatus (SCBA).
- List and describe the major components of SCBA.
- Explain the skip-breathing technique.
- Explain the safety precautions you should remember when using SCBA.
- Describe the importance of daily, monthly, and annual SCBA inspections.
- Explain the procedures for refilling SCBA cylinders.
- List the steps for donning a complete PPE ensemble.

## Skills Objectives

After completing this chapter, you will be able to perform the following skills:

- Don approved personal protective clothing.
- Doff approved personal protective clothing.
- Don an SCBA from a seat-mounted bracket.
- Don an SCBA from a side-mounted compartment.
- Don an SCBA from a storage case using the over-the-head method.
- Don an SCBA from a storage case using the coat method.
- Don a face piece.
- Doff an SCBA.
- Perform daily SCBA inspections.
- Perform monthly SCBA inspections.
- Replace an SCBA cylinder.
- Clean and sanitize an SCBA.

# You Are the Brigade Member

**Before leaving the fire station to respond to a structure fire, you don your personal protective clothing, board the apparatus, and fasten your seat belt. The fire is on the second floor. Your brigade leader tells you and your partner to mount an interior attack, so you put on your self-contained breathing apparatus (SCBA) and stretch a hose line to the entrance. Smoke fills the first floor as you make your way to the seat of the fire. You can feel the heat through your face piece and hear the sounds of breaking glass and crackling flames. You open the nozzle and direct a stream of water onto the fire to extinguish it. You continue to use your SCBA until the Safety Officer approves working without it.**

1. *How does your personal protective equipment keep you safe in this hostile environment?*
2. *What are some of the limitations of your personal protective equipment?*

## Brigade Member Qualifications

When a company or organization makes the decision to establish a facility fire brigade, it is required to establish a program that defines the mission of the organization and to establish the requirements for the organization, operation, training, and occupational safety and health of facility fire brigades.

Members of an facility fire brigade provide a vital service, the protection of life and property. A successful and proficient fire brigade member must be properly trained and supervised by a qualified brigade leader. The job requires a person who has the desire to learn, the discipline to practice, and the ability to apply skills effectively during an incident (▼ Figure 1-1). Fire brigade members are in a constant learning process; the more you know, the more effective you will be. Firefighting is manual labor, but it is the most mentally challenging manual labor anyone will encounter. Smart, well-trained, motivated brigade members are safe and proficient brigade members.

Figure 1-1 Firefighting requires a special person.

## Roles and Responsibilities of the Fire Brigade Member

The first step in understanding the organization of the facility fire brigade is to learn your roles and responsibilities as a member trained to a specific level. As you progress through this text, you will learn what to do and how to do it so that you can take your place confidently among the brigade.

Facility firefighting performance and knowledge requirements are divided into four major categories: incipient firefighting, advanced exterior firefighting, interior structural firefighting, and fire brigade leader. The training, performance, and knowledge requirements for each level of response are specified in NFPA 600, *Standard on Facility Fire Brigades*, and NFPA 1081, *Standard for facility Fire Brigade Member Qualifications.*

NFPA 1081, *Chapter 4 Incipient Facility Fire Brigade Member* is shown in (► Table 1-1). The applicable entrance requirements must be met before brigade members are qualified or certified at the incipient level. This standard also defines the job performance requirements at each successive level of response.

### Incipient Facility Fire Brigade Member

The role of the incipient facility fire brigade member is to use the knowledge, skills, and abilities received through training to safely attack incipient fires only. This offensive fire attack may include the use of portable fire extinguishers or small handlines that flow up to 125 gallons per minute. The use of personal protective equipment (PPE) and self-contained

| Table 1-1 NFPA 1081 Chapter 4 Incipient Facility Fire Brigade Member | |
|---|---|
| **4.1* General.** | This duty shall involve initiating communications, using facility communications equipment to effectively relay oral or written information, responding to alarms, returning equipment to service, and completing incident reports, according to the JPRs in 4.1.1 through 4.2.3. |
| **4.2* Manual Fire Suppression.** | This duty shall involve tasks related to the manual control of fires and property conservation activities by the incipient facility fire brigade member. |
| **4.3 Site-Specific Requirements.** | The management of the facility fire brigade shall determine the site-specific requirements that are applicable to the incipient facility fire brigade members operating on their site. The process used to determine the site-specific requirements shall be documented, and these additional JPRs added to those identified in Sections 4.1 and 4.2. |
| **4.1.1 Qualification or Certification.** | For qualification or certification at the incipient facility fire brigade level, the facility fire brigade member shall meet the entrance requirements in Chapter 1 and Sections 4.1 and 4.2, the site-specific requirements in Section 4.3 as defined by the management of the facility fire brigade, and the requirements defined in Chapter 4 of NFPA 1072. |
| **5.1.1 Qualification or Certification.** | For qualification or certification at the advanced exterior facility fire brigade member level, the facility fire brigade member shall meet the entrance requirements in Chapter 1 and Sections 4.1, 4.2, 5.1, and 5.2, the site-specific requirements in Sections 4.3 and 5.3 as defined by the management of the facility fire brigade, and the requirements defined in Chapter 5 of NFPA 1072. |
| **6.1.1 Qualification or Certification.** | For qualification or certification at the interior structural facility fire brigade member level, the member shall meet the entrance requirements in Chapter 1 and Sections 4.1, 4.2, 6.1, and 6.2, the site-specific requirements in Sections 4.3 and 6.3 as defined by the management of the facility fire brigade, and requirements defined by Chapter 5 of NFPA 1072. |
| **4.3.1*** | Attack an incipient stage fire, given a handline flowing up to 473 L/min (125 gpm), appropriate equipment, and a fire situation, so that the fire is approached safely, exposures are protected, the spread of fire is stopped, agent application is effective, the fire is extinguished, and the area of origin and fire cause evidence are preserved. |
| **4.3.2*** | Activate a fixed fire protection system, given a fixed fire protection system, a procedure, and an assignment, so that the steps are followed and the system operates. |
| **4.3.3** | Utilize master stream appliances, given an assignment, an extinguishing agent, and a master stream device, so that the agent is applied to the fire as assigned. |
| **4.3.4*** | Establish a water supply for fire-fighting operations, given an assignment, a water source, and tools, so that a water supply is established and maintained. |
| **4.3.4.1** | The incipient level is the first level of progression for the subsequent levels of progression in this standard. |
| **4.3.5** | Perform a fire safety survey in a facility, given an assignment, survey forms, and procedures, so that fire and life safety hazards are identified, recommendations for their correction are made, and unresolved issues are referred to the proper authority. |

breathing apparatus (SCBA) is not allowed when fighting incipient-stage fires.

Some brigades are trained to defensively attack exterior fires using handlines that flow up to 300 gallons per minute and/or use master stream devices for cooling of tanks and vessels.

The incipient brigade member is responsible for attending the training programs and drills that should be provided at least annually.

Incipient facility fire brigade members should have the ability to:

- Understand and correctly apply appropriate communication protocols.
- Exit a hazardous area safely as a team.
- Set up ground ladders safely and correctly.
- Conserve property with salvage tools and equipment.
- Extinguish incipient Class A, Class B, Class C, and Class D fires.
- Illuminate an emergency scene.
- Turn off utilities.
- Perform fire safety surveys.
- Clean and maintain equipment.

The facility management shall determine the site-specific requirements for the incipient-level brigade. Fire brigade leaders should refer to NFPA 600, Standard on Facility Fire Brigades, to help identify the job performance requirements, the SOPs/SOGs, and the training requirements that the brigade will need to meet to safely and efficiently perform the duties needed to mitigate the hazards found at that facility.

## Advanced Exterior Facility Fire Brigade Member

The role of the advanced exterior facility fire brigade member is to use the knowledge, skills, and abilities received through training to safely fight exterior fires that may or may not include structures.

The advanced exterior brigade member is responsible for attending the education and training programs that should be provided at least once a quarter, the semiannual drills, and the annual live fire exercise.

Advanced exterior facility fire brigade members should have the ability to:

- Don (put on) and doff (take off) PPE properly.
- Understand and correctly apply appropriate communication protocols.
- Use SCBA.
- Respond on apparatus to an emergency scene.
- Force entry into a structure.
- Exit a hazardous area safely as a member of a team.
- Set up ground ladders safely and correctly.
- Attack a passenger vehicle fire and an exterior Class A fire.
- Overhaul a fire scene.
- Conserve property with salvage tools and equipment.
- Connect a fire department engine to a water supply.
- Extinguish incipient Class A, Class B, Class C, and Class D fires.
- Extinguish an ignitable liquid fire.
- Control a flammable gas cylinder fire.
- Illuminate an emergency scene.
- Turn off utilities.
- Perform fire safety surveys.
- Clean and maintain equipment.

Every fire brigade must develop site-specific SOPs/SOGs that cover the NFPA job performance requirements (JPRs) for the emergency operations and safety considerations associated with the types of hazards that the advanced exterior facility fire brigade member can be expected to respond to at that facility.

## Interior Structural Facility Fire Brigade Member

The role of the interior structural facility fire brigade member is to use the knowledge, skills, and abilities received through training to offensively attack interior fires that involve the structure and/or its contents.

The interior structural member is responsible for attending the education and training programs that should be provided at least once a quarter, the semiannual drills, and the annual live fire exercise.

The interior structural facility brigade member should also maintain his or her physical health and agility as well as the skills to:

- Don and doff PPE properly.
- Hoist hand tools using appropriate ropes and knots.
- Understand and correctly apply appropriate communication protocols.
- Use SCBA.
- Force entry into a structure.
- Exit a hazardous area safely as a member of a team.
- Set up ground ladders safely and correctly.
- Attack an interior structure fire.
- Conduct search and rescue in a structure.
- Perform ventilation of an involved structure.
- Overhaul a fire scene.
- Conserve property with salvage tools and equipment.
- Extinguish incipient Class A, Class B, Class C, and Class D fires.
- Illuminate an emergency scene.
- Turn off utilities.
- Perform fire safety surveys.
- Clean and maintain equipment.

Every fire brigade must develop site-specific SOPs/SOGs that cover the NFPA JPRs for the emergency operations and safety considerations associated with the types of hazards that the interior structural facility fire brigade member can be expected to respond to at that facility.

# Roles Within the Department

## General Roles

A fire brigade member may have many roles during a profession, particularly in smaller brigades. Some of the more common positions that fire fighters must assume are described here:

- Brigade member: The brigade member may be assigned any task, from placing hose lines to extinguishing fires. Generally, the brigade member is not responsible for any command functions and does not supervise other personnel, except on a temporary basis when promoted to an acting officer.
- Driver/operator: Often called an engineer or technician, the driver is responsible for getting the fire apparatus to the scene safely, setting up, and running the pump or operating the aerial ladder once it arrives on the scene.
- Company officer: This individual is usually a lieutenant or captain who is in charge of a team of members that form a company. The company officer is in charge of the company both on scene and at the station. He or she is responsible for the initial firefighting strategy, personnel safety, and the overall activities of the brigade members in their company. Once command is established, the company officer focuses on tactics.
- Safety officer: The safety officer watches the overall operation for unsafe practices. He or she has the authority to stop any firefighting activity until it can be done safely and correctly. The senior ranking officer may act as safety officer until the appointed safety officer arrives or until another officer is delegated those duties.
- Training officer: The training officer is responsible for updating the training of current brigade members and for training new members. He or she must be aware of the most current techniques of firefighting and EMS.
- Incident commander: The incident commander is responsible for the management of all the incident operations. This position focuses on the overall strategy of the incident and is often filled by the brigade leader or chief.
- Public information officer: The public information officer serves as a liaison between the incident commander and the news media.
- Fire protection engineer: The fire protection engineer usually has an engineering degree, reviews plans, and works with building owners to ensure that their fire suppression and detection systems will meet the relevant codes and function as needed. Some fire protection engineers actually design these systems.

## Specialized Response Roles

Many assignments require specialized training. Many facility brigades have teams of fire fighters who respond to specific types of calls. Members of these teams are usually required to be fire fighters before they begin additional training. Specialist positions include the following:

- Hazardous materials technician: "Hazmat" technicians have training and certification in chemical identification, leak control, decontamination, and clean-up procedures.
- Technical rescue technician: A "tech rescue" technician is trained in special rescue techniques for incidents involving structural collapse, trench rescue, vehicle/machinery rescue, confined-space rescue, high-angle rescue, and other unusual situations.
- Emergency Medical Services (EMS) personnel: EMS personnel administer prehospital care to people who are sick or injured. Prehospital calls account for the majority of responses in many industrial facilities. Some members are often cross-trained as EMS personnel. EMS training levels are normally divided into categories: Advanced First Aid, CPR/AED, Emergency Medical Technician–Basic, Emergency Medical Technician–Intermediate, and Emergency Medical Technician–Paramedic.

Management of the facility fire brigade must establish the level of facility fire brigade response and is responsible for ensuring the requirements are met. Facility brigade management is also responsible for defining and documenting the site-specific requirements for each level of response.

# Brigade Member Requirements

## Age Requirements

The authority for establishing the age requirements for participation in an facility fire brigade is the responsibility of management or the authority having jurisdiction. The level of emergency response and the potential safety exposure of brigade members might affect age limitations or restrictions. Firefighting is classified as a hazardous occupation, which limits the legal minimum age to 18. Insurance liability may also be a consideration in determining age restrictions of brigade members.

## Training and Education Requirements

A training and education program must be established and maintained to ensure brigade members are able to perform their assigned response duties safely and not pose a hazard to themselves or others. All brigade members must be trained to a competency level that meets the response duties that they are expected to perform. NFPA 1081 specifies the minimum skills and knowledge requirements that brigade members must meet before participating in emergency responses.

### Medical Requirements

Firefighting is physically demanding and stressful. Brigade members are required to have a medical evaluation before being accepted for fire brigade membership when response activities are beyond the incipient level response. Medical and fitness requirements are established by reviewing the risks and tasks that will be expected, based on the fire brigade's response duties. Brigade members must have annual medical evaluations when participating in advanced exterior or interior structural firefighting.

### Physical Fitness Requirements

Because of the wide variety of industrial operations and the varied levels of fire brigade response duties, it is the responsibility of facility fire brigade management to establish job-related physical performance requirements. NFPA 600 allows fire brigade management to choose the fitness testing method that will be used for fire brigade candidates. In general, fire brigade members must have the strength and stamina needed to perform the tasks associated with firefighting and emergency operations.

### Emergency Medical Care Requirements

Delivering emergency medical care is an important function of many facility fire brigades. The level of initial medical care can vary greatly depending on the resources available during a 24-hour period. Industrial facilities operate on different shifts, and staffing levels may be minimal after the normal day shift has ended. Facilities may have the ability to provide advanced life support response during the day, but may have limited or no emergency medical capabilities at other times. NFPA 1081 states that "emergency care performance capabilities for facility fire brigade personnel shall be determined and validated by the management of the facility fire brigade." OSHA 1910.1030 mandates bloodborne pathogen training when there is potential for exposure to blood or other potentially infectious materials.

## Brigade Member Safety

Firefighting, by its very nature, is dangerous. Each individual brigade member must learn safe methods of confronting the risks presented during training exercises, on the fireground, and at other emergency scenes.

Every fire brigade must do what it can to reduce the hazards and dangers of the job and help prevent brigade member injuries and deaths. Each organization must have a strong commitment to brigade member safety and health with designated personnel to oversee these programs. Safety must be fully integrated into every activity, procedure, and job description.

Appropriate safety measures must be applied routinely and consistently. During serious incidents, safety officers are responsible for evaluating the hazards of various situations and recommending appropriate safety measures to the Incident Commander (IC). Each accident or injury must be thoroughly investigated to determine the root cause and how it can be avoided in the future.

Advances in technology and equipment require organizations to review and revise their safety policies and procedures regularly. Information reviews and research by designated safety personnel can identify new hazards as well as appropriate risk-management measures. Reports of accidents and fatalities from other fire brigades can help identify problems and develop preventive actions.

### Causes of Brigade Member Deaths and Injuries

The information and statistics in the following section involve paid and volunteer municipal fire fighters. However, facility fire brigade members are exposed to many of the same risks. Injury and fatality statistics for facility brigade members are not readily available because injuries and fatalities reported to the Occupational Safety and Health Administration (OSHA) typically don't classify "facility fire brigade" as an occupation. The information contained here is valuable in that it clearly shows what is killing and injuring the entire fire service.

Each year about 100 fire fighters are killed in the line of duty in the United States. These deaths occur not only at emergency incident scenes, but also in the station, during training, and while responding to or returning from emergency situations. Approximately the same number of fire fighter deaths occur on the fireground or emergency scene as during training or while performing other nonemergency duties. The remainder, approximately 24%, occur while responding to or returning from alarms (▶ Figure 1-2). The leading cause of fire fighter deaths is heart attacks, both on and off the fireground.

Vehicle collisions are a major cause of fire fighter fatalities. For every 1,000 emergency responses, there is one vehicle collision involving an emergency vehicle (▶ Figure 1-3). One study found that 27% of the fire fighters who died in those incidents were ejected from the vehicle, which suggests that they were not using seatbelts. Brigade members should never overlook basic safety procedures, such as always fastening seat belts, especially during emergency responses.

The NFPA estimates that 62,085 fire fighters were injured in the line of duty in 2016. Half of these injuries occurred while fighting fires, and another 17% occurred at other emergencies. The rest occurred during other on-duty activities. The most common injuries were strains, sprains, and soft-tissue injuries. Burn injuries and smoke and gas inhalation made up only a small percentage of total injuries (▶ Table 1-2).

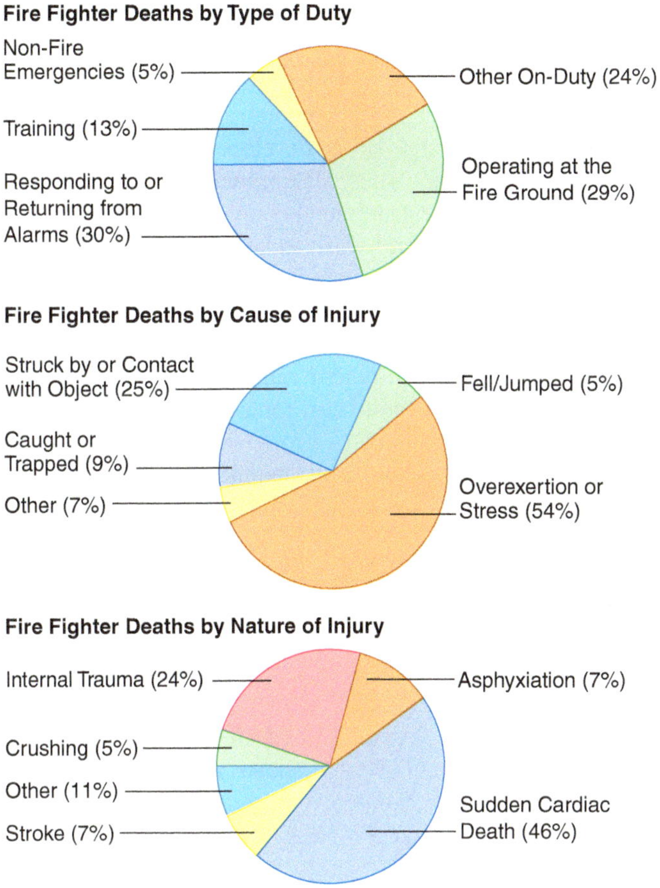


Figure 1-2 Fire fighter deaths in the United States.

Figure 1-3 Motor vehicle collisions are a major cause of death for fire fighters.

**Table 1-2 Fire Fighter Injuries**

| Type of Injury | % of total |
|---|---|
| Strains and sprains | 40 |
| Soft-tissue injuries | 22 |
| Burns | 7.9 |
| Smoke and gas inhalation | 6.2 |

## Injury Prevention

Injury prevention is a responsibility shared by each member of the firefighting team. Brigade members must always consider:

- Personal safety
- The safety of other team members
- The safety of everyone present at an emergency scene

To reduce the risks of accidents, injuries, occupational illnesses, and fatalities, a successful safety program must have four major components:

- Standards and procedures
- Personnel
- Training
- Equipment

## Standards and Procedures

Because safety is such a high priority, several organizations set standards for a safe working environment for the fire service. Although the standard doesn't apply to facility fire brigades that operate only "on property," NFPA 1500, *Standard on Fire Department Occupational Safety and Health Program,* provides a comprehensive health and safety program. Other NFPA standards focus on specific subjects directly related to health and safety. The OSHA, as well as various state and provincial health and safety agencies, develops and enforces government regulations on workplace safety. NFPA standards often are incorporated by reference in government regulations.

Every organization should have a set of standard operating procedures (SOPs) or standard operating guidelines (SOGs), which outline how to perform various functions and operations. SOPs are a requirement of NFPA 600. SOPs and SOGs cover a range of topics from uniform and grooming standards to emergency scene operations. These procedures should incorporate safe practices and policies. Each brigade member is responsible for understanding and following these procedures.

The organization's chain of command also enforces safety goals and procedures. The command structure keeps everyone working toward common goals in a safe manner. The Incident Management System (IMS) is a nationally recognized plan to establish command and control of emergency incidents. Flexible enough to meet the needs of any emergency situation, IMS should be implemented at every emergency scene, from a routine equipment accident to a major disaster involving numerous agencies.

Many brigades have a health and safety committee responsible for establishing policies on brigade member safety. Members of the committee should include representatives from every area, component, and level within the brigade, from brigade member to brigade leader. The safety officer and the brigade physician also should be members of the committee.

## Personnel

A safety program is only as effective as the individuals who implement it. Personnel selection and training in the science of safe and effective fire suppression should be a significant part of fire brigade operations and budgets.

Teamwork is an essential element of safe emergency operations. On the fireground and during any hazardous activity, brigade members must work together to get the job done. The lives of plant personnel, as well as the lives of other members of the brigade, depend on compliance with basic safety concepts and principles of operation.

An overall plan also is essential to coordinate the activities of every team, crew, or unit involved in the operation. The IMS coordinates and tracks the location and function of every individual or work group involved in an operation.

Freelancing is acting independently of a superior's orders or the organization's SOPs. Freelancing has no place on the fireground; it is a danger to both the brigade member who acts independently and every other brigade member. A brigade member who freelances can easily get into trouble by being in the wrong place at the wrong time or by doing the wrong thing. For example, a brigade member who enters a burning structure without informing a superior may be trapped by rapidly changing conditions. By the time the brigade member is missed, it may be too late to perform a rescue. Searching for a missing brigade member exposes others to unnecessary risk.

Safety officers are designated members of the brigade whose primary responsibility is safety. At the emergency scene, a safety officer reports directly to the IC and has the authority to stop any part of an action that is judged to be unsafe. Safety officers observe operations and conditions, evaluate risks, and work with the IC to identify hazards and ensure the safety of all personnel. Safety officers also determine when brigade members can work without SCBA after a fire is extinguished.

Safety officers contribute to safety in the workplace, at emergency incidents, and at training exercises. However, each member of the brigade shares the responsibility for safety, as an individual and as a member of the team.

## Training

Adequate training is essential for brigade member safety. The initial brigade member training covers the potential hazards of each skill and evolution and outlines the steps necessary to avoid injury. Brigade members must avoid sloppy practices or shortcuts that can contribute to injuries and learn how to identify hazards and unsafe conditions.

The knowledge and skills developed during training classes are essential for safety. The initial training course is only the beginning. Brigade members must continually seek out additional courses and work to keep their skills current to ensure personal and team safety.

## Equipment

A brigade member's equipment ranges from portable fire extinguishers to power and hand tools to personal protective equipment (PPE) and electronic instruments. Brigade members must know how to use equipment properly and operate it safely. Equipment also must be properly maintained. Poorly maintained equipment can create additional hazards to the user or fail to operate when needed.

Manufacturers usually supply operating instructions and safety procedures. Instructions cover proper use, limitations, and warnings of potential hazards. Brigade members must read and heed these warnings and instructions. New equipment must meet applicable standards to ensure that it can perform under difficult and dangerous conditions on the fireground.

# Safety and Health

Safety and well-being are directly related to personal health and physical fitness. Although fire organizations regularly monitor and evaluate the health of brigade members, each brigade member is responsible for personal conditioning and nutrition. Brigade members should eat a healthy diet, maintain a healthy weight, and exercise regularly.

All brigade members, whether paid or volunteer, should spend at least an hour a day in physical fitness training. Brigade members should be examined by either a personal or company physician before beginning any new workout routine. An exercise routine that includes weight training, cardiovascular workouts, and stretching with a concentration on job-related exercises is ideal. For example, many brigade members will use a stair-climbing machine and focus on the muscle groups used for firefighting. This builds strength and endurance on the fireground, but other muscle groups should not be neglected.

Hydration is an important part of every workout. A good guideline is to consume 8 to 10 ounces of water for every 5 to 10 minutes of physical exertion. Do not wait until you feel thirsty to start rehydrating. Brigade members should drink up to a gallon of water each day to keep properly hydrated. Proper hydration enables muscles to work longer and reduces the risk of injuries at the emergency scene.

Diet is another important aspect of physical fitness. A healthy menu includes fruits, vegetables, low-fat foods, whole grains, and lean protein. Pay attention to portion sizes; most

Figure 1-4 Regular exercise will help you to stay healthy and perform your job.

people eat larger portions than their bodies need. Substitute healthy choices such as fruit for high-calorie desserts.

Heart disease is the leading cause of death in the United States and among brigade members. A healthy lifestyle that includes a balanced diet, weight training, and cardiovascular exercises helps reduce many risk factors for heart disease and enables brigade members to meet the physical demands of the job ▲ Figure 1-4 .

Drug use has absolutely no place in the fire service. Many brigades have drug-testing programs to ensure that brigade members do not use or abuse drugs. The illegal use of drugs endangers your life, the lives of your team members, and the public you serve.

Everyone is subject to an occasional illness or injury. Brigade members should not try to work when ill or injured. Operating safely as a member of a team requires fitness and concentration. Do not compromise the safety of the team or your personal health by trying to work while ill or injured.

### Employee Assistance Programs

Employee assistance programs (EAPs) provide confidential help with a wide range of problems that might affect performance. Many fire brigades have EAPs so that brigade members can get counseling, support, or other assistance in dealing with a physical, financial, emotional, or substance abuse problem. A brigade leader may refer a brigade member to an EAP if the problem begins to affect job performance. Brigade members who use an EAP can do so with complete confidentiality and without fear of retribution.

## Safety During Training

During training, brigade members learn the actual skills that are later used under emergency conditions. The patterns that develop during training will continue during actual emergency incidents. Developing the proper working habits during training courses helps ensure safety later.

Many of the skills covered during training can be dangerous if they are not performed correctly. According to the NFPA, an average of nine municipal fire fighters are fatally injured during training exercises every year. Proper protective gear and teamwork are as important during training as they are on the fireground.

Instructors and veteran brigade members are more than willing to share their experiences and advice. They can explain and demonstrate every skill and point out the safety hazards involved because they have performed these skills hundreds of times and know what to do. But here, too, safety is a shared responsibility. Do not attempt anything you feel is beyond your ability or knowledge. If you see something that you feel is an unsafe practice, bring it to the attention of your instructors or a designated safety officer.

Do not freelance on the training ground. Wait for specific instructions or orders before beginning any task. Do not assume that something is safe and act independently. Follow instructions and learn to work according to the proper procedures.

Teamwork is also important during training exercises. Assignments are given to firefighting teams during most live fire exercises. Teams must stay together. If any member of the team becomes fatigued, is in pain or discomfort, or needs to leave the training area for any reason, notify the instructor or safety officer. Medical personnel should be consulted to perform an examination and refer for further treatment if necessary. A brigade member injured during training should not return until medically cleared for duty.

## Safety During Emergency Response

When a company is dispatched to an emergency, brigade members need to get to the apparatus and don appropriate PPE quickly before mounting the vehicle and proceeding to the incident. Walk quickly to the apparatus; do not run. Be careful not to slip and become injured before reaching the apparatus.

Figure 1-5 Protective clothing should be properly positioned so you can quickly don it.

Personal protective gear should be properly positioned so you can don it quickly before getting into the apparatus ▲ Figure 1-5. Be sure that seat belts are properly fastened before the apparatus begins to move. All personnel responding on fire apparatus must be seated with seat belts fastened. Seat belts should remain fastened until the apparatus comes to a complete stop. Brigade members can don SCBA while seated in some vehicles. Learn how to do this without compromising safety.

Drivers have a great responsibility. They must get the apparatus and the crew members to the emergency scene without having or causing an accident en route. They must know the plant area and any target hazards. They must be able to operate the vehicle skillfully and keep it under control at all times. They must anticipate all responses from other drivers who might not see or hear an approaching emergency vehicle or know what to do. Prompt response is a goal, but safe response is a much higher priority.

## Safety at Emergency Incidents

At the emergency scene, brigade members should never charge blindly into action. The brigade leader in command will "size-up" the situation, carefully evaluating the conditions to determine if the area is safe to enter. Wait for instructions and follow directions for the specific tasks to be performed. Do not freelance or act independently of command.

### Teamwork

On the fireground, a firefighting team should always consist of at least two brigade members who work together and are in constant communication with each other. Some brigades call this the buddy system ▼ Figure 1-6. In some cases, brigade members work directly with the brigade leader, and all brigade members function as a team. In other situations, two individual brigade members may be a team assigned to perform a specific task. In either case, the brigade leader must always know where teams are and what they are doing. Teams working in a hazardous area must maintain visual, vocal, or physical contact at all times.

Partners or assigned team members should enter together, work together, and leave together. If one member of a team must leave the fire building for any reason, the entire team must leave together, regardless if it is a two-person team or an entire brigade working as a team.

Before entering a burning building to perform interior search and rescue or fire suppression operations, brigade members must be properly equipped with approved PPE. Partners should check each other's PPE to ensure it is on and working correctly before they enter a hazardous area.

Brigade members working in a hazardous area should maintain visual, vocal, or physical contact with each other at all times. At least one member of each team should have a portable two-way radio to maintain contact with the IC or a designated individual in the chain of command who remains outside the hazardous area. The radio can be used to relay pertinent information and to summon help if the team

Figure 1-6 A firefighting team should consist of at least two members who work together.

## Brigade Member Safety Tips

The two-in, two-out rule is contained in OSHA 1910.134, *Respiratory Protection Standard.*

becomes disoriented, trapped, or injured. The IC can contact the team with new instructions or an evacuation order.

Brigade members operating in a hazardous area require back-up personnel. The back-up team must be able to communicate with the entry team, either by sight or by radio, and ready to provide assistance. During the initial stages of an incident, at least two brigade members must remain outside the hazardous area (two-in, two-out), properly equipped to respond immediately if the entry team has to be rescued.

As the incident progresses and additional crews are assigned to work in the hazardous area, a designated rapid intervention company/crew (RIC) should be established and positioned outside the hazardous area. This team's sole responsibility is to be prepared to provide emergency assistance to crews working inside the hazardous area.

### Accountability

Every fire brigade should have a personnel accountability system to track personnel and assignments on the emergency scene. The system should record the individuals assigned to each company, crew, or entry team; the assignments for each team; and the team's current activities. Several kinds of accountability systems are acceptable, ranging from paper assignments or display boards to laptop computers and electronic tracking devices.

Some brigades use a "passport" system. Each brigade leader carries a small magnetic board called a passport. Each brigade member on duty with that crew has a magnetic name tag on the board ▶ Figure 1-7. At the incident scene, the company's passport is given to a designated individual who uses it to track the assignment and location of every company at the incident.

An "accountability tag" system is often used by many organizations. Each brigade member carries a name tag and turns it in or places it in a designated location on the apparatus when the individual is on the scene. The tags are collected and used to track crews working together as a unit. This system works well when brigades are organized based on the available personnel.

Both systems provide an up-to-date accounting of everyone who is working at the incident and how they are organized. At set intervals, an accountability check is performed to account for everyone. Usually, a brigade leader reports on the status of each crew. The brigade leader should always know exactly where each brigade member is and what he or she is doing. If a crew splits into two or more teams, the brigade leader should be in contact with at least one member of each team.

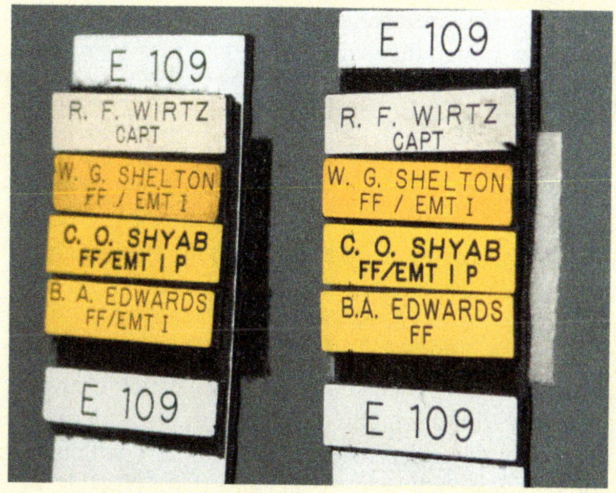


Figure 1-7 A passport lists each brigade member assigned to a crew.

An accountability check is also performed when there is a change in operational strategy or when a situation occurs that could endanger brigade members. If an accountability check is needed, the list of personnel and assignments is available at the command post.

Brigade members must learn their brigade's accountability system, how to work within it, and how it works within the IMS. Brigade members are responsible for complying with the system and staying in contact with a brigade leader or assigned supervisor at all times. Teams must stay together.

### Incident Scene Hazards

Brigade members must be aware of their surroundings when performing their assigned tasks at an emergency scene. At an incident, make a safe exit from the apparatus and look at the building or situation for safety hazards such as traffic, downed utility wires, and adverse environmental conditions. An incident on a street or facility road must first be secured with proper traffic- and scene-control devices. Flares, traffic cones, or barrier tape can keep the scene safe and bystanders at a safe distance. Always operate within established boundaries and protected work areas.

Changing fire conditions will also affect safety. During the overhaul phase and while picking up equipment, watch out for falling debris, smoldering areas of fire, and sharp objects. If a safety officer is not on the scene, another qualified person should be assigned to monitor the atmosphere for the presence of carbon monoxide (CO). CO is an odorless, colorless, tasteless gas that can cause asphyxiation, resulting in unconsciousness or death. Because the chance for injury increases when you are tired, do not let down your safety guard even though the main part of the fire is over.

### Using Tools and Equipment Safely

Brigade members must learn how to use tools and equipment properly and safely before using them at an emergency incident. Follow the proper procedures and safety precautions in training and at an incident scene. Use protective gear such as PPE, safety glasses, and hearing protection when they are required.

Proper maintenance includes sharpening, lubricating, and cleaning each tool. Equipment should always be in excellent condition and ready for use. Brigade members should be able to do basic repairs such as changing a saw blade or a handlight battery. Practice these tasks at the fire station until you can perform them quickly and safely on the emergency scene.

### Electrical Safety

Electricity is an emergency scene hazard that must always be respected. Many fires are caused by electricity, such as those ignited by faulty wiring or involving electric-powered equipment. Energized power lines may be present on the fireground. Brigade members must always check for overhead power lines when raising ladders. During any fire, the electric power supply to the building should be turned off. This is part of the fireground task called "controlling the utilities."

Fire brigades are often called to electrical emergencies, such as downed power lines, fires or arcing in transformers and switchgear, and stuck elevators. Always disconnect the power to any electrical equipment involved in an emergency incident.

Park apparatus outside the area and away from power lines when responding to a call for an electrical emergency. A downed power line should be considered energized until the power company confirms that it is dead. Secure the area around the power line and keep bystanders at a safe distance. Never drive apparatus over a downed line or attempt to move it using tools.

### Lifting and Moving

Lifting and moving objects are part of a brigade member's duties. Do not try to move something alone that is too heavy—ask for help. Never bend at the waist to lift an object; always bend at the knees and use the legs to lift. Use equipment such as handcarts, hand trucks, and wheelbarrows to move objects a long distance.

Brigade members must often move sick or injured patients. Discuss and evaluate the options before moving a patient; then proceed very carefully. If necessary, request help. Never be afraid to call for additional resources, to assist in lifting and moving a heavy patient.

**Brigade Member Safety Tips**

Never be afraid to ask for help when moving a heavy object.

### Working in Adverse Weather Conditions

In adverse weather conditions, brigade members must dress appropriately. A turnout coat and helmet can keep you warm and dry in rain, snow, or ice. Firefighting gloves and knit caps will also help retain body heat and keep you warm. If conditions are icy, make smaller movements, watch your step, and keep your balance.

### Rehabilitation

**Rehabilitation** is a systematic process to provide periods of rest and recovery for emergency workers during an incident. Rehabilitation is usually conducted in a designated area away from the hazards of the emergency scene. The rehabilitation area, or "rehab," is usually staffed by EMS personnel.

A brigade member who is sent to rehab should be accompanied by the other members of the crew. The brigade leader should inform the IC of their change in location. While in rehab, brigade members should take advantage of the opportunity to rest, rehydrate, have their vital signs checked by EMS personnel, and have minor injuries treated (► Figure 1-8). Rehab gives brigade members the chance to cool off in hot weather and to warm up in cold weather.

Rehabilitation time can be used to replace SCBA cylinders, obtain new batteries for portable radios, and make repairs or adjustments to tools or equipment. Firefighting teams can discuss recently completed assignments and plan their next work cycle. When a crew is released from rehab, they should be rested, refreshed, and ready for another work cycle. If the crew is too exhausted or unable to return to work, they should be replaced and released from the incident.

Never be afraid or embarrassed to admit you need a break when on the emergency scene. Heat exhaustion is a common condition, characterized by profuse sweating, dizziness, confusion, headache, nausea, and cramping. If a brigade member shows the signs or symptoms of heat exhaustion, the brigade leader should be notified immediately. The brigade leader will request approval for rehabilitation so the problem can be treated.

Heat exhaustion is usually not life-threatening and, if identified early, can be remedied by rehydration, cooling, and rest. Left untreated, heat exhaustion can progress quickly to heat stroke, which is a life-threatening emergency. A lack of sweating, low blood pressure, shallow breathing, and seizures are some of the signs of heat stroke.

Figure 1-8 In the rehabilitation area, brigade members can rest and rehydrate.

If you or a member of your team experience these symptoms during training or on the fireground, call for help and leave the fireground or training area to seek immediate medical attention.

A brigade member who experiences chest pain or discomfort should stop and seek medical attention immediately. Heart attacks are the leading cause of death in the fire service.

### Critical Incident Stress Debriefing

Some calls are particularly difficult and emotionally traumatic. Afterwards, brigade members who were involved may be required to attend a critical incident stress debriefing (CISD). Usually, a stress debriefing is held as soon as possible after a traumatic call. It provides a forum for personnel to discuss the anxieties, stress, and emotions triggered by a difficult call. Follow-up sessions can be arranged for individuals who continue to experience stressful or emotional responses after a challenging incident.

Many fire brigades have qualified, designated CISD staff available 24 hours a day. The initial CISD is usually a group session for all brigade members and rescuers. It can also be done on a one-on-one basis or in smaller groups. Everyone handles stress differently, and new brigade members need time to develop the personal resources to deal with difficult situations. If any call is emotionally disturbing, ask your supervisor or brigade leader for a referral to a qualified CISD counselor.

## Safety at the Fire Station

The fire station is just as much a workplace as the fireground. Be careful when working with power tools, ladders, electrical appliances, pressurized cylinders, and hot surfaces. Injuries that occur at the firehouse can be just as devastating as those that occur at an emergency incident scene.

## Safety Outside Your Workplace

Continue to follow safe practices when you are off the job as well. An accident or injury, regardless of where it happens, can end your career. For example, if you are using a ladder while off the job, follow the same safety practices that you would use at work. Use the seat belts in your personal vehicle, just as you are required to do when you are at work ▼ Figure 1-9.

### Brigade Member Safety Tips

Remember these guidelines to stay safe—on and off the job.

- **You are personally responsible for safety.** Keep yourself safe. Keep your teammates safe. Keep the citizens—your customers—safe.
- **Work as a team.** The safety of the firefighting unit depends on the efforts of each member. Become a dependable member of the team.
- **Follow orders.** Freelancing can endanger other brigade members as well as yourself.
- **Think!** Before you act, think about what you are doing. Many people are depending on you.

Figure 1-9 Always use your seat belt.

## Personal Protective Equipment

Personal protective equipment (PPE) is an essential component of a brigade member's safety system. It enables a person to survive conditions that would otherwise result in death or serious injury. Different PPE ensembles are designed for specific hazardous conditions, such as structural firefighting, hazardous materials operations, and emergency medical operations.

PPE ensembles provide specific protections, so an understanding of their designs, applications, and limitations is critical. For example, a structural firefighting ensemble will protect the wearer from the heat, smoke, and toxic gases present in building fires (▼ Figure 1-10). It cannot provide long-term protection from extreme weather conditions, and it limits range of motion. The more you know about the protection your PPE can provide, the better you will be able to judge conditions that exceed its limitations.

PPE must provide full body coverage and protection from a variety of hazards. To be effective, the entire ensemble must be worn whenever potential exposure to those hazards exists. PPE must be cleaned, maintained, and inspected regularly to ensure that it will provide the intended degree of protection when it is needed. Worn or damaged articles must be repaired or replaced.

Figure 1-10 A protective ensemble for structural firefighting provides protection from multiple hazards.

### Brigade Member Safety Tips

A structural firefighting ensemble is designed only for structural firefighting. It is not designed for other functions such as hazardous materials. Specialized functions require specific PPE. If your brigade performs specialized operations, you need the appropriate PPE for that activity.

### Standards for Personal Protective Equipment

Personal protective equipment for brigade members must be manufactured according to exacting standards. Each item must have a permanent label verifying that the particular item meets the requirements of the standard (▼ Figure 1-11). Usage limitations as well as cleaning and maintenance instructions should also be provided.

The requirements for firefighting PPE are outlined in:

- NFPA 1971, *Standard on Protective Ensembles for Structural Fire Fighting and Proximity Fire Fighting*

The requirements for self-contained breathing apparatus and personal alert safety systems are outlined in:

- NFPA 1981, *Standard on Open-Circuit Self-Contained Breathing Apparatus (SCBA) for Emergency Services*
- NFPA 1982, *Standard on Personal Alert Safety Systems (PASS)*

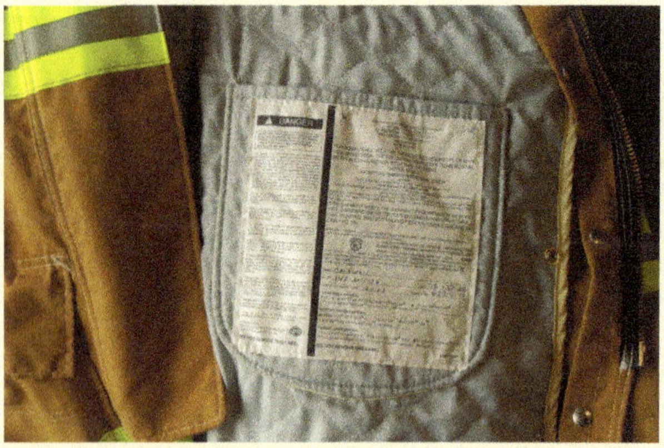

Figure 1-11 The label provides important information about each item of PPE.

Figure 1-12 The complete structural firefighting ensemble consists of a helmet, coat, trousers, protective hood, gloves, boots, SCBA, and PASS device.

## Structural Firefighting Ensemble

Structural firefighting PPE enables brigade members to enter burning buildings and work in areas with high temperatures and concentrations of toxic gases. Without PPE, brigade members would be unable to conduct search-and-rescue operations or perform fire suppression activities. A structural firefighting ensemble is designed to cover every part of the body. It provides protection from the fire, keeps water away from the body, and helps reduce trauma from cuts or falls. Structural firefighting PPE is designed to be worn with self-contained breathing apparatus (SCBA), which provides respiratory protection.

The structural firefighting ensemble consists of a protective coat, trousers, a helmet, a hood, boots, and gloves. The helmet must have a face shield, goggles, or both. The clothing is worn with SCBA and a personal alert safety system (PASS) device. All of these elements must be worn together to provide the necessary level of protection ▲ Figure 1-12.

### Protection Provided

A structural firefighting ensemble is designed for full body coverage and provides several different types of protection ► Table 1-3. The coat and trousers have tough outer shells that can withstand high temperatures, repel water, and provide protection from abrasions and sharp objects. The knees may be reinforced with pads for greater protection when crawling. Fluorescent/reflective trim adds visibility in dark or smoky environments. Insulating layers of fire-resistant materials protect the skin from high temperatures. A vapor barrier between the shell and liner keeps liquids and vapors, such as hot water or steam, from reaching the skin.

The helmet provides protection from trauma to the head and includes ear coverings. The face shield helps protect the eyes. A fire-retardant hood covers any exposed skin between the coat collar, the SCBA face piece, and the helmet. Gloves protect the hands from heat, cuts, and abrasions. Boots protect the feet and ankles from the fire, keep them dry, prevent puncture injuries, and protect the toes from crushing injuries.

**Table 1-3 Types of Protection Furnished by PPE**

Personal protective equipment:

- Provides thermal protection
- Repels water
- Provides impact protection
- Provides protection against cuts and abrasions
- Furnishes padding against injury
- Increases your visibility
- Provides respiratory protection

SCBA provides respiratory protection. An SCBA gives the brigade member an independent air supply. This protects the respiratory system from toxic products and hot gases present in the atmosphere.

### Helmet

Fire helmets are manufactured in several designs and shapes using different materials. Each design must meet the requirements specified in NFPA 1971, *Standard on Protective Ensemble for Structural Firefighting*. The hard outer shell is lined with energy-absorbing material and has a suspension system to provide impact protection against falling objects ▼ Figure 1-13. The helmet shell also repels water, protects against steam, and creates a thermal barrier against heat and cold. The shape of the helmet helps to deflect water away from the head and neck.

Figure 1-13 A helmet is constructed with multiple layers and components.

Face and eye protection can be provided by a face shield, goggles, or both. Safety goggles or glasses should meet NFPA or ANSI (American National Standards Institute) standards. These components are used when SCBA is not needed or when the SCBA face piece is not in place. A chin strap is also required and must be worn to keep the helmet in the proper position. The chin strap also helps to keep the helmet on the brigade member's head during an impact.

Fire helmets have an inner liner for added thermal protection. This liner also provides protection for the ears and neck. When entering a burning building, the brigade member should pull down the ear tabs for maximum protection.

Helmets are manufactured with adjustable inner suspension systems that hold the head away from the shell and cushion it against impacts. This suspension system must be adjusted to fit the individual, with the SCBA face piece and hood in place.

Helmet shells are often color-coded according to the brigade member's rank and function. They often carry company numbers, rank insignia, or other markings. Some organizations use a shield mounted on the front of the helmet to identify the brigade member's rank and company. Bright, reflective materials are applied to make the brigade member more visible in all types of lighting conditions.

NFPA 1971 requires that helmets approved for structural firefighting have a label permanently attached to the inside of the shell. This label lists the manufacturer, model, date of manufacture, weight, size, and recommended cleaning procedures.

## Protective Hood

Although the helmet's ear tabs cover the ears and neck, this area is still at risk for burns when the head is turned or the neck is flexed. Protective hoods provide additional thermal protection for these areas. The hood, which is constructed of flame-resistant materials such as Nomex® or PBI®, covers the whole head and neck, except for that part of the face protected by the SCBA face piece (▶ Figure 1-14). The lower part of the hood, which is called the bib, drapes down inside the turnout coat.

Protective hoods are worn over the face piece but under the helmet. After securing the face piece straps, carefully fit the hood around the face piece so that no areas of bare skin are left exposed. The hood must fit snugly around the clear area of the face piece so that vision is not compromised.

## Turnout Coat

Coats used for structural firefighting are generally called bunker coats or turnout coats (▶ Figure 1-15). Only coats that meet NFPA 1971 should be used for structural firefighting. Turnout coats have three layers. The outer layer or shell is constructed of a sturdy, flame-resistant, water-repellant material such as Nomex®, Kevlar®, or PBI®. Reflective material applied to the outer shell makes the brigade member more visible in smoky conditions and at night.

The second layer of the coat is the moisture barrier, which is a flexible membrane attached to a thermal barrier material (the third layer). The moisture barrier helps prevent the

Figure 1-14 A protective hood.

Figure 1-15 A bunker or turnout coat.

### Fire Marks

Before hoods were introduced, the skin on the neck and ears was often exposed. This often resulted in burns to the ears. Brigade members still must be careful to avoid situations where the temperature exceeds the protection provided by PPE.

transfer of water, steam, and other fluids to the skin. Water applied to a fire generates large amounts of superheated steam, which can engulf brigade members and burn unprotected skin.

The thermal barrier is a multilayered or quilted material that insulates the body from external temperatures. It enables brigade members to operate in the high temperatures generated by a fire and keeps the body warm during cold weather.

The front of the turnout coat has an overlapping flap to provide a secure seal. The inner closure is secured first, and then the outer flap is secured, creating a double seal. Several different combinations of D-rings, snaps, zippers, and Velcro can be used to secure the inner and outer closures.

The collar of the coat works with the hood to protect the neck. The collar has a closure system in front to keep it in a raised position. The coat's sleeves have wristlets that prevent hot embers from getting between the sleeves and the skin. They also prevent the sleeves from riding up the wrists, which could result in wrist burns.

Bunker coats come in a variety of lengths that will protect the body as long as the matching style of pants is also worn. The coat must be long enough to allow you to raise your arms over your head without exposing your midsection. The sleeve length should not hinder arm movement, and the coat should be large enough that it does not interfere with movements.

Pockets in the coat can be used for carrying small tools or extra gloves. Additional pockets or loops can be installed to hold radios, microphones, flashlights, or other accessories.

Figure 1-16 Bunker pants.

### Bunker Pants

Protective trousers are also called bunker pants or turnout pants (▶ Figure 1-16). They can be constructed in a waist-length design or bib style configuration. Bunker pants also must meet the NFPA 1971 and are constructed with the same multiple layers as bunker coats. The outer shell resists abrasion and repels water. The second layer is a moisture barrier to protect the skin from liquids and steam burns, and the inner layer is a quilted, thermal barrier to protect the body from elevated temperatures. Bunker pants are reinforced around the ankles and knees with leather or extra padding.

Bunker pants are manufactured with a double fastener system at the waist, similar to the front flap of a turnout coat. Reflective stripes around the ankles provide added visibility. Suspenders hold the pants up. Pants should be large enough to allow you to don them quickly. They should be big enough to allow you to crawl and bend your knees easily, but they should not be bigger than necessary.

### Boots

Structural firefighting boots can be constructed of rubber or leather and come in different lengths. Rubber firefighting boots come in a step-in style without laces (▶ Figure 1-17). Leather firefighting boots are available in a knee-length, pull-on style or in a shorter version with laces (▶ Figure 1-18).

Both boot styles must meet the same test requirements specified in NFPA 1971. The outer layer repels water and

#### Brigade Member Safety Tips

The inner thermal liner of most turnout coats can be removed while the outer shell is being cleaned, but the turnout coat must never be worn without the thermal liner. Severe injury could occur if the coat is worn without the liner.

#### Fire Marks

Bunker coats were originally made of rubber and designed primarily to repel water. Today's coats provide protection from flames, heat, abrasions, and cuts as well as water.

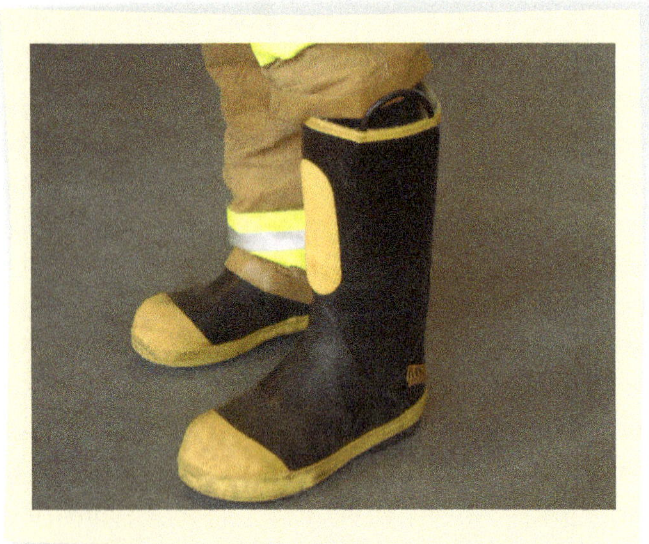

Figure 1-17 Rubber firefighting boots.

Figure 1-18 Leather firefighting boots.

must be both flame- and cut-resistant. The boots must have a heavy sole with a slip-resistant design. Boots must have a puncture-resistant sole and a reinforced toe to prevent injury from falling objects. An inner liner constructed of materials such as Nomex® or Kelvar® adds thermal protection.

### Brigade Member Safety Tips

Fire boots must be worn with approved bunker pants. Boots worn without bunker pants leave the legs unprotected and exposed to injury.

### Brigade Member Safety Tips

Plain leather work gloves or plastic-coated gloves must never be used for structural firefighting. Gloves used for structural firefighting must be labeled and meet NFPA 1971.

Boots must be the correct size for the foot. The foot should be secure within the boot to prevent ankle injuries and enable secure footing on ladders or uneven surfaces. Improperly sized boots will cause blisters and other problems.

## Gloves

Gloves are another important part of the firefighting ensemble (▼ Figure 1-19). Gloves must provide adequate protection and still enable the manual dexterity needed to accomplish tasks. NFPA 1971 specifies that gloves must be resistant to heat, liquid absorption, vapors, cuts, and penetration. Gloves must have a wristlet to prevent skin exposure during normal firefighting activities.

Firefighting gloves are usually constructed of heat-resistant leather. The wristlets are usually made of knitted Nomex® or Kevlar®. The liner adds thermal protection and serves as a moisture barrier.

Many brigade members carry a second set of gloves in their bunker gear or on the fire apparatus so they can change gloves if one pair gets wet or damaged. Do not wring or twist wet gloves because this can tear or damage the inner liners.

Although gloves furnish needed protection, they reduce manual dexterity. Brigade members should practice manual skills while wearing gloves to become accustomed to them and to adjust movement accordingly.

Figure 1-19 Firefighting gloves.

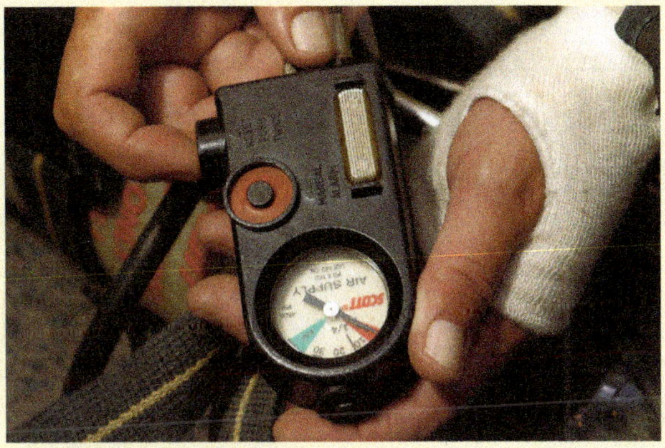

Figure 1-20 A PASS can save a brigade member's life.

Figure 1-21 Some fire apparatus are equipped with an intercom system.

### Respiratory Protection

SCBA is an essential component of the PPE used for structural firefighting. Without adequate respiratory protection, brigade members would be unable to mount an attack. The design, components, operation, use, and maintenance of this equipment are covered later in this chapter.

### Personal Alert Safety System (PASS)

A personal alert safety system (PASS) is an electronic device that sounds a loud audible signal when a brigade member becomes trapped or injured. A PASS will sound automatically if a brigade member is motionless for a set time period.

A PASS can be separate from or integrated into the SCBA unit ▲ Figure 1-20. The integrated PASS devices automatically turn on when the SCBA is activated. The separate PASS devices are worn on the SCBA harness and are required to be self-activated. Brigade members must ensure that their PASS is on and working properly before they enter a hazardous area.

The PASS combines an electronic motion sensor with an alarm system. If the user is motionless for 30 seconds, the PASS will sound a low warning tone before sounding a full alarm. The user can reset the device by moving during this warning period.

### Additional Personal Protective Equipment

Limited eye protection is provided by the face shield mounted on the helmet. When additional eye protection is needed, such as when using power saws or hydraulic rescue tools, brigade members should use NFPA or ANSI approved safety goggles or glasses.

Brigade members can also be exposed to loud noises such as sirens and engines. Because hearing loss is cumulative, it is important to limit exposure to loud sounds. An intercom system on the apparatus can provide hearing protection ▲ Figure 1-21. A small microphone is incorporated into large earmuffs located at each riding and operating position. Brigade members don the earmuffs, which reduce engine and siren noise. The microphone enables crew members to talk to each other in a normal tone of voice and to hear the apparatus radio.

Flexible ear plugs are useful in other situations involving loud sounds. Brigade members should use the hearing protection supplied by their brigades to prevent hearing loss.

A brigade member should always carry a hand light. Most interior firefighting is done in near-dark, zero-visibility conditions. A good working hand light can illuminate your surroundings, mark your location, and help you find your way under difficult conditions.

#### Brigade Member Tips

Brigade members have personal preferences about the tools and equipment they carry in their pockets. Observe what seasoned members of your brigade carry in their pockets. This will help you decide whether you need to carry similar equipment.

#### Brigade Member Tips

Practice donning your protective clothing ensemble until you can do it quickly and smoothly. Remember that each piece of the ensemble must be in place for maximum protection.

# Voices of Experience

**“ It was the aggressive training and thorough qualifications of the on-scene members that allowed the incident to be mitigated safely. ”**

Several years ago, an explosion and resulting fire at an facility facility illustrated the power of member training and qualifications. A small pipeline had ruptured at the very point it entered a process area, where there was also fire impingement on a storage tank. This fuel-fed fire was difficult to isolate because some of the isolation equipment had been compromised by the initial blast.

It was the aggressive training and thorough qualifications of the on-scene members that allowed the incident to be mitigated safely, and the team lived to fight another day with minimal equipment damage and no injuries. From a novice point of view, the incident was “big and splashy” due to a 50- to 60-foot flame plume.

Without the specific qualifications and training of the brigade, the outcome would have certainly been less desirable. Generally the formula for success in the facility emergency response arena is this: Training and education + certifications and qualifications + sound operational procedures and prefire planning = safety.

The training and educational requirements established for this brigade met the competency levels found in consensus standards and regulatory compliance rules. Without that training and education, the team would have had a much different result at this incident. Members operated according to procedures and had intimate knowledge of the operating systems of the facility and incident command. This behavior and knowledge, combined with a mastery of emergency response job performance requirements, reduced the hazards and dangers of the task at hand.

Qualifications and certifications come in many forms. It is vitally important that the required training and the maintenance of qualifications and certifications be emphasized for overall safety.

Jim Philp
Beaumont Emergency Services Training (BEST) Complex
(A division of the Facility Safety Training Council)
Beaumont, Texas

Two-way radios link the members of a firefighting team. At least one member of each team working inside a burning building or in any hazardous area should always have a radio. Some fire brigades provide a radio for every on-duty brigade member. Follow your brigade's SOP on radio use. A radio should be considered part of PPE and carried with you whenever it is appropriate.

### Limitations of the Structural Firefighting Ensemble

The structural firefighting ensemble protects a brigade member from the hostile environment of a fire. Each component must be properly donned and worn to provide complete protection. But even today's advanced PPE has limitations. Understanding those limitations will help you avoid situations that could result in serious injury or death.

There are several components that must be put on in the proper order and correctly secured. You must be able to don your equipment quickly and correctly, either at the station before you respond to an emergency, or after you arrive at the scene. Efficient donning and doffing PPE takes practice (▼ Figure 1-22).

PPE is also heavy, adding nearly 50 lb of extra weight. This increased weight means that everything you do—even walking—requires more energy and strength. Tasks such as advancing an attack line up a stairway or using an ax to create an opening can be difficult, even for a brigade member in excellent physical condition.

PPE retains body heat and perspiration, making it difficult for the body to cool itself. Perspiration is retained inside the protective clothing rather than released through evaporation to cool you. Brigade members in full protective gear can rapidly develop elevated body temperatures, even when the ambient temperature is cool. The problem of overheating is more acute when surrounding temperatures are high. This is one reason that brigade members must undergo regular rehabilitation and fluid replacement.

**Figure 1-22** It takes practice to don the full PPE.

> **Brigade Member Safety Tips**
>
> DO NOT don protective clothing inside the apparatus while en route to an emergency incident. Stay in your assigned seat, properly secured by a seat belt or safety harness while the vehicle is in motion. Don protective clothing in the firehouse before mounting the apparatus, or on the scene after you arrive.

PPE limits your mobility. Full turnout gear not only limits the range of motion, but it also makes movements awkward and difficult. This increases energy expenditure and adds stress.

Wearing PPE also decreases normal sensory abilities. The sense of touch is reduced by wearing heavy gloves. Turnout coats and pants protect skin but reduce its ability to determine the temperature of hot air. Sight is restricted when you wear SCBA. The plastic face piece reduces peripheral vision and the helmet, hood, and coat make turning the head difficult. Earflaps and the protective hood over the ears limit hearing. Speaking can become muffled and distorted by the SCBA face piece, unless it is equipped with a special voice amplification system.

For these reasons, brigade members must become accustomed to wearing and using PPE. Practicing skills while wearing PPE will help you become comfortable with its operation and limitations.

## Donning Personal Protective Clothing

Donning protective clothing must be done in a specific order to obtain maximum protection. It also must be done quickly. Although it is not a performance requirement of NFPA 1081, brigade members should be able to don personal protective clothing in two minutes or less. This requires considerable practice, but remember that your goal should first be to don PPE properly, and second to do it consistently in 120 seconds or less.

Following a set pattern of donning PPE can help reduce the time it takes to dress. Many brigade members follow the pattern described in (► Skill Drill 1-1). This exercise does not include donning SCBA. First become proficient in donning personal protective clothing, and then add the SCBA. Follow the steps in Skill Drill 1-1 to don personal protective clothing:

1. Place equipment in a logical order for donning. **(Step 1)**
   - Place the legs of your bunker pants over your boots and fold the pants down around them.
   - Lay out your coat, helmet, hood, and gloves.
2. Place your protective hood over your head and down around your neck. **(Step 2)**

### Brigade Member Safety Tips

In hot weather, remove as much of your PPE as possible when you arrive at rehab. This will help you to cool down quickly. Some fire brigades place fans and misting equipment in the rehab area to provide additional cooling.

### Brigade Member Safety Tips

Turnout clothing must be properly cleaned and maintained to provide maximum protection. Dirty turnout clothing is a sign of carelessness, not of experience.

3. Step into your boots and pull up your bunker pants. Place the suspenders over your shoulders and secure the front of the pants using the closure system. **(Step 3)**
4. Put on your turnout coat and secure the inner and outer closures. **(Step 4)**
5. Place your helmet on your head with ear flaps extended and adjust the chin strap securely. Turn up your coat collar and secure it in front. **(Step 5)**
6. Put on your gloves. **(Step 6)**
7. Check all clothing to be sure it is properly secured. Have your partner check your clothing. **(Step 7)**

## Doffing Personal Protective Clothing

To doff, or remove, your personal protective clothing, reverse the procedure used in getting dressed. Follow the steps in ▶ **Skill Drill 1-2** to doff personal protective clothing:

1. Remove your gloves. **(Step 1)**
2. Open the turnout coat collar. **(Step 2)**
3. Release the helmet chin strap and remove your helmet. **(Step 3)**
4. Remove your turnout coat. **(Step 4)**
5. Remove your protective hood. **(Step 5)**
6. Remove your bunker pants and boots. **(Step 6)**

When necessary, PPE should be properly cleaned after it is used, and then kept in a convenient location for the next response. PPE may be kept close to the apparatus, on the apparatus, or in an equipment locker. Personal protective clothing must be properly maintained, organized, and ready for the next response.

## Care of Personal Protective Clothing

Approved personal protective clothing is built to exacting standards but requires proper care to continue to afford maximum protection. Avoid unnecessary wear on turnout clothing. A complete set of approved turnout clothing (excluding SCBA) costs more than $3,000. Keep this expensive equipment in good shape for its intended use—firefighting.

Check the condition of PPE on a regular basis. Clean it when necessary; repair worn or damaged PPE at once. PPE that is worn or damaged beyond repair must be replaced immediately because it will not protect you.

Avoid unnecessary cuts or abrasions on the outer material. This material already meets NFPA standards; do not look for opportunities to test its effectiveness. If the fabric is damaged, it must be properly repaired to retain its protective qualities. Follow the manufacturer's instructions for repairing or replacing PPE.

Personal protective clothing must be kept clean to maintain its protective properties. Dirt will build up in the fibers from routine use and exposure to fire environments. Smoke particles will become embedded in the outer shell material. The interior layers will frequently be soaked with perspiration. Regular cleaning should remove most of these contaminants.

Other contaminants are formed from the by-products of burned plastics and synthetic products. These residues are combustible and can be trapped between the fibers or build up on the outside of PPE, damaging the materials and reducing their protective qualities. A brigade member who is wearing contaminated PPE is actually bringing additional fuel into the fire on the clothing.

PPE that has been badly soiled by exposure to smoke, other products of combustion, melted tar, petroleum products, or other contaminants should be cleaned as soon as possible. Items that have been exposed to chemicals or hazardous materials may have to be impounded for decontamination or disposal.

Cleaning instructions are listed on the tag attached to the clothing. Follow the manufacturer's cleaning instructions. Failure to do so may reduce the effectiveness of the garment and create an unsafe situation for the wearer.

Some fire brigades have dedicated washing machines that are approved for cleaning turnout clothing. Other brigades contract with an outside firm to clean and repair protective clothing. In either case, the manufacturer's instructions for cleaning and maintaining the garment must be followed.

Other PPE also require regular cleaning and maintenance. The outer shell of your helmet should be cleaned with a mild soap as recommended by the manufacturer. The inner parts of helmets should be removed and cleaned according to the manufacturer's instructions. The chin strap and suspension

### Brigade Member Safety Tips

Make sure your PPE is dry before using it on the fireground. If wet protective clothing is exposed to the high temperatures of a structural fire, the water trapped in the liner materials will turn into steam and be trapped inside the moisture barrier. This can result in painful steam burns.

1-1 Skill Drill

# Donning Personal Protective Clothing

Place your equipment in a logical order for donning.

Place your protective hood over your head and down around your neck.

Put on boots and pull up bunker pants. Place the suspenders over your shoulders and secure the front of the pants.

Put on your turnout coat and close the front of the coat.

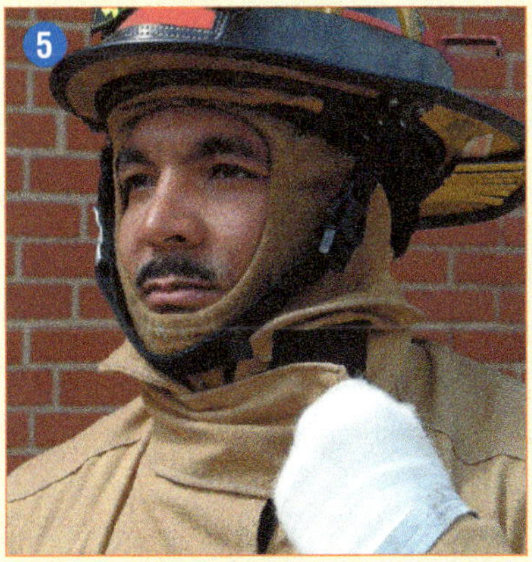

Place your helmet on your head and adjust the chin strap securely. Turn up your coat collar and secure it in front.

Put on your gloves.

Have your partner check your clothing.

1-2 Skill Drill

# Doffing Personal Protective Clothing

Remove your gloves.

Open the collar of your turnout coat.

Release the helmet chin strap and remove your helmet.

Remove your turnout coat.

Remove your protective hood.

Remove your bunker pants and boots.

system must be properly adjusted and all parts of the helmet kept in good repair.

Protective hoods and gloves get dirty quickly and should be cleaned according to the manufacturer's instructions. Most hoods can be washed with mild soaps or detergents. Repair or discard gloves or hoods that have holes in them; do not use them. A small cut or opening can result in a burn injury.

Boots should be maintained according to the manufacturer's instructions. Rubber boots should be kept in a place that does not result in damage to the boot. Leather boots must be properly maintained to keep them supple and in good repair. Boots should be repaired or replaced if the outer shell is damaged.

## Specialized Protective Clothing

### Vehicle Extrication

Due to the risk of fire at the scene of a vehicle extrication incident, most members of the emergency team will wear full turnout gear. The brigade leader may also designate one or more members to don SCBA and stand by with a charged hoseline. A structural firefighting ensemble protects against many of the hazards present at a vehicle extrication incident, such as broken glass and sharp metal objects.

There is protective clothing, such as special gloves and coveralls or jumpsuits, specifically designed for vehicle extrication. These items are generally lighter in weight and more flexible than structural firefighting PPE, although they may use the same basic materials.

Brigade members performing a vehicle extrication must always be aware of the possibility of contact with blood or other body fluids. Medical gloves should be worn when providing patient treatment. Eye protection also should be worn, due to the possibilities of breaking glass, contact with body fluids, metal debris, and accidents with tools.

# Respiratory Protection

Respiratory protection equipment is an essential component of the firefighting personal protection ensemble. Brigade members must be proficient in using SCBA before they engage in interior fire suppression activities; using one confidently requires practice.

The interior atmosphere of a burning building is considered to be IDLH (immediately dangerous to life and health). Attempting to work in this atmosphere without proper respiratory protection can cause serious injury or death. Never enter or operate in a fire atmosphere without appropriate respiratory protection.

## Respiratory Hazards of Fires

Brigade members need respiratory protection for several reasons. A fire involves a complex series of chemical reactions that can rapidly affect the atmosphere in unpredictable ways.

The most evident by-product of a fire is smoke. The visible smoke produced by a fire contains many different substances, most of which are dangerous if inhaled. In addition, smoke contains invisible, highly toxic products of combustion. The process of combustion consumes oxygen and can lower the oxygen concentration in the atmosphere below the level necessary to support life. The atmosphere of a fire may become so hot that one unprotected breath can result in fatal respiratory burns.

These respiratory hazards require brigade members to use respiratory protection in all fire environments, regardless of whether the environment is known to be contaminated, suspected of being contaminated, or could possibly become contaminated without warning. The use of SCBA allows brigade members to enter and work in a fire atmosphere with a safe, independent air supply.

### Smoke

Most fires do not have an adequate supply of oxygen to consume all of the available fuel. This results in incomplete combustion and produces a variety of by-products, which are released into the atmosphere. Many of these by-products are extremely toxic. Collectively, the airborne products of combustion are called smoke, which has three major components: solid particles, vapors, and gases.

#### Smoke Particles

Smoke particles consist of unburned, partially burned, and completely burned substances. These particles are lifted in the thermal column produced by the fire and are usually very visible. The completely burned particles are primarily ash; the unburned and partially burned smoke particles can include various substances. The concentration of unburned or partially burned particles depends on the amount of oxygen that was available to the fire.

Many smoke particles are so small that they can pass through the natural protective mechanisms of the respiratory system and enter the lungs. Some of these particles can be toxic to the body and result in severe injuries or death if they

**Fire Marks**

Modern SCBA has made interior firefighting operations safer and more effective. Before SCBA, fireground injuries and fatalities caused by smoke inhalation were common. Thousands of municipal fire fighters died from respiratory diseases, cancer, or other medical conditions that were direct consequences of repeated, unprotected exposure to smoke and other products of combustion. These exposures often had a cumulative, sometimes delayed effect. Each time fire fighters inhaled poisonous compounds, their lungs became more damaged, even though the damage was not evident until years later.

**Brigade Member Safety Tips**

Smoky environments and oxygen-deficient atmospheres are deadly! It is essential to use SCBA at all times when operating in a smoky environment.

are inhaled. They also can be extremely irritating to the eyes and digestive system.

### Smoke Vapors

Smoke also contains small droplets of liquids. These smoke vapors are similar to fog, which consists of small water droplets suspended in the air. When oil-based compounds burn, they produce small hydrocarbon droplets that become part of the smoke. If inhaled or ingested, these compounds can affect the respiratory and circulatory systems. Some of the toxic droplets in smoke can cause poisoning if they are absorbed through the skin.

Water applied to a fire creates steam and water droplets that also become part of the smoke. These water droplets can absorb some of the toxic substances contained in the smoke.

### Toxic Gases

A fire also produces several gases. The amount of oxygen available to the fire and the type of fuel being burned determine which gases are produced. A fire fueled by wood produces a different mixture of gases than one fueled by hydrocarbon-based products. (Remember, many common products such as plastics are made from hydrocarbon compounds.) The concentrations of these gases can change rapidly as the oxygen supply is consumed or as fresh oxygen is introduced to the combustion process.

Many of the gases commonly produced by commercial fires are very toxic. Carbon monoxide (CO), hydrogen cyanide, and phosgene are three of many gases often present in smoke.

CO is deadly in small quantities. When inhaled, CO quickly replaces the oxygen in the bloodstream because it combines with the hemoglobin in the blood 200 times more readily than oxygen. A small concentration of CO can quickly disable and kill a brigade member. CO is odorless, colorless, and tasteless, which adds to the danger. Because CO cannot be detected without instruments, brigade members must always assume that it is present in the atmosphere around a fire.

Hydrogen cyanide is formed when plastic products, such as the PVC pipe burns. It is a poisonous gas that is quickly absorbed by the blood and interferes with cellular respiration. A small amount of hydrogen cyanide can easily render a person unconscious.

Phosgene gas is formed from incomplete combustion of many common products, including vinyl materials.

**Table 1-4 Physiological Effects of Reduced Oxygen Concentration**

| Oxygen Concentration | Effect |
|---|---|
| 21% | Normal breathing air |
| 17% | Judgment and coordination impaired; lack of muscle control |
| 12% | Headache, dizziness, nausea, fatigue |
| 9% | Unconsciousness |
| 6% | Respiratory arrest, cardiac arrest, death |

Phosgene gas can affect the body in several ways. At low levels, it causes itchy eyes, a sore throat, and a burning cough. At higher levels, phosgene gas can cause pulmonary edema (fluid retention in the lungs) and death. Phosgene gas was used as a weapon in World War I to disable and kill soldiers.

Fires produce other gases that have different effects on the human body. Among these toxic gases are hydrogen chloride and several compounds containing different combinations of nitrogen and oxygen. Carbon dioxide, which is produced when there is sufficient oxygen for complete combustion, is not toxic, but it can displace oxygen from the atmosphere and cause asphyxiation.

## Oxygen Deficiency

Oxygen is required to sustain life. Normal outside or room air contains about 21% oxygen. A decrease in the amount of oxygen in the air will drastically affect an individual's ability to function as shown in ▲ Table 1-4. Atmosphere with an oxygen concentration of 19.5% or less is considered oxygen-deficient. If the oxygen level drops below 17%, people can experience disorientation, an inability to control their muscles, and irrational thinking, which can make escaping a fire much more difficult.

During compartment fires (fires burning within enclosed areas), oxygen deficiency occurs in two ways. First, the fire consumes large quantities of the available oxygen, decreasing the concentration of oxygen in the atmosphere. Second, the fire produces large quantities of other gases, which decrease the oxygen concentration by displacing the oxygen that would otherwise be present inside the compartment.

## Increased Temperature

Heat is also a respiratory hazard. The temperature of the gases generated during a fire varies, depending on fire conditions and the distance traveled by the hot gases. Inhaling superheated gases produced by a fire can cause severe burns of the respiratory tract. If the gases are hot enough, a single inhalation can cause fatal respiratory burns. More information about fire behavior and products of combustion is presented in Chapter 5, Fire Behavior.

## Other Toxic Environments

Not all hazardous atmospheric conditions are caused by fires. Brigade members will encounter toxic gases or oxygen-deficient atmospheres in many emergency situations. Respiratory protection is just as important in these situations as in a fire suppression operation.

Toxic gases can be released at hazardous materials incidents from leaking storage containers or industrial equipment, from chemical reactions, or from the normal decay of organic materials. CO can be produced by internal combustion engines or improperly adjusted heating systems.

Toxic gases can quickly fill confined spaces or below-grade structures. Any confined space or below-grade area must be treated as a hazardous atmosphere until it has been tested to ensure that an adequate concentration of oxygen and no hazardous or dangerous gases are present.

## Conditions That Require Respiratory Protection

We have looked at several factors that contribute to the dangers that exist in or near a fire atmosphere. Fires produce huge quantities of smoke, which contains unburned, partially burned, and completely burned poisonous particles, as well as toxic compounds and gases. Most fire deaths are caused by smoke inhalation rather than burns.

Fine liquid droplets, suspended in the smoke, contain highly toxic compounds formed from the breakdown of fuels. Smoke contains a wide variety of highly toxic gases resulting from incomplete combustion. Fires consume huge quantities of oxygen and generate huge quantities of poisonous gases, which can displace oxygen, causing an oxygen-deficient environment. Superheated gases are also produced. All of these factors contribute to classifying the atmosphere in a fire environment as IDLH.

Adequate respiratory protection is essential to your safety. The products of combustion from structure fires and commercial fires are so toxic that a few breaths can result in death. As you arrive at the scene of a fire, you do not have any way to measure the immediate danger to your life and your health posed by that fire. You must use approved breathing apparatus if you are going to enter and operate within this atmosphere.

### Brigade Member Safety Tips

Consider all below-grade areas or confined spaces to be IDLH—immediately dangerous to life and health—until they have been tested.

Anytime you are in an area where there is smoke, SCBA must be used. This includes exterior fires such as vehicle and dumpster fires. Utilize your SCBA at the fire scene until the air has been tested and proven to be safe by the safety officer. Do not remove your SCBA just because a fire has been knocked down. SCBA should be worn during overhaul until the air has been tested and deemed safe by your safety officer.

SCBA must also be used in any situation where there is a possibility of toxic gases being present or oxygen deficiency, such as a confined space. Always assume that the atmosphere is hazardous until it has been tested and proven to be safe.

## Types of Breathing Apparatus

The basic respiratory protection used by the fire service is the SCBA. The term self-contained refers to the requirement that the apparatus is the sole source of the brigade member's air supply. It is an independent air supply that will last for a predictable duration.

The two types of SCBA are <u>open-circuit breathing apparatus</u> and <u>closed-circuit breathing apparatus</u>. Open-circuit breathing apparatus is usually used for structural firefighting. A tank of compressed air provides the breathing air supply for the user (▼ Figure 1-23). Exhaled air is released into the atmosphere through a one-way valve. Approved open-circuit SCBA comes in several models, designs, and options.

**Figure 1-23** Open-circuit SCBA.

## Brigade Member Safety Tips

Always put on SCBA before entering a confined space or a below-ground structure. Continue to use respiratory protection until the air is tested and deemed safe by the safety officer. Do not attempt confined space operations without special training.

Figure 1-25 A supplied-air respirator may be needed for special rescue operations.

Closed-circuit breathing apparatus recycles the user's exhaled air. The air passes through a mechanism that removes carbon dioxide and adds oxygen within a closed system (▼ Figure 1-24). The oxygen is generated from a chemical canister. Many closed-circuit SCBAs have a small oxygen tank as well as the chemically generated oxygen. Closed-circuit breathing apparatus is more often used for extended operations, such as mine rescue work, where breathing apparatus must be worn for several hours.

A supplied-air respirator (SAR) uses an external source for the breathing air (► Figure 1-25). A hose line is connected to a breathing-air compressor or to compressed air cylinders located outside the hazardous area. The user breathes air through the line and exhales through a one-way valve, just as with an open-circuit SCBA.

Although SARs are commonly used in industrial settings, they are not used for structural firefighting. Hazardous materials teams and confined space rescue teams sometimes use SARs for specialized operations. Some fire service SCBA's can be adapted for use as SARs.

Figure 1-24 Closed-circuit SCBA.

### SCBA Standards and Regulations

In the United States, the National Institute for Occupational Safety and Health (NIOSH) sets the design, testing, and certification requirements for SCBA. NIOSH is a federal agency that researches, develops, and implements occupational safety and health programs. It also investigates fire service fatalities and serious injuries, and makes recommendations on how to prevent accidents from recurring.

OSHA and state agencies are responsible for establishing and enforcing regulations for respiratory protection programs. In some states there are individual occupational safety and health agencies to establish and enforce regulations.

The NFPA has developed three standards directly related to SCBA. NFPA 1500, *Standard on Fire Department Occupational*

## Brigade Member Tips

Do not confuse SCBA and SCUBA. Self-contained breathing apparatus, SCBA, is used by brigade members during fire suppression activities. Divers use self-contained underwater breathing apparatus (SCUBA) while swimming underwater.

## Fire Marks

Many of the occupational health and safety regulations for the fire service and other workers use the term **respirator**. A respirator is a device that provides respiratory protection for the user. There are several different types of respirators for different applications. An SCBA is a particular type of respirator.

*Safety and Health Program,* includes the basic requirements for SCBA use and program management. NFPA 1404, *Standard for Fire Service Respiratory Protection Training,* sets requirements for an SCBA training program. NFPA 1981, *Standard on Open-Circuit Self-Contained Breathing Apparatus (SCBA) for Emergency Services,* includes requirements for the design, performance, testing, and certification of open-circuit SCBA for the fire service. Each organization must follow applicable standards and regulations to ensure safe working conditions for all personnel.

## Uses and Limitations of Self-Contained Breathing Apparatus

SCBA is designed to provide pure breathing air to brigade members working in the hostile environment of a fire. It must meet rigid manufacturing specifications to properly function in the increased temperature and smoke-filled environments that brigade members encounter. When properly maintained, it will provide sufficient quantities of air for brigade members to perform rigorous tasks.

Using SCBA requires that brigade members develop unique skills, including different breathing techniques. SCBA limits normal sensory awareness; scent, hearing, and sight are all affected by the apparatus. Proficiency in the use of SCBA and other PPE requires ongoing training and practice.

SCBA also has its limitations, as with any type of equipment. Some of these limitations apply to the equipment; others apply to the user's physical and psychological abilities.

### Limitations of the Equipment

Because an SCBA carries its own air supply in a pressurized cylinder, its use is limited by the volume of air in the cylinder. SCBA for structural firefighting must carry enough air for a minimum of 30 minutes; cylinders rated for 45 minutes and 60 minutes are also available. These duration ratings, however, are based on laboratory conditions. The realistic useful life of an SCBA cylinder for firefighting operations is usually less than the rated duration, and actual use time will depend on the activities being performed and the physical condition of the person using the apparatus. An SCBA cylinder will generally have a useful life of 50% of the rated time. For example, an SCBA cylinder rated for 30 minutes can be expected to last for a maximum of 15 minutes of interior fire fighting.

**Figure 1-26** SCBA expands a user's profile, making it more difficult to get through tight spaces.

Brigade members must manage their working time while using SCBA to ensure adequate time to exit from the hazardous area before exhausting the air supply. If it takes 5 minutes to reach the work area and 5 minutes to return to fresh air, a 15-minute air supply provides about 5 minutes of working time.

The weight of an SCBA varies, based on the manufacturer and the type and size of cylinder. Generally, an SCBA weighs about 25 lb. The size of the unit also makes it more difficult for the user to fit into small places ▲ **Figure 1-26**. The added weight and bulk decrease the user's flexibility and mobility, and affects the user's center of gravity.

The design of the SCBA face piece limits visibility, particularly peripheral vision. The face piece may fog up under some conditions, further limiting visibility. SCBA also may affect the user's ability to communicate, depending on the type of face piece and any additional hardware provided, such as voice amplification and radio microphones.

### Physical Limitations of the User

Conditioning is important for SCBA users. An out-of-shape brigade member will consume the air supply more quickly and will have to exit the fire area long before a well-conditioned brigade member. Overweight or poorly conditioned brigade members are also at greater risk for heart attacks due to physical stress.

The protective clothing and SCBA that must be worn when fighting fires may weigh as much as 50 lb. Moving with this extra weight requires additional energy, which increases air consumption and body temperature. This places additional stress on a brigade members body. A person with ideal body weight will be able to perform more work per cylinder of air than a person who is overweight.

The weight and bulk of the complete PPE ensemble limits a brigade members ability to walk, climb ladders, lift weight, and crawl through restricted spaces. Brigade members must become accustomed to these limitations and learn to adjust their movements accordingly. Practice and conditioning are required to become proficient in wearing and using PPE while fighting fires.

### Psychological Limitations of the User

In addition to the physical limitations, the user must also make mental adjustments when wearing an SCBA. Breathing through an SCBA is different from normal breathing and can be very stressful. Covering your face with a face piece, hearing the air rushing in, hearing valves open and close, and exhaling against a positive pressure are all foreign sensations. The surrounding environment, which is often dark and filled with smoke, is foreign as well.

Brigade members must adjust to these stressful conditions. Practice in donning PPE, breathing through SCBA, and performing firefighting tasks in darkness helps to build confidence, not only in the equipment, but also in personal skills. Training generally introduces one skill at a time. Practice each skill as it is introduced and try to become proficient in that skill. As your skills improve, you will be able to tackle increasing levels of difficulty.

## Components of Self-Contained Breathing Apparatus

SCBA consists of four main parts: the backpack and harness, the air cylinder assembly, the regulator assembly, and the face piece assembly. There are several manufacturers and models with varying features and operations. Although the basics are similar, you need to become familiar with the specific SCBA used by your brigade.

### Backpack and Harness

The backpack provides the frame for mounting the components of the SCBA (▶ Figure 1-27). It is usually constructed of a lightweight metal or composite material. The SCBA harness consists of the straps and fasteners used to attach the SCBA to the brigade member. Most harnesses have two adjustable shoulder straps and a waist belt. Depending on the specific model of SCBA, the waist belt and shoulder straps will carry different proportions of the weight. The procedures for tightening and adjusting the straps also vary based on the model.

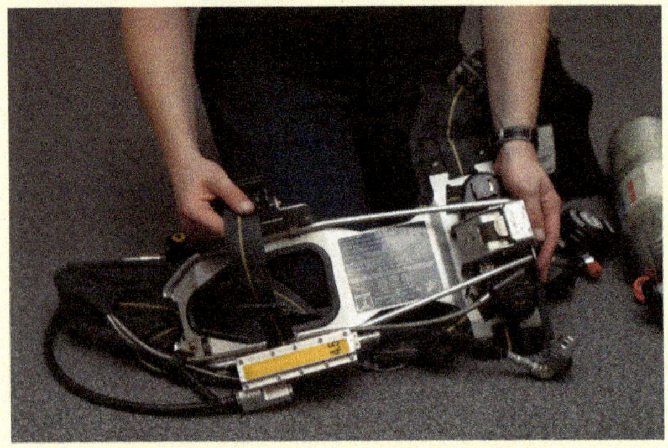

Figure 1-27 SCBA backpacks come in different models.

The harness must be secure enough to keep the SCBA firmly fastened to the user, but not so tight that it interferes with breathing or movements. The waist strap must be tight enough to keep the SCBA from moving from side to side or getting caught on obstructions.

### Air Cylinder Assembly

A compressed air cylinder holds the breathing air for an SCBA. This removable cylinder is attached to the backpack harness and can be changed quickly in the field. An experienced brigade member should be able to remove and replace the cylinder in complete darkness.

Brigade members should be familiar with the type of cylinders used in their organization. Cylinders are marked with the materials used in construction, the working pressure, and the rated duration.

The air pressure in filled SCBA cylinders ranges from 2,216 to 4,500 pounds per square inch (psi). The greater the air pressure, the higher the volume of air that can be stored in the cylinder.

Low-pressure cylinders, rated at 2,216 psi, can be constructed of steel or aluminum and are usually rated for 30 minutes of use. Composite cylinders are generally constructed of an aluminum shell wrapped with carbon, Kevlar®, or glass fibers. They are significantly lighter in weight, can be pressurized up to 4,500 psi, and are rated for 30, 45, or 60 minutes of use.

As previously noted, the rated duration times are established under laboratory conditions. A working brigade member can quickly use up the air because of exertion. Generally, the working time available for a particular cylinder is half the rated duration.

Figure 1-28 SCBA regulator.

The neck of an air cylinder is equipped with a hand-operated shut-off valve. The pressure gauge is located near the shut-off valve and indicates the pressure of the cylinder. Be careful not to damage the threads or let any dirt get into the outlet of the cylinder.

### Regulator Assembly

SCBA regulators may be mounted on the waist belt or shoulder strap of the harness or attached directly to the face piece ▲ Figure 1-28. The regulator controls the flow of air to the user. Inhaling decreases the air pressure in the face piece. This increases the regulator flow, which releases air from the cylinder into the face piece. When inhalation stops, the regulator reduces the air supply. Exhaling opens a second valve, the exhalation valve, to exhaust exhaled air into the atmosphere. SCBA regulators are capable of delivering large volumes of air to support the strenuous activities required in firefighting.

SCBA regulators will maintain a positive air pressure in the face piece in relation to the ambient air pressure outside the face piece. This feature helps to prevent the hazardous atmosphere outside the face piece from leaking into the face piece during inhalation. If there is any leakage in the area where the face piece and the face make the seal, the positive pressure breathing air inside the face piece will help to prevent the hazardous atmosphere from leaking in. Regardless of the positive pressure, a proper face piece to face seal must always be maintained. Breathing with this slight positive pressure may require some practice. New brigade members often say that it takes more energy to breathe when first using positive-pressure SCBA. This sensation gradually decreases.

Regulators are equipped for two modes of operation—the normal mode and the emergency by-pass mode. In the normal mode, described above, the regulator supplies breathing air during inhalation, decreases when inhalation stops, then opens an exhalation valve to exhaust used air into the atmosphere. To activate the normal mode, some SCBA models require that the user open a supply valve. Simply attaching the regulator to the face piece and inhaling will activate the normal mode in other models.

The emergency by-pass mode is used only if the regulator malfunctions ▼ Figure 1-29. Figure 1-29 shows a combination by-pass and purge valve on a Scott face piece-mounted regulator. Some SCBA manufacturers provide a combination by-pass and purge valve, commonly called "purge valves," on certain SCBA models. The combination valve permits the wearer to "purge" the facepiece for defogging purposes and also functions as a by-pass valve in the event of regulator failure. The by-pass valve is activated when the user turns on the red-colored emergency by-pass valve or the combination by-pass and purge valve. This releases a constant flow of breathing air into the face piece. The emergency by-pass mode uses more air, but it enables brigade members to exit a hazardous environment if the regulator malfunctions. A brigade member who must use the emergency by-pass mode must leave the hazardous area IMMEDIATELY.

Like the cylinder, the regulator has a gauge that indicates the pressure of the breathing air remaining in the cylinder. This gauge enables the user to monitor the amount of air remaining in the cylinder. The regulator pressure gauge may be mounted directly on the regulator or on a separate hose attached to a shoulder strap for easier viewing. The regulator and cylinder pressure gauges should read within 100 psi of each other.

NIOSH requires SCBA to be equipped with a low-air alarm that activates when one quarter of the air supply remains. This alarm may be a bell or whistle, a vibration, or a flashing light-emitting diode (LED). Early SCBAs (those manufactured before the 1997 edition of NFPA Standard

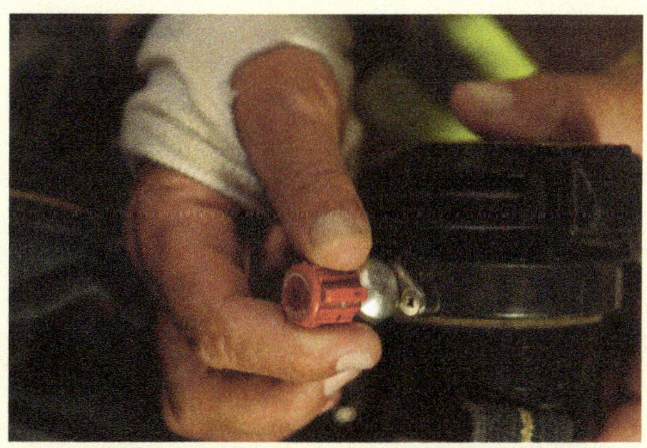

Figure 1-29 The SCBA emergency by-pass valve is used if the regulator malfunctions.

Figure 1-30 A PASS device can be integrated into an SCBA.

1981 became effective) have a single low-air supply alarm. Newer models have two different types of low-air alarms that operate independently of each other and activate different senses. For example, one alarm might ring a bell, and the second alarm might vibrate or flash an LED. Brigade members should never ignore the low-air alarm.

SCBAs meeting the current NFPA 1981 standard allows PASS devices to be stand alone or integrated. Turning on the air supply automatically activates the integrated PASS devices. This ensures that a brigade member doesn't forget to turn the PASS device on when entering a hazardous area. If your SCBA is equipped with a PASS device, learn how to activate it and how to turn it off ▲ Figure 1-30.

Although there are many different models of SCBA regulators, brigade members must learn how to operate the particular model that is used in their organization. Brigade members should be able to operate the regulator in the dark and with gloves on.

### Rapid Intervention Crew/Company Universal Air Connection

SCBAs manufactured in compliance with the 2002 edition of NFPA 1981 are equipped with a Rapid Intervention Crew/Company Universal Air Connection (RIC UAC) fitting. This fitting allows the RIC to quickly refill a low air cylinder when a brigade member has become trapped or otherwise incapacitated. The RIC UAC should not be confused with a buddy breathing system. The RIC UAC allows for the quick refilling of a low air cylinder from an air cylinder source other than another brigade member's own SCBA.

> **Brigade Member Safety Tips**
>
> Most SCBA manufacturers offer a buddy breathing system as an accessory. The NFPA does not recommend buddy breathing systems, and these systems are not part of NFPA 1981.

Figure 1-31 SCBA face pieces come in several sizes.

### Face Piece Assembly

The face piece delivers breathing air to the brigade member ▲ Figure 1-31. The face piece also protects the face from high temperatures and smoke. The face piece consists of a facemask with a clear lens, an exhalation valve, a communication system that consists of a mechanical speaking diaphragm at a minimum, and—on models with a harness-mounted regulator—a flexible low-pressure supply hose. Other models will have the regulator attached directly to the face piece.

The face piece should cover the entire face. The part in contact with the skin is made of special rubber or silicon to provide a tight seal. Exhaled air is exhausted from the face piece through the one-way exhalation valve, which has a spring mechanism to maintain positive pressure inside the face piece. Because it is hard to communicate through a face piece, some models have a voice amplification device.

Face pieces may be equipped with nose cups to help prevent fogging of the clear lens. Fogging is a greater problem in colder climates. The compressed air you breathe is dry, but the air you exhale is moist. The flow of dry air helps to prevent fogging of the lens.

The face piece is held in place with a web-like series of straps, or a net and straps. Face pieces should be stored with the straps in the extended position to make them easier to don. Pull the end of the straps toward the back of the head (not out to the sides) to tighten and ensure a snug fit.

A leak in the face piece seal can be caused by an improperly sized face piece, improper donning, or facial hair around

## Brigade Member Safety Tips

Be sure that your face piece is properly fitted and the correct size for your face.

the seal area of the face piece. A leak of any size will deplete the breathing air and reduce the amount of time available for fire fighting.

Face pieces are manufactured in several sizes. OSHA 1910.134 requires that all brigade members must have their face pieces fit-tested annually to ensure that they are wearing the proper size. Some brigades issue individual face pieces to each brigade member; others provide a selection of sizes on each apparatus. OSHA also requires that the sealing surface of the face piece must be in direct contact with the user's skin. There must not be hair, beard, or other obstruction in the seal area.

### Pathway of Air Through an SCBA

The breathing air is stored under pressure in the cylinder. The air passes through the cylinder shut-off valve into the high-pressure **air line**, or hose, that takes it to the regulator. The regulator reduces the high pressure to low pressure.

The regulator opens when the user inhales, reducing the pressure on the downstream side. In an SCBA unit with a face piece-mounted regulator, the air flows directly into the face piece. In units with a harness-mounted regulator, the air flows from the regulator through a low-pressure hose into the face piece.

When the user exhales, used air is returned to the face piece. The exhaled air is exhausted from the face piece through the exhalation valve. This cycle repeats with every breath. As the pressure in the face piece drops, the exhalation valve closes and the regulator opens.

### Skip-Breathing Technique

The skip-breathing technique helps conserve air while using an SCBA in a firefighting situation. The technique is to take a short breath, hold, take a second short breath (do not exhale in between breaths), and then relax with a long exhale. Each breath should take 5 seconds.

A simple drill can demonstrate the benefits of skip breathing. One brigade member dons PPE and an SCBA with a full air cylinder, and walks in a circle around a set of traffic cones, or, if safety permits, around the parking lot at the station. A second brigade member times how long it takes for the brigade member to completely deplete the air in the SCBA. After the first brigade member is completely rested, replace the air cylinder, and repeat the same drill using the skip-breathing technique. Compare times after completion of both evolutions.

## Brigade Member Tips

### Controlled-Breathing Technique

Controlled breathing helps extend the SCBA air supply. Controlled breathing is a conscious effort to inhale naturally through the nose and to force exhalation from the mouth. Practicing controlled breathing during training will help you to maximize the efficient use of air while you are working.

### Mounting Breathing Apparatus

SCBA should be located so that brigade members can don it quickly when they arrive at the scene of a fire. Seat-mounted brackets enable brigade members to don SCBA en route to an emergency scene, without unfastening their seat belts or otherwise endangering themselves. This enables brigade members to begin work as soon as they arrive.

There are several types of apparatus seat-mounting brackets. Some hold the SCBA with a spring clamp. Others are equipped with a mechanical hold-down device that must be released to remove the SCBA. Regardless of the mounting system used, it must hold the SCBA securely in the bracket. A collision or sudden stop should not dislodge the SCBA from the brackets. A loose SCBA can be a dangerous projectile. The brigade member who dons SCBA from a seat-mounted bracket should not tighten the shoulder straps while seated, so as not to dislodge the SCBA in a sudden stop situation. The brigade member should be secured by a seat belt or combination seat belt and shoulder harness.

Compartment-mounted SCBA units also can be donned quickly. These units are used by brigade members who arrive in apparatus without seat-mounted SCBA or whose seats were not equipped with them. The mounting brackets should be positioned high enough for easy donning. Some mounting brackets allow the brigade member to lower the SCBA without removing it from the mounting bracket. Older apparatus may have the brackets mounted on the exterior of the vehicle. An exterior-mounted SCBA should be protected from weather and dirt by a secure cover.

SCBA also may be kept in a storage case. This method is most appropriate for transporting extra SCBA units. It should not be used to transport SCBA that will be used during the initial phase of operations at a fire scene.

The SCBA should be stored on the apparatus in ready-for-use condition, with the main cylinder valve closed. After checking the SCBA, close the cylinder valve and slowly open the by-pass valve to release pressure in the system. The low-pressure alarm should sound as the pressure bleeds down. After releasing the pressure, close the by-pass valve. The SCBA is now ready to be placed in a bracket for immediate use.

## Donning Self-Contained Breathing Apparatus

Donning SCBA is an important skill. Brigade members should be able to don and activate SCBA in two minutes. Personal safety and the effectiveness of the operation depend on this skill. Brigade members must be wearing full PPE before donning SCBA. Before beginning the actual donning process, brigade members must carefully check the SCBA to ensure it is ready for operation.

- Check to be sure the air cylinder has at least 90% of its rated pressure.
- Open the cylinder valve two or three turns, listen for the low-air alarm to sound, and then open the valve fully.
- Check the pressure gauges on the regulator and on the cylinder. Both gauges should read within 100 psi of each other.
- Check all harness straps to be sure they are fully extended.
- Check all valves to be sure they are in the correct position. (An open by-pass valve will waste air.)

### Donning SCBA from an Apparatus Seat Mount

Donning SCBA while en route to an emergency can save valuable time. However, this requires that you don all of your protective clothing before mounting the apparatus. Place your arms through the shoulder straps as you sit down, and then fasten your seat belt. Or, you can fasten your seat belt first, and then slide one arm at a time through the shoulder straps of the SCBA harness. You can partially tighten the shoulder straps while you are seated.

When you arrive at the emergency scene, release your seat belt, activate the bracket release, and exit the apparatus. Face pieces should be kept in a storage bag close to each seat-mounted SCBA or attached to the harness. After exiting the apparatus, attach the waist strap, and then tighten and adjust the shoulder and waist straps.

Follow the steps in ▶ **Skill Drill 1-3** to don SCBA from a seat-mounted bracket. Before beginning this skill drill, inspect the SCBA to ensure it is ready for service.

1. Don full PPE ensemble prior to mounting the fire apparatus. Safely mount the apparatus and sit in the seat, placing arms through SCBA shoulder straps. **(Step 1)**
2. Fasten your seat belt. Partially tighten the shoulder straps. Do not fully tighten at this time. When the apparatus comes to a complete stop at the emergency scene, release your seat belt and release the SCBA from the mounting bracket. Carefully exit the apparatus. **(Step 2)**
3. Attach the waist belt and cinch down. **(Step 3)**
4. Adjust shoulder straps until they are tight. **(Step 4)**
5. Open the main cylinder valve. **(Step 5)**
6. Remove or loosen your helmet and pull back the protective hood. Don the face piece and check for leaks. Pull the protective hood up over the head, put the helmet back in place, and secure the chin strap. **(Step 6)**
7. If necessary, connect the regulator to the face piece. **(Step 7)**
8. Activate the airflow and PASS alarm. **(Step 8)**

### Donning SCBA from a Compartment Mount

To don a compartment-mounted SCBA, slide one arm through the shoulder harness strap. Slide the other arm through the other shoulder strap. Release the SCBA from the mounting bracket. Adjust the shoulder straps to carry the SCBA fairly high on your back. Attach the ends of the waist strap and tighten.

Follow the steps in **Skill Drill 1-4** to don SCBA from a side-mounted compartment or bracket. Before starting, check to be sure that the SCBA has been inspected and is ready for service. If the SCBA is mounted on an exterior bracket, remove the protective cover before beginning the donning sequence.

1. Stand in front of the SCBA bracket and fully open the main cylinder valve.
2. Turn your back toward the SCBA, slide your arms through the shoulder straps, and partially tighten the straps.
3. Release the SCBA from the bracket and step away from the apparatus.
4. Attach the waist belt and tighten.
5. Adjust the shoulder straps.
6. Remove your helmet and pull the hood back.
7. Don the face piece and check for adequate seal.
8. Pull the protective hood into position, replace your helmet, and secure the chin strap.
9. If necessary, connect the regulator to the face piece.
10. Activate the airflow and PASS alarm.

### Donning SCBA from the Ground, the Floor, or a Storage Case

Brigade members must sometimes don an SCBA that is stored in a case or on the ground. Two methods can be used: the over-the-head method and the coat method.

#### Over-the-Head Method

To don an SCBA using the over-the-head method, place the SCBA on the ground or on the floor with the cylinder valve facing away from you. Lay the shoulder straps out to each side of the backpack. Grasp the backplate with both hands and lift the SCBA over your head. Let the backpack slide down your back. The straps will slide down your arms. Balance the unit on your back. Attach and tighten the waist strap and then tighten the shoulder straps. Follow the steps in ▶ **Skill Drill 1-5** to don SCBA using the over-the-head method. Before starting, ensure that the SCBA has been inspected and is ready for service.

## 1-3 Skill Drill

# Donning SCBA from a Seat-Mounted Bracket

Don full PPE ensemble prior to mounting fire apparatus. Safely mount the apparatus and sit in the seat, placing arms through SCBA shoulder straps.

Fasten your seat belt. Partially tighten the shoulder straps. When the apparatus stops, release the seat belt and release the SCBA from its brackets. Exit apparatus.

Attach waist belt and cinch down.

Adjust shoulder straps until they are tight.

1. If necessary, open the protective case and lay out the SCBA so that the cylinder valve is away from you and the shoulder straps are to the sides. **(Step 1)**
2. Fully open the main cylinder valve. **(Step 2)**
3. Bend down and grasp the SCBA backplate with both hands. Using your legs, lift the SCBA over your head. Let the backpack slide down your back. **(Step 3)**
4. Slowly slide the pack down your back. Make sure that your arms slide into the shoulder straps. Once the SCBA is in place, tighten the shoulder straps and secure the waist strap. **(Step 4)**
5. Remove your helmet and pull the hood back. Don the face piece and check for an adequate seal. Pull your protective hood into position, replace your helmet, and secure the chin strap. **(Step 5)**
6. If necessary, connect the regulator to the face piece. Activate the airflow and PASS alarm. **(Step 6)**

### Coat Method

To don an SCBA using the coat method, place the SCBA on the ground or on the floor with the cylinder valve facing toward you. Spread out and extend the shoulder straps. Use your left hand to grasp the left shoulder strap close to the backplate. Use your right hand to grasp the right shoulder strap farther away from the backplate. Swing the SCBA over your left shoulder. Release your right arm and slide it through

## 1-3 Skill Drill (Continued)

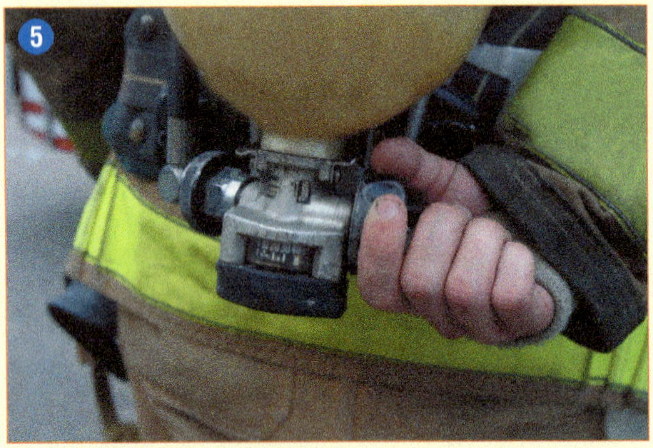


Open the main cylinder valve.

Loosen or remove helmet and pull hood back. Don face piece and check for leaks. Replace protective hood and helmet and secure chin strap.

If necessary, connect regulator to face piece.

Activate airflow and PASS alarm.

the right shoulder harness strap. Tighten both shoulder straps. Attach and tighten the waist belt.

Follow the steps in ▶ Skill Drill 1-6 to don SCBA using the coat method. Before starting, ensure that the SCBA has been inspected and is ready for service.

1. If necessary, open the protective case and lay out the SCBA so that the cylinder valve is facing you and the straps are laid out to the sides. Fully open the main cylinder valve. Place your dominant hand on the opposite shoulder strap. For safety reasons be sure to grasp the strap as close to the backplate as possible. **(Step 1)**
2. Lift the SCBA and swing it over your dominant shoulder, being careful of people or objects around you. **(Step 2)**
3. Slide your other hand between the SCBA cylinder and the corresponding shoulder strap. **(Step 3)**
4. Tighten the shoulder straps. **(Step 4)**
5. Attach the waist belt and adjust tightness. **(Step 5)**
6. Remove your helmet and pull your hood back. Don the face piece and check for an adequate seal. **(Step 6)**
7. Pull the protective hood into position, replace the helmet, and secure the chin strap. If necessary, connect the regulator to the face piece. Activate the airflow and PASS alarm. **(Step 7)**

15 Skill Drill

## Donning SCBA Using the Over-the-Head Method

1

Open the case and lay out the SCBA with the cylinder valve away from you and the shoulder straps out to the sides.

2

Fully open the main cylinder valve.

3

Bend down and grasp the SCBA backplate with both hands. Using your legs, lift the SCBA over your head.

4

Slide the SCBA down your back while your arms slide into the shoulder straps. Tighten the shoulder straps and secure the waist belt.

5

Remove your helmet and pull the hood back. Don your face piece and check for an adequate seal. Pull your protective hood into position, replace your helmet, and secure the chin strap.

6
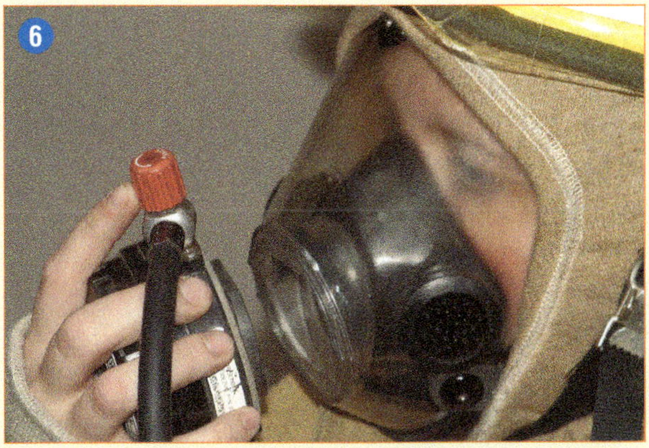

If necessary, connect the regulator to the face piece. Activate the airflow and PASS alarm.

1-6 Skill Drill

## Donning SCBA Using the Coat Method

Open the case and lay out the SCBA with the cylinder valve away from you and the shoulder straps out to the sides. Fully open the main cylinder valve. Place your dominant hand on the opposite shoulder strap.

Lift the SCBA and swing it over your dominant shoulder.

Slide your other hand between the SCBA cylinder and the corresponding shoulder strap.

Tighten the shoulder straps.

Attach the waist belt and adjust tightness.

Remove your helmet and pull your hood back. Don the face piece and check for an adequate seal.

Pull the hood into position, replace the helmet, and secure the chin strap. If necessary, connect the regulator to the face piece. Activate the airflow and PASS alarm.

These instructions will have to be modified for different SCBA units. The sequence for adjusting shoulder straps and waist belts varies with different models. Modifications must also be made for SCBAs with waist-mounted regulators. Refer to the specific manufacturer's instructions supplied with each unit. Follow the standard operating procedures for your organization.

## Donning the Face Piece

Your face piece keeps contaminated air outside and pure breathable air inside. To perform properly, it must be the correct size and it must be adjusted to fit your face. Be sure you have been tested to determine your proper size.

There must be no facial hair in the seal area. Eyeglasses that pass through the seal area cannot be worn with a face piece, because they can cause leakage between the face piece and your skin. Your face piece must match your SCBA. You cannot interchange a face piece from a different SCBA model.

Face pieces for various brands and models of SCBAs are slightly different. Some have the regulator mounted on the face piece; others have it mounted on the harness straps.

Follow the steps in ▶ **Skill Drill 1-7** to don a face piece:

1. Remove your helmet and pull the hood down over your neck.
2. Fully extend the straps on the face piece. **(Step 1)**
3. Rest your chin in the chin pocket at the bottom of the mask. **(Step 2)**
4. Fit the face piece to your face, bringing the straps or webbing over your head. **(Step 3)**
5. Tighten the lowest two straps. To tighten, pull the straps straight back, not out and away from your head. **(Step 4)**
6. Tighten the pair of straps at your temple, if any.
7. If your model has additional straps, tighten the top strap(s) last. **(Step 5)**
8. Check for a proper seal. This process depends on the model and type of face piece you use. **(Step 6)**
9. Pull the protective hood up so it covers all bare skin. Be sure it does not get under your face piece or obscure your vision.
10. Replace your helmet and secure the chin strap. **(Step 7)**
11. Install the regulator on your face piece or attach the low-pressure air supply hose to the regulator. **(Step 8)**

## Safety Precautions for Self-Contained Breathing Apparatus

As you practice using your SCBA, remember that this equipment is your protection against serious injury or death in hazardous conditions. Practice safe procedures from the beginning.

Learn to recognize the low-air alarm on your SCBA. As soon as your alarm goes off, you must exit the hazardous environment before your air supply is depleted. Never get into a situation from which you cannot escape when your low-air alarm goes off.

Before you enter a hazardous environment, make sure your PASS device is activated. Be sure you are properly logged into your accountability system. Always work in teams of two in hostile environments. Always have at least two brigade members outside at the ready whenever two brigade members are working in a hostile environment.

## Preparing for Emergency Situations

Because hostile environments are often unpredictable, brigade members must be prepared to react if an emergency situation occurs while they are using SCBA. In emergencies, follow simple guidelines. First, keep calm, stop, and think. Panic increases air consumption. Try to control your breathing by maintaining a steady rate of respirations. A calm person has a greater chance of surviving an emergency.

If the problem is with your SCBA, try to exit the hostile environment. Use the emergency by-pass valve so you can breathe if your regulator malfunctions.

If you are in danger, activate your PASS device. Use your hand light to attract attention. If you have a portable radio, call for help. These are simple but effective steps; additional emergency techniques are covered in Chapter 18, Brigade Member Survival.

## Doffing Self-Contained Breathing Apparatus

The procedure for doffing your SCBA depends on the model and whether it has a face piece-mounted regulator or a harness-mounted regulator. Follow the procedures recommended by the manufacturer and your brigade's SOPs.

### Brigade Member Tips

#### Restricted Spaces

The size and shape of SCBA may make it difficult for you to fit through tight openings. Several techniques may help you navigate these spaces.

- Change your body position: Rotate your body 45° and try again.
- Loosen one shoulder strap and change the location of the SCBA on your back.
- If you have no other choice, you may have to remove your SCBA. In this case, do not let go of the backpack and harness for any reason. Keep the unit in front of you as you navigate through the tight space. Re-attach the harness as soon as you are through the restricted space. This is an absolutely "last resort" procedure!

In general, you should reverse the steps used to don your SCBA. Follow the steps in ▶ Skill Drill 1-8 to doff your SCBA:

1. Remove your gloves. Remove the regulator from your face piece or disconnect the low-pressure air supply hose from the regulator. **(Step 1)**
2. Shut off the air-supply valve.
3. Remove your helmet and pull your protective hood down around your neck. **(Step 2)**
4. Loosen the straps on your face piece. **(Step 3)**
5. Remove your face piece. **(Step 4)**
6. Release your waist belt. **(Step 5)**
7. Loosen the shoulder straps and remove the SCBA. **(Step 6)**
8. Shut off the air-cylinder valve. **(Step 7)**
9. Bleed the air pressure from the regulator by opening the emergency by-pass valve. **(Step 8)**
10. If you have an integrated PASS device, turn it off.
11. Place the SCBA in a safe location where it will not get dirty or damaged. **(Step 9)**

1-7 Skill Drill

## Donning a Face Piece

1 Fully extend the straps on the face piece.

2 Place your chin in the chin pocket.

3 Fit the face piece to your face, bringing the straps or webbing over your head.

4 Tighten the lowest two straps.

## Putting It All Together: Donning the Entire PPE Ensemble

The complete PPE ensemble consists of both personal protective clothing and respiratory protection (SCBA). Although donning personal protective clothing and donning and operating SCBA can be learned and practiced separately, you must be able to integrate these skills to have a complete PPE ensemble. Each part of the complete ensemble must be in the proper place to provide whole-body protection.

The steps for donning a complete PPE ensemble are listed below:

- Place the protective hood over your head.
- Put on your bunker pants and boots. Adjust the suspenders and secure the front flap of the pants.
- Put on your turnout coat and secure the front.
- Open the air-cylinder valve on your SCBA and check the air pressure.
- Put on your SCBA.
- Tighten both shoulder straps.

### 1-7 Skill Drill (Continued)

5 If there are more straps, tighten the top straps last.

6 Check for a proper seal.

7 Pull your protective hood up so it covers all bare skin. Don your helmet and secure the chin strap.

8 Install the regulator on your face piece or attach the low-pressure air supply hose to the regulator.

Skill Drill 1-8

# Doffing SCBA

Remove your gloves. Remove the regulator from the face piece or disconnect the low-pressure hose from the regulator.

Remove your helmet and pull the protective hood down around your neck.

Loosen the face piece straps.

Remove your face piece.

Release your waist belt.

Loosen the shoulder straps and remove the SCBA.

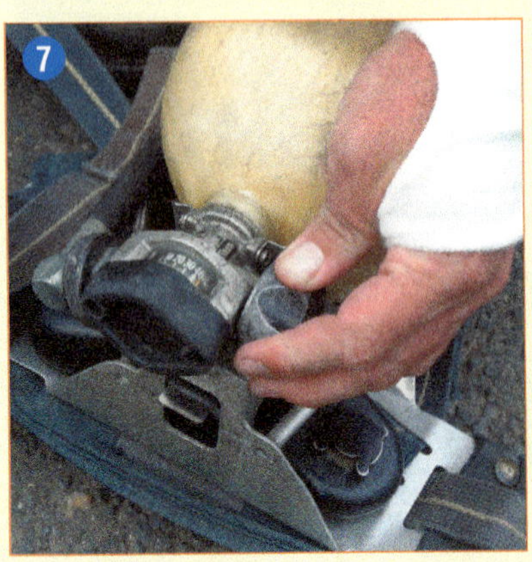


Shut off the air-cylinder valve.

Bleed the air pressure from the regulator.

Place the SCBA in a safe location.

## Brigade Member Safety Tips

Any SCBA air cylinder showing less than 90 percent of the "full" pressure should be replaced.

- Attach the waist belt and tighten it.
- Fit the face piece to your face.
- Tighten the straps, beginning with the lowest straps.
- Check the face piece for a proper seal.
- Pull the protective hood up so that it covers all bare skin, but does not obscure vision.
- Place your helmet on your head with the ear tabs extended and secure the chin strap.
- Turn up your coat collar and secure it in front.
- Put on your gloves.
- Check your clothing to be sure it is properly secured.
- Be sure your PASS device is turned on.
- Attach your regulator or turn it on to start the flow of breathing air.
- Work safely!

## SCBA Inspection and Maintenance

SCBA must be properly serviced and prepared for the next use each time it is used, whether it is an actual emergency incident or a training exercise. The air cylinder must be changed or refilled, the face piece and regulator must be sanitized according to the manufacturer's instructions, and the unit must be cleaned, inspected, and checked for proper operation. It is the user's responsibility to ensure that the SCBA is in ready condition before it is returned to the apparatus.

Each SCBA must be checked on a regular basis to ensure that it is ready for use. There are different procedures for daily, monthly, and annual inspections. The daily inspection procedure should be used when restoring a unit to service after it has been used.

If an SCBA inspection reveals any problems that cannot be remedied by routine maintenance, the SCBA must be removed from service for repair. Only properly trained and qualified personnel are authorized to repair SCBA.

### Daily Inspection

Each SCBA unit should be inspected daily or at the beginning of each shift. When stations are not staffed, SCBA should be inspected at least once a week. Follow the steps in ▶ **Skill Drill 1-9** for daily SCBA inspection:

1. Check the backpack and harness straps. Make sure these components are intact and the straps are kept lengthened. **(Step 1)**
2. Check the air-cylinder pressure. Make sure it is full. Turn on the air-cylinder valve. Check the regulator gauge or remote gauge. It should read within 100 psi of the cylinder gauge. **(Step 2)**
3. Check the condition of all hoses while they are pressurized. **(Step 3)**
4. Activate the PASS device. **(Step 4)**
5. Check the face piece. It should be clean and undamaged. Check the operation of the exhalation valve. **(Step 5)**
6. Connect the regulator to the face piece and take one or two test breaths. **(Step 6)**
7. Close the cylinder valve and open the emergency by-pass valve to bleed the pressure. **(Step 7)**
8. Check to ensure that the low-pressure alarm(s) activate at the proper pressure. Close the emergency by-pass valve and restore the SCBA to ready condition. **(Step 8)**

### Monthly Inspection

SCBA should be completely checked each month for proper operation, for leaks, and for any deterioration. Follow the steps in **Skill Drill 1-10** for monthly SCBA inspection:

1. Remove the SCBA from the apparatus and place it on the floor or on a workbench.
2. Inspect the mounting bracket for damage or wear. Lubricate if this is recommended by the manufacturer.
3. Look at the overall condition of the SCBA and note any damage.
4. Remove the air cylinder from the SCBA harness and check the hydrostatic test date.
5. Check the air cylinder for damage or wear.
6. Inspect the SCBA shoulder straps and waist belt for damage, cuts, burns, or wear.
7. Check all buckles and fasteners to ensure they work properly.
8. Examine the backplate for damage, cracks, or rust.
9. Make sure all connection points between the cylinder and the SCBA harness operate properly and are free of damage or corrosion. Lubricate them if this is recommended by the manufacturer.
10. Reattach the air cylinder to the SCBA harness.
11. Check all hoses and connection points for wear, cuts, or damage.
12. Open the air-cylinder valve. Compare readings on the cylinder and regulator gauges to ensure they match.
13. Attach the face piece and check the regulator for proper operation.
14. Activate the PASS alarm. Allow the SCBA to sit idle until the PASS alarm sounds.
15. Shut off the air-cylinder valve and open the by-pass valve to bleed the pressure. Check the low-pressure alarm as the pressure bleeds down.
16. Return the SCBA to the mounting bracket.
17. Complete all necessary paperwork.

### Annual Inspection

A complete annual inspection and maintenance must be performed on each SCBA. The annual inspection must be performed by a certified manufacturer's representative or a person who has been trained and certified to perform this work. SCBA requires regular inspection and maintenance to ensure that it will perform as intended.

## Servicing SCBA Cylinders

A pressurized SCBA cylinder contains a tremendous amount of potential energy. Not only does the air within the cylinder exert considerable pressure on its walls, but the cylinder is used under extreme conditions on the fireground. If the cylinder ruptures and suddenly releases this energy, it can cause serious injury or death. Cylinders must be regularly inspected and tested to ensure they are safe. Cylinders must be visually inspected during daily and monthly inspections. More detailed inspection is required if a cylinder has been exposed to excessive heat, come into contact with flame, exposed to chemicals, or dropped.

The U.S. Department of Transportation requires hydrostatic testing for SCBA cylinders on a periodic basis and

1-9 Skill Drill

## Daily SCBA Inspection

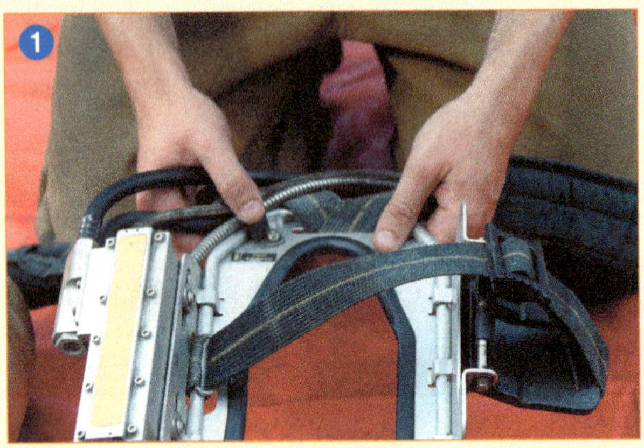

1 Check backpack and harness straps.

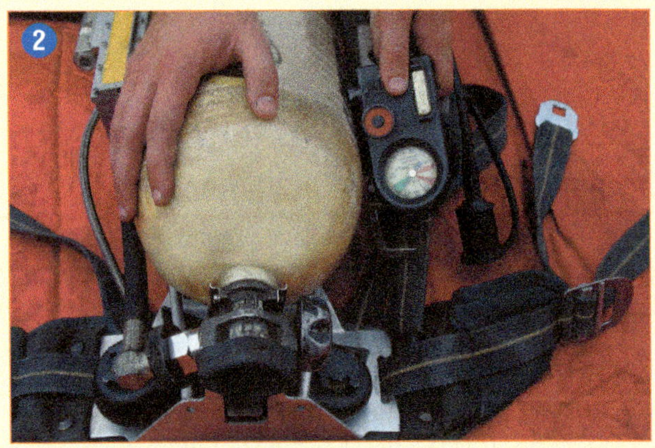

2 Check air-cylinder pressure. Turn on the air-cylinder valve and check gauge pressure.

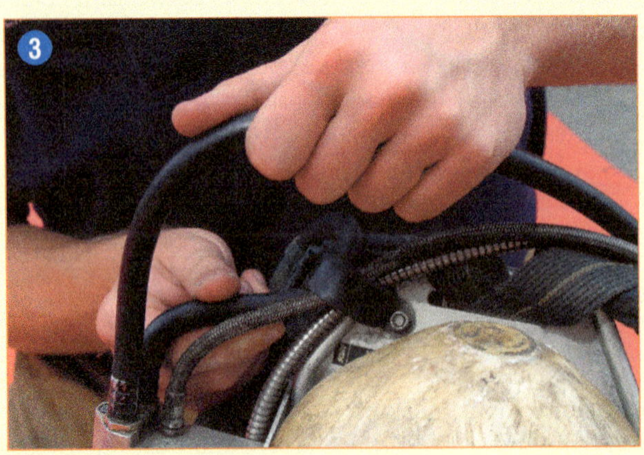

3 Check condition of all hoses while pressurized.

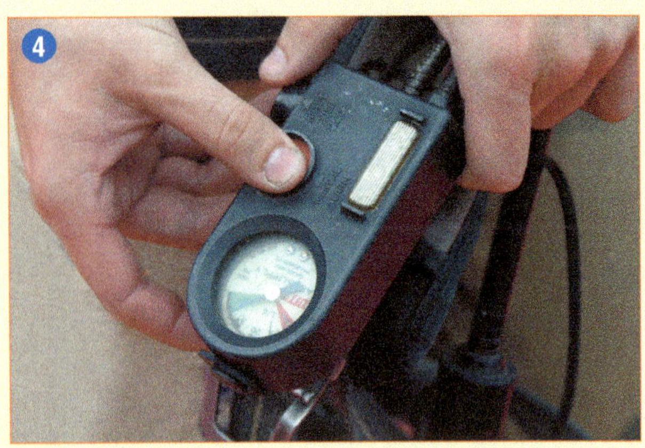

4 Activate PASS device, if present.

limits the number of years that a cylinder can be used. Hydrostatic testing identifies defects or damage that render the cylinder unsafe. Any cylinder that fails a hydrostatic test is immediately taken out of service and destroyed.

Cylinders constructed of different materials have different testing requirements. Aluminum, steel, and carbon-fiber cylinders must be hydrostatically tested every five years. Cylinders constructed of composite materials such as Kevlar-aramid or fiberglass fibers must be tested every three years. Brigade members must know what type of cylinders are used by their brigade and must check each cylinder for a current hydrostatic test date before filling.

## Replacing SCBA Cylinders

An expended air cylinder can be quickly replaced with a full cylinder in the field to enable you to continue fire-fighting activities. A single brigade member must doff SCBA to replace the air cylinder; two brigade members working together can change cylinders without removing the SCBA. The steps listed below outline how a single person makes a cylinder change. These procedures will

### 1-9 Skill Drill (Continued)

5 Check the face piece.

6 Connect the regulator to the face piece and take test breaths.

7 Close the cylinder valve and open the emergency by-pass valve to bleed the pressure.

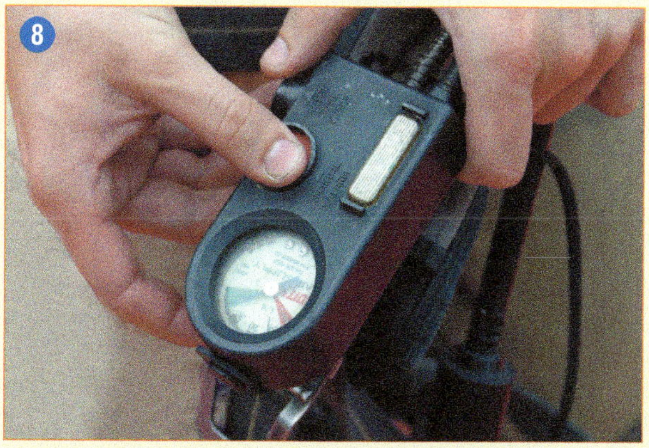

8 Check function and activation pressure of low-air alarm. Close by-pass valve and restore unit to ready condition.

change depending on the model of SCBA being used. Follow the procedures recommended by the manufacturer and by brigade SOPs.

Practice changing air cylinders until you become proficient. A brigade member should be able to change cylinders in the dark and while wearing gloves. Follow the steps in ▶ Skill Drill 1-11 to replace an SCBA cylinder:

1. Place the SCBA on the floor or a bench. **(Step 1)**
2. Turn off the cylinder valve. **(Step 2)**
3. Bleed off the pressure by opening the by-pass valve. **(Step 3)**
4. Disconnect the high-pressure supply hose. Keep the ends clean. **(Step 4)**
5. Release the cylinder from the backpack. **(Step 5)**
6. Slide a full cylinder into the backpack. Align the outlet to connect the supply hose. Lock the cylinder in place. **(Step 6)**
7. Check that the "O" ring is present and in good condition. **(Step 7)**
8. Connect the high-pressure hose to the cylinder. Hand tighten only. **(Step 8)**
9. Open the cylinder valve. Check the regulator gauge or remote gauge. It should read within 100 psi of the cylinder gauge. **(Step 9)**

To save time, someone else can replace the air cylinder while you are wearing the SCBA harness. You should not overtax yourself, however, by replacing the cylinder and going back to work without adequate rest when you need it.

## Refilling SCBA Cylinders

Compressors and cascade systems are used to refill SCBA cylinders. A compressor or a cascade system can be permanently located at a maintenance facility or at a station or they can be mounted on a truck or a trailer for mobile use. Mobile filling units are often brought to the scene of a large fire.

Compressor systems filter atmospheric air, compress it to a high pressure, and transfer it to the SCBA cylinders ▶ Figure 1-32. Cascade systems have several large storage cylinders of compressed breathing air connected by a high-pressure manifold system. The empty SCBA cylinder is connected to the cascade system, and compressed air is transferred from the storage tanks to the cylinder. The storage cylinder valves must be opened and closed, one at a time, to fill the SCBA cylinder to the recommended pressure.

Proper training is required to fill SCBA cylinders. Whether your brigade has an air compressor or a cascade system, only those brigade members who have been trained on the equipment should use it to refill air cylinders.

## Cleaning and Sanitizing SCBA

Most SCBA manufacturers will provide specific instructions for the care and cleaning of their models. The first step in cleaning the SCBA is to rinse the entire unit using a hose with clean water. The harness assembly and cylinder can

**Figure 1-32** Air compressors are installed at many fire stations to refill SCBA cylinders.

### Brigade Member Safety Tips

Refilling SCBA cylinders requires special precautions because of the high pressures that are involved. The SCBA cylinder must be in a shielded container while it is being refilled ▼ Figure 1-33. The container is designed to prevent injury if the cylinder ruptures. The hydrostatic test date must be checked before the cylinder is refilled to ensure that it has not expired. Special procedures must be followed to ensure that the air used to fill the SCBA cylinder is not contaminated.

**Figure 1-33** SCBA cylinders are refilled in a protective enclosure.

## Skill Drill 1-11

# Replacing an SCBA Cylinder

Place the SCBA on the floor or a bench.

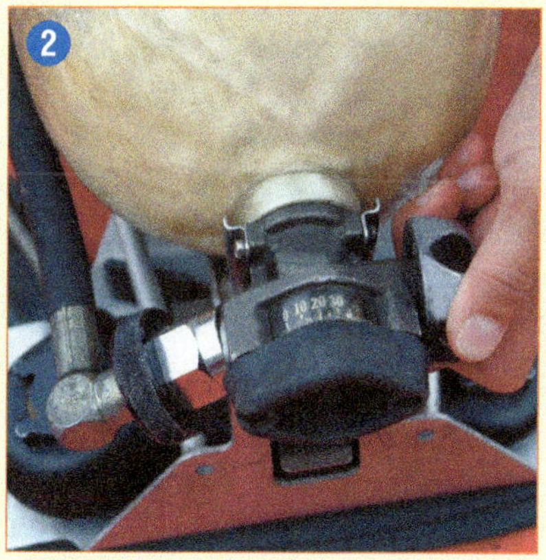


Turn off the cylinder valve.

Open the by-pass valve to bleed off pressure.

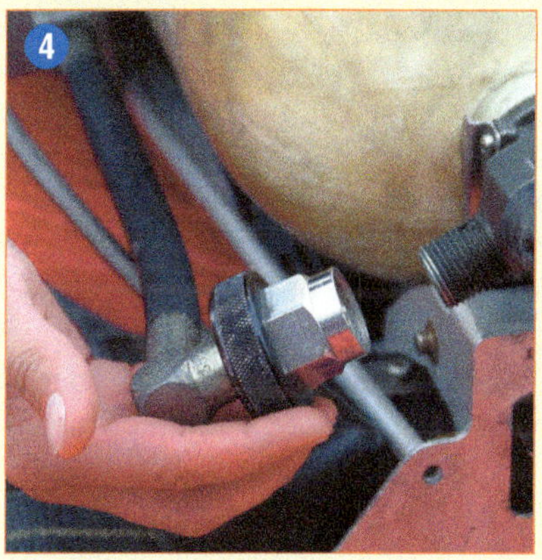


Disconnect the high-pressure supply hose.

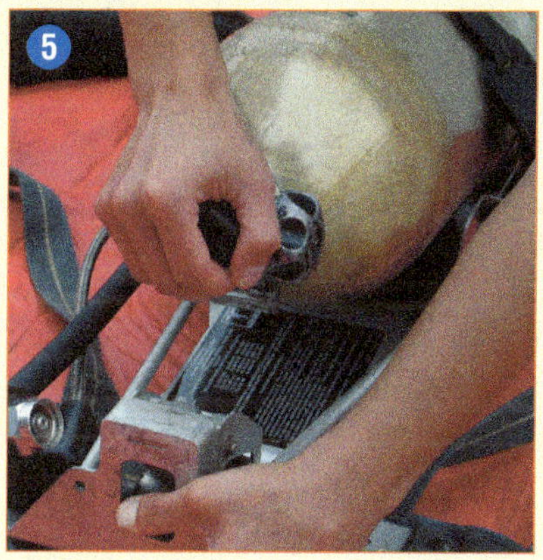


Release the cylinder from the backpack.

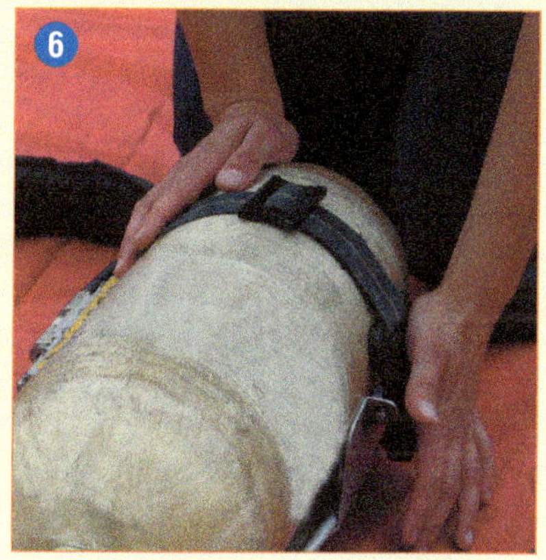


Slide a full cylinder into the backpack. Align the outlet to the supply hose. Lock the cylinder in place.

Check that the "O" ring is present and in good condition.

Connect the high-pressure hose to the air cylinder.

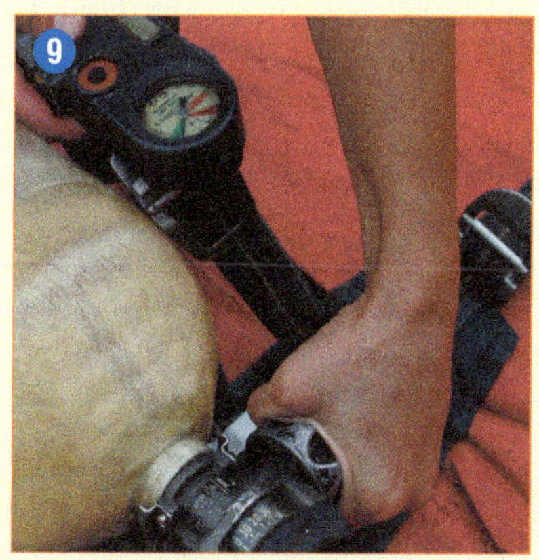


Open the cylinder valve. Check gauge reading.

be cleaned with a mild soap and water solution. If additional cleaning is needed, the unit can be scrubbed with a stiff brush. After scrubbing, the SCBA harness and cylinder should be rinsed with clean water.

After a fire, face pieces and regulators can be cleaned with a mild soap and warm water or a disinfectant cleaning solution. The face piece should be fully submerged in the cleaning solution. If additional cleaning is needed, a soft brush can be used to scrub the face piece. Avoid scratching the lens or damaging the exhalation valve. The regulator can be cleaned with the same solution but should not be submerged. The face piece and regulator should be rinsed with clean water.

Allow the SCBA time to dry completely before returning it to service. Check for any damage before returning the SCBA to service. Follow the steps in **Skill Drill 1-12** to clean and sanitize an SCBA:

1. Inspect the SCBA for any damage that may have occurred before cleaning.
2. Remove the face piece from the regulator. On some models, the regulator also can be removed from the harness.
3. Detach the SCBA cylinder from the harness.
4. Rinse all parts of the SCBA with clean water. Water from a garden hose can be used for this step.
5. Using a stiff brush, and a soap-and-water solution, scrub the SCBA cylinder and harness. Rinse and set aside to dry.
6. In a 5-gallon bucket make a mixture of mild soap and water or use the manufacturer's recommended cleaning and disinfecting solution and water.
7. Submerge the SCBA face piece into the cleaning solution. For heavier cleaning, allow the face piece to soak.
8. Clean the regulator with cleaning solution, following the manufacturer's instructions.
9. Use a soft brush, if necessary, to remove contaminants from the face piece and regulator.
10. Completely rinse the face piece and the regulator with clean water. Set them aside and allow them to dry.
11. Reassemble and inspect the entire SCBA before placing it back in service.

## Reduce Your Exposure to Contaminants

In addition, you must keep PPE clean to reduce exposure to carcinogens and other harmful, substances and to maintain its protective properties. Dirt will build up in the clothing fibers from routine use and exposure to fire environments. Smoke particles also will become embedded in the outer shell and will continue to off-gas if not properly cleaned. The interior layers will frequently be soaked with perspiration. Allowing fire by-products to stay embedded on PPE will break down the protective components of the gear, causing the potential for the material to fail.

Other contaminants are formed from the by-products of burnt plastics and synthetic products. These residues, which are flammable, can become trapped between the fibers or build up on the outside of PPE, damaging the materials and reducing their protective qualities. A fire fighter who is wearing contaminated PPE is, in essence, bringing additional fuel into the fire on the clothing.

PPE that has been badly soiled by exposure to smoke, other products of combustion, melted tar, petroleum products, or other contaminants needs to be cleaned as soon as possible. Regular and routine cleaning should remove most of these contaminants. Items that have been exposed to chemicals or hazardous materials may have to cleaned by a professional cleaning company or be impounded for decontamination or disposal.

Cleaning instructions are listed on the label or tag attached to the clothing. Always follow the manufacturer's cleaning instructions. Failure to do so may reduce the effectiveness of the garment and create an unsafe situation for the wearer.

Fire departments should have special washing machines that are approved for cleaning protective clothing. If they do not have approved cleaning devices, they can contract with an outside firm to clean and repair protective clothing. In either case, the manufacturer's instructions for cleaning and maintaining the garment must be followed. If you wash your protective clothing in house, follow the manufacturer's instructions for drying these garments properly. Make sure your PPE is dry before using it on the fire ground. If wet protective clothing is exposed to the high temperatures of a structural fire, the water trapped in the liner materials will turn into steam and be trapped inside the moisture barrier. The result may be painful steam burns. Many fire departments issue two sets of gear to each fire fighter so that they always have a set of clean, dry gear available while one set is being cleaned.

Other PPE also may require regular cleaning and maintenance. For example, the outer shell of your helmet should be cleaned with a mild soap as recommended by the manufacturer. The inner parts of helmets should be removed and cleaned according to the manufacturer's instructions. The chin strap and suspension system must be properly adjusted and all parts of the helmet kept in good repair.

Protective hoods and gloves get dirty quickly and should be cleaned according to the manufacturer's instructions. Most protective hoods can be washed with appropriate soaps or detergents. Dirty protective hoods may contain large quantities of carcinogens following a fire. Repair or discard gloves or protective hoods that have holes in them; do not use them. Even a small cut or opening in these items can result in a burn injury. Boots should be maintained according to the manufacturer's instructions. Rubber boots need to be kept in a place that does not result in damage to the boot.

Leather boots must be properly maintained to keep them supple and in good repair. Any type of boot needs to be repaired or replaced if the outer shell is damaged. Do not store these items or any other PPE in direct sunlight. All PPE that has been exposed to hazardous conditions should be cleaned and removed as soon as possible following the fire (▼ Figure 1-34). Perform a field reduction of contaminants to remove as much dirt and debris as possible. Remove all surface contaminants. Doff the PPE at the scene. Thoroughly clean the PPE as soon as you return to the station. As discussed, correctly storing your PPE is another important step in the proper care and maintenance of your PPE. PPE should be stored only in the proper PPE storage areas.

**Figure 1-34** Perform a field reduction of contaminants to remove dirt and debris before returning to the station.



## Brigade Member Safety Tips

Some recent studies have shown that protective hoods can transfer large amounts of carcinogens to your skin. When doffing PPE, *do not* leave a contaminated protective hood or SCBA mask around your neck and below your nose and mouth. You will absorb or inhale additional toxins. To remove as much dirt from your face and neck as possible, use wet wipes as soon as possible in the field. Follow this by taking a shower as soon as possible. It is also important to thoroughly clean your protective hood after each use. Follow your fire department's policies in regard to cleaning contaminated PPE. Your fire department may require that the PPE be bagged in a biohazard bag and professionally cleaned. Never clean your PPE or work uniform in home washers to prevent cross-contamination with your home laundry.

# Wrap-Up

## Ready for Review

- Industrial firefighting performance and knowledge requirements are divided into four major categories: incipient firefighting, advanced exterior firefighting, interior structural firefighting, and fire brigade leader.
- Qualifications for becoming a brigade member include age requirements, medical requirements, physical fitness requirements, and emergency medical care requirements.
- Good safety practices must be followed during training, during response, at emergency incidents, at the station, and outside the workplace.
- The PPE ensemble for structural firefighting consists of a helmet, a protective hood, a turnout coat, bunker pants, boots, gloves, SCBA, and a PASS device.
- Respiratory hazards from fires include smoke particles and vapors, toxic gases, an oxygen-deficient environment, and high temperatures.
- Understanding the uses and limitations of SCBA is essential for your safety at fire scenes.

## Hot Terms

**Air cylinder** The component of the SCBA that stores the compressed air supply.

**Air line** The hose through which air flows, either within an SCBA or from an outside source to a supplied air respirator.

**Backpack** The harness of the SCBA that supports the components worn by a brigade member.

**Brigade member** The brigade member may be assigned any task, from placing hose lines to extinguishing fires. Generally, the brigade member is not responsible for any command functions and does not supervise other personnel, except on a temporary basis when promoted to an acting officer.

**Buddy system** A system in which two brigade members always work as a team for safety purposes.

**Bunker coat** The protective coat worn by a brigade member for interior structural firefighting; also called a turnout coat.

**Bunker pants** The protective trousers worn by a brigade member for interior structural firefighting; also called turnout pants.

**Carbon monoxide (CO)** A toxic gas produced through incomplete combustion.

**Cascade system** An apparatus consisting of multiple tanks used to store compressed air and fill SCBA cylinders.

**Closed-circuit breathing apparatus** SCBA designed to recycle the user's exhaled air. The system removes carbon dioxide and generates fresh oxygen.

**Company officer** The company officer is responsible for the initial firefighting strategy, personnel safety, and the overall activities of the brigade members in their company.

**Compressor** A mechanical device that increases the pressure and decreases the volume of atmospheric air; used to refill SCBA cylinders.

**Critical incident stress debriefing (CISD)** A confidential group discussion among those who served at a traumatic incident to address emotional, psychological, and stressful issues; usually occurs within 24 to 72 hours of the incident.

**Doff** To take off an item of clothing or equipment.

**Don** To put on an item of clothing or equipment.

**Driver/operator** Often called an engineer or technician, the driver is responsible for getting the fire apparatus to the scene safely, setting up, and running the pump or operating the aerial ladder once it arrives on the scene.

**Emergency by-pass mode** Operating mode that allows an SCBA to be used even if part of the regulator fails to function properly.

**Emergency Medical Services (EMS) personnel** EMS personnel administer prehospital care to people who are sick or injured.

**Employee assistance program (EAP)** Program adopted by many organizations for brigade members to receive confidential help with problems such as substance abuse, stress, depression, or burnout that can affect their work performance.

**Face piece** Component of SCBA that fits over the face.

**Fire helmet** Protective head covering worn by brigade members to protect the head from falling objects, blunt trauma, and heat.

**Fire protection engineer** The fire protection engineer usually has an engineering degree, reviews plans, and works with building owners to ensure that their fire suppression and detection systems will meet the relevant codes and function as needed.

**Freelancing** Dangerous practice of acting independently of command instructions.

**Hand light** Small, portable light carried by brigade members to improve visibility at emergency scenes, often powered by rechargeable batteries.

**Hazardous materials technician** "Hazmat" technicians have training and certification in chemical identification, leak control, decontamination, and clean-up procedures.

**Hydrogen cyanide** Toxic gas produced by combustion of plastics and synthetics.

**Hydrostatic testing** Periodic certification test performed on pressure vessels, including SCBA cylinders.

**Immediately dangerous to life and health (IDLH)** An atmospheric concentration of any toxic, corrosive, or asphyxiant substance that poses an immediate threat to life or could cause irreversible or delayed adverse health effects. There are three general IDLH atmospheres: toxic, flammable, and oxygen-deficient.

**Incident commander** The person in charge of the incident site who is responsible for all decisions relating to the management of the incident.

**Incident management system (IMS)** The combination of facilities, equipment, personnel, procedures, and communications under a standard organizational structure to manage assigned resources effectively to accomplish stated objectives for an incident. Also known as Incident Command System (ICS).

**Incomplete combustion** A burning process in which the fuel is not completely consumed, usually due to a limited supply of oxygen.

**Kevlar®** Strong, synthetic material used in the construction of protective clothing and equipment.

**Light-emitting diode (LED)** An electronic semiconductor that emits a single-color light when activated.

**National Institute for Occupational Safety and Health (NIOSH)** A U.S. Federal agency responsible for research and development of occupational safety and health issues.

**Nomex®** A fire-resistant synthetic material used in the construction of personal protective equipment for fire fighting.

**Nose cups** An insert inside the face piece of an SCBA that fits over the user's mouth and nose.

**Occupational Safety and Health Administration (OSHA)** The federal agency that regulates worker safety and, in some cases, responder safety. OSHA is part of the Department of Labor.

**Open-circuit breathing apparatus** SCBA in which the exhaled air is released into the atmosphere and is not reused.

**Oxygen deficiency** Any atmosphere where the oxygen level is below 19.5%. Low oxygen levels can have serious effects on people, including adverse reactions such as poor judgment and lack of muscle control.

**PBI®** A fire-retardant synthetic material used in the construction of personal protective equipment.

**Personal alert safety system (PASS)** Device worn by a brigade member that sounds an alarm if the brigade member is motionless for a period of time.

**Personal protective equipment (PPE)** Gear worn by brigade members that includes helmet, gloves, hood, coat, pants, SCBA, and boots. The personal protective equipment provides a thermal barrier for brigade members against intense heat.

**Personnel accountability system** A method of tracking the identity, assignment, and location of brigade members operating at an incident scene.

**Phosgene** A chemical agent that causes severe pulmonary damage.

**Pounds per square inch (psi)** Standard unit used in measuring pressure.

**Pressure gauge** A device that measures and displays pressure readings. In an SCBA, the pressure gauges indicate the quantity of breathing air that is available at any time.

# Wrap-Up

**Protective hood** A part of a brigade member's PPE designed to be worn over the head and under the helmet to provide thermal protection for the neck and ears.

**Public information officer** The public information officer serves as a liaison between the incident commander and the news media.

**Rapid intervention company/crew (RIC)** A minimum of two fully equipped personnel on site, in a ready state, for immediate rescue of injured or trapped brigade members. In some organizations, this is also known as Rapid Intervention Team.

**Rehabilitation** A systematic process to provide periods of rest and recovery for emergency workers during an incident; usually conducted in a designated area away from the hazardous area.

**Respirator** A protective device used to provide safe breathing air to a user in a hostile or dangerous atmosphere.

**Safety officer** The position within IMS responsible for identifying and evaluating hazardous or unsafe conditions at the scene of the incident. Safety officers have the authority to stop any activity deemed unsafe.

**SCBA harness** The straps and fasteners used to attach the SCBA to the brigade member.

**SCBA regulators** Part of the SCBA that reduces the high pressure in the cylinder to a usable lower pressure and controls the flow of air to the user.

**Self-contained breathing apparatus (SCBA)** Respirator with independent air supply used by brigade members to enter toxic and otherwise dangerous atmospheres.

**Self-contained underwater breathing apparatus (SCUBA)** Respirator with independent air supply used by underwater divers.

**Smoke particles** Airborne solid material consisting of ash and unburned or partially burned fuel released by a fire.

**Standard operating procedures (SOPs)** Written rules, policies, regulations, and procedures enforced to structure the normal operations of most fire brigades.

**Supplied-air respirator (SAR)** A respirator that gets its air through a hose from a remote source, such as a compressor or compressed air cylinder.

**Technical rescue technician** A "tech rescue" technician is trained in special rescue techniques for incidents involving structural collapse, trench rescue, vehicle/machinery rescue, confined-space rescue, high-angle rescue, and other unusual situations.

**Training officer** The training officer is responsible for updating the training of current brigade members and for training new members.

**Turnout coat** Protective coat that is part of a protective clothing ensemble for structural firefighting; also called a bunker coat.

**Turnout pants** Protective trousers that are part of a protective clothing ensemble for structural firefighting; also called bunker pants.

**Two-way radio** A portable communication device used by brigade members. Every firefighting team should carry at least one radio to communicate distress, progress, changes in fire conditions, and other pertinent information.

# Brigade Member in Action

**Your fire brigade is dispatched via the plant radio system to respond to a report of hydrocarbon release and fire in a processing unit on the other side of the plant from your work area. As you walk out to your vehicle, you hear the plant alarm system sounding and see visible smoke in the sky above the involved process unit. Your arrival on the scene is followed first by other brigade members' arrival, and then by the first-due foam engines and hazmat trucks.**

**As you exit your vehicle, the on-shift fire brigade leader is walking away from the process area. He instructs you and another brigade member to "gear up" and place fixed master stream appliances on the unit perimeter in service to protect several exposures in the general fire area. The fire brigade leader quickly briefs your team on the assignment. It will require you to work in an area that is located cross-wind of the actual fire area. You will be required to activate three to four fixed monitors as well as to perform reconnaissance of the area for potential hazards.**

1. Based on the assignment your team has been given, what level of personal protective equipment would you don?
   - **A.** No specialized personal protective equipment required.
   - **B.** Normal work clothing with a self-contained breathing apparatus.
   - **C.** Full bunker gear with no respiratory protection.
   - **D.** Full bunker gear with a self-contained breathing apparatus.

2. Which of the following actions would you not perform while donning full bunker gear with a self-contained breathing apparatus?
   - **A.** Don your protective hood and then don your SCBA mask over the top of the hood.
   - **B.** Be sure your PASS device is turned on.
   - **C.** Turn up your coat collar and secure it in front.
   - **D.** Tighten your SCBA face piece, starting at the lowest straps.

3. Prior to entry into a fire site, you should always check your SCBA by-pass valve. What is the primary purpose of the by-pass valve?
   - **A.** It serves as the primary pressure relief valve for the SCBA system.
   - **B.** It provides emergency air in the event the primary regulator fails.
   - **C.** It provides a constant flow of cool air to make the SCBA easier to wear.
   - **D.** It serves as a secondary means to shut off air flow from the SCBA.

4. Prior to making entry into the process unit area, which of the following PPE inspections should you conduct jointly with your partner?
   - **A.** Check the level of air available on your SCBA.
   - **B.** Ensure no skin is visible or protruding from your PPE ensemble.
   - **C.** Ensure your PASS alarm is activated.
   - **D.** All of the above.

5. After exiting the area, which of the following actions should you not take as you doff your SCBA?
   - **A.** Remove the regulator from the SCBA or disconnect the low-pressure hose from the regulator.
   - **B.** Shut off the air cylinder valve.
   - **C.** Do not bleed off the air from the regulator.
   - **D.** Place the SCBA is a safe location.

# Standard Operating Procedures and Guidelines

# Chapter 2

## NFPA 1081 Standard

### Incipient Facility Fire Brigade Member

NFPA 1081 contains no Incipient Facility job performance requirements for this chapter.

### Advanced Exterior Facility Fire Brigade Member

NFPA 1081 contains no Advanced Exterior Facility job performance requirements for this chapter.

### Interior Structural Facility Fire Brigade Member

NFPA 1081 contains no Interior Structural Facility job performance requirements for this chapter.

## Knowledge Objectives

After completing this chapter, you will be able to:

- Define a standard operating procedure (SOP) and a standard operating guideline (SOG).
- Differentiate between an SOP and an SOG.
- Explain the procedure for developing an SOP.
- Describe the standard content of an SOP.
- Explain the need for evaluation and revision of SOPs.
- Explain how an SOP is used at a post-incident analysis.

## Skills Objectives

After completing this chapter, you will be able to perform the following skills:

- Develop an SOP for either generic or site-specific application.
- Use an SOP in the performance of either routine or emergency duties.

# You Are the Brigade Member

**You are new to the plant's facility fire brigade and have completed live fire training at a local fire training facility. As you report for work on your first day after training, many thoughts go through your head. You wonder how you would react if there were a fire during your shift. Further self-questioning asks, "How will I know what to do at the scene and who to report to? How will I even know when an incident has taken place?" You soon realize that there is nothing to worry about. You remember that the fire brigade has a published and readily available series of standard operating procedures (SOPs) for all members to follow. These SOPs guide all aspects of the brigade's administration, training, and operations.**

1. ***Why is it important to have a clearly written series of SOPs in place for facility fire brigades?***
2. ***How does an SOP and the incident action plan (IAP) relate to one another?***
3. ***How does your brigade institutionalize SOPs?***

## Introduction

There is an old saying that states "If you don't know where you are going, any road will take you there." This saying can apply to fire brigade operations when there are no standard operating procedures/guidelines (SOPs/SOGs) in place. Without procedures or guidelines, a brigade has no road map, and efficiency is reduced.

## What Are SOPs/SOGs?

SOPs/SOGs are a set of organizational directives that establish a course of action both administratively and operationally (▶ **Figure 2-1**). NFPA 1081, *Standard for Facility Fire Brigade Member Professional Qualifications*, states that SOPs are "[a] written procedure that establishes a standard course of action and documents the functional limitations of the fire brigade members in performing emergency operations." NFPA 600, *Standard on Facility Fire Brigades*, further states that an SOP is "a written organizational directive that establishes or prescribes specific operational or administrative methods to be followed routinely for the performance of designated operations or actions." The Occupational Safety and Health Administration (OSHA) standard 29 CFR 1910.156, *Fire Brigades*, when addressing hazardous materials, requires that "The employer shall develop and make available for inspection by fire brigade members, written procedures that describe the actions to be taken in situations involving the special hazards and shall include these in the training and education program."[1]

It would be nearly impossible for a brigade to operate consistently, effectively, or efficiently without SOPs/SOGs. For emergency incidents, SOPs/SOGs allow the brigade to develop a plan of action prior to the incident. SOPs/SOGs should be developed or modified as part of the pre-incident planning phase of emergency response. Established procedures allow managers and employees to make decisions more rapidly and uniformly with confidence. This improves efficiency. For operational SOPs/SOGs, the more decisions that can be made in the SOP/SOG prior to the incident, the fewer decisions that will need to be made during the incident. An SOP/SOG provides accountability and control. It lets fire units know where they should be operating and what procedures to take upon arrival at the scene of an incident.

## The Difference Between SOPs and SOGs

There has been much discussion regarding the difference between SOPs and SOGs. Some people believe that an SOP is a "must follow" document, whereas an SOG is a "guidance" document that incorporates a degree of flexibility. It is believed that with such flexibility, a brigade's liability is reduced or even avoided. With today's **litigious** society, it may not matter what you call your documents; legal proceedings can and will result. Your brigade will be held accountable based on accepted industry practice or national standards. Therefore, it is best if your SOPs/SOGs are based on nationally recognized standards. For this reason, SOPs/SOGs will be referred to only as SOPs from this point on.

SOPs should be reviewed by the legal department to ensure compliance with applicable company rules and other standards that may be required and not known by the persons writing the SOP. In some cases, the wording used in an SOP will be required to meet a company legal requirement. Other legal issues that may arise involve the use of the terms "shall," "will," "should," or "may." *Shall* and *will* indicate that these are actions that the incident commander or other persons must perform and can restrict the actions of the incident

[1]OSHA 29 CFR 1910.156, *Fire Brigades, section* 1910.156 (c) (4).

**KUPARUK RIVER UNIT**
**Kuparuk Fire Department**
**POLICY AND PROCEDURES MANUAL**

| | | | |
|---|---|---|---|
| Title: | Apparatus & Equipment<br>Foam Procedures CPF-1 | Prepared by:<br>Date: | |
| Type: | Policy [ ] Equip. O &M [x]<br>Procedure [x] Other [ ]<br>Guideline [ ] | Revised by:<br>Date: | B. K. Follett<br>7-27-97 |
| P&P No.: | 3.2.4 | Approvals: | Chief Officer |

**PURPOSE**
To establish foam procedures for diesel storage tanks at bulk fuel loading at CPF-1.

**SAFETY**
N/A

**PROCEDURES**
Make water supply connections at nearest hydrant for water supply.

Connect to the foam manifold furthest away from the tank that is being supplied with foam solution.

Supply the foam manifold with a 3" supply line for the foam solution.

Open valve number that corresponds with the tank number that will be foamed. Valves are opened by removing restraining chains and lifting the handle up.

Round diesel tank, which is tank number four, pump foam solution to the foam manifold header at 120 GPM at 75 psi. Approximately, 200 gallons of foam concentrate will be needed in order to supply the tank for 55 minutes.

Rectangular diesel tanks numbered one through three - pump the foam solution to the foam manifold header at 101 GPM at 87 psi. Approximately, 170 gallons of foam concentrate will be needed in order to supply one tank for 55 minutes.

**GENERAL**
There are a total of four (4) tanks that contain diesel, one round tank with a 420,000 gallon or 10,000 barrel capacity and three rectangular tanks with 126,000 gallon or 3000 barrel capacity. All four of these tanks are equipped with foam makers on the top of the tanks. The foam lines all share a common manifold and have separate lines to each individual tank. There are two separate foam hook-ups, one near tank #3 and the other near tank #4. Each tank is numbered, the three rectangular tanks are numbered from one through three and the round tank is number four. At the common foam manifold there is an "L" shaped valve handle with a corresponding tank number over it that indicates which fuel tank it will supply foam to. Each tank has a foam chamber at the top of the tank that has a deflector mounted inside the tank shell. The foam solution is deflected against the tank wall by the deflector.

Chief Officer

**Figure 2-1** A sample Standard Operating Procedure.

commander or allow a legal team in the courtroom to prove negligence. *Should* or *may* indicate that the action discussed is a recommendation.

## Developing SOPs

In most facility fire brigade operations, the scope of brigade functions is either established or approved by management. The scope of functions creates "side broads" that SOPs must fall within. When developing SOPs, we must start with a need. What does your brigade need SOPs for? Most SOPs have been developed because an incident that has occurred demonstrated a need to clarify a policy or procedure. Once the need is established, determine who will write the procedure. Remember that brigades are not dictatorships. Those who will be expected to follow the SOP should be part of the group or committee tasked with the SOP development. The exception to this policy would be a situation in which the SOP needs to be developed quickly because of an emergency or unsafe condition; another exception is if the SOP will be controversial.

By having a group or committee develop the SOP, a sense of ownership can be acquired, and the SOP may be more readily accepted by members of the brigade. A group or committee may also be more successful at identifying and evaluating the issues associated with the SOP's purpose and scope.

In addition, plant management should be involved during the development process. If not directly involved, they should at least be kept informed. Where an SOP addresses issues outside the fire brigade or where other departmental support is required, such as the safety or the maintenance department and possibly municipal fire departments, those departments or agencies should also be involved in the development process or at least should review the draft prior to its final signature and issue. These steps will greatly assist in the implementation and enforcement phases of the SOP process.

Some SOPs can be generic and capable of being adapted to a variety of incidents. These can be building collapse, hazardous materials response, air disaster plans, and other events that have response elements that can be incorporated into an SOP that is fairly broad in scope. Do not, however, force an SOP to work in a situation that it wasn't designed for.

## SOP Content

SOP manuals should be kept as simple as possible. They will only work if they are used. A small or compact set of procedures will accomplish more than a book of complex directions that no one can remember or follow. SOPs should instruct the user on the tasks that should be accomplished but should not dictate how the tasks should be performed. When uniformity is important in a situation, then the method of performing the task can be detailed.

SOPs should begin with the purpose and scope. The purpose outlines why the SOP has been established as well as its intent. The scope sets the boundaries for the SOP, such as who must comply with the SOP and/or in what situations this SOP applies. Terms or concepts used in the SOP should be defined in a definitions section. Responsibilities for various personnel titles or functions must also be outlined. Other standards, policies, or referenced documents should be noted in the SOP.

SOPs for facility fire brigades should reflect the general philosophy of the company as well as incorporate directions from the company's or plant's emergency response plans. A standard numbering system should be identified and used in all SOPs. This numbering system can be a standard numbering system such as numbering each section or category with a three-digit number and then subnumbering from that point. An example of this type of numbering system is as follows:[2]

| | |
|---|---|
| 100 | Rules and Regulations |
| 200 | General Administration |
| 300 | Hazardous Materials |
| 400 | Occupational Safety and Health |
| 500 | Maintenance |
| 600 | Emergency Operations |
| 700 | Emergency Medical Services |
| 800 | Communications |
| 900 | Fire Prevention |

Another variation would be an alphanumeric system, such as the following:

| | |
|---|---|
| AD-001 | Administration |
| HM-001 | Hazardous Materials |
| OPS-001 | Operations |
| EMS-001 | Emergency Medical Services |
| FP-001 | Fire Prevention |
| R&R-001 | Rules and Regulations |

## Implementation and Revision

When implementing SOPs, a time frame must be established for implementation and all personnel provided with copies and appropriate training prior to such implementation. Where possible and time allows, a pilot program should be implemented so that the application or the SOP is tested. A critique process or other method of feedback should be established; when necessary, the SOP should be revised based on lessons learned.

Once implemented, SOPs must be followed. Management should enforce the use of the SOPs so that the SOPs become

[2]Cook, J. L. (1998). *Standard Operating Procedures and Guidelines*. Saddle Brook, NJ: Fire Engineering Books and Videos.

# VOICES OF EXPERIENCE

**"It is important to identify the need for SOPs before an incident that may require them and have them developed and tested."**

Many years ago I responded to a large fire at a petroleum storage depot. A large tank was involved in a full surface fire. The storage depot had no in-plant fire brigade, and the municipality was unfamiliar with storage tank firefighting because there were only a few of these facilities located within their jurisdiction. The fire department was unsure of the correct quantities of foam solution or the correct application rates required for successful extinguishment and additionally had no policies or procedures in place for acquiring the necessary resources. The fire was successfully extinguished after a few days of intense firefighting, but the lesson learned was the need for policies and procedures to handle such an incident. Immediately after the fire, a team was formed to develop standard operating procedures (SOPs) specifically for large storage tank or other fires requiring large quantities of foam concentrate. These policies identified the units to be equipped with specialized large-volume foam delivery devices, the quantities of foam concentrate stored within the municipality, the method to acquire additional foam resources, foam solution delivery methods, and the training necessary to ensure that the procedures were relevant. It is important to identify the need for SOPs before an incident that may require them and have them developed and tested.

**Craig H. Shelley**
Advanced Fire Training Center
Saudi Aramco
Dhahran, Saudi Arabia

part of the institutional memory of the brigade. One way of having SOPs institutionalized is to include them as the subject of station and multiunit drills. Another way is to include them as subject matter on promotional exams. SOPs must be tested on a frequent basis. They must be used at every incident to be effective. How a department responds to its routine incidents will set the tone for how it responds on the major incidents. When responding and operating, a department or brigade must be organized, disciplined, and adhere to policies and procedures. Without this, the response and operations will be haphazard at best.

An SOP is a great tool to be used in the development of an incident action plan (IAP). The IAP contains the objectives for the incident strategy, tactics, risk management, and member safety. The procedures outlined in an SOP can be used to form the objectives and strategies of the IAP.

SOPs must be reviewed on a scheduled basis, such as every 3 to 5 years, and revised when necessary. An SOP that has become obsolete or unnecessary should be revoked. Those that need updating due to changes in company or department philosophy, plant expansion, changes in plant processes, new techniques, new technology, or a change in a brigade's mission or scope should be revised.

SOPs should be shared with the local municipal fire department and other agencies that may respond to an industrial facility to assist during an incident. If all parties are operating using similar policies and guidelines, it assists greatly with the interoperability of the departments during drills and incidents.

## Post-Incident Analysis

SOP review should be included as part of a brigade's post-incident critique process. This critique should reinforce positive performance and resolve problems that have occurred. If deficiencies with an SOP are noted, they should be highlighted and the SOP revised if necessary. Lessons learned after incidents as a result of the post-incident critique should be combined where necessary and used to upgrade existing or develop relevant SOPs.

## Conclusion

SOPs are a valuable and necessary tool for any brigade. These tools are just as important as any tool carried on any apparatus. They are used to define organizational policy and describe behavioral and performance expectations of the brigade's members. They provide a consistent point of reference that will allow members to perform to a measurable standard. Standard practices are required in every area of brigade operations. SOPs provide unity and ensure that brigade members work together effectively.

# Wrap-Up

## Ready for Review

- SOPs are a set of organizational directives that establish a course of action both administratively and operationally.
- SOPs allow the department or brigade to develop a plan of action prior to emergency incidents.
- SOPs should be reviewed by the company legal department to ensure compliance with applicable company rules and other standards that may be required and not known by the person writing the SOP.
- SOP manuals should be kept as simple as possible.
- SOPs will work only if they are used.
- SOPs for facility fire brigades should reflect the general philosophy of the company as well as incorporate directions from the company's or plant's emergency response plans.
- SOPs must be reviewed on a scheduled basis, such as every 3 to 5 years, and revised when necessary.

## Hot Terms

**Critique** A process that examines the overall effectiveness of a policy, drill, or emergency response and is used in the development of recommendations for improving an organization's day-to-day and emergency response procedures.

**Incident action plan (IAP)** A plan that contains objectives that reflect the incident strategy and specific control actions for the current or next operational period.

**Institutional memory** A situation in which behaviors, policies, and procedures have been structured and formalized in an organization's culture and practices.

**Litigious** Prone to engage in lawsuits.

# Brigade Member in Action

**As a member of the plant's facility fire brigade, you have been assigned to assist with development of a standard operating procedure for responding to confined-space rescue incidents.**

1. What will be the first item that should be considered when developing an SOP?
   - **A.** The person(s) who will write the SOP
   - **B.** The total number of pages required in the SOP
   - **C.** The content of the SOP
   - **D.** The need for an SOP

2. Which of the following statements is not true regarding the development of SOPs?
   - **A.** By having a group or committee develop the SOP, a sense of ownership can be acquired.
   - **B.** By having a group or committee develop the SOP the department or brigade members will be less ready to accept the SOP.
   - **C.** Plant management should be involved in the development of the SOP.
   - **D.** Some SOPs can be generic and capable of being adapted to a type of incident.

3. The SOP document should begin with which of the following?
   - **A.** Purpose and scope
   - **B.** General content
   - **C.** Definitions
   - **D.** Responsibilities

4. A time frame should be established for reviewing SOPs. This time frame should be which of the following?
   - **A.** Every 10 years
   - **B.** Annually
   - **C.** Bi-annually
   - **D.** Every 3 to 5 years

# Fire Service Communications

# Chapter 3

## NFPA 1081 Standard

### Incipient Facility Fire Brigade Member

**4.1** *General.* This duty shall involve initiating communications, using facility communications equipment to effectively relay oral or written information, responding to alarms, returning equipment to service, and completing incident reports, according to the JPRs in 5.1.1 through 5.2.3.

**4.1.2.1** Initiate a response to a reported emergency, given the report of an emergency, facility standard operating procedures, and communications equipment, so that all necessary information is obtained and communications equipment is operated properly.

**(A)** *Requisite Knowledge.* Procedures for reporting an emergency.

**(B)** *Requisite Skills.* The ability to operate facility communications equipment, relay information, and record information.

**4.1.2.2*** Transmit and receive messages via the facility communications system, given facility communications equipment and operating procedures, so that the information is promptly relayed and is accurate, complete, and clear.

**(A)** *Requisite Knowledge.* Facility communications procedures and etiquette for routine traffic, emergency traffic, and emergency evacuation signals.

**(B)** *Requisite Skills.* The ability to operate facility communications equipment and discriminate between routine and emergency communications.

**4.1.2.5*** Complete a basic incident report, given the report forms, guidelines, and incident information, so that all pertinent information is recorded, the information is accurate, and the report is complete.

**(A)** *Requisite Knowledge.* Content requirements for basic incident reports, the purpose and usefulness of accurate reports, consequences of inaccurate reports, and how to obtain necessary information.

**(B)** *Requisite Skills.* The ability to collect necessary information, proof reports, and operate facility equipment necessary to complete reports.

**4.1.3** *Activating an Emergency Call.* Activate an emergency call for assistance, given vision-obscured conditions, personal protective equipment (PPE), and department standard operating procedures (SOPs), so that the fire fighter can be located and rescued.

**(A)** *Requisite Knowledge.* Personnel accountability systems, emergency communication procedures, and emergency evacuation methods.

**(B)** *Requisite Skills.* The ability to initiate an emergency call for assistance in accordance with the AHJ's procedures, the ability to use other methods of emergency calls for assistance. [1001:4.2.4(B)]

### Advanced Exterior Facility Fire Brigade Member

NFPA 1081 contains no Advanced Exterior Facility job performance requirements for this chapter.

### Interior Structural Facility Fire Brigade Member

NFPA 1081 contains no Interior Structural Facility job performance requirements for this chapter.

## Knowledge Objectives

After completing this chapter, you will be able to:

- Describe the roles of the telecommunicator and dispatcher.
- Describe how to receive an emergency call.
- Describe how to initiate a response.
- Describe fire brigade radio communications.
- Describe radio codes.
- Describe emergency traffic and emergency evacuation signals.
- Define the content requirements for basic incident reports.
- Define how to obtain necessary information, required coding procedures, and the consequences of incomplete and inaccurate reports.
- Describe fire brigade procedures for answering nonemergency business and personal telephone calls.

## Skills Objectives

After completing this chapter, you will be able to perform the following skills:

- Transmit and receive messages via the fire brigade's radio system.
- Complete a basic incident report accurately and completely.
- Operate and answer a fire station telephone.

# You Are the Brigade Member

**Your fire brigade is practicing a table-top exercise at the fire station involving strategies and tactics. A contractor drives up in front of the station and yells out that there is a leaking flange in one of the pump houses. Your facility is a large producer of vinyl chloride.**

1. ***How would you handle this situation?***
2. ***What information do you need to get from this person?***
3. ***What role does the communication center have in this?***

## Introduction

Rapidly developing technology and advanced communications systems are having a tremendous impact on facility fire brigade communications. As a brigade member, you must be familiar with the communications systems, equipment, and procedures used in your organization. This chapter provides a basic guide to help you understand how communications systems work and how common systems are configured.

A fire brigade or emergency response team depends on a functional communications system. When there is a request for assistance or an alarm sounds, the communications center dispatches the appropriate units to the incident and maintains communication with those units throughout the incident. The communications system is the link between brigade members on the scene and the rest of the organization. The communications center monitors everything that happens at the incident scene and processes all requests for assistance or special resources.

At the scene, brigade members need to communicate with each other, so that the Incident Commander (IC) can manage the operation efficiently based on progress reports or requests for assistance. The Incident Management System depends on a functional on-site communications system. During large-scale incidents, brigade members must be able to communicate not only with each other but also with other emergency response agencies.

The communications center does more than simply dispatch units and communicate with them during emergency operations. It also must track the location and status of every other fire brigade unit. The communications center must always know which units can be dispatched to an incident, and it must be able to contact those units.

In addition to these special communications requirements, a fire brigade must have a communications infrastructure that allows it to function as an effective organization. Basic administration and day-to-day management require an efficient communications network, including telephone and data links with every fire station and work site.

## The Communications Center

The communications center is the hub of the fire brigade emergency response system. It is the central processing point for all information relating to an emergency incident and all of the information relating to the location, status, and activities of the fire brigade units. It connects and controls all of the brigade's communications systems. The communications center functions like the human brain. Information comes in via the nerves; it is processed and sent back out to be acted upon in all of the different parts of the body.

The communications center is a physical location. Its size and complexity may vary depending on the needs of the brigade. The communications center for one brigade may be a small room in the fire station; another brigade may have a specially designed, highly sophisticated facility with advanced technological equipment. The fire brigade in a small industrial facility may need a simple system, while larger more geographically spread out facilities may require a large facility. Regardless of the size, all communications centers perform the same basic functions (► **Figure 3-1**).

Many municipal fire departments operate their own "stand-alone" communications centers, serving only a single agency. Others operate or are served by a communications center that serves several fire departments and, sometimes

Figure 3-1 Regardless of the size, all communications centers perform the same basic functions.

all of the fire agencies in a county or region. In many areas, the communications center is co-located with other public safety agencies, including law enforcement and/or separate emergency medical services (EMS) providers. In a joint facility there may be separate personnel and independent systems for each service delivery agency or the entire operation may be integrated, with all employees cross-trained to receive calls and dispatch any type of emergency incident.

## Telecommunicators

In many industrial settings, communications are handled from a designated control room. In large-scale industrial facilities such as the oil, gas, and petrochemical settings, these control rooms are very sophisticated and may be staffed with designated dispatchers or telecommunicators. Telecommunicators have been professionally trained to work in a public safety communications environment. Advanced training and professional certification programs ensure that skilled, competent telecommunicators can perform their critical role in the public safety system.

Just as there are special qualities that can make an individual a good brigade member, there are qualities that a telecommunicator must possess in order to be successful. The job of a telecommunicator can be complicated, demanding, and extremely stressful. The successful telecommunicator must be able to understand and follow complicated procedures, perform multiple tasks effectively, memorize information, and make decisions quickly.

One of the telecommunicator's most important skills is an ability to communicate effectively to obtain critical information, even when the caller is highly stressed or in extreme personal danger. An emotional caller may criticize or insult the telecommunicator. The telecommunicator must respond professionally and focus on obtaining the essential information. Voice control and the ability to maintain composure under pressure are important qualities; the telecommunicator must be clear, calm, and in control.

A telecommunicator must be skilled in operating all of the systems and equipment in the communications center. He or she must understand and follow the organization's operational procedures, particularly those relating to dispatch policies and protocols, radio communications, and incident management. The telecommunicator must keep track of the status and location of each unit at all times and monitor the overall deployment and availability of resources throughout the system.

## Communications Facility Requirements

The fire brigade's communications center must be designed and operated to ensure that its critical mission can be performed with a very high degree of reliability. The communications center should be well-protected against natural threats such as floods, and it should be constructed to withstand predictable damaging forces such as severe storms and earthquakes. The communications center should be able to operate at maximum capacity, without interruption, even when other community services are severely affected. The building should be equipped with emergency generators and other systems, so that it can continue to operate for several days in the most challenging conditions.

## Communications Center Equipment

A communications center is usually equipped with several different types of communications systems and equipment. Although specific requirements depend on the size of the operation and the configuration of the local communications systems, most centers will have the following equipment:

- Dedicated 9-1-1 telephones
- Public telephones
- Direct-line phones to other agencies
- Equipment to receive alarms from public or private fire alarm systems
- Computers and/or hard copy files and maps to locate addresses and select units to dispatch
- Equipment for alerting and dispatching units to emergency calls
- Two-way radio system(s)
- Recording devices to record phone calls and radio traffic
- Back-up electrical generators

## Computer-Aided Dispatch (CAD)

Computers are used in almost all communications centers. Most large communications centers use computer-aided dispatch (CAD) systems, and many smaller centers have smaller scale versions of CAD systems.

As the name suggests, a CAD system is designed to assist a telecommunicator by performing specific functions more quickly and efficiently than they can be done manually. There are many variations in CAD systems, from very simple versions to highly sophisticated systems that perform many different functions. All of the functions performed by a CAD system can be performed manually by trained and experienced telecommunicators. The computer assists the telecommunicator by performing these functions more quickly and accurately (▼ Figure 3-2).

A well-designed CAD system can shorten the time it takes for a telecommunicator to receive and dispatch calls. A CAD system helps meet the most important objective in processing an emergency call: sending the appropriate units to the correct location as quickly as possible. The increased efficiency provided by a CAD system is particularly important in large communications centers that process a high volume of calls.

Because a computer can manage large amounts of information efficiently and accurately, a CAD system can be used to track the status of a large number of units. The CAD system will know which units are available to respond to a call, which units are assigned to incidents, and which units are temporarily assigned to cover different areas. The most advanced CAD system has global positioning system (GPS) devices in the brigade's vehicles, so it can track the exact location of each unit. A CAD system can also look up addresses and determine the closest fire stations in order of response for any location. The CAD system can then select the units that can respond quickly to an alarm, even if some of the units that would normally respond are unavailable.

Some CAD systems transmit dispatch information directly to mobile data terminals (MDTs), computers that are located in the fire station or on the apparatus (▼ Figure 3-3). CAD systems can also provide immediate access to information such as preincident plans and hazardous materials lists for an address. By linking the CAD system to other data files, brigade members have access to additional useful information such as travel route instructions, maps, and reference materials.

## Voice Recorders and Activity Logs

Almost everything that happens in a communications center is recorded, either by a voice recording system or by an activity logging system. Most communications centers

Figure 3-2 A CAD system enables a telecommunicator to work more quickly and efficiently.

Figure 3-3 Some CAD systems transmit dispatch information directly to terminals in fire stations and mobile data terminals in the apparatus.

can automatically record everything that is said over the telephone or radio, 24 hours a day. Most centers also have an instant playback unit that allows the telecommunicator to replay conversations for the previous 10 to 15 minutes at the touch of a button. This feature is particularly valuable if the caller talks very quickly, has an unusual accent, hangs up, or is disconnected. The telecommunicator can replay the message several times, if necessary, to understand exactly what was said.

The logging system keeps a detailed record of every incident and activity that occurs. The records include every call that is entered, every unit that is dispatched, and every significant event that occurs in relation to an emergency incident. Times are recorded when a call is received, when the units are dispatched, when they report that they are en route, when they arrive at the scene, when the incident is under control, and when the last unit leaves the scene. One of the advantages of a CAD system is that it automatically captures and stores every event as it occurs. Before CAD systems were developed, someone had to write down and time stamp every transaction, including every radio transmission. All hard-copy logs had to be retained for future reference.

There are several reasons for keeping voice recorders and activity logs. They serve as legal records of the official delivery of service by a fire brigade. These records may be required for legal proceedings, sometimes years after the incident occurred. They may be needed to defend the fire brigade's actions when questions are raised about an unfortunate outcome. The records provided by voice recorders and incident logs accurately document the events and can often demonstrate that the organization and its employees performed ethically, responsibly, and professionally. They also make it difficult to hide an error, if a mistake was made.

Records are also valuable in reviewing and analyzing information about brigade operations. Good recordkeeping makes it possible to examine what happened on a particular call as well as to measure workloads, system performance, activity trends, and other factors as part of planning and budget preparation. Analysts and planners can use data from CAD systems to study deployment strategies and to make the most efficient use of fire brigade resources.

## Call Response and Dispatch

Critical functions performed by most CAD systems include verifying a location and determining which units should respond to an alarm. This is usually a simple task in a small system with only one or two fire stations, but it is much more complicated in a large organization with multiple fire stations and individual units. When the telecommunicator enters an address and an incident description code into a CAD system, the system can make a recommendation on the appropriate units to dispatch in less than a second.

Figure 3-4 Telecommunicators must obtain information and relay it accurately to the appropriate responders.

Data links that provide the location of the caller and automatically enter the location can also save time. Hard copy files of the most essential information should always be maintained for back-up, so that calls can be dispatched if the system is down (out-of-service).

The telecommunicator's first responsibility is to obtain the information that is required to dispatch the appropriate units to the correct location (▲ Figure 3-4). Then the incident must be processed according to standard protocols. The telecommunicator must decide which agencies and/or units should respond and transmit the necessary information to them. The generally accepted performance objective, from the time a call reaches the communications center until the units are dispatched, is one minute or less.

Some requests for fire brigade response are made by telephone. Although some fire brigades use a regular seven-digit telephone number to report an emergency, many requests for a fire brigade response are done through an automatic alarm or by radio notification. Activation of the 9-1-1 system in North America is usually done through the internal communications center or control room for the industrial facility.

All calls to 9-1-1 are automatically directed to a designated public safety answering point (PSAP) for that community or jurisdiction. If all public safety communications are in the same facility, the call can be answered, processed, and dispatched immediately. If fire and/or EMS communications have a separate facility, those calls must be transferred to a fire/EMS-trained telecommunicator. The call-taking process is described in more detail later in this chapter.

# Communications Center Operations

Several different functions are performed in a municipal fire brigade or public safety communications center. All of these activities must be performed accurately and efficiently, even in the most challenging circumstances. For example, the activity level in a communications center can increase from calm to extreme in less than a minute during an emergency, but the chaos outside must not affect the operations inside. If the communications center fails to perform its mission, the fire brigade will not be able to deliver much-needed emergency services.

The basic functions performed in a communications center include:

- Receiving calls for emergency incidents and dispatching fire brigade units
- Supporting the operations of fire brigade units delivering emergency services
- Coordinating fire brigade operations with other agencies
- Keeping track of the status of each fire brigade unit at all times
- Monitoring the level of coverage and managing the deployment of available units
- Notifying designated individuals and agencies of particular events and situations
- Maintaining records of all emergency-related activities
- Maintaining information required for dispatch purposes

## Receiving and Dispatching Emergency Calls

Most fire brigade responses begin when a call comes in to the communications center, which then dispatches one or more units. Even if a worker reports an emergency directly to a fire station or a crew discovers a situation, the communications center is immediately notified to initiate the emergency response process.

The major steps in processing an emergency incident include:

- Call receipt
- Location validation
- Classification and prioritization
- Unit selection
- Dispatch

Call receipt refers to the process of receiving an initial call and obtaining the necessary information to initiate a response. This includes notification methods, such as automatic fire alarm systems, public fire alarm boxes, requests from other agencies, calls reported directly to fire stations, and calls initiated via radio from fire brigade units or other public safety agencies.

Location validation ensures that the information received is adequate to dispatch units to the correct location. Potential duplicate addresses, such as two streets with the same or very similar names, must be eliminated. The information must point to a valid location on a map or in a street index system. It must be within the geographic jurisdiction of the potential dispatch units. Without a valid address, a communications center will be unable to send units to the proper location.

Classification and prioritization is the process of assigning a response category, based on the nature of the reported problem. Most fire brigades respond to many kinds of situations, ranging from plant process fires to medical calls. The nature of the call dictates which units or combinations of units should be dispatched. Dispatch policies are usually spelled out in the organization's standard operating procedures (SOPs) and dispatch protocols, which state the numbers and types of units to dispatch for each type of situation.

Unit selection is the process of determining exactly which unit or units to dispatch, based on the location and classification of the incident. The usual policy is to dispatch the closest available unit that can provide the necessary assistance. When all units are available and in their fire stations, this can usually be determined quickly from preprogrammed information. Unit selection becomes more complicated when some units are out of position or unavailable. Unit selection often requires quick decision-making skills, even with a CAD system that is programmed to follow set policies for various situations.

**Dispatch** is the important step of actually alerting the selected units to respond and transmitting the information to them. Fire brigades use different dispatch systems, ranging from telephone lines to radio systems. The communications center must have at least two separate ways of notifying each fire station.

### Call Receipt

#### Telephones

The public generally uses telephones to report emergency incidents. In most communities, calls to 9-1-1 connect the caller with a PSAP. Some communities have implemented a 3-1-1 system for nonemergency calls. The PSAP can take the information immediately, or transfer the call to the appropriate agency, based on the nature of the emergency. In some systems, 9-1-1 connects the caller directly with a telecommunicator, who obtains the required information.

The communications center usually has a seven-digit telephone number as well. People who remember the seven-digit number used before their communities introduced the 9-1-1 system may continue to call that number. Even when every effort is made to encourage people to use 9-1-1 to report an emergency, some people may be reluctant to use the system because they are not sure that their particular situation is serious enough to be considered a "true emergency." Any number that is published as a fire brigade telephone number should be answered at all times, including nights and weekends. The emergency number should always be pronounced as "nine-one-one," not "nine-eleven."

The telecommunicator who takes the call must conduct a **telephone interrogation**, asking the caller questions to obtain the required information. Initially, the telecommunicator will need to know the location of the emergency and the nature of the situation. Many 9-1-1 systems can automatically provide the location of the telephone where a call originates. But this information may be unavailable or inaccurate if the call was made on a wireless phone or from someplace other than the location of the emergency incident. The exact location must be obtained, so that units can respond directly to the incident scene.

The telecommunicator must also interpret the nature of the problem from the caller's description. The caller may be distressed, excited, and unable to organize his or her thoughts. There may be language barriers; the caller may speak a different language or may not know the right words to explain the situation. Telecommunicators must follow SOPs and use active listening to interpret the information. Many communications centers use a structured set of questions to obtain and classify information on the nature of the situation. The telecommunicator must remember that the caller thinks the situation is an emergency and must treat every call as such until it is determined that no emergency exists.

Telecommunicators cannot allow gaps of silence to occur while questioning the caller. If the caller suddenly becomes silent, something may have happened to him or her; the caller may be in personal danger or extremely upset. If the telecommunicator is silent, the caller may think that the telecommunicator is no longer on the line or no longer listening. Disconnects are another problem. Callers to 9-1-1 might hang up accidentally or be disconnected after initially reaching the communications center. A telecommunicator who is unable to return the call and reach the original caller will usually dispatch a police officer to the location to determine if more help is needed.

The telecommunicator should never argue with a caller. Although callers may raise their voices, scream, or shout, the telecommunicator or brigade member must continue to speak calmly and remain professional at all times. The telecommunicator should remember that the caller is simply reacting to the emergency situation.

With just two critical pieces of information—location and nature of the problem—the telecommunicator can initiate a response. However, local SOPs may require the telecommunicator to obtain additional information. Obtaining the caller's name and contact phone number is useful in case it is necessary to call back for additional information. If the caller is in danger or distress, the telecommunicator will try to keep the line open and remain in contact with the caller until help arrives. In many communities, telecommunicators are trained to provide self-help instructions for callers, advising them on what to do until the fire brigade arrives. For example, if a building fire is being reported, the telecommunicator will advise the occupants to evacuate and wait outside; if the caller is reporting a medical incident, the telecommunicator may provide first aid instructions, based on the patient's symptoms.

## Direct Line Telephones

A **direct line** (or ring down) telephone connects two predetermined points. Picking up the phone at one end causes an immediate ring at the other end. Direct line connections often link police and fire communications centers or two fire communications centers that serve adjacent areas. Direct lines also may connect hospitals, private alarm companies, utility companies, airports, and similar facilities with the fire brigade communications center. These lines are often used in both directions, so that the communications center can receive calls for assistance or send notifications and requests for response.

The communications center may have direct lines to each fire station in its jurisdiction. A direct line also can be linked to the station's public address speakers to announce dispatch messages.

## Walk-Ins

Although most emergencies are reported by telephone, people may actually come to a fire station seeking assistance. When a walk-in occurs, the station should contact and advise the communications center of the situation immediately. The communications center will create an incident record and dispatch any needed additional assistance. Even if the units at the station can handle the situation, they should notify the communications center that they are occupied with an incident.

People should be able to come to a fire station and report an emergency at all times, even when the station is unoccupied. Many organizations install a direct line telephone to the communications center just outside each fire station. These phones should be marked with a simple sign stating, "If station is closed, pick up telephone in red box to report an emergency."

A **call box** connects a person directly to a telecommunicator. The caller can request a full range of emergency assistance—from police, fire, or emergency medical assistance to a tow truck. Call boxes can be directly connected to a wireless or hard-wired telephone network or operate on a radio system. Emergency call boxes are often located along major highways, in bridges and tunnels, and in other locations without nearby phones. In locations without electric or telephone service, call boxes are solar-powered ▶ Figure 3-5.

## Private and Automatic Fire Alarm Systems

Private and automatic fire alarm systems use several different arrangements to transmit alarms to the local fire brigade communications center. Many commercial, industrial, and residential buildings have automatic fire alarm systems, which use heat detectors, smoke detectors, or other devices to initiate an alarm. Other buildings may have manual pull-station fire alarms. These alarms may be connected to the fire

Figure 3-5 Wireless call boxes are often found in places without other telephone service.

organization or monitored from a remote location. Water flow alarms on automatic sprinkler systems also can be connected to local fire alarm systems or monitored at a remote location. Chapter 25, Fire Detection, Protection, and Suppression Systems, discusses these systems in detail.

The connection used to transmit an alarm from a private system to a public fire communications center depends on many factors. Private companies may monitor alarm systems and relay alarms to the local police or fire brigade. Many private systems have automatic telephone dialers programmed to call an emergency number with a recorded message when the system is activated. These systems may be monitored by a privately operated central station alarm service, which relays the alarm to the fire communications center. There may be a direct line between a central alarm service and the fire brigade communications center. Some communications centers provide monitoring services. Finally, private alarm systems may be connected to a public fire alarm box. No matter how the alarm reaches the communications center, the result is the creation of an incident report and the dispatch of fire brigade apparatus.

## Call Classification and Prioritization

The next step in processing an emergency incident is classifying and prioritizing each call as it is received. The nature of the problem—what is happening at the scene—determines the urgency of the call and its priority. Although most fire brigade calls are dispatched immediately, it is sometimes necessary to delay dispatch to a lower priority call if a more urgent situation arises or if several calls come in at the same time. Some calls qualify for emergency response (red lights and siren) while others are considered nonemergencies, depending on local SOPs.

Call classification determines the number and types of units that are dispatched. The standard response to different incidents is established by the fire brigade's SOPs.

## Unit Selection

After verifying the location and classifying the situation, a telecommunicator must select the specific units to dispatch to an incident. Generally, the standard assignment to each type of incident in each geographic area is stored on a system of run cards. Run cards are prepared in advance and stored in hard copy form or in computer files. Run cards list units in the proper order of response, based on response distance or estimated response time, and often specify the units that would be dispatched through several levels of multiple alarms.

The run card assignments can be used as long as all of the units that would normally respond on a call are available. If any of the units are unavailable, the telecommunicator must adjust the assignments by selecting substitute units in their order of response.

Most CAD systems are programmed to select the units for an incident automatically, based on the location, call classification, and actual status of all units. The CAD system will recommend a dispatch assignment, which the telecommunicator can accept or adjust, based on circumstances or special information. The same process is used to dispatch any additional units to an incident, whether it is a multiple alarm or a request for particular units or capabilities.

## Dispatch

The telecommunicator's next step is to transmit the dispatch information to the assigned units quickly and accurately. The information can be transmitted in many different ways, but every facility must have at least two separate methods for sending a dispatch message from the communications center to each fire station. The primary connection can be a hard-wired circuit, a telephone line, a data line, a microwave link, or a radio system.

Most fire brigades dispatch verbal messages to the appropriate fire stations. The dispatch message is broadcast over speakers in the fire station, so that everyone immediately knows the location and nature of the incident and the units that should respond. Radio transmissions are used to contact a unit that is out of the station. The fire station or each vehicle must confirm that the message was received and the units are responding. If the communications center does not receive the confirmation within a set time period, it must dispatch substitute units.

A CAD system can be programmed to alert the appropriate fire stations automatically. The system can send the dispatch information to computer terminals or printers, sound distinctive tones, turn on lights and public address speakers, turn off the stove, and perform additional functions. The dispatch message also can go directly to each individual vehicle that has a computer terminal as well as to the station. The CAD notification often is accompanied by a verbal announcement over the radio and the fire station speakers.

Fire brigades with volunteer responders must be able to reach them with a dispatch message. Some organizations issue pagers to their individual volunteers. Some pagers can receive a vocal dispatch message; others display an alphanumeric message. Volunteer fire brigades in many industrial facilities rely on outdoor sirens, horns, or whistles to notify their members of an emergency. These audible devices usually can be activated by remote control from the communications center. Volunteers call the communications center by radio or telephone to receive specific instructions.

### Operational Support and Coordination

After the communications center dispatches the units, it begins to provide incident support and coordination. Someone in the communications center must remain in contact with the responding units throughout the incident. The telecommunicator must confirm that the dispatched units actually received the alarm, record their en route times, provide any additional or updated information, and record their on-scene arrival times.

Operational support and coordination encompass all communications between the units and the communications center during an entire incident. Two-way radio or mobile computer terminals are used to exchange information. Many fire brigades also issue cellular telephones to command officers and/or individual units.

Generally, the IC will communicate with a telecommunicator operating a radio in the communications center. Progress and incident status reports, requests for additional units or release of extra units, notifications, and requests for information or outside resources are examples of incident communications. The communications center closely monitors the radio and provides any needed support for the incident.

Each part of the emergency services network—fire, EMS, and security—must be aware of what other agencies are doing at the incident. The communications center is the hub of the network that supports units operating at emergency incidents. It coordinates the fire brigade's activities and requirements with other agencies and resources. For example, the telecommunicator may need to notify nonemergency resources such as gas, electric, or telephone companies and request that they take certain actions. The communications center should have accurate, current telephone numbers and contact information for every relevant agency.

### Status Tracking and Deployment Management

The communications center must know the location and status of every fire brigade unit at all times. Units should never get "lost" in the system, whether they are available for dispatch, assigned to an incident, or in the repair shop. As previously stated, the communications center must always know which units are available and which units are not available for dispatch. The changing conditions at an incident may require frequent reassignment of units, including ambulances. In addition, units may be needed from outside the normal response area or from other districts. Status tracking is difficult if the telecommunicator must rely on radio reports and colored magnets or tags on a map or status board. CAD systems make this job much easier, because status changes can be entered through digital status units or computer terminals. GPS devices, which accurately track the location of each unit, are also helpful on large incidents.

Communications centers also must continually monitor the availability of units in each geographic area and redeploy units when there is insufficient coverage in an area. Many fire brigades list both unit relocations and multiple response units on the run cards. This information is only valid, however, if major incidents occur one at a time and when all units are available. If an organization has a large volume of routine incidents, it may need to redeploy units to balance coverage, even when there are no major incidents in progress.

Usually, a supervisor in the communications center is responsible for determining when and where to redeploy units, as well as for requesting coverage from surrounding jurisdictions. Units from many different jurisdictions can be redeployed to respond to large-scale incidents under regional or statewide plans. These plans must include a system for tracking every unit and a designated communications center for maintaining contact with all units.

## Radio Systems

Fire brigade communications systems depend on two-way radio systems. Radios link the communications center and individual units; they also link units at an incident scene. A radio system is an integral component of the Incident Management System because it links all of the units on an incident—both up and down the chain of command and across the organization chart. Usually, every fire brigade vehicle has a mobile radio, and at least one—if not every—member of a team carries a portable radio during an emergency incident. Often, a radio is the brigade member's only link to the incident organization and the only means to call for help in a dangerous situation. Radios also are used to transmit dispatch information to fire stations, to page volunteer brigade members, and to link mobile computer terminals.

Fire brigades use many different types of radios and radio systems. Technological advances are rapidly adding new features and system configurations, making it impossible to describe all of the possible features, principles, and systems. This section describes common systems and operating features, but as a brigade member, you must know how to operate your assigned radio and learn your organization's radio procedures.

## Radio Equipment

There are three types of fire service radios: the portable radio, the mobile radio, and the base station. A portable radio is a hand-held two-way radio small enough for a brigade member to carry at all times (▼ Figure 3-6). The radio body contains an integrated speaker and microphone, an on/off switch or knob, a volume control, and a "push-to-talk" (PTT) button. Most radios also have a knob or switch so the user can move between a radio channel and a talk group. A knob or button on the radio adjusts the sensitivity of the radio's signal reception. This enables the user to squelch or eliminate weak transmissions.

Figure 3-6 A portable radio can be carried by an individual brigade member.

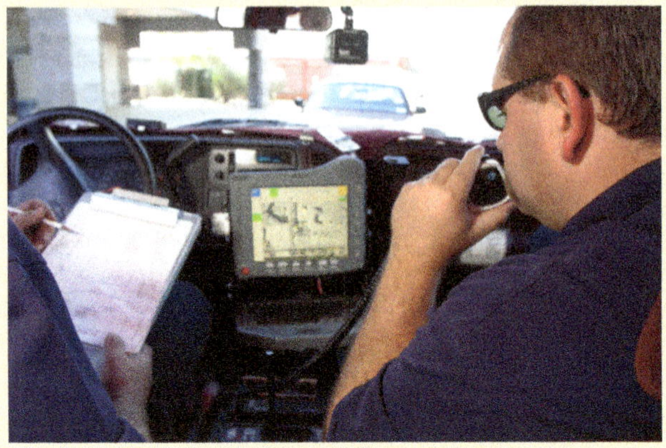

Figure 3-7 A mobile radio is permanently mounted in a vehicle.

A portable radio must have an antenna to receive and transmit signals. A frequently used optional attachment is an extension microphone/speaker unit that can be clipped to a collar or shoulder strap, while the radio remains in a pocket or pouch.

A portable radio is usually powered by a rechargeable battery, which should be checked at the beginning of each shift or prior to each use. Because the battery has a limited capacity, it must be recharged or replaced after extended operations. Battery-operated portable radios also have limited transmitting power. The signal can be heard only within a certain range and is easily blocked or overpowered by a stronger signal.

Mobile radios are more powerful two-way radios permanently mounted in vehicles and powered by the vehicle's electrical system (▲ Figure 3-7). Both mobile and portable radios have similar features, but mobile radios usually have a fixed speaker and an attached, hand-held microphone on a coiled cord. The PTT button is on the microphone, while the antenna is usually mounted on the exterior of the vehicle. Fire apparatus often include headsets with a combined intercom/radio system that enable crew members to talk to each other and hear the radio.

Base station radios are permanently mounted in a building, such as a fire station, communications center, or remote transmitter site (► Figure 3-8). Base station radios are more powerful than portable or mobile radios. The antenna is often mounted on a radio tower so the transmissions have a wide coverage area (► Figure 3-9). Emergency services radio systems often use multiple base stations at different locations to cover large geographic areas. The communications center operates these stations by remote control.

Figure 3-8 Base station radios are installed at fixed locations.

Figure 3-9 A base station radio antenna is often mounted on a radio tower to provide maximum coverage.

Mobile Data Terminals (MDTs) are computer devices that transmit data by radio. Some brigades use them to track unit status, transmit dispatch messages, and exchange different types of information. Some terminals simply use buttons to send status messages such as

Figure 3-10 Mobile computers exchange data over a radio channel.

"responding," "on the scene," and "back in service" while others are advanced laptop computers (▲ Figure 3-10).

## Radio Systems

The design, installation, and operation of two-way radio systems is closely regulated in the United States by the Federal Communications Commission (FCC). The FCC has strict limitations governing the assignment of frequencies to ensure that all users have adequate access. Every system must be licensed and operated within established guidelines.

Radios work by broadcasting electronic signals on certain frequencies. A frequency is an assigned space on the radio spectrum; only those radios tuned to that specific frequency can hear the message. The FCC licenses an agency to operate on one or more specific frequencies. A particular agency may be licensed to operate on several frequencies. The frequencies are programmed into the radio and can be adjusted only by a qualified technician.

A radio channel uses either one frequency or two frequencies. A simplex channel uses only one frequency. Each radio transmits signals and receives signals on the same frequency, so a message goes directly from one radio to every other radio that is set to receive that frequency. A duplex channel uses two frequencies, with each radio transmitting signals on one frequency and receiving on the other frequency (► Figure 3-11). Duplex channels are used with repeater systems, which are described later.

U.S. fire service frequencies are in several different ranges, including 33 to 46 MHz (VHF low band); 150 to 174 MHz (VHF high band); 450 to 460 MHz (UHF band); and the 800 MHz band. Additional groups of frequencies are allocated in specific geographic areas with a high demand for public safety agency radio channels. Each band has certain advantages and disadvantages relating to geographic coverage, topography (hills and valleys), and penetration into structures. Some

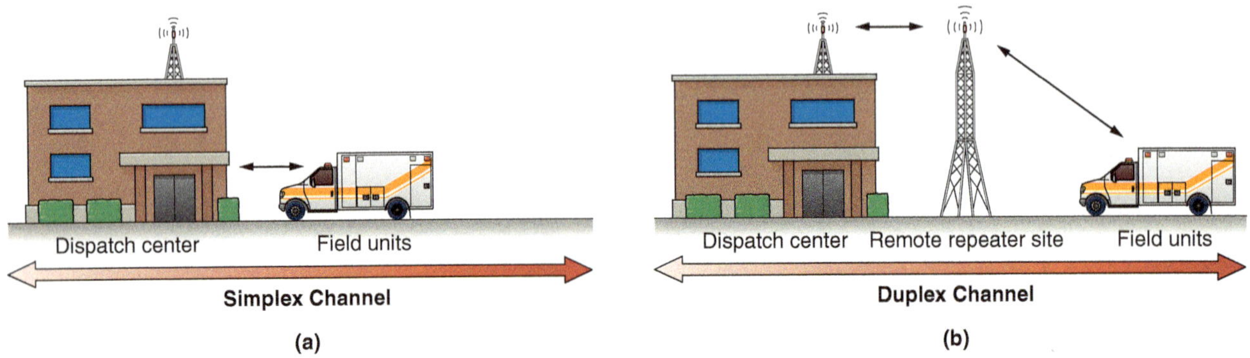


Figure 3-11 Simplex and duplex channels send and receive transmissions in different ways.

bands have a large number of different users, which can cause interference problems, particularly in densely populated metropolitan areas.

Generally, one radio can be programmed to operate on several frequencies in a particular band but cannot be used across different bands. This is a problem if neighboring fire brigades or police, fire, and EMS within the same jurisdiction are on different bands. When different agencies working at the same incident are on different bands, they must make complicated arrangements with cross band repeaters and multiple radios in command vehicles to communicate with each other.

## Repeater Systems

Radio messages can be broadcast over a limited distance. The signal weakens as it travels farther from the source. Buildings, tunnels, and topography create interference. These problems are most significant for hand-held portable radios, which have limited transmitting power. Mobile radios that operate in systems covering large geographic areas face similar problems. To compensate, fire brigades may use radio repeater systems.

In a repeater system, each radio channel uses two separate frequencies—one to transmit and the other to receive. When a low-power radio transmits over the first frequency, the signal is received by a repeater unit that automatically rebroadcasts it on the second frequency over a more powerful radio. All radios set on the designated channel receive the boosted signal on the second frequency. This enables the transmission to reach a wider coverage area.

Many public safety radio systems have multiple receivers (voter receivers), which are geographically distributed over the service area to capture weak signals. The individual receivers forward their signals to a device that selects the strongest signal and rebroadcasts it over the system's base station radio(s) and transmission tower(s). With this configuration, messages that originate from a mobile or hand-held portable radio have as strong a signal as a base station radio at the communications center. As long as the original signal is strong enough to reach one of the voter receivers, the system is effective. If the signal does not reach a repeater, no one will hear it.

Some fire brigades switch to a simplex radio channel for on-scene communications. This is sometimes called a talk-around channel, because it bypasses the repeater system. In this configuration, the radios transmit and receive on the same frequency, and the signal is not repeated. This often works well for short-distance communications, such as from the command post to inside crews at a structural fire or from one unit to another inside a building. Conversations on a talk-around channel cannot be monitored by the communications center. The IC will use a more powerful radio on a repeater channel to maintain contact with the communications center.

Mobile repeater systems also can boost signals at the incident scene by creating a localized, on-site repeater system. The mobile repeater can be permanently mounted in a vehicle or set up at the command post. It captures the weak signal from a portable radio for rebroadcast. Some fire brigades use cross-band repeaters, which boost the power from a weak signal and transmit it over a different radio band. This enables a brigade member using a UHF portable inside a building to communicate with the IC who is outside on a VHF radio. Similar systems may be used in large buildings or underground structures so crews working inside can communicate with units on the outside or with the communications center.

## Trunking Systems

Radio systems using new technology and digital communications are being developed to take advantage of the additional frequencies being allocated for public safety use. Unfortunately, the new systems are generally incompatible with older radios and require expensive infrastructure changes. They also can be more complicated to operate because the radio settings and characteristics are quite different from older systems (▶ Figure 3-12). Although the new systems will have tremendous operational advantages over conventional radio systems, certain basic requirements still must be met. The new systems demand well-designed software support, an adequate number of frequencies and transmitter/receiver sites, and well-trained users.

Figure 3-12 New technology (trunking) radios are much more advanced than conventional (VHF) radios.

The new radio technologies use a trunking system, a group of shared frequencies controlled by a computer. The computer allocates the frequencies for each transmission. The radio operator sets the radio on a talk group and communicates with the computer on a control frequency. When a user presses the transmit button, the computer assigns a frequency for that message and directs all of the radios in the talk group to receive the message on that frequency. A trunk system makes more efficient use of available frequencies and can provide coverage over extensive networks. Using digital rather than analog technology also is more efficient and allows more users to communicate at the same time in a limited portion of the radio spectrum.

## Using a Radio

As a brigade member, you must learn how to operate any radio assigned to you, and how to work with the particular radio system(s) used by your fire brigade. You must know when to use "Channel 3" or "Tac 5" or "Charlie 2," which buttons to press, and which knobs to turn. Training materials and the SOPs used by your organization should cover this information.

When a radio is assigned to you, make sure you know how to operate it and check that the battery is fully charged. On the fireground, your radio is your link to the outside world. It must work properly because your life depends on it. To use a radio, follow the steps in ▶ Skill Drill 3-1.

1. Before transmitting, determine that the channel is clear of any other traffic. Depress the PTT button and wait at least 2 seconds before speaking. This enables the system to capture the channel without cutting off the first part of the message. Some systems sound a distinctive tone when the channel is ready. **(Step 1)**
2. When you speak into the microphone, always speak across the microphone at a 45° angle and hold the microphone 1–2 inches from the mouth. Never speak on the radio if you have anything in your mouth. **(Step 2)**
3. Always know what you are going to say before you start talking. Speak clearly and keep the message brief and to the point.
4. Do not release the PTT button until you have finished speaking. This will avoid cutting off the final part of the message. **(Step 3)**

If you hold a portable radio perpendicular to the ground with the antenna pointing toward the sky, you will get better transmission and reception. Range and transmission quality also will improve if you remove the radio from the radio pocket or belt clip before you use it.

You should be familiar with all organizational SOPs governing the use of radios. All radio transmissions should be pertinent to the situation, without any unnecessary radio traffic. Remember that radio communications are automatically recorded. They also may be heard by everyone with a radio scanner, including the news media and the general public. Avoid saying anything of a sensitive nature or anything you might later regret. Radio tapes provide a complete record of everything that is said and can be admissible as evidence in a legal case.

Most fire brigades use clear speech (plain English) for radio communications, but some use codes for standard messages. A few brigades have an intricate system of radio codes to fit any situation. Ten-codes, a system of coded messages that begin with the number 10, were once widely used. They can be problematic, however, when units from different jurisdictions have to communicate with each other. To be effective, codes must be understood by all and mean the same thing to all users. NFPA standards recommend using clear speech rather than codes. If your brigade uses radio codes, you must know all the codes and how they are used.

The use of radio codes has been made redundant and is strongly discouraged since the implementation of the National Incident Management System (NIMS). All fire brigades in the United States are expected to be compliant with NIMS. All communications should be in clear text (plain English). If your brigade has a few codes or special words for special situations, they should be specified in the SOPs.

### Arrival and Progress Reports

Brigade members and telecommunicators must know how to transmit clear, accurate size-up and progress reports. The first-arriving unit at an incident—whether a brigade member, a brigade team leader, or a chief—should always give a brief initial radio report and establish command as described in the chapter on the incident management system. This initial report should convey a preliminary assessment of the situation. This report gives other units a sense of what is happening so they can anticipate what their assignments may be

## 3-1 Skill Drill

# Using a Radio

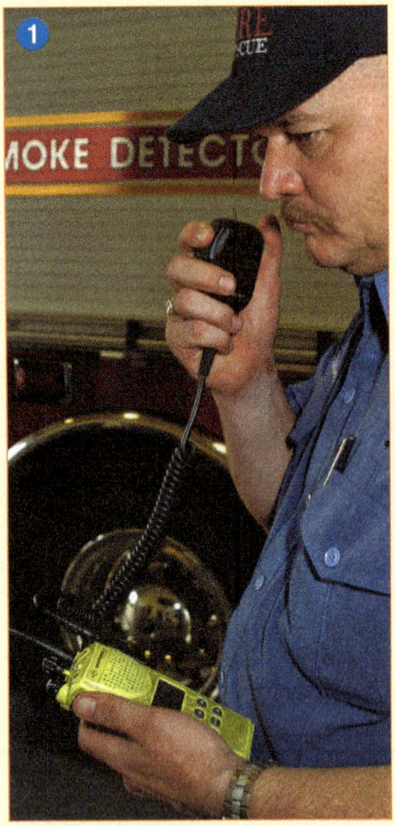

1. Before transmitting, determine that the channel is clear of any other traffic. Depress the PTT button and wait at least 2 seconds before speaking.

2. Speak across the microphone at a 45° angle and hold the microphone 1–2 inches from the mouth.

3. Speak clearly and keep the message brief. Do not release the PTT button until you have finished speaking.

when they arrive at the scene. The communications center should repeat the initial size-up information. For example, on arrival at the scene, a company may report:

*"Engine 4 to Communications. Engine 4 is on scene at the refinery terminal. We have condensate leaking from storage tank 11 and a large pool fire at the base of the tank. Engine 4 is assuming Tank 11 command. This will be an offensive operation."*

The communications center would respond by announcing:

*"Engine 4 is on the scene assuming Tank 11 command. Large pool fire at the base of storage tank, with condensate leaking from the tank. Offensive operation."*

This initial report should be followed by a more detailed report after the IC has had time to gather additional information and make initial assignments. An example of such a report follows:

IC: *"Tank 11 command to Communications."*

Communications center: *"Go ahead, Tank 11 Command."*

IC: *"All resources from Engine 4 have been committed to attacking the fire. The spill fire is estimated at 100 feet in diameter. A foam blanket has been applied for initial knock-down of the fire. All resources are engaged in an offensive operation. The fire appears to be contained. Chief 4 is in command."*

Communications center: *"Copied Tank 11 Command. You have initial knock-down of the fire, all resources are engaged in an offensive operation, and the fire appears to be contained. Chief 4 is IC."*

For the duration of an incident, the IC should provide regular updates to the communications center. Information about certain events, such as command transfer, "all clear," and "fire under control," should be provided when they occur. Reports also should be made when the situation changes significantly and at least every 10 to 20 minutes during a working incident.

In some fire brigades, the communications center prompts the IC to report at set intervals, known as time marks. Time marking allows the IC to assess the progress of the incident and decide if any changes should be made in strategy or tactics. All progress reports are noted in the activity log to document the incident.

If additional resources, such as apparatus, equipment, or personnel, are needed, the IC transmits the request to the communications center. The communications center will advise the IC of additional resources that are dispatched.

### Emergency Messages

There are times when the telecommunicator must interrupt normal radio transmissions for emergency traffic. Emergency traffic is an urgent message that takes priority over all other communications. When a unit needs to transmit emergency traffic, the telecommunicator generates a distinctive alert tone to notify everyone on the frequency to stand by, so the channel is available for the emergency communication. Once the emergency message is complete, the telecommunicator notifies all units to resume normal radio traffic. The most important emergency traffic is a brigade member's call for help.

Another emergency traffic message is "abandon the building," which warns all units inside a structure to evacuate immediately. Most fire brigades have a standard evacuation signal to warn all personnel to pull back to a safe location. The evacuation signal could be a sequence of three blasts on an apparatus air horn, repeated several times, or sirens sounded on "high-low" for 15 seconds. An evacuation warning should be announced at least three times to ensure that everyone hears it and it should be announced on the radio by the IC. Because there is no universal evacuation signal, brigade members must learn their organization's SOP for emergency evacuation.

After the evacuation, the radio airwaves should remain clear so the IC can do a roll call of all units. This ensures that all personnel have safely evacuated. After the roll call, the IC will allow the resumption of normal radio traffic.

## Incident Reports

Management and brigade SOPs commonly require that an incident report be completed for every type of emergency response. The nature and complexity of the report depend on the situation: Minor incidents generally require simple reports, whereas major incidents require extensive reports.

The National Fire Incident Reporting System (NFIRS) is a nationwide database managed by the U.S. Fire Administration that collects data related to fires and some other types of incidents. When NFIRS reports are required, they are normally submitted through the State Fire Marshal's Office or a comparable authority having jurisdiction. The requirement for facility fire brigades to submit an NFIRS varies from state to state. Typically, the first brigade leader or team leader completes the basic NFIRS information for each response, including a narrative description of the situation and the action that was taken. The full incident report may then include a supplementary report from the team leader or senior brigade member of each additional unit that responded to the incident (▶ Figure 3-13). Many fire brigades use a preferred narrative format when reporting on incidents (▶ Figure 3-14).

Some incidents require an expanded incident report narrative, in which all brigade members must submit a narrative description of their observations and activities during an incident. The brigade leader should expect to provide expanded incident reports on incidents in the following situations:

- The fire brigade was one of the first-arriving units at an incident that involved a fatality.
- The fire brigade was one of the first-arriving units at an incident that has become a crime scene or an arson investigation.
- The fire brigade participated in a rescue or other emergency scene activity that would qualify for official recognition, an award, or a bravery citation.
- The fire brigade was involved in an unusual, difficult, or high-profile activity that requires review by the brigade leader or designated authority.
- Fire brigade activity occurred that might have contributed to a death or serious injury.

### Brigade Member Safety Tips

Verifying the information received over the radio is vital. Always restate an important message or instruction to confirm that it has been received and understood.

IC: *"Command to Ladder 7. We need you to open the roof for ventilation, directly over the fire area."*

Ladder 7: *"Ladder 7 copied. We are going to the roof to ventilate directly over the fire."*

On Monday, May 12 at 1247 hours, Station 2 was alerted to respond with the Rescue Squad and the Tanker on the second alarm of a three-alarm building fire in Charles County. Chief 1 from the LaPlata VFD arrived on the scene at 8190 Port Tobacco Road, the Charles County Community Services Building, with heavy smoke showing from the rear.

While units from Station 2 were en route to the scene, the evacuation tones were sounded and all civilian county employees operating on the scene were forced to exit the building. Upon arrival, crews began cutting ventilation holes in the roof. But shortly after, they were pulled off due to the poor structural integrity of the building.

While on the scene, the crew of Rescue Squad 2 acted as RIT team number 1. Some crew members also assisted with the water supply operation. Tanker 2 was utilized in the tanker shuttle operation. All together, Tanker 2 hauled 23 loads of water to the scene which totalled 69,000 gallons by the time the operation was complete. Rescue Squad 2 was then assigned to take a transfer at LaPlata Station 1 and remained there for the remainder of the night. Both units were back in quarters at 2200 hrs.

**Figure 3-13** Supplementary report.

**Batesville Fire Department Incident Update**

Date: 09-16-2002
Time: 2349
First Company: B17
Incident Number: 101539
Address: 710 Werkcastle Lane
Incident Commander: Chief Margaret Hinkler
Dollar Loss: Still being determined
Occupancy: Residential
Displaced Occupants: 5
Adults Injured: 1
Minors Injured: 0
Civilian Fatalities: 0
Civilians Rescued: 1
Fire Fighter Injuries: 0
Fire Fighter Fatalities: 0
Total Fire Fighters on Scene: 25
Total Ambulances: 1
Time to Suppress Incident: 55 Min
Incident Duration: 7 HRS 10 MIN
Total Companies: 5

**Incident Description:**
On Saturday, September 16, 2002 at 2349 Hours (11:49 PM EST), five fire companies and one ambulance company responded to a residential fire at 710 Werkcastle Lane in Batesville. All companies were under the command of Chief Margaret Hinkler. B17 was the first company to arrive on the scene. B17 reported heavy smoke pouring from the downstairs right corner window of a two-story brick residence. Three minors and an adult female were standing outside of the residence. The adult female reported that her husband was still inside. A four-fire fighter rescue team entered the structure with a thermal imaging device. An adult male was found on his knees, coughing in the front hallway and was extracted from the residence. The fire was confined to the kitchen and family room and suppressed in 55 minutes. A dollar loss is still being determined. A grease fire on the stove was the cause of the fire.

**Figure 3-14** Sample incident report.

- Fire brigade activity occurred that might have created a liability.
- Fire brigade activity occurred that has led to an internal investigation.

The author of a narrative should describe any observations that were made en route or on the scene; he or she should also fully document his or her actions related to the incident. The narrative should provide a clear mental image of the situation and the actions that were taken. Incident reports should contain only facts—not opinion or speculation—because they may be used as part of a follow-up incident investigation or legal action. Including observations of abnormal or suspicious conditions in a report is not considered to be expressing an opinion if those observations are based on knowledge of the conditions that you would expect to encounter based on the type of incident and the equipment or process that was involved.

Incident reports should be completed as soon after the event as possible. Their prompt completion reduces the chance that important information that may have not been documented at the scene will be left out of the report. One way to retrieve information that was not written down during the event is to radio dispatch the communications center with progress reports or benchmarks such as "loss stopped" or "fuel leak is secured." Because the communications center keeps a record of all transmissions, these data can be collected later to help in completing the incident report.

The importance of accurate reports cannot be overemphasized. Inaccurate reports may leave a company or the fire brigade vulnerable to litigation. In addition, follow-up investigations and cause analysis may be hampered when an erroneous incident report is used as a tool in the investigation. For these reasons, incident reports should be consistent and completed within the period specified by company or fire brigade policy.

The specific content requirements of an incident report are determined by company management or by the fire brigade leader. At a minimum, reports should include the date and time of the incident, its location, the equipment or process involved, the conditions found on arrival, the actions taken, and the results of the brigade's actions. Reports may also include provisions for photographs, event logs or tactical worksheets, dispatch and radio logs, and witness statements. An incident report should document who, what, why, when, where, how, and how many.

Once incident reports are completed, they should be forwarded to the appropriate brigade authority for further processing and file retention in accordance with the brigade procedures or operating policy.

Accurate and timely reports serve an additional purpose: They provide a basis for lessons learned, thereby enabling fire fighters to learn from others' mistakes. Ultimately, these lessons may lead to changes or updates to training procedures or department policies.

## Taking Calls: Emergency, Nonemergency, and Personal Calls

One of the first things you should learn when assigned to a fire station is how to use the telephone and intercom systems. You must be able to use them to answer a call or to announce a response. You should keep personal calls to a minimum, so incoming phone lines are open to receive emergency calls.

A brigade member who answers the telephone in a fire station, fire brigade facility, or communications center is a representative of the fire brigade. Use your brigade's standard greeting when you answer the phone: "Good afternoon, XYZ Refinery-Emergency Response Brigade, how may I help you?" Be prompt, polite, professional, and concise.

Detailed SOPs for obtaining information and processing calls should be provided to any brigade member assigned to answer incoming emergency telephone lines. Someone may call your fire station directly instead of dialing the designated emergency number for your facility. If this happens, you are responsible for ensuring that the caller receives the appropriate emergency assistance. Your organization's SOPs should outline exactly what steps you should take in this situation, such as whether you should take the information yourself or connect the caller directly to the communications center. ▶ Skill Drill 3-2 lists the steps in obtaining essential information from a caller for an emergency response.

1. When you answer a call on any incoming line, follow your organization's SOPs. Answer promptly and professionally. Identify yourself and the brigade or location. For example, "Good afternoon. Fire Station 4, Jones speaking."
2. Determine immediately if the caller has an emergency.
3. If there is an emergency, follow your organization's SOPs. You may have to transfer the call to the communications center or take the information yourself and relay it to the communications center. **(Step 1)**
4. If you take the information from the caller, be sure to get as much information as possible, including:
   a. Type of incident/situation
   b. Location of the incident
   c. Cross streets or identifying landmarks

### Teamwork Tips

When using the radio at an incident involving multiple jurisdictions, use clear speech to avoid confusion.

# VOICES OF EXPERIENCE

**“Each of the three departments had its own frequency channel assignments, which meant that other than face-to-face communications, the three departments could not communicate with one another.”**

We had a process building fire at the refinery that was contained in a hidden space within the structure of the building. The standard response to a structure fire was an engine from the refinery, an attack engine and ladder from the city, and an engine from the neighboring fire department. Our refinery fire brigade began to initiate exterior fire control until the mutual aid departments arrived.

When the two municipal departments' personnel arrived, they were assigned to crews and began to actively fight the fire. As the crews were being assigned, they were allocated a radio frequency on which to operate. The crews could consist of fire fighters from any of the three departments on the scene.

Almost immediately, it became obvious that our lack of a communication plan was severely hampering our response. Each of the three departments had its own frequency channel assignments, which meant that other than face-to-face communications, the three departments could not communicate with one another.

In the wake of the refinery incident, a workgroup was established to identify the problems and develop a plan to correct them. The final product was a communication plan that allowed each of the departments to operate independently of one another, yet still operate seamlessly when responding mutual aid with one another.

A few months later, a crude oil leak occurred within a crude unit. During that incident, the communication plan was put to the test and passed with flying colors—all three departments were able to communicate with one another. The spill was mitigated with no injuries or property damage, and the environmental impact was minimized because of the effectiveness of our communication plan. A little planning before the incident paid large dividends during the response.

**Tyler Bones**
Flint Hills Resources Alaska
North Pole, Alaska

d. Indication of scene safety
e. Caller's name
f. Caller's location, if different from the incident location
g. Caller's call-back number

5. If your station or your unit will be responding to the call, always advise the communications center immediately before responding. Be prepared to take accurate information or messages on all calls. Include the date, time, name of the caller, caller's phone number, message, and the call taker's name. Have questions already formed and organized to maintain control of the conversation. Never leave someone on hold for any longer than necessary. Always terminate the call in a courteous manner. Let the caller hang up first. **(Step 2)**

3-2 Skill Drill

## Operating and Answering Fire Station Telephone and Intercom Systems

1 Determine immediately if the caller has an emergency. If there is an emergency, follow your organization's SOPs.

2 If you take the information from the caller, focus on obtaining vital information. If your station or your unit will be responding to the call, advise the communications center immediately. Always be prepared to take accurate information or messages for emergency, nonemergency, and personal calls. Never leave someone on hold for a long time. Always let the caller hang up first.

# Wrap-Up

## Ready for Review

- The telecommunicator works in the communications center, which may be a fire station or a separate facility.
- Brigade members must know how to use communication equipment properly, whether they are taking a request for routine or emergency assistance or using radio equipment at an emergency incident.
- The primary responsibilities of the telecommunicator are to receive requests from the public or the industrial facility, determine the appropriate level of response, notify the proper agency, and record the proper information.
- A call for help is always viewed as an emergency by the person making the call.
- Emergencies may be reported in several ways, including by telephone, wireless/cellular telephone, emergency call boxes, public alarm systems, radio, or walk-ins.
- The telecommunicator can use several systems, ranging from manual run cards to CAD systems, to determine which units should be deployed.
- The first-arriving unit on the scene—whether it is a brigade member, an engine crew, an ambulance crew, or a brigade leader—should establish command and communicate a brief initial report that conveys the initial size-up of the incident.
- In some instances, emergency traffic (an urgent message) must take precedence over normal radio communications.
- Incident reports are commonly required for every type of emergency response.

## Hot Terms

**Activity logging system** Device that keeps a detailed record of every incident and activity that occurs.

**Base station** Radios are permanently mounted in a building, such as a fire station, communications center, or remote transmitter site.

**Call box** System of telephones connected by phone lines, radio equipment, or cellular technology to communicate with a communications center or fire organization.

**Communications center** Facility that receives emergency or nonemergency reports from citizens. Many communications centers are responsible for dispatching fire brigade units as well.

**Computer-aided dispatch (CAD)** Computer-based, automated systems used by telecommunicators to obtain and assess dispatch information. The system recommends the type of response required.

**Direct line** Telephone that connects two predetermined points.

**Dispatch** A summons to fire brigade units to respond to an emergency. Also known as alerting, dispatch is performed by the telecommunicator at the communications center.

**Duplex channel** A radio system that uses two frequencies per channel. One frequency transmits and the other receives messages. The system uses a repeater site to transmit messages over a greater distance than a simplex system.

**Emergency traffic** An urgent message, such as a call for help or evacuation, transmitted over a radio that takes precedence over all normal radio traffic.

**Evacuation signal** Warn all personnel to pull back to a safe location.

**Expanded incident report narrative** A report in which all company members submit a narrative describing what they observed and which activities they performed during an incident.

**Federal Communications Commission (FCC)** United States federal regulatory authority that oversees radio communications.

**Mobile data terminals (MDTs)** Technology that allows brigade members to receive information in the apparatus or at the station.

**Mobile radio** Two-way radio that is permanently mounted in a fire apparatus.

**National Fire Incident Reporting System (NFIRS)** A nationwide database held at the National Fire Data Center under the U.S. Fire Administration that collects fire-related data so as to provide information on the national fire problem.

**Portable radio** A battery-operated, hand-held transceiver.

**Public safety answering point (PSAP)** The community's emergency response communications center.

**Radio repeater system** A radio system that automatically retransmits a radio signal on a different frequency.

**Run cards** Cards used to determine a predesignated response to an emergency.

**Simplex channel** Radio system that uses one frequency to transmit and receive all messages.

**Squelch** An electric circuit designed to cut off weak radio transmissions that are only capable of generating noise.

**Talk around channel** A simplex channel used for on-site communications.

**Telecommunicator** A trained individual responsible for answering requests for emergency and nonemergency assistance from citizens. This individual assesses the need for response and alerts responders to the incident.

**Telephone interrogation** Phase in a 9-1-1 call when a telecommunicator asks questions to obtain vital information such as the location of the emergency.

**Ten-codes** System of predetermined, coded messages, such as "What is your 10-20?" used by responders over the radio.

**Time marks** Status updates provided to the communications center every 10 to 20 minutes. This update should include the type of operation, the progress of the incident, the anticipated actions, and the need for additional resources.

**Trunking system** A radio system that uses a shared bank of frequencies to make the most efficient use of radio resources.

**Voice recording system** Recording devices or computer equipment connected to telephone lines and radio equipment in a communications center to record telephone calls and radio traffic.

# Brigade Member in Action

**You serve as the apparatus driver for the first-due quick response vehicle in your facility. During your shift, you receive an alert on your response team pager regarding a report of fire alarm activation in one of the research lab buildings in the north zone of your facility. As part of the quick response unit, it is your responsibility to respond to the alarm activation and to provide appropriate communications back to the dispatch center. You are also responsible for communicating necessary information to other responding fire brigade members.**

1. When communicating to the dispatch center, what would be your preferred type of radio to use?
   - **A.** A portable radio without an antenna should be used because it is the most compact.
   - **B.** A portable radio with a short antenna should be used because it is the lightest-weight radio.
   - **C.** The quick response radio should be used because it has more power.
   - **D.** Size up the scene and then drive to the fire station and use the base station radio.

2. Which of the following statements is true regarding your radio communications?
   - **A.** The use of radio 10 codes is strongly discouraged by the National Incident Management System.
   - **B.** You should speak very loud and give very detailed messages.
   - **C.** You should press the "push to talk" button and begin speaking immediately.
   - **D.** You should hold the microphone at least 6 inches away from your mouth.

3. What is the primary purpose of your initial on-scene radio report?
   A. It ensures that the radio of the first arriving unit is working correctly.
   B. It gives other units a sense of what is happening so they can anticipate what their assignments may be when they arrive on scene.
   C. It provides a complete listing of all tasks that will be conducted during the incident.
   D. It allows the incident commander to make all incident command assignments.

4. How often should update reports be provided during a response?
   A. Every 3 to 5 minutes
   B. Only when significant changes in the incident occur
   C. When significant changes occur and at least every 10 to 20 minutes
   D. Every 30 minutes

5. Which of the following statements is true regarding emergency radio messages?
   A. Emergency traffic is an urgent message that takes priority over all other radio traffic.
   B. The most important emergency traffic is a brigade member's call for help.
   C. An evacuation signal is used to warn all personnel to pull back to a safe location.
   D. All of the above.

# Incident Management System

# Chapter 4

## NFPA 1081 Standard

**Incipient Facility Fire Brigade Member**

**4.1.2** *Basic Incipient Facility Fire Brigade Member JPRs.* All basic facility incipient brigade members shall have a general knowledge of basic fire behavior, operation within an incident management system, operation within the emergency response operations plan for the site, the standard operating and safety procedures for the site, and site-specific hazards.

**Advanced Exterior Facility Fire Brigade Member**

**5.1.2.2*** Interface with outside mutual aid organizations, given standard operating procedures (SOPs) for mutual aid response and communication protocols, so that a unified command is established and maintained.

**(A)** *Requisite Knowledge.* Mutual aid procedures and the structure of the mutual aid organization, site SOPs, and incident management systems.

**(B)** *Requisite Skills.* The ability to communicate with mutual aid organizations and to integrate operational personnel into teams under a unified command.

**Interior Structural Facility Fire Brigade Member**

**6.2.7** Interface with outside mutual aid organizations, given SOPs for mutual aid response and communication protocols, so that a unified command is established and maintained.

**(A)** *Requisite Knowledge.* Mutual aid procedures and the structure of the mutual aid organization, site SOPs, and incident management systems.

**(B)** *Requisite Skills.* The ability to communicate with mutual aid organizations and to integrate operational personnel into teams under a unified command.

**6.3.9*** Interface with outside mutual aid organizations, given SOPs for mutual aid response and communication protocols, so that a unified command is established and maintained.

**(A)** *Requisite Knowledge.* Mutual aid procedures and the structure of the mutual aid organization, site SOPs, and incident management systems.

**(B)** *Requisite Skills.* The ability to communicate with mutual aid organizations and to integrate operational personnel into teams under a unified command.

## Knowledge Objectives

After completing this chapter, you will be able to:

- Have an understanding of operations within an incident management system.
- Have an understanding of SOPs for mutual aid response and communication protocols so that a unified command can be established and maintained.

## Skills Objectives

After completing this chapter, you will be able to perform the following skills:

- Communicate with mutual aid organizations and integrate operational personnel into teams under a unified command.
- Understand mutual aid procedures and the structure of the mutual aid organization, site SOPs, and incident management systems.
- Initiate communication using facility communication equipment to effectively relay oral or written communication for the following:
  - Responding to alarms
  - Returning equipment to service
  - Complete incident reports
- Have a basic knowledge of operation within the emergency response operations plan for the site.
- Have a basic knowledge of the standard operating and safety procedures for the site.
- Have a basic knowledge of site-specific hazards.

# You Are the Brigade Member

**You have just finished filling the pump bearing oil on the unit 1 cooling water pump when you hear a massive explosion from across the property. The fireball rising from the storage tank area seems to travel upward for well over a thousand feet when you start to feel the sudden wind of a blast wave and heat that has been somewhat buffered by the process unit located between you and the fire area. You are jolted from your focus on the rising cloud of smoke and fire by your portable radio alert tones and the site emergency siren. The control room announcement following the tones states that all emergency responders are to report to their emergency response locations. As you begin to respond to the Fire Equipment Storage Building, you hear your supervisor announce that he is assuming Incident Command and that the command post will be located next to the warehouse office area. As you are dressing out in your bunker gear, you hear the Incident Commander request mutual aid from the adjoining industrial sites and the local municipal fire brigade. Security acknowledges that off-site responders will be arriving and asks the Incident Commander where he wants to establish staging. The Incident Commander replies that the first two arriving engines should be directed to the command post, where they will establish a water and foam supply for suppression activities, and all other responders should be held at the plant entry gate, where staging will be established.**

**Based on your knowledge of the facility preplanning that you performed as a part of your fire brigade training and the Incident Management System, you understand that it will take an organized team effort to coordinate the different responders to bring about a safe conclusion to this incident.**

1. ***Why is it important to establish a well-organized command structure at a major incident?***
2. ***Why, with such an obviously large incident, are remaining responders going to be held at the plant entrance instead of being dispatched to the scene?***
3. ***Where would you expect to find the Incident Commander?***

## Introduction

The Incident Management System (IMS) is a management structure that utilizes only the components that are needed at each emergency incident. The structure of IMS is similar to that of a company. A company is organized with similar components, but different titles. The company always has a head, the CEO, who makes the final decisions and creates the general plan of action for the company. The company is then broken down into brigades (Production, Scheduling, Purchasing, Finance) to implement the CEO's plan. Each brigade has a support staff to help accomplish the organization's responsibilities. Like a company, IMS is also broken down into components. There is an overall person in charge, the Incident Commander (IC), and each brigade (Command, Operations, Planning, Logistics, and Finance/Administration) has specific responsibilities.

All fire brigade emergency operations and training exercises should be conducted within the framework of an IMS. Using an IMS ensures that operations are coordinated and conducted safely and effectively. An IMS provides a standard approach, structure, and operational procedure to organize and manage any operation, from emergencies to training sessions. The same principles apply, whether the situation involves a single brigade or hundreds of emergency responders from dozens of different agencies. They also can be applied to many nonemergency events, such as large-scale public events, that require planning, communications, and coordination.

Planning, supervision, and communications are key components of an IMS. Model procedures for incident management have been developed and widely adopted to provide a standard approach that can be used by many different agencies. Within a single fire brigade or a group of fire brigades that routinely respond together, the use of a standard system for managing an incident is essential. Enhanced coordination results when every organization that could be involved in a major incident is familiar with and has practiced with the same incident management system.

Some fire brigades use procedures that vary from the IMS model or other model systems in different respects. As a brigade member, it is essential for you to become intimately familiar with the system that is used within your jurisdiction. Where local variations exist, the overall system usually remains similar in structure and application to the IMS model.

## History of the IMS

Prior to the 1970s, each individual fire brigade had its own method for commanding and managing incidents. Often, the organization established to direct operations at an incident scene depended on the style of the leader on duty. However, this individualized approach did not work well when units from different shifts or mutual aid companies responded to a major incident. What was adequate for routine incidents was ineffective and confusing with large-scale incidents, rapidly changing situations, and personnel that did not normally work together.

Such a fragmented approach to managing emergency incidents is no longer accepted. Over the past 40 years, formal IMS, or Incident Command System (ICS), has been developed and refined. Today's IMSs include an organized system of roles, responsibilities, and standard operating procedures (SOPs) that are widely used to manage and direct emergency operations. The same basic approaches, organization structures, and terminology are used by thousands of fire brigades and emergency response agencies.

The move to develop a standard system began about 40 years ago, after several large-scale wildland fires in Southern California proved disastrous for both the fire service and residents. A number of fire-related agencies at the local, state, and federal levels decided that better organization was necessary to effectively combat these costly fires. These agencies established an organization known as FIRESCOPE (**FI**re **RES**ources of **C**alifornia **O**rganized for **P**otential **E**mergencies) to develop solutions for a variety of problems associated with large, complex emergency incidents during wildland fires. These problems included:

- Command and control procedures
- Resource management
- Terminology
- Communications

FIRESCOPE developed the first standard ICS. Originally, ICS was intended only for large, multijurisdictional/multiagency incidents involving more than 25 resources or operating units. However, it was so successful that it was applied to structural firefighting and eventually became an accepted system for managing all emergency incidents ▶ Figure 4-1.

At the same time, the Fireground Command (FGC) system was developed and adopted by many fire brigades. The concepts behind the FGC system were similar to the ICS, although there were some differences in terminology and organization structure. The FGC system was initially designed for day-to-day fire brigade incidents involving fewer than 25 fire suppression companies, but it could be expanded to meet the needs of large-scale incidents.

During the 1980s, the ICS developed by FIRESCOPE was adopted by all federal and most state wildland firefighting agencies. The National Fire Academy (NFA) also used the FIRESCOPE ICS as the model fire service IMS for all of its courses. Additional federal agencies, including the Federal Emergency Management Agency (FEMA) and the Federal Bureau of Investigation (FBI), adopted the same model for use during major disasters or terrorist events. All federal agencies could now learn and use one basic system.

Several federal regulations and consensus standards, including NFPA 1500, *Standard on Fire Department Occupational Safety and Health Program,* were adopted during the 1980s. NFPA 1081 also mandates the use of an IMS. These standards mandated the use of an ICS at emergency incidents. NFPA 1561, *Standard on a Fire Department Incident Management System,* released in 1990, identified the key components of an effective system and the importance of using such a system at all emergency incidents. Fire brigades could use either ICS or FGC system to meet the requirements of NFPA 1561.

In the following years, users of different systems from across the country formed the National Fire Service Incident Management System Consortium to develop "model procedure guides" for implementing effective IMS at various types of incidents. The resulting system, which blended the best aspects of both ICS and FGC, is now known formally as IMS. The IMS can be used at any type or size of emergency

**Figure 4-1** ICS was first developed to coordinate efforts during large-scale wildland fires.

## National Incident Management System

In 2003, President George W. Bush directed the Secretary of Homeland Security to develop and administer a National Incident Management System (NIMS). This system provides a consistent nationwide template to enable federal, state, and local governments as well as private-sector and nongovernmental organizations to work together effectively and efficiently to prepare for, prevent, respond to, and recover from domestic incidents, regardless of their cause, size, or complexity, including acts of catastrophic terrorism.

Since the terrorist attacks of September 11, 2001, much has been done to improve the prevention, preparedness, response, recovery, and mitigation capabilities and coordination processes across the United States. A comprehensive national approach to incident management, which is applicable at all jurisdictional levels and across all functional disciplines, would further improve the effectiveness of emergency response providers and incident management organizations over a full spectrum of potential incidents and hazard scenarios. Such an approach would also improve coordination and cooperation between public and private entities in a variety of domestic incident management activities. These incidents may include any of the following:

- Acts of terrorism
- Wildland and urban fires
- Floods
- Hazardous materials spills
- Nuclear accidents
- Aircraft accidents
- Earthquakes
- Hurricanes
- Tornadoes
- Typhoons
- War-related disasters

Building on the foundation provided by existing incident management and emergency response systems used by jurisdictions and functional disciplines at all levels, the NIMS integrates the practices that have proven most effective over the years into a comprehensive framework for use by incident management organizations in an all-hazards context on a national basis. To provide for interoperability and compatibility among federal, state, and local capabilities, the NIMS includes a core set of concepts, principles, terminology, and technologies addressing the following issues:

- The incident command system
- Multiagency coordination systems
- Unified command
- Training
- Identification and management of resources
- Qualifications and certification
- Collection, tracking, and reporting of incident information and incident resources

While most incidents are generally handled on a daily basis by a single jurisdiction at the local level, in some important cases successful domestic incident management operations depend on the involvement of multiple jurisdictions, functional agencies, and emergency responder disciplines. These cases require effective and efficient coordination across a broad spectrum of organizations and activities. The NIMS uses a systems approach to integrate the best of existing processes and methods into a unified national framework for incident management. This framework supports interoperability and compatibility that will, in turn, enable a diverse set of public and private organizations to conduct well-integrated and effective incident management operations.

The NIMS includes several components that work together as a system to provide a national framework for preparing for, preventing, responding to, and recovering from domestic incidents:

1. *Command and management:* The NIMS standardizes incident management for all hazards and across all levels of government. The NIMS standard incident command structures are based on three key constructs: an incident command system, multiagency coordination systems, and public information systems.
2. *Preparedness:* The NIMS establishes specific measures and capabilities that jurisdictions and agencies should develop and incorporate into an overall system to enhance operational preparedness for incident management on a steady-state basis in an all-hazards context.
3. *Resource management:* The NIMS defines standardized mechanisms to describe, inventory, track, and dispatch resources before, during, and after an incident. It also defines standard procedures to recover equipment once it is no longer needed for an incident.
4. *Communications and information management:* Effective communication, information management, and information and intelligence sharing are critical aspects of domestic incident management. The NIMS communication and information systems facilitate the essential functions needed to provide a common operating picture and interoperability for incident management at all levels.
5. *Supporting technologies:* The NIMS promotes national standards and interoperability for supporting technologies to successfully implement the NIMS and standard technologies for specific professional disciplines or incident types. It provides an architecture for science and technology support to incident management.
6. *Ongoing management and maintenance:* The Department of Homeland Security will establish a multijurisdictional, multidisciplinary NIMS Integration Center. This center will provide strategic direction for and oversight of the NIMS, supporting routine maintenance and enabling continuous improvement of the system over the long term.

incident and by any type or size organization or agency. NFPA 1561 is now called *Standard on Emergency Services Incident Management System.*

## IMS Characteristics

An IMS provides a standard, professional, organized approach to managing emergency incidents. The use of an IMS enables any type of emergency response agency to operate more safely and effectively. A standardized approach facilitates and coordinates the use of resources from multiple agencies, working toward common objectives. It also eliminates the need to develop a special approach for each situation.

All IMSs require an organizational structure to provide both a hierarchy of authority and responsibility and formal channels for communications. The specific responsibility and authority of everyone in the organization is clearly stated and the relationships are well-defined. Important characteristics of an IMS include:

- Recognized jurisdictional authority and responsibility
- Applicable to all risk and hazard situations
- Applicable to day-to-day operations as well as major incidents
- Unity of command
- Span of control
- Modular organization
- Common terminology
- Integrated communications
- Consolidated incident action plans
- Emergency response operations plan
- Designated incident facilities
- Resource management

### Jurisdictional Authority

The identification of one individual as the ultimate authority with overall responsibility for managing an incident is a key concept for the success of IMS. The issue of jurisdictional authority can become a serious concern at large-scale incidents involving multiple agencies and different levels of government. Each agency has specific responsibilities and legal authority for different situations, depending on the nature and scale of the incident, the geographic location, and other factors. Someone has to be "in charge" to ensure that operations are coordinated and safe.

Jurisdictional authority may or may not be a problem at an industrial fire, where the highest ranking or designated supervisor for the companies' fire brigade is clearly the IC. It can be more complicated when the local municipal fire organization or several organizations are involved. For example, a military aircraft crashes in an industrial complex and starts a fire that spreads across the industrial complex boundaries into an adjoining city. There, the fire causes the hazardous material to leak into a navigable waterway and approaches a jail located inside the limits of another incorporated city. Such a situation involves both military and civilian agencies, as well as multiple levels of government. Chaos would prevail if each affected jurisdiction claimed to be in charge and attempted to issue orders.

#### Establishing a Unified Command Structure

An effective IMS clearly defines the agency that will be in charge of each situation. When there are overlapping responsibilities, the IMS may employ a unified command. This brings representatives of different agencies together to work on one plan and ensures that all actions are fully coordinated. Each agency has command authority for specific, well-defined areas of responsibility.

Unified command systems should always be used whenever an incident requires mutual aid. A multiagency and/or multijurisdiction response to a mutual aid call can best be handled by using a unified command structure. This type of structure is especially important when working with local municipal first responders such as fire fighters, law enforcement personnel, and EMS responders.

The local fire service IC will want to work in tandem with the brigade team leader/IC. Working with the facility emergency response coordinator (FERC) or the team liaison officer (TLO) may be common practice in some areas, but it is preferable to work with the facility IC directly in a unified command system. In some incidents, the local fire service IC's role is to protect the community; preserving the facility may be a secondary consideration. In some cases, the fire service IC may take control of the incident.

The local ICS capabilities, including the ability to establish a unified command, will depend on the community's adoption of NIMS, the levels of ICS training completed, and the prefire planning that the local first responders have completed at the facility.

##### Integrating Brigade Members under a Unified Command

Brigade members may be integrated under a unified command only after they receive the proper level of ICS training (ICS 100–400). This training ensures that they can operate safely and proficiently when they are assigned to a division/branch, group, strike team, or task force.

Members who respond to an incident are expected to check in with the IC or with the staging areas manager, using the site-specific accountability system. At that point, they will be assigned to a team or group that has been given an assignment. Brigade members are expected to complete that assignment and then either to request another assignment or to report to the rehabilitation area.

#### Mutual Aid

Mutual aid agreements can help a facility both meet its need for resources and handle a second or cascading incident that can affect a fire brigade that is understaffed and/or underequipped.

Mutual aid agreements should be negotiated and documented before an incident occurs.

If prefire planning activities reveal that additional resources—whether in the form of staffing, equipment, or technical expertise—may be needed, then a mutual aid agreement between the local first responders, regional response teams, and even other industrial facilities may be in order. Written agreements that clearly establish the various responders' roles and responsibilities, policies for replacement of damaged equipment, and coverage for injured brigade members must be in place to protect both the giver and the receiver of mutual aid.

### Working with Mutual Aid Partners

A working relationship among the brigade leaders and members should be developed before an incident requiring mutual aid occurs. The trust engendered by knowing each member's knowledge, skills, and abilities will go a long way toward ensuring smooth operations when working with a brigade from another facility. This trust and teamwork can come from joint training sessions, exercises, and annual drills.

### Mutual Aid Response and Communications

How mutual aid is requested and which communication procedures are to be used should also be defined in the written mutual aid agreement. Some agreements call for a reconnaissance or "go" team to respond and conduct a size-up before more resources are sent in to assist the first responders. This approach eliminates the chance of losing production/maintenance workers who might otherwise serve as brigade members unnecessarily.

Phone calls, common radio frequencies, and paging systems can be used in the notification process. These systems should be tested on a regular basis.

### Structure of Mutual Aid Organizations

Mutual aid organizations usually consist of more than two facilities or agencies. In some areas of the country, these organizations include dozens or even hundreds of member facilities and/or agencies. Their activities may be coordinated by local emergency planning committees (LEPC), by other governmental agencies, or by the organization itself.

## All-Risk and All-Hazard System

IMS has evolved into an all-risk, all-hazard system that can be applied to manage resources at fires, floods, tornadoes, plane crashes, earthquakes, hazardous materials incidents, or any other type of emergency situation (▶ Figure 4-2). IMS has also been used to manage many nonemergency events, such as large-scale public events, that have similar requirements for command, control, and communications. The flexibility of the system enables the management structure to expand as needed, using whatever components are needed. Multiple agencies and organizations can be integrated in the management of the incident.

Figure 4-2 IMS can be used to manage different types of emergency incidents involving several agencies.

## NIMS and the Private Sector

Facilities should adopt the use of NIMS/ICS as a corporate policy. The FEMA fact sheet titled *Private Sector NIMS Implementation Activities* (dated November 30, 2006) provides guidance by identifying the 12 recommended activities for members of the private sector that seek to become NIMS compliant as recommended in Homeland Security Presidential Directive–5. These activities closely parallel the implementation activities that have been required of state, territorial, tribal, and local governments since 2004. Effective and consistent integration of the NIMS across federal, state, territorial, tribal, and local governments and the private sector will result in a strengthened national capacity to prevent, prepare for, respond to, and recover from any type of incident.

## Everyday Applicability

IMS can and should be used for everyday operations as well as major incidents. Someone is in command at every incident, whether it is a trash bin fire, a medical emergency, or a building fire (▶ Figure 4-3). Regular use of the system builds familiarity with standard procedures and terminology. It also increases the users' confidence in the system. Frequent use of IMS for routine situations makes it easier to apply to larger incidents.

Fire brigade members will become competent at IMS when it is used on a daily basis along with the standard operating and safety procedures as defined in the facility emergency response operations plans that are designed for the site-specific hazards of the facility.

Figure 4-3 IMS can be used effectively during day-to-day operations.

## Unity of Command

Unity of command is a management concept in which each person has only one direct supervisor. All orders and assignments come directly from that supervisor, and all reports are made to the same supervisor. This eliminates the confusion that can result when a person receives orders from more than one boss. Unity of command reduces delays in solving problems, the potential for life and property losses, and safety risks for brigade members.

IMS is not necessarily a rank-oriented system. The best qualified person should be assigned at the appropriate level for each situation, even if a lower ranking individual is temporarily assigned to a higher level position. This concept is critical for the effective application of the system and must be embraced by all participants.

## Span of Control

Span of control refers to the number of subordinates who report to one supervisor at any level within the organization. Span of control relates to all levels of IMS, from the strategic level to the operational/tactical level and to the task level (individual brigades or crews).

In most situations, one person can effectively supervise only three to seven people. Because of the dynamic nature of emergency incidents, an individual who has command or supervisory responsibilities in an IMS normally should not directly supervise more than five people. The actual span of control should depend on the complexity of the incident and the nature of the work being performed. For example, at a complex hazardous materials incident, the span of control might be only three; during less intense operations, the span of control could be as high as seven.

## Modular Organization

IMS is designed to be flexible and modular. The IMS modules—Command, Operations, Planning, Logistics, and Finance/Administration—are predefined, ready to be staffed and made operational as needed. IMS has often been characterized as an organizational toolbox—only the tools needed for the specific incident are used. In IMS, the tools are position titles, job descriptions, and an organization structure that defines the relationships between positions. Some positions and functions are used frequently; others are needed only for complex or unusual situations. Any position can be activated simply by assigning someone to it.

For example, a small structure fire can usually be managed by an IC, who directly supervises four or five brigade leaders. Each brigade leader supervises his respective brigade members. At a larger fire, with more brigades, the IMS would expand to include divisions of up to five brigades. At the same time, officers would be assigned to perform specific functions, such as safety and planning. At more complex incidents, additional levels and positions within the IMS structure would be filled in the same manner. The organization can expand as much as needed to manage the incident.

## Common Terminology

IMS promotes the use of common terminology both within an organization and among all of the agencies involved in emergency incidents. Common terminology means that each word has a single definition, and no two words used in managing an emergency incident have the same definition. Everyone uses the same terms to communicate the same thoughts, so everyone understands what is meant. Each job comes with one set of responsibilities, and everyone knows who is responsible for each duty.

Common terminology is particularly important for radio communications. For example, in many fire organizations, the term "tanker" refers to a mobile water supply apparatus. But to wildland firefighting agencies, it means a fixed-wing aircraft used to drop extinguishing agent on a fire. Without common terminology, an IC who called for a "tanker" could get an airplane instead of the truck he or she expected.

## Integrated Communications

Integrated communications ensure that everyone at an emergency can communicate with both their supervisors and their subordinates. The IMS must support communication up and down the chain of command at every level. A message must be able to move efficiently through the system from the IC down to the lowest level and from the lowest level up to the IC. Within a fire brigade, the primary means of communicating at the incident scene is by radio.

## Consolidated Incident Action Plans

An IMS ensures that everyone involved in the incident is following one overall plan. Different components of the organization may perform different functions, but all of their efforts contribute to the same goals and objectives. Everything that occurs is coordinated with everything else.

At smaller incidents, the IC develops an action plan and communicates the incident priorities, objectives, strategies, and tactics to all of the operating units or individuals. At large incidents, representatives from all participating agencies meet regularly to develop and update the plan. In both large and small incidents, those involved in the incident understand their specific roles and how they fit into the overall plan.

## Emergency Response Operations Plan

OSHA and NFPA emergency action plans require that written plans be developed in the form of a site-specific emergency response operations plan. Such a plan is designed to identify the levels of response needed for certain locations with emergency situations such as fires, chemical spills, and medical emergencies.

The site-specific plan should cover the following elements:

- Employee evacuation plans
- Evacuated-employee accountability procedure
- Procedures for reporting fires and other emergencies
- Critical shutdown procedures
- Rescue and medical response plan
- Choosing a level of response to each type of emergency

The level of response chosen must be appropriate for each type of incident, as well as the training and equipment necessary to operate at the chosen levels. By utilizing OSHA and NFPA levels of response, the brigade leaders and members understand their limitations, roles, and responsibilities when responding to an incident.

NFPA levels of response for fire brigades include the following:

- Incipient Level
- Advanced Exterior
- Interior Structural
- Advanced Exterior/Interior Structural

OSHA-regulated facilities must choose one of the following levels of response to chemical spills and leaks:

- Fire Responder Awareness Level
- First Responder Operations Level
- Hazardous Materials Technician
- Hazardous Materials Specialists

NFPA uses similar levels of response for technical rescue operations (NFPA 1670). A brigade would choose a level of response for each type of technical rescue that may be needed at the facility. At some facilities, topics such as confined-space rescue, rope rescue, and extrication from machinery are addressed at different levels based on the hazards found at the specific facility.

## Designated Incident Facilities

Designated incident facilities are assigned locations where specific functions are always performed. For example, the IC will always be based at the incident command post. The staging area, rehabilitation area, casualty collection point, treatment area, base of operations, and helispot are all examples of designated areas where particular functions take place. The facilities required for the incident are established according to the IMS plan. This plan may have been a preestablished "template" that was developed prior to an incident.

## Resource Management

Resource management is a standard system of assigning and keeping track of the resources involved in the incident. In structural firefighting, the basic units are companies. Engine companies, ladder companies, and other units are dispatched to the incident and assigned a specific function. In some cases, groups from several different companies are assigned to task forces and strike teams. The IMS keeps track of company assignments. In industry, these resources are usually referred to as "brigades" or "teams" although they may be referred to as "companies."

At large-scale incidents, units are often dispatched to a staging area and not directly to the incident location. A staging area is a location close to the incident scene where a number of units can be held in reserve, ready to be assigned if they are needed.

# The IMS Organization

The IMS structure identifies a full range of duties, responsibilities, and functions that are performed at emergency incidents. It defines the relationships among all these different components. Some components are used on almost every incident, while others apply to only the largest and most complex situations. The five major components of an IMS organization are Command, Operations, Planning, Logistics, and Finance/Administration.

The IMS organization chart may be quite simple or very complex. Each block on an IMS organization chart refers to a function area or a job description. Positions are staffed as they are needed. The only position that must be filled at every incident is the IC. The IC decides which additional components are needed for the situation and activates those positions by assigning someone to perform those tasks.

Brigade members must understand the overall structure of IMS, as well as the basic roles and responsibilities of each position. As an emergency develops, a brigade member could start in logistics, move to operations, and eventually assume a command position. Knowing how IMS works enables brigade

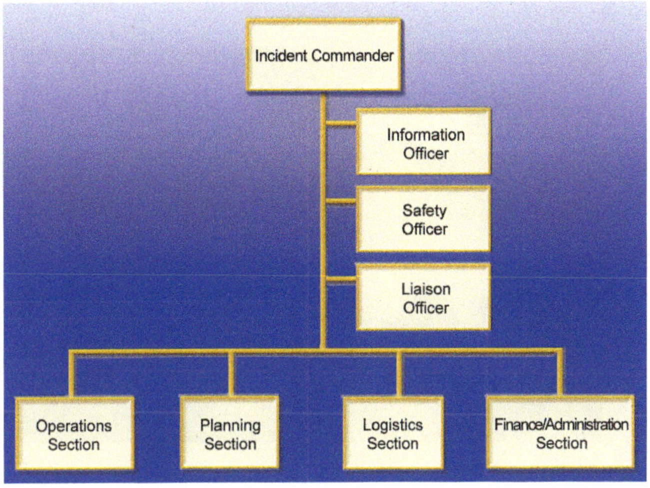


Figure 4-4 The IMS organization chart.

Figure 4-5 A unified command involves many agencies directly in the decision-making process for a large incident.

members to see how different roles and responsibilities work together and relate to each other. This makes it easier to focus on a particular assignment, instead of being overwhelmed by the incident.

## Command

On an IMS organization chart (▲Figure 4-4), the first component is Command. The IC is the only position in the IMS that must **always** be filled because there must always be someone in charge. Command is established when the first unit arrives on the scene and is maintained until the last unit leaves the scene. The IC is equivalent to the CEO of a company. The IC and staff are not full-time positions but are *functional positions*, filled by qualified responders.

In the IMS structure, one person, the IC, is ultimately responsible for managing an incident and has the necessary authority to direct all activities at the incident scene. The IC is directly responsible for:

- Determining strategy
- Selecting incident tactics
- Setting the action plan
- Developing the IMS organization
- Managing resources
- Coordinating resource activities
- Providing for scene safety
- Releasing information about the incident
- Coordinating with outside agencies

### Unified Command

Most incidents have one IC, who is directly responsible for all of the command functions. When multiple agencies with overlapping jurisdictions or legal responsibilities are involved in the same incident, a unified command is used. Representatives from each agency cooperate to share command authority. They work together and are directly involved in the decision-making process. Unified command helps ensure cooperation, avoids confusion, and guarantees agreement on goals and objectives (▲Figure 4-5).

### Command Post

The command post is the headquarters location for the incident. Command functions are centered in the command post. The IC and all direct support staff should always be located at the command post.

The command post should be in a nearby, protected location. Often, the command post for a major incident is located in a special building or vehicle. This enables the command staff to function without needless distractions or interruptions.

### Command Staff

Individuals on the command staff perform functions that report directly to the IC and cannot be delegated to other major sections of the organization (▶Figure 4-6). The Safety Officer, Liaison Officer, and Information Officer are always part of the command staff. In addition, aides, assistants, and advisors may be assigned to work directly for the IC. An aide may be a brigade member, a supervisor, or a manager who serves as a direct assistant to a command officer.

#### Safety Officer

The Safety Officer is responsible for ensuring that safety issues are managed effectively at the incident scene. The Safety Officer is the eyes and ears of the IC for safety, identifying and evaluating hazardous conditions, watching out for unsafe practices, and ensuring that safety procedures are

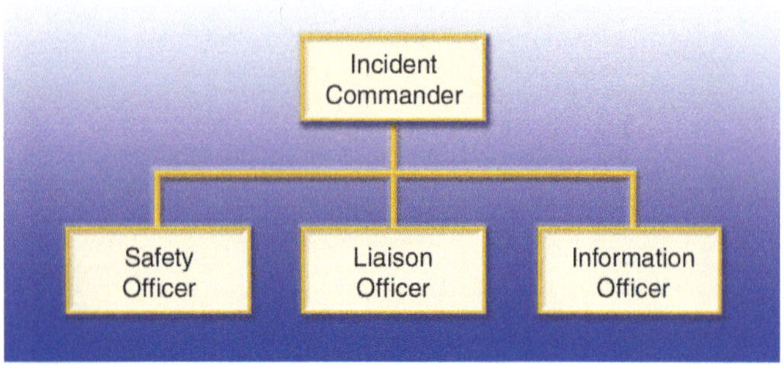


Figure 4-6 The command staff report directly to the IC.

followed. The Safety Officer is appointed early during an incident. As the incident becomes more complex and the number of resources increase, additional qualified personnel can be assigned as assistant safety officers.

The Safety Officer is an advisor to the IC but has the authority to stop or suspend operations when unsafe situations occur. This authority is clearly stated in national standards, including NFPA 1500, *Standard on Fire Department Occupational Safety and Health Program*, NFPA 1521, *Standard for Fire Department Safety Officer*, and NFPA 1561, *Standard on Emergency Services Incident Management System*. Several state and federal regulations require the assignment of a Safety Officer at hazardous materials incidents and certain technical rescue incidents.

The Safety Officer should be a qualified individual who is knowledgeable in fire behavior, building construction and collapse potential, firefighting strategy and tactics, hazardous materials, rescue practices, and departmental safety rules and regulations. A Safety Officer should also have considerable experience in incident response and specialized training in occupational safety and health. Many fire brigades have full-time Safety Officers who perform administrative functions relating to health and safety when they are not responding to emergency incidents. The NFA's *Incident Safety Officer* and *Advanced Safety Operations and Management* courses are excellent resources for people interested in serving in this capacity.

### Liaison Officer

The Liaison Officer is the IC's representative to a point of contact for representatives from outside agencies. The Liaison Officer is responsible for exchanging information with representatives from those agencies. During an active incident, the IC may not have time to meet directly with everyone who comes to the command post. The Liaison Officer position takes the IC's place, obtaining and providing information, or directing people to the proper location or authority. The Liaison area should be adjacent to, but not inside, the command post.

Figure 4-7 The Public Information Officer is responsible for gathering and releasing incident information to the media and other appropriate agencies.

### Public Information Officer

The Public Information Officer is responsible for gathering and releasing incident information to the news media and other appropriate agencies (▲ Figure 4-7). At a major incident, the public will want to know what is being done. Because the IC must make managing the incident his or her top priority, the Public Information Officer serves as the contact person for media requests. This frees the IC to concentrate on the incident. A media headquarters should be established near, but not in, the command post.

## Deputies and Assistants

Two additional staffing titles are commonly found in both the command staff and the general staff: deputy and assistant.

The position of deputy is filled by a fully qualified individual who can be delegated the authority to manage a specific task or functional operation in the absence of a superior. Because a deputy must be fully qualified, he or she can act as the relief for the superior. Deputies can be assigned to the IC, general staff functions, and branch directors.

Assistant is the title assigned to subordinates of the principal command staff positions. Assistant titles indicate qualifications, technical capability, and responsibilities that are subordinate to the primary positions. Assistants may also be assigned to unit leaders.

## General Staff Functions

The IC has the overall responsibility for the entire incident management organization, although some elements of the IC's responsibilities can be handled by the command staff. When the incident is too large or too complex for just one person to manage effectively, the IC may appoint someone

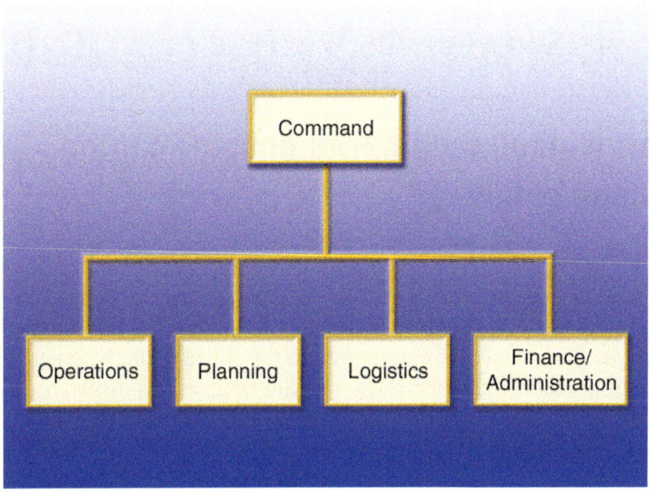


Figure 4-8 Major functional components of IMS.

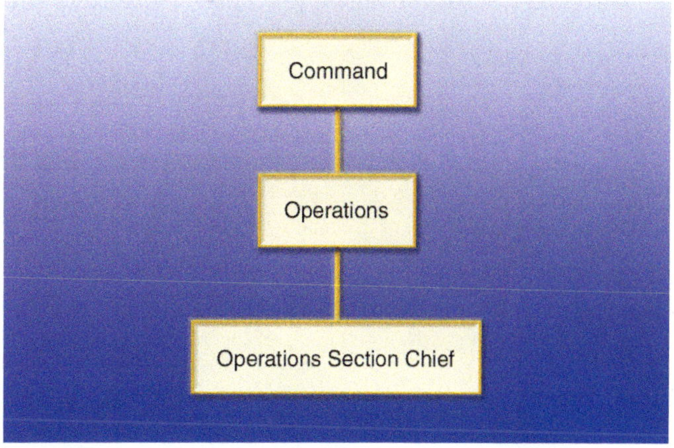


Figure 4-9 The organizational structure of the Operations Section.

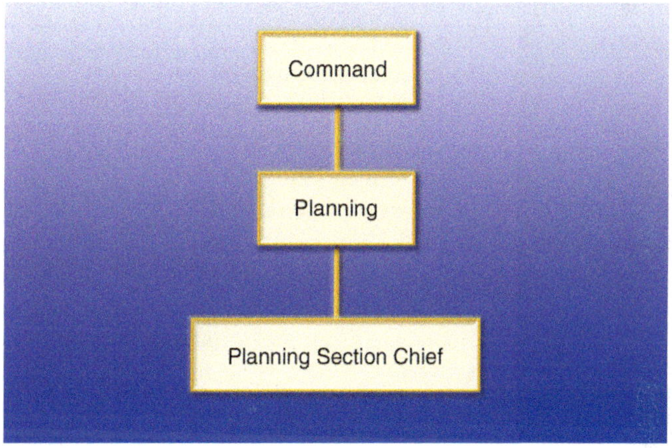


Figure 4-10 The organizational structure of the Planning Section.

to oversee parts of the operation. Everything that occurs at an emergency incident can be divided among the four major functional components within IMS (▲Figure 4-8):

- Operations
- Planning
- Logistics
- Finance/Administration

The chiefs of these four sections are known as the IMS General Staff. The IC decides which (if any) of these four positions need to be activated, when to activate them, and who should be placed in each position. (Remember that the blocks on the IMS organization chart refer to function areas or job descriptions, not positions that must always be staffed.)

The four section chiefs on the IMS General Staff, when they are assigned, may run their operations from the main command post, but this is not required. At a large incident, the four functional organizations may operate from different locations, but will always be in direct contact with the IC.

## Operations

The Operations Section is responsible for the management of all actions that are directly related to controlling the incident (►Figure 4-9). The Operations Section fights the fire, rescues any trapped individuals, treats the patients, and does whatever else is necessary to alleviate the emergency situation. At industrial facilities, this may include opening or closing valves, and securing processes and power supplies that may help resolve the emergency situation. The Operations Section is equivalent to the production department of a company. This is the department that produces results.

For most structure fires, the IC directly supervises the functions of the Operations Section. A separate Operations Section Chief is used at complex incidents so that the IC can focus on overall strategy while the Operations Section Chief focuses on the tactics that are required to get the job done.

Operations are conducted in accordance with an Incident Action Plan (IAP) that outlines what the strategic objectives are and how emergency operations will be conducted. At most incidents, the IAP is relatively simple and can be expressed in a few words or phrases. The IAP for a large-scale incident can be a lengthy document that is regularly updated and used for daily briefings of the command staff.

## Planning

The Planning Section is responsible for the collection, evaluation, dissemination, and use of information relevant to the incident (▲Figure 4-10). The Planning Section works with preincident plans, building construction drawings, maps, aerial photographs, diagrams, reference materials, and status boards. The Planning Section is also responsible for developing and updating the IAP. The Planning Section is similar to the scheduling department at a company. It plans what needs to be done by whom and what resources are needed.

# Voices of Experience

## "We all worked together to ensure that the job got done safely."

It was mid-December—and the night of the annual company party. Outside it was snowing heavily, with a 20 mile per hour wind whipping through the subzero air. When an explosion occurred at the plant, our fire brigade was dispatched to the south side of the building, to the main transformer area. Wearing our personal protective gear, we approached the main transformer area with caution. The incident commander (IC) was standing near the area; he motioned us over. He explained that the generator had shut down as designed. Unfortunately, the transformer connected to the generator had exploded and was now on fire. The IC directed two of us to stretch a hose from a nearby hydrant and prepare for a foam attack.

As we prepared for the initial attack, we noticed that the water supply line was beginning to freeze in the extreme weather conditions. At the same time, the sprinkler system around the transformer became activated and sprayed water on many of the fire fighters. Upon our retreat from the area, we discovered that much of the sprinkler system had been destroyed in the transformer explosion. Our water supply was now frozen, and the ice had broken through our deployed hoses.

The IC told us that mutual aid had been called and was expected to be on scene in about 10 minutes. In the meantime, we were directed to deploy new hose lines to replace the frozen hoses.

Once the new hoses were in place, we began our attack on the transformer fire with the foam. Mutual aid soon arrived, and members of those companies began to rotate in to relieve our brigade members as the IC directed. As the last of the fire was being extinguished, additional personnel from our sister plant arrived and began the job of planning for removal and replacement of the transformer.

Environmental specialists arrived shortly after the fire's resolution to ensure that the fire-extinguishing water and runoff were contained and properly disposed. All the while, security handled the arrival of more mutual aid responders and members of the public who arrived at the main gate. Management gave a press briefing, engineering began planning for the transformer's removal and replacement, and many other plant workers stepped forward to assist in any way they could. We all worked together to ensure that the job got done safely.

**John Shrader**
Cooper Nuclear Station
Brownville, Nebraska

## Brigade Member Tips

Strategy is the "big picture" plan of what has to be done. "Stop the fire from extending into Exposure D" is a strategic objective. Tactics are the steps that are taken to achieve the strategic objectives, such as "Make an offensive attack on the liquid fuel fire near the compressor station to prevent fire from extending into the storage area." Tasks are the specific assignments that will get the job done. "Brigade Team 1 will lay a supply line with foam and attack the fire in the compressor station area between the fire area and the storage area." Laying the supply line with foam and attacking the fire are both "tasks."

## Brigade Member Tips

NFPA 600 and 1981 and OSHA 1910 Subpart L use the term "Brigade Leader" to identify any level of brigade supervision. Facility fire brigade titles vary widely; chief, deputy chief, and captain are as common as emergency response coordinator, response team leader, crew chief, and shift supervisor. The title of the individual filling a position in an IMS is irrelevant. What is important is that that individual who assumes this role in the IMS structure be competent and qualified. Competent people, assigned to critical command functions, develop plans that make problems go away.

The IC activates the Planning Section when information needs to be obtained, managed, and analyzed. The Planning Section Chief reports directly to the IC. Individuals assigned to planning examine the current situation, review available information, predict the probable course of events, and prepare recommendations for strategies and tactics. The Planning Section also keeps track of resources at large-scale incidents and provides the IC with regular situation and resource status reports.

### Logistics

The Logistics Section is responsible for providing supplies, services, facilities, and materials during the incident (▼ Figure 4-11). The Logistics Section Chief reports directly to the IC and serves as the supply officer for the incident. Among the responsibilities of this section would be keeping apparatus fueled, providing food and refreshments for the brigade members, obtaining the foam concentrate needed to fight a flammable liquids fire, and arranging for a bulldozer to remove a large pile of debris. The Logistics Section is similar to the purchasing department of a company. It ensures that adequate resources are always available.

In many fire brigades, these logistical functions are routinely performed by support services personnel. These groups work in the background to ensure that the members of the Operations Section have whatever they need to get the job done. Resource-intensive or long-duration situations may require assignment of a Logistics Section Chief because service and support requirements are so complex or extensive that they need their own management component. Industry typically also does functions such as purchasing, warehousing, and supply for the business of the company. Some differences may exist in that industry functions as a part of the business of the company and not necessarily as a support function for emergency responders, although many industries that have fire brigades can support emergency response functions for a limited amount of time.

### Finance/Administration

The Finance/Administration Section is the fourth major IMS component under the IC (▼ Figure 4-12). This section is responsible for the accounting and financial aspects of an incident, as well as any legal issues that may arise. This

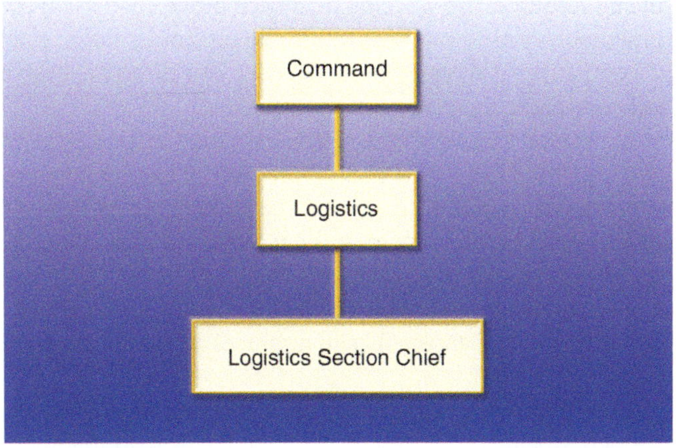


Figure 4-11 The organizational structure of the Logistics Section.

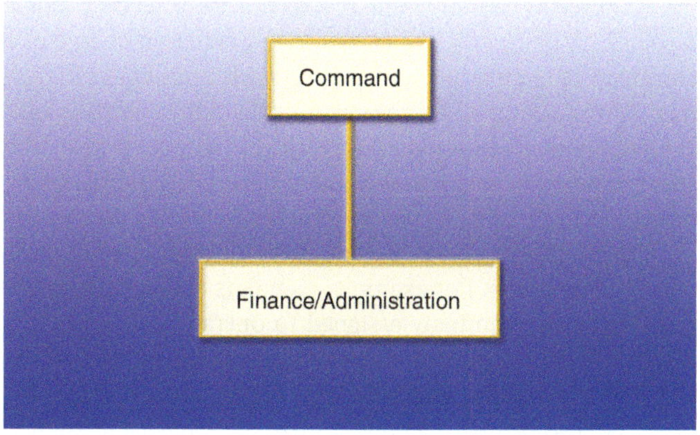


Figure 4-12 The organizational structure of the Finance/Administration Section.

function is not staffed at most incidents, because cost and accounting issues are usually addressed after the incident. A Finance/Administration Section may be needed at large-scale and long-term incidents that require immediate fiscal management, particularly when outside resources must be procured quickly. A Finance/Administration Section may also be established during a natural disaster or during a hazardous materials incident where reimbursement may come from the shipper, carrier, chemical manufacturer, or insurance company. The Finance/Administration Section is equivalent to the finance department in a company. It accounts for all activities of the company and ensures that there is enough money to keep the company running.

## Standard IMS Concepts and Terminology

Facility brigades and other emergency response agencies implement IMS to organize emergency scene activities in a standard manner. Emergency scenes tend to be chaotic, and organizing operations is often a serious challenge, particularly if the agencies involved use different terms to describe certain concepts and resources. One of the strengths of IMS is standard terminology. In IMS, specific terms apply to various parts of an incident organization. Understanding these basic concepts and terminology is the first step in understanding the system.

Some facility brigades may use slightly different terminology. Brigade members must know the terminology used by their brigade, as well as the standard IMS terminology. This section defines important IMS terms and examines their use in organizing and managing an incident.

### Single Resources and Crews

NOTE: for purposes of following the NFPA Incident Command Standard, the term "company" will be used in the following section. Your facility fire brigade may be organized using terms different than "company." A single resource is an individual vehicle and its assigned personnel . For example, an engine and its crew ▶ Figure 4-13 would be a single resource; a ladder company would be a second single resource. A company officer is the individual in charge of a company. A company operates as a work unit, with all crew members working under the supervision of the company officer. Companies are assigned to perform tasks such as search and rescue, attacking the fire with a hose line, forcible entry, or ventilation.

A crew is a group of brigade members who are working without apparatus. For example, members of an engine company or ladder company assigned to operate inside a building would be considered a crew. Additional personnel at the scene of an incident who are assembled to perform a specific task may also be called a crew. A crew must have an assigned leader or brigade leader.

Figure 4-13 A single resource is an individual company and the personnel that arrive on that unit.

### Divisions and Groups

Organizational units such as divisions and groups are established to group single resources and/or crews under one supervisor. The primary reason for establishing divisions and groups is to maintain an effective span of control.

- A division usually refers to companies and/or crews working in the same geographic area.
- A group usually refers to companies and/or crews working on the same task or objective, although not necessarily in the same location.

The flexibility of the IMS enables organizational units to be created as needed, depending upon the size and scope of the incident. In the early stages of an incident, individual companies or crews are often assigned to work in different areas or perform different tasks. As the incident grows and more companies or crews are assigned to it, the IC can establish divisions and groups. An individual is assigned to supervise each division or group as it is created.

These organizational units are particularly useful when several resources are working near each other, such as on the same floor inside a building, or are doing similar tasks, such as ventilation. The assigned supervisor can directly observe and coordinate the actions of several crews.

A division is comprised of the resources responsible for operations within a defined geographic area. This area could be a floor inside a building, the rear of the building, or an area of a refinery fire. Divisions are most often employed during routine fire brigade emergency operations. By assigning all of the units in one area to one supervisor, the IC is better able to coordinate their activities and ensure that all resources are working toward a similar strategy.

The division structure provides effective coordination of the tactics being employed by different crews working in the

Figure 4-14 The organization of a division.

same area. For example, the division supervisor would coordinate the actions of a crew that is advancing a hose line into the area, a crew that is conducting search-and-rescue operations in the same area, and a crew that is performing horizontal ventilation ▲ Figure 4-14.

For many facility fire brigades, an alternative way of organizing resources is by function rather than location. A group is comprised of resources assigned to a specific function, such as ventilation, search and rescue, or water supply. Groups are responsible for performing an assignment, wherever it may be required, and often work across division lines. For example, the officer assigned to supervise the ventilation group uses the radio designation "Ventilation Group."

An example of a local terminology preference is the term sector. A sector can be either geographically based (used instead of a division) or functionally based (used instead of a group). However, brigades that use the term sector must understand the meaning of division and group as well, because they may have to work with brigades that use those terms. The term sector is not recognized under the National Incident Management System (NIMS) and is not used by response organizations who have adopted NIMS.

Division, group, and sector supervisors all have the same rank within IMS. Divisions do not report to groups, and groups do not report to divisions. The supervisors are required to coordinate their actions and activities with each other. For example, a group supervisor must coordinate with the division supervisor when the group enters the division's geographic area, particularly if the group's assignment will affect the division's personnel, operations, or safety. The division supervisor must be aware of everything that is happening within that area. Effective communication among divisions, groups, and sectors is critical during emergency operations.

## Branches

A branch is a higher level of combined resources than divisions and groups. At a major incident, several different activities may occur in separate geographic areas or involve distinct functions. The span of control might still be a problem, even after the establishment of divisions and groups. In these situations, the IC can establish branches to place a higher level supervisor (a branch director) in charge of a number of divisions and/or groups.

## Location Designators

IMS uses a standard system to identify the different parts of a building or a fire scene. Every brigade member must be familiar with this terminology.

The exterior sides of a building are generally known as sides A, B, C, and D. The front of the building is side A, with sides B, C, and D following in a clockwise direction around the building. The companies working in front of the building are assigned to Division A and the radio designation for their supervisor is "Division A." Similar terminology is used for the sides and rear of the building.

The areas adjacent to a burning building are called exposures. Exposures take the same letter as the adjacent side of the building. A brigade member facing Side A can see the adjacent building on the left (Exposure B) and the building to the right (Exposure D). If the burning building is in a row of buildings, the buildings to the left are called Exposures B, B1, B2, and so on. The buildings to the right are Exposures D, D1, D2, and so on.

Within a building, divisions commonly take the number of the floor on which they are working. For example, brigade members working on the fifth floor would be in Division 5, and the radio designation for the brigade leader assigned to that area would be "Division 5." Crews doing different tasks on the fifth floor would all be part of this division.

► Figure 4-15 illustrates the location designators for different parts of a burning structure, exposures, or floors of a building. At some incidents where industry is the sole responder, the location name or reference is by the location in the industrial complex, such as "near tank 21 in the south complex."

### Task Forces and Strike Teams

Task forces and strike teams are groups of single resources that have been assigned to work together for a specific purpose or for a period of time. Grouping resources reduces the span of control by placing several units under a single supervisor.

A task force includes two to five single resources, such as different types of units assembled to accomplish a specific

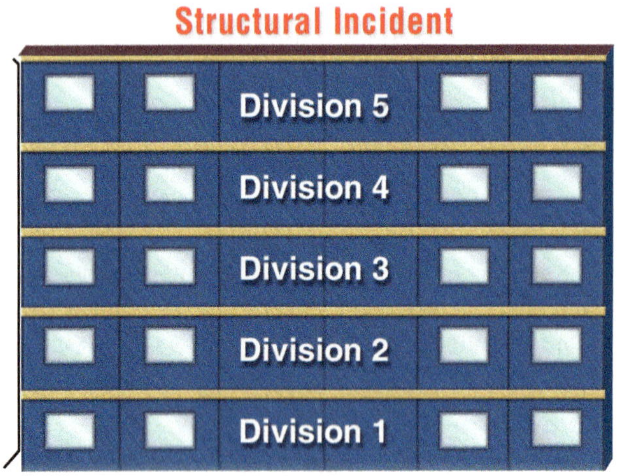


Figure 4-15 Location designators in IMS.

Figure 4-16 A strike team is five units of the same type under one leader.

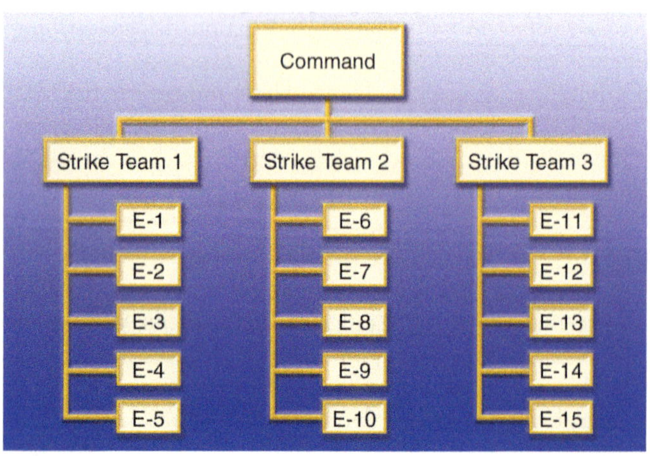


Figure 4-17 The organization of a strike team.

task. For example, a task force may be composed of two engines and one truck company, two engines and two foam units, or one rescue company and four ambulances. A task force operates under the supervision of a task force leader. All communications for the separate units in the task force are directed to the task force leader.

A strike team is usually five units of the same type with an assigned leader. A strike team could be five engines (engine strike team), five trucks (truck strike team), or five ambulances (EMS strike team) (◄ Figure 4-16). A strike team operates under the supervision of a strike team leader.

Strike teams are commonly used for wildland fires, where dozens or hundreds of companies may respond. During wildland fire seasons, many brigades establish strike teams of five engine companies that will be dispatched and work together on major wildland fires (◄ Figure 4-17). The assigned companies rendezvous at a designated location and then respond to the scene together. Each engine has an officer and brigade members, but only one officer is designated as the strike team leader. All communications for the strike team are directed to the strike team leader.

EMS strike teams, consisting of five ambulances and a supervisor, are often organized to respond to multicasualty incidents or disasters. Rather than requesting 15 ambulances and establishing an organizational structure to supervise 15 single resources, the IC can request three EMS strike teams and can coordinate with three strike team leaders.

## Implementing IMS

IMS helps to organize every incident scene in a standard, consistent manner. Fire brigades develop SOPs, then train and practice using IMS to ensure consistency at emergency incidents. This approach increases safety and efficiency. As an incident escalates in size or complexity, the IMS organization expands to fit the situation. The same system can be used at a pump seal fire or a major process fire.

The advantage of operating within a common system is that this scheme provides for effective mutual aid. When the magnitude of an incident exceeds the resources of an in-plant brigade, outside resources are required. The same applies when a municipal department lacks necessary resources, such as a foam supply to manage a fuel tanker incident. When both organizations are operating under the same management system, needed resources can quickly be made available to plug the gap and provide an effective response.

A small-scale incident can often be handled successfully by one fire brigade team or a first-alarm assignment. The organizational structure implemented at an emergency incident starts small, with the arrival of the first unit or units. When only a limited number of resources are involved, an IC can often manage the organization personally or with the assistance of an experienced aide. (► Figure 4-18) illustrates a typical command structure for an industrial structure fire.

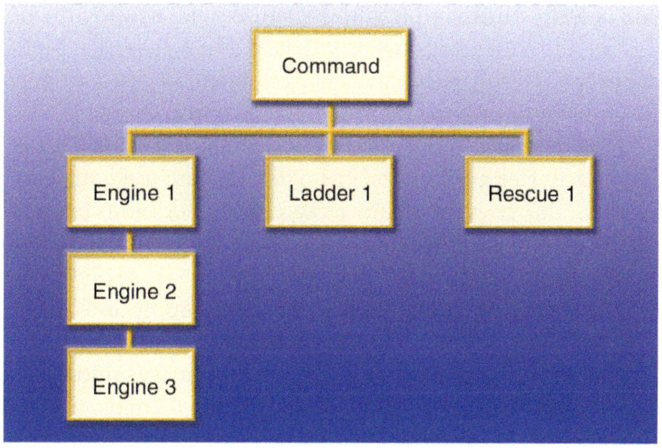


Figure 4-18 At a small-scale incident, the typical IMS command structure may consist solely of the IC and reporting resources.

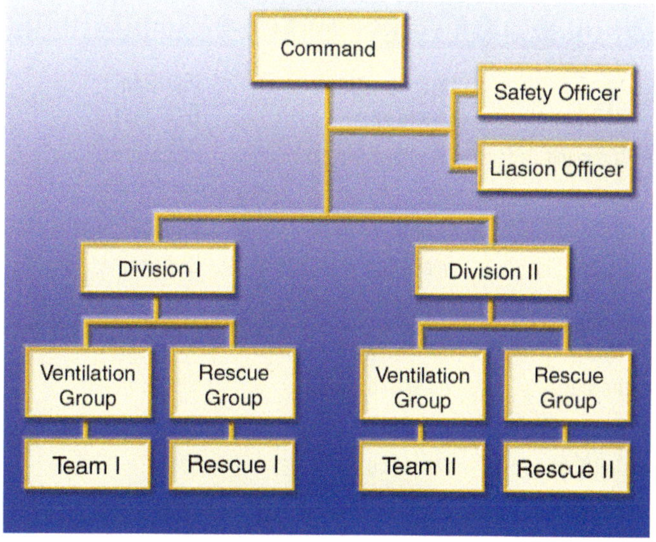


Figure 4-19 Divisions and groups are organized to manage the span of control and to supervise and coordinate units working together.

At a more complex incident, the increasing number of problems and resources places greater demands on the IC and can quickly exceed the IC's effective span of control. A larger incident requires a more complex command structure to ensure that no details are overlooked or personnel safety compromised.

The modular design of IMS allows the organization to expand, based on the needs of the incident, by activating predesignated components. The IC can delegate specific responsibilities and authority to others, creating an effective incident organization. An individual who gets an assignment knows the basic responsibilities of the job, because they are defined in advance.

The most frequently used IMS components during responses are divisions and groups (► Figure 4-19). These components place several single resources under one supervisor, effectively reducing the IC's span of control. The IC can also assign individuals to special jobs, such as Safety Officer and Liaison Officer, to establish a more effective organization for the incident.

In the largest and most complex incidents, other IMS components can be activated. The IC can delegate responsibility for managing major components of the operation, such as Operations or Logistics, to General Staff positions. Each of these positions has a wide range of responsibilities for a major component of the organization.

If necessary, branches can be established, creating an additional level within the organization (▼ Figure 4-20). For example, a structural collapse during a major fire may result

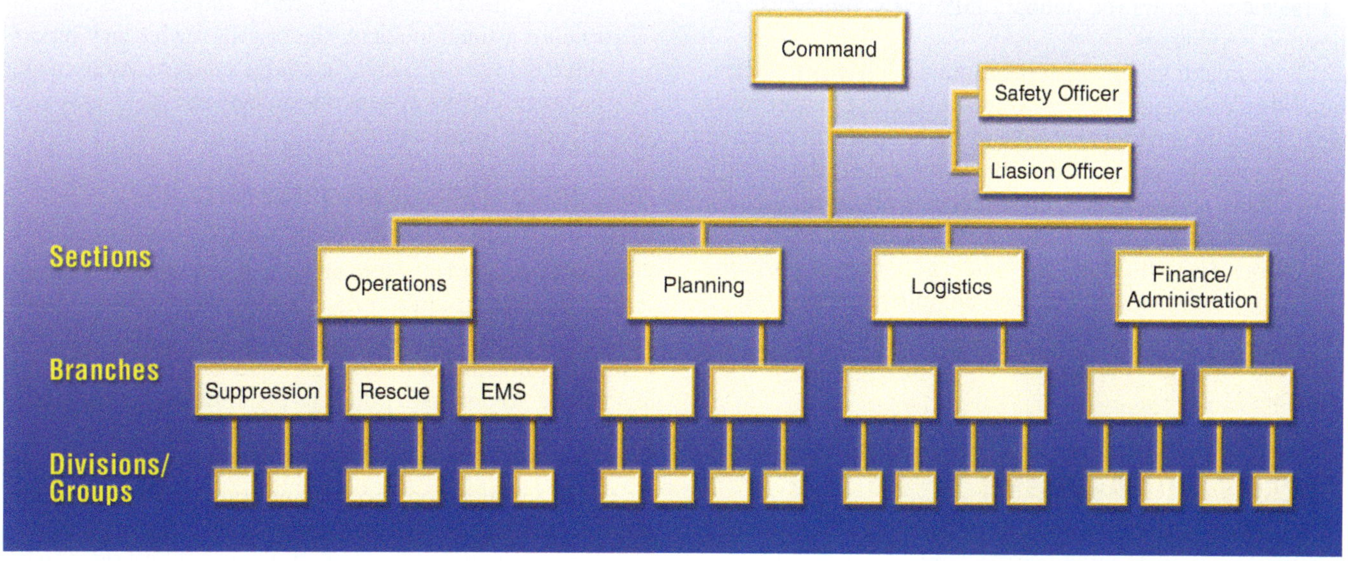


Figure 4-20 Creating branches within the Operations Section is one way to manage span of control during a large incident.

**Table 4-1 Levels of an IMS Organization**

| IMS Level | IMS Function/ Location | Position Designator |
|---|---|---|
| Command | Command and control | Incident Commander |
| Command Staff | Safety, Liaison, Information | Officer |
| General Staff | Operations, Planning, Logistics, Finance/ Administration | Section Chief |
| Branch | Varies (e.g., EMS) | Director |
| Division/Group | Varies (e.g., Div. A) | Supervisor |
| Unit/Crew/Strike | Varies (e.g., Rehab) | Leader |
| Team/Task Force | | (Brigade Leader) |

in several trapped victims and many patients who need medical treatment and transportation. In such a situation, the Operations Section Chief would have multiple responsibilities that could exceed his span of control. Activating a Fire Suppression Branch, a Rescue Branch, and a Medical Branch would address this problem. One officer would be responsible for each branch and report to the Operations Chief. Within each branch, several divisions or groups would report to the branch director.

### Standard Position Titles

To help clarify roles within the IMS organization, standard position titles are used. Each level of the organization has a different designator for the individual in charge. The position title typically includes the functional or geographic area of responsibility, followed by a specific designator. ▲ Table 4-1 shows the standard IMS levels, functions, and position designators.

Division and group supervisors are usually chief officers, but brigade leaders may be assigned to these positions as well. Typically, the brigade leader from the first unit assigned to a division or group will serve as the supervisor until a chief officer is available. A chief officer with a strong background in operations is usually assigned to the position of Operations Section Chief. The other section chief positions (Planning, Logistics, and Finance/Administration) are staffed as needed by the IC. These individuals should understand their roles and be able to meet the responsibilities of these positions.

## Working Within the IMS

To an outsider, the IMS appears to be a large, complicated organization model that involves a complex set of SOPs. But to the individual brigade member working within the system, IMS is really quite simple and uncomplicated. Brigade members should understand what IMS is and how it works. More importantly, each individual brigade member must understand his or her place and role in the system.

This section discusses how every emergency incident is conducted. Certain components of the IMS are used on every incident and at every training exercise. In addition, brigade members generally have specific responsibilities and procedures to follow in most situations.

The three basic components that always apply are:

- ***Someone is always in command of every incident,*** from the time that the first unit arrives until the last unit leaves. The identity of the IC may change, but there is always one IC in charge of the operation and responsible for everything that happens. SOPs dictate who will be the IC at any time.
- ***You always report to one supervisor.*** A brigade member's supervisor will usually be a brigade leader. The brigade leader directly supervises a small group of brigade members, such as an engine brigade or ladder brigade, who work together. At an incident scene, the brigade leader provides instructions and must always know where each brigade member is and what he or she is doing. If the brigade assigns two brigade members to work together away from the rest of the brigade, both brigade members are still under the supervision of the brigade leader. The brigade leader could be an acting brigade member (a brigade member temporarily designated as a "fill-in" brigade leader), or a brigade member could be assigned to work temporarily under the supervision of a different brigade leader.
- ***The brigade leader reports to the IC.*** If there is only one brigade on the scene, the brigade leader is the IC, until someone else arrives and assumes that role. At a small incident, the brigade leader may report directly to the IC, while at a large incident, there may be several layers of supervision between a brigade leader and the IC.

### Responsibilities of the First-Arriving Brigade Members

The first brigade members to arrive at an emergency incident are the foundation of the IMS organization structure. Rarely is a high-ranking chief sent to a fire to evaluate the situation, design the organization structure, and order the resources that will be needed. IMS builds its organization from the bottom up, around the units that take initial action.

The officer in charge of the first-arriving unit is responsible for taking initial action and becomes the IC, with all of the authority that comes with that position. This officer is responsible for managing the operation until relieved by a senior officer. If there is no officer on the first-arriving unit, the brigade member with the greatest seniority is in charge until an officer arrives and assumes command. The position of IC must have an unbroken line of succession from the moment the first unit arrives on the scene.

### Assuming Command

The IMS assumes that command is initiated when the first unit arrives and is transferred as required for the duration of the incident. The individual who is in charge of the first-arriving unit automatically assumes command of the incident until a superior takes over command.

The individual who initially assumes command must formally announce that fact over the radio. This announcement eliminates any possible confusion over who is in charge. The initial report should include the following information:

- Command designation
- Unit or individual who is assuming command
- An initial situation report
- Initial action being taken

For example, a radio report from the first-arriving unit at a fire equipment storage building fire might be as follows:

*Brigade Team 1 is on scene at the hydraulic pressure skid. We have a leak on the discharge flange of the pump. The leaking oil is on fire, and the sprinkler system has actuated and is controlling the fire but is not extinguishing the fire. After power is secured to the pump, we will attempt to extinguish the fire and secure the leaking fluid.*

Usually, the first-arriving unit at an incident scene is an engine company or some other unit that takes direct action. Regardless of the type of emergency or the type of unit, the first brigade leader on the scene is in charge. This officer must assess the situation, determine incident priorities, and provide direction for his or her own crew as well as any other units that are arriving. Any units coming in behind the first unit know that they will be taking their orders from the initial IC until a higher level officer assumes command.

The brigade leader who initially assumes command must decide whether to take action directly supervising the initial attack crew or to concentrate on managing the incident as the IC. This decision depends on many factors, including the nature of the situation, the resources that are on the scene or expected to arrive quickly, and the ability of the crew to work safely without direct supervision.

If the incident is large and complicated, the best option for the IC is to establish a command post and focus on sizing up the situation, directing incoming units, and requesting additional resources. The IC's own brigade can be assigned to work with an acting brigade leader or to join forces with another brigade leader and crew. If the situation is less critical, the IC might be able to function both as a brigade leader and as the IC at the same time, at least temporarily.

If the first-arriving unit is a chief officer, the chief officer automatically assumes command and executes command responsibilities. If a brigade leader had previously assumed command, the brigade leader would transfer command authority to the first-arriving chief officer. The officer relinquishing command should provide an assessment of the situation to the new IC, as described under Transfer of Command.

Most brigades have written procedures that specify who will assume command in certain situations. If a brigade leader arrives seconds ahead of a chief officer, the chief officer should assume command from the outset.

## Confirmation of Command

The initial announcement of command definitely confirms that command has been established at an incident. If no one announces that he or she is in command, the entire system realizes that there is no IC in place. The announcement also reinforces the IC's personal commitment to the position through a conscious personal act and a standard organizational act.

### Identification of the Incident

Individual fire brigade procedures may vary on the specific protocol for naming an incident. The first officer to assume command should establish an identity that clearly identifies the location of the incident, such as, *Engine 10 will be process unit of warehouse IC.* This reduces confusion on the radio and establishes a continuous identity for the IC, regardless of who holds that position during the incident.

There can only be one "process unit of warehouse" at a time, so there is no confusion about who is in charge of the incident. When anyone needs to talk to the IC, a call to "process unit of warehouse Command" should be answered by the individual who is in command.

### Passing Command

Passing command is an option that can be used by a first-arriving brigade leader, if there is a compelling reason that prevents that leader from assuming command of the incident. Passing command directs the next-arriving unit, whether it is a brigade team leader or chief brigade leader, to assume command.

A brigade leader is allowed to pass command only under a precise set of guidelines, in situations where his or her direct involvement in operations will have a significant impact on the outcome of the incident. For example, a four-person company might arrive at a fire and immediately need to rescue someone from a second-floor window. Because the brigade leader must help the brigade members get the ladder into position to rescue the victim, he or she would pass command to the brigade leader in charge of the next-arriving unit. The notice of passing command is simply stated: *Engine 1 passing command to Engine 2.* Engine 2 must acknowledge this message. As soon as Engine 2 arrives, the brigade leader will assume command as though he or she was the first-arriving brigade leader.

## Transfer of Command

Transfer of command occurs whenever one person relinquishes command of an incident and another individual becomes the IC. For example, the first-arriving brigade leader would transfer command to the first-arriving chief officer who would later transfer command to a higher ranking officer during a major incident. Some brigades require transfer of command when a higher ranking officer arrives at an incident; others give the higher ranking officer the option of assuming command or leaving the existing IC in charge.

When a higher ranking officer arrives at the scene of an emergency incident and takes charge, that officer assumes the moral and legal responsibility for managing the overall operation. Established procedures must be followed whenever command is transferred. One of the most important requirements of command transfer is the accurate and complete exchange of incident information ▶ Figure 4-21. The officer who is relinquishing command needs to give the new IC a current situation status report that includes:

- Tactical priorities
- Action plans
- Hazardous or potentially hazardous conditions
- Accomplishments
- Assessment of effectiveness of operations
- Current status of resources
  - Assigned or working
  - Available for assignment
  - En route
- Additional resource requirements

Figure 4-21 The new IC needs to be briefed before assuming command.

If transfer of command occurs very early during an incident, the transfer of information may be brief. For example, the first-arriving brigade leader might have been in command for only a few minutes and have little information to report when command transfers to the first-arriving chief officer. The chief may have heard all of the exchanges over the radio and know what the current situation is. However, if the brigade leader has been IC for several minutes, there may be a significant amount of information to report, such as the current assignments of all first-alarm companies. Whether the information is minimal or substantial, the transfer must be accurate and complete.

Each fire brigade establishes specific procedures for transferring command. In some cases, the transfer of command may be done via radio, but the most effective method is face-to-face communication. Standard command worksheets and a status board are valuable tools for tracking and transferring information at a command post.

Command is always maintained for the entire duration of an incident. In the later stages of an incident, after the situation is under control, command might be transferred to a lower level commander. A downward transfer of command requires the same type of briefing and exchange of information as an upward command transfer. The brigade leader in charge of the last brigade remaining on the scene would be the last IC. When that brigade leaves the scene, command is terminated.

## Command Transfer Rationale

There are important reasons for transferring command at different points during an emergency incident. A first-arriving brigade leader can usually direct the initial operations of two or three additional companies, but as situations become more complex, the problems of maintaining control increase rapidly. A brigade leader's primary responsibility is to supervise one crew and ensure that they operate safely. When three or more companies are operating at an incident, it is better to have a chief brigade leader assume command.

As more brigades are assigned to an incident, the command structure must also expand. The organization must grow to maintain an effective span of control. Additional chief brigade leaders may be assigned to the incident, and a higher ranking chief brigade leader may assume command. A command transfer may be required if the situation is beyond the training and experience of the current IC. A more experienced brigade leader may have to assume command to ensure proper management of the incident.

If multiple agencies are involved, overall command of an incident may be transferred to a different agency. At a terrorist incident, for example, the fire brigade leader could have the command responsibility until the area has been secured and fires extinguished. At that point, a law enforcement agency might assume overall command. Fire brigade units operating at the scene would maintain their internal command structure, but the IC could be a law enforcement official.

# Wrap-Up

## Ready for Review

- IMS is useful not only for large-scale operations but also for any incident, regardless of size or type. All of the functions outlined in IMS must be addressed at every incident. It is the size and/or type of incident that dictates the degree to which each function is addressed.
- IMS allows the IC to delegate responsibilities in a standard incremental manner. The IC must perform any function that is not delegated. The IC's ultimate responsibility is to ensure that all incident requirements are met.
- Several characteristics that are critical to an Incident Management System:
  - Organized approach: IMS imposes "order on chaos" on the fireground and enables a safer and more efficient operation than would be possible if personnel and units worked independently of each other.
  - Terminology: IMS uses a standard terminology for effective communications.
  - All-risk: IMS can be used at any type of emergency incident.
  - NIMS and the Private Sector: Facilities should adopt the use of NIMS/ICS as a corporate policy.
  - Jurisdictional authority: IMS enables different jurisdictions, agencies, and organizations to work cooperatively on a single incident.
  - Span of control: IMS maintains the desired span of control through flexible levels of organization.
  - Unity of command: Everyone reports to only one supervisor to avoid conflicts in giving orders.
  - Everyday applicability: IMS can and should be used on every single incident, every single time.
  - Modular: IMS is based on standard modules that are activated as needed to manage an incident.
  - Integrated communications: Everyone on the incident can communicate up and down the chain of command as needed.
  - Emergency Response Operations Plan: Designed to identify the level of response needed for certain locations with an emergency situation.
  - Incident Action Plan: Every incident has a plan that outlines the strategic objectives. Large incidents will have formal plans.
  - Facilities: Standardized facilities, such as a command post, staging area, or rehabilitation area, are established as needed.
- Five major functions are part of IMS:
  - Command: Responsible for the entire incident; this is the only function that is always staffed.
  - Operations: Responsible for most fireground functions including suppression, search and rescue, and ventilation.
  - Planning: Responsible for developing the Incident Action Plan.
  - Logistics: Responsible for obtaining the resources needed to support the incident.
  - Finance/Administration: Responsible for tracking expenditures and managing the administrative functions at the incident.
- The Command Staff assists the IC at the incident:
  - Safety Officer: Responsible for overall safety of the incident; has the authority to stop any action or operation if it creates a safety hazard on the scene.
  - Liaison: Responsible for coordinating operations between the fire brigade and other agencies that may be involved in the incident.
  - Information: Responsible for coordinating media activities and providing the necessary information to the various media organizations.
- An example of a single resource would be an engine brigade or a ladder brigade.
- Single resources can be combined into task forces or strike teams.
- Other organizational units that can be established under IMS include divisions and groups.
- IMS can be expanded infinitely to accommodate any size incident. Branches can be established to group similar functions, such as suppression, EMS, or hazardous materials.
- At every incident, someone must always assume command. As the incident grows or continues, it may be necessary to transfer command to another officer. This has to be done in a seamless manner to ensure continuity of command.

## Hot Terms

**Branch** A supervisory level established to manage the span of control above the division or group level; usually applied to operations or logistics functions.

**Branch director** Officer in charge of all resources operating within a specified branch, responsible to the next higher level in the incident organization (either a Section Chief or the Incident Commander).

**Command** The first component of the IMS system. This is the only position in the IMS system that must always be staffed.

**Command post** The location at the scene of an emergency where the Incident Commander is located and where command, coordination, control, and communications are centralized.

**Command staff** Staff positions established to assume responsibility for key activities in the incident management system; individuals at this level report directly to the IC. Command staff include the Safety Officer, Public Information Officer, and Liaison Officer.

**Company officer** Usually a lieutenant or captain in charge of a team of brigade members, both on scene and at the station. The brigade leader is responsible for firefighting strategy, safety of personnel, and the overall activities of the brigade members on their apparatus.

**Crew** An organized group of brigade members under the leadership of a brigade leader, crew leader, or other designated official.

**Designated incident facilities** Assigned locations where specific functions are always performed.

**Division** An organizational level within IMS that divides an incident into geographic areas of operational responsibility.

**Division supervisor** The officer in charge of all resources operating within a specified division, responsible to the next higher level in the incident organization, and the point-of-contact for the division within the organization.

**Emergency response operations plan** Plan designed to identify the levels of response needed for certain locations with emergency situations.

**Finance/administration section** The command-level section of IMS responsible for all costs and financial aspects of the incident, as well as any legal issues that arise.

**Fireground command (FGC)** An incident management system developed in the 1970s for day-to-day fire brigade incidents (generally handled with fewer than 25 units or companies).

**FIRESCOPE** (**FI**re **RES**ources of **C**alifornia **O**rganized for **P**otential **E**mergencies) An organization of agencies established in the early 1970s to develop a standardized system for managing fire resources at large-scale incidents such as wildland fires.

**Group** An organization level within IMS that divides an incident according to functional areas of operation.

**Group supervisor** The brigade leader in charge of all resources operating within a specified group, responsible to the next higher level in the incident organization, and the point-of-contact for the group within the organization.

**IMS general staff** The chiefs of each of the four major sections of IMS: Operations, Planning, Logistics, and Finance/Administration.

**Incident action plan (IAP)** The objectives for the overall incident strategy, tactics, risk management, and member safety that are developed by the IC. Incident Action Plans are updated throughout the incident.

**Incident command system (ICS)** The first standard system for organizing large, multijurisdictional and multiagency incidents involving more than 25 resources or operating units; eventually developed into the Incident Management System (IMS).

**Incident commander (IC)** The person in charge of the incident site who is responsible for all decisions relating to the management of the incident.

**Incident management system (IMS)** The combination of facilities, equipment, personnel, procedures, and communications under a standard organizational structure to manage assigned resources effectively to accomplish stated objectives for an incident. Also known as Incident Command System (ICS).

**Integrated communications** The ability of all appropriate personnel at the emergency scene to communicate with their supervisor and their subordinates.

**Liaison officer** The position within IMS that establishes a point of contact with outside agency representatives.

**Logistics section** The section within IMS responsible for providing facilities, services, and materials for the incident.

# Wrap-Up

**Logistics section chief** The General Staff position responsible for directing the logistics function; generally assigned on complex, resource-intensive, or long-duration incidents.

**Operations section** The section within IMS responsible for all tactical operations at the incident.

**Operations section chief** The general staff position responsible for managing all operations activities; usually assigned when complex incidents involve more than 20 single resources or when the IC cannot be involved in the details of tactical operations.

**Passing command** Option that can be used by the first-arriving brigade leader to direct the next arriving unit to assume command.

**Planning section** The section within IMS responsible for the collection, evaluation, and dissemination of tactical information related to the incident and for preparation and documentation of incident management plans.

**Planning section chief** The general staff position responsible for planning functions; assigned when the IC needs assistance in managing information.

**Public information officer** The position within IMS responsible for gathering and releasing incident information to the media and other appropriate agencies.

**Resource management** A standard system of assigning and keeping track of the resources involved in the incident.

**Safety officer** The position within IMS responsible for identifying and evaluating hazardous or unsafe conditions at the scene of the incident. Safety officers have the authority to stop any activity that is deemed unsafe.

**Single resource** An individual vehicle and the personnel that arrive on that unit.

**Span of control** The number of people that a single person supervises. The maximum number of people that one person can effectively supervise is about five.

**Staging area** A prearranged, strategically placed area where support personnel, vehicles, and other equipment can be held in an organized state of readiness for use during an emergency.

**Strike team** Five units of the same resource category, such as engines or ambulances, with a leader.

**Strike team leader** The person in charge of a strike team, responsible to the next higher level in the incident organization, and the point-of-contact for the strike team within the organization.

**Task force** Any combination of single resources assembled for a particular tactical need; has common communications and a leader.

**Task force leader** The person in charge of a task force, responsible to the next higher level in the incident organization, and the point-of-contact for the task force within the organization.

**Transfer of command** Reassignment of command authority and responsibility from one individual to another.

**Unified command** IMS option that allows representatives from multiple jurisdictions and/or agencies to share command authority and responsibility, working together as a "joint" incident command team.

**Unity of command** A characteristic of the IMS structure that has each individual reporting to a single supervisor and everyone reporting to the IC directly or through the chain of command.

# Brigade Member in Action

It is a Tuesday evening around 8:00 P.M. when the Control Center notifies you by radio that there is a smoke alarm indication in the Chemistry building. Typically at this time of the day, there is no one in the Chemistry building. As you enter the Fire Protection Equipment building, you hear on your radio a security guard who has entered the Chemistry building announce to the Control Center that there is smoke in the entry hallway of the building and that the installed fire sprinkler system in the building is locally alarming, indicating that water is flowing from the system. The guard also tells the Control Center that she is backing out of the building because the smoke is increasing. The Control Center acknowledges the security guard's transmission, and the Main Security Station also acknowledges the transmission and begins calling other members of the security force to ensure they are aware of the situation.

This information adds additional emphasis on your response as you transmit a message to the Control Center and the Main Security Station to verify that there are no Chemistry personnel on site. As you don your gear, you hear the apparatus driver open the truck door and start the apparatus. You transmit on your radio that you and the truck are en route to the Chemistry building.

As the truck arrives, you notice that there are six additional brigade members arriving at the building, and Security has arrived to control immediate scene access. There is smoke showing from the "A" side of the structure, and the majority of the smoke is coming from one of the laboratory windows. Since this is a working fire and most likely a haz-mat situation, you ask the Main Control Center to dispatch a Level I Alert with a request for Mutual Aid. You know from previous training that a Level I Alert will bring your company's haz-mat technicians from the Chemical complex. Management and engineering specialists will be dispatched to your facility to assist in the response, cleanup, and media relations. Mutual Aid will arrive from the complex a couple of miles away and will begin immediate assistance. They will also bring rehabilitation equipment and personnel, which will allow you to concentrate on the incident.

You designate a command post next to your truck and announce this over the radio so that everyone can hear who is in command and where you are located.

1. What should the command structure look like at this incident?
   - A. IC, Operations Section Chief, and Division A Supervisor
   - B. IC, fire branch director, and strike team leader
   - C. IC, PIO, rescue group supervisor
   - D. IC and team leaders

2. While working in the IMS, you abide by a management concept called unity of command. Unity of command states:
   - A. A command officer can effectively manage three to seven personnel at one time.
   - B. Each person has only one direct supervisor.
   - C. A command officer must use an accountability system in conjunction with IMS.
   - D. Each person can effectively work for three to seven managers at one time.

# Brigade Member in Action

As you brief the members of your brigade for the expected interior entry, you delegate teams and team leaders and assign them tasks to accomplish. Your driver gathers all accountability tags and transmits each name to the Control Center. A Mutual Aid brigade arrives, and you begin coordinating with that Brigade Leader on the duties you want that brigade to perform.

The Control Center calls you on the radio and tells you that no Chemistry personnel are on-site and that all personnel have been accounted for. Your Mutual Aid counterpart transfers this information to his brigade. As your teams begin entering the structure to investigate the situation, Security reports that another Mutual Aid brigade has arrived in the predesignated staging area. You ask Security to act as the Staging Officer. The first interior team reports that they have found a small fire in a chemistry sample hood and that the sprinkler system has contained the fire but they performed final extinguishment. The second interior team reports that there appears to be no fire extension to the room next to the chemistry laboratory. You acknowledge both reports and ensure that the Control Center has also copied the reports. You direct the second team to continue their primary search of the structure. You direct the team in the chemistry lab to maintain a reflash watch in case the fire returns.

3. Where in industry is a staging area typically found?
   A. At the main gate
   B. At the Control Center
   C. At the warehouse
   D. Next to the command post

4. In the IMS system with an IC, Operations, Planning, Logistics, and Finance sections established, where does the medical section typically operate?
   A. Operations
   B. Planning
   C. Logistics
   D. Finance

# Fire Behavior

Chapter 5

## NFPA 1081 Standard

**Incipient Facility Fire Brigade Member**

**4.1.2** *Basic Incipient Facility Fire Brigade Member JPRs.* All facility fire brigade members shall have a general knowledge of basic fire behavior, operation within an incident management system, operation within the emergency response operations plan for the site, the standard operating and safety procedures for the site, and site-specific hazards.

**Advanced Exterior Facility Fire Brigade Member**

**5.2.3 (A)** *Requisite Knowledge.* Principles of fire streams; types, design, operation, nozzle pressure effects, and flow capabilities of nozzles; precautions to be followed when advancing handlines to a fire; observable results that a fire stream has been correctly applied; dangerous conditions created by fire; principles of exposure protection; potential long-term consequences of exposure to products of combustion; physical states of matter in which fuels are found; the application of each size and type of attack line; the role of the backup team in fire attack situations; attack and control techniques; and exposing hidden fires.

**Interior Structural Facility Fire Brigade Member**

**6.2.1 (A)** *Requisite Knowledge.* Principles of conducting initial fire size-up; principles of fire streams; types, design, operation, nozzle pressure effects, and flow capabilities of nozzles; precautions to be followed when advancing hose lines to a fire; observable results that a fire stream has been correctly applied; dangerous building conditions created by fire; principles of exposure protection; potential long-term consequences of exposure to products of combustion; physical states of matter in which fuels are found; common types of accidents or injuries and their causes; and the application of each size and type of handlines, the role of the backup team in fire attack situations, attack and control techniques, and exposing hidden fires.

**6.2.3 (A)** *Requisite Knowledge.* The principles, advantages, limitations, and effects of horizontal and vertical ventilation; safety considerations when venting a structure; the methods of heat transfer; the principles of thermal layering within a structure on fire; fire behavior in a structure; the products of combustion found in a structure fire; the signs, causes, effects, and prevention of backdrafts; and the relationship of oxygen concentration to life safety and fire growth.

## Knowledge Objectives

After completing this chapter, you will be able to:

- Discuss the fire tetrahedron.
- Identify the physical states of matter in which fuels are found.
- Describe the methods of heat transfer.
- Define flash point, flame point, and ignition temperature as they relate to liquid fuel fires.
- Define the relationship of vapor density and flammability limits to gas fuel fires.
- Define Classes A, B, C, D, and K fires.
- Describe the phases of fire.
- Describe the characteristics of an interior structure fire.
- Describe rollover and flashover.
- Describe backdrafts.
- Describe the principles of thermal layering within a structure.

## Skills Objectives

There are no skills objectives for this chapter.

## You Are the Brigade Member

**On Monday afternoon you are paged to respond to the Administration Building to a reported fire on the third floor. As you arrive, you encounter light smoke on the ground floor. Employees are evacuating and seem to be only mildly concerned about the situation. You climb the stairs to the second floor, where the smoke is heavier. Employees are choking and coughing as they try to come down the stairs. Visibility is partially obscured, but you have no problem finding your way. When you reach the third floor, it is suddenly hot and dark. You need to use your hand light just to see the reflective material on your partner's turnout gear. After dropping down to your hands and knees, you find a layer of air that allows you to see along the floor.**

1. ***Why is visibility better near the ground?***
2. ***What concerns should you have upon reaching the third floor and it suddenly becoming hot and dark?***
3. ***What conditions are necessary for fire?***

## Introduction

Fire is a rapid, persistent chemical reaction that releases both heat and light. Before learning the methods and tactics for extinguishing fires, it is important to understand what goes on physically and chemically to make a fire occur. It is also important to understand how a fire behaves in different situations. A fire is a complex chemical process that converts one combination of substances into a different combination of substances and, at the same time, releases energy in the forms of light and heat. The understanding of fire behavior is the basis for all firefighting principles and actions.

### Brigade Member Tips

**Units of Measure**

When studying fire behavior, some basic units of measure must be used. In the United States, the British system of units is used, while in Canada and most other nations, the International System of Units, also known as the SI or metric system, is used. Either system can be used; however, it is important to use one system consistently. This book uses the British system and provides metric conversion factors or equivalent values as necessary.

In the British system, distance is measured in feet and inches, liquid volume is measured in gallons, temperature is measured in degrees Fahrenheit, and pressure is measured in pounds per square inch.

In the metric system, distance is measured in meters, liquid volume is measured in liters, temperature is measured in degrees Celsius, and pressure is measured in pascals or kilopascals.

## Fire Tetrahedron

In order to understand fire behavior, we will begin by identifying the conditions that are necessary for fire (also known as combustion) to occur. Four basic components are required to create a fire: fuel, oxygen, heat, and a sustained chemical chain reaction. We refer to these four essential components as the fire tetrahedron. First, a fuel must be present. The fuel is the material that burns. Second, oxygen must be available in sufficient quantities. Third, a source of heat must be present. The heat is essential to initiate and sustain the reaction between the fuel and the oxygen.

Adequate heat is required initially to raise the temperature of the fuel in the presence of oxygen to a point at which they will react together. After the fuel is ignited, the process of combustion releases heat energy, which is usually sufficient to keep the reaction going. If nothing interferes with the combustion process, the fuel will continue to burn until the supply of fuel or oxygen is exhausted.

The fourth component is also required to maintain a self-sustaining fire. A self-sustaining series of chemical chain reactions has to occur in order to keep the fire burning. When this component is included, the process can be represented by a four-sided geometric shape called the fire tetrahedron (► Figure 5-1). Each side of the tetrahedron depicts one of the four components that must be present for the combustion

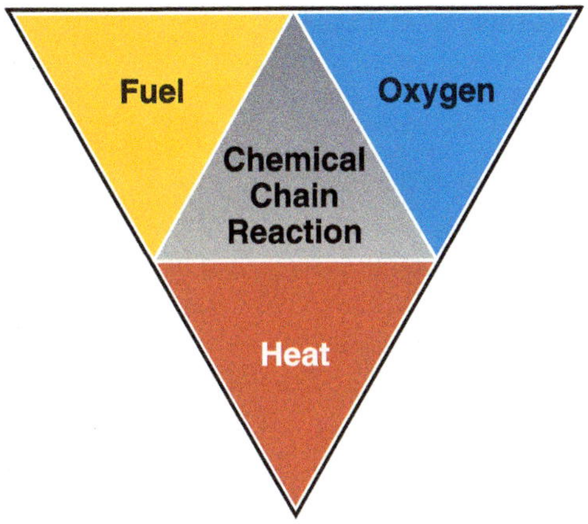


Figure 5-1 The tetrahedron model of fire (the fire tetrahedron) includes chemical chain reactions as an essential part of the combustion process.

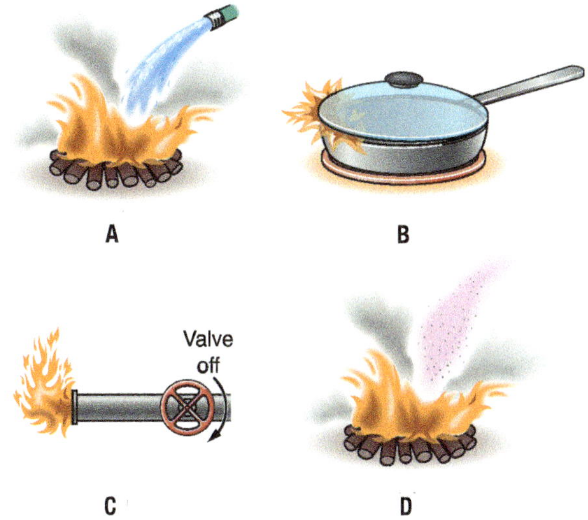


Figure 5-2 Each of the four basic methods of fire extinguishment (illustrated in relation to the fire tetrahedron). **A.** Cool the burning material. **B.** Exclude oxygen from the fire. **C.** Remove fuel from the fire. **D.** Break the chemical reaction with a chemical agent.

process to take place. A fire cannot occur without all four components.

## Methods of Extinguishment

The concept of the fire tetrahedron explains the conditions that are necessary for a fire to occur. Our objective as brigade members is to extinguish fires. Although there are many different methods that can be used to extinguish fires, they can be summarized by four main methods. The four basic methods of extinguishing fires are as follows:

- Cool the burning material.
- Exclude oxygen from the fire.
- Remove fuel from the fire.
- Break the chemical reaction.

Each method of extinguishment can be understood as removing one of the essential components of the fire tetrahedron. When the tetrahedron model of fire is considered, all four extinguishing methods can be applied (▶ Figure 5-2).

The most common method used to extinguish fires is to cool the burning material with water. The water absorbs the heat of the fire as it is turned into steam. The second method is to exclude oxygen from the fire. One example of this is placing a lid over a skillet containing flaming food. Another example is applying foam to a petroleum fire, which creates a blanket that traps the vapors and separates the oxygen from the burning fuel. The third method is to remove the fuel from a fire. The easiest way to extinguish a natural gas fire is to close the gas supply valve. The fourth method of extinguishing a fire is to interrupt the chemical chain reactions with a chemical agent.

## Fuel

Fuel is the actual material that is being consumed by a fire, allowing the fire to take place. Any material or matter that will burn is classified as a fuel. Individual fuels can be classified as solids, liquids, or gases (▶ Figure 5-3). The ability of a material to burn is regulated by its composition and its current physical state (i.e., whether it is a solid, liquid, or gas). Additional factors, such as the surface-to-mass ratio, have a strong influence on the ease of ignition and the rate of combustion for a particular fuel.

When substances are heated, they tend to change from a solid to a liquid state, and then to a gas or vapor state. When they are cooled, the reverse tends to take place. When substances are exposed to an increase in pressure, they tend to change from a gas or vapor state to a liquid, and then to a solid state.

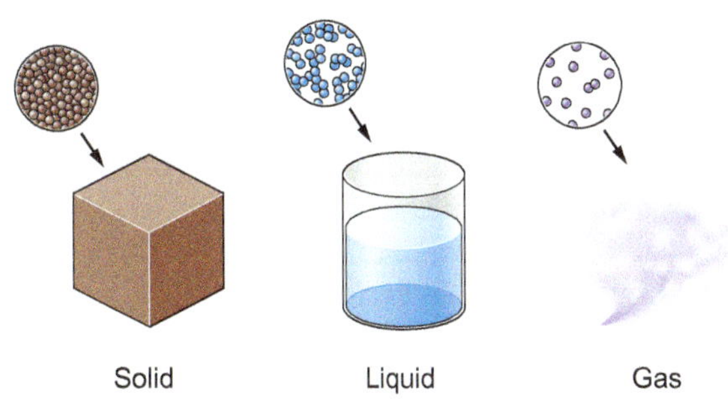


Figure 5-3 Fuels can be classified as solids, liquids, or gases.

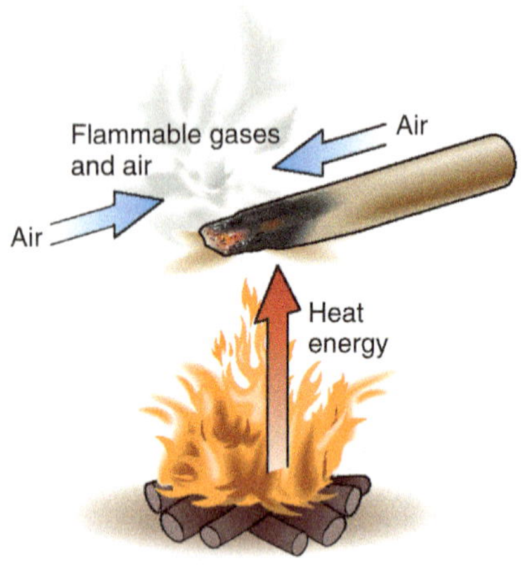


Figure 5-4 Pyrolysis.

The combustion reaction actually occurs when the fuel is in the gas or vapor state. Individual fuel molecules must be released from a solid or a liquid fuel in order to react with oxygen from the air.

## Solids

Solids have a definite shape. Most of the fuels we encounter in fires are solids. In a structure fire, the building materials and most of the contents are solids. Some of the characteristics of solids include the following:

- Solids have the ability to resist forces and retain a definite size and shape under ordinary conditions.
- Most solids expand slightly when heated and contract slightly when cooled.
- Most solids become more brittle when cooled and more flexible when heated.

When a solid fuel is burned, the actual combustion reaction does not occur within the solid itself. As the solid fuel heats up, it decomposes in a process known as pyrolysis and releases individual molecules into the atmosphere (▲ Figure 5-4). The combustion occurs when oxygen molecules in the air react with the individual fuel molecules, at the surface or slightly above the surface of the solid material.

The surface-to-mass ratio (STMR) is an important factor when looking at a solid fuel. Wood is one of the most common solid fuels. A large solid piece of wood, such as a log, will burn, but it is not easy to ignite. A log has a low surface-to-mass ratio, because it has a relatively small surface area in relation to its large mass. A large amount of energy is needed to heat a log before it will release enough molecules to sustain a fire.

If a log is cut into 2 × 4, the STMR increases, because each individual piece of wood has more surface area in relation to its total mass. Less heat energy is required to heat a 2 × 4, and there is more surface area to release molecules of fuel. Therefore, a 2 × 4 is easier to ignite and will burn more rapidly than a log.

If the wood is reduced down to sawdust, the STMR increases further. Each individual particle of sawdust has a small mass and a very large surface area in relation to its mass. A sawdust particle has a high STMR. Little energy is required to ignite the sawdust particles, and they will burn extremely quickly. Under the right conditions, the rate of combustion can be so fast that an explosion occurs.

Another consideration when discussing solids is the rate at which heat is released. One cubic foot of burning wood will generate a total heat release of approximately 156,000 BTUs (British thermal units). When the solid cubic foot of wood is divided into smaller pieces, the rate of heat release will increase. This rate will continue to climb as the surface-to-mass ratio increases. It is important to remember that the total amount of heat released does not change regardless of the surface-to-mass ratio—only the rate at which heat is released.

To cool and extinguish a wood fire with water, the volume of water must overcome the BTUs being released. A gallon of water has the ability to absorb about 9,000 BTUs. The BTUs generated from the cubic foot of wood require 17 gallons of water (156,000 ÷ 9,000 = 17.3). When the rate of heat release increases, the rate of water flow must likewise increase to absorb the BTUs that are generated. If the wood's surface-to-mass ratio was large enough to release all 156,000 BTUs in 30 seconds, the flow rate of the water would have to be doubled to deliver enough water to absorb the accelerated rate of heat release. A flow rate of 34 gpm would deliver the required 17 gallons of water in 30 seconds.

## Liquid

A liquid does not have a definite shape; it assumes the shape of the container in which it is placed. Liquids contract when cooled and expand when heated. When sufficiently heated, most liquids will turn into gases. Liquids, for all practical purposes, are not compressible; their volumes do not change significantly with changes in pressure. This characteristic allows us to pump water for long distances through pipelines or hoses.

Liquid fuels are one stage closer than solids to being in the optimum state for combustion. Vaporization occurs as a liquid fuel is heated, releasing molecules of fuel from the surface into the gaseous state. The vaporized fuel can then mix with oxygen, allowing combustion to occur. A fire involving a liquid fuel burns at the surface of the liquid.

The surface-to-volume ratio is an important factor for liquids, just as STMR is an important factor for solid fuels. As the surface area of the liquid increases, more molecules can be vaporized. The larger the surface area of a spill or container, the easier it is for combustion to occur and the more quickly the fuel is consumed.

### Gas

A gas has neither independent shape nor volume and tends to expand indefinitely. The gas we most commonly encounter is air. Air is a mix of invisible, odorless, tasteless gases that surround the earth. It consists of approximately 21% oxygen and 78% nitrogen, with small amounts of other gases like carbon dioxide making up the remaining 1%. Some fuels exist in the form of a gas such as propane or natural gas.

In the gaseous state the oxygen molecules and fuel molecules can mix freely and come into direct contact with each other. Fuels in the gaseous state need very little energy to ignite, since they are already in the optimum state for combustion. The important factor for a gaseous fuel is the ratio of fuel to air in the mixture. The mixture of fuel and air must be within a certain range for combustion to occur. The combustible range is different for each fuel. If the mixture is too lean (too much air and not enough fuel), it will not burn. If the mixture is too rich (too much fuel and not enough air), it will not burn.

## Oxygen and Oxidizing Agents

Oxygen is required to combine with the fuel for combustion to occur. The amount of oxygen required for combustion depends on the chemical composition of the solid fuel. If there is too little oxygen, the fuel will not burn. If there is an excess of oxygen, the solid fuel will generally burn faster and hotter.

An oxidizing agent can be used instead of oxygen itself in the combustion process. In a chemical reaction, an oxidizing agent behaves in the same manner as oxygen. Materials that are classified as oxidizers will support the combustion of other materials, even if no oxygen is present.

## Heat

Heat energy is required to ignite a fire. Once the fire is ignited, the combustion process produces more heat. If the fuel is in the form of a solid or a liquid, heat is required to raise its temperature in order to begin to liberate fuel molecules to a gaseous state before the fuel can burn. Additional heat is required for the gaseous fuel molecules to reach their ignition temperature, at which point the reaction between the fuel molecules and the oxygen molecules begins. Once the combustion is initiated, the extra heat energy that is produced will increase the temperature and accelerate the entire process.

Heat is one of many different forms of energy. The energy that is required to produce an ignition can come from a variety of sources. These include other forms of energy, such as mechanical, electrical, and chemical energy, that are easily converted to heat.

## Types of Energy

### Mechanical Energy

Mechanical energy is converted to heat when two materials rub against each other and create friction. For example, a fan belt rubbing against a seized pulley produces heat. Heat is also produced when mechanical energy is used to compress air in a compressor.

### Chemical Energy

Chemical energy is the energy created by a chemical reaction. Some chemical reactions produce heat (exothermic) and some absorb heat (endothermic). The combustion process is an exothermic reaction, because it releases heat energy.

Heat is produced whenever oxygen combines with a combustible material. If the reaction occurs slowly in a well-ventilated area, the heat is released harmlessly into the air. If the reaction occurs very rapidly or within an enclosed space, the mixture can be heated to its ignition temperature and can begin to burn. A bundle of rags soaked with linseed oil will begin to burn spontaneously because of the heat produced by oxidation that occurs within the mass of rags.

### Electrical Energy

Electrical energy is converted to heat energy in several different manners. Electricity produces heat when it flows through a wire, or any other conductive material. The greater the flow of electricity and the greater the resistance of the material, the greater the amount of heat produced. Examples of electrical energy that can produce enough heat to start a fire include heating elements, overloaded wires, electrical arcs, and lightning.

## Chemical Chain Reaction

The three basic ingredients of a fire are fuel, oxygen, and heat. When these three factors are present, under most circumstances, combustion will occur. The actual chemical processes of combustion involve a very complicated series of chain reactions at the molecular level. Normally, these chain reactions will continue to occur as long as there is an adequate supply of fuel and oxygen and sufficient heat. One way to extinguish a fire is to interrupt the sequence of chemical chain reactions by introducing other chemicals.

# Chemistry of Combustion

As materials decompose and burn, there are many different chemical reactions and physical changes that occur. In many cases, the chemical constituents that are present will break down and recombine several times. As described previously, reactions that result in the release of heat energy are referred to as exothermic reactions (▶ Figure 5-5). Reactions that absorb heat or require heat to be added are called endothermic reactions (▶ Figure 5-6).

The reactions that occur within most fires are complex. In addition to large quantities of heat and light, the combustion process also produces a wide variety of other by-products. The by-products change depending on the ratio of the amount of oxygen to the amount of fuel that is

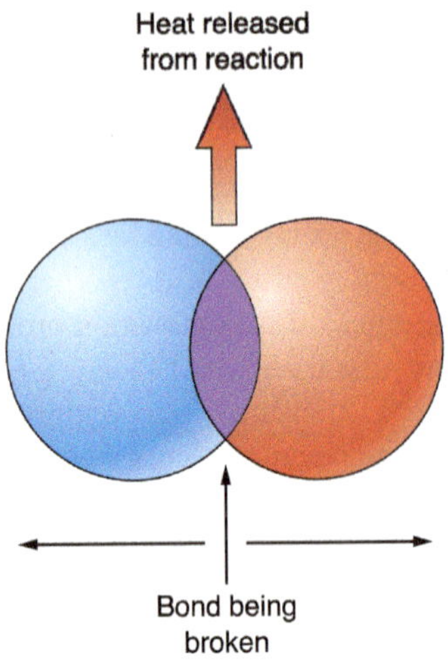


Figure 5-5 Exothermic reaction.

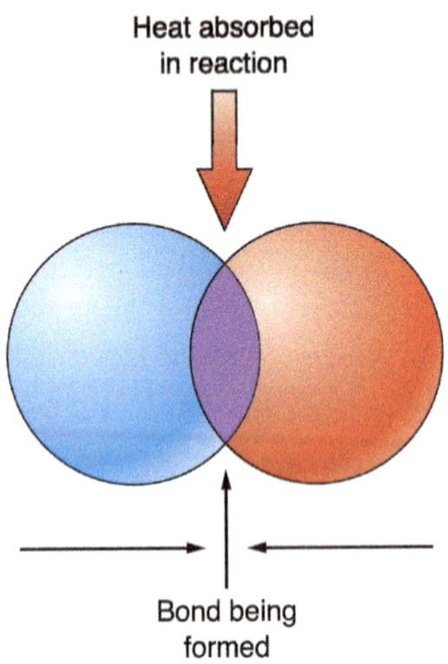


Figure 5-6 Endothermic reaction.

available to burn at any instant. This ratio can change often and quickly.

Many of the fires we encounter as brigade members burn with a limited amount of oxygen. During the early stages of a fire there is usually much more oxygen available than the fire can consume. As a fire grows inside a building, it will often consume most of the available oxygen until the rate of combustion is forced to slow down due to the shortage of oxygen. The fire can only consume oxygen at the rate that more oxygen can enter the space where the fire is burning. This tends to result in incomplete combustion, which produces large quantities of deadly toxic gases and compounds.

It is helpful to differentiate between oxidation, combustion, and pyrolysis. Oxidation is the process of chemically combining oxygen with another substance to create a new compound. The process of oxidation can be slow or fast. For example, steel that is exposed to oxygen will rust. Rusting is a very slow oxidation process.

Combustion is a rapid, self-sustaining process that combines oxygen with another substance and results in the release of heat and light. In effect, combustion is rapid oxidation. For our purposes, the terms combustion and fire can be used interchangeably.

Pyrolysis is the decomposition of a material caused by external heating. Some of the products of decomposition are gases, which can burn in the atmosphere. Eventually enough heat can be produced to cause combustion.

## Products of Combustion

Products of combustion are the substances that are produced by the combustion reaction. Fires produce a tremendous variety of products of combustion, depending on the fuel that is being burned, the temperature of the fire, and the amount of oxygen that is available to combine with the fuel. Very few fires consume all of the available fuel with maximum efficiency. This usually results in incomplete combustion and produces a variety of by-products.

We refer to the airborne products of combustion as smoke. There are three major components of smoke: particles, vapors, and gases. Many of the constituents of smoke are toxic. Inhalation of these particles, droplets, and gases can cause severe respiratory injuries. Because smoke is the direct result of fire, it is generally hot when it is created. The inhalation of superheated gases in smoke can cause severe burns of the respiratory tract. The temperature of smoke will vary depending on the conditions of the fire and the distance the smoke has traveled from the fire.

### Smoke Particles

Smoke particles are solid matter consisting of unburned, partially burned, or completely burned substances. These particles are lifted in the thermal column produced by the fire. In some cases the unburned and partially burned particles are hot enough to burn and can ignite if they come into contact with a supply of oxygen. The completely burned particles are primarily ash.

Most of the particles found in smoke are toxic. Some of these particles are small enough to get past the protective mechanisms of the respiratory system and enter the lungs.

### Vapors and Mists

Smoke often contains small droplets of liquids. When oil-based compounds burn, they produce small droplets that become part of the smoke. Oil-based or lipid compounds can be toxic and dangerous if inhaled. Some toxic droplets also cause poisoning if absorbed through the skin. Known examples of poisonous carcinogens that are readily absorbed through the skin include aromatic hydrocarbons such as xylene, toluene, and benzene.

## Brigade Member Safety Tips

Smoky environments are deadly! It is essential to use self-contained breathing apparatus any time you are operating in a smoky environment. This applies whether it is a structure fire, dumpster fire, car fire, or any other type of fire that generates smoke. Common toxic gases in smoky environments:

- Carbon monoxide
- Hydrogen chloride
- Hydrogen cyanide
- Carbon dioxide
- Nitrogen dioxide
- Phosgene
- Ammonia
- Chlorine

When water is applied to a fire, small water droplets are also suspended in the smoke or haze that forms. This is similar to the fog that is formed on a cool night. The water droplets often absorb particulate matter from the smoke.

### Gases

Smoke contains a wide variety of gases. The composition of gases in smoke varies greatly, depending on the substance being burned, the temperature, and the amount of oxygen available to the fire at a given instant. Burning wood will produce a different combination of gases than a fire fueled by petroleum-based products. (Remember that plastics are made from petroleum-based compounds.)

Most of the gases produced by a fire are toxic. Carbon monoxide, hydrogen cyanide, and phosgene are three of many toxic gases that are often present in smoke. Carbon monoxide is deadly in small quantities. Hydrogen cyanide was used to kill convicted criminals in gas chambers. Phosgene was used in World War I as a poisonous gas to disable soldiers. Fires also produce carbon dioxide, which is an inert gas. Although it is not toxic, carbon dioxide can displace oxygen and cause hypoxia.

## Heat Transfer

Heat transfer is important to a brigade member, because heat is essential for a fire to be initiated and for combustion to occur. Fires also produce large quantities of heat, which causes the fire to spread and involve additional fuel. Heat transfer is particularly important in relation to protecting the brigade member from the heat of the fire. Protective clothing is a shield that helps limit the heat that is transferred from the fire to the brigade member.

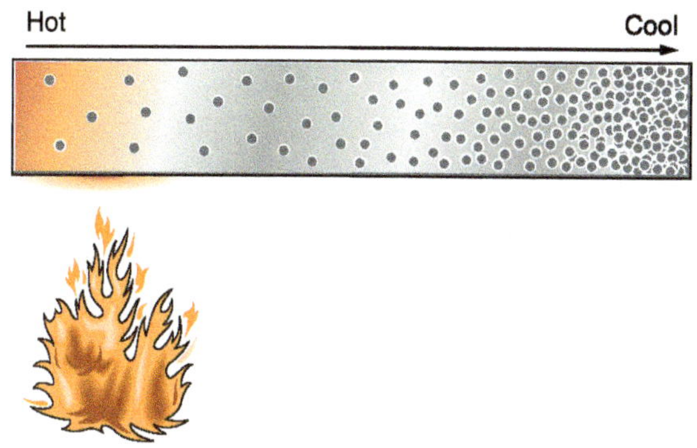


Figure 5-7 Conduction.

There are three mechanisms of heat transfer. Heat energy can be transferred from a hotter mass to a colder mass by any one of the mechanisms or by any combination of the three mechanisms working together. The fire spread we observe as brigade members is often a combination of conduction, convection, and radiation, all occurring simultaneously. In most situations one of the three mechanisms is more powerful than the other two mechanisms.

### Conduction

Conduction is the process of transferring heat energy within matter by transferring the energy directly from one molecule to another (▲ Figure 5-7). Objects vary in their ability to conduct energy. Metals generally conduct heat very easily. Insulating materials, such as fiberglass, are very poor conductors of heat.

Wood is a relatively poor conductor of heat. When wood is heated, the energy tends to remain in the area where the heat is applied, and the temperature in that area increases. Since wood has a fairly low ignition temperature, if the heat source is sustained, it will begin to burn in that localized area. The localized combustion will release more heat and cause the fire to spread to a larger area.

The same amount of heat energy applied to a steel beam will result in lower temperatures at the point where the heat is applied, because much of the heat will be dissipated by conduction to other parts of the beam. The temperature of the entire beam will increase, although the spot where the heat is applied will still be hotter than any other part of the beam.

Steel conducts heat so well that it can cause another material to ignite. Consider a steel beam that is in contact with a flame at one end and wood at the other end. By conduction, the steel could eventually transfer enough heat to the wood to cause the wood to ignite.

### Convection

Convection is the movement of heat through a fluid medium such as air or a liquid (► Figure 5-8). Heated molecules within

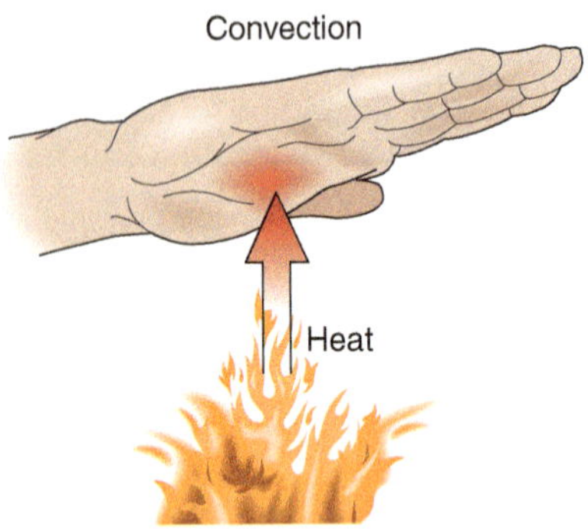


Figure 5-8 Convection.

a fluid medium will always rise. Convection currents can be observed above hot soup or a cup of coffee.

The convection currents created by a fire primarily involve gases generated by the fire. The heat of the fire warms the gases and particles in the smoke. A large fire burning in the open can generate a plume or column of heated gases and smoke that rises high into the air. This convection stream can carry the smoke and sometimes large bands of burning fuel several blocks before the gases cool and fall back to the earth. If there are winds present during a large fire, the winds can push the convection currents in different directions, causing the fire to spread.

When a fire is burning inside a room, the hot gases will rise to the ceiling. They will then travel horizontally along the ceiling until they hit a wall. If there are no openings in the room that allow the gases to escape, a layer of heated air will build up at the ceiling level and then move downward. The highest temperatures will always occur at the ceiling.

If there are openings in the fire room, convection currents can carry the fire gases outside the room of origin to other parts of the building. The superheated gases can be hot enough to ignite other materials. The hot gases will flow toward the highest level of the building, using stairways or any open shaft, and collect under the roof. Opening a ventilation hole in the roof releases the hot gases and allows energy to escape from the building by convection. This can rapidly lower temperatures inside the building, making it easier for brigade members to enter and attack the fire. We will explore the effects of convection currents on structure fires by studying how a fire spreads beyond the room of origin.

Figure 5-9 Radiation.

### Radiation

Radiation is the transfer of heat energy in the form of invisible waves (▲ Figure 5-9). The sun radiates energy to the earth through outer space. The electromagnetic radiation from the sun travels through the vacuum of space until it reaches the earth. The earth in turn absorbs the radiant heat energy from the sun. The human body absorbs radiant heat energy in the same manner, whether it comes from the sun, from a space heater, or from an open fire. Some materials reflect radiant energy more than they absorb it. A sheet of shiny aluminum foil will reflect the rays from the sun and send them in another direction.

Radiated heat energy from a fire travels in all directions. The effect of radiation is not seen or felt until the radiation strikes an object and heats the surface of the object. A large building that is fully involved in fire can radiate a tremendous amount of heat energy to the surrounding buildings. That energy can travel several hundred feet and ignite an exposed building. The radiated heat can even pass through glass windows and ignite objects inside other buildings.

## Characteristics of Liquid Fuel Fires

Fires involving liquid fuels have some different characteristics from fires that involve solid fuels. As discussed previously, solid fuels do not burn in a solid state. A fuel must be decomposed and the flammable compounds must be vaporized before the fuel can burn. The same characteristic applies to liquids. A liquid must be converted to a vapor before it will burn. In addition, the vaporized fuel molecules must be mixed with air in the proper concentration. There is a minimum and a maximum concentration that is necessary for any vapor to be ignited. These two concentrations are called the flammability limits.

There are three conditions that must be present in order for a vapor and air mixture to ignite:

- The fuel and air mixture must be within the flammable limits.
- There must be an ignition source with enough energy to ignite the vapors.
- There must be enough sustained contact between the ignition source and the fuel–air mixture to transfer the ignition energy.

As a liquid is heated, the molecules in the liquid become more active, and some of them begin to break away from the surface as vapor. The process of molecules being released from the liquid's surface is called evaporation. The rate of evaporation increases as the temperature of the liquid increases. At the boiling point, the liquid will continually give off vapors in sustained amounts and, if held at that temperature for long enough, will eventually turn into a gas. The amount of liquid that will be vaporized is also related to the volatility of the liquid. A more volatile liquid evaporates more quickly than a less volatile liquid.

As more of the liquid vaporizes, the mixture will reach a point where there is enough vapor in the air to create a flammable fuel–air mixture. Most flammable liquids can generate enough vapors to be ignited at temperatures well below their boiling point. For example, gasoline can generate flammable vapors at temperatures as low as –45°F. If the vapors are ignited, the fire will warm the liquid surface and cause vapors to be released more rapidly. This increases the rate of burning.

Because most liquid fuels are mixtures of different compounds, they do not have single boiling points. For example, gasoline is a mixture of about 100 different compounds, and the individual compounds evaporate at different rates. Gasoline vapors are composed of the individual vapors of all of the compounds mixed together.

The flammability of a liquid is defined in terms of three temperatures. These terms are used to compare characteristics of different fuels: The flash point is the lowest temperature at which a liquid produces a flammable vapor to ignite. This is measured by determining the lowest temperature at which the liquid will produce enough vapor to support a small flame for a short period of time. At this concentration the flame may go out quickly. The fire point or flame point is the lowest temperature at which a liquid produces enough vapor to sustain a continuous fire. The ignition temperature is the temperature at which the fuel will spontaneously ignite. The ignition temperature of the mixture is determined by the compound with the lowest ignition temperature.

Liquid fuels can typically be classed into two categories: hydrocarbons and polar solvents. Hydrocarbons are organic chemical compounds composed of carbon and hydrogen atoms. Examples would be crude oil, gasoline, and diesel fuel. The other type is a polar solvent. There chemicals are composed of hydrogen and oxygen atoms. Polar solvents would include chemicals that are alcohol based such as ketones, acetone, and dimethyl sulfoxide (DMSO).

One major characteristic the brigade member should know about any liquid fuel is the **specific gravity** of the substance. Specific gravity is the ratio of the density of a substance to the density of a reference substance; typically water. If the specific gravity of a chemical is a value lower than 1.0, the substance will float on the surface of water. If the value is 1.0 or more, the substance will settle or mix with the water. This factor can be valuable during size-up of the emergency and the ensuing strategy and tactics to be used. Hydrocarbons typically have a value of less than 1.0 while polar solvents typically have values of less than 1.0.

## Characteristics of Gas Fuel Fires

There are two terms that help to describe the characteristics of flammable gases and vapors. These are vapor density and flammability limits.

### Vapor Density

Vapor density refers to the weight of a gaseous fuel. Vapor density measures the weight of the gas molecules compared to air. The weight of air is assigned the value of 1.0. A gas with a vapor density of less than 1.0 will rise to the top of a confined space or rise into the open atmosphere. A gas with a vapor density greater than 1.0 is heavier than air and will settle to the ground.

In situations where a flammable gas is present, knowing the vapor density of the product allows you to predict whether the danger of ignition is at a high level or at a low level within an area. Hydrogen gas has a density of 0.07, so it is very light and will rise in the atmosphere. Propane gas has a vapor density of 1.51, so it will settle to the ground. Carbon monoxide has a density of 0.97, which is very close to air. Carbon monoxide mixes readily with all layers of the air since it has nearly the same density.

### Flammability Limits

Flammable vapors and air will ignite only when they are mixed in the proper concentrations or proportions. If the fuel vapor concentration is too low, the mixture is too lean to support combustion. Conversely, if the fuel vapor concentration is too high, the mixture is too rich to support combustion.

The range of mixtures that will burn varies from one fuel to another. Methane gas must be mixed with air in concentrations of between 4.5% and 15.0% in order to burn. These values are known as the lower and upper flammability limits. The terms flammability limits and explosive limits are used interchangeably in this book. Under most conditions, if the flammable gas and air mixture can be ignited, it has the ability to explode. Test instruments are available to measure the percent of fuels in gas and air mixtures and to determine when an emergency scene is safe.

The lower explosive limit (LEL) refers to the minimum amount of fuel vapors that must be present in a gas and air mixture for the mixture to be flammable or explosive. In the case of carbon monoxide, the LEL is 12.5%.

The upper explosive limit (UEL) of carbon monoxide is 74%. These two values tell us that carbon monoxide can burn or explode when the concentration is at least 12.5% and no greater than 74% in air.

# Voices of Experience

**"The students did learn that a small volume of water will generate a large volume of steam. They also learned that disrupting thermal balance destroys visibility and creates an even tougher environment to work in."**

There is no better way to teach firefighting techniques than to create realistic live fire training evolutions. This also provides the opportunity to demonstrate the changes in fire growth and thermal layering under controlled conditions.

During one such training session, the instructors took students into a burn room to show the changes in fire conditions from the incipient phase to the fully developed phase. Fuel for the fire was a small pile of lumber scraps and straw. The fires were kept small, and each group of students had a charged handline. The instructors also had a charged handline to provide a safety line as a back-up.

Several evolutions were conducted as the instructors demonstrated the proper use of fire streams to control the fire and maintain a good thermal balance in the room. Throughout the evolutions, the instructors stressed the potential risks of putting too much water into the heat layer at the ceiling. Prior to each evolution, steam production from hose streams was discussed, as was the proper hose stream to prevent or reduce the chance of disturbing the thermal balance in the room.

As the fire progressed, flames slowly rolled across the ceiling. One of the students thought the fire was moving too close and decided to open up a stream to push the flames back. Unfortunately, the student's nozzle was set on a fog pattern when he started flowing water. Visibility quickly went to zero, and the warm room quickly changed to a very hot room. The instructors opened the ventilation hatch and had the group retreat from the room.

No one was injured because the fire was small and the heat generated in the room was relatively low. The students did learn that a small volume of water will generate a large volume of steam. They also learned that disrupting thermal balance destroys visibility and creates an even tougher environment to work in.

Everyone left the training session with an up-close and personal lesson in fire behavior.

Scott Dornan
ConocoPhillips
Anchorage, Alaska
Kuparuk Fire Department
Kuparuk, Alaska

## Boiling-Liquid, Expanding-Vapor Explosion (BLEVE)

You must understand one potentially deadly set of circumstances involving liquid and gas fuels. This is a boiling-liquid, expanding-vapor explosion or BLEVE. A BLEVE can occur when a hydrocarbon fuel stored in a pressure vessel or atmospheric pressure storage tank is subjected to excessive heat. Flame impingement on the tank causes the internal pressure to increase. Atmospheric tanks and pressure vessels are equipped with a pressure relief device to prevent overpressure. Continued flame impingement can cause the pressure to increase at a rate higher than the design rate of the relief valve. Flame impingement below the liquid level will not damage the tank or vessel because the liquid acts as a heat sink and prevents metal fatigue.

However, flame impingement above the liquid level will cause the metal to fatigue and ultimately fail. The loss of container strength and excessive pressure is a dangerous combination. As the pressure increases, the sound from the relief valve will change: The pitch will get higher, and the sound will become louder. When the vessel can no longer contain the pressure, it will fail catastrophically. Vessel failure can be caused by metal fatigue, pressure that is in excess of the relief value capacity, or both.

When this occurs, pieces of the tank can be propelled great distances, injuring or killing personnel in the debris zone. Metal can travel a few hundred feet to over a quarter of a mile, depending on the size of the tank or vessel. The fuel that is released when the tank or vessel fails will immediately ignite, creating a fireball. Depending on the size of the tank and the fuel involved, the resulting fireball can easily reach 1,000 feet in diameter.

A common misunderstanding is that a BLEVE will occur only in a pressure vessel. Any closed container that is overpressured due to excessive heat has the potential for BLEVE. Aerosol cans, small-volume containers, and drum stock are all potential BLEVE hazards.

Over the years, dozens of emergency responders have been killed and many more injured while fighting fires in vessels and tanks because they failed to recognize the signs of impending BLEVE conditions. Understanding the characteristics of flammable and combustible liquids and gases in storage containers, tanks, and vessels and the mechanism of a BLEVE will help prevent injuries and deaths.

# Classes of Fire

Fires are classified according to the type of fuel that is burning. The classification system is designed to simplify the decisions that must be made when choosing extinguishing agents and methods for different types of fires. Fires are placed into one of five classes: Class A, Class B, Class C, Class D, and Class K. For each class of fire there are corresponding classes of extinguishing agents. Class A extinguishing agents are used to fight Class A fires and so on. We will consider the methods of attacking specific types of fires in more detail in subsequent chapters.

A fire may fit into more than one class. For example, a fire could involve a wood building (Class A) as well as petroleum products (Class B). A fire involving energized electrical circuits (Class C) might also involve Class A or Class B materials.

## Class A Fires

Class A fires involve ordinary solid combustible materials such as wood, paper, and cloth. The best method to extinguish a Class A fire is to cool the fuel to a temperature that is below its ignition temperature. Water is the most frequently used extinguishing agent for Class A fires, because it is plentiful and efficient in absorbing large quantities of heat.

A Class A fire should not be considered extinguished until the entire mass has been thoroughly cooled. If a Class A fire is not completely cooled, the remaining hot embers may rekindle, and the remaining fuel may start to burn again. Smothering a Class A fire to exclude oxygen may extinguish the fire temporarily, but it will not cool the fuel. The remaining hot embers can reignite the fire if they are exposed to air.

## Class B Fires

Class B fires involve flammable or combustible liquids such as gasoline, kerosene, oils, paints, and tar. Most Class B fires are best extinguished by creating a barrier between the fuel and the oxygen. A layer of foam is applied to the surface of the fuel to stop the production of vapors. If the liquid can no longer release vapors to mix with the air, the fire goes out. Foam also cools the surface of the liquid and heated objects in the vicinity of the fire.

Class B fuels are typically separated into two categories; hydrocarbons and polar solvents. A hydrocarbon is a compound of hydrogen and carbon, such as any of those which are the chief components of petroleum products such as gasoline, diesel fuel, naphtha, and crude oil. These products typically have a specific gravity of less than 1 indicating they will float on the surface of normal water. Polar solvents are chemicals like grain alcohol, acetic acid, propanol, ethanol, and acetone. These chemicals typically have a specific gravity of less than 1 and tend to form a solution with water.

### Brigade Member Safety Tips

When water is applied to certain combustible metals, such as magnesium, the water reacts with the metal to produce hydrogen and oxygen. The hydrogen can burn with an explosive force.

Some liquid fires can be extinguished by cooling them with water, similar to a Class A fire. This is most effective on liquids that have high flash points, in which cases the water can cool the surface of the liquids below their respective flash points. A water mist is generally used in these cases.

Alternative extinguishing agents are available for Class B fires, including dry chemicals and carbon dioxide. Dry chemical fire extinguishers attack a fire by interrupting the fire's chain reactions. Carbon dioxide is applied as a smothering agent to exclude oxygen from the area where the fire is burning.

### Class C Fires

Class C fires involve energized electrical equipment. Electricity itself doesn't burn, but it provides energy that can ignite a fire. The most common extinguishing agents for electrical fires are carbon dioxide or dry chemical agents because they do not conduct electricity. Halogenated agent extinguishers are used for special Class C fires involving electrical equipment such as computers. The fuel that is burning is often Class A or Class B; however, the special classification is required due to the particular hazards of electricity. If the electrical source is disconnected, it becomes a Class A or Class B fire.

The use of water on a Class C fire is very dangerous because water can conduct electricity. The electrical supply must be shutoff prior to applying water in any form.

### Class D Fires

Class D fires involve burning metals. Combustible metals include sodium, potassium, lithium, titanium, zirconium, magnesium, aluminum, and many of their alloys. These fires are less common, but they can be very difficult to extinguish. Most cars contain numerous components made from such alloys. The greatest ignition hazard exists when the combustible metals are present as shavings or in a molten form.

Fighting a Class D fire with water can cause a chemical reaction, and it can generate an explosive reaction. Special extinguishing powders based on sodium chloride or other salts are available. Extinguishing with clean, dry sand is another option.

### Class K Fires

Class K fires involve combustible cooking media, such as oils and grease, commonly found in commercial kitchens. The Class K designation is relatively new and coincides with a new classification of Class K extinguishing agents recognized by NFPA 10, *Standard for the Installation of Portable Fire Extinguishers*.

## Phases of Fire

When oxygen, fuel, heat, and the necessary chain reactions are all present under adequate circumstances, a fire will occur.

Figure 5-10 Incipient phase.

As a typical fire progresses, it will pass through four distinct phases, unless the process is interrupted. The four phases of fire are as follows:

- Incipient
- Growth
- Fully developed
- Decay

### Incipient Phase

The incipient phase of the fire is the starting point ▲ Figure 5-10. When the four parts of the fire tetrahedron are present and the fuel is heated to its ignition temperature, combustion occurs.

To study an example, consider a fire that begins when a burning cigarette ignites the arm of a sofa. Further consider that the sofa is sitting between an end table and a coffee table in a living room, and that there is a good supply of fuel, and the concentration of oxygen is 21% within the room. The fire begins small, with a localized flame at the point where the cigarette is in contact with the fabric.

Gradually, more of the fabric is ignited and a small plume of hot gases starts to rise from the sofa arm. This plume ignites even more of the fabric, increasing the size and intensity of the flame. At this point the convection of hot gases is the primary means of fire growth, with some additional energy being radiated to the area close to the flame.

Within a minute the sofa starts to ignite. Oxygen is drawn into the fire at floor level. At this point the fire could probably be extinguished with a portable fire extinguisher.

### Growth Phase

During the growth phase of a fire, additional fuel becomes involved in the fire. A combination of convection and radiation ignites more surfaces of the sofa and the table closest to

Figure 5-11 Growth phase.

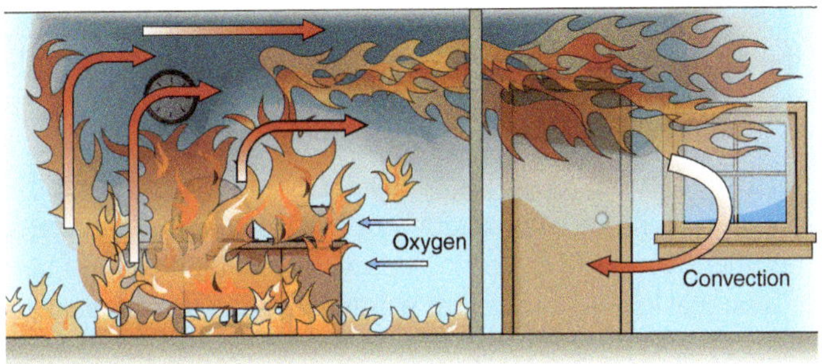


Figure 5-12 Fully developed phase.

it (▲ Figure 5-11). As more fuel is ignited, the size of the fire increases and the visible plume reaches the ceiling.

The convection flow begins to draw additional air into the fire. With this additional air and more fuel becoming involved, the fire grows in intensity. Hot gases from the plume flow across the ceiling toward the four walls of the room.

This is the point where thermal layering begins. The hotter air collects at the top of the room and banks down, while the cooler air stays closer to the floor. During this growth phase, the fire will continue to grow, as long as there is enough oxygen available and the fuel source is maintained.

### Flashover

Flashover is a the point between the growth phase and the fully developed phase where all of the combustible materials in a room become ignited.

Most fires begin small. As a fire grows in size, the combined forces of radiation, convection, and conduction cause the surface temperatures of all other combustible materials in the room or space to approach their ignition temperatures. The surface temperatures reach a point where individual items begin to ignite in rapid succession. As each additional item ignites, it adds even more heat energy to the room. At the flashover point, all of the remaining items reach their ignition temperatures at virtually the same time. Suddenly everything in the room begins to burn, releasing energy at a much faster rate. Temperatures reach the vicinity of 1,000°F in just a few seconds.

Flashover is often the most deadly to brigade members and victims because conditions can change radically in an instant. The details of a flashover are discussed in more detail in the following section.

**Brigade Member Safety Tips**

Brigade member must learn to recognize the warning signs of flashover and use the reach of their hose stream to cool the atmosphere. A brigade member even in full turnout gear has little chance of avoiding severe injury or even death if caught in a flashover.

## Fully Developed Phase

During the fully developed phase, the fire has progressed beyond flashover (◄ Figure 5-12). All combustibles have ignited and heat is being produced at the maximum rate. The fire at this point has everything it needs and produces large amounts of fire gases.

During this phase the available oxygen is consumed rapidly. The fire can burn at this rate only for as long as the oxygen supply is adequate. If the fire consumes oxygen more rapidly than it can be replaced by fresh air, the rate of combustion will slow down.

## Decay Phase

The final phase of a fire is the decay phase (▼ Figure 5-13). At this point the fire is running out of fuel. The atmosphere is still hot and heat energy is still being released; however, the

Figure 5-13 Decay phase.

rate of combustion is slowing down. The intensity of the fire decreases to the point where there is only smoldering fuel. Eventually, all of the fuel will be consumed and the fire will go out.

## Characteristics of an Interior Structure Fire

Interior fires have some special characteristics, because the fire is fully or at least partially contained within the building. The building acts as a box, keeping heat and products of combustion inside. The structure can also limit the flow of air that can reach the fire, which can change the rate of combustion and the products of combustion. In addition, the structure limits the ability of brigade members to reach the fire in order to extinguish it.

On a smaller scale, each room within a building is also a compartment where a fire can originate and grow. Sometimes the fire stays within a room and sometimes it spreads to other rooms or spaces. The following sections describe some special considerations that are directly related to fires in rooms and within buildings.

### Room Contents

Most fires in buildings involve the building's contents. A large proportion of fires in buildings do not involve the structure itself but only burn the contents within individual rooms and spaces. These fires originate in the contents, and most of the heat and products of combustion are produced by the contents. Buildings themselves can and do burn; however, the building often functions as a container, and most of the fire occurs inside it.

Fifty years ago, a typical room in an office or a residential occupancy contained many products made of wood and natural fibers. Today most rooms are heavily loaded with plastics, and many of the synthetic items in our environment are made from petroleum products. This change in room contents has changed the behavior of fires. When heated, most plastics release volatile products that are both flammable and toxic. Burning plastics generate dense smoke that is rich in flammable vapors, which makes fire suppression more difficult and more dangerous. Some of these products melt and drip at high temperatures. If the droplets are on fire, they can contribute to the spread of the fire.

The upholstered furniture that is manufactured today is generally more resistant to ignition from glowing sources such as cigarettes than in the past. However, it has very little resistance to ignition from flaming sources. After it is ignited, such furniture can be completely involved in fire in 3 to 5 minutes and reduced to a burning frame in less than 10 minutes.

Walls and ceilings are often painted using emulsions of latex, acrylics, or polyvinyl materials. When these paints are applied, they form a plastic-like coating. This coating has the characteristics of the plastic from which it was formed and can burn readily. Varnishes and lacquers are also combustible. Some paints and coatings add to the fire load in a building and can aid in the spread of the fire from one area to another. It should be apparent that the increased use of plastics in furnishings and finishes has set the stage for fires that are different from a simple campfire.

### Fuel Load and Fire Spread

Fuel load refers to the total quantity of all of the combustible products that are within a room or a space. The fuel load determines how much heat and smoke will be produced by a fire, assuming that all of the combustible fuel in that space is consumed. The size and shape of interior objects and the types of materials used to create them have a tremendous impact on the objects' ability to burn and their rate of combustion. The same factors influence the rate of fire spread to other objects. Even the arrangement of the furnishings within a room can make a difference in the way a fire burns.

Take the example of a couch on fire in a living room that is furnished in the following typical fashion: in addition to the couch, the room contains two upholstered chairs, an entertainment center, a coffee table, several small end tables, and a wastebasket. The walls are painted and the floor is covered with synthetic carpeting. Almost everything in the room is highly combustible. This situation involves many different factors that will trigger events simultaneously or in rapid succession:

- Most of the materials used to construct a couch are typically synthetics. As the couch burns, it will release tremendous quantities of heat energy within the room. If nothing is done to extinguish the fire, it will burn until all of the fuel is consumed.
- The heat will cause the synthetic materials in the couch and the other pieces of furniture in the room to release toxic and flammable gases, even before they are ignited.
- The burning synthetics will release large quantities of smoke and other products of combustion. These products are both flammable and toxic, and they will completely obscure the visibility in the room, making the atmosphere extremely hazardous to anyone who remains in the room.
- The burning synthetics in the couch will begin to drip onto the carpet. As a result, the carpet will also catch fire and burn.
- As the fire continues to burn, all of the contents of the room as well as the ceiling and wall finishes will become involved in the fire.

This is a relatively simple example, considering only the contents of a typical office or living room. A fire in a building becomes much more complicated when the heat and products of combustion begin to spread from room to room. All of these factors and many others influence the behavior of the fire.

## Special Considerations in Structure Fires

As a brigade member, you must understand the behavior of fires and work to control the power of a fire rather than try to attack it in a reckless manner. There are four conditions you must understand in order to work safely to extinguish structure fires. These conditions are flashover, rollover (flameover), backdraft, and the thermal layering of gases.

### Flashover and Rollover

Flashover was defined previously as the sudden ignition of all of the combustible objects within a room or a confined space. Flashovers can occur with explosive force, but in a different way than a backdraft. A flashover is a much more frequent occurrence than a backdraft.

A flashover occurs within a room or space when a growing fire has sufficiently heated everything around it. The hot gases produced by the fire spread across the ceiling and reach out to the walls. These gases are so hot that heat energy is radiated down from the smoke layer to the contents of the room. The ceiling and the walls are also hot and radiate heat back into the space. This intense buildup of heat causes all of the combustible objects in the space to reach their ignition temperatures almost simultaneously. As soon as the first objects begin to ignite, more heat energy is released and everything else in the room is heated even faster. The flashover occurs very quickly.

A brigade member in full protective gear cannot survive flashover conditions for more than a few seconds. The sudden increase in temperature will overwhelm the protective capabilities of protective clothing and equipment.

Brigade members often arrive at the scene of a fire when it is close to the flashover point. It is very important to be able to recognize a situation in which flashover conditions are present. The signs and conditions that indicate that a flashover may be imminent are as follows:

- Dense black smoke pushes out of a doorway or window opening under pressure. The smoke moves several feet from the opening as it is being pushed by high heat. This dense smoke with tightly packed curls is sometimes referred to as black fire.
- Dense smoke fills over one-half of a door or window opening into the fire area. Inside the room the smoke will be banked down to doorknob height. Below this level the atmosphere is often clear and the fire can be seen within the room.

Rollover or flameover is a term used to describe burning fire gases. This condition occurs in one of two ways. It can be produced when ignited fire gases leave the fuel, rise to the ceiling, and then spread out horizontally. The second way it is produced is when the fuel produces unburned combustible fire gases that rise to the ceiling and spread out horizontally. The fire gases have sufficient heat to ignite but are deficient in oxygen. Once the combustible gases reach an opening such as a door or window, they mix with oxygen-rich air and ignite into an open flame. In both situations, it is the fire gases that are burning. In order to stop rollover, the fire must be controlled. Left uncontrolled, this condition will likely lead to flashover.

The only thing that brigade members can do to prevent flashover at this point is to immediately cool the atmosphere. This situation calls for aggressive action to cool the atmosphere, to make an immediate exit, or for immediate ventilation.

### Backdraft

A backdraft is best described as an explosion that occurs when oxygen is suddenly admitted to a confined area that is very hot and contains large amounts of combustible vapors and smoke (▼ Figure 5-14). A backdraft is sometimes called a smoke explosion, because, in effect, it is the smoke produced by the fire that explodes.

Figure 5-14 Backdraft.

A backdraft usually occurs when a fire is smoldering or when the intensity of the fire has been reduced by a lack of oxygen. The typical conditions that lead to a backdraft begin with a fire inside a building that produces large amounts of smoke and carbon monoxide. These products of combustion have collected within a confined space or throughout the entire building. Because the building has relatively few openings, the products of combustion have difficulty escaping and the flow of fresh air into the building is restricted. As the fire continues to burn, it consumes most of the available oxygen, and the rate of combustion slows down. Sometimes all flaming combustion stops and the fire smolders.

The situation at this point is a building or a space within a building that is filled with carbon monoxide, which is a flammable gas, and other products of combustion. The smoke contains large quantities of material that is only partially burned. This mixture is hot enough to burn and the only thing that is keeping it from bursting into flames is the lack of oxygen. If oxygen is suddenly introduced, the mixture will ignite with explosive force. This can occur if brigade members open a door or window in a way that allows oxygen to enter the space. There have been cases where large explosions have sent glass and portions of walls toward brigade members.

Brigade members should learn to recognize the signs of an impending backdraft. These signs are as follows:

- Little or no flame is visible from the exterior of the building. (The fire is smoldering, not flaming.)
- Smoke is emanating under pressure from cracks around doors, windows, and eaves.
- Typically, there are no large openings in the building, such as open doors or windows.
- A "living fire" is visible, where smoke is puffing from the building and being drawn back in so that it looks like it is breathing.
- There is an unexplained change in the color of the smoke.
- Glass is smoke stained and blackened due to heavy carbon deposits from the smoke.
- Signs of extreme heat conditions are present.

It is important to note that these conditions could be present in an entire building or only within a space inside the building such as an attic, closet, or other concealed space. A backdraft can occur in any situation where oxygen is suddenly introduced into an enclosed space that is filled with heated products of combustion.

The best way to prevent a backdraft from occurring is to make a ventilation opening at a level above the fire so that the hot gases can escape from the interior prior to allowing fresh air to enter. This is not a simple task. It requires very close coordination between ventilation and fire attack. The most important consideration for brigade members is to recognize when potential backdraft conditions are present and to avoid triggering the backdraft by making an opening at the wrong place at the wrong time.

## Thermal Layering and Thermal Balance

During a fire in a closed room or space, the fire gases are said to be in thermal balance when they are allowed to seek their own level. Prior to flashover, the hottest gases always rise to the upper levels, and the temperature at the floor level is relatively cool compared to the ceiling (although it still could be hot enough at the floor level to burn anyone without full protective clothing and equipment). By keeping low in a normal thermal balance situation, you stay in the area where the temperature is the lowest. Your chance of survival is increased with the proper personal protective gear and by staying low.

If this normal thermal balance is upset, severe injury can occur to brigade members. When water is sprayed into the upper part of a room, the energy in the superheated gases at the top of the room converts the water to steam. Steam takes up many times the space of the same amount of water, so the steam expands rapidly. The water to steam expansion ratio is approximately 1,700 to one. The freshly produced steam becomes superheated as it absorbs more heat energy from the fire gases.

The expanding superheated steam can fill the whole room and will also displace superheated fire gases from the ceiling down toward the floor. This mixture of hot steam and fire gases can result in severe burns to brigade members, even when they are close to floor level and wearing full personal protective gear. This action also seriously jeopardizes the chances of survival for any victims trapped.

# Wrap-Up

## Ready for Review

- Matter exists in three states: solid, liquid, and gas.
- Energy exists in many forms including chemical, mechanical, and electrical.
- Conditions necessary for a fire include fuel, oxygen, heat, and a self-sustaining chemical reaction.
- Fires spread by conduction, convection, and radiation.
- The four principal methods of fire extinguishment are cooling the fuel, excluding the oxygen, removing the fuel, and inhibiting the chemical reaction.
- Fires are categorized as Class A, Class B, Class C, Class D, and Class K. These classes reflect the type of fuel that is burning and the type of hazard the fire represents.
- Solid fuel fires develop through four phases: the incipient phase, the growth phase, the fully developed phase, and the decay phase.
- The growth of room and contents fires is dependent on the characteristics of the room and the contents of the room.
- Special considerations related to room and contents fires include flameover, flashover, backdraft, and the thermal layering of gases.
- Liquid fuel fires require the proper mixture of fuel vapor and air.
- The characteristics of gas fuel fires are different from other fires.
- Vapor density reflects the weight of a gas compared to air.
- Flammability limits vary widely for different fuels.
- A BLEVE is a catastrophic explosion in a vessel containing a boiling liquid and a vapor.

## Hot Terms

**Backdraft** The sudden explosive ignition of fire gases when oxygen is introduced into a superheated space previously deprived of oxygen.

**Boiling-liquid, expanding-vapor explosion (BLEVE)** An explosion that can occur when a tank containing a liquid fuel overheats.

**British thermal unit (BTU)** The amount of heat required to raise the temperature of one pound of water by one degree Fahrenheit.

**Chemical energy** Energy created or released by the combination or decomposition of chemical compounds.

**Class A fires** Fires involving ordinary combustible materials, such as wood, cloth, paper, rubber, and many plastics.

**Class B fires** Fires involving flammable and combustible liquids, oils, greases, tars, oil-based paints, lacquers, and flammable gases.

**Class C fires** Fires that involve energized electrical equipment where the electrical conductivity of the extinguishing media is of importance.

**Class D fires** Fires involving combustible metals such as magnesium, titanium, zirconium, sodium, and potassium.

**Class K fires** Fires involving combustible cooking media such as vegetable oils, animal oils, and fats.

**Combustion** A chemical process of oxidation that occurs at a rate fast enough to produce heat and usually light in the form of either a glow or flames.

**Conduction** Heat transfer to another body or within a body by direct contact.

**Convection** Heat transfer by circulation within a medium such as a gas or a liquid.

**Decay phase** The phase of fire development where the fire has consumed either the available fuel or oxygen and is starting to die down.

**Electrical energy** Heat produced by electricity.

**Endothermic reactions** Reactions that absorb heat or require heat to be added.

**Exothermic reactions** Reactions that result in the release of energy in the form of heat.

**Fire** A rapid, persistent chemical reaction that releases both heat and light.

**Fire point (flame point)** The lowest temperature at which a liquid releases enough vapors to ignite and sustain combustion.

**Fire tetrahedron** A geometric shape used to depict the four components required for a fire to occur: fuel, oxygen, heat, and chemical chain reactions.

**Flammability limits (explosive limits)** The upper and lower concentration limits of a flammable gas or vapor in air that can be ignited, expressed as a percentage of fuel by volume.

**Flashover** The condition where all combustibles in a room or confined space have been heated to the point at which they release vapors that will support combustion, causing all combustibles to ignite simultaneously.

**Flash point** The minimum temperature at which a liquid releases sufficient vapor to form an ignitable mixture with the air.

**Fuel** All combustible materials. The actual material that is being consumed by a fire, allowing the fire to take place.

**Fully developed phase** The phase of fire development where the fire is free-burning and consuming much of the fuel.

**Gas** One of the three phases of matter. A substance that will expand indefinitely and assume the shape of the container that holds it.

**Growth phase** The phase of fire development where the fire is spreading beyond the point of origin and beginning to involve other fuels in the immediate area.

**Hydrocarbons** Organic chemical compounds composed of carbon and hydrogen atoms.

**Hypoxia** A state of inadequate oxygenation of the blood and tissue.

**Ignition temperature** The minimum temperature at which a fuel, when heated, will spontaneously ignite and continue to burn.

**Incipient phase** The phase of fire development where the fire is limited to the immediate point of origin.

**Liquid** One of the three phases of matter. A nongaseous substance that is composed of molecules that move and flow freely and assumes the shape of the container that holds it.

**Lower explosive limit (LEL)** The minimum amount of fuel vapor mixed in air that will ignite or explode.

**Mechanical energy** Heat energy created by friction.

**Oxidation** A chemical reaction initiated by combining an element with oxygen, resulting in the form of the element or one of its compounds.

**Oxidizing agent** A substance that will release oxygen or act in the same manner as oxygen in a chemical reaction.

**Plume** The column of hot gases, flames, and smoke that rises above a fire, also called a convection column, thermal updraft, or thermal column.

**Pyrolysis** The chemical decomposition of a compound into one or more other substances by heat alone; pyrolysis often precedes combustion.

**Radiation** The combined process of emission, transmission, and absorption of energy traveling by electromagnetic wave propagation between a region of higher temperature and a region of lower temperature.

**Rollover (flameover)** The condition where unburned products of combustion from a fire have accumulated in the ceiling layer of gas to a sufficient concentration (i.e., at or above the lower flammable limit) that they ignite momentarily.

**Smoke** An airborne particulate product of incomplete combustion suspended in gases, vapors, or solid and liquid aerosols.

**Solid** One of the three phases of matter. A substance that has three dimensions and is firm in substance.

**Surface-to-mass ratio (STMR)** The surface area of a material in proportion to the mass of that material.

**Surface-to-volume ratio** The surface area of a liquid in proportion to the mass.

**Thermal column** A cylindrical area above a fire in which heated air and gases rise and travel upward.

**Thermal layering** The stratification, or heat layers, that occur in a room as a result of a fire.

**Upper explosive limit (UEL)** The maximum amount of fuel vapor mixed in air that will ignite or explode.

**Vapor density** The weight of an airborne concentration (vapor or gas) as compared to an equal volume of dry air.

**Volatility** The ready ability of a substance to produce combustible vapors.

# Brigade Member in Action

**During your regular brigade training class, the instructor arranges for a tour of one of the large warehouse facilities located within your plant. During the tour, the instructor points out that the warehouse currently contains potential class A, B, C, and D fire hazards. He also points out a small room within the warehouse in which a fire occurred several years ago. This room, which was used to store documentation and files at the time of the fire, is basically isolated from the rest of the facility by four walls and separate roof. The fire started when a faulty light fixture caused the boxes of documentation to catch fire. Upon the fire brigade's arrival, the door to the room was found to be closed and the room was charged with heavy smoke, but very little, if any, fire could be seen through the window in the door. A large amount of class A materials were still present within the room at that time.**

1. Which of the following statements is incorrect regarding the different types of fire class hazards present within the warehouse?
   - A. Class C fires involve energized electrical equipment.
   - B. Class A fires involve ordinary solid combustible materials.
   - C. Class D fires involve combustible cooking media such as oil and grease.
   - D. Class B fires involve flammable or combustible liquids.

2. Based on the description of the fire provided by the instructor, what phase of fire did the responding brigade members find when they arrived at the warehouse?
   - A. Decay
   - B. Fully developed
   - C. Growth
   - D. Ignition

3. Based on the description of the incident provided by the instructor, what was the biggest potential problem in dealing with the fire situation?

   A. The fire was basically out of fuel and there were no hazards to worry about.

   B. The room was ready to reach flashover or rollover condition at any point in time.

   C. Due to the confined area and presence of combustible vapors and smoke, there was a real possibility of a backdraft if the room was opened up incorrectly.

   D. The fire had just moved from the incipient stage of the growth phase and was ready to rapidly grow in size.

4. Which of the following signs might indicate potential backdraft conditions?

   A. Little or no flame is visible from the exterior.

   B. Smoke is emanating under pressure from cracks, around doors, and around windows.

   C. Glass is smoke stained and blackened.

   D. Any or all of the above conditions.

# Building Construction

# Chapter 6

## NFPA 1081 Standard

### Incipient Facility Fire Brigade Member

NFPA 1081 contains no Incipient Facility job performance requirements for this chapter.

### Advanced Exterior Facility Fire Brigade Member

NFPA 1081 contains no Advanced Exterior Facility job performance requirements for this chapter.

### Interior Structural Facility Fire Brigade Member

**6.2.2 (A)** *Requisite Knowledge.* Basic construction of typical doors, windows, and walls within the facility; operation of doors, windows, and their associated locking mechanisms; and the dangers associated with forcing entry through doors, windows, and walls.

## Knowledge Objectives

After completing this chapter, you will be able to:

- Describe the characteristics of the following building materials: gypsum board, wood, and plastics.
- List the characteristics of each of the following types of building construction: fire-resistive construction, noncombustible construction, ordinary construction, heavy timber construction, and wood-frame construction.
- Describe how each of the five types of building construction react to fire.
- Describe the function of each of the following building components: foundations, floors, ceilings, roofs, trusses, walls, doors, windows, interior finishes, and floor coverings.

## Skills Objectives

There are no skills objectives for this chapter.

# You Are the Brigade Member

**Your fire brigade is dispatched for a building fire in the plant. The building is a single-story warehouse with smoke coming out of the loading dock area. As the engine pulls to a stop, you quickly identify the building construction and materials. As you dismount, you mentally run through the hazards presented by this construction.**

1. ***Why must brigade members understand building construction?***
2. ***What is the best way for brigade members to learn about the types of construction in their response area?***
3. ***What additional factors should be considered?***

## Introduction

Knowing the basic types of building construction is vital for brigade members because building construction affects how fires grow and spread. Brigade members must be able to identify different types of building construction quickly so that they can anticipate the fire's behavior and respond accordingly. An understanding of building construction will help determine when it is safe to enter a burning building and when it is necessary to evacuate a building. Your safety, the safety of your team members, and the safety of the building's occupants depend on a knowledge of building construction.

But construction is only one component of a complex relationship. Brigade members also must consider the occupancy of the building and the building contents. Fire risk factors vary depending on how a building is used. To establish a course of action at a fire, it is necessary to consider the interactions between three factors: the building construction, the occupancy of the building, and the contents of the building. The following sections relate to occupancy and contents, while the rest of this chapter focuses on those aspects of building construction that are important to a brigade member.

### Occupancy

The term occupancy refers to how a building is used. Based on a building's occupancy classification, a brigade member can predict who is likely to be inside the building, how many people, and what they are likely to be doing. Occupancy also suggests the types of hazards and situations that may be encountered in the building.

Occupancy classifications are used with building and safety codes to establish regulatory requirements, including the types of construction that can be used to construct buildings of a particular size, use, or location. Regulations divide building occupancies into major categories—such as residential, health care, business, and industrial—based on common characteristics.

Occupancy classifications can be used to predict the number of occupants that are likely to be at risk in a fire. An office building probably has a large population during the day, but most of the workers should be able to evacuate without assistance.

Fire hazards in different types of occupancies also vary greatly. A computer repair area has different characteristics than a warehouse. A factory could be used to produce cast-iron pipe fittings or airplanes. The materials used could be flammable, toxic, or fire resistant. Brigade members must consider each of these factors when responding to a particular building.

Building codes require that certain types of building construction be used for specific types of occupancies. In most communities, however, buildings that were built for one reason are now being used for some other purpose. Buildings constructed before modern codes were adopted were often grandfathered in during inspections. This presents hazards for fire brigade members.

### Contents

Contents also must be considered when responding to a building. Building contents vary widely but are usually closely related to the occupancy of a building. Some buildings have noncombustible contents that would not feed a fire, while others could be so dangerous that brigade members could not safely attack any fire. For example, warehouses could contain many toxic and flammable substances. Whenever possible, fire brigade members should prepare an advance plan for buildings that present special hazards.

Similar occupancies can pose different levels of risk. For example, three warehouses, identical from the exterior, could have very different contents that increase or reduce the risks to fire brigade members. One might contain ceramic floor tiles, the second might be filled with wooden furniture, and the third might be a storehouse for swimming pool chemicals. The tiles will not burn, the furniture will, and the chemicals could create toxic products of combustion and contaminated runoff.

## Types of Construction Materials

An understanding of building construction begins with the materials used. Building components are usually made of different materials. The properties of these materials and the details of their construction determine the basic fire characteristics of the building itself.

Function, appearance, cost, and compliance with building and fire codes are all considerations when selecting building materials and construction methods. Architects often place a priority on functionality and aesthetics when selecting materials; builders are often more concerned about cost and ease of construction, and building owners are usually interested in durability and maintenance expenses, as well as initial cost and aesthetics.

Fire brigade members, however, use a different set of factors to evaluate construction methods and materials. Their chief concern is the behavior of the building under fire conditions. A building material or construction method that is attractive to architects, builders, and building owners can create serious problems or deadly hazards for fire brigade members.

The most common building materials are wood, masonry, concrete, steel, aluminum, glass, gypsum board, and plastics. Within these basic categories are hundreds of variations. The key factors that affect the behavior of each of these materials under fire conditions include:

- Combustibility: Whether or not a material will burn determines its combustibility. Materials such as wood will burn when they are ignited, releasing heat, light, and smoke, until they are completely consumed by the fire. Concrete, brick, and steel are noncombustible materials that cannot be ignited and are not consumed by a fire.
- Thermal conductivity: This describes how quickly a material will conduct heat. Heat flows very quickly through metals such as steel and aluminum. Brick, concrete, and gypsum board are poor conductors of heat.
- Decrease in strength at elevated temperatures: Many materials lose strength at elevated temperatures. Steel loses strength and will bend or buckle when exposed to fire temperatures. Aluminum melts in a fire. Bricks and concrete can generally withstand high temperatures for extended periods of time.
- Thermal expansion when heated: Some materials, steel in particular, expand significantly when they are heated. A steel beam exposed to a fire will stretch (elongate); if it is restrained so that it cannot elongate, it will sag, warp, or twist.

### Brigade Member Tips

The four characteristics of building materials under fire are:

- Combustibility
- Thermal conductivity
- Decrease in strength with increased temperature
- Rate of thermal expansion

### Masonry

Masonry includes stone, concrete blocks, and brick. The individual components are usually bonded together into a solid mass with mortar, which is produced by mixing sand, lime, water, and Portland cement. A concrete-masonry unit, or CMU, is a preassembled wall section of brick or concrete blocks that contains steel reinforcing rods. A complete CMU can be delivered to a construction side, ready to be erected.

Masonry materials are inherently fire resistive ▼ Figure 6-1. They do not burn or deteriorate at the temperatures normally encountered in building fires. Masonry is also a poor conductor of heat, so it will limit the passage of heat from one side through to the other side. For these reasons, masonry is often used to construct fire walls to protect vulnerable materials. A fire wall helps prevent the spread of a fire from one side to the other side of the wall.

All masonry walls are not necessarily fire walls. A single layer of masonry may be used over a wood-framed building

Figure 6-1 Masonry materials are inherently fire resistive.

to make it appear more substantial. If there are unprotected openings in a masonry wall, a fire can often spread through them. If the mortar has deteriorated or the wall has been exposed to fire for a prolonged time, a masonry wall can collapse during a fire.

A masonry structure also can collapse under fire conditions if the roof or floor assembly collapses. The masonry falls because of the mechanical action of the collapse. Aged, weakened mortar can also contribute to a collapse. Regardless of the cause, a collapsing masonry wall is a deadly hazard to fire brigade members.

## Concrete

Concrete is also a naturally fire-resistive material. It does not burn or conduct heat well, so it is often used to insulate other building materials from fire. Concrete does not have a high degree of thermal expansion—that is, it does not expand greatly when exposed to heat and fire. Nor will it lose strength when exposed to heat and fire.

Concrete is made from a mixture of Portland cement and aggregates such as sand and gravel. Different formulations of concrete can be produced for specific building purposes. Concrete is inexpensive and easy to shape and form. It can be used for foundations, columns, floors, walls, roofs, and exterior pavement.

Under compression, concrete is strong and can support a great deal of weight, but under tension, it is weak. When used in building construction, concrete usually has embedded steel reinforcing rods to strengthen it under tension. In turn, the concrete insulates the steel reinforcing rods from heat.

Although it is inherently fire resistive, concrete can be damaged by exposure to a fire. A fire can convert trapped moisture in the concrete to steam. As the steam expands, it creates internal pressure that can cause sections of the concrete surface to break off, a process called spalling. Severe spalling in reinforced concrete can expose the steel reinforcing rods to the heat of the fire. If the fire is hot enough, the steel might weaken, resulting in a structural collapse. However, this rarely occurs.

## Steel

Steel is the strongest building material in common use. It is very strong in both tension and compression and can be produced in a wide range of shapes and sizes, from heavy beams and columns to thin sheets. Steel is often used in the structural framework of a building to support floor and roof assemblies. It is resistant to aging and does not rot; however, most types of steel will rust unless they are protected from exposure to air and moisture.

Steel is an alloy of iron and carbon; additional metals may be added to produce steel with special properties, such as stainless steel or galvanized steel.

Steel by itself is not fire resistive. Steel will melt at extremely high temperatures, but these temperatures are not normally encountered at structure fires (▲ Figure 6-2). Steel can deform and fail at temperatures common in structure fires. Other materials, such as masonry, concrete, or layers of gypsum board, are often used to protect steel from the heat of a fire. Sprayed-on coatings of mineral or cement-like materials are also used to insulate steel members. The amount of heat absorbed by steel depends on the mass of the object and the amount of protection surrounding it. Smaller, lighter pieces of steel heat more easily than larger and heavier pieces.

**Figure 6-2** Steel can deform and fail at temperatures common in structure fires.

Steel will expand as well as lose strength as it is heated. An unprotected steel roof beam directly exposed to a fire can elongate sufficiently to cause a supporting wall to collapse. Heated steel beams will often sag and twist, while columns tend to buckle as they lose strength. The bending and distortion are caused by the uneven heating that occurs in actual fire situations. Steel trusses are often used in business and industrial occupancies, particularly to support roofs. The structure of a truss enables a limited amount of material to support a heavy load. However, failure of a truss member could result in failure of the other truss.

Failure of a steel structure is dependent on three factors: the mass of the steel components, the loads placed upon them, and the methods used to connect the steel pieces. There are no accurate indicators that enable brigade members to predict when a steel beam will fail. Any sign of bending, sagging, or stretching of steel structural members is a warning of an immediate risk of failure.

## Other Metals

A variety of other metals including aluminum, copper, and zinc are used in building construction. Aluminum is often used for siding, window frames, door frames, and roof panels. Copper is used primarily for electrical wiring and piping; it is sometimes used for decorative roofs, gutters, and down spouts. Zinc is used primarily as a coating to protect metal parts from rust and corrosion.

Aluminum is occasionally used as a structural material in building construction. It is more expensive and not as strong as steel, so its use is generally limited to light-duty applications such as awnings and sunshades. It is also being used as studs in nonload-bearing walls. Aluminum expands more than steel when heated and loses strength quickly when exposed to a fire. Aluminum has a lower melting point than steel, so it will often melt and drip in a fire.

## Glass

Almost all buildings contain glass—in windows, doors, skylights, and sometimes walls. Ordinary glass breaks very easily, but glass can be manufactured to resist breakage and to withstand impacts or high temperatures.

Glass is noncombustible, but it is not fire resistive except for specifically formulated glass. Ordinary glass will usually break when exposed to fire, but specially formulated glass can be used as a fire barrier in particular situations. Thermal conductivity of glass is usually not a significant factor in the spread of fire.

There are many different types of glass:

- Ordinary window glass will usually break with a loud pop when heat exposure to one side causes it to expand and creates internal stresses that fracture the glass. The broken glass forms large shards, which usually have sharp edges.
- Tempered glass is much stronger than ordinary glass and harder to break. Some tempered glass can be broken with a spring-loaded center punch and will shatter into small pieces that do not have the sharp edges of ordinary glass.
- Laminated glass is manufactured with a thin sheet of plastic between two sheets of glass. It is much stronger than ordinary glass, difficult to break with ordinary hand tools, and will usually deform instead of breaking. When exposed to a fire, laminated glass windows are likely to crack and remain in place. Laminated glass is sometimes used in buildings to help soundproof areas.
- Glass blocks are thick pieces of glass similar to bricks or tiles. They are designed to be built into a wall with mortar so that light can be transmitted through the wall. Glass blocks have limited strength and cannot be used as part of a load-bearing wall, but they can usually withstand a fire. Some glass blocks are approved for use with fire-rated masonry walls.
- Wired glass is made by molding tempered glass with a reinforcing wire mesh (▶ Figure 6-3). When wired glass is subjected to heat, the wire holds the glass together and prevents it from breaking. It is often used in fire doors and windows designed to prevent fire spread.

Figure 6-3 Wired glass.

- Fire-rated glass Special glass formulations are available that have achieved fire ratings of up to 3 hours. This glass has much better impact rating than wired glass (in normal use) and may be found in doors or windows in commercial and industrial facilities.

## Gypsum Board

Gypsum is a naturally occurring mineral composed of calcium sulfate and water molecules, used to make plaster of Paris. Gypsum is a good insulator and noncombustible; it will not burn even in atmospheres of pure oxygen.

Gypsum board, which is also called drywall, sheetrock, or plasterboard, is commonly used to cover the interior walls and ceilings of residential living areas and commercial spaces. Gypsum board is manufactured in large sheets consisting of a layer of compacted gypsum sandwiched between two layers of specially produced paper. The sheets are nailed or screwed in place on a framework of wood or metal studs, and the edges are secured with a special tape. The nail or screw heads and tape are then covered with a thin layer of plaster.

Gypsum board has limited combustibility, because the paper covering will burn slowly when exposed to a fire. It does not conduct or release heat to an extent that would contribute to fire spread and is often used to create a firestop or to protect building components from fire. Gypsum blocks are sometimes used as protective insulation around steel members or to create fire-resistive enclosures or interior fire walls.

When gypsum board is heated, some of the water found in the calcium sulfate will evaporate, causing the board to deteriorate. If it is exposed to a fire for a long time, the gypsum board will fail. Water will also weaken and permanently damage gypsum board.

Even if the gypsum board retains its integrity after a fire, the sections that were directly exposed to the fire should be replaced. The sections that were heated will probably be unable to serve as effective fire barriers in the future.

Although gypsum board is a good finishing material and an effective fire barrier, it is not a strong structural material. It must be properly mounted on and supported by wood or steel studs. Gypsum board mounted on wood framing will protect the wood from fire for a limited time. Commercial and industrial facilities may use aluminum studs in lieu of wood framing. Also, various types of plastic and fiberglass may be used on gypsum walls to add strength or cleanability.

## Wood

Wood is probably the most commonly used building material in our environment, although less so in industrial facilities. It is inexpensive to produce, easy to use, and can be shaped into many different forms, from heavy structural supports to thin strips of exterior siding. Both soft woods such as pine and hard woods such as oak are used for building construction.

A wide variety of wood products are used in building construction:

- Solid lumber is squared and cut into uniform lengths. Examples of solid lumber include the heavy timbers used in mills and barns and the lightweight boards used for siding and decorative trim.
- Laminated wood consists of individual pieces of wood glued together. Lamination is used to produce beams that are longer and stronger than solid lumber and to manufacture curved beams.
- Wood panels are produced by gluing together thin sheets of wood. Plywood is the most common type of wood panel used in building construction. Small chips (chip board) or particles of wood (particle board) can also be used to make wood panels that are usually much weaker than plywood or solid lumber.
- Wood trusses are assemblies of pieces of wood or wood and metal, often used to support floors and roofs. The structure of a truss enables a limited amount of material to support a heavy load. However, failure of a truss member may result in failure of the entire truss.
- Wooden beams are efficient load-bearing members assembled from individual wood components. The shape of a wooden I-beam or box beam enables it to support the same load that a solid wood beam could support.

For fire brigade members, the most important characteristic of wood is combustibility. Wood adds fuel to a structural fire and can provide a path for the fire to spread. It ignites at fairly low temperatures and is gradually consumed by the fire, weakening and eventually collapsing the structure. Great quantities of heat and hot gases are created until all that remains is a small quantity of residual ash.

Wood components weaken as they are consumed by fire. A burning wood structure gets weaker every minute, making it imperative that brigade members always remember to operate on the side of safety.

How fast wood ignites, burns, and decomposes depends on several factors:

- **Ignition:** A small ignition source contains less energy and will take a longer time to ignite a fire. Using an accelerant will greatly speed up the process.
- **Moisture:** Damp or moist wood takes longer to ignite and burn. New lumber usually contains more moisture than lumber that has been in a building for many years. A higher relative humidity in the atmosphere also makes wood more difficult to ignite.
- **Density:** Heavy, dense wood is harder to ignite than lighter or less dense wood.
- **Preheating:** The more the wood is preheated, the faster it ignites.
- **Size and form:** The rate of combustion is directly related to the surface area of the wood. The combustion process occurs as high temperatures release flammable gases from the surface of the wood. A large solid beam is difficult to ignite and burns relatively slowly. A lightweight truss is more easily ignited and rapidly consumed.

Exposure to the high temperatures generated by a fire can also decrease the strength of wood through the process of pyrolysis. This chemical change occurs when wood or other materials are heated to a temperature high enough to release some of the volatile compounds in the wood, without igniting these gases. The wood begins to decompose without combustion.

### Fire-Retardant-Treated Wood

Wood cannot be treated to make it completely noncombustible. But impregnating wood with mineral salts makes it more difficult to ignite and slows the rate of burning. Fire-retardant treatment can significantly reduce the fire hazards of wood construction, although the treatment process can also reduce the strength of the wood.

In some cases, fire-retardant-treated wood can pose a danger for fire brigade members. Some fire-retardant chemicals

**Brigade Member Safety Tips**

Avoid standing on roofs that have been constructed with fire-retardant-treated wood. Atmospheric decomposition of the wood could cause these roofs to collapse, even without fire damage. Also, beware of skylights and other portions of roofs that may not be suited for walking on.

Figure 6-4 Plastic building materials come in many different forms.

used to treat plywood roofing panels will cause the wood to deteriorate and weaken. The plywood can fail, necessitating the replacement of these panels. Fire brigade members should know whether fire-retardant-treated plywood is used in the facility and avoid standing on roofs made with these panels during a fire. The extra weight could cause the plywood to collapse.

### Plastics

Plastics are synthetic materials used in many products today ▲ Figure 6-4. Plastics may be transparent or opaque, stiff or flexible, tough or brittle. They are often combined with other materials to produce building products.

Although plastics are rarely used for structural support, they can be found throughout a building. Building exteriors may include vinyl siding, plastic window frames, and plastic panel skylights. Foam plastic materials can be used as exterior or interior insulation. Plastic pipe and fittings, plastic tub and shower enclosures, and plastic lighting fixtures are commonly used today. Even carpeting and floor coverings often contain plastics.

The combustibility of plastics varies greatly. Some plastics ignite easily and burn quickly, while others will burn only while an external source of heat is present. Some plastics will withstand high temperatures and fire exposure without igniting.

Many plastics produce quantities of heavy, dense, dark smoke and release high concentrations of toxic gases as they burn. The smoke resembles smoke from a petroleum fire because most plastics are made from petroleum products.

Thermoplastic materials melt and drip when exposed to high temperatures, even those as low as 500°F. Heat is used to shape these materials into different forms, but the dripping, burning plastic can rapidly spread a fire. Thermoset materials, however, are fused by heat and will not melt. They will lose strength as the plastic burns, but will not melt.

## Types of Construction

Fire brigade members need to understand and recognize five different types of building construction. These classifications are directly related to fire protection and fire behavior in the building.

Buildings are classified based on the combustibility of the structure and the fire resistance of its components. Buildings using Type I or Type II construction are assembled with primarily noncombustible materials and limited amounts of wood and other materials that will burn. This does not mean that Type I and Type II buildings cannot be damaged by a fire. If the contents of the building burn, the fire could seriously damage or destroy the structure.

In buildings using Type III, Type IV, or Type V construction, both the structural components and the building contents will burn. Wood and wood products are used to varying degrees in these buildings. If the wood ignites, the fire will weaken and consume the structure as well as the contents. Structural elements of these building can be damaged or destroyed in a very short time.

Fire resistance refers to the length of time that a building or building component can withstand a fire before igniting. Fire resistance ratings are stated in hours, based on the time that a test assembly withstands or contains a standard test fire. For example, walls are rated based on whether they stop the progress of the test fire for periods from 20 minutes to 4 hours. A floor assembly could be rated based on whether it supports a load for 1 hour or 2 hours. Because ratings are based on a standard test fire, and an actual fire could be more or less severe than the test fire, ratings are only guidelines; a 2-hour rated assembly will not necessarily withstand a fire for 2 hours.

Building codes specify the type of construction to be used, based upon the height, area, occupancy classification, and location of the building. Additional factors, such as the presence of automatic sprinklers, will also affect the required construction classification. NFPA 220, *Standard on Types of Building Construction*, provides the detailed requirements for each type of building construction.

### Type I Construction: Fire Resistive

Type I construction is the most fire-resistive category of building construction. It is used for buildings designed for large numbers of people, buildings with a high life-safety hazard, tall buildings, large-area buildings, and buildings containing special hazards. Type I construction is commonly found in schools, hospitals, and high-rise buildings ► Figure 6-5.

Figure 6-5 Type I construction is commonly found in schools, hospitals, and high-rise buildings.

Buildings with Type I construction can withstand and contain a fire for a specified period of time. The fire resistance and combustibility of all construction materials are carefully evaluated. Each building component must be engineered to contribute to fire resistance of the entire building.

All of the structural members and components used in Type I construction must be made of noncombustible materials, such as steel, concrete, or gypsum board. In addition, the structure must be constructed or protected so that there is at least 2 hours of fire resistance. Building codes specify the fire resistance requirements for different components. For example, columns and load-bearing walls in multistory buildings could have a fire resistance requirement of 3 or 4 hours, while a floor might be required to have a fire resistance rating of only 2 hours. Some codes allow Type I buildings to use a limited amount of combustible material as interior finish.

If a Type I building exceeds specific height and area limitations, codes generally require the use of fire-resistive walls and/or floors to subdivide it into compartments. A compartment could be a single floor in a high-rise building or a part of a floor in a large-area building. A fire in one compartment should not spread to any other parts of the building. Stairways, elevators shafts, and utility shafts should be enclosed in construction that prevents fire from spreading from floor to floor or compartment to compartment.

Type I buildings usually use reinforced concrete and protected steel-frame construction. Concrete is noncombustible and provides thermal protection around the steel reinforcing rods. Reinforced concrete can fail, however, if it is subjected to a fire for a long period of time or if the building contents create an extreme fire load. Structural steel framing must be protected from the heat of a fire. In Type I construction, the structural steel members are generally encased in concrete, shielded by a fire-resistive ceiling, covered with multiple layers of gypsum board, or protected by a sprayed-on insulating material (▼ Figure 6-6). An unprotected steel beam exposed to a fire could fail and cause the entire building to collapse.

Type I building materials should not provide enough fuel by themselves to create a serious fire. It is the contents of the building that determine the severity of a Type I building fire. In theory, a fire could consume all of the combustible contents of a Type I building and leave the structure basically intact.

Although Type I construction may provide the highest degree of safety, it does not eliminate all of the risks to occupants and brigade members. Serious fires can occur in fire-resistive buildings, because the burning contents can produce copious quantities of heat and smoke. Even though the structure is designed to give brigade members time to conduct an interior attack, it could be very difficult to extinguish the fire. A fire that is fully contained within fire-resistive construction can be very hot and difficult to ventilate. For this reason, automatic sprinklers, as well as fire-resistive construction, are used in modern high-rise buildings.

Sometimes, the burning contents provide sufficient fuel and generate enough heat to overwhelm the fire-resistant properties of the construction. In these cases, the fire can escape from its compartment and damage or destroy the structure. Similarly, inadequate or poorly maintained construction can also undermine the fire-resistive properties of the building materials. Under extreme fire conditions, a Type I building can collapse, but these buildings have a much lower potential for structural failure than other buildings.

Figure 6-6 Sprayed-on fireproofing materials are often used to protect structural steel.

## Type II Construction: Noncombustible

Type II construction is also referred to as noncombustible construction (▼ Figure 6-7). All of the structural components in a Type II building must be made of noncombustible materials. The fire-resistive requirements, however, are less stringent for Type II construction. In some cases, there are no fire resistance requirements. In other cases, known as protected noncombustible construction, a fire resistance rating of 1 or 2 hours may be required for certain elements.

Noncombustible construction is most common in single-story warehouse or factory buildings, where vertical fire spread is not an issue. Unprotected noncombustible construction is generally limited to a maximum of two stories. Some multistory buildings are constructed using protected noncombustible construction.

Steel is the most common structural material in Type II buildings. Insulating materials can be applied to the steel when fire resistance is required. A typical example of Type II construction is a large-area, single-story building with a steel frame, metal or concrete block walls, and a metal roof deck. Fire walls are sometimes used to subdivide these large-area buildings and prevent catastrophic losses. Undivided floor areas must be protected with automatic sprinklers to limit the fire risk.

Fire severity in a Type II building is determined by the contents because the structural components contribute little or no fuel and interior finish materials are limited. If the building contents provide a high fuel load, a fire could destroy the structure. Automatic sprinklers should be used to protect combustible and valuable contents.

### Brigade Member Tips

Fire-resistive structures with sprinklers are a great asset in assisting brigade members in controlling the burning contents. However, when the sprinklers activate, steam is produced when the water hits the fire. As the sprinklers continue to operate, the steam is cooled and can fill an enclosed area from floor to ceiling. Without proper ventilation brigade members may find themselves with virtually no visibility.

## Type III Construction: Ordinary

Type III construction is also referred to as ordinary construction because it is used in a wide variety of buildings, ranging from commercial strip malls to small apartment buildings (▼ Figure 6-8). Ordinary construction is usually limited to buildings of no more than four stories, but it can sometimes be found in buildings as tall as six or seven stories.

Type III construction buildings have masonry exterior walls, which support the floors and the roof structure. The interior structural and nonstructural members, including the walls, floor, and roof, are all constructed of wood. In most ordinary construction buildings, gypsum board or plaster is used as an interior finish material, covering the wood framework and providing minimal fire protection.

Fire resistance requirements for interior construction of Type III buildings are limited. The gypsum or plaster coverings over the interior wood components provide some fire resistance. Key interior structural components may be required to have fire resistance ratings of 1 or 2 hours, or

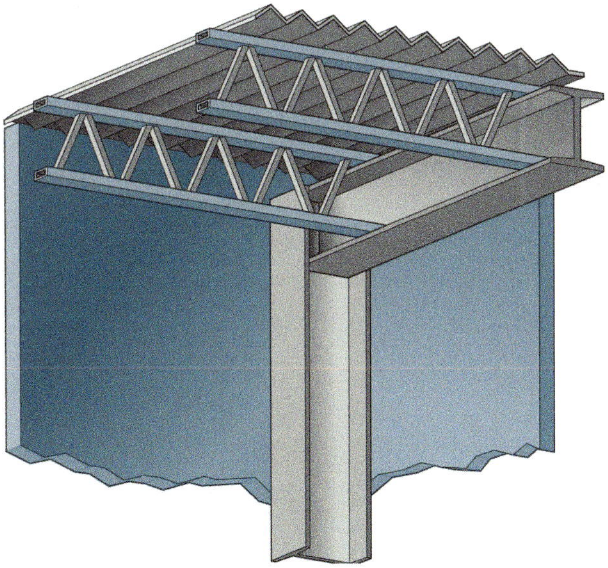

Figure 6-7 Type II construction is also referred to as noncombustible construction.

Figure 6-8 Type III construction is also referred to as ordinary construction because it is used in a wide variety of buildings.

# VOICES OF EXPERIENCE

**"The building's construction helped lead to the fire becoming a potential bomb for the unobservant."**

It was 4:30 A.M. on a Monday morning. I was the engineer for the B shift and was pulling the last of three 12-hour night shifts at the El Segundo refinery when we received a report of something burning around the facility. To investigate the report, we got into our fire patrol pickups and started checking buildings and substations. When I drove around the corner, I found the center building with smoke puffing out of every location from which it could possibly vent. It looked as if the building was being "steamed out." I reported the location to the other fire patrols, and then drove back on a Code 3 to the station to pick up the engine. Our probationary fire fighter had just walked into the station when I yelled at him to jump on the rig with his gear.

Upon our arrival back at the building, we laid a 4-inch supply line and engaged the pump. The battalion chief said that the utilities had just been controlled; the probationary fire fighter had to go with captain and the fire fighter to force the door and make entry. From the midship control deck, I saw the fire puff and then suck back in. I jumped down and told the battalion chief what I just saw. He ordered the crew to stop and break a window to ventilate the building.

When the crew used a pike pole to break the side window, it lit up and then blew all the glass out of the building like a bomb going off. This fire showed us exactly what backdraft was. It cooked everything in the main room, but the interior doors to the back rooms held back the fire. Although we managed to put the fire out with a single engine company, we had the city engine come to assist with salvage, overhaul, and investigation.

The building was an old concrete facility that had been poured in place in the 1920s; it featured a plank-and-beam roof. Luckily, our rapid and safe fire attack kept the damage confined to the main work area. During this incident, the fire-rated doors had kept the fire out of the back rooms, which allowed the fire to burn for a long period unobserved and bottled up. The building's construction helped lead to the fire becoming a potential bomb for the unobservant.

Marc J. Lawrence
Northrop Grumman Company, Former Chevron USA Fire Department
El Segundo, California

there may be no requirements at all. Some Type III buildings use interior masonry load-bearing walls to meet requirements for fire-resistant structural support.

A building of Type III construction has two separate fire loads: the contents and the combustible building materials used. Even a vacant building will contain a sufficient quantity of wood and other combustible components to produce a large fire. A fire involving both the contents and the structural components can quickly destroy the building.

Fortunately, most fires originate with the building contents and are extinguished before the flames ignite the structure itself. Once a fire extends into the structure itself and begins to consume the fuel within the walls and above the ceilings, it becomes much more difficult to control.

Type III construction presents several problems for fire brigade members. For example, an electrical fire can begin inside the void spaces within the walls, floors, and roof assemblies and extend to the contents. The void spaces also allow a fire to extend vertically and horizontally, spreading from room to room and from floor to floor. Fire brigade members will have to open the void spaces to fight the fire. An uncontrolled fire within the void spaces is likely to destroy the building.

The fire resistance of interior structural components often depends on the age of the building and on local building codes. Older Type III buildings were built from solid lumber, which can contain or withstand a fire for a limited time. Newer or renovated buildings may have lightweight assemblies that can be damaged much more quickly and are prone to early failure. Floors and roofs in these buildings can collapse suddenly and without warning.

Exterior walls also could collapse if a fire causes significant damage to the interior structure. Because the exterior walls, the floors, and the roof are all connected in a stable building, the collapse of the interior structure could make the freestanding masonry walls unstable and likely to collapse.

## Type IV Construction: Heavy Timber

Type IV construction is also known as heavy timber construction (▶ Figure 6-9). A heavy timber building has exterior walls of masonry construction, and interior walls, columns, beams, floor assemblies, and roof structure of wood. The exterior walls are usually brick and extra thick to support the weight of the building and its contents.

The wood used in Type IV construction is much heavier than that used in Type III construction. It is more difficult to ignite and will withstand a fire for a much longer time before it collapses. A typical heavy timber building could have eight-inch-square wood posts supporting 14-inch-deep wood beams and 8-inch floor joists. The floors could be constructed of solid wood planks, 2- or 3-inch thick, with top layer of wood as the finished floor surface.

Heavy timber construction has no concealed spaces or voids. This helps reduce the horizontal and vertical fire spread that often occurs in ordinary construction buildings. Many heavy timber buildings do have vertical openings for elevators, stairs, or machinery, which can provide a path for a fire to travel from one floor to another.

Figure 6-9 Type IV construction is also known as heavy timber construction.

Heavy timber construction has been used to construct buildings as tall as six to eight stories with open spaces suitable for manufacturing and storage occupancies. Similar modern buildings are usually built with either fire resistive (Type I) or noncombustible (Type II) construction. New buildings of Type IV construction are rare, except for special structures that feature the construction components as architectural elements or renovations to existing type IV structures.

The heavy, solid wood columns, support beams, floor assemblies, and roof assemblies used in heavy timber construction will withstand a fire much longer than the smaller wood members used in ordinary and lightweight combustible construction. But once involved in a fire, the structure of a heavy timber building can burn for many hours. A fire that ignites the combustible portions of heavy timber construction is likely to burn until it runs out of fuel and the building is reduced to a pile of rubble. As the fire consumes the heavy timber support members, the masonry walls will become unstable and collapse.

Mill construction was common during the 1800s, especially in northeastern states. Large mill buildings were built as factories and warehouses. In many communities, dozens of these large buildings were clustered together in industrial areas, creating the potential for huge fires involving multiple multistory buildings. The radiant heat from a major fire in one of these buildings could spread the fire to nearby buildings, resulting in the loss of several surrounding structures.

Mill construction was state of the art for its time. Automatic sprinkler systems were developed to protect buildings of mill construction; as long as the sprinklers are properly maintained, these buildings have a good safety record. Without sprinklers, a heavy timber mill building is a major fire hazard.

Only a few of the original mill buildings are still used for manufacturing. Many have been converted to small shops, galleries, office buildings, and residential occupancies. The conversions divide the open spaces into smaller compartments and create void spaces within the structure. Appropriate fire protection and life safety features, such as modern sprinkler systems, must be built into the conversions to ensure the safety of occupants.

## Type V Construction: Wood Frame

In Type V construction, all of the major components are constructed of wood or other combustible materials (▶ Figure 6-10). Type V construction is often called wood-frame construction and is the most common type of construction used today. Wood-frame construction is used not only in one- and two-family dwellings and small commercial buildings, but in larger structures such as apartment and condominium complexes and office buildings up to four stories in height. It is also found in temporary structures and outbuildings.

Many wood-frame buildings do not have any fire-resistive components. Some codes require a 1-hour fire resistance rating for limited parts of wood-frame construction buildings, particularly those of more than two stories. Plaster or gypsum board barriers are often used to achieve this rating, but brigade members should not assume that these barriers will be present or effective in all Type V construction.

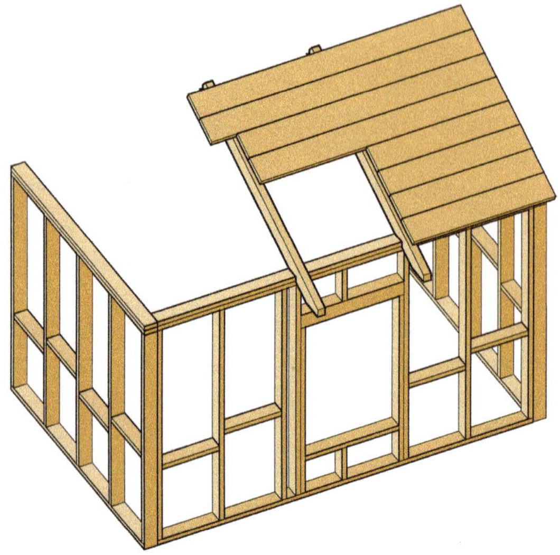

Figure 6-10 Type V, or wood-frame, construction is the most common type of construction used today.

Because all of the structural components are combustible, wood-frame buildings can rapidly become totally involved in a fire. In addition, Type V building construction usually contains voids and channels that allow a fire within the structure to spread quickly. A fire that originates in a Type V building can easily extend to other nearby buildings.

Wood-frame buildings often collapse and suffer major destruction from fires. Smoke detectors are essential to warn building occupants early if a fire occurs. Although compartmentalization can help limit the spread of a fire, automatic sprinklers are the most effective way of protecting lives and property in Type V buildings.

Wood-frame buildings are constructed in various ways. Older wood-frame construction was assembled from solid lumber, which relied on its mass to provide strength. To reduce costs and create the largest building with the least material, lightweight construction makes extensive use of wooden I-beams and wooden trusses. Structural assemblies are engineered to be just strong enough to carry the required load. As a result, there is no built-in safety margin, and these buildings can collapse early, suddenly, and completely during a fire.

Two systems are used to assemble wood-frame buildings: balloon-frame construction and platform construction. Because these systems developed in different eras, fire brigade members can anticipate the type of construction based on the age of a building.

### Brigade Member Safety Tips

Many buildings may have a brick veneer on the exterior walls of wood-frame construction. A thin layer of brick or stone might be applied to enhance the appearance of the building or to reduce the risk of fire spread. But this practice can give the impression that a building is Type III (ordinary) construction when it is really Type V (wood-frame) construction.

During a structural fire, the brick veneer is likely to collapse or peel away as the wall behind it burns or collapses. A single thickness of bricks is not usually strong enough to stand independently. Brigade members should be aware of buildings constructed in this manner and maintain a safe distance from potential falling bricks.

### Balloon-Frame Construction

Balloon-frame construction was popular between the late 1800s and the mid-1900s. A balloon-frame building has the

Figure 6-11 In a balloon-frame building, the exterior walls create channels that enable a fire to spread from the basement to the attic.

exterior walls assembled with wood studs that are continuous from the basement to the roof (▲ Figure 6-11). In a two-story building, the floor joists that support the first and second floors are nailed to these continuous studs. As a result, there is an open channel between each pair of studs that extends from the foundation to the attic. Each of these channels provides a path that enables a fire to spread from the basement to the attic without being visible on the first- or second-floor levels. Brigade members must anticipate that a fire originating on a lower level will quickly extend through these voids and should open the void spaces to check for hidden fires and to prevent rapid vertical extension.

### Platform-Frame Construction

Platform-frame construction is used for almost all modern wood-frame construction. In a building with platform construction, the exterior wall studs are not continuous. The first floor is constructed as a platform, and the studs for the exterior walls are erected on top of it. The first set of studs extends only to the underside of the second-floor platform, which blocks any vertical void spaces. The studs for the second-floor exterior walls are erected on top of the second-floor platform. At each level, the floor platform blocks the path of any fire rising within the void spaces. Platform framing prevents fire spread from one floor to another through continuous stud spaces. A fire can eventually burn through the wood platform, but the platform will slow the fire spread.

## Building Components

The construction classification system gives fire brigade members an idea of the materials used in building and an indication of how the materials will react to fire. But each building also has several different components. A brigade member who understands how these components function will have a better understanding of the risks involved with a burning building. Some building construction features are safer for brigade members, as this section shows.

The major components of a building discussed in this section are:

- Foundations
- Floors and ceilings
- Roofs
- Trusses
- Walls
- Doors and windows
- Interior finishes and floor coverings

### Foundations

The primary purpose of a building foundation is to transfer the weight of the building and its contents to the ground (▼ Figure 6-12). The weight of the building itself is called the dead load. The weight of the building's contents is called the live load. The foundation ensures that the base of the building is planted firmly in a fixed location, which helps keep all other components connected.

Modern foundations are usually constructed of concrete or masonry, although wood may be used in some areas. Foundations can be shallow or deep, depending on the type of building and soil composition. Some buildings are built on

Figure 6-12 The foundation supports the entire weight of the building.

concrete footings or piers; others are supported by steel piles or wooden posts driven into the ground.

As long as the foundation is stable and intact, it is usually not a major concern for brigade members. If the foundation shifts or is in poor repair, however, it can become a very critical concern. A burning building with a weak foundation or inadequate lateral bracing could be in imminent danger of collapse.

Most foundation problems are caused by circumstances other than a fire, such as improper construction, shifting soil conditions, or earthquakes. Fires can damage timber post and other wood foundations, but most foundations remain intact even after a severe fire has damaged the rest of the building. Fire following earthquakes can be particularly dangerous.

When examining a building, take a close look at the foundation. Look for cracks in the foundation that indicate movement. If the building has been modified or remodeled, look for areas where the support could be compromised.

## Floors and Ceilings

Floor construction is very important to brigade members for three major reasons. First, brigade members who are working inside a building must rely on the integrity of the floor to support their weight. A floor that fails could drop the brigade members into a fire on a lower level. Second, in a multistory structure, brigade members may be working below a floor (or roof), which would fall on them if it collapsed. Finally, the floor system influences whether a fire spreads vertically from floor to floor within a building or is contained on a single level. Some floor systems are designed to resist fires, while others have no fire-resistant capabilities.

### Fire-Resistive Floors

In multistory buildings, floors and ceilings are generally considered a combined structural system. The structure that supports a floor also supports the ceiling of the story below. In a fire-resistive building, this system is designed to prevent a fire from spreading vertically and to prevent a collapse when a fire occurs in the space below the floor-ceiling assembly. Fire-resistive floor-ceiling systems are rated in hours of fire resistance based on a standard test fire.

Concrete floors are common in fire-resistive construction. The concrete can be cast in place or assembled from panels or beams of precast concrete, which are made at a factory and transported to the construction site. The thickness of the concrete floor depends on the load that the floor needs to support.

Concrete floors can be self-supporting or supported by a system of steel beams or trusses. The steel can be protected by sprayed-on insulating materials or covered with concrete or gypsum board. If the ceiling is part of the fire-resistive rating, it provides a thermal barrier to protect the steel members from a fire in the area below the ceiling.

The ceiling below a fire-resistant floor can be constructed with plaster or gypsum board, or it can be a system of tiles suspended from the floor structure. In many cases, a void space between the ceiling and the floor above contains building systems and equipment such as electrical or telephone wiring, heating and air conditioning ducts, and plumbing and fire sprinkler system pipes (▲ Figure 6-13). If the space above the ceiling is not subdivided by fire-resistant partitions or protected by automatic sprinklers, a fire can quickly extend horizontally across a large area.

Figure 6-13 The space between the ceiling and the underside of the floor above is often used for electrical and communications wiring and other building systems.

### Wood-Supported Floors

Wood floor structures are common in nonfire-resistive construction. Wooden floor systems range from heavy timber construction, found in old mill buildings, to modern, lightweight construction.

Heavy timber construction can provide a huge fuel load for a fire, but it can also contain and withstand a fire for a considerable length of time without collapsing. Heavy timber construction uses posts and beams that are at least 8 inches on the smallest side and often as large as 14 inches in depth. The floor decking is often assembled from solid wood boards, 2- or 3-inch thick, which are covered by an additional 1 inch of finished wood flooring. The depth of the wood in this system will often contain a fire for an hour or more before the floor fails or burns through.

Conventional wood flooring, which was widely used for many decades, is much lighter than heavy timber, but used solid lumber as beams, floor joists, decking, and finished flooring. It burned readily when exposed to a fire, but generally took about 20 minutes to burn through or reach structural failure. This time factor is only a general, unscientific guideline that depends on many circumstances.

Modern construction uses structural elements that are much less substantial than conventional lumber. Lightweight wooden trusses or wooden I-beams are often used as supporting structures. Thin sheets of plywood are used as decking, and the top layer is often a thin layer of concrete or wood covered by carpet. This floor construction provides little resistance to fire. The lightweight structural elements can fail or the fire can burn through the decking quickly.

Unfortunately, it is impossible to tell how a floor is constructed by looking at it from above. The important information about a floor can only be observed from below, if it is visible at all. The building's age and local construction methods can provide significant clues to a floor's stability, but many older buildings have been renovated using modern lightweight systems and materials. Fire brigade members should use preincident surveys and other planning activities to gather essential structural information about buildings.

## Roofs

Safe interior firefighting operations depend on a stable roof. If the roof collapses, the brigade members inside the building may be injured or killed. Interior fires generate hot gases that accumulate under the roof. Fire brigade members performing ventilation activities on the roof to release the heated gases also must depend on the structural integrity of the roof.

The primary purpose of a roof is to protect the inside of a building from the weather. Often, roofs are not designed to be as strong as floors, especially in warmer climates. If the space under the roof is used for storage, or if extra heating, ventilating, or air conditioning equipment is mounted on the roof, the load could exceed the designer's expectations. Adding the weight of several fire brigade members to an overloaded roof could be disastrous.

Several methods and materials are used for roof construction. Roofs are constructed in three primary designs: pitched, curved, and flat. The major components of a roof assembly are the supporting structure, the roof deck, and the roof covering.

### Pitched Roofs

A pitched roof has sloping or inclined surfaces. Pitched roofs are used on many houses and some commercial buildings. The pitch or angle of the roof can vary depending on local architectural styles and climate. Variations of pitched roofs include gable, hip, mansard, gambrel, and lean-to roofs ▼ Figure 6-14.

Figure 6-14 Examples of pitched roofs include: A. A gable roof. B. A hip roof. C. A mansard roof. D. A gambrel roof.

The slope of a pitched roof can present either a minor inconvenience or a complete lack of secure footing. Fire brigade members working on a pitched roof are always in danger of falling, particularly when the roof is wet or icy, when the roof covering is not secure, or when smoke or darkness obscures vision. Roof ladders are used to provide a secure working platform for fire brigade members working on a pitched roof.

Pitched roofs are usually supported by either rafters or trusses. Rafters are solid wood joists mounted in an inclined position. Pitched roofs supported by rafters usually have solid wood boards as the roof decking. Fire brigade members can detect the impending failure of a roof with solid decking supported by rafters by a spongy feeling underfoot.

Modern lightweight construction uses manufactured wood trusses to construct most pitched roofs. The decking is usually thin plywood or a sheeting material such as wood particleboard. The trusses and decking material can burn through quickly. Fire brigade members cannot work safely on this roof if the supporting structures become involved in a fire.

Steel trusses also are used to support pitched roofs. The fire resistance of steel trusses is directly related to how well the steel is protected from the heat of the fire.

Several roof-covering materials are used on pitched roofs, usually in the form of shingles or tiles (▶ Figure 6-15). Shingles are usually made from felt or mineral fibers impregnated with asphalt, although metal and fiberglass shingles are also used. Wood shingles, often made from cedar, are popular in some areas. In older construction, wood shingles were often mounted on individual wood slats instead of continuous decking and would burn rapidly in dry weather conditions.

Shingles are generally durable, economical, and easily repaired. A shingle roof that has aged and deteriorated should be removed completely and replaced. Some older buildings may have newer layers of shingles on top of older layers, which can make it difficult to cut an opening for ventilation.

Roofing tiles are usually made from clay or concrete products. Clay tiles, which have been used for roofing since ancient times, can be flat or rounded. Rounded clay tiles are sometimes called mission tiles. Clay tiles are both durable and fire resistant.

Slate tiles are produced from thin sheets of rock. Slate is an expensive, long-lasting roofing material, but very brittle and slippery when wet.

Metal panels of galvanized steel, copper, and aluminum also are used on pitched roofs. Expensive metals such as copper are often used for their decorative appearance, while lower-cost galvanized steel is used on barns and industrial buildings. Corrugated, galvanized metal panels are strong enough to be mounted on a roof without roof decking. Metal roof coverings will not burn, but they can conduct heat to the roof decking.

Figure 6-15 Several roof-covering materials are used on pitched roofs.

## Curved Roofs

Curved roofs are often used for supermarkets, warehouses, industrial buildings, arenas, auditoriums, bowling alleys, churches, airplane hangars, and other large buildings that require large, open interiors. Curved roofs are usually supported by steel or wood bowstring trusses or arches.

The decking on curved roof buildings can range from solid wooden boards or plywood to corrugated steel sheets. The decking material must be identified before ventilation openings can be made. Often, the roof covering consists of layers that include felt, mineral fibers, and asphalt, although some curved roofs are covered with foam plastics or plastic panels. Curved roofs are extremely dangerous to operate on due to footing and collapse problems.

## Flat Roofs

Flat roofs are found on houses, apartment buildings, shopping centers, warehouses, factories, schools, and hospitals (▶ Figure 6-16). Most flat roofs have a slight slope so water can drain off. If the roof does not have the proper slope or the drains are not maintained, water can pool on the roof, overloading the structure and causing a collapse.

Figure 6-16 Most flat roofs have a slight slope so water can drain off.

The support systems for most flat roofs are constructed of either wood or steel. A wood support structure uses solid wood beams and joists; laminated wood beams may be used for extra strength or to span long distances. Lightweight construction uses wood trusses or wooden I-beams as the supporting members. Lightweight assemblies are much less fire resistant and collapse more quickly than solid wood systems.

Steel also is used to support flat roofs. Open-web steel trusses, sometimes called bar joists, can span long distances and remain stable during a fire if they are not subjected to excessive heat. Automatic sprinklers can protect the steel from the heat, but without such protection, the steel over a fire will soon weaken and sag. Heated steel support members can even elongate enough to push out the exterior walls of the building. Cracks in the outside walls at the roof level are a warning indicator.

Flat roof decks can be constructed of wood planking, plywood, corrugated steel, gypsum, or concrete. The material used depends on the building's age and size, the climate, the cost of materials, and the type of roof covering.

The roof covering is applied on top of the deck. Most flat roofs are covered with multiple-layer, built-up roofing systems that help prevent leakage. A typical built-up roof covering will have five layers: a vapor barrier, thermal insulation, a waterproof membrane, a drainage layer, and a wear course. The outside of the flat roof reveals only the top layer, which is often gravel, that serves as the wear course and protects the underlying layers from wear and tear. Cutting through all of the layers to open a ventilation hole can be a challenge.

Most flat roof coverings contain highly combustible materials, including asphalt, roofing felt, tarpaper, rubber or plastic membranes, and plastic insulation layers. These materials can be difficult to ignite, but once lit, they burn readily, releasing great quantities of heat and black smoke.

Flat roofs can present unique problems during vertical ventilation operations. A roof with a history of leaks could have several patches where additional layers of roofing material make the covering extra-thick. Sometimes a whole new roof is constructed on top of the old one. After opening a hole in the new roof, fire brigade members might discover the old roof and have to cut into that as well. Fire brigade members might discover fire burning in the space between the two roofs also.

A similar problem may be encountered in dealing with remodeled buildings or additions. There could be an old flat roof under a new pitched section, resulting in two separate void spaces. The old roof provides a large supply of fuel for the fire. Such a situation is difficult to predict unless the Incident Commander has the building plan in advance.

# Trusses

Trusses are used extensively in support systems for both floors and roofs. A truss is a structural component composed of smaller components in a triangular configuration or a system of triangles. Trusses are common in modern construction for several reasons. The triangular geometry creates a strong, rigid structure that can support a load much

**Brigade Member Safety Tips**

When working on a flat roof, avoid skylights and other openings that will not support your weight. If you are temporarily blinded by smoke, use a tool to test the surface in front of you and to your sides. Stay away from the edges of the roof.

greater than its own mass. For example, both a solid beam and a simple truss with the same overall dimensions can support the same load. The truss requires much less material than the beam, is much lighter, and can span a long distance without supports. Trusses are often prefabricated and transported to the building location.

Trusses are used in residential construction, apartment buildings, small office buildings, commercial buildings, warehouses, fast food restaurants, airplane hangars, churches, and even firehouses. They may be clearly visible, or concealed within the construction. They are widely used in new construction and often replace heavier solid beams and joists in renovated or modified older buildings.

The strength of a truss depends on both its members and the connections between them. A properly assembled and installed truss is strong. If the members or connectors begin to fail, however, the strength of the truss is compromised.

Trusses can be used for many different purposes. In building construction, trusses made of wood and steel or combinations of wood and steel are used primarily to support roofs and floors.

A parallel chord truss has two parallel horizontal members connected by a system of diagonal and sometimes vertical members. The top and bottom members are called the chords, and the connecting pieces are the web members. Parallel chord trusses are often used to support flat roofs or floors. In lightweight construction, an engineered wood truss is often assembled with wood chords and either wood or light steel web members. A steel bar joist is another example of a parallel chord truss.

A pitched chord truss is typically used to support a sloping roof. Most modern residential construction uses a series of prefabricated wood pitched chord trusses to support the roof. The roof deck is supported by the top chords, and the ceilings of the occupied rooms are attached to the bottom chords. In this way, the trusses define the shape of the attic.

A bowstring truss has the same shape as an archery bow. The top chord represents the curved bow and the bottom chord represents the straight bowstring. Bowstring trusses are usually quite large and widely spaced. They were popular in warehouses, supermarkets, and similar structures with large, open floor areas. The roof of a building with bowstring trusses has a distinctive curved shape.

## Walls

Walls are the most visible parts of a building because they shape the exterior and define the interior. Walls may be constructed of masonry, wood, steel, aluminum, glass, and many other materials.

Walls are either load-bearing or nonbearing. Load-bearing walls provide structural support (► Figure 6-17). A load-bearing wall supports a portion of both the building's weight (dead load) and its contents (live load), transmitting that load down to the building's foundation. Damaging or removing a load-bearing wall can result in a partial or total collapse of the building. Load-bearing walls can be either exterior or interior walls.

Nonbearing walls support only their own weight (▲ Figure 6-18). Most nonbearing walls can be breached or removed without compromising the structural integrity of the building. Many nonbearing walls are interior partitions

### Brigade Member Safety Tips

The United States Fire Administration (USFA) has partnered with the American Forest & Paper Association (AF&PA) to develop educational materials to enhance brigade member awareness regarding the fire performance of different types of lightweight construction components. The brigade member educational material developed under this partnership is available free of charge from http://usfa.fema.gov.

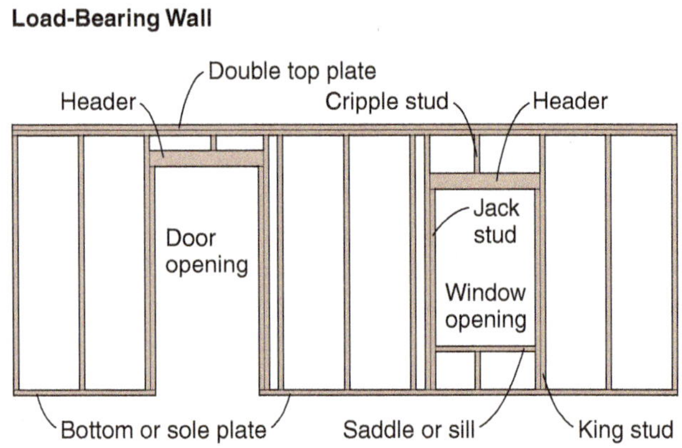


Figure 6-17 A load-bearing wall provides structural support.

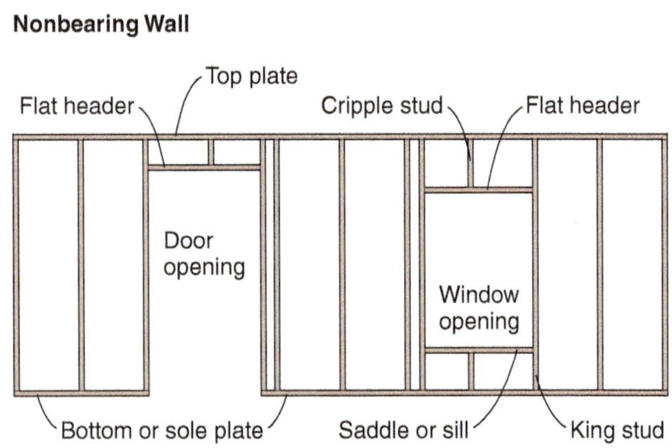


Figure 6-18 A nonbearing wall supports only its own weight.

that divide the building into rooms and spaces. The exterior walls of a building can also be nonbearing, particularly when a system of columns supports the building.

In addition to load-bearing and nonbearing walls, there are several specialized walls.

- Party walls are constructed on the line between two properties and are shared by a building on each side of the line. They are almost always load-bearing walls. A party wall is often, but not always, constructed as a fire wall between the two properties.
- Fire walls are designed to limit the spread of fire from one side of the wall to the other side. A fire wall might divide a large building into sections or separate two attached buildings. Fire walls usually extend from the foundation up to and through the roof of a building. They are constructed of fire-resistant materials and may be fire-rated.
- Fire partitions are interior walls that extend from a floor to the underside of the floor above. Fire partitions often enclose fire-rated interior corridors or divide a floor area into separate fire compartments.
- Fire enclosures are fire-rated assemblies that enclose interior vertical openings, such as stairwells, elevator shafts, and chases for building utilities. A fire enclosure prevents fire and smoke from spreading from floor to floor via the vertical opening. In multistory buildings, fire enclosures also protect the occupants using the exit stairways.
- Curtain walls are nonbearing exterior walls attached to the outside of the building. Curtain walls often serve as the exterior skin on a steel-framed high-rise building.

Solid, load-bearing masonry walls, at least 6- to 8-inch thick, can be used for buildings up to six stories high. Nonbearing masonry walls can be almost any height. Masonry walls provide a durable, fire-resistant outer covering for a building and are often used as fire walls. A well-designed masonry fire wall is often completely independent of the structures on either side. Even if the building on one side burns completely and collapses, the fire wall should prevent the fire from spreading to the building on the other side.

Older buildings often had masonry load-bearing walls several feet thick at the bottom that decreased in thickness as height increased. Modern masonry walls are often reinforced with steel rods or concrete to provide a more efficient and more durable structural system.

When properly constructed and maintained, masonry walls are strong and can withstand a vigorous assault by fire. But if the interior structure begins to collapse and exerts unanticipated forces on the exterior walls, solid masonry walls can fail during a fire.

Even though a building has an outer layer of brick or stone, brigade members should never assume that the exterior walls are masonry. Many buildings that look like masonry are actually constructed using wood-frame techniques and materials. A single layer of brick or stone, called a veneer layer, is applied to the exterior walls to give the appearance of a durable and architecturally desirable outer covering. If the wood structure is damaged during a fire, the veneer is likely to collapse.

Wood framing is used to construct the walls in most houses and many small commercial buildings. The wood framing used for exterior walls can be covered with various materials, including wooden siding, vinyl siding, aluminum siding, stucco, and masonry veneer. Moisture barriers, wind barriers, and other types of insulation are usually applied to the outside of the vertical studs before the outer covering is applied.

Vertical wooden studs support the walls and partitions inside the building. If fire resistance is critical, steel studs are used to frame walls. The wood framing is usually covered with gypsum board and a variety of interior finish materials. The space between the two wall surface coverings may be empty, it may contain thermal and sound insulating materials, or it may contain electrical wiring, telephone wires, and plumbing. These spaces often provide pathways through which fire can spread.

> **Brigade Member Tips**
>
> Manufactured trailer structures use lightweight building components throughout the structure to reduce weight. As a result, most parts of manufactured trailer structures are combustible. Manufactured trailer structures typically have few doors and small windows, making entry for fire suppression or rescue difficult. Once a fire starts, especially in an older manufactured trailer structures, it can destroy the entire structure within a few minutes. Often all that is left is the frame.

## Doors and Windows

Doors and windows are important components of any building. Although they generally have different functions—doors provide entry and exit while windows provide light and ventilation—in an emergency, doors and windows are almost interchangeable. A window can serve as an entry or an exit while a door can provide light and ventilation.

There are hundreds of door and window designs, with many different applications. Of particular concern to brigade members are fire doors and fire windows.

### Door Assemblies

Most doors are constructed of either wood or metal. Hollow-core wooden doors are often used inside buildings. A typical hollow-core door has an internal framework with its outer surfaces covered by thin sheets of wood. Because hollow-core doors can be easily opened with simple hand tools, they should not be used where security is a concern. A fire can usually burn through a hollow-core door in just a few minutes.

Solid-core wooden doors are used where a more substantial door is required. Solid-core doors are manufactured from solid panels or blocks of wood and are more difficult to force open. A solid-core door provides some fire resistance and can often keep a fire contained within a room for 20 minutes or more, giving fire brigade members a chance to arrive and mount an effective attack.

Metal doors are more durable and fire resistant than wooden doors. Some metal doors have a solid wood core covered on both sides by a thin sheet of metal. Others are constructed entirely of metal and reinforced for added security. The interior of a metal door can be hollow, or filled with wood, sound-deadening material, or thermal insulation. Although most approved fire doors are metal, not all metal doors should be considered fire doors.

## Brigade Member Safety Tips

Buildings under construction or demolition and buildings that are being renovated are extremely hazardous for brigade members.

### Window Assemblies

Fire brigade members frequently use windows as entry points to attack a fire and as emergency exits. Windows also provide ventilation during fires, allowing smoke and heat to escape and cooler, fresh air to enter.

Windows come in many shapes, sizes, and designs for different buildings and occupancies (▼ Figure 6-19). Fire brigade members must become familiar with the specific types of windows found in local occupancies, learn to recognize them, and understand how they operate. It is especially important to know if a particular type of window is very difficult to open or cannot be opened. The chapter on forcible entry contains information on window construction.

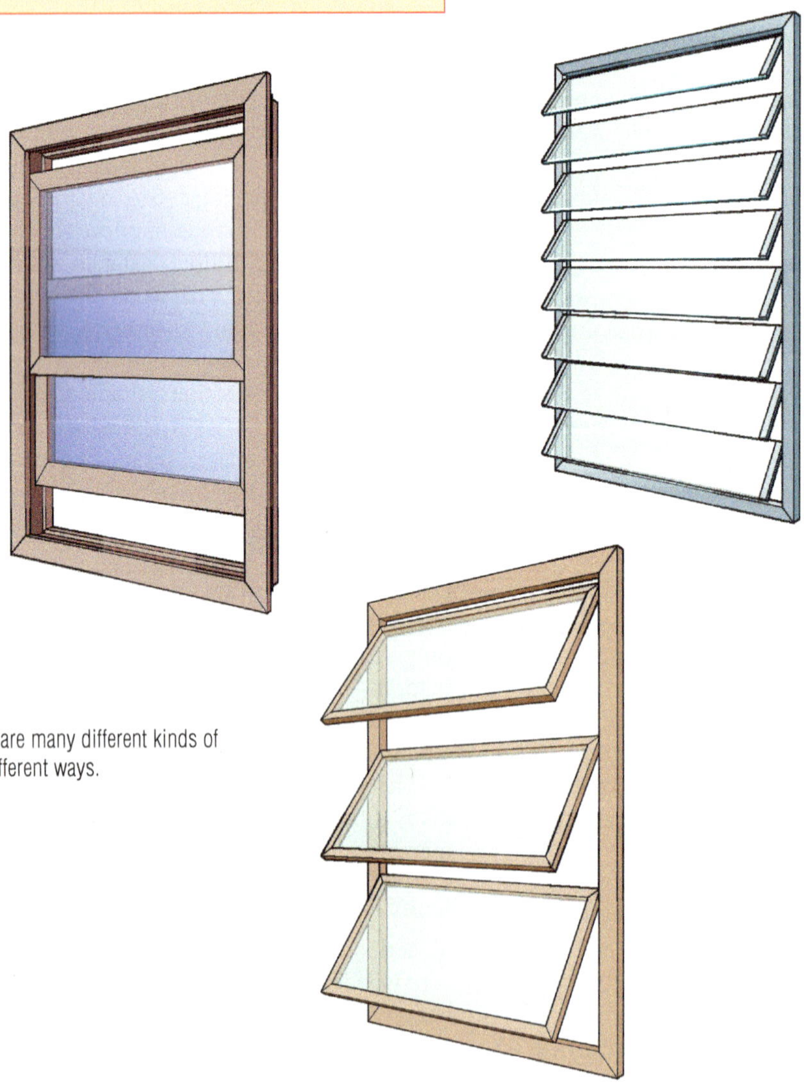

Figure 6-19 There are many different kinds of windows that open in different ways.

### Fire Doors and Fire Windows

Fire doors and fire windows are constructed to prevent the passage of flames, heat, and smoke through an opening during a fire. They must be tested and meet the standards set in NFPA 80, *Standard for Fire Doors and Fire Windows.* Fire doors and fire windows come in different shapes and sizes and provide different levels of fire resistance. For example, fire doors can swing on hinges, slide down or across an opening, or roll down to cover an opening.

The fire rating on a door or window covers the actual door or window, the frame, the hinges or closing mechanism, the latching hardware, and any other equipment that is required to operate the door or window. All of these items must be tested and approved as a combined system. All fire doors must have a mechanism that keeps the door closed or automatically closes the door when a fire occurs. Doors that are normally open can be closed by the release of a fusible link, by a smoke or heat detector, or by activation of the fire alarm system.

Fire doors and fire windows are rated for a particular duration of fire resistance to a standard test fire. This is similar to the fire resistance rating system used for building construction assemblies. A 1-hour rating, however, does not guarantee that the door will resist any fire for 60 minutes; it only establishes that the door will resist the standard test fire for 60 minutes. In any given fire situation, a door rated at 1 hour will probably last twice as long as a door rated 30 minutes.

Fire doors and windows are labeled and assigned the letters A, B, C, D, or E, based on their approved-use locations (▶ **Figure 6-20**). For example, doors designated "A" are approved for use as part of a fire wall. Doors designated "B" are approved for use in stair shafts and elevator shafts. (▶ **Table 6-1**) shows the approved uses for each of the letter designations.

**Fire windows** are used when a window is needed in a required fire-resistant wall (▶ **Figure 6-21**). Fire windows are often made of wired glass, which is designed to withstand exposures to high temperatures without breaking. Fire-resistant glass without wires is available for some applications; special steel window frames are required to keep the glass firmly in place.

Wired glass is also used to provide vision panels in or next to fire doors. Vision panels allow a person to view conditions on the opposite side of a door without opening the door, or to make sure no one is standing in front of the door before opening it. In these configurations, the entire assembly—including the door, the window, and the frame—must all be tested and approved together.

When a window is required only for light passage, glass blocks can sometimes be used instead of wired glass. Glass blocks will resist high temperatures and remain in place if properly installed. The size of the opening is limited and depends on the fire-resistive rating of the wall.

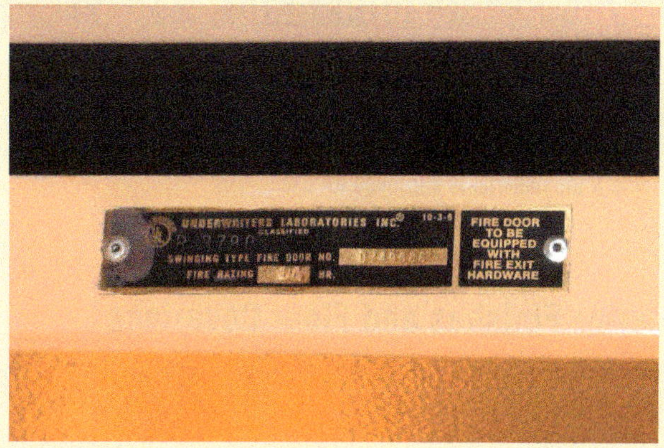


**Figure 6-20** An approved fire door has a label indicating its classification and rating.

**Table 6-1 NFPA 80 Designations for Fire Doors and Fire Windows**

| | |
|---|---|
| **Class A** | Openings in fire walls and in walls that divide a single building into fire areas |
| **Class B** | Openings in enclosures of vertical communications through buildings and in 2-hour rated partitions providing horizontal fire separations |
| **Class C** | Openings in walls or partitions between rooms and corridors having a fire resistance rating of 1 hour or less |
| **Class D** | Openings in exterior walls subject to severe fire exposure from outside of the building |
| **Class E** | Openings in exterior walls subject to moderate or light fire exposure from outside of the building |

**Figure 6-21** Fire windows are used to protect openings in walls that are required to be fire resistant.

## Brigade Member Safety Tips

Beware of materials attached to the walls or ceilings of a building to deaden sound. Some of these materials are highly flammable and can provide the fuel for a very intense, smoky fire. Check to determine if the material is approved for use.

## Interior Finishes and Floor Coverings

The term interior finish commonly refers to the exposed interior surfaces of a building. Interior finish materials will affect how a particular building or occupancy reacts when a fire occurs. Interior finish considerations include whether a material ignites easily or resists ignition, how quickly a flame will spread across the material surface, how much energy the material will release when it burns, how much smoke it will produce, and what the smoke will contain.

A room with a bare concrete floor, concrete block walls, and a concrete ceiling has no interior finishes that will increase the fire load. But if the same room had an acrylic carpet with rubber padding on the floor, wooden baseboards, varnished wood paneling on the walls, and foam plastic acoustic insulation panels on the ceiling, the situation would be different. These interior finishes would ignite quickly, flames would spread quickly across the surfaces, and significant quantities of heat and toxic smoke would be released.

Different interior finish materials contribute in various ways to a building fire. Each individual material has certain characteristics, and fire brigade members must evaluate the particular combination of materials in a room or space. Typical wall coverings include painted plaster or gypsum board, wallpaper or vinyl wall coverings, wood paneling, and many other surface finishes. Floor coverings might include different types of carpet, vinyl floor tiles, or finished wood flooring. All of these products will burn to some extent, and each one involves a different set of fire risk factors. Fire brigade members must understand the hazards posed by different interior finish materials to operate safely at the scene of a fire.

## Buildings Under Construction or Demolition

Buildings that are under construction or renovation or in the process of demolition present a variety of problems and extra hazards for fire brigade members. In many cases, the fire protection features found in finished buildings are missing. Automatic sprinklers may not yet be installed; smoke detectors may have been removed. There may be no coverings on the walls, leaving the entire wood framework fully exposed. Missing doors and windows could provide an unlimited supply of fresh air to feed the fire and allow the flames to spread rapidly. Fire-resistive enclosures could be missing, leaving critical structural components unprotected. Sprinkler and standpipe systems might be inoperative. All of these factors can contribute to a fire that spreads rapidly, burns intensely, and causes structural failure quickly.

Many fires at construction and demolition sites are inadvertently caused by workers using torches to weld or take apart pieces of the structure. Tanks of flammable gases and piles of highly combustible construction materials may be left in locations where they could add even more fuel to a fire. Buildings under construction or demolition are often unoccupied for many hours, resulting in delayed discovery and reporting of fires. It may be difficult for fire apparatus to approach the structure or for fire brigade members to access working hydrants. All of these problems must be anticipated when considering the fire risks of a construction or demolition site.

## On-Duty Fire Fighter Fatalities

The potential for fatalities or injuries due to structural collapse or failure is a risk to all interior and exterior structural fire brigade members. Although an industrial facility probably does not have a church or residential structure on-site, it may feature construction types similar to those listed in the following incidents:

**2006:** Two volunteer fire fighters die when struck by an exterior wall collapse at a commercial building during fire overhaul (Alabama)
**2005:** A career fire captain dies when trapped by a partial roof collapse in a vacant house fire (Texas)
**2004:** A volunteer chief dies and two fire fighters are injured by a collapsing church facade (Tennessee)
**2003:** A partial roof collapse in a commercial structure fire claims the lives of two career fire fighters (Tennessee)
**2003:** A career fire fighter dies from injuries received during a chimney and structural collapse after a house fire (Pennsylvania)
**2002:** Structural collapse at an auto parts store fire claims the lives of one career lieutenant and two volunteer fire fighters (Oregon)
**2002:** A volunteer lieutenant dies following structural collapse at a residential house fire (Pennsylvania)
**2002:** Parapet wall collapse at an auto body shop claims the life of a career captain and injures a career lieutenant and an emergency medical technician (Indiana)
**2002:** A career fire fighter dies after a roof collapse following roof ventilation (Iowa)
**2002:** Structural collapse at a residential fire claims the lives of two volunteer fire chiefs and one career fire fighter (New Jersey)
**2002:** One career fire fighter dies and a captain is hospitalized after a floor collapses in a residential fire (North Carolina)
**2002:** One career fire fighter dies and another is injured after a partial structural collapse (Texas)
**2002:** First-floor collapse during a residential basement fire claims the lives of two fire fighters (career and volunteer) and injures a career fire fighter captain (New York)
**2001:** A career fire fighter dies after falling through the floor fighting a structure fire at a local residence (Ohio)
**2001:** A volunteer fire fighter dies and another fire fighter is injured during a wall collapse at a fire at local business (Wisconsin)
**2000:** Roof collapse injures four career fire fighters at a church fire (Arkansas)
**1999:** A fire investigator dies after being struck by a chimney that collapsed during an origin-and-cause fire investigation (New York)
**1999:** Floor collapse claims the life of one fire fighter and injures two other fire fighters (California)
**1998:** Roof collapse in an arson church fire claims the life of a volunteer fire fighter (Georgia)
**1998:** A fire fighter dies while fighting a warehouse fire when a parapet wall collapses (Vermont)
**1996:** Sudden roof collapse at a burning auto parts store claims the lives of two fire fighters (Virginia)

Source: NIOSH, *Traumatic Occupational Injuries, Fire Fighter Fatality Investigation and Prevention Program: Fire Fighter Fatality Investigation Reports.*

# Wrap-Up

## Ready for Review

- Building materials under fire will combust, conduct heat, expand, or lose strength.
- Most construction uses masonry, concrete, steel, glass, gypsum board, wood, and plastics.
- Fire-resistive construction, noncombustible construction, ordinary construction, heavy timber construction, and wood-frame construction are the five major types of building construction.
- Each type of building construction has different strengths and hazards during a fire.
- Foundations, floors, ceilings, roofs, trusses, walls, doors, windows, interior finishes, and floor coverings are the major components of buildings.

## Hot Terms

**Balloon-frame construction** An older type of wood-frame construction in which the wall studs extend vertically from the basement of a structure to the roof without any fire stops.

**Bowstring truss** Trusses that are curved on the top and straight on the bottom.

**Combustibility** Determines whether or not a material will burn.

**Curtain walls** Nonbearing walls used to separate the inside and outside of the building, but not part of the support structure for the building.

**Curved roofs** Roofs that have a curved shape.

**Dead load** The weight of a building; the dead load consists of the weight of all materials of construction incorporated into a building, including but not limited to walls, floors, roofs, ceilings, stairways, built-in partitions, finishes, cladding, and other similarly incorporated architectural and structural items, as well as fixed service equipment, including the weight of cranes.

**Fire enclosures** Fire-rated assemblies used to enclose vertical openings such as stairwells, elevator shafts, and chases for building utilities.

**Fire partitions** Interior walls extending from the floor to the underside of the floor above.

**Fire wall** A wall with a fire-resistive rating and structural stability that separates buildings or subdivides a building to prevent the spread of fire.

**Fire window** A window or glass block assembly with a fire-resistive rating.

**Flat roofs** Horizontal roofs often found on commercial or industrial occupancies.

**Fire-rated glass** Special glass formulated to achieve a fire rating.

**Glass blocks** Thick pieces of glass similar to bricks or tiles.

**Gypsum** A naturally occurring material composed of calcium sulfate and water molecules.

**Gypsum board** The generic name for a family of sheet products consisting of a noncombustible core primarily of gypsum with paper surfacing.

**Interior finish** Any coating or veneer applied as a finish to a bulkhead, structural insulation, or overhead, including the visible finish, all intermediate materials, and all application materials and adhesives.

**Laminated glass** Glass manufactured with a thin vinyl core covered by glass on each side of the core.

**Laminated wood** Pieces of wood that are glued together.

**Live load** The weight of the building contents.

**Load-bearing wall** Walls designed for structural support.

**Masonry** Built-up unit of construction or combination of materials such as brick, clay tiles, or stone set in mortar.

**Manufactured trailer structures** A factory-assembled structure or structures transportable in one or more sections that is built on a permanent chassis and designed to be used as a dwelling without a permanent foundation when connected to the required utilities, including the plumbing, heating, air-conditioning, and electric systems contained therein.

**Nonbearing wall** Wall designed to support only the weight of the wall itself.

**Occupancy** The purpose for which a building or other structure, or part thereof, is used or intended to be used.

**Parallel chord truss** A truss in which the top and bottom chords are parallel.

**Party walls** Walls constructed on the line between two properties.

**Pitched chord truss** Type of truss typically used to support a sloping roof.

**Pitched roof** A roof with sloping or inclined surfaces.

**Platform-frame construction** Construction technique for building the frame of the structure one floor at a time. Each floor has a top and bottom plate that acts as a fire stop.

**Pyrolysis** The destructive distillation of organic compounds in an oxygen-free environment that converts the organic matter into gases, liquids, and char.

**Rafters** Joists that are mounted in an inclined position to support a roof.

**Spalling** Chipping or pitting of concrete or masonry surfaces.

**Tempered glass** Glass that is much stronger and harder to break than ordinary glass.

**Thermal conductivity** Describes how quickly a material will conduct heat.

**Thermoplastic material** Plastic material capable of being repeatedly softened by heating and hardened by cooling and, that in the softened state, can be repeatedly shaped by molding or forming.

**Thermoset material** Plastic material that, after having been cured by heat or other means, is substantially infusible and cannot be softened and formed.

**Truss** A collection of lightweight structural components joined in a triangular configuration that can be used to support either floors or roofs.

**Type I construction** Buildings with structural members made of noncombustible materials that have a specified fire resistance.

**Type II construction** Buildings with structural members made of noncombustible materials without fire resistance.

**Type III construction** Buildings with the exterior walls made of noncombustible or limited-combustible materials, but interior floors and walls made of combustible materials.

**Type IV construction** Buildings constructed with noncombustible or limited-combustible exterior walls, and interior walls and floors made of large dimension combustible materials.

**Type V construction** Buildings with exterior walls, interior walls, floors, and roof structures made of wood.

**Wired glass** Glass made by molding glass around a special wire mesh.

**Wood panels** Thin sheets of wood glued together.

**Wood trusses** Assemblies of small pieces of wood or wood and metal.

**Wooden beams** Load-bearing members assembled from individual wood components.

# Brigade Member in Action

**It is 1:00 P.M. on a Monday afternoon when your fire brigade is dispatched to a one-story warehouse with a steel bar joist roof with masonry block walls. Upon arrival you find fire in a rack storage area. Your brigade leader tells you to initiate an interior attack with a preconnected hose line.**

1. What type of construction is this structure?
   - **A.** Type I
   - **B.** Type II
   - **C.** Type III
   - **D.** Type IV
   - **E.** Type V

2. What type of roof truss system is this structure likely to have?
   - **A.** Parallel cord truss
   - **B.** Pitched cord truss
   - **C.** Bowstring truss
   - **D.** None of the above

3. You should be cautious when damaging or removing load-bearing walls because doing so:
   - A. divides the building into rooms or spaces and may encourage fire spread.
   - B. can result in a partial or total collapse of the building.
   - C. might disrupt ventilation.
   - D. may cause the veneer to collapse.

4. If drains on a flat roof are not maintained:
   - A. Roof ladders must be used to provide a secure walking platform.
   - B. The structure is subject to collapse from overload.
   - C. There is a collapse potential if Halligans are utilized.
   - D. There is a potential for horizontal fire spread.

# Portable Fire Extinguishers

# Chapter 7

## NFPA 1081 Standard

**Incipient Facility Fire Brigade Member**

**4.2** *Manual Fire Suppression.* This duty shall involve tasks related to the manual control of fires and property conservation activities by the incipient facility fire brigade member.

**4.2.1*** Extinguish incipient fires, given an incipient fire and a selection of portable fire extinguishers, so that the correct extinguisher is chosen, the fire is completely extinguished, proper extinguisher-handling techniques are followed, and the area of origin and fire cause evidence are preserved.

**(A)** *Requisite Knowledge.* The classifications of fire; risks associated with each class of fire; and the types, rating systems, operating methods, and limitations of portable fire extinguishers.

**(B)** *Requisite Skills.* The ability to select, carry, and operate portable fire extinguishers, using the appropriate extinguisher based on the size and type of fire.

**Advanced Exterior Facility Fire Brigade Member**

**5.3.6*** Extinguish an exterior fire using special extinguishing agents other than foam operating as a member of a team, given an assignment, a handline, personal protective equipment, and an extinguishing agent supply, so that fire is extinguished, reignition is prevented, and team protection is maintained.

**(A)** *Requisite Knowledge.* Methods by which special agents, such as dry chemical, dry powder, and carbon dioxide, prevent or control a hazard; principles by which special agents are generated; the characteristics, uses, and limitations of firefighting special agents; the advantages and disadvantages of using special agents; special agents application techniques; hazards associated with special agents usage; and methods to reduce or avoid hazards.

**(B)** *Requisite Skills.* The ability to operate a special agent supply for use, master various special agents application techniques, and approach and retreat from hazardous areas as part of a coordinated team.

**Interior Structural Facility Fire Brigade Member**

**6.3.11*** Extinguish a Class C (electrical) fire as a member of a team, given an assignment, a Class C fire-extinguishing appliance/extinguisher, and PPE, so that the type of Class C agent is selected for the condition, a selected agent is correctly applied to the fuel, fire is extinguished, reignition is prevented, team protection is maintained, and the hazard is faced until retreat to safe haven is reached.

**(A)** *Requisite Knowledge.* Methods by which a Class C agent prevents or controls a hazard; methods by which Class C fires are de-energized; causes of injuries from Class C fire fighting on live Class C fires with Class A agents and the Class C agents; the extinguishing agents' characteristics, uses, and limitations; the advantages and disadvantages of de-energizing using water fog nozzles on a Class A or Class B fire; and methods to reduce or avoid hazards.

**(B)** *Requisite Skills.* The ability to operate Class C fire extinguishers or fixed systems and approach and retreat from Class C fires as part of a coordinated team.

## Knowledge Objectives

After completing this chapter, you will be able to:

- State the primary purposes of fire extinguishers.
- Define Class A fires.
- Define Class B fires.
- Define Class C fires.
- Define Class D fires.
- Define Class K fires.
- Explain the classification and rating system for fire extinguishers.
- Describe the types of agents used in fire extinguishers.
- Describe the types of operating systems in fire extinguishers.
- Describe the basic steps of fire extinguisher operation.
- Explain the basic steps of inspecting, maintaining, recharging, and hydrostatic testing of fire extinguishers.

## Skills Objectives

After completing this chapter, you will be able to perform the following skills:

- Transport the extinguisher to the location of the fire.
- Select and operate a portable fire extinguisher safely to effectively extinguish an incipient fire.
- Attack a Class A fire with a stored-pressure water-type fire extinguisher.
- Attack a Class A fire with a multipurpose dry chemical fire extinguisher.
- Attack a Class B flammable liquid fire with a dry chemical fire extinguisher.
- Attack a Class B flammable liquid fire with a stored-pressure foam fire extinguisher.
- Use a wet chemical fire extinguisher.
- Use a halogenated agent-type extinguisher.
- Use dry powder fire extinguishing agents.

# You Are the Brigade Member

**You are responding to a report of a vehicle fire in the plant parking lot. The fire is in the engine compartment of a light truck. The brigade leader directs you to don your self-contained breathing apparatus (SCBA), get the dry chemical extinguisher, and back up the other brigade members who are pulling a 1 3/4-inch hose line to fight the fire. As you follow orders, you wonder:**

1. ***Are there any advantages to using a portable fire extinguisher to initially attack a car fire?***
2. ***Does it have to be a dry chemical extinguisher, or would another type of extinguisher work just as well?***
3. ***Are there instances when a dry chemical extinguisher should not be used?***

## Introduction

Portable fire extinguishers are required in many types of occupancies, as well as in commercial vehicles, boats, aircraft, and various other locations. Fire prevention efforts encourage citizens to keep fire extinguishers in their homes, particularly in their kitchens. Fire extinguishers are used successfully to put out hundreds of fires every day, preventing **millions** of dollars in property damage as well as saving lives. Most fire extinguishers are easy to operate and can be used effectively by an individual with only basic training.

Fire extinguishers range in size from models that can be operated with one hand to large, wheeled models that contain several hundred pounds of extinguishing agent (a material used to stop the combustion process) ▶ Figure 7-1. Extinguishing agents include water, water with different additives, dry chemicals, dry powders, and gaseous agents. Each agent is suitable for specific types of fires.

Fire extinguishers are designed for different purposes and involve different operational methods. As a brigade member, you must know which is the most appropriate kind of extinguisher to use for different types of fires, which kinds must not be used for certain fires, and how to use any fire extinguisher you may encounter, especially those used by your brigade and carried on your apparatus. The selection of the proper fire extinguisher builds on the information presented in Chapter 5, Fire Behavior.

Fire brigade members use fire extinguishers to control small fires that do not require the use of a hose line. A portable backpack-type fire extinguisher can be used to control and overhaul a wildland fire located beyond the reach of hoses. Special types of fire extinguishers are appropriate for situations where the application of water would be dangerous, ineffective, or undesirable. For example, applying water to a fire that involves expensive electronic equipment is both dangerous and costly. Using the correct fire extinguisher could control the fire without causing additional damage to the equipment.

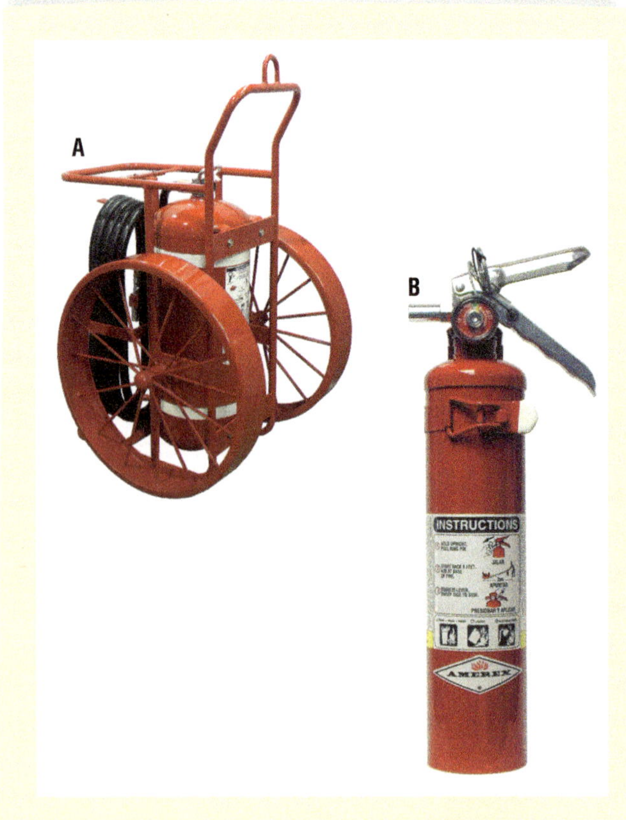


**Figure 7-1** Portable fire extinguishers can be large or small. **A.** A wheeled extinguisher. **B.** A one-hand fire extinguisher.

This chapter also covers the operation and maintenance of the most common types of portable fire extinguishers. These principles will enable you to use fire extinguishers correctly and effectively to reduce the risk of personal injury and property damage.

## Purposes of Fire Extinguishers

Portable fire extinguishers have two primary uses: to extinguish incipient fires (those that have not spread beyond the area of origin) and to control fires where traditional methods of fire suppression are not recommended.

Fire extinguishers are placed in many locations so that they will be available for immediate use on small, incipient-stage fires, such as a fire in a wastebasket. A trained individual with a suitable fire extinguisher could easily control this type of fire . As the flames spread beyond the wastebasket to other contents of the room, the fire becomes increasingly ▼ Figure 7-2 difficult to control with only a portable fire extinguisher.

Fire extinguishers are also used to control fires where traditional extinguishing methods are not recommended. For example, using water on fires that involve energized electrical equipment increases the risk of electrocution to brigade members. Applying water to a fire in a computer or electrical control room could cause extensive damage to the electrical equipment. In these cases, it would be better to use a fire extinguisher with the appropriate extinguishing agent. Special extinguishing agents are also required for fires that involve flammable liquids, cooking oils, and combustible metals. The appropriate type of fire extinguisher should be available in areas containing these hazards.

Figure 7-2 A trained individual with a suitable fire extinguisher can easily control an incipient fire.

### Brigade Member Safety Tips

Do not place yourself in a dangerous situation by trying to fight a large fire with a small fire extinguisher. You can't fight a fire or protect yourself with an empty extinguisher.

### Brigade Member Safety Tips

Proper technique in using a portable fire extinguisher is important for both safety and effectiveness. Practice using a portable fire extinguisher under the careful supervision of a trained instructor.

### Incipient Fires

Most fire brigade vehicles carry at least one fire extinguisher; many vehicles carry two or more extinguishers of different types. Brigade members often use these portable extinguishers to control incipient-stage fires quickly. At times, a brigade member may even use an extinguisher from the premises to control an incipient fire.

One advantage of fire extinguishers is their portability. It may take less time to control a fire with a portable extinguisher than it would to advance and charge a hose line. Fire brigade vehicles that are not equipped with water or fire hoses usually carry at least one multipurpose fire extinguisher. If you arrive at an incipient-stage fire in one of these vehicles, you might be able to control the flames with the portable extinguisher.

The primary disadvantage of fire extinguishers is that they are "one-shot" devices. Once the contents of a fire extinguisher have been discharged, the extinguisher is no longer effective in fighting fires until it is recharged. If the extinguisher does not control the fire, some other device or method will have to be employed. This is a serious limitation when compared to a fire hose with a continuous water supply.

### Special Extinguishing Agents

Some types of fires require special extinguishing agents. As a brigade member, you must know which fires require special extinguishing agents, what type of extinguisher should be used, and how to operate the different types of special-purpose extinguishers.

Special extinguishing agents are used for kitchen fires that involve combustible cooking oils and fats, for combustible metal fires, and for fires in electronic equipment. Using water or an unsuitable fire-extinguishing chemical to fight these types of fires can cause more damage than the fire.

Figure 7-3 Portable extinguishers containing specific extinguishing agents are sometimes helpful in overhauling a fire.

Figure 7-4 Most ordinary combustible materials are included in the definition of Class A fires.

Portable extinguishers are sometimes used in combination with other techniques. For example, with a pressurized gas fire, water may be used to cool hot surfaces and prevent reignition, while a dry chemical agent is used to extinguish the flames. Certain types of portable extinguishers can be helpful in overhauling a fire (▲Figure 7-3). The extinguishing agents contained in these devices break down the surface tension of the water, so the water can penetrate the materials and reach deep-seated fires.

## Classes of Fires

It is essential to match the appropriate type of extinguisher to the type of fire. Fires and fire extinguishers are grouped into classes according to their characteristics. Some extinguishing agents work more efficiently than others on certain types of fires. In some cases, selecting the proper extinguishing agent will mean the difference between extinguishing a fire and being unable to control it.

More importantly, in some cases it is dangerous to apply the wrong extinguishing agent to a fire. Using a water extinguisher on an electrical fire can cause an electrical shock as well as a short circuit in the equipment. A water extinguisher should never be used to fight a grease fire. Burning grease is generally hotter than 212°F (100°C), so the water converts to steam, which expands very rapidly. If the water penetrates the surface of the grease, the steam is produced within the grease. As the steam expands, the hot grease erupts like a volcano and splatters over everything and everyone nearby, resulting in burns or injuries to people and spreading the fire.

Before selecting a fire extinguisher, ask yourself, "What class of fire am I fighting?" Remember, there are five classes of fire and each will affect the choice of extinguishing equipment.

### Class A Fires

Class A fires involve ordinary combustibles such as wood, paper, cloth, rubber, rubbish, and some plastics (▲Figure 7-4). Natural vegetation, such as grass and trees, is also Class A material. Water is the most commonly used extinguishing agent for Class A fires, although several other agents can be used effectively.

### Class B Fires

Class B fires involve flammable or combustible liquids, such as gasoline, oil, grease, tar, lacquer, oil-based paints, and some plastics (►Figure 7-5). Fires involving flammable gases, such as propane or natural gas, are also categorized as Class B fires. Several different types of extinguishing agents are approved for Class B fires.

Examples of Class B fires include a fire in a pot of molten roofing tar, a fire involving splashed fuel, and burning natural gas that is escaping from a gas meter struck by a vehicle.

### Class C Fires

Class C fires involve energized electrical equipment, which includes any device that uses, produces, or delivers electrical energy (►Figure 7-6). A Class C fire could involve building wiring and outlets, fuse boxes, circuit breakers, transformers, generators, or electric motors. Power tools, lighting fixtures, household appliances, and electronic devices such as televisions, radios, and computers could be involved in Class C fires. The equipment must be plugged in

Figure 7-5 Class B fires involve flammable liquids and gases.

Figure 7-6 Class C fires involve energized electrical equipment or appliances.

or connected to an electrical source, but not necessarily operating.

Electricity does not burn, but electrical energy can generate tremendous heat that could ignite nearby Class A or B materials. As long as the equipment is energized, it must be treated as a Class C fire. Agents that will not conduct electricity, such as dry chemicals or carbon dioxide, must be used on Class C fires.

## Class D Fires

Class D fires involve combustible metals such as magnesium, titanium, zirconium, sodium, lithium, and potassium. Special techniques and extinguishing agents are required to fight combustible metals fires ► Figure 7-7. Normal extinguishing agents can react violently, even explosively, if they come in contact with burning metals. Violent reactions also can occur when water strikes burning combustible metals.

Figure 7-7 Combustible metals in Class D fires require special extinguishing agents.

Class D fires are most often encountered in industrial occupancies, such as machine shops and repair shops, as well as in fires involving aircraft and automobiles. Magnesium and titanium, both combustible metals, are used to produce automotive and aircraft parts because they combine high strength with light weight. Sparks from cutting, welding, or grinding operations could ignite a Class D fire, or the metal items could become involved in a fire that originated elsewhere.

Because of the chemical reactions that could occur during a Class D fire, it is important to select the proper extinguishing agent and application technique. Choosing the correct fire extinguisher for a Class D fire requires expert knowledge and experience.

## Class K Fires

Class K fires involve combustible cooking oils and fats ► Figure 7-8. This is a relatively new classification; cooking oil fires were previously classified as Class B combustible liquid fires. The use of high-efficiency modern cooking equipment and the trend toward using vegetable oils instead of animal fats to fry foods required the

Figure 7-8 Class K fires involve cooking oils and fats.

development of a new class of extinguishing agents. Many restaurants are still using extinguishing agents that were approved for Class B fires.

## Classification of Fire Extinguishers

Portable fire extinguishers are classified and rated based on their characteristics and capabilities. This information is important for selecting the proper extinguisher to fight a particular fire (▶ Table 7-1). It is also used to determine what type or types of fire extinguishers should be placed in a given location so that incipient fires can be quickly controlled.

In the United States, Underwriters Laboratories Inc. (UL) is the organization that developed the standards, classification, and rating system for portable fire extinguishers. This system rates fire extinguishers for both safety and effectiveness. Each fire extinguisher has a specific rating that identifies the class or classes of fires for which it is both safe and effective.

The classification system for fire extinguishers uses letters and numbers. The letters indicate the class or classes of fire for which the extinguisher can be used, and the numbers indicate its effectiveness. Fire extinguishers that are safe and effective for more than one class will be rated with multiple letters. For example, an extinguisher that is safe and effective for Class A fires will be rated with an "A;" one that is safe and effective for Class B fires will be rated with a "B;" and one that is safe and effective for both Class A and Class B fires will be rated with both an "A" and a "B."

Class A and Class B fire extinguishers also include a number, indicating the relative effectiveness of the fire extinguisher in the hands of a nonexpert user.

On Class A extinguishers, the number is related to an amount of water. An extinguisher that is rated 1-A contains the equivalent of 1.25 gallons of water. A typical Class A extinguisher contains 2.5 gallons of water and has a 2-A rating. The higher the number, the greater the extinguishing capability of the extinguisher. An extinguisher that is rated 4-A should be able to extinguish approximately twice as much fire as one that is rated 2-A.

The effectiveness of Class B extinguishers is based on the approximate area (measured in square feet) of burning fuel they are capable of extinguishing. A 10-B rating indicates that a nonexpert user should be able to extinguish a fire in a pan of flammable liquid that is 10 square feet in surface area. An extinguisher rated 40-B should be able to control a flammable liquid pan fire with a surface area of 40 square feet.

Numbers are used to rate an extinguisher's effectiveness only for Class A and Class B fires. If the fire extinguisher can also be used for Class C fires, it contains an agent proven to be nonconductive to electricity and safe for use on energized electrical equipment. For instance, a fire extinguisher that carries a 2-A:10-B:C rating can be used on Class A, Class B, and Class C fires. It has the extinguishing capabilities of a

### Brigade Member Tips

The safest and surest way to extinguish a Class C fire is to turn off the power and treat it like a Class A or B fire. If you are unable to turn off the power, you should be prepared for reignition. The electricity could reignite the fire after it is extinguished.

**Table 7-1 Types of Fires**

| | |
|---|---|
| Class A | Ordinary combustibles |
| Class B | Flammable or combustible liquids |
| Class C | Energized electrical equipment |
| Class D | Combustible metals |
| Class K | Combustible cooking media |

## Brigade Member Tips

Brigade members must understand both the Underwriters Laboratories classification and rating system and the labeling system for fire extinguishers.

2-A extinguisher when applied to Class A fires, the capabilities of a 10-B extinguisher for Class B fires, and can be used safely on energized electrical equipment.

Standard test fires are used to rate the effectiveness of fire extinguishers. The testing may involve different agents, amounts, application rates, and application methods. Fire extinguishers are rated for their ability to control a specific type of fire as well as for the extinguishing agent's ability to prevent rekindling. Some agents can successfully suppress a fire but are unable to prevent the material from reigniting. A rating is given only if the extinguisher completely extinguishes the standard test fire and prevents rekindling.

## Labeling of Fire Extinguishers

Fire extinguishers that have been tested and approved by an independent laboratory are labeled to clearly designate the class or classes of fire the unit is capable of extinguishing safely. The traditional lettering system has been used for many years and is still found on many fire extinguishers. Recently, however, a universal pictograph system, which does not require the user to be familiar with the alphabetic codes for the different classes of fires, has been developed.

### Traditional Lettering System ▶ Figure 7-9

The traditional lettering system uses the following labels:

- Extinguishers suitable for use on Class A fires are identified by the letter A on a solid green triangle.
- Extinguishers suitable for use on Class B fires are identified by the letter B on a solid red square.
- Extinguishers suitable for use on Class C fires are identified by the letter C on a solid blue circle.
- Extinguishers suitable for use on Class D fires are identified by the letter D on a solid yellow five-pointed star.
- Extinguishers suitable for use on Class K (combustible cooking oil) fires are identified by a pictograph showing a fire in a frying pan. Because the Class K designation is new, there is no traditional-system alphabet graphic for it.

### Pictograph Labeling System

The pictograph system, such as described for Class K fire extinguishers, uses symbols rather than letters on the labels. This system also clearly indicates if an extinguisher is inappropriate for use on a particular class of fire. The pictographs are all square icons that are designed to represent each class of fire ▲ Figure 7-10. The icon for Class A fires is a burning trash can beside a wood fire. The Class B fire extinguisher icon is a flame and a gasoline can; the Class C icon is a flame and an electrical plug and socket. There is no pictograph for Class D extinguishers. Extinguishers rated for fighting Class K fires are labeled with an icon showing a fire in the frying pan.

Under this pictograph labeling system, the presence of an icon indicates that the extinguisher has been rated for that class of fire. A missing icon indicates that the extinguisher has not been rated for that class of fire. A red slash across an icon indicates that the extinguisher must not be used on that type of fire, because doing so would create additional risk.

An extinguisher rated for Class A fires only would show all three icons, but the icons for Class B and Class C would have a red diagonal line through them. This three-icon array signifies that the extinguisher uses a water-based extinguishing agent, making it unsafe to use on flammable liquid or electrical fires.

Certain extinguishers labeled for Class B and Class C fires do not include the Class A icon but may be used to put

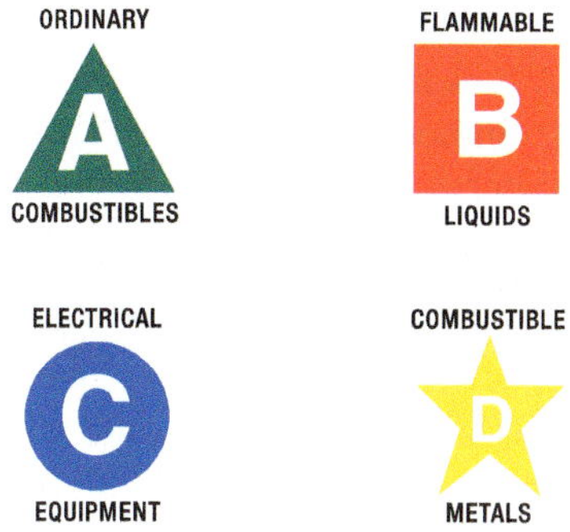


Figure 7-9 Traditional letter labels on fire extinguishers often incorporated a shape as well as a letter.

Figure 7-10 The icons for Classes A, B, C, and K fires.

out small Class A fires. The fact that they have not been rated for Class A fires indicates that they are less effective in extinguishing a common combustible fire than a comparable Class A extinguisher would be.

## Fire Extinguisher Placement

Fire codes and regulations require the installation of fire extinguishers in many areas so that they will be available to fight incipient fires. *NFPA 10 Standard for Portable Fire Extinguishers* lists the requirements for placing and mounting portable fire extinguishers as well as the appropriate mounting heights.

The regulations for each type of occupancy specify the maximum floor area that can be protected by each extinguisher, the maximum travel distance from the closest extinguisher to a potential fire, and the types of fire extinguishers that should be provided. Two key factors must be considered when determining which type of extinguisher should be placed in each area: the class of fire that is likely to occur and the potential magnitude of an incipient fire.

Extinguishers should be mounted so they are readily visible and easily accessed (▼ Figure 7-11). Heavy extinguishers should not be mounted high on a wall. If the extinguisher is mounted too high, a smaller person might be unable to lift it off its hook or could be injured in the attempt.

Figure 7-11 Extinguishers should be mounted in locations with unobstructed access and visibility.

According to *NFPA 10*, the recommended mounting heights for the placement of fire extinguishers are:

- Fire extinguishers weighing up to 40 lb (18.14 kg) should be mounted so that the top of the extinguisher is not more than five feet (1.53 m) above the floor.
- Fire extinguishers weighing more than 40 lb (18.14 kg) should be mounted so that the top of the extinguisher is not more than three feet (1.07 m) above the floor.
- The bottom of an extinguisher should be at least four inches (10.2 cm) above the floor.

### Classifying Area Hazards

Areas are divided into three risk classifications—light, ordinary, and extra hazard—according to the amount and type of combustibles that are present, including building materials, contents, decorations, and furniture. The quantity of combustible materials present is called a building's fire load and is measured as the average weight of combustible materials per square foot or per square meter of floor area. The larger the fire load, the larger the potential fire.

Occupancy use category does not necessarily determine the appropriate hazard classification. The recommended hazard classifications for different types of occupancies are guidelines based on typical situations. The hazard classification for each area should be based on the actual amount and type of combustibles that are present.

#### Light or Low Hazard

Light (or low) hazard locations are areas where the majority of materials are noncombustible or arranged so that a fire is not likely to spread. Light hazard environments usually contain limited amounts of Class A combustibles, such as wood, paper products, cloth, and similar materials. A light hazard environment might also contain some Class B combustibles (flammable liquids and gases), such as copy machine chemicals or modest quantities of paints and solvents, but all Class B materials must be kept in closed containers and stored safely. Examples of common light hazard environments are most offices and meeting rooms. (► Figure 7-12).

#### Ordinary or Moderate Hazard

The ordinary (or moderate) hazard locations contain more Class A and Class B materials than light hazard locations. Examples of ordinary hazard locations include light manufacturing, light vehicle repair shops, research labs, and electronic plants (► Figure 7-13).

Ordinary hazard areas also include warehouses that contain Class I and Class II commodities. Class I commodities include noncombustible products stored on wooden pallets or in corrugated cartons that are shrink-wrapped or wrapped in paper. Class II commodities include noncombustible products stored in wooden crates or multilayered corrugated cartons.

Figure 7-12 Light hazard areas include offices, conference rooms, and training rooms.

Figure 7-13 Light manufacturing, mobile equipment repair shops, and electronic plants are classified as ordinary hazard areas.

### Extra or High Hazard

Extra (or high) hazard locations contain more Class A combustibles and/or Class B flammables than ordinary hazard locations. Examples of extra hazard areas include woodworking shops, mobile equipment repair facilities, plastics processing areas, and combustible fluid use areas ▶ Figure 7-14. In addition, areas used for manufacturing processes such as painting, dipping, or coating, and facilities used for storing or handling flammable liquids are classified as extra hazard environments. Warehouses containing products that do not meet the definitions of Class I and Class II commodities are also considered extra hazard locations.

Figure 7-14 Mobile equipment repair facilities, plastics processing areas, and combustible fluid use areas are classified as extra hazard locations.

## Determining the Appropriate Class of Fire Extinguisher

Several factors must be considered when determining the number and types of fire extinguishers that should be placed in each area of an occupancy. Among these factors are the types of fuels found in the area and the quantities of those materials.

Some areas may need extinguishers with more than one rating or more than one type of fire extinguisher. Environments that include Class A combustibles require an extinguisher rated for Class A fires; those with Class B combustibles require an extinguisher rated for Class B fires; and areas that have both Class A and Class B combustibles require either an extinguisher that is rated for both types of fires or a separate extinguisher for each class of fire.

Most buildings require extinguishers that are suitable for fighting Class A fires because ordinary combustible materials, such as furniture, partitions, interior finish materials, paper, and packaging products, are so common. Even where other classes of products are used or stored, there is still a need to defend the facility from a fire involving common combustibles.

A single multipurpose extinguisher is generally less expensive than two individual fire extinguishers and eliminates the problem of selecting the proper extinguisher for a particular fire. However, it is sometimes more appropriate to install Class A extinguishers in general use areas and to place extinguishers that are especially effective in fighting Class B or Class C fires near those hazards.

Some facilities present a variety of conditions. In these occupancies, each area must be individually evaluated so that extinguisher installation is tailored to the particular circumstances.

## Methods of Fire Extinguishment

Understanding the nature of fire is key to understanding how extinguishing agents work and how they differ from each other. All fires require four elements: fuel, sufficient heat, oxygen, and a chemical chain reaction. Scientifically, burning is called rapid oxidation. It is a chemical process that occurs when a fuel is combined with oxygen, resulting in the formation of ash or other waste products and the release of energy as heat and light.

The combustion process begins when the fuel is heated to its ignition point—the temperature at which it begins to burn. The energy that initiates the process can come from many different sources, including a spark or flame, friction, electrical energy, or a chemical reaction. Once a substance begins to burn, it will generally continue burning as long as there are adequate supplies of oxygen and fuel to sustain the chemical chain reaction, unless something interrupts the process.

Most extinguishers stop the burning by cooling the fuel below its ignition point or kindling temperature, by cutting off the supply of oxygen, or by combining these two techniques. Some extinguishing agents interrupt the complex system of molecular chain reactions that occur between the heated fuel and the oxygen. Modern portable fire extinguishers contain agents that use one or more of these methods.

### Cooling the Fuel

If the temperature of the fuel falls below its kindling temperature, the combustion process will stop. Water extinguishes a fire using this method.

### Cutting Off the Supply of Oxygen

Creating a barrier that interrupts the flow of oxygen to the flames will also extinguish a fire. Putting a lid on a pan of burning food is an example of this technique (▼Figure 7-15). Applying a blanket of foam to the surface of a burning liquid is another example. Surrounding the fuel with a layer of carbon dioxide can also cut off the supply of oxygen necessary to sustain the burning process.

Figure 7-15 Covering a pan of burning food with a lid will extinguish a fire by cutting off the supply of oxygen.

### Interrupting the Chemical Chain

Some extinguishing agents work by interrupting the reaction required to sustain combustion. In some cases, a very small quantity of the agent can accomplish the objective.

## Types of Extinguishing Agents

An extinguishing agent is the substance contained in a portable fire extinguisher that puts out a fire. The best extinguishing agent for a particular hazard depends on several factors, including the types of materials involved and the anticipated size of the fire. Portable fire extinguishers use seven basic types of extinguishing agents:

- Water
- Dry chemicals
- Carbon dioxide
- Foam
- Wet chemicals
- Halogenated agents
- Dry powder

## Water

Water is an efficient, plentiful, and inexpensive extinguishing agent. When water is applied to a fire, it quickly converts from liquid into steam, absorbing great quantities of heat in the process. As the heat is removed from the combustion process, the fuel cools below its ignition temperature and the fire stops burning.

Water is an excellent extinguishing agent for Class A fires. Many Class A fuels will absorb water, which further lowers the temperature of the fuel. This also prevents rekindling.

Water is a much less effective extinguishing agent for other fire classes. Applying water to hot cooking oil can cause splattering, which can spread the fire and possibly endanger the extinguisher operator. Because water conducts electricity, it is dangerous to apply a stream of water to any fire that involves energized electrical equipment. If water is applied to a burning combustible metal, a violent reaction can occur. Because of these limitations, plain water is only used in Class A fire extinguishers.

One disadvantage of water is that it freezes at 32°F (0°C). In areas that are subject to freezing, loaded-stream extinguishers can be used. These extinguishers combine an alkali metal salt with water. The salt lowers the freezing point of water, so the extinguisher can be used in much colder areas.

Wetting agents can also be added to the water in a fire extinguisher. These agents reduce the surface tension of the water, allowing it to penetrate more effectively into many fuels, such as baled cotton or fibrous materials.

## Dry Chemical

Dry chemical fire extinguishers deliver a stream of very fine particles onto a fire. Different chemical compounds are used to produce extinguishers of varying capabilities and characteristics. The dry chemicals interrupt the chemical chain reaction that occurs within the combustion process.

Dry chemical extinguishing agents offer several advantages over water extinguishers:

- They are effective on Class B (flammable liquids and gases) fires.
- They can be used on Class C (energized electrical equipment) fires, because the chemicals are nonconductive.
- They are not subject to freezing.

The first dry chemical extinguishers were introduced during the 1950s and were rated only for Classes B and C fires. The industry term for these B:C-rated units is "ordinary dry chemical" extinguishers.

During the 1960s multipurpose dry chemical extinguishers were introduced. These extinguishers are rated for Classes A, B, and C fires. The chemicals in these extinguishers form a crust over Class A combustible fuels to prevent rekindling ▼Figure 7-16.

Figure 7-16 Multipurpose dry chemical extinguishers can be used for Classes A, B, and C fires.

Multipurpose dry chemical extinguishing agents are in the form of fine particles and are treated with other chemicals to help maintain an even flow when the extinguisher is being used. Additional additives prevent them from absorbing moisture, which could cause caking and interfere with the discharge.

One disadvantage of dry chemical extinguishers is that the chemicals, particularly the multipurpose dry chemicals, are corrosive and can damage electronic equipment, such as computers, telephones, and copy machines. The fine particles are carried by the air and settle like a fine dust inside the equipment. Over time, the residue can corrode metal parts, causing considerable damage. If electronic equipment is exposed to multipurpose dry chemical extinguishing agents, it should be cleaned professionally within 48 hours after exposure.

The five primary compounds used as dry chemicals extinguishing agents are:

1. Sodium bicarbonate (rated for Classes B and C fires only)
2. Potassium bicarbonate (rated for Classes B and C fires only)
3. Urea-based potassium bicarbonate (rated for Classes B and C fires only)
4. Potassium chloride (rated for Classes B and C fires only)
5. Ammonium phosphate (rated for Classes A, B, and C fires)

Sodium bicarbonate is often used in small household extinguishers. Potassium bicarbonate, potassium chloride, and urea-based potassium bicarbonate all have greater fire-extinguishing capabilities (per unit volume) for Class B fires than sodium bicarbonate. Potassium chloride is more corrosive than the other dry chemical extinguishing agents.

Ammonium phosphate is the only dry chemical extinguishing agent rated as suitable for use on Class A fires.

Which dry chemical extinguisher to use depends on the compatibility of different agents with each other and with products they might contact. Some dry chemical extinguishing agents can be used in combination with some types of foam.

## Carbon Dioxide

Carbon dioxide ($CO_2$) is a gas that is 1.5 times heavier than air. When carbon dioxide is discharged on a fire, it forms a dense cloud that displaces the air surrounding the fuel. This interrupts the combustion process by reducing the amount of oxygen that can reach the fuel.

In portable carbon dioxide fire extinguishers, carbon dioxide is stored under pressure as a liquid. It is colorless and odorless. It is discharged through a hose and expelled on the fire through a horn. When it is released, the carbon dioxide is very cold and forms a visible cloud of "dry ice" because moisture in the air will freeze when it comes into contact with the carbon dioxide.

Carbon dioxide is rated for Classes B and C fires only. It does not conduct electricity and has two significant advantages over dry chemical agents: it is not corrosive and it does not leave any residue.

Figure 7-17 Carbon dioxide extinguishers are heavy due to the weight of the container and the quantity of agent needed. They also have a large discharge nozzle, making them easily identifiable.

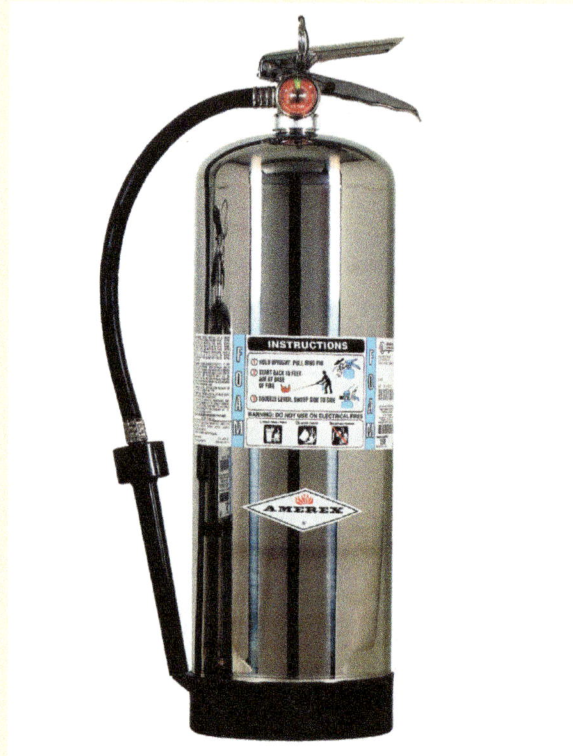


Figure 7-18 An AFFF extinguisher produces an effective foam for use on Class B fires.

Carbon dioxide also has several limitations and disadvantages. These include:

- Weight: Carbon dioxide extinguishers are heavier than similarly rated extinguishers that use other extinguishing agents (▲ Figure 7-17).
- Range: Carbon dioxide extinguishers have a short discharge range, which requires the operator to be close to the fire, increasing the risk of personal injury.
- Weather: Carbon dioxide does not perform well at temperatures below 0°F (-18°C) or in windy or drafty conditions, because it dissipates before it reaches the fire.
- Confined spaces: When used in confined areas, carbon dioxide dilutes the oxygen in the air. If enough oxygen is displaced, people in the space can begin to suffocate.
- Suitability: Carbon dioxide extinguishers are not suitable for use on fires involving pressurized fuel or on cooking grease fires.

## Foam

Foam fire extinguishers discharge a water-based solution with a measured amount of foam concentrate added. The nozzles on foam extinguishers are designed to introduce air into the discharge stream, thus producing a foam blanket. Foam extinguishing agents are formulated for use on either Class A or Class B fires.

Class A foam extinguishers for ordinary combustible fires extinguish fires in the same way that water extinguishes fires. The foam concentrate reduces the surface tension of the water, allowing for better penetration into the burning materials.

Class B foam extinguishers discharge a foam solution that floats across the surface of a burning liquid and prevents the fuel from vaporizing. The foam blanket forms a barrier between the fuel and the oxygen, extinguishing the flames and preventing reignition. These agents are not suitable for Class B fires that involve pressurized fuels or cooking oils.

The most common Class B additives are aqueous film-forming foam (AFFF) and film-forming fluoroprotein (FFFP) foam (▲ Figure 7-18). Both concentrates produce very effective foams. Which should be used depends on the product's compatibility with a particular flammable liquid and other extinguishing agents that could be used on the same fire.

Some Class B foam extinguishing agents are approved for use on polar solvents, which are water-soluble flammable liquids such as alcohols, acetone, esters, and ketones. Only extinguishers that are specifically labeled for use with polar solvents should be used if these products are present.

Although they are not specifically intended for Class A fires, most Class B foams can also be used on ordinary combustibles. The reverse is not true, however; Class A foams are not effective on Class B fires. Foam extinguishers are not suitable for use on Class C fires and cannot be stored or used at freezing temperatures.

### Wet Chemical

Wet chemical extinguishers are the only type of extinguisher to qualify under the new Class K rating requirements. They use wet chemical extinguishing agents, which are chemicals applied as water solutions. Before Class K extinguishing agents were developed, most fire extinguishing systems for kitchens used dry chemicals.

All new fixed extinguishing systems in restaurants and commercial kitchens now use wet chemical extinguishing agents. These agents are specifically formulated for use in commercial kitchens and food-product manufacturing facilities, especially where food is cooked in a deep fryer. The fixed systems discharge the agent directly over the cooking surfaces. There is no numeric rating of their efficiency in portable fire extinguishers.

The Class K wet chemical agents include aqueous solutions of potassium acetate, potassium carbonate, and potassium citrate, either singly or in various combinations. The wet agents convert the fatty acids in cooking oils or fats to a soap or foam, a process known as saponification.

When wet chemical agents are applied to burning vegetable oils, they create a thick blanket of foam that quickly smothers the fire and prevents it from reigniting while the oil cools. The agents are discharged as a fine spray, which reduces the risk of splattering.

### Halogenated Agents

Halogenated extinguishing agents are produced from a family of liquefied gases, known as halogens, that includes fluorine, bromine, iodine, and chlorine. Hundreds of different formulations can be produced from these elements with many different properties and potential uses. Although several of these formulations are very effective for extinguishing fires, only a few of them are commonly used as extinguishing agents.

Halogenated extinguishing agents are called clean agents because they leave no residue and are ideally suited for areas that contain computers or sensitive electronic equipment. Per pound, they are approximately twice as effective at extinguishing fires as carbon dioxide.

There are two categories of halogenated extinguishing agents: halons and halocarbons. A 1987 international agreement, known as the Montreal Protocol, limits halon production because these agents damage the earth's ozone layer. Halons have been replaced by a new family of extinguishing agents, halocarbons.

The halogenated agents are stored as liquids and are discharged under relatively high pressure. They release a mist of vapor and liquid droplets that interrupt the chemical chain reaction of the combustion process to extinguish the fire.

These agents dissipate rapidly in windy conditions, as does carbon dioxide, so their effectiveness is limited in outdoor locations.

Halon 1211 (bromochlorodifluoromethane) should be used judiciously and only in situations where its clean properties are essential. Small Halon 1211 extinguishers are rated for Classes B and C fires, but are unsuited for use on fires involving pressurized fuels or cooking grease. Larger halon extinguishers are also rated for Class A fires.

Currently, four types of halocarbon agents are used in portable extinguishers: hydrochlorofluorocarbon (HCFC), hydrofluorocarbon (HFC), perfluorocarbon (PFC), and fluoroiodocarbon (FIC).

### Dry Powder

Dry powder extinguishing agents are chemical compounds used to extinguish fires involving combustible metals (Class D fires). These agents are stored in fine granular or powdered form and are applied to smother the fire. They form a solid crust over the burning metal to exclude oxygen and absorb heat.

The most commonly used dry powder extinguishing agent is formulated from finely ground sodium chloride (table salt) plus additives to help it flow freely. A thermoplastic material mixed with the agent binds the sodium chloride particles into a solid mass when they come into contact with a burning metal.

Another dry powder agent is produced from a mixture of finely granulated graphite powder and compounds containing phosphorus. This agent cannot be expelled from fire extinguishers; it is produced in bulk form and applied by hand, using a scoop or a shovel. When applied to a metal fire, the phosphorus compounds release gases that blanket the fire and cut off its supply of oxygen. The graphite absorbs heat from the fire, allowing the metal to cool below its ignition point. Other specialized dry powder extinguishing agents are available for fighting specific types of metal fires. For details, see *NFPA's Fire Protection Handbook*.

Class D agents must be applied very carefully so that the molten metal does not splatter. No water should come in contact with the burning metal. Even a trace quantity of moisture can cause a violent reaction.

## Fire Extinguisher Design

All portable fire extinguishers use pressure to expel their contents. Many portable fire extinguishers rely on pressurized gas to expel the extinguishing agent. The gas can be stored with the extinguishing agent in the body of the extinguisher, or externally, in a separate cartridge. The extinguishing agent is put under pressure only as it is used.

Some extinguishing agents, such as carbon dioxide, are called self-expelling agents. These agents are normally gases, which are stored as liquids under pressure. When the pressure is released, the agent rapidly expands, causing it to self-discharge.

## Portable Fire Extinguisher Components

There are two basic styles of handheld portable fire extinguishers: stored-pressure and cartridge-operated. As the name implies, the stored-pressure extinguisher shell holds both the extinguishing agent and a pressurized gas to expel the agent. In a cartridge-operated extinguisher, the extinguishing agent is stored at atmospheric pressure. This kind of extinguisher does not become pressurized until the gas cartridge is punctured, which releases the charging gas into the extinguisher shell. Stored-pressure handheld portable fire extinguishers have seven basic parts (▼ Figure 7-19):

- A cylinder or shell that holds the extinguishing agent
- A carrying handle
- A nozzle or horn
- A discharge hose
- A trigger and discharge valve assembly
- A locking mechanism (ring pin) to prevent accidental discharge
- A pressure indicator

A cartridge-operated handheld portable fire extinguisher has nine parts (▼ Figure 7-20):

- A cylinder or shell that holds the extinguishing agent
- A fill cap
- A gas cartridge containing either $CO_2$ or nitrogen
- A cartridge guard
- A carry handle
- A nozzle and discharge valve assembly
- A discharge hose
- A hose retainer to prevent accidental discharge
- A gas tube to distribute the pressure into the extinguisher shell

### Cylinder or Shell

The body of the extinguisher, known as the cylinder or shell, holds the extinguishing agent. Nitrogen, compressed air, or carbon dioxide are used to pressurize the cylinder to expel the agent. Stored-pressure extinguishers store both the extinguishing agent, in wet or dry form, and the gas under pressure in the cylinder. Cartridge-operated extinguishers rely on an external cartridge of pressurized gas, which is released only when the extinguisher is charged.

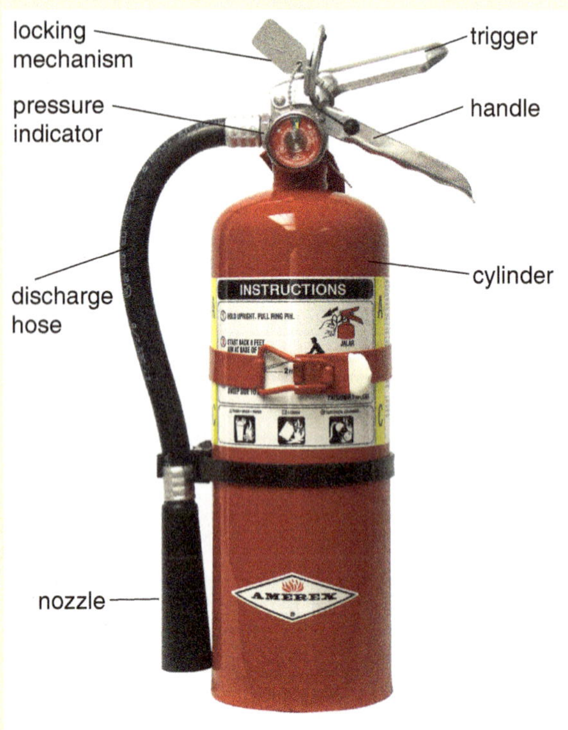


Figure 7-19 The stored-pressure portable fire extinguisher.

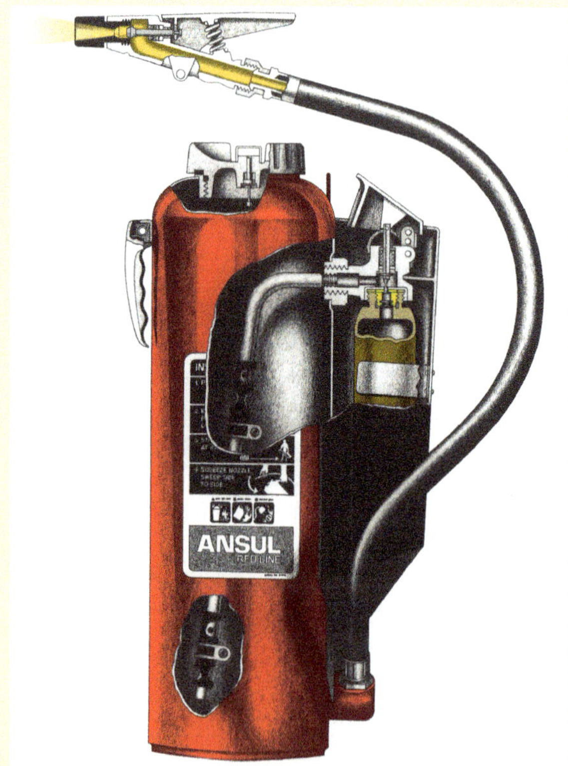


Figure 7-20 The cartridge-operated portable fire extinguisher.

### Handle

The handle is used to carry a portable fire extinguisher and, in many cases, to hold it during use. The actual design of the handle varies from model to model, but all extinguishers that weigh more than 3 lb have handles. In many cases, the handle is located just below the trigger mechanism.

### Nozzle or Horn

The extinguishing agent is expelled through a nozzle or horn. In some extinguishers, the nozzle is attached directly to the valve assembly at the top of the extinguisher. In other models, the nozzle is at the end of a short hose.

Foam extinguishers have a special aspirating nozzle that introduces air into the extinguishing agent, creating the foam. Carbon dioxide extinguishers have a tubular or conical horn, which is often mounted at the end of a short hose.

Pump tank extinguishers, which are nonpressurized, manually operated water extinguishers, usually have a nozzle at the end of a short hose. The manually operated pump may be mounted directly on the cylinder or it may be part of the nozzle assembly.

### Trigger

The trigger is the mechanism that is squeezed or depressed to discharge the extinguishing agent. In some models, the trigger is a button positioned just above the handle. Most of the time, the trigger is a lever located above the handle. The operator lifts the extinguisher by the handle and simultaneously squeezes down on the discharge lever.

Cartridge-operated extinguisher models usually have a two-step operating sequence. First, a handle or lever is pushed to pressurize the stored agent, then a trigger-type mechanism incorporated in the hose/nozzle assembly is used to control the discharge.

### Locking Mechanism

The locking mechanism is a simple quick-release device that prevents accidental discharge of the extinguishing agent. The simplest form of locking mechanism is a stiff pin, which is inserted through a hole in the trigger to prevent it from being depressed. The pin usually has a ring at the end so that it can be removed quickly.

A special plastic tie, called a tamper seal, is used to secure the pin. The tamper seal is designed to break easily when the pin is pulled. Removing the pin and tamper seal is best accomplished with a twisting motion. The tamper seal makes it easy to see whether the extinguisher has been used and not recharged. The seal also discourages people from playing or tinkering with the extinguisher.

### Pressure Indicator

The pressure indicator or gauge shows that a stored-pressure extinguisher has sufficient pressure to operate properly. Over time, the pressure in an extinguisher may dissipate. Checking the gauge first will tell you whether the extinguisher is ready for use.

Pressure indicators vary in design. Most extinguishers use a needle gauge. Pressure may be shown in pounds per square inch (psi) or on a three-part scale (too low, proper range, too high). Pressure gauges are usually color-coded; a green area indicates the proper pressure zone. Carbon dioxide extinguishers do not have pressure indicators or gauges, but are weighed to determine the remaining agent.

### Wheeled Fire Extinguishers

Wheeled fire extinguishers are large units mounted on wheeled carriages. Wheeled extinguishers typically contain between 150 and 350 lb of extinguishing agent. The wheeled design lets one person transport the extinguisher to the fire. If a wheeled extinguisher is intended for indoor use, doorways and aisles must be wide enough to allow passage to every area where it could be needed.

Wheeled fire extinguishers usually have long delivery hoses, so the unit can stay in one spot as the operator moves around to attack the fire. A separate cylinder containing nitrogen or some other compressed gas provides the pressure necessary to operate the extinguisher.

# Fire Extinguisher Characteristics

Portable fire extinguishers vary according to their extinguishing agent, capacity, effective range, and the time it takes to completely discharge their agent. They also have different mechanical designs. This section describes the basic characteristics of several types of extinguishers, organized by type of extinguishing agent.

The seven types of extinguishers described include:

- Water extinguishers
- Dry chemical extinguishers
- Carbon dioxide extinguishers
- Class B foam extinguishers
- Wet chemical extinguishers
- Halogenated-agent extinguishers
- Dry powder extinguishing agents

## Water Extinguishers

Water extinguishers are used to cool the burning fuel below its ignition temperature. Water extinguishers are intended for use primarily on Class A fires. Class B foam extinguishers are intended for flammable liquids fires. Water extinguishers include stored-pressure, loaded-stream, and wetting-agent models.

### Stored-Pressure Water-Type Extinguishers

The most common stored-pressure water-type extinguisher is the 2.5 gallons model with a 2-A rating ► Figure 7-21. This extinguisher expels water in a solid

**Figure 7-21** Stored-pressure water-type extinguisher.

stream with a range of 35–40 feet through a nozzle at the end of a short hose. The discharge time is approximately 55 seconds if the extinguisher is used continuously. A full extinguisher weighs about 30 lb.

Because the contents of these extinguishers can freeze, they should not be installed in areas where the temperature is expected to drop below 32°F (0°C). Antifreeze models of stored-pressure water-type extinguishers, called loaded-stream extinguishers, are available.

The recommended procedure for operating a stored-pressure water extinguisher is to set it on the ground, grasp the handle with one hand, and pull out the ring pin or release the locking latch with the other hand. Now the extinguisher can be lifted and used to extinguish the fire. Use one hand to aim the stream at the fire, and squeeze the trigger with the other hand. The stream of water can be made into a spray by putting a thumb at the end of the nozzle; this technique is often used after the flames have been extinguished to thoroughly soak the fuel.

Stored-pressure water-type extinguishers can be recharged at any location that provides water and a source of compressed air. Follow the manufacturer's instructions to ensure proper and safe recharging.

### Loaded-Stream Water-Type Extinguishers

Loaded-stream water-type extinguishers discharge a solution of water containing an alkali metal salt that prevents freezing at temperatures as low as -40°F (-40°C). The most common model is the 2.5 gallon unit, which is identical to a typical stored-pressure water extinguisher. Handheld models are available with capacities of 1–2.5 gallons of water.

### Wetting-Agent and Class A Foam Water-Type Extinguishers

Wetting-agent water-type extinguishers expel water that contains a solution to reduce its surface tension (the physical property that causes water to bead or form a puddle on a flat surface). Reducing the surface tension allows water to spread over the fire and penetrate more efficiently into Class A fuels.

Class A foam extinguishers contain a solution of water and Class A foam concentrate. This agent has foaming properties as well as the ability to reduce surface tension.

Both wetting-agent and Class A foam extinguishers are available in the same configurations as water extinguishers, including handheld stored-pressure models and wheeled units. These extinguishers should not be exposed to temperatures below 40°F (4°C).

### Pump Tank Water-Type Extinguishers

Pump tank water-type extinguishers come in sizes ranging from 1-A rated, 1.5 gallon units to 4-A rated, 5 gallon units. The water in these units is not stored under pressure. The pressure to expel the water is provided by a hand-operated, double-acting, vertical piston pump, which moves water out through a short hose on both the up and the down strokes. This type of extinguisher sits upright on the ground during use. A small bracket at the bottom allows the operator to steady the extinguisher with one foot while pumping.

Pump tank extinguishers can be used with antifreeze. The manufacturer should be consulted for details because some antifreezes (such as common salt) can corrode the extinguisher or damage the pump. Extinguishers with steel shells corrode more easily than those with copper or nonmetallic shells.

## Dry Chemical Extinguishers

Dry chemical extinguishers contain a variety of chemical extinguishing agents in granular form. Handheld dry chemical extinguishers are available with capacities ranging

> **Fire Marks**
>
> Stored-pressure water extinguishers have replaced soda-acid extinguishers, which had to be inverted (turned upside down) to trigger their discharge. The soda-acid models are no longer manufactured and should have been replaced in all installations.

from 1 to 30 lb of agent. Wheeled fire extinguishers are available with capacities up to 350 lb of agent. Ordinary dry chemical models can be used to extinguish Class B and C fires. Multipurpose dry chemical models are rated for use on Class A, B, and C fires.

All dry chemical extinguishing agents can be used on Class C fires that involve energized electrical equipment; however, the residue left by the dry chemical can be very damaging to computers, electronic devices, and electrical equipment.

Stored-pressure units expel the dry chemical agent in the same manner as a stored-pressure water extinguisher. The dry chemical agent in a cartridge-operated extinguisher is not stored under pressure. These extinguishers have a sealed, pressurized cartridge connected to the storage cylinder. They are activated by pushing down on a lever that punctures the cartridge and pressurizes the cylinder.

Most small, handheld dry chemical extinguishers are designed to discharge completely in 8–20 seconds. Larger units may discharge for as long as 60 seconds. All handheld dry chemical extinguishers are designed to be carried and operated simultaneously.

Depending on the extinguisher's size, the range of the discharge stream can be from 5 to 30 feet. Some models have special nozzles that allow for a longer range. The long-range nozzles are useful when the fire involves burning gas or a flammable liquid under pressure or working in a strong wind.

The trigger allows the extinguisher to be discharged intermittently, starting and stopping the agent flow. Releasing the trigger stops the flow of agent; however, this does not mean that the extinguisher can be put aside and used again later. These extinguishers do not retain their internal pressure for extended periods.

Anytime a dry chemical extinguisher has been activated or partially used, the extinguisher must be serviced and recharged to replenish the extinguishing agent and restore the unit's pressure. Disposable models are not refillable. Low temperature rated, dry chemical fire extinguishers can be stored and used in areas with temperatures below freezing.

### Ordinary Dry Chemical Extinguishers

Ordinary dry chemical extinguishers are available in handheld models with ratings up to 160-B:C. Larger, wheeled units carry ratings up to 640-B:C.

## Brigade Member Safety Tips

Discharging a dry chemical fire extinguisher in a confined space can create a cloud of very fine dust that can impair vision and cause difficulty breathing. SCBA should be used to protect brigade members from both toxic gases from the fire and the dry chemical dust discharged from the fire extinguisher.

## Brigade Member Tips

A dry chemical extinguisher will often continue to lose pressure after a partial discharge. Pressure loss can occur even when only a very small amount of agent has been discharged and the pressure gauge still indicates that the extinguisher is properly charged. This occurs because the agent can leave residue in the valve assembly and allow the stored pressure to leak out slowly.

### Multipurpose Dry Chemical Extinguishers

Ammonium phosphate is commonly called a multipurpose dry chemical agent because it can be used on Class A, B, and C fires. When it is used on Class A fires, the chemical coats the surface of the fuel to prevent continuing combustion.

Multipurpose dry chemical extinguishers are available in handheld models with ratings ranging from 1-A to 20-A and from 10-B:C to 120-B:C. Larger, wheeled models have ratings ranging from 20-A to 40-A and from 60-B:C to 320-B:C.

Multipurpose dry chemical extinguishers should never be used on cooking oil (Class K) fires. The ammonium phosphate-based extinguishing agent is acidic and will not react with cooking oils to produce the smothering foam needed to extinguish the fire. The acid will counteract the foam-forming properties of any alkaline extinguishing agent that is applied to the same fire.

## Carbon Dioxide Extinguishers

Carbon dioxide extinguishers are rated to fight Class B and C fires. Carbon dioxide extinguishes a fire by enveloping the fuel in a cloud of inert gas, which reduces the oxygen content of the surrounding atmosphere and smothers the flames. Because the carbon dioxide discharge is very cold, it also helps to cool the burning materials and surrounding areas.

Carbon dioxide gas is 1.5 times heavier than air, colorless, odorless, nonconductive, and inert. It is also noncorrosive and does not leave any residue. These factors are important where costly electronic components or computer equipment must be protected. Carbon dioxide is also used around food preparation areas and in laboratories.

Carbon dioxide is both an expelling agent and an extinguishing agent. The agent is stored under a pressure of 823 psi, which keeps the carbon dioxide in liquid form. When the pressure is released, the liquid carbon dioxide rapidly converts to a gas. The expanding gas forces the agent out of the container.

The carbon dioxide is discharged through a siphon tube that reaches to the bottom of the cylinder. It is forced through a hose to a horn or cone-shaped applicator that is used to direct the flow of the agent. When discharged, the agent is very cold and contains a mixture of carbon dioxide gas and solid carbon dioxide, which quickly converts to gas.

Compared to other types of extinguishers, carbon dioxide extinguishers have relatively short discharge ranges of 3–8 feet. This presents a safety issue because the operator must be close to the fire. Depending on size, carbon dioxide extinguishers can discharge completely in 8–30 seconds.

Carbon dioxide fire extinguishers are not recommended for outdoor use with strong air currents. The carbon dioxide will be rapidly dissipated and will not be effective in smothering the flames.

Carbon dioxide extinguishers have a trigger mechanism that can be operated intermittently to preserve the remaining agent. The pressurized carbon dioxide will remain in the extinguisher, but the extinguisher must be recharged after use. The extinguisher is weighed to determine the amount of agent left in the cylinder.

The smaller carbon dioxide extinguishers contain from 2 to 5 lb of agent. These units are designed to be operated with one hand. The horn is attached directly to the discharge valve on the top of the extinguisher by a hinged metal tube. In larger models, the horn is attached at the end of a short hose.

The horns of some older carbon dioxide fire extinguishers were constructed of metal, which conducts electricity. These extinguishers do not carry a Class C rating and must not be used around fires involving energized electrical equipment. Metal horns are no longer made for carbon dioxide fire extinguishers, but some of these units may still be in service.

## Class B Foam Extinguishers

Class B foam extinguishers are very similar in appearance and operation to water extinguishers. They discharge a solution of water and either AFFF or FFFP foam concentrate. The agent is discharged through an aspirating nozzle, which mixes air into the stream. They create a foam blanket that will float over the surface of a flammable liquid.

Class B foam agents are also very effective in fighting Class A fires. They are not suitable for Class C fires or fires involving flammable liquids or gases under pressure. They are not intended for use on cooking oil fires, and only alcohol resistant foam extinguishers can be used on fires involving polar solvents. Detailed information on the use of AFFF and FFFP is available in *NFPA 11 Standard for Low-, Medium-, and High-Expansion Foam*.

Foam extinguishing agents are not effective at freezing temperatures. Consult the extinguisher manufacturer for information on using foam agents effectively at low temperatures.

## Wet Chemical Extinguishers

**Wet chemical extinguishers** are used to protect Class K installations, which include cooking oils, deep fryers, and grills. Many commercial cooking installations use fixed, automatic fire extinguishing systems as their first line of defense. Portable Class K wet chemical extinguishers are currently available in two sizes, 1.5 gallons and 2.5 gallons. There are no numerical ratings for these extinguishers.

### Brigade Member Safety Tips

Do not aim carbon dioxide extinguisher discharge at anyone or allow it to come in contact with exposed skin. Frostbite could result. Carbon dioxide discharged into a confined space will reduce the oxygen level in that space. Self-contained breathing apparatus must be used by anyone entering the confined area.

## Halogenated-Agent Extinguishers

Halogenated-agent extinguishers include both halon agents and halocarbon agents. Because halon agents can destroy the earth's protective ozone layer, their use is strictly controlled. The halocarbons are not subject to the same environmental restrictions. Both types of agents are available in handheld extinguishers rated for Classes B and C fires. Larger capacity models are also rated for use on Class A fires.

The agent is discharged as a streaming liquid, which can be directed at the base of a fire. The discharges from these extinguishers have a horizontal stream range of 9–15 feet.

The halogenated agents are nonconductive and leave no residue. These agents are relatively expensive but they perform better than carbon dioxide models in most applications, particularly in windy conditions.

Halon 1211 (bromochlorodifluoromethane) is available in handheld stored-pressure extinguishers with capacities that range from 1 lb, rated 1-B:C, to 22 lb, rated 4-A:80-B:C. Wheeled Halon 1211 models are available with capacities up to 150 lb with a rating of 30-A:160-B:C. The wheeled fire extinguishers use a nitrogen charge from an auxiliary cylinder to expel the agent.

## Dry Powder Extinguishing Agents

**Dry powder extinguishing agents** are intended for fighting Class D fires involving combustible metals. The extinguishing agents and the techniques required to extinguish Class D fires vary greatly, and depend on the specific fuel, the quantity involved, and the physical form of the fuel, such as grindings, shavings, or solid objects. The agent and the application method must be suited to the particular situation.

### Brigade Member Tips

Dry chemical agents and dry powder agents have very different meanings in relation to fire extinguishers. Dry powder fire extinguishers are designed for use in suppressing Class D fires. Dry chemical fire extinguishers are rated for Classes B and C fires or for Classes A, B, and C fires. These two terms must not be confused.

## Brigade Member Safety Tips

Before deciding to use a fire extinguisher, size up the fire to ensure that the extinguisher is adequately sized and has the proper extinguishing agent.

Some dry powder extinguishers are listed for use on specific combustible metal fires. This information and recommended application methods are printed on the container. Consult the manufacturer's recommendations for information about Class D agents and extinguishers, as well as *NFPA's Fire Protection Handbook*.

### Dry Powder Fire Extinguishers

Dry powder fire extinguishers using sodium chloride-based agents are available with 30 lb capacity in either stored-pressure or cylinder/cartridge models. Wheeled models are available with 150 and 350 lb capacities.

Dry powder extinguishers have adjustable nozzles that allow the operator to vary the flow of the extinguishing agent. When the nozzle is fully opened, the handheld models have a range of 6–8 feet. Extension wand applicators are available to direct the discharge from a more distant position.

### Bulk Dry Powder Agents

Bulk dry powder agents are available in 40- and 50-lb pails and 350-lb drums. The same sodium chloride-based agent that is used in portable fire extinguishers can be stored in bulk form and applied by hand.

Another dry powder extinguishing agent for Class D fires, graded granular graphite mixed with compounds containing phosphorus, cannot be expelled from a portable fire extinguisher. This agent must be applied manually from a pail or other container using a shovel or scoop.

## Use of Fire Extinguishers

Fire extinguishers should be simple to operate. An individual with basic training should be able to use fire extinguishers safely and effectively. Every portable extinguisher should be labeled with printed operating instructions.

There are six basic steps in extinguishing a fire with a portable fire extinguisher. They are:

1. Locate the fire extinguisher.
2. Select the proper classification of extinguisher.
3. Transport the extinguisher to the location of the fire.
4. Activate the extinguisher to release the extinguishing agent.
5. Apply the extinguishing agent to the fire for maximum effect.
6. Ensure your personal safety by having an exit route.

Although these steps are not complicated, practice and training are essential for effective fire suppression. Tests have shown that the effective use of Class B portable fire extinguishers depends heavily on user training and expertise. A trained expert can extinguish a fire twice as large as a non-expert, using the same extinguisher.

### Selecting the Proper Fire Extinguisher

Selecting the proper extinguisher requires an understanding of the classification and rating system for fire extinguishers. Knowing the different types of agents, how they work, the ratings of the fire extinguishers, and which extinguisher is appropriate for a particular fire situation is also important.

Brigade members should be able to assess a fire quickly, determine if the fire can be controlled by an extinguisher, and identify the appropriate extinguisher. Using an extinguisher with an insufficient rating may not completely extinguish the fire, which can place the operator in danger of being burned or otherwise injured. If the fire is too large for the extinguisher, you will have to consider other options such as obtaining additional extinguishers or making sure that a charged hose line is ready to provide backup.

Understanding the fire extinguisher rating system and the different types of agents will enable a brigade member who must use an unfamiliar extinguisher to determine if it is suitable for a particular fire situation. A quick look at the label should be all that is needed.

Brigade members should also be able to determine the most appropriate type of fire extinguisher to place in a given area, based on the types of fires that could occur and the hazards that are present. In some cases, one type of extinguisher might be preferred over another. An extinguishing agent such as carbon dioxide or a halogenated agent is better for a fire involving electronic equipment because it leaves no residue. A dry chemical extinguisher is generally more appropriate than a carbon dioxide extinguisher to fight an outdoor fire, because wind will quickly dissipate carbon dioxide. A dry chemical extinguisher would also be the best choice for a fire involving a flammable liquid leaking under pressure from a pipe, but foam is better for a liquid spill on the ground. ▸ Table 7-2 provides information on different types of extinguishers.

### Transporting a Fire Extinguisher

The best method of transporting a handheld portable fire extinguisher depends on the size, weight, and design of the extinguisher. Handheld portable fire extinguishers can weigh as little as 1 lb to as much as 50 lb. The ability to handle the heavier extinguishers depends on an individual operator's strength.

Extinguishers with a fixed nozzle should be carried in the favored or stronger hand. This enables the operator to depress

## Table 7-2 Fire Extinguisher Types

| Extinguishing Agent | Method of Operation | Capacity | Horizontal Range of Stream | Approximate Time of Discharge | Protection Required below 40°F (4°C) | UL or ULC Classifications[a] |
|---|---|---|---|---|---|---|
| Water | Stored-pressure | 6L | 30–40 ft | 40 sec | Yes | 1-A |
| | Stored-pressure or pump | 2 1/2 gal | 30–40 ft | 1 min | Yes | 2-A |
| | Pump | 4 gal | 30–40 ft | 2 min | Yes | 3-A |
| | Pump | 5 gal | 30–40 ft | 2 to 3 min | Yes | 4-A |
| Water (wetting agent) | Stored-pressure | 1 1/2 gal | 20 ft | 30 sec | Yes | 2-A |
| | | 25 gal (wheeled) | 35 ft | 1 1/2 min | Yes | 10-A |
| | | 45 gal (wheeled) | 35 ft | 2 min | Yes | 30-A |
| | | 60 gal (wheeled) | 35 ft | 2 1/2 min | Yes | 40-A |
| Loaded stream | Stored-pressure | 2 1/2 gal | 30–40 ft | 1 min | No | 2 to 3-A:1-B |
| | | 33 gal (wheeled) | 50 ft | 3 min | No | |
| AFFF, FFFP | Stored-pressure | 2 1/2 gal | 20–25 ft | 50 sec | Yes | 3-A:20 to 40-B |
| | Stored-pressure | 6L | 20–25 ft | 50 sec | Yes | 2A:10B |
| | Nitrogen cylinder | 33 gal | 30 ft | 1 min | Yes | 20-A:160-B |
| Carbon dioxide[b] | Self-expelling | 2 1/2–5 lb | 3–8 ft | 8–30 sec | No | 1 to 5-B:C |
| | Self-expelling | 10–15 lb | 3–8 ft | 8–30 sec | No | 2 to 10-B:C |
| | Self-expelling | 20 lb | 3–8 ft | 10–30 sec | No | 10-B:C |
| | Self-expelling | 50–100 lb (wheeled) | 3–10 ft | 10–30 sec | No | 10 to 20-B:C |
| Regular dry chemical (sodium bicarbonate) | Stored-pressure | 1–2 1/2 lb | 5–8 ft | 8–12 sec | No | 2 to 10-B:C |
| | Cartridge or stored-pressure | 2 3/4–5 lb | 5–20 ft | 8–25 sec | No | 5 to 20-B:C |
| | Cartridge or stored-pressure | 6–30 lb | 5–20 ft | 10–25 sec | No | 10 to 160-B:C |
| | Stored-pressure | 50 lb (wheeled) | 20 ft | 35 sec | No | 160-B:C |
| | Nitrogen cylinder or stored-pressure | 75–350 lb (wheeled) | 15–45 ft | 20–105 sec | No | 40 to 320-B:C |
| Purple K dry chemical (potassium bicarbonate) | Cartridge or stored-pressure | 2–5 lb | 5–12 ft | 8–10 sec | No | 5 to 30-B:C |
| | Cartridge or stored-pressure | 5 1/2–10 lb | 5–20 ft | 8–20 sec | No | 10 to 80-B:C |
| | Cartridge or stored-pressure | 16–30 lb | 10–20 ft | 8–25 sec | No | 40 to 120-B:C |
| | Cartridge or stored-pressure | 48–50 lb (wheeled) | 20 ft | 30–35 sec | No | 120 to 160-B:C |
| | Nitrogen cylinder or stored-pressure | 125–315 lb (wheeled) | 15–45 ft | 30–80 sec | No | 80 to 640-B:C |
| Super K dry chemical (potassium chloride) | Cartridge or stored-pressure | 2–5 lb | 5–8 ft | 8–10 sec | No | 5 to 10-B:C |
| | Cartridge or stored-pressure | 5–9 lb | 8–12 ft | 10–15 sec | No | 20 to 40-B:C |
| | Cartridge or stored-pressure | 9 1/2–20 lb | 10–15 ft | 15–20 sec | No | 40 to 60-B:C |
| | Cartridge or stored-pressure | 19 1/2–30 lb | 5–20 ft | 10–25 sec | No | 60 to 80-B:C |
| | Cartridge or stored-pressure | 125–200 lb (wheeled) | 15–45 ft | 30–40 sec | No | 160-B:C |

## Table 7-2—continued

| Extinguishing Agent | Method of Operation | Capacity | Horizontal Range of Stream | Approximate Time of Discharge | Protection Required below 40°F (4°C) | UL or ULC Classifications[a] |
|---|---|---|---|---|---|---|
| Multipurpose/ABC dry chemical (ammonium phosphate) | Stored-pressure | 1–5 lb | 5–12 ft | 8–10 sec | No | 1 to 3-A[c] and 2 to 10-B:C |
| | Stored-pressure or cartridge | $2\frac{1}{2}$–9 lb | 5–12 ft | 8–15 sec | No | 1 to 4-A and 10 to 40-B:C |
| | Stored-pressure or cartridge | 9–17 lb | 5–20 ft | 10–25 sec | No | 2 to 20-A and 10 to 80-B:C |
| | Stored-pressure or cartridge | 17–30 lb | 5–20 ft | 10–25 sec | No | 3 to 20-A and 30 to 120-B:C |
| | Stored-pressure or cartridge | 45–50 lb (wheeled) | 20 ft | 25–35 sec | No | 20 to 30-A and 80 to 160-B:C |
| | Nitrogen cylinder or stored-pressure | 110–315 lb (wheeled) | 15–45 ft | 30–60 sec | No | 20 to 40-A and 60 to 320-B:C |
| Dry chemical (foam compatible) | Cartridge or stored-pressure | $4\frac{3}{4}$–9 lb | 5–20 ft | 8–10 sec | No | 10 to 20-B:C |
| | Cartridge or stored-pressure | 9–27 lb | 5–20 ft | 10–25 sec | No | 20 to 30-B:C |
| | Cartridge or stored-pressure | 18–30 lb | 5–20 ft | 10–25 sec | No | 40 to 60-B:C |
| | Nitrogen cylinder or stored-pressure | 150–350 lb (wheeled) | 15–45 ft | 20–150 sec | No | 80 to 240-B:C |
| Dry chemical (potassium bicarbonate urea based) | Stored-pressure | 5–11 lb | 11–22 ft | 18 sec | No | 40 to 80-B:C |
| | Stored-pressure | 9–23 lb | 15–30 ft | 17–33 sec | No | 60 to 160-B:C |
| | | 175 lb (wheeled) | 70 ft | 62 sec | No | 480-B:C |
| | | 3L | 8–12 ft | 30 sec | No | K |
| Wet chemical | Stored-pressure | 6L ($2\frac{1}{2}$ gal) | 8–12 ft | 35–45 sec | No | 2A:1-B:C:K |
| | | | 8–12 ft | 75–85 sec | No | 2-A:1-B:C:K |
| Halon 1211 (bromochlorodi-fluoromethane) | Stored-pressure | 0.9–2 lb | 6–10 ft | 8–10 sec | No | 1 to 2-B:C |
| | | 2–3 lb | 6–10 ft | 8–10 sec | No | 5-B:C |
| | | $5\frac{1}{2}$–9 lb | 9–15 ft | 8–15 sec | No | 1-A:10-B:C |
| | | 13–22 lb | 14–16 ft | 10–18 sec | No | 2 to 4-A and 20 to 80-B:C |
| | | 50 lb | 35 ft | 30 sec | No | 10-A:120-B:C |
| | | 150 lb (wheeled) | 20–35 ft | 30–44 sec | No | 30-A:160 to 240-B:C |
| Halon 1211/1301 (bromochlorodi-fluoromethane bromotrifluoro-methane) mixtures | Stored-pressure or self-expelling | 0.9–5 lb | 3–12 ft | 8–10 sec | No | 1 to 10-B:C |
| | Stored-pressure | 9–20 lb | 10–18 ft | 10–22 sec | No | 1-A:10-B:C to 4-A:80-B:C |
| Halocarbon type | Stored-pressure | 1.4–150 lb | 6–35 ft | 9–23 sec | No | 1B:C to 10A:80-B:C |

Note: Halon should be used only where its unique properties are deemed necessary.

[a] UL and ULC ratings checked as of July 24, 1987. Readers concerned with subsequent ratings should review the pertinent lists and supplements issued by these laboratories: Underwriters' Laboratories Inc., 333 Pfingsten Road, Northbrook, IL 60062, or Underwriters' Laboratories of Canada, 7 Crouse Road, Scarborough, Ontario, Canada M1R 3A9.

[b] Carbon dioxide extinguishers with metal horns do not carry a C classification.

[c] Some small extinguishers containing ammonium phosphate-based dry chemical do not carry an A classification.

*Source:* National Fire Protection Association. *Fire Protection Handbook* Vol. 2, 19th ed. Quincy, MA, 2003, pp 11-121–11-122.

the trigger and direct the agent easily. Extinguishers that have a hose between the trigger and the nozzle should be carried in the weaker or less-favored hand so that the favored hand can grip and aim the nozzle.

Heavier extinguishers may have to be carried as close as possible to the fire and placed upright on the ground. The operator can depress the trigger with one hand, while holding the nozzle and directing the agent with the other hand.

Follow the steps in ▶ **Skill Drill 7-1** to transport a fire extinguisher:

1. Locate the extinguisher and remove it from the mounting bracket. **(Step 1)**
2. When lifting an extinguisher from a low bracket or off the floor, always lift with your legs and not your back. **(Step 2)**
3. If the extinguisher is very heavy, use both hands to carry it. Grasp the handle with your strong hand and support the bottom of the extinguisher in your weak hand. **(Step 3)**
4. Walk briskly toward the fire. Never run when carrying a fire extinguisher.
5. If the extinguisher has a fixed nozzle, hold the extinguisher at arm's length in your favored or stronger hand.
6. If the extinguisher has a hose with a nozzle, carry the extinguisher in your less-favored hand and grip the nozzle with your other hand. **(Step 4)**

## Basic Steps of Fire Extinguisher Operation

Activating a fire extinguisher to apply the extinguishing agent is a single operation in four steps. The P-A-S-S acronym is a helpful way to remember these steps:

- Pull the safety pin.
- Aim the nozzle at the base of the flames.
- Squeeze the trigger to discharge the agent.
- Sweep the nozzle across the base of the flames.

Practice discharging different types of extinguishers in training situations to build confidence in your ability to use them properly and effectively.

## Ensure Your Personal Safety

When using a fire extinguisher, always approach the fire with an exit behind you. If the fire suddenly expands or the extinguisher fails to control it, you must have a planned escape route. Never let the fire get between you and a safe exit. After suppressing a fire, do not turn your back on it. Always watch and be prepared for a rekindle until the fire has been fully overhauled.

If you must enter an enclosed area where an extinguisher has been discharged, wear full PPE and use SCBA. The atmosphere within the enclosed area will probably contain a mixture of combustion products and extinguishing agents. The oxygen content within the space may be dangerously depleted.

Follow the steps in ▶ **Skill Drill 7-2** to operate a carbon dioxide extinguisher:

1. Size up the fire to determine what is burning, if there is energized electrical equipment involved or nearby, and if there are any other hazards present. Select the proper extinguisher.
2. Be sure the rating of the extinguisher matches the size of the fire. If the fire is too large for a portable fire extinguisher, back away until other suppression methods are available. If possible, close off the area where the fire is located to limit the spread of smoke and fire.
3. Pull the pin on the handle. **(Step 1)**
4. Remove the horn or nozzle from the secured position on the extinguisher and aim in the direction of your approach. **(Step 2)**
5. Give the trigger a quick squeeze to ensure that the extinguisher is operational and the agent discharges properly. **(Step 3)**
6. Approach the fire upwind with an exit to your back. Never let the fire get between you and the exit. Always have a safe exit path in case you have to evacuate. **(Step 4)**
7. Aim the nozzle of the fire extinguisher at the base of the fire and squeeze the trigger. **(Step 5)**
8. Sweep the extinguishing agent from side to side, continuing to aim at the base of the flames. Continue to use the extinguisher until the fire is out or the extinguisher is empty. **(Step 6)**
9. If the extinguisher empties before the fire is completely suppressed, back away to a safe location and wait for assistance.
10. Back away from the fire. Never turn your back on the fire. Be prepared in case it reignites.
11. Overhaul the fire to ensure that it is completely extinguished. Have additional fire extinguishers or a hose line available in case the fire flares up again. **(Step 7)**

Follow the steps of ▶ **Skill Drill 7-3** to attack a Class A fire with a stored-pressure water-type fire extinguisher.

1. Begin to attack the fire upwind from a safe distance to take advantage of the reach of the stream. **(Step 1)**
2. Aim the stream directly at the base of the flames.
3. Sweep the nozzle back and forth, moving closer as the fire goes out. **(Step 2)**
4. After the flames are out, position your finger in front of the nozzle to create a spray and soak the fuel. **(Step 3)**
5. If the fire is deep-seated (burning below the surface, as in a tightly packed trash barrel or a brush pile), break the fuel apart with a long-handled tool. Apply the extinguishing agent to any smoldering, smoking, or glowing surfaces. **(Step 4)**
6. Apply additional water spray to prevent the fire from reigniting. **(Step 5)**

## Brigade Member Safety Tips

Always test a fire extinguisher before using it. Activate the trigger long enough to ensure that the agent discharges properly.

Follow the steps of Skill Drill 7-4 to attack a Class A fire with a multipurpose dry chemical fire extinguisher.

1. Begin to attack the fire upwind from a safe distance, taking advantage of the reach of the dry chemical discharge stream.
2. Aim the stream directly at the base of the flames.
3. Sweep the nozzle back and forth, moving closer as the fire is extinguished.
4. After the flames are out, apply additional agent in short bursts to ensure that all hot surfaces are coated with dry chemical.
5. If the fire is deep-seated (burning below the surface, as in a tightly packed trash barrel or a brush pile), break the fuel apart with a long-handled tool. Apply additional extinguishing agent to any smoldering, smoking, or glowing surfaces.
6. Watch for indications of reignition and apply additional agent in short bursts as required. If necessary, use a fine water spray to soak the fuel.

Follow the steps in ▶ Skill Drill 7-5 to attack a Class B flammable liquid spill or pool fire with a dry chemical extinguisher.

1. Check the pressure gauge to ensure that the extinguisher is properly charged **(Step 1)** and pull the pin on the handle. **(Step 2)**
2. Begin fighting the fire upwind and uphill from a safe distance, as specified on the extinguisher's label. **(Step 3)**
3. Do not aim the initial discharge directly into the liquid from close range. The high-velocity stream of extinguishing agent could splash and spread the burning fuel. **(Step 4)**
4. Discharge the stream at the base of the flame, starting at the near edge of the fire and working toward the back. **(Step 5)**
5. Sweep the nozzle back and forth across the surface of the flammable liquid. **(Step 6)**
6. After the flames are extinguished, watch for indications of reignition and be prepared to apply additional agent. Look for hot objects that could provide a source of reignition. **(Step 7)**

Follow the steps in ▶ Skill Drill 7-6 to attack a Class B flammable liquid spill or pool fire with a stored-pressure fire extinguisher containing AFFF or FFFP extinguishing agent.

1. Begin fighting the fire upwind and uphill from a safe distance, as specified on the extinguisher's label. **(Step 1)**
2. Aim the discharge stream high over the top of the fire. Aiming directly into the liquid could cause the burning fuel to splash and spread the fire. **(Step 2)**
3. Discharge the stream in an arc over the top of the fire, so that the foam drops gently onto the surface of the burning liquid. Lob the foam to create a blanket that floats on the surface. **(Step 3)**
4. Slowly sweep the stream back and forth above the flammable liquid to widen the foam blanket. Use the stream to carefully push the foam blanket toward the back of the liquid surface.
5. If the fire is a spill on the ground, aim the stream at the ground in front of the fire and let the foam bounce onto the front part of the fire. Let the foam blanket flow across the surface of the burning liquid.
6. After the flames are extinguished, continue to apply agent until the surface of the flammable liquid is fully covered by a foam blanket. Look for hot objects that could reignite the liquid and apply the agent directly on them. **(Step 4)**
7. Apply additional agent as needed to maintain the foam blanket and control any reignition. **(Step 5)**

Follow the steps in ▶ Skill Drill 7-7 to apply a wet (Class K) extinguishing agent on a fire in a deep fryer.

1. Begin to apply the agent from a safe distance. **(Step 1)**
2. Do not direct the agent stream directly into the burning liquid. Avoid splashing the burning liquid.
3. Lob the extinguishing agent, expelled as a fine spray, lightly onto the burning surface. It will form a thick foam blanket to smother the fire. **(Step 2)**
4. Continue discharging the extinguisher until the foam blanket has extinguished all flames and secured the entire surface of the oil. **(Step 3)**
5. Do not disturb the foam blanket after the flames die down. The surrounding metal and other materials may still be hot enough to reignite the oil if the foam blanket is broken.
6. Be prepared for a possible flare-up until the cooking oil and the surrounding area have adequately cooled. If reignition occurs, repeat these steps. **(Step 4)**

Follow the steps in Skill Drill 7-8 to apply a halogenated extinguishing agent to a fire in an electrical equipment room.

1. If possible, turn off or disconnect the electrical power from a remote location before attacking the fire.
2. Stand back at least 8 feet (2.4 m) when you begin to discharge the agent.
3. Direct the stream at the base of the flames, sweeping the agent slowly from side to side.

# Voices of Experience

**"Many of the employees didn't know how to use an extinguisher or which one to use."**

On a brisk fall day in my previous employer's facility, the plant shift supervisor was informed by radio of a fire in a warehouse on the north end of the facility. The plant fire alarms had not been pulled, and the building did not have an automatic system. The warehouse was an older building made of lightweight metal bowstring construction with corrugated metal siding. It was used as a carpentry shop and for lumber storage.

The supervisor called the dispatcher and had fire rescue activated. Upon my arrival on scene, he notified me that smoke was showing and that several employees were operating fire extinguishers inside and outside the building, including into open windows. I assumed command and immediately sounded a general alarm, as fire was issuing out of several windows and some whirlybird vents. Once all employees were accounted for, the fire crew began our attack.

We laid several LD hose lines and began an aggressive attack with 2.5-inch hoses and master streams to do a quick knockdown. Because the roof had self-ventilated, we stayed in a defensive posture for quite some time. Once the major body of fire was knocked down, we began an interior attack and secondary search. Only when the loss was stopped and we determined the fire was out did we begin surveying the scene. We soon discovered that the entire back shop—which occupied half of the building—was a total loss, including a forklift that had been inside the facility. But what amazed me the most was that at least 15 used fire extinguishers were piled up by the front door.

When I spoke with the foreman, he told me that workers in the shop had tried to fight the fire with ABC extinguishers for at least 10 minutes before they called for help. The water extinguishers were still in their brackets on a post. Each person thought the others had called for help. Many of the employees didn't know how to use an extinguisher or which one to use.

This experience provides some valuable lessons:

1. Call for help early, and any time you use an extinguisher. If the fire is out, fire fighters can always go home after checking for extension.
2. Don't bring a knife to a gun fight: If one extinguisher isn't effective, and two don't solve the problem, then the third extinguisher probably won't help either.
3. What the employees do in the first five minutes of a fire determines what fire fighters will be doing for the next five hours. Your brigade will benefit if you implement a thorough fire extinguisher training program for all employees. You never know—they may be able to use the training at home as well.

Ronald K. Sommers
Chevron Phillips Chemical Company
Pasadena Plastics Complex Fire and Rescue
Pasadena, Texas

4. Continue to apply the extinguishing agent for a short time after the flames have gone out. This allows the fuel to cool, preventing reignition.
5. Watch the fire area for reignition. Repeat the agent application, if necessary.

Follow the steps in ▶ Skill Drill 7-9 to apply a dry powder extinguishing agent to an incipient-stage fire involving combustible metal filings or shavings (Class D fire).

1. If the fire is very hot, begin to discharge the agent with the nozzle fully opened at the maximum range of 6–7 feet (1.7–2.4 m) away from the fire. Direct the agent so that it falls onto the top of the burning material. **(Step 1)**
2. As the fire comes under control, close the nozzle valve to produce a soft, heavy flow. Move in closer and cover the fire area completely. **(Step 2)**
3. If you are using a scoop or a shovel instead of an extinguisher, apply a thick, even coat of extinguishing agent over the entire fire to smother it.
4. If the burning metal is on a combustible surface, first cover the fire with the dry powder extinguishing agent. Then spread a 1–2 inches (2.5–5.0 cm) layer of dry powder extinguishing agent nearby. Carefully shovel the burning metal onto this layer of powder and top with more extinguishing agent as needed.
5. If hot spots develop, apply more extinguishing agent to cover them.
6. Allow the remains of the fire to cool undisturbed.
7. Do not disturb the blanket of extinguishing agent until the fire has cooled completely. **(Step 3)**
   NOTE: Do not use water or any other agent to cool the fuel or the area around it! Combustible metals react violently with water and many other agents.

## The Care of Fire Extinguishers: Inspection, Maintenance, Recharging, and Hydrostatic Testing

Fire extinguishers must be regularly inspected and properly maintained so they are available for use in an emergency. Records must be kept to ensure that the required inspections and maintenance have been performed on schedule. The individuals assigned to perform these functions must be properly trained and must always follow the manufacturer's recommendations for inspecting, maintaining, recharging, and testing the equipment.

### Inspection

According to *NFPA 10,* an inspection is a monthly "quick check" to verify that a fire extinguisher is available and ready for immediate use. Fire extinguishers on fire apparatus should be inspected as part of the regular equipment check as mandated by your standard operating procedures (SOPs). Monthly inspections reinforce a brigade member's familiarity with the equipment and its location. The brigade member charged with inspecting the extinguishers should:

- Ensure that tamper seals are intact.
- Examine all parts for signs of physical damage, corrosion, or leakage.
- Check the pressure gauge to be sure it is in the operable range.
- Ensure that the extinguisher is properly identified by type and rating.
- Check the nozzle for damage or obstruction by foreign objects.

If an inspection reveals any problems, the extinguisher should be removed from service until the required maintenance procedures are performed. Spare extinguishers should be used until the problem is corrected.

The pressure gauge on a stored-pressure extinguisher indicates whether the pressure is sufficient to expel the entire agent. Some stored-pressure extinguishers use compressed air while others use compressed nitrogen. The weight of the extinguisher and the presence of an intact tamper seal should indicate that it is full of extinguishing agent.

Cartridge-type extinguishers contain a predetermined quantity of a pressurizing gas that expels the agent. The gas is released only when the cartridge is punctured. A properly charged extinguisher will be full of extinguishing agent and will have a cartridge that has not been punctured.

Self-expelling agents, such as carbon dioxide and some halogenated agents, do not require a separate gas cartridge. The only way to determine if a carbon dioxide extinguisher is properly charged is to weigh it. The proper weight should be indicated on the extinguisher label.

### Maintenance

The maintenance requirements and intervals for various types of fire extinguishers are outlined in NFPA 10. Maintenance includes an internal inspection as well as any repairs that may be required. Maintenance procedures must be performed periodically, depending on the type of extinguisher. An inspection may also reveal the need to perform maintenance procedures.

Maintenance procedures must always be performed by a qualified person. Some procedures can be performed only at a properly licensed facility. The specific qualifications and training requirements are determined by the manufacturer and the jurisdictional authority. Untrained personnel should never be allowed to perform fire extinguisher maintenance.

7-1 Skill Drill

## Transporting a Fire Extinguisher

Locate the extinguisher and remove it from its mounting bracket.

When lifting an extinguisher, always lift with your legs and not your back.

Grasp the handle with the strong hand and support the bottom of the extinguisher in the weak hand.

If the extinguisher has a fixed nozzle, hold the extinguisher at arm's length in your stronger hand. If the extinguisher has a hose with a nozzle, carry the extinguisher in your weak hand and use the other hand to grip the nozzle.

7-2 **Skill Drill**

## Operating a Carbon Dioxide Extinguisher

Pull the pin on the handle.

Remove the horn or nozzle from the secured position on the extinguisher and aim in the direction of your approach.

Give the trigger a quick squeeze to ensure that the extinguisher is operational and the agent discharges properly.

Approach the fire upwind with an exit to your back. Never let the fire get between you and the exit.

Aim the nozzle of the fire extinguisher at the base of the fire and squeeze the trigger.

Sweep the extinguishing agent from side to side, continuing to aim at the base of the flames. Continue to use the extinguisher until the fire is out or the extinguisher is empty.

Back away from the fire. Overhaul the fire to ensure that it is completely extinguished.

Skill Drill 7-3

## Attacking a Class A Fire with a Stored-Pressure Water-Type Fire Extinguisher

1. Begin to attack the fire upwind from a safe distance.

2. Aim the stream directly at the base of the flames. Sweep the nozzle back and forth, moving closer as the fire goes out.

3. After the flames are out, position your finger in front of the nozzle to create a spray and soak the fuel.

4. Break apart the fuel with a tool and apply the extinguishing agent to any smoldering, smoking, or glowing surfaces.

5. Apply additional water spray to prevent the fire from reigniting.

7-5 Skill Drill

# Attacking a Class B Flammable Liquid Fire with a Dry Chemical Fire Extinguisher

1. Check the pressure gauge to ensure that the extinguisher is properly charged.

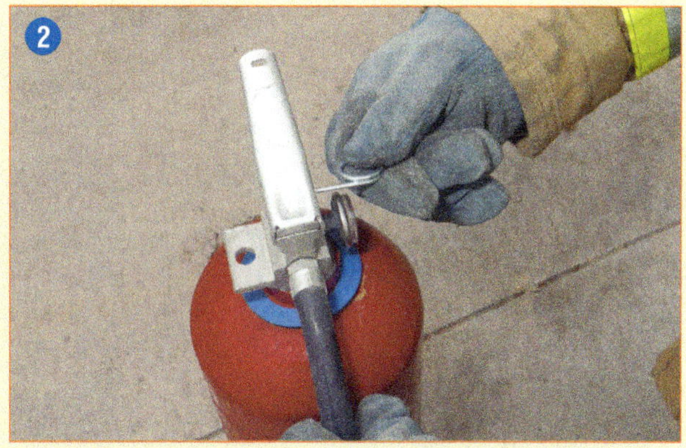

2. Pull the pin on the handle.

3. Begin fighting the fire upwind and uphill from a safe distance.

4. Do not aim the initial discharge into the liquid at close range.

5. Discharge the stream at the base of the flame, starting at the near edge of the fire and working toward the back.

6. Sweep the nozzle back and forth across the surface of the flammable liquid.

7. Look for hot or smoldering objects that could reignite the liquid.

7-6 Skill Drill

# Attacking a Class B Flammable Liquid Fire with a Stored-Pressure Foam Fire Extinguisher (AFFF or FFFP)

Begin fighting the fire upwind and uphill from a safe distance.

Do not discharge stream directly into the liquid.

Lob the foam in an arc over the fire to create a blanket that floats on the surface.

Slowly sweep the stream back and forth above the flaming surface to widen the foam blanket. Let the foam blanket flow across the surface of the burning liquid. After the flames are extinguished, continue to apply agent until the liquid surface is fully covered by the foam blanket.

Apply additional agent as needed to maintain the foam blanket and control any reignition.

7-7 Skill Drill

# Use of Wet Chemical Fire Extinguishers

Begin to apply the agent from a safe distance.

Do not direct the agent stream directly into the burning liquid. Lob the extinguishing agent, expelled as a fine spray, lightly onto the burning surface to create a foam blanket.

Continue to discharge the extinguisher until the foam blanket has extinguished all flames.

Do not disturb the foam blanket even after all flames have died down. If reignition occurs, repeat these steps.

*Courtesy of Ansul Incorporated training materials.

7-9 Skill Drill

# Use of Dry Powder Fire Extinguishing Agents

Open the nozzle completely and direct the agent so that it falls onto the top of the burning material.

Close the nozzle valve to produce a soft, heavy flow and move closer to cover the fire area completely.

Do not disturb the blanket of extinguishing agent until the fire has cooled completely.

Common indications that an extinguisher needs maintenance include:

- The pressure gauge reading is outside the normal range.
- The inspection tag is out-of-date.
- The tamper seal is broken, especially in extinguishers with no pressure gauge.
- There is any indication that the extinguisher is not full of extinguishing agent.
- The hose and/or nozzle assembly is obstructed.
- There are signs of physical damage, corrosion, or rust.
- Signs of leakage around the discharge valve or nozzle assembly can be seen.

## Recharging

A fire extinguisher must be recharged after each use, even if it was not completely discharged. The only exceptions are nonrechargeable extinguishers, which should be replaced after any use. All performance standards assume that an extinguisher will be fully charged when it is ready for use. Immediately after use, an extinguisher should be taken out of service until it has been properly recharged. This also applies to any extinguisher that leaks or has a pressure gauge reading above or below the proper operating range. Extinguishers must be recharged by qualified or licensed personnel.

When an extinguisher is recharged, the extinguishing agent is refilled and the system that is provided to expel the agent is fully charged. Both the quantity of the agent and the pressurization must be verified. A tamper seal is installed after recharging. The tamper seal provides assurance that the extinguisher has not been charged.

Typical 2.5 gallon stored-pressure water extinguishers can be recharged by brigade members using water and a source of compressed air. Before the extinguisher is refilled, all remaining stored pressure must be discharged so that the valve assembly can be safely removed. Water is added up to the water-level indicator and the valve assembly is replaced. Compressed air is then introduced to raise the pressure to the level indicated on the gauge.

## Hydrostatic Testing

Fire extinguishers are pressure vessels, designed to hold a steady internal pressure. The ability of an extinguisher to withstand this internal pressure is assured by periodic hydrostatic testing. The hydrostatic testing requirements for fire extinguishers are established by the U.S. Department of Transportation and can be found in *NFPA 10* ▶ Table 7-3. The test is conducted in a special test facility and involves filling the extinguisher with water and applying a specified pressure.

Each extinguisher has an assigned maximum interval between hydrostatic tests, usually 5 or 12 years, depending on the construction material and type ▶ Table 7-4. The date of the most recent hydrostatic test must be indicated on the outside of the extinguisher. An extinguisher may not be refilled if the date of the most recent hydrostatic test is not within the prescribed limit. Any extinguisher that is out-of-date should be removed from service and sent to the appropriate maintenance facility for hydrostatic testing.

**Table 7-3 Hydrostatic Test Pressure Requirements**

| | |
|---|---|
| Carbon dioxide extinguishers<br>Carbon dioxide and nitrogen cylinders (used with wheeled extinguishers) | $^5/_3$ service pressure stamped on cylinder |
| Carbon dioxide extinguishers with cylinder specification ICC3 | 3,000 psi (20 685 kPa) |
| All stored-pressure and bromochlorodifluoromethane (1211) | Factory test pressure not to exceed 3 times the service pressure |
| Carbon dioxide hose assemblies | 1,250 psi (8,619 kPa) |
| Dry chemical and dry powder hose assemblies | 300 psi (2,068 kPa) or at service pressure, whichever is higher |

Note: The factory test pressure is the pressure at which the shell was tested at time of manufacture. This pressure is shown on the nameplate.

The service pressure is the normal operating pressure as indicated on the gauge and nameplate.

**Table 7-4 Hydrostatic Test Interval for Extinguishers**

| Extinguisher Type | Test Interval (Years) |
|---|---|
| Stored-pressure water-loaded stream and/or antifreeze | 5 |
| Wetting agent | 5 |
| AFFF (aqueous film-forming foam) | 5 |
| FFFP (film-forming fluoroprotein foam) | 5 |
| Dry chemical with stainless-steel shells | 5 |
| Carbon dioxide | 5 |
| Wet chemical | 5 |
| Dry chemical, stored-pressure, with mild steel shells, brazed brass shells, or aluminum shells | 12 |
| Dry chemical, cartridge- or cylinder-operated, with mild steel shells | 12 |
| Halogenated agents | 12 |
| Dry powder, stored-pressure, cartridge- or cylinder-operated, with mild steel shells | 12 |

# Wrap-Up

## Ready for Review

- Portable fire extinguishers are effective for fighting fires only if the extinguisher is suitable for the class of fire.
- There are five classes of fire:
  - Class A: ordinary combustibles such as wood, paper, and cloth.
  - Class B: flammable or combustible liquids.
  - Class C: energized electrical equipment.
  - Class D: combustible metals such as magnesium.
  - Class K: combustible cooking oils and fats.
- Extinguishers are rated by independent testing laboratories according to the class or classes of fire the units can extinguish.
- Portable fire extinguishers that are rated for use on Class A and Class B fires are also rated numerically according to the size of the fire they can extinguish.
- The environment where a fire is likely to take place is an important factor in choosing the proper extinguisher.
- Environments are classified in three major hazard groups: light, ordinary, and extra.
- The basic steps of fire extinguisher operation are: locate, select, transport, activate, and apply.
- The common types of fire extinguishers are: water, dry chemical, carbon dioxide, foam, halogenated agents, dry powder, and wet chemical.
- Selecting the best fire extinguisher depends on the type of material burning, the size of the fire, the location of the fire, the effectiveness of the extinguisher, the ambient temperature, and wind conditions.
- Extinguishers differ in how their extinguishing agents are stored and expelled.
- Fire extinguishers must be regularly inspected, maintained, recharged, and hydrostatically tested.

## Hot Terms

**Ammonium phosphate** An extinguishing agent used in dry chemical fire extinguishers that can be used on Classes A, B, and C fires.

**Aqueous film-forming foam (AFFF)** A water-based extinguishing agent used on Class B fires that forms a foam layer over the liquid and stops the production of flammable vapors.

**Carbon dioxide ($CO_2$) fire extinguisher** A fire extinguisher that uses carbon dioxide gas as the extinguishing agent.

**Cartridge-operated extinguisher** A fire extinguisher that has the expellant gas in a cartridge separate from the extinguishing agent storage shell. The storage shell is pressurized by a mechanical action that releases the expellant gas.

**Class A fires** Fires involving ordinary combustible materials such as wood, cloth, paper, rubber, and many plastics.

**Class B fires** Fires involving flammable and combustible liquids, oils, greases, tars, oil-based paints, lacquers, and flammable gases.

**Class C fires** Fires that involve energized electrical equipment where the electrical conductivity of the extinguishing media is of importance.

**Class D fires** Fires involving combustible metals such as magnesium, titanium, zirconium, sodium, and potassium.

**Class K fires** Fires involving combustible cooking media such as vegetable oils, animal oils, and fats.

**Clean agent** Gaseous fire extinguishing agent that does not leave a residue when it evaporates. Also known as halogenated agents.

**Cylinder** The body of the fire extinguisher where the extinguishing agent is stored.

**Dry chemical fire extinguisher** An extinguisher that uses a mixture of finely divided solid particles to extinguish fires. The agent is usually sodium bicarbonate-, potassium bicarbonate-, or ammonium phosphate-based, with additives to provide resistance to packing and moisture absorption and to promote proper flow characteristics.

**Dry powder extinguishing agent** Extinguishing agent used in putting out Class D fires. The common dry powder extinguishing agents include sodium chloride and graphite-based powders.

**Dry powder fire extinguisher** A fire extinguisher that uses an extinguishing agent in powder or granular form, designed to extinguish Class D combustible metal fires by crusting, smothering, or heat-transferring means.

**Extinguishing agent** Material used to stop the combustion process; extinguishing agents may include liquids, gases, dry chemical compounds, and dry powder compounds.

**Extra hazard locations** Occupancies where the total amounts of Class A combustibles and Class B flammables are greater than expected in occupancies classed as ordinary (moderate) hazard.

**Film-forming fluoroprotein (FFFP)** A water-based extinguishing agent used on Class B fires that forms a foam layer over the liquid and stops the production of flammable vapors.

**Fire load** The amount of combustibles in a fire area or on a floor in buildings and structures, including either contents or building parts, or both.

**Halogenated-agent extinguisher** An extinguisher that uses halogenated extinguishing agents.

**Halogenated extinguishing agent** A liquefied gas extinguishing agent that puts out fires by interrupting the chemical chain reaction.

**Halon 1211** Bromochlorodifluoromethane ($CBrClF_2$), a halogenated agent that is effective on Classes A, B, and C fires.

**Handle** The grip used for holding and carrying a portable fire extinguisher.

**Horn** The tapered discharge nozzle of a carbon dioxide-type fire extinguisher.

**Hydrostatic testing** Periodic testing of an extinguisher to verify it has sufficient strength to withstand internal pressures.

**Ignition point** The minimum temperature at which a substance will burn.

**Incipient** The initial stage of a fire.

**Light hazard locations** Occupancies where the total amount of combustible materials is less than expected in an ordinary hazard location.

**Loaded-stream extinguisher** A water-based fire extinguisher that uses an alkali metal salt as a freezing point depressant.

**Lob** A method of discharging extinguishing agent in an arc to avoid splashing or spreading the burning fuel.

**Locking mechanism** A device that locks an extinguisher's trigger to prevent accidental discharge.

**Multipurpose dry chemical extinguishers** Extinguishers rated to fight Class A, B, and C fires.

**Nozzle** The discharge orifice of a portable fire extinguisher.

**Ordinary hazard locations** Occupancies that contain more Class A and Class B materials than are found in light hazard locations.

**P-A-S-S** Acronym used for operating a portable fire extinguisher: Pull pin, Aim nozzle, Squeeze trigger, Sweep the nozzle across burning fuel.

**Polar solvent** A water-soluble flammable liquid such as alcohol, acetone, ester, and ketone.

**Pressure indicator** A gauge on a pressurized portable fire extinguisher that indicates the internal pressure of the expellant.

**Pump tank extinguishers** Nonpressurized, manually operated water extinguishers, usually have a nozzle at the end of a short hose.

**Pump tank water-type extinguisher** A nonpressurized portable water fire extinguisher. Discharge pressure is provided by a hand-operated double-acting piston pump.

**Rapid oxidation** Chemical process that occurs when a fuel is combined with oxygen, resulting in the formation of ash or other waste products and the release of energy as heat and light.

**Saponification** The process of converting the fatty acids in cooking oils or fats to soap or foam.

**Self-expelling** A fire extinguisher in which the agents have sufficient vapor pressure at normal operating temperatures to expel themselves.

**Stored-pressure extinguisher** A fire extinguisher in which both the extinguishing agent and the expellant gas are kept in a single container; generally equipped with a pressure indicator or gauge.

**Stored-pressure water-type extinguisher** A fire extinguisher in which water or a water-based extinguishing agent is stored under pressure.

**Tamper seal** A retaining device that breaks when the locking mechanism is released.

**Trigger** The button or lever used to discharge the agent from a portable fire extinguisher.

# Wrap-Up

**Underwriters Laboratories Inc. (UL)** The U.S. organization that tests and certifies that fire extinguishers (among many other products) meet established standards.

**Wet chemical extinguisher** A fire extinguisher for use on Class K fires that contains wet chemical extinguishing agents.

**Wet chemical extinguishing agent** An extinguishing agent for Class K fires; commonly uses solutions of water and potassium acetate, potassium carbonate, potassium citrate, or any combination thereof.

**Wetting-agent water-type extinguisher** An extinguisher that expels water combined with a chemical or chemicals to reduce its surface tension.

**Wheeled fire extinguisher** A portable fire extinguisher equipped with a carriage and wheels intended to be transported to the fire by one person.

# Brigade Member in Action

**Your brigade is dispatched to respond to a report of a small fire located on a scaffold in a construction area of the facility. Upon your arrival, the fire brigade leader informs you that the wooden scaffolding decking boards have been ignited by sparks from a welding job in the area. The brigade leader says that it appears as if there is a very small fire contained to the scaffold area but some active piping systems are also located in the area. He requests that you and your partner obtain an extinguisher from the first-due pumping apparatus and extinguish the fire. The pumping apparatus carries a 2.5-gallon water extinguisher, a 30-pound ABC dry chemical extinguisher, a 20-pound carbon dioxide extinguisher, and a 30-pound class D extinguisher.**

1. Based on the initial information provided by the fire brigade leader, what would your first choice of fire extinguisher from the pumping apparatus be?
   - **A.** The 30-pound class D extinguisher.
   - **B.** The 20-pound carbon dioxide extinguisher.
   - **C.** The 2.5-gallon water extinguisher.
   - **D.** The 30-pound ABC dry chemical extinguisher.

2. The fire brigade leader reminds you to use the P-A-S-S method to place the extinguisher into operation. What does P-A-S-S stand for?
   - **A.** Pull the pin, Aim the nozzle, Squeeze the trigger, and Sweep the nozzle.
   - **B.** Pull the pin, Activate the agent, Sweep the nozzle, and Suffocate the fire.
   - **C.** Press the button, Aim the nozzle, Sweep the nozzle, and Suffocate the fire.
   - **D.** Press the plunger, Aerate the agent, Squeeze the trigger, and Sweep the nozzle.

3. What is the primary advantage of using a stored-pressure water extinguisher rather than a ABC dry chemical extinguisher for a typical class A fire?
   - **A.** The water extinguisher lowers the temperature and thus prevents rekindling.
   - **B.** The water extinguisher has a longer effective range (35–40 feet) than the dry chemical extinguisher (5–30 feet).
   - **C.** Both A and B.
   - **D.** There is no real advantage to using a water extinguisher over an ABC dry chemical extinguisher.

4. Which of the following is an incorrect statement regarding the general use of extinguishers?
   - **A.** Always ensure that the extinguisher is operating correctly prior to attacking the fire.
   - **B.** Aim the stream at the base of the fire.
   - **C.** Attack the fire from uphill and upwind whenever possible.
   - **D.** If the extinguisher empties prior to full extinguishment of the fire, turn and immediately run from the area.

5. What are the requirements for recharging the extinguisher after you are finished using it?
   - **A.** The extinguisher must be recharged only if it is completely empty.
   - **B.** The extinguisher must be recharged after every use, even if it was not completely discharged.
   - **C.** The extinguisher can be recharged by fire brigade members only after it is verified to be empty by the fire brigade leader.
   - **D.** Extinguishers cannot be recharged; they must be replaced with new extinguishers.

# Brigade Member Tools and Equipment

# Chapter 8

## NFPA 1081 Standard

### Incipient facility Fire Brigade Member

**4.1.2.4*** Return equipment to service, given an assignment, policies, and procedures, so that the equipment is inspected, damage is noted, the equipment is clean, and the equipment is placed in a ready state for service or is reported otherwise.

**(A)** *Requisite Knowledge.* Types of cleaning methods for various equipment, correct use of cleaning materials, and manufacturer's or facility guidelines for returning equipment to service.

**(B)** *Requisite Skills.* The ability to clean, inspect, and maintain equipment and to complete recording and reporting procedures.

### Advanced Exterior Facility Fire Brigade Member

**5.3.10*** Utilize tools and equipment assigned to the facility fire brigade, given an assignment and specific tools, so that tools are selected and correctly used under adverse conditions in accordance with manufacturer's recommendations and the policies and procedures of the facility fire brigade.

**(A)** *Requisite Knowledge.* Available tools and equipment, their storage locations, and their correct use in accordance with recognized practices, and selection of tools and equipment given different conditions.

**(B)** *Requisite Skills.* The ability to select and use the correct tools and equipment for various tasks, follow guidelines, and restore tools and equipment to service after use.

### Interior Structural Facility Fire Brigade Member

**7.3.7*** Utilize tools and equipment assigned to the facility fire brigade, given an assignment and specific tools, so that tools are selected and correctly used under adverse conditions in accordance with manufacturer's recommendations and the policies and procedures of the facility fire brigade.

**(A)** *Requisite Knowledge.* Available tools and equipment, their storage locations, and their correct use in accordance with recognized practices; and selection of tools and equipment given different conditions.

**(B)** *Requisite Skills.* The ability to select and use the correct tools and equipment for various tasks, follow guidelines, and restore tools and equipment to service after use.

## Knowledge Objectives

After completing this chapter, you will be able to:

- Describe the general purposes of tools and equipment.
- Describe the safety considerations for the use of tools and equipment.
- Describe why it is important to use tools and equipment effectively.
- Describe why it is important for you to know where tools are stored.
- List and describe tools and equipment that are used for rotating.
- List and describe tools and equipment that are used for pushing or pulling.
- List and describe tools and equipment that are used for prying or spreading.
- List and describe tools and equipment that are used for striking.
- List and describe tools and equipment that are used for cutting.
- Describe the tools used in response and scene size-up activities.
- Describe the tools used in a forcible entry.
- Describe the tools used during an interior attack.
- Describe the tools used in search-and-rescue operations.
- Describe ventilation tools.
- Describe the hand tools needed during an overhaul assignment.
- Describe the importance of properly maintaining tools and equipment.
- Describe how to clean and inspect hand tools.
- Describe how to clean and maintain power tools.

## Skills Objectives

After completing this chapter, you will be able to perform the following skills:

- Properly select and use the correct tools and equipment.
- Properly inspect and clean tools.
- Follow guidelines and restore tools and equipment to service after use.
- Complete recording and reporting procedures.

# You Are the Brigade Member

**Your brigade has just been told the company has approved the purchase of a new pumping apparatus for the facility. This new apparatus will be the primary response vehicle for all emergencies that occur within the facility. The fire brigade leader has assigned you and four other brigade members to a committee to identify the types of tools and equipment that will be carried on the new vehicle. You have been asked to identify the tools necessary to conduct routine forcible entry, interior firefighting, rapid intervention, ventilation, and salvage and overhaul operations. You have limited space on the apparatus and are told to pick your equipment wisely.**

1. ***With what type of tools would you choose to equip the new apparatus?***
2. ***What would be the primary use for each tool that you have chosen?***
3. ***What would be your justification for the purchase of the specific tools chosen?***
4. ***What type of maintenance requirements would be needed for the tools?***

## Introduction

Brigade members use tools and equipment to perform a wide range of activities. A brigade member must know how to use tools effectively, efficiently, and safely, even when it is dark or visibility is limited. This chapter provides an overview of the general functions of the most commonly used tools and equipment and discusses how they are used during fire suppression and rescue operations. The same tools may be used in different ways during each phase of fire suppression or rescue operations.

## General Considerations

Tools and equipment are used in almost all fire suppression and rescue operations. As you progress through your training, you will learn how to use and operate the different types of tools and equipment carried by your brigade.

Hand tools are used to extend or multiply the actions of your body and increase your effectiveness in performing specific functions. Most hand tools operate using simple machine principles. A pike pole extends your reach, allows you to penetrate through a ceiling, and enables you to apply force to pull down ceiling material. An axe multiplies the cutting force you can exert on a given area. Power tools and equipment use an external source of power, such as an electric motor or an internal combustion engine, and are faster and more efficient than hand tools.

### Safety

Safety is a prime consideration when using tools and equipment. You need to operate the equipment so that you, fellow brigade members, victims, and bystanders are not accidentally injured. Personal safety also requires the use of proper personal protective equipment (PPE) that includes:

- Approved helmet
- Firefighting hood
- Eye protection
- Face shield
- Approved firefighting gloves
- Turnout coat
- Bunker pants
- Boots
- Self-contained breathing apparatus (SCBA)
- Personal alert safety system

### Conditions of Use/Operating Conditions

The best way to learn how to use tools and equipment properly is under optimal conditions of visibility and safety. In the beginning, you must be able to see what you are doing and practice without endangering yourself and others. As you become more proficient, you should practice using tools and equipment under more realistic working conditions. Eventually, you must be able to use tools and equipment safely and effectively when darkness or smoke decreases visibility. You must be able to work safely in hazardous areas, where you are surrounded by noise and other activities, while wearing all of your protective clothing and using your SCBA. Many brigades require brigade members to practice certain skills and evolutions in total darkness or with their face masks covered to simulate the darkness of actual fires.

### Effective Use

Effective, efficient use of tools and equipment means using the least amount of energy to accomplish the task. Being effective means you achieve the desired goal. Being efficient means that you produce the desired effect without wasting time or energy. When assigned a task on the fireground, your objective is to complete that task safely and quickly. If you waste energy by working inefficiently, you will not be able to perform additional tasks. However, if you know which tools and equipment are needed for each phase of firefighting, you will be able to achieve the desired objective quickly and have the energy needed to complete the remaining tasks.

New brigade members are often surprised by the strength and energy required to perform many tasks. An aggressive, continuous program of physical fitness will enable you to maintain your body in the optimal state of readiness. It does little good to practice for hours if you are not in prime condition.

As your training continues, you will learn which tools and equipment are used during different phases of fireground operations. For example, the tools needed for forcible entry are different from the tools usually needed for overhaul. Knowing which tools are needed for the work that must be done will help you prepare for the different tasks that unfold on the scene of a fire. The specific steps for properly using tools and equipment will be presented in this chapter, throughout the textbook, and by your fire brigade.

Most fire brigades have standard operating procedures or guidelines that specify the tools and equipment needed in various situations. As a brigade member, you must know where every tool and piece of equipment is located. Knowing how to use a piece of equipment does you no good if you cannot find it quickly. Your brigade leader is responsible for telling you which tools to bring along for different situations.

Some brigade members carry a selection of small tools and equipment in the pockets of their turnout coats or bunker pants. Check to see if your brigade requires you to carry certain tools and equipment at all times. Ask senior brigade members for recommendations about what tools and equipment you should carry.

## Functions

Response apparatus carries a number of tools and different types of equipment. Often, the easiest way to learn and remember these tools is to group them by the function each performs (► Tables 8-1 and 8-2). Most of the tools used by fire brigades fit into the following functional categories:

- Rotating (assembly or disassembly)
- Pushing or pulling
- Prying or spreading
- Striking
- Cutting
- Multiple function

### Rotating Tools

Rotating tools apply a rotational force to make something turn. The most common rotating tools are screwdrivers, wrenches, and pliers, which are used to assemble (fit together) or disassemble (take apart) parts that are connected with threaded fasteners (► Table 8-3).

**Table 8-1 Tool Functions**

Rotate, assemble, or disassemble (► Figure 8-1).

Pry or spread (► Figure 8-2).

Push or pull (► Figure 8-3).

Strike (► Figure 8-4).

Cut (► Figure 8-5).

A combination of functions (► Figure 8-6).

Assembling and disassembling are basic mechanical skills that are routinely employed by brigade members to solve problems. Most fire response vehicles carry a tool kit with a selection of screwdrivers with different heads, open-end wrenches, box wrenches, socket wrenches, pliers, adjustable wrenches, and pipe wrenches.

There are various sizes and types of screw heads, including slotted head, Phillips head, Roberts head, and others. A screwdriver with interchangeable heads is sometimes more useful than a selection of different screwdrivers. Pliers and wrenches also come in various shapes and sizes for different applications.

A spanner wrench is used to tighten or loosen fire hose couplings (one set or pair of connection devices that allow a fire hose to be interconnected with additional lengths of hose). If the couplings are not too tight, they can be loosened by hand, without the use of a spanner wrench. A hydrant wrench is used to open or close a hydrant by rotating the valve stem, or to remove the caps from the hydrant outlets.

Common examples of rotating tools include (► Figure 8-7):

- Box-end wrench: A hand tool with a closed end that is used to tighten or loosen bolts; some styles of box-end wrenches have ratchets for easier use.
- Gripping pliers: A hand tool with a pincer-like working end that can also be used to bend wire or hold smaller objects.
- Hydrant wrench: A hand tool used to operate the valves on a hydrant; some are plain wrenches and others have a ratchet (a mechanism that will only rotate one way) feature (► Figure 8-8).

**Brigade Member Safety Tips**

Learn to operate tools and equipment in areas without hazards and with adequate light. Gradually, increase the amount of PPE worn and decrease the light to levels typically found on the fireground. Eventually, you should be so familiar with your tools and equipment that you can use them under adverse conditions while wearing PPE.

- Open-end wrench: A hand tool with an open end that is used to tighten or loosen bolts.
- Pipe wrench: A wrench with one fixed grip and one movable grip that can be adjusted to fit securely around pipes and other tubular objects.
- Screwdriver: A tool that is used to turn screws.
- Socket wrench: A wrench that fits over a nut or bolt and uses the action of an attached handle to tighten or loosen the nut or bolt.
- Spanner wrench: A special wrench used to tighten or loosen hose couplings; spanners come in several sizes.

Figure 8-1 Rotate, assemble, or disassemble.

Figure 8-2 Pry or spread.

Figure 8-3 Push or pull.

Figure 8-4 Strike.

Figure 8-5 Cut.

Figure 8-6 A combination of functions.

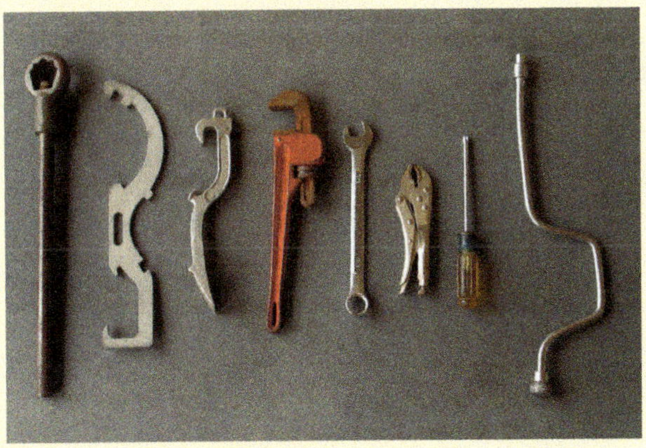

Figure 8-7 Hydrant wrench, spanner wrench, pipe wrench, open-end wrench, box-end wrench, gripping pliers, screwdriver, socket wrench.

## Table 8-2 Tool Functions

| Function | Tool |
|---|---|
| Rotate | Box-end wrench<br>Gripping pliers<br>Hydrant wrench<br>Open-end wrench<br>Pipe wrench<br>Screwdriver<br>Socket wrench<br>Spanner wrench |
| Push or pull | Ceiling hook<br>Clemens hook<br>Drywall hook<br>K tool<br>Multipurpose hook<br>Pike pole<br>Plaster hook<br>Roofman's hook<br>San Francisco hook |
| Pry or spread | Claw bar<br>Crowbar<br>Flat bar<br>Halligan tool<br>Hux bar<br>Hydraulic spreader<br>Kelly tool<br>Pry bar<br>Rabbit tool |
| Strike | Battering ram<br>Chisel<br>Flat-head axe<br>Hammer<br>Mallet<br>Maul<br>Sledgehammer<br>Spring-loaded center punch |
| Cut | Axe<br>Bolt cutter<br>Chain saw<br>Cutting torch<br>Hacksaw<br>Handsaw<br>Hydraulic shears<br>Reciprocating saw<br>Rotary saw<br>Seatbelt cutter |

## Table 8-3 Common Tools for Assembly and Disassembly

| | |
|---|---|
| Box-end wrenches | Pipe wrenches |
| Gripping pliers | Screwdrivers |
| Hydrant wrenches | Socket wrenches |
| Open-end wrenches | Spanner wrenches |

Figure 8-8 Hydrant wrench in action.

## Fire Marks

The word "pike" means sharp point or spike. In ancient times, a pike was the tip of a spear, and pike poles were used for very different purposes than current pike poles.

Figure 8-9 A pike pole in action.

## Pushing/Pulling Tools

A second type of tool is used for pushing or pulling. These tools can extend the reach of the brigade member as well as increase the power that can be exerted upon an object (▼ Table 8-4). These tools have many different uses in fire brigade operations.

An example of a tool that extends reach is a pike pole (► Figure 8-9). A pike pole consists of a wood or fiberglass pole with a metal head attached to one end. A pike pole is used primarily to pull down a ceiling to get to the seat of a fire burning above. The metal head has a sharpened point that can be punched through the ceiling and a hook that can grab and pull it down. The proper use of a pike pole will be discussed in depth in Chapter 19, Salvage and Overhaul.

Pike poles come in several different sizes and with a variety of heads (► Figure 8-10). The most common length of 4–6 feet enables a brigade member to stand on a floor and pull down a 10-foot high ceiling. Closet hooks, intended for use in tight spaces, are commonly 2- to 4-foot long. Some pike poles are equipped with handles as long as 12 or 14 feet for use in rooms with very high ceilings; others may have a "D"-type handle for better pulling power.

The different head designs are intended for different types of ceilings and come in a variety of configurations. Many fire brigades use pike poles that are specialized or specific to the types of applications typically found in their response areas.

Common types of pulling poles and hooks include:

- Ceiling hook: A tool with a long wood or fiberglass pole and a metal point with a spur at right angles that can be used to probe ceilings and pull down plaster lath material.
- Clemens hook: A multipurpose tool that can be used for forcible entry and ventilation applications because of its unique head design.
- Drywall hook: A specialized version of a pike pole designed to remove drywall.
- Multipurpose hook: A long pole with a wooden or fiberglass handle and a metal hook.
- Plaster hook: A long pole with a pointed head and two retractable cutting blades on the side.

**Table 8-4 Common Tools for Pushing and Pulling**

| | | |
|---|---|---|
| Ceiling hook | Multipurpose hook | Roofman's hook |
| Clemens hook | Pike pole | San Francisco hook |
| Drywall hook | Plaster hook | |

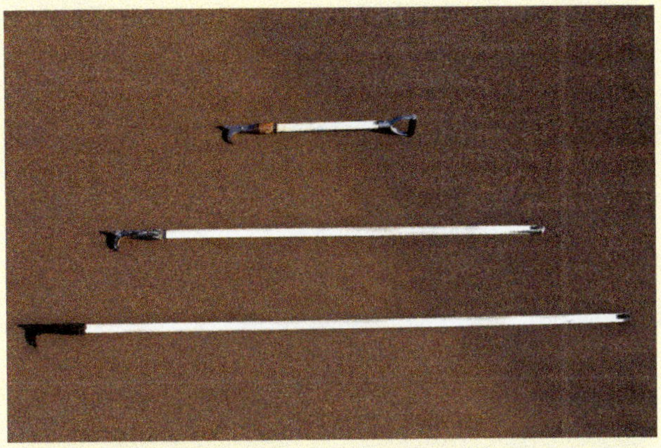

Figure 8-10 Pike poles come in several different sizes.

Figure 8-11 Components of a K tool.

- Roofman's hook: A long pole with a solid metal hook.
- San Francisco hook: A multipurpose tool used for forcible entry and ventilation applications because it has a built-in gas shut-off and directional slot.

A K tool is another type of pushing or pulling tool ▶ Figure 8-11. This tool is used to pull the lock cylinder out of a door, exposing the locking mechanism so it can be unlocked easily.

## Prying/Spreading Tools

A third type of tool is used for prying or spreading. These tools may be as simple as a pry bar or as mechanically complex as a hydraulic spreader ▶ Table 8-5. They also come in several variations for different applications.

A simple pry bar consists of a hardened steel rod with a tapered end that can be inserted into a small area. The bar acts as a lever to multiply the force that a person can exert to bend or pry objects apart. A properly positioned pry bar can apply an enormous amount of force.

Fire brigades use a wide range of prying tools in different sizes and with different features. One of the most popular is a Halligan tool, which was designed in the 1940s by Hugh Halligan of the New York Fire Department. This tool incorporates a sharp pick, a flat prying surface, and a forked claw. It can be used for forcible entry applications. Several other prying tools can also be used for forcible entry. The specific uses and applications of these tools will be covered in the chapters on the phases of fire suppression.

Tools used for prying include ▶ Figure 8-12:

- Claw bar: A tool with a pointed claw-hook on one end and a forked- or flat-chisel pry on the other end that can be used for forcible entry.
- Crowbar: A straight bar made of steel or iron with a forked-like chisel on the working end.
- Flat bar: A specialized prying tool made of flat steel with prying ends suitable for performing forced entry.

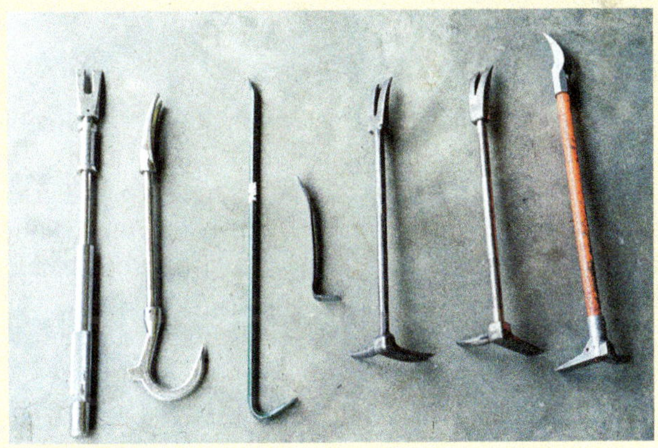

Figure 8-12 Common tools used for prying and spreading.

**Table 8-5 Common Tools for Prying and Spreading**

| | | |
|---|---|---|
| Claw bar | Halligan tool | Kelly tool |
| Crowbar | Hux bar | Pry bar |
| Flat bar | Hydraulic spreader | Rabbit tool |

### Teamwork Tips

Always try before you pry! One person should check to see if a door or window is already unlocked while another readies the right forcible entry tool or tools.

- Halligan tool: A prying tool designed for use in the fire service that also incorporates a pick and a claw; sometimes known as a Hooligan tool.
- Hux bar: A multipurpose tool that can be used for forcible entry and ventilation applications because of its unique design; also can be used as a hydrant wrench.
- Kelly tool: A steel bar with two main features: a large pick and a large chisel or claw.
- Pry bar: A specialized prying tool made of a hardened steel rod with a tapered end that can be inserted into a small area ▶ Figure 8-13.

Hydraulic spreaders are an example of machine-powered rescue tools ▶ Figure 8-14. The use of hydraulic power enables you to apply several tons of force on a very small area. You need to have special training to operate these machines safely. Fire and rescue brigades most commonly use them for extrication from vehicles ▶ Figure 8-15.

Hand-powered hydraulic tools are also used for prying and spreading. One of these, called a rabbit tool, is designed for quickly opening doors ▶ Figure 8-16.

## Striking Tools

Striking tools are used to apply an impact force to an object ▶ Table 8-6. They are often used to gain entrance to a building or a vehicle or to make an opening in a wall or roof. They can also be used to force the end of a prying tool into a small opening. Specific use of these tools will be covered in Chapter 11, Forcible Entry. Striking tools include:

- Hammer: A hand tool constructed of solid material with a long handle and a head affixed to the top of the handle with one side of the head used for striking and the other side used for prying.
- Mallet: A short-handled hammer with a round head.
- Sledgehammer: A long, heavy hammer that requires the use of both hands.
- Maul: A specialized striking tool (weighing six pounds or more) with an axe head on one side and a sledgehammer head on the other side ▶ Figure 8-17.
- Flat-head axe: A tool with an axe head blade on one side and a flat head on the opposite side ▶ Figure 8-18.
- Pick-head axe: A tool with an axe head blade on one side and a pointed "pick" on the opposite side ▶ Figure 8-19.

### Brigade Member Tips

Many fire brigades carry a striking tool and a prying bar strapped together. This combination of tools is sometimes referred to as the irons. A flat-head axe and a Halligan tool are carried together on many vehicles and are always taken into a fire building by one of the crew members.

Figure 8-13 A pry bar in action.

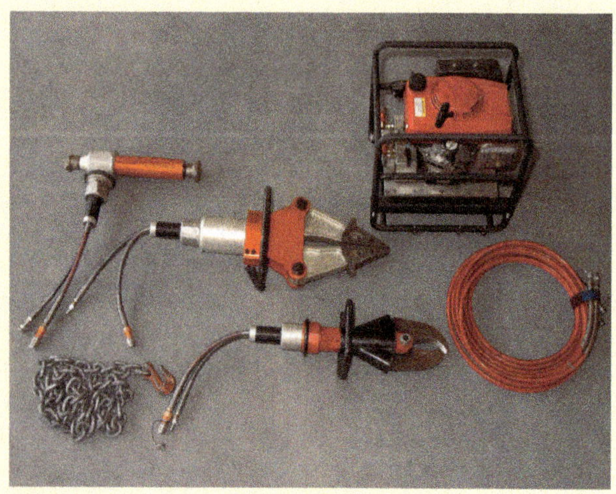

Figure 8-14 The components of a hydraulic rescue tool.

**Table 8-6 Common Striking Tools**

| | | |
|---|---|---|
| Battering ram | Hammer | |
| Chisel | Mallet | Sledgehammer |
| Flat-head axe | Maul | Spring-loaded center punch |

Figure 8-15 A hydraulic spreader is often used for vehicle extrication.

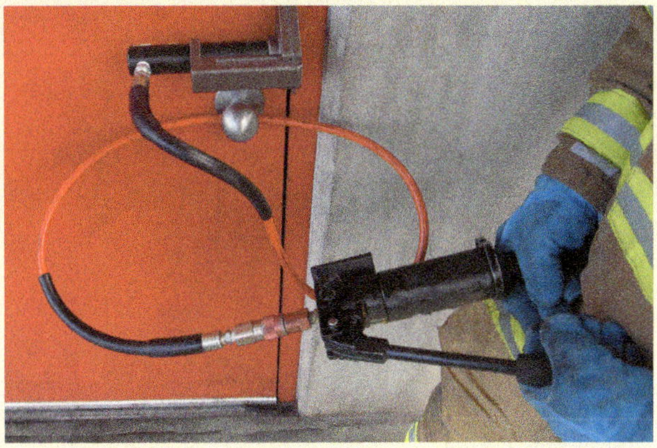

Figure 8-16 A rabbit tool can open doors quickly.

Figure 8-17 Striking tools (from top): hammer, mallet, maul, sledgehammer.

## Brigade Member Safety Tips

Some brigade members carry knives, combination tools, a small length of rope, or wooden wedges in their turnout gear. The tools and equipment that you carry will depend on your brigade requirements, the type of area served by your brigade, and your personal preference.

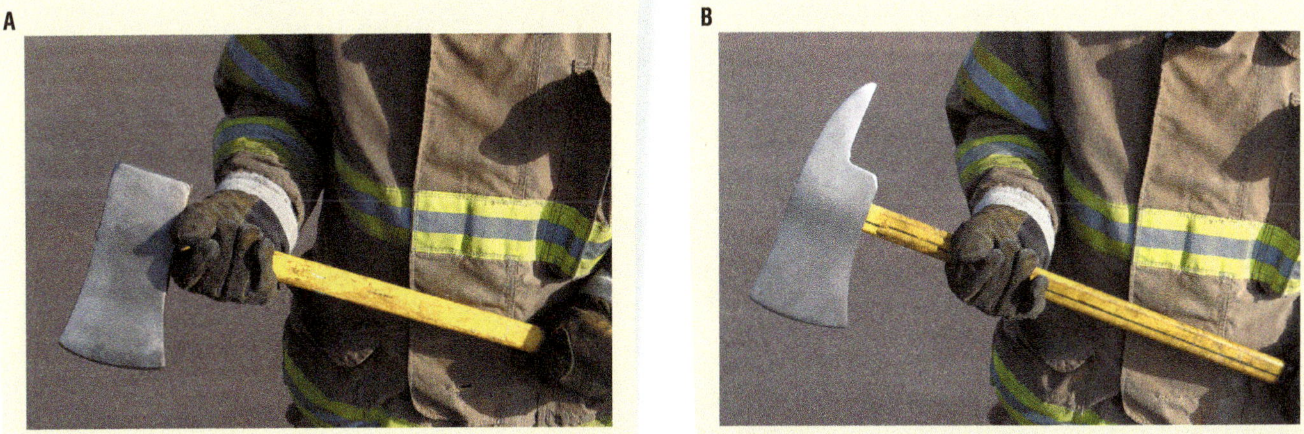

Figure 8-18 Two types of axes: **A.** Flat-head axe and **B.** A pick-head axe.

## Brigade Member Safety Tips

Some tools require a considerable amount of movement and room to operate. Always check to be sure no one is in danger of being injured before you use these tools. Look around and make sure you can operate an axe or sledgehammer safely and effectively.

- Battering ram: A heavy metal bar used to break down doors ▶ Figure 8-19.
- Chisel: A metal tool with one sharpened end that can be used to break apart material in conjunction with a hammer, mallet, or sledgehammer ▶ Figure 8-20.
- Spring-loaded center punch: A spring-loaded punch used to break tempered automobile glass.

One of the most frequently used tools in the fire service is the axe. Both flat-head axes and pick-head axes are used. Both types of axes have a wide cutting blade that can be used to chop into a wall, roof, or door. A flat-head axe also can be used as a striking tool for forcible entry, usually in combination with a prying tool, such as a Halligan tool. Together, the flat-head axe and the Halligan tool are sometimes referred to as "the irons" and are very effective for most forcible entry situations ▶ Figure 8-21. A pick-head axe has a point or pick that can be used for puncturing, pulling, and prying ▶ Figure 8-22. Forcible entry will be covered in depth in Chapter 11, Forcible Entry.

The spring-loaded center punch is a striking tool used primarily on cars ▶ Figure 8-23. This tool can exert a large amount of force on a pinpoint-size portion of tempered automobile glass. This disrupts the integrity of the glass and causes the window to shatter into small, uniform-sized pieces. A spring-loaded center punch is often used in vehicular crashes to gain access to a patient who needs care.

## Cutting Tools

Cutting tools have a sharp edge that severs an object. They come in several forms and are used to cut a wide variety of substances ▶ Figure 8-24. Cutting tools used by brigade members range from knives or wire cutters carried in the pockets of turnout coats to seatbelt cutters, bolt cutters (a scissors-like tool used to cut through items such as chains or padlocks), saws, cutting torches (a torch that produces a high temperature flame capable of melting through metal), and hydraulic shears ▶ Table 8-7. Each is designed to work on certain types of materials. Brigade members can be injured and cutting tools can be ruined if tools are used incorrectly.

Bolt cutters are most often used to cut through chains or padlocks to open doors or gates. By concentrating the cutting force on a small area, you can break through a chain in just a few seconds.

Figure 8-19 Using a battering ram.

Figure 8-20 Using a chisel with a hammer.

Figure 8-21 A flat-head axe can be used with a Halligan tool to force open a door.

Figure 8-22 A pick-head axe can be used to pry up boards.

Figure 8-23 A spring-loaded center punch can be used to break a car window safely.

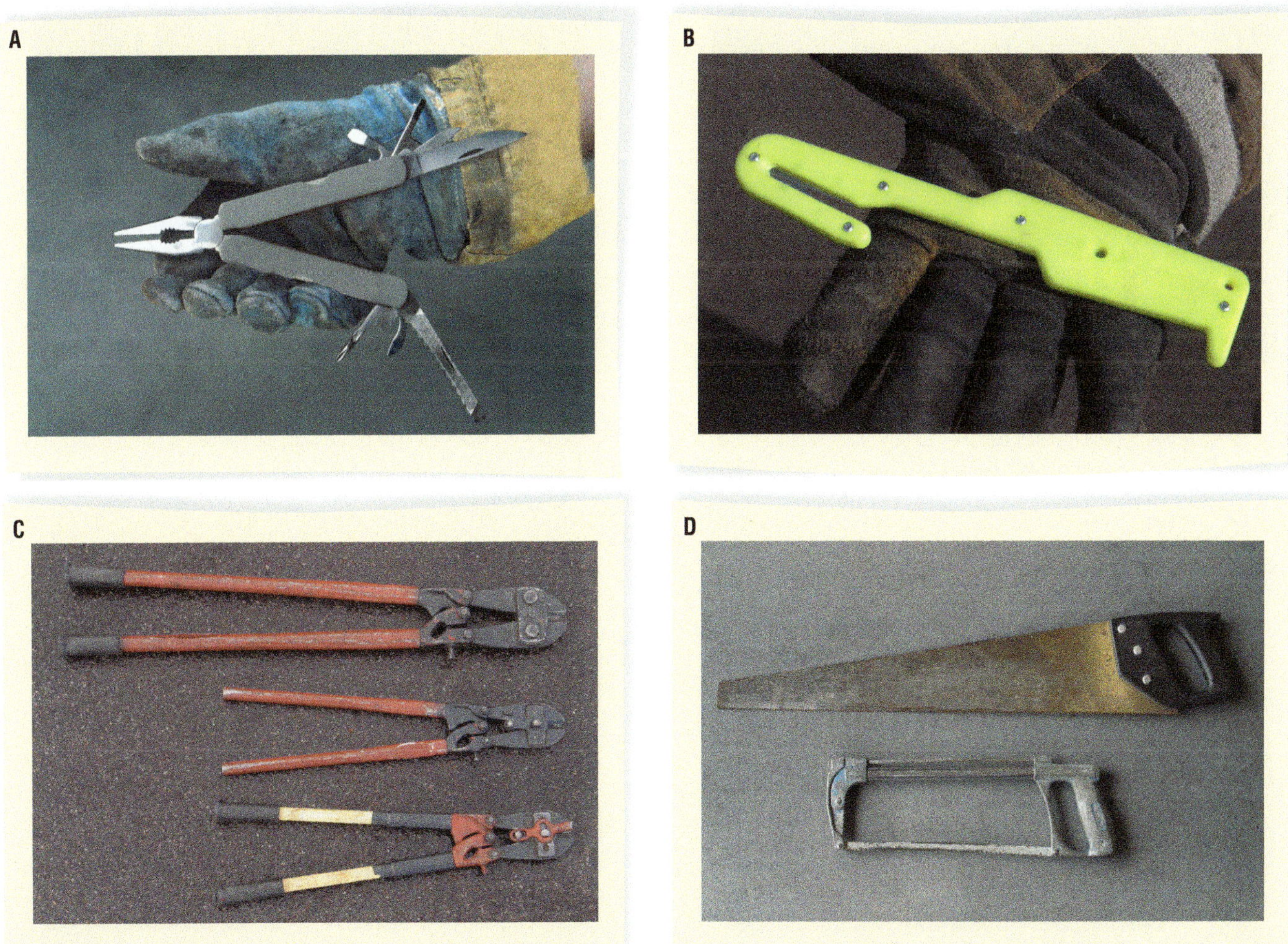

Figure 8-24 Cutting tools. A. Combination tool. B. Seatbelt cutter. C. Bolt cutters. D. Various handsaws.

**Table 8-7 Common Cutting Tools**

| | |
|---|---|
| Axes | Bolt cutters |
| Chain saws | Cutting torches |
| Hacksaws | Handsaws |
| Hydraulic shears | Reciprocating saws |
| Rotary saws | Seatbelt cutter |

Fire brigades often carry several different types of saws. Saws can be divided into two main categories, based on the power source. **Handsaws** are manually powered and **mechanical saws** are usually powered by electric motors or gasoline-powered engines. Handsaws include hacksaws, carpenter's handsaws, keyhole saws, and coping saws.

**Hacksaws** are designed to cut metal ▼ Figure 8-25. Different blades can be used, depending on the type of metal being cut. Hacksaws are useful when metal needs to be cut under closely controlled conditions.

**Carpenter's handsaws** are designed for cutting wood ▼ Figure 8-26. Saws with large teeth are effective in cutting large timbers or tree branches. These saws are useful at auto crashes where tree limbs may hamper the rescue effort. Saws with finer teeth are designed for cutting finished lumber. A **coping saw** is used to cut curves in wood. It is a handsaw with a narrow blade set between the ends of a U-shaped frame. A **keyhole saw**, a specialty saw, is narrow and slender and can be used to cut keyholes in wood.

There are three primary types of mechanical saws: chain saws, rotary saws, and reciprocating saws. Although handsaws have a valuable role, power saws have the advantage of accomplishing more work in a shorter period of time. They also enable brigade members to conserve energy, resulting in less fatigue. Because mechanical saws are powerful, they must only be used by trained operators. But there are

**Figure 8-25** A brigade member using a hacksaw to cut through metal.

**Figure 8-27** A brigade member can use a chain saw to cut through a roof.

**Figure 8-26** A carpenter's handsaw is designed for cutting wood.

**Figure 8-28** Brigade member using a rotary saw.

## Brigade Member Safety Tips

When operating any mechanical saw, it is important to use a saw blade or saw disk designed specifically for the material being cut. Always use the proper eye protection as well as a helmet shield when using power saws.

some disadvantages to using power saws. They are heavy to carry and sometimes can be difficult to start. They may also require an electrical connection, although cordless models are becoming more available.

Most people are familiar with gasoline-powered or electric chain saws commonly used to cut wood, particularly trees. Brigade members often use saws with special chains to cut ventilation openings in roofs constructed of wood, metal, tar, gravel, or insulating materials ◄ Figure 8-27.

Rotary saws are powered by electric motors or gasoline engines ◄ Figure 8-28. In some rotary saws, the cutting part of the saw is a round metal blade with teeth. Different blades are used depending on the type of material being cut. Other rotary saws use a flat, abrasive disk for cutting. The disks are made of composite materials and are designed to wear down as they are used. It is important to match the appropriate saw blade or saw disk to the material being cut.

Reciprocating saws are powered by an electric or battery motor that rapidly pulls a saw blade back and forth ► Figure 8-29. As with rotary saws, reciprocating saws use different blades to cut different materials. Reciprocating saws are most commonly used to cut metal during a vehicle extrication.

The cutting tools that require the most training are hydraulic shears and the cutting torch. Hydraulic shears are often used with hydraulic spreaders and rams in automobile extrication. The same hydraulic power source can be used with all three types of tools. Hydraulic shears can cut quickly through metal posts and bars.

Cutting torches produce an extremely high temperature flame and are capable of heating steel until it melts, thereby cutting through the object ► Figure 8-30. Because these torches produce such high temperatures (5,700°F), operators must be specially trained before using this tool. Cutting torches are sometimes used for rescue situations and for cutting through heavy steel objects.

### Multiple Function Tools

Certain tools are designed to perform multiple functions, thus reducing the total number of tools needed to achieve a goal. One example is a flat-head axe because it can be used as either a cutting or a striking tool. There are combination tools that can be used to cut, to pry, to strike, and to turn off utilities.

Figure 8-29 **A.** The components of a reciprocating saw. **B.** A reciprocating saw being used during a vehicle extrication.

Figure 8-30 A cutting torch can be used to cut through a metal door.

# Voices of Experience

**"I'm forever indebted to my partner for his actions that day."**

We were investigating the report of smoke at one of our facilities that elevates dry product up approximately 80 feet vertically and then transfers the product to horizontal belts for movement to one of several large bulk storage silos. The vertical elevating mechanism consisted of a web belt with several hundred "buckets" attached along the length of the web. The dry product is dropped into a below-ground bin, and the buckets on the belt pass through the product in the bin, fill with product, and the belt then passes up the length of the "head house" to the top, where the buckets pass over the top pulley wheel and dump their contents onto one of the horizontal belts for transfer to a silo.

As I stood at the bottom of the head house and looked up, I noticed a significant amount of smoke near the top. I was assigned to carry a Halligan tool in case it was needed to pry open any number of covered boxes or bin openings. Suddenly I heard a large roar like a jet engine was passing by. My partner pushed me away from the bucket belt with enough force that I dropped the Halligan, just as I realized the belt had severed and was crashing down. Several hours later, after the fire was extinguished, we noted that the bucket belt was piled higher than I stood upright and completely covered the area. I would certainly have been under that pile of buckets. As we pulled the pile apart, there it was—the Halligan, somewhat tarnished but just fine otherwise. Afterward at the equipment storage location, I finished cleaning the Halligan, and as I placed it back in its storage location I again realized how fortunate I was. I'm forever indebted to my partner Bill for his actions that day.

**John Shrader**
Cooper Nuclear Station
Brownville, Nebraska

### Special Use Tools

Some fire situations require special use tools that perform other functions. For example, fire brigades located in areas where brush and ground fires occur frequently may need to carry rakes, brooms, shovels, and combination tools that can be used for raking, chopping, cutting, and leaf blowing.

Fire brigade rescue squads also use specialized equipment such as jacks and air bags for lifting heavy objects, come alongs or lever blocks (small, hand-operated winches) for dragging heavy objects, and tripods ( ▶ Figure 8-31 ). You can learn more about the proper use of this special equipment by taking special rescue courses or during in-service training.

Figure 8-31 Heavy-duty air bags can be used to lift vehicles in rescue situations.

## Phases of Use

The process of extinguishing a fire usually involves a sequence of steps or stages. Each phase of a fireground operation may require the use of certain types of tools and equipment. The basic steps of fire suppression include:

- Response/Size-Up: This phase begins when the emergency call is received and continues as the units travel to the incident scene. The last part of this phase involves the initial observation and evaluation of factors used to determine the strategy and tactics that will be employed.
- Forcible Entry: This phase begins when entry to buildings, vehicles, aircraft, or other confined areas is locked or blocked, requiring brigade members to use special techniques to gain access.
- Interior Attack: During this phase, a team of brigade members is assigned to enter a structure and attempt fire suppression.
- Search and Rescue: As its name suggests, this phase involves a search for any victims trapped by the fire and their rescue from the building.
- Rapid Intervention: When a rapid intervention company/crew (RIC) provides immediate assistance to injured or trapped brigade members.
- Ventilation: This step involves changing air within a compartment by natural or mechanical means.
- Overhaul: The final phase is to ensure that all hidden fires are extinguished after the main fire has been suppressed.

### Response/Size-Up

The response and size-up phase enables you to anticipate emergency situations. At this time, you should consider the information from the dispatcher along with preincident plan information about the location. This information can provide you with an idea of the nature and possible gravity of the situation, as well as the types of problems that might arise. For example, an automobile fire in the parking lot may present different problems and require different tools than a call for smoke coming from a warehouse. A different kind of thinking process occurs when you respond to a pump seal fire that may have maintenance personnel injured or trapped than when you respond to a report of smoke alarms near an outdoor metering control pit. Even though information is limited, this is the time to start thinking about the types of tools and equipment that you might need.

Most fire brigades have standard operating procedures or guidelines that specify the tools and equipment required for different types of fires. Each crew member is expected to bring specific tools and equipment from the apparatus. These requirements take into account the roles that different units have within the fire brigade. The tools carried for an interior fire attack are different from those carried for an outside or defensive attack.

Upon arrival at the scene, the brigade leader will size-up the situation and develop the action plans for each company, following standard operating procedures and guidelines. The brigade leader will use their senses of sight, sound, and smell, as well as any other available information, to make a determination of how to best attack the fire. As a crew member, you are responsible for following the directions of your brigade leader.

### Forcible Entry

Gaining entrance to a locked building or structure can present a challenge to even the most seasoned brigade member. Buildings are often equipped with security devices designed to keep unwanted people out. However, these same devices can make it very difficult for brigade members to gain access to the building. Forcible entry is the process of entering a building by overcoming these barriers. The skills involved will be discussed in depth in Chapter 11, Forcible Entry.

Several types of tools can be used for forcible entry, including an axe, a prying tool, or a K tool. A flat-head axe and a Halligan tool are often used in combination to quickly pry open a door, although they may permanently damage both the door and the frame. Prying tools used for gaining access include pry bars, crowbars, Halligan tools, hux bars, and the hydraulic-powered rabbit tool.

A K tool can be used to pull out a cylinder lock mounted in a wood or heavy metal door, so that the lock can be released ▶ Figure 8-32. This is a comparatively nondestructive process that leaves the door and most of the locking mechanism undamaged. The building owner can have the lock cylinder replaced at a relatively low cost.

Various striking tools can be used for forcible entry when brute force is needed to break into a building. These include flat-head axes, hammers, sledgehammers, and battering rams.

Sometimes the easiest or only way to gain access is to use cutting tools. An axe can be used to cut out a door panel. A power saw can be used to cut through a wood wall. Bolt cutters can be used to remove a padlock. Cutting torches or power saws can be used to cut through metal security bars. Although cutting is a destructive process, it is justified to save lives or property.

Many techniques and types of tools can be used to gain entry into secured structures ▼ Table 8-8. The exact tool needed will depend upon the method of entry and the type of obstacle. Because experience usually determines the best way to gain entry in each situation, rely on the orders and advice of your brigade leader and coworkers.

Figure 8-32 Use a K tool to pull out and release a cylinder lock.

## Interior Firefighting Tools and Equipment

The process of fighting a fire inside a building involves several tasks that are usually performed simultaneously or in rapid succession by teams of brigade members. While one crew is advancing a hose line to attack the fire, another crew may be searching for occupants and another may be performing ventilation tasks. An additional crew may be standing by as an RIC, in case a brigade member needs to be rescued.

Some basic tools and equipment should be carried by every crew working inside a burning building. Crews may also carry specialized tools and equipment needed for their particular assignment. The basic tools enable them to solve problems they may encounter while performing interior operations. For example, the crew may encounter obstacles

**Table 8-8 Forcible Entry Tools**

| Type of tool | Use in forcible entry |
|---|---|
| Prying tools | These can include Halligan tools, a flat bar, a crowbar, and other prying tools. They can be used to break windows or force open doors. |
| Axe, flat-head axe, pick-head axe | An axe can be used to cut through a door or break a window. A flat-head axe can be used in conjunction with a Halligan tool to force open a door. |
| Sledgehammer | A sledgehammer can be used to breach walls or break a window. It can also be used in conjunction with other tools, such as a Halligan tool, to force open a door or break off a padlock. |
| Hammer, mallet | These tools may be used in conjunction with other tools such as chisels or punches to force entry through windows or doors. |
| Chisel, punch | These tools can be used to make small openings through doors or windows. |
| K tool | The K tool provides a "through the lock" method of opening a door, minimizing damage to the door. |
| Rotary saw, chain saw, reciprocating saw | Power saws can cut openings through various obstacles, including doors, walls, fences, gates, security bars, and other barriers. |
| Bolt cutter | A bolt cutter can be used to cut off padlocks or cut through obstacles such as fences. |
| Battering ram | A battering ram can be used to breach walls. |
| Hydraulic door opener | A hydraulic door opener can be used to force open a door. |
| Hydraulic rescue tool | A hydraulic rescue tool can be used in a variety of situations to break, breach, or force openings in doors, windows, walls, fences, or gates. |

such as locked doors, or they may need to open an emergency escape route. They may need to establish **horizontal ventilation** by forcing, opening, or breaking a window. They may have to gain access to the space above the ceiling by using a pike pole or making a hole in a wall or floor with an axe. A powerful light is important, because smoke can quickly reduce interior visibility to just a few inches.

The basic set of tools for interior firefighting includes:

- A prying tool, such as a Halligan tool
- A striking tool, such as a flat-head axe or sledgehammer
- A cutting tool, such as an axe
- A pushing/pulling tool, such as a pike pole
- A strong **hand light** or portable light

The specific tools that must be carried by each crew are usually defined in a fire brigade's training manuals and standard operating procedures. The requirements are based on local conditions and preferences and may differ, depending on the type of company and the assignment.

The interior attack team is responsible for advancing a hose line, finding the fire, and applying water to extinguish the flames. They need the basic tools that will allow them to reach the seat of the fire.

## Search and Rescue Tools and Equipment

Search and rescue needs to be carried out quickly, shortly after arrival on the fireground. A search team should carry the same basic hand tools as the interior attack team:

- Pushing tool (short pike pole)
- Prying tool (Halligan tool)
- Striking tool (sledgehammer or flat-head axe)
- Cutting tool (axe)
- Hand light

In addition to being equipped for forcible entry and emergency exit, a search-and-rescue team may also use tools to probe under obstructions or collapse areas for unconscious victims. A short pike pole or closet hook is relatively light and reduces the time needed to search an area by extending the brigade member's reach. An axe handle can also be used for this purpose.

Figure 8-34 Brigade member using a thermal imaging device.

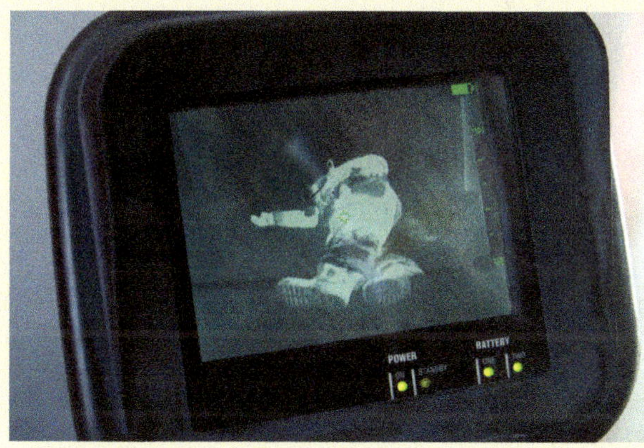

Figure 8-35 An image produced by a thermal imaging device.

Figure 8-33 A thermal imaging device.

Other types of tools used for search and rescue include **thermal imaging devices** (▶ Figure 8-33, ▲ 8-34, ▲ 8-35), portable lights, and **lifelines**. The techniques of search and rescue are discussed in Chapter 13, Search and Rescue.

## Rapid Intervention Tools and Equipment

An RIC is designated to stand by to provide immediate assistance to any brigade members who become lost, trapped, or injured during an incident or training exercise. The RIC team should have the standard set of tools for interior firefighting as well as extra tools and equipment particularly important for search-and-rescue tasks. The extra tools and equipment

**Brigade Member Tips**

Industrial use of thermal imaging cameras has expanded to include identification of gas leaks in vessels and piping, liquid levels in noninsulated tanks, fuel spills under snow, and similar applications.

should help them find and gain access to a brigade member who is in trouble, extricate a brigade member who is trapped under debris, provide breathing air for a brigade member who has experienced an SCBA failure or run out of air, and remove an injured or unconscious brigade member from the building. All of this equipment should be gathered and staged with the RIC where it will be immediately available if it is needed.

The special equipment that an RIC should carry includes:

- Thermal imaging device
- Additional portable lighting
- Lifelines
- Prying tools
- Striking tools
- Cutting tools, including a power saw
- SCBA and spare air cylinders

### Ventilation Tools and Equipment

The objective of ventilation is to provide openings so that fresh air can enter and hot gases and products of combustion can escape. Many of the same tools used for forcible entry are also used to provide ventilation. Power saws and axes are commonly used to cut through roofs and vent combustion by-products.

Fans are often used either to remove smoke from a building or to introduce fresh air into a structure. With positive pressure ventilation, fresh air is blown into a building through selected openings to force contaminated air out through other openings. Positive pressure ventilation is often used to allow brigade members to work quickly, safely, and effectively to locate and suppress the fire.

Negative pressure ventilation uses fans placed at selected openings to draw contaminated air out of a building. It is used when there are no suitable openings for positive pressure ventilation or when introducing pressurized air could accelerate the fire. Negative pressure ventilation may also be the best option when positive pressure could spread the products of combustion.

Ventilation fans can be powered either by electric or gasoline motors or by water pressure ▶ Figure 8-36. A gasoline-powered fan may not be suitable in some situations, because it can introduce carbon monoxide into a structure if an exhaust hose is not available. If there are potentially dangerous levels of gases in the building atmosphere, a water-powered fan may be the best choice.

Figure 8-36 Different types of fans used in ventilation.

Horizontal ventilation usually involves opening outer doors and windows to allow fresh air to enter and to remove contaminated air. When ventilation fans are used, they are placed at these openings. Positive pressure fans blow in fresh air, and negative pressure fans draw out smoke and contaminants. Unlocked or easily released windows and doors should be opened normally. Locked or jammed windows and doors may have to be broken or forced open using basic interior firefighting tools.

It may also be necessary to make interior openings within the building so that contaminated air can reach the exterior openings. Opening interior doors is the easiest, quickest, and often most effective way to ventilate interior spaces. A series of electrically powered fans may be placed throughout a large structure to help move smoke in a desired direction.

Vertical ventilation requires openings in the roof or the highest part of a building to allow smoke and hot gases to escape. Whenever possible, existing openings such as doors, windows, roof hatches, and skylights should be used for vertical ventilation. It may be necessary to force them open or to break them using forcible entry tools.

In some circumstances it may be necessary to cut through a roof to make an effective vertical ventilation opening. Cutting tools such as axes and power saws are used to make these openings. Pike poles will also be needed to pull down ceilings after the roof covering is opened.

The special equipment needed for ventilation includes:

- Positive pressure fans
- Negative pressure (exhaust) fans
- Pulling and pushing tools (long pike poles)
- Cutting tools (power saws and axes)

## Overhaul Tools and Equipment

The purpose of overhaul is to examine the fire scene carefully and ensure that all hidden fires are located and extinguished. Burned debris must be removed and potential hot spots in

## Brigade Member Safety Tips

There is a valuable saying in the fire service. "Hand tools—never leave your vehicle without them."

enclosed spaces behind walls, above ceilings, and under floors must be exposed. Both tasks can be accomplished using simple hand tools.

Pike poles are commonly used for pulling down ceilings and opening holes in walls. Axes, and sometimes power saws, are used to open walls and floors. Prying and striking tools are also used to open closed spaces. Shovels, brooms, and rakes are used to clear away debris.

The increasing use of infrared thermal imaging devices has made it possible to "see" hot spots behind walls without physically cutting into them (▼ Figure 8-37). This technology has reduced the risk of missing dangerous hot spots as well as the time and effort it takes to overhaul a fire scene. The process of overhauling a fire scene and the tools used to accomplish this will be covered more fully in Chapter 19, Salvage and Overhaul.

Tools used during overhaul operations include:

- Pushing tools (pike poles of varying lengths)
- Prying tools (Halligan tool)
- Striking tools (sledgehammer, flat-head axe, hammer, mallet)
- Cutting tools (axes, power saws)
- Debris-removal tools (shovels, brooms, rakes, carryalls)
- Water-removal equipment (water vacuums)
- Ventilation equipment (electric, gas, or water-powered fans)
- Portable lighting
- Thermal imaging device

Figure 8-37 Thermal imaging devices can "see" hot spots.

# Tool Staging

Many fire brigades have standard operating procedures for staging necessary equipment nearby during a fire or rescue operation. This often involves placing a salvage cover on the ground at a designated location and laying out commonly used tools and equipment where they can be accessed readily. A similar procedure may be used for rescue operations, where tools that are likely to be needed can be laid out ready for use. This saves valuable time because brigade members do not have to return to their own apparatus or search several different vehicles to find a particular tool.

A brigade's standard operating procedure usually specifies the types of tools and equipment to be staged. The tool-staging area could be outside the building or, in the case of a high rise or large building, at a safe and convenient location inside. Additional personnel may be directed to bring particular items to the tool staging location at working fires.

# Maintenance

Tools and equipment must be properly maintained so that they will be ready for use when they are needed. Keep equipment clean and free from rust. Keep cutting blades sharpened and fuel tanks filled. Every tool and piece of equipment must be ready for use before you respond to an emergency incident.

Use power equipment only after you have been instructed on its use. Read and heed the instructions supplied with the equipment. Test power equipment frequently and have it serviced regularly by a qualified shop. Keep records of all inspections and maintenance performed on power tools. Preventive maintenance will help ensure that equipment will operate properly when it is needed. Treat tools and equipment as if someone's life depends on them, because it will!

Use equipment only for its intended purposes. For example, a pike pole is made for pushing and pulling; it is not a lever and will break if used inappropriately. Use the right tool for the job.

## Cleaning and Inspecting Salvage, Overhaul, and Ventilation Equipment and Tools

Clean tools used in ventilation according to the manufacturer's instructions. Power tools are used frequently during

## Brigade Member Safety Tips

Keep the manufacturers' manuals to all of the brigade's tools and equipment in a safe and easily accessible location.

overhaul and should also be cleaned according to the manufacturer's instructions. After returning from a fire, clean and inspect, and record all of your overhaul tools to ensure that they are in a "ready state."

## Cleaning and Inspecting Hand Tools

All hand tools should be completely cleaned, inspected, and recorded after use, as shown in ▶ Table 8-9. Remove all dirt and debris. If appropriate, use water streams to remove the debris and soap to clean the equipment thoroughly. To prevent rust, metal tools must be dried completely, either by hand or by air, before being returned to the apparatus. Cutting tools should be sharpened after each use. Before any tool is placed back into service, it should be inspected for damage.

Avoid painting tools, because this will hide any possible defects or visible damage. Keep the number of markings on a tool to a minimum.

## Cleaning and Inspecting Power Tools

All power equipment should be left in a "ready state" for immediate use at the next incident. This means that:

- All debris should be removed and the tool should be clean and dry.
- All fuel tanks should be filled completely with fresh fuel.
- Any dull or damaged blades/chains should be replaced.
- Belts should be inspected to ensure they are tight and undamaged.
- All guards should be securely in place.
- All hydraulic hoses should be cleaned and inspected.
- All power cords should be inspected for damage.
- All hose fittings should be cleaned, inspected, and tested to ensure tight fit.
- The tools should be started to ensure that they operate properly.
- Tanks on water vacuums should be emptied, washed, cleaned, and dried.
- Hoses and nozzles on water vacuums should be cleaned and dried.

It is very important to read the manufacturer's manual and follow all instructions on the care and inspection of power tools and equipment. Keep all manufacturers' manuals in a safe and easily accessible location. Refer to them when cleaning and inspecting the tools and equipment. Remember, your safety depends on the quality of your tools and equipment.

**Table 8-9 Cleaning and Inspecting Hand Tools**

| | |
|---|---|
| **Metal parts** | All metal parts should be clean and dry. Remove rust with steel wool. Do not oil the striking surface of metal tools because this may cause them to slip. |
| **Wood handles** | Inspect for damage such as cracks, splinters, etc. Sand if necessary. Do not paint or varnish. Apply a coat of boiled linseed oil. Check that the tool head is tightly fixed to the handle. |
| **Fiberglass handles** | Clean with soap and water. Inspect for damage. Check that the tool head is tightly fixed to the handle. |
| **Cutting edges** | Inspect for nicks or other damage. File and sharpen as needed. *Note:* Power grinding may weaken some tools. Hand sharpening may be required. |

# Wrap-Up

## Ready for Review

- Tools and equipment extend your reach and multiply the force you can apply to an object.
- Your safety and the safety of others are of paramount importance when using tools and equipment.
- Learn to use tools and equipment under adverse conditions, while wearing motion-limiting PPE gear.
- Strive for effective and efficient use of tools and equipment.
- Know where tools and equipment are stored in the fire station and on the apparatus.
- The primary functions of tools and equipment are:
  - Rotate—assemble or disassemble
  - Push or pull
  - Pry or spread
  - Strike
  - Cut
  - A combination
- Use dispatch information and size-up information to anticipate the types of tools you may need.
- Learn which tools and equipment are used during the following phases of fire suppression:
  - Response/Size-up
  - Forcible Entry
  - Interior Attack
  - Search and Rescue
  - Rapid Intervention Company/Crew
  - Ventilation
  - Overhaul
- Proper selection and maintenance of tools and equipment are essential.

## Hot Terms

**Battering ram** A tool made of hardened steel with handles on the sides used to force doors and to breach walls. Larger versions are used by up to four people; smaller versions are made for one or two people.

**Bolt cutter** A cutting tool used to cut through thick metal objects such as bolts, locks, and wire fences.

**Box-end wrench** A hand tool used to tighten or loosen bolts. The end is enclosed, as opposed to an open-end wrench. Each wrench is a specific size and most have ratchets for easier use.

**Carpenter's handsaw** A saw designed for cutting wood.

**Ceiling hook** A tool with a long wooden or fiberglass pole that has a metal point with a spur at right angles at one end. It can be used to probe ceilings and pull down plaster lath material.

**Chain saw** A power saw that uses the rotating movement of a chain equipped with sharpened cutting edges. Typically used to cut through wood.

**Chisel** A metal tool with one sharpened end used to break apart material in conjunction with a hammer, mallet, or sledgehammer.

**Claw bar** A tool with a pointed claw-hook on one end and a forked- or flat-chisel pry on the other end. It is often used for forcible entry.

**Clemens hook** A multipurpose tool that can be used for several forcible entry and ventilation applications because of its unique head design.

**Closet hook** A type of pike pole intended for use in tight spaces, commonly 2–4 feet in length.

**Come along** A hand-operated tool used for dragging or lifting heavy objects. Sometimes known as lever blocks.

**Coping saw** A saw designed to cut curves in wood.

**Coupling** One set or a pair of connection devices attached to a fire hose that allows the hose to be interconnected to additional lengths of hose.

**Crowbar** A straight bar made of steel or iron with a forked-like chisel on the working end suitable for performing forcible entry.

**Cutting torch** A torch that produces a high temperature flame capable of heating metal to its melting point, thereby cutting through an object. Because of the high temperatures (5,700ºF) that these torches produce, the operator must be specially trained before using this tool.

**Drywall hook** A specialized version of a pike pole that can remove drywall more effectively because of its hook design.

**Flat bar** A specialized type of prying tool made of flat steel with prying ends suitable for performing forcible entry.

**Flat-head axe** A tool that has an axe head on one side and a flat head on the opposite side.

**Forcible entry** Techniques used by brigade members to gain entry into buildings, vehicles, aircraft, or other areas when normal means of entry are locked or blocked.

**Gripping pliers** A hand tool with a pincer-like working end that can be used to bend wire or hold smaller objects.

**Hacksaw** A cutting tool designed for use on metal. Different blades can be used for cutting different types of metal.

**Halligan tool** A prying tool that incorporates a pick and a claw, designed for use in the fire service. Sometimes known as a Hooligan tool.

**Hammer** A striking tool.

**Hand light** Small, portable light carried by brigade members to improve visibility at emergency scenes, often powered by rechargeable batteries.

**Handsaw** A manually powered saw designed to cut different types of materials. Examples include hacksaws, carpenter's handsaws, keyhole saws, and coping saws.

**Horizontal ventilation** The process of making openings so that smoke, heat, and gases can escape horizontally from a building through openings such as doors and windows.

**Hux bar** A multipurpose tool that can be used for several forcible entry and ventilation applications because of its unique design. Also can be used as a hydrant wrench.

**Hydrant wrench** A hand tool is used to operate the valves on a hydrant; may also be used as a spanner wrench. Some are plain wrenches, and others have a ratchet feature.

**Hydraulic shears** A lightweight hand-operated tool that can produce up to 10,000 pounds of cutting force.

**Hydraulic spreader** A lightweight hand-operated tool that can produce up to 10,000 pounds of prying and spreading force.

**Interior attack** The assignment of a team of brigade members to enter a structure and attempt fire suppression.

**Irons** A combination tool, normally the Halligan tool and the flat-head axe.

**K tool** Used to remove lock cylinders from structural doors so the locking mechanism can be unlocked.

**Kelly tool** A steel bar with two main features: a large pick and a large chisel or claw.

**Keyhole saw** A saw designed to cut keyholes in wood.

**Lifeline** A rope secured to a brigade member that enables him or her to retrace his or her steps out of a structure.

**Mallet** A short-handled hammer.

**Maul** A specialized striking tool, weighing six pounds or more, with an axe head on one side and a sledgehammer head on the other side.

**Mechanical saw** Usually powered by electric motors or gasoline engines. The three primary types are chain saws, rotary saws, and reciprocating saws.

**Multipurpose hook** A long pole with a wooden or fiberglass handle and a metal hook on one end used for pulling.

**Negative pressure ventilation** Ventilation that relies upon electric fans to pull or draw the air from a structure or area.

**Open-end wrench** A hand tool used to tighten or loosen bolts. The end is open, as opposed to a box-end wrench. Each wrench is a specific size.

**Overhaul** Examination of all areas of the building and contents involved in a fire to ensure that the fire is completely extinguished.

**Personal protective equipment (PPE)** Gear worn by brigade members that includes helmet, gloves, hood, coat, pants, SCBA, and boots. The PPE provides a thermal barrier for brigade members against intense heat.

**Pick-head axe** A tool that has an axe head on one side and a pointed "pick" on the opposite side.

**Pike pole** A pole with a sharp point, or pike, on one end coupled with a hook. Used to make openings in ceilings, walls, etc.

**Pipe wrench** A wrench having one fixed grip and one movable grip that can be adjusted to fit securely around pipes and other tubular objects.

**Plaster hook** A long pole with a pointed head and two retractable cutting blades on the side.

**Positive pressure ventilation** Ventilation that relies upon fans to push or force clean air into a structure.

**Pry bar** A specialized prying tool made of a hardened steel rod with a tapered end that can be inserted into a small area.

**Rabbit tool** Hydraulic spreading tool designed to pry open doors that swing inward.

# Wrap-Up

**Rapid intervention company/crew (RIC)** A minimum of two fully equipped personnel on site, in a ready state, for immediate rescue of injured or trapped brigade members. In some organizations, this is also known as a Rapid Intervention Team.

**Reciprocating saw** Powered by electric or battery motors, a reciprocating saw's blade moves back and forth.

**Response** Activities that occur in preparation for an emergency and continue until the arrival of emergency apparatus at the scene.

**Roofman's hook** A long pole with a solid metal hook used for pulling.

**Rotary saw** Powered by electric motors or gasoline engines, a rotary saw uses a large rotating blade to cut through material. The blades can be changed depending upon the material that is being cut.

**San Francisco hook** A multipurpose tool that can be used for several forcible entry and ventilation applications because of its unique design, which includes a built-in gas shut-off and directional slot.

**Screwdriver** A tool used for turning screws.

**Search and rescue** The process of searching a building for a victim and extricating the victim from the building.

**Seatbelt cutter** A specialized cutting device that cuts through seatbelts.

**Size-up** The ongoing observation and evaluation of factors that are used to develop objectives, strategy, and tactics for fire suppression.

**Sledgehammer** A long, heavy hammer that requires the use of both hands.

**Socket wrench** A wrench that fits over a nut or bolt and uses the ratchet action of an attached handle to tighten or loosen the nut or bolt.

**Spanner wrench** A type of tool used in coupling or uncoupling hoses by turning the rocker lugs on the connections.

**Spring-loaded center punch** A spring-loaded punch used to break automobile glass.

**Thermal imaging device** High-tech device that uses infrared technology to find objects giving off a heat signature.

**Ventilation** The process of removing smoke, heat, and toxic gases from a burning structure and replacing them with clean air.

**Vertical ventilation** The process of making openings so that the smoke, heat, and gases can escape vertically from a structure.

# Brigade Member in Action

**You have recently completed your initial training as an facility fire brigade member for your company. Because the new member to the team is typically assigned the duties of forcing entry or carrying the tools necessary for team support, you check out the location of all the tools and equipment available to your brigade at this site.**

1. One of the first things that you should do is:
   - **A.** check the warranty for all of the power tools carried on the apparatus.
   - **B.** make a mental note of where each tool is carried so you can find it quickly when you need it.
   - **C.** sharpen the blades on all of the cutting tools.
   - **D.** make sure that the striking tools and the prying tools are kept in separate compartments.

2. The brigade leader tells you that your assignment for the day is to carry "the irons." You understand from this assignment that your first responsibility at the scene of a fire will be:
   - **A.** search and rescue.
   - **B.** forcible entry.
   - **C.** ventilation.
   - **D.** rapid intervention.

3. "The irons" refers to two specific tools that are usually used together. These tools are:
   - **A.** pick-head axe and pike pole.
   - **B.** K tool and rabbit tool.
   - **C.** flat-head axe and Halligan tool.
   - **D.** battering ram and bolt cutters.

4. You respond on the second alarm for a fire in a three-story site administration building. While the brigade leader reports to the Incident Commander, you observe that smoke is coming from the top floor and escaping from under the eaves of the building. When the officer returns, he tells you and the other crew members to bring pike poles inside the building. Your assignment will most likely be:
   - **A.** vertical ventilation.
   - **B.** rapid intervention crew.
   - **C.** horizontal ventilation.
   - **D.** opening ceilings to expose hidden fire.

5. While working with a pike pole inside an office on the third floor, you observe smoke coming from behind the wooden baseboards, just above the floor. The brigade leader tells you to open this area and see if there is fire in the wall. You should:
   - **A.** use your pike pole to pry the baseboard away from the wall.
   - **B.** use a sledgehammer to break the baseboard.
   - **C.** get a screwdriver from the tool kit and remove the screws that secure the baseboard to the wall.
   - **D.** get a pick-head axe and use it to cut the baseboard away from the wall.

6. The brigade leader tells you and another brigade member to set up positive pressure ventilation to clear the smoke out of the area where you will be working. You should place a positive pressure ventilation fan:
   - **A.** outside a doorway to blow fresh air into the building.
   - **B.** on the roof of the building to suck smoke out through a hole.
   - **C.** in an open window to blow smoke out.
   - **D.** inside the building to blow smoke toward an open door.

7. Late one night, your brigade responds to a call for smoke coming from a below-grade transfer tunnel. When you arrive, you see that there is smoke exiting the main tunnel entrance above the normally closed and locked door. The door lock consists of a cylinder lock in the metal frame, metal door. No one seems to have a key for the lock. Which tool are you most likely to use to open the door?
   - **A.** Pick head axe
   - **B.** Spring-loaded center punch
   - **C.** K tool
   - **D.** Halligan tool

# Ropes and Knots

# Chapter 9

## NFPA 1081 Standard

**Incipient Facility Fire Brigade Member**

NFPA 1081 contains no Incipient Facility job performance requirements for this chapter.

**Advanced Exterior Facility Fire Brigade Member**

NFPA 1081 contains no Advanced Exterior Facility job performance requirements for this chapter.

**Interior Structural Facility Fire Brigade Member**

NFPA 1081 contains no Interior Structural Facility job performance requirements for this chapter.

## Knowledge Objectives

After completing this chapter, you will be able to:

- Describe the differences between life safety rope and utility rope.
- List the three most common synthetic fiber ropes used for fire brigade operations.
- Describe the construction of a kernmantle rope.
- Describe how to use rope to support response activities.
- Describe how to clean and check ropes.
- Describe how to record rope maintenance.
- List the reasons for placing a life safety rope out of service.
- Describe the knot types and their usage in the fire service.
- Describe how to tie safety, half hitch, clove hitch, figure eight, figure eight on a bight, figure eight with a follow-through, bowline, and sheet bend or Becket bend knots.
- Describe the types of knots to use for given tools, ropes, or situations.
- Describe hoisting methods for tools and equipment.

## Skills Objectives

After completing this chapter, you will be able to perform the following skills:

- Place a life safety rope into a rope bag.
- Tie the following knots:
  Safety (overhand)
  Half hitch
  Clove hitch
  Figure eight
  Figure eight on a bight
  Figure eight follow-through
  Bowline
  Sheet bend or becket bend
- Hoist the following tools using the correct knots:
  Axe
  Pike pole
  Ladder
  Charged hose line
  Uncharged hose line
  Exhaust fan

# You Are the Brigade Member

**You are a new brigade member responding to your first structure fire at the plant. Flames are showing from the third floor in the rear of the building. Two brigade members go into the building to perform search and rescue operations while the rest of the team sets up a rapid intervention crew and an attack line to begin fire suppression. As the attack crew enters the building, you see the rescue team at an upstairs window. The brigade leader tells you that they need another axe and that they are going to drop a rope so you can send it up to them.**

1. ***As the rescue team lowers a rope from the upstairs window, what factors do you need to consider?***
2. ***How do you safely tie the axe to the rope?***
3. ***What kind of knot will you use?***
4. ***How can you be sure that the axe will not slip out of the rope?***

## Introduction

In the fire service, ropes are widely used to hoist or lower tools, appliances, or people; to pull a person to safety; or to serve as a lifeline in an emergency. A rope may be your only means of accessing a trapped person or your only way of escaping from a fire.

Learning about ropes and knots is an important part of your training as a brigade member. This chapter will give you a basic understanding of the importance of ropes and knots. You can then build on this foundation as you develop skills in handling ropes and tying knots. You must be able to tie simple knots accurately without hesitation or delay regardless of the conditions.

This chapter discusses different types of rope construction and the materials used in making ropes. It covers the care, cleaning, inspection, and storage of ropes. It also shows how to tie eight essential knots and how to secure tools and equipment so they can be raised or lowered using ropes. Finally, it discusses additional uses for ropes in various rescue situations.

## Types of Rope

There are two primary types of rope used in the fire service, each dedicated to a distinct function. Life safety rope is used solely for supporting people. Life safety rope must be used anytime a rope is needed to support a person, whether during training or during firefighting, rescue, or other emergency operations ▶ Figure 9-1. Utility rope is used when it is NOT necessary to support the weight of a person, such as when hoisting or lowering tools or equipment.

### Life Safety Rope

The life safety rope is a critical tool used only for life-saving purposes. It must never be used for utility purposes. Life safety rope must be used in every situation where the rope must support the weight of one or more persons. In these situations, rope failure could result in serious injury or death.

Because a brigade member's equipment must be extremely reliable, the criteria for design, construction, and performance of life safety rope and related equipment are specified in *NFPA 1983, Standard on Fire Service Life Safety Rope and System Components*. Life safety ropes are rated for either one

Figure 9-1 A safety rope is a critical tool for brigade members.

person or two persons. A two-person rope must be used in rescue operations where both the rescued individual and the rescuer require support.

NFPA 1983 lists very specific standards for the construction of life safety rope. The *Standard* also requires the rope manufacturer to include detailed instructions for the proper use, maintenance, and inspection of the life safety rope, including the conditions for removing the rope from service. The manufacturer must also supply a list of criteria that must be reviewed before a life safety rope that has been used in the field can be used again. If the rope does not meet all of the criteria, it must be retired from service.

### Types of Life Safety Ropes

The two primary types of life safety ropes are the one-person rope and the two-person rope. In NFPA 1983, the one-person rope is classified as a light duty life safety rope and the two-person rope is classified as a general duty life safety rope.

A

B

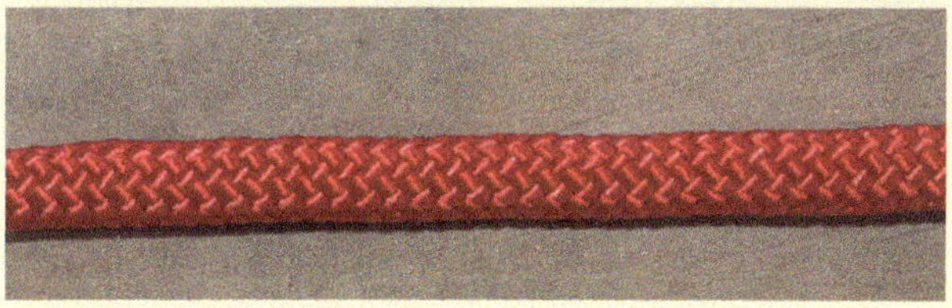

Figure 9-2 **A.** A one-person rope is a light duty life safety rope. **B.** A two-person rope is a general duty life safety rope.

A one-person life safety rope is designed to bear the weight of a single person (300 lb) ◄ Figure 9-2A. A two-person life safety rope is designed to bear the weight of two people (600 lb) ◄ Figure 9-2B. After each use, these ropes must be inspected according to the criteria provided by the manufacturer before they can be used again. If a life safety rope has been damaged or overstressed, or if it does not meet the inspection criteria, it cannot be reused as a life safety rope.

## Personal Escape Ropes

A personal escape rope is intended to be used by a brigade member only for self-rescue from an extreme situation. This rope is designed to carry the weight of only one person and to be used only one time ► Figure 9-3. Its purpose is to provide the brigade member with a method of escaping from a life-threatening situation. After one use, the personal escape rope should be replaced by a new rope.

When you are fighting a fire, you should always have a safe way to get out of a situation and to a safe location. You may be able to go back through the door that you entered, or you may have another exit route, such as through a different door, through a window, or down a ladder. If conditions suddenly change for the worse, having an escape route can save your life.

Sometimes, you may find yourself in a situation where conditions deteriorate so quickly that you cannot use your planned exit route. For example, the stairway you used collapses behind you, or the room you are in suddenly flashes over (a phase in the development of a contained fire in which exposed surfaces reach ignition temperature more or less simultaneously and fire spreads rapidly throughout the space), blocking your planned route out. In such a situation, you may need to take extreme measures to get out of the building. The personal escape rope was developed specifically for this type of emergency self-rescue situation. A personal escape rope can support the weight of one person and fits easily in a small packet or pouch ▼ Figure 9-4.

Figure 9-3 A personal escape rope is designed to be used only once.

Figure 9-4 A personal escape rope can be easily carried by a brigade member.

Figure 9-5 Utility ropes are used for hoisting and lowering tools.

Figure 9-6 Some utility ropes are made from natural fibers.

Because these personal escape ropes are so important, they can be used only once. After they are used once, they are discarded. The rope may have been damaged in some way and might fail if it ever has to be used again for rescue. You cannot take that chance; your life may depend on the quality and strength of your personal escape rope.

## Utility Rope

Utility rope is used when it is NOT necessary to support the weight of a person. Fire brigade utility rope is used for hoisting or lowering tools or equipment, for ladder halyards (rope used on extension ladders to raise a fly section), for marking off areas, and for stabilizing objects ▲ Figure 9-5. Utility ropes also require regular inspection.

### Brigade Member Safety Tips

Many fire brigades use color-coding or other visible markings to identify different types of rope. This allows a brigade member to very quickly determine if a rope is a one-person or two-person life safety rope or a utility rope. The length of each rope should also be clearly marked by a tag or a label on the rope bag.

Utility ropes must not be used in situations where life safety rope is required. A brigade member must be able to instantly recognize the category of a rope from its appearance and markings.

# Rope Materials

Ropes are made from different types of materials. The earliest ropes were made from naturally occurring vines or fibers that were woven together. Ropes are now made of synthetic materials such as nylon or polypropylene. Because ropes have many different uses, different materials may work better than others, depending on the situation.

## Natural Fibers

In the past, fire brigades used ropes made from natural fibers, such as manila, because there were no alternatives ▲ Figure 9-6. The natural fibers are twisted together to form strands. A strand may contain hundreds of individual fibers of different lengths. Today, ropes made from natural fibers are still used as utility ropes but are no longer acceptable as life safety ropes ► Table 9-1. Natural fiber ropes can be weakened by mildew and deteriorate with age, even when properly stored. A wet manila rope can absorb 50% of its weight in water, making it very susceptible to deterioration. A natural fiber rope is very difficult to dry.

## Synthetic Fibers

Since nylon was first manufactured in 1938, synthetic fibers have been used to make ropes. In addition to nylon, several newer synthetic materials such as polyester, polypropylene, and polyethylene are used in rope construction ► Figure 9-7. Synthetic fibers have several advantages over natural fibers ► Table 9-2. Synthetic fibers are generally stronger than natural fibers, so it may be possible to use a smaller diameter rope without sacrificing strength. Synthetic materials can also produce very long

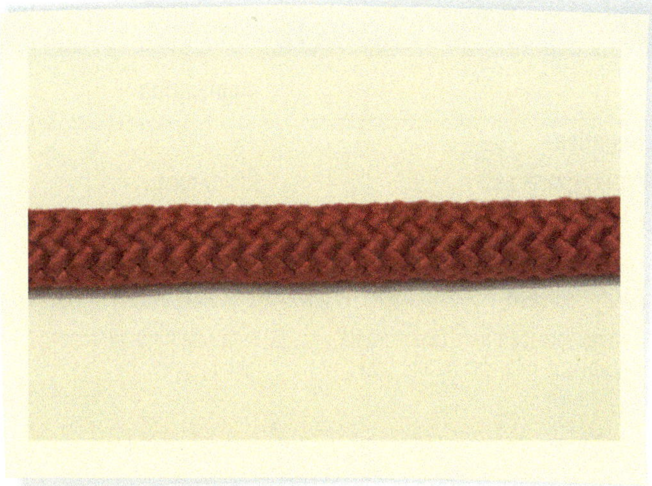

Figure 9-7 Synthetic fibers are generally stronger than natural fibers.

Figure 9-8 Polypropylene rope is often used in water rescues.

fibers that run the full length of a rope to provide greater strength and added safety.

Synthetic ropes are more resistant to rot and mildew than natural fiber ropes and do not age or degrade as quickly. They also absorb much less water and can be easily washed and dried. Some types of synthetic rope can float on water, which is an advantage in water rescue situations.

However, ropes made from synthetic fibers do have some drawbacks (▶ Table 9-3). Prolonged exposure to ultraviolet light as well as exposure to strong acids or alkalis can damage a synthetic rope and decrease its life expectancy. In addition, synthetic materials may be highly susceptible to abrasion or cutting.

Life safety ropes are always made of synthetic fibers. Before any rope can be used for life safety purposes, it must meet the requirements outlined in *NFPA 1983*. These standards specify that life safety rope must be made of continuous filament virgin fiber and woven of block creel construction (without knots or splices in the yarns, ply yarns, strands, braids, or rope). Rope of any other material or construction may not be used as a life safety rope.

The most common synthetic fiber used in life safety ropes is nylon. It has a high melting temperature with good abrasion resistance and is strong and lightweight. Nylon ropes are also resistant to most acids and alkalis. Polyester is the second most common synthetic fiber used for life safety ropes. Some life safety ropes are made of a combination of nylon and polyester or other synthetic fibers.

Polypropylene is the lightest of the synthetic fibers. Because it does not absorb water and floats, polypropylene rope is often used for water rescue situations (▶ Figure 9-8). However, it is not as suitable as nylon for life safety uses because it is not as strong, it is hard to knot, and it has a low melting point.

### Table 9-1 Drawbacks to Using Natural Fiber Ropes

- Lose their load-carrying ability over time
- Subject to mildew and rot
- Absorb 50% of their weight in water
- Degrade quickly

### Table 9-2 Advantages to Using Synthetic Fiber Ropes

- Thinner without sacrificing strength
- Less absorbent than natural fiber ropes
- Greater resistance to rotting and mildew
- Longer-lasting than natural fiber ropes
- Greater strength and added safety
- More burn resistant than natural fiber ropes

### Table 9-3 Drawbacks to Using Synthetic Fiber Ropes

- Can be damaged by prolonged exposure to ultraviolet light
- Can be damaged by exposure to strong acids or alkalis
- Susceptible to abrasion

**Table 9-4 Properties of Rope Materials**

| Type | Material | Properties: Positive | Properties: Negative | Application |
|---|---|---|---|---|
| Natural | Manila | No real advantages over synthetic rope | Absorbs water easily<br>Cannot bear as much weight as synthetic rope<br>Noncontinuous fibers<br>Easily degraded | Utility rope |
| Synthetic | Nylon<br>Polyester | High melting temperature<br>Good abrasion resistance<br>Less absorbent than natural ropes<br>Greater resistance to rotting and mildew<br>Lasts longer than natural ropes | Can be damaged from exposure to sunlight, oils, gas, acids, bases, or fumes | Life safety rescue rope<br>Utility rope |
| | Polypropylene | Does not absorb water<br>Floats | Hard to knot<br>Low melting point | Water rescue rope<br>Utility rope |

## Rope Construction

There are several different types of rope construction. The best choice of rope construction depends on the specific application (▲ Table 9-4).

### Twisted and Braided Rope

Twisted rope, which is also called laid rope, is made of individual fibers twisted into strands. The strands are then twisted together to make the rope (► Figure 9-9 A). This method of rope construction has been used for hundreds of years. Both natural and synthetic fibers are used to make twisted rope.

This method of construction exposes all of the fibers to the outside of the rope where they are subject to abrasion. Abrasion can damage the rope fibers and may reduce rope strength. Twisted ropes tend to stretch and are prone to unraveling when a load is applied.

Braided rope is constructed by weaving or intertwining strands together (► Figure 9-9 B). This method of construction also exposes all of the strands to the outside of the rope where they are subject to abrasion. Most braided rope is constructed from synthetic fibers. Braided rope will stretch under a load, but it is not prone to twisting. A double braided rope has an inner braided core covered by a protective braided sleeve, so that only the fibers in the outer sleeve are exposed. The inner core is protected from abrasion.

### Kernmantle Rope

Kernmantle rope consists of two distinct parts: the kern and the mantle. The kern is the center or core of the rope; it provides about 70% of the strength of the rope. The mantle or sheath is a braided covering that protects the core from dirt and abrasion. Only about 30% of the strength of the rope comes from the mantle. Both parts of a kernmantle rope are made with synthetic fibers, but different fibers may be used for the kern and the mantle.

Each fiber in the kern extends for the entire length of the rope without knots or splices. This block creel construction is required under NFPA 1983 for all life safety ropes. The continuous filaments produce a core that is stronger than one constructed of shorter fibers that are twisted or braided together.

**Brigade Member Tips**

To understand how a kernmantle rope gets its strength, imagine a small, monofilament fishing line rated at 15 lb. Putting 100 identical strands of this filament side-by-side would create a cable capable of supporting 1,500 lb. When covered with a sheath, the fishing line cable would create a static kernmantle rope.

Figure 9-9 A. Twisted rope. B. Braided rope.

Figure 9-10 A. Suspension bridge cables use the same type of construction as kernmantle ropes. B. This close-up shows the core and mantle construction.

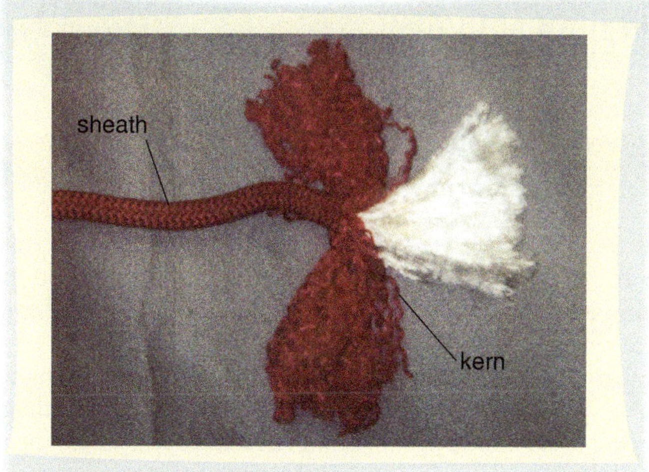


Figure 9-11 The parts of a kernmantle rope.

Kernmantle construction is also used for the cables that support a suspension bridge ▲ Figure 9-10. Thin strands of steel are laid from one end of the bridge to the other. Although each strand only supports a small weight, multiple strands are strong enough to support both the bridge and the load carried by the bridge. The cables are then wrapped with a protective covering, similar to the mantle portion of a rope.

Kernmantle construction produces a very strong and flexible rope that is relatively thin and lightweight ► Figure 9-11. This construction is well suited for rescue work and is very popular for life safety rope.

## Dynamic and Static Rope

A rope can be either dynamic or static, depending on how it reacts to an applied load. A dynamic rope is designed to be elastic and will stretch when it is loaded. A static rope will not stretch under load. The differences between dynamic and static ropes result from both the fibers used and the construction method.

Dynamic rope is usually used in safety lines for mountain climbing, because it will stretch and cushion the shock if a climber falls a long distance. A static rope is more suitable for most fire rescue situations, where falls from great heights are not anticipated. Teams that specialize in rope rescue often carry both static and dynamic ropes for use in different situations.

### Dynamic and Static Kernmantle Ropes

Kernmantle ropes can be either dynamic or static. A dynamic kernmantle rope is constructed with overlapping or woven fibers in the core. When the rope is loaded, the core fibers are pulled tighter, which gives the rope its elasticity.

The core of a static kernmantle rope has all of the fibers laid parallel to each other. A static kernmantle rope has very little elasticity and limited elongation under an applied load. Most fire brigade life safety ropes use static kernmantle construction. It is well suited for lowering a person and can be used with a pulley system for lifting individuals. It can also be used to create a bridge between two structures.

# Rope Strength

Life safety ropes are rated to carry a specific amount of weight under the minimum requirements of NFPA 1983 ▼ Table 9-5. The required minimum breaking strength for

**Table 9-5 Required Strength of Life Safety Ropes**

| Classification | Rated Load (Persons) | Rated Load (Weight) | Minimum Breaking Strength | Safety Factor |
|---|---|---|---|---|
| Personal escape rope | One | 300 lb | 3,000 lbf (13.34 kN) | 10:1 |
| Light use life safety rope | One | 300 lb | 4,500 lbf (20 kN) | 15:1 |
| General use life safety rope | Two | 600 lb | 9,000 lbf (40 kN) | 15:1 |

SOURCE: NFPA 1983, *Standard on Fire Service Life Safety Rope and System Components.*

## Fire Marks

Although synthetic fiber ropes were introduced to the fire service in the 1950s, they were not widely adopted until the 1980s.

a life safety rope is based on an assumed loading of 300 lb per person with a safety factor of 15:1. The safety factor allows for reductions in strength due to knots, twists, abrasion, or any other cause. The safety factor also allows for shock loading if a weight is applied very suddenly. For example, shock loading could occur if the person who is tied to the rope falls and then is stopped by the rope. A personal escape rope is also expected to support a weight of 300 lb, representing one person, with a safety factor of 10:1.

The actual breaking strength of a rope depends on the material, the diameter, and the type of construction. The rope manufacturer should be consulted for detailed specifications.

## Technical Rescue Hardware

During technical rescue incidents, ropes are often used to access and extricate individuals. In addition to the rope, several hardware components may also be used. The one most commonly used by brigade members is a carabiner, or a snap link (▶ Figure 9-12). This device is used to connect one rope to another rope, to a harness, or to itself. There are different types of carabiners, and you should know how to operate the type used by your organization.

Figure 9-12 A carabiner.

### Harnesses

A harness is a piece of rescue safety equipment made of webbing and worn by a person. It is used to secure the person to a rope or to a solid object. Three different types of harnesses—belts, seats, and chest harnesses—are used by rescuers, depending upon the type of circumstances encountered.

- The ladder belt harness (NFPA Class I harness) is used to keep a brigade member in place on a ladder. Generally, the most accurate way to differentiate a Class I harness from a Class II harness is to inspect the harness label.
- The seat harness (NFPA Class II harness) is used to support a brigade member, particularly in rescue situations (▶ Figure 9-13).
- The chest harness (NFPA Class III harness) is the most secure type of harness and is used to support a brigade member who is being raised or lowered on a life safety rope (▶ Figure 9-14).

Harnesses must be cleaned and inspected regularly, just as you do for life safety ropes. Follow the manufacturer's instructions for cleaning and inspecting harnesses.

Figure 9-13 Class II harness (seat harness).

### Rope Rescue

Rope rescue involves raising and lowering rescuers to access injured or trapped individuals, as well as raising or lowering those rescued so they can be given appropriate medical treatment. An approved rope rescue course is required to attain proficiency in rope rescue skills. However, this chapter will provide you with the basics of rope rescue and give you a foundation for learning the more complex parts of rope rescue.

## Brigade Member Tips

Harnesses that are labeled as NFPA Class III and independently certified to meet American National Standards Institute (ANSI) Z359.1 are OSHA compliant for fall-arrest protection.

Figure 9-14 Class III harness (chest harness).

Figure 9-15 Rope rescues require intense technical training.

Rope rescue courses cover the technical skills needed to raise or lower people using mechanical advantage systems and to remove someone from a rock ledge or a confined space ▲ Figure 9-15. They also cover the equipment and skills needed to accomplish these rescues safely.

Figure 9-16 Ropes are invaluable when a person is trapped in an inaccessible area, such as a high platform.

## Rope Rescue Incidents

Most rope rescue incidents involve people who are trapped in normally inaccessible locations such as a tower or platform ▲ Figure 9-16. Rescuers often have to lower themselves using a system of anchors, webbing, ropes, carabiners, and other devices to reach the trapped person. Once rescuers reach the person, they then have to stabilize him or her and determine how to get the person to safety. Sometimes the person will have to be lowered or raised to a safe location. Extreme cases could involve more complicated operations, such as transporting the person in a basket lowered by a helicopter.

The type and number of ropes used in a rope rescue will depend upon the situation. There is almost always a primary rope that will bear the weight of the rescuer (or rescuers) in order to reach the person. The rescuers will often have a second safety line attached to them, which serves as a backup if the main line fails. Additional lines may be needed to raise or lower the trapped individual, depending upon the circumstances.

## Trench Rescue

Rescues in collapsed trenches often are complicated and involve a number of different skills, such as shoring, air quality monitoring, confined space operations, and rope rescue. Ropes are often used to remove the trapped person. After the rescuers shore the walls of the trench and remove the dirt covering the person, they will place the person in a Stokes basket or on a backboard and lift him or her to the surface. If the trench is deep, ropes may be used to raise the patient to the surface.

## Confined Space Rescue

A confined space rescue can take place in locations such as tanks, towers, underground electrical vaults, storm drains,

# Voices of Experience

**"Ropes and proper knots can be life savers, and I encourage the members of my team to keep a piece of knot-tying rope handy in their personal cars to practice while stopped at traffic lights."**

Technical rescue including rope work is a key technique for facility brigades owing to the vast range of vessels, racks, cranes, process plants, confined spaces, and constant construction activities they may encounter. Rope use can be called upon frequently for rescue and equipment movement.

For example, two summers ago, our team was called upon to assist with an extremely difficult and hazardous rescue. A bulldozer operator had become trapped in his cab after a coal pile at a power plant shifted, causing the dozer to become submerged in a 40-foot pile of coal. The initial rescuers faced a daunting task, similar to trying to keep the walls of a sandy hole dug at the beach from collapsing. The coal pile was so unstable that the only possible way to extricate the victim was vertically, without disturbing the coal pile.

Luckily, a screw conveyor was located directly over the sinkhole, at an elevation of some 100 feet in the air. Our team of rescue technicians, who were highly skilled in ropes and knots, made quick work of setting up a lowering system with multiple anchor points and a 4:1 mechanical advantage system for raising the victim. A rescuer was lowered into the hole. Gingerly digging by hand, he removed the coal and freed the cab door of the bulldozer. The operator, who had suffered a broken arm, was packaged by the rescuer and outfitted with a rescue harness. Rescue technicians on the conveyor were ready and waiting with the haul system and gently raised the victim and then the rescuer from the hole. The victim was transported to a regional trauma center, where he underwent surgery; he eventually made a full recovery.

Our rescuer was quoted as saying, "A lot goes through your mind—first about the victim, and then about yourself and the safety of the operation, and then all the things you care about in life rush through your entire being." In this case, the coordinated efforts of individuals who continuously practice rope and knot work made the completion of this extremely hazardous rescue flawless and safe.

I cannot stress enough how important a piece of rope and a proper knot are in a fire fighters' chest of tools. It may, indeed, save your life one day in a bailout situation or another hazardous condition. Practice, practice, practice, and more practice will keep your knot skills sharp and ready. Ropes and proper knots can be life savers, and I encourage the members of my team to keep a piece of knot-tying rope handy in their personal cars to practice while stopped at traffic lights.

**John A. Welling, III**
Bristol-Myers Squibb Company
New Brunswick, New Jersey

and similar spaces. It is often very difficult to extricate an unconscious or injured person from these locations because of the poor ventilation and limited entry or exit area. For this reason, ropes are often used to remove an injured or unconscious person ▼ Figure 9-17.

### Water Rescue

Ropes can be used in a variety of ways during water rescue operations. The simplest situation involves a rescuer on the shore throwing a rope to a person in the water and pulling the person to shore ▼ Figure 9-18. A more complicated situation may involve a rope stretched across a stream or river. A boat is tethered to the rope, and rescuers on shore maneuver the boat using a series of ropes and pulleys.

Figure 9-17 Ropes are often used to remove an injured or unconscious person from a confined space.

Figure 9-18 Ropes ensure rescuer safety during water rescues.

## Rope Maintenance

All ropes, especially life safety ropes, need proper care to perform in an optimal manner. Maintenance is necessary for all kinds of equipment and all types of rope, and it is absolutely essential for life safety ropes. Your life and the lives of others depend on the proper maintenance of your life safety ropes.

There are four parts to the maintenance formula:

- Care
- Cleaning
- Inspection
- Storage

### Care

You must follow certain principles to preserve the strength and integrity of rope ▼ Table 9-6:

- Protect the rope from sharp and abrasive surfaces. Use edge protectors when the rope must pass over a sharp or unpadded surface.
- Protect the rope from heat, chemicals, and flames.
- Protect the rope from rubbing against another rope or webbing. Friction generates heat, which can damage or destroy the rope.
- Protect the rope from prolonged exposure to sunlight. Ultraviolet radiation can damage rope.
- Never step on a rope! Your footstep could force shards of glass, splinters, or abrasive particles into the core of the rope, damaging the rope fibers.
- Follow the manufacturer's recommendations for rope care.

### Cleaning

Many ropes made from synthetic fibers can be washed with a mild soap and water. A special rope washer can be attached to a garden hose ► Figure 9-19. Some manufacturers recommend placing the rope in a mesh bag and washing it in a frontloading washing machine.

Use a mild detergent. Do not use bleach because it can damage rope fibers. Follow the manufacturer's recommendations for specific care of your rope. Do not pack or store wet

**Table 9-6 Principles to Preserve Strength and Integrity of Rope**

- Protect from sharp abrasive surfaces.
- Protect from heat, chemicals, and flames.
- Protect from rubbing against another rope.
- Protect from prolonged exposure to sunlight.
- Never step on a rope.
- Follow manufacturer's recommendations.

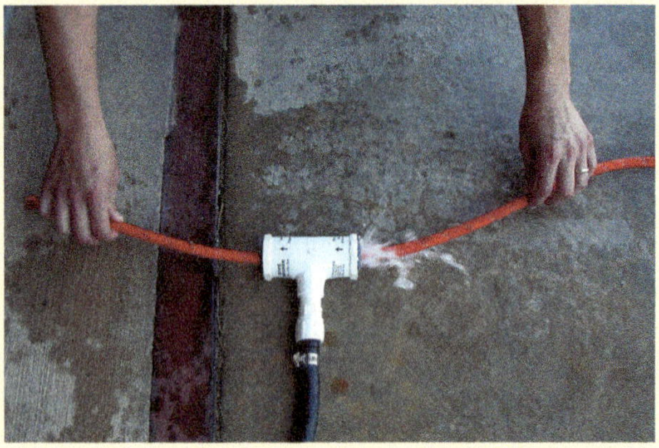

Figure 9-19 Some fire brigades use a rope washer to clean their ropes.

or damp rope. Air-drying is usually recommended, but rope should not be dried in direct sunlight. The use of mechanical drying devices is not usually recommended.

## Inspection

Life safety ropes must be inspected after each use, whether the rope was used for an emergency incident or in a training exercise (► Table 9-7). Unused rope should be inspected on a regular schedule. Some organizations inspect all rope, including life safety and utility ropes, every 3 months. Obtain the inspection criteria from the rope manufacturer.

Inspect the rope visually, looking for cuts, frays, or other damage, as you run it through your fingers. Because you cannot see the inner core of a kernmantle rope, feel for any **depressions** (flat spots or lumps on the inside). Examine the sheath for any discolorations, abrasions, or flat spots

**Table 9-7 Questions to Consider When Inspecting Life Safety Ropes**

- Has the rope been exposed to heat or flame?
- Has the rope been exposed to abrasion?
- Has the rope been exposed to chemicals?
- Has the rope been exposed to shock loads?
- Are there any depressions, discoloration, or lumps in the rope?

**Table 9-8 Signs of Possible Rope Deterioration**

- Discoloration
- Shiny markings from heat or friction
- Damaged sheath
- Core fibers poking through the sheath

(▼ Figure 9-20). If you have any doubt about whether the rope has been damaged, consult with your brigade leader (▲ Table 9-8).

A life safety rope that is no longer usable must be destroyed or downgraded. In some cases, a used life safety rope can be downgraded and used as a utility rope. A downgraded rope must be clearly marked so that it cannot be confused with a life safety rope.

### Rope Record

Each piece of rope must be marked for identification. A **rope record** must be kept for each piece of life safety rope.

**Brigade Member Safety Tips**

A **shock load** can occur when a rope is suddenly placed under unusual tension. This could occur if someone attached to a life safety rope falls until the length of the rope or another rescuer stops the drop. A utility rope can be shock-loaded in a similar manner if a piece of equipment that is being raised or lowered suddenly drops.

Any rope that has been shock-loaded must be inspected and may have to be removed from service. Although there may not be any visible damage, shock-loading may cause damage that is not immediately apparent. Repeated shock loads can severely weaken a rope so that it can no longer be used safely. Accurate rope records will help identify potentially damaged rope. Whenever a life safety rope is shock loaded, it should be removed from service or downgraded to utility rope.

Figure 9-20 Rope inspection is a critical step.

Figure 9-21 Ensure that ropes are stored safely.

This record should include a history of when the rope was purchased, each time it was used, how it was used, and the types of loads applied to it. Each inspection should also be recorded. Many fire brigades maintain records for both utility ropes and life safety ropes.

## Storage

Proper care will ensure a long life for your rope and reduce the chance of equipment failure and accident. Store ropes away from temperature extremes, out of sunlight, and in areas where there is some air circulation. Avoid placing ropes where fumes from gasoline, oils, or hydraulic fluids can damage the rope. Apparatus compartments used to store ropes should be separated from compartments used to store any oil-based products or machinery powered by gasoline or diesel fuel. Do not place any heavy objects on top of the rope (▲ Figure 9-21).

A rope bag is used to protect and store ropes. Each bag should contain only one rope. Rope may also be coiled for storage (► Figure 9-22). Very long pieces of rope are sometimes stored on reels (► Figure 9-23). Follow the steps in (► Skill Drill 9-1) to place a life safety rope into a rope bag:

1. Tie a figure eight knot on a bight in the first end of the life safety rope to be placed into the rope bag. **(Step 1)**
2. Load the life safety rope into the rope bag carefully. **(Step 2)**
3. Do not try to coil the rope in the bag, because this will cause it to kink and become tangled when it is pulled out. **(Step 3)**

### Brigade Member Tips

Rope identification is facilitated by labeling each end of the rope and protecting the label with clear heat-shrink tubing.

Figure 9-22 Ropes may be coiled for storage.

Figure 9-23 Long sections of rope are sometimes stored on reels.

## Knots

Knots are prescribed ways of fastening lengths of rope or webbing to objects or to each other. As a brigade member, you must know how to tie and when to use certain knots. Knots can be used for multiple purposes. Hitches, such as the clove hitch, are used to attach a rope around an object. Knots, such as the figure eight and the bowline, are used to form loops. Bends, such as the sheet bend or becket bend, are used to join two ropes together. Safety knots, such as the overhand knot, are used to secure the ends of ropes to prevent them from coming untied.

Any knot will reduce the load-carrying capacity of the rope by a certain percentage (► Table 9-9). You can avoid an unnecessary reduction in rope strength if you know what type of knot to use and how to tie it correctly.

**Table 9-9 Effect of Knots on Rope Strength**

| Group | Knot | Reduction in Strength (%) |
|---|---|---|
| Loop knots | Figure eight on a bight | 20 |
| | Figure eight with a follow-through | 19 |
| | Bowline | 33 |

## Terminology

Specific terminology is used to refer to the parts of a rope in describing how to tie knots ▶ Figure 9-24.

- The working end is the part of the rope used for forming the knot.
- The running end is the part of the rope used for lifting or hoisting.
- The standing part is the rope between the working end and the running end.
- A bight is formed by reversing the direction of the rope to form a "U" bend with two parallel ends ▶ Figure 9-25.

9-1 Skill Drill

## Placing a Life Safety Rope into a Rope Bag

1 Tie a figure eight knot on a bight in the first end of the rope to go into the bag.

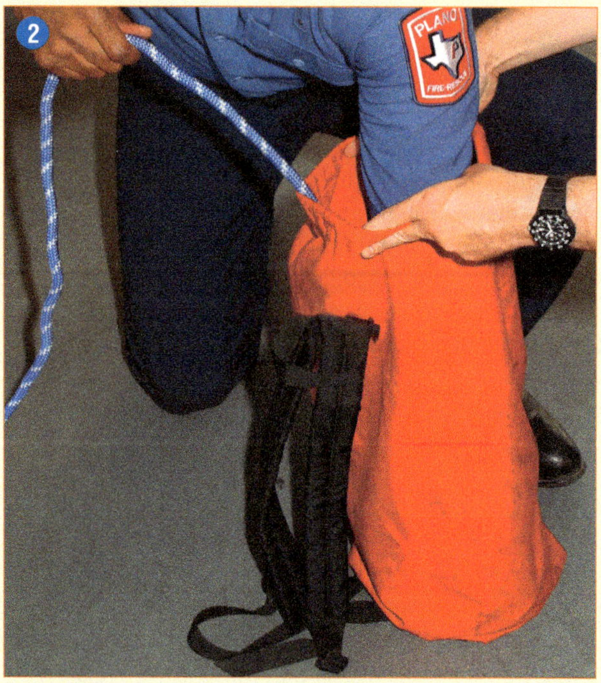


2 Load the rope into the bag.

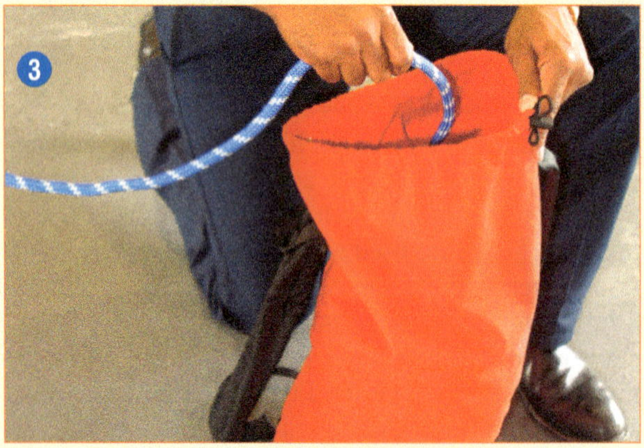

3 Do not coil the rope in the bag.

Figure 9-24 The sections of a rope used in tying knots.

Figure 9-26 A loop.

Figure 9-25 A bight.

Figure 9-27 A round turn.

- A loop is formed by making a circle in the rope ▶ Figure 9-26.
- A round turn is formed by making a loop and then bringing the two ends of the rope parallel to each other ▶ Figure 9-27.

A brigade member must know how to tie eight basic knots or hitches and how to use them properly. The eight basic knots are:

- Safety knot (overhand knot)
- Half hitch
- Clove hitch
- Figure eight
- Figure eight on a bight
- Figure eight with a follow-through
- Bowline
- Bend (sheet or becket bend)

## Safety Knot

A safety knot (also referred to as an overhand knot or a keeper knot) is used to secure the leftover working end of the rope. It provides a degree of safety to ensure that the primary knot will not become undone. A safety knot should always be used to finish the other basic knots.

A safety knot is simply an overhand knot in the loose end of the rope that is made around the standing part of

### Teamwork Tips

In a rescue situation, always have your partner check your knots. Taking a few extra seconds to be sure knots are tied correctly can save lives.

9-2 Skill Drill

## Safety Knot

Take the loose end of the rope, beyond the knot, and form a loop around the standing part of the rope.

Pass the loose end of the rope through the loop.

Tighten the safety knot by pulling on both ends at the same time.

the rope. This secures the loose end and prevents it from slipping back through the primary knot. Follow the steps in ▲ **Skill Drill 9-2** to tie a safety knot:

1. Take the loose end of the rope, beyond the knot, and form a loop around the standing part of the rope. **(Step 1)**
2. Pass the loose end of the rope through the loop. **(Step 2)**
3. Tighten the safety knot by pulling on both ends at the same time. To test whether you've tied a safety knot correctly, try sliding it on the standing part of the rope. A knot that is tied correctly will slide. **(Step 3)**

## Hitches

Hitches are knots that wrap around an object, such as a pike pole or a fencepost. They are used to secure the working end of a rope to a solid object or to tie a rope to an object before hoisting it.

### Half Hitch

The half hitch is not a secure knot by itself. It is used only in conjunction with other knots. For example, when hoisting an axe or pike pole, you will use the half hitch to keep the hoisting rope aligned with the handle. On long objects,

93 Skill Drill

## Half Hitch

Make a round turn in the rope around the object.

Pass the standing end of the rope under the round turn on the opposite side from the direction of the pull.

you may need to use several half hitches. Follow the steps in ▲ Skill Drill 9-3 to tie a half hitch:

1. Make a round turn in the rope around the object you are hoisting. **(Step 1)**
2. Pass the standing end of the rope under the round turn on the opposite side from the direction of the pull. **(Step 2)**

### Clove Hitch

A clove hitch is used to attach a rope firmly to a round object, such as a tree or a fence post. It can also be used to tie a hoisting rope around an axe or pike pole. A clove hitch can be tied anywhere in a rope and will hold equally well if tension is applied to either end of the rope or both ends simultaneously. There are two different methods of tying this knot.

A clove hitch tied in the open is used when the knot can be formed and then slipped over the end of an object, such as an axe or pike pole. It is tied by making two consecutive loops in the rope. Follow the steps in ► Skill Drill 9-4 to tie a clove hitch in the open:

1. The first loop is made with the left hand and has the running part of the rope passing over the working part. **(Step 1)**
2. The second loop is made with the right hand and has the running part of the rope passing under the working part. **(Step 2)**
3. The right-hand loop is then placed on top of the left-hand loop so that the openings are aligned. **(Step 3)**
4. The two loops are held together, and the knot is slipped over the object. **(Step 4)**
5. Tighten the clove hitch by simultaneously pulling both ends of the rope in opposite directions. Tie a safety knot in the working end of the rope. **(Step 5)**

If the object is too large or too long to slip the clove hitch over one end, the same knot can be tied around the object. Follow the steps in ► Skill Drill 9-5 to tie a clove hitch around an object:

1. First, loop the rope completely around the object with the working end below the running end. **(Step 1)**
2. Loop the working end around the object a second time to form a second loop, slightly above the first loop. **(Step 2)**
3. Pass the working end under the second loop, just above the point where the second loop crosses over the first loop. **(Step 3)**
4. Secure the knot by pulling on both ends. **(Step 4)**
5. Tie a safety knot in the working end of the rope. **(Step 5)**

## Loop Knots

Loop knots are used to form a loop in the end of a rope. These loops may be used for hoisting tools, for securing a person during a rescue, for securing a rope to a fixed object, or for identifying the end of a rope stored in a rope bag. When tied properly, these knots will not slip and are easy to untie.

94 Skill Drill

# Clove Hitch Tied in the Open

Make a loop using your left hand, with the running part of the rope over the working part of the rope.

Make a second loop using your right hand, with the running part of the rope under the working part of the rope.

Bring the right-hand loop on top of the left-hand loop.

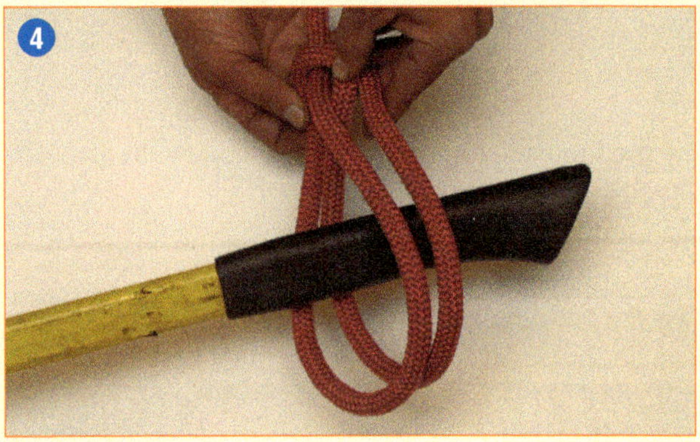


Slide both loops over the object.

Pull in opposite directions to tighten the clove hitch. Tie a safety knot in the working end of the rope.

95 Skill Drill

# Clove Hitch Tied Around an Object

Make a complete loop around the object, working end down.

Make a second loop around the object a short distance above the first loop.

Now pass the working end of the rope under the second loop, above the point where the second loop crosses over the first loop.

Tighten the knot and secure it by pulling on both ends.

Tie a safety knot in the working end of the rope.

96 Skill Drill

## Figure Eight Knot

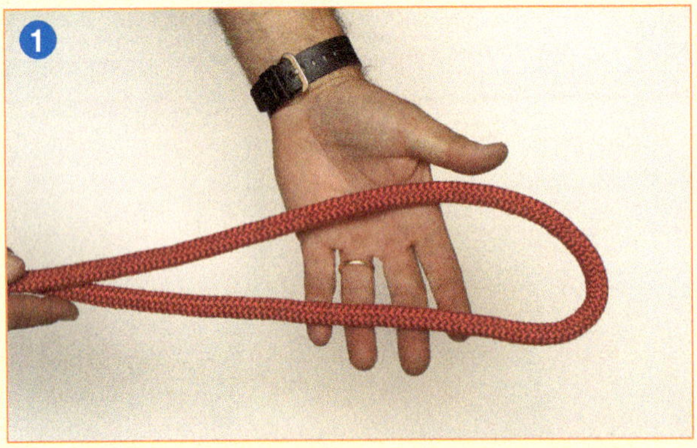

Form a bight in the rope.

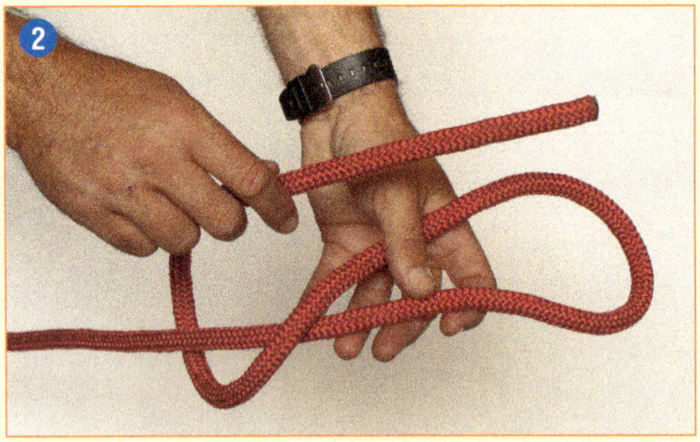

Loop the working end of the rope completely around the standing part of the rope.

Thread the working end back through the bight.

Tighten the knot by pulling on both ends simultaneously.

### Figure Eight Knot

A figure eight is a basic knot used to produce a family of other knots, including the figure eight on a bight and the figure eight with a follow-through. A simple figure eight knot is seldom used alone. Follow the steps in ▲ **Skill Drill 9-6** to tie a figure eight knot:

1. Form a bight in the rope. **(Step 1)**
2. Loop the working end of the rope completely around the standing end of the rope. **(Step 2)**
3. Thread the working end through the opening of the bight. **(Step 3)**
4. Tighten the knot by pulling on both ends simultaneously. When you pull the knot tight, it will have the shape of a figure eight. **(Step 4)**

### Figure Eight on a Bight

The figure eight on a bight knot creates a secure loop at the working end of a rope. The loop can be used to attach the end of the rope to a fixed object or a piece of equipment, or to tie a life safety rope around a person. The loop may be any size—from an inch to several feet in diameter. Follow the steps in ► **Skill Drill 9-7** to tie a figure eight on a bight:

1. The figure eight on a bight is tied in a section of the rope that has been doubled over to form a bight. The closed end of the bight becomes the working end of the rope. **(Step 1)**
2. Hold the two sides of the bight as if they were one rope. Form a loop in the doubled section of the rope. **(Step 2)**

9-7 Skill Drill

## Figure Eight on a Bight

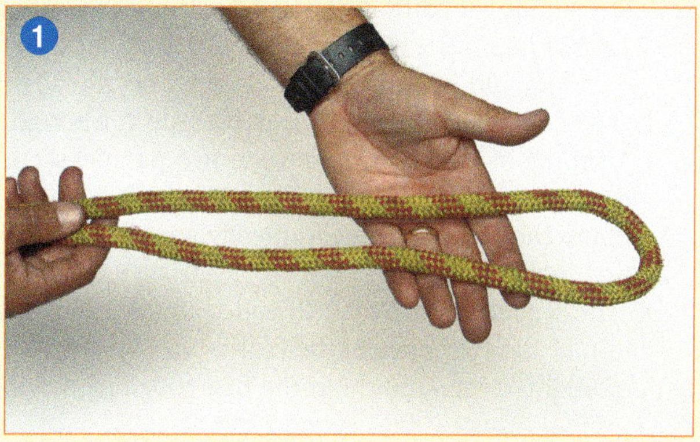


Form a bight and identify the end of the bight as the working end.

Holding both sides of the bight together, form a loop.

Feed the working end of the bight back through the loop.

Pull the knot tight.

Secure the loose end of the rope with a safety knot.

3. Pass the working end of the bight through the loop. **(Step 3)**
4. Pull the knot tight. Pulling the knot tight locks the neck of the bight and forms a secure loop. **(Step 4)**
5. Use a safety knot to tie the loose end of the rope to the standing part. **(Step 5)**

### Figure Eight Follow-Through

A figure eight follow-through knot creates a secure loop at the end of the rope when the working end must be wrapped around an object or passed through an opening before the loop can be formed. It is very useful for attaching a rope to a fixed ring or a solid object with an "eye." Follow the steps in ▶ **Skill Drill 9-8** to tie a figure eight follow-through:

1. The first step in a figure eight follow-through is to tie a simple figure eight in the standing part of the rope, about 2 feet from the working end. Leave this knot loose. **(Step 1)**
2. Thread the working end through the opening or around the object and bring it back to the figure eight knot.
3. Secure the working end by threading it back through the figure eight, tracing the path of the original knot in the opposite direction. **(Step 2)**
4. Once the working end has been threaded through the knot, you can pull the knot tight. **(Step 3)**
5. Secure the loose end with a safety knot. **(Step 4)**

This knot can also be used to tie the ends of two ropes together securely. To do this:

1. Tie the simple figure eight near the end of one rope.
2. Thread the end of the second rope completely through the knot in the opposite direction.
3. Pull the knot tight.
4. Secure the loose end of each rope to the standing part of the second rope.

### Bowline

A bowline knot also can be used to form a loop. It is frequently used to secure the end of a rope to an object or anchor point. Follow the steps in ▶ **Skill Drill 9-9** to tie a bowline:

1. To tie a bowline, form a loop and bring the working end of the rope back to the standing part. **(Step 1)**
2. Make a small loop in the standing part of the rope so that the section closer to the working end passes on top of the section closer to the running end. **(Step 2)**
3. Thread the working end of the rope through the opening in the small loop from below. **(Step 3)**
4. Pass the working end over the loop, around and under the standing part of the rope, and back down through the same opening. **(Step 4)**
5. Tighten the knot by holding the working end and pulling the standing part of the rope backward. **(Step 5)**
6. Use a safety knot to secure the working end of the rope. **(Step 6)**

## Bends

Bends are used to join two ropes together. The sheet bend or becket bend can be used to join two ropes of unequal size. A sheet bend knot also can be used to join rope to a chain. Follow the steps in ▶ **Skill Drill 9-10** to tie a sheet or becket bend:

1. This knot is tied by forming a bight in the working end of one rope. If the ropes are of unequal size, the bend should be made in the larger rope. **(Step 1)**
2. The end of the second rope is passed up through the opening of the bight, under the two parallel sections of the first rope. **(Step 2)**
3. Loop the second rope completely around both sides of the bight. **(Step 3)**
4. The second rope is then threaded under itself and on top of the two sides of the bight. **(Step 4)**
5. To tighten the knot, hold the first rope firmly while pulling back on the second rope. **(Step 5)**
6. A safety knot should be used to secure the loose ends of both ropes. **(Step 6)**

There are many ways to tie each of these knots. Find one method that works for you and use it all the time. In addition, your brigade may require that you learn how to tie other knots. It is important to become proficient in tying knots. With practice, you should be able to tie these knots in the dark, with heavy gloves on, and behind your back.

A knot should be properly "dressed" by tightening and removing twists, kinks, and slack from the rope. The finished knot is firmly fixed in position. The configuration of a properly dressed knot should be evident so that it can be easily inspected. All loose ends should be secured by safety knots to ensure that the primary knot cannot be released accidentally.

Knot-tying skills can be quickly lost without practice. Practice tying knots while you are on the telephone or watching TV ▼ **Figure 9-28**. You never know when you will need to use your skills in an emergency situation.

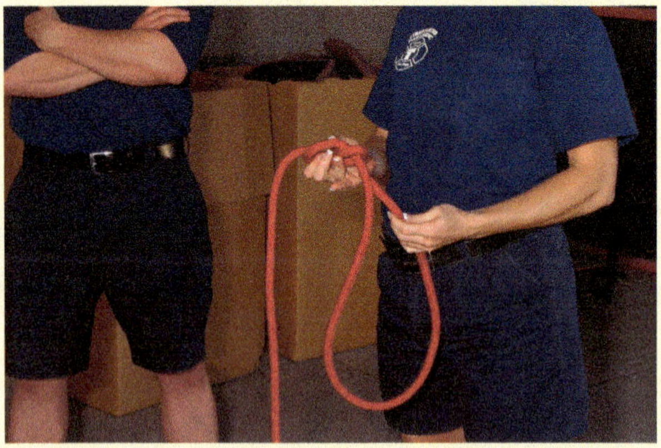

**Figure 9-28** To maintain your knot-tying skills, practice tying different knots frequently.

98 Skill Drill

# Figure Eight Follow-Through

1 Tie a regular figure eight leaving approximately 2 feet of rope at the working end.

2 Loop the working end around or through the object to be secured. Thread the working end back through the original figure eight in the opposite direction.

3 Pull the knot tight.

4 Tie a safety knot in the working end.

99 Skill Drill

# Bowline

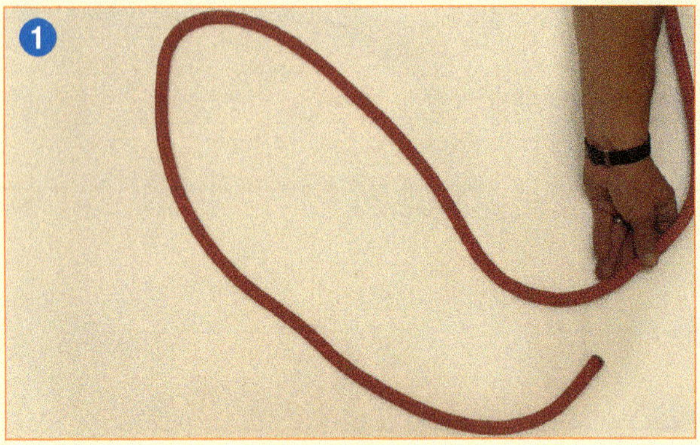


Make the desired size loop and bring the working end back to the standing part.

Form another small loop in the standing part of the rope with the section closest to the working end on top.

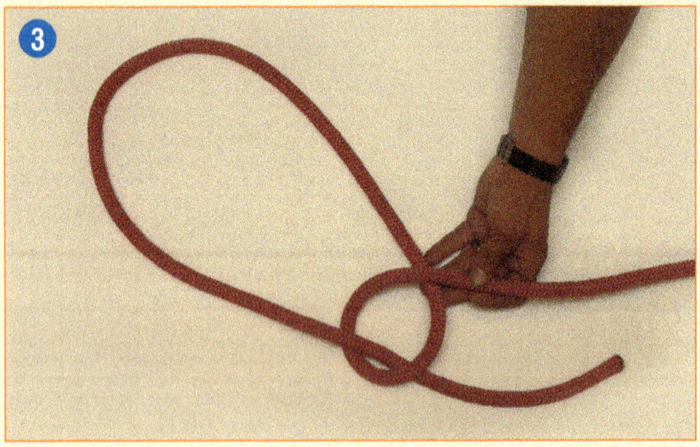


Thread the working end up through this loop from the bottom.

Pass the working end over the loop, around and under the standing part, and back down through the same opening.

Tighten the knot by holding the working end and pulling the standing part of the rope backward.

Tie a safety knot in the working end of the rope.

Skill Drill 9-10

## Sheet or Becket Bend

Using your left hand, form a bight at the working end of the first (larger) rope.

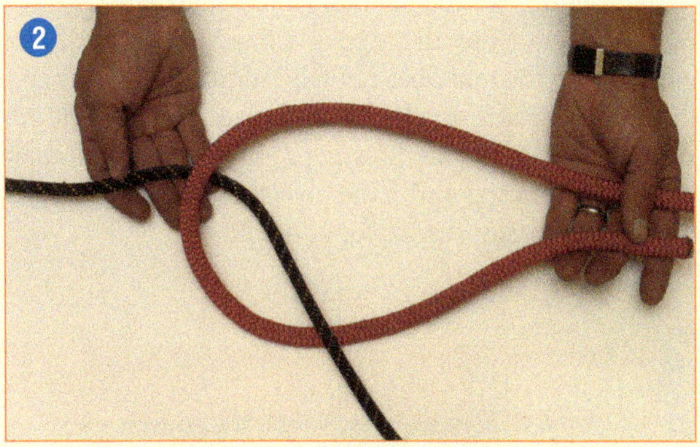

Thread the working end of the second (smaller) rope up through the bight.

Loop the second (smaller) rope completely around both sides of the bight.

Pass the working end of the second (smaller) rope between the original bight and under the second rope.

Tighten the knot.

Tie a safety knot in the working end of each rope.

## Hoisting

Tying knots is not an idle exercise but a practical skill that you will use on the job. In emergency situations, you may have to raise or lower a tool to other brigade members. It is important for you to learn how to raise and lower an axe, a pike pole, a ladder, a hose line, and an exhaust fan.

You must ensure that the rope is tied securely to the object being hoisted so the tool does not fall. Additionally, your crew members must be able to remove and place the tool into service quickly. When you are hoisting or lowering a tool, make sure no one is standing under the object. Keep the scene clear of people to avoid any chance for accident.

### Hoisting an Axe

An axe should be hoisted in a vertical position with the head of the axe down. Follow the steps in ▼ **Skill Drill 9-11** to hoist an axe:

1. The team that needs the axe should lower a rope with enough extra rope to tie the required knot around the axe and for the tag line. **(Step 1)**
2. Tie the end of the hoisting rope around the handle near the head using either a figure eight on a bight or a clove hitch. **(Step 2)**
3. Slip the knot down the handle from the end to the head. **(Step 3)**

**9-11 Skill Drill**

## Hoisting an Axe

The team that needs the axe should lower a rope with enough extra rope to tie the required knot around the axe.

Tie a figure eight on a bight to make a small loop.

Place the loop over the axe handle near the head.

Pass the standing part of the rope around the head of the axe.

4. Loop the standing part of the rope under the head. **(Step 4)**
5. Place the standing part of the rope parallel to the axe handle. **(Step 5)**
6. You can use one or two half hitches along the axe handle to keep the handle parallel to the rope. **(Step 6)**
7. Prepare to raise the axe. **(Step 7)**

To release the axe, hold the middle of the handle, release the half hitches, and slip the knot up and off.

## Hoisting a Pike Pole

A pike pole should be hoisted in a vertical position with the head at the top, so it can be used immediately. Follow the steps in ▶ Skill Drill 9-12 to hoist a pike pole:

1. The team that needs the pike pole should lower a rope with enough extra rope available for the required knot and the tag line. **(Step 1)**
2. Place a clove hitch over the handle and secure it near the butt. **(Step 2)**
3. Place a half hitch mid point on the handle. **(Step 3)**
4. Place a second half hitch on the head, securing it under the hook. **(Step 4)**
5. Prepare to raise the pike pole. **(Step 5)**

## Hoisting a Ladder

A ladder should be hoisted in a vertical position. A tag line should be attached to the bottom to keep it under control as it is hoisted. If it is a roof ladder, the hooks should be in the

## 9-11 Skill Drill (Continued)

Place the standing part of the rope parallel to the axe handle.

Tie one or two half hitches along the axe handle.

Prepare to raise the axe.

9-12 Skill Drill

## Hoisting a Pike Pole

The team that needs the pike pole should lower a rope with enough extra rope available for the required knot and the tag line.

Place a clove hitch over the handle and secure it near the butt.

Place a half hitch mid point on the handle.

Place a second half hitch on the head, securing it under the hook.

Prepare to raise the pike pole.

retracted position. Follow the steps in ▶ Skill Drill 9-13 to hoist a ladder:

1. The team that needs the ladder should lower a rope with enough extra rope to tie onto the ladder. **(Step 1)**
2. A figure eight on a bight should be tied to create a loop, approximately 3 or 4 feet in diameter, that is large enough to fit around both ladder beams. **(Step 2)**
3. The rope should be passed between two rungs of the ladder, three or four rungs from the top. **(Step 3)**
4. The end of the loop is then pulled under the rungs toward the top of the ladder.
5. Place the loop around the top of the ladder. **(Step 4)**
6. Remove the slack from the rope and allow the loop to slide down the ladder. **(Step 5)**
7. Attach a tag line from below to control the ladder as it is hoisted. The size and weight of a ladder require that a tag line be attached to the lower part of the ladder. This enables a brigade member on the ground to keep the ladder under control as it is raised. **(Step 6)** Prepare to raise the ladder. When the hoisting rope is pulled up, the ladder will be securely tied to the rope. **(Step 7)**

The ladder can be easily released by pulling the loop back over the top. When lowering a ladder, reverse the loop down through the ladder. This will pull the bottom of the ladder away from the building, preventing damage.

## Brigade Member Tips

Raise a pike pole in the same orientation it will be used, pick first. This eliminates the need to turn the pole in close confines.

## Hoisting a Charged Hose Line

It is almost always preferable to hoist a dry hose line, because water adds considerable weight to a charged line. Water weighs 8.33 pounds per gallon, which can make hoisting much more difficult. Follow the steps in ▶ **Skill Drill 9-14** to hoist a charged hose line:

1. The team that needs the hose should lower a rope with enough extra rope available to tie onto the hose. **(Step 1)**
2. Make sure that the nozzle is completely closed and secure. A charged hose line should have the nozzle secured in a closed position as it is hoisted. An unsecured handle (known as the bale) could hang up as the hose is being hoisted. The nozzle could open suddenly, resulting in an out-of-control hose line suspended in mid-air, and the potential for serious injuries and damage. **(Step 2)**
3. Use a clove hitch, 1 or 2 feet behind the nozzle, to tie the end of the hoisting rope around the hose. A safety knot should be used to secure the loose end of the rope below the clove hitch. **(Step 3)**
4. A bight is made in the rope. **(Step 4)**
5. Insert the bight through the handle opening and slip it over the end of the nozzle. When the bight is pulled tight, it will create a half hitch and secure the handle in the off position while the charged hose line is hoisted. **(Step 5)**
6. Prepare to hoist the hose. **(Step 6)**

The knot can be released after the line is hoisted by removing the tension from the rope and slipping the bight back over the end of the nozzle.

## Hoisting an Uncharged Hose Line

Before hoisting a dry hose line, you should fold the hose back on itself and place the nozzle on top of the hose. This ensures that water will not reach the nozzle if the hose is accidentally charged while being hoisted. It also eliminates any unnecessary stress on the couplings by ensuring that the rope pulls on the hose and not directly on the nozzle or coupling. Follow the steps in ▶ **Skill Drill 9-15** to hoist an uncharged hose line:

1. The team that needs the hose should lower enough rope so the ground crew can tie on the hose. **(Step 1)**
2. Fold about 6 feet of hose back on itself and place the nozzle on top of the hose. **(Step 2)**
3. Tie a clove hitch securely around both the nozzle and the hose. **(Step 3)**
4. Tie a hitch half way between the nozzle and the fold, and tie a second half hitch about 6 inches from the fold. **(Step 4)**
5. Hoist the hose with the fold at the top and the nozzle pointing down. Before releasing the rope, the brigade members at the top must pull up enough hose so that the weight of the hanging hose does not drag down the hose. **(Step 5)**

## Hoisting an Exhaust Fan or Power Tool

Several different types of tools and equipment, including an exhaust fan, a chain saw, or circular saw, or any other object that has a strong closed handle, can be hoisted using the same technique. The hoisting rope is secured to the object by passing the rope through the opening in the handle. A figure eight with a follow-through knot is used to close the loop.

Some types of equipment require that you use additional half hitches to balance the object in a particular position while it is being hoisted. Power saws are hoisted in a level position to prevent the fuel from leaking out.

You should practice hoisting the actual tools and equipment used in your organization. You should be able to do them automatically and in adverse conditions.

Remember, you always use utility rope for hoisting tools. You do not want to get oil or grease on designated life safety ropes. If a life safety rope gets oily or greasy, it should be taken out of service and destroyed so that it will not be mistakenly used again as a life safety rope. It can be cut into short lengths and used for utility rope.

Follow the steps in ▶ **Skill Drill 9-16** to hoist an exhaust fan:

1. The team that needs the equipment lowers enough rope so the ground crew can tie on the exhaust fan. **(Step 1)**
2. Tie a figure eight knot in the rope about 3 feet from the working end of the rope. **(Step 2)**
3. Loop the working end of the rope around the fan handle and back to the figure eight knot. **(Step 3)**
4. Secure the rope by tying a figure eight with a follow-through. Thread the working end back through the first figure eight in the opposite direction. **(Step 4)**
5. Attach a tag line to the fan for better control. **(Step 5)**
6. Prepare to hoist the fan. **(Step 6)**

## Brigade Member Safety Tips

When hoisting tools:

- Wear a helmet and gloves.
- Keep the scene clear.
- Use the hand-over-hand method.
- Use an **edge roller** (a device used to prevent damage to a rope from jagged edges or friction) to protect the rope.
- Avoid electric wires.

9-13 Skill Drill

## Hoisting a Ladder

The team that needs the ladder should lower a rope with enough extra rope to tie onto the ladder.

Tie a figure eight on a bight to make a loop 3 or 4 feet in diameter.

Pass the rope between the rungs of the ladder, three or four rungs from the top. Pull the loop under the rungs toward the top of the ladder.

Place the loop around the top of the ladder.

Remove the slack from the rope and allow the loop to slide down the ladder.

Attach a tag line from below to control the ladder as it is hoisted.

Prepare to raise the ladder.

9-14 Skill Drill

## Hoisting a Charged Hose Line

1 The team that needs the hose should lower a rope with enough extra rope available to tie onto the hose.

2 Make sure that the nozzle is completely closed and secure.

3 Tie a clove hitch around the hose, 1 or 2 feet behind the nozzle.

4 Make a bight in the rope.

5 Pass the bight through the nozzle handle (bale) and slip the bight over the nozzle tip. This creates a half hitch that will help keep the nozzle closed while the hose is being raised.

6 Prepare to hoist the hose.

9-15 Skill Drill

# Hoisting an Uncharged Hose Line

1 The team that needs the hose should lower enough rope so the ground crew can tie on the hose.

2 Fold about 6 feet of hose back on itself and place the nozzle on top of the hose.

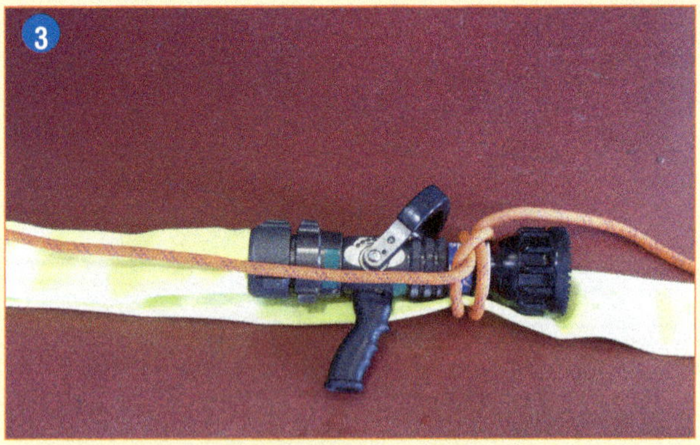

3 Tie a clove hitch near the end of the rope, wrapping the rope around both the nozzle and the hose.

4 Tie a half hitch and slip it over the nozzle. Move the half hitch along the hose and secure it about 6 inches from the fold.

5 Prepare to hoist the hose line with the fold at the top and the nozzle pointing down.

Skill Drill 9-16

## Hoisting an Exhaust Fan

The team that needs the equipment should lower enough rope so the ground crew can tie on the exhaust fan.

Tie a figure eight knot in the rope about 3 feet from the working end of the rope.

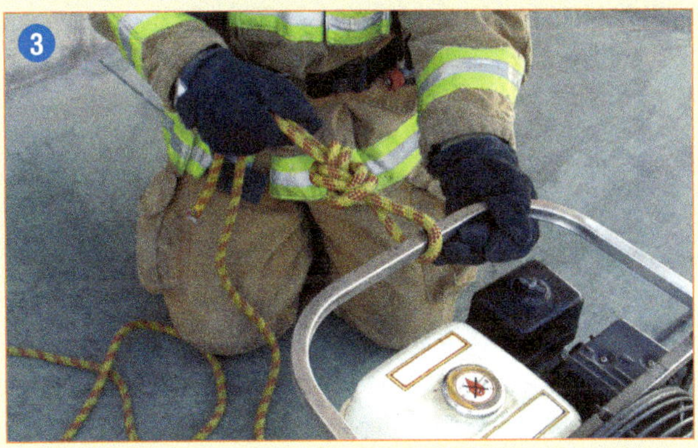


Loop the working end of the rope around the fan handle and back to the figure eight knot.

Secure the rope by tying a figure eight with a follow-through. Thread the working end back through the first figure eight in the opposite direction.

Attach a tag line to the fan for better control.

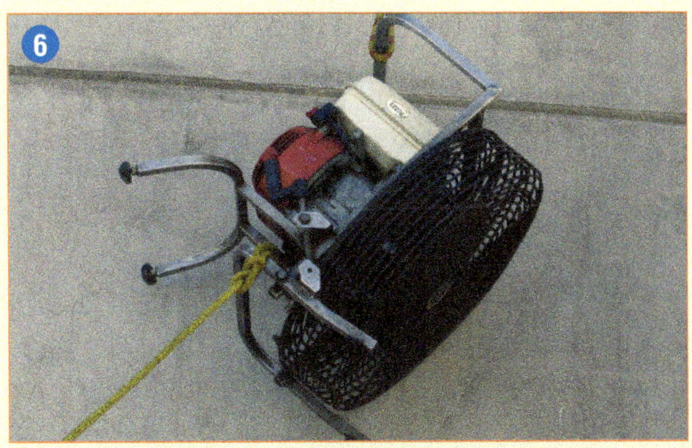


Prepare to hoist the fan.

# Wrap-Up

## Ready for Review

- Life safety ropes are used for supporting a person (or persons) during rescue operations and should be used only for that purpose.
- Utility ropes are used when it is not necessary to support the weight of a person.
- Life safety ropes must be made of continuous filament virgin fiber and woven of block creel construction.
- Proper maintenance consists of preventing damage through proper care, proper cleaning, regular inspection, and proper storage.
- Knots have characteristics that make them good for selected uses. Basic knots used by brigade members include the overhand safety knot, hitches, loop knots, and bends.
- Brigade members must be able to select and tie the appropriate knots to hoist axes, pike poles, ladders, hose lines, and exhaust fans.

## Hot Terms

**Bends** Knots used to join two ropes together.

**Bight** A U-shape created by bending a rope with the two sides parallel.

**Block creel construction** Rope constructed without knots or splices in the yarns, ply yarns, strands, braids, or rope.

**Braided rope** Rope constructed by intertwining or weaving strands together.

**Carabiner** A piece of metal hardware used extensively in rope rescue operations. It is generally an oval-shaped device with a spring-loaded clip that can be used for connecting together pieces of rope, webbing, or other hardware. Also referred to as a "snap link."

**Depressions** Indentations felt on a kernmantle rope that indicate damage to the interior core, or kern, of the rope.

**Dynamic rope** A rope generally made out of synthetic materials that is designed to be elastic and stretch when loaded. Used often by mountain climbers.

**Edge roller** A device used to prevent damage to a rope from jagged edges or friction.

**Harness** A piece of equipment worn by a rescuer that can be attached to a life safety rope.

**Hitches** Knots that attach to or wrap around an object.

**Kernmantle rope** Rope made of two parts—the kern (interior core) and the mantle (the outside sheath).

**Knots** A fastening made by tying together lengths of rope or webbing in a prescribed way, used for a variety of purposes.

**Ladder halyards** Rope used on extension ladders to raise and lower a fly section.

**Life safety rope** Rope used solely for the purpose of supporting people during firefighting, rescue, other emergency operations, or during training exercises.

**Loop** A piece of rope formed into a circle.

**One-person rope** A rope rated to carry the weight of a single person (300 lb).

**Personal escape rope** An emergency use rope designed to carry the weight of only one person and to be used only once.

**Rope bag** A bag used to protect and store rope so that the rope can be easily and rapidly deployed without kinking.

**Rope record** A record for each piece of rope that includes a history of when the rope was placed in service, when it was inspected, when and how it was used, and what types of loads were placed on it.

**Round turn** A piece of rope looped to form a complete circle with the two ends parallel.

**Running end** The part of a rope used for lifting or hoisting.

**Safety knot** A knot used to secure the leftover working end of the rope. Also known as an overhand knot or keep knot.

**Shock load** An instantaneous load that places a rope under extreme tension, such as when a falling load is suddenly stopped when the rope becomes taut.

**Standing part** The part of a rope between the working end and the running end.

**Static rope** A rope generally made out of synthetic material that stretches very little under load.

**Twisted rope** Rope constructed of fibers twisted into strands, which are then twisted together.

**Two-person rope** A rope rated to carry the weight of two people (600 lb).

**Utility rope** Rope used for securing objects, for hoisting equipment, or for securing a scene to prevent bystanders from being injured. It is never to be used in life safety operations.

**Working end** The part of the rope used for forming the knot.

# Brigade Member in Action

**You and other members of your brigade arrive on the scene of a smoldering fire in an overhead conveyor belt structure that carries boxes of finished products from a packing building to a warehouse. The fire is relatively small, but will require extinguishment with a hose. Afterward, salvage and overhaul operations will be conducted. The brigade leader instructs you and your crew to access the conveyor belt structure and complete the extinguishment and salvage/overhaul. The only access to the 25-foot-high structure is via a fixed ladder system. The structure does not contain any type of standpipe system.**

1. You decide that you and your crew will access the structure and then use a rope to raise the necessary equipment to complete your tasks. You return to the apparatus to retrieve the proper rope to complete the task. What type of rope should you choose?
   - **A.** A 20-foot personal webbing you carry in your bunker gear.
   - **B.** A 200-foot life safety rope.
   - **C.** A 100-foot utility rope.
   - **D.** A 50-foot piece of wire rope.

2. You decide that the first piece of equipment that will be raised to the structure will be a hose line. Which of the following statements is correct regarding hoisting a hose line up into a structure?
   - **A.** Whenever possible, it is always preferable to hoist a dry hose line.
   - **B.** A charged hose line should have the nozzle secured in a closed position as it is hoisted.
   - **C.** Before hoisting a dry hose line, fold the hose back on itself and place the nozzle on top of the hose.
   - **D.** All of the above are correct.

3. A pike pole will be required to complete the overhaul operation. What is the proper technique to rig a pike pole to be hoisted?
   - **A.** Place a clove hitch over the handle near the butt, place two to three half hitches along the handle 2–3 feet apart, and then hoist the pike pole vertically.
   - **B.** Tie a bowline in the center of the handle, and then hoist the pike pole horizontally.
   - **C.** Tie a series of clove hitches along the pike pole handle every 6 inches, and then hoist the pike pole vertically.
   - **D.** Tie a bowline around the hook of the pike pole, and then hoist the pike pole vertically.

4. Which of the following statements is *incorrect* regarding the hoisting of tools?
   - **A.** You should always wear a helmet and gloves.
   - **B.** Another brigade member should always stand below the hoisting operation to watch the progress of the operation.
   - **C.** You should use the hand-over-hand method to hoist the equipment.
   - **D.** Beware of and avoid electric wires in the area.

5. How should rope be stored when you no longer need to use it?
   - **A.** Rope bags are the only acceptable method to store rope.
   - **B.** Ropes should be stored in "daisy-chained" fashion or coiled.
   - **C.** Ropes should be stored in a coiled fashion or should be wrapped around a piece of wood.
   - **D.** Ropes should be stored in rope bag, coiled, or stored on a reel.

# Response and Size-Up

# Chapter 10

## NFPA 1081 Standard

**Incipient Facility Fire Brigade Member**

**4.1.2.3** Respond to a facility emergency, given the necessary equipment and facility response procedures, so that the team member arrives in a safe manner.

**(A)** *Requisite Knowledge.* Facility layout, special hazards, and emergency response procedures.

**(B)** *Requisite Skills.* The ability to recognize response hazards and to safely use each piece of response equipment provided.

**Advance Exterior Facility Fire Brigade Member**

**5.3.12*** Respond on apparatus to an emergency scene, given personal protective clothing and other necessary PPE, so that the apparatus is correctly mounted and dismounted, seat belts are used while the vehicle is in motion, and other personal protective equipment is correctly used. [1001:4.3.2]

**(A)** *Requisite Knowledge.* Mounting and dismounting procedures for riding fire apparatus, hazards and ways to avoid hazards associated with riding apparatus, prohibited practices, and types of department PPE and the means for usage. [1001:4.3.2(A)]

**(B)** *Requisite Skills.* The ability to use each piece of provided safety equipment. [1001:4.3.2(B)]

**Interior Structural Facility Fire Brigade Member**

**6.2.1 (A)** *Requisite Knowledge.* Principles of conducting initial fire size-up; principles of fire streams; types, design, operation, nozzle pressure effects, and flow capabilities of nozzles; precautions to be followed when advancing hose lines to a fire; observable results that a fire stream has been correctly applied; dangerous building conditions created by fire; principles of exposure protection; potential long-term consequences of exposure to products of combustion; physical states of matter in which fuels are found; common types of accidents or injuries and their causes; and the application of each size and type of handlines, the role of the backup team in fire attack situations, attack and control techniques, and exposing hidden fires.

## Knowledge Objectives

After completing this chapter, you will be able to:

- Describe your role in ensuring safe and efficient response to an emergency scene.
- Describe how to ride an emergency vehicle safely.
- Describe how to dismount an emergency vehicle safely.
- Describe how to shut off utilities.
- Define and describe size-up.

## Skills Objectives

After completing this chapter, you will be able to perform the following skills:

- Mount an apparatus safely.
- Dismount from an apparatus safely.
- Transport equipment safely.

# You Are the Brigade Member

**Your fire brigade is dispatched to a report of fire in a building that processes chemicals. As you get ready to respond to the scene, the emergency dispatcher indicates that she has received several alarm activations and two calls from employees reporting active fires burning inside the structure.**

1. ***What should be some of your primary response concerns as you travel to the scene in the first-due apparatus?***
2. ***What should be your first actions as a fire brigade member when you arrive on the scene?***
3. ***What should be the key elements of your incident size-up?***
4. ***What are your basic fireground objectives for controlling the emergency?***

## Introduction

Response involves a series of actions that begin when a crew is dispatched to an alarm and end with their arrival at the emergency incident. Response actions include receiving the alarm, donning protective clothing and equipment, mounting the apparatus, and transporting equipment and personnel to the emergency incident quickly and safely.

Because brigade members must be ready to react immediately to an alarm, preparations for response should begin even before the alarm is sounded. These preparations include checking personal equipment, ensuring the fire apparatus is ready, and making sure that all equipment carried on the apparatus is ready for use. Brigade members should also be familiar with their response area, know the buildings under their protection, and understand their brigade's standard operating procedures (SOPs).

Response actions for the apparatus driver also include considering road and traffic conditions, determining the best route to the incident, identifying nearby hydrant locations or water sources, and selecting the best position for the apparatus at the incident scene.

Size-up is a systematic process of information gathering and situation evaluation that begins when an alarm is received. Size-up continues during response and includes the initial observations made upon arrival at the incident scene. Size-up information is essential for determining the appropriate strategy and tactics for each situation and is ongoing throughout the entire incident.

The Facility Fire Brigade Leader is ultimately responsible for obtaining the necessary information to manage the emergency incident. Each brigade member should also be involved in the process of gathering and processing information. The observations made by individual brigade members will help them anticipate necessary actions and provide brigade leaders with needed information.

## Response

Trained brigade members must be ready to respond to an emergency at any time during their tour of duty. This process begins by ensuring that personal protective equipment (PPE) is complete, ready for use, and in good condition. This includes checking that gloves, protective hood, and flashlight are ready for immediate use. At the beginning of each tour of duty, place PPE in its designated location, which may be your assigned riding position on the apparatus (▶ Figure 10-1).

Conduct an inspection of the self-contained breathing apparatus (SCBA) at the beginning of each tour of duty or during drill days (▶ Figure 10-2). The air supply should be 90% of full pressure, the face piece clean, and the personal alert safety system (PASS) operable. Also check the availability and operation of your hand light and any hand tools you might require, based on your assigned position on the team (▶ Figure 10-3). Recheck personal protective equipment and tools thoroughly when you return from each response. This will help ensure that your gear and equipment will be functional for the next alarm.

### Alarm Receipt

The emergency response process begins when an alarm is received by your dispatch center. Brigade members should be familiar with the dispatch method or methods used by their brigade. Brigades may be alerted by radio, pager, plant alarm, or some other method.

**Figure 10-1** Check your PPE and place it in the designated location for your riding position or work area.

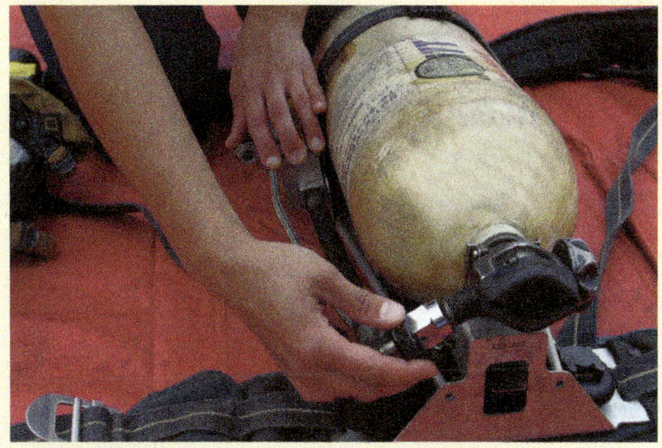

**Figure 10-2** Conduct an inspection of your SCBA at the beginning of each tour of duty.

**Figure 10-3** Check the tools assigned to your riding position or function.

## Brigade Member Tips

Volunteer brigade members who are not assigned to specific tours of duty or riding positions should check PPE, SCBA, and associated tools and equipment on a regular basis so that they are ready for use whenever there is an emergency response. After each use, all PPE should be carefully checked before it is put away.

Radio, telephone, or public address systems are often used to transmit information to fire brigades, and the use of computer terminals and printers to transmit dispatch messages is increasing. Some fire brigades may use outdoor sirens or horns to summon brigade members to an emergency.

Dispatch messages contain varying amounts of useful information, based on what the dispatcher learns from the caller. At a minimum, the dispatch information will include the location of the incident, the type of emergency (vehicle fire, structure fire, medical call), and the units that are due to respond. Computer-aided dispatch systems often provide additional information about the building or premises on the dispatch printout.

As additional information becomes available, the telecommunicator will include it in dispatch messages to later-responding units or advise the responding units by radio while they are en route. Each additional piece of information can help responding brigade members plan strategies and prepare themselves for the incident.

When an alarm is received, the response should be prompt and efficient. Responding brigade members should walk briskly to the apparatus. There is no need to run; the objective is to respond quickly, without injuring anyone or causing any damage. Follow established procedures to ensure that stoves, faucets, and other equipment and appliances at the station or work area are shut off or tended to by other workers. Wait until the apparatus doors are fully open before leaving the station.

10-1 Skill Drill

## Mounting Apparatus

When mounting (climbing aboard) fire apparatus, always have at least one hand firmly grasping a handhold and at least one foot firmly placed on a foot surface. Maintain the one hand and one foot placement until you are seated.

Fasten your seatbelt and then don any other required safety equipment for response, such as hearing protection. Eye and face protection are required for seating areas that are not fully enclosed.

## Riding the Apparatus

Don PPE before mounting the apparatus. Do not attempt to dress while the apparatus is on the road. Wait until you dismount at the incident scene to don any protective clothing that is not donned prior to mounting the apparatus. Don SCBA only after the apparatus stops at the scene, unless the SCBA is seat-mounted.

All equipment should be properly mounted, stowed, or secured on the fire apparatus. Any unsecured equipment in the crew compartment can be dangerous if the apparatus must stop or turn quickly. A flying tool, map book, or PPE can seriously injure a brigade member.

Be careful when mounting and dismounting apparatus. The steps on fire apparatus are often high and can be slippery. Follow the steps in ▲ Skill Drill 10-1 to mount an apparatus properly.

1. When mounting (climbing aboard) fire apparatus, always have at least one hand firmly grasping a handhold and at least one foot firmly placed on a foot surface. Maintain the one hand and one foot placement until you are seated. **(Step 1)**
2. Fasten your seatbelt and then don any other required safety equipment for response, such as hearing protection. Eye and face protection are required for seating areas that are not fully enclosed. **(Step 2)**

All brigade members must be seated in their assigned riding positions with seatbelts and/or harnesses fastened before the apparatus begins to move. To minimize the potential for injury, brigade members should be in their seats, with seatbelts secured, whenever the vehicle is in motion. Do not unbuckle a seatbelt to don any clothing or equipment while the apparatus is en route to an incident.

The noise produced by sirens and air horns can have long-term, damaging effects on your hearing. If your brigade provides hearing protection for personnel riding on fire apparatus, use it. These devices often include radio and intercom capabilities so that brigade members can talk to each other and hear information from the dispatcher or Incident Commander (IC).

During transport, limit conversation to the exchange of pertinent information. Listen for instructions from the IC, your brigade leader, and for additional information about the incident over the radio. The vehicle operator's attention should be focused on driving the apparatus safely to the scene of the incident.

The ride to the incident is a good time to consider any relevant factors that could affect the situation. These factors could include the time of day or night, the temperature, the presence of precipitation or wind, as well as the location and type of incident. Using the time to think ahead will help you mentally prepare for various possibilities.

## Emergency Response

The fire apparatus operator must always exercise caution when driving to an incident. Fire apparatus can be very large, heavy, and difficult to maneuver. Operating an emergency vehicle without the proper regard for safety can endanger the lives of both the brigade members on the vehicle and civilian drivers and pedestrians. Although the impulse to respond as speedily as possible is understandable, never compromise safety for a faster response time.

Brigade members who drive emergency vehicles must have special driver training, be familiar with company policy, SOPs, and know the laws and regulations that apply to emergency response. Many jurisdictions require a special driver's license to operate fire apparatus. The rules that apply to emergency vehicles are very specific, and the driver is legally responsible for the safe operation of the vehicle at all times.

An emergency vehicle must always be operated with due regard for the safety of everyone on the road. Although most states and provinces permit drivers of emergency vehicles to disregard specific traffic regulations when responding to emergency incidents, apparatus operators must consider the actions of other drivers before making such a decision. For example, traffic laws require other drivers to yield the right of way to an emergency vehicle. There is no assurance, however, that other drivers will do so when an emergency vehicle approaches. The apparatus driver must also anticipate what routes other units responding to the same incident will take.

Brigade members who respond to emergency incidents in personal vehicles must follow the specific laws and regulations of their state or province and follow departmental SOPs and regulations. Brigade members driving their privately owned vehicles to the plant in response to an emergency must follow all applicable state and local regulations.

## Prohibited Practices

For safety, SOPs often prohibit specific actions during response. As noted previously, brigade members must remain seated with seatbelts securely fastened while the emergency vehicle is in motion. Never unfasten your seatbelt to retrieve or don equipment. Do not dismount the apparatus until the vehicle comes to a complete stop.

Never stand up while riding on apparatus. Do not hold on to the side of a moving vehicle or stand on the rear step. When a vehicle is in motion, everyone aboard must be seated and belted in an approved riding position.

## Dismounting a Stopped Apparatus

When the apparatus arrives at the incident scene, the driver will park it in a location that is both safe and functional. Wait until the vehicle comes to a complete stop before dismounting. Always check for traffic before opening doors or stepping out of the apparatus. During the dismount, watch for other hazards, such as ice and snow, downed power lines, or hazardous materials, that could be present.

Be careful when dismounting apparatus. The increased weight of PPE and adverse conditions can contribute to slips, strains, and sprains. Use handrails when mounting or dismounting the apparatus. Follow the steps in ▶ Skill Drill 10-2 to dismount an apparatus safely.

1. Become familiar with your riding position and the safest way to dismount. **(Step 1)**
2. One hand should always be grasping a handhold, and one foot should always be placed firmly on a flat surface when leaving the apparatus, especially on wet or potentially icy roadway surfaces. **(Step 2)**

## Traffic Safety on the Scene

An emergency incident scene presents several risks to brigade members in addition to the hazards of fighting fires and performing other duties. One of these dangers is traffic, particularly when the incident scene is on a street or highway. Traffic safety should be a major concern for the first-arriving units because approaching drivers might not see emergency workers or realize how much room brigade members need to work safely.

The first unit or units to arrive at the incident scene have a dual responsibility. Not only must the brigade members focus on the emergency situation facing them, but they must also consider approaching traffic, including other emergency vehicles, and other, less obvious hazards. Always check for traffic before opening doors and dismounting from the apparatus. Watch out for traffic when working in the street. Follow departmental SOPs to close streets quickly and to block access to areas where emergency operations are being conducted.

### Brigade Member Safety Tips

- Do not attempt to mount or dismount a moving vehicle.
- Do not remove your seatbelt until the apparatus comes to a complete stop.
- Do not stand directly behind an apparatus that is backing up. Stand off to one side where the driver can see you in the rear-view mirror. All fire apparatus should have working, audible alarms when in reverse gear.

10-2 Skill Drill

## Dismounting Apparatus

1 Become familiar with your riding position and the safest way to dismount.

2 Maintain the one hand and one foot placement when leaving the apparatus, especially on wet or potentially icy roadway surfaces.

One of the most dangerous work areas for brigade members is on the scene of a highway incident, where traffic can be approaching at high speeds. Place traffic cones and other warning devices far enough away from the incident to slow approaching traffic and direct it away from the work area. All brigade members working at highway incidents should wear high-visibility safety vests, as well as their normal PPE. Many fire brigades have specific SOPs covering required safety procedures for these incidents.

## Arrival at the Incident Scene

From the moment that brigade members arrive at an emergency incident scene, SOPs and the structured incident management system must guide all actions. Brigade members should always work in assigned teams (companies or crews) and be guided by a strategic plan for the incident. Teamwork and disciplined action are essential for the safety of all brigade members and the effective, efficient conduct of operations.

The command structure plans, coordinates, and directs operations. Brigade members responding to an incident on an apparatus comprise the crew assigned to that vehicle and take direction from their brigade leader, who ensures that their actions coordinate with the overall plan. A brigade member who arrives at the scene independent of an apparatus must report to the IC and receive an assignment to a company or crew under the supervision of a brigade leader.

Brigade members who take action on their own, without regard to SOPs, the command structure, or the strategic plan, are **freelancing**. This type of activity, whether it is done by individual brigade members, groups of brigade members, or full companies, is unacceptable. Freelancing cannot be tolerated at any emergency incident. The safety of every person on the scene can be compromised by brigade members who do not work within the system.

Brigade members should not respond to an emergency incident scene unless they have been dispatched or have an assigned duty to respond. Unassigned units and individual personnel arriving on the scene can overload the IC's ability to manage the incident effectively. Individuals who simply show up and find something to do are likely to compromise their own safety and create more problems for the command

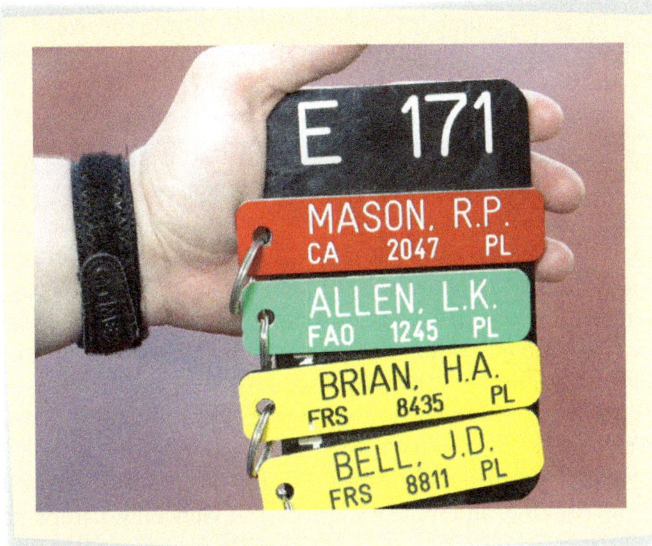


Figure 10-4 There should be a designated location on the fire apparatus for brigade members to deposit their PATs.

staff. All personnel must operate within the established system, reporting to a designated supervisor under the direction of the IC.

## Personnel Accountability System

A personnel accountability system should be used to track every brigade member at every incident scene. The system maintains an updated list of the brigade members assigned to each vehicle or crew and tracks each crew's assignment at the fire scene. If any brigade members are reported missing, lost, or injured, the accountability system can identify who is missing and their last assignment.

Different fire brigades use various types of accountability systems. Consult your SOPs for more specific information about the system that is being used and always follow the required procedures.

Many fire brigades use a personnel accountability tag (PAT) to track individual brigade members. An accountability tag generally includes the brigade member's name, ID number, and photograph and may include additional information, such as an individual's medical history, medications, allergies, and other factors. Brigade members who respond to an incident on an apparatus deposit their PATs in a designated location on the vehicle (▲ Figure 10-4). The PATs are then collected from each vehicle and taken to the command post. Brigade members who respond directly to the scene must report to the command post to receive an assignment and deposit their PATs.

## Controlling Utilities

Controlling utilities is one of the first tasks that must be accomplished at many major event working fires. Operating

### Teamwork Tips

Teamwork is a hallmark of effective fireground operations. Training and teamwork produce effective, coordinated operations. Unassigned individual efforts and freelancing can disrupt the operations and endanger the lives of both brigade members and civilians.

control or process valves is one of the important tasks that brigade members may perform. Most brigades have written SOPs that define when the utilities are to be shut off. Although this responsibility is often assigned to a particular company, crew or plant maintenance utilities personnel, all brigade members should know how to shut off an operating area's electrical, gas, and water service. When shutting down utilities, brigade members must be aware of potential impacts to any plant or facility processes that may be operating and must follow facility lockout/tagout procedures.

There are several reasons to disconnect electrical and gas utilities to a burning building or process area. If faulty electrical equipment or a gas service caused the fire, shutting off the supply will help alleviate the problem. Disconnecting electrical service can also prevent problems such as short circuits and electrical arcing that could result from fire or water damage. Shutting down gas service eliminates the potential for explosion due to damaged, leaking, or ruptured gas piping.

Controlling utilities is particularly important if brigade members need to open walls or ceilings to look for fire in the void spaces, to cut ventilation holes in roofs, or to penetrate through floors. Because these spaces often contain both electrical lines and gas pipes, the danger of electrocution or explosion exists unless these utilities are disconnected.

For example, a brigade member using an axe to open a void space could be electrocuted if the tool contacts energized electrical wires. Brigade members operating a hose could be injured if the water conducts electricity back to the nozzle. Brigade members can also be electrocuted by coming in contact with wires that have been pulled loose or exposed by a structural collapse.

Gas pipes in walls, ceilings, roofs, or floors can also be inadvertently damaged by brigade members. A power saw slicing through a gas pipe can cause a rapid release of gas and create the potential for an explosion. A structural collapse can rupture gas lines or create leaks in the stressed piping.

Controlling utilities also alleviates the risk of an additional fire or explosion. Before any utility service is shut off, it should be coordinated with facility operations or maintenance personnel to prevent an upset condition that may create another hazardous situation. Shutting off electrical service eliminates potential ignition sources that could cause an explosion if leaking gas has accumulated. Shutting off the gas supply will prevent any further leakage. The electrical

Figure 10-5 The electric service usually has an exterior meter connected to above-ground utility wires.

supply must be disconnected at a location outside the area where gas might be present. Interrupting the power at a remote location alleviates the risk of an electrical arc that could cause a gas explosion.

If there is a serious water leak inside the building, shut off the water supply to prevent electrical problems and to help minimize additional water damage to the structure and contents. Shutting off the water supply to a fire-damaged building is often necessary to control leaking pipes, which could cause more water damage than the water that was used to fight the fire.

## Electrical Service

Electric delivery service arrangements depend on the providing utility company (which may be an in-plant generated electrical service) and the age of the system. The most common installation is a service drop from above-ground utility wires to the electric meter, which is mounted on the outside of the building (▲ Figure 10-5). Other electric service installations include underground connections, inside meters, and multiple services or separate electric meters for different tenants or areas inside a building.

Fire brigades must work with their site maintenance utility personnel to identify the different types of electric services and arrangements used and the proper procedures for dealing with each type. Often, a main disconnect switch can be operated to interrupt the power. Most plants will train brigade members to identify and operate shut-off devices on typical installations. Shutting off large systems that involve high-voltage equipment should be done by personnel qualified to do so with the proper PPE.

A utility company representative should be called to interrupt power from a remote location, such as a utility pole or substation. This may be necessary if the outside wires have been damaged by fire, if brigade members are working with ladders or aerial apparatus, or if an explosion is possible. In many facilities, qualified technicians are dispatched automatically to all working fires. Some utility companies can interrupt the service to an entire plant from a remote location in response to a fire brigade request. Be aware that emergency generators or other back-up power sources may be feeding power into a building or area when the main power sources are shut off.

## Gas Service

Natural gas or liquefied propane (LP) gas are used for heating, cooking, and industrial processes. Generally, natural gas is delivered through a network of underground pipes. LP gas is usually delivered by a tank truck and stored in a container on the premises. In some areas, LP gas is distributed from one large storage tank through a local network of underground pipes to several facility buildings or process areas.

A single valve usually controls the natural gas supply to a building, although large industrial buildings may have several valves. This valve is generally located outside the building at the entry point of the gas piping. Natural gas service has a distinctive piping arrangement. In older buildings, the shut-off valve for a natural gas system may be in the basement.

The shut-off valve for a natural gas system is usually a quarter-turn valve with a locking device so it can be secured in the off position (▼ Figure 10-6). When the handle is in line with the pipe, the valve is open; when the handle is at a right angle to the pipe, the valve is closed.

The valve for an LP gas system is usually located at the storage tank. The more common type of LP gas valve, however, has

Figure 10-6 A typical natural gas service includes a pressure-reducing valve and a gas meter along with the shut-off valve.

a distinctive handle that indicates the proper rotation direction to open or close the valve. To shut off the flow of gas, rotate the handle to the fully closed position.

A gas valve that has been shut off must not be reopened until the system piping has been inspected by a qualified person. Air must be purged from the system and any pilot lights must be reignited to prevent gas leaks.

### Water Service

Water service to a building can usually be shut off by closing one valve at the entry point. Many systems are piped to allow the water service to be turned off at the connection between the utility pipes and the building's system. This underground valve is outside the building and can be operated with a special wrench or key. When shutting off water service, ensure that this will not effect the supply to the fire protection system.

# Size-Up

Size-up is the process of evaluating an emergency situation to determine what actions need to be taken. It is always the first step in making plans to bring the situation under control. The initial size-up of an incident is conducted when the first unit arrives on the scene and determines the appropriate actions for that unit or units. Conducting a 360-degree walk-around of the incident during the initial size-up provides a crucial big picture view and can reduce the chance of surprises as the event progresses.

Initial size-up is often conducted by the first-arriving brigade leader, who serves as the IC until a higher ranking officer arrives and assumes command. The IC uses the size-up to develop an initial plan and to set the stage for the actions that follow.

As more complete and detailed information becomes available, the IC will improve both the initial size-up and the initial plan. At a major incident, the size-up process might continue through several stages. The ongoing size-up must consider the effectiveness of the initial plan, the impact brigade members are having on the problem, and any changing circumstances at the incident.

Although brigade leaders usually perform size-up, brigade members must understand how to formulate an operational plan, how to gather and process information, and how this information can change plans during the operation. If there is no brigade leader on the first arriving unit, a brigade member could be responsible for assuming command and conducting the preliminary size-up until an officer arrives. Individual brigade members are often asked to obtain information or to report their observations for ongoing size-up. Brigade members should routinely make observations during incidents to maintain their personal awareness of the situation and to develop their personal competence.

## Managing Information

The size-up process requires a systematic approach to managing information. Emergency incidents are often complicated, chaotic, and rapidly changing. The IC must look at a complicated situation, identify the key factors that apply, and develop an action plan based on known facts, observations, realistic expectations, and certain assumptions. As the operation unfolds, the IC must continually revise, update, supplement, and reprocess the size-up information to ensure that the plan is still valid or to identify when the plan needs to be changed.

Size-up is based on two basic categories of information: facts and probabilities. Facts are data elements that are accurate and based on prior knowledge, a reliable source of information, or an immediate, onsite observation. Probabilities are factors that can be reasonably assumed, predicted, or expected to occur, but are not necessarily accurate.

Often, the initial size-up is based on a limited amount of factual information and a larger amount of probable information. It is refined as more factual information is obtained and as probable information is either confirmed or revised. Effective size-up requires a combination of training, experience, and good judgment.

### Facts

Facts are bits of accurate information obtained from various sources. For example, the communications center will provide some facts about the incident during dispatch. A preincident plan will contain many facts about the structure. Maps, manuals, and other references provide additional information. Plant or facility operations personnel should be utilized as early into the incident as possible to provide access condition information that may affect firefighting operations. Individuals with specific training, such as the building engineer or a utility representative, can add specific information. A seasoned brigade leader will build a bank of information based on experience, training, and direct observations.

The initial dispatch information will contain facts such as the location and nature of the situation. The dispatch information could be very general *(a building reported on fire in the vicinity of Central Shops)* or very specific *(a smoke alarm activation in room 3102 of the Chemical Process Building)*.

The time of day, temperature, and weather conditions are other factors that can easily be determined and incorporated into the initial size-up. Based on these basic facts, a brigade leader might have certain expectations about the incident. For example, whether a building is likely to be occupied or unoccupied, whether processes are running, and whether traffic will delay the arrival of additional units may be inferred from the time of day. Weather conditions such as snow and ice will delay the arrival of fire apparatus, create operational problems with equipment, and require additional brigade members to perform basic functions such as

# VOICES OF EXPERIENCE

**"A call came in as a reported wall fire in the south end of one of our warehouses."**

When responding to incidents, the facility fire brigade may have several advantages. The first is a better than average response time, and another benefit is an intimate knowledge of the buildings and their fire protection systems.

A call came in as a reported wall fire in the south end of one of our warehouses. As we responded to this event, each of us knew exactly the location and potential hazards associated with the warehouse. It was built pre-WWI, and its structural elements were wood frame with metal sheeting on the exterior.

As the officer in charge, I responded with the command vehicle and was first on location. Fire and smoke were visible from a louvered attic vent on the south side of the building. I conducted a 360-degree walkaround of the building as part of my size-up. I reported the situation to the incoming responding units and let them know that we had a wall fire that had extended into the attic space and were commencing an offensive attack on the fire. I passed command to the next incoming unit, and with one handline attacking the fire at street level, another brigade member and I took a handline into the attic space and started pulling the wall to get ahead of the fire and prevent it from spreading further. We were successful and actually stopped the fire from spreading into the attic space before it generated enough heat to fuse a sprinkler head.

We were successful because of our quick response time. We had water on this fire in about three minutes. We conduct training with all of our local fire departments and are integrated into the IMS system and communicate using the same terminology. This relationship and prework aided tremendously in this response. While inside fighting fire and communicating our actions to the IC, the incoming city brigade vented the roof and provided support to our personnel fighting the fire.

As expressed, the facility fire brigade needs to leverage very fast response times and knowledge of their structures to operate efficiently and effectively. Our value to those agencies is to minimize injuries, damage, and production interruption to the best of our abilities. The warehouse fire is a great example of this.

**Randy J. Krause**
Boeing Fire Department
Seattle, Washington

## Brigade Member Tips

Maintenance personnel can often provide valuable information about a structure and its mechanical systems.

stretching hose lines and raising ladders. Strong winds can cause rapid extension or spread of a fire to exposed buildings. High heat and humidity will affect brigade members' performance and may cause heat casualties.

An experienced brigade leader will also have a basic knowledge of the facility and the brigade's available resources. The brigade leader may remember the types of occupancies, construction characteristics of typical buildings, and specific information about particular buildings from previous incidents or preincident planning visits.

A preincident plan can be especially helpful during size-up because it contains significant information about a structure. A preincident plan generally provides details about a building's construction, layout, contents, special hazards, and fire protection systems. Chapter 22, Preincident Planning, contains specific information about the contents and development of such plans.

Basic facts about a building's size, layout, construction, and occupancy can often be observed upon arrival, if they are not known in advance. The brigade leader must consider the size, height, and construction of the building during the size-up. The action plan for a single-story, wood-frame temporary office on an acre of land will be quite different than that for a multistory chemical processing facility in the middle of a large plant.

The age of the building is often an important consideration in size-up because building and fire safety codes change over time. Older wooden buildings may have balloon-frame construction, which can provide a path for fire to spread rapidly into uninvolved areas. Floor and roof systems in newer buildings use trusses for support, which are susceptible to early failure and collapse. The weight of air conditioning or heating units on the roof may also contribute to structural collapse.

A plan for rescuing occupants and attacking a fire must consider information about the building layout, such as the number, locations, and construction of stairways. The plan for a building with an open stairway that connects several floors might have keeping the fire out of the stairway as a priority. This enables brigade members to use the stairway for rescue and prevents the fire from spreading to other floors. Ladder placement, use of aerial or ground ladders, and emergency exit routes all depend on the building layout.

Any special factors that will assist or hinder operations must be identified during size-up. For example, bars on the windows will limit access and complicate the rescue of trapped victims, but firewalls and sprinkler systems will help to confine or extinguish a fire.

The occupancy of a building is also critical information. An apartment building, a cafeteria, maintenance garage, an office building, a warehouse filled with concrete blocks, and a chemical manufacturing plant all present brigade members with different sets of problems that must be addressed.

The size of the fire and its location within the building help determine hose line placement and ventilation sites. The fire's location is also critical in determining which occupants are in immediate need of rescue. For example, someone directly above a burning office should be rescued immediately, but a person at the opposite end of the building, far from the fire, is probably not in immediate danger.

Direct visual observations will give the best information about the size and location of a fire, particularly when combined with information about the building. Visible flames indicate where the fire is located and how intensely it is burning but might not tell the whole story. Flames issuing from only one window suggest that the fire is confined to that room, but it could also be spreading through void spaces to other parts of the building.

Often, the location of a fire within a building cannot be determined from the exterior, particularly when visibility is obscured by smoke (▼ Figure 10-7). An experienced brigade member will observe where smoke is visible, how much and

Figure 10-7 It can be difficult to determine the location and extent of the fire inside a building.

what color it is, how it moves, and even how it smells. A burning pot of food, an overheated electrical motor, or burning wood all have distinctive odors.

Inside a building, brigade members can use their observations and sensations to help them work safely and effectively. If smoke limits visibility, a crackling sound or the sensation of heat coming from one direction may indicate the seat of the fire, which is the main area of the fire. Blistering paint and smoke seeping through cracks can lead brigade members to a hidden fire burning within a wall. Advanced technology, such as thermal imaging cameras, can be even more effective in identifying hidden fires.

The IC needs to gather as much factual information as possible about a fire. Because the IC often is located at a command post outside the building, brigade leaders are requested to report their observations from different locations. A company that is operating a hose line inside the building can report on interior conditions, while the ventilation crew on the roof will have a different perspective. The IC may request that a brigade leader or a brigade member prepare a reconnaissance report. The reconnaissance report is the inspection and exploration of a specific area to gather information for the IC. The IC assembles, interprets, and bases decisions on this information.

Regular progress reports from companies working in different areas update information about the situation. Progress reports enable an IC to judge whether an operational plan is effective or needs to be changed.

## Probabilities

Probabilities refer to events and outcomes that can be predicted or anticipated, based on facts, observations, common sense, and previous experiences. Brigade members frequently use probabilities to anticipate or predict what is likely to happen in various situations. The attack plan is also based on probabilities, predicting where the fire is likely to spread and anticipating potential problems.

An IC must be able to identify the probabilities that apply to a given situation quickly. For example, a late night fire in a warehouse that is only occasionally occupied during the night shift may involve occupants who need to be rescued. Therefore, the IC would assign additional crews to search for potential victims, even if no factual information indicates that any occupants need to be rescued. Similarly, a fire burning on the top floor of a structure in a row of attached buildings has a high probability of spreading to adjoining storage. In this case, the IC's plan would include opening the roof above the fire and sending crews into the exposed areas to check for fire extension.

The concepts of convection, conduction, and radiation enable an IC to predict how a fire will extend in a particular situation. By observing a particular combination of smoke and fire conditions in a particular type of building, the IC can identify a range of possibilities for fire extension within the structure or to other exposed buildings. Using these probabilities, the IC can predict what is likely to happen and develop a plan to control the situation effectively.

The IC must also evaluate the potential for collapse of a burning structure. The building's construction, the location and intensity of the fire, and the length of time the structure has been burning are all factors that must be considered. If the possibility of collapse exists, risk management dictates that the IC order brigade members out of the building before it occurs.

## Resources

Resources include all of the means that are available to fight a fire or conduct emergency operations at any other type of emergency incident. Resource requirements depend on the size and type of incident. Resource availability depends on the capacity of a fire brigade to deliver brigade members, fire apparatus, equipment, water, and other items that can be used at the scene of an incident.

A fire brigade's basic resources are personnel and equipment. Firefighting resources are usually defined as the numbers of crews, equipment, and command officers required to control a particular fire. An IC should be able to request the required number of crews and know that each unit will arrive with the appropriate equipment and the necessary brigade members to perform a standard set of functions at the emergency scene. The IC usually will have a good idea of the number and types of crews available to respond, how they are staffed and equipped, and how long they should take to arrive. This information might have to be updated at the incident, particularly in facilities served by volunteer brigade members, because the number of brigade members available to respond may vary at different times.

Water supply is another critical resource. Water supply is generally not a problem in most plants with brigades. Water supply could limit operations in a plant without hydrants or with a limited or malfunctioning water supply. It takes time to establish a water supply from a static source, and the amount of water that can be delivered by a shuttle is limited. In this situation, the IC would need to call for additional tankers or engine companies with large-diameter hoses.

Resources include more than brigade members, fire equipment, and a water source. A fire in a flammable liquids storage facility will require large quantities of foam and the equipment to apply it effectively. A hazardous materials incident might require special monitoring equipment, chemical protective clothing, and bulk supplies to neutralize or absorb a spilled product. A building collapse might require heavy equipment to move debris and a structural engineer to determine where brigade members can work safely. The fire brigade must have these supplies available or be able to obtain them quickly when they are needed. Resources for a large-scale incident must include food and fluids for the rehydration of brigade members, fuel for the apparatus, and other supplies.

The size-up process enables the IC to determine what resources will be needed to control the situation and to ensure their availability. An action plan to control an incident can be effective only if the necessary resources can be assembled on a timely basis. If there is a delay, the IC must anticipate how much the fire will grow and where it will spread. If the desired resources cannot be obtained, the IC must develop a realistic plan using available resources to gain eventual control of the situation.

Ideally, a fire brigade will be able to dispatch enough brigade members and equipment to control any situation within its jurisdiction. Most brigades have mutual aid agreements with surrounding jurisdictions to assist each other if a situation requires more resources than the plant brigade or local community can provide. In some areas, hundreds of brigade members or municipal fire fighters and equipment can be assembled for a large-scale incident. In more remote areas, the resources available to fight a fire can be very limited.

If resources are insufficient or delayed, a fire can become too large to be controlled by available personnel. For example, if only 20 brigade members and two apparatus are available, they will have to bring the fire under control before it gets too big or wait until the fire burns itself down to a manageable size.

Resources must be organized to support efficient emergency operations. Brigade members must be properly equipped and organized in crews. Equipment and procedures must be standardized. The incident management system must be designed to manage all of the resources that could be used at a large-scale incident, and the communications system must enable the IC to coordinate operations effectively.

## Incident Action Plan

The IC develops an incident action plan that outlines the steps needed to control the situation. The incident action plan is based on information gathered during size-up. The initial IC develops a basic plan for beginning operations. If the situation expands or becomes more complicated, this plan is revised and expanded as additional information is obtained, more resources become available, and the incident management structure grows.

The incident action plan should be based on the five basic fireground objectives. In order of priority, these objectives are:

1. Rescue any victims.
2. Protect exposures.
3. Confine the fire.
4. Extinguish the fire.
5. Salvage property and overhaul the fire.

This system of priorities clearly establishes that the highest priority in any emergency situation is saving lives. The remaining priorities involve saving property. Making sure that the fire does not spread to any **exposure** (an area adjacent to the fire that may become involved if not protected) is a higher priority than confining the fire within the burning building. After the fire is confined, the next priority is to extinguish it. The final priority is to protect property from additional damage and make sure the fire is completely out.

These priorities are not separate and exclusive. One objective does not have to be accomplished before brigade members tackle the next one. Often, more than one objective can be addressed simultaneously, and certain activities help achieve more than one objective. For example, if a direct attack on the fire will bring it under control very quickly, the objectives of protecting exposures, confining the fire, and extinguishing the fire might all be accomplished together. Extinguishing the fire might also be the best way to protect the lives of building occupants. Frequently, salvage crews may be working on lower floors while fire suppression crews are still attacking a fire on an upper floor.

These priorities guide the IC in making difficult decisions, particularly if not enough resources are available to address every priority. If the decision is between saving lives and saving property, saving lives always comes first. After rescue is completed and if the fire is still spreading, the IC should place exposure protection ahead of salvage and overhaul.

A similar set of priorities can be established for any emergency situation. Saving lives is always more important than protecting property. The IC must always place a higher priority on bringing the problem under control than on cleaning up after the problem.

### Rescue

Protecting lives is the first consideration at a fire or any other emergency incident. The need for rescue depends on many circumstances. The number of people in danger varies based on the type of occupancy and the time of day. A commercial building that is crowded during the workday might have few, if any, occupants at night.

The degree of risk to the lives of occupants must also be evaluated. A fire that involves one office on the 10th floor of a multistory building could threaten the lives of the occupants of that floor and those directly above the fire. However, workers below the fire are probably not in any danger, and workers several floors above the fire might be safer staying in their office until the fire is extinguished, instead of walking down smoke-filled stairways.

Often, the best way to protect lives is to extinguish the fire quickly, so efforts to control the fire are usually initiated at the same time as rescue. Hose lines may have to be used to protect exit paths and keep the fire away from victims during search-and-rescue operations. The chapter on search and rescue discusses specific tactics and techniques for these operations.

## Exposure Protection

Within the secondary objective of protecting property, there are multiple priorities. The first priority in protecting property is to keep the fire from spreading beyond the area of origin or involvement when the fire brigade arrives. The IC must start by making sure the fire is not expanding. If the fire is burning in only one room, the objective should be to ensure that it does not spread beyond that room. If more than one room is involved, the objective might be to contain the fire to one area or one floor level. Sometimes the objective is to confine the fire to one building, particularly if multiple buildings are attached or closely spaced.

The IC has to look ahead of the fire and identify a place to stop its spread. If flames are extending quickly through the loft in a mill construction building, the IC will place crews with hose lines ahead of the fire to stop its progress.

The IC must sometimes weigh potential losses when deciding where to attack a fire. If a fire in a vacant building threatens to spread to an adjacent occupied building, the IC will usually assign crews to protect the exposure before attacking the main body of fire. If, however, a fire in an occupied building could spread to a vacant building, the IC's decision might be to attack the fire first and address controlling the spread later.

## Confinement

After ensuring that the fire is not extending into any exposed areas, the IC will focus on confining it to a specific area. The IC will define a perimeter and plan operations so that the fire does not expand beyond that area. Brigade members on the perimeter must be alert to any indications that the fire is extending.

Thermal imaging devices are electronic cameras that can detect sources of heat. They are valuable tools for finding fires in void spaces. The principles and use of thermal imaging devices are covered more fully in Chapter 13, Search and Rescue.

## Extinguishment

Depending on the size of the fire and the risk involved, the IC will mount either an offensive attack or a defensive strategy to extinguish a fire. An offensive attack is used with most small fires. Brigade members advance on or into the fire area with hose lines or other extinguishing agents and overpower the fire. If the fire is not too large and the attacking brigade members can apply enough extinguishing agent, the fire can usually be extinguished quickly and efficiently. Extinguishing the fire in this way often resolves several priorities at the same time, including exposure protection and confinement.

At regular intervals during an offensive attack, the IC must evaluate the progress being made. If not enough progress is being made, the IC must either adjust the tactics or alter the overall strategy. For example, the fire may produce more heat than the water from hose lines can absorb or brigade members may be unable to penetrate far enough to make a direct attack on the fire.

Sometimes the problem can be resolved by adding resources and intensifying the offensive attack. Larger hose lines or more brigade members with additional hose lines might be able to extinguish the fire successfully. Coordinating ventilation with an offensive attack might enable brigade members to get close enough to extinguish the fire. Special extinguishing agents may be used to extinguish the fire.

When the fire is too large or too dangerous to extinguish with an offensive attack, the IC will implement a defensive attack. In these situations, all brigade members are ordered out of the building, and heavy streams are operated from outside the fire building. A defensive strategy is required when the IC determines that the risk to brigade members' lives is excessive, as in situations where structural collapse is possible. The IC who adopts a defensive strategy has determined that there is no property left to save or that the potential for saving property does not justify the risk to brigade members. Sometimes a defensive strategy is effective in extinguishing the fire; at other times, it simply keeps the fire from spreading to exposed properties (▶ Figure 10-8).

There are situations in which all brigade members are withdrawn from the area and the fire is allowed to burn itself out. These situations generally involve potentially explosive or hazardous materials that pose an extreme danger to brigade members.

## Salvage and Overhaul

Salvage operations are conducted to save property by preventing avoidable property losses. Salvage is the removal or protection of property that could be damaged during firefighting or overhaul operations (▶ Figure 10-9). Salvage operations are often aimed at reducing smoke and water damage to the structure and contents once the fire is under control.

The overhaul process is conducted after a fire is under control to completely extinguish any remaining pockets of fire (▶ Figure 10-10). The IC is responsible for ensuring that the fire is completely extinguished before terminating operations. Floors, walls, ceilings, and attic spaces should be checked for signs of heat, smoke, or fire. Window casings, wooden door jambs, baseboards, electrical outlets, and heating/air conditioning vents can often hide small, smoldering fires. Debris from the burned contents of the structure should be removed and thoroughly doused to reduce the potential for rekindle, or a reignition of the fire. For large fires, fire brigade personnel may be assigned as a fire watch or a fire brigade crew may be assigned to return to the scene every few hours to check for indications of residual fire. These operations are covered in depth in Chapter 19, Salvage and Overhaul.

Figure 10-8 A defensive strategy involves an exterior fire attack with heavy streams that emphasizes protecting the exposures to the fire building.

Figure 10-9 Salvage operations are conducted to save property by preventing avoidable property losses.

Figure 10-10 Overhaul is conducted after a fire is under control to completely extinguish any remaining pockets of fire.

# Wrap-Up

## Ready for Review

- Preparations for an emergency response begin long before an alarm is received. Each brigade member must ensure that all personal protective equipment, apparatus, and tools are ready for response. Adhering to simple procedures enables crews to arrive at an emergency quickly and safely.
- Size-up is the ongoing mental evaluation of an emergency situation. The size-up process begins when information about an emergency incident is received. Size-up involves both an evaluation of the known facts of the situation, including the incident location, time, weather, type of structure, exposures, available resources, and life hazard, and a consideration of any probabilities that could alter the situation.
- Brigade members must understand the need to receive dispatch information properly.
- As a brigade member, your job is to respond safely to a scene. Follow your brigade's SOPs when riding in apparatus to the scene.
- Carefully mount and dismount all apparatus.
- Be aware of your surroundings at all times, especially when responding to an incident on the highway. Motorists might be distracted and not see you in the road.
- All information received at the communication center must be relayed to fire brigade units. What may seem insignificant to the dispatcher may be essential to the incident commander.
- Size-up is critical to the successful outcome of an emergency incident.
- The incident action plan should be based on the five basic fireground objectives:
  - Rescue any victims.
  - Protect exposures.
  - Confine the fire.
  - Extinguish the fire.
  - Salvage property and overhaul the fire.

## Hot Terms

**Balloon-frame construction** An older style of wood-frame construction in which the wall studs extend vertically from the basement to the roof without any fire stops.

**Defensive attack** Exterior fire suppression operations directed at protecting exposures.

**Exposure** Any person or property that may be endangered by flames, smoke, gases, heat, or runoff from a fire.

**Extension** Fire that moves into areas not originally involved, including walls, ceilings, and attic spaces; also the movement of fire into uninvolved areas of a structure.

**Freelancing** Dangerous practice of acting independently of command instructions.

**Offensive attack** An advance into the fire building by brigade members with hose lines or other extinguishing agents to overpower the fire.

**Overhaul** Examination of all areas of the building and contents involved in a fire to ensure that the fire is completely extinguished.

**Personal alert safety system (PASS)** Device worn by a brigade member that sounds an alarm if the brigade member is motionless for a period of time.

**Personal protective equipment (PPE)** Gear worn by brigade members that includes helmet, gloves, hood, coat, pants, SCBA, and boots. The PPE provides a thermal barrier against intense heat.

**Personnel accountability system** A method of tracking the identity, assignment, and location of brigade members operating at an incident scene.

**Personnel accountability tag (PAT)** Identification card used to track the location of a brigade member on an emergency incident.

**Preincident plan** A written document resulting from the gathering of general and detailed information to be used by public emergency response agencies and private industry for determining the response to reasonable anticipated emergency incidents at a specific facility.

**Reconnaissance report** The inspection and exploration of a specific area in order to gather information for the incident commander.

**Rekindle** A situation where a fire, which was thought to be completely extinguished, reignites.

**Response** Activities that occur in preparation for an emergency and continue until the arrival of emergency apparatus at the scene.

**Salvage** Removing or protecting property that could be damaged during firefighting or overhaul operations.

**Seat of the fire** The main area of the fire origin.

**Size-up** The ongoing observation and evaluation of factors that are used to develop objectives, strategy, and tactics for fire suppression.

**Thermal imaging devices** Electronic devices that detect differences in temperature based on infrared energy and then generate images based on those data. Commonly used in obscured environments to locate victims.

# Brigade Member in Action

**You are dispatched as a member of the personnel accompanying the first-due apparatus to a vehicle fire located inside a maintenance shop. You are riding in the back jump seat of the apparatus and will more than likely be assigned to deploy an attack line upon your arrival at the scene.**

1. Which of the following actions is *not* correct regarding your response inside the apparatus?
   - **A.** Retrieve your bunker gear and don it while en route to the scene.
   - **B.** Don an SCBA after the apparatus stops at the scene unless the SCBA is seat mounted.
   - **C.** Always wear your seat belt while riding in a moving apparatus.
   - **D.** During the response, limit your conversation to the exchange of pertinent information.

2. Your brigade's SOPs require the use of a personnel accountability system at every scene. What is the primary purpose of a personnel accountability system?
   - **A.** The system is used to identify who is responsible for any mistakes.
   - **B.** The system is used to track only the location of the attack team.
   - **C.** The system is used to track every brigade member at every incident scene.
   - **D.** The system is used to indicate who is in charge of the incident.

3. Upon arrival at the scene, the IC is met by an earlier-arriving fire brigade leader, who provides the IC with a reconnaissance report. What is the purpose of the "recon report"?
   - **A.** The recon report provides information regarding past responses to the facility.
   - **B.** The recon report provides information from an actual inspection and exploration of the area of concern.
   - **C.** The recon report provides information given by witnesses in the area.
   - **D.** The recon report provides information from the prefire planning.

4. Upon arrival, the IC implements a basic incident action plan based on the five major fireground objectives. What are those objectives, in order?
   - **A.** Rescue victims, protect exposures, confine the fire, extinguish the fire, conduct salvage and overhaul.
   - **B.** Rescue victims, extinguish the fire, control the utilities, check for extension, conduct salvage and overhaul.
   - **C.** Confine the fire, rescue victims, establish a water supply, extinguish the fire, conduct ventilation.
   - **D.** Rescue victims, establish a water supply, conduct ventilation, extinguish the fire, conduct salvage and overhaul.

# Forcible Entry

# Chapter 11

## NFPA 1081 Standard

### Incipient Facility Fire Brigade Member

NFPA 1081 contains no Incipient Facility job performance requirements for this chapter.

### Advanced Exterior Facility Fire Brigade Member

**5.3.2*** Gain access to facility locations, given keys, forcible entry tools (e.g., bolt cutters, small hand tools, and ladders), and an assignment, so that areas are accessed and remain accessible during advanced exterior facility fire brigade operations.

**(A)** *Requisite Knowledge.* Site drawing reading, access procedures, forcible entry tools and procedures, and site-specific hazards, such as access to areas restricted by railcar movement, fences, and walls. Procedures associated with special hazard areas such as electrical substation, radiation hazard areas, and other areas specific to the site if needed.

### Interior Structural Facility Fire Brigade Member

**6.2.2** Force entry into a structure, given PPE, tools, and an assignment, so that the tools are used, the barrier is removed, and the opening is in a safe condition and ready for entry.

**(A)** *Requisite Knowledge.* Basic construction of typical doors, windows, and walls within the facility; operation of doors, windows, and their associated locking mechanisms; and the dangers associated with forcing entry through doors, windows, and walls.

**(B)** *Requisite Skills.* The ability to transport and operate site-specific tools to force entry through doors, windows, and walls using assorted methods and tools.

## Knowledge Objectives

After completing this chapter, you will be able to:

- Force entry into a structure, given personal protective equipment, tools, and an assignment, so that the tools are used, the barrier is removed, and the opening is in a safe condition and ready for entry.
- Understand basic construction of typical doors, windows, and walls within the facility; operation of doors, windows, and locks; and the dangers associated with forcing entry through doors, windows, and walls.

## Skills Objectives

After completing this chapter, you will be able to perform the following skills:

- Force entry through an inward-opening door.
- Force entry through an outward-opening door.
- Use the triangle method to open an overhead door.
- Force entry through a double-hung window.
- Force entry through a casement window.
- Force entry through a projected window.
- Force entry by unscrewing a lock.
- Breach masonry and metal walls.
- Breach a wood floor.

# You Are the Brigade Member

**You are responding to multiple reports of fire in the plant's administrative building. Even though it is late evening, you know that some of the engineering and management staff are working late. There is also janitorial staff in the building. You also know that some of the internal areas of the building are locked, including the normal access doors and sensitive design and document storage areas. Your portable radio sounds with a report of people trapped on the second floor.**

1. ***What tools will you need to force entry in this situation?***
2. ***What obstacles might you encounter inside the building?***
3. ***What are some of the dangers associated with forcible entry?***

## Introduction

One of a brigade member's most dynamic and challenging tasks, forcible entry is defined as gaining access to a structure when the normal means of entry are locked, secured, obstructed, blocked, or unable to be used for some reason. The term forcible entry usually refers to structures; extrication is the term used for entry to a vehicle. This chapter examines forcible entry into structures.

Forcible entry requires strength, knowledge, proper techniques, and skill. Because forcible entry often causes damage to the property, brigade members must consider the results of using different forcible entry methods and select the one appropriate for the situation. If rapid entry is needed to save a life or prevent a more serious loss of property, it is appropriate to use maximum force. When the situation is less urgent, brigade members can take more time and use less force.

Another factor to consider when making forcible entry is the need to secure the premises after operations are completed. Brigade members must never leave the premises in a condition that would allow unauthorized entry. If a door or a lock is destroyed during an urgent forcible entry, arrangements should be made to board up or repair the opening afterward. When the situation is not urgent, consider using entry methods that result in less damage and can be more easily repaired.

A brigade member who is skilled in forcible entry should be able to get the job done quickly, with as little damage as possible. The brigade member must consider the type of building construction, possible entry points, the types of securing device(s) that are present, and the best tools and techniques for each situation. Selecting the right tool and using the proper technique initially saves valuable time and could save lives.

Doors, windows, locks, and security devices can be combined in countless variations to prevent easy entry. Brigade members must keep up with technology, including the new styles of windows, doors, locks, and security devices that are common in the company area and how they operate. Learn to recognize different types of doors, windows, and locks. The best time to examine these components is during inspection and preincident planning visits. Touring buildings under construction or renovation is also an excellent way to learn about building construction and to examine different devices. Talking with construction workers and locksmiths can provide valuable information about how to best attack a particular lock, window, or door.

This chapter will review door, window, and lock construction, as well as discuss the tools and techniques used for forcible entry in several types of situations. The skill drills cover the most common forcible entry techniques.

## Forcible Entry Situations

Forcible entry is usually required at emergency incidents where time is a critical factor. A rapid forcible entry, using the right tools and methods, might result in a successful rescue or allow brigade members to make an interior attack and control a fire before it extends. Brigade members must study and practice forcible entry techniques so that proficiency increases with every experience.

Brigade leaders usually select both the point of entry and the method to be used. They ensure that the efforts of different teams are properly coordinated for safe, effective operations. For example, forcible entry actions must be coordinated with hose teams because the entry must be made before brigade members can get inside to attack a fire. Opening a door before the hose lines are in place and ready to advance could allow fresh air to enter, resulting in a possible fire spread or backdraft. Making an opening at the wrong location could undermine a well-planned attack.

Before beginning forcible entry, remember to "try before you pry." Always check doors or windows to ensure that forcible entry is needed. An unlocked door requires no force. A window that can be opened does not need to be broken. Checking first takes only a few seconds and could save several minutes of effort and unnecessary property damage. It is equally important to look for alternative entry points. Do not spend time working on a locked door when a nearby window provides easy access to the same room.

Unusually difficult forcible entry situations may require the expertise of specialty equipment and specially trained individuals. These individuals may have training and experience working with specialized tools and equipment to gain access to almost any area.

## Forcible Entry Tools

Choosing the right forcible entry tool can be a very important decision. Fire brigades use several forcible entry tools, ranging from basic cutting, prying, and striking tools to sophisticated mechanical and hydraulic equipment. Brigade members must know:

- What tools the brigade uses
- The uses and limitations of each tool
- How to select the proper tool for the job
- How to safely operate each tool
- How to carry the tools safely
- How to inspect and maintain each tool, to ensure readiness for service and safe operation

### General Tool Safety

Any tool that is used incorrectly or is maintained improperly can be dangerous to both the operator and other persons. Anyone who uses a tool should understand the proper operating and maintenance procedures before beginning the task. Always follow the manufacturer's recommendations for operation and maintenance.

General safety tips for using tools include:

1. Always wear the appropriate protective equipment. When conducting forcible entry during fire suppression operations, this includes a full set of structural firefighting personal protective equipment (PPE).
   - Goggles or approved eye protection are required when working with cutting or striking tools to prevent eye injuries from pieces of metal, glass, or other materials.
   - Gloves provide protection from sharp cutting blades as well as sharp edges and broken glass. A good glove can easily make the difference between a minor injury and a severed finger.
   - A helmet provides essential protection from falling debris.
   - Turnout coat and pants help to protect skin from sharp metal edges or broken glass.
   - Boots protect the feet from nails or other sharp objects.
2. Learn to recognize the materials used in building and lock construction and the appropriate tools and techniques for each. For example, many locks are made of case-hardened steel, which combines carbon and nitrogen to produce a very hard outer coating that cannot be penetrated by most ordinary cutting tools. Using the wrong tool to cut case-hardened steel could break the tool and potentially injure the user and other brigade members.
3. Keep all tools clean, properly serviced according to the manufacturer's guidelines, and ready for use. Immediately report any tool that is damaged or broken and take it out of service.
4. Do not leave tools lying on the ground or floor. Return them to a tool staging area or the apparatus. Tools that are left lying around could cause injuries.

### General Carrying Tips

Tools can cause injuries even when not in use. Carrying a tool improperly can result in muscle strains, abrasions, or lacerations. General carrying tips that apply to all tools include:

- Do not try to carry a tool or piece of equipment that is too heavy or designed to be used by more than one person (▼Figure 11-1). Request assistance from another brigade member.
- Always use your legs—not your back—when lifting heavy tools.

Figure 11-1 Request assistance to help carry a heavy tool.

Figure 11-2 All tools should be kept in a ready state.

- Keep all sharp edges and points away from your body at all times. Cover or shield them with a gloved hand to protect those around you.
- Carry long tools with the head down toward the ground. Be aware of overhead obstructions and wires, especially when using pike poles.

## General Maintenance Tips

All tools should be kept in a "ready state." This means that the tool is in proper working order, in its proper storage place, and ready for immediate use. Forcible entry tools are generally stored in a designated compartment on the apparatus. Hand tools should be clean, and cutting blades should be sharp. Power tools should be completely fueled and treated with a fuel stabilizing product, if necessary, to ensure easy starting. Every brigade member should be able to locate the right tool immediately, confident that it is ready for use ( ▲ Figure 11-2 ).

All tools require regular maintenance and cleaning to ensure that they will be ready for use in an emergency. Thorough, conscientious daily or weekly checks should be performed, particularly with infrequently used tools. Always follow the manufacturer's instructions and guidelines, and store maintenance manuals in an easily accessible location. Keep proper records to track maintenance, repairs, and any warranty work that is performed.

## Types of Forcible Entry Tools

Forcible entry tools include hand tools and power tools. They can be categorized as:

- Striking tools
- Prying/Spreading tools
- Cutting tools
- Lock tools

Many of the basic fire fighting tools were described and pictured in Chapter 8, Brigade Member Tools and Equipment.

### Brigade Member Tips

The forcible entry tools covered in this chapter are not the only ones available. Different fire brigades use different tools, and there are often regional preferences for a particular piece of equipment. Become familiar with the tools used in your brigade.

### Striking Tools

Striking tools are used to generate an impact force directly on an object or another tool. Striking tools are generally hand tools powered by human energy. The head of a striking tool is usually made of hardened steel.

#### Flat-head Axe

The flat-head axe was one of the first tools developed by humans and is still widely used. One side of the axe head is a cutting blade, and the other side is a flat striking surface. Brigade members often use the flat side to strike a Halligan tool and drive a wedge into an opening. Most fire apparatus carry both flat-head axes and pick-head axes. The pick-head axe is a cutting tool, not a striking tool.

#### Battering Ram

The battering ram is another tool used to force doors and breach walls. Originally, it was a large log used to smash through enemy fortifications. Today's battering rams are usually made of hardened steel and have handles; two to four people are needed to use a battering ram.

#### Sledgehammer or Maul

Sledgehammers (sometimes called mauls) come in various weights and sizes. The head of the hammer can weigh from 2 to 20 lb. The handle may be short like a carpenter's hammer or long like an axe handle. A sledgehammer can be used by itself to break down a door or with other striking tools such as the Halligan.

### Prying/Spreading Hand Tools

Hand tools designed for prying and spreading are often used by brigade members to force entry into buildings. This section describes the most commonly used forcible entry tools.

### Brigade Member Safety Tips

Before swinging a striking tool, make sure there is a clear area of at least the length of the tool handle. Position another brigade member outside the swing area as a safety person to keep others from entering the zone.

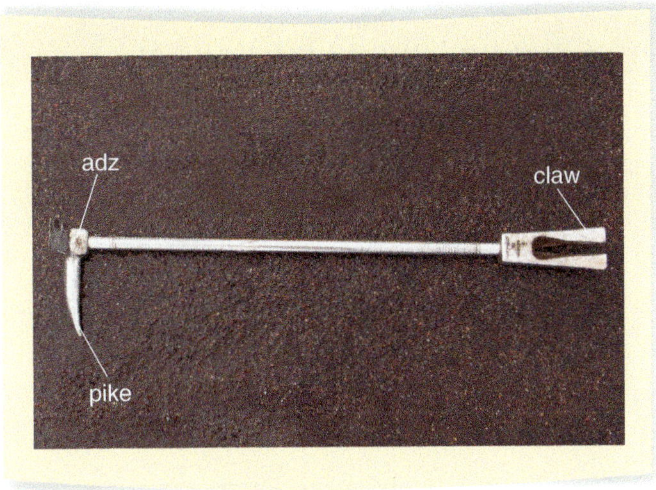


Figure 11-3 A Halligan tool has multiple purposes.

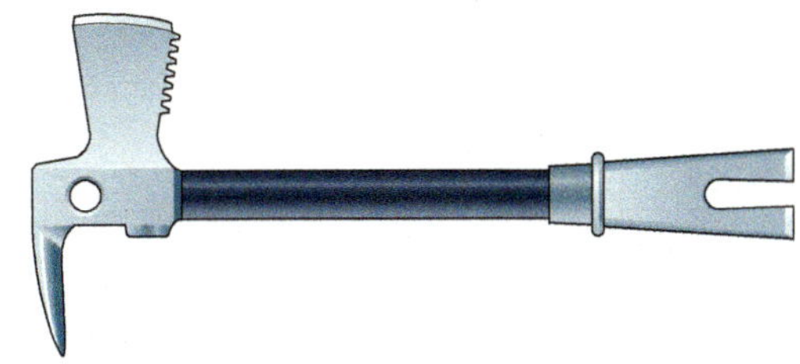

Figure 11-4 The pry axe is a multipurpose tool that can be used to cut and pry.

### Halligan Tool (Bar)

The Halligan tool or bar is widely used by the fire service (▲Figure 11-3). Pairing a Halligan tool with a flat-head axe creates a tool often referred to as "the irons." These two tools are commonly used to perform forcible entry. The Halligan tool incorporates three different tools—the adz, pick, and claw. The adz end is used to pry open doors and windows; the pick end is used to make holes or break glass; and the claw is used to pull nails and to pry apart wooden slats. A brigade member can use a flat-head axe to strike the Halligan into place and create a better bite (a small opening that allows better tool access in forcible entry) between a door or window and its frame.

### Pry Bar/Hux Bar/Crowbar

Pry bars, Hux bars, and crowbars are made from hardened steel in a variety of shapes and sizes. They are most commonly used to force doors and windows, to remove nails, or to separate building materials. The various shapes allow brigade members to exert different amounts of leverage in diverse situations.

### Pry Axe

The pry axe is a multipurpose tool that can be used both to cut and to force open doors and windows (►Figure 11-4).

## Fire Marks

The Halligan bar, designed by New York City fire fighter Hugh Halligan, is one of the most innovative tools ever created by a fire fighter. It is used by thousands of fire departments and brigades.

The pry axe includes an axe head, a pick, and a claw. The tool consists of two parts, the body, which has the axe head and pick and looks similar to a miniature pick axe. It also includes a handle with a claw at the end, which slides into the body. The handle can be extended to provide extra leverage when prying or it may be removed and inserted into the head of the axe to provide rotational leverage. Extreme caution should be used when handling this tool. Over time the mechanism that locks the handle into position may become worn, allowing the handle to slip out and potentially injure brigade members.

### Hydraulic Tools

Hydraulic-powered tools such as spreaders, cutters, and rams are often used for forcible entry. These tools require hydraulic pressure, which can be provided by a high-pressure, motor-operated pump or a hand pump.

Hydraulic cutters and spreaders are usually used in vehicle extrication but can also be used in some forcible entry situations (►Figure 11-5). Hydraulic rams come in different lengths and sizes and can be used to apply a powerful force in one direction. The operation of these tools is covered in the chapter on vehicle rescue and extrication.

A rabbit tool is a small hydraulic spreader operated by a hand-powered pump. This tool is designed with teeth that will fit into a door jamb or rabbit (a type of door frame that has the door stop cut into the frame). As the spreader opens, it applies a powerful force that can open many doors.

## Cutting Tools

Cutting tools are primarily used for cutting doors, roofs, walls, and floors. Although not as fast as power tools, hand-operated cutting tools are proven and reliable in many situations. Because they do not require a power source, they often can be deployed more quickly than power tools.

Figure 11-5 A hydraulic spreader.

## Brigade Member Safety Tips

If hydraulic fluid under pressure begins to leak, it can cause serious injuries. Do not try to disconnect a pressurized hose because this can cause serious injury or splash hydraulic fluid into the eyes. If the fluid enters the body, it can cause tissue necrosis (death). Always relieve hydraulic system pressure before disconnecting a hose or fitting. Avoid wiping your face or eyes if there is hydraulic fluid on your hands.

Several power cutting tools are available for different applications. These tools can be powered by batteries, electricity, gasoline, or hydraulics, depending on the amount of power required. Each type of tool and power source has advantages and disadvantages. For example, battery-powered tools are portable and can be placed in operation quickly, but have limited power and operating times.

### Axe

There are many different types of axes, including flat-head, pick-head, pry, and multipurpose axes. The cutting edge of an axe is used to break into plaster and wood walls, roofs, and doors (▶ Figure 11-6).

Although a flat-head axe has been described as a striking tool, it is generally classified as a cutting tool. The pick-head axe is similar to a flat-head axe, but it has a pick instead of a striking surface opposite the blade. This pick can be used to make an entry point or a small hole if needed.

Specialty axes such as the pry axe and the multipurpose axe can be used for purposes other than cutting. The multipurpose axe, for example, can be used for cutting, striking, or prying, and includes a pick, a nail puller, a hydrant wrench, and a gas main shut-off wrench.

Figure 11-6 An axe is used to break into plaster, wood walls, roofs, and doors.

### Bolt Cutters

Bolt cutters are used to cut metal components such as bolts, padlocks, chains, and chain-link fences. Bolt cutters are available in several different sizes, based on the blade opening and the handle length. The longer the handles, the greater the cutting force that can be applied. Bolt cutters may not be able to cut into some heavy-duty padlocks that are made with case-hardened metal.

## Brigade Member Tips

An axe and Halligan are often carried together as a set and are referred to as "the irons."

### Circular Saw

Gasoline-powered circular saws are used by most fire brigades both for forcible entry and for cutting ventilation holes. They are light, powerful, and easy-to-use, with blades that can be changed quickly. The different blades enable the saw to cut a variety of materials.

## Brigade Member Tips

Most power cutting tools have built-in safety features. Always take advantage of and use the safety features on every tool. To avoid injury, never remove safety guards from tools, not even for "just one cut."

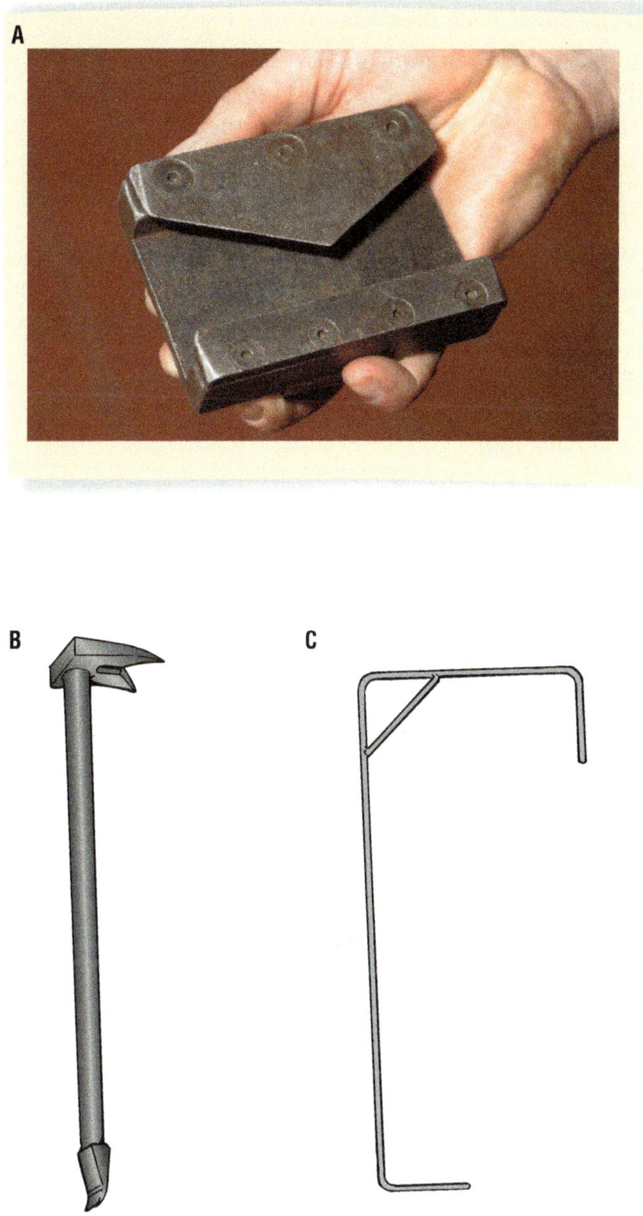


Figure 11-7A–C A. K tool. B. A tool. C. J tool.

Fire brigades generally carry several different circular saw blades so that the proper blade can be used in different situations.

- Carbide-tipped blades are specially designed to cut through hard surfaces or wood. They stay sharp for long periods, so more cuts can be made before the blade needs to be changed.
- Metal-cutting blades are a composite material made with aluminum oxide. They are used to cut metal doors, locks, or gates.
- Masonry-cutting blades are abrasive and made of a composite material that includes silicon carbide or steel. They can cut concrete, masonry, and similar materials. Because they resemble metal-cutting blades, the operator must check the label on the blade before using it. Blades with missing labels should be discarded. Do not store masonry-cutting blades near gasoline because the gasoline vapors will cause the composite materials in the blade to decompose. When the blade is later used, it could disintegrate.

## Lock Tools/Specialty Tools

The lock and specialty tools are used to disassemble the locking mechanism on a door. These devices will cause minimal damage to the door and the door frame. If they are used properly, the door and frame should be undamaged, although the lock will generally need to be replaced. An experienced user can usually gain entry in less than a minute with these tools.

### K Tool, A Tool, and J Tool

The K tool is designed to cut into a lock cylinder ► Figure 11-7A. To shear the cylinder from a lock, place the cutting edge of the K tool on the top of the lock, insert a pry bar into the K tool, and strike the pry bar with a flat-head axe. If the lock has a protective ring, the K tool may not be able to cut through it. Once the lock cylinder is removed, another tool is then used to open the locking mechanism.

The A tool is similar to the K tool except that it has the pry bar built into the cutting part of the tool ► Figure 11-7B. This tool gets its name from its A-shaped cutting edges. To use an A tool, put the cutting head on the lock cylinder and use a striking tool to force it down into the cylinder until the lock can be forced out of the door. Once the lock cylinder is removed, use another tool to open the locking mechanism.

A J tool will fit between double doors that have panic bars ▲ Figure 11-7C. Slide the J tool between the doors and pull to engage the panic bars.

### Shove Knife

A shove knife is an old tool frequently used. Slipping the knife between the door and the frame at the latch forces the latch back and opens the door. Most new doors have latches that will not respond to a tool like this.

### Duck-Billed Lock Breakers

To open a padlock using a duck-billed lock breaker, drive the point into the shackles (the U-shaped part) of the lock. The increasing size of the point forces the shackles apart until they break.

### Locking Pliers and Chain

The locking pliers and chain are used to clamp a padlock securely in place so that the shackles can be cut safely with a circular saw or cutting torch. Clamp the pliers to the lock body while a second brigade member maintains a steady tension on the chain as the lock is being cut.

### Bam-Bam Tool

The bam-bam tool has a case-hardened screw, which is inserted and secured in the keyway of a lock. After the screw is in place, drive the handle of the tool into the keyway and pull the tumbler out of the lock.

Figure 11-8 The parts of a door.

# Doors

Usually, the best point to attempt forcible entry to a structure is the door or window. Doors and windows are constructed as entry points and are generally of weaker materials than walls or roofs. An understanding of the basic construction of doors will help you select the proper tool and increase the likelihood of successfully gaining entry.

## Basic Door Construction

Doors can be categorized by their construction material and by the way they open. Both interior and exterior doors have the same basic components ▶ Figure 11-8:

- Door (the entryway itself)
- Jamb (the frame)
- Hardware (the handles, hinges, etc.)
- Locking device

## Construction Material

Doors are generally constructed of wood, metal, or glass. The design of the door and the construction material used will determine the difficulty in forcing entry.

### Wood

Wood doors may be found in some commercial buildings. There are three types of wood swinging doors: slab, ledge, and panel. Slab doors come in solid-core or hollow-core designs and are attached to wood-frame construction with normal hardware ▶ Figure 11-9.

Solid-core doors are constructed of solid wood core blocks covered by a face panel. Solid-core doors may have an Underwriters Laboratories or Factory Mutual fire rating. They are usually used for entrance doors. Solid-core doors are heavy and may be difficult to force, but their construction enables them to contain fire better than hollow-core doors.

Figure 11-9 A slab door may have either solid-core or hollow-core construction.

Hollow-core doors have a lightweight, honeycomb interior, covered by a face panel. They are often used as interior doors, such as for bedrooms. Hollow-core doors are easy to force and will burn through quickly.

Ledge doors are simply wood doors with horizontal bracing. These doors, which are often constructed of

## Brigade Member Safety Tips

Opening a door can allow fresh air to enter, resulting in a possible fire spread or backdraft. Be careful and listen to instructions.

tongue-and-groove boards, may be found on warehouses, sheds, and barns.

Panel doors are solid wood doors that are made from solid planks to form a rigid frame with solid wood panels set into the frame. Panel doors are used as both exterior and interior doors and may be made from a variety of types of wood. These doors resist fire longer than hollow-core slab doors and are typically easier to breach than solid-core slab doors if entry is attempted at the panels.

### Metal

Metal doors are utilitarian for warehouses and factories. They also may have either a hollow-core or solid-core construction. Hollow-core metal doors have a metal framework interior, so they are as lightweight as possible. Solid-core metal doors have a foam or wood interior to reduce weight without affecting strength. Metal doors may be set in either a wood frame or a metal frame.

### Glass

Glass doors generally have a steel frame with tempered glass; or are simply tempered glass, and do not require a frame, but have metal supports to attach hardware. Glass doors are easy to force, but can be dangerous due to the large amount of small broken pieces that are produced when glass is broken.

## Types of Doors

Doors also can be classified by how they open. The five most common ways that doors open are: inward, outward, sliding, revolving, and overhead. Overhead, sliding, and revolving doors are the most easily identifiable. Inward-opening and outward-opening doors can be differentiated by whether or not the hinges are visible. If you can see the hardware, the door will swing toward you (outward); if the hinges are not visible, the door will swing away from you (inward) ▶ Figure 11-10.

Door frames may be constructed of either wood or metal. Wood-framed doors come in two styles: stopped and rabbit. Stopped door frames have a piece of wood attached to the frame to stop the door from swinging past the latch. Rabbited door frames are constructed with the stop cut built into the frame so it cannot be removed.

Metal-framed doors are more difficult to force than wood-framed doors. Metal frames have little flexibility; when a metal frame is used with a metal door, forcing entry

A

B

Figure 11-10 A. An outward-opening door will have hinges showing. B. A door with the hinges not showing will open inward.

will be very difficult. Metal frames look like rabbited door frames.

## Inward-Opening Doors

### Design

Inward-opening doors of wood, steel, or glass can be found in most structures. They have an exterior frame with a stop or rabbit that keeps the door from opening past the latch. The locking mechanisms may range from standard door knob locks to deadbolt locks or sliding latches.

### Forcing Entry

A simple solution to gaining entry through these doors may be to break a small window on the door or adjacent to it, reach inside, and operate the locking mechanism. Remember to "try before you pry."

If no window is available and the door is locked, stronger measures might be required. To force an inward-opening door, first determine what type of frame it has. If the door has a stopped frame, use a prying tool near the locking mechanism to pry the stop away from the frame. After removing the stop, reinsert the prying tool near the latch and pry the door away from the frame. Once the latch clears the strike plate, push the door inward. Use a striking tool to force the prying tool further into the jamb. To force entry into an inward-opening door, follow the steps in ▶ Skill Drill 11-1.

1. Look for any safety hazards as you evaluate the door. Inspect the door for the location and number of locks and their mechanisms. **(Step 1)**
2. Place the forked end of the Halligan tool into the door frame, between the door jamb and the door stop. Place the tool near the lock, with the beveled end against the door. **(Step 2)**
3. Once the Halligan tool is in position, have your partner, on your command, drive the tool further into the gap between the rabbited jamb or stop and the door. Make sure that the tool is not driven into the door jamb itself. **(Step 3)**
4. Once the tool is past the stop, and between the door and the jamb, push the Halligan toward the door to force it open. If more leverage is needed, your partner can slide the axe head between the bevel of the Halligan bar and the door. It may be necessary to push in the door. **(Step 4)**
5. Secure the door to prevent it from closing behind you.

## Outward-Opening Doors

### Design

Outward-opening doors are used in commercial occupancies and for most exits ▶ Figure 11-11. They are designed so that people can leave a building quickly during an emergency. Outward-opening doors may be constructed of wood, metal, or glass. They usually have exposed hinges, which may present an entry opportunity. More frequently, however, these hinges will be sealed so that the pins cannot be removed. Several types of locks, including handle-style locks and deadbolts, may be used with these doors.

**Figure 11-11** An outward-opening door enables rapid exit.

### Forcing Entry

Before forcing entry to an outward-opening door, check the hinges to see if they can be disassembled or the pins removed. If that would take too long or cannot be done, place the adz end of a prying tool into the door frame near the locking mechanism. Use a striking tool to drive it further into the door jamb and get a good bite on the door. Then leverage the tool to force the door outward away from the jamb. To force entry into an outward-opening door, follow the steps in ▶ Skill Drill 11-2.

1. Size-up the door looking for safety hazards. Inspect the door for the location and number of locks and mechanisms. **(Step 1)**
2. Place the adz end of the Halligan tool between the door and the frame, near the locking mechanism, or between the mechanism and a secondary lock. **(Step 2)**
3. Once the Halligan tool is in position, give the command to your partner to strike the Halligan and drive the adz end further into the gap.

11-1 **Skill Drill**

# Forcing Entry into an Inward-Opening Door

Size-up the door looking for any safety hazards. Inspect the door for the location and number of locks and mechanisms.

Place the forked end of the Halligan tool into the door frame between the door jamb and the door stop. Insert the tool near the lock, with the beveled end of the tool against the door.

Once the Halligan tool is in position, have your partner, on your command, drive the tool further into the gap between the rabbited jamb or stop and the door. Make sure that the tool is not driven into the door jamb itself.

Once the tool is past the stop, and between the door and the jamb, push the Halligan toward the door to force it open. If more leverage is needed, your partner can slide the axe head between the bevel of the Halligan bar and the door. It may be necessary to push in on the door. Secure the door to prevent it from closing behind you.

11-2 Skill Drill

## Forcing Entry into an Outward-Opening Door

1 Size-up the door looking for safety hazards. Inspect the door for the location and number of locks and mechanisms.

2 Place the adz end of the Halligan tool between the door and the frame, near the locking mechanism, or between the mechanism and a secondary lock.

3 Pry in a downward direction with the fork end of the tool and then force the door outward. Always secure the door to prevent it from closing behind you.

4. Pry down with the fork end of the tool and then force the door outward.
5. Always secure the door to prevent it from closing behind you. **(Step 3)**

## Sliding Doors

### Design

Most sliding doors are constructed of tempered glass in a wooden or metal frame. They are commonly found in residences and hotel rooms that open onto balconies or patios (▶ **Figure 11-12**). Sliding doors generally have two sections and a double track; one side is fixed in place while the other side slides. A weak latch on the frame of the door secures the movable side. Many people prop a wood or metal rod in the track to provide additional security. If this has been done, look for another means of entry if possible to avoid destroying the door and leaving the premise open.

### Forcing Entry

Forcing open sliding glass doors may be either very easy or very difficult. If the doors are not reinforced with a rod in the track, they should be easy to force. The locking mechanisms are not strong, and any type of prying tool can be used. Place the tool into the door frame near the locking mechanism and

Figure 11-12 A reinforced sliding glass door may be difficult to open without breaking the glass.

Figure 11-13 Building codes may require outward-opening doors next to a revolving door.

force the door away from the mechanism. If there is a rod in the track that prevents the door from moving and there is no other way to enter, the only forcible entry choice is to break the glass. Remember to clean all broken glass from the opening.

## Revolving Doors

### Design

Revolving doors are most commonly found in upscale buildings and buildings in large cities (▶ Figure 11-13). They are usually made of four glass panels with metal frames. The panels are designed to collapse outward with a certain amount of pressure to allow rapid emergency escape. Revolving doors are generally secured by a standard cylinder lock or slide latch lock.

Because of *Life Safety Code*®, fire prevention, and building code requirements, there are often standard, outward-opening doors adjacent to the revolving doors. It may be easier to attempt to force entry through these doors instead of through the revolving doors. Repairing any damage to the side doors will be less expensive for the building owner, too.

### Forcing Entry

Forced entry through revolving doors should be avoided whenever possible. Even if the doors can be forced open, the opening created will not be large enough to allow many people to exit easily. Forced entry through a revolving door can be done by attacking the locking mechanism directly or by breaking the glass.

## Overhead Doors

### Design

Overhead doors come in many different designs—from standard residential garage doors to high-security commercial roll-up doors (▶ Figure 11-14). Most residential garage doors have three or four panels that may include windows. Some residential overhead garage doors come in a single section and tilt rather than roll-up. These doors may be made of wood or metal. They usually have a hollow-core filled with insulation or foam. Commercial security overhead doors are made of

**Brigade Member Safety Tips**

Always secure doors in the open position before entering a building. Prop open overhead doors with a pike pole. Overhead doors could come down on brigade members or their attack lines, trapping them or cutting off their water supply. This could complicate the attack and lead to injuries or deaths.

Figure 11-14 A. Commercial overhead doors are made of metal panels or steel rods. B. Lightweight overhead garage doors may tilt or roll-up.

metal panels or hardened steel rods. They may be solid-core or hollow-core, depending on the amount of security needed. Overhead doors can be secured with cylinder-style locks, padlocks, or automatic garage door openers.

### Forcing Entry

Before forcing entry through an overhead door, make a careful size-up of the door. Most residential garage doors are not very sturdy; breaking a window or panel and manually operating the door lock or pulling the emergency release on the automatic opener from the inside may be all that is needed. If the fire is behind the overhead door, it may have weakened the door springs, making it impossible to raise the door. In such a case, the door must be cut or another means of entry found.

Secure the opened door with a pike pole or support to ensure that it does not close. Put the prop under the door near the track.

The quickest way to force entry through a security roll-up door is to cut the door with a torch or saw. Make the cut in the shape of a triangle, with the point at the top. Pad the opening to prevent injury to those entering and exiting through the opening. To open an overhead garage door using the triangle method, follow the steps in Skill Drill 11-3.

1. Before cutting, check for any safety hazards during the size-up of the garage door.
2. Select the appropriate tool to make the cut. The best choice will probably be a power saw with a metal-cutting blade.
3. Wearing full turnout gear and eye protection, start the saw and ensure it is in proper working order.
4. Be aware of the environment behind the door. If necessary, cut a small inspection hole, large enough to insert a hose nozzle.
5. Starting at a center high point in the door, make a diagonal cut to the right, down to the bottom of the door.
6. From the same starting point, make a second diagonal cut to the left, down to the bottom of the door, forming a large triangle.
7. Pad or protect the cut edges of the triangle and the bottom panel to prevent injuries as brigade members enter or leave.

## Windows

Windows provide airflow and light to the inside of buildings; they also can provide emergency entry or exits. Windows are often easier to force than doors. Understanding how to force entry into a window requires an understanding of both window-frame construction and glass construction. Window frames are made of the same materials used in doors—wood, metal, vinyl, or combinations of these materials—and will often match the door construction of a building.

This section reviews window construction, glass construction, and forcible entry techniques. As always, the goal is to force entry with minimal or no damage to the window. As with doors, try to open the window before using any force. Although breaking the glass is the easiest way to force entry, it is also dangerous. Glass may be the least expensive part of the window, but replacement costs are increasing because energy-efficient windows are being used more widely.

### Safety

Always take proper protective measures, including wearing full PPE, when forcing entry through windows. Ensure that the area around the window, both inside and outside the building and below the window, is clear of other personnel. Broken pieces of wood, shattered glass, and sharp metal can cause serious injury.

Remember that forcing entry through windows during a fire situation may cause the fire to spread. Do not attempt forcible entry through a window unless a proper fire attack is in place. Always stand to the windward side, with your hands higher than the breaking point, when breaking windows. This ensures that broken glass will fall away from the hands and body. Placing the tip of the tool in the corner of the window will give you more control in breaking the window. After the window is broken, clear glass from the entire frame, so that no glass shards will stick out and cause injury. This also allows safe passage for those entering or exiting through the window.

## Glass Construction

The glazed (transparent) part of the window is most commonly made of glass. Window glass comes in several configurations: regular glass, double-pane glass, plate glass (for large windows), laminated glass, and tempered glass. Other substances such as Plexiglas may also be used. The window may contain one or more panes of glass; insulated glass usually has two or more pieces of glass in the window and will be discussed in more detail later.

### Regular or Annealed Glass

Single-pane, regular, or annealed glass is normally used in construction because it is relatively inexpensive; larger pieces are called plate glass. It is easily broken with a pike pole. When broken, plate glass creates long, sharp pieces called shards, which can penetrate helmets, boots, and other protective gear, causing severe lacerations and other injuries.

### Double-Pane Glass (Insulated Windows)

Double-pane glass is being used in many structures because it improves insulation by using two panes of glass with an air pocket between them. Some double-pane windows may have an inert gas such as argon between the panes for additional insulation value. These windows are sealed units and are more expensive to replace. However, replacing the glass alone is less expensive than replacing the entire window assembly. Forcing entry through these windows is basically the same as for single-pane windows except that the two panes may need to be broken separately. These windows also will produce dangerous glass shards.

### Plate Glass

Commercial plate glass is a stronger, thicker glass used in large window openings. Although it is being replaced by tempered glass for safety reasons, commercial plate glass can still be found frequently in older large buildings, storefronts, and residential sliding doors. It can easily be broken with a sharp object such as a Halligan tool or a pike pole. When broken, commercial plate glass will create many large, sharp pieces.

### Laminated Glass

Laminated glass, also known as safety glass, is used to prevent windows from shattering and causing injury. Laminated glass is molded with a sheet of plastic between two sheets of glass. Laminated glass is most commonly used in vehicle windshields, but may also be found in other applications such as doors or building windows.

### Tempered Glass

Tempered glass is specially heat-treated, making it four times stronger than regular glass. It is commonly found in side and rear windows in vehicles, in commercial doors, in newer sliding glass doors, and in other locations where a person might accidentally walk into the glass and break it. Tempered glass breaks into small pellets without sharp edges to help prevent injury during accidents. The best way to break tempered glass is by using a sharp, pointed object in the corner of the frame. During a vehicle extrication, a center punch is often used to break a vehicle window.

#### Wired Glass

Wired glass is tempered glass with wire reinforcing. It may be clear or frosted and is often used in fire-rated doors that require a window or sight-line from one side of the door to the other. This glass is difficult to break and force.

## Frame Designs

Window frames come in many styles. This chapter covers the most common ones, but brigade members must be familiar with both the types in their response area and forced entry techniques for each type.

### Double-Hung Windows

#### Design

Double-hung windows have two movable sashes, usually of wood or vinyl, that move up and down (▼ Figure 11-15). They are most common in residences and have wood, plastic, or metal runners. Newer double-hung window sashes may be removed or swung in for cleaning. They may have one locking

Figure 11-15 Double-hung windows allow the inner and outer sashes to move freely up and down.

# Voices of Experience

**"Mike checked the door for heat by pulling down part of his glove and placing the back of his exposed hand onto the door."**

The control room announced over my radio that multiple smoke detectors were activated in Warehouse 3. As I drove my response vehicle toward the facility, I recalled from a prefire planning inspection that Warehouse 3 is a maze of partitioned rooms. The materials stored in this particular facility require separation to prevent their cross-contamination.

When I arrived at the warehouse, I saw the fire brigade assembling and an entry team setting up for interior operations. I got into my PPE and reported to the incident commander (IC) near his vehicle. The IC told me and two other fire brigade members to gather the forcible-entry tools because he needed us to be able to force open several doors.

As we moved into the structure through the already-opened main entry door, I remembered that all of the interior doors in Warehouse 3 were made of heavy metal and encased in metal frames. The good news was that all of the interior doors would open toward us—that is, their hinges would be exposed. Most of the doors used a nonlocking latch with a padlock for security. The padlocks would make matters pretty simple once we determined which areas we needed to enter.

Mike carried the "irons," I carried the "rabbit tool," and Devin had a thermal camera. As we encountered moderate smoke conditions, Devin indicated to Mike the door that needed to be forced; he also stepped around us to motion the hose team toward the door. As the members of the hose team positioned themselves, Mike placed the Halligan forks over the padlock. I set the rabbit tool next to the door and picked up the flat-head axe. I placed the axe's striking end on the Halligan forks and made a "check swing" to ensure that there was enough clearance for my swing. I looked to Mike, who nodded; I then looked to Devin, who was looking through the thermal camera and who also nodded in acknowledgment. The smoke was increasing and I pointed to the fire fighter on the hose team nozzle. He gave me a "thumbs up." I then raised the flat-head axe and swung it, striking the Halligan head. The padlock broke.

Mike checked the door for heat by pulling down part of his glove and placing the back of his exposed hand onto the door. Mike started near the bottom of the door and worked his way up, indicating that the door was hot about halfway up the door. He pulled his glove back up and pointed to the hose team operator, who immediately began to open his nozzle slightly, ensuring that water was available. I stepped to the side as Mike pulled the door open.

The hose team immediately went to work as heavy smoke flowed out of the top half of the door opening. Periodic discharges of water into the upper areas of the room made the hallway warmer at first and then much cooler as the team knocked the fire down.

**John Shrader**
Cooper Nuclear Station Fire Brigade
Brownville, Nebraska

mechanism in the center of the window, or two locks on each side of the lower sash that prevent the sashes from moving up or down.

### Forcing Entry

Forcing entry through double-hung windows involves opening or breaking the locking mechanism. Place a prying tool under the lower sash and force it up to break the lock or remove it from the track. This technique must be done carefully because it may cause the glass to shatter. Because forcing the lock(s) causes extensive damage to the window, breaking the glass and opening the lock(s) may be a less expensive method of gaining entry. To force entry through a wooden double-hung window, follow the steps in ▼ Skill Drill 11-4.

1. Size-up the window for any safety hazards and find the locking mechanism(s). **(Step 1)**
2. Place the pry end of the Halligan tool under the bottom sash in line with the locking mechanism. **(Step 2)**
3. Pry the bottom sash upward to displace the locking mechanism.
4. Secure the window so that it does not close. **(Step 3)**

11-4 **Skill Drill**

## Forcing Entry Through a Wooden Double-Hung Window

1

Size-up the window for any safety hazards, and find the locking mechanism(s).

2

Place the pry end of the Halligan tool under the bottom sash in line with the locking mechanism.

3

Pry the bottom sash upward to displace the locking mechanism. Secure the window so that it does not close.

## Brigade Member Tips

When forcing entry through older, double-hung windows, try sliding a hacksaw blade or similar blade up between the sashes. The teeth on the blade will grip the locking mechanism and can often move it into the unlocked position. This will not work on newer double-hung windows, because of the newer latch design.

## Single-Hung Windows

### Design

Single-hung windows are similar to double-hung windows, except that the upper sash is fixed, and only the lower sash moves (▼ Figure 11-16). The locking mechanism is the same as on double-hung windows as well. It may be difficult to distinguish between single-hung and double-hung windows from the exterior of a building.

### Forcing Entry

Forcing entry through a single-hung window uses the same technique as forced entry through a double-hung window. Place a prying tool under the lower sash and force it up, which will usually break the lock or remove the sash from the track. Again, be careful because the glass will probably shatter. Breaking the glass and opening the window is generally easier with a single-hung window.

## Jalousie Windows

### Design

A jalousie window is made of adjustable sections of tempered glass in a metal frame that overlap each other when closed (▲ Figure 11-17). This type of window is often found in mobile homes and is operated by a small hand-wheel or crank located in the corner of the window.

Figure 11-16 A single-hung window has only one movable sash. Used with permission of Andersen Corporation.

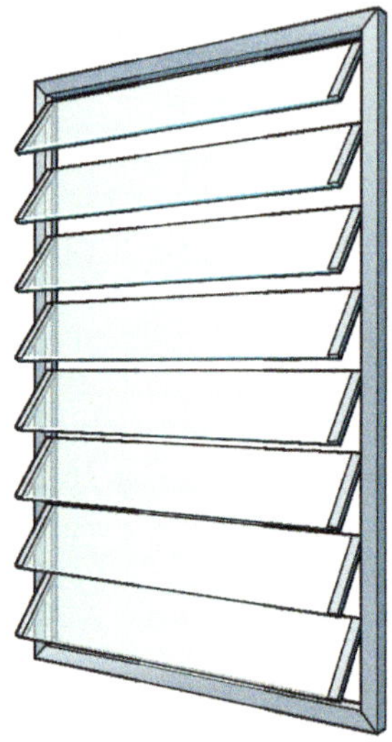

Figure 11-17 Jalousie windows are opened and closed with a small hand-crank.

### Forcing Entry

Forcing entry through jalousie windows can be difficult because they are made of tempered glass and have several panels. Forcing open the panels or breaking a lower one to operate the crank is possible but doesn't leave a big enough opening to enter. Removing or breaking the panels one at a time is time-consuming and does not always leave a clean entry point with the framing system still in place. The best strategy is to avoid these windows.

## Awning Windows

### Design

Awning windows are similar in operation to jalousie windows, except that they usually have one large or two medium-sized glass panels instead of many small ones (► Figure 11-18). Awning windows are operated by a hand crank located in the corner or in the center of the window. Awning windows can be found in residential, commercial, or industrial structures. Residential awning windows may be framed in wood, vinyl, or metal, while commercial and industrial windows will usually be metal framed.

Commercial and industrial awning windows often use a lock and a notched bar to hold the window open, rather than a crank. These are called projected windows and will be discussed later.

### Forcing Entry

Forcing entry through an awning window is the same as for jalousie windows. Break or force the lower panel and operate the crank, or break out all the panels. Depending on the size of the panels and the window frame, it may be easier

Figure 11-18 An awning window has larger panels than a jalousie window.

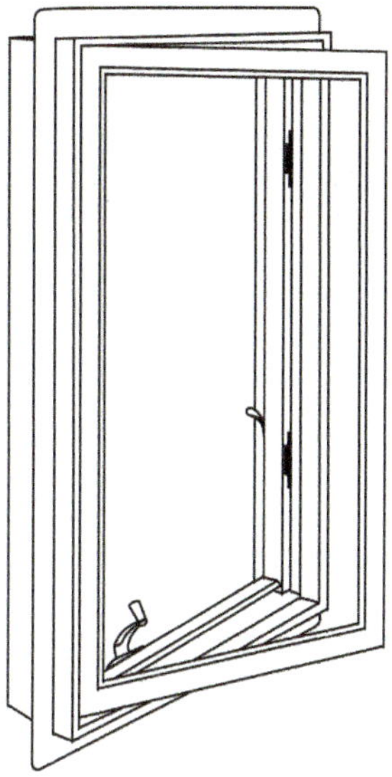

Figure 11-20 Casement windows open to the side with a crank mechanism. Used with permission of Andersen Corporation.

Figure 11-19 Horizontal-sliding windows work like sliding doors. Used with permission of Andersen Corporation.

to access an awning window, and the larger opening makes entry easier as well.

## Horizontal-Sliding Windows

### Design

Horizontal-sliding windows are similar to sliding doors (▲ Figure 11-19). The latch is similar as well and attaches to the window frame. People often place a rod or pole in the track to prevent break-ins. Newer sliding windows have latches between the windows, similar to those on double-hung windows.

### Forcing Entry

Forcing entry through sliding windows is just like forcing entry through sliding doors. Place a pry bar near the latch to break the latch or the plate. If there is a rod in the track, look for another entry point or break the glass, although this should be the last resort.

## Casement Windows

### Design

Casement windows have a steel or wood frame and open away from the building with a crank mechanism (▲ Figure 11-20). Although they are similar to jalousie or awning windows, casement windows have a side hinge, rather than a top hinge. Several types of locking mechanisms can be used. Like jalousie windows, these windows should be avoided because they are difficult to force open.

### Forcing Entry

The best way to force entry through a casement window is to break out the glass from one or more of the panes, and operate the locking mechanisms and crank manually. To force entry through a casement window, follow the steps in (Skill Drill 11-5).

1. Size-up the window to check for any safety hazards and locate the locking mechanism.
2. Select an appropriate tool to break out a windowpane.
3. Stand to the windward side of the window and break out the pane closest to the locking mechanism, keeping the tool lower than your hands.
4. Remove all of the broken glass in the pane to prevent injuries.
5. Reach in and manually operate the window crank to open the window.

### Projected Windows

#### Design

Projected windows, also called factory windows, are usually found in older warehouse or commercial buildings (▼ Figure 11-21). They can project inward or outward on an upper hinge. Screens are rarely used with these windows, but forcing entry may not be easy, depending on the integrity of the frame, the type of locking mechanism used, and their distance off the ground. These windows may have fixed, metal-framed wire glass panes above them.

#### Forcing Entry

Avoid forcing entry through a projected window when possible. They are often difficult to force open and may be difficult to enter. To force entry, access one pane, unlock the mechanism, and open the window by hand. If the opening created is not large enough, break out the entire window assembly. To force entry through a projected or factory window, follow the steps in (Skill Drill 11-6).

1. Size-up the window to check for any safety hazards and locate the locking mechanism.
2. Select an appropriate tool, such as a pike pole, to break a pane.
3. Stand to the windward side of the window and break the pane closest to the locking mechanism, keeping the tool head lower than your hands.
4. Remove all of the broken glass in the frame to prevent injuries.
5. Reach in and manually open the locking mechanism and the window.
6. A cutting tool, such as a torch, can be used to remove the window frame and enlarge the hole.

## Locks

Locks have been used for hundreds of years. They range from very simple push button locks found in most homes to complex computer-operated locks found in banks and high security areas.

### Parts of a Door Lock

All door locks have the same basic parts. Gaining entry will be easier with an understanding of these parts and how they operate (▼ Figure 11-22). The major parts of a door lock include:

- Latch—The part of the lock that "catches" and holds the door frame.
- Operating lever—The handle, doorknob, or keyway that turns the latch to lock it or unlock it.
- Deadbolt—A second, separate latch that locks and reinforces the regular latch from the inside of the door.

### Parts of a Padlock

All padlocks, like all door locks, have similar parts (► Figure 11-23). The basic parts of a padlock include:

- Shackles—The U-shaped top of the lock that slides through a hasp and locks in the padlock itself.
- Unlocking device—The key way, combination wheels, or combination dial used to open the padlock.
- Lock body—The main part of the padlock that houses the locking mechanisms and the retention part of the lock.

Figure 11-21 Projected windows may open inward or outward.

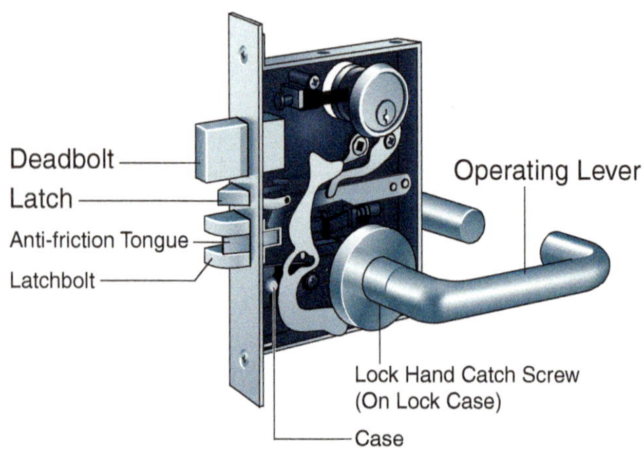


Figure 11-22 All door locks have the same basic parts.

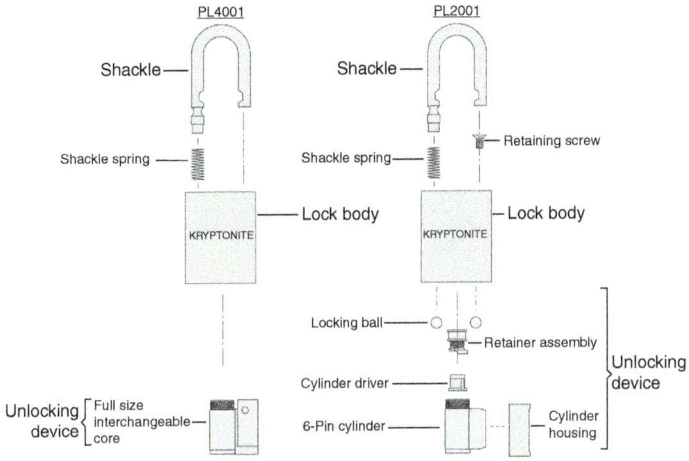


Figure 11-23 All padlocks have three basic parts.

## Safety

Tool safety is important when executing forcible entry through a door or padlock. The tools used for cutting the cylinders must be kept sharp, and brigade members must be careful to avoid being cut. Care also must be taken when using the striking tools to avoid pinching or crushing fingers or hands. Brigade members must wear proper protective equipment such as gloves and use tools carefully to minimize the risk of injury.

## Types of Locks

The four major lock categories are: cylindrical locks, padlocks, mortise locks, and rim locks.

### Cylindrical Locks

#### Design

Cylindrical locks are the most common fixed lock in use today (▶ Figure 11-24). They are relatively inexpensive and are standard in most doors. The locks and handles are set into predrilled holes in the doors. One side of the door usually has a key-in-the-knob lock; the other side will have a keyway, a button, or some type of locking/unlocking device.

#### Forcing Entry

Forcing entry into a cylindrical lock is usually very easy because most mechanisms are not very strong. Place a pry bar near the locking mechanism and lever it to force the lock.

### Padlocks

#### Design

Padlocks are the most common locks on the market today and come in many designs and strengths. Both regular-duty and heavy-duty padlocks are available.

Padlocks come with various unlocking devices, including keyways, combination wheels, or combination dials. Operating the unlocking device opens one side of the lock to release the shackle and allow entry. Shackles for regular padlocks generally have a diameter of 1/4 inch or less and are not made of case-hardened metal. Shackles for heavy-duty padlocks are 1/4 inch or larger in diameter and made of case-hardened steel or some other case-hardened metal. The design of some padlocks hides the latch, making forced entry difficult.

#### Forcing Entry

Several ways can be used to force entry through padlocks without causing extensive property damage. Before breaking the padlock, consider cutting the shackle or hasp first. This saves the lock and makes securing the building easier. Breaking the shackle is the best method for forcing entry through padlocked doors. If the padlock is made of case-hardened steel, many conventional methods of breaking the lock will be ineffective. The most common tools used to force entry through a padlock are bolt cutters, duck-billed lock breakers, a bam-bam tool, locking pliers and chain, and a torch.

- Bolt cutters—Bolt cutters can quickly and easily break regular-duty padlocks, but cannot be used on heavy-duty, case-hardened steel padlocks. To use bolt cutters on a padlock, open the jaws as wide as possible. Close the jaws around one side of the lock shackles to cut

Figure 11-24 A cylindrical lock.

through the shackle. Once one shackle of a regular-duty lock is cut, the other side will spin freely and allow access.

- Duck-billed lock breakers —The duck-billed lock breaker has a large metal wedge attached to a handle. Place the narrow end of the wedge into the center of the shackle and force it through with another striking tool. The wedge will spread the shackle until it breaks, freeing the padlock and allowing access.
- Bam-bam tool—The bam-bam tool can pull the keyway from the lock cylinder, but will not work on higher end padlocks that add security rings to hold the cylinder in place ▼ Figure 11-25. This tool has a case-hardened screw that is placed in the keyway. Once the screw is set, the sliding hammer will pull the tumblers out of the cylinder so that the trip-lever mechanism inside the lock can be opened manually.
- Locking pliers and chain—The locking pliers and chain are attached to a padlock to secure it in a fixed position so that it can be cut. Use a rotary saw with a metal-cutting blade or a torch to cut the padlock after it has been secured. Securing the padlock is necessary because it is too dangerous to hold the lock with a gloved hand or to cut into it while it is loose.

## Mortise Locks

### Design

Mortise locks, like cylindrical locks, are designed to fit in predrilled openings inside a door, and are commonly found in hotel rooms ► Figure 11-26. They have both a latch and a bolt built into the same mechanism, which operate independently of each other. The latch will lock the door, but the bolt can also be deployed for added security.

### Forcing Entry

The design and construction of mortise locks make them difficult to force. A door with a mortise lock may be forced with conventional means but will probably require the use of a through-the-lock technique.

## Rim Locks/Deadbolts

### Design

Rim locks and deadbolts are different types of locks that can be surface mounted on the interior of the door frame ▲ Figure 11-27. They can be identified from the outside by the keyway that has been bored into the door. These locks have a bolt that extends at least 1 inch into the door frame, making the door more difficult to force.

### Forcing Entry

Rim locks and deadbolts are the hardest of the locks to break. They can be very difficult to force with conventional methods, so that through-the-lock methods may be the only option. Skill Drills 11-7 through 11-9 present various through-the-lock techniques. To force entry using a K tool, follow the steps in ► Skill Drill 11-7.

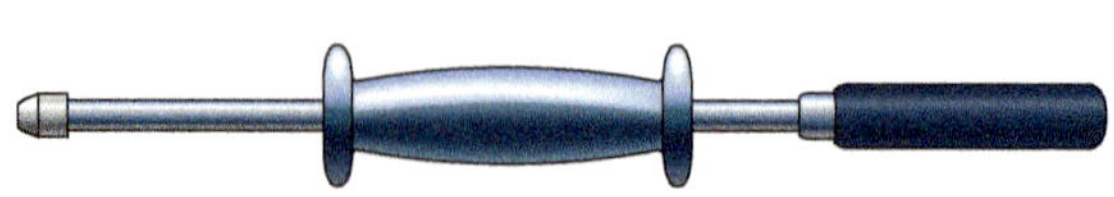

Figure 11-25 A bam-bam tool removes the keyway from the lock cylinder.

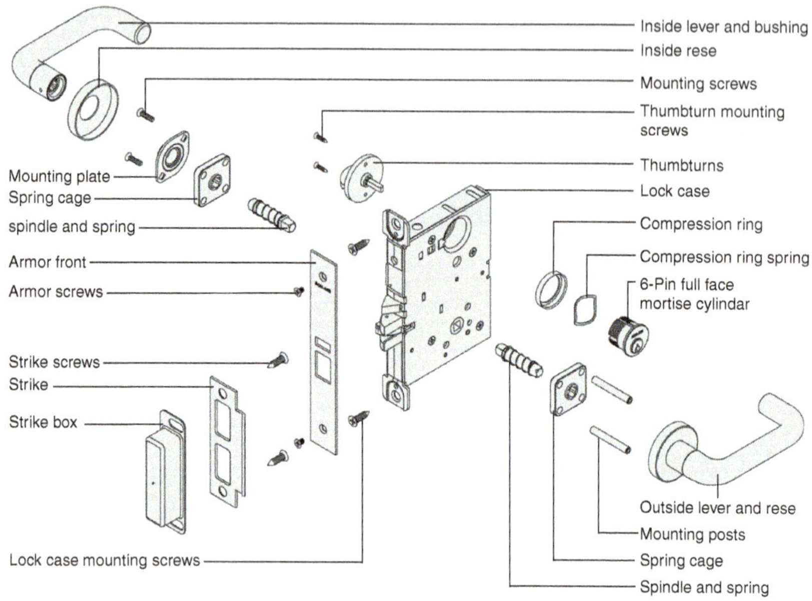


Figure 11-26 Mortise locks have a latch and a bolt, which operate independently of each other.

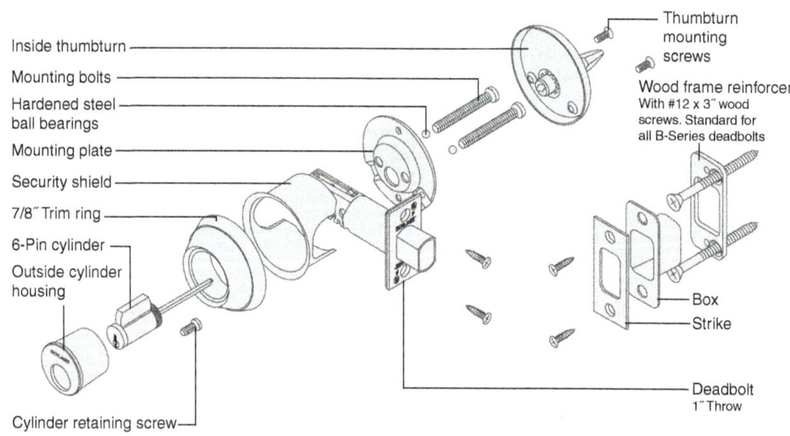


Figure 11-27 Rim locks and deadbolts are mounted on the inside of a door.

1. Size-up the lock area to check for any safety hazards. Determine what type of lock is used, whether it is regular or heavy-duty construction, and whether the cylinder has a case-hardened collar, which may hamper proper cutting.
2. Place the K tool over the face of the cylinder, noting the location of the keyway. **(Step 1)**
3. Using a Halligan or similar pry tool, tap the K tool down into the cylinder. **(Step 2)**
4. Place the adz end of the Halligan tool into the slot on the K tool, and strike the end of the Halligan tool with a flat-head axe to drive the K tool further into the lock cylinder. **(Step 3)**
5. Pry up on the Halligan to pull out the lock and expose the locking mechanism.
6. Using the small tools that come with the K tool, turn the mechanism to open the lock. **(Step 4)**

To force entry using an A tool, follow the steps in ► **Skill Drill 11-8**.

1. Size-up the door and lock area to check for any safety hazards.
2. Place the cutting edges of the A tool at the top of the cylinder, between the cylinder and the door frame. **(Step 1)**
3. Using a flat-head axe or similar tool, drive the A tool into the cylinder.

**11-7 Skill Drill**

## Forcing Entry Using a K Tool

Place the K tool over the face of the cylinder, noting the location of the keyway.

Using a Halligan or similar pry tool, tap the K tool down into the cylinder.

Place the adz end of the Halligan tool into the slot on the K tool, and strike the end of the Halligan tool with a flat-head axe to drive the K tool further into the lock cylinder.

Pry up on the Halligan tool to pull out the lock and expose the locking mechanism. Using the small tools that come with the K tool, turn the mechanism to open the lock.

11-8 Skill Drill

## Forcing Entry Using an A Tool

Place the cutting edges of the A tool at the top of the cylinder, between the cylinder and the door frame.

Using a flat-head axe or similar tool, drive the A tool into the cylinder. Pry up on the A tool to remove the cylinder from the door. Insert a key tool into the hole to manipulate the locking mechanism and open the door.

4. Pry up on the A tool to remove the cylinder from the door.
5. Insert a key tool into the hole to manipulate the locking mechanism and open the door. **(Step 2)**

To force entry by unscrewing the lock, follow the steps in ► **Skill Drill 11-9**.

1. Size-up the door and lock area to check for any safety hazards.
2. Using a set of vise grips, lock the pliers onto the outer housing of the lock with a good grip. **(Step 1)**
3. Unscrew the housing until it can be fully removed.
4. Manipulate the locking mechanism with lock tools to disengage the arm, then open the door. **(Step 2)**

## Breaching Walls and Floors

On occasion, forced entry through a window, lock, or door may not be possible or may take too long. In these cases, consider the option of breaching a wall or floor. An understanding of basic construction concepts is required to safely and quickly execute this technique.

### Load-bearing/Nonbearing Walls

Before breaching a wall, first consider whether it is a load-bearing wall. A load-bearing wall supports the building's ceiling and/or rafters. Removing or damaging a load-bearing wall could cause the building or wall to collapse. A nonbearing wall can be removed safely and without danger (► **Figure 11-28**). Nonbearing walls are also called partition walls or simply partitions. Refer to the chapter on building construction for a review of wall construction.

### Exterior Walls

Exterior walls can be constructed of one or more materials. Many residences have both wood and brick, aluminum siding, or masonry block construction (► **Figure 11-29**). Commercial buildings usually have concrete, masonry, or metal exterior walls. Commercial buildings also may have steel I-beam construction; older commercial structures may have heavy timber construction. Because there are exceptions to every rule, it is important to know the buildings in the specific response area.

Whether to attempt to breach an exterior wall is a difficult decision. Masonry, metal, and brick are formidable materials, and breaking through them can be very difficult. The best tools to use in breaching a concrete or masonry exterior wall are a battering ram, a sledgehammer, or a rotary saw with a concrete blade.

### Interior Walls

Interior walls in light construction structures are usually constructed of wood or metal studs covered by plaster, gypsum,

## Skill Drill 11-9

# Forcing Entry by Unscrewing the Lock

1. Using a set of vise grips, lock the pliers onto the outer housing of the lock with a good grip.

2. Unscrew the housing until it can be removed. Manipulate the locking mechanism with lock tools to disengage the arm, then open the door.

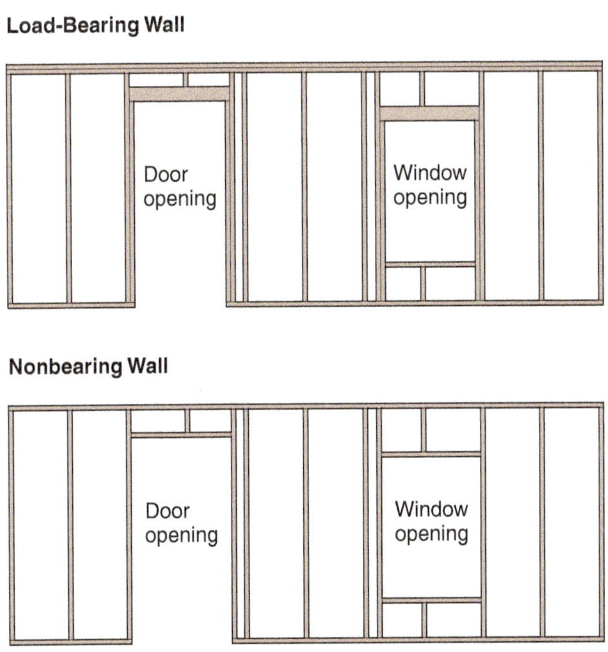


Figure 11-28 A load-bearing wall supports the building's ceiling and/or rafters; a nonbearing wall simply partitions the space.

or sheetrock. Commercial buildings may have concrete block interior walls. Breaching an interior wall can be dangerous to brigade members for several reasons. Many interior walls contain electrical wiring, plumbing, cable wires, and phone wires, which present various hazards. Interior walls may also be load-bearing. Although load-bearing walls may be breached, extreme care should be taken, particularly if studs are removed.

Figure 11-29 Structures such as offices and control rooms may have exterior walls of multiple materials such as metal and masonry block.

After determining whether the wall is load-bearing or not, sound it to locate a stud away from any electrical outlets or switches. Tap on the wall; the area between studs will make a hollow sound compared to the solid sound directly over the stud.

After locating an appropriate site, make a small hole to check for any obstructions. If the area is clear, expand the opening to reveal the studs. Walls should be breached as close as possible to the studs because this makes a large opening, and cutting is easier. If possible, enlarge the opening by removing at least one stud to enable quick escapes. To breach a wall frame, follow the steps in Skill Drill 11-10.

1. Size-up the wall looking for safety hazards such as electrical outlets, wall switches, or any signs of plumbing. Inspect the overall scene to ensure that the wall is not load-bearing.
2. Using a striking or cutting tool, sound the wall to locate any studs, and make a hole between the studs.
3. Cut as close to the studs as possible.
4. Enlarge the hole by extending it from stud to stud and as high as necessary.

To breach a masonry wall, you will create an upside-down V in the wall. Follow the steps in Skill Drill 11-11.

1. Size-up the wall for any safety hazards such as electrical outlets, wall switches, and plumbing. Inspect the overall area to ensure that the wall is not load-bearing.
2. Select a row of masonry blocks that is at or near floor level.
3. Using a sledgehammer, knock two holes each in five masonry blocks along the selected row. Each hole should pierce into the hollow core of the masonry blocks.
4. Repeat the process on four masonry blocks above the first row.
5. Repeat the process on three masonry blocks above the second row.
6. Repeat the process on two masonry blocks above the third row.
7. Repeat the process on one masonry block above the fourth row. An upside down V has now been created.
8. Begin knocking out the remaining portion of the masonry blocks by hitting the masonry blocks parallel to the wall rather than perpendicular to it.
9. Clear any reinforcing wire using bolt cutters.
10. Enlarge the hole as needed.

To breach a metal wall, follow the steps in Skill Drill 11-12.

1. Size-up the wall for any safety hazards such as electrical outlets, wall switches, or plumbing. Inspect the overall area to ensure that the wall is not load-bearing.
2. Select an appropriate tool such as a rotary saw with a metal-cutting blade or a torch.
3. Open the wall using a diamond-shaped or V-cut.
4. Protect or pad the metal edges of the cut to ensure that no one who goes through the opening is cut.

## Floors

The two most popular floor materials found in light offices construction and commercial buildings are wood and poured concrete. Both are very resilient and may be difficult to breach. Conduct a thorough size-up; breaching a floor should be a last resort measure ▶ Figure 11-30. A rotary saw with an appropriate blade is the best tool to breach a floor. To breach a floor, follow the steps in Skill Drill 11-13.

1. Size-up the floor area to ensure there are no hazards in the area to be cut.
2. Sound the floor with an axe or similar tool to locate the floor joists.
3. Use an appropriate cutting tool to cut one side of the hole first, then cut the opposite side.
4. Remove any flooring such as carpet, tile, or floorboards that loosens.
5. Cut a similar opening into the subfloor until the proper size hole is achieved.
6. Secure the area around the hole to prevent others working in the area from falling through the hole.

Figure 11-30 Breaching a floor should be a last resort.

# Forcible Entry and Salvage

## Before Entry

The decision of which type of forcible entry to use will have a direct impact on salvage operations. In a nonemergency, through-the-lock methods will keep damage to a minimum, thus salvaging the property from excessive repairs. In an emergency situation, more urgent methods may cause massive amounts of destruction, requiring expensive repairs to a door, window, or wall.

## After Entry

When the operation is over, and normal salvage and overhaul have begun, the structure must be secured before brigade members leave the scene. The use of conventional forced-entry techniques and specialized tools requires that doors and windows be secured to prevent break-ins or theft after the call. If through-the-lock methods are used, a hasp and padlock may be all that is needed to secure a site. Broken windows or doors can be boarded with sheets of plywood. Police or company security will often increase patrols around a damaged building when needed to deter potential thieves.

Remember that the fire brigade is charged not only with saving lives but also with protecting property. Do everything possible to reduce the amount of damage caused by both the fire and your own actions. Management should be informed of efforts by the fire brigade to protect property at the fire scene and efforts after extinguishment to preserve evidence, protect investigators, cleanup personnel, and secure the scene. It does not serve as justification for a fire brigade, its equipment and training, if the actions of the fire brigade cause much more damage than the fire has already caused.

# Wrap-Up

## Ready for Review

- Forcing entry into a structure can be performed with a basic understanding of forcible entry tools as well as window, door, wall, and lock construction.
- Using proper safety techniques is the most important part of using forcible entry tools.
- Always use full PPE when performing forcible entry.
- Striking tools are used for driving other tools or for hitting doors, windows, walls, or other objects.
- Prying tools are used to spread doors and windows away from their frames.
- Cutting tools are used for cutting into doors, windows, or locks to gain entry.
- Lock tools are specialized tools used for attacking the lock itself, rather than a door or window.
- An understanding of door construction, composition, and opening techniques will facilitate successful forcible entry.
- An understanding of window construction, composition, and glazing will facilitate successful forcible entry.
- An understanding of how a lock works, what it is made of, and how it secures a door or window will facilitate successful forcible entry.
- Floors and walls can be used for forcible entry if precautions are taken and all other methods have failed. Due to potential safety problems, this should be the last type of entry considered.
- Conventional forcible entry has been used for many years and is still the most common form of forcible entry used. With hand tools and leverage, most doors and windows can be forced open without problems.
- Through-the-lock forcible entry is an option when a brigade has the correct tools and conventional methods have not been successful.

## Hot Terms

**A tool** A cutting tool with a pry bar built into the cutting part of the tool.

**Adz** The prying part of the Halligan tool.

**Annealed** The process of forming standard glass.

**Awning window** Windows that have one large or two medium panels operated by a hand crank from the corner of the window.

**Bam-bam tool** A tool with a case-hardened screw, which is secured in the keyway of a lock, to remove the keyway from the lock.

**Battering ram** A tool made of hardened steel with handles on the sides used to force doors and to breach walls. Larger versions are used by up to four people; smaller versions are made for one or two people.

**Bite** A small opening made to enable better tool access in forcible entry.

**Bolt cutter** A cutting tool used to cut through thick metal objects, such as bolts, locks, and wire fences.

**Case-hardened steel** A process that uses carbon and nitrogen to harden the outer core of a steel component, while the inner core remains soft. Case-hardened steel can only be cut with specialized tools.

**Casement window** Windows in a steel or wood frame that open away from the building via a crank mechanism.

**Claw** The forked end of a tool.

**Cutting tool** Tools that are designed to cut into metal or wood.

**Cylindrical lock** The most common fixed lock in use today. The locks and handles are placed into predrilled holes in the doors. One side of the door will usually have a key-in-the-knob lock, and the other will have a keyway, a button, or some other type of locking/unlocking device.

**Deadbolt** A secondary locking device used to secure a door in its frame.

**Door** An entryway; the primary choice for forcing entry into a vehicle or structure.

**Double-hung window** Windows that have two movable sashes that can go up and down.

**Double-pane glass** Window design that traps air or inert gas between two pieces of glass to improve insulation.

**Duck-billed lock breaker** A tool with a point that can be inserted in the shackles of a padlock. As the point is driven further into the lock, it gets larger and forces the shackles apart until they break.

**Exterior wall** A wall—often made of wood, brick, metal, or masonry—that makes up the outer perimeter of a building. Exterior walls are often load-bearing.

**Forcible entry** Gaining access to a structure or vehicle when normal means of entry have been blocked or cannot be used.

**Glazed** Transparent glass.

**Hardware** The parts of a door or window that enable it to be locked or opened.

**Hollow-core** A door made of panels that are honeycombed inside, creating an inexpensive and lightweight design.

**Horizontal-sliding window** Windows that slide open horizontally.

**Interior wall** A wall inside a building that divides a large space into smaller areas. Also known as a partition.

**Irons** A combination tool, normally the Halligan tool and the flat-head axe.

**Jalousie window** Windows made of small slats of tempered glass that overlap each other when closed. Jalousie windows are held by a metal frame and operated by a small hand wheel or crank found in the corner of the window.

**Jamb** The part of a doorway that secures the door to the studs in a building.

**J tool** A tool that is designed to fit between double doors equipped with panic bars.

**K tool** A tool designed to remove the face plate of a lock cylinder.

**Laminated glass** Also known as safety glass; the lamination process places a thin layer of plastic between two layers of glass, so that the glass does not shatter and fall apart when broken.

**Latch** The part of the door lock that secures into the jamb.

**Load-bearing wall** A wall that supports structural members or upper floors of a building.

**Lock body** The part of a padlock that holds the main locking mechanisms and secures the shackles.

**Mortise lock** Door locks with both a latch and a bolt built into the same mechanism; the two locking devices operate independently of each other. This is a common lock found in places such as hotel rooms.

**Nonbearing wall** A wall that does not support a ceiling or structural member, but simply divides a space. See *interior wall* and *partition*.

**Operating lever** The handle of a door that turns the latch to open it.

**Padlock** The most common lock on the market today, built to provide regular-duty or heavy-duty service. Several types of locking mechanism are available, including a keyway, combination wheels, or combination dials.

**Partition** A wall that does not support a ceiling or structural member, but simply divides a space. See *interior* or *nonbearing wall*.

**Pick** The pointed end of a pick axe that can be used to make a hole or bite in a door, floor, or wall.

**Plate glass** A type of glass with additional strength so it can be formed in larger sheets, but will still shatter upon impact.

**Projected window** Also called factory windows. Usually found in older warehouse or commercial buildings. They project inward or outward on an upper hinge.

**Pry axe** A specially designed hand axe that serves multiple purposes. Similar to a Halligan bar, it can be used to pry, cut, and force doors, windows, and many other types of objects. Also called a multipurpose axe.

**Rabbit** A type of door frame that has the stop for the door cut into the frame.

**Rim lock** Surface or interior mounted locks on or in a door with a bolt that provide additional security.

**Shackles** The U-shaped part of a padlock that runs through a hasp and secures back into the lock body.

**Solid-core** A door design that consists of wood filler pieces inside the door. This creates a stronger door that may be fire-rated.

**Striking tool** Tools designed to strike other tools or objects such as walls, doors, or floors.

**Tempered glass** A type of safety glass that is heat-treated so that it will break into small pieces that are not as dangerous.

**Unlocking device** A key way, combination wheel, or combination dial.

# Brigade Member in Action

**It is 3:00 A.M. and your fire brigade has been dispatched to assist the adjoining power plant fire brigade to fight a warehouse fire. As you drive through the plant's security gate, you are informed that no personnel are in the warehouse but there definitely is a fire inside. You remember from a joint training walk-down that the warehouse contained many interior rooms and secured areas. As you approach the building on the "C" side, you notice that heavy smoke is issuing from the vents near the roof.**

1. Basic safety rules that must be followed when using striking tools include:
   - **A.** wear bunker gear and gloves.
   - **B.** wear eye protection.
   - **C.** ensure there is a clear area prior to swinging a striking tool.
   - **D.** swing the striking tool with short and controlled swings.
   - **E.** All of the above

2. Your brigade leader instructs you to take the irons and force entry. The irons consists of a(n):
   - **A.** pick-head axe and bolt cutters.
   - **B.** axe and a Halligan tool.
   - **C.** pry bar and a circular saw.
   - **D.** a tool and a bam-bam tool.

**The power plant fire brigade is deploying hose lines preparing for an interior attack, and the IC directs your team leader to force open the door to the building because the key cannot be located. You notice that the door is heavy metal construction and is in a metal frame, and the building wall is masonry block. The door has a single deadbolt and a lockable handle, indicating that there are two locking mechanisms.**

3. Deadbolts are the most difficult type of lock to break into. What forcible entry method should you use?
   - A. Through-the-lock-technique
   - B. Conventional
   - C. Specialty
   - D. None of the above

4. Once the deadbolt has been pulled and unlocked, you are now ready to force the cylindrical lock that is in the door handle. Based on this scenario, what would be the best forcible entry method?
   - A. Pull the hinge pins
   - B. Use the battering ram
   - C. Use the irons to force the door
   - D. Use a cutting torch to cut the hinges

# Ladders

# Chapter 12

## NFPA 1081 Standard

**Incipient Facility Fire Brigade Member**

NFPA 1081 contains no Incipient Facility job performance requirements for this chapter.

**Advanced Exterior Facility Fire Brigade Member**

**5.3.11** Setup and use portable ladders, given an assignment, single and extension ladders, and team members as appropriate, so that hazards are assessed, the ladder is stable, the angle is correct for climbing, extension ladders are extended to the correct height with the fly locked, the top is placed against a reliable structural component, and the assignment is accomplished.

**(A)** *Requisite Knowledge.* Parts of a ladder, hazards associated with setting up ladders, what constitutes a stable foundation for ladder placement, different angles for various tasks, safety limits to the degree of angulation, and what constitutes a reliable structural component for top placement.

**(B)** *Requisite Skills.* The ability to carry ladders, raise ladders, extend ladders and lock flies, determine that a wall and roof will support the ladder, judge extension ladder height requirements, and place the ladder to avoid obvious hazards.

**Interior Structural Facility Fire Brigade Member**

**6.3.8** Setup and use portable ladders, given an assignment, single and extension ladders, and team members as appropriate, so that hazards are assessed, the ladder is stable, the angle is correct for climbing, extension ladders are extended to the correct height with the fly locked, the top is placed against a reliable structural component, and the assignment is accomplished.

**(A)** *Requisite Knowledge.* Parts of a ladder, hazards associated with setting up ladders, what constitutes a stable foundation for ladder placement, different angles for various tasks, safety limits to the degree of angulation, and what constitutes a reliable structural component for top placement.

**(B)** *Requisite Skills.* The ability to carry ladders, raise ladders, extend ladders and lock flies, determine that a wall and roof will support the ladder, judge extension ladder height requirements, and place the ladder to avoid obvious hazards.

## Knowledge Objectives

After completing this chapter, you will be able to:

- List and describe the parts of a ladder.
- Describe the different types of ladders.
- Describe how to clean and inspect ladders.
- Describe the hazards with ladders.
- Describe how to deploy a ladder.
- Describe how to work on a ladder.

## Skills Objectives

After completing this chapter, you will be able to perform the following skills:

- One-person carry
- Two-person shoulder carry
- Three-person shoulder carry
- Two-person suitcase carry
- Three-person suitcase carry
- Three-person flat carry
- Four-person flat carry
- Three-person flat-shoulder carry
- Four-person flat-shoulder carry
- One-person flat raise for ladders under 14 feet
- One-person flat raise for ladders over 14 feet
- Two-person beam raise
- Two-person flat raise
- Three-person flat raise
- Four-person flat raise
- Tie the halyard
- Climb a ladder
- Work from a ladder
- Deploy a roof ladder

# You Are the Brigade Member

**You are paged out to a report of a fire alarm in the Administration Building. The building is protected by a sprinkler system, but the system has been shut down for maintenance. When you arrive, you find heavy smoke showing from the third level. As another crew stretches handlines into the entrance, you are assigned to ladder the roof and prepare to ventilate.**

1. ***What size ladder do you need to access the roof?***
2. ***What type of carry and raise will your crew use to set up the ladder?***
3. ***What are your safety considerations when placing the ladder?***

## Introduction

Despite many technological advances in the fire service, one fundamental piece of equipment has not changed much. The ladder remains one of the brigade member's basic tools, carried on nearly every piece of apparatus.

The portable ladder is one of the most functional, versatile, durable, inexpensive, easy-to-use, and rapidly deployable tools used by brigade members. No advanced technology can substitute for a ladder as a means of rapid, safe vertical access for fire suppression and rescue operations. Every brigade member must be proficient in the basic skills of working with ladders.

## Functions of a Ladder

Ladders provide a vertical path, either up or down, from one level to another. Ladders can provide access to an area or egress (a method to exit or escape) from an area ▶ Figure 12-1. At times, a ladder can be used as a work platform so that a brigade member can perform various functions in locations that cannot be reached otherwise.

### Fire Marks

On March 16, 1993, the Chicago Fire Department responded to a fire in a four-story 1930s vintage hotel. The fire rapidly made the interior stairways and hallways impassable. Because the building was occupied by approximately 160 people, fire fighters were faced with a massive rescue situation.

Although 19 people lost their lives, fire fighters made over 100 successful rescues. Many of those rescues resulted from the quick and effective ladder work by the first-arriving ladder companies.

Ladders are most often used to provide access to and egress from areas above grade (the level at which the ground intersects the foundation of a structure). Used outside a structure, ladders can enable brigade members to reach the roof for ventilation operations, to enter a window for an interior search, or to rescue a victim. Exterior ladders also provide an emergency exit for crews working inside a structure. Within the structure, ladders can be used to access the attic or to provide a safe path between floors, avoiding a damaged stairway. A response team can use a ladder to access and exit from the top of a rail car during a hazardous materials incident.

Brigade members also use ladders to access and exit areas below grade. Lowering a ladder into a trench or manhole allows brigade members to reach the level of an incident or to escape safely. A ladder can be used to reach an injured person in a ravine or to access a highway accident from an overpass.

**Figure 12-1** Ladders can provide access to or exit from a structure.

Ladders are useful even at grade level. A ladder can be used to create a bridge across a small opening or to enable brigade members to climb over a fence or obstruction to reach an emergency scene.

Ladders also can be used as work platforms. For example, a brigade member who has climbed a ladder to a window can stand on the ladder to open the window for ventilation or to direct a hose stream into the building. Ladders also can be used during overhaul operations, providing a platform for a brigade member to remove exterior trim in a wood-frame building.

### Secondary Functions of Ladders

Ladders serve several other functions in the fire service. Roof ladders provide stable footing and distribute the weight of brigade members during pitched roof operations. A two-section ladder, rigged with rope and webbing, forms a makeshift lift called a ladder gin. Ladder gins are used to lower a rescuer into a trench or manhole or to raise a victim safely from a below-grade site ▶ Figure 12-2.

Ladders provide elevated platforms for equipment as well as for brigade members. They are especially useful when elevation would extend the reach or enhance the operations of the equipment. For example, a hose line attached to a ladder section can protect exposures or confine a fire. A smoke ejector may be mounted on a ladder to provide ventilation. Portable lighting also may be secured to ladders to illuminate working areas.

Ladders can be used as ramps for moving equipment or for hoisting or lowering victims. They can be used to help shore up a damaged wall, to support a hose line over an opening, or to guide a rope up, down, or over an obstacle. With a tarp or salvage cover, brigade members can turn a ladder into a channel or chute for water and debris.

Figure 12-2 Rescuers can use a ladder gin to remove a fall victim from an underground vault.

## Ladder Construction

In its most basic design, a ladder consists of two beams connected by a series of parallel rungs. Fire service ladders, however, are specialized tools with several parts. To use and maintain ladders properly, brigade members must be familiar with the different types of ladders, as well as with the different parts and terms used to describe them.

### Basic Ladder Components

The basic components of a straight ladder are used in most other types of ladders as well ▶ Figure 12-3.

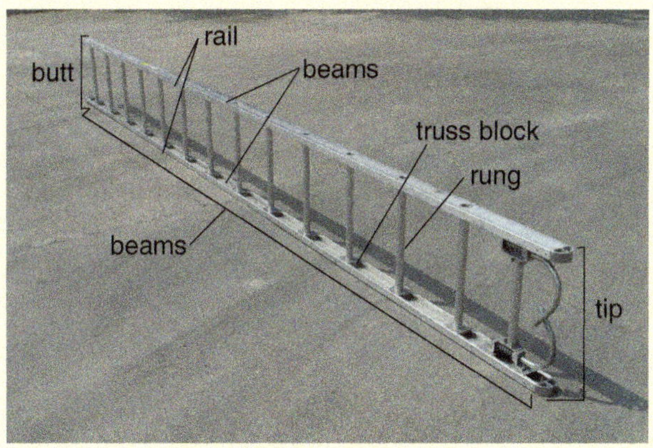


Figure 12-3 Basic components of a straight ladder.

#### Beams

A beam is one of the two main structural components that run the entire length of most ladders or ladder sections. The beams support the rungs and carry the load of a person from the rungs down to the ground.

There are three basic types of beam construction: trussed beam, I-beam, and solid beam.

- Trussed beam: A trussed beam has a top and a bottom rail, joined by a series of smaller pieces called truss blocks. The rungs are attached to the truss blocks. Trussed beams are usually constructed of aluminum or wood.
- I-beam: An I-beam has thick sections at the top and bottom, connected by a thinner section. The rungs are

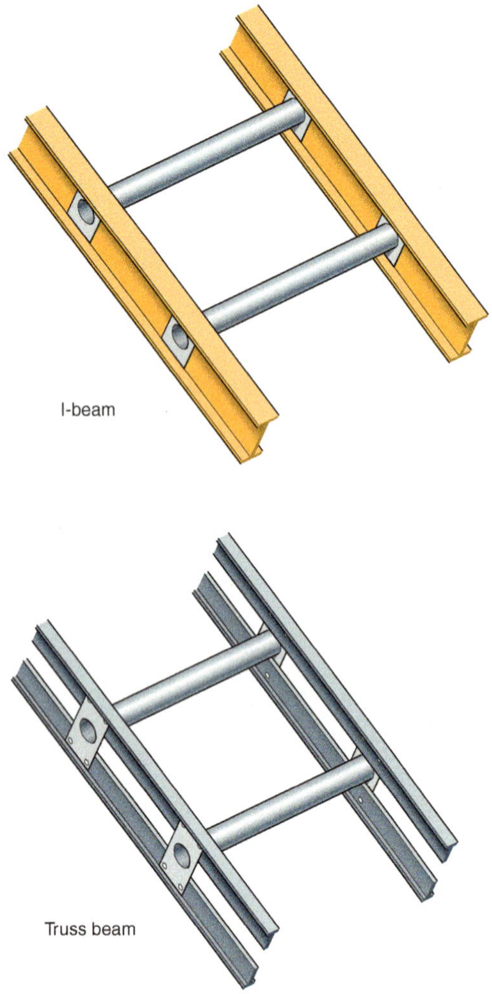


Figure 12-4 Two types of beam construction are truss beam and I-beam.

attached to the thinner section of the beam. This type of beam is usually made from fiberglass (▲ Figure 12-4).

- Solid beam: A solid beam has a simple rectangular cross section. Many wooden ladders have solid beams. Rectangular aluminum beams, which are usually hollow or C-shaped, are also classified as solid beams.

### Rail

The rail is the top or bottom section of a trussed beam. Each trussed beam has two rails. The term rail also can be used to refer to the top and bottom surfaces of an I-beam.

### Truss Block

A truss block is a piece or assembly that connects the two rails of a trussed beam. The rungs are attached to the truss blocks. Truss blocks can be made from metal or wood.

### Rung

A rung is a crosspiece that spans the two beams of a ladder. The rungs serve as steps and transfer the weight of the user to the beams. Most portable ladders used by fire brigades have aluminum rungs, but wooden ladders are still constructed with wood rungs.

### Tie Rod

A tie rod is a metal bar that runs from one beam of the ladder to the other to keep the beams from separating. Tie rods are typically found in wood ladders.

### Tip

The tip is the very top of the ladder.

### Butt

The butt is the end of the ladder that is placed against the ground when the ladder is raised. It is sometimes called the heel or base.

### Butt Spurs

Butt spurs are metal spikes attached to the butt of a ladder. The spurs prevent the butt from slipping out of position.

### Butt Plate or Footpad

A butt plate or footpad is an alternative to a simple butt spur. It is a swiveling plate attached to the butt of the ladder and incorporates both a spur and a cleat or pad.

### Roof Hooks

Roof hooks are spring-loaded, retractable, curved metal pieces attached to the tip of a roof ladder. The hooks are used to secure the tip of the ladder to the peak of a pitched roof.

### Heat Sensor Label

A heat sensor label identifies when the ladder has been exposed to specific heat conditions that could damage the structural integrity of the ladder. The label changes colors when exposed to a particular temperature. Ladders that have been exposed to excessive heat must be removed from service and tested before being used again. A ladder that fails the structural stability test should be removed from service permanently.

### Protection Plates

Protection plates are reinforcing pieces placed on a ladder at chaffing and contact points to prevent damage from friction or contact with other surfaces.

## Extension Ladder Components

An extension ladder is an assembly of two or more ladder sections that fit together and can be extended or retracted to adjust the length. Extension ladders have additional parts (► Figure 12-5).

### Bed Section

The bed section is the widest section of an extension ladder. It serves as the base; all other sections are raised from the bed

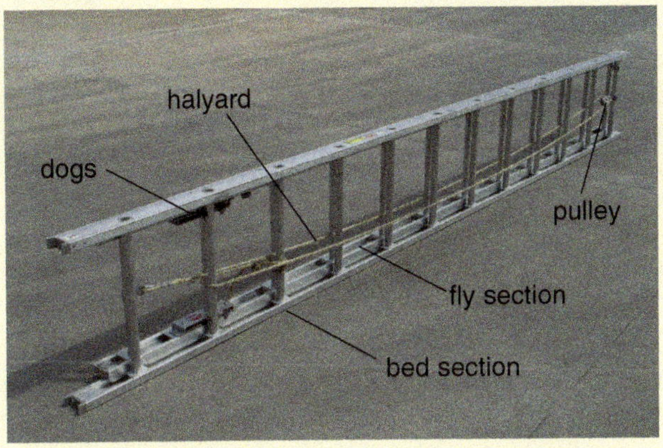


Figure 12-5 Extension ladders have additional parts not found in simple ladders.

section. The bottom of the bed section rests on the supporting surface.

### Fly Section

A fly section is the part of an extension ladder that is raised or extended from the bed section. Extension ladders often have more than one fly section, each of which extends from the previous section.

### Dogs

Dogs are mechanical locking devices used to secure the extended fly section(s). Dogs are also referred to as pawls, ladder locks, or rung locks.

### Guides

Guides are strips of metal or wood that guide a fly section as it is being extended. Channels or slots in the bed or fly section may also serve as guides.

### Halyard

The halyard is the rope or cable used to extend or hoist the fly section(s) of an extension ladder. The halyard runs through the pulley.

### Pulley

The pulley is a small grooved wheel used to change the direction of the halyard pull. A downward pull on the halyard creates an upward force on the fly section(s), extending the ladder.

### Stops

Stops are pieces of wood or metal that prevent the fly section(s) of a ladder from overextending and collapsing the ladder. Stops are also referred to as stop blocks.

### Staypole

Staypoles (or tormentors) are long metal poles attached to the top of the bed section. They are used to help stabilize the ladder as it is being raised and lowered. One pole is attached to each beam of a long (40 feet or longer) extension ladder with a swivel joint. Each pole has a spur on the other end. Ladders with staypoles are typically referred to as Bangor ladders.

# Types of Ladders

Ladders used in the fire service can be classed into two broad categories. Aerial ladders are permanently mounted and operated from a piece of fire apparatus. Portable ladders are carried on fire apparatus but are designed to be removed and used in other locations.

## Aerial Apparatus

Aerial apparatus vary greatly in design and function. This discussion is limited to a brief overview of the basic styles of aerial apparatus. Detailed requirements for aerial apparatus are documented in NFPA 1901 *Standard for Automotive Fire Apparatus*. This standard sets minimum performance requirements for apparatus referred to as either "aerial ladders" or "elevating platforms."

Aerial ladders are permanently mounted, power-operated ladders with a working length of at least 50 feet. They have at least two sections (▼ Figure 12-6). Some aerial ladders have a permanently mounted waterway and monitor nozzle. Aerial ladders are often referred to as "straight-stick" aerials.

An elevating platform apparatus includes a passenger-carrying platform (bucket) attached to the tip of a ladder or boom (► Figure 12-7). The ladder or boom must have at least two sections, which may be telescoping or articulating (jointed). An elevating platform apparatus that is 110 feet or less in length must also have a prepiped waterway and a

Figure 12-6 An aerial ladder can be used to access the roof of a commercial structure.

Figure 12-7 An elevated platform serves as a secure working platform for brigade members.

permanently mounted monitor nozzle. An elevated platform supported by a boom may have an aerial ladder for continuous access to the platform.

## Portable Ladders

Portable ladders, often called ground ladders, are carried on most fire apparatus. Portable ladders are designed to be removed from the apparatus and used in different locations. They may be general purpose ladders, such as straight or extension ladders, or specialized ladders. Specialized ladders have names that indicate their particular function.

The number and lengths of portable ladders used by a fire brigade depend on the maximum height of buildings in the response area. Most fire brigade engines carry 24- or 28-foot ladders, which can reach the roof of a typical two-story building. Ladder trucks usually carry 35- or 40-foot ladders, which can reach the roofs of most three-story buildings.

**Brigade Member Safety Tips**

Fire service portable ladders should be constructed and certified as compliant with the most recent edition of the NFPA 1931, *Standard for Manufacturer's Design of Fire Department Ground Ladders.*

Generally, fire service portable ladders are limited to a maximum length of 50 feet. If building heights exceed the capabilities of portable ladders, aerial apparatus must be used. ▶ Table 12-1 shows the various types of portable ladders, their construction, and their weights.

Figure 12-8 Roof ladders are commonly 12- to 18-foot long.

### Straight Ladder

A straight ladder is a single-section, fixed-length ladder. Straight ladders may also be called wall ladders or single ladders. They are lightweight, can be raised quickly, and can reach windows and roofs of one- and two-story structures. Straight ladders are commonly 12- to 20-foot long, but can be up to 30 feet and longer. Longer straight ladders are rarely used by the fire service because they are difficult to store and to handle.

### Roof Ladder

A roof ladder (sometimes called a hook ladder) is a straight ladder equipped with retractable hooks at one end. The hooks secure the tip of the ladder to the peak of a pitched roof, so that the ladder lies flat on the roof. A roof ladder provides stable footing and distributes the weight of brigade members and their equipment. This helps reduce the risk of structural failure in the roof assembly. Although roof ladders are available in lengths from 10–30 feet, they are usually 12- to 18-foot ▲ Figure 12-8.

Figure 12-9 An extension ladder can be used at any length, from fully retracted to fully extended.

**Table 12-1 Lengths and Weights of Various Ladders**

| Length (ft) | Type | No. of Sections | Construction | Material | Weight (lb) |
|---|---|---|---|---|---|
| 20 | Straight | 1 | Solid beam | Aluminum | 45 |
| 20 | Straight | 1 | Solid beam | Fiberglass | 50 |
| 20 | Extension | 2 | Solid beam | Aluminum | 60 |
| 20 | Extension | 2 | Truss beam | Aluminum | 76 |
| 20 | Extension | 2 | Solid beam | Fiberglass | 95 |
| 24 | Extension | 2 | Solid beam | Aluminum | 72 |
| 24 | Extension | 2 | Truss beam | Aluminum | 90 |
| 24 | Extension | 2 | Solid beam | Fiberglass | 110 |
| 30 | Extension | 2 | Solid beam | Aluminum | 107 |
| 30 | Extension | 3 | Solid beam | Aluminum | 108 |
| 30 | Extension | 2 | Truss beam | Aluminum | 131 |
| 30 | Extension | 2 | Solid beam | Fiberglass | 140 |
| 30 | Extension | 3 | Truss beam | Aluminum | 164 |
| 30 | Extension | 3 | Solid beam | Fiberglass | 170 |
| 35 | Extension | 2 | Solid beam | Aluminum | 114 |
| 35 | Extension | 2 | Solid beam | Aluminum | 129 |
| 35 | Bangor | 2 | Truss beam | Aluminum | 153 |
| 35 | Extension | 2 | Solid beam | Fiberglass | 160 |
| 35 | Bangor | 3 | Truss beam | Aluminum | 179 |
| 35 | Extension | 3 | Solid beam | Fiberglass | 195 |
| 40 | Bangor | 2 | Solid beam | Aluminum | 171 |
| 40 | Bangor | 3 | Solid beam | Aluminum | 193 |
| 40 | Bangor | 2 | Truss beam | Aluminum | 210 |
| 40 | Bangor | 3 | Truss beam | Aluminum | 245 |
| 45 | Bangor | 2 | Solid beam | Aluminum | 182 |
| 45 | Bangor | 2 | Truss beam | Aluminum | 255 |
| 45 | Bangor | 3 | Truss beam | Aluminum | 265 |
| 45 | Bangor | 3 | Solid beam | Aluminum | 280 |
| 50 | Bangor | 2 | Truss beam | Aluminum | 229 |
| 50 | Bangor | 3 | Truss beam | Aluminum | 297 |

### Extension Ladder

An extension ladder is an adjustable-length ladder with multiple sections. The bed or base section supports one or more fly sections. Pulling on the halyard extends the fly sections along a system of brackets or grooves ◄ Figure 12-9. The fly sections lock in place at set intervals so that the rungs are aligned to facilitate climbing.

An extension ladder is usually heavier than a straight ladder of the same length and requires more than one person to set up. Because the length is adjustable, an extension ladder

**Brigade Member Safety Tips**

Roof ladders are not free-hanging ladders. The roof hooks will not support the full weight of the ladder and anyone on it when the ladder is in a vertical position.

Figure 12-10 Staypoles are used to stabilize a Bangor ladder while it is being raised or lowered.

Figure 12-11 Combination ladders can be used in several different configurations.

can replace several straight ladders and can be stored in places where a longer straight ladder would not fit.

## Bangor Ladder

Bangor ladders are extension ladders with staypoles for added stability during raising and lowering operations. Staypoles are required on ladders of 40 feet or greater length and can be found on 35-foot ladders. The poles help keep these heavy ladders under control during maneuvers ▲ Figure 12-10. When the ladder is positioned, the staypoles are planted in the ground on either side for additional stability.

## Combination Ladder

A combination ladder can be converted from a straight ladder to a stepladder configuration (A-frame), or from an extension ladder to a stepladder configuration. Combination ladders are convenient for indoor use and for maneuvering in tight spaces. Combination ladders are generally 6–10 feet in the A-frame configuration and 10–15 feet in the extension configuration ► Figure 12-11.

## Folding Ladder

A folding ladder (or attic ladder) is a narrow, collapsible ladder designed to allow access to attic scuttle holes and confined areas ► Figure 12-12. The two beams fold in for portability. Folding ladders are commonly available in 8- to 14-foot lengths.

## Fresno Ladder

A Fresno ladder is a narrow, two-section extension ladder, designed to provide attic access. The Fresno ladder is generally short, just 10 feet to 14 inches, so it has no halyard; it is extended manually ► Figure 12-13. A Fresno ladder can be used in tight space applications, such as bridging over a damaged section of an interior stairway.

## Pompier Ladder

A pompier ladder, also known as a scaling ladder, is a lightweight, single-beam ladder. Municipal fire fighters once used pompier ladders to climb up the outside of a

### Brigade Member Safety Tips

It is very easy to pinch your hands when opening and closing a folding ladder. Always wear gloves when working with folding ladders.

Figure 12-12 Folding ladders are designed to be used in tight spaces.

## Fire Marks

Chris Hoell, a St. Louis Fire Department Lieutenant, is credited with developing the pompier or scaling ladder in the late 1800s. The device, originally called the Hoell Life Saving Appliance, caught the attention of the Fire Department of New York City (FDNY), which hired Hoell to train its members to use the device. The pompier ladder was used by FDNY companies from the early 1880s until 1996.

building, one floor at a time, to rescue a trapped person. Today, pompier ladders are used only in situations where no other option is available.

The beam of a pompier ladder has a large hook at the tip. The rungs extend through the beam, creating steps and handholds on both sides. The fire fighter would hook the ladder over a windowsill or solid object one or two floors above ground level, and climb up ▶ Figure 12-14. The fire fighter would then dismount, pull the ladder up, and secure the hook at a higher level, repeating the process up to the desired level. The fire fighter would then reverse direction and carry the person down to safety. The use of pompier ladders has almost vanished in municipal fire department and has virtually no application in industrial fire and emergency operations.

# Inspection, Maintenance, Cleaning, and Service Testing of Portable Ladders

The portable ladders used by the fire service must be able to withstand extreme conditions. NFPA 1931, *Standard on Design of and Design Verification Tests for Fire Department Ground Ladders* establishes requirements for the construction of new ladders. The requirements are based on probable conditions during emergency operations.

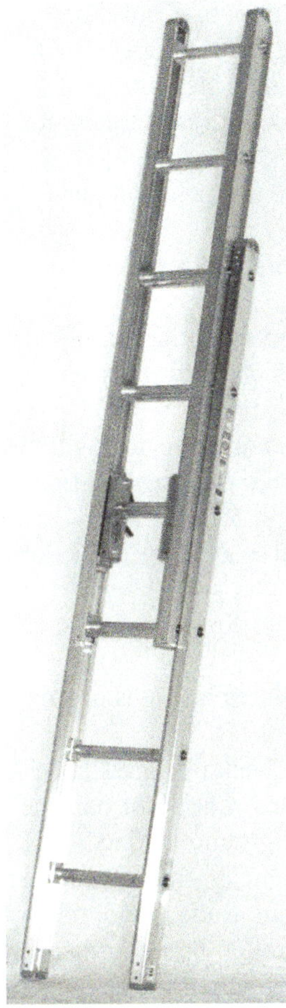

Figure 12-13 A Fresno ladder is extended manually.

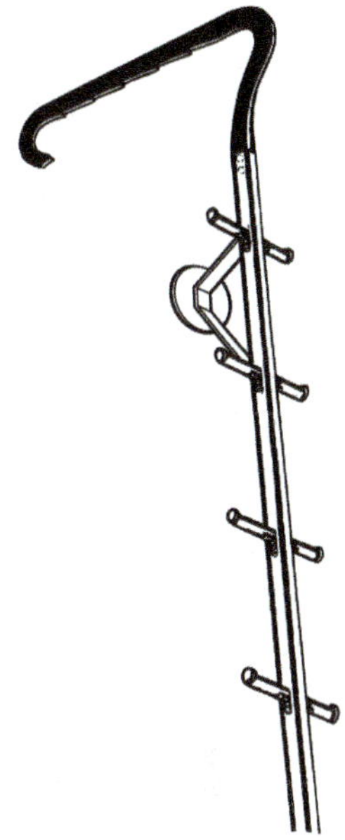

Figure 12-14 The pompier ladder enables a brigade member to scale a vertical wall.

Ladders can be dropped, overloaded, or exposed to temperature extremes during use. NFPA 1932, *Standard on Use, Maintenance, and Service Testing of In-Service Fire Department Ground Ladders,* provides general guidance for ground ladder users. Portable ladders must be regularly inspected, maintained, and service-tested, following the NFPA Standard. In addition, ladders should always be inspected and maintained in accordance with the manufacturer's recommendations.

## Inspection

Ground ladders must be visually inspected at least monthly. A ladder should also be inspected after each use. A general visual inspection should, at a minimum, note the following items:

- Beams: Check the beams for cracks, splintering, breaks, gouges, checks, wavy conditions, or deformations.
- Rungs: Check all rungs for snugness, tightness, punctures, wavy conditions, worn serrations, splintering, breaks, gouges, checks, or deformations.
- Halyard: Check the halyard for fraying or kinking; ensure that it moves smoothly through the pulleys.
- Wire-Rope Halyard Extensions: Check the wire-rope halyard extensions on three- and four-section ladders for snugness. This check should be performed when

the ladder is in the bedded position, to ensure that upper sections will align properly during operation.
- Ladder Slides: Check the ladder slide areas for chaffing. Also check for adequate wax, if the manufacturer requires wax.
- Dogs: Check the dog (pawl) assemblies for proper operation.
- Butt Spurs: Check the butt spurs for excessive wear or other defects.
- Heat Sensors: Check the labels to see if the sensors indicate that the ladder has been exposed to excessive heat.
- Bolts and Rivets: Check all bolts and rivets for tightness; bolts on wood ladders should be snug and tight without crushing the wood.
- Welds: Check all welds on metal ladders for cracks or apparent defects.
- Roof Hooks: Check the roof hooks for sharpness and proper operation.
- Metal Surfaces: Check metal surfaces for signs of surface corrosion.
- Fiberglass and Wood Surfaces: Check fiberglass ladders for loss of gloss on the beams. Check for damage to the varnish finish on wooden ground ladders.

If the inspection reveals any deficiencies, the ladder must be removed from service until repairs are made. Minor repairs that require simple maintenance can often be performed by properly trained brigade members at the fire station. Repairs involving the structural or mechanical components of a ladder must be performed only by qualified personnel at a properly equipped repair facility.

## Maintenance

All brigade members should be able to perform routine ladder maintenance. Maintenance is simply the regular process of keeping the ladder in proper operating condition. Fundamental maintenance tasks include:
- Clean and lubricate the dogs, following the manufacturer's instructions (▶ Figure 12-15).
- Clean and lubricate the slides on extension ladders in accordance with the manufacturer's recommendations.
- Replace worn halyards and wire rope on extension ladders when they fray or kink (▶ Figure 12-16).
- Clean and lubricate hooks. Remove rust and other contaminants and lubricate the folding roof hook assemblies on roof ladders to keep them operational (▶ Figure 12-17).
- Check the heat sensor labels. Remove a ladder that has been exposed to high temperatures from service for testing.
- Maintain the finish on fiberglass and wooden ladders in accordance with the manufacturer's recommendations.
- Portable ladders should NOT be painted except for the top and bottom 18 feet of each section. Paint can hide structural defects in the ladder. The tip and butt are painted for purposes of identification and visibility.

Ladders that are in storage should be placed on racks or in brackets and protected from the weather. Fiberglass ladders can be damaged by prolonged exposure to direct sunlight (▶ Figure 12-18).

### Brigade Member Safety Tips

Brigade members should perform routine ladder maintenance. Repairs to portable ladders should be performed ONLY by trained repair personnel.

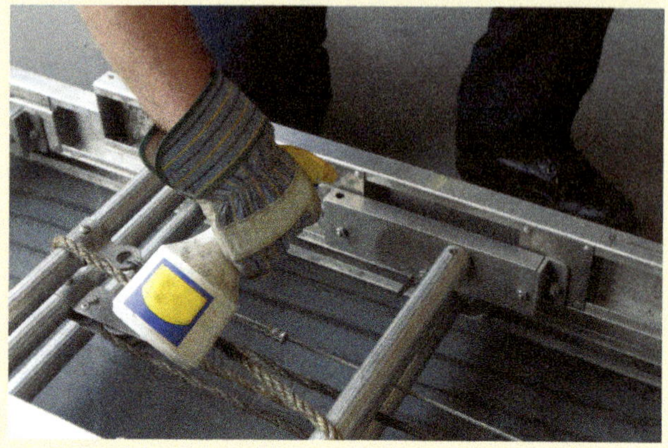

Figure 12-15 Extension ladder dogs must operate smoothly.

Figure 12-16 Replace the halyard if it is worn or damaged.

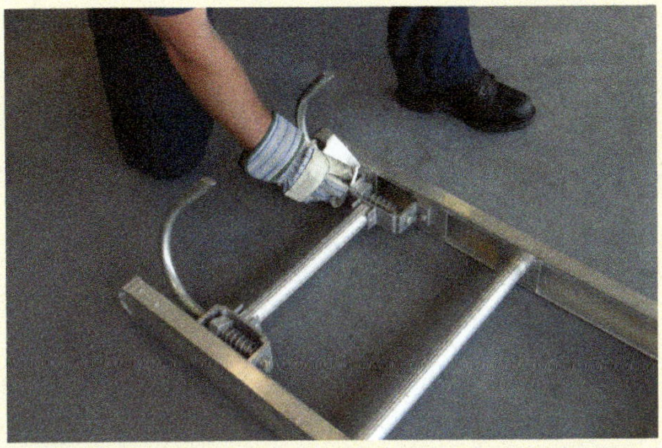

Figure 12-17 Roof hooks must operate smoothly.

Figure 12-18 Portable ladders should be stored on racks or in brackets, out of the weather or direct sunlight.

## Cleaning

Ladders must be regularly cleaned to remove road grime and dirt that build up during storage on the apparatus. Ladders should be cleaned before each inspection to ensure that any hidden faults can be observed. Ladders should also be cleaned after each use to remove dirt and debris.

Use a soft-bristle brush and water to clean ladders. A mild, diluted detergent may be used, if allowed by the manufacturer's recommendations ▶ Figure 12-19. Remove any tar, oil, or grease deposits with a safety solvent as recommended by the manufacturer.

Rinse the cleaned ladder before replacing it on the apparatus.

## Service Testing

Service testing is performed annually to evaluate the continued usefulness of a ladder during its life. Service tests should be performed on new ladders as well as on ladders that have been in use for some time. Service testing is different from design verification testing. Design verification tests are conducted by testing laboratories to ensure that new ladders are constructed in compliance with manufacturing specifications. Service tests measure the structural integrity of a portable ladder.

Service testing of portable ladders must follow NFPA 1932, *Standard on Use, Maintenance, and Service Testing of In-Service Fire Department Ground Ladders*.

### Brigade Member Safety Tips

Do not get any solvent on the halyard of an extension ladder. Contact with solvents can damage halyard ropes.

Figure 12-19 Ladders should be cleaned regularly, as well as after each use, with a mild detergent.

Service tests should be conducted before a new ladder is used and annually while it is in service. A ladder that has been exposed to extreme heat, overloaded, impact- or shock-loaded, visibly damaged, or is suspected of being unsafe for any other reason must be removed from service until it has passed a service test. A repaired ladder must also pass a service test before it can be returned to service, unless a halyard replacement was the only repair.

The horizontal-bending test evaluates the structural strength of a ladder. The ladder is placed in a horizontal position across a set of supports. A weight is then put on the ladder, and the amount of deflection or bending caused by

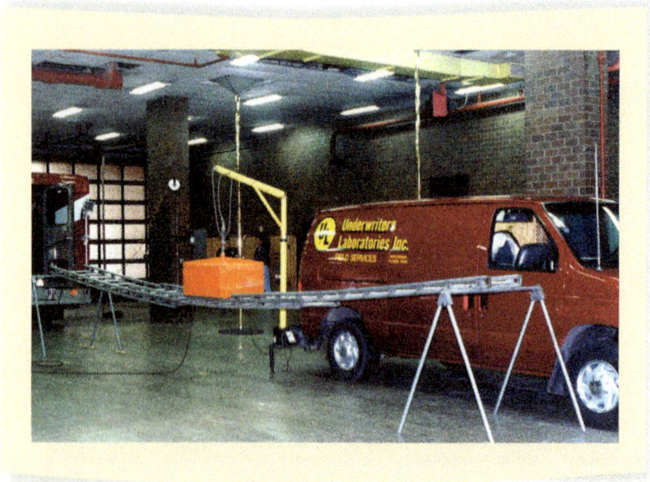


Figure 12-20 Horizontal bend test evaluates the structural strength of a ladder.

the weight is measured to evaluate the strength of the ladder (▲ Figure 12-20). Additional tests are performed on the extension hardware of extension ladders. The hooks on roof ladders are also tested.

Service testing of ground ladders requires special training and equipment and must be conducted only by qualified personnel. Many fire brigades use outside contractors to perform these tests. The results of all service tests must be recorded and kept for future reference.

# Ladder Safety

Several potential hazards are associated with ladder use. These risks are easily overlooked during emergency operations. Many brigade members have been seriously injured or killed in ladder accidents, both on the fireground and during training sessions.

Ladders must be used with caution; follow standard procedures and regularly reinforce your skills through training. Many safety precautions should be followed from the time a ladder is removed from the apparatus until it is returned to the apparatus. Basic safety issues include:

- general safety
- lifting and moving ladders
- placement of ground ladders
- working on the ladder
- rescue operations
- ladder damage

These issues are explained in general here and will be revisited under the specifics of using ladders.

## General Safety Requirements

In any firefighting operation, full turnout clothing and protective equipment are essential for working with ground ladders. Turnout gear provides protection from mechanical injuries, as well as from fire and heat injuries. Helmets, coats, pants, gloves, and protective footwear provide protection from falling debris, impact injuries, or pinch injuries. Brigade members must be able to work with and on ladders while wearing self-contained breathing apparatus. You should practice working with ladders and tying knots while wearing all of your protective equipment, including gloves and eye protection.

## Lifting and Moving Ladders

Teamwork is essential when working with ladders. Some ladders are heavy and can be very awkward to maneuver, particularly when they are extended. Crew members must use proper lifting techniques and coordinate all movements.

Never attempt to lift or move a ladder that weighs more than you are capable of lifting safely. Because of their shape, ladders may be awkward to carry, especially over uneven terrain, through gates, or over snow and ice. It is better to ask for help to move a ladder, rather than risk injury or delay the placement of the ladder.

## Placement of Ground Ladders

Brigade members working on the fireground should survey the area where a portable ladder is going to be used before placing the ladder or beginning to raise it. If possible, inspect the area before retrieving the ladder from the apparatus. If you note any hazards, consider changing the position of the ladder. Work in the safest available location.

Sometimes a ladder has to be placed in a potentially hazardous location. If you are aware of the hazard, you can take corrective action and special precautions to avoid accidents.

The most important check is the location of overhead utility lines. If a ladder comes in contact with an electrical wire, the brigade members who are handling it can be electrocuted. If the line falls, it can electrocute other brigade members as well. Avoid placing a ladder against a surface that has been energized by a damaged or fallen power line.

Do not assume that it is safe to use wood or fiberglass ladders around power lines. Metal ladders will conduct electricity, but a wet or dirty wood or fiberglass ladder can also conduct electricity.

If possible, do not raise ladders anywhere near overhead wires. At a minimum, keep ladders at least 10 feet away from any power lines while using or raising them. If a power line is nearby, make sure that there are enough brigade members to keep the ladder under control as it is raised. One brigade member should watch to make sure that the ladder does not come too close to the line. If the ladder falls into a power line, the results can be deadly.

Other types of overhead obstructions can also be hazards during a ladder raise. If the ladder hits something as it is being raised, the weight and momentum of the ladder will shift. This can cause the brigade members to lose control and drop the ladder.

Ladders must be placed on stable and relatively level surfaces (► Figure 12-21). The shifting weight of a climber

Figure 12-21 Ladders must be placed on stable and relatively level surfaces.

Figure 12-22 To prevent a ladder from slipping, brigade members should secure the base by heeling the ladder.

can easily tip a ladder placed on a slope or unstable surface. Always evaluate the stability of the surface before placing the ladder, and check it again during the course of the firefighting operations. A ladder placed on snow or mud may become unstable as the snow melts or as the base of the ladder sinks into the mud. Water from the firefighting operations can make soft ground even more unstable.

Ladders should not be exposed to direct flames or extreme temperatures. Excessive heat can cause permanent damage or catastrophic failure. Check the heat sensor labels immediately after a ladder has been exposed to high temperatures.

## Working on a Ladder

Before climbing a ladder, be sure it is set at the proper climbing angle (approximately 75°) for maximum load capacity and safety. The proper angle can be quickly determined by placing the butt of the ladder one foot away from the wall, for every 4 feet of vertical distance. Dogs must be locked and the halyard tied before anyone climbs an extension ladder.

To prevent slipping, another brigade member should secure the base by heeling (footing) the ladder (▶ Figure 12-22). This technique uses the brigade member's weight to keep the base of the ladder from slipping. The base of the ladder also can be mechanically secured to a solid object. An alternative to securing the base is to use a rope or strap to secure the tip of the ladder to the building.

All ladders have weight limits. Most portable ladders are designed to support a weight of 750 lb. In a rescue situation, this weight limit is equivalent to one person and two brigade members with full protective clothing and equipment. This weight limit would also accommodate two brigade members climbing or working on the ladder along with their equipment.

The weight should be distributed along the length of the ladder. Only one brigade member should be on each section of an extension ladder at any time.

While climbing the ladder, be prepared for falling debris, misguided hose streams, or people jumping from the building

### Brigade Member Safety Tips

Always maintain a minimum 10 feet clearance between a ladder and utility lines to prevent electrocution.

## Brigade Member Tips

Operators of aerial equipment are trained to extend the bucket or tip of the aerial ladder above a trapped individual during a rescue. They then lower it down to the person. This generally prevents the individual from jumping onto the ladder.

that could knock you off the ladder. This is another reason why brigade members should wear full turnout gear when working on ladders. It will help protect a brigade member who is knocked to the ground. It is even more important that brigade members working from ladders near the fire wear protective clothing and equipment.

If fire conditions should change rapidly while you are working on a ladder, you must be prepared to climb down quickly. For example, if the fire suddenly flashes over or if flames break out through a window near the ladder, you must be prepared to move quickly out of danger. Turnout gear will not protect you from direct exposure to the flames for more than a few seconds.

A brigade member working from a ladder is in a less stable position than one working on the ground. You must constantly adjust your balance, especially when swinging a hand tool or reaching for a trapped occupant. There is the danger of falling as well as the risk to people below if something falls or drops. Brigade members who are working from a ladder should use a safety belt or a leg lock to secure themselves to the ladder.

### Rescue

Rescue is a brigade member's most important and unpredictable duty. Portable ladders are often used to reach and remove trapped occupants from the upper stories of a building. However, brigade members must address several important safety concerns before going up a ladder to rescue someone.

A person who is in extreme danger may not wait to be rescued. Jumpers risk their own lives and may endanger the brigade members trying to rescue them. Several brigade members have been seriously injured by persons who jumped before a rescue could be completed.

A trapped person might try to jump onto the tip of an approaching ladder, or to reach out for anything or anyone nearby. You might be pulled or pushed off the ladder by the person you are trying to rescue. If several people are trapped, they might all try to climb down the ladder at the same time.

It is important to make verbal contact with any person you are trying to rescue. You must remain in charge of the situation and not let the individual panic. Tell the person to remain calm and wait to be rescued. If there are enough brigade members, one could maintain contact with the person while others raise the ladder.

After you reach the person, you still have to get him or her down to the ground safely. Try to make the person realize that any erratic movements could tip over the ladder. It is often helpful to have one brigade member assist the person, while a second brigade member acts as a guide and back-up to the rescuer. Modern fire ladders are designed to support this load, but a ladder that is overloaded during a rescue operation must be removed from service for inspection and service testing. See Chapter 13 for specific ladder rescue techniques.

### Ladder Damage

Ladders may easily be damaged while in use. A ladder might be overloaded or used at a low angle during a rescue. An unexpected shift in fire conditions could bring the ladder in direct contact with flames. Whenever a ladder is used outside of its recommended limits, it should be taken out of service for inspection and testing, even if there is no visible damage.

## Using Portable Ladders

Portable ladders are often urgently needed during emergency incidents. Because an accident or error in handling or using a ladder can result in death or serious injury, all brigade members must know how to work with ladders.

Using a ladder requires that brigade members complete a series of consecutive tasks. The first step is to select the best ladder for the job from those available. Brigade members must then remove the ladder from the apparatus and carry it to the location where it will be used. The next step is to raise and secure the ladder. At the end of the operation, the ladder must be lowered and returned to the apparatus. Each of these tasks is important to the safe and successful completion of the overall objective.

### Ladder Selection

The first step in using a portable ladder is to select the appropriate ladder from those available. Brigade members must be familiar with all of the ladders carried on their apparatus. Engine and ladder company apparatus usually carry several portable ladders of various lengths. Many other types of apparatus, such as tankers (water tenders) and rescue units, often carry additional portable ladders. NFPA 1901, *Standard for Automotive Fire Apparatus,* provides minimum requirements for portable ladders for each type of apparatus (▶ Table 12-2).

Selecting an appropriate ladder requires that brigade members estimate the heights of windows and rooflines. Facility preplanning will improve ladder selection efficiency and identify potential safety hazards.

Ladder placement will also affect the size and length of the ladder needed. When the ladder is used to access a roof, the tip of the ladder should extend several feet above the roofline. This provides a handhold and footing for brigade members as they mount and dismount. The extra length also makes the tip of the ladder visible to brigade members working on the

**Table 12-2 Minimum Ladder Complement for Apparatus as Specified in NFPA 1901, *Standard for Automotive Fire Apparatus* (2003 ed.)**

**Pumper**
1 Attic Ladder
1 Roof Ladder
1 Extension Ladder

**Quick Attack**
1 Extension Ladder must be 12 feet or longer

**Aerial/Ladder**
Portable ladders that have a total length of 115 feet or more and contain a minimum of:

- 1 Attic Ladde
- 2 Roof Ladders
- 2 Extension Ladders

**Quint**
Portable ladders that have a total length of 85 feet or more and contain a minimum of:

- 1 Attic Ladder
- 1 Roof Ladder
- 1 Extension Ladder

roof (▶ Figure 12-23). A common rule of thumb is to be sure at least five ladder rungs show above the roofline.

Ladders used to provide access to windows must be longer than those used in rescue operations. During access operations, the ladder is placed next to the window, with the ladder tip even with the top of the window opening (▶ Figure 12-24). However, during rescue operations, the tip of the ladder should be immediately below the windowsill. This prevents the ladder from obstructing the window opening while a trapped occupant is removed (▶ Figure 12-25).

The final factor in determining the correct length to use is the angle formed by the ladder and the placement surface (ground). A portable ladder should be placed at an angle of approximately 75° for maximum strength and stability. This means that the ladder will have to be slightly longer than the vertical distance between the ground and the target point. Generally, a ladder requires an additional 1 feet in length for every 15 feet of vertical height.

To reach a window 30 feet above grade level, the ladder would have to be at least 32-foot long. Because ladders used in roof operations need to extend at least above the roofline, accessing a roof that is 30-foot above grade level requires a ladder at least 35-foot long.

## Removing the Ladder from Apparatus

Ladders are mounted on apparatus in various ways. Ladders should be mounted in locations where they will not be

Figure 12-23 The tip of the ladder should extend above the roofline during roof operations.

Figure 12-24 The tip of the ladder should be even with the top of the window opening to provide access through the window.

Figure 12-25 For rescue operations, the tip of the ladder should be just below the windowsill.

# Voices of Experience

**"If they had not placed their feet in the proper position against the ladder, one or both could have received a serious crushing injury."**

During a "rookie" academy, brigade members were practicing raising and lowering extension ladders. One member was having trouble getting a good grip on the halyard because his gloves were wet. To improve his grip, he wrapped the halyard around his hand. The instructor told him to unwrap his hand to prevent an injury. When the student unwrapped the halyard from his hand, he lost his grip and the two sections of the ladder slammed into the ground.

The two students who were keeping the ladder steady had a foot against the outside of the beam on each side of the ladder. If they had not placed their feet in the proper position against the ladder, one or both could have received a serious crushing injury.

It is important to remember that long ladders can be difficult to control. Proper raising and lowering techniques should always be followed. Maintain control at all times and follow safety rules. A crushed foot or serious hand injury is a bad way to end a training session.

**Scott Dornan**
ConocoPhillips
Anchorage, Alaska
Kuparuk Fire and Rescue
Kuparuk, Alaska

exposed to excessive heat, engine exhaust, or mechanical damage. Brigade members must be familiar with how ladders are mounted on their apparatus and practice removing them safely and quickly.

Ladders are often nested one inside another and mounted on brackets on the side of a pumper. Brigade members should note the nesting order and location of the ladders relative to the brackets when removing or replacing them. Ladders that are not needed should not be placed on the ground in front of an exhaust pipe because they could be damaged by the hot exhaust.

Ladders can also be stored on overhead hydraulic lifts (▶ Figure 12-26). The hydraulic mechanism keeps them out of the way until they are needed, when they can be lowered to a convenient height.

On some vehicles, portable ladders are stored in compartments under the hose bed or aerial device. The ladders may lie flat or vertically on one beam (▶ Figure 12-27). Brigade members slide the ladders out the rear of the vehicle to remove or replace them.

## Lifting Ladders

Many ladders are heavy and awkward to handle. As noted in Table 12-1, some ladders weigh more than 200 lb. To prevent lifting injuries while handling ladders, brigade members must work together to lift and carry long or heavy ladders.

When brigade members are working as a team to lift or carry a ladder, one brigade member must act as the leader, providing direction and coordinating the actions of all team members. The lead brigade member needs to call out the intended movements clearly, using standard commands and terminology. For example, when lowering a ladder, the lead should say "Prepare to lower," followed by the command "Lower." Team members have to communicate in a clear and concise fashion, using the same terminology for commands. There should be no confusion about the specific meaning of each command.

A prearranged method should exist for determining the lead brigade member for ladder lifts. In some fire brigades, the brigade member on the right side at the butt end of the ladder is the standard leader. In other brigades, the brigade member at the tip of the ladder on the right side may be the leader. The departmental policy should be consistent for all ladder operations.

Additionally, brigade members must use good lifting techniques when handling ladders. When bending to pick up a ladder, bend at the knees and keep the back straight (▶ Figure 12-28). Lift and lower the load with the legs rather than the back. Take care to avoid twisting motions during lifting and lowering, because these motions often lead to back strains.

## Carrying Ladders

Once the ladder has been removed from the apparatus or lifted from the ground, it must be carried to the placement site. Ladders can be carried at shoulder height or at arm's

**Figure 12-26** This hydraulic mechanism lowers the ladders to a convenient height when they are needed.

**Figure 12-27** Ladders in rear compartments can be stored flat or vertically.

**Figure 12-28** Bend the knees and keep the back straight when lifting or lowering a ladder.

length, and on edge or flat. The most common carries are explained in the following sections.

### One-Person Carry

Most straight ladders and roof ladders less than 18-foot long can be safely carried by a single person. The steps involved in the one-person carry are outlined in ▼ Skill Drill 12-1.

1. Start with the ladder mounted in a bracket or standing on edge (on one beam). Locate the center of the ladder. **(Step 1)**
2. Place an arm between two rungs of the ladder just to one side of the middle rung. **(Step 2)**
3. The top beam of the ladder rests on the brigade member's shoulder as it is carried. **(Step 3)**

A straight ladder is carried with the butt end first and pointed slightly downward. A roof ladder should be carried tip first. Most ladders are carried with the butt end forward, because the placement of the butt determines the positioning of the ladder. Roof ladders are carried with the hooks toward the front. A roof ladder usually must be carried up another ladder to reach the roof. Then, the roof ladder is usually pushed up the pitched roof slope from the butt end until the hooks engage the peak.

### Two-Person Shoulder Carry

The two-person shoulder carry is generally used with extension ladders up to 35-foot long. It can also be used to carry straight ladders or roof ladders that are too long for one

## 12-1 Skill Drill

## One-Person Carry

Locate the center of the ladder.

Place one arm through two rungs, just to one side of the middle rung.

Bring the top beam to rest on the brigade member's shoulder.

person to handle. To perform the two-person shoulder carry, follow the steps in ▼Skill Drill 12-2.

1. Start with the ladder mounted in a bracket or standing on edge. Both brigade members stand on the same side of the ladder, facing the butt, one near the butt and one near the tip. **(Step 1)**
2. Facing the butt end of the ladder, each brigade member places an arm between two rungs and lifts the ladder onto the shoulder. The ladder is carried butt end first. **(Step 2)**
3. The brigade member closest to the butt covers the butt spur with a gloved hand to prevent injury to other brigade members in the event of a collision. **(Step 3)**

## Brigade Member Safety Tips

Industrial structures with metal roofs may not provide a safe and secure placement of roof ladder hooks. Careful consideration of the risk versus gain should be made before operating on metal roofs.

### Three-Person Shoulder Carry

Three brigade members may be needed to carry a heavy ladder. This carry is similar to the two-person shoulder carry, with the additional brigade member at the middle of the

## 12-2 Skill Drill

### Two-Person Shoulder Carry

1 Both brigade members approach the ladder from the same side, facing the butt. One brigade member stands near the butt and the other near the tip.

2 Each brigade member places an arm between two rungs and lifts the ladder onto the shoulder.

3 The butt spurs are covered with a gloved hand while the ladder is transported.

ladder. All three brigade members stand on the same side of the ladder. To perform the three-person shoulder carry, follow the steps in ▼ **Skill Drill 12-3**.

1. Start with the ladder mounted in a bracket or standing on edge. All three brigade members stand on the same side of the ladder, facing the butt, one near each end and one at the middle. **(Step 1)**
2. Each brigade member places an arm between two rungs and hoists the ladder onto the shoulder. **(Step 2)**
3. The brigade member closest to the butt covers the butt spur with a gloved hand to prevent injury to other brigade members in the event of a collision. **(Step 3)**

### Two-Person Straightarm Carry

The two-person straightarm carry is commonly used with straight and extension ladders. The ladder is carried at arm's length. To perform the two-person straightarm carry, follow the steps in ► **Skill Drill 12-4**.

1. Begin with the ladder resting on the ground on one beam. Both brigade members stand on the same side of the ladder, at opposite ends, and face the butt. **(Step 1)**
2. The brigade members reach down and grasp the upper beam of the ladder. **(Step 2)**
3. Pick the ladder up from the ground and carry it with the butt end forward. **(Step 3)**

12-3 **Skill Drill**

## Three-Person Shoulder Carry

All three brigade members approach the ladder from the same side and face the butt end. Two brigade members stand at each end and one in the middle.

Each brigade member places an arm between two rungs and lifts the ladder onto the shoulder.

The butt spurs are covered with a gloved hand while the ladder is transported.

### Three-Person Straightarm Carry

A three-person straightarm carry can be used for heavier ladders. This carry is similar to the two-person straightarm carry, with the addition of a third brigade member at the center of the ladder. All three brigade members remain on the same side of the ladder. To perform the three-person straightarm carry, follow the steps in ▼ Skill Drill 12-5.

12-4 Skill Drill

## Two-Person Straightarm Carry

1 Both brigade members face the butt of the ladder, at opposite ends.

2 The upper beam of the ladder is grasped.

3 Pick up the ladder using good lifting techniques.

12-5 Skill Drill

## Three-Person Straightarm Carry

1 Three brigade members stand on one side of the ladder, facing the butt. All three brigade members grasp the upper beam.

2 Pick up the ladder using good lifting techniques.

1. The carry begins with the ladder resting on the ground on one beam. One brigade member stands near the butt of the ladder, one near the center, and the third near the tip. All three brigade members stand on the same side and face the butt of the ladder. All three brigade members reach down and grasp the upper beam of the ladder. **(Step 1)**
2. Pick the ladder up from the ground and carry it at arm's length with the butt forward. **(Step 2)**

### Three-Person Flat Carry

The three-person flat carry is typically used with extension ladders up to 35-foot long. To perform the three-person flat carry, follow the steps in ▼ Skill Drill 12-6.

1. The carry begins with the bed section of the ladder flat on the ground. Two brigade members stand on one side of the ladder, one at the butt and one at the tip. The third brigade member stands on the opposite side of the ladder near the center. **(Step 1)**
2. All three brigade members kneel down, facing the butt end, and grasp the closer beam at arm's length. **(Step 2)**
3. The brigade members rise to a standing position lifting the ladder to arm's length. **(Step 3)**

### Four-Person Flat Carry

This carry is similar to the three-person flat carry described above except that two brigade members are positioned on each side of the ladder. On each side, one brigade member is at the tip and the other is at the butt end of the ladder. To perform the four-person flat carry, follow the steps in ► Skill Drill 12-7.

1. The carry begins with the bed section of the ladder flat on the ground. Two brigade members stand on each side of the ladder, one at the butt and one at the tip. **(Step 1)**
2. All four brigade members kneel down, facing the butt end of the ladder and grasp the closer beam at arm's length. **(Step 2)**
3. The brigade members rise to a standing position lifting the ladder at arm's length and walk with the butt end forward. **(Step 3)**

### Three-Person Flat Shoulder Carry

The three-person flat shoulder carry is similar to the three-person flat carry. The difference is that the ladder is carried on the shoulders instead of at arm's length. The brigade members face the tip of the ladder as they lift it, then pivot into the ladder as they raise it to shoulder height. This carry is useful when brigade members must carry the ladder over short obstacles. Because raising the ladder to shoulder height increases the potential for back strain, brigade members must follow proper lifting techniques. To perform the three-person flat shoulder carry, follow the steps in ► Skill Drill 12-8.

1. The carry begins with the bed section of the ladder flat on the ground. On one side of the ladder, a brigade

12-6 Skill Drill

## Three-Person Flat Carry

Two brigade members stand at the ends of the ladder on one side, and the third stands in the middle on the opposite side.

All three face the butt of the ladder and kneel and grasp the closer beam at arm's length.

The brigade members rise and carry the ladder at arm's length.

12-7 Skill Drill

## Four-Person Flat Carry

Two brigade members stand on each side of the ladder facing the butt, one at the butt and one at the tip.

All four kneel down, facing the butt end, and grasp the closer beam at arm's length.

The ladder is raised to arm's length.

12-8 Skill Drill

## Three-Person Flat Shoulder Carry

Two brigade members are positioned on the same side at the ends of the ladder, facing the tip. The third is on the opposite side at the middle. All three brigade members kneel and grasp the closer beam.

As the ladder approaches chest height, the brigade members all pivot into the ladder.

The ladder is carried on the shoulders facing the butt.

**12-9 Skill Drill**

## Four-Person Flat Shoulder Carry

Two brigade members are positioned on each side of the ladder at the ends. All four face the tip.

All four brigade members kneel and grasp the closer beam.

As the ladder approaches chest height, the brigade members all pivot toward the ladder, bringing it to rest on the shoulder. The ladder is carried on the shoulders facing the butt.

member is located at the butt and another is located at the tip. On the other side of the ladder, a brigade member is positioned near the center of the ladder. All three brigade members face the tip of the ladder. The brigade members kneel and grasp the closer beam. **(Step 1)**

2. The brigade members stand, raising the ladder. As the ladder approaches chest height, the brigade members all pivot into the ladder. **(Step 2)**
3. The ladder rests on the shoulders of the brigade members, with all three facing the butt of the ladder. The ladder is carried in this position with the butt moving forward. **(Step 3)**

### Four-Person Flat Shoulder Carry

This carry is similar to the three-person flat shoulder carry described above, except that there are two brigade members on each side of the ladder. On each side, one brigade member is at the tip and the other at the butt end of the ladder. To perform the four-person flat shoulder carry, follow the steps in **▲ Skill Drill 12-9**.

1. The carry begins with the bed section of the ladder flat on the ground. On each side of the ladder a brigade member is located at the butt and another is located at the tip. All four brigade members face the tip of the ladder. **(Step 1)**
2. The brigade members kneel and grasp the closer beam. **(Step 2)**
3. The brigade members stand, raising the ladder. As the ladder approaches chest height, the brigade members all pivot toward the ladder. The ladder ends at rest on the shoulders of the brigade members, with all four facing the butt of the ladder. The ladder is carried at this position with the butt moving forward. **(Step 3)**

## Placing a Ladder

The first step in raising a ladder is selecting the proper location for the ladder. Generally, a brigade leader or senior brigade member will select the general area for ladder placement, and the brigade member at the butt of the ladder will determine the exact site. Both the brigade leader ordering the ladder and the brigade member placing the ladder need to consider several factors in their decisions.

A raised ladder should be at an angle of approximately 75° to provide the best combination of strength, stability, and vertical reach. This angle also creates a comfortable climbing angle. When a brigade member is standing on a

**Brigade Member Safety Tips**

Do not place chocks or wood blocks under one beam of a ladder to position it on sloping ground. The ladder could slip off the blocks and overturn.

Figure 12-29 A ladder placed on an uneven surface can easily tip over.

rung, the rung at shoulder height will be about an arm's length away. The ratio of ladder height (vertical reach) to distance from the structure should be 4:1. For example, if the vertical reach of a ladder is 20 feet, the butt of the ladder should be placed 5 feet out from the wall. Most new ladders have an inclination guide on the beams so that the ladder will be set at the proper angle.

When calculating vertical reach, remember to add additional footage for rooftop operations. If the butt of the ladder is placed too close to the building, the climbing angle will be too steep. The tip of the ladder could pull away from the building as the brigade member climbs, making the ladder unstable. If the butt is too far from the structure, the climbing angle will be too shallow, reducing the load capacity of the ladder and increasing the risk that the butt could slip out from under the ladder.

The ladder should be placed on a stable and relatively level surface. Ladders placed on uneven surfaces are prone to tipping (▲ Figure 12-29). As a brigade member climbs the ladder, the moving weight will cause the load to shift back and forth between the two beams. Unless both butt ends are firmly placed on solid ground, this can start a rocking motion that will tip the ladder over.

Ladders should not be placed on top of manholes or trap doors. The weight of the ladder, brigade members, and equipment could cause the cover or door to shift or fail, injuring the brigade member(s).

Ladders should only be placed where there are no overhead obstructions. If a ladder comes into contact with overhead utility lines, particularly electric power lines, the brigade members working on it, as well as those stabilizing it, could be injured or killed. This is true for all ladders, regardless of their composition (fiberglass, wood, or metal).

A ladder can be energized even if it does not actually touch an electric line. A ladder that enters the electromagnetic field surrounding a power line can become energized. Ladders should stay at least 10 feet away from energized power lines.

Finally, portable ladders should not be placed in high traffic areas unless no other alternative is available. For example, because the main entrance to a structure is heavily used, a ladder should not be placed where it would obstruct the door.

## Raising a Ladder

Once the position has been selected, the ladder must be raised. Two common techniques for raising portable ladders are the beam raise and the flat raise. A beam raise is usually used when the ladder must be raised parallel to the target surface. A flat raise is often used when the ladder can be raised from a position perpendicular to the target surface.

The number of brigade members required to raise a ladder depends on the length and weight of the ladder, as well as on the available clearance from obstructions. A single brigade member can safely raise many straight ladders and lighter extension ladders. Two or more brigade members are required for longer and heavier ladders.

### One-Person Flat Raise

There are two variations of a one-person flat raise. One is used by a single brigade member to raise a small, straight ladder, typically 14 feet or less in length. To perform this raise, follow the steps in (▶ Skill Drill 12-10).

1. Start with the brigade member carrying a ladder using the one-person carry described earlier. Check for overhead hazards. **(Step 1)**
2. The brigade member places the ladder flat on the ground. The heel of the ladder should be positioned approximately where it will be when the ladder is in the raised position.
3. Standing at the tip, the brigade member raises the ladder to chest level.
4. The brigade member walks hand-over-hand down the rungs until the ladder is vertical. **(Step 2)**
5. The brigade member places one foot against the beam of the ladder and leans it into place. **(Step 3)**

The other variation of the one-person flat raise is generally used with straight ladders longer than 14 feet and with extension ladders that can be safely handled by the brigade member.

**Brigade Member Safety Tips**

Brigade members must recognize their individual limits. If you need assistance, do not try to raise the ladder alone. The fireground is no place to prove your strength or daring. If you overexert and injure yourself, your team will be a member short, making it more difficult to perform necessary tasks.

12-10 Skill Drill

## One-Person Flat Raise for Ladders Under 14 feet

Carry the ladder to the structure. Check for overhead hazards before raising the ladder.

By the tip, raise the ladder to chest level. Walk hand-over-hand down the rungs until the ladder is vertical.

Heel the ladder and lean it into place.

Each brigade member will have a different safety limit, depending on his or her strength and the weight of the ladder. To perform this variation of the one-person flat raise, follow the steps in ▶ **Skill Drill 12-11**.

1. Start with the brigade member carrying a ladder using the one-person carry described earlier. Check for overhead hazards. **(Step 1)**
2. The brigade member places the butt of the ladder on the ground directly against the structure and rotates the ladder so both spurs contact the ground and structure. Then the brigade member lays the ladder on the ground. If the ladder is an extension ladder, the base section should be against the ground (fly section up). **(Step 2)**
3. The brigade member takes hold of a rung near the tip, brings that end of the ladder to chest height and then steps beneath the ladder and pushes upward on the rungs. **(Step 3)**
4. The ladder is raised using a hand-over-hand motion as the brigade member walks toward the structure until the ladder is vertical and against the structure. **(Step 4)**
5. If an extension ladder is being used, the brigade member holds the ladder vertical against the structure and extends the fly section by pulling the halyard

Skill Drill 12-11

# One-Person Flat Raise for Ladders Over 14 feet

1 Carry the ladder to the structure. Check for overhead hazards before raising the ladder.

2 Place the butt against the base of the structure and rotate ladder so both spurs contact the ground and structure. Lay the ladder flat on the ground.

3 Grasp a rung near the tip, bring that end of the ladder to chest height, step beneath the ladder, and push upward on the rungs.

4 Walk toward the structure, lifting the rungs hand-over-hand until the ladder is vertical against the structure.

5 Pull the butt away from the structure.

12-12 Skill Drill

## Tying the Halyard

1 Wrap the excess halyard around two rungs and pull it tight over the upper rung.

2 Tie a clove hitch around the upper rung.

3 Pull the knot tight and add a safety knot.

smoothly, with a hand-over-hand motion, until desired height is reached and the dogs are locked.

6. The butt of the ladder is then pulled out from the structure to create the proper climbing angle. To move the butt away from the structure, the brigade member grips a lower rung and lifts slightly while pulling outward. At the same time, pressure should be applied to an upper rung to keep the tip of the ladder against the structure. **(Step 5)**
7. If the ladder is an extension ladder, it will be necessary to rotate the ladder so the fly section is out. The halyard should be tied as described in Skill Drill 12-12.

### Tying the Halyard

The halyard of an extension ladder should always be tied after the ladder has been extended and lowered into place. A tied halyard stays out of the way and provides a safety back-up to the dogs for securing the fly section. To tie the halyard, follow the steps in ▲ Skill Drill 12-12.

1. The brigade member wraps the excess halyard rope around two rungs of the ladder and pulls the rope tight across the upper of the two rungs. **(Step 1)**
2. The brigade member ties a clove hitch around the upper rung and the vertical section of the halyard. Refer to Chapter 9, Ropes and Knots, to review how to tie a clove hitch. **(Step 2)**
3. Pull the clove hitch tight and place an overhand safety knot as close to the clove hitch as possible to prevent slipping. **(Step 3)**

### Two-Person Beam Raise

The two-person beam raise is used with midsized extension ladders up to 35-foot long. To perform the two-person beam raise, follow the steps in ► Skill Drill 12-13.

1. The two-person beam raise begins with a shoulder or straightarm carry. One brigade member is near the butt of the ladder and one near the tip. The brigade members check for overhead hazards. **(Step 1)**
2. The brigade member at the butt of the ladder places the butt of the lower beam on the ground, while the brigade member at the tip of the ladder holds the other end. The brigade member at the butt of the ladder places a foot on the butt of the beam that is in contact with the ground and grasps the upper beam. **(Step 2)**
3. The brigade member at the tip of the ladder begins to walk toward the butt, while raising the beam in a hand-over-hand fashion until the ladder is vertical. **(Step 3)**
4. The two brigade members pivot the ladder into position as necessary. **(Step 4)**
5. The brigade members face each other, one on each side of the ladder, and heel the ladder by each placing

#### Brigade Member Safety Tips

Pull the halyard of the extension ladder straight down in line with the ladder to avoid pulling the ladder over.

12-13 Skill Drill

# Two-Person Beam Raise

Begin with one brigade member at the butt of the ladder and one at the tip. Check for overhead hazards.

The brigade member at the butt lowers the ladder until one beam is on the ground. The brigade member at the butt of the ladder places a foot on the butt of the beam that is in contact with the ground and grasps the upper beam.

The brigade member at the tip walks toward the butt of the ladder, raising the beam hand-over-hand until it is vertical.

The two brigade members stand on opposite sides and pivot the ladder into position, as necessary.

Each brigade member places one foot against the butt of a beam to brace the ladder. The brigade member with his back to the building extends and locks the fly section.

The brigade member on the outside heels the ladder while both lean it into place. Secure the halyard before climbing the ladder.

the toe of one boot against the opposing beams of the ladder.

6. One brigade member extends the fly section by pulling the halyard smoothly with a hand-over-hand motion until the fly section is the height desired and the dogs are locked. The other brigade member stabilizes the ladder by holding the outside of the two beams—so that if the fly comes down suddenly it will not strike the brigade member's hands. **(Step 5)**
7. The brigade member facing the structure places one foot against one beam of the ladder and then both brigade members lean the ladder into place. The halyard is tied as described previously. **(Step 6)**

### Two-Person Flat Raise

The two-person flat raise is also commonly used with midsized extension ladders up to 35-foot long. To perform the two-person flat raise, follow the steps in ▼ **Skill Drill 12-14**.

1. The two-person flat raise begins from a shoulder carry or straightarm carry, with one brigade member near the butt of the ladder and one near the tip. Check for overhead hazards. The brigade member at the butt of the ladder places the butt of the lower beam on the

**12-14 Skill Drill**

## Two-Person Flat Raise

1 Begin with one brigade member near the butt of the ladder and one at the tip. Check for overhead hazards. The brigade member at the butt places the butt of the lower beam on the ground. The brigade member at the tip rotates the ladder until both beams are in contact with the ground. The brigade member at the butt places both feet on the bottom rung, grasps a higher rung, crouches, and leans backward. The brigade member at the tip swings under the ladder and walks toward the butt, raising the rungs hand-over-hand until the ladder is vertical. On opposite sides, the brigade members pivot the ladder into position, if necessary.

2 Each brigade member places one foot against the butt of a beam to brace the ladder. The brigade member with his back to the structure extends and locks the fly section.

3 The brigade member on the outside heels the ladder while both lean it into place. Secure the halyard before climbing the ladder.

## Brigade Member Safety Tips

When raising the fly section of an extension ladder, do not wrap the halyard around your hand. If the ladder falls or the fly section unexpectedly retracts, your hand could be caught in the halyard, and you could be seriously injured. Also, do not place your foot under the fly sections. If the rope slips, your foot will be crushed.

ground, while the brigade member at the tip holds the other end.
2. The brigade member at the tip rotates the ladder so that both beams are in contact with the ground.
3. The brigade member at the butt of the ladder stands on the bottom rung, grasps a higher rung with both hands, crouches down and leans backward.
4. The brigade member at the tip of the ladder swings under the ladder and walks toward the butt, advancing down the ladder and lifting the rungs in a hand-over-hand fashion until the ladder is vertical. The two brigade members stand on opposite sides of the ladder and pivot it into position as necessary. **(Step 1)**
5. The brigade members face each other, one on each side of the ladder, and heel the ladder by each placing the toe of one boot against the opposite beams of the ladder. If using an extension ladder, one brigade member extends the fly section by pulling the halyard smoothly with a hand-over-hand motion until the tip is at the desired height and the dogs are locked. The other brigade member stabilizes the ladder by holding the outside of the base section's beams—so that if the fly comes down suddenly it will not strike the brigade members hands. **(Step 2)** The brigade member facing the structure places one foot against one beam of the ladder and then both brigade members lean the ladder into place. The halyard is tied as described in Skill Drill 12-12. **(Step 3)**

### Three- and Four-Person Flat Raises

The three- and four-person flat raises are used for very heavy ladders. The basic steps in these raises are similar to the two-person flat raise. In a three-person flat raise, the third brigade member assists with the hand-over-hand raising of the ladder.

When four brigade members are raising the ladder, two anchor the butt of the ladder while the other two raise the ladder. The two brigade members at the butt each place their inside foot on the bottom rung of the ladder and bend over, grasping a rung in front of them. To perform the three-person flat raise, follow the steps in ▶ **Skill Drill 12-15**.

1. The three-person flat raise begins from a shoulder carry or straightarm carry, with one brigade member near the butt of the ladder, one in the middle, and one near the tip. **(Step 1)**
2. The brigade members check for overhead hazards. **(Step 2)**
3. The brigade member at the butt of the ladder places the butt of the lower beam on the ground, while the brigade member at the tip holds the other end. The brigade member in the middle moves to the tip. The brigade members at the tip rotates the ladder so that both butts are in contact with the ground. **(Step 3)**
4. The brigade member at the butt of the ladder stands on the bottom rung, grasps a higher rung with both hands, crouches down and leans backward. **(Step 4)**
5. The brigade members at the tip of the ladder begin to walk toward the butt, advancing down the ladder and lifting the rungs in a hand-over-hand fashion until the ladder is vertical. **(Step 5)**
6. The three brigade members pivot the ladder into position as necessary. **(Step 6)**
7. Two brigade members face each other, one on each side of the ladder, heel the ladder by each placing the toe of one boot against the opposite beams of the ladder, and grasp the outsides of the beams. **(Step 7)**
8. One brigade member extends the fly section by pulling the halyard smoothly with a hand-over-hand motion until the tip is at the desired height and the dogs are locked. **(Step 8)**
9. One brigade member heels the ladder while the other two lean the ladder into place.
10. The halyard is tied as described previously. **(Step 9)**

To perform the four-person flat raise, follow the steps in **Skill Drill 12-16**.

1. The four-person flat raise begins with a four-person flat carry. Two brigade members are at the butt of the ladder and two brigade members are at the tip. Check for overhead hazards.
2. The brigade members at the butt of the ladder place both butts on the ground while the brigade members at the tip of the ladder hold the other end.
3. The brigade members at the butt of the ladder stand side-by-side, facing the ladder. Each brigade member places the inside foot on the bottom rung and the other foot on the ground outside the beam. Both then crouch down, grab a rung and the beam, and lean backward.
4. The two brigade members at the tip of the ladder begin to walk toward the butt of the ladder, advancing down the rungs in a hand-over-hand fashion until the ladder is vertical.
5. The brigade members pivot the ladder into position as necessary.
6. Two brigade members heel the ladder by placing a boot against each beam. Each brigade member places the toe of one boot against one of the beams. The third

Skill Drill 12-15

## Three-Person Flat Raise

1 Begin with one brigade member at the butt of the ladder, one in the middle, and one at the tip.

2 Check for overhead hazards.

3 The brigade member at the butt lowers the ladder until the butt of one beam is on the ground. The brigade member in the middle moves to the tip. The brigade members at the tip rotate the ladder until both butts are on the ground.

4 The brigade member at the butt of the ladder places both feet on the bottom rung, grasps a higher rung, crouches, and leans backward.

5 The brigade members at the tip walk toward the butt of the ladder, raising the rungs hand-over-hand until the ladder is vertical.

## 12-15 Skill Drill (Continued)

The three brigade members pivot the ladder into position, as necessary.

Two brigade members stabilize the ladder, each placing one foot against the butt of a beam and grasping the outsides of the beams with both hands.

One brigade member extends and locks the fly section.

The brigade member on the outside heels the ladder while two brigade members lean it into place. The halyard is secured before climbing the ladder.

brigade member stabilizes the ladder by holding it on the outside of the rails.

7. The fourth brigade member extends the fly section by pulling the halyard smoothly with a hand-over-hand motion until the tip reaches the desired height and the dogs are locked.
8. The two brigade members facing the structure each place one foot against one beam of the ladder while the other two brigade members lower the ladder into place.
9. The halyard is tied as described previously.

### Fly Section Orientation

When raising extension ladders, brigade members must know whether the fly section should be placed toward the building (fly in) or away from the building (fly out). The ladder manufacturer and brigade standard operating procedures will specify whether the fly should be facing in or out. In general, manufacturers of fiberglass and metal ladders recommend that the fly sections be placed away from the structure. Wood fire-service ladders are often designed to be used with the fly in.

## Securing the Ladder

There are several ways to prevent a ladder from moving once it is in place. One option is to have a brigade member stand between the ladder and the structure, grasp the beams, and lean back to pull the ladder into the structure (▶ Figure 12-30). This is called heeling the ladder. A brigade member(s) on the outside of the ladder, facing the structure, can also heel the ladder by placing a boot against the beam(s) of the ladder (▶ Figure 12-31).

A rope, a rope-hose tool, or webbing also can be used to secure a ladder in place. The lower part of the ladder can be tied to any solid object to keep the base from kicking out. The base should always be secured if the ladder is used at a low angle. The tip of the ladder can be tied to a secure object near the top to keep it from pulling away from the building.

## Climbing the Ladder

Climbing a ladder at the fire or emergency scene should be done in a deliberate and controlled manner. Always make sure the ladder is secure (tied or heeled). Before climbing an extension ladder, verify that the dogs are locked and the halyard is secured.

The proper climbing angle should be checked as well. Some ladder beams have level guide stickers placed there by the manufacturer. The bottom line of the indicator will be parallel to the ground when the ladder is positioned properly. Standing on the bottom rung of the ladder and extending the arms straight out is another way to check the climbing angle (▶ Figure 12-32). The hands should comfortably reach the beams or rungs if the angle is appropriate. Finally, the angle can be measured by ensuring that the butt of the ladder is one-fourth of the working height out from the base of the structure.

**Figure 12-30** The brigade member's weight pulls the ladder against the structure to secure the ladder.

Keep bounce and shifting to a minimum while climbing. Eyes should be focused forward with only occasional glances upward. This will prevent debris such as falling glass from injuring the face or eyes. Lower the protective face shield for additional protection.

Use a hand-over-hand motion on the rungs of the ladder, or slide both hands along the underside of the beams while climbing.

If tools must be moved up or down, it is better to hoist them by ropes than to carry them up a ladder. Carrying tools on a ladder reduces the brigade member's grip and increases the potential for injury if a tool slips or falls. To climb a ladder while holding a hand tool, the brigade member should hold the tool against one beam with one hand and maintain contact with the opposite beam with the other hand. Tools such as pike poles can be hooked onto the ladder and moved up every few rungs. To climb a ladder while carrying a tool, follow the steps in (▶ Skill Drill 12-17).

1. The brigade member prepares to climb the ladder by placing the tool in one hand and holding it against the beam of the ladder. **(Step 1)**

Figure 12-31 A brigade member can push against the ladder to heel it.

Figure 12-32 The climbing angle can be checked by standing on the bottom rung and holding the arms straight out.

12-17 Skill Drill

## Climbing the Ladder While Carrying a Tool

1 Hold the tool in one hand and place it against the beam.

2 Slide the tool up the beam, while sliding the opposite hand up the other beam.

Figure 12-33 Check the stability of the roof before dismounting from the ladder.

Figure 12-34 A ladder belt has a large hook that is designed to secure a brigade member to a ladder.

2. The brigade member wraps the other hand around the opposite beam and begins climbing. Contact is maintained between the free hand and the beam by sliding the tool along the beam while climbing. **(Step 2)**

Be sure not to overload the ladder. There should be no more than two brigade members on a ladder at one time. A properly placed ladder should be able to support two rescuers (with their protective clothing and equipment) and one victim.

## Dismounting the Ladder

When a ladder is used to reach a roof or a window entry, the brigade member will have to dismount the ladder. Brigade members can minimize the risk of slipping and falling by making sure that the surface is stable before dismounting. Test the stability with a tool before dismounting. For example, brigade members who dismount from a ladder onto a roof typically sound the roof with an axe before stepping onto the surface (▲ Figure 12-33).

Try to maintain contact with the ladder at three points when dismounting. For example, a brigade member who steps onto a roof should keep two hands and one foot on the ladder while checking the footing. This is particularly important if the surface slopes or is covered with rain, snow, or ice. Do not shift your weight onto the roof until you have tested the footing.

It is also important to check the footing when entering a window from a ladder positioned to the side of the window. Before stepping into the window, be sure that the interior surface is structurally sound and offers secure footing. The three points of contact rule also applies in this situation.

Under heavy fire and smoke conditions, brigade members sometimes dismount by climbing over the ladder tip and sliding over the windowsill into the building. Sound the floor inside the window before entering to be sure it is solid and stable. To remount a ladder under these conditions, back out the window feet first and rest your abdomen on the sill until you can feel the ladder under your feet. Under better conditions, sit on the windowsill with legs out and roll onto the ladder.

## Working from a Ladder

Brigade members must often work from a ladder. To avoid falling while working from a ladder, the brigade member must be secured to the ladder. Two different methods are used by brigade members to secure themselves to a ladder.

The first method is a ladder belt. A ladder belt is specifically designed to secure a brigade member to a ladder or elevated surface (▲ Figure 12-34). Brigade members must be sure to use only ladder belts designed and certified to

12-18 Skill Drill

## Working from a Ladder

1 The brigade member climbs the ladder to the desired work height and then one rung higher.

2 The brigade member notes the side where the work will be performed. The opposite leg is extended between the rungs.

3 The knee is bent around the rung and the foot is brought back under the rung.

4 The foot is secured around the lower rung or the beam. The brigade member moves the other leg down one rung.

5 The brigade member is now free to work with both hands.

NFPA 1983, *Standard on Fire Service Life Safety Rope and System Components*. Utility belts designed to carry tools should never be used as ladder belts. The alternate method is to apply the leg lock. The leg lock is simple, secure, and requires no special equipment. To work from a ladder, follow the steps of ◄ Skill Drill 12-18.

1. The brigade member climbs to the desired work height and steps up to the next higher rung. **(Step 1)**
2. The brigade member notes the side of the ladder where the work will be performed. The leg on the opposite side is extended between the rungs. **(Step 2)**
3. Once the leg is between the rungs, the brigade member bends the knee and brings that foot back under the rung and through to the climbing side of the ladder. **(Step 3)**
4. The foot is secured against the next lower rung or the beam of the ladder. The brigade member uses the thigh for support and steps down one rung with the opposite foot. **(Step 4)**
5. The brigade member is now free to lean out to the side of the ladder and work with two hands on the tool. **(Step 5)**

## Brigade Member Safety Tips

Working with ladders requires attention to basic safety precautions. Brigade members should always check these important safety items:

- Choose the proper ladder for the job.
- Wear protective gear, including gloves, when working with ladders.
- Use the proper number of brigade members for each raise.
- Use leg muscles, not back or arm muscles, when lifting ladders.
- Never raise ladders into or near electrical wires.
- Check the ladder for proper angle.
- Check the ladder locks to be sure they are seated over the rungs.
- Check staypoles to be sure they are set properly.
- Make sure the ladder is secure at the top or the bottom before climbing.
- Climb smoothly and rhythmically.
- Always tie in with a leg lock or a ladder belt when working from a ladder.
- Do not overload the ladder.
- Inspect ladders for damage and wear after each use.

Mastering the leg lock and using a ladder belt enable brigade members to accomplish several advanced tasks while on a portable ladder. Among these tasks are placing a roof ladder for ventilation or using a hose stream to apply water to a hard-to-reach location.

### Placing a Roof Ladder

Several methods can be used to properly position a roof ladder on a sloping roof. The most common method is described in Skill Drill 12-19.

1. The brigade member carries the roof ladder to the base of a ladder that is already in place to provide access to the roofline.
2. The brigade member places the roof ladder on the ground and rotates the hooks of the roof ladder to the open position.
3. The brigade member uses a one-person beam or flat raise to lean the roof ladder against one beam of the other ladder with the hooks oriented outward.
4. The brigade member climbs the lower ladder until reaching the mid point of the roof ladder. The brigade member then slips one shoulder between two rungs of the roof ladder and shoulders the ladder.
5. The brigade member climbs to the roofline of the structure carrying the roof ladder on one shoulder.
6. The brigade member then uses a ladder belt to secure to the ladder or applies a ladder leg lock (described in Skill Drill 12-18).
7. The brigade member places the roof ladder on the roof surface with hooks down. The ladder is pushed up toward the peak of the roof with a hand-over-hand motion.
8. Once the hooks have passed the peak, the brigade member pulls back on the roof ladder to set the hooks and checks to ensure they are secure.
9. A roof ladder is removed from the roof by reversing the process described above. After releasing the hooks from the peak, it may be necessary to turn the ladder on one of its beams so that it can slide down the roof without catching the hooks on the roofing material.

# Wrap-Up

## Ready for Review

- The primary function of ladders is to provide safe access to and egress from otherwise inaccessible areas.
- Ladders can be used for several auxiliary purposes including channeling debris, serving as a lift point, and holding other firefighting equipment.
- To inspect and maintain ladders, the brigade member must become familiar with ladder construction and the terminology used to describe ladders.
- Ladders must be regularly inspected, cleaned, and maintained.
- To select the appropriate ladder for a job, the brigade member must know what types of ladders are available as well as their uses and limitations.
- Ladders used in the fire service can be classed into two broad categories: aerial ladders and portable ladders.
- Communication is key to coordinating efforts when working with ladders.
- To deploy a portable ladder, brigade members must be able to carry, place, raise, and climb the ladder safely, using common techniques.
- Brigade members must be able to work safely from ladders.

## Hot Terms

**Aerial ladder** A power-operated ladder permanently mounted on a piece of apparatus.

**Bangor ladder** A ladder equipped with tormentor poles or staypoles that stabilize the ladder during raising and lowering operations.

**Beam** One of the two main structural pieces running the entire length of each ladder or ladder section. The beams support the rungs.

**Bed section** The lowest and widest section of an extension ladder. The fly sections of the ladder extend from the bed section.

**Butt** Often called the heel or base, the butt is the end of the ladder that is placed against the ground when the ladder is raised.

**Butt plate** (also referred to as footpad) An alternative to a simple butt spur; a swiveling plate with both a spur and a cleat or pad that is attached to the butt of the ladder.

**Butt spurs** The metal spikes attached to the butt of a ladder. The spurs help prevent the butt from slipping out of position.

**Combination ladder** A ladder that converts from a straight ladder to a step ladder configuration (A-frame) or from an extension ladder to a step ladder configuration.

**Dogs** (also referred to as pawls, ladder locks, and rung locks) A mechanical locking device used to secure the fly section(s) of a ladder after they have been extended.

**Egress** A method of exiting from an area or a building.

**Extension ladder** An adjustable-length, multiple-section ladder.

**Fly section** A section of an extension ladder that is raised or extended from the base section or from another fly section. Some extension ladders have more than one fly section.

**Folding ladder** A ladder that collapses by bringing the two beams together for portability. Unfolded, the folding ladder is narrow and used for access to attic scuttle holes and confined areas.

**Fresno ladder** (also referred to as attic ladder) A narrow, two-section extension ladder that has no halyard. Because of its limited length, it can be extended manually.

**Grade** The level at which the ground intersects the foundation of a structure.

**Guides** Strips of metal or wood that serve to guide a fly section during extension. Channels or slots in the bed or fly section may also serve as guides.

**Halyard** The rope or cable used to extend or hoist the fly section(s) of an extension ladder.

**Heat sensor label** A piece of heat-sensitive material on each section of a ladder that identifies when the ladder has been exposed to high heat conditions.

**I-beam** A ladder beam constructed of one continuous piece of I-shaped metal or fiberglass to which the rungs are attached.

**Ladder belt** A belt specifically designed to secure a brigade member to a ladder or elevated surface.

**Ladder gin** An A-shaped structure formed with two ladder sections. It can be used as a makeshift lift when raising a trapped person. One form of the device is called an A-frame hoist.

**Pompier ladder** (scaling ladder) A lightweight, single beam ladder.

**Portable ladder** Ladder carried on fire apparatus, but designed to be removed from the apparatus and deployed by brigade members where needed.

**Protection plates** Reinforcing material placed on a ladder at chaffing and contact points to prevent damage from friction and contact with other surfaces.

**Pulley** A small, grooved wheel through which the halyard runs. The pulley is used to change the direction of the halyard pull, so that a downward pull on the halyard creates an upward force on the fly section(s).

**Rail** The top or bottom piece of a trussed-beam assembly used in the construction of a trussed ladder. The term rail is also sometimes used to describe the top and bottom surfaces of an I-beam ladder. Each beam will have two rails.

**Roof hooks** The spring-loaded, retractable, curved metal pieces that allow the tip of a roof ladder to be secured to the peak of a pitched roof. The hooks fold outward from each beam at the top of a roof ladder.

**Roof ladder** (hook ladder) A straight ladder equipped with retractable hooks so that the ladder can be secured to the peak of a pitched roof. Once secured, the ladder lies flat against the surface of the roof, providing secure footing for brigade members.

**Rung** A ladder crosspiece that provides a climbing step for the user. The rung transfers the weight of the user out to the beams of the ladder or back to a center beam on an I-beam ladder.

**Solid beam** A ladder beam constructed of a solid rectangular piece of material, typically wood, to which the ladder rungs are attached.

**Staypole** (tormentor) A long piece of metal attached to the top of the bed section of an extension ladder and used to help stabilize the ladder during raising and lowering. The pole attaches to a swivel point and has a spur on the other end. One pole is attached to each beam of long (40 feet or longer) extension ladders.

**Stop** A piece of material that prevents the fly section(s) of a ladder from overextending and collapsing the ladder.

**Tie rod** A metal rod that runs from one beam of the ladder to the other to keep the beams from separating. Tie rods are typically found in wood ladders.

**Tip** The very top of the ladder.

**Truss block** A piece of wood or metal that ties the two rails of a trussed beam ladder together and serves as the attachment point for the rungs.

**Trussed beam** A ladder beam constructed of top and bottom rails joined by truss blocks that tie the rails together and support the rungs.

# Brigade Member in Action

**Your fire brigade is dispatched to a processing building for a report of a fire inside a filter system. The filter system is mounted to the flat roof of the $2\frac{1}{2}$-story industrial building. When your first-due engine arrives on scene, you find heavy smoke coming from the filter system. The best area in which to locate the ladder for access is in a narrow alley, where the adjacent building is less than 10 feet from the fire building. You and your crew are assigned to the task of placing the ladder and stretching a line to the roof area. Your engine carries a 24-foot extension ladder, a 35-foot extension ladder, and a 16-foot roof ladder.**

1. Based on your assignment, what would be your first ladder of choice for the operation?
   A. 16-foot roof ladder
   B. 24-foot extension ladder
   C. 35-foot extension ladder
   D. Special-call mutual aid for a 75-foot aerial truck

2. Based on the location in which the ladder must be raised, which of the following techniques would be the most useful?
   A. One-person flat raise
   B. Two-person flat raise
   C. Two-person beam raise
   D. Three-person flat raise

3. Which of the following angles should the ladder be set to before you climb it?
   A. 45º
   B. 75º
   C. 90º
   D. 180º

4. Which of the following maneuvers should be conducted when dismounting the ladder at the roof level?
   A. Sound the roof area with a tool prior to dismounting.
   B. Attempt to maintain three points of contact with the ladder when dismounting.
   C. Do not shift your weight from the ladder until you check your footing.
   D. All of the above.

# Search and Rescue

# Chapter 13

## NFPA 1081 Standard

**Incipient Facility Fire Brigade Member**

NFPA 1081 contains no Incipient Facility job performance requirements for this chapter.

**Advanced Exterior Facility Fire Brigade Member**

**5.2.4** Conduct search and rescue operations as a member of a team, given an assignment, obscured vision conditions, PPE, scene lighting, forcible entry tools, handlines, and ladders when necessary, so that all equipment is correctly used, all assigned areas are searched, all victims are located and removed, team integrity is maintained, and team members' safety, including respiratory protection, is not compromised.

**(A)** *Requisite Knowledge.* Use of appropriate tools and equipment, psychological effects of operating in obscured conditions and ways to manage them, methods to determine if an area is tenable, primary and secondary search techniques, team members' roles and goals, methods to use and indicators of finding victims, victim removal methods, and considerations related to respiratory protection.

**(B)** *Requisite Skills.* The ability to use SCBA to exit through restricted passages, use tools and equipment for various types of rescue operations, rescue a facility fire brigade member with functioning respiratory protection, rescue a facility fire brigade member whose respiratory protection is not functioning, rescue a person who has no respiratory protection, and assess areas to determine tenability.

**Interior Structural Facility Fire Brigade Member**

**6.2.8** Conduct search and rescue operations as a member of a team, given an assignment, obscured vision conditions, PPE, a flashlight, forcible entry tools, handlines, and ladders when necessary, so that all equipment is correctly used, all assigned areas are searched, all victims are located and removed, team integrity is maintained, and team members' safety, including respiratory protection, is not compromised.

**(A)** *Requisite Knowledge.* Use of appropriate tools and equipment, psychological effects of operating in obscured conditions and ways to manage them, methods to determine if an area is tenable, primary and secondary search techniques, team members' roles and goals, methods to use and indicators of finding victims, victim removal methods, and considerations related to respiratory protection.

**(B)** *Requisite Skills.* The ability to use SCBA to exit through restricted passages, use tools and equipment for various types of rescue operations, rescue a facility fire brigade member whose respiratory protection is not functioning, rescue a person who has no respiratory protection, and assess areas to determine tenability.

## Knowledge Objectives

After completing this chapter, you will be able to:

- Define search and rescue.
- Describe the importance of scene size-up in search and rescue.
- Describe search techniques.
- Describe the primary search.
- Describe search patterns.
- Describe the secondary search.
- Describe how to ensure brigade member safety during a search.
- Describe ladder rescue techniques.
- Describe industrial hazards and issues, which may affect search-and-rescue operations.

## Skills Objectives

After completing this chapter, you will be able to perform the following skills:

- Demonstrate the one-person walking assist.
- Demonstrate the two-person walking assist.
- Demonstrate the two-person extremity carry.
- Demonstrate the two-person seat carry.
- Demonstrate the two-person chair carry.
- Demonstrate the cradle-in-arms carry.
- Demonstrate the clothes drag.
- Demonstrate the blanket drag.
- Demonstrate the webbing sling drag.
- Demonstrate the brigade member drag.
- Demonstrate the one-person emergency drag from a vehicle.
- Demonstrate the long backboard rescue.
- Demonstrate rescuing a conscious person from a window.
- Demonstrate rescuing an unconscious person from a window.
- Demonstrate rescuing an unconscious small adult from a window.
- Demonstrate rescuing a large adult from a window.

# You Are the Brigade Member

**As you enter the classroom of your fire brigade training session, several fire brigade members are discussing a search-and-rescue operation at a chemical processing facility similar in size to your plant. The incident occurred in a large indoor processing building. When the facility fire brigade arrived on scene, the area foreman indicated that one of the process technicians who was working within the building was not present at the evacuation site and was presumed to still be inside the building. A team of fire brigade members, including a member familiar with the process building, was assigned to conduct search-and-rescue operations. It took several teams of search-and-rescue personnel to find the process technician inside the large open-area building, which was a maze of piping, pumps, and other process equipment.**

1. ***What techniques are used to search large open-area buildings?***
2. ***What hazards may be encountered by search teams within a process-area type of environment?***
3. ***What would be the quickest and safest method to remove an employee from this type of environment?***

## Introduction

The mission of the fire brigade is to save lives and protect property. Saving lives is the highest priority at a fire scene. That is why search and rescue are so important at any incident. The first brigade members to arrive must always consider the possibility that lives could be in danger and act accordingly.

Saving lives remains the highest priority until it is determined that everyone who was in danger has been found and moved to a safe location, or until it is no longer possible to rescue anyone successfully. The first situation occurs when a thorough search is done and there is no one remaining to be rescued. The second situation might occur if fire conditions or other factors make it likely that no one could still be alive to be rescued.

## Search and Rescue

Search and rescue are almost always performed in tandem. A search is done to look for victims who need assistance to leave a dangerous area. Rescue is the physical removal of a person from confinement or danger, such as when a brigade member leads an occupant to an exit or carries an unconscious victim out of a burning building and down a ladder. Both are examples of rescues because the victim is physically removed from imminent danger through the actions of a rescuer.

Although any fire brigade unit could be assigned to search-and-rescue operations, many fire brigades assign this responsibility to brigade members with specialized rescue training (high angle, confined space, etc.). Additionally, fire brigade members must be trained to the proper level of hazmat response to enter fire areas that contain hazardous materials. All fire brigade members must be trained and prepared to perform basic search-and-rescue functions.

Search-and-rescue operations must be conducted quickly and efficiently. A systematic approach will ensure that everyone who can possibly be saved is successfully located and removed from danger. Brigade members should practice the specific search-and-rescue procedures used by their brigades to sharpen their skills and increase their efficiency.

Often, the first-arriving brigade members may not know if anyone is inside a burning building, near a process area, or how many people could be in the area or in a structure. Brigade members should never assume that an area is unoccupied. Industrial facilities routinely maintain evacuation plans and roll call procedures to account for workers and visitors in an operating area or a structure. Searches are not generally necessary if all personnel have been accounted for. If there are doubts or questions regarding the accountability of personnel in the affected area, a thorough search is necessary. Due to the hazards associated with many industrial operations, a search may not be possible because of the risk to brigade members when weighed against the possibility of survival of missing employees or guests. There are times when a search cannot be conducted immediately. Depending on the circumstances, the search could be delayed or limited to a specific area because of the risk to brigade members and the low probability of survivors.

Figure 13-1 To support search-and-rescue operations, brigade members may first have to position hose lines to protect the means of egress.

Figure 13-2 The IC must weigh the risk to brigade members against the possibility of saving anyone inside before authorizing an interior search.

## Coordinating Search and Rescue with Fire Suppression

Although search and rescue are always the first priorities at a fire, they are never the only actions taken by first-arriving brigade members. Brigade members must plan and coordinate all other activities to support the search-and-rescue priority. When search-and-rescue operations are finished, fire brigade members can focus totally on fire suppression and other incident stabilization operations.

Often, brigade members must take action to confine or control the fire before search-and-rescue operations can begin. It might be necessary to position hose lines to keep the fire away from potential victims or to protect the entry and exit paths, so that the victims can be found and safely removed (▲ Figure 13-1). In some cases, the best way to save lives is to control the fire and eliminate the danger quickly.

Other fire scene activities also must be coordinated with search and rescue. Forcible entry might be needed to provide entrances and exits for search-and-rescue teams. Well-placed ventilation can reduce interior temperatures and improve visibility, enabling search teams to locate victims more rapidly. Portable lighting can provide valuable assistance to interior search crews. Isolation of hazardous conditions (sources of electricity, hazardous materials, etc.) can make areas more tenable for victims and safer for fire brigade members conducting search operations.

Often, the search for potential victims also provides valuable information about the location and extent of the fire and other hazardous conditions within a building and/or exterior process operation area. The searchers act as a reconnaissance team to determine which areas are involved and where the fire or hazards might spread. They report this information to the Incident Commander (IC) who is developing an overall plan for the situation.

## Search-and-Rescue Size-Up

The size-up process at every fire should include a specific evaluation of the critical factors for search and rescue. These include the number of occupants in the building, their location, the degree of risk to their lives, and their ability to evacuate by themselves. Usually, this information is not immediately available, so actions have to be based on a combination of observations and expectations. The type of facility; the size, construction, complexity, and condition of the building/exterior process area; and the apparent smoke, fire, and hazardous conditions, as well as the time of day and day of week, are important observations that could indicate whether and how many people may need to be rescued.

Brigade members can then develop a search-and-rescue plan based on their conclusions. This plan identifies the areas to be searched, the priorities for searching different areas, the number of search teams required, and any additional actions needed to support the search-and-rescue activities. One search-and-rescue team with standard firefighting gear may be sufficient for a small industrial building, but multiple teams with thermal imaging cameras and air monitoring equipment might be necessary to search a large, multihazard industrial facility.

### Risk-Benefit Analysis

Plans for search and rescue must take into consideration the risks and benefits of the operation. In some situations, search-and-rescue operations must be limited or cannot be performed. For example, if a building is exhibiting potential backdraft conditions, or eminent overpressure or boiling liquid, expanding vapor explosion (BLEVE) conditions exist, the risk to fire brigade members might be too high and the possibility of saving anyone inside might be too low to justify sending a search-and-rescue team into the building/exterior process area (▲ Figure 13-2). A similar decision might be

## Brigade Member Safety Tips

### Search-and-Rescue Size-Up Considerations

- Occupancy
- Size of building
- Construction of building
- Time of day and day of week
- Number of occupants
- Degree of risk to the occupants presented by the fire
- Ability of occupants to exit on their own

made if the fire is in an abandoned building or a building in danger of structural collapse. In other situations, it might not be possible to conduct a search until the fire has been extinguished and/or other hazards have been controlled.

### Occupancy Factors

When a building is known or believed to be occupied, brigade members should first rescue the occupants who are in the most immediate danger, followed by those who are in less danger. Search teams should be assigned on the basis of these priorities.

Several factors determine the level of risk faced by the occupants of a burning building. These include the location of the fire within the building, the direction of spread, the volume and intensity of the fire, and smoke conditions in different areas.

In establishing search priorities, brigade members should also consider where occupants are most likely to be located. Occupants who are close to the fire, above the fire, or in the path of spread are usually at greater risk than those who are farther away from the fire. A person in the immediate area where the fire started is probably in immediate danger, whereas the occupants of a lower floor or a remote area of a large building or facility are relatively safe.

Workers who are at windows or other elevated areas calling for help obviously realize they are in danger and want to be rescued. But there may be other employees within the facility or area who cannot be immediately seen. These people could be unconscious, incapacitated, or trapped. A search has to be conducted to locate them before they can be rescued.

The occupancies of industrial facilities can very greatly. During a day shift, large facilities may have an occupancy of hundreds of production and support personnel. If the facility is running multiple production shifts, there may be large populations on each shift. During the nonproduction shifts, a small group of maintenance and janitorial personnel may inhabit the facility. During turnaround or retooling periods, the facility populations may even exceed the normal levels.

Another type of occupancy factor that must be considered within industrial facilities is the type and quantity of hazards within the facility. Refineries and chemical plants will contain a vast array of petrochemicals in extremely large quantities. A power generation facility will contain high-voltage distribution systems as well as hazards associated with the type of fuel used (coal dust, fuel oil, nuclear, etc.). Steel mills and foundries will contain molten metal and heavy machinery hazards. Other industrial types of industrial facilities will house other unique hazards. These hazards must be considered during the search-and-rescue size-up process.

### Observations

Brigade members should never assume that a building is completely unoccupied. An observant brigade member will notice clues that indicate whether or not a building is occupied and how many people are likely to be present. The initial size-up at an incident can provide valuable information on the probability of finding occupants, the number of occupants, and the most likely location of any occupants.

Unlike many municipal fire service agencies, facility fire brigade members have the requirement of having a good working knowledge of the facility in which they work. The brigade members will typically have an understanding of what portions of the facility are occupied due to their routine work activities. They are more than likely very aware of the areas of the facility that are populated during any given time. They may not only know what areas are populated, but may also know the general areas where the occupants may be located based on routine work habits.

Much like search-and-rescue operations in the municipal sector, key observations at the incident site can also give more information regarding the potential for victims to be present in the area (► Figure 13-3). Vehicles adjacent to a building or plant area may indicate a potential for victims. Personal effects such as lunch boxes, hard hats, coats, tool belts, and so forth could indicate that personnel could be in the area. Other observations that could indicate potential victims within the area include operating process equipment (which is not normally left unattended), unsecured facilities (which are normally secured when unoccupied), and paperwork (work permits, loading papers, etc.) that indicates work was under way in the area.

### Occupant Information

Occupants who have already escaped and workers who know the other personnel in the work area can provide valuable information about how many people might still be inside the facility and where they might be. Although this information may be valuable, it is not always accurate. Employees standing outside a burning building or process area is a good sign, but this does not necessarily mean that everyone is safe. An important question to ask is "Are you sure that no one else is left inside the area?"

**Figure 13-3** Exterior observations can often provide a good indication if a building is occupied. **A.** A single vehicle parked outside the building may indicate minimal occupancy. **B.** Industries may retire unused facilities and identify their out-of-service status.

It is often difficult to obtain accurate information from people who have just escaped from a burning industrial facility, especially if they believe a coworker or close friend may still be inside. People may be too emotional too speak or even think clearly. If someone says they think workers are still inside, brigade members should obtain as much information as possible. They should ask specific questions such as "Where was the employee last seen?" and "What was the employee doing at the time of the emergency?"

Other important sources of information available to facility fire brigade members include a knowledge of the facility evacuation plans and an understanding of the area worker accountability process. Almost all industrial facilities have emergency evacuation plans with predesignated evacuation sites. These sites must be checked during emergency situations to help account for any personnel. Brigade members must also use the facility accountability systems (area sign-in sheets, accountability/ID tags, etc.) to help verify potential missing personnel. These resources actually provide facility brigade members an advantage over their municipal counterparts, who routinely respond to residential fires with no idea of how many personnel occupy the building.

### Building Size and Arrangement

The size and arrangement of the building are other important factors to consider in planning and conducting a search. A large building, with many rooms, must be searched on a systematic basis. Access to an interior layout or floor plan is often helpful when planning and assigning teams to search a building. To ensure that each area is searched promptly and that no areas are omitted, there must be a systematic way of assigning specific areas to different teams. Teams must report when they have completed searching each area. Assignments are often based on the stairway locations, corridor arrangements, building segments (areas of a building partitioned by firewalls and/or grouped by work areas), or room numbering systems. Because it is difficult to determine this information in the midst of an emergency, fire brigades often conduct preincident surveys of facilities. Fire brigades assemble detailed information about the facilities during the survey and use it to prepare preplans so they are ready when an emergency occurs.

Preincident plans for industrial facilities can include valuable information such as:

- Corridor layouts
- Exit locations
- Stairway locations
- Process/manufacturing equipment locations
- Special function rooms or areas
- Locations of specialized hazards
- Evacuation site locations
- Safe shelter/places of refuge locations
- Typical population of areas by shift

Brigade members should note how the floors of a building are numbered. Buildings constructed on a slope may appear to have a different number of levels when viewed from different sides. For example, a fire may appear to be on the third floor to a brigade member standing in front of the building, while a brigade member at the rear of the building might report that it is on the fifth floor. Without knowing how floors are numbered, brigade members can be sent to work above a fire instead of below it.

## Outdoor Process Area Size and Arrangement

Industrial facilities such as refineries, chemical processing plants, and similar large-scale operations have large outdoor processing areas. These areas include a maze of different types of processing equipment such as tanks, pressure vessels, reactors, pumps, and similar equipment. The equipment may be located at ground level or on elevated structures that may be accessed by regular stairwells or vertical ladder cages. Confined-space areas and buildings may be dispersed throughout the outdoor process areas dependent on the facility.

Fire brigades conduct a preincident survey of the outdoor process area very similar to the surveys conducted for a building. The information gathered during the survey is used to develop a preplan (or plot plan) of the processing area. Preplans of outdoor processing areas typically include the following information:

- Stairwell/vertical ladder location
- Process/manufacturing equipment locations
- Locations of specialized hazards
- Building locations within process areas
- Heights of process equipment
- Evacuation site locations
- Safe shelter/places of refuge locations
- Typical population of areas by shift

## Search Coordination

The overall plan for the incident must focus on the life-safety priority as long as search-and-rescue operations are still underway. As soon as all searches are complete, the priority can shift to controlling the incident, whether that be extinguishing the fire or controlling other hazardous situations.

The IC makes search assignments and serves as the search coordinator. As the teams complete their search of each area, they must notify the IC of the results. An "all clear" report indicates that an area has been searched and all victims have been removed.

If brigade members who have been assigned to perform some other task discover a victim or come upon a critical rescue situation, they must notify the IC immediately. Because life safety is always the highest priority, the IC will adjust the overall plan to support the rescue situation. For example, another group might be assigned to perform the first team's task and to assist with rescue.

Another aspect of search coordination is keeping track of everyone who was rescued or escaped without assistance. This information should be tracked at the command post so that reports of missing persons can be matched to reports of rescued persons. Brigade members should also conduct an exterior search for any missing people. An employee who escaped from the emergency location may be lying unconscious on the ground nearby or in the care of coworkers or may have reported to an area evacuation site. Someone who jumped from an elevated location could be injured and unable to move.

During emergencies at industrial facilities, it is extremely important for the IC to coordinate the search operation in unison with the emergency evacuation plans that are implemented by facility employees. The IC must understand the locations of the evacuation sites and safe shelters for the affected areas and interface with the personnel conducting accountability operations at those sites. Being able to obtain a rapid status of the area evacuation and the possible number and location of missing personnel can greatly improve the efficiency of search operations. The coordination between fire brigade search operations and employee evacuation operations cannot be emphasized enough and needs to be routinely practiced to make the communications second nature.

## Search Priorities

A search begins with the areas where victims are at the greatest risk. One or two well-trained search teams can usually go through all rooms in a smaller industrial building (the size of a typical single-family dwelling) in less than five minutes. Multiple teams and a systematic division of the building are needed in larger structures such as warehouses and manufacturing areas. Outdoor processing areas will also require large-scale search resources as well as the potential of specialized personnel and equipment for rescue operations. Area search assignments should be based on a system of priorities:

- The first priority is to search the area immediately around the fire, then the rest of the fire floor.
- The second priority is to search the area directly above the fire and the rest of that floor.
- The next priority is higher level floors, working from the top floor down, because smoke and heat are likely to accumulate in these areas.
- Generally, areas below the fire floor are a lower priority.

At an emergency involving an elevated structure or building, the IC might assign two or more search-and-rescue teams to each floor. The teams must work together closely and coordinate their searches to ensure that all areas are covered. For example, one team might search all areas/rooms on the right side of the corridor, while the other searches areas/rooms on the left side.

For emergencies involving hazardous materials or other highly hazard situations or large-scale search areas, initial search priorities may need to be reevaluated. Initial searches may need to be conducted from the fringe areas of the hot zone, and search-and-rescue personnel may only be requested to enter the area if victims can be confirmed from the fringe area search. This type of search prioritization provides greater safety for brigade members and a more rapid coverage of wide areas.

# Search Techniques

Brigade members should employ standard techniques to search assigned areas quickly, efficiently, and safely. Searchers

Figure 13-4 Search teams consist of at least two members.

must always operate in teams of two and should always stay together ▲ Figure 13-4. The partners must remain in direct visual, voice, or physical contact with each other.

At least one member of each search team must also have a radio to maintain contact with the command post or someone outside the building. The team uses the radio to call for help if they become disoriented, trapped by fire, or need assistance. If the search team finds a victim, they must notify the IC so that help is available to remove the victim from the building and provide medical treatment. The team must also notify the IC when the search of each area has been completed so the IC can make informed decisions. A backup team is required for these searches.

Two types of searches are performed in buildings. A primary search is a quick attempt to locate any potential victims who are in danger. The primary search should be as thorough as time permits and should cover any place where victims are likely to be found. Fire conditions might make it impossible to conduct a primary search in some areas or may limit the time available for an exhaustive search.

A secondary search is conducted after the situation is under control. During this follow-up search, brigade members should take the time to look everywhere and ensure that no one is missing. If possible, the secondary search should be conducted by a different team, so that each area of the building is examined with a fresh set of eyes.

## Primary Search

During the primary search, brigade members rapidly search the accessible areas of a burning structure to locate any potential victims. The objective of the primary search is to find any potential victims as quickly as possible and remove them from danger. When brigade members complete the primary search, they have gone as far as they could and have removed anyone that could be rescued. The phrase "primary all clear" is used to report that the primary search has been completed.

### Brigade Member Tips

Primary search teams are usually the first brigade members to enter a burning building. As they search for possible victims, team members will often obtain valuable information about the location and spread of the fire. This reconnaissance information should be communicated to the IC to help in managing the overall operation.

By necessity, the primary search is conducted quickly and gives priority to the areas where victims are most likely to be located. Time is always a critical concern, because brigade members must reach potential victims before they are burned, overcome by smoke and toxic gases, or trapped by a structural collapse. Search teams may have only a few minutes to conduct a primary search. In that limited time, brigade members must try to find anyone who could be in danger and remove them to a safe area. Often, active fire conditions may limit the areas that can be searched quickly as well as the time that can be spent in each area.

Fire brigade members should try to check all areas where victims might be, such as areas where workers are normally assigned during specific work hours, areas where they were last seen by coworkers, or areas where they were assigned as outlined in the facility accountability process. Personnel who try to escape on their own are often found near doors, windows, stairwells, and other means of egress. In some cases, employees may seek places of refuge such as office/control rooms, levels of buildings/process areas above or below the actual fire area, or other areas that provide relief from heat or smoke conditions.

The primary search is frequently conducted in conditions that expose both brigade members and victims to the risks presented by heavy smoke, heat, structural collapse, and entrapment by the fire. Search teams must often work in conditions of zero visibility and may have to crawl along the floor to stay below layers of hot gases. Because the beam from a powerful hand light might be visible for only a few inches inside a smoke-filled building, search-and-rescue teams must practice searching for victims in total darkness. They must know how to keep track of their location and how to get back to their entry point. Practicing these skills in a controlled environment will enable brigade members to perform confidently under similar conditions at an emergency incident.

Brigade members must rely on their senses when they search a building. The three most important senses during a search are:

- Sight—Can you see anything?
- Sound—Can you hear someone calling for help, moaning, or groaning?
- Touch—Do you feel a victim's body?

Figure 13-5 Use your tools to extend your reach and sweep the area in front of you.

If smoke and fire restrict visibility, searchers must use sound and touch to find victims. Every few seconds a member of the search team should yell out, "Is anyone in here? Can anyone hear me?" Then listen. Hold your breath to quiet your breathing regulator and stop moving. People who have suffered burns or are semi-conscious may not be able to speak intelligibly, so you will need to listen for guttural sounds, cries, faint voices, groans, and moans. Focus on the direction of any sounds you hear.

As you search, feel around for hands, legs, arms, and torsos of potential victims. Use hand tools to extend your reach, sweeping ahead and to the sides with the handle as you crawl. Use the tool to reach toward the center of the room as you move along the walls (▲ Figure 13-5). Practice until you can tell the difference between a piece of furniture or industrial equipment and a human body. In many industrial environments, the presence of "softer" objects such as furniture, clothing, and so forth is very limited, thus making the determination of a victim within a hostile environment much easier.

Zero-visibility conditions can be very disorienting. Searchers must follow walls and note turns and doorways to avoid becoming lost. After locating a victim or completing a search of an area, the search team must be able to retrace its path and return to its entry point. Search teams must also have a secondary escape route in case fire conditions change and block their planned exit. During the search, brigade members should note the locations of stairways, doors, and windows. Searchers should always be aware of the nearest exit and an alternate exit.

If the situation deteriorates rapidly, a window could become the emergency exit from a room. Searchers should keep track of the exterior walls and window locations, even reaching up to feel for the windows if conditions force them to crawl. In industrial areas with multiple levels with open hand rail systems, open stairwells, and ladder cage systems, fire brigade members must be extremely careful to note their position during heavy smoke conditions in regard to the openings and elevated areas. Disorientation or careless movements in such unguarded areas can be disastrous.

Searchers must remain in contact with someone on the outside and must be able to state their location at all times, particularly if they need assistance. The search team must be able to give their location, including the building section and floor, to the IC. If the team needs a ladder for the rescue, the IC will need to know where the ladder should be placed (which side of the building) and how long a ladder will be needed (which floor). Searchers should use standard incident management system terminology to describe their location over the radio. In process equipment areas, fire brigade members must have a basic understanding of the process equipment in the area. In these large open-space areas, an understanding of the facility layout and/or plot plan will be essential to a successful search process. Frequent review of facility layouts and changes to the facility will provide a great advantage to fire brigades.

## Search Patterns

Each room should be searched using a standard pattern. If the rooms are relatively small, searchers should follow the walls around the perimeter of each room and reach toward the middle to feel for victims. In larger rooms, one team member can maintain contact with the wall, while the other maintains contact with the first, and the two work their way around the room in tandem.

Some fire brigades use a clockwise pattern to search a room while others use a counterclockwise pattern. In a clockwise search (also known as a left-hand search), brigade members turn left at the entry point, keep the left hand in contact with the wall, and use the right arm to sweep the room. At each corner the searchers make a right turn, eventually returning to the entry door (► Figure 13-6). A counterclockwise search moves around the room in the opposite direction (► Figure 13-7). Brigade members should regularly practice and use the standard system adopted by their brigade.

Brigade members can use the handle of a hand tool to extend their reach while sweeping across the floor and under furniture. Searchers should always check under desks, in chairs, and in other locations where victims may hide, such as closets, bathrooms, and enclosed stairwells.

The same clockwise/counterclockwise search pattern also applies to search areas that are divided into small spaces, such as office cubicle areas or warehouse/maintenance areas. Searchers should work their way around the area from partition to partition or aisle to aisle in a standard direction, searching each area in the same manner. At the end of the search, they should be back at the entry point.

Figure 13-6 To use a clockwise search pattern, turn left when you enter, then right at each corner around the room.

Figure 13-7 In a counterclockwise search pattern, turn right when you enter, then left at each corner around the room.

If a door is closed, brigade members should check the temperature of the door to determine if there is an active fire on the other side. Brigade members should follow their brigade standard procedures for performing such a check. Some organizations permit a brigade member to remove a glove just long enough to feel the temperature of a door. Other organizations consider this to be a risky procedure. A thermal imaging device, which shows heat images instead of visual images, can be used to determine if there is a fire behind the door. A hot door should not be opened unless there is a hose line ready to douse the fire. It is usually better to leave a hot door closed and move on to search adjacent rooms.

Figure 13-8 In large buildings, doors should be marked after the rooms are searched.

Searchers must keep track of their position in relation to the entry door so they can find their way out of the room. To keep the door open during the search of a small room, one team member should remain at the door while the second team member performs the search. A chock can be used to keep the door from closing or a flashlight can be placed in the doorway to serve as a beacon and to help searchers maintain their orientation. Searchers should always exit through the same door that they entered.

The search room or process area should be marked so that other personnel will know that it has been searched. Chalk, crayons, felt tip markers, or masking tape can be used to mark the door. Some fire brigades use a two-part marking system to indicate when a search is in progress and when it has been completed. A "/" indicates a search in progress, and an "X" indicates that the search has been completed (▲ Figure 13-8). Other brigades place an object in the doorway or attach a tag or latch strap to the doorknob to indicate that the room has been searched.

## Thermal Imaging Devices

A thermal imaging device is a valuable tool for conducting a primary search in a smoke-filled building. A thermal imaging device is similar to a television camera, except that it captures

## Brigade Member Tips

Learn the features of your thermal imager and practice with it regularly to get the maximum advantage.

heat images instead of visible light images. The images appear on a display screen and show the relative temperatures of different objects. The device can be set to distinguish small temperature differences, enabling brigade members to conduct a search quickly and thoroughly.

The major benefit of using a thermal imaging device is that it can "see" the image of a person in conditions of total darkness or through smoke that totally obscures normal vision (▼ Figure 13-9). Brigade members will be able to identify the shape of a human body with a thermal image scan because the body will be either warmer or cooler than its surroundings.

Temperature differences also enable the thermal imaging device to show furniture, walls, doorways, and windows. This enables a brigade member to navigate through the interior of a smoke-filled building. The thermal imaging device can even be used to locate a fire in a smoke-filled building or behind walls or ceilings (► Figure 13-10). A thermal image scan of the exterior of a building can be used to locate the fire source and the direction of fire spread. Scanning a door before opening it can indicate whether the room is safe to enter.

There are several different types of thermal imaging devices. Some thermal imagers are handheld devices, similar to a handheld computer with a built-in monitor. Others are helmet-mounted and produce an image in front of the user's self-contained breathing apparatus (SCBA) facepiece. Brigade members need training and practice to become proficient in using these devices, interpreting the images, and maneuvering confidently through a building while wearing a thermal imager. The use of a thermal imaging device in an industrial facility takes additional training and interpretation of the camera's images since there can be many heat sources within the facility. Brigade members must understand both the camera's capabilities and the type of reaction the camera will have to the different type of heat sources within their facility.

Figure 13-10 A thermal imaging device can be used to check for hidden fire.

Figure 13-11 The use of search ropes will be essential when searching large open-area environments such as warehouses.

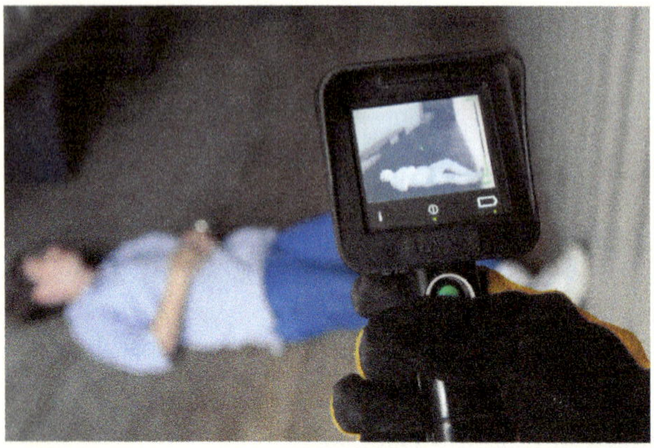

Figure 13-9 The thermal imaging device can capture an image of a person and hot processing equipment in the same area.

### Search Ropes

Special techniques must be used to search large, open areas, such as warehouses or indoor production areas, when visibility is limited (▲ Figure 13-11). Search ropes should be used when it is impossible to cover the interior by following the walls. They also should be used in areas with interconnected rooms or spaces, or with multiple aisles created by production equipment or storage racks. Without a search rope, the search teams might not be able to find their way out of the area.

Search ropes should be anchored at the entry point to provide a direct path to this location for each searcher. One brigade member should remain at the anchor point to tend the ropes and maintain accountability for the brigade members who enter. One brigade member stretches a large diameter rope down the center of a room. The other searchers then attach their individual search ropes to the main line and branch off to cover the area. The search ropes always provide a reliable return path to the entry point. If the brigade member assigned to the anchor point of a search rope is part of the rapid intervention team (two-in/two-out rule), that individual cannot be utilized as a member of the interior search team.

### Secondary Search

A secondary search is conducted after the fire is under control or fully extinguished. During the secondary search, brigade members locate any victims that might have been missed in the primary search and search any areas not included in the primary search. Conditions in the building are usually better during the secondary search because the fire is under control and there is adequate ventilation. Brigade members will be able to see better and move around without having to crawl under layers of hot gases (▼ Figure 13-12). After a secondary search is completed, brigade members should notify the IC by reporting, "Secondary all clear."

Even though the fire is under control, safety remains a primary consideration during a secondary search. SCBA must be used until the air is tested and pronounced safe to breathe. The levels of carbon monoxide and other poisonous gases often remain high for a long time after a fire. In burned areas, brigade members must have a hose line available in case the fire **rekindles** or starts up again. The structural stability of the building also must be evaluated before beginning a secondary search. Portable lighting should be used, and holes in floors and other possible hazards should be marked.

Figure 13-12 A secondary search is conducted after the fire is controlled.

The secondary search should be started as soon as the fire is under control and sufficient resources are available. It is conducted slowly and methodically to ensure that no areas are overlooked. Whenever possible, a different team of brigade members should perform the secondary search. They should check all places where employees may try to exit the fire area and/or seek refuge from the fire, including closets, behind doors, in hallways, under windows, and under desks or work benches.

The secondary search must include areas that were involved in the fire. In these areas, the secondary search should be conducted as a body recovery process. Brigade members should move carefully and look closely at the debris to identify the remains of a deceased victim.

## Search Safety

Search-and-rescue operations often present the highest risk to brigade members during emergency situations. During these operations, brigade members are exposed to the same risks that endanger the lives of potential victims. Even though brigade members have the advantages of protective clothing, protective equipment, training, teamwork, and standard operating procedures, they can still be seriously or fatally injured. Safety must be an essential consideration in all search-and-rescue operations.

### Risk Management

Search-and-rescue situations require a very special type of risk management. Every emergency operation involves a degree of unavoidable, inherent risk to brigade members. During search-and-rescue operations, brigade members will probably encounter situations that involve a significantly higher degree of personal risk. This level of risk is acceptable only if there is a reasonable probability of saving a life.

The IC is responsible for managing the level of risk during emergency operations. The IC must determine which actions are taken in each situation. Actions that present a high level of risk to the safety of brigade members are justified only if lives can be saved. Only a limited risk level is acceptable to save property. When there is no possibility of saving either lives or property, no risk is acceptable.

The acceptance of unusual risk can be justified only when victims are known or believed to be in immediate danger, and there is a reasonable probability that they can be rescued. It is not acceptable to risk the lives of brigade members when there is no possibility of rescuing any victims. For example, if a building is fully involved in flames, there is no possibility that any occupants would still be alive, so there is no reason to send brigade members inside to search for them. The risk involved in conducting search-and-rescue operations must always be weighed against the probability of finding someone who is still alive to be rescued.

A primary search is conducted during the active stages of a fire, when time is a critical factor. During a primary search, it may be necessary to accept a high level of risk to save a life. A secondary search, conducted after the fire is extinguished, should not expose brigade members to any avoidable risks.

The IC must decide whether to conduct a primary search, based on a risk–benefit evaluation of the situation. In making this decision, the IC must consider the stage of the fire, the condition of the building, and the presence of any other hazards. These risks must be weighed against the probability of finding any occupants who could be rescued. This on-scene evaluation is often guided by established policies. For example, many fire brigade have policies that prohibit brigade members from entering abandoned structures known to be in poor structural condition. This policy is based on the probability that such a building has a high risk of collapse and the low probability that anyone would be inside.

The IC may decide not to conduct a primary search because the risk to brigade members is too great or the possibility of making a successful rescue is too remote. Such a decision might be based on advanced fire conditions, potential backdraft conditions, imminent structural collapse, or other circumstances ▶ Figure 13-13. The IC may be able to identify these conditions from the exterior or may learn of them from a team assigned to conduct a search. A search team that encounters conditions that make entry impossible should report their findings back to the IC. In these situations, a secondary search is conducted when it is feasible to enter.

Figure 13-13 A primary search should not be conducted in a process area that is fully involved in fire.

## Search-and-Rescue Equipment

To perform search and rescue properly, brigade members must have the appropriate equipment ▶ Figure 13-14. Search-and-rescue equipment includes:

- Personal protective equipment
- Portable radio
- Hand light or flashlight
- Forcible entry tools
- Hose lines
- Thermal imaging devices
- Ladders
- Long rope(s)
- A piece of tubular webbing or short rope (16–24 feet)
- Multigas analyzer
- Rapid Intervention Team (RIT) breathing apparatus system
- Stokes basket
- Emergency rescue drag harness or rescue drag tarp

Figure 13-14 Fire brigade members must be thoroughly trained in the use of search-and-rescue equipment.

## Brigade Member Safety Tips

During search-and-rescue operations, brigade members should remember to:

- Work from a single plan.
- Maintain radio contact with the IC, through the chain of command and portable radios.
- Monitor fire conditions during the search.
- Coordinate ventilation with search-and-rescue activities.
- Maintain your accountability system.
- Stay with a partner.

## Brigade Member Tips

### Risk-Benefit Analysis

The IC must always balance the risks involved in an emergency operation with the potential benefits.

- Actions that present a high level of risk to the safety of brigade members are justified only if there is a potential to save lives.
- Only a limited level of risk is acceptable to save valuable property.
- It is not acceptable to risk the safety of brigade members when there is no possibility to save lives or property.

### NFPA 1500 Requirements

Specific safety requirements for search-and-rescue operations are defined in NFPA 1500, *Standard on Fire Department Occupational Safety and Health Program,* and in regulations enforced by the Occupational Safety and Health Administration 29CFR1910.134. The NFPA requirements state that a team of at least two brigade members must enter together, and at least two other brigade members must remain outside the danger area, ready to rescue the brigade members who are inside the building. This is sometimes called the **two-in/two-out rule**.

An exception to this rule is permitted in an imminent life-threatening situation, where immediate action can prevent the loss of life or serious injury. Only under such specific circumstances are brigade members allowed to take actions at a higher level of risk. The initial IC must evaluate the situation and determine if the risk is justified. The IC also must be prepared to explain his or her decision.

Full personal protective equipment, which is always required for structural fire fighting, is essential for interior search and rescue. The proper attire includes helmet, protective hood, bunker coat, turnout pants, boots, and gloves. Each brigade member must use SCBA and carry a flashlight or hand light. Before entering the building, brigade members must activate their personal alert safety system (PASS) devices.

At least one member of each search team should be equipped with a portable radio. If possible, each individual should have a radio. If a brigade member gets into trouble, the radio is the best means to obtain assistance.

Each brigade member assigned to search and rescue should carry at least one forcible entry hand tool, such as an ax, Halligan tool, or short ceiling hook. These tools can be used both to open an area for a search and, if necessary, to open an emergency exit path. A hand tool can also be used to extend the brigade member's reach during a sweep for unconscious victims. Due to the high potential for hazardous materials within industrial occupancies, many fire brigades require search teams to carry a multigas analyzer to constantly monitor the level of contaminants in the air. Another very important piece of rescue equipment in the industrial environment is a RIT breathing apparatus system. This type of system can be used to deliver breathing air to a victim who has no respiratory protection or to assist a person who is wearing respiratory protection but is low on air.

A search-and-rescue team working close to the fire should have a hose line or be accompanied by another team with a hose line. A hose line can protect the brigade members and enable them to search a structure more efficiently. A hose line is essential when a search team is working close to or directly above the fire. The hose line can be used to knock down the fire, to protect a means of egress (stairway or corridor), or to protect the victims as they are escaping.

A hose line also can be used to guide searchers out of a structure. Searchers can follow the hose line to a coupling, feel the coupling to determine the male and female ends, and then follow the hose in the direction of the male coupling. If the team is working without a hose line, it should have search ropes, particularly if it would be difficult for brigade members to find their way out of a search area. More information on emergency exit procedures is presented in Chapter 18, Brigade Member Survival.

Brigade members also must pay attention to their air supplies during search-and-rescue operations. Brigade members must have adequate air to make a safe exit. This limits the time that can be spent searching and the distance that brigade members can penetrate into a building.

## Methods to Determine if an Area Is Tenable

Brigade members must make rapid, accurate, and ongoing assessments about the safety of the building while working at a structure fire. When they arrive, brigade members must quickly determine the type of structure involved, the possibility of collapse, and the life safety risk involved. They also must evaluate the stability of the structure and the potential for backdraft or flashover. Sagging walls, chipped or cracked mortar or cement, warped or failing structural steel, and any partial collapse are significant indicators of an impending collapse. Dark black smoke, blackened windows, the appearance of a "breathing" building, and intense heat are signs of possible flashover or backdraft. Fire brigade members must also be able to evaluate other hazards associated with process-area fire search-and-rescue operations. This could include hazards such as the release of hazardous materials, potential BLEVE of pressurized containers, presence of high-voltage systems, and/or the presence of operating machinery. Fire brigade members must be aware of the special hazards common to their industrial sites and be able to quickly evaluate the search locations to determine if the areas can be searched safely.

Even after the decision has been made to enter a burning structure, brigade members must continue to reevaluate the safety of the operation. Brigade members working inside the building must rely on other brigade members and the IC outside the structure to notify them of changing fire conditions. The IC may know or see things that brigade members working inside may not know and may call for an evacuation as the fire situation changes. Conditions inside the structure must constantly be evaluated. Brigade members should check and recheck the surface they are working on. Floors that have been burnt through or feel soft or spongy should be avoided. A rapid rise in the amount of heat or flame "rollover" may indicate a potential flashover.

## Rescue Techniques

Rescue is the removal of a person who is unable to escape from a dangerous situation. Brigade members rescue people not only from fires but also from a wide variety of accidents and mishaps. Although this section refers primarily to rescuing occupants from burning buildings, some of the techniques can be used for other types of situations. As a brigade member, you must learn and practice the assists, carries, drags, and other techniques used to rescue people from fires. After mastering these techniques, you will be able to rescue victims from a life-threatening situation.

Rescue is the second component of search and rescue. When you locate a victim during a search, you must direct, assist, or carry the person to a safe area. Rescue can be as basic and simple as verbally directing an occupant toward an exit. Or it can be as demanding as extricating a trapped, unconscious victim and physically carrying that person out of the building. The term rescue is generally applied to situations where the rescuer physically assists or removes the victim from the dangerous area.

### Brigade Member Safety Tips

To exit from a fire, follow the hose line in the direction of the male coupling.

Most people who realize that they are in a dangerous situation will attempt to escape on their own. Elderly persons, physically or developmentally handicapped persons, and ill or injured persons, however, may be unable to escape and will need to be rescued. People who are sleeping or under the influence of alcohol or drugs may not become aware of the danger in time to escape on their own. Toxic gases may incapacitate even healthy individuals before they can reach an exit. Victims also may be trapped when a rapidly spreading fire, an explosion, or a structural collapse cuts off potential escape routes.

Because fires are life-threatening to both victims and rescuers, the first priority is to remove the victim from the fire building or dangerous area as quickly as possible. It is usually better to move the victim to a safe area first, and then provide any necessary medical treatment. The assists, lifts, and carries described in this chapter should not be used if you suspect that the victim has a spinal injury, unless there is no other way to remove him or her from the life-threatening situation.

Always use the safest and most practical means of egress when removing a victim from a dangerous area. A building's normal exit system, such as interior corridors and stairways, should be used if it is open and safe. If the regular exits cannot be used, an outside fire escape, a ladder, or some other method of egress must be found. Ladder rescues, which are covered in this chapter, can be both difficult and dangerous, whether the victim is conscious and physically fit or unconscious and injured (▶ Figure 13-15).

### Shelter-in-Place

In some situations, the best option is to shelter the occupants in place instead of trying to remove them from a fire building. This option should be considered when the occupants are conscious and in a part of the building that is adequately protected from the fire by fire-resistive construction and/or fire suppression systems. If smoke and fire conditions block the exits, they might be safer staying in the sheltered location than attempting to evacuate through a hazardous environment.

Such a situation could occur in a high-rise office or processing building with a fire contained to one room or area within the building. The stairways and corridors could be filled with smoke, but the employees who are remote from the fire would be very safe in the protected rooms within the facility with the windows open to provide fresh air. They would be exposed to more risk if they attempted to exit than if they remained in their location until the fire is extinguished. This decision must be made by the IC.

### Exit Assist

The simplest rescue is the exit assist. The victim is responsive and able to walk without assistance or with very little assistance. The brigade member may only need to guide the person to safety or to provide a minimal level of

Figure 13-15 Bringing an unconscious victim or a person who needs assistance down a ladder can be difficult and dangerous.

physical support. Even if the victim can walk without assistance, the brigade member should take the person's arm or use the one-person walking assist (see below) to make sure that the victim does not fall or become separated from the rescuer.

The following assists can be used to help responsive victims exit a fire situation:

- One-person walking assist
- Two-person walking assist

### One-Person Walking Assist

The one-person walking assist can be used if the person is capable of walking.

To perform a one-person walking assist, follow the steps in ▶ Skill Drill 13-1.

1. Help the person to stand next to you, facing the same direction. **(Step 1)**
2. Have the person place his or her arm behind your back and around your neck. Hold the person's wrist as it drapes over your shoulder. **(Step 2)**
3. Put your free arm around the person's waist and help the victim to walk. **(Step 3)**

### Two-Person Walking Assist

The two-person walking assist is useful if the victim cannot stand and bear weight without assistance. The two rescuers completely support the victim's weight. It may be difficult to walk through doorways or narrow passages using this type of assist. To perform a two-person walking assist, follow the steps in ▶ Skill Drill 13-2.

1. Two brigade members stand facing the victim, one on each side of the victim. **(Step 1)**
2. Both brigade members assist the victim to a standing position. **(Step 2)**
3. Once the victim is fully upright, place the victim's right arm around the neck of the brigade member on the right side. Place the victim's left arm around the neck of the brigade member on the left side. The victim's arms should drape over the brigade members' shoulders. The brigade members hold the person's wrist in one hand. **(Step 3)**
4. Both brigade members put their free arms around the person's waist, grasping each other's wrists for support and locking arms together behind the victim. **(Step 4)**
5. Both brigade members slowly assist the victim to walk. Brigade members must coordinate their movements and move slowly. **(Step 5)**

## Simple Victim Carries

Four simple carries can be used to move a victim who is conscious and responsive, but incapable of standing or walking:

- Two-person extremity carry
- Two-person seat carry
- Two-person chair carry
- Cradle-in-arms carry

### Two-Person Extremity Carry

The two-person extremity carry requires no equipment and can be performed in tight or narrow spaces, such as mobile home corridors, small hallways, and narrow spaces between buildings. The focus of this carry is on the victim's extremities. To perform a two-person extremity carry, follow the steps in ▶ Skill Drill 13-3.

1. Two brigade members help the victim to sit up. **(Step 1)**
2. One brigade member kneels behind the victim, reaches under the victim's arms, and grasps the victim's wrists. **(Step 2)**
3. The second brigade member backs in-between the victim's legs, reaches around, and grasps the victim behind the knees. **(Step 3)**
4. At the command of the first brigade member, both brigade members stand up and carry the victim away, walking straight ahead. Brigade members must coordinate their movements. **(Step 4)**

### Two-Person Seat Carry

The two-person seat carry is used with victims who are disabled or paralyzed. This carry requires two brigade members, and moving through doors and down stairs may be difficult. To perform a two-person seat carry, follow the steps in ▶ Skill Drill 13-4.

1. Two brigade members kneel near the victim's hips, one on each side of the victim. **(Step 1)**
2. Both brigade members raise the victim to a sitting position and link arms behind the victim's back. **(Step 2)**
3. The brigade members' remaining free arms are then placed under the victim's knees and linked. **(Step 3)**
4. If possible, the victim puts his or her arms around the brigade members' necks for additional support. **(Step 4)**

### Two-Person Chair Carry

The two-person chair carry is particularly suitable when a victim must be carried through doorways, along narrow corridors, or up or down stairs. Two rescuers use a chair to transport the victim. A folding chair cannot be used, and the chair must be strong enough to support the weight of the victim while being carried. The victim should feel much more secure with this carry than with the two-person seat carry. The victim should be encouraged to hold on to the chair. To perform a two-person chair carry, follow ▶ Skill Drill 13-5.

1. Tie the victim's hands together or have the victim grasp hands together.
2. One brigade member stands behind the seated victim, reaches down, and grasps the back of the chair. **(Step 1)**
3. The brigade member tilts the chair slightly backward on its rear legs so that the second brigade member can step back between the legs of the chair and grasp the tips of the chair's front legs. The victim's legs should be between the legs of the chair. **(Step 2)**
4. When both brigade members are correctly positioned, the brigade member behind the chair gives the command to lift and walk away.
5. Because the chair carry may force the victim's head forward, watch the victim for airway problems. **(Step 3)**

13-1 **Skill Drill**

## One-Person Walking Assist

1. Help the person to stand.
2. Have the person place his or her arm around your neck, and hold on to the person's wrist, which should be draped over your shoulder.
3. Put your free arm around the person's waist and help the victim to walk.

## 13-2 Skill Drill

# Two-Person Walking Assist

1. Two brigade members stand facing the victim, one on each side of the victim.

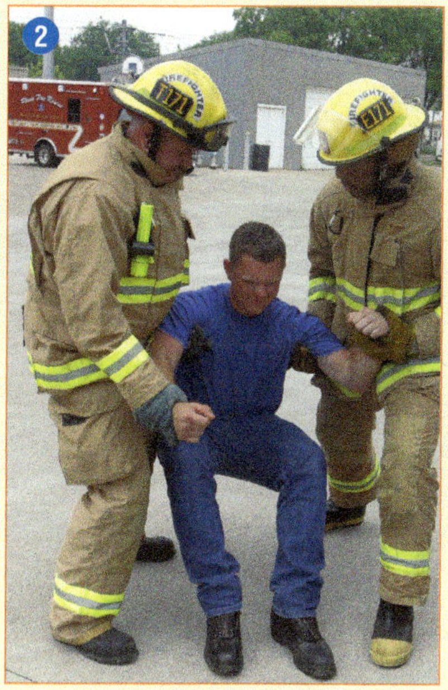

2. The brigade members assist the victim to a standing position.

3. Once the victim is fully upright, drape the victim's arms around the necks and over the shoulders of the brigade members, who each hold one of the victim's wrists.

4. Both brigade members put their free arm around the person's waist, grasping each other's wrists for support and locking their arms together behind the victim.

5. Assist walking at the victim's speed.

13-3 Skill Drill

## Two-Person Extremity Carry

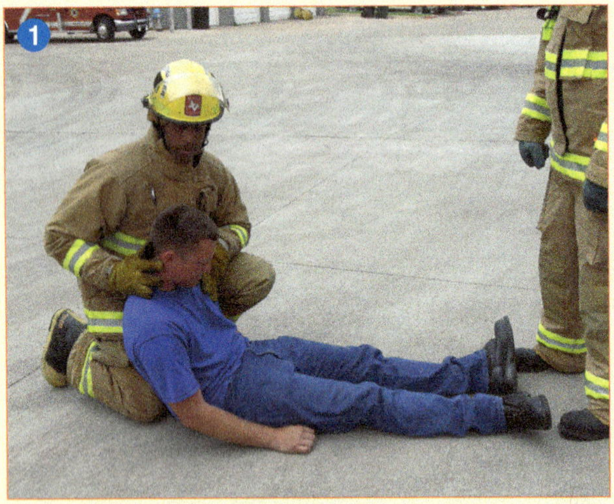

1 Two brigade members help the victim to sit up.

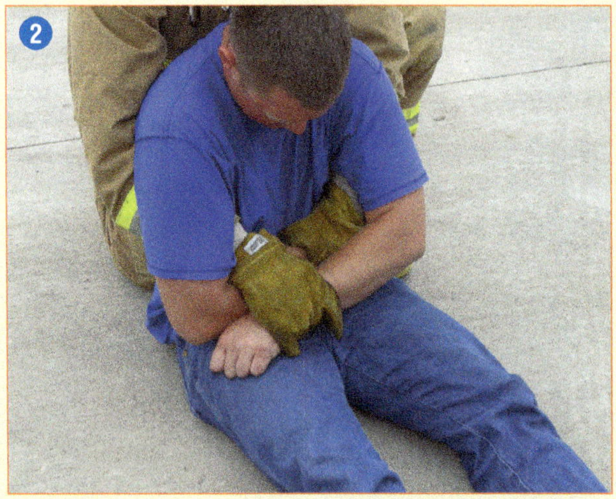

2 The first brigade member kneels behind the victim, reaches under the victim's arms, and grasps the victim's wrists.

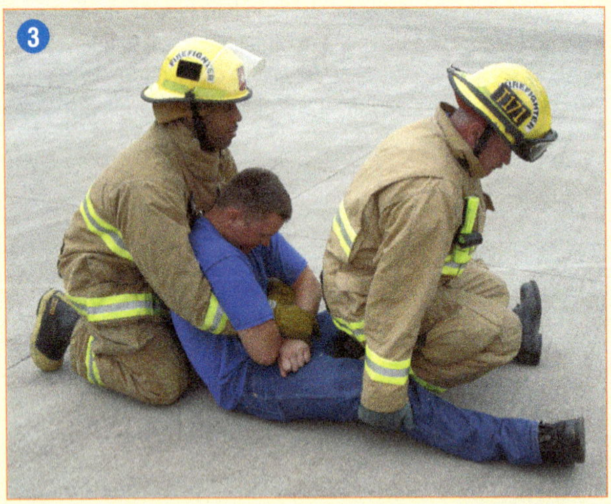

3 The second brigade member backs in-between the victim's legs, reaches around, and grasps the victim behind the knees.

4 The first brigade member gives command to stand and carry the victim away, walking straight ahead. Brigade members must coordinate their movements.

## 13-4 Skill Drill

# Two-Person Seat Carry

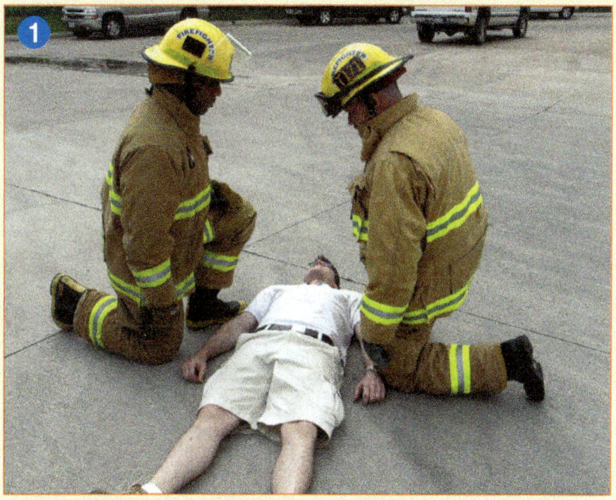

Kneel beside the victim near the victim's hips.

Raise the victim to a sitting position and link arms behind the victim's back.

Place your free arms under the victim's knees and link arms.

If possible, the victim puts his or her arms around your necks for additional support.

13-5 Skill Drill

## Two-Person Chair Carry

One brigade member stands behind the seated victim, reaches down, and grasps the back of the chair.

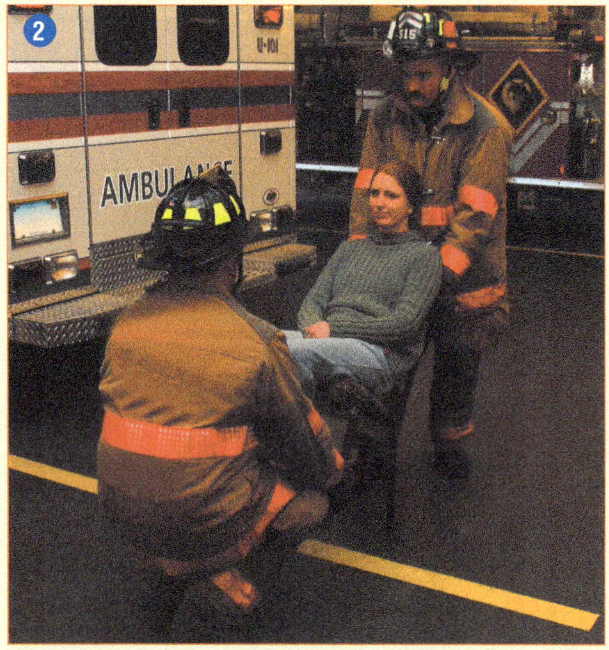


The brigade member tilts the chair slightly backward on its rear legs so that the second brigade member can step back between the legs of the chair and grasp the tips of the chair's front legs. The victim's legs should be between the legs of the chair.

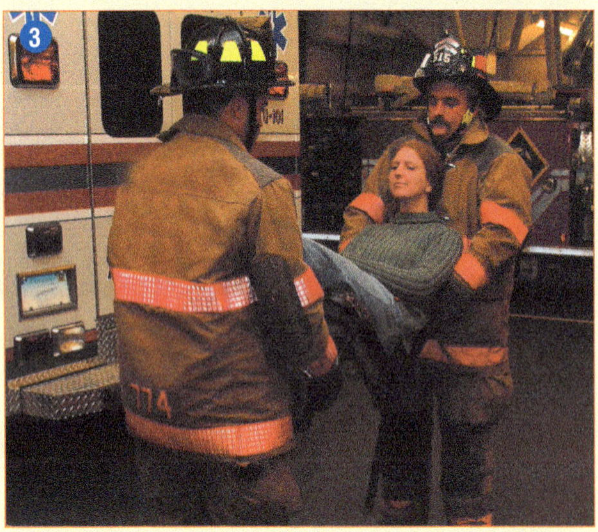


When both brigade members are correctly positioned, the brigade member behind the chair gives the command to lift and walk away. Because the chair carry may force the victim's head forward, watch the victim for airway problems.

## Brigade Member Safety Tips

Keep your back as straight as possible and use the large muscles in your legs to do the lifting!

### Cradle-in-Arms Carry

The cradle-in-arms carry can be used by one brigade member to carry a small adult ▼ Figure 13-16. The brigade member should be careful of the victim's head when moving through doorways or down stairs. To perform the cradle-in-arms carry, follow the steps in Skill Drill 13-6.

1. Kneel beside the child and place one arm around the child's back and the other arm under the thighs.
2. Lift slightly and roll the child into the hollow formed by your arms and chest.
3. Be sure to use your leg muscles to stand.

## Emergency Drags

The most efficient method to remove an unconscious or unresponsive victim from a dangerous location is a drag. Five emergency drags can be used to remove unresponsive victims from a fire situation:

- Clothes drag
- Blanket drag
- Webbing sling drag
- Brigade member drag
- Emergency drag from a vehicle

When using an emergency drag, the rescuer should make every effort to pull the victim in line with the long axis of the body to provide as much spinal protection as possible. The victim should be moved head first to protect the head.

Figure 13-16 Cradle-in-arms carry.

### Clothes Drag

The clothes drag is used to move a victim who is on the floor or ground and is too heavy for one rescuer to lift and carry. The rescuer drags the person by pulling on the clothing in the neck and shoulder area. The rescuer should grasp the clothes just behind the collar, use the arms to support the victim's head, and drag the victim away from danger. To perform the clothes drag, follow the steps in ► Skill Drill 13-7.

1. Crouch behind the victim's head, grab the shirt or jacket around the collar and shoulder area, and support the head with your arms. **(Step 1)**
2. Lift with your legs until you are fully upright. Walk backwards, dragging the victim to safety. **(Step 2)**

### Blanket Drag

The blanket drag can be used to move a victim who is dressed in clothing that is too flimsy for the clothes drag, or it can be used to provide additional protection from the area conditions. This procedure requires the use of a large blanket, curtain, rug, salvage cover, or specialized rescue tarp. Place the item on the floor and roll the victim onto it, then pull the victim to safety by dragging the sheet or blanket.

To perform a blanket drag, follow the steps in ► Skill Drill 13-8.

1. Lay the victim supine (face up) on the ground. Stretch out the material for dragging next to the victim. **(Step 1)**
2. Roll the victim onto the right or left side. Neatly bunch one-third of the material against the victim so the victim will lie approximately in the middle of the material. **(Step 2)**
3. Lay the victim back down (supine) on the material. Pull the bunched material out from underneath the victim and wrap it around the victim. **(Step 3)**
4. Grab the material at the head and drag backwards to safety. **(Step 4)**

The standing drag is performed by following the steps in Skill Drill 13-9:

1. Stand at the head of the supine victim. Then kneel at the victim's head.
2. Raise the victim's head and torso 90°. The victim is leaning against you.
3. Reach under the victim's arms, wrap your arms around the victim's chest, and lock your arms.
4. Stand straight up using your legs.
5. Drag the victim out.

### Webbing Sling Drag

The webbing sling drag provides a secure grip around the upper part of a victim's body, for a faster removal from the dangerous area. In this drag, a sling is placed around the victim's chest and under the armpits, and used to drag the victim. The webbing sling helps support the victim's head and neck. A webbing sling can be rolled and kept in a turnout coat pocket.

13-7 Skill Drill

## Clothes Drag

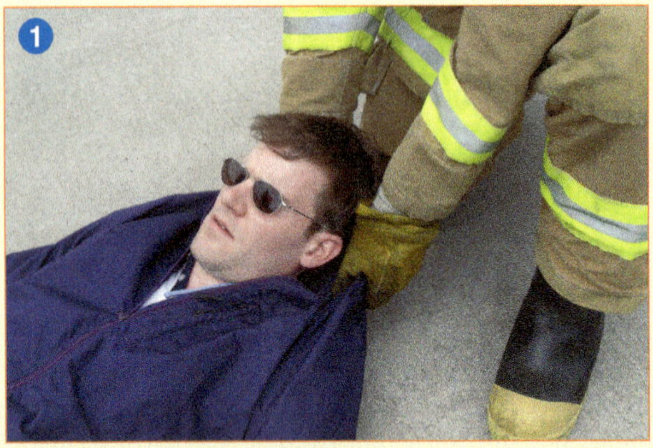
1 Crouch behind the victim's head and grab the shirt or jacket around the collar and shoulder area.

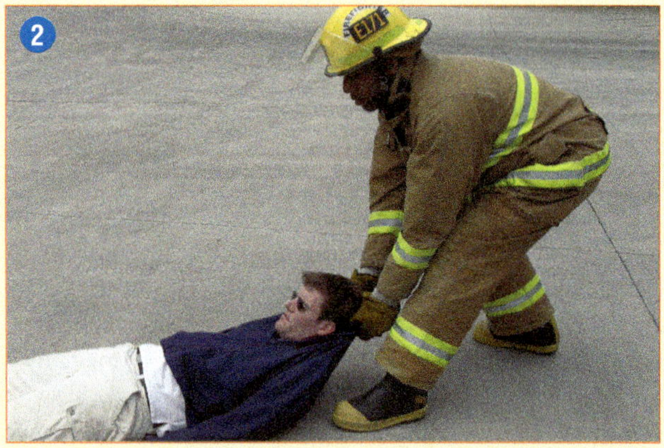

2 Lift with your legs until you are fully upright. Walk backwards, dragging the victim to safety.

A carabiner can be attached to the sling to secure the straps under the victim's arms and provide additional protection for the head and neck.

To perform the webbing sling drag, follow the steps in ▶ Skill Drill 13-10.

1. Using a prepared webbing sling, place the victim in the center of the loop so the webbing is behind the victim's back in the area just below the armpits. **(Step 1)**
2. Take the large loop over the victim and place it above the victim's head. Reach through, grab the webbing behind the victim's back, and pull through all the excess webbing. This creates a loop at the top of the victim's head and two loops around the victim's arms. **(Step 2)**
3. Adjust hand placement to protect the victim's head while dragging. **(Step 3)**

### Brigade Member Drag

The brigade member drag can be used if the victim is heavier than the rescuer because it does not require lifting or carrying the victim. To perform the brigade member drag, follow the steps in ▶ Skill Drill 13-11.

1. Tie the victim's wrists together with anything that is handy: a cravat (a folded triangular bandage), gauze, belt, or necktie. **(Step 1)**
2. Get down on hands and knees and straddle the victim. **(Step 2)**
3. Pass the victim's tied hands around your neck, straighten your arms, and drag the victim across the floor by crawling on your hands and knees. **(Step 3)**

### Emergency Drag from a Vehicle

An emergency drag from a vehicle is performed when the victim must be quickly removed from a vehicle to save his or her life. The drags described below might be used if the vehicle is on fire or if the victim requires cardiopulmonary resuscitation (CPR).

#### One Rescuer

There is no effective way for one person to remove a victim from a vehicle without some movement of the neck and spine. Preventing excess movement of the victim's neck, however, is important. To perform an emergency drag from a vehicle with only one rescuer, follow the steps in ▶ Skill Drill 13-12.

1. Grasp the victim under the arms and cradle the head between your arms. **(Step 1)**
2. Gently pull the victim out of the vehicle. **(Step 2)**
3. Lower the victim down into a horizontal position in a safe place. **(Step 3)**

#### Long Backboard Rescue

If four or more brigade members are present, one brigade member can support the victim's head and neck, while the second and third brigade members move the victim by lifting under the arms. The victim can then be moved in line with

13-8 Skill Drill

## Blanket Drag

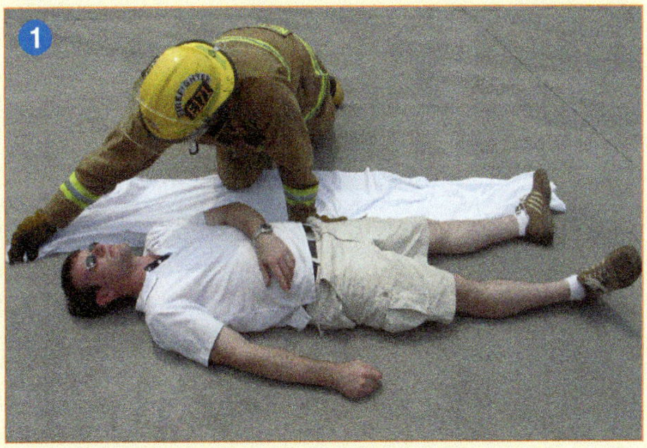

1 Stretch out the material you are using next to the victim.

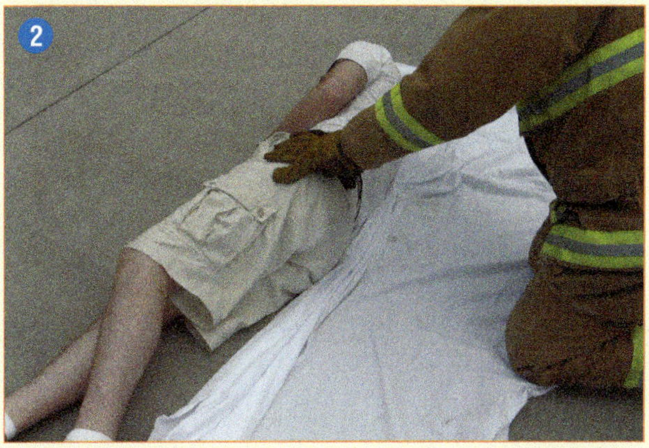

2 Roll the victim onto one side. Neatly bunch one-third of the material against the victim's body.

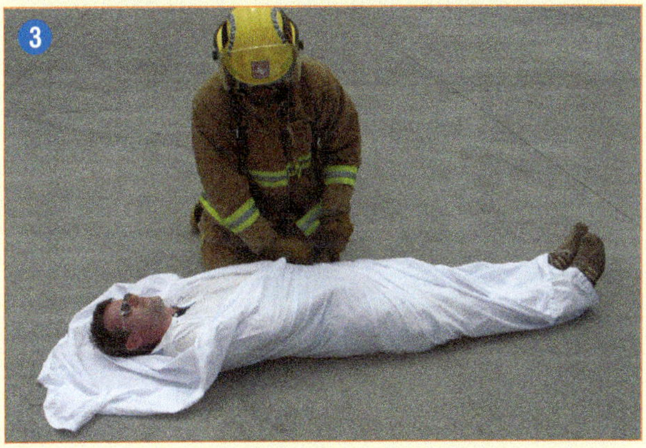

3 Lay the victim back down (supine). Pull the bunched material out from underneath the victim and wrap it around the victim.

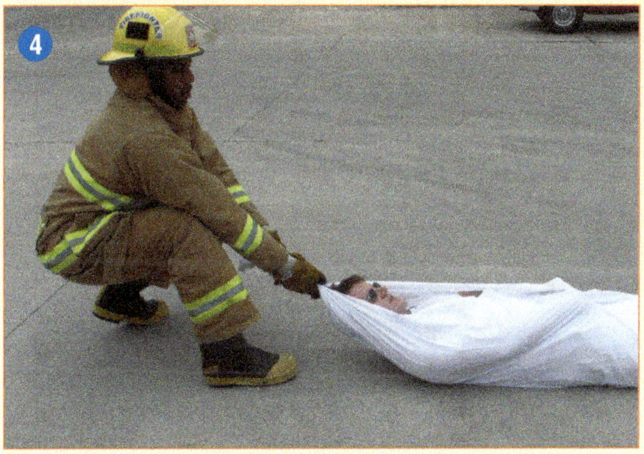

4 Grab the material at the head and drag backwards to safety.

the long axis of the body, with the head and neck stabilized in a neutral position. Whenever possible, a long backboard should be used to remove a victim from the vehicle. Follow the steps in ▶ Skill Drill 13-13 to perform this skill.

1. The first brigade member supports the victim's head and cervical spine from behind. Support may be applied from the side, if necessary, by reaching through the driver's side doorway. **(Step 1)**
2. The second brigade member serves as team leader and, as such, gives the commands until the patient is supine on the backboard. Because the second brigade member lifts and turns the victim's torso, he or she must be physically capable of moving the patient. The second brigade member works from the driver's side doorway. If the first brigade member is also working from that doorway, the second brigade member should stand closer to the door hinges toward the front of the vehicle. The second brigade member applies a cervical collar. **(Step 2)**
3. The second brigade member provides continuous support of the victim's torso until the victim is supine on the backboard. Once the second brigade member takes control of the torso, usually in the form of a body hug, he or she should not let go of the victim for any reason. Some type of cross-chest shoulder hug usually works well, but you will have to decide what method works best for you on any given victim. You must remember that you cannot simply reach into the

Skill Drill 13-10

## Webbing Sling Drag

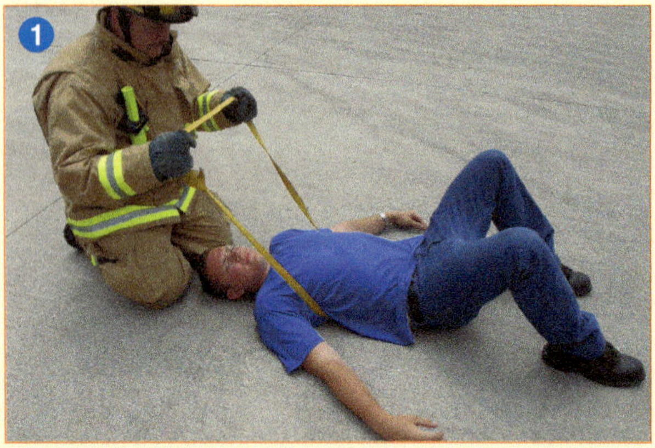

1. Place the victim in the center of the loop so the webbing is behind the patient's back.

2. Take the large loop over the victim and place it above the victim's head. Reach through, grab the webbing behind the victim's back, and pull through all the excess webbing. This creates a loop at the top of the victim's head and two loops around the victim's arms.

3. Adjust hand placement to protect the victim's head while dragging.

car and grab the victim; this will only twist the victim's torso. You must rotate the victim as a unit.

4. The third brigade member works from the front passenger's seat and is responsible for rotating the victim's legs and feet as the torso is turned, ensuring that they are free of the pedals and any other obstruction. With care, the third brigade member should first move the victim's nearer leg laterally without rotating the victim's pelvis and lower spine. The pelvis and lower spine rotate only as the third brigade member moves the second leg during the next step. Moving the nearer leg early makes it much easier to move the second leg in concert with the rest of the body. After the third brigade member moves the legs together, they should be moved as a unit. **(Step 3)**
5. The victim is rotated 90° so that the back is facing out the driver's door and the feet are on the front passenger's seat. This coordinated movement is done in three or four short, quick "eighth turns." The second brigade member directs each quick turn by saying, "Ready, turn" or "Ready, move." Hand position changes should be made between moves.
6. In most cases, the first brigade member will be working from the back seat. At some point, either because the doorpost is in the way or because he or she cannot reach farther from the back seat, the first brigade

13-11 Skill Drill

## Brigade Member Drag

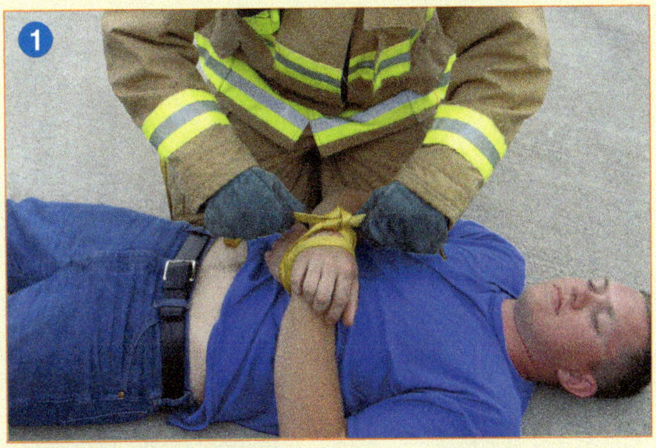

1 Tie the victim's wrists together with anything that is handy.

2 Get down on hands and knees and straddle the victim.

3 Pass the victim's tied hands around your neck, straighten your arms, and drag the victim across the floor by crawling on your hands and knees.

member will be unable to follow the torso rotation. At that time, the third brigade member should assume temporary support of the head and neck until the first brigade member can regain control of the head from outside the vehicle. If a fourth brigade member is present, he or she stands next to the second brigade member. The fourth brigade member takes control of the head and neck from outside the vehicle without involving the third brigade member. As soon as the change has been made, the rotation can continue. **(Step 4)**

7. Once the victim has been fully rotated, the backboard should be placed against the victim's buttocks on the seat. Do not try to wedge the backboard under the victim. If only three brigade members are present, be sure to place the backboard within arm's reach of the driver's door before the move so that the board can be pulled into place when needed. In such cases, the far end of the board can be left on the ground. When a fourth brigade member is available, the first brigade member exits the rear seat of the car, places the backboard against the patient's buttocks, and maintains pressure in toward the vehicle from the far end of the board. (Note: When the door opening allows, some brigade members prefer to insert the backboard onto the car seat before the victim is rotated.)

13-12 Skill Drill

## One-Person Emergency Drag from a Vehicle

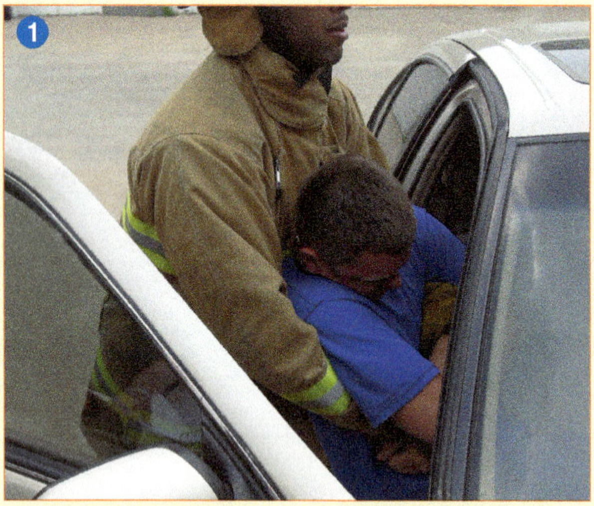

Grasp the victim under the arms and cradle the head between your arms.

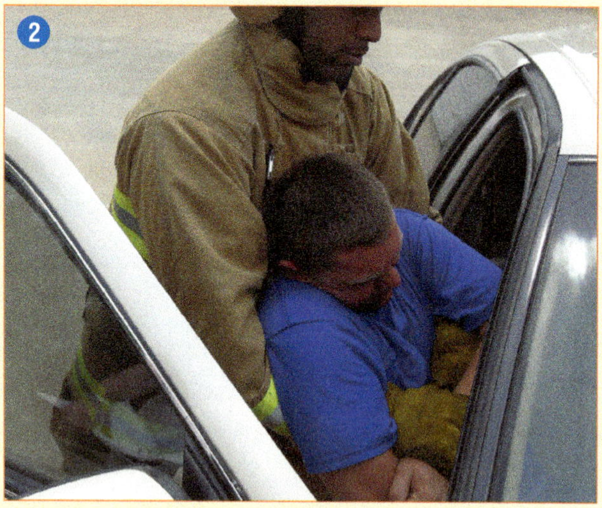

Gently pull the victim out of the vehicle.

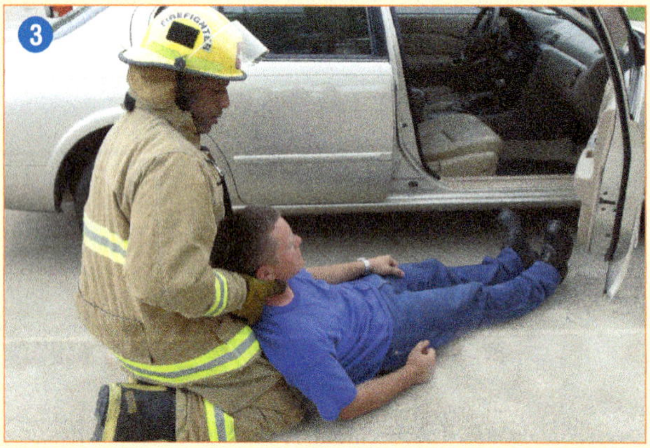


Lower the victim down into a horizontal position in a safe place.

8. As soon as the victim has been rotated and the backboard is in place, the second brigade member and third brigade member lower the patient onto the board while supporting the head and torso so that neutral alignment is maintained. The first brigade member holds the backboard until the victim is secured. **(Step 5)**
9. Next, the third brigade member must move across the front seat to be in position at the victim's hips. If the third brigade member stays at the victim's knees or feet, he or she will be ineffective in helping to move the body's weight. The knees and feet follow the hips.
10. The fourth brigade member maintains support of the head and now takes over giving the commands. The second brigade member maintains direction of the extrication. The second brigade member stands with his or her back to the door, facing the rear of the vehicle. The backboard should be immediately in front of the third brigade member. The second brigade member grasps the patient's shoulders or armpits. Then, on command, the second brigade member and the third brigade member slide the victim 8–12 inches along the backboard, repeating this slide until the victim's hips are firmly on the backboard. **(Step 6)**

13-13 Skill Drill

## Long Backboard Rescue

The first brigade member provides in-line manual support of the head and cervical spine.

The second brigade member gives commands and applies a cervical collar.

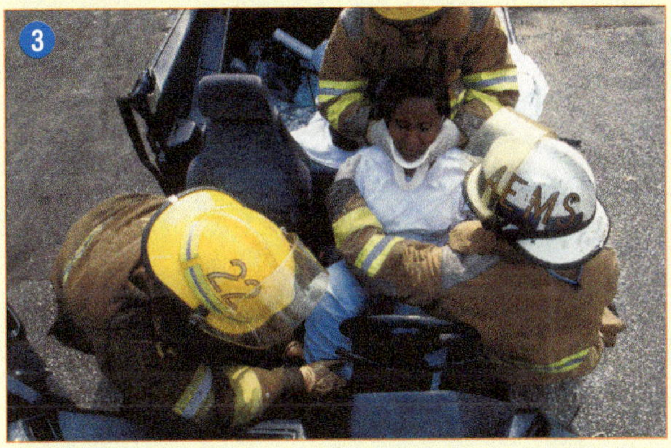

The third brigade member frees the victim's legs from the pedals and moves the legs together without moving the pelvis or spine.

The second and third brigade members rotate the victim as a unit in several short, coordinated moves. The first brigade member (relieved by the fourth brigade member as needed) supports the head and neck during rotation (and later steps).

11. At that time, the third brigade member gets out of the vehicle and moves to the opposite side of the backboard, across from the second brigade member. The third brigade member now takes control at the shoulders, and the second brigade member moves back to take control of the hips. On command, these two brigade members move the victim along the board in 8–12 inches slides until the victim is placed fully on the board. **(Step 7)**
12. The first (or fourth) brigade member continues to maintain support of the head. The second brigade member and third brigade member now grasp their side of the board, and then carry it and the victim away from the vehicle onto the prepared cot nearby. **(Step 8)**

These steps must be considered a general procedure to be adapted as needed. Two-door cars differ from four-door models. Larger cars differ from smaller compact models, pickup trucks, and full-size sedans and four-wheel-drive vehicles. You will handle a large, heavy adult differently from a small adult or child. Every situation will be different—a different car, a different patient, and a different crew. Your resourcefulness and ability to adapt are necessary elements to successfully perform this technique.

## 13-13 Skill Drill (Continued)

The first (or fourth) brigade member places the backboard on the seat against the victim's buttocks. The second and third brigade members lower the victim onto the long backboard.

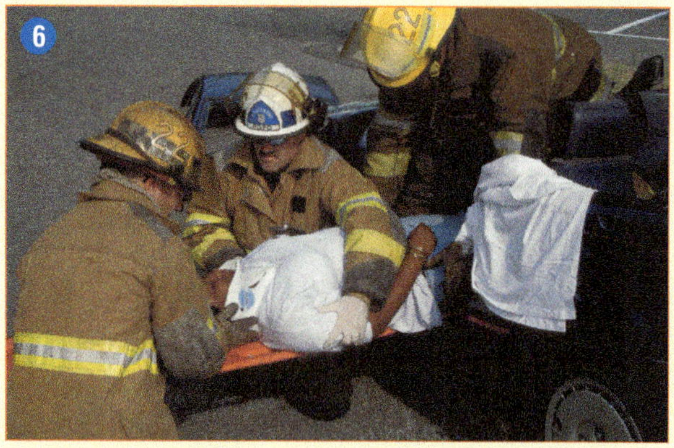


The third brigade member moves to an effective position for sliding the victim. The second and third brigade members slide the victim along the backboard in coordinated, 8–12 inches moves until the hips rest on the backboard.

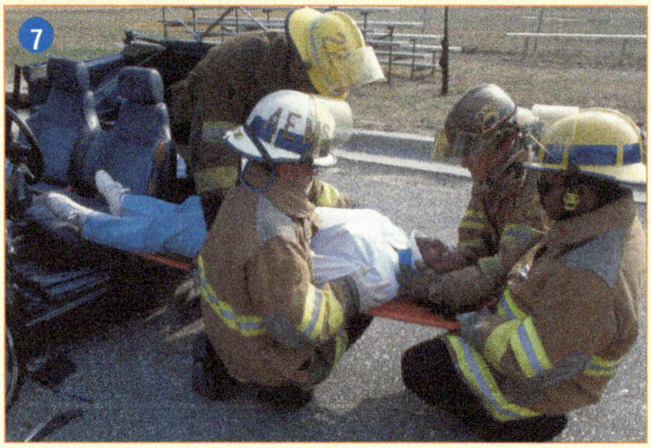


The third brigade member exits the vehicle, moves to the backboard opposite the second brigade member, and together they continue to slide the victim until the victim is fully on the backboard.

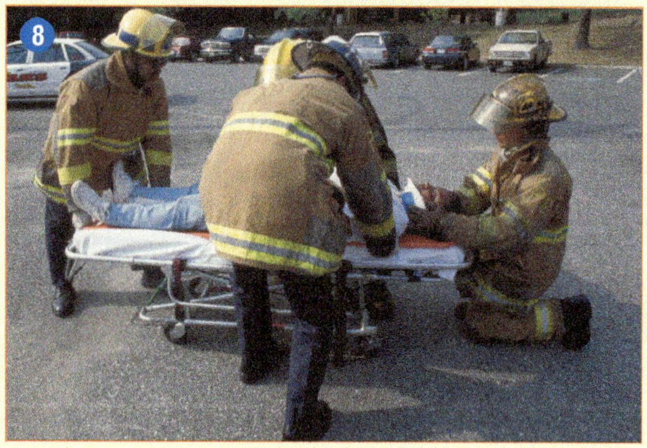


The first (or fourth) brigade member continues to stabilize the victim's head and neck while the second, third, and fourth brigade members carry the victim away from the vehicle.

## Assisting a Person down a Ground Ladder

Using a ground ladder to rescue a trapped occupant is one of the most critical, stressful, and demanding tasks performed by brigade members. Assisting someone down a ladder involves a considerable risk of injury to both brigade members and occupants. The brigade member must use the proper technique to safely accomplish a ladder rescue. In addition, the brigade member must have the physical strength and stamina to accomplish the rescue without injury to anyone involved. Although circumstances could require an individual brigade member to work alone, at least two brigade members should work as a team to rescue a person whenever possible.

Time is a critical factor in many rescue situations. Someone waiting to be rescued is often in immediate danger and may be preparing to jump. Brigade members may have only a limited time to work. They must quickly and efficiently raise a ladder and assist the occupant to safety.

Ladder rescue begins with proper placement of the ladder. A ladder used to rescue a person from a window should have its tip placed just below the windowsill. This positioning makes it easier for the person to mount the ladder. If possible, one or more brigade members in the interior should help the person onto the ladder, and one brigade member should stay on the ladder to assist the individual down.

A ladder used for rescue must be heeled or tied in. The weight of an occupant and one or two brigade members, all moving on the ladder at the same time, can easily destabilize a ladder that is not adequately secured.

### Rescuing a Conscious Person from a Window

When a rescue involves a conscious person, brigade members should establish verbal contact as quickly as possible to reassure the individual that help is on the way. Many people have jumped to their deaths seconds before a ladder could be raised to a window. All ladder rescues should be performed with two brigade members whenever possible; Skill Drill 13-14 presents a technique that could, if necessary, be performed by a single brigade member.

1. The rescue team will place the ladder into the rescue position with the tip of the ladder just below the windowsill, and secure the ladder in place.
2. The first brigade member should climb the ladder, make contact with the victim, and climb inside the window to assist the victim. Contact should be made as soon as possible to calm the victim. The victim should be encouraged to stay at the window until the rescue can be performed.
3. The second brigade member should climb up to the window, standing one rung below the windowsill. This leaves at least one rung available for the victim. When ready, the brigade member should advise the victim to slowly come out onto the ladder, feet first and facing the ladder.
4. The brigade member should form a semi-circle around the victim, with both hands on the beams of the ladder.
5. The brigade member and victim should then proceed slowly down the ladder, one rung at a time, with the brigade member one rung below the victim.
6. If the victim slips or loses footing, the brigade member's legs should keep the victim from falling.
7. The brigade member can take control of the victim at any time by leaning in toward the ladder and squeezing the victim against the ladder.

### Rescuing an Unconscious Person from a Window

If the trapped person is unconscious, one or more brigade members will have to climb inside the building and pass the person out of the window to a brigade member on the ladder. Caution should be used when lowering an unconscious person down a ladder, because it is very easy for the person's arms or legs to get caught in the ladder. To rescue an unconscious victim from a window and down a ladder, follow the steps in ► Skill Drill 13-15.

1. The rescue team will place and secure the ladder in the rescue position with the tip of the ladder just below the windowsill. **(Step 1)**
2. One brigade member should climb up the ladder and into the window to assist from the inside. The second brigade member climbs up to the window opening and waits for the victim. **(Step 2)**
3. The brigade member on the ladder should have a firm grip on the ladder with both hands on the rungs. One leg should be straight and the other should be bent so that the thigh is horizontal to the ground, with the knee at a 90° angle. The foot of the straight leg should be one rung below the foot of the bent leg.
4. When both brigade members are ready, the interior brigade member will pass the victim out through the window and onto the ladder. The victim's back should be toward the ladder, so the victim is face-to-face with the brigade member on the ladder. **(Step 3)**
5. The victim should be lowered so that the groin of the victim will rest on the horizontal leg of the brigade member. The brigade member's arms should be under the arms of the victim and holding onto the rungs. It is important to keep the balls of both feet on the rungs of the ladder. It is much more difficult to move your feet in this position if the heels are close to the rungs. **(Step 4)**
6. The brigade member can now climb down the ladder slowly one rung at a time. The victim is always supported at the groin by one of the brigade member's legs. The brigade member's arms are under the victim's arms to support the upper torso. As an option the interior brigade member may tie the victim's hands together and place them over the neck of the brigade member on the ladder. **(Step 5)**

### Rescuing an Unconscious Small Adult from a Window

Small adults can be cradled across a brigade member's arms during a rescue. The victim must be light enough that the brigade member can descend safely, using only arm strength to support the victim. To carry an unconscious small adult from a window and down a ladder, follow the steps in ► Skill Drill 13-16.

1. The rescue team will set up and secure the ladder in the rescue position, with the tip of the ladder just below the windowsill. **(Step 1)**
2. The first brigade member should climb the ladder and enter the window to assist from the interior. The second brigade member will climb the ladder to the window opening and wait for the victim. Both arms should be level with hands on the beams. **(Step 2)**
3. When ready, the interior brigade member should pass the victim to the brigade member on the ladder so the victim is cradled across the brigade member's arms. **(Step 3)**
4. The brigade member can now climb down the ladder slowly with the victim being held in his or her arms. The brigade member's hands should slide down the beams. **(Step 4)**

# VOICES OF EXPERIENCE

**"Search-and-rescue operations in a process unit environment required an open area search operation which is much different than the right or left-hand search operations commonly used inside of structures."**

As part of our regular emergency response training program, we conduct monthly drills with our on-shift emergency response personnel. One of our experienced shift supervisors was coordinating the normal monthly drill, which included a simulated response to one of our newer processing units.

Prior to starting the drill, the shift supervisor review the proposed drill scenario with the process operators and the shift supervisor from the process are where the drill was going to take place. The shift supervisor obtained a rescue mannequin and staged the mannequin on the deck of a compressor platform. After the drill location was properly set up and all involved parties were briefed, the drill, response was initiated by the dispatch center.

The on-shift fire crew personnel responded to the process area and were met by a process operator who informed the responders that a gas leak had occurred on the unit and that a process operator was unaccounted for. Due to the unstable situation, the other process operators had not entered the area and they could contact the operator via radio.

The IC immediately assigned a search-and-rescue team to conduct a primary search of the area and a backup team to standby to assist as necessary. The search team initiated a rapid open area search of the outdoor process area and after a period of time located the "victim" on the compressor deck.

The search team attempted to remove the 150 lb training mannequin from the deck and soon found that it was much more difficult than they anticipated. The backup team was requested to assist and the two teams worked together to "man handle" the victim through the tight confines of the compressor deck, down the stairwell and out to the plot edge of the process area.

After the rescue drill was completed, an on-scene review of the drill operations was completed. The drill was deemed successful but all personnel involved agreed that the search-and-rescue operations were extremely physically demanding. Search-and-rescue operations in a process unit environment required an open area search operation which is much different than the right- or left-hand search operations commonly used inside of structures. Moving patients on and off the grating decks of the process area and through congested areas was found to be very difficult. This drill was also conducted during an evening shift so limited lighting also played a factor in the response.

Since this drill was conducted several years ago, our refinery has conducted additional search-and-rescue training for fire crew members and have also added extra search-and-rescue equipment to the response apparatus to be better prepared for potential search-and-rescue operations. Multigas analyzers are now carried on each front line response apparatus. Rescue drag harnesses are carried on each foam engine and "drag bags" are carried on the hazmat and rescue response vehicles. A SCBA "RIT" bag is also carried on the primary foam engine. All of these changes will hopefully better prepare our team to respond to an actual search-and-rescue operation if it is ever needed.

**Rick Haase**
ConocoPhillips Wood River Refinery
Roxana, Illinois

Skill Drill 13-15

## Rescuing an Unconscious Victim from a Window

The tip of the ladder is placed just below the windowsill.

One brigade member enters to assist the victim. The second brigade member climbs to the window.

The brigade member waiting on the ladder places both hands on the rungs with one leg straight and the other horizontal to the ground with the knee bent at 90°. The interior brigade member will then pass the victim through the window onto the ladder with the victim's back toward the ladder.

Lower the victim to straddle the brigade member's leg. The brigade member's arms should be under the victim's arms holding onto the rungs.

Step down one rung at a time, transferring the victim's weight from one leg to the other. The victim's arms can be secured around the brigade member's neck.

13-16 Skill Drill

# Rescuing an Unconscious Small Adult from a Window

The ladder is placed in rescue position, with the tip below the windowsill.

One brigade member enters the window to assist the victim. The second brigade member stands on the ladder to receive the victim, with both arms level and hands on beams.

The victim is placed in the brigade member's arms.

The brigade member descends, keeping arms level and sliding the hands down the beams.

## Brigade Member Tips

If you are removing a deceased person, first place the victim in a body bag, then place the body bag on a backboard or basket stretcher. Carrying a body bag without a rigid support is difficult and awkward.

### Rescuing a Large Adult

Three brigade members using two ladders may be needed to rescue very tall or heavy adults. To rescue a large adult using a ladder, follow the steps in Skill Drill 13-17.

1. The rescue team will place and secure two ladders, side-by-side, in the rescue position. The tips of the two ladders should be just below the windowsill.
2. Multiple brigade members may be required to enter the window to assist from the inside.
3. Two brigade members, one on each ladder, should climb up to the window opening and wait for the victim.
4. When ready, the victim should be lowered down across the arms of the brigade members, with one supporting the victim's legs and the other supporting the victim's arms. Once in place, the brigade members can slowly descend the ladder, using both hands to hold onto the ladder rungs.

## Removal of Victims by Ladders

Ladders should be used to remove victims only when it is not possible to use interior stairways or fire escapes. A ladder rescue is often frightening to conscious victims. Rescuing an unconscious victim by ladder is dangerous and difficult, but may be the best way to save a life.

***Aerial ladders and platforms:*** An aerial ladder or platform often is used for rescue operations. The same basic rescue techniques are used with both aerial and ground ladders, but aerial ladders have several advantages over ground ladders. Aerial ladders are much stronger and have a longer reach. They also are wider and more stable, with side rails for additional security ► Figure 13-17.

Figure 13-17 An aerial ladder is stronger and more stable than a ground ladder.

Aerial platforms are even more suitable for rescue operations. These devices reduce the risk of slipping and falling because the victim is lowered to the ground mechanically. An aerial platform is usually preferred for rescue work if one is available.

***Ground ladders:*** Before a ground ladder can be used in a rescue, it must be properly positioned and secured. Positioning and securing ground ladders are covered in Chapter 12, Ladders. Additional personnel will be needed to secure the ladder and to assist in bringing the victim down the ladder.

# Wrap-Up

## Ready for Review

- Search and rescue are the highest priorities at a fire scene and may be assigned to any type of fire company.
- Search and rescue must be integrated with other firefighting functions.
- The initial overview and observations of the fire scene provide valuable information for making decisions about search-and-rescue functions.
- The IC must be able to make an informed decision about whether it is safe to begin a search.
- Search-and-rescue priorities start with the fire floor and then move to the floor above the fire.
- Brigade members must be properly dressed and equipped for search and rescue.
- The primary search should be as thorough as possible in the time available.
- The secondary search is made after the fire is under control.
- Brigade members must use the senses of sight, touch, and hearing when searching.
- Thermal imaging equipment can improve the effectiveness and efficiency of a search in a smoke-filled building.
- Brigade members must always be aware of possible escape routes.
- Rescue techniques include assists, drags, and carries.

## Hot Terms

**Primary search** An initial search conducted to determine if there are victims who must be rescued.

**Rekindle** A situation where a fire, which was thought to be completely extinguished, reignites.

**Rescue** Those activities directed at locating endangered persons at an emergency incident, removing those persons from danger, treating the injured, and providing for transport to an appropriate healthcare facility.

**Search** The process of looking for victims who are in danger.

**Search rope** A guide rope used by brigade members that allows them to maintain contact with a fixed point.

**Secondary search** A more thorough search undertaken after the fire is under control. This search is done to ensure that there are no victims still trapped inside the building.

**Thermal imaging devices** Electronic devices that detect differences in temperature based on infrared energy and then generate images based on those data. Commonly used in obscured environments to locate victims.

**Two-in/two-out rule** A safety procedure that requires a minimum of two personnel to enter a hazardous area and a minimum of two backup personnel to remain outside the hazardous area during the initial stages of an incident.

# Brigade Member in Action

**During the second shift on a Tuesday, the plant fire alarm at your vehicle manufacturing facility sounds for a report of working fire in a dust collection area. Brigade members and apparatus begin to arrive at the fire building and are met by moderate to heavy smoke but with no fire showing. The area foreman indicates that two maintenance technicians were conducting welding repairs to a ductwork system at the time of the fire, and only one technician was present at the area evacuation site. The on-shift fire brigade leader tells you and two other brigade members to prepare for a primary search-and-rescue operation. You enter the buildings and are met with heavy smoke and heat conditions in an area that is highly congested with multiple pieces of ductwork, blower fans, and conduit banks.**

1. The following statements are true about search and rescue risk-benefit except:
   - A. It is not acceptable to risk the safety of brigade members when there is no possibility of saving lives or property.
   - B. Pets are considered family members and should have the same risk consideration as human loss of life.
   - C. Only a limited level of risk is acceptable to save valuable property.
   - D. Actions that present a high level of risk to the safety of brigade members are justified only if there is a potential to save lives.

2. You use the thermal imaging device to scan the work area. The thermal imaging device captures:
   - A. infrared light rather than visible light images.
   - B. light using a photocathode to convert photon energy to electrons and to strike a phosphor screen that emits an image you can see.
   - C. heat images rather than visible light images.
   - D. None of the above

**With the aid of the thermal imaging device, you and your partners find the victim within 10 minutes. He is unconscious but breathing. You need to remove the victim fast. The smoke is heavy and visibility is poor.**

3. What would be the most appropriate rescue technique?
   - A. A webbing sling drag
   - B. A two-person walking assist
   - C. An exit assist
   - D. A cradle-in-arms carry

4. When removing a victim from a dangerous area, you should:
   - A. use the safest and most practical means of egress.
   - B. use the building's normal exit system if clear and safe.
   - C. use a ladder rescue whenever possible.
   - D. Both A and B

# Ventilation

# Chapter 14

## NFPA 1081 Standard

### Incipient Facility Fire Brigade Member

NFPA 1081 contains no Incipient Facility job performance requirements for this chapter.

### Advanced Exterior Facility Fire Brigade Member

NFPA 1081 contains no Advanced Exterior Facility job performance requirements for this chapter.

### Interior Structural Facility Fire Brigade Member

**6.2.3*** Perform ventilation on a structure operating as a member of a team, given an assignment, PPE, and tools, so that a sufficient opening is created, all ventilation barriers are removed, structural integrity is not compromised, and products of combustion are released from the structure.

**(A)** *Requisite Knowledge.* The principles, advantages, limitations, and effects of horizontal and vertical ventilation; safety considerations when venting a structure; the methods of heat transfer; the principles of thermal layering within a structure on fire; fire behavior in a structure; the products of combustion found in a structure fire; the signs, causes, effects, and prevention of backdrafts; and the relationship of oxygen concentration to life safety and fire growth.

**(B)** *Requisite Skills.* The ability to transport and operate tools and equipment to create an opening and implement ventilation techniques.

## Knowledge Objectives

After completing this chapter, you will be able to:

- Define ventilation as it relates to fire suppression activities.
- List the effects of properly performed ventilation on fire and fire suppression activities.
- Describe how fire behavior principles affect ventilation.
- Describe how building construction features within a structure affect ventilation.
- List the principles, advantages, limitations, and effects of horizontal ventilation.
- List the principles, advantages, limitations, and effects of natural ventilation.
- List the principles, advantages, limitations, and effects of mechanical ventilation.
- List the principles, advantages, limitations, and effects of negative-pressure and positive-pressure ventilation.
- List the principles, advantages, limitations, and effects of hydraulic ventilation.
- List the principles, advantages, limitations, and effects of vertical ventilation.
- List safety precautions for ventilating roofs.
- List the basic indicators of roof collapse.
- Explain the role of ventilation in the prevention of backdraft and flashover.

## Skills Objectives

After completing this chapter, you will be able to perform the following skills:

- Break glass with a hand tool.
- Break a window with a ladder.
- Break windows on upper floors using the Halligan toss.
- Establish negative-pressure ventilation.
- Establish positive-pressure ventilation.
- Sound a roof.
- Operate a power saw.
- Perform a rectangular or square cut.
- Perform a louver cut.
- Perform a triangular cut.
- Perform a peak cut.
- Perform a trench cut.

# You Are the Brigade Member

You are a fire brigade member who is working the second shift at your industrial complex. Your shift assignment, should an alarm occur, is building ventilation.

At 19:05 hours, the plant's coded alarm system activates for building 12. You respond directly to the building, knowing the plant's fire apparatus will also be en route. After checking in and donning your PPE, you and your partner enter the building's mechanical room, where you ensure that the HVAC system is shut down and the building is compartmentalized. During your investigation, you discover that smoke detectors in the HVAC system have shut it down. You report this fact to the incident commander (IC), who assigns you and your partner to access the roof and open roof vents in section 4 of the building. The IC notes that an extension ladder has been raised to side C of the building, and that a hose team is preparing for fire suppression operations on the second floor of the site.

As you and your partner make your way to the roof, you are contacted by your backup team, which consists of two members at the extension ladder on the C side of the building. While approaching the roof, you experience light smoke, so you and your partner don and activate your SCBA. You access the roof through a roof door, closing the door behind you. You step over the fire walls to what you know is section 4 of the building, open the roof vents, and report the completion of this task to the IC. As you exit the roof via the extension ladder and give the thumb's-up sign to your stand-by crew, you can see the thermal column rise from the vents and you know the hose team is applying water to the fire.

1. *Why was it important to ensure that the building's HVAC system was in the off position?*
2. *Why was it important to shut the doors behind you on your way to the roof?*
3. *Was it necessary to use the buddy system, to have a backup team, and to identify a second exit from the roof for this operation?*

## Introduction

Fire service ventilation is the process of removing smoke, heat, and toxic gases from a burning building and replacing them with cooler, cleaner, more oxygen-rich air. When ventilation is coordinated with fire attack, it can save lives and reduce property damage. Proper ventilation assists in the location and rescue of victims, enables hose teams to advance and locate the source of the fire, and prevents fire spread. The lack of ventilation or improper ventilation techniques can spread the fire, injure both brigade members and occupants, and increase property damage.

During its normal progression and growth, a fire gives off smoke, heat, and toxic gases. As long as there is fuel and oxygen, the fire will continue to burn and produce these products of combustion. When the fire is inside a building, the structure acts as a container or box, trapping the products of combustion. As the fire grows and develops, the smoke, heat, and toxic gases spread throughout the structure, presenting a direct risk to the lives of occupants and brigade members, reducing visibility, and increasing property damage. In some cases, the trapped products of combustion create a potential for explosion. Ventilation removes these products from the interior atmosphere or allows them to escape in a controlled manner.

The primary principle that controls the spread of smoke, heat, and toxic gases within a room or a building is convection. See Chapter 5, Fire Behavior, for more information. Heated gases expand, becoming less dense than cooler gases. As a result, the hot gases produced by a fire in a closed room will rise to the ceiling and spread outward, displacing cooler air and pushing it toward the floor (▶ Figure 14-1). As the fire continues to burn, the hot layer of gas banks (curves) down closer to the floor.

If the heated products of combustion escape from the room, the same principles will apply as they spread throughout the structure. Smoke, heat, and toxic gases will spread horizontally, along the ceiling, until they find a path such as a stairway, elevator shaft, or pipe chase (an open space within a wall where wires and pipes can run) that allows them to reach

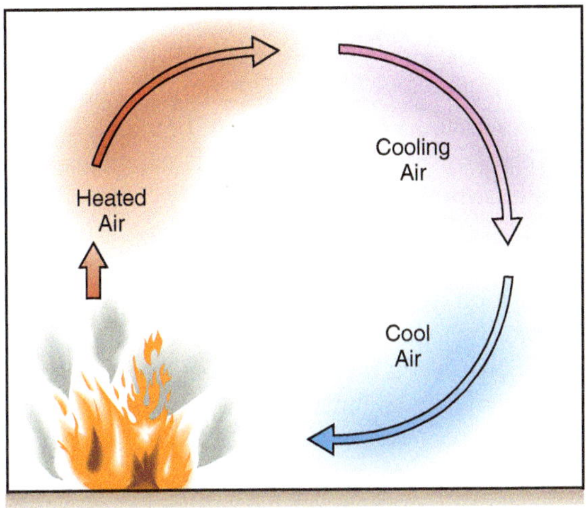


Figure 14-1 Convection currents cause heated products of combustion to rise within a room and spread along the ceiling.

Figure 14-3 Mushrooming occurs as the heated products of combustion flow outward and downward.

Figure 14-2 The heated products of combustion spread horizontally along the ceiling until they find a path that allows them to flow upward.

Figure 14-4 Self-contained breathing apparatus is essential when entering an area contaminated by products of combustion.

a higher level. They will then flow upward through the vertical opening until they reach another horizontal obstruction, such as the ceiling of the highest floor or the underside of the roof (▲ Figure 14-2). At that point, they will again spread out horizontally and bank down as they accumulate. This process of spreading out and banking down is called **mushrooming** (► Figure 14-3).

As long as the products of combustion are trapped within the structure, they present a series of risks and dangers to the occupants and to brigade members. Most of the gases produced by a fire are toxic. The contaminated atmosphere they create poses a life-threatening condition to occupants as well as to anyone who enters without self-contained breathing apparatus (SCBA) (► Figure 14-4). The gases may be so hot that simply breathing them can cause fatal respiratory burns.

Figure 14-5 Fire spreads where there is fuel and wherever the hot products of combustion flow or accumulate.

As the fire continues to consume oxygen from the atmosphere, an additional respiratory hazard may be created. The particulate matter in smoke can severely obscure visibility, making it impossible for occupants to find their way to an exit or for rescuers to locate trapped occupants. Smoke is also irritating to the eyes and the mucous membranes of the respiratory system.

Convection, the flow of heated gases produced by the fire, is one of the primary mechanisms of fire spread. The gases may be hot enough to ignite combustible materials along their path. The fire spreads along the paths taken by the heated gases and in the areas where they accumulate. If the hot gases spread into additional areas, through vertical or horizontal openings, there is a high probability that those areas will also become involved in the fire (▲ Figure 14-5).

In addition to igniting combustible materials, the hot gases, smoke, and other products of combustion can ignite or explode. In many cases, they may include a rich supply of partially burned fuels that are hot enough to ignite, but lack sufficient oxygen to support combustion. If these products are mixed with clean air, the atmosphere itself can ignite or, in an extreme situation, explode in a backdraft (sudden explosive ignition of fire gases when oxygen is introduced into a superheated space). Even if they do not ignite, the products of combustion are often hot enough to cause thermal burns to exposed skin.

Yet another concern is the property damage caused by exposure to smoke and heat. Soot and other residues of smoke often cause as much property damage as flames. Significant heat damage can occur in areas that were exposed only to products of combustion, but not to the fire. Even when a fire is contained and under control, prompt and effective ventilation of the accumulated heat and smoke can be an important factor in limiting property losses.

## Benefits of Proper Ventilation

Ventilation is the process of controlling the flow of smoke, heat, and toxic gases so that they are released safely and effectively from a building. Ventilation must be closely coordinated with all other activities being carried out at a fire scene.

The life safety benefits of ventilation are of primary importance. Ventilation allows the smoke to rise so that brigade members can locate trapped occupants more rapidly. In addition, ventilation can provide clean air to occupants who may be overcome by toxic products of combustion.

Ventilation's role in removing heat is important for brigade members advancing an initial attack hose line. As the brigade members advance, ventilation can relieve the pressure and intense heat of the fire and create a much safer environment. A vent downwind or above the attack hose line pulls away the steam that is created when water cools the flames. It also cools the atmosphere and prevents the steam from banking down on top of the attack line. The ventilation opening also causes the smoke to lift and allows brigade members to aim the hose stream directly at the seat of the fire (the main area of the fire). By reducing the potential for a backdraft or a flashover (the sudden ignition of all combustible objects in a room), effective ventilation also increases the safety of the attack team.

A

B

Figure 14-6 Proper ventilation enables brigade members to control a fire rapidly. A. Unvented structure. B. Vented structure.

Proper ventilation also helps to limit fire spread within the structure. Ventilating near the source of the fire can limit the area of involvement and reduce the spread of heat and toxic gases throughout the structure (◄ Figure 14-6). Initially, this may allow more oxygen to reach the fire, causing a flare-up. A hose line should be in place and ready to hit the seat of the fire prior to ventilation. Careful coordination of ventilation will enable brigade members to confine and extinguish a fire more effectively.

Releasing trapped heat and smoke from the upper level of a building will often prevent the fire from spreading horizontally into adjoining areas or neighboring buildings. A well-placed opening will release heat and smoke to the outside so that brigade members can operate effectively inside the building.

After the fire is under control, ventilation should be maintained until all of the heat, smoke, and toxic products of combustion have been removed from the building. Complete ventilation creates a clear atmosphere for overhauling the fire and limits property losses caused by smoke damage.

# Factors Affecting Ventilation

In order to ventilate a structure successfully, brigade members must first consider how fire behavior dictates the movement of the products of combustion. With this knowledge, brigade members can develop a plan of where and when to create openings to release these products and limit smoke and fire spread. Fire behavior is directly related to the principles of heat transfer, which involve convection, conduction, and radiation. The movement of smoke and heat within a structure is primarily related to convection.

## Convection

Convection refers to the transfer of heat through a circulating medium of liquid or gas. Heated air naturally expands and rises, carrying smoke and gases as it flows up and out from a fire. This heated mixture flows across the underside of ceilings, through doorways, and up stairways, elevator shafts, and vertical concealed spaces such as pipe chases. Unless it is released to the exterior, the heat will accumulate in any accessible space in the upper levels of a building.

As the heated air rises, it displaces cooler air, which is drawn toward the seat of the fire. The cooler air brings oxygen to support combustion and causes the fire to intensify. As the fire continues to burn, the products of combustion spread throughout the building, eventually banking down to lower levels. Within any room or space, the hottest gases are at the highest level and cooler gases are stratified below.

Convection currents can carry smoke and superheated gases to uninvolved areas within the structure or into adjoining spaces, if there is an opening available. The flow of heated gases will always follow the path of least resistance and can be altered by making an opening that offers less resistance.

Brigade members use this basic principle by making ventilation openings that will cause the convective flow to draw the heated products out of the building. Doorways, windows, or roof openings can be used for ventilation. To accomplish this, brigade members must identify the most appropriate type of openings as well as the most feasible locations for openings in the structure.

## Mechanical Ventilation

In addition to making or controlling openings to influence convective flow, brigade members can also use mechanical ventilation to direct the flow of combustion gases. Fans can be used to draw or pull smoke through openings (negative-pressure ventilation) or to introduce fresh air to displace smoke and other products of combustion (positive-pressure ventilation). Hose streams can be used to create air currents to ventilate an area. Some buildings have ventilation systems designed to remove smoke or prevent smoke from entering certain areas.

## Wind and Atmospheric Forces

Wind and other atmospheric forces can play a significant role in ventilation. The wind should always be considered when determining where and how to ventilate a structure. Even a slight breeze can have an impact on the effectiveness of a ventilation opening.

For example, a wind blowing against the upwind side of a building will prevent smoke, heat, or other products of combustion from escaping through an opening on that side. If a door or window on that side of the building is opened, a strong current of oxygen-rich air could enter the structure and accelerate the fire. The wind would push the heat and other products of combustion throughout the structure, creating serious life safety risks for those inside the building.

Conversely, opening a ventilation outlet on the downwind side of a building could be especially effective on a windy day. As the wind blows around the building, it creates a negative-pressure zone that literally removes smoke and heat out of the building. After the fire has been extinguished, opening additional doors and windows on the upwind side will rapidly clear residual smoke (► Figure 14-7).

Temperature and humidity also can affect ventilation, particularly in tall buildings and large-area structures. On a cold day, there will probably be a strong updraft through a heated multifloor building. A downdraft is more likely on a hot day in an air-conditioned building. Because humidity makes the atmosphere denser, removing cool smoke from a building could be less efficient on a humid day.

Wind, temperature, and humidity are powerful, persistent forces of nature. It is more efficient and more effective to work with them than against them.

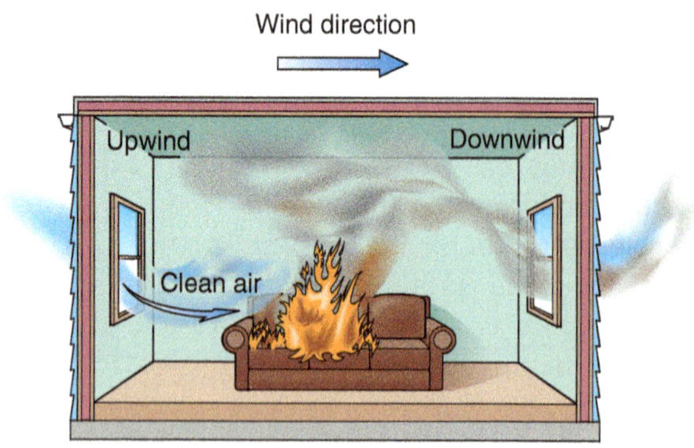


Figure 14-7 On a windy day, proper ventilation will remove smoke and heat from a structure.

# Building Construction Considerations

The way a building is constructed will affect ventilation operations. Each construction type presents its own set of challenges. More information on building construction can be found in Chapter 6, Building Construction.

## Fire-Resistive Construction

Fire-resistive construction (Type I) refers to a building in which all of the structural components are made of noncombustible or limited-combustible materials. This type of construction generally has spaces divided into compartments, which limit potential fire spread. A fire is most likely to involve the combustible contents of an interior space within the building.

Although fire-resistive construction is intended to confine a fire, there are potential avenues for fire spread. Openings for mechanical systems, such as heating and cooling ducts, plumbing and electrical chases, elevator shafts, and stairwells all provide paths for fire spread. A fire can also spread from one floor to another through exterior windows, a phenomenon called leap-frogging (or auto-exposure).

Smoke can travel through a fire-resistive building using the same routes. Heating, ventilation, and air conditioning (HVAC) systems, stairways, and elevator shafts are common pathways. Because the windows in these buildings are often sealed and are not broken easily, there may be limited opportunities for creating ventilation openings. In some cases, one stairway can be used as an exhaust shaft, while other stairways are cleared for rescue and access. Ventilation fans are often required to direct the flow of smoke.

The roof on a fire-resistive building is usually supported by either a steel or a concrete roof deck. It may be difficult or impossible to make vertical-ventilation holes in such a roof. Similar roofing materials are used in buildings with noncombustible construction.

## Ordinary Construction

Ordinary construction (Type III) buildings have exterior walls made of noncombustible or limited-combustible materials, which support the roof and floor assemblies. The interior walls and floors are usually wood construction; the roof usually has a wood deck and a wood structural support system. The roof can be cut with power saws or axes to open vertical-ventilation holes. Ordinary construction buildings usually have windows and doors that can be used for horizontal ventilation.

This type of construction often includes numerous openings within the walls and floors for plumbing and electrical chases. If a fire breaks through the interior finish materials (plaster or drywall), it can spread undetected within the void spaces to other portions of the structure. Interior stairwells generally allow a fire to extend vertically within the building.

The vertical openings in ordinary construction often provide a path for fire to extend into an attic or cockloft (open space between the ceiling of the top floor and the underside of the roof). Heat and smoke can accumulate and spread laterally within these spaces between sections of a large building or into attached buildings. Vertical ventilation is essential when fire is burning in an attic or cockloft ▼ Figure 14-8.

## Wood-Frame Construction

Wood-frame construction (Type V) has many of the same features as ordinary construction. The primary difference is that the exterior walls in a wood-frame building are not required to be constructed of masonry or noncombustible materials.

Wood-frame buildings often have many void spaces where fire can spread, including attics or cocklofts. Wood

Figure 14-8 Vertical ventilation is essential when a fire is burning in an attic or cockloft.

truss roofs and floors, which can fail quickly under fire conditions, are common in these buildings.

Older buildings of wood-frame construction were often assembled with balloon-frame construction. This type of construction has direct vertical channels within the exterior walls, so a fire can spread very quickly from a lower level to the attic or cockloft. Heated smoke and gases can accumulate under the roof, requiring rapid vertical roof ventilation.

Modern wood-frame construction uses platform-frame techniques. In these buildings, the structural frame is built one floor at a time. Between each floor, there is a plate at the floor and the ceiling that acts as a fire stop. This prevents the fire from spreading up and contains it on a single floor.

## Tactical Priorities

Ventilation is directly related to the three major priorities in firefighting operations: life safety, fire containment, and property conservation.

### Venting for Life Safety

Life safety is our primary goal whether it is rescuing occupants or protecting fire brigade personnel. Ventilation helps to clear smoke, heat, and toxic gases from the structure, which gives occupants a better chance to survive. It provides firefighting crews increased visibility and makes the structure more tenable, enabling more rapid searches for victims.

Ventilation also limits fire spread and allows brigade members to advance hose lines more safely and rapidly to attack the fire. Controlling the fire reduces the risk to brigade members as well as occupants.

A dedicated ventilation team can often have a significant impact on life safety. Interior search-and-fire attack teams should be prepared to create ventilation openings as needed.

### Venting for Fire Containment

A brigade member's second priority is to contain the fire and gain control of the situation. Fire spread can often be controlled or limited through effective ventilation. Releasing smoke and superheated gases to the exterior prevents them from spreading throughout the interior of a building or into adjoining spaces.

Attack teams can be more effective when there is less smoke and heat. This enables attack lines to be advanced more easily to extinguish the fire. A coordinated fire attack and ventilation effort should consider the timing and the direction of the attack as well as the type and location of possible ventilation openings.

### Venting for Property Conservation

The third priority is property conservation, which involves reducing losses caused by means other than direct involvement in the fire. Ventilation can play a significant role in limiting property damage. If a structure is ventilated rapidly and correctly, the damage caused by smoke, heat, water, and overhaul operations can all be reduced.

## Location and Extent of Smoke and Fire Conditions

A brigade member must be able to recognize when ventilation is needed and where it should be provided, based on the circumstances of each fire situation. There are many factors that must be considered, including the size of the fire, the stage of combustion, the location of the fire within a building, and the available ventilation options. An experienced brigade member learns to recognize these significant factors and quickly interpret each situation.

When feasible, ventilation should be as close to the fire as possible. The most desirable options are to provide a ventilation opening directly over the seat of the fire or to open a door or window that will release heat and smoke from the fire directly to the exterior. If it is not possible to create a ventilation opening in the immediate area of the fire, brigade members must be able to predict how ventilation from another location will affect the fire. Usually, fire will travel toward a ventilation opening, following the flow of heat and smoke. To make the fire travel in the opposite direction, brigade members could use the wind or a fan to push fresh air through a ventilation opening.

The color, location, movement, and amount of smoke can provide valuable clues about the fire's size, intensity, and fuel ▼ Figure 14-9. Thin, light-colored smoke, moving lazily out of the building, usually indicates a small fire involving ordinary combustibles. Thick, dark gray smoke "pushing" out of a structure suggests a larger, more intense fire. A fire involving petroleum products will produce large quantities of black, rolling smoke that rises in a vertical column.

Figure 14-9 The color, location, and amount of smoke can provide valuable clues.

Smoke movement is a good indicator of the fire's temperature. A very hot fire will produce smoke that moves quickly, rolling and forcing its way out through an opening. The hotter the fire is, the faster the smoke will move. Cooler smoke moves more slowly and gently. On a cool, damp day with very little wind, this type of smoke might hang low to the ground (a phenomenon known as **smoke inversion**).

An unusual phenomenon occurs in buildings with automatic sprinkler systems. The water discharged from the sprinklers cools the smoke, so that it hardly moves within the building. This type of smoke can fill a large warehouse space from floor to ceiling with an opaque mixture that behaves like fog on a damp day. Mechanical ventilation is needed to clear this cold smoke from the building. A similar situation occurs when smoke is trapped within a building long enough to cool to ambient temperature.

## Types of Ventilation

To remove the products of combustion and other airborne contaminants from a structure, brigade members use two basic types of ventilation: horizontal and vertical. **Horizontal ventilation** utilizes the doors and windows on the same level as the fire, as well as any other horizontal openings that are available. In some cases, brigade members might make additional openings in a wall to provide horizontal ventilation.

**Vertical ventilation** involves openings in roofs or floors so that heat, smoke, and toxic gases escape from the structure in a vertical direction. Pathways for vertical ventilation can include stairwells, exhaust vents, and roof openings such as skylights, scuttles, or monitors. Additional openings can be created by cutting a hole in the roof or the floor.

Ventilation can be either natural or mechanical. Natural ventilation depends on convection currents and other natural forces, such as the wind, to move heat and smoke out of a building and allow clean air to enter. Mechanical ventilation uses fans or other powered equipment to exhaust heat and smoke and/or to introduce clean air.

When referring to ventilation techniques, brigade members use the term contaminated atmosphere to describe the products of combustion that must be removed from a building and the term clean air to refer to the outside air that replaces them. The contaminated atmosphere can include any combination of heat, smoke, and toxic gases produced by combustion. A contaminated atmosphere also may be any dangerous or undesirable atmosphere not caused by a fire. For example, natural gas, carbon monoxide, ammonia, and many other products can contaminate the atmosphere as a result of a leak, spill, or similar event. The same basic techniques of ventilation can be applied to many of these situations.

## Horizontal Ventilation

Horizontal ventilation uses horizontal openings in a structure, such as windows and doors, and can be employed in many situations, particularly in small fires (▲ Figure 14-10). Horizontal ventilation is commonly used in room-and-contents fires, and fires that can be controlled quickly by the initial attack team.

Figure 14-10 Windows are frequently used in horizontal ventilation.

Horizontal ventilation can be a rapid, generally easy way to clear a contaminated atmosphere. Often, outlets for horizontal ventilation can be made simply by opening a door or window. In other cases brigade members may need to break a window or to use forcible entry techniques. Search-and-attack teams operating inside a structure, as well as brigade members operating outside the building, use horizontal ventilation to reduce heat and smoke conditions.

Horizontal ventilation is most effective when the opening goes directly into the room or space where the fire is located. Opening an exterior door or window allows heat and smoke to flow directly outside and eliminates the problems that can occur when these products travel through the interior spaces of a building.

Horizontal ventilation is more difficult if there are no direct openings to the outside or if the openings are inaccessible. In these situations, it may be necessary to direct the air flow through other interior spaces or to use vertical ventilation. Horizontal ventilation may have to be used in situations where an inaccessible or damaged roof makes vertical ventilation too dangerous or impossible.

Horizontal ventilation also can be used in less urgent situations if the building can be ventilated without additional structural damage. Low urgency situations might include residual smoke caused by a small fire or a small natural gas leak.

Horizontal ventilation tactics include both natural and mechanical methods. Mechanical ventilation involves the use of fans or other powered equipment.

## Natural Ventilation

Natural ventilation depends on convection currents, wind, and other natural air movements to allow a contaminated atmosphere to flow out of a structure. The heat of a fire creates convection currents that move smoke and gases up toward the roof or ceiling and out away from the fire source. Opening or breaking a window or door will allow these products of combustion to escape through natural ventilation.

Natural ventilation can be used only when the natural air currents are adequate to move the contaminated atmosphere out of the building and replace it with fresh air. Mechanical devices can be used if natural forces do not provide adequate ventilation.

Natural ventilation is often used when quick ventilation is needed, such as when attacking a room-and-contents fire or a first-floor office building fire with people trapped on the second floor. For search-and-attack teams to enter and act quickly, smoke and heat must be ventilated immediately.

Wind speed and direction play an important role in natural ventilation. If possible, windows on the downwind side of a building should be opened first so the contaminated atmosphere flows out. Openings on the upwind side can then be used for cross ventilation, bringing in clean air. Opening a window on the upwind side first could push the fire into uninvolved areas of the structure.

### Breaking Glass

Natural ventilation uses preexisting or created openings in a building. The methods used to create horizontal openings employ a variety of tools and techniques. For example, some windows can simply be opened by hand. But if the window cannot be opened and the need for ventilation is urgent, brigade members should not hesitate to break the glass. Breaking a window is a fast and simple way to create a ventilation opening.

When breaking glass, the brigade member should always use a hand tool (Halligan tool, axe, or pike pole), work from the upwind side and keep hands above or to the side of the falling glass. This will prevent pieces of glass from sliding down the tool and potentially injuring the brigade member. Then the tool should be used to clear the entire opening of all remaining pieces of glass. This creates the largest opening possible and provides a way for brigade members to enter or exit through the window in the event of an emergency.

Clearing the broken glass also reduces the risk of injury to anyone who may be in the area and eliminates the danger of sharp pieces falling out later. Before breaking glass, particularly if the window is above the ground floor, brigade members must look out to ensure that no one will be struck by the falling glass. To break glass on the ground floors with a hand tool, follow the steps in ► Skill Drill 14-1.

1. The brigade member must wear full personal protective equipment (PPE), including eye protection and gloves.
2. Select a hand tool and position yourself upwind and to the side of the window. **(Step 1)**
3. With back against the wall, swing backward forcefully with the tip of the tool striking the top $1/3$ of the glass. **(Step 2)**
4. Clear remaining glass from the opening with the hand tool. **(Step 3)**

### Breaking a Window from a Ladder

A similar technique can be used to break a window on an upper floor, with the brigade member working from a ladder. The ladder should be positioned upwind and to the side of the window. The brigade member should climb to a position level with the window and lock in to the ladder for safety. Using a hand tool, the brigade member strikes and breaks the window, then clears the opening completely. The brigade member must not be positioned below the window, where glass could slide down the handle of the tool ▼ Figure 14-11.

> **Brigade Member Tips**
>
> Opening windows on the upwind side of a building before opening downwind side windows may force the fire into uninvolved areas of the structure.

Figure 14-11 When breaking an upper-story window, the brigade member should place the ladder to the side of the window.

### Breaking a Window with a Ladder

Brigade members on the ground can use the tip of a ladder to break and clear a window when immediate ventilation is needed on an upper floor. Dropping the tip of the ladder into the upper half of the glass will break it. Moving the tip of the ladder back and forth inside the opening will usually clear the remaining glass.

This technique requires proper ladder selection. Usually, second-floor windows can be reached by 16- or 20-foot roof ladders, as well as by bedded 24- or 28-foot extension ladders. Extension ladders are needed to break windows on the third or higher floors.

The ladder can be raised directly into the top half of the window or it can be raised next to the window to determine the proper height for the tip. The ladder is then rolled into the window, the tip is drawn back, and forcibly dropped into the top third of the window. The objective is to push the broken glass into the window opening, but there is always a risk that some of the glass will fall outward. Brigade members working below the window must wear full PPE to shield themselves from falling glass. A word of caution—the higher the window, the greater the danger of falling glass.

14-1 Skill Drill

## Breaking Glass with a Hand Tool

1 Position yourself upwind and to the side of the window.

2 With back against the wall, swing backward forcefully with the tip of the tool striking the top 1/3 of the glass.

3 Clear remaining glass from the opening with the hand tool.

When a ladder is used to break a window, there may be shards of glass left hanging from the edges of the opening. If the opening will be used for access, it will need to be completely cleared with a tool before anyone enters through it. To break a window with a ladder, follow the steps in Skill Drill 14-2.

1. Wear full PPE, including eye protection.
2. Select the proper size ladder for the job.
3. Check for overhead power lines. Use standard procedures for performing a ladder raise utilizing one, two, or three brigade members, depending on the size of the ladder.
4. If a roof ladder is used, extend the hooks toward the window.
5. Raise the ladder next to the window. Extend the tip so it is even with the top third of the window.
6. Position the ladder in front of the window.
7. The ladder is forcibly dropped into the window.
8. Exercise caution; falling glass can cause serious injury. Glass may slide down the beams of the ladder.
9. Raise the ladder from the window and move it to the next window to be ventilated. Either carry the ladder vertically or pivot the ladder on its feet.

## Opening Doors

Door openings also can be used for natural ventilation operations. Doorways have certain advantages when used for ventilation. They are large openings, often twice the size of an

## Brigade Member Tips

### Insulated Thermopane™ Windows

Concerns for energy savings have increased the use of insulated thermal glass both in new construction and as a replacement for conventional glass in older windows. Thermopane windows have multiple layers of glass sealed together, with a small airspace between the layers. These windows are much more difficult to break than conventional glass windows. A brigade member may have to use multiple strikes and additional force to break a thermopane window.

Thermopane windows are used to increase the energy efficiency of a structure. The windows fit tightly so there is little air leakage, and the glass resists heat transfer through it. The tight fitting of the window is not a feature unique to thermopane windows, but rather generally to the age of the installation. Older window installations leak more air. Thermopane windows may have a more substantial frame, but if they do not fit properly, they will leak air as will any window. This can cause potentially dangerous fire conditions to develop in some situations. Any tight fitting window can cause potentially dangerous fire conditions to develop.

Conventional windows often break from the heat of a fire, providing immediate ventilation. Thermopane windows are more likely to stay intact as the fire develops to an advanced stage because the fire would have to break multiple layers of glass. By this time, the fire may consume most of the available oxygen. When the window is finally broken, the sudden rush of fresh air can cause instantaneous burning of these gases with explosive forces being created ▼ Figure 14-2.

Brigade members should evaluate the structure and fire conditions for signs of backdraft before creating any type of ventilation opening. Refer to Chapter 5, Fire Behavior, for a list of indications.

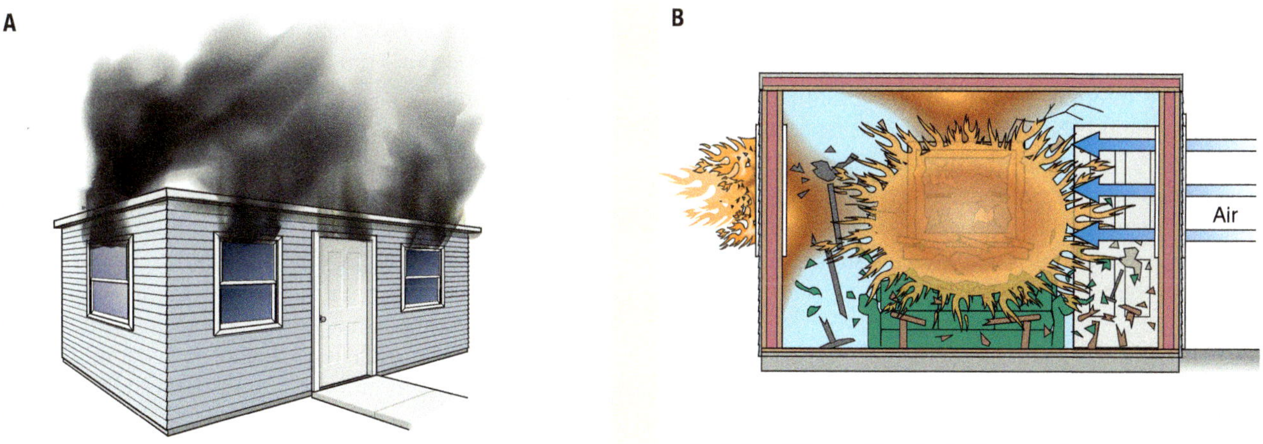


Figure 14-12 A. Building structure and fire conditions should be evaluated for signs of backdraft before any type of ventilation opening is made. B. Improperly ventilating a fire in an oxygen-deficient atmosphere can cause a backdraft.

average window. A locked door can be opened using forcible entry techniques.

The problem with using doorways as ventilation openings is that this compromises entry and exit for interior fire attack teams, as well as search-and-rescue teams. If heat and smoke are pouring through a doorway, brigade members cannot enter or exit through that doorway. Because protecting exit paths is a priority, doorways are more suitable as entries for clean air. The contaminated atmosphere should be discharged through a different opening.

Brigade members will approach a fire through the interior of a building to attack from the unburned side. Opening an exterior door that leads directly to the fire creates a vent and pushes the heat and smoke out of the building. This plan also creates a natural clean air entrance, as clear, cool air flows toward the fire with the attack team.

Doorways are also good places to set up mechanical devices when natural ventilation needs to be augmented. Fans can be used to push clean air in (positive-pressure ventilation) or to pull contaminated air out (negative-pressure ventilation).

## Mechanical Ventilation

Mechanical ventilation uses mechanical means to augment natural ventilation. There are three different methods of mechanical ventilation. Negative-pressure ventilation uses fans called smoke ejectors to exhaust smoke and heat from a structure. Positive-pressure ventilation uses fans to introduce clean air into a structure and push the contaminated atmosphere out. Hydraulic ventilation moves air by using a fog-pattern fire stream to create a pressure differential a behind and in front of the nozzle.

Most industrial structures have HVAC systems that can provide at least some mechanical ventilation. To be used effectively, such a system must be configured so that zones can be controlled correctly. Some HVAC systems are specifically designed to vent smoke or to supply fresh air if a fire occurs; other HVAC systems must be shut down immediately if they interfere with emergency ventilation objectives. Consideration should be given to using a building's or structure's HVAC system as the primary means of ventilation wherever possible. The facility engineer or maintenance staff may have the specific knowledge required to assist in this effort. Information about HVAC systems should be obtained during prefire planning surveys.

### Negative-Pressure Ventilation

The basic principles of air flow are used in negative-pressure ventilation. Brigade members locate the source of the fire, and with a fan or blower, exhaust the products of combustion out through a window or door. The fan draws the heat, smoke, and fire gases out by creating a slightly negative pressure inside the building. In turn, fresh air is drawn into the structure, replacing the contaminated air.

Figure 14-13 A smoke ejector pulls products of combustion out of a structure.

Smoke ejectors are generally 16–24 inches in diameter, although larger models are available. Ejectors can be powered by electricity, gasoline, or water (▲ Figure 14-13).

Negative-pressure ventilation can be used to clear smoke out of a structure after a fire, particularly if natural ventilation would be too slow or if no natural cross ventilation exists.

In large buildings, negative-pressure ventilation may be used to pull heat and smoke out while brigade members are working, although it is not generally used before or during fire attack. Negative-pressure ventilation is usually associated with horizontal ventilation, but an ejector can be used in a roof-vent operation to pull heated gases out of.

Negative-pressure ventilation has several limitations, including positioning, power source, maintenance, and air flow control. Often, brigade members must enter the heated and smoke-filled environment to set up the ejector, and they may need to use braces and hangers to position it properly. Because most ejectors run on electricity, a power source and a cord are required.

Air flow control is also difficult. The fan must be completely sealed so that the exhausted air is not immediately drawn back into the building. This effect is called churning and reduces the effectiveness of the ejector. Churning can be eliminated by completely blocking the opening around the fan.

Smoke ejectors usually have explosion-proof motors, which makes them excellent choices for venting flammable or combustible gases and other hazardous environments. To perform negative-pressure ventilation, follow the steps in (► Skill Drill 14-3).

1. Determine the area to be ventilated and the outside wind direction.
2. If possible, place the smoke ejector to exhaust on the downwind side of the building.

3. The fan should be hung as high in the selected opening as possible.
4. Remove any obstructions from the area used to ventilate smoke, including curtains, screens, and other debris.
5. Use a salvage cover to prevent churning if window is not intact. **(Step 1)**
6. Provide an opening on the upwind side of the structure to provide cross ventilation. **(Step 2)**

## Positive-Pressure Ventilation

Positive-pressure ventilation uses large, powerful fans to force fresh air into a structure. The fans create a positive pressure inside the structure, displace the contaminated atmosphere, and push the heat and products of combustion out. Positive-pressure ventilation can be used to reduce interior temperatures and smoke conditions in coordination with an initial attack, or to clear a contaminated atmosphere after a fire.

Positive-pressure fans are usually set up at exterior doorways, often at the same opening used by the attack team ▶ **Figure 14-14**. If the fan is placed several feet away from the entry point, it will produce a cone of air that completely covers the opening. This pushes clean air in the same direction as the advancing hose line, so brigade members move toward the fire in a stream of clean, cool air. Heat and smoke are pushed back and out of the building through an opening on the opposite side of the building.

For this system to work effectively, the integrity of the building must be intact and ventilation openings kept to a minimum. Positive-pressure ventilation will not work properly if there are too many openings. There must be an opening near the seat of the fire for the heat and smoke to exhaust ▶ **Figure 14-15**. The entry and exit openings should be approximately the same size to create the desired positive pressure within the structure. If there is no exhaust opening, or if it is too small, the heat and smoke will migrate into other areas of the building or back toward the attack team. If the exhaust opening is too large, it will not create a buildup of pressure inside the structure that creates the velocity to effectively remove the smoke.

To increase the efficiency of positive-pressure ventilation, brigade members should close doors to unaffected areas of the structure. Because smaller areas can be ventilated rapidly, venting one room at a time or one section of the building at a time can quickly clear trapped smoke out of a building after a fire. In a multistory building, positive-pressure fans can blow fresh air up through a stair shaft. By opening the doors, one floor at a time can be cleared. Very large structures can be ventilated by placing multiple fans side-by-side or one behind the other.

**Figure 14-14** Positive-pressure fans are usually set up at exterior doorways, often at the same opening used by the attack team.

**Figure 14-15** There must be an opening near the seat of the fire to allow the heat and products of combustion to exhaust.

### Brigade Member Tips

Doors used for ventilation must be kept open to ensure that ventilation is sustained. Depending on the time available and the type of door and lock, brigade members may remove the door, disable the closure device, or place a wedge in the hinge side or under the door.

Overhead doors must be blocked in the open position, so they cannot accidentally roll down and close.

14-3 Skill Drill

## Negative-Pressure Ventilation

Hang the fan in the upper part of the opening. Use salvage covers to prevent churning.

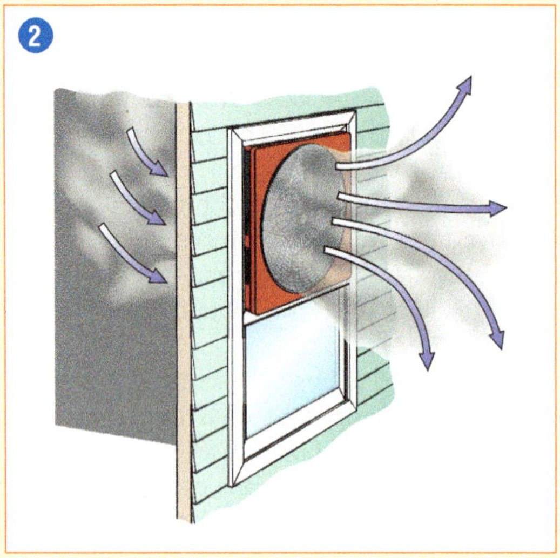

Provide openings on the upwind side for cross ventilation.

Positive-pressure ventilation has several advantages over negative-pressure ventilation. A single brigade member can set up a fan very quickly. Because the fan is positioned outside the structure, the brigade member does not have to enter a hazardous environment, and the fan does not interfere with interior operations. Positive pressure is both quick and efficient, because the entire space within the structure is under pressure. When used properly, positive-pressure ventilation can help confine a fire to a smaller area and increase safety for interior operating crews, as well as any building occupants, by reducing interior temperatures. Positive-pressure fans do not require as much cleaning and maintenance as negative-pressure fans, because the products of combustion never move through the fan.

Positive-pressure ventilation does have its disadvantages, however. The most important concern is that improper use of positive pressure can spread a fire. If the fire is burning in a structural space such as in a pipe chase or above a ceiling, positive pressure can push it into unaffected areas of the structure. If the fire is in structural void spaces, positive-pressure ventilation should not be used until access to these spaces is available and attack crews are in place. Similar problems can occur if positive pressure is used without an adequate exhaust opening for the heat and smoke.

14-4 Skill Drill

## Positive-Pressure Ventilation

1

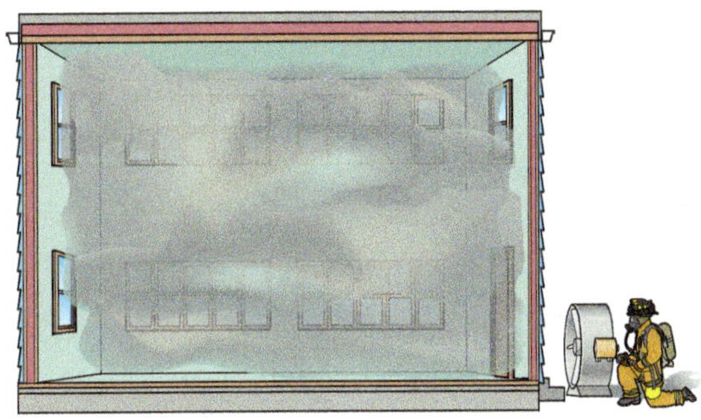

Place the fan in front of the opening to be used for attack.

2

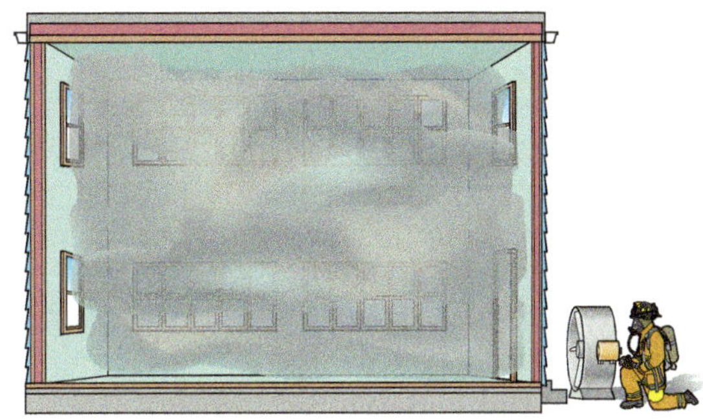

Provide an exhaust opening at or near the fire.

3

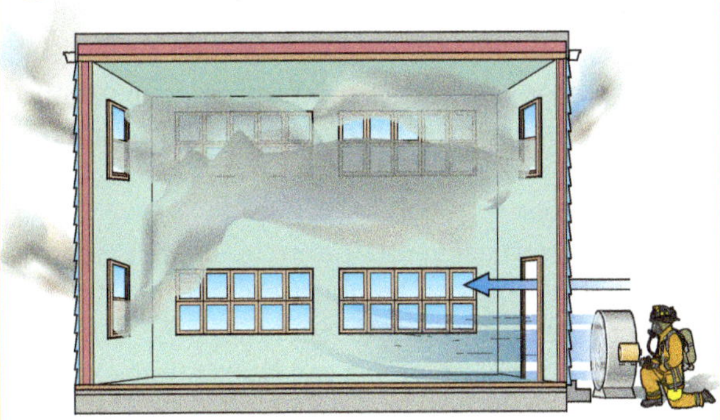

Start fan and allow smoke to clear.

Positive-pressure fans operate at high velocity and can be very noisy. Most are powered by internal combustion engines and can increase carbon monoxide levels if they are run for significant periods of time after the fire is extinguished. Natural ventilation should be used after the structure is cleared to prevent carbon monoxide build-up. Because positive-pressure fan motors can get hot, they are unsafe to use if combustible vapors are present. ▲ **Skill Drill 14-4** outlines the steps in performing positive-pressure ventilation.

1. Determine the location of the fire within the building and the direction of attack.
2. Place the fan 4–10 feet in front of the opening to be used for attack. **(Step 1)**
3. Provide an exhaust opening at or near the fire. This opening can be made before starting the fan or when the fan is started. **(Step 2)**
4. Check for interior openings that could allow the products of combustion to be pushed into unwanted areas.
5. Start the fan and check the cone of air produced. It should completely cover the opening. This can be checked by running a hand around the door frame to feel the direction of air currents.

6. Allow smoke to clear—usually 30 seconds to 1 minute depending on the size of the area to be ventilated and smoke conditions. **(Step 3)**

### Hydraulic Ventilation

Hydraulic ventilation uses the water stream from the hose line to exhaust smoke and heated gases from a structure. The brigade member working the hose line directs a fog stream out of the building through an opening, such as a window or doorway. The contaminated atmosphere is drawn into a low-pressure area behind the nozzle. An induced draft created by the high-pressure stream of water pulls the smoke and gas out through the opening (▼ Figure 14-16). A well-placed fog stream can move a tremendous volume of air through an opening.

Hydraulic ventilation is most useful in clearing smoke and heat out of a room after the fire is under control. To perform hydraulic ventilation, a brigade member must enter the room and remain close to the ventilation opening. The brigade member places the nozzle 2–4 feet inside the opening and opens the nozzle to a narrow fog or broken spray pattern. The brigade member keeps directing the stream outside and backs into the room until the fog pattern almost fills the opening.

The brigade member operating the nozzle must stay low, out of the heat and smoke, or to one side to keep from partially obstructing the opening. This technique is covered more fully in Chapter 16, Fire Hose, Nozzles, and Streams.

Hydraulic ventilation can move several thousand cubic feet of air per minute and is effective in clearing heat and smoke from a fire room. Because it does not require any specialized equipment, it can be performed by the attack team, using the same hose line that was used to control the fire.

There are some disadvantages to using hydraulic ventilation. Brigade members must enter the heated, toxic environment and remain in the path of the products of combustion as they are being exhausted. If the water supply is limited, using water for ventilation must be balanced with using water for fire attack. This technique can cause excessive water damage if it is used improperly or for long periods of time. Also, in cold climates, ice build-up can occur on the ground, creating a safety hazard.

Ideally, the ventilation opening should be created before the hose line advances into the fire area. The hose stream also can be used to break a window before it is directed on the fire. The steam created when the water hits the fire will push the heat and smoke out through the vent. Hydraulic ventilation can then be used to clear the remaining heat and smoke from the room.

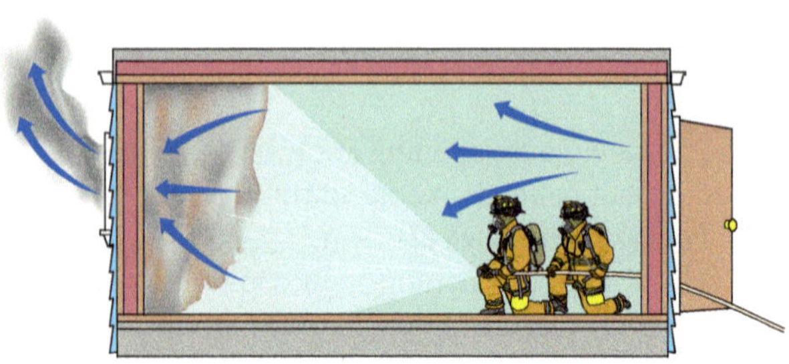

**Figure 14-16** Hydraulic ventilation can be used to draw smoke out of a building after the fire is controlled.

## Brigade Member Safety Tips

If the fire is in hidden spaces, do not use positive-pressure ventilation until access to these spaces is available and attack crews are in place.

## Vertical Ventilation

Vertical ventilation refers to the release of smoke, heat, and other products of combustion into the atmosphere in a vertical direction. Vertical ventilation will occur naturally, due to convection currents, if there is an opening available above a fire. Convection causes the products of combustion to rise and flow through the opening. In some situations, vertical ventilation can be assisted by mechanical means such as fans or hose streams.

Although vertical ventilation refers to any opening that allows the products of combustion to travel up and out, it is most often applied to operations on the roof of a structure. The roof openings can be existing features, such as skylights or bulkheads, or created by brigade members who cut through the roof covering. The choice of roof openings depends primarily on the building's roof construction.

A vertical-ventilation opening should be made as close as possible to the seat of the fire (► Figure 14-17). Smoke issuing from the roof area, melted asphalt shingles, or steam coming from the roof surface are all signs that brigade members can use to identify the hottest point.

## Brigade Member Tips

Remember, ventilation openings allow products of combustion, as well as steam from the attack lines, to escape from the fire site. If ventilation is not coordinated with attack, the smoke, heat, steam, and embers will be forced back at the attack team.

Figure 14-17 The vertical-ventilation opening should be as close as possible to the seat of the fire.

## Safety Considerations in Vertical Ventilation

Vertical ventilation should be performed only when it is necessary and can be done safely. Rooftop operations involve significant inherent risks. If horizontal ventilation can sufficiently control the situation, vertical ventilation is unnecessary. Before performing vertical ventilation, brigade members must evaluate all the pertinent safety issues and avoid unnecessary risks.

The most obvious risk in a rooftop operation is the possibility that the roof will collapse. Many municipal fire fighters have died during vertical ventilation attempts because the fire had compromised the structural integrity of the roof support system. If the structural system supporting the roof fails, the roof and the personnel standing on it will fall into the fire. Personnel also can fall through an area of the roof deck that has been weakened by the fire.

Falling from a roof—either off of the building or through a ventilation opening, open shaft, or skylight into the building—is another hazard. Smoke can cause poor visibility and disorientation, increasing the risk of falling off the edge of a roof. Factors such as darkness, snow, ice, or rain can create hazardous situations. A sloping roof presents greater risks than a flat roof.

Effective vertical ventilation often results in rapid fire control and reduces the risks to brigade members operating inside the building. In some situations, prompt vertical ventilation can save the lives of building occupants by relieving interior conditions, opening exit paths, and allowing brigade members to enter and perform search-and-rescue operations.

Vertical ventilation should always be performed as quickly and efficiently as possible. If they provide an adequate opening, skylights, ventilators, bulkheads, and other existing openings should be opened first. If brigade members must cut through the roof covering to create a hole, the location should be identified and the operation should be performed promptly and efficiently.

### Brigade Member Safety Tips

Whenever possible, brigade members conducting vertical ventilation should operate from the safety of roof ladders or aerial apparatus ▼ Figure 14-18. This reduces the danger of falling from the roof or into the building if the roof collapses.

Figure 14-18 Brigade members operating from the safety of aerial apparatus.

Brigade members working on the roof should always have two safe exit routes. A second ground ladder or aerial device should be positioned to provide a quick alternative exit, in case conditions on the roof deteriorate. The two escape routes should be separate from each other, preferably in opposite directions from the operation site. The team working on the roof should always know where these exit routes are located, and plan ventilation operations to work toward these locations.

The ventilation opening should never be made between the crew and their escape route. A ventilation opening will release smoke and hot gases from the fire. If this column of smoke (and possibly fire) gets between the crew and their exit from the roof, they could be in serious danger. A charged hose line should be ready for use on the roof to protect brigade members and exposures, but a hose stream should never be directed into a ventilation opening ▶ Figure 14-19.

Once the ventilation opening has been made, the brigade members should withdraw to a safe location. There is no good reason to stand around the vent hole and look at it; this simply exposes the brigade members to an unnecessary risk.

Before brigade members climb or walk onto any roof, they should test the roof to ensure that it will support their weight. If the roof is spongy or soft, it may not support a crew and their equipment.

Sounding is a way to test the stability of a roof with a tool such as an axe or pike pole. Simply strike the surface

Figure 14-19 Be prepared for smoke and flames that may be released when the roof is opened.

of the roof with the blunt end of the tool. If the roof is in good condition, the impact should produce a firm rebound and a reassuring sound. If the impact does not produce this sound or if the tool penetrates the roof, the structure is not safe. During ventilation operations, brigade members should continually sound the roof, probing ahead and around themselves with the handle of an axe or a pike pole to ensure that the area is solid.

Modern lightweight construction (not common in industrial structures) frequently uses thin plywood for roof decking, with a variety of waterproof and insulating coverings. A relatively minor fire under this type of roof can cause the plywood to delaminate, without any visible indication of damage or weakness from above. The weight of a brigade member on a damaged area could be enough to open a hole and cause the brigade member to fall through the roof.

Sounding is not a foolproof method to test the condition of a roof and may give brigade members a false sense of security on some types of roofs. Sounding is most commonly used on roofs with wood-frame rafters and trusses. Sounding is generally not reliable when operating on roofs with a steel support system. Also, roofs finished with slate or tile can appear solid even if the supporting structure is ready to fail, due to the rigid surface of the tiles. Always look for other visible indicators of possible collapse and take the roof construction into account. To sound a roof for safety, follow the steps in ► Skill Drill 14-5.

1. Before stepping onto a roof, use a hand tool (such as an axe, pike pole, sledge, etc.), to strike the roof decking with considerable force. Listen and feel for rebound. **(Step 1)**
2. Sound the area ahead and to each side of your path. The material should sound solid and remain intact. If the tool penetrates, the roof is not safe.
3. Locate support members by listening to the sound and watching the bounce of the tool. Support members create a solid sound and make the tool bounce from the roof. The spaces between supports will sound hollow, and the tool will not bounce.
4. Continue to sound the area around you periodically to monitor conditions. **(Step 2)**
5. Remember to sound the roof as you leave the area as well. You are not completely safe until your feet are back on solid ground. **(Step 3)**

A brigade member's path to a proposed vent hole site should follow the areas of greatest support and strength. These include the roof edges, which are supported by bearing walls, and the hips and valleys where structural materials are doubled for strength. These areas should be stronger than interior sections of the roof.

Brigade members should always be aware of their surroundings when working on a roof. Knowing where the roof's edge is located, for example, can help avoid accidentally walking or falling off the roof. This is especially important, although difficult, in darkness, adverse weather, or heavy smoke conditions. It is equally important to watch for openings in the roof, such as skylights and ventilation shafts.

The order of the cuts in making a ventilation opening should be carefully planned. Brigade members should be upwind, have a clear exit path, and be standing on a firm section of the roof or using a roof ladder. If brigade members are upwind from the open hole, the wind will push the heat and smoke away from them. The escaping heat and smoke should not block their exit path. To prevent an accidental fall into the building, brigade members should stand on a portion of the roof that is firmly supported, particularly when making the final cuts.

## Basic Indicators of Roof Collapse

Roof collapse is the greatest risk to brigade members performing vertical ventilation. Some roofs, particularly truss roofs, give little warning that they are about to collapse.

Brigade members assigned to vertical-ventilation tasks should always be aware of the condition of the roof. They should immediately retreat from the roof if they notice any of the following signs:

- Any spongy feeling or indication that the roof is not as solid as when the venting operation began
- Any visible indication of sagging roof supports
- Any indication that the roof assembly is separating from the walls, such as the appearance of fire or smoke near the roof edges
- Any structural failure of any portion of the building, even if it is some distance from the ventilation operation
- Any sudden increase in the intensity of the fire from the roof opening

## Roof Construction

As we learned in Chapter 6, Building Construction, roofs can be constructed of many different types of materials in several configurations. All roofs have two major components—a support structure and a roof covering. The roof support system provides the structural strength to hold the roof in place.

It must also be able to bear the weight of any rain or snow accumulation, any rooftop machinery or equipment, and any other loads placed on the roof, such as brigade members or other people walking on the roof.

The support system can be constructed of solid beams of wood, steel, or concrete, or a system of trusses. Trusses are produced in several configurations and are made of wood, steel, or a combination of wood and steel.

The roof covering is the weather-resistant surface of the roof and may have several layers. Roof covering materials include shingles and composite materials, tar and gravel, rubber, foam plastics, and metal panels. The roof decking is a rigid layer made of wooden boards, plywood sheets, or metal panels. In addition to the decking, the roof covering generally includes a waterproof membrane and insulation to retain heat in winter and limit solar heating in the summer. Fire will very quickly burn through some roofs, but others will retain their integrity for long periods.

Vertical-ventilation operations often involve cutting a hole through the roof covering. The selection of tools, the technique used, the time required to make an opening, and the personnel requirements all depend on the roofing materials and the layering configurations.

Each type of material used in roof construction is affected differently by fire. These reactions will either increase or decrease the time available to perform roof operations. For

14-5 Skill Drill

## Sounding a Roof

Use a hand tool to check the roof before stepping onto it.

Use the tool to sound ahead and to both sides as you walk. Locate support members by sound and rebound. Check conditions around your work area periodically.

Sound the roof along your exit path.

example, solid beams are generally more fire resistant than truss systems, because they are larger and heavier. Wooden beams are affected by fire more quickly than concrete beams. Steel elongates and loses strength when heated.

Most roofs will eventually fail as a result of fire exposure, some very quickly and others more gradually. In some situations, the fire will weaken or burn through the supporting structure so that it collapses while the roof covering remains intact. Structural failure usually results in sudden and total collapse of the roof. A structural failure involving other building components could also cause the roof to collapse.

In other cases, the fire may burn through the roof covering, while the structure is still sound. This type of roof failure usually begins with a "burn through" close to the seat of the fire or at a high point above the fire. As combustible products in the roof covering become involved in the fire, the opening spreads, eventually causing the roof to collapse.

The inherent strength and fire resistance of a roof is often determined by local climate conditions. In areas that receive winter snow loads, roof structures can usually support a substantial weight and may include multiple layers of insulating material. A serious fire could burn under this type of roof with very little visible evidence from above. Heavily constructed roofs, supported by solid structural elements, can burn for hours without collapsing.

In warmer climates, roof construction is often very light. These roofs may simply function as umbrellas. The supporting structure may be just strong enough to support a thin deck with a waterproof, weather-resistant membrane. This type of roof may burn through very quickly, and the supporting structure may collapse after a short fire exposure.

## Solid-Beam Versus Truss Construction

The two major structural support systems for roofs use either solid-beam or truss construction. It may be impossible to determine which system is used simply by looking at the roof from the exterior of the building, particularly when the building is on fire. This information should be obtained during preincident planning surveys.

The basic difference between solid-beam and truss construction is in the way individual load-bearing components are made. Solid-beam construction uses solid components, such as girders, beams, and rafters. Trusses are assembled from smaller, individual components. In most cases, it makes no difference whether solid-beam or truss construction is used. For brigade members, however, there is an important difference: A truss system can collapse quickly and suddenly when it is exposed to a fire.

Trusses are constructed by assembling relatively small and lightweight components in a series of triangles. The resulting system can efficiently span long distances while supporting a load. In most cases, a solid beam that spanned an equal distance and carried an equivalent load would be much larger and heavier than a truss (▲ Figure 14-20). The disadvantage of truss construction is that some trusses can fail completely when only one of the smaller components is weakened or when one of the connections between components fails.

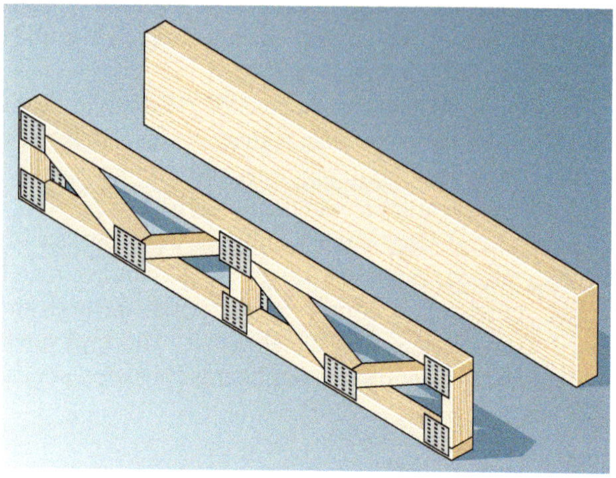

Figure 14-20 A lightweight truss can carry as much weight as a much heavier solid beam.

A truss is not necessarily bad or inherently weak construction. Some trusses are very strong and are assembled with substantial components that can resist fire as well as solid beams. However, many trusses are made of lightweight materials and are designed to carry as much load as possible with as little mass as possible. These trusses can be expected to fail much more quickly when exposed to fire.

Truss construction can be used with almost any type of roof or floor. Trusses can support flat roofs, pitched roofs, arched roofs, or overhanging roofs. Brigade members should assume that any modern construction uses truss construction for the roof support system until proven otherwise.

The individual components of a lightweight truss used for roof support are often wood 2- × 4-foot sections or a combination of wood and lightweight steel bars or tubes. Because these components are small, they can be weakened quickly by a fire. Even more critical, however, are the points where the individual components join together. These connection points are often the weakest portion of the truss and the most common point of failure.

For example, the components of a truss constructed with 2- × 4-inch wood pieces can be connected by heavy-duty staples or by gusset plates (connecting plates made of wood or lightweight metal) (► Figure 14-21). These connections can fail quickly in a fire. When one connection fails, the truss loses its ability to support a load.

If only one truss fails, other trusses in the system may be able to absorb the additional load. However, the overall

Figure 14-21 Gusset plates can fail in a fire, weakening the truss.

strength of the system will be compromised. A fire that causes one truss to fail will probably weaken additional trusses in the same area. These trusses also can fail, sometimes as soon as the additional load is transferred. In most cases, a series of trusses will fail in rapid succession, resulting in a total collapse of the roof. Sometimes the weight of a brigade member walking on the roof will be enough to trigger a collapse.

Trusses also can be made of metal, usually individual steel bars or angle sections that are welded together. Lightweight steel trusses, known as bar joists, often support flat roofs on commercial or industrial buildings. When exposed to the heat of a fire, these metal trusses will expand and lose strength. A roof supported by bar joists will probably sag before it collapses, a warning that brigade members should heed. The expansion can stretch the trusses, which may cause the supporting walls to collapse suddenly. Horizontal cracks in the upper part of a wall indicate that the steel roof supports are pushing outward.

## Roof Designs

Brigade members must be able to identify the types of roofs and the materials used in their construction. Creating ventilation openings may require the use of assorted tools and techniques, depending on the type of roof decking.

### Flat Roof

Flat roofs can be constructed with many different support systems, roof decking systems, and materials. Although they are classified as flat, most roofs have some slope so water can flow to roof drains or scuppers.

Flat roof construction is generally very similar to floor construction. The roof structure can be supported by solid components, such as beams and rafters (the elements that support the roof), or by trusses. The horizontal beams or trusses often run from one exterior bearing wall (wall that supports the weight of a floor or roof) to another bearing wall. In some buildings, the roof is supported by a system of vertical columns and/or interior bearing walls.

The roof deck is usually constructed of multiple layers, beginning with wooden boards, plywood sheets, or metal decking. Then come one or more layers of roofing paper, insulation, tar and gravel, rubber, gypsum, lightweight concrete, or foam plastic. Some roof decks have only a single layer, such as metal sheets or precast concrete sections.

Flat roofs often have vents, skylights, scuttles (small openings or hatches with movable lids), or other features that penetrate the roof deck. Removing the covers from these openings will provide vertical ventilation without cuts through the roof deck.

Flat roofs also may have parapet walls, freestanding walls that extend above the normal roofline. A parapet can be an extension of a firewall, a division wall between two buildings, or a decorative addition to the exterior wall.

### Pitched Roofs

Pitched roofs have a visible slope for rain, ice, and snow runoff. The pitch or angle of the roof usually depends on both local weather conditions and the aesthetics of construction. A pitched roof can be supported either by trusses or by a system of rafters and beams. The rafters usually run from one bearing wall up to a center ridge pole and back to another bearing wall.

Most pitched roofs have a layer of solid sheeting, which can be metal, plywood, or wooden boards. This layer is often covered by a weather-resistant membrane and an outer covering such as shingles, slate, or tiles. Some pitched roofs have a system of laths (thin, parallel strips of wood) instead of solid sheeting to support the outer covering.

As with flat roofs, the type of roof construction material will dictate how to ventilate a pitched roof. For example, a slate or tile pitched roof may be opened by breaking the tiles and pushing them through the supporting laths. A tin roof can be cut and "peeled" back like the lid on a can. Opening a wooden roof usually requires cutting, chopping, or sawing.

Pitched roofs may have steep or gradual slopes. Roof ladders should be used to provide a stable support while working on a pitched roof. A ground or aerial ladder is used to access the lower part of the roof; the roof ladder is placed on the sloping surface and hooked over the center ridge or roof peak.

### Arched Roofs

Arched roofs are generally found in commercial structures because they create large open spans without columns

# VOICES OF EXPERIENCE

**"We noticed that this was "cold smoke" and that there was no heat whatsoever."**

During a very violent thunderstorm, several of us were in the engine room watching a spectacular lightning show when we observed a brilliant flash and heard a very loud clap of thunder. With that, our tones dropped for a reported building fire at a recently vacated meat-packing plant.

The address was well known to all of us because we had used the large industrial building as a preplan and drill site many times in the past. The building was the largest in the district—a one-story, 15,000-square-foot structure with 16-foot ceilings that could accommodate many pieces of heavy industrial machinery. We also knew that the company occupying it had ceased operations as recently as last week.

As we dressed, we heard the report from the first-due police officer. He had observed "a huge volume of smoke" in the front parking lot of the building and noted that this incident was a "working fire." We responded with two engines, a rescue unit, and an ambulance; we also sought mutual aid units from two other stations.

Upon our arrival on the A side of the building, I observed a large amount of thick black smoke pouring from the building. At this point, however, we saw no fire. The chief ordered the first engine in through the A side of the building. My engine was sent to the C side of the building to make entry via the tractor-trailer-sized overhead door on the loading dock. We positioned the apparatus accordingly and gathered our forcible entry tools. When we opened the overhead door, a huge volume of smoke rushed out from the door. We noticed that this was "cold smoke" and that there was no heat whatsoever. I reported this fact to the IC, and he confirmed that the interior crews were reporting the same conditions on the A side of the building.

We knew that the only ladder truck on the assignment was coming from another run that had taken place on the opposite side of the response district. Given that fact, we would have limited ability to vertically ventilate this fire. Furthermore, owing to the cold smoke conditions, we had no idea where the fire was or what was on fire.

Our engine had a line ready to find the seat of the fire. The IC told us to vent any windows and to open all interior doors to lighten the smoke. He reported that the first engine was doing the same from the A side. Both crews worked with limited visibility.

Once the crews had opened the interior and horizontally vented the building, we could see across the rooms inside the building. Our first-due engine was then able to find the seat of the fire, which was located in an electrical room. Once the utility company had secured the power supply to the building, we were able to control the fire quickly.

If we had not used horizontal ventilation, this fire would have spread and likely destroyed the building. The use of horizontal ventilation allowed us to locate the seat of the fire and control it rapidly.

The cause of the fire was determined to be a direct lightning strike to a gang of transformers that had formerly supplied the building. The strike caused the power room to explode owing to the massive overload caused by the bolt of lightening. In the end, the damage to the building was minor.

**Louis N. Molino, Sr.**
LNM Emergency Consulting Services (LNMECS)
Bryan, Texas

Figure 14-22 An arched roof is commonly used for warehouses, supermarkets, and bowling alleys.

▲ Figure 14-22. Arched roofs are common in warehouses, supermarkets, bowling alleys, and similar buildings. Arched roofs can be supported by trusses or by other types of construction, including large wood, steel, or concrete arches.

Arched roofs are often supported by bowstring trusses, which give the roof its distinctive curved shape. Bowstring trusses are usually constructed of wood and spaced 6–20 feet apart. They support a roof deck of wooden boards or plywood sheeting and a covering of waterproof membrane. Although these roofs are quite distinctive when seen from above or outside, they may not be evident from inside the building because a flat ceiling is attached to the bottom chords of the trusses. This creates a huge attic space that is often used for storage. A hidden fire within this space can severely and quickly weaken the bowstring trusses.

The collapse of a bowstring truss roof is usually very sudden. A large area may collapse due to the long spans and wide spaces between trusses. Brigade members should be wary of any fire involving the truss space in these buildings. Many municipal fire fighters have been killed when a bowstring truss roof collapsed.

Buildings with this type of roof construction should be identified and documented during preincident planning surveys. Parapet walls, smoke, and other obstructions may make it difficult to recognize the distinctive arched shape of a bowstring truss roof during an incident.

# Vertical-Ventilation Techniques

## Roof Ventilation

Among the different roof openings that can provide vertical ventilation are built-in roof openings, inspection openings to locate the optimal place to vent, primary expandable openings directly over the fire, and defensive secondary openings to prevent fire spread ▲ Table 14-1.

**Table 14-1 Types of Vertical-Ventilation Openings**

- Built-in roof openings
- Inspection openings
- Primary (expandable) openings
- Secondary (defensive) openings

The objective of any roof ventilation operation is simple: to provide the largest opening in the appropriate location, using the least amount of time and the safest technique.

Before starting any vertical-ventilation operation, brigade members must make an initial assessment. Construction features and indications of possible fire damage should be noted, safety zones and exit paths established, and built-in roof openings that can be used immediately should be identified. The brigade member should ensure appropriate automatic fire vents are opened, and fire dampers are closed. The sooner a building is ventilated, the safer it will become for hose teams working inside.

Ventilation operations must not be conducted in unsafe locations. A less efficient opening in a safe location is better than an optimal opening in a location that jeopardizes the lives of the ventilation team. In some cases, it may not be safe to conduct any rooftop operations and brigade members will have to rely on horizontal ventilation techniques.

Vertical ventilation is most effective when the opening is at the highest point, directly over the fire. Information on the fire's location, as well as visible clues from the roof, can be used to pinpoint the best location to vent. The IC or the interior crews may be able to provide some direction, and the ventilation team should look for signs such as smoke issuing through the roof, tar bubbling up, or other indications of heat. If the roof is wet, rising steam could indicate the hottest area.

Examination holes can be made to evaluate conditions under the roof and to verify the proper location for a ventilation opening. A triangular examination hole can be created very quickly by making three small cuts in the roof. A power saw can also be used to make a kerf cut, which is only as wide as the saw blade. Examination holes also are used to determine how large an area is involved, whether a fire is spreading, and which direction it is moving ► Figure 14-23.

Once the spot directly over the fire is located, the ventilation team should determine the most appropriate type of opening to make. Built-in rooftop openings provide readily available ventilation openings ► Figure 14-24. Skylights, rooftop stairway exit doors, louvers, and ventilators can quickly become

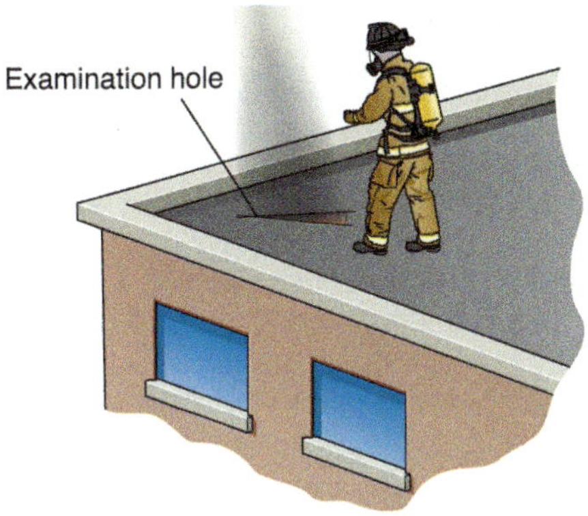


Figure 14-23 A triangular cut can be used as an examination hole.

Figure 14-24 Built-in rooftop openings can often provide vertical ventilation very quickly.

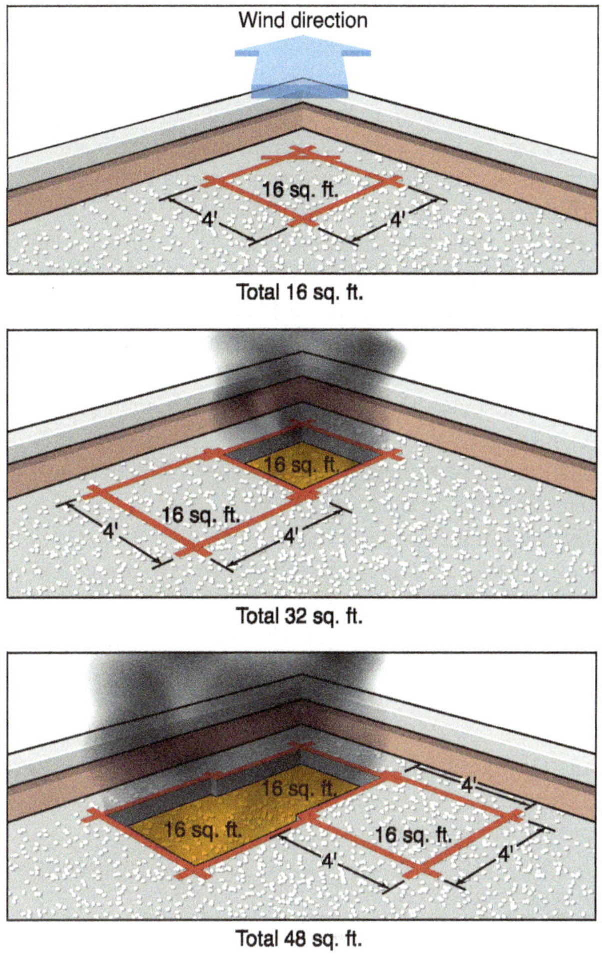


Figure 14-25 A ventilation hole can be quickly expanded by extending the cuts.

ventilation openings simply by removing a cover or an obstruction. This also results in less property damage to the building.

If the roof must be cut to provide a ventilation opening, cutting one large hole is better than making several small ones. The original hole should be rectangular and at least 4-foot long by 4-foot wide. If necessary, this hole can be expanded by continuing the cuts to make a larger opening ▶ Figure 14-25. If the crews inside the building report that smoke conditions are lifting and temperature levels are dropping, the vent is effective and probably large enough. If interior crews do not see a difference, the opening might be obstructed or need to be expanded.

Once the roof opening is made, a hole of the same size should be made in the ceiling material below to allow heat and smoke to escape from the interior of the building. If the ceiling is not opened, only the heat and smoke from the space between the ceiling and the roof will escape through the roof opening. The blunt end of a pike pole or hook can be used to push down as much of the ceiling material as possible. A roof opening will be much less effective if there is no ceiling hole or if the ceiling hole is too small.

## Tools Used in Vertical Ventilation

Several tools can be used in roof ventilation. Power saws are commonly used to cut the vent openings, but many hand tools are also used. Axes, Halligan tools, pry bars, tin cutters, pike poles, and other hooks can be used to remove coverings from existing openings, cut through the roof decking, remove sections of the roof, and punch holes in the interior ceiling. The specific tools that will be needed can usually be determined by looking at the building and the roof construction features.

Ventilation team members should always carry a standard set of tools, as well as a utility rope for hauling additional equipment if needed. All personnel involved in roof operations must use breathing apparatus and wear full protective clothing. Ground ladders or an aerial apparatus can provide access to the roof.

Power saws will effectively cut through most roof coverings. A rotary saw with a wood- or metal-cutting blade or a chain saw can be used. Special carbide-tipped saw blades can cut through typical roof construction materials. To operate a power saw, follow the steps in Skill Drill 14-6.

1. The saw should be checked during apparatus inspections.
2. Be sure the cutting device or blade is appropriate for the material anticipated. If it is not, put the correct blade on before going to the roof.
3. Briefly inspect the blade or chain for obvious damage.
4. Ensure that your proper protective gear is in place, including eye protection, SCBA, and full PPE.
5. The saw should be started to ensure that it runs properly before going to the roof.
6. Stay clear of any moving parts of the saw, use a foot or knee to anchor the saw to the ground, and pull the starter cord as recommended to start the saw.
7. Run the saw briefly at full throttle to verify proper operation.
8. Shut the saw down, wait for the blade to stop completely, then carry the saw to the roof.
9. If possible, always work off of a roof ladder or aerial platform for added safety.
10. Start the saw in an area slightly away from where you intend to cut.
11. Always run a saw at maximum throttle when cutting. The saw should be running at full speed before the blade touches the roof decking. Keep the throttle fully open while cutting and removing the blade from the cut to reduce the tendency for the blade to bind.

## Types of Roof Cuts

Roof construction is a major consideration in determining the type of cut to use. Some roofs are thin and easy to cut with an axe or power saw; others have multiple layers that are difficult to cut. Familiarity with roof types and the best methods for opening them will increase your efficiency and effectiveness in ventilation operations.

### Rectangular or Square Cut

A 4- × 4-foot rectangular or square cut is the most common vertical-ventilation opening. It requires four cuts completely through the roof decking, using an axe or power saw. When using a power saw, the brigade member must carefully avoid cutting through the structural supports. After the four cuts are completed, a section of the roof deck can be removed.

The brigade member should stand upwind of the opening, with an unobstructed exit path. The first and last cuts should be made parallel to and just inside the roof supports. The brigade member making the cuts must always stand on a solid portion of the roof. A triangular cut in one corner of the planned opening can be used as a starting point for prying the deck up with a hand tool.

Depending on the roof construction, it may be possible to lift out the entire section at once. If there are several layers of roofing material, brigade members may have to peel them off in layers. The decking could be plywood sheets or individual boards that have to be removed one at a time. To perform the rectangular cut, follow the steps in ► Skill Drill 14-7.

1. Sound the roof with a tool to locate the roof supports.
2. Make the first cut parallel to a roof support. The cut should be made so that the support will be outside the area that will be opened. **(Step 1)**
3. Make a small triangular cut at the first corner of the opening. **(Step 2)**
4. Follow with two cuts perpendicular to the roof supports. Do not cut through roof supports. Rock the saw over them to avoid damaging the integrity of the roof structure. Then make the final cut parallel to another roof support. **(Step 3)**
5. Make the final cut parallel to and slightly inside a roof support. Be sure to stand on the solid portion of the roof when making this cut.
6. Use a hand tool to pull out the corner triangle or push it through to create a small starter hole. **(Step 4)**
7. If possible, use the hand tool to pull the entire roof section free and flip it over onto the solid roof. It may be necessary to pull the decking out, one board at a time.
8. After opening the roof deck, use a pike pole to punch out ceiling below. This hole should be the same size as the opening in the roof decking. Be careful for a sudden updraft of hot gases or flames. **(Step 5)**

### Louver Cut

Another common cut is the louver cut, which is particularly suitable for flat or sloping roofs with plywood decking.

Using power saws or axes, brigade members make two parallel cuts, approximately 4-foot apart, perpendicular to the roof supports. Then the brigade members make cuts parallel

## Brigade Member Safety Tips

Always carry and handle tools and equipment in a safe manner. When carrying tools up a ladder, hold the beam of the ladder instead of the rungs and run the hand tool up the beam. This technique will work for most hand tools, including axes, pike poles and hooks, and tin roof cutters. Hoist tools using a rope with an additional tagline to keep them away from the building. Carry power saws and equipment in a sling across your back. Always TURN OFF power equipment before you carry it.

147 Skill Drill

# Rectangular or Square Cut

Locate the roof supports by sounding. Make the first cut parallel to the roof support.

Make a triangle cut at the first corner.

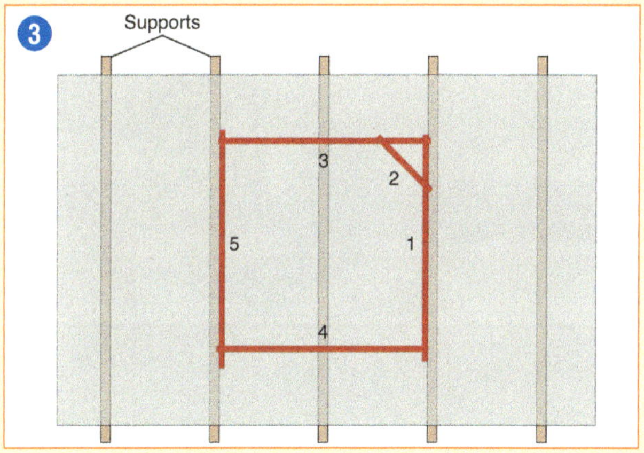


Make two cuts perpendicular to the roof supports (3 and 4). Then make the final cut parallel to another roof support (5).

Pull out or push in the triangle cut.

Punch out the ceiling below. Be careful for a sudden updraft of hot gases or flames.

Skill Drill 14-8

## Louver Cut

Locate the roof supports by sounding.

Make two parallel cuts perpendicular to the roof supports.

Cut parallel to the supports and between pairs of supports in a rectangular pattern.

Tilt the panel to a vertical position.

to the roof supports, approximately half way between each pair of supports. Using each roof support as a fulcrum, the cut sections are tilted to create a series of louvered openings.

Louver cuts can quickly create a large opening. Continuing the same cutting pattern in any direction creates additional louver sections. To make a louver cut, follow the steps in ▲ Skill Drill 14-8.

1. Sound the roof with a tool to locate the roof supports. **(Step 1)**
2. Make two parallel cuts, perpendicular to the roof supports. Do not cut through the roof supports. Rock the saw over them to avoid damaging the integrity of the roof structure. **(Step 2)**
3. Make cuts parallel to the supports and between pairs of supports in a rectangular pattern. **(Step 3)**
4. Strike the nearest side of each section of the roofing material with an axe or maul, pushing it down on one side; use the support at the center of each panel as a fulcrum. Tilt the panel over the middle support. **(Step 4)**
5. Moving horizontally along the roof, make additional louver openings.

## Skill Drill 14-9

# Triangular Cut

Locate the roof supports.

The first cut is made from just inside a support member in a diagonal direction toward the next support member.

The second cut begins at the same location as the first, and is made in the opposite diagonal direction, forming a "V" shape.

The final cut is made along the support member and connects the first two cuts. Cutting from this location allows brigade members the full support of the member directly below them while performing ventilation.

6. Open the interior ceiling area below the opening by using the butt end of a pike pole. This hole should be the same size as the opening made in the roof decking.

### Triangular Cut

The triangular cut works well on metal roof decking because it prevents the decking from rolling away as it is cut. Using saws or axes, the brigade member removes a triangle-shaped section of decking. Smaller triangular cuts can be made between supports, so that the decking falls into the opening, or over supports to create a louver effect. Because triangular cuts are generally smaller than other types of roof ventilation openings, several may be needed to create an adequately sized vent. To make a triangular cut, follow the steps in ▲ Skill Drill 14-9.

1. Locate the roof supports. **(Step 1)**
2. The first cut is made from just inside a support member in a diagonal direction toward the next support member. **(Step 2)**
3. The second cut begins at the same location as the first and is made in the opposite diagonal direction, forming a "V" shape. **(Step 3)**

14-10 Skill Drill

## Trench Cut

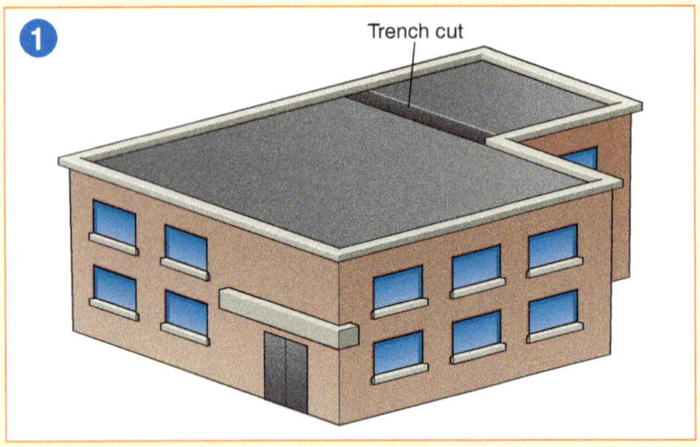


Make two parallel cuts, 2- to 4-foot apart, across the entire roof, starting at the ridge pole (for peaked roofs) or a bearing wall (for flat roofs). Cut between the two long cuts to make a row of rectangular sections. Remove the rectangular panels to open the trench.

4. The final cut is made along the support member and connects the first two cuts. Cutting from this location allows brigade members the full support of the structural member directly below them while performing ventilation. **(Step 4)**

### Trench Cut

Trench cut ventilation is used to stop fire spread in long narrow buildings. The trench cut creates a large opening ahead of the fire, by removing a section of roof and letting heated smoke and gases flow out of the building. Essentially, it is a firebreak in the roof.

Trench cuts are a defensive ventilation tactic to stop the progress of a large fire, particularly one that is advancing through an attic or cockloft. The IC who chooses this tactic is "writing off" part of the building and identifying a point where crews will be able to stop the fire.

A trench cut is made from one exterior wall across to the other. It begins with two parallel cuts, spaced 2- to 4-foot apart. About every 4 feet, brigade members make short perpendicular cuts between the two parallel cuts. They can then lift the roof covering out in sections, completely opening a section of the roof. On a pitched roof, the trench should run from the peak down. On a sloping roof, it should start at the higher end and work toward the lower end.

A trench cut is a secondary cut, used to limit the fire spread, rather than a primary cut located over the seat of the fire. A primary vent should still be made before crews start working on the trench cut.

A trench cut must be made far in advance of the fire. The ventilation crew must be able to complete the cut, and the interior crew needs time to get into position before the fire passes the trench. Inspection holes should be made to ensure that the fire has not already passed the chosen site before the trench cut is completed. Crew lives will be in danger and the tactic is useless if the fire advances beyond the trench before it is opened.

Although trench cuts are effective, they require both time and personnel. As with all types of vertical ventilation, careful coordination between the ventilation crew and the interior attack crew is essential. While the trench cut is being made, hose teams should be deployed inside the building to defend the area in front of the cut (▼ **Figure 14-26**). To make a trench cut, follow the steps in (▲ **Skill Drill 14-10**).

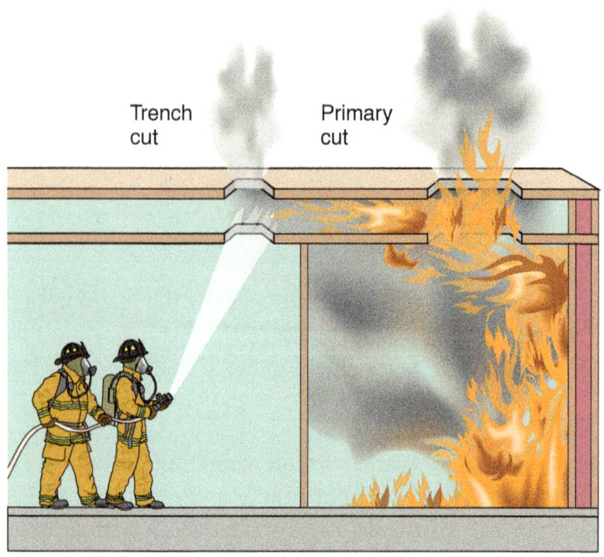


**Figure 14-26** Ventilation and attack teams work together to use a trench cut effectively.

1. After a primary cut has been made over the seat of the fire, the crew cuts a number of small inspection holes to identify a point sufficiently ahead of the fire travel.
2. Make two parallel cuts, 2- to 4-foot apart, across the entire roof, starting at the ridge pole (for peaked roofs) or a bearing wall (for flat roofs).
3. Cut between the two long cuts to make a row of rectangular sections.
4. Remove the rectangular panels to open the trench. **(Step 1)**

## Special Considerations

Many obstacles can be encountered during ventilation operations. Brigade members must be creative and remember the basic objectives of ventilation—to create high openings as rapidly as possible so that hose lines can be advanced into the building.

Poor access or obstructions such as fences, or tight exposures can prevent brigade members from getting close enough to place ladders. Many industrial roofs have multiple layers; some may even have a new roof built on top of an older one. Window openings in abandoned buildings may be boarded or sealed.

Steel bars, shutters, and other security features can hamper ventilation efforts. A building that requires forcible entry is also likely to present ventilation challenges.

### Ventilating a Concrete Roof

Some industrial structures have concrete roofs. Concrete roofs can be constructed with "poured-in-place" concrete, with precast concrete sections of roof decking placed on a steel or concrete supporting structure, or with T-beams. Concrete roofs are generally flat and difficult to breach. The roof decking is usually very stable, but fire conditions underneath could weaken the supporting structural components or bearing walls, leading to failure and collapse.

There are few options for ventilating concrete roofs. Even special concrete-cutting saws are generally ineffective. Brigade members should use alternative ventilation openings such as vents, skylights, and other roof penetrations or horizontal ventilation.

### Ventilating a Metal Roof

Metal roofs and metal roof decks also present many challenges. Because metal conducts heat more quickly than other roofing materials, discoloration and warping may indicate the seat of the fire. Tin-cutter hand tools can be used to slice through thin metal coverings; special saw blades may be needed to cut through metal roof decking. In many cases, the metal is on the bottom and supports a built-up or composite roof covering.

Metal roof decking is often supported by lightweight steel bar joists, which can sag or collapse when exposed to a fire. Because the metal decking is lightweight, the supporting structure may be relatively weak, with widely spaced bar joists. The resulting assembly may fail quickly with only limited fire exposure.

As the fire heats the metal deck, the tar roof covering can melt and leak through the joints into the building, where it can release flammable vapors. This can quickly spread the fire over a wide area under the roof decking. Brigade members should look for indications of dripping or melting tar, and begin rapid ventilation to dissipate the flammable vapors before they can ignite. Hose streams should be used to cool the roof decking from below to stop the tar from melting and producing vapors.

When a metal roof deck is cut, the metal can roll down and create a dangerous slide directly into the opening. The triangular cut prevents the decking from rolling away as easily, so it is the preferred option, even though several cuts may be needed to create an adequately sized vent.

### Ventilating a Basement

Basement fires are especially difficult to ventilate. Basements generally have just a few small windows, if they have any at all. Basement stairways may lead only to the ground floor interior, with no exterior exit.

If a basement fire occurs, windows or exterior doorways into the basement should be opened or broken to provide as much ventilation as possible. If the basement has few or no exterior windows or doors, the interior stairways and other vertical openings will act as chimneys. Heat, smoke, and gases will rise up them into the rest of the structure. In buildings with balloon-frame construction, a basement fire can travel through unprotected wall spaces directly up to the underside of the roof.

A combination of vertical and horizontal ventilation can sometimes be used in attacking a basement fire. A direct path is created from the basement to the first floor, where the heat and smoke are pushed out through a door or window. A stairway or some other opening can be used for vertical ventilation, or a hole can be cut in the floor directly over the fire. The ventilation cut in the floor must be large enough to effectively move the smoke from the basement area to the ground floor.

The opening from the basement to the ground floor should be near a window or door that can be used for horizontal ventilation. This double ventilation must be coordinated to ensure that the smoke and heat ventilated from the basement is exhausted out of the building to avoid excessive smoke or heat on the floor level above. The heat and smoke can be exhausted using a hose line (hydraulic ventilation) or a negative-pressure smoke ejector. A hose line must be ready in case the fire spreads into the ground floor.

A basement fire presents problems to brigade members working in the basement and above it. To attack the fire, brigade members must enter the basement. Smoke and hot gases moving up the stairway as the brigade members descend can make entry difficult or impossible. Brigade members operating above the fire are in danger if the floor or stairway collapses.

The preferred method of attacking a basement fire is to create as many ventilation openings as possible on one side of the basement. This draws the heat and smoke in that direction. Brigade members can then enter the basement from the opposite side, along with clean air. The direction of the attack must be determined and the entry and ventilation points must be clearly identified before this operation begins.

Ventilation openings in basement fires help push heat and smoke away from the attacking brigade members. As the products of combustion rise, brigade members can sometimes crawl beneath them to reach the basement. Because the basement floor is usually much cooler than the stairway, brigade members can attack the fire while the heat and smoke exit over their heads. However, if the entrance stairway is the only ventilation opening, when the water hits the fire, the steam that is produced will push the heat and smoke back toward the attack team. A ventilation opening through the floor into the basement provides a second exit for the products of combustion.

## Ventilating Windowless Buildings

Many industrial structures do not have windows ▶ Figure 14-27. Some buildings are designed without windows; others have bricked-up or covered windows. These buildings pose two significant risks to brigade members: Heat and products of combustion are trapped, and brigade members have no secondary exit route.

Windowless buildings are similar to basements. Any ventilation will need to be as high as possible and will probably require mechanical assistance. Using existing rooftop openings, cutting openings in the roof, reopening boarded-up windows or doors, and making new openings in exterior walls are all possible ways to ventilate windowless structures.

Figure 14-27 Structures without windows pose significant problems in ventilation.

## Ventilating Large Buildings

Providing adequate ventilation is more difficult in large buildings than in smaller ones. A ventilation hole in the wrong location can draw the fire toward the opening, spreading the fire to an area that had not been involved. This underscores the importance of coordinating ventilation operations with the overall fire attack strategy.

Smoke will cool as it travels into unaffected portions of a large building. A sprinkler suppression system will also cool the smoke, causing it to stratify. As the cold smoke fills the area, it becomes more difficult to clear.

If possible, brigade members should use interior walls and doors to create several smaller compartments in the building. This can limit the spread of heat and smoke. The smaller areas can be cleared one at a time with positive-pressure fans. Several fans can be used in a series or in parallel to clear smoke from a large area.

## Backdraft and Flashover Considerations

Ventilation is a major consideration in two significant fireground phenomena: backdraft and flashover. Both can be deadly situations, and brigade members should exercise great caution when conditions indicate that either is possible.

### Backdraft

A backdraft can occur when a building is charged with hot gases, and most of the available oxygen has been consumed. There may be few flames, but the hot gases contain rich amounts of unburned or partially burned fuel. If clean air is introduced to the mixture, the fuel can ignite and explode. To reduce the danger, brigade members must release as much heat and unburned products of combustion as possible, without allowing clean air to enter.

A ventilation opening as high as possible within the building or area can help to eliminate potential backdraft conditions. A roof opening will draw the hot mixture up and relieve the interior pressure. As the mixture rises into the open atmosphere, it may ignite. The ventilation crew should have a charged line ready to protect themselves and nearby exposures.

The attack crews should charge their hose lines outside the building. They should not force entry or begin to apply water until the structure is ventilated. Once they see flaming combustion inside the structure, they can open their hose streams to cool the interior atmosphere as quickly as possible.

### Flashover

Both ventilation and cooling are needed to relieve potential flashover conditions. Flashover can occur when the air in a room is very hot, and exposed combustibles in the space are near their ignition point. Applying water will cool the upper atmosphere, while ventilation draws the heated smoke and gases out of the space. Ventilation openings should be placed to draw the heat and flames away from the hose crew. Simultaneously, the water stream should be used to push the products of combustion toward the opening. Although flame conditions may intensify briefly, the heat will be directed away from the hose crew.

# Wrap-Up

## Ready for Review

- Ventilation is a process that helps remove heat, smoke, and toxic gases from a burning building.
- Ventilation is a critical component of fire attack and must be coordinated with the advancement of attack hose lines.
- Ventilation saves lives and enhances brigade member safety.
- Horizontal ventilation utilizes the doors and windows on the same level as the fire.
- Vertical ventilation involves openings in roofs or floors.
- Brigade members should know and recognize the warning signs of flashover and backdraft.
- Brigade members should be able to recognize differences in building construction that will enhance or inhibit ventilation efforts.
- Brigade members should be able to identify the hazards associated with ventilation operations.
- Brigade members should be able to identify when ventilation is necessary and when it may be an unacceptable risk.
- Brigade members should understand natural and mechanical ventilation techniques and when each should be used.

## Hot Terms

**Arched roof** A rounded roof usually associated with a bow-truss.

**Backdraft** The sudden explosive ignition of fire gases when oxygen is introduced into a superheated space previously deprived of oxygen.

**Balloon-frame construction** An older type of wood-frame construction in which the wall studs extend vertically from the basement of a structure to the roof without any fire stops.

**Bearing wall** A wall that is designed to support the weight of a floor or roof.

**Chase** Open space within walls for wires and pipes.

**Churning** Recirculation of exhausted air that is drawn back into a negative-pressure fan in a circular motion.

**Cockloft** The concealed space between the top floor ceiling and the roof of a building.

**Convection** Heat transfer by circulation within a medium such as a gas or a liquid.

**Ejectors** Electrical fans used in negative-pressure ventilation.

**Fire-resistive construction** Buildings that have structural components of noncombustible materials with a specified fire resistance. Materials can include concrete, steel beams, and masonry block walls. Type I building construction, as defined in NFPA 220, *Types of Building Construction.*

**Flashover** The condition where all combustibles in a room or confined space have been heated to the point at which they release vapors that will support combustion, causing all combustibles to ignite simultaneously.

**Flat roofs** Horizontal roofs often found on commercial or industrial occupancies.

**Gusset plates** The connecting plate made of wood or lightweight metal used in trusses.

**Horizontal ventilation** The process of making openings so that smoke, heat, and gases can escape horizontally from a building through openings such as doors and windows.

**Hydraulic ventilation** Ventilation that relies upon the movement of air caused by a fog stream.

**Kerf cut** A cut that is only the width and depth of the saw blade. It is used to make inspection holes.

**Laths** Thin strips of wood used to make the supporting structure for roof tiles.

**Leap-frogging** A fire spread from one floor to the other through exterior windows (auto-exposure).

**Louver cut** A cut that is made using power saws and axes to cut along and between roof supports so that the sections created can be tilted into the opening.

**Mechanical ventilation** Ventilation that uses mechanical devices to move air.

**Mushrooming** The process that occurs when rising smoke, heat, and gases encounter a horizontal barrier such as a ceiling and begin to move out and back down.

**Natural ventilation** Ventilation that relies upon the natural movement of heated smoke and wind currents.

**Negative-pressure ventilation** Ventilation that relies upon electric fans to pull or draw the air from a structure or area.

**Ordinary construction** Buildings where the exterior walls are noncombustible or limited-combustible, but the interior floors and walls are made of combustible materials. Also known as Type III building construction, as defined in NFPA 220, *Types of Building Construction.*

**Parapet walls** Walls on a flat roof that extend above the roof line.

**Pitched roof** A roof with sloping or inclined surfaces.

**Platform-frame techniques** Subflooring is laid on the joists, and the frame for the first floor walls is erected on the first floor.

**Positive-pressure ventilation** Ventilation that relies upon fans to push or force clean air into a structure.

**Primary cut** The main ventilation opening made in a roof to allow smoke, heat, and gases to escape.

**Products of combustion** Heat, smoke, and toxic gases.

**Rafters** Solid structural components that support a roof.

**Roof covering** The material or assembly that makes up the weather-resistant surface of a roof.

**Roof decking** The rigid component of a roof covering.

**Seat of the fire** The main area of the fire.

**Secondary cut** An additional ventilation opening to create a larger opening, or to limit fire spread.

**Smoke inversion** Smoke hanging low to the ground due to the cold air.

**Sounding** The process of striking a roof with a tool to determine if the roof is solid enough to support the weight of a brigade member.

**Trench cut** A cut that is made from bearing wall to bearing wall to prevent horizontal fire spread in a building.

**Triangular cut** A triangle-shaped ventilation cut in the roof decking that is made using saws or axes.

**Truss** A collection of lightweight structural components joined in a triangular configuration that can be used to support either floors or roofs.

**Ventilation** The process of removing smoke, heat, and toxic gases from a burning structure and replacing them with clean air.

**Vertical ventilation** The process of making openings so that the smoke, heat, and gases can escape vertically from a structure.

**Wood-frame construction** Buildings with exterior walls, interior walls, floors, and roof made of combustible wood material. Type V building construction, as defined in NFPA 220, *Types of Building Construction.*

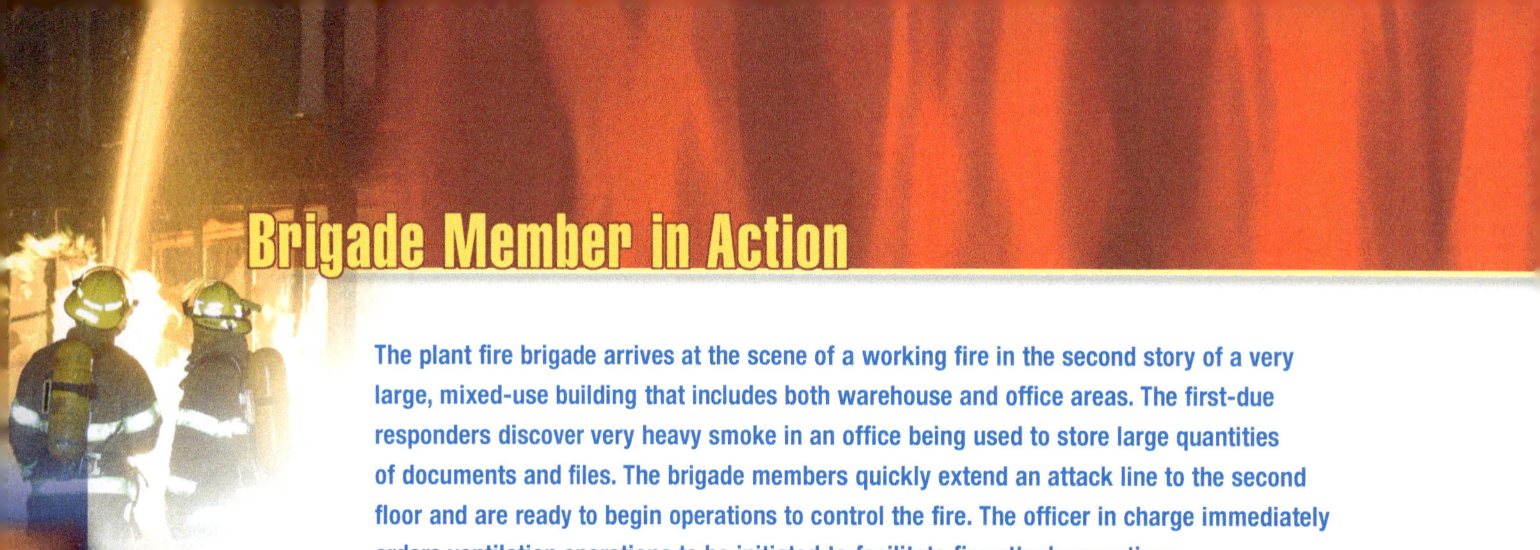

# Brigade Member in Action

**The plant fire brigade arrives at the scene of a working fire in the second story of a very large, mixed-use building that includes both warehouse and office areas. The first-due responders discover very heavy smoke in an office being used to store large quantities of documents and files. The brigade members quickly extend an attack line to the second floor and are ready to begin operations to control the fire. The officer in charge immediately orders ventilation operations to be initiated to facilitate fire attack operations.**

1. The prefire plan indicates that the room of fire origin has two large windows. The roof of the building above the room of fire origin is a metal truss construction with two to three layers of roofing materials. What would be the quickest method to initiate effective ventilation operations?
   - **A.** Send a crew to the roof to initiate vertical ventilation on the opposite end of the roof from the room of fire origin.
   - **B.** Initiate negative-pressure ventilation operations on the first floor of the building.
   - **C.** Initiate hydraulic ventilation at the doorway of the first-floor entrance door.
   - **D.** Place a ladder on the side of the building and break the windows of the room of fire origin.

2. The attack team makes a very quick knock of the fire, but heavy smoke lowers the visibility within the room of fire origin to near zero. What can the attack team do to initiate additional ventilation operations?
   - **A.** Conduct hydraulic ventilation operations by placing a narrow fog stream through the window and allowing it to carry the smoke and heat out of the room of fire origin.
   - **B.** Conduct hydraulic ventilation operations by placing a straight stream through the door of the room of fire origin, thereby allowing fresh air from the windows to move into the room of fire origin.
   - **C.** Use a pike pole to pull the ceiling and vent smoke and heat into the attic area.
   - **D.** Request a ladder truck to place fans outside the windows of the room of fire origin.

3. If the IC requested a trench-cut-type ventilation operation, to which of the following operation would he or she be referring?
   A. A trench cut creates a large opening in the roof ahead of the fire, essentially establishing a fire break.
   B. "Trench cut" is another term for "triangular cut," which is essentially a three-sided cut on a roof.
   C. A trench cut creates a very narrow opening behind the fire, which allows hose lines to be inserted into the room of fire origin.
   D. A trench cut creates a large opening in the wall of the building, which allows water streams to access the seat of the fire.

4. Which of the following statements is *false* regarding roof ventilation operations?
   A. You should always "sound" the roof with a tool prior to walking on the roof.
   B. You should start the saw at ground level and keep it running as it is carried up the ladder to the roof.
   C. You should work off a roof ladder or aerial ladder whenever possible.
   D. Ventilation teams should always carry standard hand tools as well as a utility rope.

# Water Supply

# Chapter 15

## NFPA 1081 Standard

### Incipient Facility Fire Brigade Member

**4.3.4*** Establish a water supply for firefighting operations, given an assignment, a water source, and tools, so that a water supply is established and maintained.

**(A)** *Requisite Knowledge.* Water sources, operation of site water supply components, hydraulic principles, and the effect of mechanical damage and temperatures on the operability of the water supply source.

**(B)** *Requisite Skills.* The ability to operate the site water supply components and to identify damage or impairment.

### Advanced Exterior Facility Fire Brigade Member

**5.2.7*** Establish a water supply for firefighting operations, given a water source and tools, so that a water supply is established and maintained.

**(A)** *Requisite Knowledge.* Water sources, correct operation of site water supply components, hydraulic principles, and the effect of mechanical damage and temperatures on the operability of the water supply source.

**(B)** *Requisite Skills.* The ability to operate the site water supply components and identify damage or impairment.

### Interior Structural Facility Fire Brigade Member

**6.2.6*** Establish a water supply for firefighting operations, given a water source and tools, so that a water supply is established and maintained.

**(A)** *Requisite Knowledge.* Water sources, correct operation of site water supply components, hydraulic principles, and the effect of mechanical damage and temperatures on the operability of the water supply source.

**(B)** *Requisite Skills.* The ability to operate the site water supply components and take action to address damage or impairment.

## Knowledge Objectives

After completing this chapter, you will be able to:

- Understand the basic principles of hydraulics.
- Identify the components of an industrial water distribution system.
- Identify the components of various tools used in establishing and maintaining water supplies.
- Know how to operate fire hydrants.
- Conduct a hydrant flow test.

## Skills Objective

After completing this chapter, you will be able to perform the following skill:

- Operate the site water system

# You Are the Brigade Member

**You are a plant operator and a member of the plant's fire brigade team. You are called to an industrial fire in another part of the plant. As you arrive, the fire is spreading rapidly and involving several other pieces of process equipment. Other fire brigade members are beginning to deploy hydrant-mounted monitors and hose reels in an attempt to begin cooling the exposures. Fixed deluge systems have also been activated and the fire brigade's fire trucks are now arriving at the scene and begin to arrange for a water supply.**

**Plant operators are beginning to shut down parts of the process area, and plant management has been notified and is beginning to assemble at the command post. On the radio, you hear additional brigade members activating additional fire pumps and checking the level of water in the fire water tanks. In the middle of this fire and the chaos, you begin to wonder if the plant's fire pumping capacity and volume of water are sufficient to control this large industrial fire.**

1. ***If you activate too many monitors, will this affect other cooling operations?***
2. ***What will happen if the additional fire pumps fail to operate?***

## Introduction

Hydraulics is the science of applied mechanics that specifically relates to the properties of fluids and the study of the physical characteristics of fluids both at rest and in motion. The term fluid applies to gas or liquid form. For this chapter, we will deal only with liquids, either at rest or in motion.

The science of hydraulics began thousands of years ago with the need to control and manipulate water for irrigation purposes. This was particularly relevant in ancient Egypt, Mesopotamia, and India. Although a more formal understanding of hydraulics did not occur until much later in civilization, these earlier cultures did understand the nature of flow and the use of channels.

The ability of water to control and suppress fires comes from the water's ability to absorb heat. Water has the highest heat-absorbing ability of most common substances. Most fires are suppressed and extinguished by cooling, so water is very efficient for this function, due to its physical characteristics. This heat- or energy-absorption characteristic of water combined with its abundance make it an ideal substance for firefighting.

## Water Supply

The importance of a dependable and adequate water supply for fire-suppression operations is self-evident. The hose line or monitor is not only the primary weapon for fighting fire, it is also the brigade member's primary defense against being burned or driven out of a burning process fire or to provide cooling to the surrounding structure to prevent the further spread of fire. The basic plan for fighting most fires depends on having an adequate supply of water to confine, control, and extinguish the fire.

If the water supply is interrupted while crews are working inside an industrial complex, the brigade members can be trapped, injured, or killed. Brigade members entering a process area need to be confident that their water supply is both reliable and adequate to operate hose lines for their protection and to extinguish the fire.

Ensuring a dependable water supply is a critical fireground operation that must be accomplished as soon as possible. A water supply should be established at the same time as other initial fireground operations. At many fire scenes, size-up, forcible entry, raising ladders, search and rescue, ventilation, and establishing a water supply all occur concurrently.

The brigade can obtain water from one of two sources. The plant's private water systems furnish water via a dedicated fire water system or from a combined fire and process water system (► Figure 15-1). Some remote industrial facilities may depend on static water sources such as ponds, lakes, and rivers. They serve as either main sources of water supply for the plant or drafting sites for fire brigade apparatus to obtain and deliver water to the fire scene.

Often, the water that is carried in a tank on one of the first-arriving apparatus is used in the initial attack. Although many fires are successfully controlled using tank water, this tactic does not ensure the adequacy and reliability of the

Figure 15-1 The water that comes from a hydrant is provided by a municipal or private water system.

Figure 15-2 Mobile water supply apparatus can deliver large quantities of water to the scene of a fire.

water supply. The establishment of an adequate, continuous water supply then becomes the primary objective to support the fire attack ► Figure 15-2. The operational plan must ensure that an adequate and reliable water supply is available before the tank is empty.

## Static Sources of Water

Several potential static water sources can be used for fighting fires in remote areas. Both natural and man-made bodies of water such as rivers, streams, lakes, ponds, oceans, reservoirs, and cisterns can be used to supply water for fire suppression ► Figure 15-3. Some areas have many different static sources, while others have few or none at all.

Water from a static source can be used to fight a fire directly, if it is close enough to the fire scene. Otherwise, it must be transported to the fire using long hose lines, engine relays, or mobile water supply tankers.

Static water sources must be accessible to a fire engine or portable pump. If there is a road or hard surface adjacent to the water source, a fire engine can drive close enough to draft water directly into the pump through a hard suction hose. Some fire brigades construct special access points so engines can approach the water source.

Dry hydrants also provide quick and reliable access to static water sources. A dry hydrant is a pipe with a strainer on one end and a connection for a hard suction hose on the other end. The strainer end should be at least two pipe diameters below the water's surface to avoid creating a vortex and away from any silt or potential obstructions. The other end of the pipe should be accessible to fire apparatus, with the connection at a convenient height for an engine hook-up ► Figure 15-4. When a hard suction hose is connected to the dry hydrant, the engine can draft water from the static source.

Figure 15-3 Any accessible body of water can be used as a static water source.

Dry hydrants are often installed in lakes and rivers and close to clusters of buildings where there is a recognized need for fire protection. In some areas, dry hydrants are used to enable brigade members to reach water under the frozen surface of a lake or river. NFPA 1142, *Standard on Water Supplies for Suburban and Rural Fire Fighting* has more information about dry hydrants.

The portable pump is another alternative for areas that are inaccessible to fire apparatus ► Figure 15-5. The portable pump can be hand-carried or transported by an off-road vehicle to the water source. Portable pumps can deliver up to 500 gallons of water per minute.

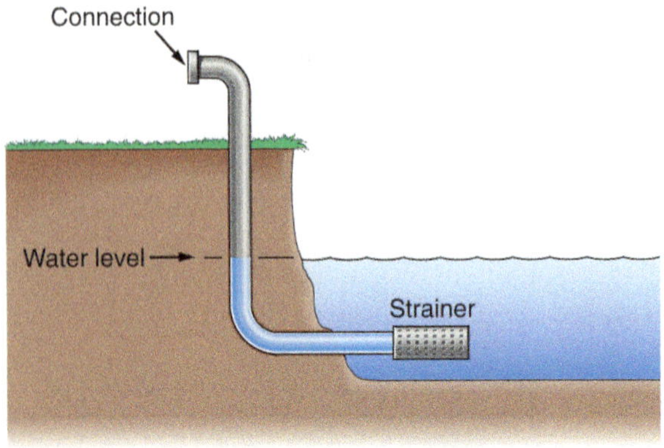


Figure 15-4 A dry hydrant can be placed at an accessible location near a static water source.

Figure 15-5 A portable pump can be used if the water source is inaccessible to a fire brigade engine.

In cold climates, static water sources may freeze over. Dry hydrants and equipment may also ice over or freeze solid. These events will affect the operation both during the incident and after the incident (when it is time to break down the operation).

One key move to keep water flowing once it starts is to crack a nozzle or pumper discharge port and let the water move continuously. SOPs should describe how to handle this issue when working with any type of water supply system.

Mechanical damage from rocks or mud, damage from vehicles, or actions of vandals (such as placement of bottles or cans in the piping) can also affect static water systems.

## Industrial Water Systems

Similar to municipal water systems, industrial water systems provide fire water for hydrants, monitors, and fixed fire protection systems. In some cases, the facility fire water system is combined with the process water system of the facility. Most facility fire water systems protect a single site or process plant, but they can protect an entire geographical area such as an industrial park that contains multiple industrial facilities.

Although industrial and municipal fire water systems are similar in the types of components used, many of these facility fire water systems can deliver extremely large amounts of water, sometimes in excess of 15,000 gpm for dedicated fire water systems and over 100,000 gpm for fire and process water combined systems.

There are two parts to an industrial fire water system: the water source and the distribution system. Whether there is a water treatment plant depends on the particular facility and its proximity to other cities and water sources. In general, industrial water systems take an already treated supply from the municipal water system and store this volume of water in large tanks. In other cases, industrial facilities take a supply from a source that's not treated to a "drinking" level standard. At best only mild treatment such as anti-corrosion and microbiological chemicals are added to prevent build-up and as a form of preventive maintenance.

## Water Sources

Industrial water systems can draw water from a wide variety of sources such as wells, rivers, lakes, oceans, or man-made storage facilities such as retention ponds or reservoirs. In most instances, only a single source is used, but multiple sources also can be utilized. Multiple sources may be found at larger facilities where process expansion or increased production dictated the need for either process water or fire water.

The water source used for industrial facilities needs to be extremely reliable. This can be attributed to the fact that a combination process and fire water system, if disabled, not only will isolate the fire protection systems but also will shut down production. Although having a fire water system out of service is never a great idea, if a process or other industrial facility loses water necessary for production, it could lead to catastrophic events and millions of dollars in lost production and equipment failures.

## Water Distribution System

The distribution system delivers water from the source (river, ocean, or storage tank) to the fire hydrants, hose reel, and sprinkler systems through a complex network of underground or aboveground water mains. In most industrial applications, this includes the use of fire pumps and other components to ensure that the required water volume and pressure can be delivered where and when they are needed. In very limited applications, a gravity feed system will accomplish this by using elevated water

Figure 15-6 A gravity-feed system can deliver water to a low-lying plant without the need for pumps.

Figure 15-7 Water that is stored in an elevated tank can be delivered to the end users under pressure.

storage tanks where large pressure requirements are not needed ▲ Figure 15-6 and Figure 15-7.

Water pressure requirements differ from facility to facility, depending on how the water is used. Generally, static water pressures in industrial facilities range from 85 psi to 175 psi or higher, depending on the type of facility and fire protection demands. Many of the pressure requirements are dictated by the distance from the source to the facility, the size of the facility, or the need to supply water to high elevations such as those found in many of the oil and gas process industries. Residual pressures in most industrial applications rarely fall below 40 psi, but as with most pumping operations, it is possible to operate residual pressures at or below 20 psi under certain circumstances.

As previously stated, most facility fire water systems rely on an arrangement of pumps to provide the required volume and pressure. These pumps can be either electric or diesel engine driven. Most industrial systems have redundant power supplies for the electric drivers to reduce the risk of service interruption. In addition to the redundant power supply, many facilities use diesel drivers as secondary or back-up water supply pumps. This adds reliability because of the independent power supply.

In some limited applications, a combination of pumps and gravity tanks may be used to deliver water, similar to a municipal system. In this case, pumps may be used to deliver the water from the treatment plant to elevated water storage tanks or towers or to reservoirs located on hills or high ground areas. The elevated storage tanks maintain the desired water pressure in the plant distribution system, so that water can be delivered under pressure. In many cases, a dual connection between the municipal and industrial distribution system exists. This dual connection might consist of a connection to an elevated tank providing adequate pressure and volume. In the case of a large fire, a connection via the municipal distribution system can be opened to provide additional volume.

The water mains that deliver water to the fire hydrants, hose reels, and fire protection systems come in various sizes. Large mains, known as primary feeders, carry large quantities of water from the source, such as a river or storage tank, to the distribution feeder. These smaller mains, called secondary feeders, distribute water to plant areas or buildings. The smallest pipes, called distributors, deliver water to the individual device such as a sprinkler system, fire hydrant, or hose reel.

The size of water mains in the distribution system depends on the amount of water needed for fire protection or process demand in combination systems. Most industrial primary feeders can be as large as 36–42 inches in diameter, but generally range between 14 and 18 inches. Secondary feeders supplying water to process units range in size from 10 to 12 inches in diameter, while distributors can range from 6 to 8 inches in diameter.

Water distribution systems in a well-designed industrial facility will follow a grid pattern. A grid arrangement provides water flow to a fire hydrant from two or more directions and establishes multiple paths from the source to each area. This helps ensure an adequate flow of water for firefighting. The grid design also helps to minimize downtime for the other portions of the system if a water main breaks or needs maintenance work. With a grid, the water flow can be diverted around the affected sections. The grid layout promotes good hydraulics and reduces dead-ends, multiple bends, and other hydraulically challenging situations that can increase friction loss, thus reducing available flow.

Figure 15-8 A shut-off valve controls the water supply to an individual user or fire hydrant.

Figure 15-9 A wet-barrel hydrant has a separate valve for each outlet.

Shut-off valves are installed at intervals throughout an industrial distribution system and allow different sections to be turned off or isolated (▲ Figure 15-8). These valves are used when a water main breaks or when work must be performed on a section of the system. In general, control valves that isolate sections of the primary or secondary feeders are called sectional valves. Sectional valves isolate large sections of pipe and up to four or five devices on it, such as a fire hydrant, hose reel, or fire protection system, but the remainder of the system remains in service. Smaller valves, known as isolation or system control valves, isolate a single device such as a hose reel or sprinkler system.

In either case, it is important to notify affected groups in the event it is necessary to close a sectional or isolation valve. Many plants implement a "tagging" system similar to hot work (welding), cold work, or lockout/tagout permitting systems that must be followed whenever a portion of a fire water system is placed in isolation.

## Types of Fire Hydrants

Fire hydrants provide water for firefighting purposes. Hydrants can be installed on private water systems supplied by the municipal water system or from a separate source. The water source as well as the adequacy and reliability of the supply to private hydrants must be identified to ensure that they will be sufficient in fighting fires.

Most fire hydrants consist of an upright steel casing (barrel) attached to the underground water distribution system. The two main types of hydrants are the dry-barrel hydrant and the wet-barrel hydrant. Hydrants are equipped with one or more valves to control the flow of water through the hydrant. One or more outlets are provided to connect fire hoses to the hydrant. These outlets are sized to fit $2\frac{1}{2}$ inches or larger fire hoses.

### Wet-Barrel Hydrants

Wet-barrel hydrants are used in locations where temperatures do not drop below freezing. Wet-barrel hydrants always have water in the barrel and do not have to be drained after each use.

Wet-barrel hydrants usually have separate valves controlling the flow to each individual outlet (▲ Figure 15-9). The brigade member can hook up one hose line and begin flowing water, and later attach a second hose line and open the valve for that outlet, without shutting down the hydrant.

A third type of water delivery device that is similar to a wet-barrel hydrant found in industrial facilities is called a manifold. Manifolds are large water delivery devices that feature a large pipe with generally up to six 6-inch outlets. They are commonly found in industrial facilities where large volumes of water are needed, such as tank farms, or they can be used in lieu of a hydrant. Manifolds usually are connected to large diameter (larger than 14 inches) primary or secondary mains and have an inlet diameter of 12 inches. Some manifolds can be as large as a 24-inch inlet from the primary main and have eight 6-inch outlets.

### Dry-Barrel Hydrants

Dry-barrel hydrants are used in climates where temperatures can be expected to fall below freezing. The valve that controls the flow of water into the barrel of the hydrant is located at the base, below the frost line, to keep the hydrant from freezing (► Figure 15-10). The length of the barrel depends on the climate and the depth of the valve. Water enters the barrel of the hydrant only when it is going to be used. Turning the nut on

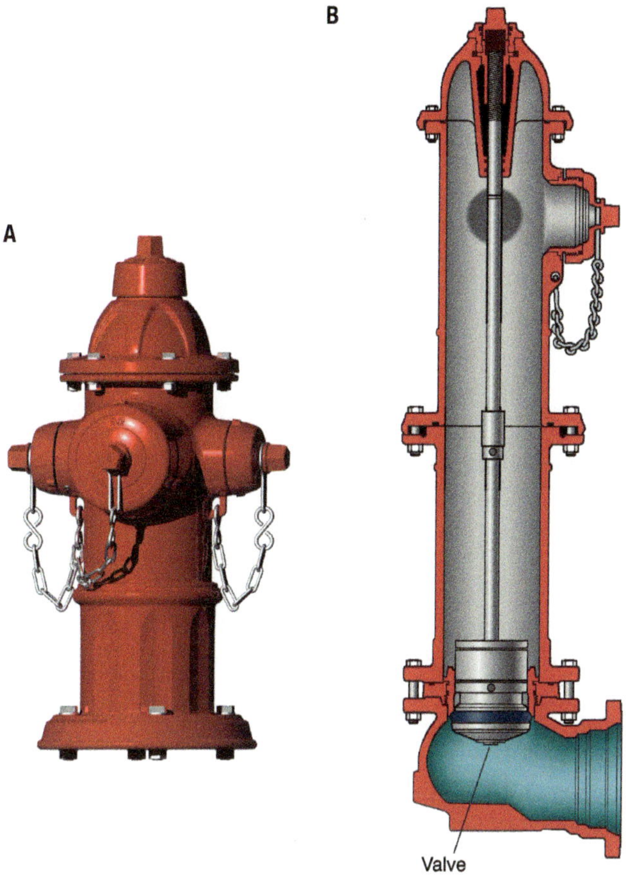


Figure 15-10 A dry-barrel hydrant (A) is controlled by an underground valve (B).

the top of the hydrant rotates the operating stem, which opens the valve so that water flows up into the barrel of the hydrant.

Whenever the hydrant is not in use, the barrel must remain dry. If the barrel contains standing water, it will freeze in cold weather and render the hydrant inoperable. After each use, the water runs out through a drain at the bottom of the barrel. This drain is fully open when the hydrant valve is fully shut. When the hydrant valve is opened, the drain closes. This prevents water from being forced out of the drain when the hydrant is under pressure.

In many areas of the United States, dry-barrel hydrants do not have drains in the bottom of the barrels owing to concerns about the potential for contaminated water to enter the water supply system. The potential for a cross-connection is small, but most municipal water supplies that follow American Water Works Association Standards (AWWA) have plugged this drain. As a consequence, the hydrant barrel must be pumped after each use and weekly inspections must be performed during the winter months to assure that water is not leaking into the barrel through the valve.

A partially opened valve means that the drain is also partially open, and pressurized water can flow out. This can erode (undermine) the soil around the base of the hydrant and may damage the hydrant. For this reason a hydrant should always be either fully opened or fully closed. A fully opened hydrant also makes the maximum flow available to fight a fire.

Most dry-barrel hydrants have only one large valve controlling the flow of water (▲ Figure 15-11). Each outlet must be connected to a hose or an outlet valve, or have a hydrant cap firmly in place before the valve is turned on. Many fire brigades use special hydrant valves so additional connections can be made after water is flowing through the first hose. If additional outlets are needed later, separate outlet valves can be connected before the hydrant is opened. For more information on hoses, see Chapter 16, Fire Hose, Nozzles and Streams.

Figure 15-11 Most dry-barrel hydrants have only the one large valve at the bottom of the barrel controlling the flow of water.

## Fire Marks

The large opening on a fire hydrant is called the steamer port (or steamer connection). Its name goes back to the days when horse-drawn steam-powered engines were used. Large-diameter outlets on hydrants allowed as much water as possible to flow directly into the engine.

## Fire Marks

Hydrants are sometimes called fire plugs. The term "fire plug" dates to a time when water mains were made of hollow wooden logs that were drilled to allow for water connections in the event of a fire.

Dry-barrel hydrants are also used in applications other than cold weather environments. Dry-barrel hydrants are used in process areas that are at risk from explosions such as vapor cloud explosions (VCEs). A VCE is defined as an explosion that produces damaging overpressures by the unplanned release of a large quantity of flammable vaporizing liquid or high pressure gas from a process vessel or pipe. Dry-barrel hydrants are used in these areas so that in the event of a VCE, the resulting damaged hydrant is sheared off at the top and the valve remains closed. This prevents unnecessary and uncontrollable loss of water during emergencies.

## Fire Hydrant Locations

Fire hydrants are located according to local standards, facility requirements, and nationally recommended practices. Fire hydrants may be placed a certain distance apart, perhaps every 300 feet in industrial areas.

In some cases, the requirements for locating hydrants are based on the occupancy, construction, and size of a building, and insurance requirements.

Knowing the plan for installing fire hydrants makes them easier to find in emergency situations. Brigade members who perform fire inspections or develop preincident plans should identify the locations of nearby fire hydrants for each building or group of buildings.

## Fire Hydrant Operation

### Operating a Fire Hydrant

Brigade members must be proficient in operating a fire hydrant. ▶ Skill Drill 15-1 outlines the steps in getting water from a dry-barrel hydrant efficiently and safely. These same steps, with the modifications noted, apply to wet-barrel hydrants as well.

1. Remove the cap from the outlet you will be using. **(Step 1)**
2. Quickly look inside the hydrant opening for any objects that may have been thrown into the hydrant. **(Step 2)** (Omit this step for a wet-barrel hydrant.)
3. Check to ensure that the remaining hydrant caps are snugly attached. **(Step 3)** (Omit this step for a wet-barrel hydrant.)
4. Place the hydrant wrench on the stem nut. Check the top of the hydrant for the arrow indicating which direction to turn the nut to open the hydrant valve. **(Step 4)**
5. Open the hydrant just enough to determine that there is a good flow of water and to flush out any objects that may have been put into the hydrant. **(Step 5)** (Omit this step for a wet-barrel hydrant.)
6. Shut off the flow of water. **(Step 6)** (Omit this step for a wet-barrel hydrant.)
7. Attach the hose or valve to the hydrant outlet. **(Step 7)**

> **Brigade Member Tips**
>
> Before you leave the fire scene, make sure that dry-barrel hydrants are completely drained, even if the weather is warm. By winter, any water left in the hydrant could freeze. Brigade members will lose valuable time connecting a hose to a frozen hydrant only to discover that it will not operate. If this happens, brigade members will be without water until they can locate a working hydrant and can reposition and reconnect the hose lines.

8. When instructed by your brigade leader or the pump operator, start the flow of water. Turn the hydrant wrench to fully open the valve. This may take 12–14 turns, depending on the type of hydrant. **(Step 8)**
9. Open the hydrant slowly to avoid a pressure surge. Once the flow of water has begun, you can open the hydrant valve more quickly. Make sure that you open the hydrant valve completely. If the valve is not fully opened, the drain hole will remain open. **(Step 9)**

NOTE: This skill drill applies to a dry-barrel hydrant. For a wet-barrel hydrant, omit steps 2, 3, 5, and 6. Open the valve for the particular outlet that will be used.

Individual fire brigades may vary these procedures. For example, some brigades specify that the wrench be left on the hydrant. Other brigades require that it be removed and returned to the fire apparatus so that an unauthorized person cannot interfere with the operation. Always follow the standard operating procedures for your brigade.

### Shutting Down a Hydrant

Shutting a hydrant down properly is just as important as opening a hydrant properly. Improper closing of a fire hydrant, such as closing the valve too quickly, can result in a water hammer or hydraulic shock. A water hammer is a sudden and momentary increase in pressure that occurs in a water system (either municipal or industrial); it is caused by a sudden change in velocity or direction of the water. The pressure "wave" can travel throughout the water distribution system, causing damage to pipe, hydrants, and valves. If the hydrant is damaged during shutdown, it cannot be used until it has been repaired. Following the steps in ▶ Skill Drill 15-2 will enable you to shut down a hydrant efficiently and safely.

1. Turn the hydrant wrench slowly until the valve is closed. **(Step 1)**
2. Allow the hose to drain by opening a drain valve or disconnecting a hose connection downstream. Slowly disconnect the hose from the hydrant outlet, allowing any remaining pressure to escape. **(Step 2)**
3. On dry-barrel hydrants, leave one outlet open until the water drains from the hydrant. **(Step 3)**
4. Replace the hydrant cap. **(Step 4)**

15-1 Skill Drill

# Operating a Fire Hydrant

Remove the cap from the outlet you will be using.

Quickly look inside the hydrant opening for foreign objects. (Dry-barrel hydrant only.)

Check to ensure that the remaining caps are snugly attached. (Dry-barrel hydrant only.)

Attach the hydrant wrench to the stem nut. Check for an arrow indicating the direction to turn to open.

Open the hydrant enough to verify flow and flush hydrant. (Dry-barrel hydrant only.)

Shut off the flow of water. (Dry-barrel hydrant only.)

Attach hose or valve to the hydrant outlet(s).

When instructed, turn the hydrant wrench to fully open the valve.

Open slowly to avoid pressure surge.

NOTE: Do not leave or replace the caps on a dry-barrel hydrant until you are sure that the water has completely drained from the barrel. If you feel suction on your hand when you place it over the opening, the hydrant is still draining. In very cold weather, you may have to use a hydrant pump to remove all of the water and prevent freezing.

## Maintaining Fire Hydrants

### Inspecting Fire Hydrants

Because hydrants are essential to fire suppression efforts, brigade members must understand how to inspect and maintain them. Hydrants should be checked on a regular schedule, no less than once a year, to ensure that they are in proper operating condition. During inspections, brigade members may encounter some common problems and should know how to correct them.

The first factors to check when inspecting hydrants are visibility and accessibility. Hydrants should always be visible from every direction, so they can be easily spotted. A hydrant should not be hidden by tall grass, brush, fences, debris, dumpsters, or any other obstructions (▶ Figure 15-12). In winter, hydrants must be clear of snow. No vehicles should be allowed to park in front of a hydrant.

15-2 Skill Drill

## Shutting Down a Hydrant

1 Turn the wrench slowly to close the hydrant valve.

2 Drain the hose line. Slowly disconnect the hose from the hydrant outlet.

3 Leave one hydrant outlet open until the hydrant is fully drained.

4 Replace the hydrant cap.

Figure 15-12 Hydrants should not be hidden or obstructed.

In many facilities, hydrants are painted in bright reflective colors for increased visibility. The bonnet (the top of the hydrant) may also be color-coded to indicate the available flow rate of a hydrant. Colored reflectors are sometimes mounted next to hydrants or placed in the pavement in front of them to make them more visible at night.

Hydrants should be installed at an appropriate height above the ground. The outlets should not be so high or so low that brigade members have difficulty connecting hose lines to them. NFPA 24, *Standard for the Installation of Private Fire Service Mains and Their Appurtenances,* requires a minimum of 18 inches from the center of a hose outlet to the finished grade. Hydrants should be positioned so that the connections, especially the large steamer connection, are facing the facility road or apparatus access.

During a hydrant inspection, check the exterior for signs of damage. Open the steamer port of dry-barrel hydrants to ensure that the barrel is dry and free of debris. Make sure that all caps are present and that the outlet hose threads are in good working order ▶ Figure 15-13.

The second part of the inspection ensures that the hydrant works properly. Open the hydrant valve just enough to ensure that water flows out and flushes any debris out of the barrel. After flushing, shut down the hydrant. Leave the cap off dry-barrel hydrants to ensure that they drain properly. A properly draining hydrant will create suction against a hand placed over the outlet opening ▶ Figure 15-14. When the hydrant is fully drained, replace the cap.

If the threads on the discharge ports need cleaning, use a steel brush and a small triangular file to remove any burrs in the threads. Also check the gaskets in the caps to make sure they are not cracked, broken, or missing. Replace worn gaskets with new ones. Follow the manufacturer's recommendations for any parts that require lubrication.

Figure 15-13 All hydrants should be checked at least annually.

Figure 15-14 You should feel suction to indicate the hydrant is draining.

# Voices of Experience

**"Without a reliable and sufficient water supply and distribution system, controlling an incident is highly unlikely."**

One day, while leaving my office in the plant, I noticed a large black cloud of smoke emanating from what appeared to be a location behind the asphalt plant, near our fire training ground. Upon closer examination, I discovered that the fire was, in fact, coming from the asphalt plant and not the training ground. I immediately raised the alarm and notified the in-plant fire brigade.

Upon my arrival, I discovered that a section of the unit had caught fire, and the fire was spreading rapidly. I began to open the valves to the hydrant-mounted monitors in an attempt to impede the fire's progress. As I opened additional monitors, I could sense that the intensity of the flow was decreasing as additional monitors began to flow.

The fire brigade and trucks soon arrived and established a water supply from nearby hydrants. This method, although correct, began to affect the monitors I had previously opened. It became readily apparent that the large amount of flow was beginning to exceed the online fire-pumping capacity.

A quick radio call was made to the utility department to activate the diesel-driven backup fire pumps so as to increase the flow and pressure in the dedicated fire water system. This did the trick. The pressure and flow immediately increased and gave the fire brigade the much-needed water supply to control and extinguish the fire. It was evident to all that without a reliable and sufficient water supply and distribution system, controlling an incident is highly unlikely.

Anthony R. Cole
Saudi Aramco
Dhahran, Saudi Arabia

## Brigade Member Tips

Flow rates for hydrants that are color-coded in accordance with NFPA 291, *Recommended Practice for Fire Flow Testing and Marking of Hydrants.*

| | | |
|---|---|---|
| Class C | Less than 500 GPM | Red |
| Class B | 500 to 999 GPM | Orange |
| Class A | 1,000 to 1,499 GPM | Green |
| Class AA | 1,500 GPM and above | Light blue |

GPM = gallons per minute.

## Testing Fire Hydrants

The amount of water available to fight a fire at a given location is a crucial factor in planning an attack. Will the hydrants deliver enough water at the needed pressure to enable brigade members to control a fire? If not, what can be done to improve the water supply? How can brigade members obtain additional water if a fire does occur?

Fire brigade teams may be assigned to test the flow from hydrants in their plants. The procedures for testing hydrants are relatively simple, but a basic understanding of the concepts of hydraulics and careful attention to detail are required. This section explains some of the basic theory and terminology of hydraulics, and describes how the tests are conducted and the results are recorded.

### Flow and Pressure

To understand the procedures for testing fire hydrants, brigade members must understand the terminology that is used. The flow or quantity of water moving through a pipe, hose, or nozzle is measured by its volume, usually in gallons (or liters) per minute. Water pressure refers to an energy level and is measured in pounds per square inch (psi) (or kilopascals). Volume and pressure are two different, but mathematically related, measurements.

Water that is not moving has potential energy. When the water is moving, it has a combination of potential energy and kinetic energy. Both the quantity of water flowing and the pressure under a specific set of conditions must be measured as part of testing any water system, including hydrants.

Static pressure is the pressure in a system when the water is not moving. Static pressure is potential energy, because it would cause the water to move if there were some place the water could go. Static pressure causes the water to flow out of an opened fire hydrant. If there were no static pressure, nothing would happen when brigade members opened a hydrant.

Static pressure is generally created by elevation pressure and/or pump pressure. An elevated storage tank creates elevation pressure in the water mains. Gravity also creates elevation pressure in a water system as the water flows from a hilltop reservoir to the water mains in the valley below. Pumps create pressure by bringing the energy from an external source into the system.

Static pressure in a water distribution system can be measured by placing a pressure gauge on a hydrant port and opening the hydrant valve. There cannot be any water flowing out of the hydrant when static pressure is measured.

Static pressure measured in this way assumes that there is no flow in the system. Because municipal water systems deliver water to hundreds or thousands of users, there is almost always water flowing within the system. In most cases, a static pressure reading is actually measuring the normal operating pressure of the system.

Normal operating pressure refers to the amount of pressure in a water distribution system during a period of normal consumption. In an industrial plant, for example, water is constantly used for process needs. In an industrial or commercial area, normal consumption occurs during a normal business day as water is used for various purposes. The system uses some of the static pressure to deliver this water. A pressure gauge connected to a hydrant during a period of normal consumption will indicate the normal operating pressure of the system.

Brigade members need to know how much pressure will be in the system when a fire occurs. Because the regular users of the system will be drawing off a normal amount of water even during a fire fight, the normal operating pressure is sufficient for measuring available water.

Residual pressure is the amount of pressure that remains in the system when water is flowing. When brigade members open a hydrant and start to draw large quantities of water out of the system, some of the potential energy of still water is converted to the kinetic energy of moving water. However, not all of the potential energy turns into kinetic energy; some of it is used to overcome friction in the pipes. The pressure remaining while the water is flowing is residual pressure.

Residual pressure is important because it provides the best indication of how much more water is available in the system. The more water that is flowing, the less residual pressure there is. In theory, when the maximum amount of water is flowing, the residual pressure is zero, and there is no more potential energy to push more water through the system. In reality, 20 psi is considered the minimum usable residual pressure, to reduce the risk of damage to underground water mains or pumps.

At the scene of a fire, a pump operator uses the difference between static pressure and residual pressure to determine how many more attack lines or appliances can be operated from the available water supply. The operator can refer to a set of tables to calculate the maximum amount of available water. The tables are based on the static and residual pressure readings taken during hydrant testing. Flow pressure measures the quantity of water flowing through an opening during a hydrant test. When a stream of water flows out through an opening (known as an orifice), all of the pressure is converted to kinetic energy. To calculate the volume of water flowing, brigade members measure the pressure at the center of the water stream as it passes through the opening, and factor in the size and flow characteristics of the orifice. A Pitot gauge is used to measure flow pressure in pounds per square inch (or kilopascals), and to calculate the flow in gallons per minute (or liters per minute).

Knowing the static pressure, the flow in gallons (liters) per minute, and the residual pressure enables brigade members to calculate the amount of water that can be obtained from a hydrant or a group of hydrants on the same water main. The test procedure is described in the following section.

## Hydrant Testing Procedure

The procedure for testing hydrant flows requires two adjacent hydrants, a Pitot gauge, and an outlet cap with a pressure gauge. Brigade members will measure static pressure and residual pressure at one hydrant and will open the other to let water flow out. The two hydrants should be connected to the same water main and preferably at about the same elevation ▼ Figure 15-15.

The cap gauge is placed on one of the outlets of the first hydrant. The hydrant valve is then opened to allow water to fill the hydrant barrel. The initial pressure reading on this gauge is recorded as the static pressure.

At the second hydrant, brigade members remove one of the discharge caps and open the hydrant. They put the Pitot gauge into the middle of the stream and take a reading. This is recorded as the Pitot pressure. At the same time, brigade members at the first hydrant record the residual pressure reading.

Using the size of the discharge opening (usually $2\frac{1}{2}$ inches) and the Pitot pressure, brigade members can calculate the flow in gallons per minute or look it up in a table. The table usually incorporates factors to adjust for the shape of the discharge opening. Brigade members can use special graph paper or computer software to plot the static pressure and the residual pressure at the test flow rate. The line defined by these two points shows the number of gallons (liters) per minute

### Brigade Member Tips

You have learned the steps needed to establish a water supply. Now you can understand how important this task is for the whole firefighting team, and how identifying the location of fire hydrants and static water sources during preincident planning can save time and lives.

Figure 15-15 Testing hydrant flow requires two hydrants on the same water main.

that are available at any residual pressure. The flow available for fire suppression is usually defined as the number of gallons (liters) per minute available at 20 psi residual pressure.

Several special devices are available to simplify taking accurate Pitot readings. Some outlet attachments have smooth tips and brackets that hold the Pitot gauge in the exact required position (▼ Figure 15-16). The flow can also be measured with an electronic flow meter instead of a Pitot gauge (► Figure 15-17).

Figure 15-16 The Pitot gauge.

Jones & Bartlett Learning. Photographed by Glen E. Ellman.

Figure 15-17 Pitot gauge (Cap gauge).

Jones & Barlett Learning. Photographed by Glen E. Ellman.

# Wrap-Up

## Ready for Review

- Industrial water systems consist of a water source and a water distribution system.
- Industrial water supply systems deliver larger volumes of pressure than municipal systems.
- Fire hydrants enable brigade members to obtain water from the municipal water supply system. Brigade members must know how to operate and hook up hoses to a fire hydrant.
- The two main types of hydrants are the dry-barrel hydrant and the wet-barrel hydrant.
- Maintenance of fire hydrants includes inspection and testing.

## Hot Terms

**Distributors** Relatively small-diameter underground pipes that deliver water to systems or devices.

**Dry-barrel hydrant** A type of hydrant used in areas subject to freezing weather. The valve that allows water to flow into the hydrant is located underground, and the barrel of the hydrant is normally dry.

**Dry hydrant** A permanent piping system that provides access to draft water from a static water source.

**Elevated water storage tank** An aboveground water storage tank that is designed to maintain pressure on a water distribution system.

**Elevation pressure** The amount of pressure created by gravity.

**Flow pressure** The amount of pressure created by moving water.

**Gravity feed system** A water distribution system that depends on gravity to provide the required pressure. The system storage is usually located at a higher elevation than the end users.

**Industrial water system** A water distribution system that is designed to deliver water to either fire protection pumps or both process and fire systems.

**Normal operating pressure** The observed static pressure in a water distribution system during a period of normal demand.

**Pitot gauge** A type of gauge that is used to measure the velocity pressure of water that is being discharged from an opening. Used to determine the flow of water from a hydrant.

**Primary feeder** The largest diameter pipes in a water distribution system, carrying the greatest amounts of water.

**Private water system** A privately owned water system operated separately from the municipal water system.

**Reservoir** A water storage facility.

**Residual pressure** The pressure remaining in a water distribution system while water is flowing. The residual pressure indicates how much more water is potentially available.

**Secondary feeder** Smaller diameter pipes that connect the primary feeders to the distributors.

**Shut-off valve** Any valve that can be used to shut down water flow to a water user or system.

**Static pressure** The pressure in a water pipe when there is no water flowing.

**Static water source** A water source such as a pond, river, stream, or other body of water that is not under pressure.

**Steamer port** The large diameter port on a hydrant.

**Water main** The generic term for any aboveground or underground water supply pipe.

**Water supply** A source of water.

**Wet-barrel hydrant** A hydrant used in areas that are not susceptible to freezing. The barrel of the hydrant is normally filled with water.

# Brigade Member in Action

**Shortly after lunchtime, a loud explosion emanates from a section of the plant. The plant sirens begin to alert plant personnel and fire brigade members, while automatic sprinkler deluge systems begin to activate. As plant responders and fire brigade team members arrive on the scene, hydrant-mounted monitors are flowing to cool the surrounding equipment. The first-in fire engine is about to arrive on the scene and radios to you that they plan to lay a line from a hydrant to the involved process unit. You arrive at the hydrant and await the fire engine. You prepare for a forward hose lay and to hook up to the dry-barrel hydrant.**

1. You should open and flow water out of the hydrant prior to hooking up to the supply hose to determine:
   - A. that air is not trapped in the barrel.
   - B. that all foreign matter is flushed out.
   - C. that the valve stem rotates.
   - D. the color of the water.

2. Opening the hydrant partially will:
   - A. cause the valve to prematurely fail.
   - B. undermine the soil around the hydrant.
   - C. allow maximum flow of water
   - D. not obstruct the flow of water.

3. You can check to see if a dry-barrel hydrant is draining water by:
   - A. using a flashlight and looking through an open outlet.
   - B. attaching a static gauge and checking for a vacuum.
   - C. holding your hand over the open outlet to feel for suction.
   - D. looking at the drain gauge indicator.

4. The purpose of draining a dry-barrel hydrant is to prevent:
   - A. freezing in cold temperatures.
   - B. corrosion.
   - C. potential contaminating algae bloom.
   - D. water hammer that could damage the industrial water system.

# Fire Hose, Nozzles, and Streams

# Chapter 16

# Fire Hose, Nozzles, and Streams

## NFPA 1081 Standard

### Incipient Facility Fire Brigade Member

**4.3.1*** Attack an incipient stage fire, given a handline flowing up to 473 L/min (125 gpm), appropriate equipment, and a fire situation, so that the fire is approached safely, exposures are protected, the spread of fire is stopped, agent application is effective, the fire is extinguished, and the area of origin and fire cause evidence are preserved.

**(A)** *Requisite Knowledge.* Types of handlines used for attacking incipient fires, precautions to be followed when advancing handlines to a fire, observable results that a fire stream has been properly applied, dangerous building conditions created by fire, principles of exposure protection, and dangers such as exposure to products of combustion resulting from fire condition.

**(B)** *Requisite Skills.* The ability to recognize inherent hazards related to the material's configuration; operate handlines; prevent water hammers when shutting down nozzles; open, close, and adjust nozzle flow; advance charged and uncharged hose; extend handlines; operate handlines; evaluate and modify water application for maximum penetration; assess patterns for origin determination; and evaluate for complete extinguishment.

### Advanced Exterior Facility Fire Brigade Member

**5.2.3 (A)** *Requisite Knowledge.* Principles of fire streams; types, design, operation, nozzle pressure effects, and flow capabilities of nozzles; precautions to be followed when advancing handlines to a fire; observable results that a fire stream has been correctly applied; dangerous conditions created by fire; principles of exposure protection; potential long-term consequences of exposure to products of combustion; physical states of matter in which fuels are found; the application of each size and type of attack line; the role of the backup team in fire attack situations; attack and control techniques; and exposing hidden fires.

**5.2.3 (B)** *Requisite Skills.* The ability to prevent water hammers when shutting down nozzles; open, close, and adjust nozzle flow and patterns; apply water using direct, indirect, and combination attacks; advance charged and uncharged 38 mm ($1^1/_2$ inch) diameter or larger handlines; extend handlines; replace burst hose sections; operate charged handlines of 38 mm ($1^1/_2$ inch) diameter or larger; couple and uncouple various handline connections; carry hose; attack fires; and locate and suppress hidden fires.

### Interior Structural Facility Fire Brigade Member

**6.2.1 (A)** *Requisite Knowledge.* Principles of conducting initial fire size-up; principles of fire streams; types, design, operation, nozzle pressure effects, and flow capabilities of nozzles; precautions to be followed when advancing hose lines to a fire; observable results that a fire stream has been correctly applied; dangerous building conditions created by fire; principles of exposure protection; potential long-term consequences of exposure to products of combustion; physical states of matter in which fuels are found; common types of accidents or injuries and their causes; and the application of each size and type of handlines, the role of the backup team in fire attack situations, attack and control techniques, and exposing hidden fires.

**6.2.1 (B)** *Requisite Skills.* The ability to prevent water hammers when shutting down nozzles; open, close, and adjust nozzle flow and patterns; apply water using direct, indirect, and combination attacks; advance charged and uncharged 38 mm ($1^1/_2$ inch) diameter or larger handlines; extend handlines; replace burst hose sections; operate charged handlines of 38 mm ($1^1/_2$ inch) diameter or larger; couple and uncouple various handline connections; carry hose; attack fires; and locate and suppress hidden fires.

## Knowledge Objectives

After completing this chapter, you will be able to:

- Describe how to prevent water hammers.
- Describe how a hose is constructed.
- Describe the types of hoses used in the fire service.
- Describe how to clean and maintain a hose.
- Describe how to inspect a hose.
- Describe how to note a defective hose.
- Describe how to roll a hose.
- Describe how to lay a supply line.
- Describe how to load a hose.
- Describe how to connect a hose to a water supply.
- Describe how to carry and advance a hose.
- Describe the types and designs of nozzles.
- Describe pressure effects and flow capabilities of nozzles.

## Skills Objectives

After completing this chapter, you will be able to perform the following skills:

- Replace the swivel gasket.
- Perform the one-person foot-tilt method of coupling a fire hose.
- Perform the two-person method for coupling a fire hose.
- Perform the one-person knee-press method of uncoupling a fire hose.
- Perform the two-person stiff-arm method.
- Uncouple a hose with spanners.
- Connect two lines with damaged coupling.
- Clean hoses.
- Mark a defective hose.
- Perform the straight hose roll.
- Perform the single-donut hose roll.
- Perform the double-donut hose roll.
- Perform the self-locking double-donut hose roll.
- Perform the forward lay.
- Perform the four-way hydrant valve.
- Perform the reverse lay.
- Perform the split hose lay.
- Perform a flat hose load.
- Perform a horseshoe load.
- Perform an accordion hose load.
- Attach a soft sleeve hose to a fire hydrant.
- Attach a hard suction hose to a fire hydrant.
- Load the minuteman hose load.
- Advance the minuteman hose load.
- Loading a preconnected flat load.
- Advance the preconnected flat hose load.
- Load the triple layer hose load.
- Advance the triple layer hose load.
- Unload and advance the wyed lines.
- Perform a working hose drag.
- Perform a shoulder carry.
- Advance an accordion load.
- Advance a hose line up a stairway.
- Advance a hose line down a stairway.
- Advance an uncharged hose line up a ladder.
- Use a hose stream from a ladder.
- Connect to a standpipe system.
- Advance from a standpipe.
- Replace a defective hose section.
- Drain and carry a hose.
- Operate a smooth bore nozzle.
- Operate a fog nozzle.

# You Are the Brigade Member

**You and your crew are responding on the first-due foam engine to the report of a fire in a quality assurance laboratory within your petrochemical facility. Based on what you know about the location of the fire within the lab building, you anticipate that you will have to fight either a class A or class B fire. You arrive on the scene to find a fire in a storage room on the second floor of the building. The incident commander (IC) instructs you and your crew to stretch a line to combat the fire.**

1. *What would be the easiest method to extend the attack line to the second floor?*
2. *If you want the capability of providing either water or foam, what would your nozzle of choice be on the attack line?*
3. *What are the advantages of using a 1¾-inch line for fire attack operations?*

## Introduction

### Fire Hydraulics

Fire hydraulics deal with the properties of energy, pressure, and water flow as related to fire suppression. When operating hose lines at a fire, it is important to understand some basic principles of hydraulics. You need to understand the basic concepts of friction loss in different size hose lines, changes in pressure due to elevation, and water hammer. Brigade members who advance to the position of pump operator will learn more about fire service hydraulics.

#### Flow

Flow refers to the volume of water that is being moved through a pipe or hose. In fire hydraulics, flow is measured in gallons per minute (gpm).

#### Pressure

The amount of energy in a body or stream of water is measured as pressure. In fire hydraulics, pressure is measured in pounds per square inch (psi). Pressure is required to push water through a hose, to expel water through a nozzle, or to lift water up to a higher level. A pump adds energy to a water stream, causing an increase in pressure.

#### Friction Loss

Friction loss is a loss of pressure as water moves through a pipe or hose. This loss of pressure represents the energy required to push the water through the hose. Friction loss is influenced by the diameter of the hose, the volume of water traveling through the hose, and the distance the water travels. In a given size hose, a higher flow rate produces more friction loss. In a 2½-inch hose, a 300-gpm flow causes much more friction loss than a 200-gpm flow. At a given flow rate, the smaller the diameter of the hose, the greater the friction loss. At a flow of 500 gpm, the friction loss in a 2½-inch hose is much greater than the friction loss in a 4-inch hose. With any combination of flow and diameter, the friction loss is directly proportional to the distance. At a flow of 250 gpm in a 2½-inch hose, the friction loss in 200 feet is double the friction loss in 100 feet.

#### Elevation Pressure

Elevation affects water pressure. An elevated water tank supplies pressure to a water system because of the difference in height between the water in the water tank and the underground delivery pipes. If a fire hose is laid down a hill, the water at the bottom will have additional pressure due to the change in elevation. If a fire hose is advanced upstairs to the third floor of a building, it will lose pressure due to the energy required to lift the water. The fire pump operator has to take elevation changes into account when setting the discharge pressure.

#### Water Hammer

Water hammer is a surge in pressure caused by suddenly stopping the flow of a stream of water. A fast-moving stream

## Brigade Member Tips

Open and close hydrants and nozzles slowly to prevent water hammer.

## Fire Marks

OSHA 1910.155 Subpart L, limits incipient firefighting to the use of Class II standpipe systems or hose systems flowing 125 gpm or less.

of water has a large amount of kinetic energy. If the water suddenly stops moving when a valve is closed, all of the kinetic energy is converted to an instantaneous increase in pressure. Because water cannot compress, the additional pressure is transmitted along the hose or pipe as a shock wave. Water hammer can rupture a hose, cause a coupling to separate, or damage the plumbing on a piece of fire apparatus. Severe water hammer can even damage an underground piping system. Brigade members have been injured by equipment that was damaged by a water hammer.

A similar situation can occur if a valve is opened too quickly and a surge of pressurized water suddenly fills a hose. The surge in pressure can damage the hose or cause the brigade member at the nozzle to lose control of the stream.

To prevent water hammer, always open and close fire hydrant valves slowly. Pump operators also need to open and close the valves on fire engines slowly. When you are operating the nozzle on an attack line, open the nozzle slowly. Most importantly, when you close the shut-off valve on an attack line, do it slowly.

# Fire Hoses

## Functions of Fire Hoses

Fire hoses are used for two main purposes. The hoses used to discharge water from an attack engine onto the fire are called attack hoses, or attack lines. Most attack hoses carry water directly from the attack engine to a nozzle that is used to direct the water onto the fire. In some cases, an attack line is attached to a deck gun, ground monitor, or some other type of master stream appliance. Attack lines can also be used to deliver water to a fire department connection that supplies a standpipe or sprinkler system inside a building.

Hoses used to deliver water to an attack engine are called supply hoses, or supply lines. The water can come directly from a hydrant or it can come from another engine that is being used to provide a water supply for the attack engine. Supply line sizes range from 3 to 12 inches. Attack hoses usually operate at higher pressures than supply lines. Supply hoses are designed to carry larger volumes of water at lower pressures.

## Sizes of Hose

Fire hoses range in size from 1 to 12 inches in diameter (▶ Figure 16-1). The nominal hose size refers to the inside diameter of the hose when it is filled with water. The smaller diameter hoses are used as attack lines and the large-diameter hoses (LDHs) are almost always used as supply lines. Medium diameter hose sizes can be used as either attack lines or supply lines. Because there are differing definitions of hose terms throughout the country, hose sizes can be grouped into three general categories.

Small diameter hose (SDH) lines range in size from 1 to 2 inches in diameter. Some fire apparatus are equipped with a reel of hard rubber hose called a Booster hose, or booster line, which is used for small fires such as a dumpster fire. Booster lines are still in use around the United States, but have continued to lose ground in favor of larger hose. Small hose standpipes (Class II) are equipped with $1^{1}/_{2}$-inch rigid rubber or flat hose on reels. The rubber hose is often used in areas subject to harsh environments. A lightweight collapsible 1-inch hose, known as forestry hose, is often used to fight brush fires.

The hoses that are most commonly used to attack incipient and interior fires are either $1^{1}/_{2}$ or $1^{3}/_{4}$ inches in diameter. These hoses are usually connected directly to a handline nozzle. Some fire brigades also use 2-inch attack lines. Each section of attack hose is typically 50-inch long.

Hoses $2^{1}/_{2}$ to 4-inches in diameter are called medium diameter hose (MDH). Hoses in this size range can be used as either supply lines or attack lines. Large handline nozzles are often

Figure 16-1 Fire hose comes in a wide range of sizes for different uses and situations.

used with $2^1/_2$-inch hose to attack larger fires. When used as an attack hose, the 3-inch hose is more often used to deliver water to a master stream device or a department connection and the 4-inch hose is usually the smallest hose that is used to supply apparatus. These hose sizes typically come in 50-foot lengths.

NFPA 1962 defines hoses that are $3^1/_2$ inches or larger in diameter as large-diameter hoses (LDH). Most industrial brigades refer to LDH as hose that is 5 inches or larger. Five- and six-inch hoses are commonly used to provide apparatus water supply. Larger supply hose (up to $7^1/_4$ inches) is used by many fire brigades that require high-volume water flow. LDH up to 12" is used in some industrial facilities when protection of process plant operations requires an extremely large-volume water supply. Standard lengths of 50 and 100 feet are available for LDH.

Fire hose is designed to be used as either attack hose or supply hose. Attack hose must withstand higher pressures and is designed to be used in a fire environment where it may be subjected to high temperatures, sharp surfaces, abrasion and other potentially damaging conditions. LDH is constructed to operate at lower pressures than attack hose and in less severe operating conditions; however, it must still be durable and resistant to external damage. Attack hose can be used as supply hose, but LDH must never be used as attack hose.

Fire hose must be tested annually. Testing requirements can be found in NFPA 1962, *Inspection, Care, and Use of Fire Hose, Couplings, and Nozzles, and the Service Testing of Fire Hose.*

## Hose Construction

Most fire hose is constructed with an inner waterproof liner surrounded by either one or two outer layers. The outer layers provide the strength to withstand the high pressures that are exerted by the water inside the hose. The strength is provided by a woven mesh made from high strength synthetic fibers such as nylon that are resistant to high temperatures, mildew, and many chemicals. These fibers can also withstand some mechanical abrasion.

Double-jacket hose is constructed with two layers of woven fibers. The outer layer serves as a protective covering, while the inner layer provides most of the strength. The tightly woven outer jacket can resist abrasion, cutting, hot embers, and other external damage. The woven fibers are treated to resist water and provide added protection from many common hazards that are likely to be encountered at the scene of a fire.

Instead of a double jacket, some fire hoses are constructed with a durable rubber-like compound as the outer covering. This material is bonded to a single layer of strong woven fibers that provides the strength to keep the hose from rupturing under pressure. This type of construction is called rubber–covered hose, or rubber-jacket hose ► Figure 16-2.

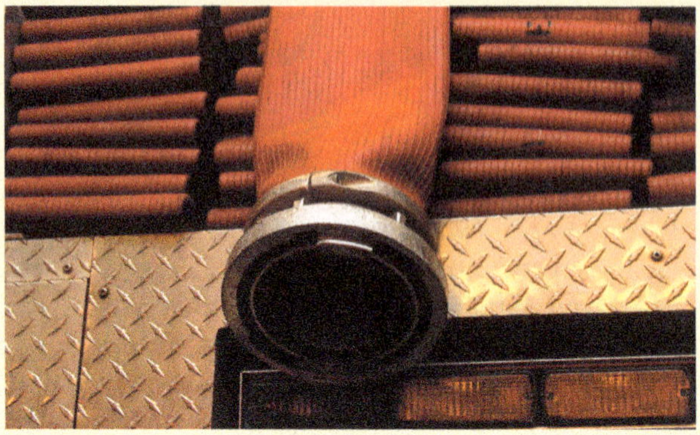

Figure 16-2 Rubber-covered hose.

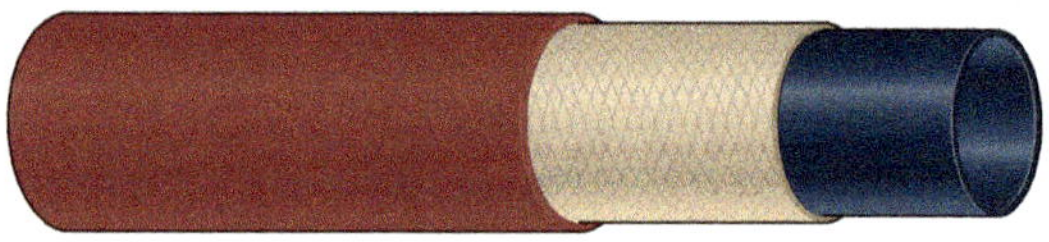

Figure 16-3 The liner inside a fire hose can be made from synthetic rubber or a variety of membrane materials.

Both types of hose are designed to be stored flat and to fold easily. This allows a much greater lengths of hose to be stored in the hose bed on fire apparatus.

The hose liner, or hose inner jacket, is the inner part of the hose ▲ Figure 16-3. This liner prevents the water from leaking out of the hose and provides a smooth inside surface for the water to move against. Without this smooth surface, there would be excessive friction between the moving water and the inside of the hose, reducing the amount of pressure that could reach the nozzle. The inner liner is usually made of a synthetic rubber compound or a thin flexible membrane material that can be flexed and folded without developing leaks. In double-jacket hose, the liner is bonded to the inner woven jacket. In a rubber-covered hose, the inner and outer layers are usually bonded together and the woven fibers are contained within.

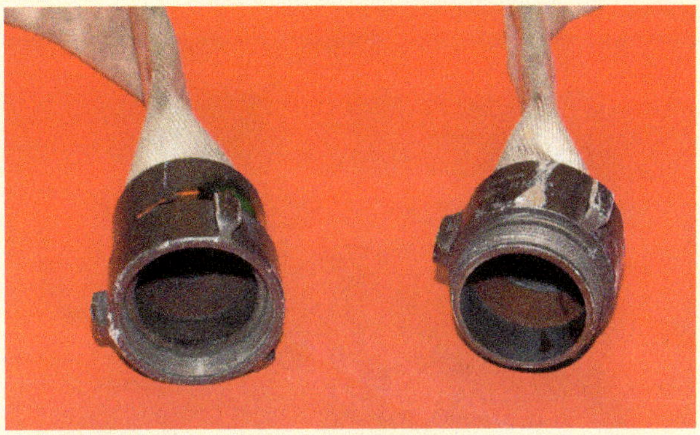

Figure 16-4 A set of threaded couplings includes one male and one female coupling. The male coupling has exposed threads, while the threads on the female coupling are inside the swivel.

Figure 16-5 A spanner wrench is used to tighten or loosen a hose coupling.

Figure 16-6 Higbee indicators show the position where the threads on a pair of couplings are properly aligned with each other.

## Hose Couplings

Couplings are used to connect individual lengths of fire hose together. Couplings are also used to connect a hose line to a hydrant; to an intake or discharge valve on a engine; or to a variety of nozzles, fittings, and appliances. A coupling is permanently attached to each end of a section of fire hose. The two most common types of fire hose couplings are threaded hose couplings or nonthreaded (Storz-type couplings).

### Threaded Couplings

Threaded couplings are used on most hoses up to 3" in diameter and on soft sleeve hose and hard suction hose. A set of threaded couplings consists of a male coupling, which has the threads on the outside, and a female coupling, which has matching threads on the inside (▲ Figure 16-4). The female coupling has a swivel, so the male and female ends can be attached together without twisting the hose. A length of fire hose has a male coupling on one end and a female coupling on the other end.

When connecting fire hoses with threaded couplings, make sure the threads are properly aligned so the male and female couplings will engage fully. When the couplings are properly aligned, the two ends should attach together with minimal resistance. The swivel on the female coupling should be turned until the connection is snug, but only hand tight, so the couplings can be easily disconnected.

If there is any leakage after the hose is filled with water, further tightening may be needed. Use a spanner wrench to gently tighten the couplings until the leak is stopped. Spanner wrenches are used to connect and disconnect hose couplings (► Figure 16-5). Normally two spanners are used together to rotate the two couplings in opposing directions.

Couplings are constructed with either rocker lugs, or pin lugs, to engage a spanner wrench. Using a wrench to tighten couplings on an empty hose or overtighten couplings on a filled hose can damage the gaskets and cause them to leak. A spanner wrench may be needed to uncouple the hose after it has been pressurized with water.

Higbee indicators (sometimes called a Higbee notch) show the position where the ends of the threads on a pair of couplings are properly aligned with each other. Using the Higbee indicators will help you to couple hose more quickly. When the indicators on the male and female couplings are aligned, the two couplings should connect quickly and easily (▲ Figure 16-6).

An important part of a threaded coupling is the rubber gasket. The gasket is an O-shaped piece of rubber that sits inside the swivel section of the female coupling. When the male coupling is tightened down against it, a seal is formed that stops water from leaking. If the gasket is damaged or missing, the coupling will leak. These gaskets can deteriorate with time and can also be damaged by overtightening the coupling.

16-1 Skill Drill

## Replacing the Swivel Gasket

1 Fold the new gasket, bringing the thumb and forefinger together, creating two loops.

2 Place either of the two loops into the coupling and against the gasket seat.

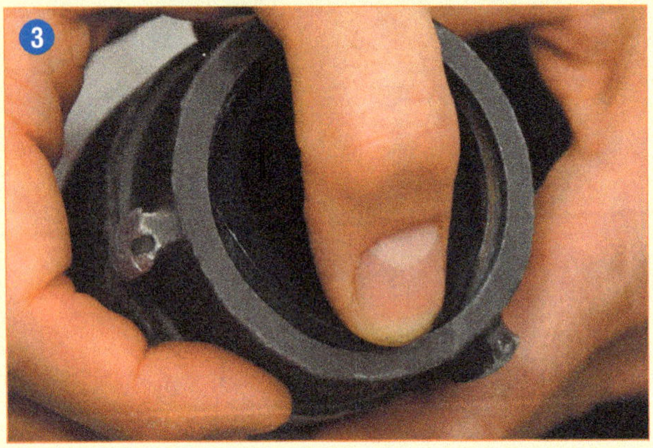

3 Using the thumb, push the remaining unseated portions into the coupling until the entire gasket is properly positioned against the coupling seat.

Periodically, the gaskets must be changed as part of the maintenance on the hose.

While a leaking coupling is not a critical problem during most firefighting operations, it can result in unnecessary water damage. The best way to prevent leaks is to make sure the gaskets are in good condition and replace any gaskets that are missing or damaged. To replace the swivel gasket, follow the steps in ▲ Skill Drill 16-1.

1. Fold the new gasket, bringing the thumb and forefinger together, creating two loops. **(Step 1)**
2. Place either of the two loops into the coupling and against the gasket seat. **(Step 2)**
3. Using the thumb, push the remaining unseated portions into the coupling until the entire gasket is properly positioned against the coupling seat. **(Step 3)**

### Storz-Type Couplings

Storz-type couplings are designed so that the couplings on both ends of a length of hose are the same. There is no male or female end to the hose. When this system is used, each coupling can be attached to any other coupling of the same diameter ▶ Figure 16-7. Storz-type couplings are made for all hose sizes; however, in North America they are most commonly used on an LDH.

Figure 16-7 Storz-type couplings are designed so that the couplings on both ends of a length of hose are the same.

Storz-type couplings are connected by mating the two couplings face-to-face and then turning clockwise one-third of a turn. To disconnect a set of couplings, the two parts are rotated counterclockwise one-third of a turn. A spanner wrench can be used to tighten a leaking coupling or to release a connection that cannot be loosened by hand.

Adaptors are used to connect Storz-type couplings to threaded couplings or to connect couplings of different sizes together. Many fire brigades use LDH with Storz-type couplings as a supply line between a hydrant and an engine.

## Coupling and Uncoupling Hose

There are several techniques for coupling and uncoupling. Depending upon the circumstances, each is effective. A brigade member should learn how to perform each. To perform the one-person foot-tilt method of coupling fire hose, follow the steps in ▶ Skill Drill 16-2.

1. Place one foot on the hose behind the male coupling.
2. Push down with your foot to tilt the male coupling upward. (Step 1)
3. Place one hand behind the female coupling and grasp the hose. (Step 2)
4. Place the other hand on the coupling swivel.
5. Bring the two couplings together and align the Higbee indicators. Rotate the swivel in a clockwise direction to connect the hoses. (Step 3)

To perform the two-person method for coupling a fire hose, follow the steps in ▶ Skill Drill 16-3.

### Brigade Member Safety Tips

NEVER attempt to uncouple charged hose lines.

1. Pick up the male end of the coupling. Grasp it directly behind the coupling and hold it tightly against the body. (Step 1)
2. The second brigade member holds the female coupling firmly with both hands. (Step 2)
3. The second brigade member brings the female coupling to the male coupling. (Step 3)
4. The second brigade member aligns the female coupling with the male coupling. Use the Higbee indicators for easy alignment. (Step 4)
5. The second brigade member turns the female coupling counterclockwise until it clicks. This indicates the threads are aligned. (Step 5)
6. Turn the female coupling clockwise to couple the hoses. (Step 6)

Charged hose lines should never be disconnected while the hose is under pressure. The loosened couplings can flail around wildly and cause serious injury to personnel or bystanders in the vicinity. Always shut off the water supply and bleed off the pressure before uncoupling. The water pressure will make it difficult to uncouple a charged hose line. If the coupling resists an attempt to uncouple, check to make sure the pressure is relieved before using spanner wrenches to loosen the coupling.

To perform the one-person knee-press method of uncoupling a fire hose, follow the steps in ▶ Skill Drill 16-4.

1. Pick up the connection by the female coupling end. (Step 1)
2. Turn the connection upright, resting the male coupling on a firm surface. (Step 2)
3. Place a knee on the female coupling and with body weight press down (this compresses the gasket).
4. Turn the female swivel counterclockwise and loosen the coupling. (Step 3)

To perform the two-person stiff-arm method of uncoupling a hose, follow the steps of ▶ Skill Drill 16-5.

1. Two brigade members face each other and firmly grasp their respective coupling. (Step 1)
2. With elbows locked straight, they push toward each other. (Step 2)
3. While pushing toward each other, the brigade members turn the coupling counterclockwise, loosening the coupling. (Step 3)

To uncouple a hose with spanners, follow the steps in ▶ Skill Drill 16-6.

1. With the connection on the ground, straddle connection above the female coupling. (Step 1)

16-2 Skill Drill

## Performing the One-Person Foot-Tilt Method of Coupling a Fire Hose

Place one foot on the hose behind the male coupling. Push down with your foot to tilt the male coupling upward.

Place one hand behind the female coupling and grasp the hose.

Place the other hand on the coupling swivel. Bring the two couplings together and align the Higbee indicators. Rotate the swivel in a clockwise direction to connect the hoses.

2. Place one spanner wrench on the female coupling with handle to the left. **(Step 2)**
3. Place the second spanner wrench on the male coupling with the handle to the right. **(Step 3)**
4. Push both spanner handles down toward the ground, loosening the connection. **(Step 4)**

To connect two lines with damaged coupling, follow the steps of **Skill Drill 16-7**.

1. Using a hose jacket, open the hose jacket and place the damaged coupling in one end.
2. Place the second coupling in the other end of the jacket.
3. Close the hose jacket, ensuring that the latch is secure.

Slowly bring the hose line up to pressure, allowing the gaskets to seal around the hose ends.

## Attack Hose

Attack hose is designed to be used for fire suppression where it can be exposed to heat and flames, hot embers, broken glass, sharp objects, and many other potentially damaging conditions. It must be tough, but flexible and light in weight.

Fire brigades typically use two sizes of hose as attack lines for fire suppression. The smaller size is usually a $1\frac{1}{2}$ or $1\frac{3}{4}$ inches in diameter, while $2\frac{1}{2}$-inch hose is most often used for heavy interior attack lines.

16-3 Skill Drill

## Performing the Two-Person Method of Coupling a Fire Hose

Pick up the male end of the coupling. Grasp it directly behind the coupling and hold it tightly against the body.

The second brigade member holds the female coupling firmly with both hands.

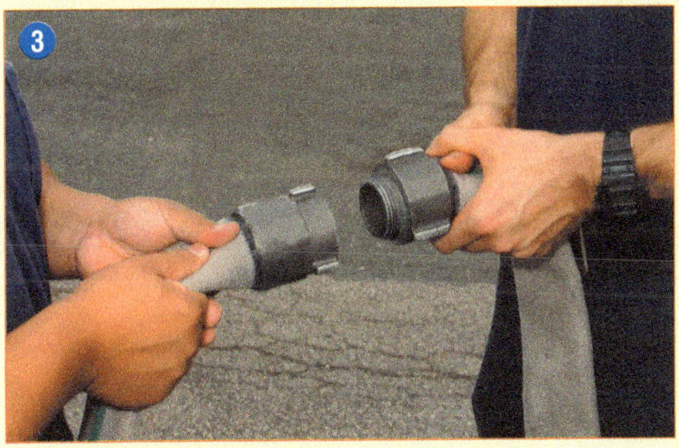

The second brigade member brings the female coupling to the male coupling.

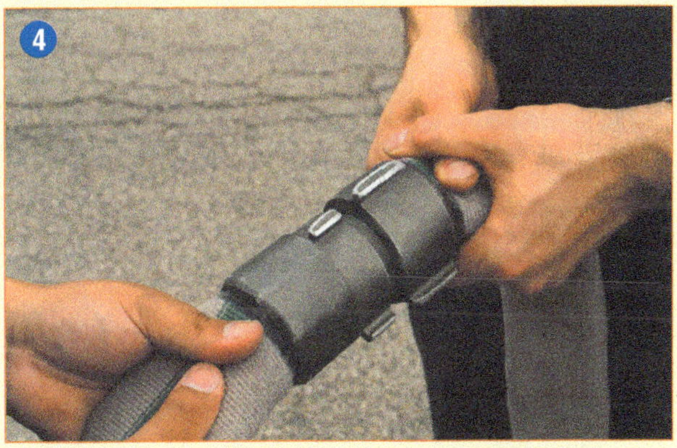

The second brigade member aligns the female coupling with the male coupling. Use the Higbee indicators for easy alignment.

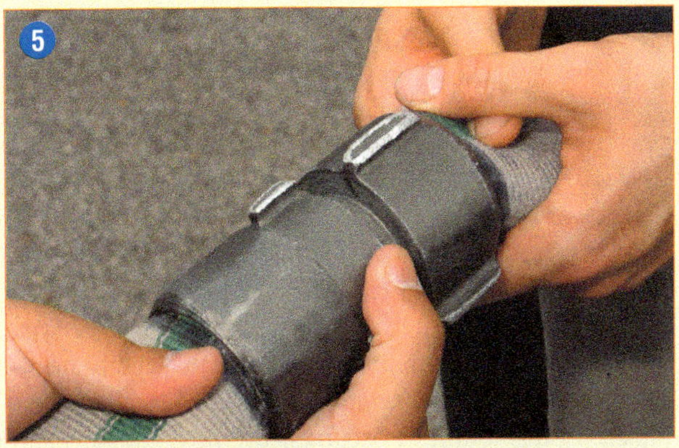

The second brigade member turns the female coupling counterclockwise until it clicks. This indicates the threads are aligned.

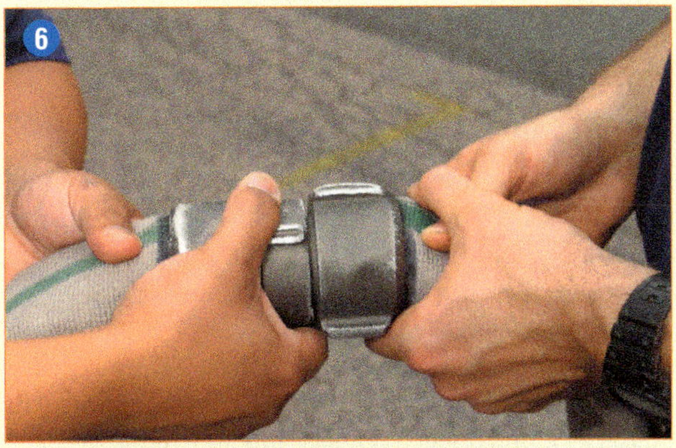

Turn the female coupling clockwise to couple the hoses.

16-4 Skill Drill

# Performing the One-Person Knee-Press Method of Uncoupling a Fire Hose

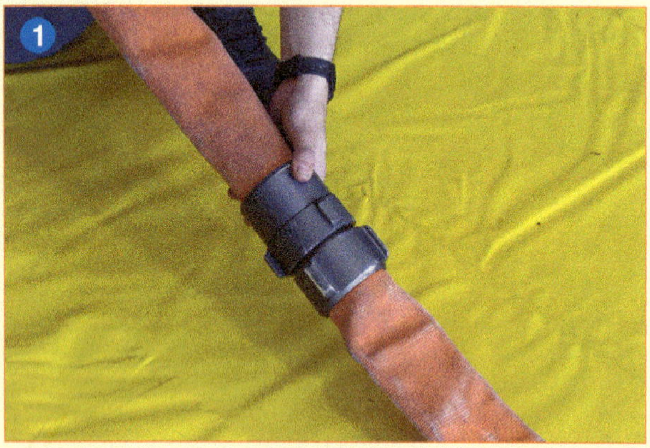

1 Pick up the connection by the female coupling end.

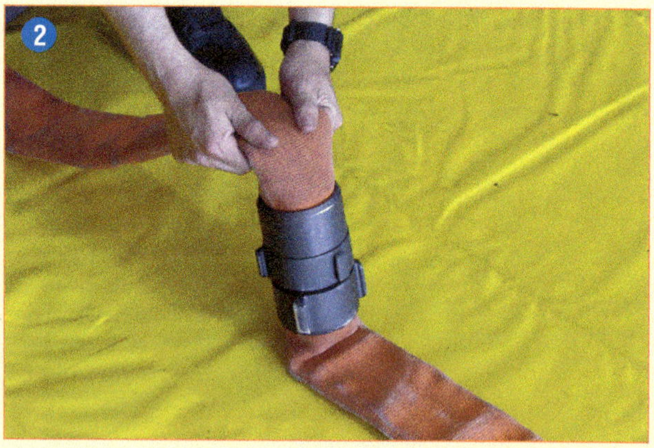

2 Turn the connection upright, resting the male coupling on a firm surface.

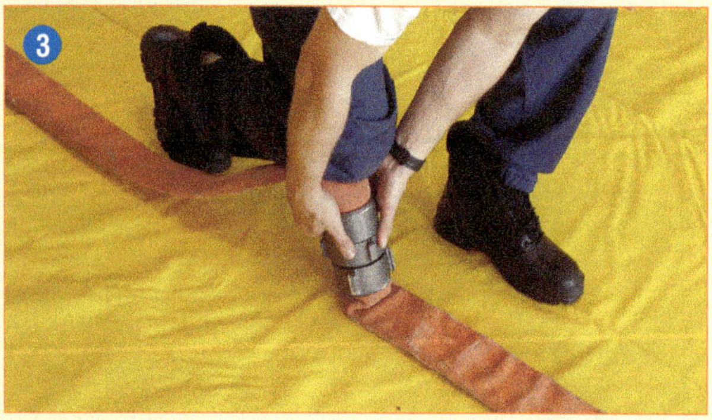

3 Place a knee on the female coupling and with body weight press down. Turn the female swivel counterclockwise and loosen the coupling.

## $1\frac{1}{2}$- and $1\frac{3}{4}$-inch Attack Hose

Most fire brigades use either $1\frac{1}{2}$- or $1\frac{3}{4}$-inch hose as the primary attack line for most fires. Both sizes of hose use the same $1\frac{1}{2}$-inch couplings. This is also the attack hose that is used most often during basic fire training. Handlines of this size can usually be operated by two brigade members, although a third person on the line makes it much easier to advance and control. This hose is often stored on fire apparatus as a preconnected attack line in lengths of 150 to 350 feet, ready for immediate use.

The primary difference between $1\frac{1}{2}$ and $1\frac{3}{4}$-inch hose is the amount of water that can flow though the hose. Depending on the pressure in the hose and the type of nozzle that is used, a $1\frac{1}{2}$-inch hose can generally flow between 60 and 125 gal1ons of water per minute. An equivalent $1\frac{3}{4}$-inch hose can flow between 120 and 200 gpm. This is important because the amount of fire that can be extinguished is directly related to the amount of water that is applied to it. A $1\frac{3}{4}$-inch hose can deliver much more water and is only slightly heavier and more difficult to advance than a $1\frac{1}{2}$-inch hose line. A $1\frac{1}{2}$-inch attack hose is found in occupant-use hose cabinets and small hose standpipe systems (Class II). Flat hose is found in hose cabinets and horizontal racks. Both rigid rubber hose and flat hose are found in hose reels.

16-5 Skill Drill

## Performing the Two-Person Stiff-Arm Method of Uncoupling a Hose

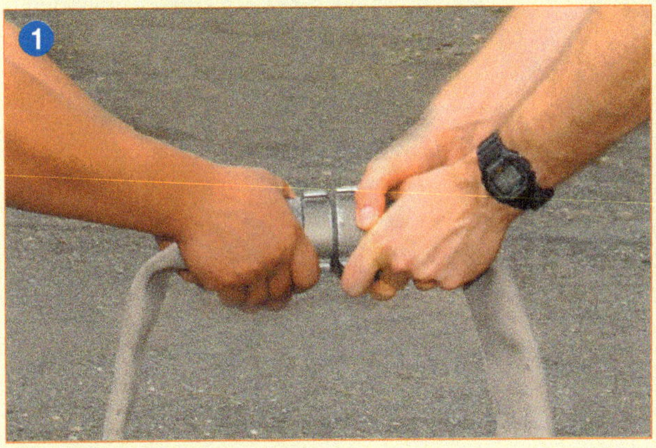

Two brigade members face each other and firmly grasp their respective coupling.

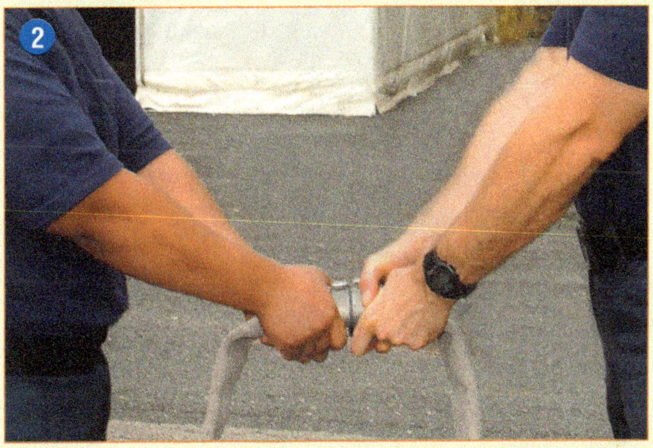

With elbows locked straight, the brigade members push toward each other.

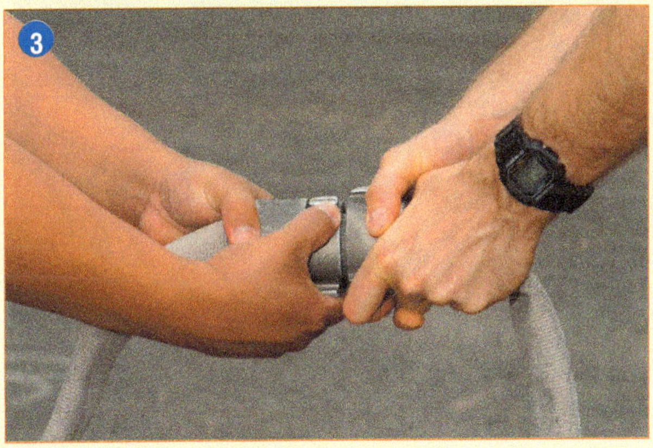

While pushing toward each other, the brigade members turn the coupling counterclockwise, loosening the coupling.

### $2^1/_2$-Inch Attack Hose

A $2^1/_2$-inch hose is used as an attack line for fires that are too large to be controlled by a $1^1/_2$- or $1^3/_4$-inch hose line. A $2^1/_2$-inch handline hose is generally considered to flow about 250 gallons of water per minute. It takes at least two brigade members to safely control a $2^1/_2$-inch handline hose due to the weight of the hose and the water and the nozzle reaction force. A 50-foot length of dry 3-inch hose weighs up to 30 lb. When the hose is charged and filled with water it can weigh as much as 150 lb per length.

Higher flows, up to approximately 350 gpm, can be achieved with higher pressures and larger nozzles; however, it is difficult to operate a handline hose at these high flow rates. These flows are more likely to be used to supply a master stream device.

### Booster Hose

A booster hose is usually carried on a hose reel that holds 150 or 200 feet of rubber hose. Booster hose contains a steel wire that gives it a rigid shape. The rigid shape of this hose allows it to flow water without pulling all the hose off the reel. It is light in weight and can be advanced quickly by one person.

The disadvantage of booster hose is its limited flow. The normal flow from a 1-inch booster hose is 40–50 gpm, which

Skill Drill 16-6

## Uncoupling Hose with Spanners

1 With the connection on the ground, straddle connection above the female coupling.

2 Place one spanner wrench on the female coupling with handle to the left.

3 Place the second spanner wrench on the male coupling with the handle to the right.

4 Push both spanner handles down toward the ground, loosening the connection.

is not an adequate flow for structure fires. The use of booster hose is usually limited to small outdoor fires and trash dumpsters. Booster hose should not be used for structural firefighting.

### Supply Hose

Supply hose is used to deliver water to an attack engine from a pressurized source, which could be a hydrant or another engine working in a relay operation. Supply lines range from 3 to 12 inches in diameter. The choice is made based on the preferences and operating requirements of each fire brigade. This depends on the amount of water needed to supply the attack engine, the distance from the source to the attack engine, and the pressure that is available at the source.

Engines are normally loaded with at least one bed of hose that can be laid out as a supply line. When threaded couplings are used, this hose can be loaded to lay out from the hydrant to the fire (known as a forward lay) or from the fire to the hydrant (known as a reverse lay). Sometimes engines are loaded with two beds of hose so they can easily drop a supply line in either direction. If Storz-type couplings are used or the necessary adaptors are provided, hose from the same bed can be laid in either direction.

Figure 16-8 Soft sleeve hose.

Figure 16-9 A hard suction hose.

When 2½-inch hose is used as a supply line, it is usually the same type of hose used for attack lines. This size hose has a limited flow capacity, but it can be effective at low to moderate flow rates and over short distances. Sometimes two parallel lines of 2½-inch hose are used to provide a more effective water supply.

Large-diameter supply lines are much more efficient than 2½-inch hose for moving larger volumes of water over longer distances. Many fire brigades use 5-inch hose or larger as their standard supply line. A single 5-inch supply line can deliver flows exceeding 1,500 gpm under some conditions. LDH is heavy and difficult to move after it has been charged with water. This hose comes in 50 and 100-foot lengths. A typical fire engine may carry 750 to 1250 feet of supply hose on it.

### Soft Sleeve Hose

A soft sleeve hose (also known historically as soft sleeve hose) is a short section of LDH that is used to connect a fire department pumper directly to the large streamer outlet on a hydrant or other pressurized water source (▶ Figure 16-8). Use of this type of hose allows as much water as possible to flow from the hydrant or pressurized water source to the pump through a single hose. A soft sleeve hose may have Storz or similar connections on both ends or a female connection on each end. If it has two Storz connections, an adapter is required to connect the large-diameter inlet to the engine apparatus and the hydrant or pressurized water source. If the hose uses two female threaded connections, one end should match the local hydrant threads, and the other end should match the threads on a large-diameter inlet to the engine apparatus. The couplings may have large handles to allow for quick tightening by hand. The hose can be from 4 to 6 inches in diameter and is usually between 10 and 25 feet in length.

### Hard Suction

A hard suction hose is a special type of supply hose used to draft water from a static source such as a river, lake, or portable drafting basin (▶ Figure 16-9). The water is drawn through this hose into the pump on an engine or into a portable pump. It is called a hard suction hose because it is designed to remain rigid and will not collapse when a vacuum is created in the hose to draft the water into the pump.

Hard suction hose typically comes in 10 or 20 feet sections. The diameter is based on the capacity of the pump and can be as large as 6 inches. The hose can be made from either rubber or plastic; however, the newer plastic versions are much lighter and more flexible.

Long handles are provided on the female couplings of hard suction hose to assist in tightening the hose. In order to draft water, it is essential to have an airtight connection at each coupling. Sometimes it may be necessary to gently tap these handles with a rubber mallet to tighten the hose or to disconnect it. Tapping these handles with anything metal could cause damage to the handles or the coupling.

## Hose Care, Maintenance, and Inspection

Fire hose should be regularly inspected and tested following the procedures in NFPA 1962, *Inspection, Care and Use of Fire Hose, Couplings, and Nozzles and the Service Testing of Fire Hose*. Hoses that are not properly maintained can deteriorate over time and eventually fail. The gaskets in female couplings need to be checked regularly and replaced when they are worn or damaged.

### Causes and Prevention of Hose Damage

The fire hose is a lifeline for brigade members. Every time brigade members fight a fire, they have to rely on fire hose to deliver the water needed to attack the fire and protect themselves from the fire. Fire hose is a highly engineered product designed to perform well under adverse conditions. We must be careful to prevent damage to the hose that could result in premature or unexpected failure. The most common factors that can cause damage to fire hose include mechanical causes, chemicals, heat, cold, and mildew.

### Mechanical Damage

Mechanical damage can occur from many sources. Hose that is dragged over rough objects or along a roadway can be damaged by abrasion. Sharp objects can cut through the hose. Be especially careful if you need to place a hose line through a broken window; remove any protruding sharp edges of glass first. Particles of grit caught in the fibers can damage the jacket or puncture holes in the liner. Reloading dirty hose can cause damage to the fibers in the hose jacket.

Fire hose is likely to be damaged if it is run over by a vehicle. Hose ramps should be used if traffic has to drive over a hose. Hose couplings can also be damaged by other mechanical forces, such as dropping them on the ground. The exposed threads on male couplings are easily damaged if they are dropped. Avoid dragging hose couplings, since this can cause damage to the threads and to the swivels.

### Heat and Cold

Hoses can be damaged by heat and cold and prolonged exposure to sunlight. Heat is an obvious concern when fighting a fire. A hose that is directly exposed to a fire can burn through and burst quickly. Burning embers can also damage the hose, causing small leaks or weakening the hose so that it is likely to burst under pressure. Always visually inspect any hose that has been in direct contact with a fire.

Avoid storing a hose in places where it will be in contact with hot surfaces, such as a heating unit or the exhaust pipe on a vehicle. If the apparatus is parked outside, use a hose cover to protect the hose from sunlight.

In cold weather, freezing is a threat to hoses. Freezing can rupture the inner liner and break fibers in the hose jacket. When working in below freezing temperatures, water should be kept flowing through the hose to prevent freezing. If a line has be shut down temporarily, the nozzle should be left partly open to keep the water moving and the stream should be directed to a location where it will not cause additional water damage. When a line is no longer needed, the hose should be drained and rolled before it freezes.

Hose that is frozen or encased in ice can often be thawed out with a steam generator. Another option is to carefully chop the hose out using an axe, being careful not to cut the hose. The hose can then be transported back to the fire station to thaw. Do not attempt to bend a section of frozen hose. In situations where the hose is frozen solid, it may be necessary to transport the hose back to the fire station on a flatbed truck.

### Chemicals

Many chemicals can cause damage to fire hoses. These chemicals can be encountered at incidents in facilities where chemicals are manufactured, stored, or used and in locations where their presence is not anticipated. Most vehicles contain a wide variety of chemicals that can damage fire hose, including battery acid, gasoline, diesel fuel, antifreeze, motor oil, and transmission fluid. The hose can come in contact with these chemicals at vehicle fires or at the scene of a collision where chemicals are spilled. Supply hoses often come in contact with residue from these chemicals when lines are laid in the roadway. It is important to remove chemicals from the hose as soon as possible and to wash the hose with an approved detergent, thoroughly rinse, and let dry thoroughly.

**Brigade Member Safety Tips**

Any time a fire hose has suffered possible damage, it should be thoroughly inspected and tested according to NFPA 1962, *Inspection, Care and Use of Fire Hose, Couplings, and Nozzles and the Service Testing of Fire Hose,* before it is returned to service.

### Mildew

**Mildew** is a type of fungus that can grow on fabrics and materials in warm, moist conditions. A fire hose that has been packed away while it is still wet and dirty is a natural breeding ground for mildew. Mildew feeds on nutrients found in many natural fibers and can cause them to rot and deteriorate. In the days when cotton fibers were used in fire hose jackets, mildew was a large problem. Hose had to be washed and completely dried after every use before it could be placed back on the apparatus.

Modern fire hose is made from synthetic fibers that are resistant to mildew, and most types can be repacked without drying. Mildew can still grow on exposed fibers if they are soiled with contaminants that will provide mildew with the necessary nutrients. The fibers in rubber-covered hose are protected from mildew.

## Cleaning and Maintaining Hoses

Hose that is dirty or contaminated should be cleaned, following the steps in Skill Drill 16-8.

1. Lay the hose out flat.
2. Rinse the hose with water.
3. Gently scrub the hose with a soft bristle brush and mild detergent, paying attention to soiled areas.
4. Turn over the hose and repeat steps two and three.
5. Give a final rinse to the hose with water.
6. Hang the hose and allow it to dry before properly storing it.

## Hose Inspections

Each length of hose should be tested at least annually, according to the procedures listed in NFPA 1962, *Inspection, Care and Use of Fire Hose, Couplings, and Nozzles and the Service Testing of Fire Hose.* The hose testing equipment must be operated according to the manufacturer's instructions.

A visual inspection should also be performed after each use, either while the hose is being cleaned and dried or when it is reloaded onto the apparatus. If any defects are found, that length of hose should be immediately removed from service and tagged with a description of the problem. The appropriate notifications must be made to have the hose repaired.

To clearly mark a defective hose, follow the steps in Skill Drill 16-9.

1. Inspect the hose for defects.
2. Upon finding a defect, mark the area on the hose and remove the hose from service.
3. Tag the hose as defective with a description of the defect, take it out of service, and notify your superiors.

### Hose Records

Hose records are important documents. A hose record is a written history of each individual length of fire hose. Each length of hose should be identified with a unique number stenciled or painted on it. A hose record will contain information such as:

- Hose size, type, and manufacturer
- Date the hose was manufactured
- Date the hose was purchased
- Dates when the hose was tested
- Any repairs that have been made to the hose

## Hose Appliances and Tools

A hose appliance is used to connect to a fire hose apparatus or hydrant for the purpose of delivering water. You should be familiar with appliances such as a wye, water thief, Siamese connection, double-male and double-female adaptors, reducers, and tools such as hose clamps, hose jackets, and hose rollers. It is important for you to learn how to use the hose appliances and tools required by your fire brigade. You should understand the purpose of each device and be able to utilize it properly.

### Wyes

A wye is a device that splits one hose stream into two hose streams. The word wye refers to a Y-shaped part or object. When threaded couplings are used, a wye has one female connection and two male connections.

The wye that is most commonly used in the fire service splits one supply hose line into two attack hose lines. A gated wye is equipped with two quarter-turn ball valves so that the flow of water to each of the lines can be controlled independently (▶ Figure 16-10). A gated wye enables you to initially attach and operate one hose line and then add a second hose later. The use of a gated wye avoids the need to shut down the hose line supplying the wye in order to attach the second line.

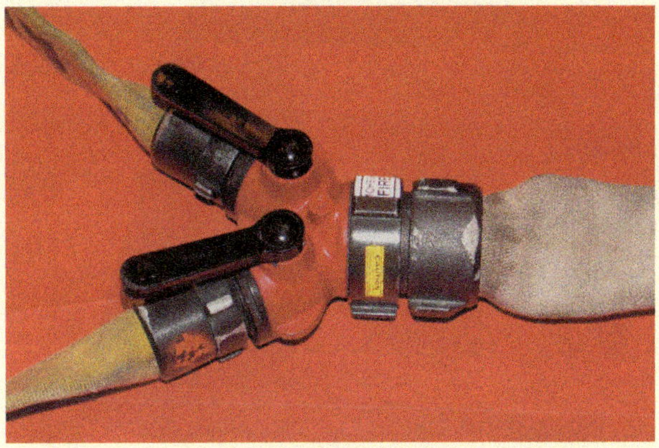

Figure 16-10 A gated wye is used to split one 2 ½-inch hose line into two lines.

Figure 16-11 A water thief.

### Water Thief

A water thief is similar to a gated wye, with an additional outlet (▲ Figure 16-11). The water that comes from a single inlet can be directed to multiple outlets and one outlet the same size as the inlet. Under most conditions, it will not be possible to supply all three outlets at the same time because the capacity of the supply hose is limited.

A water thief can be placed near the entrance to a building to provide the water for interior attack lines. One or two attack lines can be used and, if necessary, can be shut down and a larger line can be substituted. Sometimes the larger line is used to knock down a fire and the two smaller lines are used for overhaul.

# Voices of Experience

**"Although the fire was not a large-scale event . . . the challenge came in trying to reach the fire with our fire streams."**

As a chief officer for the Refinery Terminal Fire Company, I recently had the opportunity to respond to a very challenging elevated fire at a local refinery. Although the fire was not a large-scale event—it was a medium-size fire surrounding the flange of a reactor—the challenge came in trying to reach the fire with our fire streams.

The flange was located near the top of the reactor and was estimated to be approximately 200 feet above the ground. The reactor was surrounded by steel I-beams and catwalks as well as several gauges and instruments that contained radioactive sources. It was determined that a cooling operation, geared toward the surrounding structures and radioactive shielding materials and instruments, would be necessary to protect their integrity while the refinery employees de-inventoried product from the reactor (which would stop the leaking product and extinguish the fire). The process of de-inventorying the reactor was expected to take several hours.

To accomplish the objective of cooling these structures and instruments, which was located 200 feet above the ground, our company elected to use two pieces of fire apparatus. The main apparatus was FT-2, a 4,000-gpm industrial pumper that has a remote-controlled 80-foot articulating boom and a 4,000-gpm fog nozzle at the tip of the boom. The secondary apparatus was F-6, another 4,000-gpm industrial pumper that has two rear-mounted remote-controlled deck guns, each with a 2,000-gpm fog nozzle.

Our objective was to set up the primary apparatus (FT-2) on the battery limits of the unit (about 150 feet from the base of the reactor) and directly across from the side of the reactor where the fire was burning. This would allow us to unfold the 80-foot boom and use the wind direction to help propel a water stream up to the fire and surrounding structure. The secondary apparatus (F-6) would be set up perpendicular to FT-2 at the side of the reactor (also about 150 feet from the base of the reactor) and would use the cross winds to propel a water stream into the same area of the fire and surrounding structure. Given that this procedure was anticipated to be a lengthy cooling operation (while the reactor was de-inventoried), the purpose of F-6 would be to cover any areas of the structure that FT-2 could not reach and to allow for refueling of FT-2 throughout the incident, while maintaining a water stream onto the structure during these periods.

Upon setting up and flowing both apparatus, even when using the stiff winds to our advantage, we discovered that we could not able to reach the 200-foot level of the fire with our water streams. The apparatus crews immediately realized the shortfall of the streams, and both pumpers discontinued their operations so as to replace the apparatus's fog nozzles with smooth-bore nozzles. A 4-inch smooth-bore nozzle was placed on FT-2's articulating boom and two 1-inch stack-tip nozzles were placed on F-6's rear-mounted deck guns.

The apparatus crews resumed their water flows using the smooth-bore nozzles and were able to successfully cool the fire and the surrounding structures throughout the duration of the incident. In the postincident analysis, it was determined that having the added reach of the smooth-bore nozzles allowed us to successfully cool the structure, which resulted in very minimal damage to the surrounding structure and no compromise of the sensitive radiological instruments.

**John David Lowe**
Refinery Terminal Fire Company
Corpus Christi, Texas

Figure 16-12 A typical Siamese connection has two female inlets and a single outlet.

Figure 16-13 Double-male and double-female adaptors are used to join two couplings of the same sex.

## Siamese Connections

A Siamese connection is a hose appliance that combines two hose lines into one (▲ Figure 16-12). This increases the flow of water on the outlet side of the Siamese. A Siamese connection that is used with threaded couplings has two female connections on the inlets and one male connection on the outlet. Water can be flowed through a Siamese connection even if only one supply line exists. In this case, the pressure from the supplied side of the Siamese would cause the flapper to close over the unused intake opening. If and when another supply line is connected to the Siamese, the flapper valve would then remain centered to allow the water to flow from both supply lines.

A Siamese connection is sometimes used on a engine inlet to allow water to be received from two different supply lines. Siamese connections are also used to supply master stream devices and ladder pipes. Siamese connections are commonly installed on the fire department connections that are used to supply water to standpipe and sprinkler systems.

## Adaptors

Adaptors are used for connecting hose couplings of the same diameter that have dissimilar threads. Dissimilar threads could be encountered when different fire brigades are working together or in industrial settings where the hose threads do not match the threads of the municipal fire department. This is not a common problem because the most widely used coupling is national standard thread (also called national hose thread). Adaptors are also used to connect threaded couplings to Storz-type couplings.

Adaptors can also be used when it is necessary to connect two female couplings or two male couplings. A double-female adaptor is used to join two male hose couplings. A double-male adaptor is used to join two female hose couplings (► Figure 16-13).

Figure 16-14 A reducer is used to connect a smaller hose line to the end of a larger line.

## Reducers

A reducer is used to attach a smaller hose to a larger hose (▲ Figure 16-14). Usually the larger end has a female connection and the smaller end has a male connection. A common type of reducer is used to reduce a $2^1/_2$-inch hose thread to a $1^1/_2$ inches.

## Hose Jacket

A hose jacket is a device that is placed over a leaking section of hose to stop a leak (► Figure 16-15). The best way to handle a leak in a section of hose is to replace the defective section of hose. A hose jacket can provide a temporary fix until the section of hose can be replaced. A hose jacket should be used only in cases where it is not possible to quickly replace the leaking section of hose.

Figure 16-15 A hose jacket is used to repair a leaking hose line.

The hose jacket consists of a split metal cylinder that fits tightly over the outside of a hose line. The cylinder is hinged on one side to allow it to be placed over the leak; then a fastener is used to clamp the cylinder tightly around the hose. Gaskets on each end of the hose jacket prevent water from leaking out the ends of the hose jacket.

## Hose Roller

A hose roller is used to protect a hose line that is being hoisted over the edge of a roof or over a windowsill ▶ Figure 16-16. The hose roller keeps the hose from chafing or kinking at the sharp edge. A hose roller is sometimes called a hose hoist because it makes it easier to raise or hoist a hose over the edge of the building. Hose rollers can also be used to protect ropes when hoisting an object over the edge of a building and during rope rescue operations.

## Hose Clamp

A hose clamp is used to temporarily stop the flow of water in a hose line. Hose clamps are often used on supply lines, so that the hydrant can be opened before the line is hooked up to the intake of the attack engine. The brigade member at the hydrant does not have to wait for the pump operator to connect the line to the pump intake before opening the hydrant. As soon as the intake line is connected, the clamp is released. A hose clamp can also be used to stop the flow in a line if a hose ruptures or it has to be connected to a different appliance hose ▶ Figure 16-17.

### Brigade Member Safety Tips

A hose clamp should always be opened slowly to prevent water hammer.

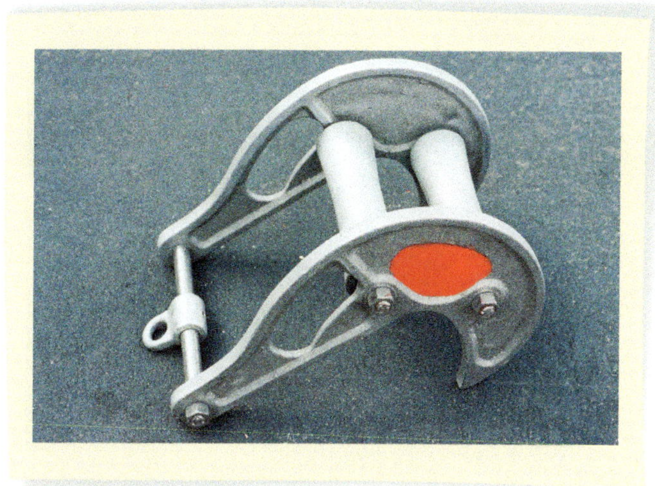

Figure 16-16 A hose roller is used to protect a hose when it is hoisted over a sharp edge of a roof or a windowsill.

Figure 16-17 A hose clamp is used to temporarily interrupt the flow of water in a hose line.

## Master Stream Devices

A master stream device is a large-capacity nozzle that can be supplied by two or more hose lines or a single LDH. Master stream devices include deck guns and portable ground monitors. A deck gun is usually attached to the top of an engine and may be supplied by a direct pipe connection from the pump ▶ Figure 16-18. A ground monitor can be removed from the apparatus and placed on the ground. When it is placed on the ground, the water is supplied by $2\frac{1}{2}$ inches or larger hose lines ▶ Figure 16-19. Some devices can be used as either a deck gun or removed from

Figure 16-18 A deck gun is mounted on top of an engine or other apparatus.

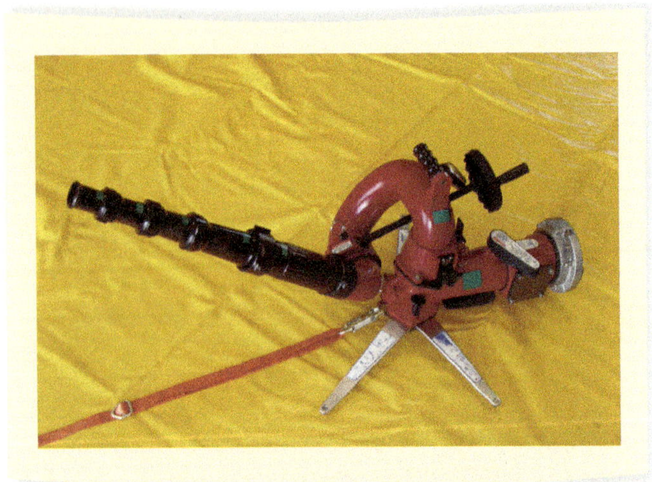

Figure 16-19 A ground monitor can be removed from the apparatus and placed on the ground.

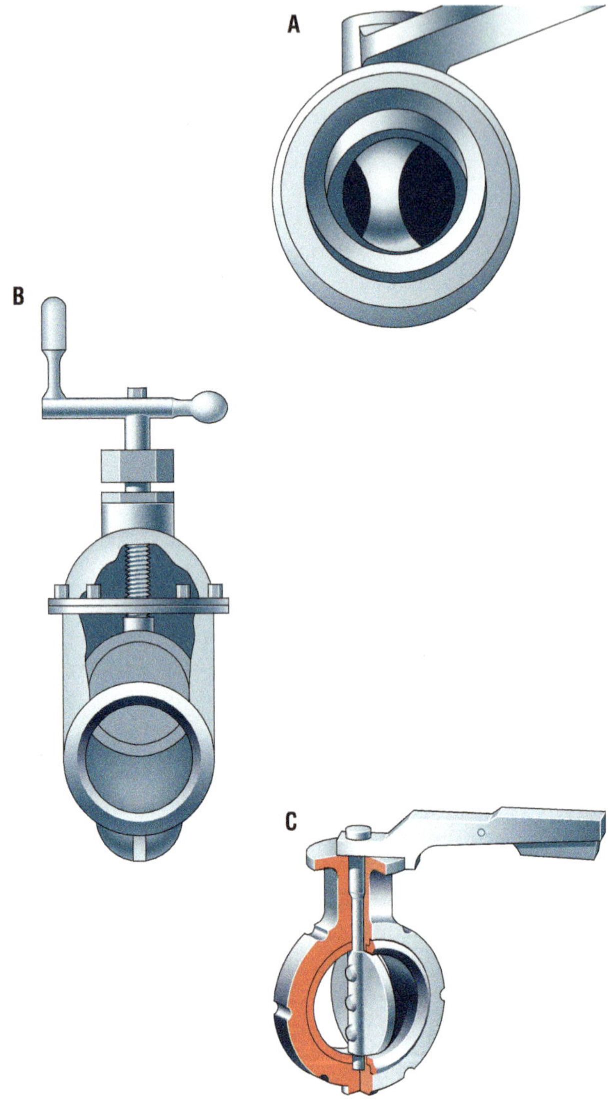


Figure 16-20 A. Ball valve. B. Gate valve. C. Butterfly valve.

the apparatus and used as a ground monitor. Master stream devices are used during defensive firefighting operations.

## Valves

Valves are used to control the flow of water in a pipe or hose line. Several different types of valves are used on fire hydrants, fire apparatus, standpipe and sprinkler systems, and hose lines. The important thing to remember when opening and closing any valve or nozzle is to do it S-L-O-W-L-Y to prevent water hammer.

Some of the common valves that you will encounter include:

- Ball Valves: These valves are used on nozzles, gated wyes, and engine discharge gates. Ball valves are made up of a ball with a hole in the middle. When the hole is in line with inlet and outlet, water flows through it. As the ball is rotated, the flow of water is gradually reduced until it is shut off completely (▲ Figure 16-20A).
- Gate Valves: These valves are found on hydrants and on sprinkler systems. Rotating a spindle causes a gate to move slowly across the opening. The spindle is rotated by turning it with a wrench or a wheel-type handle (▲ Figure 16-20B).
- Butterfly valves: These valves are often found on the large pump intake connections where a hard suction hose or soft sleeve hose is connected. They are opened or closed by rotating a handle one-quarter turn (▲ Figure 16-20C).

# Hose Rolls

## Overview of Hose Rolls

An efficient way to transport a single section of fire hose is in the form of a roll. Rolled hose is compact and easy to manage. A fire hose can be rolled many different ways, depending on how it will be used. Follow the standard operating procedures (SOPs) of your brigade.

### Straight or Storage Hose Roll

The straight roll is a simple and frequently used hose roll. It is used for general handling and transporting of hose. This roll is also used for rack storage of hose. To roll a straight hose roll, follow the steps of ▶ Skill Drill 16-10.

1. Lay the length of hose to be rolled flat and straight. **(Step 1)**
2. Begin by rolling the male coupling over on top of the hose. **(Step 2)**
3. Roll the hose to the female coupling. **(Step 3)**
4. Lay the hose roll on its side and tap any protruding hose flat with a foot. **(Step 4)**

### Single-Donut Hose Roll

The single-donut roll is used when the hose will be put into use directly from the rolled state. The hose has both couplings on the outside of the roll. The hose can be connected and extended by one brigade member. As the hose is extended, it unrolls. To perform a single-donut hose roll, follow the steps in ▶ Skill Drill 16-11.

1. Place the hose flat and in a straight line. **(Step 1)**
2. Locate the mid-point of the hose. **(Step 2)**
3. From the mid-point, move 5 feet toward the male coupling end.
4. Start rolling the hose toward the female coupling. **(Step 3)**
5. At the end of the roll, wrap the excess hose of the female end over the male coupling to protect the threads. **(Step 4)**

### Double-Donut Hose Roll

The double-donut roll is used primarily to make a small compact roll that can be carried. To perform the double-donut hose roll, follow the steps in ▶ Skill Drill 16-12.

1. Lay the hose flat and in a straight line. **(Step 1)**
2. Bring the male coupling alongside the female coupling. **(Step 2)**
3. Fold the far end over and roll toward the couplings, creating a double roll. **(Step 3)**
4. The roll can be carried by hand, by a rope, or strap. **(Step 4)**

### Self-Locking Double-Donut Hose Roll

The self-locking double donut is similar to the double donut with the exception that it forms its own carry loop. To perform the self-locking double-donut roll, follow the steps in ▶ Skill Drill 16-13.

1. Lay the hose flat and bring the couplings alongside each other. **(Step 1)**
2. Cross one side of the hose over the other, creating a loop. This loop creates the carrying shoulder loop. **(Step 2)**
3. Bring the loop back toward the couplings to the point where the hose crosses. **(Step 3)**
4. From the point where the hose crosses, begin to roll the hose toward the couplings with the loop as its center. This creates a loop on each side of the roll. **(Step 4)**
5. On completion of the rolling, position the couplings on the top of the rolls. **(Step 5)**
6. Position the loops so one is larger than the other. Then pass the larger loop over the couplings and through the smaller loop. This secures the rolls together and forms the shoulder loop. **(Step 6)**

> **Brigade Member Safety Tips**
>
> When performing a forward lay, the brigade member who is connecting the supply line to the hydrant must not stand between the hose and the hydrant. When the apparatus starts to move off, the hose could become tangled and suddenly be pulled taut. Anyone standing between the hose and the hydrant could be seriously injured.

# Fire Hose Evolutions

Fire hose evolutions are standard methods of working with fire hose to accomplish different objectives in a variety of situations. Most fire brigades set up their equipment and conduct regular training in order to be prepared to perform a set of standard hose evolutions. Hose evolutions involve specific actions that are assigned to each member of a crew. Every brigade member should know how to perform all of the standard evolutions quickly and proficiently. When a brigade leader calls for a particular evolution to be performed, each crew member should know exactly what to do.

Hose evolutions are divided into supply line operations and attack line operations. Supply line operations involve laying hose lines and making connections between a water supply source and an attack engine. Attack line operations involve advancing hose lines from an attack engine to apply water onto the fire.

## Supply Line Operations

The objective of laying a supply line is to deliver water from a hydrant or an alternative water source to an attack engine. In most cases, this involves laying a hose line with

Skill Drill 16-10

# Rolling a Straight Hose Roll

Lay the length of hose to be rolled flat and straight.

Begin by rolling the male coupling over on top of the hose.

Roll the hose to the female coupling.

Lay the hose roll on its side and tap any protruding hose flat with a foot.

Skill Drill 16-11

# Performing a Single-Donut Hose Roll

1 Place the hose flat and in a straight line.

2 Locate the mid-point of the hose.

3 From the mid-point, move 5 feet toward the male coupling end. Start rolling the hose toward the female coupling.

4 At the end of the roll, wrap the excess hose of the female end over the male coupling to protect the threads.

16-12 Skill Drill

## Performing a Double-Donut Hose Roll

Lay the hose flat and in a straight line.

Bring the male coupling alongside the female coupling.

Fold the far end over and roll toward the couplings, creating a double roll.

The roll can be carried by hand, rope, or strap.

16-13 Skill Drill

## Performing a Self-Locking Double-Donut Hose Roll

1 Lay the hose flat and bring the couplings alongside each other.

2 Cross one side of the hose over the other, creating a loop. This loop creates the carrying shoulder loop.

3 Bring the loop back toward the couplings to the point where the hose crosses.

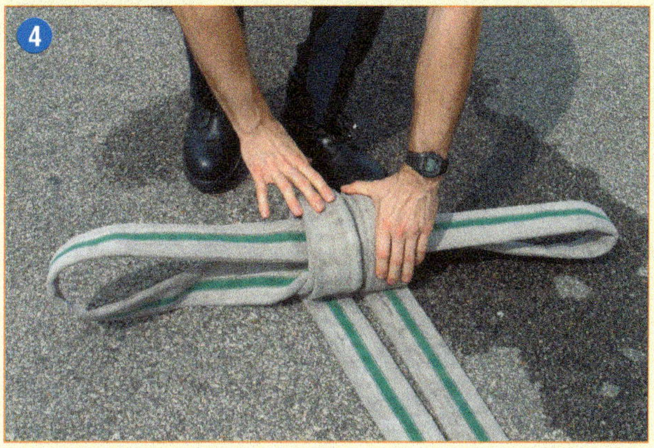

4 From the point where the hose crosses, begin to roll the hose toward the couplings with the loop as its center. This creates a loop on each side of the roll.

5 On completion of the rolling, position the couplings on the top of the rolls.

6 Position the loops so one is larger than the other. Then pass the larger loop over the couplings and through the smaller loop.

a moving vehicle or dropping a continuous line hose out of a bed as the vehicle moves forward. This can be done using either a forward lay or reverse lay. A forward lay starts at the hydrant and proceeds toward the fire; the hose is laid in the same direction as the water flows, from the hydrant to the fire. A reverse lay involves laying the hose from the fire to the hydrant; the hose is laid in the opposite direction to the water flow. Each fire brigade determines its own preferred methods and procedures based on apparatus, water supply, and regional considerations.

## Brigade Member Tips

When connecting a supply line to a hydrant, wait until you get the signal from the driver/operator to charge the line. If the hydrant is opened prematurely, the hose bed could become charged with water or a loose hose line could discharge water at the fire scene. Either situation will disrupt the operation and could cause serious injuries. Be sure that you know your brigade's signal to charge a hose line and do not become so excited or rushed that you make a mistake.

### Forward Hose Lay

The forward hose lay is most often used by the first arriving engine company at the scene of a fire (▼ Figure 16-21). This method allows the engine company to establish a water supply without assistance from an additional company. A forward lay also places the attack engine close to the fire, allowing access to additional hoses, tools, and equipment that are carried on the apparatus.

A forward hose lay can be performed using MDH lines (2½- or 3-inch hose) or with LDH (3½ inches and larger). The larger the diameter of the hose, the more water can be delivered to the attack engine through a single supply line. When MDH is used and the beds are arranged to lay dual lines, a company can lay two parallel lines from the hydrant to the fire.

To perform a forward hose lay, follow the steps in (Skill Drill 16-14).

1. The fire apparatus should stop within about 10' of the hydrant.
2. Grasp enough hose to reach to the hydrant and to loop around the hydrant. Step off the apparatus with the hydrant wrench and all necessary tools.
3. Loop the end of the hose around the hydrant or secure the hose as specified in the SOP. Ensure that you are not between the hose and the hydrant. Never stand on the hose.
4. Signal the driver/operator to proceed to the fire once the hose is secured.
5. Once the apparatus has moved off and a length of supply line has been removed from the apparatus and is lying on the ground, the brigade member should remove the appropriate hydrant cap. Follow local SOP for checking the operating condition of the hydrant.
6. Attach the supply hose to the outlet. An adaptor may be needed if an LDH with Storz-type couplings is used.
7. Attach the hydrant wrench to the hydrant.
8. The driver/operator should uncouple the hose and attach the end of the supply line to the pump inlet or clamp the hose close to the pump, depending on local SOPs.
9. The driver/operator should signal to charge the hose by prearranged hand signal, radio, or air horn.
10. Slowly open the hydrant completely.
11. Follow the hose back to the engine and remove any kinks from the supply line.

### Four-Way Hydrant Valve

In cases where long supply lines are needed or when MDH is used, it is often necessary to place a supply engine at the hydrant. The supply engine pumps water through the supply line in order to increase the flow to the attack engine. Some brigades use a four-way hydrant valve to connect the supply line to the hydrant, so that the supply line can be charged

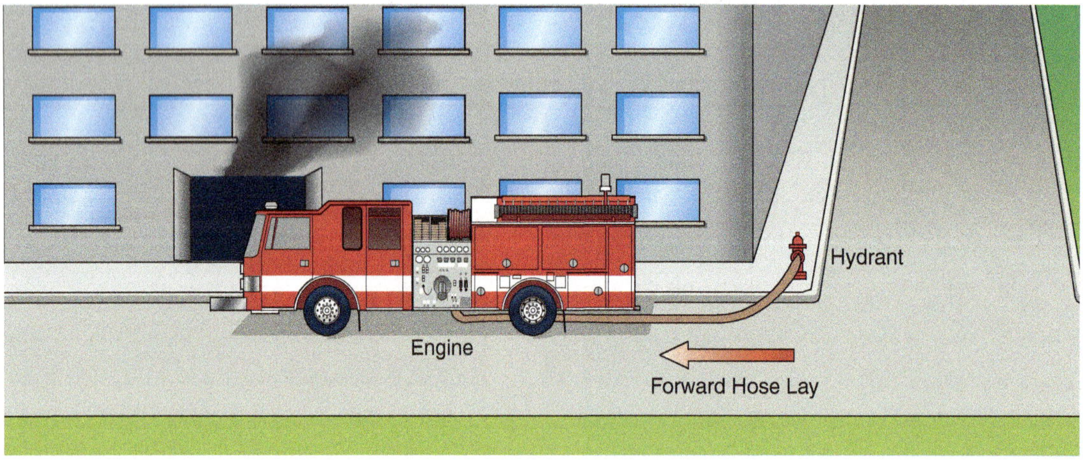


Figure 16-21 Forward hose lay.

## Brigade Member Tips

When laying out supply hose with threaded couplings, you may find that the wrong end of the hose is on top of the hose bed. Double-male adaptors and double-female adaptors will allow you to attach a male coupling to a male discharge or to attach a female coupling to a female discharge. A set of adaptors (one double male and one double female) should be easily accessible for these situations. Some brigades place a set of adaptors on the end of the supply hose for this purpose.

## Brigade Member Tips

### Split Hose Beds

A **split hose bed** is a hose bed that is divided into two or more sections. This is done for several purposes.

- One compartment in a split hose bed can be loaded for forward lay (female coupling out), and the other side can be loaded for a reverse lay (male coupling out). This allows a line to be laid in either direction without adaptors.
- Two parallel hose lines can be laid at the same time. (This is sometimes called laying dual lines.) Dual lines are beneficial if the situation requires more water than one hose line can supply.
- The split beds can be used to store different sized hoses. For example, one side of the hose bed could be loaded with 3-inch hose that can be used as a supply line or as an attack line. The other side of the hose bed could be loaded with 5-inch hose for use as a supply line. This setup enables the use of the most appropriate sized hose for a given situation.
- All of the hose from both sides of the hose bed can be laid out as a single hose line. This is done by coupling the end of the hose in one bed to the beginning of the hose in the other bed.

A variation of the split hose bed is known as a combination load, where the last coupling in one bed is normally connected to the first coupling in the other bed. When one long line is needed, all of the hose plays out of one bed first; then the hose continues to play out from the second bed. To lay dual lines, the connection between the two hose beds is uncoupled and the hose can play out of both beds simultaneously. When the two sides of a split bed are loaded with the hose in opposite directions, either a double-female or a double-male adaptor is used to make the connection between the two hose beds.

with water immediately and still allow for a supply engine to connect into the line later.

When the four-way valve is placed on the hydrant, the water flows initially from the hydrant through the valve to the supply line, which delivers the water to the attack engine. The second engine can hook up to the four-way valve and redirect the flow by changing the position of the valve. The water then flows from the hydrant to the supply engine. The supply engine boosts the pressure and discharges the water into the supply line, boosting the flow of water to the attack engine. This can be accomplished without uncoupling any lines or interrupting the flow. To use a four-way valve, follow the steps in Skill Drill 16-15.

1. The attack engine should stop about 10' past the hydrant to be used.
2. Grasp the four-way valve, the attached hose, and enough hose to reach to and loop around the hydrant. Remove the four-way valve off the apparatus along with the hydrant wrench and any other needed tools.
3. Loop the end of the hose around the hydrant or secure the hose with a rope as specified in the local SOPs. DO NOT stand between the hydrant and the hose.
4. Signal the driver/operator to proceed to the fire.
5. Once enough hose has been removed from the apparatus and is lying on the ground, the hydrant person should remove the cap from the fire hydrant. Follow local SOPs for checking the operating condition of the hydrant.
6. Attach the four-way valve to the hydrant outlet (an adaptor may be needed).
7. Attach the hydrant wrench to the hydrant.
8. The driver/operator uncouples the hose and attaches the end of the supply line to the pump inlet.
9. The driver/operator signals you by prearranged hand signal, radio, or air horn to charge the supply line.
10. Slowly open the hydrant completely.
11. Initially, the attack engine is supplied with water from the hydrant. When the supply engine arrives at the fire scene, the driver/operator should stop at the hydrant with the four-way valve.
12. The driver/operator should attach a hose from the four-way valve outlet to the intake side of the engine.
13. Attach a second hose to the inlet side of the four-way valve and connect the other end to the pump discharge.
14. Change the position of the four-way valve to direct the flow of water from the hydrant through the supply engine and into the supply line.

### Reverse Hose Lay

The reverse hose lay is the opposite of the forward lay ► Figure 16-22. In the reverse lay, the hose is laid out from the fire to the hydrant, in the direction opposite to the flow of the water. This evolution can be used when the attack engine arrives at the fire scene without a supply line. This could be a standard tactic in areas where there are sufficient hydrants available and additional companies that can assist in establishing a water supply will be arriving quickly. One of these companies will be assigned to lay a supply line from the attack engine to a hydrant.

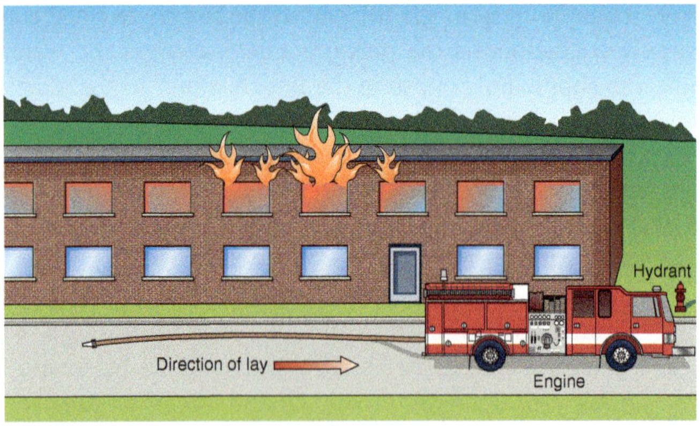


Figure 16-22 Reverse hose lay.

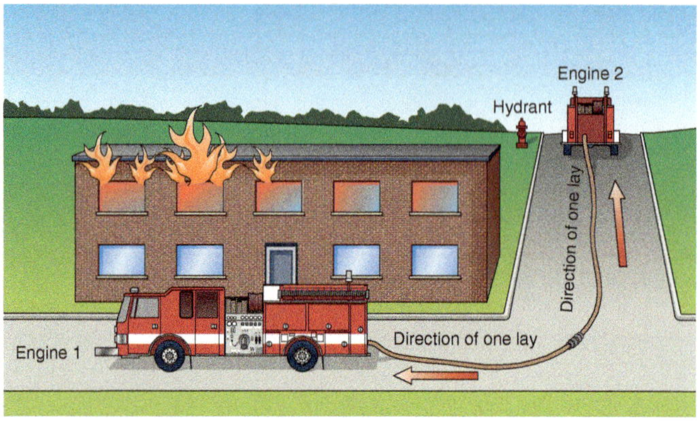


Figure 16-23 Split hose lay.

In this case, the attack engine will focus on immediately attacking the fire using water from the on-board tank. The supply engine stops close to the attack engine, and hose is pulled from the bed of the supply engine to an intake on the attack engine. The supply engine then drives to the hydrant (or alternative water source) and pumps water back to the attack engine. Usually the supply engine parks so that hose can be pulled from the supply engine to the inlet to the attack engine. To perform a reverse lay, follow the steps in Skill Drill 16-16.

1. Pull sufficient hose to reach from the supply engine to the inlet of the attack engine.
2. Anchor the hose to a stationary object if possible.
3. The driver/operator of the supply engine drives away, laying out hose from the attack engine to the fire hydrant or other static water source.
4. Connect the supply line to the inlet of the attack engine.
5. The driver/operator of the supply engine should uncouple the supply hose from the hose remaining in the hose bed and attach the supply hose to the discharge side of the supply engine.
6. The driver/operator will then connect the supply engine to the hydrant or water source. The four-way hydrant valve can be used here if needed to boost the supply pressure.
7. Upon signal from the attack engine, the supply engine driver/operator should charge the supply line.

## Split Hose Lay

A split hose lay is performed by two engine companies in situations where hose must be laid in two different directions to establish a water supply ▶ Figure 16-23. This evolution could be used when the attack engine has to approach a fire along a dead-end access road with no hydrant or down a long roadway. To perform a split hose lay, the attack engine drops the end of its supply hose and performs a forward lay toward the fire. The supply engine stops at the end of the forward lay, pulls off enough hose to connect to the end of the supply line, and then performs a reverse lay to the hydrant or water source. When the two lines are connected together, the supply engine can provide water to the attack engine.

A split lay often requires coordination by two-way radio, because the attack engine must advise the supply engine of the plan and where the end of the supply line is being dropped. In many cases, the attack engine is out of sight when the supply engine arrives at the split point. NOTE: A split hose lay does not necessarily require split hose beds. It can be performed with or without split beds if the necessary adaptors are used. To perform a split hose lay, follow the steps in Skill Drill 16-17.

1. The driver/operator of the first arriving engine company will stop at a point away from the fire.
2. Remove the supply line from the hose bed and anchor it in a kneeling position.
3. The driver/operator should proceed slowly toward the fire.
4. Either proceed by foot to the fire or wait for the supply engine. Check with local SOPs.
5. If the supply engine returns to the fire, the supply engine needs to stop and connect the supply hose to the hose end laid by the attack engine. If threaded couplings are used, a double-male adaptor may be required.
6. If the supply engine remained at the start of the forward lay, anchor the supply line from the second engine company. After the hose is laid to a hydrant, connect the two lines to form one supply line.
7. The driver/operator of the attack engine will start pumping water from the booster tank and also connect the supply hose to the pump intake.
8. The driver/operator of the supply engine should position the apparatus at the hydrant according to local SOPs.
9. Then, the driver/operator of the supply engine should pull off hose from the hose bed until the next coupling. The hose is broken at this connection and is

## Fire Marks

If the supply line from the water source to the fire is long, it may be necessary to use a series of engines to relay water to the attack engine. In a relay operation, each engine boosts the pressure enough to move the water on to the next engine.

## Brigade Member Tips

When referring to the hosebed, the end closest to the cab is called the front of the hosebed. The end closest to the tailboard is called the rear of the hosebed.

connected to the pump discharge or the four-way hydrant valve. Check with local SOPs.

10. Upon signal from the attack engine, the supply engine driver/operator will charge the supply line.

## Loading Supply Hose

This section describes the basic procedures for loading supply line hose into the hose beds on a fire apparatus. Hose can be loaded in several different configurations, depending on the way the hose is planned to be laid out. The hose must be easily removable from the hose bed, without kinks or twists and without the possibility of becoming caught or tangled. The ideal hose load would be easy to load, avoid wear and tear on the hose, have few sharp bends, and allow the hose to play out of the hose bed smoothly and easily.

You must learn the specific hose loads used by your fire brigade. When loading hose, always remember that the time and attention that goes into loading the hose properly will become worthwhile when it is time to use the hose at a fire.

There are three basic hose loads that are commonly used for supply line hose: the flat load, the horseshoe load, and the accordion load. Any one of these methods can be used to load hose for either a forward lay or for a reverse lay.

### Flat Hose Load

The flat hose load is the easiest to load and can be used for any size of hose, including LDH. Because the hose is placed flat in the hose bed, it should lay out flat without twists or kinks. Wear and tear on the edges of the hose from the movement and vibration of the vehicle during travel are limited. The flat hose load can be used with a single hose bed or a split bed. There are many variations of the flat hose load. Follow the SOPs of your brigade. To load hose in a flat load, follow the steps in ▶ Skill Drill 16-18.

1. If you are loading hose with threaded couplings, determine whether the hose will be used for a forward or reverse lay.
2. To set up the hose for a forward lay, place the male hose coupling in the hose bed first. To set up the hose for a reverse lay, place the female hose coupling in the hose bed first. **(Step 1)**
3. Start the hose load with the coupling at the front end of the hose bed. **(Step 2)**
4. Fold the hose back on itself at the rear of the hose bed. **(Step 3)**
5. Run the hose back to the front end on top of the previous length of hose.
6. Fold the hose back on itself so the top of the hose is on the previous length. **(Step 4)**
7. While laying the hose back to the rear of the hose bed, angle the hose to the side of the previous fold. **(Step 5)**
8. Continue to lay the hose in neat folds until the whole hose bed is covered with a layer of hose.
9. To make this hose load neat, make every other layer of hose slightly shorter or alternating the folds. This keeps the ends from getting too high at the folds.
10. Continue to load the layers of hose until the required amount of hose is loaded. **(Step 6)**

### Horseshoe Hose Load

The horseshoe hose load is accomplished by standing the hose on its edge and placing it around the perimeter of the hose bed in a U-shape. At the completion of the first U-shape, the hose is folded inward to form another U in the opposite direction. This continues until a complete layer is filled; then another layer is started on top of the first. When the hose load is completed, the hose in each layer is in the shape of a horseshoe. A major advantage of the horseshoe load is that it contains fewer sharp bends than the other hose loads.

The horseshoe load is not normally used for LDH because the hose tends to fall over when it stands on edge. The horseshoe load causes more wear on the hose because the weight of the hose is supported by the edges. There are many variations of the horseshoe hose load. Follow the SOPs of your brigade. To perform a horseshoe hose load, follow the steps in ▶ Skill Drill 16-19.

1. If you are loading threaded hose, determine whether you want to load the hose for a forward lay or a reverse lay. For a forward lay, start with the male coupling in the rear corner of the hose bed. For a reverse lay, start with the female coupling in the rear corner of the hose bed. **(Step 1)**
2. Lay the first length of hose on its edge against the right or left wall of the hose bed. **(Step 2)**
3. At the front of the hose bed, lay the hose across the width of the bed and continue down the opposite side toward the rear. **(Step 3)**

16-18 Skill Drill

# Performing a Flat Hose Load

1 To set up the hose for a forward lay, place the male hose coupling in the hose bed first. To set up the hose for a reverse lay, place the female hose coupling in the hose bed first.

2 Start the hose lay with the coupling at the front end of the hose bed.

3 Fold the hose back on itself at the rear of the hose bed.

4 Run the hose back to the front end on top of the previous length of hose. Fold the hose back on itself so the top of the hose is on the previous length.

5 While laying the hose back to the rear of the hose bed, angle the hose to the side of the previous fold.

6 Continue to lay the hose in neat folds until the whole hose bed is covered with a layer of hose. Continue to load the layers of hose until the required amount of hose is loaded.

16-19 **Skill Drill**

# Performing a Horseshoe Hose Load

For a forward lay, start with the male coupling in the rear corner of the hose bed. For a reverse lay, start with the female coupling in the rear corner of the hose bed.

Lay the first length of hose on its edge against the right or left wall of the hose bed.

At the front of the hose bed, lay the hose across the width of the bed and continue down the opposite side toward the rear.

When the hose reaches the rear of the hose bed, fold the hose back on itself and continue laying it back toward the front of the hose bed. Keep the hose tight to the previous row of hose around the hose bed until it is back to the rear on the starting side. Fold the hose back on itself again and continue packing the hose tight to the previous row.

Continue to pack the hose on the first layer. Each fold of hose will decrease the amount of space available inside of the horseshoe. Once the center of the horseshoe is filled in, begin a second layer by bringing the hose from the rear of the hose bed and laying it around the perimeter of the hose bed. Complete additional layers using the same pattern as you did for the first layer. Finish the hose load with your adaptors or appliances.

## Brigade Member Tips

When a split hose bed is loaded for a combination load, the end of the last length of hose in one bed is coupled to the beginning of the first length in the opposite bed. Begin loading the first bed with the initial coupling hanging out. This coupling will be the last to be deployed from the first bed when the hose is laid out. Leave enough of the end of the hose hanging down to reach the other hose bed. Load the second bed in the normal manner. When both beds have been loaded, connect the hanging coupling from the bottom of the first hose bed to the end coupling on the top of the adjoining bed.

## Brigade Member Tips

The following tips will help you to do a better job when loading hose:

- Drain all of the water out of the hose before loading.
- Rolling the hose first will result in a flatter hose load, because there will be no air in the hose.
- Do not load hose too tightly. Leave enough room so that you can slide a hand between the folds of hose. If hose is loaded too tightly, it may not lay out properly.
- Load hose so that couplings do not have to turn around as the hose is pulled out of the hose bed. Make a short fold in the hose close to the coupling to keep the hose properly oriented. This short fold is called a **Dutchman** (▼ **Figure 16-24**).
- Couple sections of hose with the flat sides oriented in the same direction.
- Check gaskets before coupling hose.
- Tighten couplings hand tight only. With a good gasket, the hose should not leak.

**Figure 16-24** A Dutchman is used so that a coupling will not have to turn around and possibly become stuck as the hose is laid out.

4. When the hose reaches the rear of the hose bed, fold the hose back on itself and continue laying it back toward the front of the hose bed. Keep the hose tight to the previous row of hose around the hose bed until it is back to the rear on the starting side. Fold the hose back on itself again and continue packing the hose tight to the previous row. **(Step 4)**
5. Continue to pack the hose on the first layer. Each fold of hose will decrease the amount of space available inside of the horseshoe. Once the center of the horseshoe is filled in, begin a second layer by bringing the hose from the rear of the hose bed and laying it around the perimeter of the hose bed. Complete additional layers using the same pattern as you did for the first layer. Finish the hose load with your adaptors or appliances. **(Step 5)**

### Accordion Hose Load

The accordion hose load is performed with the hose placed on its edge. The hose is laid side-to-side in the hose bed. One advantage of the accordion load is that it is easy to load in the hose bed. One layer is loaded from left to right, and then the next layer is loaded above it from right to left.

There are disadvantages to the accordion hose load. Because the hose is stacked on its side, there is more wear on the hose than with a flat load. The accordion hose load is not recommended for LDH because LDH tends to collapse when placed on its side. There are many variations of the accordion hose load. Follow the SOPs of your brigade. To perform the accordion hose load, follow the steps in (► **Skill Drill 16-20**).

1. Determine whether the hose will be used for a forward or reverse lay.
2. To set up the hose for a forward lay, place the male hose coupling in the hose bed first. To set up the hose for a reverse lay, place the female hose coupling in the hose bed first.
3. Start the hose lay with the coupling at the front end of the hose bed.
4. Lay the first length of hose in the hose bed on its edge against the side of the hose bed. **(Step 1)**
5. Double the hose back on itself at the rear of the hose bed. Leave the female end extended so that the two hose beds can be cross-connected. **(Step 2)**
6. Lay the hose next to the first length and bring it to the front of the hose bed.
7. Fold the hose at the front of the hose bed so the bend is even to the edge of the hose bed. Continue to lay folds of hose across the hose bed. **(Step 3)**
8. Alternate the length of the hose folds at each end to allow more room for the folded ends.
9. When the bottom layer is completed, angle the hose upward to begin the second tier.
10. Continue the second layer by repeating the steps you used to complete the first layer. **(Step 4)**

16-20 Skill Drill

## Performing an Accordion Hose Load

Lay the first length of hose in the hose bed on its edge against the side of the hose bed.

Double the hose back on itself at the rear of the hose bed. Leave the female end extended so that the two hose beds can be cross-connected.

Lay the hose next to the first length and bring it to the front of the hose bed. Fold the hose at the front of the hose bed so the bend is even to the edge of the hose bed. Continue to lay folds of hose across the hose bed.

Alternate the length of the hose folds at each end to allow more room for the folded ends. When the bottom layer is completed, angle the hose upward to begin the second tier. Continue the second layer by repeating the steps you used to complete the first layer.

## Connecting an Engine to a Water Supply

When an engine is located at a hydrant, a supply hose must be used to deliver the water from the hydrant to the engine. This is a special type of supply line that is intended to deliver as much water as possible over a short distance. In most cases, a soft sleeve hose is used to connect directly to a hydrant. This connection can also be made with a hard suction hose or with a short length of large-diameter supply hose.

### Attaching a Supply Hose to a Hydrant

The large soft sleeve hose is normally used to connect an engine directly to a hydrant. To attach a soft sleeve hose to a hydrant, follow the steps in ▶ Skill Drill 16-21.

1. The driver/operator positions the apparatus so that its inlet is the correct distance from the hydrant. **(Step 1)**
2. Remove the hose, any needed adaptors, and the hydrant wrench. **(Step 2)**
3. Attach the soft sleeve hose to the inlet of the engine. In some organizations this end of the hose is preconnected. **(Step 3)**
4. Unroll the hose. **(Step 4)**
5. Remove the large hydrant cap. **(Step 5)**
6. Attach the soft sleeve hose to the hydrant. Sometimes it will be necessary to use an adaptor. **(Step 6)**
7. Ensure that there are no kinks or sharp bends in the hose that will restrict the flow of water. **(Step 7)**
8. Open the hydrant slowly when indicated by the pump operator. Check all connections for leaks. Tighten if necessary. **(Step 8)**
9. Place chafing blocks under the hose where it contacts the ground to prevent abrasion. **(Step 9)**

### Attaching a Hard Suction Hose to a Hydrant

Although it is not commonly done, a hard suction hose is sometimes used to connect an engine to a hydrant. This can be a difficult task because the hard suction is heavy and has limited flexibility, so additional personnel will often be needed to lift and attach the hose. The apparatus must be carefully positioned to make this connection properly. To attach a hard suction hose to a hydrant, follow the steps in ▶ Skill Drill 16-22.

1. The driver/operator will position the apparatus so that the intake on the apparatus is located the correct distance from the hydrant. **(Step 1)**
2. Remove the pump inlet cap on the apparatus.
3. Remove the hydrant steamer outlet cap.
4. Place an adaptor on the hydrant if needed.
5. With your partner's help, remove a section of hard suction hose from the engine. **(Step 2)**
6. Connect the hard suction hose to the intake on the engine. **(Step 3)**
7. Connect the opposite end to the hydrant, using a double-female adaptor with the hydrant thread on one side and the suction hose thread on the other side.
8. The driver/operator may need to move the apparatus slightly to position the apparatus to make the final connection.
9. Slowly open the hydrant when instructed by the driver/operator. **(Step 4)**

> **Fire Marks**
>
> A transverse preconnected hose located over the pump is sometimes called a "Mattydale Hose Bed" or a "Mattydale lay" because this setup was first used by the fire brigade in Mattydale, New York.

## Attack Line Evolutions

Attack lines are the hose lines used to deliver water from an attack engine to a nozzle, which discharges the water onto the fire. Any hose line that is used to discharge water onto the fire without going through another pump is defined as an attack line. Attack lines can use several different hose sizes and any length of hose.

Attack lines are usually stretched from an engine or an apparatus that is functioning as an attack engine to the fire. The attack engine is usually located close to the fire, and attack lines are stretched manually by brigade members. In some situations, an engine will drop an attack line at the fire and drive from the fire to a hydrant or water source. This is similar to a reverse lay evolution as described in the supply line section, except that the hose will be used as an attack line.

Most engines are equipped with preconnected attack lines, which provide a predetermined length of attack hose that is already equipped with a nozzle and connected to a pump discharge outlet. An additional supply of attack hose is usually carried in a hose bed or compartment that is not preconnected. To create an attack line with this hose, the desired length of hose is removed from the bed; then a coupling is disconnected and attached to a pump discharge outlet. This hose can also be used to extend a preconnected attack line or to attach to a wye or a water thief.

The attack hose is loaded so that it can be quickly and easily deployed. There are many ways to load attack lines into a hose bed. This section presents some of the most common hose loads. Your brigade may use a variation of one of these loads. It is important for you to master the hose loads used by your brigade.

### Preconnected Attack Lines

Preconnected hose lines are intended for immediate use as attack lines. A preconnected hose line has a predetermined

16-21 Skill Drill

# Attaching a Soft Sleeve Hose to a Fire Hydrant

1 The driver/operator positions the apparatus so that its inlet is the correct distance from the hydrant.

2 Remove the hose, any needed adaptors, and the hydrant wrench.

3 Attach the soft sleeve hose to the inlet of the engine.

4 Unroll the hose.

5 Remove the large hydrant cap.

6 Attach the soft sleeve hose to the hydrant.

## 16-21 Skill Drill (Continued)

Ensure that there are no kinks or sharp bends in the hose that will restrict the flow of water.

Open the hydrant slowly when indicated by the driver/operator. Check all connections for leaks. Tighten if necessary.

Place chafing blocks under the hose where it contacts the ground to prevent abrasion.

length of hose with the nozzle already attached and is connected to a discharge outlet on the engine. The most commonly used attack lines are 1³/₄-inch hose, generally from 150' to 200' in length. Many engines are also equipped with a preconnected 2¹/₂- or 3-inch hose line for quick attack on larger fires.

Attack lines should be loaded in the hose bed so they can be quickly stretched from the attack engine to the fire. It should be possible for one or two brigade members to quickly remove the hose from the hose bed and advance the hose to the fire. In many cases, an attack line is stretched in two stages. First, the hose from the attack engine to the building entrance or to a location close to the fire will be laid out. From that point, the hose is advanced into the building to reach the fire. Extra hose should be deposited at the entrance to the fire building. For exterior defensive operations, master stream attack, and exposure protection, hose lines should be laid in a position that provides the most effective use of attack lines.

Several different hose loads can be used. The hose should not get tangled as it is being removed from the bed and advanced. Laying out the hose should not require multiple trips between the engine and the fire and it should be easy to lay the hose around obstacles and corners. It should

16-22 Skill Drill

## Attaching a Hard Suction Hose to a Fire Hydrant

1

The driver/operator will position the apparatus so that the intake is located the correct distance from the hydrant.

2

Remove the pump inlet cap. Remove the hydrant steamer outlet cap. Place an adaptor on the hydrant if needed. With your partner's help, remove a section of hard suction hose from the engine.

3

Connect the hard suction hose to the intake on the engine.

4

Connect the opposite end to the hydrant. Slowly open the hydrant when instructed to do so by the driver/operator.

also be possible to repack the hose quickly and with minimal personnel. There is no perfect hose load that works well for every situation. The three most common hose loads for preconnected attack lines are the minuteman load, the flat load, and the triple layer load. Your circumstances will make one type of hose load preferable for your operation.

Because of different situations, most brigades load attack lines of different lengths. An engine could be equipped with a 150' preconnected line and a 250' preconnected line. Brigade members should pull an attack line that is long enough to reach the fire, but not so long that there will be an excess of hose to slow down the operation and become tangled.

Preconnected hose lines can be placed in several different locations on an engine. A section of a divided hose bed at the rear of the apparatus can be loaded with a preconnected attack line. Transverse hose beds are installed above the pump on many engines and loaded so the hose can be pulled off from either side of the apparatus. Preconnected lines can be loaded into special trays that are mounted on the side of fire apparatus.

Some engines have a special compartment in the front bumper that can store a short preconnected hose line. This

**Brigade Member Safety Tips**

When loading hose on an apparatus, you should always wear gloves and use caution in climbing up and down on the apparatus. If you are loading hose at a fire scene, watch out for wet, slippery surfaces, ice, or other hazards. Also, wet hose can be heavy and you may need to reach, stretch, or lift the hose to get it into the hose bed. Use caution!

shorter preconnected hose line often is used for vehicle fires and dumpster fires, where the apparatus can drive up close to the incident and a longer hose line is not needed. Booster hose is another type of preconnected attack line. Booster reels holding ¾- or 1-inch diameter hose can be mounted in a variety of locations on fire apparatus. Booster lines are rarely found on newer industrial apparatus. A preconnected handline can be put into service as quickly as a hose reel and provides higher flow rates.

### The Minuteman Load

To prepare a 200-foot minuteman hose load, connect the female coupling to the preconnect discharge. Lay the first two sections in the hose bed using a flat load. After the first loop of hose, make short pulling handles at the end of the hose bed. Leave the male end of the second section of hose to the side of the hose compartment. Attach the remaining two sections of hose together, and place the nozzle on the male end of the last section of hose. Place these sections of hose in the hose bed, starting with the nozzle end. Leave long pulling loops in the first loops attached to the nozzle. When you reach the female end of the last section of hose, attach it to the male end of the second section of hose placed into the hose bed. To prepare the minuteman load, follow the steps in ► Skill Drill 16-23.

1. Connect the female end of the first length of hose to the discharge outlet. **(Step 1)**
2. Lay the hose flat to the edge of the hose bed and leave the remaining hose with the male coupling out of the front of the bed to be connected later. Do not connect to the rest of the load. **(Step 2)**
3. Assemble the remaining hose sections and attach the nozzle. **(Step 3)**
4. Place the nozzle on the first length already in the hose bed. **(Step 4)**
5. Load the remaining hose flat into the bed, alternating the folds from the left to right sides of the bed. **(Step 5)**
6. Connect the last section loaded to the first section placed in the bed. **(Step 6)**
7. Lay the remaining loose hose on top the load. **(Step 7)**

To advance the minuteman hose load, follow the steps in ► Skill Drill 16-24.

1. Grasp the nozzle and the folds next to it. **(Step 1)**
2. Pull the load approximately one-third out of the bed. **(Step 2)**
3. Turn away from the hose bed and place the load on the shoulder. Walk away from the apparatus until all hose is clear of the hose bed. **(Step 3)**
4. Continue walking away, pulling the remaining hose from the hose bed and then allow the hose to deploy from the top of the load on the shoulder. **(Step 4)**

### The Preconnected Flat Load

To prepare the preconnected flat load, attach the female end of the hose to the preconnect discharge. Begin placing the hose flat in the hose bed. When about one-third of the hose is in the bed, make an 8-inch loop at the end of the hose bed. This loop will be used as a pulling handle. When two-thirds of the hose is loaded, make a second pulling loop about twice the size of the first loop. Finish loading the hose, attach the nozzle, and place it on top of the hose bed. The preconnected flat load is now ready for use. To prepare the preconnected flat hose load, follow the steps in ► Skill Drill 16-25.

1. Attach the female end of the hose to the preconnect discharge. **(Step 1)**
2. Begin laying the hose flat in the hose bed. **(Step 2)**
3. When about one-third of the hose is in the bed, make an 8-inch loop at the end of the hose bed. This loop will be used as a pulling handle.
4. When two-thirds of the hose is loaded, make a second pulling loop about twice the size of the first loop. **(Step 3)**
5. Finish loading the hose, attach the nozzle, and place it on top of the hose bed. The preconnected flat load is now ready for use. **(Step 4)**

To advance the preconnected flat hose load, follow the steps in ► Skill Drill 16-26.

1. Place arm through the larger lower loop.
2. Grasp the smaller loop with the same hand.**(Step 1)**
3. Grasp the nozzle with the opposite hand. **(Step 2)**
4. Pull the load from the bed. **(Step 3)**
5. Walk away from the vehicle. **(Step 4)**
6. As the load deploys, drop the small loop.
7. Extend the remaining hose to length. **(Step 5)**

### The Triple Layer Load

To prepare the triple layer load, attach the female end of the hose to the preconnect discharge. Connect the sections of hose together. Extend the hose directly from the hose bed. Pick up the hose two-thirds of the distance from the discharge to the hose nozzle. Carry the hose back to the apparatus, forming a three-layer loop. Pick up the entire length of folded hose (this will take several people.) Lay the tripled folded hose in the hose bed in an S-shape with

16-23 Skill Drill

## Loading the Minuteman Hose Load

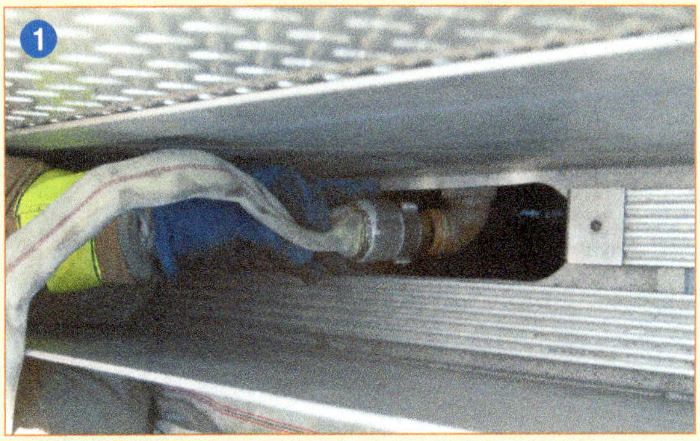

Connect the female end of the first length of hose to the discharge outlet.

Lay the hose flat to the edge of the hose bed and leave the remaining hose with the male coupling out of the front of the bed to be connected later. Do not connect to the rest of the load.

Assemble the remaining hose sections and attach the nozzle.

Place the nozzle on the first length already in the hose bed.

the nozzle on top. To prepare the triple layer hose load, follow the steps in ► **Skill Drill 16-27**.

1. To make the triple layer load, attach the female end of the hose to the preconnect discharge. **(Step 1)**
2. Connect the sections of hose together. **(Step 2)**
3. Extend the hose directly from the hose bed. Pick up the hose two-thirds of the distance from the discharge to the hose nozzle. **(Step 3)**
4. Carry the hose back to the apparatus, forming a three-layer loop. **(Step 4)**
5. Pick up the entire length of folded hose. (This will take several people.) **(Step 5)**
6. Lay the tripled folded hose in the hose bed in an S-shape with the nozzle on top. **(Step 6)**

To advance the triple layer hose load, follow the steps in ► **Skill Drill 16-28**.

1. Grasp the nozzle and the top fold. **(Step 1)**
2. Turn away from the hose bed and place the hose on the shoulder. **(Step 2)**
3. Walk away from vehicle until the entire load is out of the bed. **(Step 3)**
4. When the load is out of the bed, drop the fold. **(Step 4)**
5. Extend the nozzle to the remaining distance. **(Step 5)**

16-23 Skill Drill (Continued)

Load the remaining hose flat into the bed, alternating the folds from the left to right sides of the bed.

Connect the last section loaded to the first section placed in the bed.

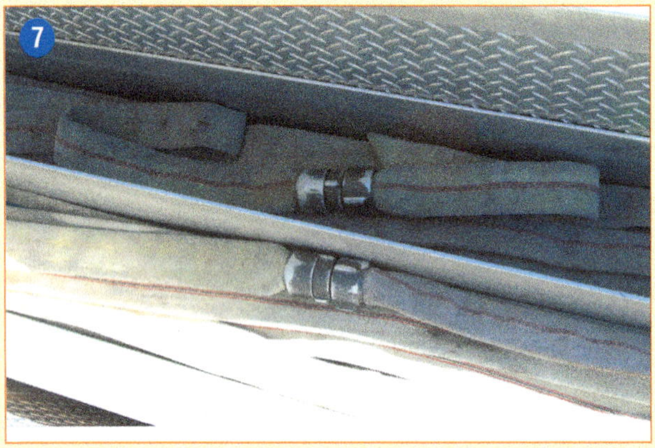


Lay the remaining loose hose on top of the load.

## Wyed Lines

In order to reach a fire that may be some distance from the engine, it may be necessary to first advance a larger diameter line, such as $2\frac{1}{2}$- or 3-inch hose line, and then split it into two $1\frac{3}{4}$-inch attack lines. This is accomplished by attaching a gated wye or a water thief to the end of the supply line and then attaching the two attack lines to the gated outlets. To unload and advance the wyed lines, follow the steps of Skill Drill 16-29.

1. Grasp one of the attack lines and pull it from the bed.
2. Pull the second attack line from the bed and place it far enough from the first line so that you can walk between the hose lines. This will keep the lines from becoming entangled.
3. Grasp the gated wye and pull it from the bed.
4. The apparatus will deploy the remaining hose from the wye back to a water source. Place the gated wye so that one attack line is on the other side.
5. The individual attack lines can now be extended to desired positions.

16-24 Skill Drill

## Advancing the Minuteman Hose Load

1 Grasp the nozzle and the folds next to it.

2 Pull the load approximately one-third out of the bed.

3 Turn away from the hose bed and place the load on the shoulder. Walk away from the apparatus until all hose is clear from the hose bed.

4 Continue walking away, pulling the remaining hose from the hose bed and then allow the hose to deploy from the top of the load on the shoulder.

## Hose Carries and Advances

Several different techniques are used to carry and advance fire hose. The best technique for a particular situation will depend on the size of the hose, the distance it must be moved, and the number of brigade members available to perform the task. The same techniques can be used for supply lines or attack lines.

Whenever possible, a hose line should be laid out and positioned as close as possible to the location where it will be operated before it is charged with water. A charged line is much heavier and more difficult to maneuver than a dry hose line. A suitable amount of extra hose should be available to allow for maneuvering and advancement after the line is charged.

### Advancing Standpipe and Occupant Use Hose

Hose reels (▶ Figure 16-25) and hose cabinets are designed for quick deployment of a hose line by trained occupants or fire brigade members. Hose cabinets and reels or racks in a Class II standpipe are equipped with 1½-inch outlets and are intended to deliver a maximum of 125 gpm. This is the maximum flow rate that can be used by trained occupants and incipient-level brigade members.

16-25 Skill Drill

## Loading the Preconnected Flat Hose Load

Attach the female end of the hose to the preconnect discharge.

Begin laying the hose flat in the hose bed.

When about one-third of the hose is in the bed, make an 8-inch loop at the end of the hose bed. This loop will be used as a pulling handle. When two-thirds of the hose is loaded, make a second pulling loop that is about twice the size of the first loop.

Finish loading the hose, attach the nozzle, and place it on top of the hose bed. The preconnected flat load is now ready for use.

A single brigade member can usually advance a hose line from an occupant-use hose cabinet (▶ Figure 16-26). A second brigade member may be needed to help remove kinks from the hose or to assist in advancing the hose around a corner or other obstacle.

To advance a hose from an occupant-use hose cabinet, follow the steps in (Skill Drill 16-30):

1. Open the cabinet door.
2. Swing the hose rack out of the cabinet.
3. Grasp the nozzle and pull it free from the retaining clamp.
4. Pull the hose from the rack and extend it until all of the hose is free.
5. Ensure that the hose is free of kinks.
6. Turn on the water supply.
7. Adjust the spray pattern and advance the hose toward the fire.

(*Note:* Some occupant-use hose cabinets are equipped with an automatic valve that opens the water supply when the last section of hose is pulled free of the rack.)

To advance a hose from a hose reel, follow the steps in (Skill Drill 16-31):

Figure 16-25 A hose reel.

Figure 16-26 An occupant-use hose cabinet.

1. Grasp the nozzle and a short section of hose.
2. Pull the hose straight out of the hose reel.
3. Extend the hose until the entire length has been pulled from the reel.
4. Ensure that the hose is free of kinks.
5. Turn on the water supply.
6. Adjust the nozzle pattern and advance the hose toward the fire.

When using a hose reel that is equipped with rigid rubber hose, the water can be turned on before the hose is pulled from the reel. Because rigid rubber hose is not subject to kinking, it eliminates the need to pull the entire length of hose from the reel.

A second brigade member may be needed to advance a long length of hose. A charged rack or reel hose may be difficult for a single brigade member to maneuver and advance because of the weight of the charged hose or the need to move around corners or other obstacles.

Industrial facilities utilize a variety of hose stations or enclosures to protect fire hose, rack systems, hose reels, and other firefighting equipment. Enclosures such as yard boxes or hose houses (▼ **Figure 16-27**) may also store extra hose and tools such as hydrant and valve wrenches, contain appliances such as spare nozzles, and be designed to protect the water supply valves or hydrant.

## Working Hose Drag

The working hose drag technique is used to deploy hose from a hose bed and advance the line a relatively short distance to the desired location. Depending on the size and length of the hose, several brigade members may be required to perform this task. To perform a working hose drag, follow the steps of (► **Skill Drill 16-32**).

1. Place the end of the hose over your shoulder. **(Step 1)**
2. Hold onto the coupling with your hand. **(Step 2)**
3. Walk in the direction you want to advance the hose. **(Step 3)**
4. As the next hose coupling is ready to come off the hose bed, have a second brigade member grasp the coupling and place the hose over the shoulder. **(Step 4)**
5. Continue this process until you have enough hose out of the hose bed. **(Step 5)**

Figure 16-27 A hose house.

16-26 Skill Drill

## Advancing the Preconnected Flat Hose Load

1 Place the arm through the larger lower loop. Grasp the smaller loop with the same hand.

2 Grasp the nozzle with the opposite hand.

3 Pull the load from the bed.

4 Walk away from the vehicle.

5 As the load deploys, drop the small loop. Extend the remaining hose to length.

16-27 Skill Drill

## Loading the Triple Layer Hose Load

1 Attach the female end of the hose to the preconnect discharge.

2 Connect the sections of hose together.

3 Extend the hose directly from the hose bed. Pick up the hose two-thirds of the distance from the discharge to the hose nozzle.

4 Carry the hose back to the apparatus, forming a three-layer loop.

5 Pick up the entire length of folded hose.

6 Lay the tripled folded hose in the hose bed in an S-shape with the nozzle on top.

16-28 Skill Drill

## Advancing the Triple Layer Hose Load

Grasp the nozzle and the top fold.

Turn away from the hose bed and place the hose on the shoulder.

Walk away from the vehicle until the entire load is out of the bed.

When the load is out of the bed, drop the fold.

Extend the nozzle to the remaining distance.

16-32 Skill Drill

# Performing a Working Hose Drag

1 Place the end of the hose over your shoulder.

2 Hold onto the coupling with your hand.

3 Walk in the direction you want to advance the hose.

4 As the next hose coupling is ready to come off the hose bed, a second brigade member grasps the coupling and places the hose over the shoulder.

5 Continue this process until you have enough hose out of the hose bed.

## Shoulder Carry

The shoulder carry is used to transport full lengths of hose over a longer distance than is practical to drag the hose. The shoulder carry is also useful when a hose line has to be advanced around obstructions. For example, this technique could be used to stretch an attack line from the front of a structure, around to the rear entrance, and up to the second floor. This technique could also be employed to stretch an additional supply line to an attack engine in a location where the hose cannot be laid out by another engine.

This technique requires practice and good teamwork in order to be successful. By working together to complete tasks efficiently, we achieve our goal of extinguishing the fire in the shortest period of time. To perform a shoulder carry, follow the steps in **Skill Drill 16-33**.

1. Stand at the tailboard of the engine. Grasp the end of the hose and place it over your shoulder so the coupling is at chest height.
2. Have a second brigade member place additional hose on your shoulder so that the ends of the folds reach to about knee level. Continue to place enough folds on your shoulder that you can safely carry.
3. Hold the hose to prevent it from falling off your shoulder. Continue to hold the hose and move forward about 15'.
4. A second brigade member should stand at the tailboard to receive a load of hose.
5. When enough brigade members have received hose loads, the hose can be uncoupled from the hose bed.
6. To move the hose, all brigade members must coordinate their movements. The driver/operator should connect this coupling to the engine.
7. All of the brigade members should start walking toward the fire. The last brigade member should not start off-loading hose from his or her shoulder until it has all been laid out. The next brigade member in line then starts laying out hose from his or her shoulder.
8. Each brigade member lays out his or her supply of hose until the entire length is laid out.

To advance an accordion load using a shoulder carry, follow the steps in **► Skill Drill 16-34**.

1. Find the end of the accordion load, whether it is a nozzle or coupling.
2. Using two hands, grasp the end of the load and the number of folds it will take to make an adequate shoulder load. **(Step 1)**
3. Pull the accordion load about one-third of the way off the apparatus. **(Step 2)**
4. The folds must then be twisted so they become flat with the end of the accordion load (nozzle or coup-ling) on the bottom of the now flat shoulder load. **(Step 3)**
5. Transfer the hose to the opposite shoulder while turning so you face in the direction you will walk. **(Step 4)**
6. Place the shoulder load over your shoulder and grasp tightly with both hands. **(Step 5)**
7. Walk away from the apparatus, pulling the shoulder load out of the hose bed. **(Step 6)**
8. If additional brigade members are needed, they may follow steps 1-7 to assist in removing the required amount of hose.

## Advancing an Attack Line

In order to attack an interior fire, an attack line is usually advanced in two stages. The first stage involves laying out the hose to the building entrance. The second stage is to advance the line into the building to the location where it will be operated.

When the attack line has been laid out to the entry point, the extra hose that will be advanced into the building should be flaked out in a serpentine pattern so that it will not become tangled when it is charged. The hose should be flaked out with the lengths of hose running parallel to the front of the fire building so that it can be easily advanced into the building **▼ Figure 16-28**. The flaked hose should be set back from the doorway so that it does not obstruct the entry and exit path.

You should make sure you flake out the hose BEFORE it is charged with water. Once the line is charged, the hose becomes much more difficult to maneuver and advance. It is also important to be sure that there is enough hose to reach the location where it will be needed inside the building.

While you are flaking out the hose and preparing to enter the building, your brigade leader will be completing the size-up. Other members of your team should be carrying out other tasks to support the operation. They may be forcing entry, getting into position for ventilation, establishing a water supply, and performing search and rescue. All of these tasks must be performed in sequence to maintain a safe environment and to efficiently extinguish the fire.

**Figure 16-28** Hose should be flaked out in a serpentine pattern outside the building entrance.

16-34 Skill Drill

## Advancing an Accordion Load

1 Using two hands, grasp the end of the load and the number of folds it will take to make an adequate shoulder load.

2 Pull the accordion load about one-third of the way off the apparatus.

3 The folds must then be twisted so they become flat with the end of the accordion load (nozzle or coupling) on the bottom of the now flat shoulder load.

4 Transfer the hose to the opposite shoulder while turning so you face in the direction you will walk.

5 Place the shoulder load over your shoulder and grasp tightly with both hands.

6 Walk away from the apparatus, pulling the shoulder load out of the hose bed.

Once the hose is flaked out, signal the driver/operator to charge the line. Open the nozzle slowly to bleed out any trapped air and to make sure the hose is operating properly. If you are using an adjustable nozzle, make sure the nozzle is set to deliver the appropriate stream. Once this is done, slowly close the nozzle.

When you are given the command to advance the hose, keep safety as your number one priority. Make sure the other members of the nozzle team are ready. Do not stand in front of the door as it is opened. You do not know what may happen when the door is opened.

As you move inside, stay low to avoid the greatest amount of heat and smoke. If you cannot see because of the dense smoke, use your hands to feel the pathway in front of you. Feel in front of you so you do not fall into a hole or opening. Look for the glow of fire, and check for the sensation of heat coming through your face piece. Communicate with the other members of the nozzle team as you advance.

As you advance the hose line, you need to have enough hose to enable you to move forward. A good hose line crew consists of at least two members at the nozzle and a third member outside the door. As resistance is encountered in advancing the line, the second brigade member at the nozzle can help to pull more hose, while the brigade member at the door is responsible for feeding more hose into the building. If necessary, the second brigade member can retrace the hose back to relieve an obstruction that prevents the hose from advancing. Charged hose lines are not easy to advance through a structure. It is only with good teamwork that efficient hose line advancement can occur.

### Advancing a Hose Line up a Stairway

When advancing a hose line up stairs, arrange an adequate amount of extra hose close to the bottom of the stairs. Make sure all members of the team are ready to move on command. It is hard to move a charged hose line up a set of stairs while flowing water through the nozzle. Shutting down the hose line while moving up the stairs will often allow you to get to the top of the stairs more quickly and safely. Follow the directions of your brigade leader. It is easiest to position an uncharged hose line. To advance an uncharged hose line up a stairway, follow the steps in ▶ Skill Drill 16-35.

1. Use a shoulder carry to advance up the stairs. **(Step 1)**
2. When ascending the stairway, lay the hose against the outside of the stairs to avoid sharp bends and kinks and to reduce tripping hazards. **(Step 2)**
3. Arrange excess hose so that it is available to brigade members entering the fire floor. **(Step 3)**

### Advancing a Hose Line Down a Stairway

Advancing a charged hose line down a stairway is also difficult. The smoke and flames from the fire tend to travel up the stairway. You want to get down the stairway and position yourself below the heat and smoke as quickly as possible. Keep as low as possible to avoid the worst of the heat and smoke. The one thing you have going for you as you advance a hose line down a stairway is that gravity is working with you to bring the hose line down the stairs. You should never advance toward a fire unless your hose line is charged and ready to flow water.

Wearing personal protective equipment (PPE) and SCBA changes your center of gravity. If you try to crawl down a stairway headfirst, you are likely to find yourself tumbling head over heels. Move down the stairway feet first, using your feet to feel for the next step. Move carefully, but as quickly as possible to get below the worst of the heat and smoke. To advance a hose line down a stairway, follow the steps in Skill Drill 16-36.

1. Advance a charged hose line.
2. Descend stairs feet first.
3. Position brigade members at areas where hose lines could snag.

### Advancing a Hose Line up a Ladder

If a hose line has to be advanced up a ladder, this should be done before the line is charged. To do this, place the hose on the side of the ladder. Place the hose across your chest with the nozzle draped over your shoulder. Climb up the ladder with the uncharged hose line in this position. If the line is mistakenly charged while you are climbing, it will not push you away from the ladder.

Additional brigade members should pick up the hose about every 25 feet and help to advance it up the ladder. The nozzle is passed over the top rung of the ladder and into the fire building. Additional hose should be fed up the ladder until sufficient hose is inside the building to reach the fire. The hose should be secured to the ladder with a hose strap to keep it from becoming dislodged. To advance an uncharged hose line up a ladder, follow the steps in ▶ Skill Drill 16-37.

1. If a hose line needs to be advanced up a ladder, it should be advanced before it is charged.
2. Advance the hose line to the ladder. **(Step 1)**
3. Pick up the nozzle; place the hose across your chest with the nozzle draped over your shoulder. **(Step 2)**
4. Climb up the ladder with the uncharged hose line. **(Step 3)**
5. Once the first brigade member reaches the first fly section of the ladder, a second brigade member should shoulder the hose to assist advancing the hose line up the ladder. To avoid overloading of the ladder, limit one brigade member per fly section. **(Step 4)**
6. The nozzle is placed over the top rung of the ladder and advanced into the fire area. **(Step 5)**
7. Additional hose can be fed up the ladder until sufficient hose is in position.
8. The hose can be secured to the ladder with a hose strap to support its weight and keep it from becoming dislodged. **(Step 6)**

16-35 Skill Drill

## Advancing an Uncharged Hose Line Up a Stairway

1

Use a shoulder carry to advance up the stairs.

2

When ascending the stairway, lay the hose against the outside of the stairs to reduce tripping hazards. Avoid sharp bends.

3

Arrange excess hose so that it is available to brigade members entering the fire floor.

### Operating a Hose Stream from a Ladder

A hose stream can be operated from a ladder and directed into a building through a window or other opening. To operate a fire hose from a ladder, follow the steps in Skill Drill 16-38.

1. Climb the ladder with a hose line to the height at which the line will be operated.
2. Apply a leg lock or use a ladder belt.
3. Place the hose between two rungs and secure the hose to the ladder with a rope hose tool, rope, or piece of webbing.
4. Carefully operate the hose stream from the ladder. Be careful when opening and closing nozzles and redirecting the stream because of the nozzle back-pressure. This force could destabilize the ladder.

### Extending an Attack Line

When choosing a preconnected hose line or assembling an attack line, it is better to have too much hose than a line that is too short. With a hose that is longer than necessary, you can flake out the excess hose. With a hose line that is too short, you cannot advance it to the seat of the fire without shutting it down and taking the time to extend it.

There may be circumstances where it is necessary to extend a hose line by adding additional lengths of hose. This could occur if the fire is further from the apparatus than

16-37 Skill Drill

## Advancing an Uncharged Hose Line Up a Ladder

Advance the hose line to the ladder.

Pick up the nozzle; place the hose across your chest with the nozzle draped over your shoulder.

Climb up the ladder with the uncharged hose line.

Once the first brigade member reaches the first fly section of the ladder, a second brigade member should shoulder the hose to assist advancing the hose line up the ladder. To avoid overloading of the ladder, limit one brigade member per fly section.

The nozzle is placed over the top rung of the ladder and advanced into the fire area.

Additional hose can be fed up the ladder until sufficient hose is in position. The hose can be secured to the ladder with a hose strap to support its weight and keep it from becoming dislodged.

initially estimated. More hose line could be needed to reach the burning area or to complete extinguishment.

There are two basic ways to extend a hose line. The first is to disconnect the hose from the discharge on the attack engine and add the extra hose at that location. This requires advancing the full length of the attack line to take advantage of the extra hose, which could take time and considerable effort. The alternative is to add the hose to the discharge end of the hose. This provides fast extension of an attack line by one or two brigade members and eliminates the need to drag water-filled hose. The attack line can either be shut down at the apparatus or a hose clamp can be utilized on the last section of hose to shut off the flow. The advantage of using a hose clamp is that the crew can reestablish flow without relying on the engineer to open the discharge at the apparatus.

## Connecting Hose Lines to Standpipe and Sprinkler Systems

Fire department connections on buildings are provided so that the fire brigade can pump water into a standpipe or sprinkler system. The function of the hose line is to provide either a primary or secondary water supply for the sprinkler or standpipe system. The same basic techniques are used to connect the hose lines to either type of system.

Standpipe systems are used to provide a water supply for attack lines that will be operated inside the building (▼ Figure 16-29). Outlets are provided inside the building where brigade members can connect attack lines. The brigade members inside the building have to depend on brigade members outside to supply the water to the fire department connection.

There are two types of standpipe systems. A dry standpipe system depends on the fire brigade to provide all of the water. A wet standpipe system has a built-in water supply, but the fire department connection is provided to deliver a higher flow or to boost the pressure. The pressure requirements for standpipe systems depend on the height where the water will be used inside the building.

Figure 16-29 A standpipe connection.

### Brigade Member Safety Tips

Some brigades use LDH with Storz-type couplings to connect to fire department connections. Hose that is rated for use as "attack hose" should be used to connect to a standpipe system.

Many fire brigades SOPs require two hose lines to be connected to a fire department connection. If one line breaks or the flow is interrupted, water will still be delivered to the system through the other line.

The fire department connection for a sprinkler system is also used to supplement the normal water supply. The required pressures and flows for different types of sprinkler systems can vary significantly. As a guideline, sprinkler systems should be fed at 150 psi unless there is more specific information available. To connect a hose line to supply a fire department connection, follow the steps in (Skill Drill 16-39).

1. Locate the fire department connection to the standpipe or sprinkler system.
2. Extend a hose line from the engine discharge to the fire department connection using the size hose required by the fire brigade's SOPs. Some brigades use a single hose line, while others call for two or more lines to be connected.
3. Remove the caps on the fire department connection. Some caps are threaded into the connections and have to be unscrewed. Other caps are designed to break away when struck with a tool such as a hydrant wrench or spanner.
4. Visually inspect the interior of the connection to ensure that there is no debris that could obstruct the water flow.
5. Attach the hose line(s) to the connection(s).
6. Notify the driver/operator when the connection has been completed.

### Advancing an Attack Line from a Standpipe Outlet

The standpipe outlets inside a building are provided for brigade members to connect attack hose lines. This eliminates the need to advance hose lines all the way from the attack engine to an upper floor or a fire deep inside a large area building. In tall structures, it would be impossible to advance hose lines up the stairways in a reasonable time and to supply sufficient pressure to fight a fire on an upper floor.

The standpipe outlets are often located in stairways and SOPs generally require attack lines to be connected to an outlet one floor below the fire. The working space in the stairway and around the outlet valve is usually limited. Before opening the door, it is important to properly flake out the hose line so it will be ready to advance into the fire floor. Before charging the hose line, the hose should be flaked out on the stairs going up from the fire floor. When the hose line is charged and advanced into the fire floor, gravity will help to move the line forward. This is much easier than having to pull the charged hose line up the stairs.

To connect and advance an attack line from a standpipe outlet, follow the steps in Skill Drill 16-40.

1. Carry a standpipe hose bundle to the standpipe connection below the fire. Remove the cap from the standpipe.
2. Attach the proper adaptor or appliance such as a gated wye.
3. Flake the hose up the stairs to the floor above the fire.
4. Extend the hose to the fire floor and prepare for your fire attack.

## Replacing a Defective Section of Hose

With proper maintenance and testing, the risk of fire hose failure should be low, but it is always possible. Every brigade member should know how to quickly replace a length of defective hose and restore the flow.

A burst hose line should be shut down as soon as possible. If the line cannot be shut down at the pump or at a control valve, a hose clamp can be used to stop the flow in an undamaged section of hose upstream from the problem. After the water flow has been shut off, quickly remove the damaged section of hose and replace it with two sections of hose. Using two sections of hose will ensure that the replacement hose is long enough to replace the damaged section. To replace a defective hose section, follow the steps in Skill Drill 16-41.

1. Shut down or clamp off damaged line.
2. Remove damaged section of hose.
3. Replace with two sections to ensure length will be adequate.
4. Restore water flow.

## Draining and Picking Up Hose

In order to put the hose back into service, the hose must be drained of water. That is accomplished by laying the hose straight on a flat surface. Then lift one end of the hose to shoulder level. Gravity will allow the water to flow to the lower portion of the hose and eventually out of the hose. As you proceed down the length of hose, fold the hose back and forth over your shoulder. When you reach the end of the section, you will have the whole section of hose on your shoulder. To drain the hose, follow the steps in ▶ Skill Drill 16-42.

1. Lay the section of hose straight on a flat surface. **(Step 1)**
2. Start at one end of the section and lift the hose to shoulder level. **(Step 2)**
3. Moving down the length of hose, fold it back and forth over your shoulder. **(Step 3)**
4. Continue down the length until the entire section is on your shoulder. **(Step 4)**

### Brigade Member Safety Tips

Always open and close nozzles slowly to prevent water hammer.

### Brigade Member Tips

Some nozzles are made so that the tip of the nozzle can be separated from the shut-off valve. This is a **breakaway nozzle**. This allows you to shut off the flow, unscrew the nozzle tip, and then add additional lengths of hose to extend the hose line without shutting off the valve at the engine.

## Unloading Hose

There are times other than a fire where you will need to unload the hose from an engine. Hose should be unloaded and reloaded on a regular basis to place the bends in different portions of the hose. Leaving bends in the same locations for long periods of time is likely to cause weakened areas. Hose might have to be unloaded to change out apparatus. It could also be necessary to offload the hose for annual testing to be conducted.

The following procedure should be used to unload a hose bed:

- A large area such as a parking lot should be used for this procedure.
- Disconnect any gate valves or nozzles from the hose before you begin.
- Grasp the end of the hose, and pull it off the engine in a straight line.
- When a coupling comes off the engine, disconnect the hose and pull off the next section of hose.
- When all of the hose has been removed from the hose bed, use a broom to brush off any dirt or debris on both sides of the hose jacket.
- Sweep out any debris or dirt from the hose bed.
- Roll all of the hose into doughnut rolls.
- Store hose rolls off the floor on a rack, in a cool dry area.

# Nozzles

Nozzles are attached to the discharge end of attack lines to give fire streams shape and direction. Nozzles are used on all sizes of handlines as well as on master stream devices. Nozzles can be classified into three groups. Low volume nozzles flow 40 gpm or less. These are primarily used for booster hoses, and their use is limited to small outside fires.

16-42 Skill Drill

## Drain and Carry a Hose

1 Lay the section of hose straight on a flat surface.

2 Start at one end of the section and lift the hose to shoulder level.

3 Move down the length of hose, folding it back and forth over your shoulder.

4 Continue down the length until the entire section is on your shoulder.

**Handline nozzles** are used on hose lines ranging from $1\frac{1}{2}$ to $2\frac{1}{2}$ inches in diameter. Handline streams usually flow between 60 and 350 gpm. **Master stream nozzles** are used on deck guns, portable monitors, and ladder pipes that flow more than 350 gpm.

Low volume and handline nozzles incorporate a shut-off valve that is used to control the flow of water. The control valve for a master stream is usually separate from the nozzle. All nozzles have some type of mechanism to direct the water stream into a certain shape. Some nozzles also incorporate a mechanism to automatically adjust the flow based on the water volume and pressure that are available.

### Nozzle Shut Offs

The **nozzle shut off** enables a person at the nozzle to start or stop the flow of water. The most common nozzle shut off mechanism is a quarter-turn valve. The handle that controls this valve is called a bale. Some nozzles incorporate a rotary control valve operated by rotating the nozzle in one direction to open and the opposite direction to shut off the flow of water.

Two different types of nozzles are manufactured for the fire service. These are the **smooth bore nozzles** and **fog stream nozzles**. Smooth bore nozzles produce a solid stream of water. Fog stream nozzles separate the water into

Figure 16-30 Smooth bore nozzle.

droplets. The size of the water droplets and the discharge pattern can be varied by adjusting the nozzle setting. Nozzles must have an adequate volume of water and an adequate pressure in order to produce a good fire stream. The volume and pressure requirements vary according to the type and size of nozzle.

## Smooth Bore Nozzles

The simplest smooth bore nozzle consists of a shut-off valve and a smooth bore tip that gradually decreases the diameter of the stream to a size smaller than the hose diameter ▲ Figure 16-30. Smooth bore nozzles are manufactured to fit both handlines and master stream devices. Smooth bore nozzles that are used for master stream and ladder pipes often consist of a set of stacked tips. Each successive tip in the stack has a smaller diameter opening. Tips can be quickly added or removed to provide the desired stream size. This allows different sizes of streams and volumes to be produced under different conditions.

There are several advantages of using a smooth bore nozzle. A good smooth bore has a longer reach than a combination fog nozzle operating at a straight stream setting. A smooth bore is capable of deeper penetration into burning materials, resulting in quicker knock-down and extinguishment. Most smooth bore nozzles are designed to operate at 50 psi on handline nozzles and 80 psi on master stream nozzles. Adjustable stream nozzles such as fog nozzles are typically designed to operate at 100 psi whether found on a handline or a master stream device. However, some newer models are being designed to operate at lower pressures while allowing similar volumes of water to flow. Lower nozzle pressure makes it easier for a brigade member to handle the nozzle.

A straight stream extinguishes a fire with less air movement and less disturbance of the thermal balance than a fog stream. This makes the heat conditions less intense for brigade members during an interior attack. It is also easier for the operator to see the pathway of a solid stream than a fog stream.

There are also disadvantages with smooth bore nozzles. Smooth bore nozzles do not absorb heat as readily as fog streams and are not effective for hydraulic ventilation. You cannot change the setting of a smooth bore nozzle to produce a fog pattern; however, a fog nozzle can be set to produce a straight stream. To operate a smooth bore nozzle, follow the steps in ► Skill Drill 16-43.

1. Select the desired tip size and attach to nozzle shut-off valve. **(Step 1)**
2. Obtain a stable stance. **(Step 2)**
3. Slowly open the valve, allowing water to flow. **(Step 3)**
4. Open the valve completely to achieve maximum effectiveness. (Failure to fully open the valve will result in reduced flow, depriving you of the necessary 1 gpm.) **(Step 4)**
5. Direct the stream to the desired location. **(Step 5)**

## Fog Stream Nozzles

Fog stream nozzles produce fine droplets of water ► Figure 16-31 and Figure 16-32. The advantage of creating these droplets of water is that they absorb heat much more quickly and efficiently than a solid stream of water. This is an important characteristic when immediate reduction of room temperature is needed to avoid a flashover. Fog nozzles can produce a variety of stream patterns from a straight stream to a narrow fog cone of less than 45° to a wide-angle fog pattern that is close to 120°.

The straight streams produced by fog nozzles have openings in the center. Therefore, a fog nozzle can produce a straight stream, but not a solid stream. The straight stream from a fog stream nozzle will break up faster and will not have the reach of a solid stream. A straight stream will be affected more by wind than a solid stream.

There are several advantages of using fog stream nozzles. First, fog stream nozzles can be used to produce a variety of stream patterns by rotating the tip of the nozzle. Fog streams are effective at absorbing heat and can be used to create a water curtain to protect brigade members from extreme heat.

Fog nozzles move large volumes of air along with the water. This can be an advantage or a disadvantage, depending

16-43 Skill Drill

# Operating a Smooth Bore Nozzle

Select the desired tip size and attach to nozzle shut-off valve.

Obtain a stable stance.

Slowly open the valve, allowing water to flow.

Open the valve completely to achieve maximum effectiveness.

Direct the stream to the desired location.

Figure 16-31 Fog stream nozzle.

Figure 16-32 Brass occupant use fog stream nozzle.

on the situation. A fog stream can be used to exhaust smoke and gases through hydraulic ventilation. This air movement can also result in sudden heat inversion in a room that pushes hot steam and gases down onto the brigade members. If used incorrectly, a fog pattern can push the fire into unaffected areas of a building.

In order to produce an effective stream, nozzles must be operated at the pressure recommended by the manufacturer. For many years, the standard operating pressure for fog stream nozzles was 100 psi. In recent years, some manufacturers have produced low-pressure nozzles that are designed to operate at 50 or 75 psi. The advantage of low-pressure nozzles is that they produce less reaction force, which makes them easier to control and advance. Lower nozzle pressure also decreases the risk that the nozzle will get out of control. To operate a fog nozzle, follow the steps in ► Skill Drill 16-44.

1. Select the desired nozzle.
2. Obtain a stable stance. **(Step 1)**
3. Slowly open the valve, allowing water flow. **(Step 2)**
4. Open the valve completely. (Failure to open the valve fully will restrict water flow reducing the necessary gpm). **(Step 3)**
5. Select the desired water pattern by rotating the bezel of the nozzle.
6. Apply water where needed. **(Step 4)**

### Types of Fog Stream Nozzles

There are three types of fog stream nozzles. The difference between the types is in the water delivery capability. A fixed gallonage fog nozzle will deliver a preset flow in gpm at the rated discharge pressure. The nozzle could be designed to flow 30, 60, or 100 gpm.

An adjustable gallonage fog nozzle allows the operator to select a desired flow from several settings. This is done by rotating a selector bezel to adjust the size of the opening. For example, a nozzle could have the options of flowing 60, 95, or 125 gpm. Once the setting is chosen the nozzle will deliver the rated flow only as long as the rated pressure is provided at the nozzle.

An automatic adjusting fog nozzle can deliver a wide range of flows. The nozzle has an internal spring-loaded piston. As the pressure at the nozzle increases or decreases, its piston moves in or out to adjust the size of the opening. The amount of water flowing through the nozzle is adjusted to maintain the rated pressure and produce a good stream. A typical automatic nozzle could have an operating range of 90 to 225 gpm while maintaining adequate discharge pressure.

## Other Types of Nozzles

There are other types of nozzles that are used for special purposes. Piercing nozzles are used to make a hole in sheet metal, aircraft, or building walls, in order to extinguish fires behind these surfaces ► Figure 16-33.

16-44 Skill Drill

# Operating a Fog Nozzle

1 Obtain a stable stance (if standing).

2 Slowly open the valve, allowing water flow.

3 Open the valve completely.

4 Select the desired water pattern, by rotating the bezel of the nozzle. Apply water where needed.

Figure 16-33 Piercing nozzle.

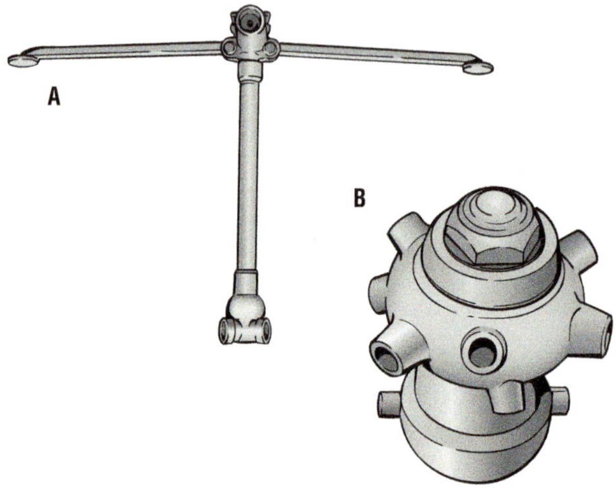


Figure 16-34 Cellar nozzle (A) and Bresnan distributor nozzle (B).

Cellar nozzles and Bresnan distributor nozzles are used to fight fires in cellars, voids, and other inaccessible places ► Figure 16-34. These nozzles discharge water in a wide circular pattern as the nozzle is lowered vertically through a hole into the cellar. They work like a large sprinkler head.

Water curtain nozzles are used to deliver a flat screen of water to form a protective sheet of water on the surface of an exposed building ► Figure 16-35. The water curtains must be directed onto the exposed building because radiant heat can pass through the water curtain. If your fire brigade has other types of specialty nozzles, you need to become proficient in the use and operation of them.

## Nozzle Maintenance and Inspection

Nozzles should be inspected on a regular basis, along with all of the equipment on the apparatus. A nozzle should be checked after each use before being placed back on the apparatus. It should be kept clean and clear of debris. Debris

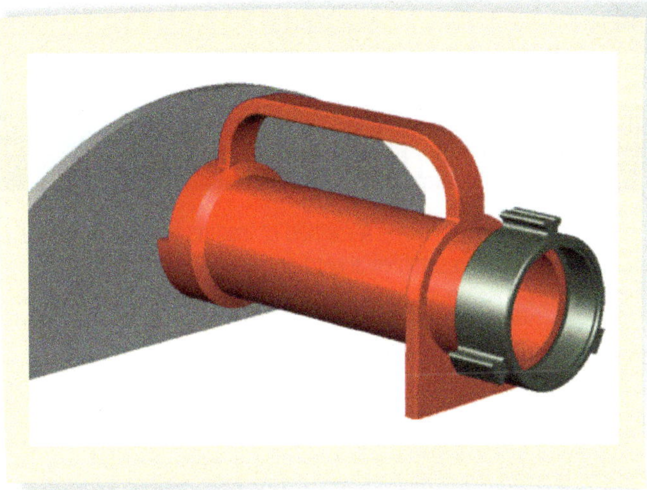

Figure 16-35 One type of a water curtain nozzle.

inside the nozzle will affect the performance of the nozzle, possibly reducing flow. Dirt and grit can interfere with the valve operation and prevent opening and closing fully. A light lubricant on the valve ball will keep it operating smoothly.

On fog nozzles, inspect the teeth on the face of the nozzle. Make sure all teeth are present and the ring can spin freely. Any missing teeth or failure of the ring to spin will drastically affect the fog pattern. Any problems noted should be referred to a competent technician for repair.

# Wrap-Up

## Ready for Review

- Fire hose is the most fundamental firefighting tool.
- Fire hose comes in different sizes for a variety of uses.
- Attack hose is designed to be used for fire suppression.
- Supply hose is used to deliver water to an attack engine from a pressurized source such as a hydrant or another engine.
- Through the use of hose appliances, brigade members have many options to choose from for configuring fire hose to meet the requirements of the situation.
- There are many different ways to carry, load, and transport hose on fire apparatus. The method chosen is usually determined by the advantages of the method and the needs of the individual fire brigade.
- Fire hose and nozzles need proper care and maintenance just as any other firefighting equipment. Properly cared for equipment will have a long service life.
- Fire hydraulics influences hose selection and configurations.
- Nozzle selection is determined by the effects desired. Examples are fog stream and straight stream.

## Hot Terms

**Accordion hose load** A method of loading hose on a vehicle whose appearance resembles accordion sections. It is achieved by standing the hose on its edge, then placing the next fold on its edge and so on.

**Adaptor** A device that joins hose couplings of the same type, such as male to male or female to female.

**Adjustable gallonage fog nozzle** A nozzle that allows the operator to select a desired flow from several settings.

**Attack engine** The engine from which the attack lines have been pulled.

**Attack hose** (attack line) The hose that delivers water from a fire pump to the fire. Attack hoses range in size from 1 to 2$^1/_2$ inches.

**Automatic adjusting fog nozzle** A nozzle that can deliver a wide range of water stream flows. It operates by an internal spring-loaded piston.

**Ball valves** Valves used on nozzles, gated wyes, and engine discharge gates. Made up of a ball with a hole in the middle of the ball.

**Booster hose** (booster lines) A rigid hose that is $^3/_4$ or 1 inch in diameter. This hose delivers only 30 to 60 gpm, but can do so at high pressures. It is used for small fires.

**Breakaway nozzle** A nozzle with a tip that can be separated from the shut-off valve.

**Bresnan distributor nozzle** A device that can be placed in confined spaces. The nozzle spins, spreading water over a large area.

**Butterfly valves** Valves that are found on the large pump intake valve where the hard or soft sleeve hose connects.

**Cellar nozzles** Nozzles used to fight fires in cellars and other inaccessible places. These devices work by spreading water in a wide pattern as the nozzle is lowered by a hole into the cellar.

**Double-female adaptor** A hose adaptor that is equipped with two female connectors. It allows two hoses with male couplings to be connected together.

**Double-jacket hose** A hose constructed with two layers of woven fibers.

**Double-male adaptor** A hose adaptor that is equipped with two male connectors. It allows two hoses with female couplings to be connected together.

**Dutchman** A term used for a short fold placed in a hose when loading it into the bed. This fold prevents the coupling from turning in the hose bed.

**Fire hydraulics** The physical science of how water flows through a pipe or hose.

**Fixed gallonage fog nozzle** A nozzle delivers a set number of gallons per minute that the nozzle was designed for, no matter what pressure is applied to the nozzle.

**Flat hose load** A method of putting a hose on apparatus in which the hose is laid flat and stacked on top of the previous section.

**Fog stream nozzle** Device placed at the end of a fire hose that separates water into fine droplets to aid in heat absorption.

**Forward lay** A method of laying a supply line where the line starts at the water source and is laid toward the fire.

**Four-way hydrant valve** A specialized type of valve that can be placed on a hydrant that allows another engine to increase the supply pressure without interrupting flow.

**Friction loss** The reduction in pressure due to the water being in contact with the side of the hose. This contact requires force to overcome the drag the wall of the hose creates.

**Gate valves** Valves found on hydrants and sprinkler systems.

**Gated wye** A valved device that splits a single hose into two separate hoses, allowing each hose to be turned on and off independently.

**Handline nozzles** Used on hoses ranging from 1½- to 2½-inch hose lines, usually flow between 90 and 350 gpm.

**Hard suction hose** A hose designed to prevent collapse under vacuum conditions so that it can be used for drafting water from below the pump (lakes, rivers, wells, or sea water, etc.).

**Higbee indicators** An indicator on the male and female threaded couplings that indicates where the threads start. These indicators should be aligned before starting to thread the couplings together.

**Horseshoe hose load** A method of loading hose where the hose is laid into the bed along the three walls of the bed, resembling a horseshoe.

**Hose appliance** Any device used to connect to a fire hose for the purpose of delivering water.

**Hose clamp** A device used to compress a fire hose to stop water flow.

**Hose jacket** A device used to stop a leak in a fire hose or to join hoses that have damaged couplings.

**Hose liner** (hose inner jacket) The inside portion of a hose that is in contact with the flowing water.

**Hose roller** A device that is placed on the edge of a roof and is used to protect hose as it is hoisted up and over the roof edge.

**Large-diameter hose (LDH)** Hose in the 4–12 inches range.

**Low volume nozzles** Nozzles that flow 40 gpm or less.

**Master stream device** A large-capacity nozzle that can be supplied by two or more hose lines or a single LDH. Can flow more than 350 gpm. Includes deck guns and portable ground monitors.

**Master stream nozzles** A nozzle used on deck guns, portable monitors, and ladder pipes that flows more than 350 gpm.

**Medium diameter hose (MDH)** Hose of 2½ or 3-inch size.

**Mildew** A condition that can occur on hose if it is stored wet. Mildew can damage the jacket of a hose.

**Nozzle shut off** Device that enables the person at the nozzle to start or stop the flow of water.

**Nozzles** Attachments to the discharge end of attack hoses to give fire streams shape and direction.

# Wrap-Up

**Piercing nozzle** A nozzle that can be driven through sheet metal or other material to deliver a water stream to that area.

**Reducer** A device that can join two hoses of different sizes.

**Reverse lay** A method of laying a supply line where the line starts at the fire and ends at the water source.

**Rocker lug** (pin lug) Fittings on threaded couplings that aid in coupling the hoses.

**Rubber-covered hose** (rubber-jacket hose) Hose whose outside covering is made of rubber, said to be more resistant to damage.

**Siamese connection** A device that allows two hoses to be connected together and flow into a single hose.

**Small diameter hose (SDH)** Hose in the 1–2 inches range.

**Smooth bore nozzle** Nozzles that produce a solid stream of water.

**Smooth bore tip** A nozzle device that is a smooth tube, used to deliver a solid stream of water.

**Soft Sleeve Hose** A LDH that is designed to be connected to the large port on a hydrant (steamer connection) and into the engine.

**Spanner wrench** A type of wrench used in coupling or uncoupling hoses by turning the lugs on the connections.

**Split hose bed** A hose bed that is divided into two or more sections.

**Split hose lay** A scenario where the attack engine will lay a supply line from a point away from the fire, and the supply engine will lay a supply line from the hose left by the attack engine to the water source.

**Storz-type coupling** A hose coupling that has the property of being both the male and female coupling. It is connected by engaging the lugs and turning the coupling one-third of a turn.

**Supply hose** (supply line) The hose used to deliver water from a source to a fire pump.

**Threaded hose couplings** A type of coupling that requires a male and female fitting to be screwed together.

**Triple layer load** A hose loading method that utilizes folding the hose back onto itself to reduce the overall length to one-third before loading in the bed. This load method reduces deployment distances.

**Water curtain nozzles** Nozzles used to deliver a flat screen of water to form a protective sheet of water.

**Water hammer** An event that occurs when flowing water is suddenly stopped; the velocity force of the moving water is transferred to everything it is in contact with. These can be tremendous forces that can damage equipment and cause injury.

**Water Thief** A device with an inlet and an outlet of the same size and several additional outlets of smaller size.

**Wye** A device used to split a single hose into two separate lines.

# Brigade Member in Action

**You are paged out to a fire in a three-story building, which is a combination of nonhazardous warehouse storage and offices. When you arrive at the scene, the structure is 60% involved with fire showing out of the windows. The IC has decided on a defensive strategy and has ordered your brigade leader to establish water supply and operate a ground monitor attack on the west side of the structure.**

1. Which of the following is the correct method of laying a supply line from the hydrant to the fire?
   - **A.** Forward hose lay
   - **B.** Reverse hose lay
   - **C.** Split hose lay
   - **D.** Flat hose lay

2. What type of hose should supply a ground monitor?
   - **A.** Multiple booster hose lines
   - **B.** Multiple $1\frac{1}{2}$- or $1\frac{3}{4}$-inch hose lines
   - **C.** Multiple $2\frac{1}{2}$ inches or larger hose lines
   - **D.** Multiple forestry hose lines

**The fire is successfully knocked down without spreading to exposures. There are hot spots that need to be extinguished. Your brigade leader orders you to disconnect the ground monitor and connect a gated wye with two 200 foot $1\frac{3}{4}$-inch hose lines for overhaul. Once the overhaul is completed, you reload them on your apparatus.**

3. A gated wye:
   - **A.** Combines two hose lines into one.
   - **B.** Splits one hose stream into two.
   - **C.** Joins two female hose couplings.
   - **D.** Joins two hose lines of similar diameter but differing threads.

4. LDH should be loaded using a/an:
   - **A.** Horseshoe load.
   - **B.** Accordion load.
   - **C.** Dutchman load.
   - **D.** Flat load.

# Foam

# Chapter 17

## NFPA 1081 Standard

**Incipient Facility Fire Brigade Member**

NFPA 1081 contains no Incipient Facility job performance requirements for this chapter.

**Advanced Exterior Facility Fire Brigade Member**

**5.3.4*** Extinguish an ignitible (or simulated ignitible) liquid fire operating as a member of a team, given an assignment, a handline, PPE, a foam proportioning device, a nozzle, foam concentrates, and a water supply, so that the correct type of foam concentrate is selected for the given fuel and conditions, a correctly proportioned foam stream is applied to the surface of the fuel to create and maintain a foam blanket, the fire is extinguished, reignition is prevented, and team protection is maintained.

**(A)** *Requisite Knowledge.* Methods by which foam prevents or controls a hazard; principles by which foam is generated; causes for poor foam generation and corrective measures; difference between hydrocarbon and polar solvent fuels and the concentrates that work on each; the characteristics, uses, and limitations of firefighting foams; the advantages and disadvantages of using fog nozzles versus foam nozzles for foam application; foam stream application techniques; hazards associated with foam usage; and methods to reduce or avoid hazards.

**(B)** *Requisite Skills.* The ability to prepare a foam concentrate supply for use, assemble foam stream components, master various foam application techniques, and approach and retreat from fires and spills as part of a coordinated team.

**Interior Structural Facility Fire Brigade Member**

**6.3.4*** Extinguish an ignitable liquid fire operating as a member of a team, given an assignment, a handline, PPE, a foam proportioning device, a nozzle, foam concentrates, and a water supply, so that the correct type of foam concentrate is selected for the given fuel and conditions, a correctly proportioned foam stream is applied to the surface of the fuel to create and maintain a foam blanket, fire is extinguished, reignition is prevented, and team protection is maintained.

**(A)** *Requisite Knowledge.* Methods by which foam prevents or controls a hazard; principles by which foam is generated; causes for poor foam generation and corrective measures; difference between hydrocarbon and polar solvent fuels and the concentrates that work on each; the characteristics, uses, and limitations of firefighting foams; the advantages and disadvantages of using fog nozzles versus foam nozzles for foam application; foam stream application techniques; hazards associated with foam usage; and methods to reduce or avoid hazards.

**(B)** *Requisite Skills.* The ability to prepare a foam concentrate supply for use, assemble foam stream components, master various foam application techniques, and approach and retreat from fires and spills as part of a coordinated team.

## Knowledge Objectives

After completing this chapter, you will be able to:

- Understand foam terms.
- Describe how foam works.
- Understand the foam tetrahedron.
- Describe expansion rates.
- Explain the different types of foam concentrate.
- Describe foam characteristics.
- Describe foam percentages and their importance.
- Explain foam guidelines and limitations.
- Identify various foam proportioning devices.
- Recognize the causes of poor foam quality delivery.
- Calculate the application rates for a spill fire.
- Calculate the application rates for a diked fire.
- Calculate the application rates for a tank fire.

## Skills Objectives

After completing this chapter, you will be able to perform the following skills:

- Assemble the correct foam stream components.
- Perform the roll-on method of applying foam.
- Perform the bounce-off method of applying foam.
- Perform the rain-down method of applying foam.

# You Are the Brigade Member

**You are responding to a plant alarm with a report of a running spill in a partially diked area. When you arrive, the area operator tells you that the source of the leak hasn't been determined; the liquid spill could be a light hydrocarbon blend, methanol, or a combination of both. You are carrying aqueous film-forming foam (AFFF) in your foam tank. A second engine carrying "alcohol foam" is 5 minutes away. You decide to begin foaming operations to prevent the vapors from reaching an ignition source. A few minutes after foam application has started, you observe that the foam blanket is remaining on the fuel's surface and spreading out, reducing the release of vapors.**

1. ***What are the six categories of foam concentrates that are commonly used in the fire service?***
2. ***What type of foam concentrate is required for polar solvent fuels?***
3. ***Was the fuel spill hydrocarbon or methanol?***

## Introduction

Brigade members are faced with a wide variety of flammable and combustible liquid risks. Successful control and extinguishment requires not only the proper application of foam on the fuel surface but also an understanding of the physical characteristics of foam production. A thorough knowledge of the chemistry of the variety of foam concentrates that are available today is necessary to ensure brigade member safety and fire control.

Firefighting foam is divided into two basic classifications: Class A or Class B. Class A foams are used to fight ordinary combustible material (wood, textiles, and paper). Often referred to as "wetting agents," Class A foams are very effective because they improve the penetrating effect of water and allow for greater heat absorption. Class A foams are most commonly used by municipal, rural, and wildland fire departments, but are being increasingly used by facility fire brigades. Class B foam is used on Class B flammable and combustible liquid fires. This chapter will discuss Class B foams that are widely used in the petrochemical industry.

Foam is not a "one size fits all" extinguishing agent. The liquid fuel involved will determine the type of foam concentrate required as well as the volume and the duration necessary to extinguish the fire and control the vapors.

Large-volume flammable or combustible liquid fires are spectacular and can have devastating effects. They occur less frequently than smaller incidents handled every day by facility fire organizations and fire brigades. Understanding the capabilities and limitations of foam will increase the probability of a safe, efficient, and effective response.

## How Foam Works

Hydrocarbon fuels have a lower surface tension than water. When fuel and water are mixed, the two fluids quickly separate; the fuel rises to the top and the water remains on the bottom. When foam concentrate is mixed with water, the surface tension is reduced, allowing the foam/water mix to float on the surface of the fuel.

Foam extinguishes flammable or combustible liquid fires in four ways (▼ Figure 17-1):

- It prevents air from mixing with the vapors on the fuel surface. The foam blanket provides a physical barrier on the fuel surface.

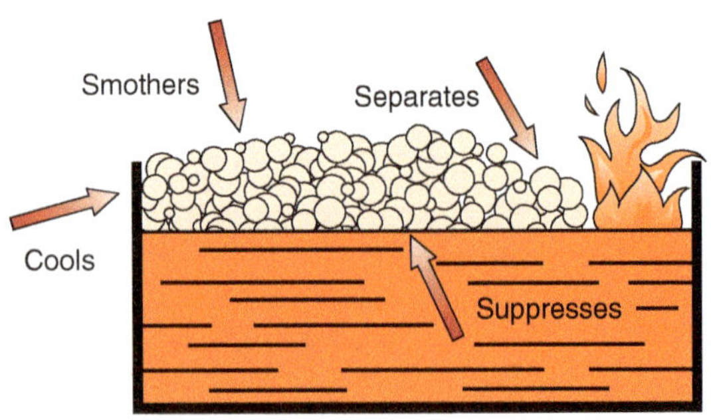


Figure 17-1 How foam works.

- It eliminates vapor release from the fuel surface. The effective time for vapor suppression depends on the vapor pressure of the fuel and the quality of the foam blanket.
- It separates the flames from the fuel surface. Because the combustion process is taking place in the vapors just above the fuel surface, foam flowing on the fuel surface interrupts flame production.
- It cools the fuel surface and surrounding metal. Foam solution is mostly water in the form of bubbles. The increased surface area provides a greater amount of heat absorption than water droplets.

## Foam Tetrahedron

Just as the elements of the fire tetrahedron must exist in the correct proportions to sustain the combustion process, the elements of the foam tetrahedron must also be in the correct proportions to produce an effective foam solution ► Figure 17-2.

Firefighting foam (finished foam) is produced by mechanical agitation and mixing of air into a foam solution. Too little foam concentrate in water will produce a foam solution that is too thin to be effective and may quickly dissipate into the fuel. Too much concentrate will produce foam that may be too thick to be properly expanded or aspirated when mixed with air. The expansion of foam solution is dependent on good mechanical agitation and air aspiration. When an insufficient amount of air is introduced into the solution stream, the solution is poorly aspirated. This results in foam with few bubbles, and fewer bubbles mean the foam will break down quickly and will not be able to suppress vapors. Poorly aspirated foam will also break down quickly when exposed to heat and flame.

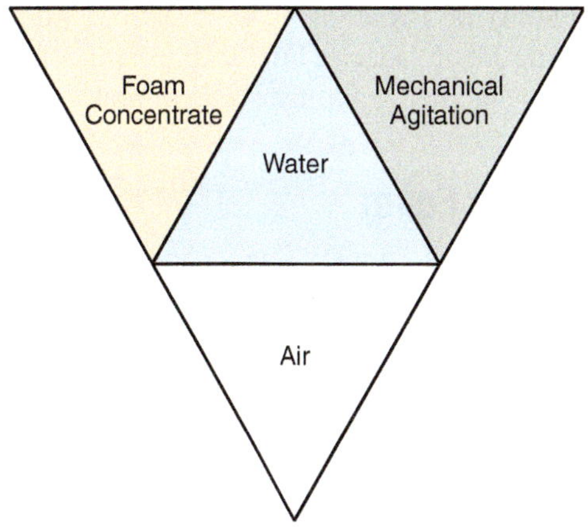


Figure 17-2 Foam tetrahedron.

## Expansion Rates

The expansion rate is the ratio of finished foam produced from foam solution after being agitated and aspirated through a foam-making appliance. NFPA 11 divides foam concentrates into three expansion ranges:

- Low expansion
- Medium expansion
- High expansion

### Low Expansion

These foams, which have an expansion ratio up to 20:1, are primarily designed for flammable and combustible liquids. Low expansion foam is effective in controlling and extinguishing most Class B fires. Special low expansion foams are also used on Class A fires where the penetrating and cooling effect of the foam solution is important.

### Medium Expansion

Medium expansion foams, which have an expansion ratio from 20:1 to 200:1, are used primarily to suppress vapors from hazardous chemicals. Mid-range expansion ratios of 30:1 and 55:1 produce an effective foam blanket for vapor suppression on low boiling point organics and chemicals that are highly water reactive.

### High Expansion

High expansion foams, which have an expansion ratio from 200:1 to approximately 1000:1, are used for a variety of confined space firefighting situations. The synthetic, detergent type foam is used in basements, ships, aircraft hangars, and mines.

# Foam Concentrates

Foam concentrates are divided into six categories commonly used in the fire service:

- Protein foam
- Fluoroprotein foam
- Film-forming fluoroprotein foam (FFFP)
- Aqueous Film-Forming Foam (AFFF)
- Alcohol-resistant aqueous film-forming foam (AR-AFFF)
- Synthetic detergent foam

## Protein Foam

Protein foams are limited to use on hydrocarbon fires only. They form a tough stable foam blanket with excellent heat and burnback resistance as well as good drainage rate characteristics. Although protein foams provide slower knockdown than other concentrates, they provide a long-lasting foam blanket after the fire is extinguished. Protein foams can be used with salt water or fresh water. These foams require good air aspiration through a foam nozzle; they cannot be properly aspirated through a structural fog nozzle.

Mechanical protein foams were first developed between the late 1930s and mid 1940s, and came into nonmilitary use after World War II. These foams are produced from hydrolyzed keratin protein (such as hoof and horn meal or chicken feathers) with stabilizing additives and inhibitors to prevent corrosion and control viscosity.

## Fluoroprotein Foam

Fluoroprotein foams contain fluorochemical surfactants, which improve the performance with better resistance to fuel pickup, faster knockdown, and compatibility with dry chemical agents. These foams are used on hydrocarbon fuels and some oxygenated fuel additives. Fluoroprotein foams have excellent heat and burnback resistance and maintain a good foam blanket after extinguishment. The addition of surfactants makes the foam more fluid, which increases the knockdown rate and provides better fuel tolerance than protein foam.

## Film-Forming Fluoroprotein Foam

A derivative of fluoroprotein and AFFF, this foam has performance characteristics similar to protein and fluoroprotein foams. Knockdown performance is improved because this foam releases an aqueous film on the surface of the hydrocarbon fuel. The overall performance of FFFP lies between fluoroprotein foam and AFFF. The foam does not have the quick knockdown of AFFF on a spill fire. When used on a fuel in depth fire, FFFP does not have the burnback resistance of fluoroprotein foam.

## Aqueous Film-Forming Foam

Commonly referred to as AFFF, this foam has the fastest knockdown on hydrocarbon fuels. Since AFFF is very fluid, it quickly flows around obstacles and across the fuel surface (▼ Figure 17-3). AFFF can be used as a premixed solution; it is compatible with dry chemical agents and can be used with fresh or salt water. Although AFFF can be used through nonaspirating nozzles, maximum performance can be achieved only through aspirating foam nozzles.

AFFF is composed of synthetic foaming agents and fluorochemical surfactants. AFFF extinguishes fire by forming an aqueous film on the fuel surface. The film is a thin layer of foam solution, which quickly spreads across the surface of a hydrocarbon fuel, creating an extremely fast fire knockdown. Surfactants reduce the surface tension of the foam solution, which allows it to remain on the surface of the hydrocarbon fuel. The aqueous film is formed by the action of the foam solution draining from the foam blanket.

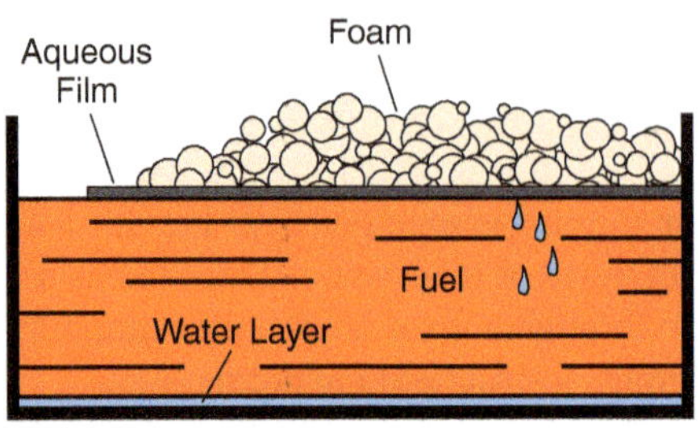


Figure 17-3 Aqueous Film-Forming Foam.

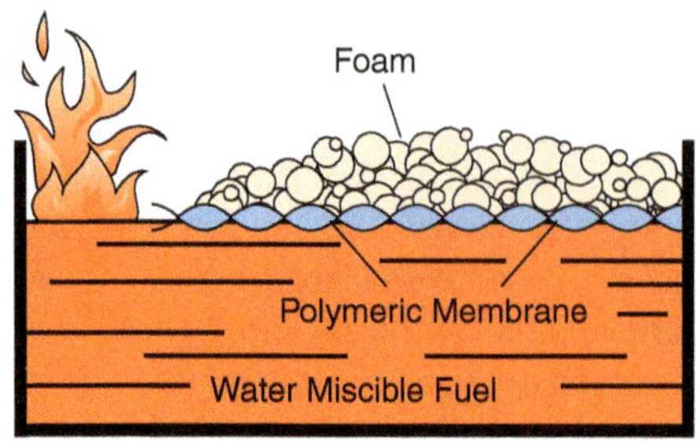


Figure 17-4 Alcohol-Resistant Aqueous Film-Forming Foam.

## Alcohol-Resistant Aqueous Film-Forming Foam

AR-AFFFs are a combination of synthetic detergents, fluorochemicals, and high molecular weight polymers (▲ Figure 17-4). Polar solvents (water miscible fuels) are not compatible with non-AR foams. Common polar solvents include alcohols (isopropyl, methanol, ethanol), esters (butyl acetate) amines, ketones (methyl ethyl ketone), and aldehydes. When non-AR foam is applied to the surface of a polar solvent, the foam blanket quickly breaks down into a liquid and mixes with the fuel. AR-AFFF performs as a conventional AFFF on hydrocarbon fuels, forming an aqueous film on the fuel surface. When applied to polar solvents, the foam solution forms a polymeric membrane on the fuel surface. This tough membrane separates the fuel from the foam and reduces destruction of the foam blanket.

AR-AFFF is one of the most versatile types of foam. It provides good knockdown and burnback resistance and has a high fuel tolerance on polar solvent and hydrocarbon fires.

## Synthetic Detergent Foam (High Expansion)

This foam group is most commonly used on Class A fires. High expansion foam is highly effective in confined space firefighting or in areas where access is limited or entry is dangerous to brigade members. These areas include basements, shipboard compartments, warehouses, aircraft hangars, and mine shafts. High expansion foams can be used in fixed generating systems and portable foam generators.

Rapid smothering and cooling achieve fire control and extinguishment. High expansion foams have a tremendous smothering and steam generation effect because the water is divided into such fine particles (bubbles), which enhance the heat absorption quality of the water. Care must be taken with

## Brigade Member Safety Tips

It is important to remember that foam is mostly water and presents the same electrical shock hazard potential to brigade members as water application.

regard to electrical power sources in the area when foam is applied.

## Foam Characteristics

Good foam must contain the right combination of physical characteristics to be effective. Knockdown speed and flow are the time required for a foam blanket to spread out across a fuel surface. The foam must also be able to flow around obstacles in order to achieve complete extinguishment and vapor suppression.

Foam must have good heat resistance to avoid breakdown from the effects of direct flame contact of burning fuel vapors or heat generated from metal objects. Fuel resistance is foam's ability to minimize fuel pickup. This oleophobic quality reduces the amount of fuel saturation in the foam.

Foam must produce a good vapor-suppressing blanket. A vapor-tight foam blanket reduces the generation of flammable or combustible vapors above the fuel surface and minimizes reignition.

When used on polar solvent fuels, foam must be alcohol resistant. Because alcohol readily mixes with water and since foam is mostly water, a foam blanket that is not alcohol resistant will quickly dissolve into the fuel and be destroyed.

A comparison of the properties of the various foam types is shown in ▼ Table 17-1.

## Foam Percentages

Foam concentrates are designed to be mixed with water at specific ratios. Foam concentrate ratios vary from 1 to 6%. The amount of concentrate varies depending on the manufacturer, the type of application, and the type of fuel.

A 3% concentrate is mixed at a ratio of 97 parts water to 3 parts foam concentrate. For example, each 100 gallons of foam solution would contain 97 gallons of water and 3 gallons of foam concentrate. A 6% solution would require 6 gallons of concentrate mixed with 94 gallons of water to produce the same 100 gallons of foam solution. When mixed with water (proportioned), the 100 gallons of solution have virtually the same performance characteristics. A 3% concentrate is twice as concentrated as a 6% concentrate.

It is important to understand that foam concentrates must be proportioned at the percentage listed by the manufacturer. If you want to produce a 3% foam solution, you cannot use half the amount of a 6% concentrate, you must use the concentrate at the percentage listed on the container. Foam concentrates are manufactured at different percentages for a variety of reasons. The chemical make-up of the concentrate, freeze protection additives, military use specifications (Mil-Specs), and cost are some of the basic factors that determine the percentage.

The trend in industry is to reduce foam concentrate percentages as low as possible. Lower proportioning rates mean less bulk storage for the user. Lower percentage rates also means that you can increase your firefighting capacity by carrying the same volume of foam concentrate or you can reduce your foam supply without reducing your suppression capabilities. Lower proportioning rates can also reduce the cost of fixed foam system components and concentrate transportation costs. Historically, foam concentrates were manufactured at 3 and 6%. Today, foam concentrates are produced for use at rates as low as 1% and as high as 6%, depending on the liquid fuel and how the foam is to be used.

AR-AFFFs are effective on both hydrocarbon and polar solvent fuel. The most common concentrate in use is labeled "AR-AFFF 3–3%". This means that the foam can be used at 3% concentration on hydrocarbons and polar solvents. There are also concentrates on the market that are used at different proportioning percentage rates depending on the type of fuel. Concentrates labeled as 3–6% are still common and have been used for many years. Concentrates labeled as 1–3% are seeing increased use as the chemical technology improves.

**Table 17-1 Comparison of the Properties of Various Foam Types**

| Property | Protein | Fluoroprotein | AFFF | FFFP | AR-AFFF |
|---|---|---|---|---|---|
| Knockdown | Fair | Good | Excellent | Good | Excellent |
| Heat resistance | Excellent | Excellent | Fair | Good | Good |
| Fuel resistance | Fair | Excellent | Moderate | Good | Good |
| Vapor suppression | Excellent | Excellent | Good | Good | Good |
| Alcohol resistance | None | None | None | None | Excellent |

The lower percentage rate is used for most hydrocarbons. It is important to review manufacturers' data sheets when determining which foam best meets your needs. A concentrate that is labeled 1% for hydrocarbons may require proportioning at 3% for some blended gasoline; this information should be contained in the product data sheet. The higher percentage rate is used for polar solvents. The higher proportioning percentage must be used to produce the polymeric membrane on the fuel surface. If the lower percentage rate is used on a polar solvent fuel, the foam will quickly be destroyed. An important consideration when using foams at 1% concentration is that the foam proportioner must be extremely accurate in order to ensure that a true 1% solution is being delivered to the fuel surface.

Because so many types of foam concentrates are available, selecting the right concentrate can be a challenge. The key is to identify the type of exposure that is to be protected and the type of foam delivery system that will be used. A foam concentrate that provides excellent performance characteristics for hydrocarbon storage tank and containment dike fires using monitors may not be as effective for warehouse protection utilizing a foam sprinkler system. Good hazard evaluation as a part of the overall preplanning process will ensure that the correct foam concentrate is selected.

## Foam Production

Finished foam is a combination of water, foam concentrate, air, and mechanical agitation. When these four elements are brought together in the correct proportions, foam is produced. The simplified diagram in ▼ Figure 17-5 shows how foam is produced through a typical proportioning system.

## Foam Proportioners

Foam proportioners are designed to supply the correct percentage of foam concentrate into the water stream. A variety of proportioning devices and systems are available to the industrial fire service today. Proportioning equipment ranges from simple in-line eductors used in hose systems to "around-the-pump" and "balanced pressure" systems, found on mobile fire apparatus. This equipment is discussed in detail later in the chapter.

## Foam Guidelines

Proper storage is critical to foam shelf life. Foam concentrates have temperature limitations that prevent degradation. The concentrates are stored in sealed containers to prevent air contact that causes evaporation and chemical breakdown. Manufacturers' guidelines will list the storage requirements to ensure concentrates are ready for service after many years of storage.

Foam concentrates in general tend to be more stable when used with moderate water temperatures. Although foam liquids will perform with water temperatures that exceed 100°F, preferred water temperatures are 35 to 80°F. Foam concentrates can be used with either fresh water or seawater. Water that contains contaminants such as detergents, certain corrosion inhibitors, or oil residues may adversely affect foam quality.

Ideal nozzle pressures range between 50 and 200 pounds per square inch (psi). When a proportioner is used, proportioner pressure should not exceed 200 psi. Higher pressures will cause foam quality to deteriorate, while lower pressures will reduce the reach of foam streams.

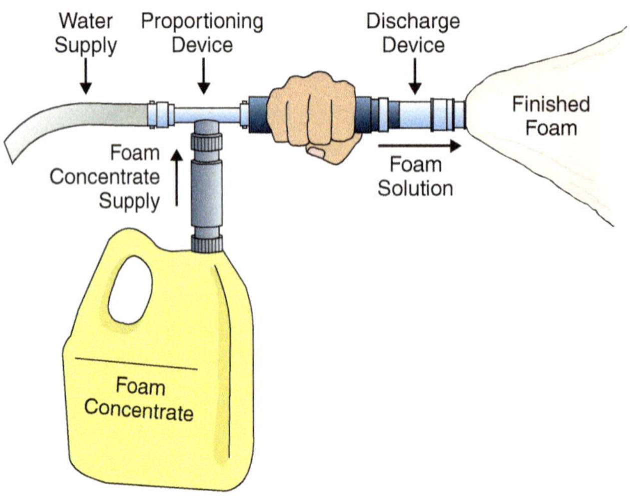


Figure 17-5 Foam proportioning system.

When flammable or combustible liquids have been spilled, prompt coverage with a foam blanket can prevent ignition. Additional foam application may be required periodically to maintain the blanket for extended periods until the spill has been cleaned up.

Foam should be considered the same as water when used on or near electrical fires and is not generally recommended in those circumstances. Electrical systems should be de-energized before applying foam.

Foam is not recommended for use on products that are stored as liquids but are normally vapor or gas at ambient conditions, such as propane, butane, and vinyl chloride. Foam is not recommended on water-reactive materials such as magnesium, titanium, lithium, potassium, and other combustible metals.

## Foam Equipment

A basic foam equipment system consists of a water supply hose, a foam proportioner to mix foam concentrate into the water stream, a foam source, and a hose line with either a standard fog stream nozzle or an air aspirating foam nozzle ► Figure 17-6. There are a variety of proportioning systems; the appropriate type and size of the system required is determined by the anticipated size of the exposure or fire risk and the capacity of the water supply. Small volume, low flow portable equipment is suitable for small Class B spill fires or for Class A fires in relatively small structures. Large spill fires, tank fires, and large structures are more effectively protected by fixed systems or proportioning systems integrated into industrial fire apparatus.

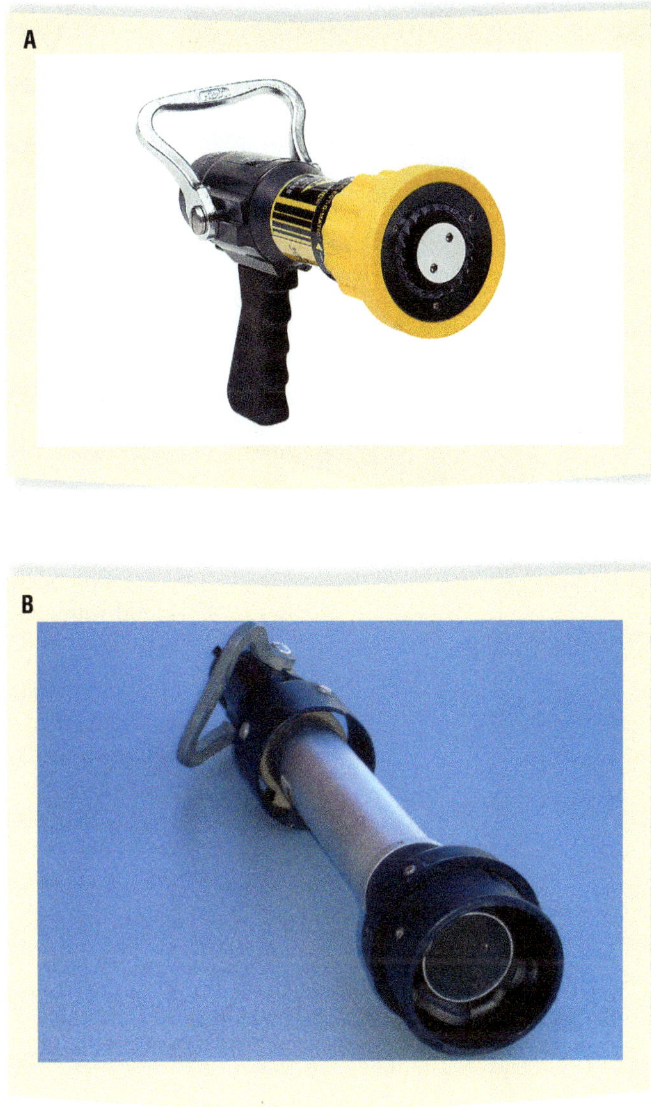


Figure 17-6 Fog stream nozzle and air aspirating foam nozzle.

### Foam Proportioning Systems

A foam proportioner is the device that mixes foam concentrate into a water stream in the correct percentage. Foam eductors and injectors are the two types of proportioners. Proportioners are manufactured in a range of proportioning percentages from 0.5 to 6% and vary in delivery capacity from as little as 60 gpm to over 14,000 gpm ► Figure 17-7.

Figure 17-7 4,000 to 14,000 gpm variable gallonage foam proportioner and master stream nozzle.

#### Foam Eductors

A foam eductor functions by flowing water through a venturi, which causes an increase in the velocity of the water,

**Brigade Member Safety Tips**

Extinguishing flammable or combustible liquid fires is not listed as an Incipient Facility Fire Brigade Member job performance requirement in NFPA 1081. A liquid fire that cannot be extinguished with either portable fire extinguishers or $1\frac{1}{2}$-inch foam hand lines is beyond incipient-stage firefighting.

# Voices of Experience

**"I quickly completed the calculations for the required foam extinguishment and noted there were insufficient water supplies to support both exposure protection and firefighting operations."**

Our refinery's emergency dispatch center was contacted by the assistant chief of the municipal department, who stated that his department was fighting a fire in a 120-foot asphalt tank in a terminal facility on the outskirts of the city. The municipal department was requesting additional foam resources. The assistant chief informed me that they had used all available on-site foam resources (numerous 5-gallon buckets) but that the fire was continuing to grow. Our refinery agreed to respond with a foam engine, a foam tender, and two quick-attack vehicles. Each vehicle was staffed with a full crew.

Upon our arrival at the incident, we noted that the tank was fully involved and that only exposure protection operations were being conducted. I quickly completed the calculations for the required foam extinguishment and noted there were insufficient water supplies to support both exposure protection and firefighting operations. This information was relayed to the command staff; they informed our crews that the tactical units would remain in a defensive operation and that they planned to allow the tank to burn out.

Our apparatus was released from staging and returned to the refinery. Approximately 8 hours later, the municipal department again contacted the refinery and requested that our resources return to the scene. The department staff had decided they would attempt extinguishment of the fire before daybreak. We contacted the command post and informed personnel there of the water supply we would require to complete extinguishment. The command staff informed us they had contacted additional resources and that they would have additional large-diameter hose lays in place by the time the refinery apparatus arrived on the scene.

A refinery quick-attack truck with a 2,000-gpm monitor was set up on the northwest corner of the tank, and dual 5-inch supply lines for the monitor were stretched to Foam Engine 3 (a 3,000-gpm foam pumper with a 1,850-gallon tank). Engine 3 was set up 150 feet north of Quick Attack 1. Two 5-inch supply lines and one 4-inch supply line were redirected from other operations to supply Engine 3. Foam Tender 1 (a 4,000-gallon tender) was set up adjacent to Engine 3, and a foam resupply system was established using a portable foam-transfer pump.

Foam extinguishment operations were started using a 3% foam application from a rapidly sweeping stream from the 2,000-gpm deck gun on Quick Attack 1. Within 15 minutes of the application, significant fire and smoke knockdown could be seen. The west side of the tank area could not be reached owing to an overhang from the failed sidewall. A portable monitor (1,250 gpm) with a foam tube was fed with two 3-inch supply lines to supplement extinguishment in the western portion of the tank. After 45 minutes, no visible flame or smoke could be seen. Foam operations were slowed, such that they used intermittent flows from the 2,000-gpm and 1,250-gpm monitors. The fire was deemed secure, and cooling operations were turned back over to the municipal department.

This incident highlighted the importance of proper foam-application rates and techniques. Ironically, this same tank caught fire approximately 1 month later as the terminal staff was trying to de-inventory the tank. The refinery resources were again requested, and they extinguished the fire in an even shorter time frame using the same tactical plan.

Rick Haase
ConocoPhillips Wood River Refinery
Roxana, Illinois

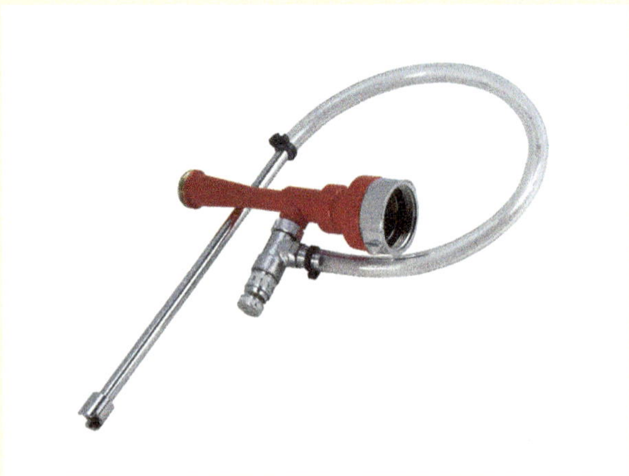

Figure 17-8 Portable in-line eductor.

creating a low-pressure area on the discharge side of the venturi. The low pressure creates a vacuum, drawing foam concentrate through the pickup tube into the water stream. Eductors are manufactured with either fixed percentage or adjustable percentage.

A commonly used portable eductor is the in-line eductor (also referred to as a line proportioner, inductor, or ratio controller) ▲ Figure 17-8. Portable eductors are a common choice when limited use is expected, when flammable liquid fires are relatively small in size, or when it is difficult to justify the expense of high-capacity proportioning systems. Because in-line proportioners are portable, they are easy to set up and can be operated some distance from the apparatus or fixed water source.

Most eductor systems have operating requirements and limitations. The flow rate of the eductor must be matched with a nozzle of the same flow rate. A 95-gpm eductor must be operated with a 95-gpm nozzle in order to deliver an effective foam stream. Mismatching nozzles and eductors is a common cause of proportioning problems. Mismatches can cause a poor quality foam solution or cause the foam concentrate pickup to shut down.

Eductors typically require a fairly high inlet pressure. Most eductors develop their rated flow at 200-psi inlet pressure, although some eductors are designed to operate at lower pressures. In-line eductors will not operate properly if there is excessive back pressure. Total back pressure cannot exceed 65% of the inlet pressure. An eductor operating at 200 psi is limited to a total of 130-psi back pressure. Excessive back pressure may be caused by:

- Nozzle flow that is rated lower than the eductor.
- Too much hose (friction loss) or hose kinks between the eductor and the nozzle.

## Brigade Member Safety Tips

It is important to verify the manufacturer's recommendation for the length of time that a foam concentrate can remain premixed with water.

- Nozzle elevation that is too high above the eductor.
- A nozzle shutoff valve that is not fully open.

### Foam Injectors

Foam injectors, which are typically found on apparatus, provide foam to the water stream under pressure. A metering system senses the pressure and flow rate of the water and adjusts the injection rate to ensure the correct percentage of foam concentrate is supplied to the water stream. Injection proportioners are effective over a broad range of water flow rates and pressures.

To place a foam line in service, use the following steps Skill Drill 17-1:

1. Ensure all foam system components are available. This includes the water supply hose, the foam eductor, foam concentrate, attack hose line, and an air aspirating nozzle or foam nozzle.
2. Connect the water supply line to the water source and to the inlet side of the eductor.
3. Connect the attack line to the discharge side of the eductor.
4. Place the foam concentrate container(s) next to the eductor.
5. Set the metering device on the eductor to match the concentrate percentage on the container.
6. Place the eductor pickup tube into the foam concentrate container.
7. Charge the hose line with water, ensuring the minimum required inlet pressure at the eductor.
8. Flow water through the hose line until foam solution begins to flow from the nozzle.
9. Apply foam from the upwind side using the appropriate application technique (bank in, bounce off, or rain-down).

### Batch Mixing

Foam concentrate can be poured directly into the booster tank on apparatus that is not equipped with a foam proportioning system. Batch mixing can be used either with foam concentrates that are designed for premixed systems or when the manufacturer states that the concentrate can be batch mixed for a short period of time. A booster tank with a volume of 1,000 gallons would require 30 gallons of 3% concentrate or 60 gallons of 6% concentrate. Once the concentrate has been added, the solution should be mixed by

Figure 17-9 Trailer-mounted twin agent system.

Figure 17-10 Vehicle-mounted twin agent system.

circulating it through the water pump. The solution should be circulated at low pump speed and pressure to prevent excessive foaming of the solution.

### Premixing

Premixed foam is a solution of either AFFF or FFFP. Premixed foam is commonly used in 2½-gallon hand portable extinguishers and 33-gallon wheeled extinguishers. These portable extinguishers are effective on small spill fires. A 2½-gallon extinguisher is rated for a spill fire area up to 40 sq ft, while the 33-gallon extinguisher is rated for up to 160 sq ft. Large-volume mobile foam systems are used on skid- or trailer-mounted twin agent systems (foam and dry chemical) and on vehicle-mounted systems (▲ Figures 17-9 and 17-10).

Premixed systems are discharged from stored pressure in the extinguisher or from an inert gas source, which pressurizes a pressure-rated tank. The systems are easy to use and provide quick delivery of the agent on the fire. They are limited to a single use and must be emptied and recharged afterward.

# Foam Tactics

Successful foam operations require an effective size-up before firefighting starts. It is not enough to arrive on scene and start applying foam on a flammable or combustible liquids fire. You have to know the type of fuel you are dealing with, the size of the area involved, the required application rate, and the required application duration. Once you begin foam operations, you must be able to sustain foam application until the fire is out.

## Spill Fires

Spill fires are those where the average depth of the fuel is 1 inch or less. NFPA 11 defines a spill fire as "nondiked spill areas where a flammable or combustible liquid spill might occur, uncontained by curbing, dike walls, or walls of a room or building. A spill fire is bounded only by the contour of the surface on which it is lying."

### Spill Fire Tactical Elements

Because a spill fire is not contained, an important tactical consideration is the topography of the spill area. You must take into consideration the path that the liquid will follow and any additional hazards caused by migrating fuel. Additional firefighting resources may be needed to protect exposures from potential spill migration. All spill fires should be fought from the uphill, upwind side. This provides safety for personnel and helps carry the foam across the fuel surface.

The type of fuel involved is another tactical consideration. Polar solvent fuels cannot be extinguished with protein, fluoroprotein, or AFFF. AR foam must be used. The foam concentrate must be matched to the liquid fuel involved.

The next tactical consideration is spill size. The size of the spill area must be calculated to ensure an adequate foam supply. Estimate the length and width of the spill to determine the square footage (► Figure 17-11). Spills tend to be irregular in shape, so it is easier to estimate the area at its largest point. This allows for quicker calculations, and although the actual spill area is smaller than what is calculated, it provides a margin of safety and ensures an adequate foam application rate.

The minimum application rate will vary depending on the type of foam concentrate used. In all cases, the minimum application time for a spill fire is 15 minutes. The minimum application rates and discharge times for

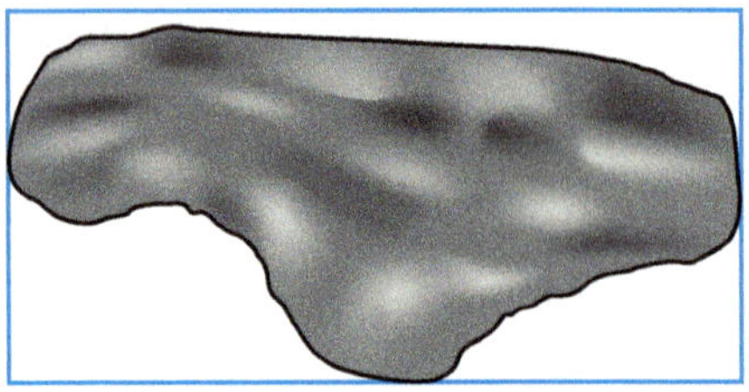

Figure 17-11 Determine a spill area by estimating the length and width at the largest points.

portable foam nozzles and monitors are shown in ▼ Table 17-2.

Based on the table, a 1,000 square foot hydrocarbon spill using AFFF requires an application rate of 0.1 gallons per minute (gpm)/square foot (sq ft), which equates to a 100-gpm foam solution flow. The 15-minute application time requires a total solution flow of 1,500 gallons. Foam proportioned at 3% would require a total of 45 gallons of concentrate and 1,455 gallons of water.

It is important to remember that before flammable or combustible liquid firefighting begins, adequate foam concentrate supply, a sustained water supply, and adequate personnel and equipment must be staged at the scene. If foam firefighting cannot be sustained, it should be delayed until the required resources are available. If foam operations begin on a spill fire and must be shut down before the fire is extinguished, the fire will break down and consume the foam that has been applied. Much of the fire area that was extinguished before the operation was shut down will have to be extinguished again when operations are resumed. The end result is that foam concentrate is wasted, additional firefighting time is required, and little was gained by starting the operation without adequate resources.

## Three-Dimensional Fires

Three-dimensional fires involve liquid fuel dripping, pouring, or running from one or more horizontal surfaces. These fires can be challenging because the vertical burning liquid provides a constant ignition source to the liquids on each level below. Successful extinguishment requires proper foam application in the correct sequence. Three-dimensional fires can be relatively small, such as those involving with piping that operates at low pressure, or they can be large and difficult to fight, such as when a burning storage tank is pouring fuel into a ground fire. The fuel source supply must be shut off as soon as possible to limit the surface area and depth of the fuel.

### Three-Dimensional Fire Tactical Elements

A key tactic is to extinguish the fire at the lowest level first. If the fire at this lower level is not extinguished, it provides a continuous ignition source, in the form of heat and flame to the fuel at the upper level. Even when the fire at the lower level has been completely extinguished, there is the strong probability that a small fire will persist where the burning liquid falls from above and breaks up the foam blanket. The size of the fire at the lower level and the rate at which that size increases over time will determine when and if additional foam application is required to keep the fire area to a minimum while the fire at the upper level is being extinguished.

After the fire is extinguished at the lowest level, foam can be applied to the next horizontal liquid fire. The foam flow should reach the fuel surface as gently as possible. Gentle application will minimize the amount of fuel that splashes onto the areas below. If the foam flow volume is excessive and the containment volume is minimal, a large amount of burning fuel may be pushed over the side of the containment and increase the size of the fire on the lower level.

**Table 17-2 Minimum Application Rates and Discharge Times for Nondiked Spills Using Portable Foam Nozzles or Monitors**

| Foam Type | Minimum Application Rate gpm/ft² | Minimum Discharge Time (min) | Anticipated Product Spill |
|---|---|---|---|
| Protein and fluoroprotein | 0.16 | 15 | Hydrocarbon |
| AFFF, FFFP, AR-AFFF, and AR-FFFP | 0.10 | 15 | Hydrocarbon |
| Alcohol-resistant foams | Consult manufacturer (typically 0.2) | 15 | Polar solvent |

As each successive horizontal fire is extinguished, the volume of vertical burning fuel should decrease. A large-volume, foam-compatible dry chemical extinguisher may be effective in extinguishing these vertical fuel fires. Success in extinguishment depends on the volume and volatility of the fuel and fire fighters' ability to reach the burning fuel with an adequate amount of dry chemical.

When the source of the fuel is from a flange, valve, or piping break, fire fighters can partially control the fire by "capturing" it with a power cone or modified fog water stream. The water stream pattern should be adjusted so that the fire and fuel are contained inside the water pattern. Capturing will reduce both the amount of fire at the source and the amount of ground fire below it. This technique also allows crews to keep the fire away from piping, tanks, and vessels that might otherwise create an exposure hazard.

Small leaks near the ground can be controlled by handlines when such an approach does not put brigade members in danger of operating from a position in the fuel on the ground or under an elevated leak source. Ground monitors or aerial monitors can be used to control large leaks that cannot be controlled by handlines or that are too far away for handline streams to be effective. A spotter with a clear view of the leak and fire stream may be needed to direct the positioning and shape of the stream so as to provide optimal control and capture.

When foam and water operations are conducted in the same location, the streams must be closely coordinated to minimize the destruction or disturbance of the foam blanket that is protecting unburned fuel. Reduction of the ground fire to a manageable size may allow foam operations to be run intermittently. This strategy saves foam supplies, extends the amount of time crews can operate foam lines, and facilitates the logistics required to replenish the foam concentrate.

## Diked Fires

Diked or contained fires are defined by NFPA 11 as "areas bounded by contours of land or physical barriers that retain a depth of fuel greater than 1 inch." Diked fires, also referred to as spill fires in depth, require greater resources and present potential tactical challenges that may not exist in spill fires. Fixed discharge outlets and fixed or portable monitors commonly protect large diked areas. Diked areas can also be protected by foam hose lines, but this is often impractical due to the high application rates and extended discharge times that are required.

### Diked Fire Tactical Elements

Successful extinguishment of a diked fire depends on good preplanning, applying the correct amount of foam for the required time, and proper application technique. The area of a dike can vary greatly depending on the volume of the tank or tanks inside the dike. A properly designed dike will hold the volume of the largest tank plus 10%. Because a dike has a known area, preplanning will ensure that the required resources, application rate, and application duration are identified. The worst possible time to start making calculations is when your world is on fire.

### Preplanning

Preplanning provides the opportunity to identify exposures and potential problems that are unique to the area. Exposed piping, flanges, fittings, pumps, and meters can create additional problems that must be addressed before an event if the tactical plan is to be successful.

Protection of these exposures should be an early tactical consideration. Heat from a fire can quickly cause flanges and fittings to fatigue. Flange gaskets can fail causing additional fuel to leak into the diked area. Any mechanical connection inside the dike has a high potential for failure and is a potential source of uncontrolled fuel flowing into the dike.

In diked areas that provide containment to more than one tank, multiple failures in piping connections may cause the release of fuel volumes that are in excess of the capacity of the dike. If the dike overflows, you are faced with the potential of a running spill fire that could put personnel and equipment in danger. Fuel burning outside of the dike may also cut off access routes and prevent personnel from operating foam equipment from the most effective position. Applying foam to the fuel surface under pipe fittings will help to maintain their integrity.

### Application Rate and Duration

Application rates and discharge times for fixed foam application on hydrocarbon liquids are shown in ▶ Table 17-3. Class I hydrocarbons are those fuels with a flash point of less than 100°F; class II hydrocarbons are fuels with a flash point greater than 100°F. Application rates and discharge times are not shown for fuels (polar solvent group) that require AR foams. The characteristics of the fuels within this group vary widely. The manufacturers' recommendations should be followed. These recommendations are based on listings or approvals for specific products and foam-making devices. In all cases, the minimum application time is 30 minutes.

### Application Techniques

The minimum application rates and discharge times assume that the foam discharge is reaching the fuel's surface. The key to successful foam firefighting is putting foam on the fuel surface in the most efficient and effective manner. Gentle foam application is very important. One of the reasons that foam application rates are higher for handline and monitor application is the potential for foam being plunged into the fuel, which reduces its effectiveness. Foam applied from handlines and monitors can also be carried away by wind and thermal columns from the fire. Foam that does not reach the fuel surface cannot extinguish the fire. It is important to

**Table 17-3 Minimum Application Rates and Discharge Times for Fixed Foam Application on Diked Areas Involving Hydrocarbon Liquids**

| Type of Foam Discharge Outlets | Minimum Application Rate gpm/ft² | Minimum Discharge Time (min) Class I Hydrocarbon | Minimum Discharge Time (min) Class II Hydrocarbon |
|---|---|---|---|
| Low-level foam discharge outlets | 0.10 | 30 | 20 |
| Foam monitors | 0.16 | 30 | 20 |

understand that the minimum discharge times may not be adequate because wind, thermal updraft, or plunging into the fuel will reduce the amount of foam reaching the fire. Foam not only extinguishes the fire but also prevents or reduces reignition by suppressing fuel vapors.

Water streams that are used for cooling operations must be coordinated to prevent disrupting the foam blanket. Personnel should avoid walking in areas where foam has been applied. This breaks up the foam blanket, creating the potential for reignition, and puts fire crews at risk. Small areas of fuel may continue to burn where the foam stream breaks up the foam blanket. This can be corrected by moving the foam stream away and allowing the blanket to reseal, thus extinguishing the remaining fire. Foam should be reapplied as necessary to maintain an effective foam blanket depth.

Three methods of applying foam to the fuel surface are roll on, bounce off, and rain-down. Their effectiveness depends on the size of the area obstructions and the amount of time that the fuel has been burning.

### Roll On

Roll on, also referred to as bank in and sweep, is most effective on spill fires and small dike fires (▼Figure 17-12). In this technique the foam is applied to the ground in front of the spill, allowing it to build up at the edge of the spill. The energy of the stream pushes the foam blanket out across the fuel surface. Gentle sweeping of the nozzle from side to side in a horizontal motion will cause the foam blanket to spread out across the fuel surface. It may be necessary to change location to ensure the entire area is covered.

To perform the roll-on method, follow the steps in (Skill Drill 17-2):

1. Open the nozzle away from the spill until foam is flowing.
2. Move into a safe distance on the upwind side of the spill and open the nozzle.
3. Direct the foam stream onto the ground just in front of the edge of the spill.
4. Allow the foam to pile up and roll out across the top of the fuel until the area is completely covered.
5. Change position as necessary to ensure the foam has covered the entire area.

### Bounce Off

Bounce off, also referred to as bank shot and bank down, is an effective method of gently applying foam when an object such as piping, a vessel, or a wall is available (▼Figure 17-13). The bounce-off method allows foam to strike the object and run down onto the fuel surface and spread out. This method is especially effective when using

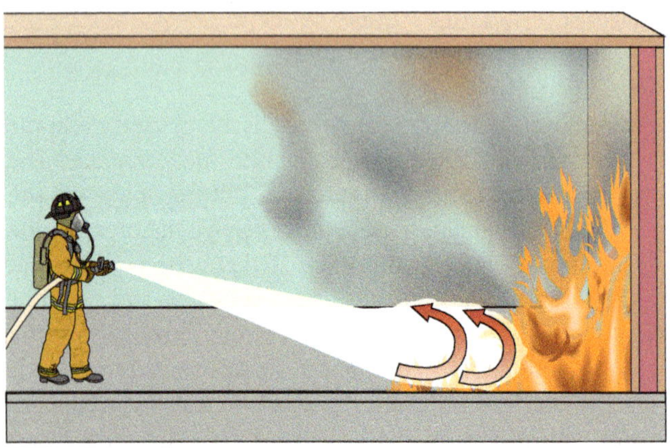

Figure 17-12 Roll-on method.

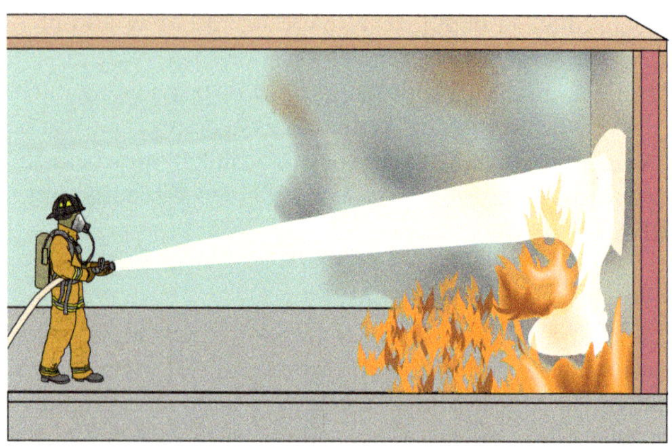

Figure 17-13 Bounce-off method.

a straight stream to apply foam. It may be necessary to bounce the foam off several points to ensure complete foam coverage.

To perform the bounce-off method, follow the steps in Skill Drill 17-3:

1. Open the nozzle away from the spill until foam is flowing.
2. Move into a safe distance on the upwind side of the spill and open the nozzle.
3. Direct the foam stream onto an object such as a wall or tank so that the foam is directed off the object and onto the fuel surface.
4. Allow the foam to flow across the fuel surface until the area is completely covered. The foam stream may have to be bounced off several areas of the object to ensure complete coverage and extinguishment.

#### Rain-Down

In the rain-down method the nozzle is raised up at a sharp angle so that the foam stream is directed into the air (▼ Figure 17-14). This allows the foam stream to reach a height where the stream will break into smaller drops and fall gently back onto the fuel surface. The nozzle angle may have to be adjusted to ensure that the foam pattern effectively reaches the fuel area. The rain-down method can provide an effective and fast knockdown. However, the foam stream may be carried away from the fuel surface in high wind conditions or if the fuel has had an extended preburn, creating a strong thermal column. When these conditions exist, this technique may be ineffective.

To perform the rain-down method, follow the steps in Skill Drill 17-4:

1. Open the nozzle away from the spill until foam is flowing.
2. Move into a safe distance on the upwind side of the spill and open the nozzle.
3. Direct the foam stream into the air so that the foam stream breaks up and falls onto the fuel surface.
4. Allow the foam to flow across the fuel surface until the area is completely covered.
5. Adjust your position and the elevation of the nozzle as necessary to ensure complete coverage of the fuel surface.

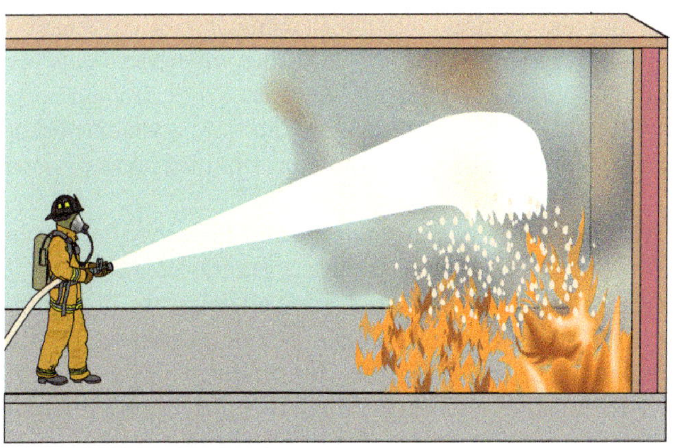

Figure 17-14 Rain-down method.

## Tank Fires

Tank fires are some of the most dramatic and intense incidents that can occur. Tank fires also require a great amount of preplanning and resource management. A thorough knowledge of tank construction and the fire effects on different construction types will help reduce the number of surprises during an incident. Effective protection systems, resources, and proper strategy and tactics will result in more efficient and effective fire ground operations. The foam requirements for tanks vary depending on the type of tank, the size of the area involved, and the type of product in the tank. Typical application rates are:

- 0.10 gpm/sq ft for hydrocarbon fixed systems
- 0.30 gpm/sq ft for fixed system seal protection
- 0.10 gpm/sq ft for subsurface injection
- 0.10 to 0.16 gpm/sq ft for hydrocarbon spills—portable equipment
- 0.16 gpm/sq ft hydrocarbon storage tanks—portable equipment
- 0.20 gpm/sq ft polar solvent storage tanks—portable equipment

NOTE: Industry recommends an application rate of 0.18 to 0.20 gpm/sq ft for tanks larger than 140 feet.

### Preplanning

Tank emergencies and fires require a level of preplanning that exceeds the elements of spill fires and smaller dike fires because of the volume of fuel, the size of tanks, and the volume of foam and water resources required to bring the incident under control.

There are different issues and challenges to be addressed. Fires in diesel tanks, gasoline tanks, and crude oil tanks are not the same. Extinguishing times are different, and they pose different hazards to responding personnel.

The geography of the area is a critical element of the preplan. Access and egress points determine how and if you can get close enough to a tank to operate effectively. Egress addresses the question, can I evacuate quickly enough if the incident gets worse and puts personnel in danger? Dirt roads that will support the weight of fire equipment may be inadequate after heavy rains. Facility drainage is another factor to be considered. It is important to evaluate not only drainage capacity for the dike system around the tanks but also drainage direction if the dike should overflow and the flow direction of the tank contents if the tank should fail.

How is the facility staffed? If the brigade is fully staffed during the day shift, but only a skeleton crew is in the plant at other times, the response is going to be delayed until adequate personnel are on scene. Delays in starting firefighting operations could result in a longer firefight.

Tank data in the preplan is essential. Critical questions can be addressed if the information is in the preplan. The type of tank construction, tank capacity, height, diameter, and the capacity of piping systems should not be left to memory at 2 o'clock in the morning.

Knowing the characteristics of the dike will prevent unwanted surprises. Having the capacity of the dike, drain valve locations, and the drain volume rates readily available in the preplan will reduce the chance of overfilling the dike when foam and water are added to the volume released by the tank or process piping.

The exposures at the scene will determine the type and capacity of support activities and personnel required. Tanks, piping, process units, and structures pose different challenges. These exposures require different strategies and tactics and affect the priority order of the incident.

Tank fires are large and complex. Water requirements may exceed the capacity of a plant water system. Water is needed for extinguishment and exposure protection. If the water requirements can't be met, and additional water cannot be brought to the scene, the tank will burn out or exposures will be lost, possibly both.

Large tanks or multiple tanks can require tremendous volumes of foam. Foam flow calculations for every tank and the associated dike should be calculated and easily retrievable from the preplan. The question "do I have enough foam on hand?" is answered by having this information before the incident. Preplanning of foam requirements will also identify the need for mutual aid or regional foam supplies.

## Application Rate and Duration

The firefighting tactics determine application rates and application duration. Over-the-top foam application requires higher rates and longer application times than fixed systems. Foam application from ground monitors and apparatus also requires more personnel. A 120-foot cone roof diesel storage tank has a total surface area of 11,500 sq ft. Over-the-top or topside foam stream operations require 0.16 gpm/sq ft or 1,850 gpm application rate for a minimum of 50 minutes. The same tank protected by a subsurface system requires an application rate of 0.10 gpm/sq ft or 1,150 gpm for a minimum of 30 minutes. Detailed listings for foam application rates and durations for various tank designs, fuel products, and discharge types can be found in NFPA 11.

To calculate the application rate for storage tanks, use the following steps in Skill Drill 17-5:

1. Determine the area: 3.14 × radius squared (for quick field calculations use 0.8 × diameter squared).
2. Determine the foam application rate: area × rate.
3. Determine the required foam solution: rate × duration.
4. Determine the foam concentrate quantity: solution × foam percentage.

## General Tank Fire Foam Tactics

Extinguish dike fires first. If the tank is extinguished before the dike fire, there is a very good chance that the dike fire will reignite the tank. Extinguishing the dike fire also reduces the chance of damage to process piping and their fittings located within the dike.

Apply cooling streams to tank shells above the liquid level. The product in the tank will provide cooling and act as a heat sink. Cooling streams applied above the liquid level will help reduce distortion and help prevent reignition. Be aware of the volume of cooling water and foam collecting in the dike and make provisions to pump out the dike to prevent overflow.

Apply foam in the most effective manner. Foam that doesn't reach the fuel surface can't extinguish the fire. Whenever possible, foam should be applied from the upwind side of the tank. Allow the wind to help carry foam to the tank's surface rather than blowing it away. Foam should be applied from outside of the dike. It is dangerous to put personnel inside the dike of a storage tank. A sufficient volume of foam must be applied to overcome the effects of the thermal column that is generated by the burning liquid. The intense heat at and above the fuel surface will destroy some of the initial foam. Initial foam application must overcome this heat before the foam can spread out across the fuel surface and extinguish the fire.

Adequate foam supplies and an uninterrupted water source must be available before operations begin. The minimum application rate and application duration assume that all the foam is reaching the fuel surface continuously. The actual extinguishing time for a tank fire can vary greatly. Although tanks have been extinguished in under an hour, other tanks have required days to extinguish because of their size, inadequate water or foam supplies, and the time required to establish enough delivery capability.

## Crude Oil Tanks

Crude oil storage tanks present unique challenges and hazards that are not present in other flammable or combustible liquid storage tanks. One of the major differences between crude oil and other products is water. Crude storage tanks contain a layer of water in the tank bottom, and some crude oils may contain small and fairly thin stratified areas of water at various levels.

Crude oil is a mixture of light and heavy hydrocarbons that burn off and separate as the fire progresses. As the crude oil continues to burn, the cold crude oil below the fire warms up; this causes more of the light product to rise to the surface and feed the fire. The heavier products in the crude oil start to build up and begin to fall down through the crude oil below. This heavy layer, called a **heat wave**, warms up the crude oil and sends more light fuel to the surface. As this process continues, more fuel is warmed up, creating a layer of superheated fuel just below the surface. The heat wave gets thicker and heavier as it moves toward the bottom of the tank. Three different events can occur during crude oil firefighting.

As the crude oil burns, the layer of preheated fuel below the surface is hot enough to boil water and can cause water droplets to flash to steam. This causes a **frothover** of hot burning crude oil. If foam is properly applied, the froth will usually stay in the tank. If foam or cooling water is not properly applied, the froth may float burning fuel over the top of the tank. The burning crude oil will spread out in the dike and down ditches. This is another reason to avoid plunging foam into the fuel, to stay out of dike areas and, whenever possible, to operate uphill.

As the heat wave continues to fall and contacts small layers or pockets of water, a small steam explosion creates a **slopover**. Slopovers tend to be relatively small in intensity. The steam explosion creates a burp of burning crude oil. The frequency and size of slopovers are determined by the volume and number of water layers in the crude oil column. There could be many of these water pockets or none at all, depending on the crude oil.

A **boilover** is the most serious and potentially deadly incident at a crude tank fire. During the burning of the crude oil, the heat wave continues to build up, getting thicker and heavier. During this process the fire and smoke on the fuel surface show little or no change in appearance. If the fire has not been extinguished or if it has not been possible to drain the water bottom in the tank, a boilover will occur. When the heat wave comes in contact with the water bottom, the water is immediately heated well above 212°F. The water flashes to steam with an expansion ratio of 1,700 to 1 and the resulting steam explosion ejects the entire contents of the tank in seconds. Without knowing how much water is in the bottom of the tank, it's not possible to determine how big the fireball will be. The fireball can easily be up to 10 times the tank's diameter. A relatively small 120-ft diameter tank could create a fireball over 1,000 ft.

Playing a water stream up and down the side of the tank can indicate the level of the heat wave because the water will either change to steam or rapidly evaporate. Proper tactics, adequate resources, and properly trained personnel can help avert a crude tank fire disaster.

# Wrap-Up

## Ready for Review

- Foam is a fundamental flammable and combustible liquids firefighting tool.
- Understanding the chemistry of foam helps ensure brigade member safety and fire control.
- Foam is an effective tool for vapor suppression and fire control of hydrocarbon and polar solvent liquids.
- Effective firefighting foam requires the correct mix of concentrate, water, air, and mechanical agitation.
- The correct foam type must be matched to the hazard.
- Foam characteristics differ with the type of foam concentrate and the performance required.
- Correct foam proportioning must be matched to the type of hazard.
- Correct foam application and duration is necessary to extinguish a fire or suppress vapors.
- Safety must be the top priority; equipment and tanks can be replaced.

## Hot Terms

**Batch mixing** Pouring foam concentrate directly into the booster tank of an apparatus.

**Boilover** Violent ejection of fuel from a tank when hot heavy hydrocarbons contact water in a tank bottom, causing a steam explosion.

**Bounce off** Foam application utilizing an object to bounce foam off of to gently flow onto the fuel surface.

**Burnback resistance** The ability of a foam blanket to resist direct flame impingement.

**Drainage rate** The rate at which solution drains from the foam blanket. For foam quality test purposes, it is the time it takes for 25% of the solution to drain from the foam blanket.

**Eductor** A foam proportioner that operates as a venturi to draw foam concentrate into the water stream.

**Finished foam** The homogeneous blanket obtained by mixing water, foam concentrate, and air.

**Fluorochemical surfactant** A chemical compound containing fluorine that is used to reduce surface tension when dissolved in a solution.

**Foam concentrate** The foaming agent that is mixed with the appropriate amounts of water and air to produce mechanical foam.

**Foam proportioner** The device that mixes foam concentrate into a water stream in the correct percentage.

**Foam solution** A homogeneous mixture of water and foam concentrate in the correct proportions.

**Frothover** A frothing of burning crude oil caused when water contacts superheated fuel and flashes to steam.

**Heat wave** A build-up of heavy hydrocarbons that collect as crude oil burns.

**Hydrolyzed** Decomposition of a chemical compound by reaction with water.

**Miscible** Readily mixes with water.

**Oleophobic** Oil hating; having the ability to shed hydrocarbon liquids.

**Oxygenated** To treat, combine, or infuse with oxygen.

**Polymer** A naturally occurring or synthetic compound consisting of large molecules made up of a linked series of repeated simple monomers.

**Premixed foam** Mixed foam and water used in portable extinguishers and dual agent systems.

**Rain-down** Foam application method to apply a raised foam stream to allow the foam to gently fall onto the fuel surface.

**Roll on** Foam application method of applying foam at the front edge of the fuel and allowing the foam to flow across the fuel surface.

**Slopover** Burps of crude oil caused by steam explosions when the heat wave contacts small areas of water in the fuel column below the surface.

**Surface tension** The attractive force exerted upon the surface molecules of a liquid by the molecules beneath.

**Thermal column** Heated air that rises above a burning fuel.

**Vapor pressure** The pressure exerted by a vapor.

**Venturi** A tube with a constricted throat that causes an increase in the velocity of water, creating a low-pressure area.

**Viscosity** The degree to which a fluid resists flow under an applied force. The lower the viscosity, the easier a fluid will flow.

# Brigade Member in Action

**You arrive at the scene of a contained methanol spill fire. Another engine has already begun to apply foam to the fire. You observe that the fire crew is not making any progress in knocking the fire down even though you see a good foam discharge. The brigade leader orders your crew to start applying foam and shuts down the other engine crew's operations. After a few minutes the size of the fire is reduced and your foam blanket is flowing out across the fuel surface, extinguishing the fire.**

1. What is the likely cause of the other engine's lack of success?
   - A. The wrong type of foam concentrate was being used.
   - B. The water pressure was too low.
   - C. The proportioning percentage was too high.
   - D. The chief didn't give the other crew enough time.

2. When applied to polar solvents, what type of foam forms a polymeric membrane on the fuel surface?
   - A. Synthetic foam
   - B. AR-AFFF
   - C. AFFF
   - D. Fluoroprotein foam

3. What is the proportioning percentage required for this fire using AR-AFFF 3–6% concentrate?
   - A. 1%
   - B. 3%
   - C. 6%
   - D. You can use either; AR-AFFF is effective at any percentage.

4. What type of foam concentrate is most commonly used on Class A fires?
   - A. Protein foam
   - B. Synthetic detergent foam
   - C. Fluoroprotein foam
   - D. FFFP

# Brigade Member Survival

# Chapter 18

## NFPA 1081 Standard

### Incipient Facility Fire Brigade Member

**4.2.3** Exit hazardous area, given that the fire has progressed beyond the incipient stage, so that a safe haven is found and the team members' safety is maintained.

**(A)** *Requisite Knowledge.* Communication procedures, emergency evacuation methods, what constitutes a safe haven, and elements that create or indicate a hazard.

**(B)** *Requisite Skills.* The ability to follow facility evacuation routes, evaluate areas for hazards, and identify a safe haven.

### Advanced Exterior Facility Fire Brigade Member

**5.2.8*** Exit a hazardous area as a team, given vision-obscured conditions, so that a safe haven is found before exhausting the air supply, others are not endangered, and the team integrity is maintained.

**(A)** *Requisite Knowledge.* Personnel accountability systems, communication procedures, emergency evacuation methods, what constitutes a safe haven, elements that create or indicate a hazard, and emergency procedures for loss of air supply.

**(B)** *Requisite Skills.* The ability to operate as a team member in vision-obscured conditions, locate and follow a guideline, conserve air supply, evaluate areas for hazards, and identify a safe haven.

**6.2.9*** Operate as a member of a rapid intervention crew (RIC), given size-up information, basic rapid intervention tools and equipment, and an assignment, so that strategies to effectively rescue the facility brigade member(s) are identified and implemented; hazard warning systems are established and understood by all participating personnel; incident-specific personal protective equipment (PPE) is identified, provided, and utilized; physical hazards are identified; and confinement, containment, and avoidance measures are discussed.

**(A)** *Requisite Knowledge.* Identification and care of PPE; specific hazards associated with the facility; strategic planning for rescue incidents; communications and safety protocols; atmospheric monitoring equipment needs; identification, characteristics, expected behavior, type, causes, and associated effects of personnel becoming incapacitated or trapped; and recognition of, potential for, and signs of impending building collapse.

**(B)** *Requisite Skills.* The ability to use PPE, determine resource needs, select and operate basic and specialized tools and equipment, implement communications and safety protocols, and mitigate specific hazards associated with rescue of trapped or incapacitated personnel.

### Interior Structural Facility Fire Brigade Member

**6.2.5*** Exit a hazardous area as a team, given vision-obscured conditions, so that a safe haven is found before exhausting the air supply, others are not endangered, and the team integrity is maintained.

**(A)** *Requisite Knowledge.* Personnel accountability systems, communication procedures, emergency evacuation methods, what constitutes a safe haven, elements that create or indicate a hazard, and emergency procedures for loss of air supply.

**(B)** *Requisite Skills.* The ability to operate as a team member in vision-obscured conditions, locate and follow a guideline, conserve air supply, and evaluate areas for hazards and identify a safe haven.

**6.2.10*** Operate as a member of a RIC, given size-up information, basic rapid intervention tools and equipment, and an assignment, so that strategies to effectively rescue the brigade member(s) are identified and implemented; hazard warning systems are established and understood by all participating personnel; incident-specific PPE is identified, provided, and utilized; physical hazards are identified; and confinement, containment, and avoidance measures are discussed.

**(A)** *Requisite Knowledge.* Identification and care of PPE; specific hazards associated with the facility; strategic planning for rescue incidents; communications and safety protocols; atmospheric monitoring equipment needs; identification, characteristics, expected behavior, type, causes, and associated effects of personnel becoming incapacitated or trapped; and recognition of, potential for, and signs of impending building collapse.

**(B)** *Requisite Skills.* The ability to use PPE, determine resource needs, select and operate basic and specialized tools and equipment, implement communications and safety protocols, and mitigate specific hazards associated with rescue of trapped or incapacitated personnel.

## Knowledge Objectives

After completing this chapter, you will be able to:

- Describe the procedure for making an appropriate risk–benefit analysis.
- Describe the procedures for the personnel accountability system.
- Describe the role of the RIC.
- Define self-rescue techniques.
- Describe how to conserve self-contained breathing apparatus (SCBA) air supply.
- Describe the critical incident stress management process.

## Skills Objectives

After completing this chapter, you will be able to perform the following skills:

- Operate as a team member.
- Evaluate areas for hazards.
- Exit a structure following a hose line or guideline.
- Initiate an emergency traffic procedure.
- Respond as part of a RIC to locate a missing brigade member.

# You Are the Brigade Member

**While assigned to the night shift, your engine is dispatched to an alarm for smoke in the support maintenance shop. The shop is a three-story wood and metal frame structure containing radio communication, instrumentation, and electrical repair shops. The shops are normally occupied during the day shift only. Upon arrival, no smoke or fire is visible from the outside, and a security guard is standing at the entrance. You approach the entrance with your brigade leader, carrying a pressurized water extinguisher.**

**The security guard informs you that there is light smoke in the second floor near the electric shop. You start to go inside to investigate but your brigade leader stops you. He tells you to wait until the rest of the crew members are ready to go in with a hose line. He tells the pump engineer to be ready to charge the line and to collect the accountability tags from the other arriving brigade members. He transmits an initial report on the radio and tells the second engine to provide the rapid intervention crew (RIC).**

1. ***Why does your brigade leader stop you from going inside?***
2. ***What is the purpose of the accountability tags?***
3. ***What does a company assigned to RIC do in this situation?***

## Introduction

Brigade member survival is the primary desired outcome in any successful fire brigade operation. Brigade members often perform their duties in environments that are inherently dangerous. Survival is a positive outcome that can only be achieved by working in a manner that recognizes risks and hazards and consistently applies safe operating policies and procedures. This chapter will introduce you to the actions, attitudes, and systems that are important in achieving that goal.

Brigade members often encounter situations where survival depends on making the right decisions and taking the appropriate actions. Many of the factors that cause fire service deaths and injuries each year appear again and again in post–incident studies. These findings identify risks that are often present at fires and other emergency incidents. Brigade members must learn to recognize dangerous situations and to take actions appropriate for the situation.

### Risk–Benefit Analysis

A **risk–benefit analysis** approach to emergency operations can limit the risk of brigade member deaths and injuries. Risk–benefit analysis is based upon comparing the positive results that can be achieved with the probability and severity of potential negative consequences. A standard approach to risk–benefit analysis should be incorporated into the standard operating procedures (SOPs) for all fire brigades. NFPA 1500, *Standard on Fire Department Occupational Safety and Health Program,* incorporates this approach. This standard is an excellent reference guide to fire brigade operations.

In the fire service, risk–benefit analysis has to be practiced at several different levels. The incident commander (IC) is responsible for the high-level risk–benefit analysis that determines the strategy and tactics to be implemented at every incident scene. A brigade leader processes risk and benefits to ensure the safety of a group of brigade members working as a team. As a brigade member, you should perform a similar risk–benefit analysis of situations you encounter and any actions you are involved in performing.

The IC must always assess the risks and benefits before committing crews to the interior of a burning structure. If the potential benefits are too low and the risks are too high, brigade members should not be committed to conduct an interior attack. The IC must also reassess the risks and benefits periodically during an operation. There should be no hesitation to terminate an interior attack and move to an exterior defensive operation at any time if the risks outweigh the benefits.

A simply stated risk–benefit philosophy for a fire brigade can be expressed as follows:

- We will not risk our lives at all for persons or property that are already lost.
- We will accept a limited level of risk, under measured and controlled conditions, to save property of value.
- We will accept a higher level of risk only where there is a reasonable and realistic possibility of saving lives.

This method attempts to predict and recognize the potential risks that brigade members would face when entering a

Figure 18-1 If a fire has already reached the stage where no occupants could survive and no property of value can be saved, there is no justification for risking the lives of brigade members.

burning structure and weighs them against the results that could be achieved. If a building is known to be unoccupied after normal day shift hours or if the fire has reached the stage where no occupants could survive, there is no justification for risking the lives of brigade members to save it. Similarly, if a fire has already reached the stage where no property of value can be saved, there is no justification for risking the lives of brigade members (▲Figure 18-1).

In a situation where no lives are at stake, but there is a reasonable possibility that property could be saved, this policy allows for brigade members to be committed to an aggressive offensive attack. This policy recognizes that a standard set of hazards is anticipated in an offensive fire attack situation, but the combination of personal protective clothing and equipment, training, and SOPs is designed to allow brigade members to work in relative safety in this type of environment. All standard approaches to operational safety must be followed without compromise. No property is worth the life of a brigade member.

It is permissible to risk the life of a brigade member only in a situation where there is a reasonable and realistic possibility of saving a life. Even in these situations, the actions must be conducted in a manner that is as safe as possible. The determination that a risk is acceptable in a particular situation does not justify taking unsafe actions. It only justifies taking actions that involve a higher level of risk.

Brigade leaders and safety officers are involved in risk–benefit analysis on a continuous basis. If there is a change in the balance of risk to benefit, from their perspective, it is their responsibility to report their observations and recommendations to the IC. The IC reevaluates the risk–benefit balance whenever conditions change and decides if the strategy must be changed. The International Association of Fire Chiefs has developed a "Risk Assessment/Rules of Engagement" Model (▼Table 18-1).

In addition, each individual brigade member involved in the operation should make a risk–benefit analysis from his or her perspective. A report from an observant brigade member to his or her brigade leader can be crucial to the safety of an operation.

## Hazard Indicators

As you progress as a brigade member, it will become evident that fire suppression and other types of emergency operations involve many different types of hazards. Brigade members must be capable of working safely in an environment that includes a wide range of inherent hazards. While the danger of firefighting should never be taken for granted or thought of as routine, a brigade member has to learn to routinely follow safety SOPs. Some of these hazards are encountered at almost every incident, while others are rare occurrences. A brigade member should recognize many

**Table 18-1 Risk Assessment/Rules of Engagement**

| Fire Fighter Injury/ Life Safety Risk | High Probability of Success | Marginal Probability of Success | Low Probability of Success |
|---|---|---|---|
| Low Risk | Initiate offensive operations. Continue to monitor risk factors. | Initiate offensive operations. Continue to monitor risk factors. | Initiate offensive operations. Continue to monitor risk factors. |
| Medium Risk | Initiate offensive operations. Continue to monitor risk factors. Employ all available risk control options. | Initiate offensive operations. Continue to monitor risk factors. Be prepared to go defensive if risk increases. | Do not initiate offensive operations. Reduce risk to fire fighters and actively pursue risk control options. |
| High Risk | Initiate offensive operations only with confirmation of realistic potential to save endangered lives. | Do not initiate offensive operations that will put fire fighters at risk for injury or fatality. | Initiate defensive operations only. |

Figure 18-2 The NFPA 704 diamond indicates that hazardous materials are present.

different types of hazardous conditions and react to them appropriately.

An example of a common hazard would be the presence of smoke. All fires produce smoke, and it has long been known that breathing smoke is dangerous. The proper response to this hazard is to always wear a breathing apparatus where smoke is in the atmosphere. This is a situation when an obvious hazard is recognized and a standard solution is applied.

Hazardous conditions may or may not be evident by observation. For example, smoke and flames are usually visible, while hazards related to the integrity of process piping, vessels, or control systems might not be easily detectable. You should be aware of the indicators of hazards that present themselves in a less obvious manner. This requires knowledge of what can be hazardous, given a particular set of circumstances. Hazard recognition becomes easier through study and experience.

Some examples are:

- **Construction:** Knowledge of pipe rack construction can help in anticipating fire behavior in different types of steel support systems and recognizing the potential for metal fatigue and structural collapse. Are the supports protected by a fire-resistive coating that is designed to withstand a fire for some period of time, or are they likely to weaken quickly? Has the pipe support system been modified from its original use? Does the structure have unprotected major load-bearing components that would allow a fire to weaken the metal quickly?
- **Weather conditions:** Inclement weather might make an otherwise routine incident hazardous. Rain and snow make operations on a roof extra dangerous. Raising a ladder close to power lines during windy weather involves an increased level of risk.
- **Occupancy:** A name such as "Welding Shop" on a building should cause brigade members to anticipate that hazardous materials could be present. A warning placard outside would provide a general hazardous materials warning indicator (◄ Figure 18-2).
- **Right to know:** Hazard communication training, identifying your workplace hazards, including chemicals.

These are just a few examples of observable factors that might indicate a hazard. At each and every incident, you should observe and learn what could be a critical indication of a hazard.

## Safe Operating Procedures

Safe operating procedures define the manner in which a fire brigade conducts operations at an emergency incident. Many of the SOPs are based on brigade member health and safety. The most important component of brigade member survival is to consistently follow safe operating procedures.

Safe operating procedures must be learned and practiced before they can be implemented. Training is the only way for individuals to become proficient at performing any set of skills. In order to be effective, the same skills must be applied routinely and consistently at every training session and at every emergency incident. It has been shown that when under great pressure, most people will revert to habit. When self-survival is at stake, well-practiced habits can be a lifesaver.

### Team Integrity

Teamwork is an essential requirement in firefighting (▼ Figure 18-3). A team can be defined as two or more individuals attempting to achieve a common goal. Due to the intense physical labor necessary to pull hose or raise ladders, team efforts are imperative. The most important

Figure 18-3 A team attempts to achieve a common goal.

reason for team integrity in emergency operations is for safety; team members keep track of each other and provide immediate assistance to each other when needed. The standard operational team in firefighting operations is a company, usually consisting of three to five brigade members working under the direct supervision of a brigade leader.

Individual companies are assigned as a group to perform tasks at the scene of an emergency incident. The IC keeps track of companies and communicates with brigade leaders. Every brigade leader must have a portable radio to maintain contact with the IC. The brigade leader is responsible for knowing the location of every company member at all times. The brigade leader should also know exactly what each individual brigade member is doing at all times.

Team integrity means that a company arrives at a fire together, works together, and leaves together. The members of a crew should always be oriented to each other's location, activities, and condition. In a hazardous atmosphere, such as inside a structure, you should always be able to contact your company members by voice, sight, or touch. When air cylinders have to be replaced, all of the company members leave the building together, change cylinders together, and return to work together. If the members require rehabilitation, the full crew goes to rehabilitation together and returns to action together.

In some cases a crew will be divided into smaller teams to perform specific tasks. For example, a ladder crew could be split into a roof team and an interior search team. The initial attack on a room and contents fire in a plant office might be made by two brigade members, while two other brigade members remain outside, following the two-in/two-out rule. For safety and survival reasons, brigade members should always use a buddy system, working in teams of at least two. If you should become incapacitated, your partner can provide help immediately and call for assistance. If you are alone, no one will know that you need help. The two team members must remain in direct contact with each other. At least one member of every team must have a portable radio to maintain contact with the IC; in some fire brigades, every member carries a portable radio.

## Personnel Accountability System

A personnel accountability system can be defined as a systematic way to keep track of the location and function of all personnel operating at the scene of an incident (▶ Figure 18-4). The IC has the responsibility to account for each brigade member involved in an emergency incident. A personnel accountability system is designed to establish standard procedures for performing this function and to manage the necessary information.

An accountability system has to keep track of every company or team of brigade members working at the scene of the incident from the time they arrive until the time they are released. In addition, the system must be able to identify each individual brigade member who is assigned to that company or team. The system assumes that all of the crew members will be together, working as a unit under the supervision of the brigade leader, unless the IC has been advised of a change. At any point in the operation, the IC should be able to identify the location and function of each crew and, within seconds, know the individual brigade members assigned to that crew.

Figure 18-4 Personnel accountability system.

The personnel accountability systems used by fire brigades can take different forms. Whatever type of system is used, the only way for it to be effective is if every member takes responsibility for its use at all incidents. This includes even the smallest of incidents that appear to pose no immediate hazards. Some fire brigades start with a written roster of the team members while others use a computer database.

One common system uses a Velcro™ or magnetic nametag for each member. The tags for the members are affixed to a special board carried in the cab. This board is sometimes called a passport. These boards should be kept up to date throughout the shift and denote any change in the team.

At the scene of an incident, these boards are physically left with a designated person at the command post or at the entry point to a hazardous area. Depositing the board and the tags indicates that the company and those individual brigade members have entered the hazardous area. When a crew leaves the hazardous area or is released from the scene, the brigade leader must retrieve the board. Retrieving the board indicates that the crew has safely left the hazardous area.

A Personnel Accountability Report (PAR) is a roll call taken by each supervisor at an emergency incident. When the IC requests a PAR, each brigade leader physically verifies that all assigned members are present and confirms this information to the IC. This should occur at regular, predetermined intervals throughout the incident. A PAR should also be requested at tactical benchmarks, such as a change from offensive strategy to defensive strategy.

The brigade leader must be in visual or physical contact with all crew members in order to verify their status. The brigade leader then communicates with the IC by radio to report a PAR. Any time a brigade member cannot be accounted for, he or she is considered missing until proven otherwise. A report of a missing brigade member always becomes the highest priority at the incident scene.

If unusual or unplanned events occur at an incident, a PAR should always be performed. For example, if there is a report of an explosion, a structural collapse, or a brigade member missing or in need of assistance, a PAR should immediately take place so it can be determined how many personnel might be missing, identify the missing individuals, and establish their last known location and assignment. This last location would be the starting point for any search and rescue crews.

## Emergency Communications Procedures

Communication is a critical part of any emergency operation (▼ Figure 18-5). This is particularly important in relation to brigade member safety and survival. Breakdown in the communication process is a major contributing cause of injury and death during incidents. Good communications can save lives in a dangerous situation.

Chapter 3, Fire Service Communications, discussed the proper radio communications procedures. The following is a model of good radio communication:

"Ladder 3 to Central IC."

"This is Central IC. Go ahead, Ladder 3."

"Ladder 3 has completed exposure protection on tank 103 and is available for another assignment."

"IC copies. Tank 103 exposure protection completed and Ladder 3 available for reassignment."

**Figure 18-5** Communication is a critical part of any emergency operation.

### Brigade Member Safety Tips

**Standard Radio Terminology**

EMERGENCY TRAFFIC is used to report a brigade member in trouble and requiring immediate assistance or it is used to report a hazardous condition or situation. The term MAYDAY is another term commonly used to indicate that a brigade member is in need of immediate assistance.

These standard communications procedures ensure that the message is clearly stated and then repeated back as confirmation that it has been received and understood.

When a brigade member is missing or in trouble, or an imminent safety hazard is detected, it is even more important to communicate clearly and ensure that the message is understood. There are standard methods to transmit emergency information, standard information items to convey, and standard responses to emergency messages. Reserved phrases, sounds, and signals for emergency messages should be a part of your brigade's SOPs. These phrases should be known and practiced by everyone in the brigade. The same procedures and terminology should be used by all of the fire brigades that can be expected to work together at emergency incidents.

The term "Emergency Traffic" is used to indicate that a brigade member is in trouble and requires immediate assistance. It can also be used if a brigade member is lost, trapped, has a SCBA failure, or is running out of air. A brigade member can transmit Emergency Traffic to request assistance, or another brigade member could transmit a message to report that a brigade member is missing or in trouble.

The term "Emergency Traffic" is also used to indicate an imminent fire ground hazard, such as a potential explosion or structural collapse. An Emergency Traffic message would also be used to order brigade members to immediately withdraw from interior offensive attack positions and switch to defensive strategy. An Emergency Traffic message takes precedence over all other radio communications.

Any use of this term should bring an immediate response. The IC will interrupt any other communications and direct the person reporting to proceed. That person will then state the emergency, including all pertinent information; for instance, "This is brigade member Jones with Engine 12. I'm running low on air and I am separated from my crew. I'm on the third floor and I believe that I'm in the northwest section of the building." The IC will repeat this information and then initiate procedures to rescue brigade member Jones. This could include committing the RIC, calling for additional resources, and redirecting other teams to support a search and rescue operation.

In many fire brigades, the communications center will sound a special tone over the radio to alert all members that an emergency situation is in progress and then repeat the information to be certain that everyone at the incident scene heard the message correctly.

Similar procedures are followed for hazardous situations. All imminent hazards and emergency instructions should be announced in a manner that captures the attention of everyone at the incident scene and ensures that the message is clearly understood.

### Initiating Emergency Traffic

An Emergency Traffic message, as described previously, is used to indicate that a brigade member is in trouble and needs immediate assistance. The analysis of fire fighter fatalities and serious injuries has shown that many wait until it is too late to call for assistance. Instead of initiating Emergency Traffic when first getting into trouble, too often a fire fighter will wait until the situation is absolutely critical before requesting assistance. This could be due to a fear of embarrassment if the situation turns out to be less severe than anticipated, combined with a hope that the fire fighter will be able to resolve the problem without assistance.

The failure to act promptly can be fatal in many situations. Systems have been designed to do everything possible to protect brigade members and to rescue brigade members from dangerous situations. These systems can be effective only if you act appropriately when you find yourself in trouble. Do not hesitate to call for help when you think you need it.

To initiate Emergency Traffic, follow the SOPs for your fire brigade. When acknowledged by the IC, clearly state your identity, the nature of the problem, and your location or approximate location. Manually activate your personal alert safety system device so that other brigade members can hear the signal and come to your location. Also activate the emergency alert button on your portable radio, if it has this feature. All of these actions will increase the probability that the RIC or other brigade members will be able to locate you.

## Rapid Intervention Company/Crew (RIC)

Rapid intervention companies/crews (RIC) are established for the sole purpose of rescuing brigade members who are operating at emergency incidents. In some fire brigades, the term rapid intervention team is used to describe this assignment. A RIC is a company or crew that is assigned to stand by at the incident scene, fully dressed and equipped for action, and be ready to deploy immediately when assigned by the IC.

The U.S. federal regulation that establishes requirements for using respiratory protection, OSHA 29 CFR 1910.134, mandates the use of back-up teams (two in/two out) at incidents where operations are conducted in an unsafe atmosphere (potentially IDLH).

A RIC is an extension of the two-in/two-out rule. During the early stages of an interior fire attack, a minimum of two brigade members is required to establish an entry team and a minimum of two additional brigade members is required to remain outside the hazardous area. Only one of the two brigade members who remain outside the hazardous area can perform other functions, but they must have full personal protective equipment (PPE) clothing and SCBA, so that they can immediately enter and assist the entry team if the interior brigade members need to be rescued. The two brigade members who remain outside the hazardous area are the first stage of a RIC. The second stage is a dedicated RIC, specifically assigned to stand by for brigade member rescue assignments.

NFPA 1500, *Standard on Fire Department Occupational Safety and Health Program,* includes a complete description of RIC procedures and examples of their use at typical scenes. A number of options are available for establishing and organizing RIC crews or teams, including dispatching one or more additional companies to incidents with this specific assignment. Some small fire brigades may depend on an automatic response from an adjoining facility brigade or municipal department to provide a dedicated RIC at working incidents.

### RIC Assignments and Accountability

SOPs should dictate when a RIC is required, how it is assigned, and where it should be positioned at an incident scene. A RIC should be in place at any incident where brigade members are operating in conditions that are immediately dangerous to life and health (IDLH), which includes any structure fire where brigade members are operating inside a building and using SCBA.

Air-monitoring equipment should be used whenever brigade members might be working in a potentially IDLH atmosphere. Many fire brigades carry a standard "four-gas monitor," which is capable of detecting levels of oxygen, carbon monoxide, methane, and hydrogen sulfide. Structure fires consume oxygen and produce large volumes of carbon monoxide; hydrogen sulfide and methane are commonly found in petrochemical operations. Other atmospheric monitors are used to detect industry-specific hazards such as volatile organic compounds, sulfur dioxide, hydrogen, hydrogen chloride, chlorine, ammonia, and a wide variety of other atmospheric hazards.

For these instruments to provide a reliable warning to brigade members, they must be calibrated on a regular basis, maintained in accordance with the manufacturer's instructions, and used correctly by trained personnel.

The assignments of the RIC team members are set up to position the crew at the point where they will be most effective. RIC locations and team size may change as the incident progresses or becomes more complicated. Although the minimum required crew size is two, this number is rarely

# Voices of Experience

**"The crews advised that the heat was too intense to move any further into the drilling rig and the only way to provide safe access was to utilize foam master streams through the opening above the rig floor."**

We were paged out to a report of a drilling rig fire. When we arrived, we found heavy smoke showing throughout the drilling rig and a large amount of fire in the mud pits, which were located about 15-foot aboveground level.

I met with the drilling rig supervisor, who advised that the fire had originated in the mud pit area and was spreading throughout the rest of the structure. A drilling rig is composed of multiple levels, with narrow passageways and narrow stairs. The power had been shut off, so the interior was completely dark. We reviewed the facility preplan and determined that we could put crews into three different areas of the drilling rig to knock down the fires. Drilling rigs are steel and hydraulics—the longer they burn, the more structurally unstable they become. As the fire crews worked their way through the rig, they continually reported their progress and verified their crew headcount.

The three crews were making progress in all areas, until they tried to advance their hose lines up to the drilling rig "floor," which is the highest elevation. The crews advised that the heat was too intense to move any further into the drilling rig and the only way to provide safe access was to utilize foam master streams through the opening above the rig floor. This would be similar to flowing water through a roof opening on a structure fire with crews still inside. All crews were ordered to retreat to a safe area away from the rig floor and report when their crews were in a safe haven. The crew leaders reported that they were clear and all members were accounted for.

In less than 5 minutes of operating an elevated master stream at 1,500 gpm, the fire was knocked down and the area was cooled to the point that it was safe for the interior crews to continue. Although these kinds of fires are dangerous, experienced crew leaders, working with well-trained and experienced brigade members, resulted in a successful and effective operation.

There was a limited level of risk because the leaders working interior maintained a strong and measured control of the conditions that the crews were exposed to. It is important to remember that we should not risk the safety and lives of our personnel for property that is already lost or cannot be saved.

Scott Dornan
ConocoPhillips
Anchorage, Alaska
Kuparuk Fire & Rescue
Kuparuk, Alaska

adequate. RIC staffing should be based on incident-specific needs, rather than on some arbitrary minimum.

Each RIC position should have a positional name for accountability purposes. A RIC may be identified by a team number (such as "RIC 1") or by location (such as "North RIC"). During a search and rescue operation, the RIC officer must be accountable for all RIC personnel *at all times*. Doing so requires a PAR. Under this system, the RIC officer calls for a PAR, either by voice or by radio, and expects a return call from each of the RIC personnel.

The IC should immediately deploy the RIC to any situation where a brigade member needs immediate assistance. A situation that calls for the use of the RIC would include a lost or missing brigade member, an injured brigade member who has to be removed from a hazardous location, or a brigade member trapped by a structural collapse.

## Incapacitated Personnel

The notification or witnessing of a fire brigade member being lost, missing, trapped, or injured can be a very emotional event. Responding to a major incident in the plant can be difficult enough without added complications such as explosions, flammable liquid spills, pressurized vapor release, and employee evacuation or rescues. The additional element of losing a fellow fire brigade member who is a coworker and a friend can challenge your training and ability to remain objective. It is at that point when the true importance of a RIC crew is realized. The intensive and repetitive training undertaken before any such event occurs will determine the RIC team's ability to start a successful rescue effort without becoming part of the escalating problem.

Weighing how much risk should be taken against how much benefit might be realized from performing a rescue is one of the most difficult decisions that fire fighters will ever have to make. The essence of safety for any facility fire brigade is to reduce the risk and increase the benefits of their efforts. In other words, evacuate the building, reduce property loss, and clean up so that everyone can return to work without any type of injury. Unfortunately, unforeseen events can occur that will increase the risk dramatically, and in those instances well-equipped and highly trained personnel are the most important asset. The most valuable resources of any facility fire brigade are quality training and practice. When the brigade applies those previously learned lessons and exhibits discipline, the risk can be calculated and safely dealt with.

## Identity and Implementation of Rescue Strategies

RIC rescue strategies for facility fire brigades are driven by the type and magnitude of the event and the resources that are available. An effective size-up will determine whether the in-house resources are adequate or whether mutual aid is required. Some facility fire brigades may not have adequate staffing or equipment to accomplish more complex RIC rescue strategies. It is important to identify the required RIC resources before they are needed.

Knowledge of building construction as it relates to fire behavior, internal vessel and tank configurations, metal fatigue, and toxic gas or vapor exposure are a few of the considerations that must be addressed to develop an effective strategy and assign RIC operations.

The RIC rescue strategy should address three major components:

- **Confinement:** The RIC crew can use a backup hose line to confine the fire from an exterior or interior position until the trapped fire attack personnel can be rescued or successfully escape on their own.
- **Containment:** The RIC crew can contain the fire by closing interior doors to stop or slow fire spread until the fire attack personnel can be successfully rescued.
- **Avoidance measures:** The RIC crew must be aware of potential hazards and ensure that their actions do not make the situation worse. If a fire team is trapped on the downwind side of a major fuel spill fire, for example, the RIC crew should ensure that its foam streams do not push fuel, heat, and smoke toward the trapped team.

Depending on the location of the fire, recognition of the facility-specific hazards and potential for structural collapse are best identified by a comprehensive prefire planning program that is integrated with the fire brigade's regularly scheduled training classes.

Once a fire has occurred, the RIC crew members can use their training, experience, and the acronym "HEAT" to address potential hazards:

### RIC "HEAT" Acronym

**Hazards:** Obvious hazards may be related to construction features, type of occupancy, hazardous materials, entry restrictions, high-security systems, and electrical hazards.

**Entry/egress:** All points of entry will serve as priority considerations for rescue entry and exit limitations.

**Access:** Access to all sides and floors of a structure are to be reviewed. Knowing which areas are inaccessible will help determine the type of strategy to be employed for that area of a structure or enclosure, in addition to informing the decision to request additional fire department mutual aid.

**Tactics:** Are we winning or losing the fight? Should offensive tactics change to defensive tactics? Where is the fire? Where are the crews? Where should the RIC crew be assigned?

## RIC Tools and Equipment

Each member of a RIC crew must optimally be equipped with approved NFPA PPE and self-contained breathing apparatus (SCBA). Because many fire brigades confront not only structural fires but also fires that involve hazardous materials, vessels, tanks, and pressurized fuels, it is important that all facility fire brigade personnel recognize and understand the types of incidents that might occur in their facility. The nature of the incident will determine the type of PPE required. For example, a structure fire requires full PPE and SCBA, whereas a tank rescue in a nonhazardous atmosphere may require only a fire-resistant jumpsuit or coveralls, plus head, and eye protection. Regardless of the type of incident, the appropriate level of PPE must be provided, including training in its proper use and limitations.

Once the RIC has been established on the scene, crew members should assemble the tools and equipment that may be needed based on the type of incident:

*Portable radio.* A portable radio should be issued to all RIC members. It serves the following purposes:

- A portable radio allows the RIC officer to communicate with the victims in need of help.
- It allows the RIC officer to communicate with the IC.
- The RIC officer can maintain accountability for positioning and accomplished tasks.
- The RIC officer can be warned of additional hazards or ordered to evacuate by the IC.
- The RIC officer or any personnel on the RIC crew can communicate with each other.

*Available backup hose line.* In a structural fire scenario, a hose line may be needed to provide a safe entry for the RIC to reach downed brigade members. Depending on the situation, the backup hose line may also take the place of a search rope.

*Large hand light.* All personnel should have a durable and bright hand light. Although bulky compared to a smaller personal hand light, the large hand light is imperative to see and to be seen. These lights may also be useful if the victim requires disentanglement or requires a shared SCBA procedure. The large hand light can also be used as another means of communication to attract personnel to a certain location or to point to dangerous areas.

*Search rope.* The search rope not only helps in keeping the crew together but also allows additional crews outside to locate the RIC crew and the victims.

*Thermal imaging device.* A thermal imaging device allows the RIC crew to "see" through darkness and smoke while conducting a search operation. Although expensive, it is an essential piece of equipment when the crew is conducting search rope operations and extended fire fighter rescue tasks.

*Hand tools.* One of the most versatile firefighting rescue hand tools is the Halligan bar. It is capable of cutting, prying, lifting, and striking, yet is small enough to use while conducting search operations. Other hand tools commonly carried by RIC crews include pick-head or flat-head axes, closet hooks, pry axes, wire cutters, and utility knives.

*Additional SCBA air.* An additional SCBA air supply can best be provided by a "RIC pack"—that is, an SCBA air cylinder without a harness assembly, which is equipped with a special air regulation system that will supply air to a victim.

*Miscellaneous tools and equipment.* Special rescue situations may require the use of saws, hydraulic rescue tools, pulleys and carabiners, individual ropes and webbing, and ladders of various sizes.

It is important that each member of the fire brigade be thoroughly trained in the use of each piece of equipment and tool to be employed during their RIC operations. Brigade leaders should understand not only the ability of the fire brigade members to perform RIC operations but also their limitations. These limitations may include lack of tools and equipment, insufficient training, limited personnel staffing, or the need to call for mutual aid to support RIC operations.

## Brigade Member Survival Procedures

Some of the most important procedures you need to learn are directly related to your personal safety and the safety of the brigade members who will be working with you. These procedures are designed to keep you from getting into dangerous situations and to guide you in situations where you could be in immediate danger. Your personal safety will depend on learning, practicing, and consistently following these procedures.

### Maintaining Orientation

As you practice working with SCBA, you will find out that it is very easy to become disoriented in a dark, smoke-filled building. Sometimes the visibility inside a burning building can be measured in inches. In these conditions, it is extremely important to stay oriented. If you get lost, you could run out of air before you find your way out. The atmosphere outside your facemask could be fatal in one or two breaths. Even if you call for assistance, rescuers might not be able to find you. Safety and survival inside a fire building can be directly related to remaining oriented within the building.

Several methods can be used to stay oriented inside a smoke-filled building. Before entering a building, look at it from the outside to get an idea of the size, shape, arrangement, and number of stories. After entering, follow walls and pay attention to where you go, whether you are moving toward the front or the rear, toward the left side or right side, and upstairs or downstairs.

One of the most basic methods to remain oriented is to always stay in contact with a hose line. This can be accomplished by always keeping a hand or foot on the hose line to remain oriented. To find your way out, follow the hose line back from the nozzle to the entry point.

## Teamwork Tips

At a minimum, one member of every company must remain oriented to their location in the building.

Team integrity is an important factor in maintaining orientation. Everyone works as a team to stay oriented. When the team members cannot see each other, they have to stay in direct physical contact or within the limits of verbal contact. Standard procedures are employed to work efficiently as a team, such as having one member remain at the doorway into an office area or control room while the other members perform a systematic search inside. The searchers work their way around the area in one direction until they get back to the brigade member at the door.

A guideline can be used for orientation when inside a structure. This is a rope attached to an object on the exterior or a known fixed location. The guideline is stretched out as a team enters the structure. They can follow the guideline back out or another team can follow the guideline in to find them. Some fire brigades use a series of knots to mark distances along the rope to indicate how far it is stretched. Do not use guidelines if they might cause entanglement, if they might be cut by sharp corners or edges, or when their use would compromise brigade member safety. The guideline technique requires intense practice.

Training sessions in conditions with no visibility will build confidence. This type of training can be conducted inside any building, while using an SCBA facemask with the lens covered. More realistic conditions can be simulated by using smoke machines inside training facilities. Brigade members should practice navigating around obstructions, following charged hose lines, climbing and descending stairs, and fitting through narrow spaces. Distracting noises, such as operating fans and alarms, can add a realistic feel. Team integrity must always be maintained in these practice sessions.

### Self-Rescue

In the event that you become separated from your team, or lost, disoriented, or trapped inside a structure, several techniques can be used for self-rescue. The first thing to do is to call for assistance. Do not wait until it is too late for anyone to find you before making it known that you are in trouble. You need to initiate the process as soon as you think you are in trouble, not when you are absolutely sure that you are in grave danger.

If you are simply separated from your team, your best option is to take a normal route of egress to the exterior. Following a hose line back to an open doorway, descending a ladder, or climbing out through a ground floor window can all provide a direct exit and require no special tools or techniques. You must immediately communicate with your brigade leader or the IC to inform them that you have exited safely.

There are many more complicated techniques that brigade members can use to escape from dangerous predicaments. These include some standard methods, such as breaching a wall or using a rescue line and harness to rappel down to the ground. These methods must be learned and practiced regularly so that you will be ready to perform them without hesitation if you are ever in a situation that requires them.

Disentanglement is an important skill that needs to be learned and practiced. When crawling in a smoke-filled building, a brigade member can easily become entangled in wires, cables, debris, or whatever may be hidden in the smoke. Many brigade members carry small tools in the pockets of their PPE that can be used to cut through wires or small cables. This can be very dangerous and difficult if visibility does not allow the entangling material to be seen and identified.

### Safe Havens

A safe haven is a temporary location that provides refuge, while awaiting rescue or finding a method of self-rescue from an extremely hazardous situation. The term safe in this case is relative; it may not be particularly safe, but it is less dangerous than the alternatives.

Safe havens are important when situations become critical and few alternatives are available. When a critical situation occurs, brigade members should know where to look for a safe haven and recognize one when it presents itself.

## Brigade Member Safety Tips

Some proven actions to assist in self-rescue are:

- If you have a portable radio, use it. Follow the procedures for relaying critical information. Activate the emergency button on your radio, if it has one.
- Manually activate the alarm mode of your personal alert safety system.
- Turn on your flashlight. In darkness, this could help lead rescuers to your location.
- Conserve air. Remain as calm as possible, regulate your rate of breathing, and take shallow breaths or skip breaths when possible.
- Locate a window or a door. Place yourself as close as possible to locations where rescuers are likely to enter. Try to make it easy for them to find you.
- If possible, close the door to the room you are in before opening or breaking the window.
- Break a window to make yourself visible and to reach fresh air.
- Look for a safe haven where you can wait for help.

Remember: The only way for other brigade members to know you are in trouble is to tell them. Call for assistance as soon as you realize you are in trouble. Do not wait until it is too late!

If brigade members are trapped inside a burning building, a room with a door and a window could be a safe haven. Closing the door could keep the fire out of the room for at least a few minutes, and breaking the window could provide fresh air and an escape route. The safe haven provides time for a rescue team to reach the brigade members, a ladder to be raised to the window, or another group of brigade members to control the fire.

A roof or floor collapse often leaves a void adjacent to an exterior wall. If trapped brigade members can reach this void, it could be the best place to go. Breaching the wall might be the only way out of the void or the only way for a rescue team to reach them.

Maintaining team integrity is important when escaping to a safe haven. If your partner exits an area, you must do so as well. Brigade members have survived in safe havens by buddy breathing with SCBA and by working together to create an opening. The ability to use a radio to call for help and to describe the location where brigade members are trapped can be critical.

As with all of the emergency procedures described in this chapter, these activities require good instruction and practice and must follow your brigade's operating guidelines.

## Air Management

Air management is important to all brigade members when using an SCBA. Air management relates to the basic fact that air equals time. The time that you can survive in a dangerous atmosphere depends on how much air is in your SCBA and how quickly you use it. You have to manage your air supply in order to manage your time in the hazardous area.

Your time in the hazardous atmosphere has to include the time it takes to enter and get to the location where you plan to operate and the time it will take to exit safely. Your work time is limited by the amount of air that you have in your SCBA cylinder, after allowing for entry and exit time. If it will take 6 minutes for you to reach a safe atmosphere and you only have enough air left for 3 minutes, the results could be disastrous.

Remember that the time rating for an SCBA is based on a standard rate of consumption for a typical adult under low exertion conditions in a test laboratory. Brigade members typically consume air much more rapidly than the standard test. A brigade member can often consume a 30-minute rated air supply in 10–12 minutes.

The actual rate of air consumption varies significantly among individual brigade members and depends on the activities that are being performed. Chopping holes with an axe or pulling ceilings with a pike pole will cause you to consume air more rapidly than conducting a survey with a thermal image camera. Even given the same workload, no two individuals will use exactly the same amount of air from their SCBA cylinder. Differences in body size, fitness levels, and even heredity factors cause air consumption rates to be faster or slower than the average.

Air management has to be a team effort as well as an individual responsibility. You will always be working as part of a team when using SCBA. The team members will use air at different rates. The member who uses the air supply most rapidly determines the working time for the team.

A good way to determine your personal air usage rate is to participate in an SCBA consumption exercise. This drill puts brigade members through a series of exercises and obstacles to learn how quickly each brigade member uses air when performing different types of activities. Repeated practice can help you learn to use air more efficiently by building your confidence and learning to control your breathing rate. Keeping physically fit results in a more efficient use of oxygen and allows you to work longer with a limited air supply.

Practice sessions should be done in a nonhazardous atmosphere. A typical skill-building exercise might include the following:

- While wearing full PPE and breathing air from your SCBA, perform typical activities such as carrying hose packs up stairs, hitting an object with an axe or sledgehammer, or crawling along a hallway.
- Record the following: How much air is in the cylinder when you start; how long you work; whether the physical labor is light or heavy; the time when the low air alarm sounds; and the time when the cylinder is completely empty.

This type of exercise will allow you to determine your personal rate of consumption and work on improving your times with practice. With practice you will be able to predict how long your air supply should last under a given set of conditions.

Knowledge of your team members' physical conditions and their workloads can help you look out for their safety as well. Since many tasks are divided up in any given operation, one member may be working to his or her maximum level while another may be acting only in a supportive role. This could cause your team member to use up his or her air supply much faster without realizing it.

When using SCBA at an emergency incident, be aware of the limitations of the device. Do not enter a hazardous area unless your air cylinder is full. (OSHA regulations do not allow a cylinder to be used unless it is at least 90% full.) Keep track of the time you are using air and check the gauge regularly to know how much air you have left at all times. Do not wait until the low-pressure alarm sounds to start thinking about leaving the hazardous area; it may not provide enough time to make a safe exit.

When operating SCBA in restricted spaces, the size and shape of SCBA may make it difficult for you to fit through tight openings. Several techniques may help you navigate these spaces:

- Change your body position: Rotate your body by 45 degrees and try again.
- Loosen one shoulder strap and change the location of the SCBA on your back.
- As a last resort, remove your SCBA. In this case, do not let go of the backpack and harness for any reason. Keep the unit in front of you as you navigate through the tight space. Don the harness as soon as you have passed through the restricted space. You should not remove your SCBA unless there is no other way to move through a restrictive area.

Even using the best of air management practices, emergency situations can occur. An SCBA can malfunction or brigade members can be trapped inside a building by a structural collapse or an unexpected situation. These are the times when remaining calm and knowing how to use all of the emergency features of an SCBA can be critically important. Remaining calm, controlling your breathing rate, and taking shallow breaths can slow air consumption. All of these techniques require practice to build confidence.

## Rescuing a Downed Brigade Member

As discussed, one of the most critical and demanding situations in the fire service is when brigade members have to rescue another crew member who is in trouble. Air management has particular implications in a brigade member rescue situation. Air management has to be considered for the rescuers as well as the brigade member who is in trouble.

If you reached a downed crew member, the first thing to do is assess the brigade member's condition. Is the brigade member conscious and breathing? Does the brigade member have a pulse? Is the brigade member trapped or injured? Make a rapid assessment and notify the IC of your situation and location. Have the RIC deployed to your location and quickly determine the additional resources you will need.

The most critical decision at this point is whether the brigade member can be moved out of the hazardous area quickly and easily or if considerable time and effort will be required. If it will take more than a minute or two to remove the brigade member from the building, air supply will be an important consideration. You need to provide enough air to keep the brigade member breathing for the duration of the rescue operation.

Brigade members must know and understand all aspects of their SCBA in this type of situation. If the brigade member has a working SCBA, determine how much air remains in the cylinder. A brigade member who is breathing and has an adequate air supply is not in immediate, life-threatening danger. If there is very little air or no air in the SCBA, this is a critical priority. You need to move the brigade member out of the hazardous area immediately or provide an additional air supply. Can a spare air cylinder be brought to your location and connected to the downed brigade member's SCBA? Can the RIC bring in a whole replacement SCBA?

If the brigade member's SCBA is not working, check to see if a simple fix can make it operational. Did a hose or regulator come off? Was the cylinder valve accidentally closed? If regulator failure is a possibility, open the by-pass valve.

Many newer SCBA units are designed with an additional hose or hose connections exclusively for buddy breathing. Buddy breathing is a method for two brigade members to breathe air from the same SCBA cylinder. If you have this equipment, you need to know the policies and procedures for their use in your fire brigade.

As discussed in Chapter 13, the RIC breathing apparatus system can also be used to deliver breathing air to a victim who has no respiratory protection or to assist a person who is wearing respiratory protection but is low on air.

# Rehabilitation

Chapter 20 covers Brigade Member Rehabilitation. The purpose of **rehabilitation** is to reduce the effects of fatigue during an emergency operation (▼ Figure 18-6). Firefighting involves very demanding physical labor. When combined with the extremes of weather and the mental stresses of emergency incidents, it can challenge the strength and endurance of brigade members who are physically fit and well prepared. Rehabilitation helps brigade members retain the ability to perform at the current incident and restores their capacity to work at later incidents.

At the simplest of incidents, rehabilitation can be set up on the tailboard of an apparatus with a water cooler. At larger incidents, a complete rehabilitation operation should be established, including personnel to monitor the vital signs of brigade members and provide first aid. In whatever size or form it takes, rehabilitation is integral to brigade

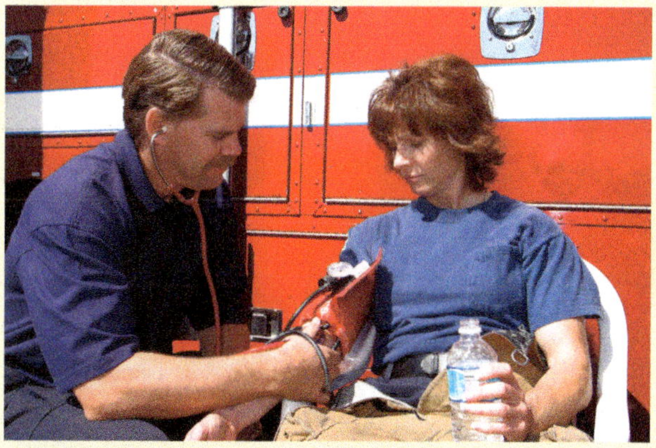

Figure 18-6 Rehabilitation reduces the effects of fatigue caused by an operation.

member safety and survival. The personnel accountability system must continue to track brigade members who are assigned to report to rehabilitation and when they are released from rehabilitation for another assignment.

## Critical Incident Stress

Brigade members may be exposed to stressful situations. There are times when brigade members are involved in situations that are described as critical incidents. These incidents challenge the capacity of most individuals to deal with stress. Examples of critical incidents include:

1. Line-of-duty deaths (police, fire/rescue, emergency medical services)
2. Suicide of a colleague
3. Serious injury to a colleague
4. Situations that involve a high level of personal risk to brigade members
5. Events in which the victim is known to the brigade members
6. Multicasualty/disaster/terrorism incidents
7. Events involving death or life-threatening injury/illness to a victim, especially a child
8. Events that are prolonged or end with a negative or unexpected outcome

This list is not complete, nor is it a fact that any of these situations will seriously trouble every individual. Normal coping mechanisms help many brigade members to handle many situations. Some individuals have a high capacity to deal with stressful situations through exercise, talking to friends and family, or turning to their religious beliefs. These are healthy, nondestructive ways to deal with the pressures of being exposed to a critical incident.

Sometimes individuals react to critical incidents in ways that are not positive—such as alcohol abuse, depression, the inability to function normally, or a negative attitude toward life and work. These symptoms can occur in anyone, even individuals who normally have healthy coping skills. Reactions will vary from one individual to the next, both in type and severity. Many times brigade members do not realize they are affected in a deeply negative manner. A somewhat routine incident can trigger negative reactions from a critical incident that occurred in the past. Critical incident stress can also be cumulative. Sometimes it is called burnout and cannot be traced to any one incident.

The aim of critical incident stress management (CISM) is to prevent these reactions from having a negative impact on the brigade member's work and life, both short-term and long-term (▶ Figure 18-7). Brigade members should understand the CISM system and how they can access it. The concept of CISM is a simple but important part of brigade member survival.

The recognized stages of emotional reaction experienced by brigade members and other rescue personnel after a critical incident can include:

1. Anxiety
2. Denial/disbelief
3. Frustration/anger
4. Inability to function logically
5. Remorse
6. Grief
7. Reconciliation/acceptance

These stages can occur within minutes or hours, or they can take several days or months. Not all of the steps will occur for every event, and they do not necessarily occur in order.

Figure 18-7 Critical incident stress debriefings are used to alleviate the stress reactions caused by high-stress emergency situations.

CISM helps brigade members recognize and deal with these reactions in the most positive manner possible. There are some variations in the way this is accomplished, but the purpose remains the same. In-house, peer-driven programs guided and assisted by mental healthcare professionals are a proven method for CISM. Team members who have faced the same situations and experienced similar feelings can be understanding and nonjudgmental when providing assistance with stress management.

Most CISM programs operate in a similar manner. To begin with, there is an informal process where a trained CISM team member will have a conversation with an individual or small group that has been involved in a critical incident. This is a simple and quick approach to see if any further action is required for the group or for any individual brigade member. This might take place on-scene, at the fire station, or at another location away from the critical incident scene.

If an incident is recognized as requiring further intervention, a series of steps might follow. These steps range from a semi-formal group discussion of the incident to the inclusion of healthcare professionals. The most common form of CISM is peer defusing, when a group informally discusses events that they experienced together. Critical incident stress debriefings are meetings between brigade members and specially trained leaders. The purpose of a debriefing is to allow an open discussion of feelings, fears, and reactions to the situation that occurred. A debriefing is not an investigation or an interrogation. Debriefings are usually held within 24 to 72 hours after a major incident. The leaders offer suggestions and information on overcoming stress.

Your fire brigade should have some type of CISM program in place. All crew members should know how it is activated and be supportive of its intent and function. Brigade members should realize that emotional and mental health must be protected, just as much as physical health and safety.

# Wrap-Up

## Ready for Review

- Risk–benefit analysis should be done continuously by all members when operating at hazardous incidents.
- Hazards of some kind are always present. They should be identified and used to make a risk–benefit analysis.
- The personnel accountability system is designed to keep track of the location and function of all personnel operating at an incident.
- The PAR is a roll call taken at an emergency incident.
- Communication procedures should be known and used in a consistent manner at all incidents. This process is especially important when brigade members are in trouble.
- RICs are in place on the fireground for one central purpose: to rescue brigade members.
- Brigade members should learn, by whatever means possible, the important procedures that are directly related to personal safety and the safety of fellow brigade members.
- Air management is a critical skill that takes training and practice.
- Emergency evacuation methods are used when normal means of egress are blocked, and include the use of normal tools and techniques in nontraditional ways.
- Critical incident stress is a known hazard that emergency personnel can learn to deal with. Organizational help is usually available in the form of in-house programs or referral to professionals. This hazard deals with long-term health, safety, and survival.

## Hot Terms

**Air management** The way in which an individual utilizes a limited air supply to be sure it will last long enough to enter a hazardous area, accomplish needed tasks, and return safely.

**Critical incident stress management (CISM)** A system to help personnel deal with critical incident stress in a positive manner; its aim is to promote long-term mental and emotional health after a critical incident.

**Guideline** A rope used for orientation when inside a structure when there is low or no visibility; it is attached to a fixed object outside the hazardous area.

**Immediately dangerous to life and health (IDLH)** Any condition that poses an immediate or delayed threat to life, might cause irreversible adverse health effects, or might interfere with an individual's ability to escape unaided from a hazardous environment.

**Personnel Accountability Report (PAR)** A roll call report confirming that all members of a company are present.

**Personnel accountability system** A method of tracking the identity, assignment, and location of brigade members operating at an incident scene.

**Rapid intervention company/crew (RIC)** A minimum of two fully equipped personnel on site, in a ready state, for immediate rescue of injured or trapped brigade members. In some brigades, this is also known as a rapid intervention team.

**Rehabilitation** A systematic process to provide periods of rest and recovery for emergency workers during an incident; usually conducted in a designated area away from the hazardous area.

**Risk–benefit analysis** The process of weighing predicted risks against potential benefits and making decisions based on the outcome of that analysis.

**Safe haven** A temporary place of refuge to await rescue.

**Self-rescue** The activity of a brigade member using techniques and tools to remove himself or herself from a hazardous situation.

# Brigade Member in Action

**It is 4:00 P.M. on a summer day, and you are preparing for your afternoon work activities when you are dispatched to a control room structure fire. While responding to the fire, the dispatcher radios an update and states that an employee is missing and was last seen inside the break area. Your engine is first on the scene, and your brigade leader tells you to start a search of the second floor.**

1. What are the risk–benefit philosophies that should be considered in this incident?
   A. Accept a higher level of risk when there is a higher possibility to save lives
   B. Accept limited risk to save property of value
   C. Deny risk for persons or property that are already lost
   D. All of the above

2. What risk factors should be considered in this scenario?
   A. Smoke, heat, building construction, length of burn time, location of fire, and utilities
   B. Radioactive materials, snow, and smoke
   C. Industrial equipment, weather, and large heating, ventilation, and air conditioning systems on the roof
   D. High fire load, hazardous materials, and high–voltage lines

**You and your partner enter the control room and begin to do a left wall search from the main entry over to the break room. While performing your search, you become disoriented and lose your partner and your hose line.**

3. As soon as you realize that you are lost, you should:
   A. try to find your way out of the structure.
   B. wait until the situation becomes critical to save yourself some embarrassment.
   C. immediately seek assistance by calling for help.
   D. Both A and B.

4. The IC should call for a PAR because there are brigade members missing. A PAR is a(n):
   A. Personnel Accountability Report.
   B. a temporary place of refuge.
   C. team of brigade members designated for brigade member rescue.
   D. emergency signal that is signaled through your portable radio.

# Salvage and Overhaul

Chapter 19

## NFPA 1081 Standard

**Incipient Facility Fire Brigade Member**

**4.2.2*** Conserve property, given special tools and equipment and an assignment within the facility, so that the facility and its contents are protected from further damage.

**(A)** *Requisite Knowledge.* The purpose of property conservation and its value to the organization, methods used to protect property, methods to reduce damage to property, types of and uses for salvage covers, and operations at properties protected with automatic sprinklers or special protection systems.

**(B)** *Requisite Skills.* The ability to deploy covering materials, control extinguishing agents, and cover building openings, including doors, windows, floor openings, and roof openings.

**Advanced Exterior Facility Fire Brigade Member**

**5.2.5*** Conserve property operating as a member of a team, given special tools and equipment and an assignment within the facility, so that exposed property and the environment are protected from further damage.

**(A)** *Requisite Knowledge.* The purpose of property conservation and its value to the organization, methods used to protect property, methods to reduce damage to property, operations at properties protected with automatic sprinklers or special protection systems, understanding the impact of using master streams and multiple hose streams on property conservation, particularly as it can relate to the impact on outside facilities.

**(B)** *Requisite Skills.* The ability to deploy covering materials, control extinguishing agents, and cover openings and equipment such as doors, windows, floor openings, and roof openings related to the impact of outside facilities.

**5.2.6** Overhaul a fire scene, given PPE, a handline, hand tools, scene lighting, and an assignment, so that structural integrity is not compromised, all hidden fires are discovered, fire cause evidence is preserved, and the fire is extinguished.

**(A)** *Requisite Knowledge.* Types of fire handlines and water application devices most effective for overhaul, water application methods for extinguishment that limit water damage, types of tools and methods used to expose hidden fire, dangers associated with overhaul, obvious signs of area of origin or signs of arson, and reasons for protection of fire scene.

**(B)** *Requisite Skills.* The ability to deploy and operate a handline, expose void spaces without compromising structural integrity, apply water for maximum effectiveness, expose and extinguish hidden fires, recognize and preserve obvious signs of area of origin and fire cause, and evaluate for complete extinguishment.

**Interior Structural Facility Fire Brigade Member**

**6.2.4*** Overhaul a fire scene, given PPE, attack line, hand tools, a flashlight, and an assignment, so that structural integrity is not compromised, all hidden fires are discovered, fire cause evidence is preserved, and the fire is extinguished.

**(A)** *Requisite Knowledge.* Types of fire handlines and application devices most effective for overhaul, application methods for extinguishing agents that limit damage, types of tools and methods used to expose hidden fire, dangers associated with overhaul, obvious signs of area of origin and signs of arson, and reasons for protection of fire scene.

**6.2.9*** Conserve property operating as a member of a team, given special tools and equipment and an assignment within the facility, so that exposed property and the environment are protected from further damage.

**(A)** *Requisite Knowledge.* The purpose of property conservation and its value to the organization, methods used to protect property, methods to reduce damage to property, types of and uses for salvage covers, operations at properties protected with automatic sprinklers or special protection systems, and understanding the impact of using master streams and multiple hose streams on property conservation, particularly as it can relate to the impact on outside facilities.

**(B)** *Requisite Skills.* The ability to deploy covering materials, control extinguishing agents, and cover building openings, including doors, windows, floor openings, and roof openings.

## Knowledge Objectives

After completing this chapter, you will be able to:

- Describe the importance of adequate lighting at the fire scene and in the fire building.
- Describe the safety precautions that need to be considered when performing salvage.
- List the tools that are used for salvage.
- Describe some general steps that can be taken to limit water damage.
- Describe the steps needed to stop the flow of water from activated sprinkler heads.
- Describe how brigade members can limit losses from smoke and heat.
- Describe the steps needed to protect building contents using a salvage cover.
- Describe overhaul.
- List the safety concerns that must be addressed to ensure safety for brigade members performing overhaul.
- Describe how to preserve structural integrity during overhaul.
- List the tools that are used for overhaul.

## Skills Objectives

After completing this chapter, you will be able to perform the following skills:

- Inspect and run a generator.
- Stop water from a sprinkler head using a sprinkler stop.
- Stop water from a sprinkler head using sprinkler wedges.
- Close and open main control valve (outside screw-and-yoke).
- Close and open main control valve (postindicator).
- Construct a water chute.
- Construct a water catch-all.
- Perform the one-person salvage cover fold.
- Perform the two-person salvage cover fold.
- Fold and roll a salvage cover.
- Perform a one-person salvage cover roll.
- Perform a shoulder toss.
- Perform a balloon toss.
- Open a ceiling to check for fire with a pike pole.
- Open an interior wall.

# You Are the Brigade Member

**Your fire brigade is dispatched to a reported fire in a warehouse at your plant. When you arrive, you and your crew are assigned to salvage operations. The incident safety officer has confirmed safe levels of carbon monoxide in the warehouse and your PPE is subsequently modified. You and your partner take salvage covers into the warehouse area, move cartons of finished products, gather computers, and then pile them. You cover the pile with a salvage cover to protect everything from water damage, using water control and debris control techniques. As you feel the rush of "fresh air" from the forced-air ventilation in use, you know that smoke, water, and debris damage will be minimal and that this area will be back in operation quickly.**

1. ***What are the most important objects to try to save in a commercial facility?***
2. ***Should you concentrate your salvage operations on certain areas?***
3. ***Aside from salvage covers, what additional salvage tools may be needed?***

## Introduction

The basic goals of firefighting are first to save lives, second to control the fire, and third to protect property and the environment. Salvage and overhaul operations help meet that third goal by limiting and reducing property losses resulting from a fire. Salvage efforts protect property and belongings from damage, particularly from the effects of smoke and water. Overhaul ensures that a fire is completely extinguished by finding and exposing any smoldering or hidden pockets of fire in an area that has been burned.

Salvage and overhaul are usually conducted in close coordination with each other and have a lower priority than search and rescue or controlling the fire. If a brigade has enough personnel on scene to address more than one priority, salvage can be done concurrent with fire suppression. Overhaul, however, can begin only when the fire is under control.

Throughout the salvage and overhaul process, brigade members must attempt to preserve evidence related to the cause of a fire, particularly when working near the suspected area of fire origin and when the fire cause is not obvious. In general, disturb as little as possible in or near the area of origin until a fire investigator determines the cause of the fire and gives permission to clear the area. Knowing what caused a fire can help prevent future fires and enable companies to recoup losses under their insurance coverage. This subject is covered in greater depth in Chapter 26, Fire Cause Determination.

## Lighting

Lighting is an important concern at many fires and other emergency incidents. Because many emergency incidents occur at night, lighting is required to illuminate the scene and enable safe, efficient operations. Inside fire buildings, heavy smoke can completely block out natural light, and electrical service is often interrupted or disconnected for safety purposes, making interior lighting essential for brigade members to conduct search and rescue, ventilation, fire suppression, salvage, and overhaul.

Most brigades have different types of lighting equipment for various situations. Spotlights project a narrow concentrated beam of light. Floodlights project a more diffuse light over a wide area. Lights can be portable or permanently mounted on fire apparatus.

Lighting equipment can be mounted on any apparatus, and some brigades equip special vehicles with high-powered lights to illuminate incident scenes.

### Safety Principles and Practices

The lighting and power equipment used at an incident generally operates on 110-volt alternating current (AC). Some systems require higher voltage. All cords, junction boxes, lights, and power tools must be maintained properly and handled carefully to avoid electrical shocks. All electrical equipment must be properly grounded. Electrical cords must be properly insulated, without cuts or defects, and the wire gauze sized to handle the required amperage.

Outlets on mobile or portable generators with more than 5 kilowatts must have ground fault circuit interrupters (GFCI)

> **Brigade Member Safety Tips**
>
> Never operate lighting at an emergency scene without GFCI protection—electrocutions can occur!

to prevent a brigade member from receiving a potentially fatal electric shock. A GFCI senses when there is a problem with an electrical ground and interrupts the current flow, shutting off the power to the equipment it is feeding.

Some portable generators are equipped with a grounding rod that must be inserted into the ground. Always use a grounding device if one is provided. Avoid areas of standing or flowing water when placing power cords and junction boxes at a fire scene, and put electrical equipment on higher ground whenever possible.

## Lighting Equipment

Portable lights can be taken into buildings to illuminate the interior. They can also be set up outside to illuminate the fire or emergency incident scene (▶ Figure 19-1). Portable lights usually range from 300 watts to 1500 watts and can use several types of bulbs, including quartz and halogen bulbs.

Portable light fixtures are connected to the generator with electrical cords. Electrical cords should be stored neatly coiled or on permanently mounted reels. The cord is pulled from the reel to where the power is needed. Portable reels can be taken from the apparatus into the fire scene.

Junction boxes are used as mobile power outlets. They are placed in convenient locations so that cords for individual lights and electrical equipment can be attached. Junction boxes used by brigades are protected by waterproof covers and are often equipped with small lights so that they can be easily located.

The connectors and plugs used for fire brigade lighting may have special connectors that attach with a slight clockwise twist. This keeps the power cords from becoming unplugged during operations.

Lights also can be permanently mounted to illuminate incident scenes (▶ Figure 19-2). The vehicle or trailer operator can immediately illuminate the emergency scene simply by pressing the generator starter button. Some vehicle-mounted lights can be manually raised to illuminate a larger area. Mechanically operated light towers, which can be raised and rotated by remote control, can create nearly daylight conditions at an incident scene.

## Battery-Powered Lights

Battery-powered lights are generally used by individual brigade members to find their way in dark areas or to illuminate their immediate work area. Battery-powered lights are most often used during the first few critical minutes of an incident because they are lightweight, easily transported, do not require power cords, and can be used immediately (▶ Figure 19-3). These lights are powered by either disposable or rechargeable batteries and have a limited operating time before the batteries need to be recharged or replaced.

Powerful lights are needed to pierce the smoke inside a unit. Large hand lights that project a powerful beam of light are preferred for search-and-rescue and interior fire suppression operations when brigade members must penetrate smoke-filled areas quickly. These lights can be equipped with a shoulder strap for easy transport. Every crew member entering a fire building should be equipped with a high-powered hand light.

Figure 19-1 Portable emergency lights come in various sizes and levels of brightness.

Figure 19-2 An apparatus-mounted light tower can provide near daylight conditions at an incident scene.

Figure 19-3 Hand lights manufactured for use by brigade members have light outputs ranging from 3,500 to 75,000 candlepower.

A personal flashlight is another type of battery-operated light used by brigade members. Always carry a flashlight with your personal protective equipment (PPE); it could be a lifesaver if your primary light source fails. Flashlights are not as powerful as the larger hand lights, and they will not operate for as long on one set of batteries. A brigade member's flashlight should be rugged and project a strong light beam.

## Electrical Generators

More powerful lighting equipment requires a separate source of electricity, generally a 110-volt AC delivered through power cords. The electricity can be supplied by a generator, an inverter, or a building's electrical system, if the power has not been interrupted.

Power inverters convert 12-volt DC (direct current) from a vehicle's electrical system to 110-volt AC power. An inverter can provide a limited amount of AC current and is usually used to power one or two lights. Some vehicles have additional outlets for operating a portable light or a small electrically powered tool. Most power inverters do not produce enough power to operate high-intensity lighting equipment, large power tools, or ventilation fans. Connecting devices that draw too much current to an inverter can seriously damage it.

Electrical generators are powered by either gasoline or diesel engines and can be portable or permanently mounted. Portable generators, which are small enough to be carried to the fire scene, come in various sizes and produce up to 6,000 watts (6 kilowatts) of power. A portable generator has a small gasoline- or diesel-powered motor with its own fuel tank.

Apparatus-mounted generators have much larger capacities, sometimes exceeding 20 kilowatts. The smaller units, which are similar to portable generators, can be permanently mounted in a compartment on the apparatus. Larger generators can be mounted directly on the vehicle, often have diesel engines, and draw fuel directly from the vehicle's fuel tank. Permanently mounted generators can also be powered by a vehicle engine through a power take-off and a hydraulic pump.

Another possible power source for lighting is a building's normal power supply. This is usually not an option if there is a serious fire in the building because the current will be disconnected for safety reasons. If the fire is relatively minor, the power might not have to be interrupted and can be used for lighting. Sometimes power can be obtained from a nearby building.

## Lighting Methods

Effective lighting improves the efficiency and safety of all crews at an emergency incident but requires practice to set up quickly and efficiently. Departmental standard operating procedures (SOPs) or the incident commander (IC) will dictate responsibility for setting up lighting.

### Exterior Lighting

Exterior lighting should be provided at all incident scenes during hours of darkness so that brigade members can see what they are doing, recognize hazards, and find victims who need rescuing. Scene lighting also makes brigade members more visible to drivers approaching the scene or maneuvering emergency vehicles. Powerful exterior lights can often be seen through the windows and doors of a dark, smoke-filled building and can guide disoriented brigade members or occupants to safety.

Apparatus operators should turn on their apparatus-mounted floodlights and position them to illuminate as much of the area as possible. The entire area around a fire building should be illuminated. If some areas cannot be covered with apparatus-mounted lights, use portable lights.

### Interior Lighting

Portable lights can provide interior lighting at a fire scene. During interior operations, quickly extend a power cord and set up a portable light at the entry point. Brigade members looking for the building exit can use the light at the door as a beacon. Extend additional lights into the building to illuminate interior areas as needed and as time permits.

Interior operations often progress through the search-and-rescue and fire suppression stages before effective

**Brigade Member Safety Tips**

Never run a portable generator in a confined space or area without adequate ventilation or wearing SCBA. The exhaust contains carbon monoxide, which can incapacitate you.

## Brigade Member Safety Tips

A good rule to remember is: Light early, light often, and light safely.

interior lighting can be established. If interior operations are lengthy, portable lighting should be used to provide as much light as possible in the areas where brigade members are working.

When operations reach the salvage and overhaul phase, take extra time to provide adequate interior lighting in all areas so that crews can work safely. Proper lighting enables brigade members to see what they are doing and to observe any dangerous conditions that need to be addressed.

### Cleaning and Maintenance

Portable electrical equipment should be cleaned and properly maintained to ensure that it will work when it is needed. Test and run generators to ensure they start, run smoothly, and produce power as specified by the manufacturer. Run gasoline-powered generators for 15–30 minutes to reduce any deposit build-up that could foul the spark plugs and make the generator hard to start.

At the same time, inspect and test other electrical equipment. Take junction boxes out and check them for cracked covers or broken outlets. Plug them in to make sure the GFI works properly. Check power cords for tears in the protective covering, mechanical damage, fraying, heat damage, or burns. Inspect plugs for loose or bent prongs. Inspect and test run all power tools and equipment on the apparatus.

Follow the manufacturer's recommendations for generator engine maintenance, including regular oil changes. To conduct a weekly/monthly generator test, follow the steps in **Skill Drill 19-1**.

1. Place the generator outside (if required) or open all doors as needed and install grounding rod (as needed).
2. Check oil and fuel levels and start generator.
3. Connect the power cord or junction box to the generator.
4. Connect a load (fan or lights) and listen as the engine revs up to proper speed. Check the voltage and amperage gauges to confirm efficient operation.
5. Run the generator under load for 15–30 minutes.
6. Turn off the load and listen as the generator slows down to idle speed.
7. Allow generator to idle for approximately 2 minutes, then turn it off.
8. Disconnect all power cords and junction boxes, and remove grounding rod (as needed).
9. Clean all power cords and tools and replace them in proper storage areas.
10. Allow the generator to cool for 5 minutes before refueling and checking oil.
11. Return the generator to its storage location.

## Salvage Overview

Salvage operations are conducted to save property from a fire and to reduce the damage that results from a fire. Often, the damage caused by smoke and water can be more extensive and costly to repair or replace than the property that is burned. Salvage operations are also conducted in other situations that endanger property, such as floods or water leaking from burst pipes.

Salvage efforts usually are aimed at preventing or limiting secondary losses that result from smoke and water damage, fire suppression efforts, and other causes. They must be performed promptly to preserve the building and protect the contents from avoidable damage. Salvage operations include expelling smoke, removing heat, controlling water runoff, removing water from the building, securing a building after a fire, covering broken windows and doors, and temporarily patching ventilation openings in the roof to protect the structure and contents.

Brigade members must help protect property from avoidable loss or damage as part of their responsibilities. At the scene of a fire or any other type of emergency, the plant property is entrusted to the care of brigade members who are expected to protect that property to the best of their ability.

Salvage priorities at commercial and industrial fires are considerably different from the operations conducted at residential structure fires. At industrial fires, the preincident plan should be used to identify critical items that should be preserved if possible and enables fire brigades to focus their salvage efforts appropriately. For example, hospitals typically protect medical records, whereas some other industries must protect critical areas such as process controls. A preincident plan should document these critical items requiring protection. Fire brigade team leaders should work with their company's management staff to determine the priorities prior to an incident.

### Safety Considerations During Salvage Operations

Safety during salvage operations is a primary concern. ICs must be fully aware of where every crew member is working but, most importantly, must ensure they are performing in a safe environment. That environment is not only the structure around them but also the PPE required to properly protect team members operating on the site during salvage operations. The required PPE will depend on the site of the salvage operations, the stage of the fire, the proximity of the crews to the actual fire, and any other hazards that are present.

## Brigade Member Safety Tips

At the emergency scene, the IC should know where every crew is working and what they are doing at all times. The IC relies on your brigade leader to know your location and activities and to relay this information to the command post.

Salvage operations often begin while the fire is still burning and can continue for several hours after the fire has been extinguished. You must use PPE identified for your brigade for salvage operations. Once the fire is extinguished, the fire must be ventilated, and the atmosphere should be tested and determined to be safe to operate in by the safety officer or designated industrial hygienist, if the facility has one.

Salvage operations being conducted in different parts of the building, such as several floors below the fire floor or in an adjacent building, may be conducted by incipient fire brigade members with industry-approved PPE such as gloves and eye protection. Salvage operations conducted in areas of a building where hazardous atmospheres may be present must be conducted by interior structural fire brigade members, using full thermal protective clothing and self-contained breathing apparatus (SCBA). Hazards in these areas will be quite different from those in the fire area. Fire brigade team members conducting salvage operations in these remote areas can wear industry-approved PPE such as safety boots, helmets, gloves, and eye protection, with the approval of the safety officer.

Fire brigade team members performing salvage and overhaul must be able to see where they are going, what they are doing, and what potential hazards are present. Although the smoke may have cleared, electrical power may still be unavailable. Portable lights should be set up to fully illuminate areas during salvage and overhaul operations. Care must be taken when operating portable generators for lighting and other purposes during salvage operations. Exhaust from the generators must be properly ventilated.

Structural integrity is always a major concern during all firefighting phases. ICs must continue to assess the structural integrity of the building even during salvage operations. Structural collapse is always a possibility during salvage operations and must be continuously monitored. Fire brigade team members should immediately leave the area if the integrity of the structure is in question, and report their findings to the IC. Lightweight trusses can collapse quickly when damaged by a fire. Heavy objects, such as heating, ventilation, and air conditioning (HVAC) units, might crash through roofs or ceilings that have been weakened by the fire. The water that is used to extinguish the fire can contribute to structural collapse. Each gallon of water weighs 8.3 lb. If enough water is discharged onto a floor or roof without an adequate drainage system to remove the water, or if the drains are clogged, the extra weight could overload the structure and cause a catastrophic failure of the structure with the potential of seriously injuring or killing team members.

In certain instances, the serious danger of excess pooled water may not be obvious. This is particularly true in commercial or industrial areas that contain materials that could absorb the water, such as fabric warehouses, cardboard and paper mills, or other finished products with absorbent characteristics. Two principal hazards exist in this situation. First, the added weight of the water absorbed into the product could overload the structure just as if it were pooled or standing water. In this case, if the hazard isn't recognized, it won't be considered by personnel. Second, if the bottom materials absorbing or wicking the water become waterlogged and unable to support the upper load, the stacked material may collapse onto fire brigade team members operating in the area.

Water can create potential hazards for fire brigade teams conducting salvage operations on floors below the fire. Water used to extinguish a fire, whether from a sprinkler system or handlines, can leak down through the building and can cause major damage. Water trapped in the void space above a ceiling can cause sections of the ceiling to collapse. Fire brigade teams should watch for warning signs such as sagging ceiling tiles and water leaking through ceiling joints. Water that accumulates in a basement or other pit could conceal a stairway and present additional hazards if utility connections are also located in the basement. Finally, water creates environments for mold and bacterial growth, and its rapid removal precludes the formation and growth of these contaminants.

Damage to the building's utility systems presents significant risks to fire brigade team members conducting salvage operations. Gas and electrical services should always be shut off to eliminate the potential hazards of electrocution or explosion stemming from leaking gas.

### Salvage Tools

Brigade members conducting salvage operations use special tools and equipment as listed in ▶ Table 19-1. Salvage covers and other protective items shield and cover building contents. Other salvage tools remove smoke, water, and other products that can damage property ▶ Figure 19-4.

## Fire Marks

For many years, salvage operations were conducted by insurance-funded fire patrols. These organizations worked with fire departments, preventing fire damage to valuable items. The last insurance services fire patrol was disbanded in New York City in 1996.

**Table 19-1 Salvage Tools**

- Salvage covers; treated canvas or plastic
- Box cutter for cutting plastic
- Floor runners
- Wet/dry vacuums
- Squeegees
- Submersible pumps and hose
- Sprinkler shut-off kit
- Ventilation fans, power blowers
- Small tool kit

**Figure 19-4** Salvage tools.

# Using Salvage Techniques to Prevent Water Damage

The best way to prevent water damage at a fire scene is to limit the amount of water used to fight the fire. Use only enough water to knock down the fire quickly. Do not use more water than is needed. Do not spray water blindly into smoke or continue to douse a fire that is already out; this only creates unnecessary water damage.

Turn off hose nozzles when they are not in use. If a nozzle must be left partly open to prevent freezing during cold weather, direct the flow through a window. Take time to tighten any leaky hose couplings after the fire has been knocked down. This prevents additional water damage during overhaul. Look for leaking water pipes in the building; if you find any, shut off the building's water supply.

## Master Streams and Property Conservation

The use of master streams and/or multiple hose streams pours thousands of gallons of water each minute into a structure and onto outside structures.

The rule of thumb regarding the weight of water is 8.5 pounds per gallon. A master stream device with a $1\frac{1}{2}$-inch tip is flowing 500 gallons per minute into the structure. Thus the total weight added to the structure and its contents by this flow is 4250 pounds per minute. Not surprisingly, this additional stress on the structure and its contents can cause a structural collapse.

On the outside of a facility, such an intense water flow can cause erosion of the structural components and piping that support the parts and pieces of the facility. This can also lead to collapse of the structure.

SOPs for the removal of this water, the removal of contents that may absorb this water, and consideration of the outside of the facilities that may suffer from erosion or water weight issues must be a part of the salvage operation.

## Controlling Extinguishing Agents

The methods used to control extinguishing agents depend on the type of agent being used. An agent that is expelled from a fire extinguisher, such as a dry chemical, will require cleanup of the agent from the area. The decontamination of the floor and items exposed to this agent should follow the recommendations found in the Material Safety Data Sheet for that extinguishing agent. Fire fighters should not use any more of this agent than is necessary to extinguish the fire.

The amount of water needed to extinguish a fire can be calculated based on the needed fire flow (in gallons per minute) relative to the BTUs that the burning material is expected to generate. A hose line flowing 95 gallons per minute, for example, may need to be operated for only 3 minutes to extinguish the fire.

The amount of agent needed to extinguish a fire can be specified as part of the prefire planning for a specific hazard. Appropriate control of these agents will make the loss-control function be easier, be less time-consuming, and do less property damage.

## Deactivating Sprinklers

Buildings equipped with sprinkler systems require special steps to limit water damage. Sprinklers can control a fire efficiently but will keep flowing until the activated sprinklers are shut off or the entire system is shut down. This can cause water damage.

Sprinklers should be shut down as soon as the IC declares that the fire is under control. Do not shut down sprinklers prematurely; the fire may rekindle, causing major damage. Between the time that sprinklers are shut down and overhaul is completed, a brigade member with a portable radio

# VOICES OF EXPERIENCE

**"Numerous openings existed in the wall allowing smoke, heat, and water to enter the sales floor area."**

One of the most basic skills taught to all fire brigade members is salvage operations. Although a basic skill, it can be the most important skill used on a fire ground to minimize the damages caused by smoke, water, and heat.

While commanding firefighting operations at a fully involved fire in a warehouse section of a military installation commissary (supermarket), the decision was made to stop the fire at the rated wall between the warehouse and the sales floor. Numerous openings existed in the wall allowing smoke, heat, and water to enter the sales floor area. Two crews were dispatched to the front of the store to begin salvage operations to protect as much of the inventory and equipment in the store as possible. As more personnel arrived on scene, they were assigned to the salvage team. Salvage operations continued well after the fire was extinguished.

Salvage operations were such a success that although the warehouse contents were a total loss, the store area was reopened three days later for a "fire sale." Seven days later, operating with a temporary receiving dock with "just in time logistics," the store was completely restocked and open to the community for full business.

Frederick J. Knipper
Kiaserslautern Military Community Fire Brigade
Kiaserslautern, Germany

Figure 19-5 Most fires are controlled by the activation of only one or two sprinkler heads.

should stand by and be ready to reactivate the sprinklers if necessary.

Sprinkler systems are usually designed so that only those sprinkler heads directly over or close to the fire will activate. Most fires can be contained or fully extinguished by the release of water from only one or two sprinkler heads ▲ Figure 19-5. Although movies may show all sprinkler heads activating at once, this does not occur in real life. More information on sprinkler systems is presented in Chapter 25, Fire Detection, Protection, and Suppression Systems.

If only one or two sprinkler heads have been activated, inserting a sprinkler wedge or a sprinkler stop can quickly stop the flow. This also keeps the rest of the system operational during overhaul.

A sprinkler wedge is a simple triangular piece of wood. Inserting two wedges on opposite sides, between the orifice and the deflector of the sprinkler, and pushing them together effectively plugs the opening, stopping the flow from standard upright and pendant sprinkler heads.

A sprinkler stop is a more sophisticated mechanical device with a rubber stopper that can be inserted into a sprinkler head. Several types of sprinkler stops are available, including some that work only with specific sprinkler heads. Not all sprinkler heads, however, can be shut off with a wedge or a sprinkler stop. Recessed sprinklers, which are often installed in buildings with finished ceilings, are usually difficult to shut off. If the individual heads cannot be shut off, stop the water flow by closing the sprinkler control valve.

To stop the flow of water from a sprinkler head using a sprinkler stop, follow the steps in ► Skill Drill 19-2.

1. Have a sprinkler stop in hand. **(Step 1)**
2. Place the flat-coated part of the sprinkler stop over the sprinkler head orifice and between the frame of the sprinkler head. **(Step 2)**
3. Push the lever to expand the sprinkler stop until it snaps into position. **(Step 3)**

To stop the flow of water from a sprinkler head using a pair of sprinkler wedges, follow the steps in ► Skill Drill 19-3.

1. Hold one wedge in each hand. **(Step 1)**
2. Insert the two wedges, one from each side, between the discharge orifice and the sprinkler head deflector. **(Step 2)**
3. Bump the wedges securely into place to stop the water flow. **(Step 3)**

If individual heads cannot be plugged or if several sprinklers have been activated, stop the flow by closing the sprinkler control valve. After the valve is closed, the water trapped in the piping system will drain out through the activated sprinkler heads for several minutes. If there is a drain valve near the control valve, open it to drain the system quickly. This also directs the water flow to a location where it will not cause additional damage.

The main control valve for a sprinkler system is usually an outside stem and yoke (OS&Y) valve or a postindicator valve (PIV). An OS&Y valve is typically found in a mechanical room in the basement or on the ground floor level of a building. PIVs are located outside the building or on an exterior wall. Some sprinkler systems also have zone valves that control the flow of water to sprinklers in different areas of the building. Multistory buildings usually have a zone valve for each floor. Closing a zone valve stops the flow to sprinklers in the zone; in the rest of the building, sprinklers remain operational. The locations of the main control valves and zone valves should be identified during preincident planning visits to a building.

Any sprinkler control valves that are not supervised must always be locked in the open position, usually with a chain and padlock. Supervised control valves are equipped with a tamper alarm that is tied into the fire alarm system, but they may be locked for added protection. Brigade members who are sent to shut off a sprinkler control valve should take a pair of bolt cutters to remove the lock if the key is not available.

To close and reopen an OS&Y valve, follow the steps in ► Skill Drill 19-4.

1. Locate the OS&Y valve as indicated on the preincident plan. It may be inside the building or close by outside. The stem of an OS&Y valve protrudes from the valve handle in the open position.
2. Identify the valve that controls sprinklers in the fire area. Smaller systems will have only one OS&Y valve. Large buildings may have multiple sprinkler systems or one main control valve and additional zone valves controlling the flow to different areas.
3. If the valve is locked in the open position with a chain and padlock, and the key is readily available, unlock and remove the chain. If no key is available, cut the lock or the chain with a pair of bolt cutters. Cut a link close to the padlock so that the chain can be reused. **(Step 1)**

19-2 Skill Drill

## Using Sprinkler Stops

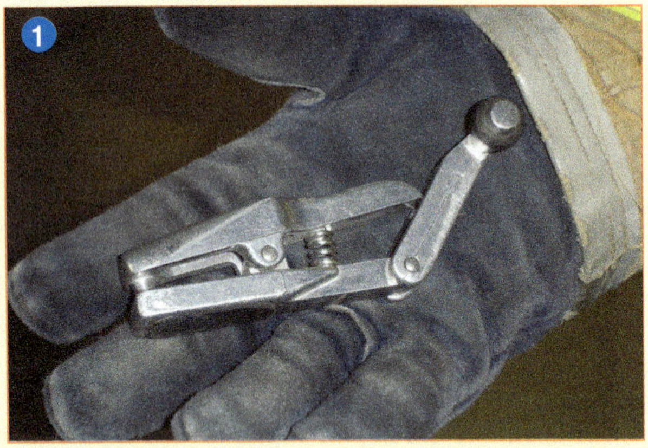

Have a sprinkler stop in hand.

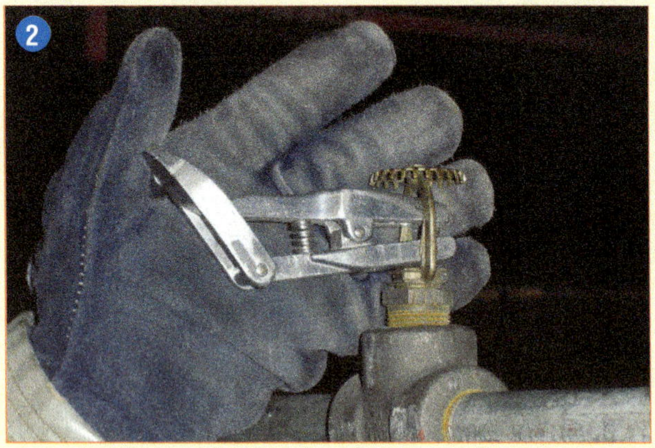

Place the flat-coated part of the sprinkler stop over the sprinkler head orifice and between the frame of the sprinkler head.

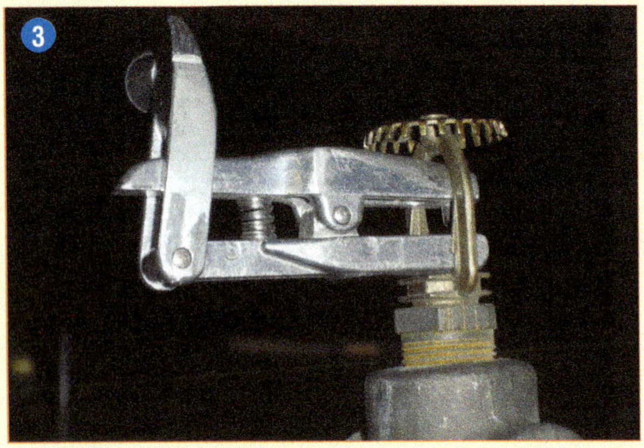

Push the lever to expand the sprinkler stop until it snaps into position.

4. Turn the valve handle clockwise to close the valve. Keep turning until resistance is strong and little of the stem is visible. **(Step 2)**
5. To open the OS&Y valve, turn the handle counterclockwise until resistance is strong and the stem is visible again. Lock the valve in the open position. **(Step 3)**

To close and open a post indicator valve (PIV), follow the steps in **▶ Skill Drill 19-5**.

1. Locate the PIV (usually located outside the building). The location of the valve should be indicated on a preincident plan.
2. Most PIVs will be locked in the open position by a padlock. Unlock the padlock if the key is readily available. If no key is available, cut the lock with a pair of bolt cutters. **(Step 1)**
3. Remove the handle from its storage position on the PIV and place it on top of the valve, similar to a hydrant wrench.
4. Turn the valve stem in the direction indicated on top of the valve to close the valve. Keep turning until resistance is strong and the visual indicator changes from "OPEN" to "SHUT." **(Step 2)**
5. To reopen the PIV, turn in the opposite direction until resistance is strong and the indicator changes back to "OPEN." The valve should then be locked in the open position. **(Step 3)**

19-3 Skill Drill

## Using Sprinkler Wedges

1 Hold one wedge in each hand.

2 Insert the two wedges, one from each side, between the discharge orifice and the sprinkler head deflector.

3 Bump the wedges securely into place to stop the water flow.

Before a sprinkler system can be restored to normal operation, the sprinkler heads that have been activated must be replaced. Every sprinkler system, when installed, has spare heads stored somewhere, usually near the main control valves. Some brigades may carry a selection of different heads in a sprinkler kit. An activated sprinkler head must be replaced with another head of the same design, size, and temperature rating.

The main sprinkler control valve or the appropriate zone valve must be closed, and the system must be drained before a sprinkler head can be changed. Special wrenches must be used when replacing sprinkler heads to prevent damage to the operating mechanism.

After the activated heads are replaced, the sprinkler system can be restored to service. Restoring a sprinkler system to service takes special training and must be performed only by qualified individuals.

### Removing Water

Water that accumulates within a building or drips down from higher levels should be channeled to a drain or to the outside of the building to prevent or limit water damage. A pump may be needed to help remove the water in some cases.

Some buildings have floor drains that will funnel water into the belowground sewer system. Floor drains should be kept free from debris so that the water can drain freely.

19-4 Skill Drill

## Close and Reopen Main Control Valve (OS&Y)

Locate the OS&Y valve as indicated on the preincident plan. It may be inside the building or close by outside. The stem of an OS&Y valve protrudes from the valve handle in the open position. Identify the valve that controls sprinklers in the fire area. Smaller systems will have only one OS&Y valve. Large buildings may have multiple sprinkler systems or one main control valve and additional zone valves controlling the flow to different areas. If the valve is locked in the open position with a chain and padlock, and the key is readily available, unlock and remove the chain. If no key is available, cut the lock or the chain with a pair of bolt cutters. Cut a link close to the padlock so that the chain can be reused.

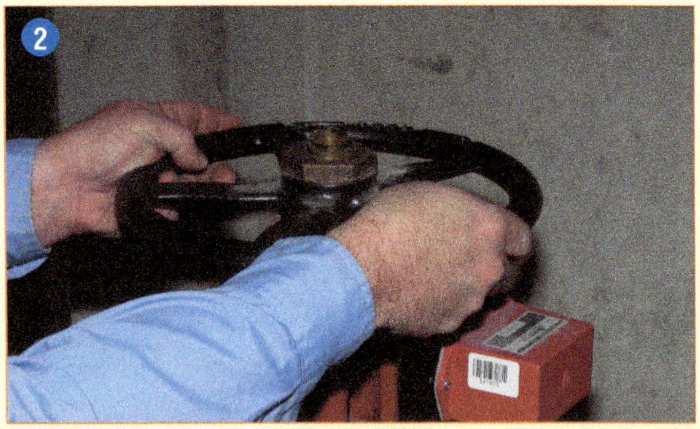

Turn the valve handle clockwise to close the valve. Keep turning until resistance is strong and little of the stem is visible.

To open the OS&Y valve, turn the handle counterclockwise until resistance is strong and the stem is visible again. Lock the valve in the open position.

Brigade members can use squeegees to direct the water into the drain.

In buildings without floor drains, it may be possible to shut off the water supply to a toilet, remove the nuts that hold the toilet bowl to the floor, and remove the toilet from the floor flange. This creates a large drain capable of handling large quantities of water, as long as someone keeps the opening from becoming clogged with debris.

Water on a floor at grade level can often be channeled to flow outside through a doorway or other opening. Water on a floor aboveground level can sometimes be drained to the outside by making an opening at floor level in an exterior wall of the building. This type of opening is called a scupper.

Water chutes or water catch-alls are often used to collect water leaking down from firefighting operations on higher floor levels and to help protect property on calls involving burst pipes or leaking roofs. Water chutes and catch-alls can be constructed quickly using salvage covers.

A water chute catches dripping water and directs it toward a drain or to the outside through a window or doorway. A water catch-all is a temporary pond that holds dripping water in one location. The accumulated water must then be drained to the outside of the building.

To construct a water chute, follow the steps in ► Skill Drill 19-6. A water chute can be constructed with a single salvage cover, or with a cover and two pike poles for support.

19-5 Skill Drill

## Close and Open Main Control Valve (PIV)

Locate the PIV. The PIV is usually located outside the building as indicated on a preincident plan. Most PIVs will be locked in the open position with a padlock. If the key is readily available, unlock the padlock. If no key is available, cut the lock with a pair of bolt cutters.

Remove the handle from its storage position on the PIV and place it on top of the valve. Turn the valve stem in the direction indicated on top of the valve to close the valve. Keep turning until resistance is strong and the visual indicator changes from "OPEN" to "SHUT."

To reopen the PIV, turn in the opposite direction until resistance is strong and the indicator changes back to "OPEN." The valve should then be locked in the open position.

1. Fully open a large salvage cover flat on the ground. **(Step 1)**
2. Roll the cover tightly from one edge toward the middle. If using pike poles, lay one pole on the edge and roll the cover around the handle. Roll the opposite edge tightly toward the middle in the same manner. Stop when the rolls are 1' to 3' apart. **(Step 2)**
3. Turn the cover upside down. Position the chute so it collects the dripping water and channels it toward a drain or outside opening. The chute can be placed on the floor, with one end propped-up by a chair or other object.
4. Use a stepladder or other tall object to support chutes constructed with pike poles. **(Step 3)**

To construct a water catch-all, follow the steps in ► **Skill Drill 19-7**.

1. Fully open a large salvage cover flat on the ground. **(Step 1)**
2. Roll two edges inward from the opposite sides, approximately 3' on each side. **(Step 2)**
3. Fold each of the four corners over at a 90° angle, starting each fold approximately 3' in from the edge. **(Step 3)**
4. Roll the remaining two edges inward approximately 2'. **(Step 4)**
5. Lift the rolled edge over the corner flaps, and tuck it in under the flaps to lock the corners in place. **(Step 5)**

19-6 Skill Drill

## Construct a Water Chute

1

Fully open a large salvage cover flat on the ground.

2

Roll the cover tightly from one edge toward the middle. If using pike poles, lay one pole on the edge and roll the cover around the handle. Roll the opposite edge tightly toward the middle until the two rolls are 1–3 feet apart.

3

Turn the cover upside down. Position the chute so it collects the dripping water and channels it toward a drain or outside opening. The chute can be placed on the floor, with one end propped-up by a chair or other object. Use a stepladder or other tall object to support chutes constructed with pike poles.

### Water Vacuum

Special vacuum cleaners that suck up water also can be used during salvage operations. There are two types of water vacuums: a small capacity backpack-type and a larger wheeled unit. The backpack vacuum cannot be used by someone who is wearing SCBA. Additionally, wet/dry shop vacs may also be used as a low-cost alternative.

### Drainage Pumps

Drainage pumps remove water that has accumulated in basements or belowground level. Portable electric submersible pumps can be lowered directly into the water to pump it out of a building. Gasoline-powered portable pumps are placed outside because they exhaust poisonous carbon monoxide gas. These pumps use a hard-suction hose lowered into the basement through a window to draft water out of the building.

#### Brigade Member Safety Tips

When a gasoline-powered pump is used inside a structure it may generate poisonous carbon monoxide gas. Subsequently, ventilation, air monitoring, or the use of SCBA may be required.

Skill Drill 19-7

# Construct a Water Catch-All

Fully open a large salvage cover flat on the ground.

Roll two edges inward from the opposite sides, approximately 3 feet on each side.

Fold each of the four corners at a 90° angle, starting each fold approximately 3 feet in from the edge.

Roll the remaining two edges inward approximately 2 feet.

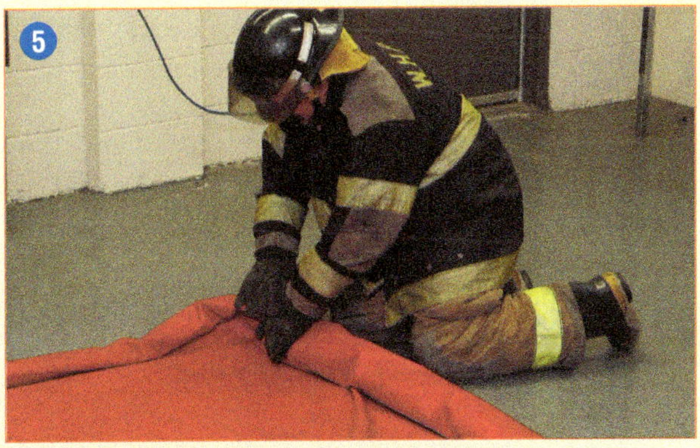

Lift the rolled edge over the corner flaps, and tuck it in under the flaps, to lock the corners in place.

## Using Salvage Techniques to Limit Smoke and Heat Damage

The visible components of smoke are mostly soot particles and other products of combustion, including corrosive chemicals that will settle on any horizontal surface when the smoke cools. Effectively channeling the smoke out of the building is a better option than allowing contaminants to settle on furniture, contents, or equipment. All members of the fire suppression team, not just the salvage crew, should recognize opportunities to prevent smoke and heat damage. One of the most effective methods to reduce the property loss due to smoke and heat damage is to close doors to unaffected areas as soon as possible. Fire brigade team members should remember to close doors of unaffected areas once they have been searched. Other effective methods of removing smoke, gases, and heat are natural ventilation and mechanical ventilation. These methods of ventilation and others are discussed in detail in Chapter 14, Ventilation.

### Salvage Covers

The most common method of protecting building contents is to cover them with salvage covers. Salvage covers are large square or rectangular sheets of heavy canvas or plastic material used to protect items from water runoff, falling debris, soot, and particulate matter in smoke residue.

Salvage crews usually begin on the floor immediately below the fire to prevent water damage to contents on lower floors. If the fire is in an attic, brigade members may have enough time to spread covers over the items in the rooms directly below the fire before pulling the ceilings to attack the flames.

The most efficient way to protect a room's contents is to move all the items to the center of the room, away from the walls. This reduces the total area that must be covered, enabling one or two brigade members to cover the pile quickly and move on to the next room (▶ Figure 19-6).

Some brigades use rolls of construction-grade polyethylene film instead of salvage covers. This material comes in rolls as long as 120 feet so it can be unrolled over the building contents and cut to the correct length with a box cutter. It is useful for covering long surfaces such as retail display counters. This material is disposable and can be left behind after the fire, unlike traditional salvage covers, which must be picked up, washed, dried, and properly folded for storage. Occasionally, salvage covers may be left behind, as protection for the building contents, but should be picked up when they are no longer needed.

Special folding and rolling techniques are used to store salvage covers so that they can be deployed quickly by one or two brigade members. Follow your organization's SOPs when folding or rolling a salvage cover.

To fold a salvage cover to prepare it for one person to deploy, follow the steps in (▶ Skill Drill 19-8).

**Figure 19-6** Move furniture to the center of the room so that a single salvage cover can protect all of the contents.

1. Spread the salvage cover flat on the ground. Stand at one end, facing your partner standing at the other end. **(Step 1)**
2. You and your partner each place one hand on the outer edge of the cover and the other hand one quarter of the way in from the edge. **(Step 2)**
3. Together, flip the outside edge in 3 inches from the middle of the cover, creating a fold at the quarter point. **(Step 3)**
4. Flip the outside fold in to the same point of the cover, creating a second fold. Repeat steps 2, 3, and 4, from the opposite side of the cover. The folded edges should meet at the middle of the cover, with the folds 6-inch apart. **(Step 4)**
5. Fold the two halves of the salvage cover together. **(Step 5)**
6. Starting from the middle of the cover, use a broom to brush the air out of the cover. **(Step 6)**
7. Move to the newly created narrow end of the salvage cover. **(Step 7)**
8. Fold the narrow end of the salvage cover 3 inches from the middle of the cover, creating a fold at the quarter point.
9. Flip the outside fold of the narrow end in to the same point of the cover, creating a second fold. The folded edge should meet at the middle of the cover, with the folds 6-inch apart. **(Step 8)**
10. Fold the two halves of the salvage cover together. **(Step 9)**

To fold a salvage cover to prepare it for two brigade members to deploy, follow the steps in (▶ Skill Drill 19-9).

1. Spread the salvage cover flat on the ground. Stand at one end, facing your partner standing at the other end.
2. You and your partner each place one hand on the outer edge of the cover and the other hand one quarter of the way in from the edge. **(Step 1)**
3. Together, fold the cover in half. **(Step 2)**

19-8 Skill Drill

# One-Person Salvage Cover Fold

Spread the salvage cover flat on the ground. Stand at one end, facing your partner standing at the other end.

You and your partner each place one hand on the outer edge of the cover and the other hand one quarter of the way in from the edge.

Together, flip the outside edge in 3 inches from the middle of the cover, creating a fold at the quarter point.

Flip the outside fold in to the same point of the cover, creating a second fold. Repeat steps 2, 3, and 4, from the opposite side of the cover. The folded edges should meet at the middle of the cover, with the folds 6-inch apart.

Fold the two halves of the salvage cover together.

Starting from the middle of the cover, use a broom to brush the air out of the cover.

*continued*

19-8 Skill Drill (Continued)

## One-Person Salvage Cover Fold—continued

Move to the newly created narrow end of the salvage cover.

Fold the narrow end of the salvage cover 3 inches from the middle of the cover, creating a fold at the quarter point. Flip the outside fold of the narrow end in to the same point of the cover, creating a second fold. The folded edge should meet at the middle of the cover, with the folds 6-inch apart.

Fold the two halves of the salvage cover together.

4. Together, grasp the unfolded edge and fold the cover in half again. **(Step 3)**
5. Starting from the middle of the cover, use a broom to brush the air out of the cover. **(Step 4)**
6. Move to the newly created narrow end of the salvage cover.
7. Fold the salvage cover in half length-wise. **(Step 5)**
8. Fold the salvage cover in half length-wise again. Make certain that the open end is on top.
9. Fold the cover in half a third and final time. **(Step 6).**

To fold and roll a salvage cover, follow the steps in ▶ **Skill Drill 19-10**.

1. Spread the salvage cover flat on the ground. Stand at one end, facing your partner standing at the other end. **(Step 1)**
2. You and your partner each place one hand on the outer edge of the cover and the other hand one quarter of the way in from the edge. **(Step 2)**
3. Together, flip the outside edge in to the middle of the cover, creating a fold at the quarter point. **(Step 3)**
4. Flip the outside fold in to the middle of the cover, creating a second fold. Repeat steps 2, 3, and 4, from the opposite side of the cover. **(Step 4)**

19-9 Skill Drill

## Two-Person Salvage Cover Fold

Spread the salvage cover flat on the ground. Stand at one end, facing your partner standing at the other end. You and your partner each place one hand on the outer edge of the cover and the other hand one quarter of the way in from the edge.

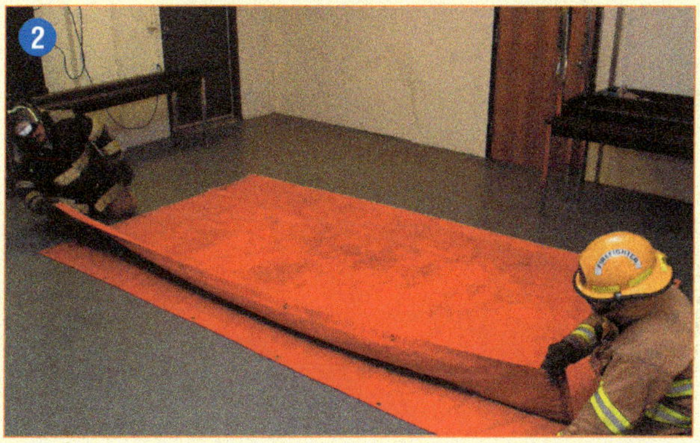


Together, fold the cover in half.

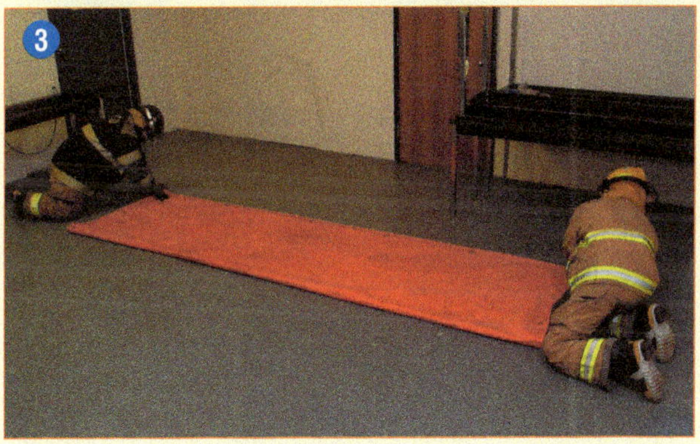

Together, grasp the unfolded edge and fold the cover in half again.

Starting from the middle of the cover, use a broom to brush the air out of the cover.

Move to the newly created narrow end of the salvage cover. Fold the salvage cover in half length-wise.

Fold the salvage cover in half length-wise again. Make certain that the open end is on top. Fold the cover in half a third and final time.

19-10 Skill Drill

# Fold and Roll a Salvage Cover

Spread the salvage cover flat on the ground. Stand at one corner, facing your partner standing at the other end.

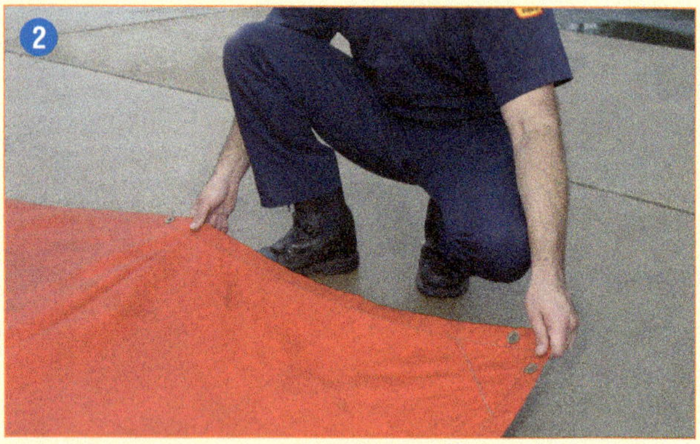

You and your partner each place one hand on the outer edge of the cover and the other hand one quarter of the way in from the edge.

Together, flip the outside edge in to the middle of the cover, creating a fold at the quarter point.

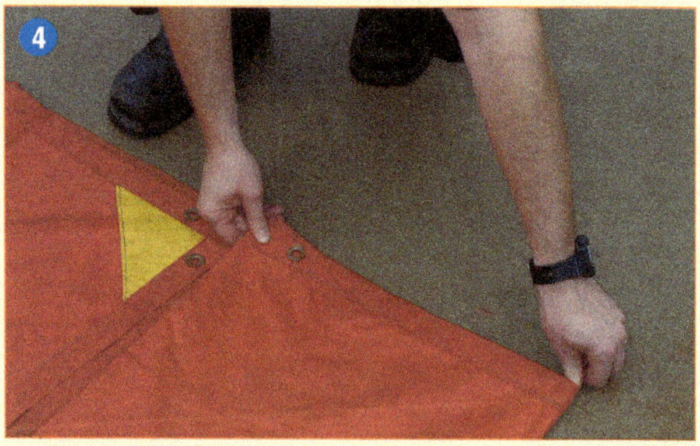

Flip the outside fold in to the middle of the cover, creating a second fold. Repeat steps 2, 3, and 4 from the opposite side of the cover.

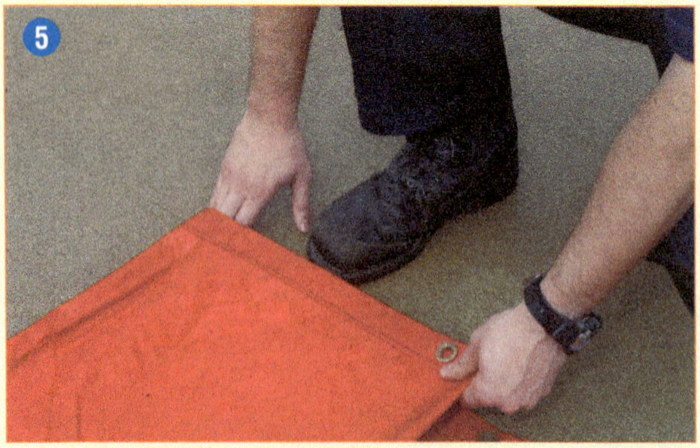

The folded edges should meet at the middle of the cover, with the folds touching, but not overlapping.

Tightly roll up the folded salvage cover from the end.

5. The folded edges should meet at the middle of the cover, with the folds touching, but not overlapping. **(Step 5)**
6. Tightly roll up the folded salvage cover from the end. **(Step 6)**

Although it takes two people to fold and roll a salvage cover for storage, this technique enables one person to unroll the cover easily and quickly. To perform the one-person salvage cover roll, follow the steps in **▶ Skill Drill 19-11**.

1. Stand in front of the end of the object that you are going to cover. **(Step 1)**
2. Stand to unroll the cover over one end of the object. **(Step 2)**
3. Continue unrolling the cover over the top of the object. Allow the remainder of the roll to settle at the other side of the object. **(Step 3)**
4. Spread the cover, unfolding each side outward over the object to the first fold. **(Step 4)**
5. Unfold the second fold on each side to drape the cover completely over the object. **(Step 5)**
6. Tuck in all loose edges of the cover around the object. **(Step 6)**

A single brigade member can also use a shoulder toss to spread a salvage cover. Follow the steps in **▶ Skill Drill 19-12** to do a shoulder toss.

1. Place a folded salvage cover over one arm. **(Step 1)**
2. Toss the cover over the salvaged object with a straight-arm movement. **(Step 2)**
3. Unfold the cover until it completely drapes the object. **(Step 3)**

Two brigade members can use a balloon toss to cover a pile of building contents quickly. The balloon toss is shown in **▶ Skill Drill 19-13**.

1. Place the cover on the ground beside the object. **(Step 1)**
2. Unfold the cover so that it runs along the entire base of the object. Each person grabs one edge of the cover and brings it up to waist height. **(Step 2)**
3. Together, lift the cover quickly so that it fills with air like a balloon. **(Step 3)**
4. Move quickly to the other side of the item, using the air to help support the cover, and spread the entire cover over the object. **(Step 4)**

### Floor Runners

A floor runner is a long section of protective material used to cover a section of flooring or carpet. Floor runners protect floor coverings and finishes from water, debris, brigade members' boots, and firefighting equipment. Brigade members entering an area for salvage operations should unroll the floor runner ahead of themselves and stay on the floor runner while working in the area.

### Other Salvage Operations

The best way to protect the contents of a building may be to move them to a safe location. The IC will make this determination. Any items removed from the building should be placed in a dry, secure area, preferably a single location. In some cases, salvaged items can be moved to a suitable location within the same building. If items are moved outside, they should be protected from further potential damage caused by firefighting operations or the weather.

There are times when the building contents can provide clues to the cause or spread of the fire. In these situations, fire investigators should be consulted and supervise the removal process.

Salvage operations sometimes extend outside the building to include valuable property such as vehicles or machinery. Use a salvage cover to protect outside property or move vehicles if this can be done without compromising the fire suppression efforts.

## Overhaul Overview

Overhaul is the process of searching for and extinguishing any pockets of fire that remain after a fire has been brought under control. The risk of a new fire remains even if 99% of a fire is out and just 1% is smoldering. A single pocket of embers can rekindle after brigade members leave the scene and cause even more damage and destruction than the original fire. A fire cannot be considered fully extinguished until the overhaul process is complete.

The process of overhaul begins after the fire is brought under control. Overhaul can be a time-consuming, physically demanding process. The greatest challenge during overhaul is to identify and open any void spaces in a building where the fire might be burning undetected. If the fire extends into any void spaces, brigade members must open the walls and ceilings to expose the burned area. Any materials that are still burning must be soaked with water or physically removed from the building. This process must continue until all of the burned material is located and unburned areas are exposed. Overhaul is also required for non-structure fires such as automobiles, bulk piles (tires or woodchips), vegetation, and even garbage.

### Safety Considerations During Overhaul

Many injuries can occur during overhaul. Overhaul is strenuous work conducted in an area already damaged by fire. Brigade members who were involved in the fire suppression efforts may be physically fatigued and overlook hazards. Many organizations summon fresh crews to conduct overhaul, giving the crews that fought the fire time to recover.

Several hazards may be present in the overhaul area. The structural safety of the building could be compromised. Heavy objects could collapse roofs or ceilings, debris could litter the area, and there could be holes in the floor. Visibility is often limited, so brigade members may have to depend on portable lighting. Wet or icy surfaces make falls more likely. Smoldering areas may burst into flames, and

19-11 **Skill Drill**

# One-Person Salvage Cover Roll

Stand in front of the end of the object that you are going to cover.

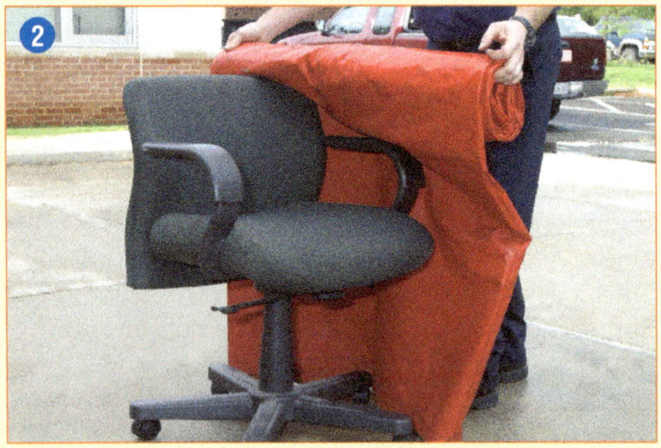


Start to unroll the cover over one end of the object.

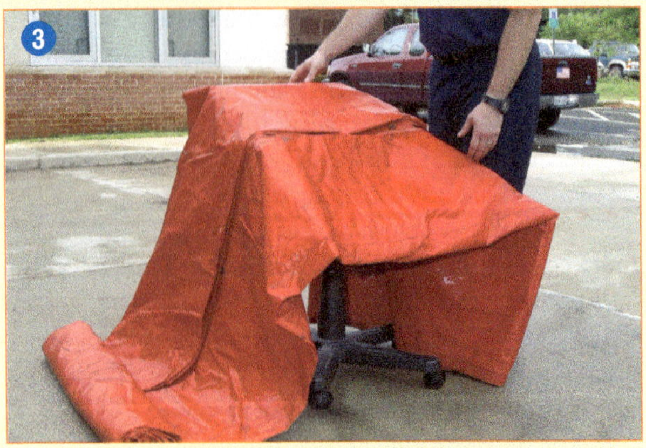


Continue unrolling the cover until you reach the top of the object. Allow the remainder of the cover to unroll and settle at the end of the object.

Spread the cover, unfolding each side outward over the object to the first fold.

Unfold the second fold on each side and drape the cover completely over the object.

Tuck in all loose edges of the cover around the object.

19-12 Skill Drill

## Shoulder Toss

Place a folded salvage cover over one arm.

Toss the cover over the salvaged object with a straight-arm movement.

Unfold the cover until it completely drapes the object.

19-13 Skill Drill

## Balloon Toss

1 Place the cover on the ground beside the object.

2 Unfold the cover so that it runs along the entire base of the object. Each person grabs one edge of the cover and brings it up to waist height.

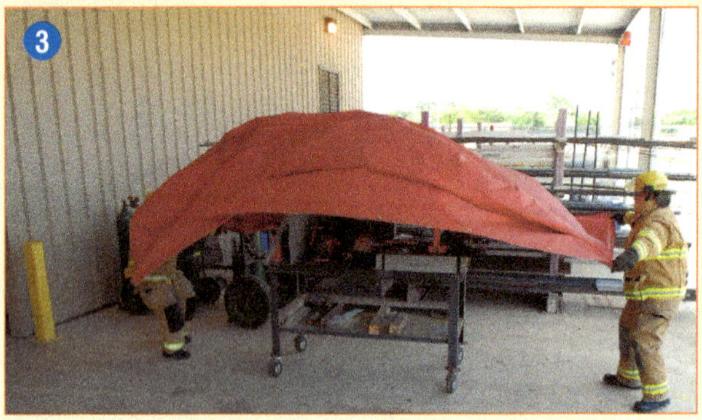

3 Together, lift the cover quickly so that it fills with air like a balloon.

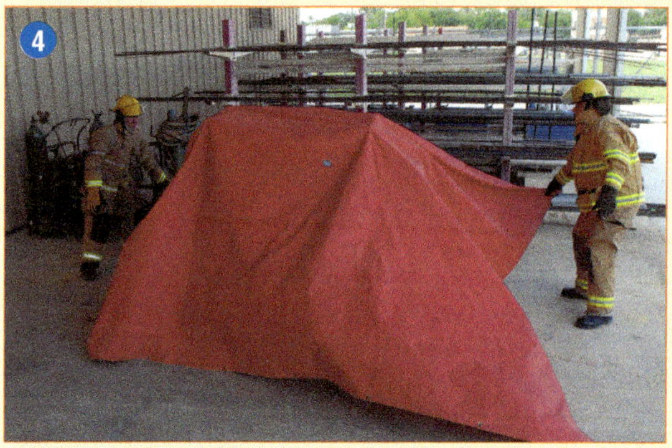

4 Move quickly to the other side of the item, using the air to help support the cover, and spread the entire cover over the object.

the air is probably not safe to breath. Dangerous equipment, including axes, pike poles, and power tools, are used in close quarters.

Brigade members must be aware of these hazards and proceed with caution. When necessary, take extra time to evaluate the hazards and determine the safest way to proceed. There is no need to rush during overhaul, and there is no excuse for risking injury during this phase of operations.

A charged hose line must always be ready for use during overhaul operations because flare-ups and explosions may occur. A smoldering fire in a void space could flare-up suddenly when brigade members open the space and allow fresh air to enter. Such a fire could quickly become intense and dangerous. Finely grained combustible materials such as sawdust can smolder for a long time and ignite explosively when they are disturbed and oxygenated.

Brigade members must wear PPE typical to their organization during overhaul. SCBA and thermal protective clothing may be necessary for interior structural fire brigades. Fresh crews or crews that have been properly rehabbed can replace fatigued crews.

A safety officer should always be present during overhaul operations to note any hazards and ensure that operations are conducted safely. Brigade leaders should supervise operations, look for hazards, and make sure that all crew members work carefully. The work should proceed at a cautious pace. Too many people working in a small area creates chaos.

Always evaluate the structural condition of a building before beginning overhaul. Look for the following indicators of possible structural collapse:

- Lightweight and/or truss construction
- Cracked walls, walls out of alignment, sagging floors
- Heavy mechanical equipment on the roof
- Overhanging cornices or heavy signs
- Accumulations of water

Do not compromise the structural integrity of the building. When opening walls and ceilings, remove only the outer coverings and leave the structural members in place. If more invasive overhaul is required, be careful not to compromise the structure's load-bearing members. Avoid cutting lightweight wood tresses, load-bearing wall studs, and floor or ceiling joists, especially when using power tools.

If fire damages the structural integrity of a building, the IC can call for a "hydraulic overhaul" rather than a standard overhaul. In this situation, large-caliber hose streams are used to completely extinguish a fire from the exterior. This strategy is appropriate if there are excessive risks to brigade members and such extensive damage to the property that it has no salvage value. Heavy mechanical equipment may be used to demolish unsafe buildings and expose any remaining hot spots.

If a complete overhaul cannot be conducted, the IC can establish a fire watch. The fire watch team remains at the fire scene and watches for signs of rekindling. The team can request additional help if needed.

## Coordinating Overhaul with Fire Investigators

Overhaul crews must work with fire investigators to ensure that important evidence is not lost or destroyed. Ideally, a fire investigator should examine the area before overhaul operations begin, identifying evidence and photographing the scene before it is disturbed.

If an investigator is not immediately available, the overhaul crews should make careful observations and report to the investigator later. When performing overhaul in or near the suspected area of fire origin, note burn patterns on the walls or ceilings that could indicate the exact site of origin or the path of fire travel. When moving electrical items, note whether they were plugged in or turned on. Always look for evidence that the fire investigator can use to determine the cause of the fire.

If you observe anything suspicious, particularly indications of arson, delay overhaul until an investigator can examine the scene. Ensure that the fire will not rekindle, but do not go any further until the investigator arrives.

Chapter 26, Fire Cause Determination has additional information on evidence preservation.

## Where to Overhaul

Determining when, where, and how much property needs to be overhauled requires good judgment. Overhaul must be thorough and extensive enough to ensure that the fire is completely out. At the same time, brigade members should try not to destroy any more property than necessary.

### Brigade Member Safety Tips

A charged hose line must always be available during overhaul.

### Brigade Member Safety Tips

Before overhaul operations begin, the safety officer should confirm that electric power and gas service in the overhaul area are shut off. Beware—even though the gas supply may be turned off, there could still be gas under pressure in the lines that must be bled off.

Generally, it is better to make sure that the fire is definitely out than to be too careful about damaging property.

The area that must be overhauled depends on the building's construction, its contents, and the size of the fire. All areas directly involved in the fire must be overhauled. If the fire was confined to a single room, all of the contents in that room must be checked for smoldering fire. The overhaul process also must ensure that the fire did not extend into any void spaces in the walls, above the ceiling, or into the floor. If there is any indication that the fire spread into the structure itself or into the void spaces, all suspect areas must be opened to expose any hidden fire.

If the fire involved more than one room, overhaul must include all possible paths of fire extension. The paths available for a fire to spread within a building are directly related to the type of construction. Learn to anticipate where and how a fire is likely to spread in different types of buildings to find any hidden pockets of fire.

Fire-resistive construction can help contain a fire and keep it from spreading within a building, although this is not guaranteed. Look for any openings that would allow the fire to spread, including utility shafts, pipe chases, and fire doors or dampers that failed to close tightly.

Wood-frame and ordinary-construction buildings may have several areas where a hidden fire could be burning. These structures often have void spaces under the wall covering materials. When a serious fire strikes a building of ordinary or wood-frame construction, brigade members may need to open every wall, ceiling, and void space to check for fire extension. Neighboring buildings may also need to be overhauled, if the fire could have spread into them.

If a building has been extensively remodeled, overhaul presents special challenges. Fire can hide in the space between a dropped ceiling and the original ceiling or extend into a different section of the building through doors and windows covered by new construction. Some buildings have two roofs, one original and one added later, with a void space between them. A fire in this void space presents very difficult overhaul problems.

The cause of the fire also can indicate the extent of necessary overhaul. A break room stove fire will probably extend into the stove exhaust duct and could ignite combustible materials in the immediate vicinity. Follow the path of the duct and open around it to locate any residual fire. A lightning strike releases enormous energy through the wiring and piping systems, which can start multiple fires in different parts of the building. Overhaul after this incident must be quite extensive.

### Using Your Senses

Efficient and effective overhaul requires the use of all your senses. Look, listen, and feel to detect signs of potential burning.

Look for:

- Smoke seeping from cracks or from around doors and windows
- Fresh or new smoke
- Red, glowing embers in dark areas
- Burnt areas
- Discolored material
- Peeling paint or cracked plaster

Listen for:

- Crackling sounds that indicate active fire
- Hissing sounds that indicate water has touched hot objects

Feel for:

- Heat, using the back of the hand (only if it is safe to remove your glove)

Experience is a valuable trait during overhaul situations. By using the senses in a systematic manner, an experienced brigade member can often determine which parts of the structure need to be opened and which areas remain untouched by the fire.

### Thermal Imager

The thermal image device is a valuable, high-tech tool used during overhaul. The same type of thermal imager used for search and rescue can also be used to locate hidden hot spots or residual pockets of fire. It can quickly differentiate between unaffected areas and areas that need to be opened. Using a thermal imager can decrease the time needed to overhaul a fire scene and reduce the amount of physical damage to the building.

The thermal imager distinguishes between objects or areas with different temperatures and displays hot areas and cold areas as different colors on the video screen. Because it is very sensitive, the thermal imager can "see" a hot spot, even if the heat source is behind a wall. The imager also can show the pattern of fire within a wall, even if there is only a few degrees' difference in the wall's surface temperature.

Figure 19-7 Thermal imaging devices can help locate hot spots.

Interpreting the readings from a thermal image device requires practice and training (▲ Figure 19-7). The imager displays relative temperature differences, so an object will appear to be warmer or cooler than its surroundings. If the room has been superheated by fire, all of the contents, including the walls, floor, and ceiling, will appear hot. An extra-hot spot behind a wall could be difficult to distinguish. It might be better to look at the wall from the side that was not exposed to the fire to identify any hot spots.

The thermal imager, however, is only a tool; it cannot completely rule out the possibility of concealed fire. The only way to ensure that no hidden fires exist is to open a ceiling or wall and do a direct inspection.

## Overhaul Techniques

The objective of overhaul is to find and extinguish any fires that could still be burning after a fire is brought under control. Any fire uncovered during overhaul must be thoroughly extinguished. Overhaul operations should continue until the IC is satisfied that no smoldering fires are left.

During overhaul operations, a charged hose line should be available to extinguish any sudden flare-ups or exposed pockets of fire. If necessary, use a direct stream from the hose line but avoid unnecessary water damage. Extinguish small pockets of fire or smoldering materials with the least possible amount of water by using a short burst from the hose line or simply drizzling water from the nozzle directly onto the fire.

Extinguish smoldering objects that can be safely picked up by dropping them into a bucket filled with water. Remove materials prone to smoldering, such as cushioned furniture, from the building and thoroughly soak them outside. Densely packed materials such as newspapers, fabric, and the like must be opened and thoroughly separated to ensure there are no deep-seated hot spots.

Place debris that is moved outside far enough away from the building to prevent any additional damage if it reignites. Do not allow debris to block entrances or exits. In some cases, a window opening can be enlarged so that debris can be removed easily. Heavy machinery, such as a front-end loader, may be used to remove large quantities of debris.

## Overhaul Tools

Tools used during overhaul are designed for cutting, prying, and pulling so brigade members can access spaces that might contain hidden fires. Many of the tools used for overhaul are also used for ventilation and forcible entry. The tools used for overhaul operations include:

- Pike poles and ceiling hooks for pulling ceilings and removing gypsum wallboard
- Crowbars and Halligan-type tools for removing baseboards and window or door casings
- Axes for chopping through wood, such as floor boards and roofing materials
- Power tools such as battery-powered saws for opening up walls and ceilings
- Pitchforks and shovels for removing debris
- Rubbish hooks and rakes for pulling things apart

Because overhaul situations usually do not require high pressures or large volumes of water, a 1½- or 1¾-inch hose line is usually sufficient to extinguish hot spots. After the fire has been extinguished by using the proper-size hose line, water may still be needed during overhaul because of hot spots in the walls, smoldering materials, or rekindling of a small fire. Some brigades use garden hose lines, pressurized water extinguishers, and even spray bottles as methods of extinguishment in an effort to reduce further water damage. Follow your brigade's SOPs.

Buckets, tubs, wheelbarrows, and carryalls (rubbish carriers) are used to remove debris from a building (▶ Figure 19-8). A carryall is a 6-square feet piece of heavy canvas material with rope or flat webbing handles.

## Opening Walls and Ceilings

Pike poles are used to open ceilings and walls to expose hidden fire. To pull down a ceiling with a pike pole, follow the steps in (▶ Skill Drill 19-14).

### Brigade Member Tips

Overhaul is physically demanding work. Pace yourself and use proper technique to increase your efficiency. Ask experienced members of your team for tips on improving your technique.

Figure 19-8 A carryall is used to remove debris during overhaul.

1. Select the appropriate length pike pole based on the height of the ceiling. For most office area applications, a 6-foot pole is sufficient; longer poles are needed for higher ceilings. **(Step 1)**
2. Determine what area of the ceiling will be opened. In most cases, the officer in charge makes this determination. Typically, the most heavily damaged areas will be opened first, followed by the surrounding areas. **(Step 2)**
3. Position yourself to begin work with your back toward a door, so the debris you pull down will not block your access to the exit. **(Step 3)**
4. Using a strong, upward-thrusting motion, penetrate the ceiling with the tip of the pike pole. Face the hook side of the tip away from you. **(Step 4)**
5. Pull down and away from your body, so the ceiling material falls away from you. **(Step 5)**
6. Continue pulling down sections of the ceiling until the desired area is open. Pull down any insulation, such as rolled fiberglass, found in the ceiling. **(Step 6)**

Use the pike pole to break through and pull down large sections of ceilings made with gypsum board. Pulling down laths and breaking through plaster ceilings requires more force. Power saws may be required to cut through ceilings made with plywood or solid boards.

Pike poles, axes, power saws, and handsaws can be used to open a hole in a wall. Use the same technique to open a wall with a pike pole as you do to open a ceiling. When using an axe, make vertical cuts with the blade of the axe, then pull the wallboard away from the studs by hand or with the pick end of the axe. A power saw also can be used to make vertical cuts. Pull the wall section away with another tool or by hand.

To open an interior wall with a pick-head axe, follow the steps in (▶ Skill Drill 19-15).

19-14 **Skill Drill**

# Pull a Ceiling Using a Pike Pole

Select the appropriate length pike pole based on the height of the ceiling. For most office area applications, a 6-foot pole is sufficient; longer poles are needed for higher ceilings.

Determine what area of the ceiling will be opened. Typically, the most heavily damaged areas are opened first, followed by the surrounding areas.

Position yourself to begin work with your back toward a door, so the debris you pull down will not block your access to the exit.

Using a strong, upward-thrusting motion, penetrate the ceiling with the tip of the pike pole. Face the hook side of the tip away from you.

Pull down and away from your body, so the ceiling material falls away from you.

Continue pulling down sections of the ceiling until the desired area is opened. Pull down any insulation, such as rolled fiberglass, found in the ceiling.

NOTE: Use the pike pole to break through and pull down large sections of ceilings made with gypsum board. Pulling down laths and breaking through plaster ceilings will require more force. Power saws may be required to cut through ceilings made of plywood or solid boards.

Skill Drill 19-15

## Open an Interior Wall Using a Pick-Head Axe

Determine what area of the wall will be opened up. The brigade leader usually makes this determination. Typically, the areas most heavily damaged by the fire are opened first, followed by the surrounding areas, working outward.

Use the axe blade to begin cutting near the top of the wall. Cut downward between wall studs. Survey the wall for electrical switches or receptacles because they are evidence of electric wires behind the wall.

After making two vertical cuts, use the pick end of the axe to pull the wall material away from the studs and open the wall. Work from top to bottom.

If necessary, remove items such as baseboards or window and door trim with a Halligan tool or axe.

Continue opening sections of the wall until the desired area is open. Pull out any insulation, such as rolled fiberglass, found behind the wall.

1. Determine what area of the wall will be opened. The officer in charge will usually make this determination. Typically, the areas most heavily damaged by the fire are opened first, followed by the surrounding areas, working outward. **(Step 1)**
2. Use the axe blade to begin cutting near the top of the wall. Cut downward between wall studs. Survey the wall for electrical switches or receptacles because they are evidence of electric wires behind the wall. **(Step 2)**
3. After making two vertical cuts, use the pick end of the axe to pull the wall material away from the studs and open the wall. Work from top to bottom. **(Step 3)**
4. If necessary, remove items such as baseboards or window and door trim with a Halligan tool or axe. **(Step 4)**
5. Continue opening additional sections of the wall until the desired area is open. Pull out any insulation, such as rolled fiberglass, found behind the wall. **(Step 5)**

## Covering Openings on Walls, Ceilings, or Windows

Building openings into walls, ceilings, or windows should be covered as soon as practical. This helps to protect the structure and contents from weather. It is also done for security reasons.

The brigade members can use tarpaulins, sheets of plastic, or rolled roofing as a temporary fix until maintenance crews or contractors arrive with plywood to cover the openings. Tools needed may include staple guns, hammers and nails, and sheetrock screws. To cover openings on walls, ceilings, or windows, follow the steps in **Skill Drill 19-16**.

1. Set up a temporary barrier.
2. Clear the opening of all glass and/or debris.
3. Measure the opening.
4. Cut the covering to overlap the opening.
5. Staple, nail, or screw the covering over the opening.

## Covering Floor and Roof Openings

Openings in a floor or a roof represent serious safety hazards. A brigade member should be posted at any such opening as soon as possible to prevent a brigade member from falling through the opening to a floor below; such falls often lead to fatalities.

Temporary fixes include blocking the opening with furniture, scene tape, portable safety nets, and barriers. A barrier built of 2-inch by 4-inch boards (2 × 4) may be used as a temporary barrier. Plywood can be used as a temporary cover for the opening as long as it need not support a static load.

To cover openings in floors or roofs, follow the steps in Skill Drill 19-17.

1. Set up a temporary barrier.
2. Clear the opening of debris and rough edges.
3. Measure the opening.
4. Cut the covering to overlap the opening.
5. Nail or screw the covering over the opening.

# Wrap-Up

## Ready for Review

- The objective of salvage is to protect property and belongings from damage.
- Primary loss refers to the damage caused by the fire. Secondary losses are caused by smoke, heat, and water damage as well as the damage from the firefighting operations.
- Lighting is needed at emergency scenes to improve safety and efficiency.
- Salvage is a property conservation issue. It should be started as soon as life safety and incident stabilization are addressed.
- It may be necessary to pump water out of a building or to rechannel water to remove it from a building.
- Good ventilation helps to reduce the damage caused by smoke.
- Objects can be removed or covered.
- Proper safety practices must be followed when engaging in salvage and overhaul.
- The objective of overhaul is to locate and extinguish any remaining fire.
- Overhaul may be quickly completed for a minor fire or may be extensive and time-consuming for a major fire.
- Use your senses of sight, hearing, and touch to determine where overhaul is needed.

## Hot Terms

**Carryall** A piece of heavy canvas with handles, which can be used to tote debris, ash, embers, and burning materials out of a structure.

**Floodlight** A light that can illuminate a broad area.

**Floor runner** A piece of canvas or plastic material, usually 3- to 4-foot wide and in various lengths, used to protect flooring from dropped debris and/or dirt from shoes and boots.

**Generator** An engine-powered device that provides electricity.

**Inverter** A device that converts the direct current from an apparatus electrical system into alternating current.

**Junction box** A device that attaches to an electrical cord to provide additional outlets.

**Outside stem and yoke (OS&Y) valve** A sprinkler control valve with a valve stem that moves in and out as the valve is opened or closed.

**Overhaul** Examination of all areas of the building and contents involved in a fire to ensure that the fire is completely extinguished.

**Post indicator valve (PIV)** A sprinkler control valve with an indicator that reads either open or shut depending on its position.

**Rekindle** A situation where a fire thought to be out reignites.

**Salvage** Removing or protecting property that could be damaged during firefighting or overhaul operations.

**Salvage cover** Large square or rectangular sheets made of heavy canvas or plastic material spread over furniture and other items to protect them from water run-off and falling debris.

**Scupper** An opening through which water can be removed from a building.

**Secondary loss** Property damage that occurs due to smoke, water, or other measures taken to extinguish the fire.

**Spotlight** A light designed to project a narrow, concentrated beam of light.

**Sprinkler stop** A mechanical device inserted between the deflector and the orifice of a sprinkler head to stop the flow of water.

**Sprinkler wedge** A piece of wedge-shaped wood placed between the deflector and the orifice of a sprinkler head to stop the flow of water.

**Thermal imaging devices** Electronic devices that detect differences in temperature based on infrared energy and then generate images based on that data. Commonly used in obscured environments to locate victims and hot spots.

**Water catch-all** A salvage cover folded to form a container to hold water until it can be removed.

**Water chute** A salvage cover folded to direct water flow out of a building or away from sensitive items or areas.

**Water vacuum** A device similar to a household vacuum cleaner, but with the ability to pick up liquids. Used to remove water from buildings.

# Brigade Member in Action

**It is 5:30 A.M., when you receive a phone call from your company asking you to assist with salvage and overhaul operations following a structure fire in a large industrial warehouse. Upon arrival at the scene, you are assigned to Division A. The Division A supervisor tells your brigade leader to salvage the second floor. He advises that utilities have been shut off to that area of the warehouse. Your brigade leader orders you and the crew to retrieve the salvage equipment and begin salvaging the second floor. Another crew is on the floor above you overhauling the fire.**

1. What salvage equipment will you potentially need?
   - **A.** Submersible pump, chainsaw, pike pole, and shovel
   - **B.** Box knife, ladder, circular saw, and Halligan bar
   - **C.** Salvage cover, small tool kit, floor runners, and a wet/dry vacuum
   - **D.** Floor runner, sprinkler shut-off kit, rope, and clothespins

2. The purpose of salvage is to:
   - **A.** find and extinguish hidden fires.
   - **B.** prevent or limit secondary losses and preserve the building and protect the contents.
   - **C.** remove potential fuel for the fire so that it will slow the spread of fire.
   - **D.** remove smoke from furniture and rugs.

**You reach the second floor, which has several offices, a conference room, a break room area, large display cases, and a bathroom. The offices and the conference room are nicely decorated and appear to have expensive furniture. It is a dark area and a light odor of smoke remains.**

3. What items should be the highest priority to salvage?
   - **A.** The computers and projection equipment
   - **B.** The furniture
   - **C.** The carpets and rugs
   - **D.** Break room items

4. What are some safety concerns associated with salvage?
   - **A.** Breathing in products of combustion
   - **B.** Structural collapse
   - **C.** Electrocution
   - **D.** All of the above

# Brigade Member Rehabilitation

# Chapter 20

## NFPA 1081 Standard

### Incipient Facility Fire Brigade Member

NFPA 1081 contains no Incipient Facility Job Performance Requirements for this chapter.

### Advanced Exterior Facility Fire Brigade Member

NFPA 1081 contains no Advanced Exterior Facility Job Performance Requirements for this chapter.

### Interior Structural Facility Fire Brigade Member

NFPA 1081 contains no Interior Structural Facility Job Performance Requirements for this chapter.

## Knowledge Objectives

After completing this chapter, you will be able to:

- Define emergency incident rehabilitation.
- Describe why brigade members need emergency incident rehabilitation.
- List and describe the types of extended fire incidents where brigade members need emergency incident rehabilitation.
- Describe four other types of incidents where brigade members would benefit from emergency incident rehabilitation.
- Describe the seven functions of a rehabilitation center.
- List four parts of revitalization.
- Describe the types of fluids that are well suited for brigade members to drink during emergency incident rehabilitation.
- Describe the types of food that are well suited for brigade members to eat during emergency incident rehabilitation.
- Describe the personal responsibilities related to emergency incident rehabilitation.

## Skills Objectives

There are no skill objectives for this chapter.

# You Are the Brigade Member

**You are a process operator who works at a small refinery. Your brigade has been fighting a pump seal fire for the past four hours without much rest. Due to your rural nature, mutual aid has still not arrived. A large pool fire is beginning to spread and is impinging on some high pressure lines. Your team has been ordered by the incident commander to extinguish the pool fire. Your team applies a foam blanket and successfully extinguishes the pool fire. Despite the success in containing the pool fire, two members of your team succumb to heat exhaustion and need urgent medical attention. Incident command orders your team to rehabilitation.**

1. ***What is rehabilitation, and why is it important for facility fire brigade members?***
2. ***Why is the incident commander assigning your entire team to rehabilitation?***
3. ***What steps will you take to rehabilitate yourself?***

## Introduction

In most industrial settings, facility fire brigade members are part of an emergency response team. Responding to and fighting industrial fires are not typically the primary roles of these brigade members. Some brigade members may not be accustomed to the physical, mental, and emotional stresses in an industrial fire response. Small fires can quickly escalate into much larger incidents; teams can quickly become tired and exhausted (▼ Figure 20-1). Incident commanders must be cognizant of the need to consider back-up teams, and due consideration needs to be made for fire brigade members to rehabilitate.

Emergency incident rehabilitation is part of the overall emergency effort. Brigade members and other emergency workers who are exhausted, thirsty, hungry, ill, injured, or emotionally upset can take a break for rest, fluids, food, medical evaluation, and treatment of illnesses and/or injuries (▼ Figure 20-2). Without the opportunity to rest and recover, you may develop physical symptoms such as fatigue, headaches, or gastrointestinal problems.

Rehabilitation enables brigade members to perform more safely and effectively at an emergency scene. The effort that is required to rescue a collapsed or injured brigade member takes time and resources away from fire

**Figure 20-1** Firefighting often involves extreme physical exertion and results in rapid fatigue.

**Figure 20-2** Rehabilitation provides brigade members an opportunity to take a break for rest, fluids, food, medical evaluation, and treatment of illnesses and injuries.

## Brigade Member Tips

The amount of rest needed to recover from physical exertion is directly related to the intensity of the work performed. Brigade members who have exerted tremendous efforts will require a longer recovery period than those who have performed moderate work.

suppression activities. Rehabilitation is essential to fight fires and perform rescues safely and effectively.

## Factors, Cause, and Need for Rehabilitation

Many conditions come together during a fire fight to produce a stressful environment. Consider the stresses involved in a middle-of-the-night callout at your home for a working fire at your facility. The loud, jarring sound of the pager alarm jolts your sleeping body awake. In response, you must get up immediately and without hesitation, get dressed, drive to the facility, and put on your personal protective equipment (PPE). As soon as you arrive at the facility, you may have to drive an emergency vehicle, haul hoses, position a ladder, and set up a master stream. All of these tasks require a significant amount of energy and concentration. You must be able to move into action quickly with no time to warm up your muscles as athletes do before an event.

You may be called to a fire on the hottest day of the year, the coldest day of the year, and under all types of adverse circumstances. Because you know that lives and property are at risk, you may feel an added emotional stress, which affects the body. Brigade members may end up working in an unfamiliar, smoke-filled environment, which makes the job more difficult and stressful.

### Personal Protective Equipment

Personal protective equipment (PPE), the protective clothing and breathing apparatus used by brigade members to reduce and prevent injuries, adds heat stress on the body ▶ Figure 20-3. PPE can weigh 40 lb or more, and the extra weight increases the amount of energy needed simply to move around. PPE creates a protective envelope around a brigade member that protects the body from the smoke, flames, heat, and steam of a fire. But at the same time, it traps almost all body heat inside the protective envelope.

Normally, evaporating perspiration helps cool the body so that it does not overheat. PPE acts as a vapor barrier that keeps harmful liquids and vapors out, but it also prevents most of a brigade member's perspiration from evaporating. When a brigade member is wearing PPE, perspiration soaks

Figure 20-3 The PPE that is worn to protect a brigade member also contributes to heat stress.

the inner clothing, but the evaporative cooling does not occur.

### Dehydration

Dehydration is a state in which fluid losses are greater than fluid intake into the body, leading to shock and even death if untreated. Fighting fires is a very strenuous activity, and the large amounts of muscular energy required can produce a significant amount of heat. During this exertion, the body loses a substantial amount of water through perspiration. Brigade members in action can lose up to 2 quarts of fluid in less than 1 hour ▶ Figure 20-4. Dehydration reduces strength, endurance, and mental judgment. Replenishing fluids during rehabilitation is essential to correct this imbalance in the body.

### Energy Consumption

Food provides the fuel the body needs to do muscular work. In times of strenuous activity, the body burns carbohydrates and fats for energy. These energy sources then need to be replenished. Without a sufficient supply of the right food for energy, the body cannot continue to perform at peak levels for extended periods. During rehabilitation, it is essential to refuel the body with nutritious food.

### Brigade Member Safety Tips

A tired or dehydrated brigade member is more likely to be injured and runs the risk of collapsing. Rehabilitation is essential to correct imbalances in the body that, if left untreated, could endanger the brigade member, coworkers, and others.

Figure 20-4 When perspiring, brigade members can lose up to 2 quarts of fluid in less than 1 hour.

Figure 20-5 A well-conditioned brigade member will have a greater tolerance for the stresses encountered when fighting fires.

Figure 20-6 Taking short breaks to rehabilitate reduces the risks of injury and illness.

## Tolerance for Stress

Each individual has a different tolerance level for the stresses encountered when fighting fires. For example, younger individuals tend to have greater endurance and can tolerate higher levels of stress. A person who is well-rested and well-conditioned will have more endurance than one who is tired and in poor condition. Carrying extra weight and performing strenuous tasks will strain the heart and increase the risk of heart attack.

Conditioning plays a significant role in a brigade member's level of endurance. A well-conditioned person with good cardiovascular capacity, good flexibility, and well-developed muscles will be better able to tolerate the stresses of fighting fires than a person who is out of shape (▲ Figure 20-5). Because very few industrial facilities provide exercise facilities for their employees, many brigade members work out at private health clubs to maintain their physical conditioning. However, even the most impressive conditioning will not keep a brigade member from becoming exhausted under physically stressful situations.

## The Body's Need for Rehabilitation

Rehabilitation provides periods of rest and time to recover from the fatigue and stresses of fighting fires and participating in emergency operations. Studies have shown that proper rehabilitation is one way to prevent brigade members from collapsing or suffering injuries during fire suppression activities. Taking short breaks, replacing fluids, and obtaining energy from food reduces the risks of injury and illness ◄ Figure 20-6. Rehabilitation even helps to improve the quality of decision-making, because people who are tired tend to make poor decisions.

## Types of Incidents Affecting Brigade Member Rehabilitation

The concept of rehabilitation needs to be addressed at all incidents, but it will not be necessary to implement all the components of a rehabilitation center for every incident. For example, fire brigades should always have fluids and high-quality energy foods available for rehabilitation. Brigade members putting out a small fire in a single room might require only water for rehydration while those involved in extinguishing a major plant process fire may need a full rehabilitation station.

### Extended Fire Incidents

#### Structure Fires

Large emergency incidents require full-scale rehabilitation efforts. Major structure fires that involve extended time on the scene will be hard on crews ▼ Figure 20-7. Crews working on the interior will become dehydrated and fatigued quickly due to the intense heat and stressful conditions. They will require rehabilitation so they can continue working at a healthy and safe level. Rotating crews off the fireground and bringing in fresh crews also promotes health and safety and helps get the job done in an efficient manner.

Figure 20-7 Major fires often require a strong effort for an extended duration. Rehabilitation and crew rotation are important to limit the risk of exhaustion and injuries to brigade members.

**Fire Marks**

The introduction of structured rehabilitation procedures has had a very positive impact on reducing injuries due to heat stress and exhaustion. The old philosophy was to "fight hard until you drop," and it was not unusual to see exhausted municipal fire fighters being carried out of burning buildings to waiting ambulances. However, this disrupted effective firefighting operations because the fire fighters who could still function were busy rescuing their injured and exhausted crew members.

**Brigade Member Tips**

Various activities can help develop optimal stair-climbing endurance. At the health club, the brigade member may participate in step aerobics classes or use the stair-climbing devices (with or without gear) to develop leg muscles and lung capacity. Outside of the health club, a brigade member can use step boxes or stairwells to enhance climbing capacity.

## Tank Farm Fires and Flammable Liquid or Gas Fires

Large-scale incidents involving flammable liquids or gases may require significant amounts of manpower to extinguish, and in the event of a storage tank fire, these campaigns can invariably extend from several hours to several days ► Figure 20-8. Emergency incidents that are complicated both from a technical perspective and from sheer magnitude of size will create additional demands on fire brigade members. The mental and physical requirements of firefighting and other emergency operations combined with the environmental factors of extreme cold or extreme heat and humidity create conditions that can adversely affect the health and safety of the individual fire brigade member. When brigade members become fatigued, their ability to operate safely is impaired. This impacts reaction time and their ability to make critical decisions. Rehabilitation in large-scale incidents involving storage tanks and other flammable liquid or gas fires is essential to prevent more serious conditions such as heat exhaustion or heat stroke from occurring.

### Other Types of Incidents Requiring Rehabilitation

Other incidents may also require extensive rehabilitation efforts to maintain the health and safety of brigade members. Hazardous materials incidents that require brigade members to wear fully encapsulated suits (a protective suit that fully covers the responder, including the breathing apparatus) are

Figure 20-8 Tank fires are often prolonged events involving many brigade members working for days. Rehabilitation may be conducted on a large scale at these incidents.

Figure 20-9 Brigade members wearing fully encapsulated suits must be carefully monitored for symptoms of heat stress.

especially draining (► Figure 20-9). Hazardous materials incidents can expose brigade members to strenuous conditions for extended periods of time. These incidents may also require that the staging area be located at a distance from the hazardous materials, so brigade members will have to walk long distances while wearing the heavy PPE. As a result, hazardous materials incidents require adequate rehabilitation for brigade members.

At times, the facility fire brigade is involved in long-duration search-and-rescue activities or other incidents that require the presence of public safety agencies for extended periods of time. These situations can be both mentally and physically stressful. The establishment of a rehabilitation center that brigade members can use during these incidents is essential.

The need for rehabilitation is not limited to emergency situations. Training exercises and even stand-by assignments may also require rehabilitation. Whenever brigade members are required to be ready for action for an extended period of time, some provision should be made for providing nourishing foods and replenishing fluids.

Figure 20-10 The same type of rehabilitation procedures should be implemented for training activities as those used for actual emergency incidents.

## Teamwork Tips

During active operations, brigade members may be very reluctant to admit that they need a rest. Asking an obviously exhausted brigade member if he or she needs to go to rehabilitation almost invariably brings the response, "No—I'm OK." All brigade members need to watch out for one another, and brigade leaders must monitor their teams for indications of fatigue. It is better to go to rehabilitation a few minutes early than to wait too long and risk the consequences of injury or exhaustion. The incident commander should always plan ahead so that a fresh or rested team is ready to rotate with a team that needs rehabilitation.

Large-scale training activities, including live burn exercises, involve the same concerns as major fire incidents, and rehabilitation should be incorporated in the planning for these activities (▲ Figure 20-10). Training exercises may be conducted over a full day and involve a series of activities, as well as time to set up each exercise. As part of the process, time should be set aside for the participants to go to rehabilitation between strenuous activities.

Weather conditions can have a significant impact on the need for rehabilitation. Brigade members should always dress appropriately for the weather, and plans for rehabilitation procedures should take into account the anticipated environmental conditions. Whether it is hot or cold, returning the body's temperature back to normal is one of the primary goals of rehabilitation.

Emergencies that occur when the temperature is very hot will increase the need to rotate teams and allow extensive rehabilitation. Teams working on the interior fire attack may not notice much difference, because their environment during the attack will be hot regardless of the temperature outside. However, hot weather will affect teams working on the outside. These brigade members will tend to become dehydrated and fatigued much faster than they would if temperatures were in a more comfortable range. Another factor that must be considered is the humidity of the air. Humidity plays an important role in evaporative cooling. High humidity reduces evaporative cooling, making it more difficult for the body to regulate its internal temperature.

Cold weather also increases the need for rehabilitation and crew rotation. The cold can be just as dangerous as the heat. Hypothermia, a condition in which the internal body temperature falls below 95°F (35°C), can lead to loss of coordination, muscle stiffness, coma, and death. Even in cold weather, the weight of PPE and the physical exertion of fighting a fire will cause the body to sweat inside the protective clothing. Damp clothing and cold temperatures can quickly lead to hypothermia.

## How Does Rehabilitation Work?

One way to understand an emergency incident rehabilitation center is to look at the functions it is designed to perform. The most common model has seven different functions ▼ Table 20-1:

1. Physical assessment
2. Revitalization
3. Medical evaluation and treatment
4. Regular monitoring of vital signs
5. Transportation to a hospital
6. Critical incident stress management
7. Reassignment

**Table 20-1 Seven Functions of Rehabilitation**

| |
|---|
| 1. Physical assessment |
| 2. Revitalization |
| 3. Medical evaluation and treatment |
| 4. Regular monitoring of vital signs |
| 5. Transportation to a hospital |
| 6. Critical incident stress management |
| 7. Reassignment |

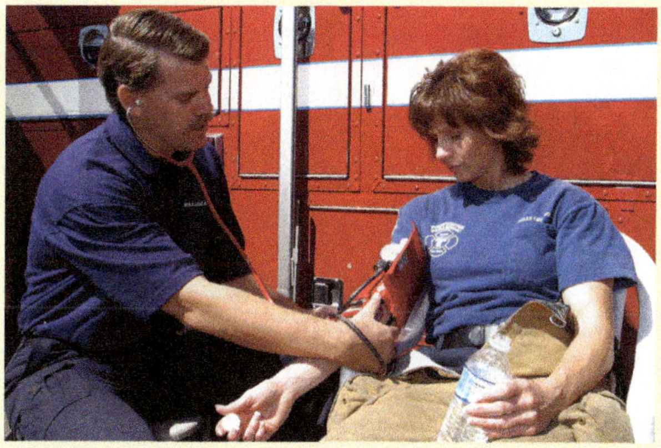

Figure 20-11 The physical condition of each brigade member who arrives at the rehabilitation center should be evaluated, and vital signs should be recorded by assigned medical personnel.

### Physical Assessment

Whenever possible, entire teams should be assigned to rehabilitation and stay together in the rehabilitation center. All brigade members should be identified as they enter and leave the rehabilitation center to maintain accountability. When brigade members are ready to be released, the team should be released as a unit.

The first function of emergency incident rehabilitation is physical assessment. A brigade member's vital signs, including pulse, respiration, blood pressure, and temperature, should be taken and recorded by assigned medical personnel when the brigade member arrives at the rehabilitation center ▲ Figure 20-11. Many signs and symptoms of fatigue will indicate the need for rehabilitation and are a function of many factors, including poor nutrition and mental, physical, and emotional exhaustion. Each team member should be questioned and observed for indications of emotional stress. From here, the brigade member may be referred for medical treatment or sent for rest, rehydration, and food before being reassigned to operational activities.

### Revitalization

Revitalization is the main part of the rehabilitation process. The four components of revitalization are ▼ Table 20-2:

**Table 20-2 Four Components of Revitalization**

| |
|---|
| 1. Rest |
| 2. Fluid replacement |
| 3. Nutrition |
| 4. Temperature stabilization |

Figure 20-12 In the rehabilitation center, brigade members should be able to remove personal protective clothing and get some rest before returning to action.

Figure 20-13 The rehabilitation center should have plenty of fluids to rehydrate brigade members. Plain water or diluted sports drinks are preferred.

1. Rest
2. Fluid replacement
3. Nutrition
4. Temperature stabilization

## Rest

Rest begins as soon as the brigade member arrives for rehabilitation. Rehabilitation should be located away from the central activity of the emergency, so the brigade member can disengage from all other stressful activities and remove personal protective clothing (▲ Figure 20-12). Many times brigade personnel who are in rehabilitation do not remove all of their gear, including bunker pants. Doing so achieves maximum cooling and facilitates rest and rehydration. Rest continues as the brigade member goes through the other parts of revitalization. Rehabilitation centers are equipped with chairs or cots so brigade members can sit or lie down and relax.

## Fluid Replacement

When the body becomes overheated, it sweats so that evaporative cooling can reduce body temperature. Perspiration is composed of water and other dissolved substances such as salt. During fire suppression activities, a brigade member can perspire enough to lose 2 quarts of water in the time it takes to go through 2 bottles of air. Because 1 pint of water weighs 1 pound, losing 2 quarts of water is equivalent to a 2% loss in body weight for a 200-lb brigade member. The loss of that much body fluid can result in impaired body temperature regulation.

Further dehydration can result in reduced muscular endurance, reduced strength, and heat cramps. Severe cases of dehydration can contribute to heat stroke. Because a brigade member can lose so much water during fire suppression activities, it is important to replace fluids before severe dehydration takes place (▲ Figure 20-13).

When the body perspires, it loses **electrolytes** (certain salts and other chemicals that are dissolved in body fluids and cells) as well as water. If a brigade member loses large amounts of water through perspiration, the electrolytes must be replenished as well as the water. This is why sport-type drinks are often used in rehabilitation. Sport drinks supply water, balanced electrolytes, and some sugars. These drinks can be mixed with twice the recommended amount of water to dilute the sugars.

Rehydrating your body is not as easy as just drinking lots of fluids quickly. Drinks like colas, coffee, and tea should be avoided because they contain caffeine. Caffeine acts as a diuretic that causes the body to excrete more water. Sugar-rich carbonated beverages are not tolerated or absorbed as well as straight water or diluted sports drinks. Too much sugar is difficult to digest and causes swings in the body's energy levels. Drinks that are too cold or too hot may be hard to consume and may prevent you from ingesting enough liquids.

## Brigade Member Tips

The sensation of thirst is not a reliable indicator of the amount of water the body has lost. Thirst develops only after the body is already dehydrated. Do not wait until you feel thirsty to be concerned with your hydration. Start to drink before you get thirsty. Remember to drink early and often.

Try to drink enough fluids to keep the body properly hydrated at all times, particularly during hot weather. This will decrease the risk of dehydration when an emergency incident occurs. Adequate hydration also ensures peak physical and mental performance.

The best indicator of proper hydration is the color of urine. If urine is dark or amber in color, the body is dehydrated. If urine is light or clear, the body is properly hydrated. Fatigue and heat sensitivity are also signs of dehydration. Brigade members should strive to drink a quart of water an hour during periods of work.

An additional concern is the rate at which fluids can be absorbed from the stomach. Drinking too much too quickly can cause bloating, a condition in which air fills the stomach. This causes a feeling of fullness and can lead to discomfort, nausea, and even vomiting.

Studies demonstrate that the stomach can absorb 1–1.5 quarts of fluid per hour. However, the body can lose up to 2 quarts of fluid per hour. Thus, the body loses fluids much more rapidly than they can be replaced. Once the initial 2 quarts of water are lost, the body will require 1–2 hours to recover.

### Nutrition

A brigade member performs more physical tasks and exerts more energy than the average worker. Like an engine that runs on diesel fuel, the body runs on glucose. Glucose, also known as blood sugar, is carried throughout the body by the bloodstream and is needed to burn fat efficiently and release energy.

In order for the body to work properly, the glucose (blood sugar) levels need to be in balance. If blood sugar drops too low, the body becomes weak and shaky. If blood sugar is too high, the body becomes sluggish. Blood sugar levels can be balanced by eating a proper diet of carbohydrates, proteins, and fats.

Carbohydrates are a major source of fuel for the body and can be found in grains, vegetables, and fruits ► Figure 20-14. The body converts carbohydrates into glucose, making "carbs" an excellent energy source. A common dietary myth is that carbohydrates are fattening and should be avoided. In fact, carbohydrates should make up 55–65% of calories in a balanced diet. Carbohydrates have the same number of calories per gram as proteins and fewer calories than fat. Additionally, carbohydrates are the only fuel that the body can readily use during high-intensity physical activities such as fighting fires.

Figure 20-14 Carbohydrates are a major source of fuel for the body.

Figure 20-15 Proteins perform many vital functions in the body.

Proteins perform many vital functions within the body ▲ Figure 20-15. Most protein comes from meats and dairy products. Smaller amounts of protein are found in grains, nuts, legumes, and vegetables. Proteins (amino acids) are

# Voices of Experience

**"Our facility is located on the Gulf Coast, and the training usually takes place during hot humid weather."**

At our facility, we train all new process operation hires for emergency response roles. In fact, our emergency response brigade consists mainly of process operators. With each class of new hires, we conduct approximately 120 hours of emergency response training. Our facility is located on the Gulf Coast, and the training usually takes place during hot humid weather.

One July, as we were planning the brigade training for August, we were discussing past training classes. The discussion quickly turned to a familiar topic: "We spend a lot of time dealing with the effects of the weather on the brigade rather than the actual emergency response training procedures and techniques." With that, we decided to undertake a proactive approach to responder safety and rehabilitation.

We now include the subject of rehabilitation in our safety classes at the start of each day. In the field during physical training activities, we ensure that everyone stays hydrated. In addition to providing fluids, we provide nutritious foods, scheduled periods of rest, and means of body temperature stabilization.

Managing the four components of revitalization is very important. The one component that has shown to be the greatest benefit to our program is body temperature stabilization. We provide cold or cool neck wraps to be used during training activities. Our goal is to keep body temperature near normal levels during physical exertion. By controlling body temperature, the rest periods can be more manageable, allowing more time to train.

We also realized we were having trouble with the same issues during incidents at the plant, so we now practice the same procedures during emergency response activities. Recognizing the importance of rehabilitation to protect our greatest resource has improved the safety, capability, and functions of our brigade.

Roy Robichaux, Jr.
ConocoPhillips—Alliance Refinery
Belle Chasse, Louisiana

## Brigade Member Safety Tips

PPE is designed to provide protection from hazards encountered in emergency operations. Brigade members who are overheated should remove their PPE as soon as possible to permit evaporative cooling. PPE must only be removed in a safe place, outside the hazard area. The incident commander or a safety officer may approve working without full PPE in situations where the risks have been fully evaluated.

used by the body to grow and repair tissues and are only used as a primary fuel source in extreme conditions such as starvation. Like other nutrients, excess proteins are converted and stored in the body as fat. A balanced diet has 10–12% of calories from proteins.

Fats are also essential for life. Fats are used for energy, for insulating and protecting organs, and for breaking down certain vitamins. Some fats, such as those found in fish and nuts, are healthier and more beneficial to the body than others, such as the fats found in margarine. However, no more than 25–30% of the diet should come from fats. Excess fat consumption, particularly of saturated fats (which come mostly from animal products), is linked to high cholesterol, high blood pressure, and cardiovascular disease.

Candy and soft drinks contain sugar. The body can quickly absorb and convert these foods to fuel. But simple sugars also stimulate the production of insulin, which reduces blood glucose levels. That's why eating a lot of sugar can actually result in lower energy levels.

For a brigade member to sustain peak performance levels, it is necessary to refuel during rehabilitation. During short incidents, low-sugar, high-protein sports bars can be used to keep the glucose balance steady. During extended incidents, a brigade member should eat a more complete meal. The proper balance of carbohydrates, proteins, and fats will maintain energy levels throughout the emergency. To ensure peak performance, the meal should include complex carbohydrates such as whole grain breads, whole grain pasta, rice, and vegetables. It is also better to eat smaller meals; larger meals can increase glucose levels and slow down the body.

Healthy, balanced eating should be a lifestyle. Proper nutrition reduces stress, improves health, and provides more energy. It is just as important to keep blood sugar levels balanced throughout the day as it is to remain hydrated. Then, you will be ready to react to an emergency at any time.

### Temperature Stabilization

The fourth part of revitalization is stabilizing body temperature. Body temperature must return to a normal range before a brigade member resumes strenuous activities. Brigade members exposed to hot or cold temperatures need to have a place where their body temperature can return to normal levels before resuming further activities.

As mentioned previously, fighting fires in full turnout gear can result in some degree of heat stress (▼ Figure 20-16). The heat generated by the body during the intense physical exertion, coupled with the increased temperature from the fire, can increase the body's internal temperature and produce profuse sweating. In most situations, turnout gear should be completely removed as soon as possible to allow the body to cool. If necessary, additional steps such as the use of cold compresses should be taken to reduce body temperature. When ambient temperatures are low, damp clothes should be removed and blankets should be used to stabilize body temperature.

Stabilizing body temperature is further complicated during periods of hot weather or high humidity. At these times, brigade members cannot cool off even when they take a break and remove their turnout gear. For this reason, rehabilitation centers need to be climate-controlled so that brigade members can achieve a normal body temperature before resuming active duties. Some organizations have air-conditioned vehicles for rehabilitation. Others use buses or establish rehabilitation centers in an already cooled area such as a control room.

Figure 20-16 Fighting fires in complete turnout gear can result in varying degrees of heat stress.

Industrial fire fighters, whether at the incipient, advanced exterior, or interior structural level, may be subject to heat-related disorders. This kind of situation may evolve rapidly as temperatures in the firefighting environment rise or when barriers to the body's ability to cool itself are used, such as fire fighters' turnout gear. Additional physiological activities, including increased work rates or breathing against a resistance (e.g., in SCBA), increase the chance that one of these thermal disorders will occur. Older brigade members or those with circulatory system problems are particularly susceptible to heat-related concerns. All fire brigade members must be cognizant of the two heat-related events described next, recognize their signs and symptoms, and take appropriate actions if they occur.

The symptoms of *heat stress,* sometimes known as heat cramps or heat exhaustion, include profuse sweating, weakness, nausea, headache, and fainting associated with loss of fluids. During heat stress events, the body continues to try to cool itself, but symptoms are present. Individuals who present with these symptoms should immediately be removed from firefighting activities and placed in a cool atmosphere to rest. Water or electrolyte fluids should be provided.

*Heat stroke* is a true medical emergency. It can occur if heat stress is not properly treated and can lead to brain damage or death. Symptoms include red/blue dry skin (because sweating has ceased), confusion, and loss of consciousness. In heat stroke, the body loses its ability to cool itself and typically the core body temperature exceeds 104°F. First aid for heat stroke requires dosing the body continually with cool liquid and summoning emergency medical assistance immediately. Brigade members who are experiencing heat stroke need to be taken to a medical facility as soon as possible.

Cold temperatures cause problems as well. Brigade members responding to incidents during cold weather are subject to hypothermia and frostbite (damage to tissues resulting from prolonged exposure to cold) ▶ Figure 20-17. In these cases, the rehabilitation center needs to be warm enough so they can get warmed up before returning to the chilly environment. Brigade members who are wet or severely chilled should be wrapped in warming blankets and moved into a well-heated area before they remove their turnout gear. As soon as possible, all wet clothing should be removed and replaced with warm, dry clothing.

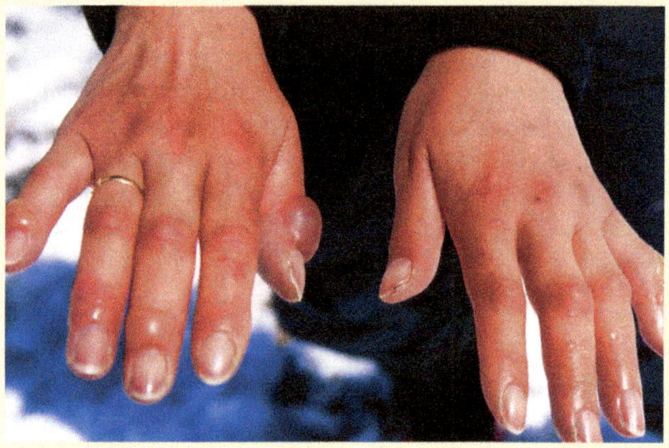

Figure 20-17 Prolonged exposure to freezing weather can result in severe frostbite.

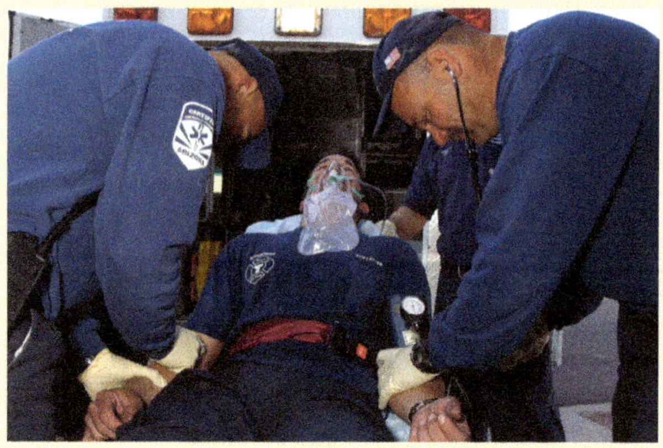

Figure 20-18 A medical team should be part of rehabilitation to check vital signs and treat any injuries that occur.

## Medical Evaluation and Treatment

A brigade member who has abnormal vital signs, is suffering pain, or is injured needs to have further medical evaluation. A medical evaluation and treatment team should be assigned to the rehabilitation center ▶ Figure 20-18. Signs of illness or injury should be checked in the rehabilitation center before the brigade member resumes activities. Identifying problems in the rehabilitation center can prevent later disruption on the fire scene. A brigade member who collapses or becomes incapacitated while in action endangers the entire team.

## Regular Monitoring of Vital Signs

Pulse, respiration, blood pressure, and temperature should be monitored at regular intervals during rehabilitation. These readings should be compared with readings taken when the brigade member first enters the rehabilitation center. Vital signs must be taken repeatedly to ensure that they return to normal limits before the brigade member is reassigned. If vital signs do not return to normal levels, a brigade member may need to spend more time in the rehabilitation center or be further evaluated by medical personnel.

## Transportation to a Hospital

Another function of the rehabilitation center is to transport ill or injured brigade members to the hospital ▶ Figure 20-19.

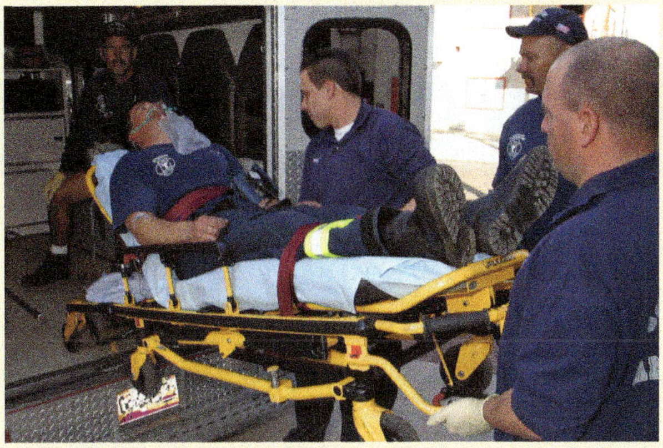

Figure 20-19 An ambulance should always be ready near the rehabilitation center to transport an ill or injured brigade member to a hospital.

## Brigade Member Safety Tips

Stress management skills are crucial for high-quality performance, regardless of the type of stressor involved or what role the person plays in emergency assistance. A brigade member needs to know basic stress management skills to deliver high-quality performance. Exercise, proper nutrition, and emotional support are all helpful in managing stress.

## Brigade Member Tips

You must take care of yourself first to be able to help others.

An ambulance should be available to ensure that ill or injured brigade members receive the best possible care.

### Critical Incident Stress Management

When emergencies involve stressful situations, particularly mass casualty incidents or brigade member fatalities, rehabilitation should include staff trained in critical incident stress management (CISM). CISM is a process that confronts the responses to critical incidents, defuses them, and directs the brigade member toward physical and emotional balance. The CISM team members should talk to each team in rehabilitation to assess the emotional state of brigade members. Talking with a CISM member in the rehabilitation center is the first step in the CISM system. After the incident, CISM team members may meet with brigades or individual brigade members to discuss the event and work through it. CISM is discussed further in Chapter 18, Brigade Member Survival.

### Reassignment

Once rested, rehydrated, refueled, and rechecked to make certain that they are fit for duty, brigade members can be released from rehabilitation and reassigned to active duty. Reassignments may be to the same job performed before or to a different task, depending on the decision of the officer in charge. Generally, teams stay together during the rehabilitation process because this creates a more effective work unit and helps to ensure accountability at all times. However, it is sometimes necessary to temporarily reassign members to different companies. This allows brigade members who require additional rest or medical treatment to remain in rehabilitation and releases fully staffed teams for additional duty.

## Personal Responsibility in Rehabilitation

The goal of a facility fire brigade is to save lives first and property second. To achieve this goal, however, you need to take care of yourself first, take care of the rest of your team second, and then take care of the people involved in the incident. Another way of looking at this is to remember that safety begins with you. The other members of your team depend on you to bear your share of the load. To do this, you need to maintain your body in peak condition.

Part of your responsibility is to know your own limits. No one else can know what you ate, whether you are lightheaded or dehydrated, whether you are feeling ill, or whether you need a breather. You are the only person who knows these things. Therefore, you may be the only person who knows when you need to request rehabilitation. It may be difficult to say "I need a break" while your team members are still hard at work. But it is better to break when you need it than to push yourself too far and have to be rescued by other members of your brigade.

Regular rehabilitation enables brigade members to accomplish more work during a major incident. It decreases the risk of stress-related injuries and illness. Remember, safety begins and ends with you.

# Wrap-Up

## Ready for Review

- Rehabilitation is a specially designated area where emergency personnel can rest, receive fluids and nourishment, and be evaluated for medical and CISM problems.
- Rehabilitation centers help prevent injuries and illnesses, enable brigade members to accomplish more during an emergency incident, and reduce the need for rescuing exhausted or collapsed brigade members from the fire scene.
- Types of incidents that may necessitate rehabilitation centers include major structure fires, storage tank fires, and flammable liquid and gas fires. Rehabilitation is also needed for hazardous materials incidents, search-and-rescue operations and training activities.
- The seven parts of rehabilitation are: physical assessment, revitalization, medical evaluation and treatment, regular monitoring of vital signs, transport, critical incident stress management, and reassignment. Revitalization consists of rest, fluid replacement, nutrition, and temperature stabilization.
- Your responsibilities in rehabilitation are to know your limits, to listen to your body, and to use the rehabilitation facilities when needed.

## Hot Terms

**Critical incident stress management (CISM)** A system to help personnel deal with critical incident stress in a positive manner; its aim is to promote long-term mental and emotional health after a critical incident.

**Dehydration** A state in which fluid losses are greater than fluid intake into the body, leading to shock and even death if untreated.

**Electrolytes** Certain salts and other chemicals that are dissolved in body fluids and cells.

**Emergency incident rehabilitation** A function on the emergency scene that cares for the well-being of the brigade members. It includes physical assessment, revitalization, medical evaluation and treatment, and regular monitoring of vital signs.

**Frostbite** Damage to tissues as the result of exposure to cold; frozen or partially frozen body parts.

**Fully encapsulated suits** A protective suit that completely covers the brigade member, including the breathing apparatus, and does not let any vapor or fluids enter the suit.

**Glucose** The source of energy for the body. One of the basic sugars, it is the primary fuel, along with oxygen.

**Hypothermia** A condition in which the internal body temperature falls below 95°F, usually a result of prolonged exposure to cold or freezing temperatures.

**Personal protective equipment (PPE)** Gear worn by brigade members that includes helmet, gloves, hood, coat, pants, SCBA, and boots. The PPE provides a thermal barrier for brigade members against intense heat.

**Rehabilitate** To restore to a condition of health or to a state of useful and constructive activity.

# Brigade Member in Action

**Your facility fire brigade team is dispatched to a fire in the canteen. It is a hot, humid afternoon in July. The temperature is 94°F (35°C), and relative humidity is 98%. When you arrive, the structure is completely involved and the fire has extended into a nearby garage storage facility. The Incident Commander orders your team leader to establish a rehabilitation center. Your team leader tells you to establish rehab at the designated staging area, which is at the entrance of the plant.**

1. What equipment would you NOT bring with you?
   - A. Medical equipment bag
   - B. Drinking water and ice
   - C. Self-contained breathing apparatus
   - D. Clipboards and pens

2. Your brigade leader tells you and your partner to check in the brigade members as they report to rehabilitation. This means that you will:
   - A. write down the identity of each team reporting to rehabilitation.
   - B. write down the time each team reports to rehabilitation.
   - C. record the name of each individual brigade member reporting to rehabilitation.
   - D. determine if anyone reporting to rehabilitation is injured.
   - E. All of the above

3. Should the brigade members go back into the building after rehabilitation?
   - A. Yes—the fire is still burning and they are needed urgently.
   - B. Yes—they should be fine for another work period after a quick drink.
   - C. No—they cannot leave rehabilitation until you have written down all of their names.
   - D. No—they need more time to cool down and rehydrate.

4. What should they do to revitalize?
   - A. Remove protective clothing, drink a carbonated soda, and eat a donut.
   - B. Keep protective clothing on and drink cold water or sports beverage.
   - C. Remove protective clothing, eat a protein bar, and drink cold water or diluted sports beverage.
   - D. Remove protective clothing, drink coffee, and eat a donut.

# Fire Suppression

# Chapter 21

# NFPA 1081 Standard

### Incipient Facility Fire Brigade Member

**4.3.1*** Attack an incipient stage fire, given a handline flowing up to 473 L/min (125 gpm), appropriate equipment, and a fire situation, so that the fire is approached safely, exposures are protected, the spread of fire is stopped, agent application is effective, the fire is extinguished, and the area of origin and fire cause evidence are preserved.

**(A)** *Requisite Knowledge.* Types of handlines used for attacking incipient fires, precautions to be followed when advancing handlines to a fire, observable results that a fire stream has been properly applied, dangerous building conditions created by fire, principles of exposure protection, and dangers such as exposure to products of combustion resulting from fire condition.

**(B)** *Requisite Skills.* The ability to recognize inherent hazards related to the material's configuration; operate handlines; prevent water hammers when shutting down nozzles; open, close, and adjust nozzle flow; advance charged and uncharged hoses; extend handlines; operate handlines; evaluate and modify water application for maximum penetration; assess patterns for origin determination; and evaluate for complete extinguishment.

**4.3.3*** Utilize master stream appliances, given an assignment, an extinguishing agent, and a master stream device, so that the agent is applied to the fire as assigned.

**(A)** *Requisite Knowledge.* Safe operation of master stream appliances, uses for master stream appliances, tactics using fixed master stream appliances, and property conservation.

**(B)** *Requisite Skills.* The ability to put into service a fixed master stream appliance, and to evaluate and forecast a fire's growth and development.

### Advanced Exterior Facility Fire Brigade Member

**5.2.3*** Attack an exterior fire operating as a member of a team, given a water source, a handline, PPE, tools, and an assignment, so that team integrity is maintained, the attack line is correctly deployed for advancement, access is gained into the fire area, appropriate application practices are used, the fire is approached in a safe manner, attack techniques facilitate suppression given the level of the fire, hidden fires are located and controlled, the correct body posture is maintained, hazards are avoided or managed, and the fire is brought under control.

**(A)** *Requisite Knowledge.* Principles of fire streams; types, design, operation, nozzle pressure effects, and flow capabilities of nozzles; precautions to be followed when advancing handlines to a fire; observable results that a fire stream has been correctly applied; dangerous conditions created by fire; principles of exposure protection; potential long-term consequences of exposure to products of combustion; physical states of matter in which fuels are found; the application of each size and type of attack line; the role of the backup team in fire attack situations; attack and control techniques; and exposing hidden fires.

**(B)** *Requisite Skills.* The ability to prevent water hammers when shutting down nozzles; open, close, and adjust nozzle flow and patterns; apply water using direct, indirect, and combination attacks; advance charged and uncharged 38 mm (1 1/2 inches) diameter or larger handlines; extend handlines; replace burst hose sections; operate charged handlines of 38 mm (1 1/2 inches) diameter or larger; couple and uncouple various handline connections; carry hose; attack fires; and locate and suppress hidden fires.

**5.3.3** Utilize master stream appliances, given an assignment, an extinguishing agent, and a master stream device and supply hose, so that the appliance is set up correctly and the agent is applied as assigned.

**(A)** *Requisite Knowledge.* Correct operation of master stream appliances, uses for master stream appliances, tactics using master stream appliances, selection of the master stream appliance for different fire situations, the effect of master stream appliances on search and rescue, ventilation procedures, and property conservation.

**(B)** *Requisite Skills.* The ability to correctly put in service a master stream appliance and evaluate and forecast a fire's growth and development.

**5.3.5*** Control a flammable gas fire operating as a member of a team, given an assignment, a handline, PPE, and tools, so that crew integrity is maintained, contents are identified, the flammable gas source is controlled or isolated, hazardous conditions are recognized and acted upon, and team safety is maintained.

**(A)** *Requisite Knowledge.* Characteristics of flammable gases, components of flammable gas systems, effects of heat and pressure on closed containers, boiling liquid expanding vapor explosion (BLEVE) signs and effects, methods for identifying contents, water stream usage and demands for pressurized gas fires, what to do if the fire is prematurely extinguished, alternative actions related to various hazards, and when to retreat.

**(B)** *Requisite Skills.* The ability to execute effective advances and retreats, apply various techniques for water application, assess gas storage container integrity and changing conditions, operate control valves, and choose effective procedures when conditions change.

**5.3.9*** Extinguish a Class C (electrical) or simulated Class C fire as a member of a team, given an assignment, a Class C fire-extinguishing appliance/extinguisher, and PPE, so that the proper type of Class C agent is selected for the condition, the selected agent is correctly applied to the fuel, the fire is extinguished, reignition is prevented, team protection is maintained, and the hazard is faced until retreat to safe haven is reached.

**(A)** *Requisite Knowledge.* Methods by which Class C agent prevents or controls a hazard; methods by which Class C fires are de-energized; causes of injuries from Class C firefighting on live Class C fires with Class A agents and the Class C agents; the extinguishing agents' characteristics, uses, and limitations; the advantages and disadvantages of de-energizing as using water fog nozzles on a Class A or Class B fire; and methods to reduce or avoid hazards.

**(B)** *Requisite Skills.* The ability to operate Class C fire extinguishers or fixed systems and approach and retreat from Class C fires as part of a coordinated team.

**5.3.13** Attack a vehicle fire operating as a member of a team, given PPE, an attack line, and hand tools, so that hazards are avoided, leaking flammable liquids are identified and controlled, protection from flash fires is maintained, all vehicle compartments are overhauled, and the fire is extinguished.

**(A)** *Requisite Knowledge.* Principles of fire streams as they relate to fighting fires; precautions to be followed when advancing hose lines; observable results that a fire stream has been properly applied; identifying alternative fuels and the hazards associated with them; dangerous conditions created during a fire; common types of accidents or injuries related to fighting fires and how to avoid them; how to access locked passenger, trunk, and engine compartments; and methods for overhauling.

**(B)** *Requisite Skills.* The ability to identify fuel type; assess and control fuel leaks; open, close, and adjust the flow and pattern on nozzles; apply water for maximum effectiveness while maintaining flash fire protection; advance 38 mm (1 1/2 inches) or larger-diameter attack lines; and expose hidden fires by opening all compartments.

**Interior Structural Facility Fire Brigade Member**

**6.2** Manual Fire Suppression

**6.2.1*** Attack an interior structural fire operating as a member of a team, given a water source, a handline, PPE, tools, and an assignment, so that team integrity is maintained, the handline is deployed for advancement, access is gained into the fire area, correct application practices are used, the fire is approached safely, attack techniques facilitate suppression given the level of the fire, hidden fires are located and controlled, the correct body posture is maintained, hazards are avoided or managed, and the fire is brought under control.

**(A)** *Requisite Knowledge.* Principles of conducting initial fire size-up; principles of fire streams; types, design, operation, nozzle pressure effects, and flow capabilities of nozzles; precautions to be followed when advancing hose lines to a fire; observable results that a fire stream has been correctly applied; dangerous building conditions created by fire; principles of exposure protection; potential long-term consequences of exposure to products of combustion; physical states of matter in which fuels are found; common types of accidents or injuries and their causes; and the application of each size and type of handlines, the role of the backup team in fire attack situations, attack and control techniques, and exposing hidden fires.

**(B)** *Requisite Skills.* The ability to prevent water hammers when shutting down nozzles; open, close, and adjust nozzle flow and patterns; apply water using direct, indirect, and combination attacks; advance charged and uncharged 38 mm (1½ inches) diameter or larger handlines; extend handlines; replace burst hose sections; operate charged handlines of 38 mm (1½ inches) diameter or larger; couple and uncouple various handline connections; carry hose; attack fires; and locate and suppress hidden fires.

**6.2.4 (B)** *Requisite Skills.* The ability to deploy and operate handlines, expose void spaces without compromising structural integrity, apply extinguishing agents for maximum effectiveness, expose and extinguish hidden fires, recognize and preserve obvious signs of area of origin and fire cause, and evaluate for complete extinguishment.

**6.3.3** Utilize master stream appliances, given an assignment, an extinguishing agent, a master stream device, and supply hose, so that the appliance is set up correctly and the agent is applied as assigned.

**(A)** *Requisite Knowledge.* Correct operation of master stream appliances, uses for master stream appliances, tactics using master stream appliances, selection of the master stream appliances for different fire situations, and the effect of master stream appliances on search and rescue, ventilation procedures, and property conservation.

**(B)** *Requisite Skills.* The ability to correctly put in service a master stream appliance and to evaluate and forecast a fire's growth and development.

**6.3.5*** Control a flammable gas fire operating as a member of a team, given an assignment, a handline, PPE, and tools, so that team integrity is maintained, contents are identified, the flammable gas source is controlled or isolated, hazardous conditions are recognized and acted upon, and team safety is maintained.

**(A)** *Requisite Knowledge.* Characteristics of flammable gases, components of flammable gas systems, effects of heat and pressure on closed containers, BLEVE signs and effects, methods for identifying contents, water stream usage and demands for pressurized gas fires, what to do if the fire is prematurely extinguished, alternative actions related to various hazards, and when to retreat.

**(B)** *Requisite Skills.* The ability to execute effective advances and retreats, apply various techniques for water application, assess gas storage container integrity and changing conditions, operate control valves, and choose effective procedures when conditions change.

**6.3.6*** Extinguish a fire using special extinguishing agents other than foam operating as a member of a team, given an assignment, a handline, PPE, and an extinguishing agent supply, so that fire is extinguished, reignition is prevented, and team protection is maintained.

**(A)** *Requisite Knowledge.* Methods by which special agents, such as dry chemical, dry powder, and carbon dioxide, prevent or control a hazard; principles by which special agents are generated; the characteristics, uses, and limitations of firefighting special agents; the advantages and disadvantages of using special agents; special agents application techniques; hazards associated with special agents usage; and methods to reduce or avoid hazards.

**(B)** *Requisite Skills.* The ability to operate a special agent supply for use, master various special agents application techniques, and approach and retreat from hazardous areas as part of a coordinated team.

**6.3.11*** Extinguish a Class C (electrical) fire as a member of a team, given an assignment, a Class C fire-extinguishing appliance/extinguisher, and PPE, so that the type of Class C agent is selected for the condition, a selected agent is correctly applied to the fuel, fire is extinguished, reignition is prevented, team protection is maintained, and the hazard is faced until retreat to safe haven is reached.

**(A)** *Requisite Knowledge.* Methods by which Class C agent prevents or controls a hazard; methods by which Class C fires are de-energized; causes of injuries from Class C firefighting on live Class C fires with Class A agents and the Class C agents; the extinguishing agents' characteristics, uses, and limitations; the advantages and disadvantages of de-energizing using water fog nozzles on a Class A or Class B fire; and methods to reduce or avoid hazards.

**(B)** *Requisite Skills.* The ability to operate Class C fire extinguishers or fixed systems and approach and retreat from Class C fires as part of a coordinated team.

## Knowledge Objectives

After completing this chapter, you will be able to:

- Describe offensive versus defensive operations.
- Describe how to operate hose lines.
- Describe how to attack an interior structure fire.
- Describe exposure protection.
- Describe how to attack a vehicle fire.
- Describe how to extinguish a flammable gas cylinder fire.
- Describe a BLEVE.
- Describe how to attack fires involving electricity.

## Skills Objectives

After completing this chapter, you will be able to perform the following skills:

- Apply water using the direct attack.
- Apply water using an indirect attack.
- Apply water using the combination attack.
- Use a large handline using the one-person method.
- Use a large handline using the two-person method.
- Operate deck guns.
- Operate portable monitors.
- Locate and suppress interior wall and subfloor fires.
- Extinguish a fire in a flammable gas cylinder.

## You Are the Brigade Member

**During your shift at your plant, you are dispatched to a reported fire in the warehouse area. You are well aware of the hazards associated with this warehouse, so you don your personal protective equipment (PPE) and your self-contained breathing apparatus (SCBA). As you arrive with your team at the designated area, you notice heavy smoke coming from the warehouse bay. You pull the appropriate hose line from the hose cabinet and make your way to the seat of the fire. As your brigade turns the corner into the aisle, the situation changes; you see a fully involved forklift. On the rear of the forklift is a compressed gas cylinder. The radiant heat from the forklift fire is beginning to heat the combustible materials to the point that they are starting to burn.**

1. *What are the hazards at this fire?*
2. *How should you attack this fire?*

## Introduction

The term fire suppression refers to all of the tactics and tasks that are performed on the fire scene to achieve the final goal of extinguishing the fire. Fire suppression can be accomplished through a variety of methods that will stop the combustion process. All of these methods involve removal of one of the four components of the fire tetrahedron. A fire can be extinguished by removing the oxygen, the fuel, or the heat from the combustion process. Interrupting the chemical chain reactions will also stop the combustion process and extinguish the fire.

This chapter presents the methods that are most frequently used by brigade members for extinguishing fires. The apparatus and equipment used by most fire brigades is designed to apply large volumes of water to a fire in order to cool the fuel below its ignition temperature. Although fire brigades typically use a variety of extinguishing agents for different situations, water is used more often than any other agent.

Water is a very effective extinguishing agent for many different types of fires, because tremendous quantities of heat energy are required to convert water into steam. When water is applied to a fire, all of the heat that is used to create steam is removed from the combustion process. If a sufficient quantity of water is applied, the fuel is cooled below its ignition temperature and the fire is extinguished.

## Offensive versus Defensive Operations

All Facility fire suppression operations can be classified as either offensive or defensive operations. When fire brigade team members attack a fire directly either using portable fire extinguishers or by advancing hose lines into a facility to attack the seat of a fire, the strategy is offensive. Defensive operations are conducted from the exterior of a facility or structure. Defensive operations are usually conducted by directing water streams at a fire from the exterior from a safe distance.

Offensive operations expose brigade members directly to the heat and smoke generated by a fire inside a facility and may require the use of **thermal protective clothing**. Offensive operations expose brigade members to additional risk factors, such as the possibility of being trapped by a collapse of warehouse storage units, structural collapse, or failure of structural components. The objective in an offensive operation is for the brigade members to get close enough to the fire to apply extinguishing agents directly to the seat of the fire at close range where it can be most effective. When an offensive attack is undertaken and completed successfully, the fire can be controlled with the least amount of agent and the least amount of property damage.

Facility brigade members usually undertake offensive strategies when the fire is reported quickly and is usually still in the **incipient stage**. Offensive operations are accomplished by applying water from hose lines or using other extinguishing agents such as portable fire extinguishers. Large handlines are often used to conduct interior operations on larger fires, although they are much heavier, require additional personnel, and are more difficult to maneuver inside a facility. Large handlines provide a greater volume of water and operating pressure and should be used only by properly trained fire brigade members.

Large handlines and master streams are more often used in defensive operations. Defensive strategy is used in situations where the fire is too large to be controlled by an offensive attack. It is also used in situations where

## Brigade Member Safety Tips

Offensive (interior attack) and defensive (exterior attack) operations must NEVER be performed simultaneously.

## Brigade Member Safety Tips

A brigade leader should never risk the lives of fire brigade team members when there are no lives to save.

the integrity of the structure is in question and the level of risk to brigade members conducting an interior attack would be unacceptable. The primary objective in a defensive operation is to prevent the fire from spreading. Water or other suppression agents are directed into the structure through doorways, windows, and other openings, or onto exposures to keep a fire from extending from one area to another, while brigade members remain outside the building and operate from safe positions.

The decision to conduct offensive or defensive operations must be made by the brigade leader at the beginning of fire suppression operations and is periodically reevaluated throughout the incident. The decision to employ offensive or defensive operations must be made before operations begin and must be clearly communicated and understood by everyone involved in the operation. There is no room for confusion between offensive and defensive operations. Master streams from defensive operations will push fire, heat, and smoke back into the structure. It would be extremely dangerous if one group of brigade members was operating defensive master streams outside a facility while another brigade crew was inside the facility conducting an offensive attack. Retired Phoenix Fire Department Chief Alan Brunacini said it best, "Combining strategic modes (offensive/defensive) is like ordering artillery on yourself."

If the decision is made to switch from offensive to defensive or from defensive to offensive strategy at any point during an operation, the change must be clearly communicated and understood. The change from offensive to defensive operations could occur if an interior attack is unsuccessful or the risk factors are determined to be too great to justify having fire brigade team members inside the structure. Prior to transitioning from offensive to defensive strategy, the brigade leader should complete a full personnel accountability check to ensure all team members are accounted for and outside the structure. Sometimes the strategy is switched from defensive to offensive after an exterior attack has reduced the volume of fire inside a building to the point that fire brigade team members can enter the building and complete the extinguishment with handlines. An offensive fire attack requires well-planned coordination between crews performing different tasks, such as ventilating, operating hose lines, and conducting aggressive search and rescue.

Figure 21-1 Exposures should be cooled to prevent equipment failure.

A defensive strategy should be implemented when the brigade leader determines that it would be impossible to enter the fire area or structure and control the fire with handlines, as well as in situations where the risk of injury or death of a team member is excessive. Defensive operations involve the use of large hose lines and master stream devices from the exterior of the structure to confine the fire. Exposure protection should be a high priority during defensive operations, and equipment should be cooled to prevent equipment failure (▲ Figure 21-1).

## Command Considerations

The incident commander (IC) has to evaluate a wide range of factors to decide if offensive strategy (interior attack) or defensive strategy (exterior attack) should be used at a particular fire. If the risk factors are too great, an exterior attack is the only acceptable option. If the decision is made to launch an interior attack, the IC has to determine where and how to attack, considering both safety and the effectiveness of the operation. The factors to be evaluated when considering whether to enter the structure to mount an attack include the following:

- What are the risks versus the potential benefits?
- Is it safe to send brigade members into the building?
- Are there any structural concerns?
- Are there any lives at risk?
- Does the size of the fire prohibit entry?

**Brigade Member Safety Tips**

Incipient fire brigade members are limited to operating handlines up to $1\frac{1}{2}$ inches and master streams up to 300 gpm.

- Are there enough brigade members on the scene to mount an interior attack? (Remember the two-in/two-out rule.)
- Is there an adequate water supply?
- Can proper ventilation be performed to support offensive operations?

After sizing up the situation, the IC must determine which type of attack is appropriate. As a new brigade member, you are not responsible for determining the type of fire attack that will be used, but you should understand the factors that go into making these decisions and why the IC orders different types of fire attacks for different types of fires.

## Operating Hose Lines

The most basic fire brigade team member skills that must be mastered by every team member involve the use of hose lines to apply water onto a fire for extinguishment. As a brigade member, you must learn how to advance and operate a hose line effectively to extinguish a fire. Understanding the dangers of hose lines is essential for your protection, the protection of your brigade members, and the protection of other brigades that may be operating nearby (▶ Figure 21-2). One general rule that must always be observed is to know where your stream is going. Care must always be taken to avoid opposing hose lines, where two crews are operating at the same time "against" each other from opposite directions.

Fire brigade members must learn to operate both large and small hose lines, as well as master stream devices. Small hose lines are generally up to 2 inches in diameter with the most common being $1\frac{1}{2}$ or $1\frac{3}{4}$ inches. Hose lines discharge large volumes of water under operating pressures ranging from 100 to 200 pounds per square inch (psi), so brigade members must be physically capable of controlling the nozzles at all times. Generally, one brigade member can usually operate the nozzle while a second brigade member can provide added support and assistance while advancing and maneuvering hose lines.

**Brigade Member Safety Tips**

Care must be taken to avoid opposing hose lines—two teams operating at the same time "against" each other.

**Figure 21-2** Understanding the dangers of hose lines is essential for your brigade's protection, and for the protection of other brigades that may be operating nearby.

Large hose lines are defined as hoses that are $2\frac{1}{2}$ inches in diameter or larger. The most common large handline used for fire attack is $2\frac{1}{2}$ inches. Water can flow through such a hose at a rate of 250 gpm and a pressure of 100 pounds per square inch (psi). Compared to smaller handlines, these large hose lines are much heavier and less maneuverable in a confined space, such as when operating inside a facility. One brigade member can control a properly anchored hose line; however, at least three brigade members are required to effectively advance and maneuver a large hose line.

Master streams are used when even larger quantities of water are required to control a large fire or when protecting exposures. Master stream devices generally deliver in excess of 350 gallons of water per minute (gpm) with some nozzles having the ability to deliver more than 10,000 gpm. The most commonly used devices generally flow between 1500 and 4500 gpm. Because of their weight, flow rates, and higher operating pressures, master stream devices are operated from fixed positions either on the ground, on top of a piece of fire apparatus, or from some other elevating device.

### Fire Streams

Several different types of fire streams can be produced by different types of nozzles. The nozzle defines the pattern and form of the water that is discharged onto the fire. A fire stream can be produced with either a smooth bore nozzle or an adjustable fog nozzle. Brigade policies and standard operating procedures (SOPs) will usually determine the type of nozzle that is used with different types of hose lines. The nozzle operator must know which type of nozzle should be used in different situations. When an adjustable nozzle is used, the nozzle operator must know how to set the discharge pattern to produce different kinds of streams.

## Brigade Member Tips

Although every effort should be made to minimize water damage, remember, property can be dried out but it cannot be unburned.

## Brigade Member Safety Tips

It is imperative to extinguish all fire as you proceed to its seat. Failure to do so may allow the fire to grow behind you, entrapping you by cutting off your escape route.

The first major distinction in nozzle discharge patterns is between a fog stream and a straight stream ► Figure 21-3 . A fog stream divides water into droplets, which have a very large surface area and can absorb heat efficiently. When heat levels in a building need to be lowered quickly, a fog stream is usually the pattern of choice. A fog stream can also be used to protect brigade members from the heat of a large fire. Most adjustable nozzles can be adjusted from a straight stream, to a narrow fog pattern, to a wide fog pattern, depending on the reach that is required and how the stream will be used.

A straight stream provides more reach than a fog stream, so it can hit the fire from a greater distance. A straight stream also keeps the water concentrated in a small area, so it can penetrate through a hot atmosphere to reach and cool the burning materials. A straight stream is produced by setting an adjustable nozzle to the narrowest pattern it can discharge. This type of stream is made up of a highly concentrated pattern of droplets that are all discharged in the same direction.

A solid stream is produced by a smooth bore nozzle ► Figure 21-4 . A solid stream has more reach and penetrating power than a straight stream, because it is discharged as a continuous column of water.

One important point that must be remembered when selecting and operating nozzles is the amount of air that is moved along with the water. A fog stream naturally moves a large quantity of air along with the mass of water droplets. This air flows into the fire area along with the water. When this air movement is combined with steam production as the water droplets encounter a heated atmosphere, the thermal balance is likely to be disrupted quickly. The hot fire gases and steam can be displaced back down toward the nozzle operator or pushed into a different area of the building. Straight and solid streams move little air in comparison with a fog stream, so there are fewer concerns with displacement and disruption of the thermal balance.

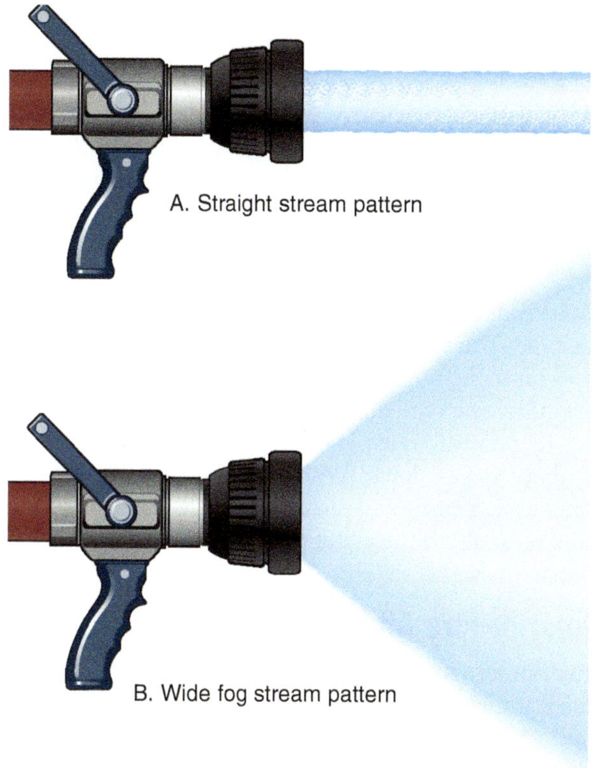


Figure 21-3 A straight stream and a fog stream are both produced using adjustable fog nozzles.

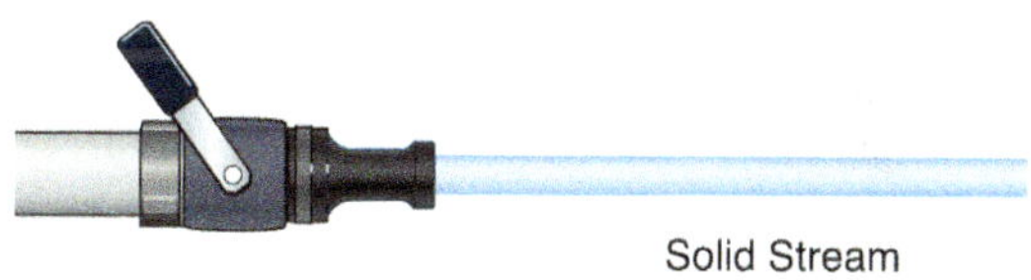


Figure 21-4 A solid stream is produced by a smooth-bore nozzle.

The air movement created by a fog stream can be used for ventilation. Discharging a fog stream out through a window or doorway will draw smoke and heat out in the same manner as an exhaust fan. This operation must be performed carefully to prevent accidentally drawing hidden fire toward the nozzle operator.

## Water Hammer

Water hammer is the surge of pressure caused when a high-velocity flow of water is abruptly shut off. A fast-moving stream of water has a large amount of kinetic energy. If the stream is suddenly stopped by closing a valve, all of its kinetic energy is converted to an instantaneous increase in pressure behind the valve. Because water cannot be compressed, the additional pressure is transmitted along the hose or pipe as a shock wave (water hammer). The resulting water hammer can rupture a hose, cause a coupling to separate, or damage the plumbing on a piece of fire apparatus. The faster a nozzle or valve is closed, the greater the shock-wave-generated effect will be. Severe water hammer can even damage an underground water distribution system. Damage generated by a water hammer can injure brigade team members and cause additional damage to firefighting equipment.

A similar situation can occur if a valve is opened too quickly and a surge of pressurized water suddenly fills a hose or pipe. The surge in pressure can damage the hose or pipe and cause the brigade member at the nozzle to lose control of the nozzle or appliance.

To prevent water hammer, always open and close fire hydrant valves slowly. When operating nozzles on an attack line, always open and close the nozzle slowly. Pump operators must also be aware of the need to open and close all fire apparatus valves slowly.

## Interior Fire Attack

Interior structure fires occur inside a building, structure, or facility that is usually completely enclosed and has a roof structure. The fuel can be the contents, structural members, or other items stored in the structure. Interior structural firefighting involves the physical activity of suppression, rescue, or both inside of structures that are involved in a fire situation beyond the incipient stage. An interior fire attack is an offensive operation that requires brigade members to enter a structure and discharge an extinguishing agent (usually water) onto the fire. The larger the fire, the greater the challenge the fire brigade teams experience and the greater the risks that are encountered.

Interior fire attack can be conducted on many different scales. In many cases an interior attack is conducted on a fire that is only burning in one room, and the fire can be controlled quickly by one attack hose line. Larger fires require more water, which could be provided by two or more small handlines working together or one or more larger handlines. Fires that involve multiple rooms, large spaces, or concealed spaces are more complicated and require much more coordination; however, the basic techniques for attacking the fire are similar.

Direct attack and indirect attack are two different methods of discharging water onto a fire. A combination attack is performed in two stages, beginning with an indirect attack and then continuing with a direct attack.

### Direct Attack

The most effective means of fire suppression in most situations is the direct attack. The direct attack uses a straight or solid hose stream to deliver water directly onto the base of the fire (► Figure 21-5). The water cools the fuel until it is below its ignition temperature. The water is directed into the fire in short bursts and in a controlled method. Brigade members should not apply more water than necessary to extinguish the fire, in order to keep damage from excess water to a minimum. To perform a direct attack, follow the steps in (Skill Drill 21-1).

1. Select the proper hose line used to fight the fire depending on the fire's size, location, and type.
2. Advance the hose line to the entry point of the structure on the unburned side.
3. Don face piece, activate SCBA and Personal Alert Safety System (PASS) device prior to entering the building.
4. Signal the operator/driver that you are ready for water.
5. Open the nozzle to purge air from the system and make sure water is flowing.
6. Make sure ventilation is completed or in progress.
7. Enter into the structure and locate the seat of the fire.
8. Apply water in either a straight or solid stream onto the base of the fire in short bursts.
9. Watch for changes in fire condition; use only enough water to extinguish the fire.
10. Locate and extinguish hot spots.

### Indirect Attack

Indirect application of water is used by interior structural brigades in situations where the temperature is increasing and it appears that the room or space is ready to flashover. A short burst of water is aimed at the ceiling to cool the super-heated gases in the upper levels of the room or space. This action can prevent or delay flashover long enough for brigade members to apply water directly to the seat of the fire or to make a safe exit. Follow the SOP of your organization regarding the application of water.

The objective of an indirect attack is to quickly remove as much heat as possible from the fire atmosphere. An indirect attack is particularly effective at preventing flashover from occurring. This method should be used when a fire has produced a layer of hot gases at the ceiling level. When the water is applied into the hot fire gases, it is converted to steam, absorbing tremendous quantities of heat in the process. The atmosphere is cooled quickly down. The heat that is used to convert the water to steam is removed from the combustion process.

Figure 21-5 In a direct attack, a straight or solid hose stream is used to deliver water directly onto the base of the fire.

Figure 21-6 In an indirect attack, brigade members direct a fog stream at the ceiling of the intensely heated area to create steam.

An indirect attack can be performed by brigade members using a straight stream, a solid stream, or a narrow fog stream. The brigade members direct their water at the ceiling of the intensely heated area where the hot gases are layered, in order to create steam ▲ Figure 21-6. This is often referred to as "painting the ceiling" with the water stream. The water is distributed over a large surface area to absorb heat as quickly as possible. Once the temperature has been reduced and the area has been properly ventilated, the brigade members can switch to a direct attack to complete extinguishment.

As soon as enough steam has been produced to reduce the fire, the fire stream should be shut down so that the thermal layering of the superheated gases is disturbed as little as possible. When water is converted to steam, it expands to occupy a volume 1700 times greater than the volume of the water. This expansion tends to displace the hot gases that were near the ceiling and push them down toward the floor. This mixture of steam and hot gases is capable of causing serious steam burns to brigade members through their PPE. Serious injuries can occur if brigade members put too much water into the upper atmosphere and the hot gases are forced down on top of them.

To perform an indirect attack, follow the steps in Skill Drill 21-2.

1. Don PPE, including SCBA.
2. Select the correct hose line to be used to attack the fire depending on the type of fire, its location, and size.
3. Advance the hose line from the apparatus to the opening in the structure where the indirect attack will be made; ideally from the unburned side.
4. Don face piece; activate SCBA and PASS device.
5. Notify the operator/driver that you are ready for water.
6. Open the nozzle to be sure that air is purged from the hose line and that water is flowing. If using a fog nozzle, ensure proper nozzle pattern for entry. Shut down the nozzle until you are in a position to apply water.
7. Advance with a charged hose line to the location where you are going to apply water.
8. Direct the water stream toward the upper levels of the room and ceiling into the heated area overhead and move the stream back and forth. This should be done in short 3-4 second bursts. Flow water until the room begins to darken. Then shut the nozzle off and assess conditions.
9. Watch for changes and reduction in the amount of fire. Once the fire is reduced, shut down the nozzle.
10. Check to make sure ventilation has been completed.
11. Attack remaining fire and hot spots until extinguished.

### Combination Attack

The combination attack employs both the indirect and direct attack methods sequentially. This method should be used when a room's interior has been heated to the point where it is nearing a flashover condition. The brigade members should first use an indirect attack method to cool the fire gases down to safer temperatures and prevent flashover from occurring. This is followed with a direct attack on the main body of fire.

When using this method, the brigade member who is operating the nozzle should be given plenty of space to maneuver it. Only enough water as needed to control the fire should be used, thus avoiding unnecessary water damage.

To perform a combination attack, follow the steps in Skill Drill 21-3.

1. Wearing full PPE and SCBA, select the correct hose line to accomplish the suppression task at hand.
2. Stretch the hose line to the entry point of the structure and signal the operator/driver that you are ready to receive water.

## Brigade Member Safety Tips

Indicators of possible building collapse
- Leaning walls
- Smoke emitting from cracks
- Creaking sounds
- Sagging roofs or floors

Considerations for possible building collapse
- Exposure to serious fire
- Water loading due to master streams
- Age of building

## Brigade Member Tips

When using the indirect or combination attack, watch for droplets of water raining down. This indicates lowered ceiling temperatures. If no droplets fall, the ceiling is still too hot.

3. Open the nozzle to get the air out and be sure that water is flowing.
4. Enter the structure and locate the area of origin of the fire.
5. Aim your nozzle at the upper left corner of the fire and make either a "T," "O," or "Z" pattern with the nozzle. Remember to start high and work the pattern down to the fire level.
6. Use only enough water to darken down the fire without upsetting the thermal layering.
7. Once the fire has been reduced, find the remaining hot spots, and complete extinguishments using a direct attack.

## Large Handlines

Large handlines can be used for either offensive fire attack or for defensive operations. In an offensive attack situation, a 2½-inch attack line can be advanced into a building to apply a heavy stream of water onto a large volume of fire. The same direct and indirect attack techniques as were described for small hose lines can also be used with large handlines. A 2½-inch handline can overwhelm a substantial interior fire if it can be discharged directly into the involved area. The extra reach of the stream can also be valuable for an interior attack in a large building.

It is more difficult for brigade members to advance and maneuver a large handline inside a building, particularly in tight quarters or around corners. At least three brigade members are usually needed to advance and maneuver a 2½-inch handline inside a building. The brigade members have to contend with the nozzle reaction force, as well as the combined weight of the hose and the water. In situations where the hose line has to be advanced a considerable distance into a building, additional brigade members will be required to move the line. The extra effort is balanced by the powerful fire suppression capabilities of a large handline.

Large handlines are often used in defensive situations to direct a heavy stream of water onto a fire from an exterior position. In these cases the nozzle is usually positioned to be operated from a single location by one or two brigade members. The stream can be used to attack a large exterior fire or to protect exposures. The stream can also be directed into a building through a doorway or window opening to knock down a large volume of fire inside. If the exterior attack is successful in reducing the volume of fire, the IC might make the decision to switch to an offensive (interior) attack to complete extinguishment.

### One-Person Method

One brigade member can control a large attack hose by forming a large loop of hose about 2 feet behind the nozzle. By placing the loop over the top of the nozzle, the weight of the hose stabilizes the nozzle and reduces the nozzle reaction. To add more stability, lash the hose loop to the hose behind the nozzle where they cross. This reduces the energy needed to control the line if it is necessary to maintain the water stream for a long period of time. This method does not allow the hose to be moved while water is flowing, but it is good for protecting exposures when operating in a defensive attack mode. To perform the one-person method for operating a large handline, follow the steps in ▶ Skill Drill 21-4.

1. Select the correct size fire hose for the task to be performed.
2. In full PPE and SCBA, advance the hose into the position from which you plan to attack the fire.
3. Signal the operator/driver that you are ready for water.
4. Open the nozzle to allow air to escape the hose and to ensure that water is flowing.
5. Close the nozzle and then make a loop with the hose, ensuring that the nozzle is UNDER the hose line that is coming from the fire apparatus. **(Step 1)**
6. Lash the hose together at the section of hose where they cross or use your body weight to kneel or sit on the hose line at the point where the hose crosses itself. **(Step 2)**
7. Be sure to allow enough hose to extend past the section where the line crosses itself for maneuverability. **(Step 3)**
8. Open the nozzle and direct water onto the designated area. **(Step 4)**

### Two-Person Method

When two brigade members are available to operate a large handline, one should be the nozzle operator, while the other provides a back-up. The nozzle operator grasps the nozzle with one hand and holds the hose behind the nozzle with the other hand. The hose should be cradled across the brigade member's hip for added stability. The back-up brigade member should be positioned about 3 feet behind the nozzle operator. This person grasps the hose with both hands and holds the hose against a leg or hip. A hose strap can also be used to provide a better hand grip on a large hose line. When the line is operated from a fixed position, the second brigade member can kneel on the hose with one knee to stabilize it against the ground. ▶ Skill Drill 21-5 demonstrates this skill.

1. Don all PPE and SCBA.
2. Select the correct hose line for the task at hand.
3. Stretch the hose line from the fire apparatus into position. **(Step 1)**

> **Brigade Member Safety Tips**
>
> Master streams should NEVER be directed into a building while brigade members are operating inside the structure. The force of the stream can push the heat, smoke, and fire onto the brigade members.

21-4 Skill Drill

## One-Person Method for Operating a Large Handline

Close the nozzle and then make a loop with the hose, ensuring that the nozzle is UNDER the hose line that is coming from the fire apparatus.

Lash the hose together at the section of hose where they cross, or use your body weight to kneel or sit on the hose line at the point where the hose crosses itself.

Be sure to allow enough hose to extend past the section where the line crosses itself for maneuverability.

Open the nozzle and direct water onto the designated area.

4. Signal the pump operator that you are ready for water.
5. Open the nozzle a small amount to allow air to escape and to ensure that water is flowing.
6. Advance the hose line as needed. **(Step 2)**
7. Before attacking the fire, the brigade member on the nozzle should cradle the hose on his or her hip while grasping the nozzle with one hand and supporting the hose with the other hand. **(Step 3)**
8. The second brigade member should stay approximately 3 feet behind the brigade member on the nozzle. The second brigade member should grasp the hose with two hands and, if necessary, use a knee to stabilize the hose against the ground.
9. A hose strap can be used to provide a better grip on the hose.
10. Open the nozzle in a controlled fashion and direct water onto the fire or designated exposure. **(Step 4)**

If you need to advance a flowing $2^1/_2$-inch handline a short distance and only two brigade members are available, be aware of the large reaction force exerted by the flowing water. It is much easier to shut down the nozzle momentarily and move it to the new position than to relocate a flowing line.

21-5 Skill Drill

## Two-Person Method for Operating a Large Handline

1

Stretch the hose line from the fire apparatus into position.

2

Advance the hose line as needed.

3

Before attacking the fire, the brigade member on the nozzle should cradle the hose on his or her hip while grasping the nozzle with one hand and supporting the hose with the other hand.

4

The second brigade member should stay approximately 3 feet behind the brigade member on the nozzle. The second brigade member should grasp the hose with two hands and may use a knee to stabilize the hose against the ground if necessary. Open the nozzle in a controlled fashion and direct water onto the fire or designated exposure.

If the line must be moved while water is flowing, both brigade members must brace the hose against their bodies to keep it under control. Three brigade members can stabilize and advance a large handline more comfortably and safely than two.

## Master Stream Devices

Master stream devices are used to produce high-volume water streams for large fires. There are several different types of master stream devices, including portable monitors, deck guns, ladder pipes, and other elevated stream devices. Most master streams discharge between 350 and 1500 gallons of water per minute, although much larger capacities are available for special applications. In addition, the stream that is discharged from a master stream device has a greater range than a handline, so it can be effective from a greater distance.

A master stream device can be either manually operated or directed by remote control. Many master stream devices can be set up and then left to operate unattended. This can be extremely valuable in a high-risk situation, because there is no need to leave a brigade member in an unsafe location or hazardous environment to operate the device (► Figure 21-7).

Master streams are used only during defensive operations. A master stream should *never* be directed into a building while brigade members are operating inside the structure. Master streams may be placed in position during search and rescue activities if it appears that offensive operations will not be able to stop the progression of the fire. Nevertheless, the master stream must never be turned on until all search and rescue team members have evacuated the building and all members are accounted for, because it can push heat, smoke, or fire onto the brigade team members.

Master streams generate strong air current movement, so they will have a major effect on any ventilation efforts that may have been initiated. When using a master stream through an opening in a structure, it is essential to ensure that additional openings exist through which the fire, hot gases, and smoke can escape. Failure to have adequate openings could result in trapping the fire and by-products inside the structure.

Master stream devices create water streams that have excessive force. The force and impact of the stream can dislodge loose materials, cause extensive damage to interior property, or lead to structural collapse. Because master streams are considered to be the last defense against a fire, property conservation is usually a low priority when they are used.

When a master stream is properly positioned, brigade members should see a change in the fire growth. A properly positioned master stream device should be angled so that the stream is deflected off the ceiling or other overhead equipment. Deflecting the stream causes a raindrop effect and provides the best extinguishing characteristics. If brigade team members do not see a decrease in the fire's progression, an evaluation should be conducted to determine whether the stream is being properly applied and whether a sufficient quantity of water is being applied to the fire.

**Figure 21-7** Master streams are effective when long reach and large cooling streams are required.

**Figure 21-8** A deck gun is permanently mounted on a vehicle and equipped with a piping system that delivers water to the device.

## Deck Guns

A deck gun is permanently mounted on a vehicle and equipped with a piping system that delivers water to the device. These devices are sometimes called turret pipes or wagon pipes (▲ **Figure 21-8**). If the vehicle is equipped with a pump, the pump operator can usually open a valve to start the flow of water. Sometime a hose must be connected to a special inlet to deliver water to the deck gun. If your fire apparatus is equipped with a deck gun, you need to learn your role when placing it in operation. To operate a deck gun, follow the steps in **Skill Drill 21-6**.

1. Make sure that all firefighting personnel are out of a structure before using a deck gun.
2. Place the deck gun in position.
3. Aim the deck gun at the fire or at the intended exposure.
4. Signal the operator/driver that you are ready for water.
5. Once water is flowing, adjust the angle, pattern, or water flow as necessary.

## Portable Monitor

A portable monitor is a master stream device that can be positioned wherever a master stream is needed and placed on the ground (► **Figure 21-9**). Hose lines are connected to the portable monitor to supply the water. Most portable monitors are supplied with either two or three $2^{1}/_{2}$-inch inlets or with one large-diameter hose inlet. Some monitors can be used as a deck gun or taken off the fire apparatus and used as a portable monitor.

To deploy a portable monitor, remove it from the apparatus and carry it to the location where it will be used.

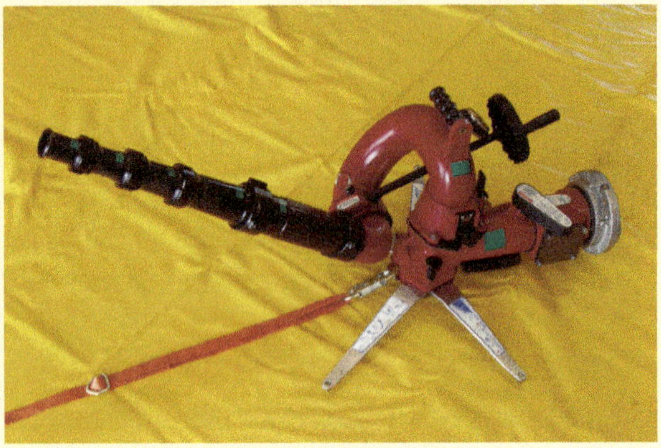

Figure 21-9 A portable monitor is placed on the ground and supplied with water from one or more hose lines.

Advance an adequate number of hose lines from the engine to the monitor. The number of hose lines needed depends on the volume of water needed and the size of the hose lines. Form a large loop in the end of each hose line in front of the monitor and then attach the male coupling to the inlets of the monitor. The loops serve to counteract the force created by the flow of the water through the nozzle.

To set up and operate a portable monitor, follow the steps in ▼ Skill Drill 21-7.

1. Remove the portable monitor from the fire apparatus and move it into position.
2. Attach the necessary hose lines to the monitor as per SOPs or manufacturer's instructions. **(Step 1)**
3. Be sure to loop the hose lines in front of the monitor to counteract the force created by water flowing out of the nozzle. **(Step 2)**
4. Signal the operator/driver that you are ready for water.
5. Aim the water stream at the fire or onto the designated exposure and adjust as necessary.

The nozzle reaction force can cause a portable monitor to move from the position where it was placed if it is not adequately secured. A moving portable monitor can be extremely dangerous to anyone in its path. Many portable monitors are equipped with a strap or chain that must be secured to a fixed object to prevent the monitor from moving. Pointed feet on the base also help to keep a portable monitor from moving. If the stream is operated at a low angle, the reaction force will tend to make the monitor unstable. A safety lock is usually provided to prevent the monitor from being lowered beyond a safe limit of 35°. When setting up any portable monitor, always follow the manufacturer's instructions and the SOPs of your brigade for safe and effective operation.

### Elevated Master Streams

Elevated master stream devices can be mounted on aerial ladders, aerial platforms, or special hydraulically operated booms (▶ Figure 21-10). A ladder pipe is an elevated stream device that is mounted at the tip of an aerial ladder. On many

21-7 Skill Drill

## Set Up and Operate a Portable Monitor

1 Attach the necessary hose lines to the monitor as per SOPs or manufacturer's instructions.

2 Be sure to loop the hose lines in front of the monitor to counteract the force created by water flowing out of the nozzle.

Figure 21-10 Aerial ladder master stream.

aerial ladders the ladder pipe is only attached to the top of the ladder when it is needed, and a hose is run up the ladder to deliver water to the device. Most newer aerial ladders and tower ladders are equipped with a fixed piping system to deliver water to a permanently mounted master stream device at the top. This saves valuable set-up time at a fire scene. If your apparatus is equipped with a ladder pipe, you need to learn how to assist in its set-up.

## Protecting Exposures

Protecting exposures refers to actions that are taken to prevent the spread of a fire to areas that are not already burning. Exposure protection is a consideration at every fire; however, it is a much more important consideration when the fire is large. If the fire is relatively small and contained within a limited area, the best way to protect exposures is usually to extinguish the fire; when the fire is extinguished, the exposure problem ceases to exist. In cases where the fire is too large to be controlled by an initial attack, exposure protection becomes a priority. In some cases the best outcome you can hope to accomplish is to stop the fire from spreading.

The IC has to consider the size of the fire and the risk to exposures in relation to the amount of firefighting capability that is available and how quickly it can be assembled. In some cases the IC will direct the first brigades to protect exposures while a second group prepares to attack the fire.

**Brigade Member Safety Tips**

Any time a floor feels hot through your PPE, consider the possibility of a fire burning in the level below you. Beware of a weakened floor caused by a hidden basement fire!

Figure 21-11 Protecting an exposure from radiant heat.

Sometimes the IC has to identify a point where the progress of the fire can be stopped and direct all firefighting efforts toward that objective.

Protecting exposures involves very different tactics from offensive fire attack. At a large fire, the first priority is to protect exposed buildings and property from a combination of radiant heat, convective heat, and burning embers (▲ Figure 21-11). The best option is usually to direct the first hose streams at the exposures rather than the fire itself. Wetting the exposures will keep the fuel from reaching its ignition temperature. Because radiant heat can travel through a water stream, directing water onto the exposed surface is more effective than aiming a stream between the fire and the exposure. Master stream devices such as deck guns, portable monitors, and elevated master streams are excellent tools for protecting exposures. Large volumes of water can be directed onto the exposures from a safe distance.

### Ventilation

Before any interior attack is initiated, it is important that the structure be ventilated. Ventilation must be coordinated with the suppression efforts to ensure that both events occur simultaneously and in a manner that supports the attack plan. Proper ventilation allows for the hot gases and smoke to be removed from the building, improving visibility and tenability in the building for any trapped victims and brigade members. Improper ventilation can create conditions that allow a fire to burn more aggressively and make it more difficult for brigade members to enter and attack the fire. Coordination is essential to ensure that the hose lines will be ready to attack when the ventilation openings are made. The ventilation openings must be located so that the hot smoke and gases will be drawn away from the attack crews.

### Concealed-Space Fires

Fires in ordinary and wood frame construction can burn in combustible void spaces behind walls and under subfloors. In order to prevent the fire from spreading, these fires must be found and suppressed (► Figure 21-12).

# VOICES OF EXPERIENCE

**"As I saw this, I called for the units inside the structure to back out."**

Fighting a fire offensively or defensively is a key decision that must be made by the command function at the start of operation. Once that decision is made, it must be clearly communicated to everyone on the fireground.

While fighting a very large fire in a warehouse, I made the decision to attack the fire offensively using $1^{3}/_{4}$-inch attack lines from two sides of the warehouse. Defensive lines were also put in place to be used in the event the fire began to spread into the issue counter area, and a ladder truck with a master stream device was placed at the rear of the facility to be used only if the offensive operation failed. During the initial attack, the advancing firefighting crews pushed the fire out the rear windows. The operator of the master stream device thought the fire had gotten out of control and opened up the nozzle and began to discharge water into the structure. As I saw this, I called for the units inside the structure to back out, and was quickly able to get the operator of the master stream to shut down. Once the master stream was shut down, the offensive crews were advanced back into the structure and were able to extinguish the fire.

It is imperative that everyone operating on the fire ground is aware of the command decisions that have been made and control their impulse to act independently based on their own observations. Combining offensive and defensive operations can be very dangerous. It is essential that everyone knows when a change from offensive to defensive operations is going to occur, and all crews operating in the interior must be accounted for before defensive operations are undertaken.

Frederick J. Knipper
Duke University—Duke Health System
Durham, North Carolina

Figure 21-12 Fires may be hidden behind walls.

To locate and suppress fires behind walls and under subfloors, follow the steps in Skill Drill 21-8.

1. Locate the area of the fire building where a hidden fire is believed to be.
2. Look for signs of fire such as smoke coming from cracks or openings in walls, charred areas with no outward evidence of fire, and peeling or bubbled paint or wallpaper. Listen for cracks and pops or hissing steam.
3. If available, use a thermal imager to look for areas of heat that may indicate a hidden fire.
4. Using the back of your hand, feel for heat coming from a wall or floor.
5. If a hidden fire is suspected, use a tool such as an axe or Halligan to remove the building material over the area.
6. If fire is located, expose the area as well as possible and extinguish the fire using conventional firefighting methods. Be sure to expose as much area as needed without causing unnecessary damage.

For more information on evaluating the structure for complete extinguishment, see Chapter 19, Salvage and Overhaul.

## Basement Fires

Fires in basements or below-grade level are rare in most industries and present several different challenges. Basements are difficult and dangerous to enter, and they have limited routes of egress. Basements are usually difficult to ventilate, which means that an interior attack often has to be made in conditions of high heat and low visibility. This also means that it will be difficult to remove the fire gases and steam produced by the attack lines. It can be difficult to see even after ventilation has been performed. Basements are often used for storage, and it may be hard to keep your sense of orientation in the narrow disorganized spaces.

Figure 21-13 Fire below-grade.

Brigade members should identify the safest means of entry and exit into the area where firefighting operations will be conducted. An exterior access point allows brigade members to enter a basement without passing through the hot gas layers at the basement ceiling level. If the only point of entry is an interior stairway, brigade members must protect that stairway opening to keep the fire from extending to the upper floors ▲ Figure 21-13. Ventilation must be planned and conducted early. If ventilation is not managed properly, the interior stairwell will act as a chimney and bring heat and smoke up from the basement.

Always consider the possibility of a fire burning below the ground floor level when you enter a fire building. A fire that appears to be on the ground floor could have originated in the cellar and weakened the floor. Cellar fires can also spread to upper floors in houses with balloon construction.

## Fires Aboveground Level

Advancing charged hose lines up stairs and along narrow hallways requires much more physical effort than fighting a fire on the ground level in an open area. It is always important to protect stairways and other vertical openings between floors when fighting a fire in a multiple-level structure. Hose lines must be placed to keep the fire from extending vertically and to ensure that exit paths are always available ► Figure 21-14.

When advancing a hose above the ground floor, advance them uncharged until you reach the fire floor and have extra hose available. This allows for easier advancement of attack lines.

Interior fire crews must always look for a secondary exit path in case their entry route is blocked by the fire or by a structural collapse. The secondary exit could be a second interior stairway, an outside fire escape, a ground ladder placed to a window, or an aerial device.

Be aware of the risk of structural instability and collapse. Check the floor that you are working on and listen for the sounds associated with a failing ceiling or roof. Do not use

Figure 21-14 With fires above grade, stairways in structures must be protected from fire.

more water than is needed to extinguish the fire. Additional water adds weight to the structure and could lead to structural collapse.

In multistory structures, the standpipe system may be used to supply water for hose lines. Brigade members must practice connecting hose lines to standpipe outlets and extending lines from stairways into remote floor areas. Additional hose, tools, air cylinders, and emergency medical service equipment should be staged one or two floors below the fire.

### Fires in Large Buildings

Many large buildings have floor plans that can cause brigade members to become lost or disoriented while working inside, particularly in low-visibility or zero-visibility conditions. The use of guide lines may be necessary to keep brigade members from becoming lost and running out of air. A well-organized preincident plan of the structure can be essential when fighting this type of fire. Knowing the occupancy and the other hazards beforehand will help in determining the best strategy and tactics.

### Fires in Buildings During Construction, Renovation, or Demolition

Buildings that are under construction, renovation, or demolition are all at increased risk for destruction by fire. These buildings often have large quantities of combustible materials exposed, without the fire-resistant features of a finished building. If the building lacks windows and doors, an almost unlimited supply of oxygen could be available to fuel a fire. Fire detection, fire alarm, and automatic fire suppression systems might not be operational. Construction workers may be using torches and other flame-producing devices. These buildings are often unoccupied and could be targets for arson. The fire suppression systems are many times nonoperational in these buildings.

Under these conditions a fire in a large building could be impossible to extinguish. If there are no life safety hazards involved, brigade members should use a defensive strategy for this type of fire. No brigade members should enter the building, and a collapse zone should be established. A defensive exterior operation should be conducted using master streams and large handlines to protect exposures.

### Fires in Stacked or Piled Materials

Fires occurring in stacked or piled materials can present a variety of hazards. The greatest danger to brigade members is the possibility that a stack of heavy material such as rolled paper or baled rags will collapse without warning. This can occur if a fire has damaged the stacked materials or if water has soaked into them. Absorbed water can greatly increase the weight of many materials and also weaken cardboard and paper products. The water discharged by automatic sprinklers alone can be sufficient to make some stacked materials unstable. A tall stack of material that falls on top of a team of brigade members can cause injury or death.

Fires in stacked materials should be approached cautiously. All brigade members must remain outside potential collapse zones. Mechanical equipment should be used to move material that has been partially burned or water-soaked.

Conventional methods of fire attack can often be used to gain control of the fire; however, water must penetrate into the stacked material in order to fully extinguish the residual combustion. Class A foam can often be used to extinguish smoldering fire in tightly packed combustible materials. Overhaul will require the materials to be separated to expose any remaining deep-seated fire. This can be a labor intensive process unless mechanical equipment can be used to dig through the material.

### Trash Container and Rubbish Fires

Trash container (dumpster) fires usually occur outside of any structure and appear to present fewer challenges than a fire inside a building. Brigade members still must be vigilant in wearing full PPE and using SCBA when fighting trash container fires. There is no way of knowing what might be inside a trash container.

If the fire is deep seated, the trash will have to be overhauled. Manual overhaul involves pulling the contents of a trash container apart with pike poles and other hand tools so that water can reach the burning material. This can be

## Brigade Member Safety Tips

Fires that occur in confined spaces generate deadly levels of toxic gases or vapors and are very dangerous. Entry should be made only by brigade members who are specially trained in interior structural operations and who use PPE and SCBA.

labor intensive and involves considerable risk to brigade members. The brigade members are exposed to whatever contaminants are in the container, as well as the risks of injury from burns, smoke, or other causes. Considering the value of the contents of a trash container, it is difficult to justify any risk to brigade members.

Class A foam is useful for many trash container fires, because it allows water to soak into the materials. This can eliminate the need for manual overhaul. Some fire brigades use the deck gun on the top of an engine to extinguish large trash container fires and then complete extinguishments by filling the container with water. This is done by pointing the deck gun at the dumpster and slowly opening the discharge gate to lob water into the container (▶ Figure 21-15).

**Figure 21-15** Some brigades use a deck gun on top of an engine to extinguish large trash container fires.

Trash containers are often placed behind, or immediately adjacent to, large buildings and businesses. If the container is close to the structure, be sure to check for extension. Also look above the container for telephone, cable, and power lines that might have been damaged by the fire.

### Confined Spaces

Fires and other emergencies can occur in confined spaces. Fires in underground vaults and utility rooms such as transformer vaults are too dangerous to enter. Brigade members should summon the plant utility group and keep the area around manhole covers and other openings clear while awaiting the arrival of the utility personnel. For emergencies in these areas, the Occupational Safety and Health Administration requires specially trained entry teams. You should be familiar with your fire brigade's operational procedures for these incidents. It is also important to be familiar with confined spaces throughout your facility. These areas should be toured with plant personnel so that preincident plans can be developed.

Other hazards exist in confined spaces, including oxygen deficiencies, toxic gases, and standing water. Conditions might appear to be safe, but there might not be enough oxygen in the confined space to sustain life. It is common for an employee to enter these spaces and pass out, usually followed by another employee who comes in to assist and also passes out. If entering these spaces to attempt rescue, an SCBA must be worn, and brigade members should be attached to a lifeline. An additional lifeline should be lowered or brought with the brigade member to tie to the victim, harness, or stretcher line on a bight (rescue knot), which allows surface personnel to raise the victim out of the hole. Due to the lack of ventilation in most confined spaces, brigade members may notice an increased amount of heat once in the space. Brigade members will tire quickly and must recognize the signs of heat exhaustion and heat stroke.

Confined spaces may have low oxygen levels and high levels of combustible or flammable gases, such as methane. Brigade members who must enter a confined space should take an air monitoring device. Air quality must be checked constantly, looking for the buildup of explosive gases as well as decreased oxygen levels.

A strict accountability system must be adhered to when brigade members enter into a confined space. This ensures that only those with proper training and equipment enter the confined space. It is important for a safety officer to track the movement of personnel and the time that they are in the confined space.

Fire suppression in confined spaces must not begin until all associated utilities and industrial processes have been turned off. Fire suppression may require the use of hose streams, high and low expansion foams, carbon dioxide ($CO_2$) flooding systems, or built-in sprinkler systems.

## Vehicle Fires

Vehicle fires are common occurrences in most areas. Vehicles contain hundreds of pounds of plastics, which give off toxic smoke when they burn. For this reason it is important to wear SCBA when fighting a vehicle fire. A $1\frac{1}{2}$ or $1\frac{3}{4}$-inch hose line should always be used to attack a vehicle fire, in order to provide sufficient protection from a sudden flare-up. Many modern vehicles have shock absorbers in the bumpers, and trunk or hatchback components that are gas-filled. Be aware that when these cylinders are heated they can burst and send pieces flying from the vehicle at high velocity.

If the operator or driver is present, ask about any specific hazards that may be present in the vehicle, such as a gas cylinder, cans of spray paint, or any other hazardous materials. If no driver or operator is around, do not assume that the vehicle is safe; always be cautious as you approach a vehicle fire.

## Attacking Vehicle Fires

### Fires Under the Hood

Fighting fires under the closed hood of a vehicle can not only be very difficult, but also very dangerous. When approaching a burning vehicle with fire under the hood, approach from the uphill and upwind side at a 45° angle (▼ Figure 21-16). Using the full reach of the stream, direct the stream of water into the wheel well and through the front grill. This technique gets some water onto the fire quickly to cool the engine area, the shock absorbers, and the pistons of shock-absorbing bumpers. Shock absorbers and shock-absorbing bumpers are enclosed cylinders that when heated are capable of exploding. An exploding cylinder can travel distances up to 20 feet and pack enough force to break bones, and they can cause severe burns. If the cylinders in a front (or rear) shock-absorbing bumper explode, they can throw the bumper up to 3 feet with enough force to break a fire brigade team member's legs. Vehicles contain other small pistons that could also be dangerous, such as the hold-open pistons sometimes used on the front hood to keep the hood raised. These are sealed cylinders that can explode and shoot a spear-like rod through the front or rear of a vehicle.

Once firefighting operations have begun under the hood area, have another fire brigade team member stabilize the vehicle to keep it from rolling. The fire brigade team member entering the vehicle should open the driver's door and pull the hood release latch, which is usually located at the bottom of the "A" post. Pulling the hood latch may be ineffective if the fire has already damaged the release cable; however, it is a worthwhile first step. Always communicate with the team fighting the fire before releasing the hood. Always make sure the shock absorbers are cool before attempting to open the hood. If the shock absorbers (hold-open cylinders) are still hot, the cylinder could be over-pressurized and cause the hood to spring open with tremendous force.

After the vehicle has been cooled and some of the fire has been darkened, try to open the hood. If the hood release operated, a gloved hand should be used to activate the secondary latch, which is usually located in the middle of the front of the hood. If the release cable is damaged, a Halligan bar can be used to force the corner of the hood up enough to allow a nozzle to be inserted into the engine compartment. Another method is to break the front grill out and cut the release cable close to the latch mechanism. Using a pair of vise-grip pliers, grasp the remaining piece of cable at the release mechanism and pull. Once the hood is raised, extinguish the remaining fire, being careful not to splash battery acid. In most engine compartment fires, the plastic case of the battery will have melted and leaked battery acid onto the frame, engine, and ground.

It is also important to be aware of any leaking fluids on the ground. These could possibly ignite, spreading the fire beyond the vehicle. In addition, care should be taken to prevent these fluids from entering into storm drains or waterways, which could create an environmental problem.

After the main body of the fire has been extinguished, it is important to perform a complete overhaul just as with a structure fire. All of the seating material should be removed or exposed to ensure that the fire has not spread into this combustible material. Other concealed areas of the vehicle should also be checked for fire extension. Because the vehicle has probably already been extensively damaged from the fire and firefighting operations, the liberal use of water will not cause any undue damage beyond what has already occurred.

### Fires in the Passenger Area

Fires in the passenger area should be approached from the upwind side at a 90° angle from the car. Using the reach of the 1¾-inch hose line, start approximately 50 feet from the vehicle, with the nozzle set on straight stream, and darken the fire down by slowly sweeping the stream back and forth in a

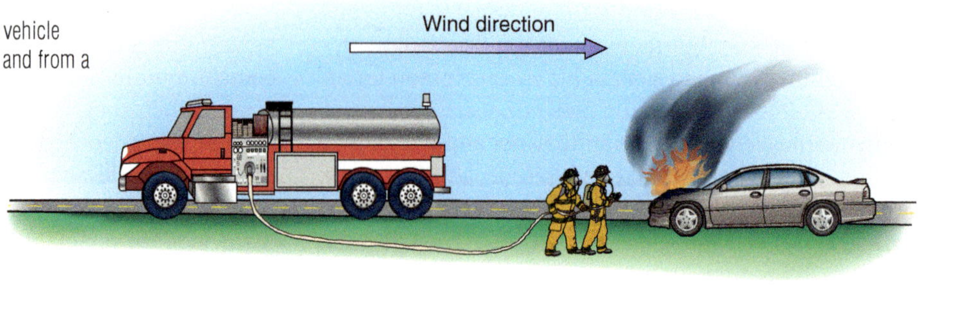


Figure 21-16 When possible, approach vehicle fires in the hood or engine area from upwind and from a 45º angle.

horizontal motion. Extinguish all visible fire while walking toward the car. Observe the area under the vehicle during the approach for any sign of leaking flammable liquids. If burning flammable liquids are present, widen the spray pattern on the nozzle. Foam can be used to extinguish the burning liquid and provide a vapor barrier to prevent reignition.

After all visible fire has been knocked down, allow a few minutes for the steam and smoke to dissipate before starting overhaul. This will allow visibility to improve so that overhaul can be completed safely. During overhaul of interior fires, remember that air bags can deploy without warning in a burning automobile. Never place any part of your body in the path of a front or side air bag.

### Fire in the Trunk

A fire in the trunk area of an automobile can be accessed by using the pike of the Halligan bar to force the lock into the trunk, then using a screwdriver or key tool (K-tool lock puller) to turn the lock cylinder in a clockwise direction. A charged hose line must be ready when the trunk lid is raised.

Fires in the rear of light trucks and vans must always be approached cautiously. Vans are often used for maintenance and support functions and could contain hazardous waste, chemicals, and radioactive material.

### Alternative Fuel Vehicles

Always be alert for signs that a burning vehicle could be powered by an alternate fuel, such as compressed natural gas (CNG) or liquefied propane gas (LPG). Fully involved fires in vehicles powered by either type of fuel should be fought with an unmanned master stream to prevent injuries from exploding gas cylinders.

Vehicles powered by CNG contain high pressure storage cylinders. These cylinders are usually located in the trunk or truck bed and contain CNG at high pressures. They must be cooled and protected like any gas cylinder. CNG is a lighter-than-air gas that will rise and dissipate if it is released into the atmosphere.

Propane is stored in pressure cylinders similar to those used for heating or cooking purposes. Propane is heavier than air, so the vapors will pool or collect in low areas.

Hybrid automobiles have small gasoline-powered engines and large battery banks. The batteries power electric motors that drive the wheels, much like a train locomotive. There are two noteworthy hazards in these vehicles:

- The nickel metal hydride batteries are very hazardous when burning and may explode. The runoff from water used in firefighting will also be hazardous and should be avoided.
- High-voltage direct current cables connect the batteries to the electric motors that power the wheels. Cutting these orange cables can cause serious injury or death. The cables usually run from the battery bank to the front of the car via the undercarriage. They are usually placed so that they will pass directly under the center of the driver's chair. Extra care should be used when using hydraulic metal cutters or spreaders on these types of autos.

# Flammable Liquids Fires

Flammable liquids fires can be encountered in almost any type of occupancy. Most fires involving a vehicle (plane, train, ship, car, truck) are also likely to involve a combustible or flammable liquid. Special tactics must be used when attempting to extinguish a flammable liquids fire. Special extinguishing agents such as foam or dry chemicals may be needed.

## Hazards

Fires involving flammable liquids such as gasoline require special extinguishing agents. Most flammable liquids can be extinguished using either foam or dry chemicals. Class B extinguishing agents are approved for use on Class B (flammable liquids) fires.

Flammable liquids fires can be classified as either two-dimensional or three-dimensional. A two-dimensional fire refers to a spill, pool, or open container of liquid that is burning only on the top surface. A three-dimensional fire refers to a situation where the burning liquid is dripping, spraying from the source, or flowing over the edges of a container.

A two-dimensional flammable liquid fire can usually be controlled by applying the appropriate Class B foam onto the burning surface. There are several different formulations of Class B foams that are suitable for different liquids and situations. The foam will flow across the surface and create a seal that stops the fuel from vaporizing. This separates the fuel from the oxygen and extinguishes the fire. Foam will also cool the liquid and reduce the possibility of reignition.

Brigade members should be aware of hot surfaces or open flames that could cause the vapors to reignite after a fire has been extinguished. It is important to determine the identity of the liquid that is involved in order to select the appropriate extinguishing agent and to determine whether the vapors are lighter or heavier than air.

A three-dimensional fire is much more difficult to extinguish with foam, because the foam cannot establish an effective seal on the fuel surface. Dry extinguishing agents may be more effective than foam in controlling three-dimensional fires. Dry chemical agents can also be used to extinguish two-dimensional fires; however, it does not provide a seal between the fuel and the oxygen. In some cases a fire can be extinguished with a dry chemical, and then the surface can be covered with foam to prevent reignition.

Brigade members should avoid standing in pools or contaminated runoff from flammable liquids. Brigade members' PPE could absorb the product and become contaminated. If it is seriously contaminated, the PPE itself could become flammable.

### Brigade Member Safety Tips

Some vehicle components (engines or body) may be made out of magnesium or metal alloys that can react violently when water is applied during suppression. A Class D extinguishing agent should be used instead of water.

### Suppression

The skills for suppressing flammable liquids fires are presented in the Foam chapter, Chapter 17.

## Flammable Gas Characteristics

A flammable gas is defined as a substance that is in a gaseous form at a temperature of 68°F (20°C) or less. It has a pressure of 14.7 psi (101.3 kPa) and either is ignitable at 14.7 psi (101.3 kPa) when in a mixture of 13 percent or less volume with air or possesses a flammable range at 14.7 psi (101.3 kPa) with air at least 12% regardless of the lower limit. The U.S. Department of Transportation (DOT) categorizes flammable gases within Division 2.1, meaning that containers holding these substances are placarded/labeled using the DOT 2.1 placards and labels.

Most flammable gases are stored in cylinders or tanks and are used as fuels in a flammable gas system. The remainder of the flammable gas system usually consists of a regulator (which controls the flow and the pressure of the gas), valves (which are used to turn on and shut off the flow of the gas), and hoses, pipes, tubes and other devices (which direct the gas to its intended point of ignition or use).

### Compressed-Gas Cylinders

Flammable gases are stored in compressed-gas cylinders that are designed to DOT specifications. Compressed-gas cylinders are designed to withstand temperature increases to approximately 130°F (54°C). Most cylinders are equipped with pressure-relief valves that will open when the pressure inside the cylinder increases above the designed pressure. The pressure-relief valve will continue to relieve the excess pressure until either the cylinder is cooled and its pressure reduced or the cylinder's contents and pressure are completely emptied. When sufficient heat is continuously applied to the tank but its pressure does not decrease, the internal pressures can reach a level that could cause a catastrophic failure of the cylinder.

### Brigade Member Safety Tips

Do not attempt to extinguish a propane gas fire until the fuel flow is shut off.

### Boiling-Liquid, Expanding-Vapor Explosion (BLEVE)

Flammable gas cylinders that are exposed to direct heating from an external source are subject to catastrophic failure, an incident commonly called a **boiling-liquid, expanding-vapor explosion (BLEVE)**. A BLEVE can occur when the liquid fuel in a compressed-gas cylinder or tank is stored under pressure. A propane tank is an example: Such a vessel is partly filled with the liquid propane and the rest of the vessel is occupied by propane in the form of a vapor.

The most common cause of a BLEVE is a fire that impinges upon the tank. The fire heats the liquid in the tank, causing it to generate more vapors. This reaction increases the internal pressure of the tank and causes the pressure-relief valve to open in an attempt to siphon off the excess pressure. If the pressure cannot be relieved fast enough, the pressure can increase to a point where the tank will rupture catastrophically. When this happens, large pieces of the tank can be propelled over significant distances, potentially injuring or even killing brigade team members. As the tank ruptures, any remaining flammable liquid is immediately released from the tank. Because the temperature of the liquid is at or above its boiling point, the liquid immediately turns into vapor and creates a rapidly expanding cloud. The fire can then ignite the escaping vapors, creating a fireball. All of this can happen quickly, in a matter of a few seconds.

The best method to prevent a BLEVE is to direct heavy streams of water onto the tank using unmanned master streams from a safe distance. The water should be directed at the area where the tank is being heated. The brigade members operating these streams should work from protected positions or use remote-controlled or unmanned monitors. An adequate water supply is a must for pressurized-gas fires. To control a flammable gas or liquid container fire, a minimum of 500 gpm is needed at each point of flame impingement for tankers. For smaller fires, two 1¾-inch lines should be used to cool the container.

Horizontal tanks are designed to fail at the ends if a catastrophic failure occurs. Nevertheless, the tank failure will likely create a fireball, with fragments flying in all directions. There is no "safe side" when a pressurized tank fails.

Over the years, dozens of emergency services personnel have been killed by BLEVEs that occurred while they were trying to fight fires that involved tanks of liquefied gaseous

### Brigade Member Safety Tips

A propane tank with the relief valve operating should not be approached. Remotely directing water onto the tank allows it to cool until the relief valve resets. Then the tank may be approached and the valve turned off.

fuels. By understanding the characteristics of flammable gas fuels and the mechanism underlying a BLEVE, we can help to prevent injuries or deaths in emergency situations.

## Propane Gas

The popularity of propane gas has caused these cylinders to become commonplace in many locations. In addition, propane is used as an alternative fuel for vehicles and is often stored to power emergency electrical generators. Brigade members should be familiar with the basic hazards and characteristics of propane and procedures for fighting propane fires.

Propane or LPG exists as gas in its natural state at temperatures above –44°F (–42.2°C). When it is placed into a storage cylinder under pressure, it is changed into a liquid. Storing propane as a liquid is very efficient, because it has an expansion ratio of 270 to 1. (One cubic foot of liquid propane will convert into 270 cubic feet of gas when it is released into the atmosphere.) A large quantity of fuel can be stored in a small container.

Inside a propane container, there is a space filled with propane gas above the level of the liquid propane. As the contents of the cylinder are used, the liquid level becomes lower and the vapor space increases. The internal piping is arranged to draw product from the vapor space.

Propane gas containers come in a variety of sizes and shapes, with capacities ranging from a few ounces to thousands of gallons. The cylinder itself is usually made of steel or aluminum. A discharge valve keeps the gas inside the cylinder from escaping into the atmosphere and controls the flow of gas into the system where it is used. This valve should be easily visible and accessible. In the event of a fire, closing this valve should stop the flow of the product and extinguish the fire. The valve should be clearly marked to indicate the direction it should be turned or moved to reach the closed position.

A connection to a hose, tubing, or piping allows the propane gas to flow from the cylinder to its destination. In the case of portable tanks, this connection is often the most likely place for a leak to occur. If the gas is ignited, this area could become involved in fire.

A propane cylinder is always equipped with a relief valve to allow excess pressure to escape in order to prevent an explosion if the tank is overheated. Propane cylinders must be stored in an upright position so that the relief valve remains within the vapor space. If a propane cylinder is placed on its side, the relief valve could be below the liquid level. If a fire were to heat the tank and cause an increase in pressure, the relief valve would release liquid propane, which would then expand by the 270 to 1 ratio. This would create a huge cloud of potentially explosive propane gas.

### Brigade Member Safety Tips

Fighting a fire involving any pressurized tank is dangerous. If there is any question regarding the integrity of the tank or the relief valve, stay away! Cool the tank from a safe distance and let it burn out. Don't risk lives for a tank.

## Propane Hazards

Propane is highly flammable. It is nontoxic, but it can displace oxygen and cause asphyxiation. By itself propane is odorless, and leaks could not be detected by a human sense of smell. Mercaptan is added to propane to create a distinctive odor. Propane gas is heavier than air, so it will flow along the ground and accumulate in low areas.

When responding to a reported LPG leak, brigade members and their apparatus should stage uphill and upwind of the scene. Brigade members should be aware that an explosion could happen at any time, so full PPE and SCBA must be worn. When using meters to check for LPG, be sure to check storm drains, basements, and other low-lying areas for concentrations of the gas. Life safety should be the highest priority; depending on the type and size of the leak, an evacuation might be necessary.

To protect the container from rupture, the relief valve will open to release some of the pressure. The relief valve should exhaust vapor until the pressure drops to a preset level, and then the valve should close or reseat. As the heating and relieving cycles continue, the liquid can begin to boil within the container. If the flame is impinging directly on the tank, the container can weaken and fail somewhere above the liquid line. When this happens, the container will rupture and release its contents with explosive speed. The boiling liquid will expand, vaporize, and ignite in a giant fireball, accompanied by flying fragments of the ruptured container. Brigade members and municipal fire fighters have been killed in these explosions.

### Propane Fire Suppression

Fighting fires involving LPG or other flammable gas cylinders requires careful analysis and logical procedures. If the gas itself is burning because of a pipe or regulator failure, the best way to extinguish the fire is to shut off the main discharge valve at the cylinder. If the fire is extinguished and the fuel continues to leak, there is a high probability that it will reignite explosively. Do not attempt to extinguish the flames unless the source of the fuel can be shut off or all of the fuel has been consumed. If the fire is heating the storage tank, hose streams should be used to cool the cylinder, being careful not to extinguish the fire.

Unless there is a remote shutoff valve, the flow can only be stopped if it is safe to approach the cylinder. The integrity of the cylinder should be inspected from a distance before any attempt is made to approach and shut off the valve. If the container is damaged or the valve is missing, the fuel should be allowed to burn off, while hose streams continue to cool the tank from a safe distance.

Approach the fire with two hose lines working together. The nozzles should be set on a wide fog pattern, with the discharge streams interlocked to create a protective curtain. The brigade leader should be located between the two nozzle operators. On the commands of the brigade leader, the crew should move forward, remaining together and never turning their backs to the burning product. Upon reaching the valve, the brigade member in the center turns off the valve, stopping the flow of gas. Any remaining fire is then extinguished by normal means. Continue the flow of water as a protective curtain and to reduce sources of ignition.

Should the fire be extinguished prematurely, the valve should still be turned off as soon as the team reaches it. Always approach and retreat from these types of fires facing the objective with water flowing, in case of reignition.

Unmanned master streams should be used to protect flammable gas containers that are exposed to a severe fire. Direct the water to cool the point of direct flame impingement and the vapor space or upper part of the container (▶ Figure 21-17). If the container is next to a fully involved building or fire that is too large to control, evacuate the area and do not fight the fire. If there is nothing to save, risk nothing.

Keep in mind that if the relief valve is open, the container is under stress. Exercise extreme caution when this is occurring. As the gas pressure is relieved, it will generate a shrill whistle; if the sound is rising in frequency, an explosion could be imminent and evacuation should be ordered.

To suppress a flammable gas cylinder fire, follow the steps in (Skill Drill 21-9).

1. Cool the tank from a distance until the relief valve resets.
2. Wearing full PPE, two teams of brigade members using a minimum of two 1¾-inch hose lines advance using an interlocking 90°-wide fog pattern for protection. Do not approach the cylinder from the ends. The brigade leader is located between the two nozzle persons and coordinates and directs the advance upon the cylinder.
3. When the cylinder is reached, the two nozzle teams isolate the shutoff valve from the fire with their fog streams while the leader closes the tank valve, eliminating the fuel source.
4. As cooling continues, brigade members slowly back away from the cylinder, never turning their backs to it.

### Shutting Off Gas Service

There are many industrial applications for natural gas and propane. If a gas line inside a structure is compromised during a fire, the escaping gas can add fuel to the fire. The method in which the gas is supplied to the structure must be located to stop the flow.

Figure 21-17 Using a master stream to protect a flammable gas container exposed to fire.

Gas supplies are delivered through a gas meter connected to an underground utility network or from a storage tank located outside the building. If the gas is supplied by an underground distribution system, the flow can be stopped by closing a quarter-turn valve on the gas meter. If there is an outside LPG storage cylinder, closing the cylinder valve will stop the flow. After the gas service has been shut off, use a lockout tag to be sure that it is not turned back on. Only qualified personnel should reestablish the flow of gas to a structure.

## Fires Involving Electricity

Class C fires involve energized electrical equipment, which includes any device that uses, produces, or delivers electrical energy. A Class C fire could involve plant wiring and outlets, fuse boxes, circuit breakers, transformers, generators, or electric motors. Power tools, lighting fixtures, household appliances, and electronic devices such as televisions, radios, and computers could be involved in Class C fires as well. The equipment must be plugged in or connected to an electrical source for the incident to be considered a Class C fire, although the equipment does not need to be switched on to be energized.

Electricity does not burn, but electrical energy can generate tremendous heat that could ignite nearby Class A or B materials. As long as the equipment is energized, the incident must be treated as a Class C fire. To effectively and safely extinguish fires involving charged electrical equipment, extinguishing agents that will not conduct electricity, such as dry chemicals or carbon dioxide, must be used. Chapter 17 provides detailed information on extinguishing agents.

The primary hazard for facility fire brigade team members during Class C fires is the possibility of electrical shock. Whenever possible, the electrical circuit to the equipment should be disconnected or de-energized before the brigade makes an attack on the fire. De-energizing the circuit does not always guarantee that the equipment is de-energized, however. Care must be taken that the equipment does not store charged electrical power in transformers, capacitors, or any other power sources. If possible, the IC should confer with a responsible person from the electrical utility division of the plant regarding the best means to deal with this threat.

De-energizing electrical equipment can be as easy as unplugging the equipment from the wall or as complicated as shutting down an entire high-voltage distribution grid. Other methods of de-energizing electrical circuits and equipment include removing and covering fuses, switching circuit breakers off, and terminating power at utility poles; only qualified electrical personnel should perform the last task. Brigade members must ensure that they are familiar with the systems they are working on prior to operating during an emergency and should never attempt to disrupt power by accessing areas or using devices they are not trained on.

All electrical equipment should be considered as potentially energized until the power utility department or a qualified electrician confirms that the power is off. Once the electrical service has been disconnected, most fires in electrical equipment can be controlled using the same tactics and agents as are used for a Class A fire or Class B fire, depending on the fuels involved. It is imperative when dealing with Class C incidents that all LOTO (Lock-out, Tag-Out) procedures for the facility be followed.

## Injuries

The danger of electrocution when using hose streams is minimal. With the exception of power plants, most buildings generally do not produce voltages sufficient to be hazardous to brigade team members. Tests have shown that airspaces between the water droplets help to dissipate enough of the voltage that a brigade team member would receive only a low-order shock, which could cause an involuntary reflex action. Although it is a minimal shock, a low-order shock could still be sufficient to cause a brigade team member to lose control of the nozzle, which then becomes a greater hazard.

Brigade team members must keep in mind that the use of hose streams around charged electrical equipment differs from the use of pressurized water extinguishers. Pressurized water extinguishers should never be used on or around charged electrical circuits or equipment. Most electrocution-related injuries that occur on the fire ground have been the result of either the brigade team member coming in direct contact with charged equipment or personnel working with ground ladders near electrical wires. To maximize safety on the fire ground, it is advisable to de-energize all electrical circuits whenever possible.

When it is not possible to turn the electricity off at the breaker or another readily available isolation point, it will be necessary to have a qualified electrician who has been trained in handling high voltage disconnect the power outside the building.

## Class C Fire Extinguishing Agents

Certain situations may dictate that fire suppression efforts cannot wait until electrical equipment is de-energized before the brigade begins attacking a fire. When these tactics become necessary, it is very important that brigade team members know the types of agents available, the procedures for properly applying these agents, and the limitations of these agents. Class C fire extinguishing agents include dry chemicals, carbon dioxide ($CO_2$), and halogenated agents.

### Dry Chemicals

Dry chemical fire extinguishers deliver a stream of finely ground particles into the base of the fire. Different chemical compounds are used to produce extinguishers of varying capabilities and characteristics.

The dry chemical extinguishing agents work in two ways. First, the dry chemicals interrupt or terminate the chemical chain reactions that occur during the combustion process. Once the chemical chain reaction is interrupted, the fire will go out. Second, the tremendous surface area of the finely ground particles allows them to absorb large quantities of heat. Absorbing the heat lowers the overall temperature at the fire area and helps to extinguish the fire and prevent its reignition.

### Carbon Dioxide

Carbon dioxide is a gas that is 1.5 times heavier than air. When $CO_2$ is discharged on a fire, it forms a dense cloud that displaces the air surrounding the fuel. This interrupts the combustion process by reducing the amount of oxygen that can reach the fuel. A blanket of carbon dioxide over the surface of a liquid fire can also disrupt the fuel's ability to vaporize.

### Halogenated Agents

Halogenated extinguishing agents are produced from a family of liquefied gases, known as halogens, that includes fluorine, bromine, iodine, and chlorine. Halogenated extinguishing agents are also called "clean agents" because they leave no residue; this characteristic makes them ideal agents for fire suppression in areas that contain computers or other sensitive electronic equipment. A 1987 international agreement, known as the Montreal Protocol, limits Halon production because these agents damage the earth's ozone layer. Halons have since been largely replaced by halocarbons and other clean agent gaseous agents.

The halogenated agents are stored as liquids and are discharged under relatively high pressure. They release a mist of vapor and liquid droplets that disrupts the molecular chain reactions within the combustion process. Just like dry chemical agents, the large surface area offered by the vapor and droplets absorbs large amounts of heat and helps to reduce the temperature.

## Limitations

Class C fire extinguishing agents do have certain limitations that might become a factor when determining which type of agent should be used on certain fires. One common factor that affects almost all Class C agents is wind. Class C agents are composed of fine particulate materials or gaseous clouds that can easily be swept away by even a mild breeze. Brigade team members should position themselves so the wind provides the best advantage; keeping the wind at your back usually provides the greatest advantage.

Dry chemical extinguishers are generally limited to a single use. Such extinguishers tend to bleed off pressure even if they are only partially discharged. Fine particulates may lodge in the seal assembly and often cause the extinguisher to bleed off air pressure even if only a very small amount of agent is used.

Other limitations include corrosive effects (certain dry chemicals), toxic effects (Halons), oxygen displacement (Halons), and possibly thermal shock ($CO_2$) when used on certain products and circuits.

## Suppression

Fire suppression methods for fires involving electrical equipment depend on the type of equipment and the power supply. In many cases, the best approach is to wait until the power is disconnected and then use the appropriate extinguishing agents to control the fire. If power cannot be disconnected or the situation requires immediate action, only Class C extinguishing agents, such as those described previously and in Chapter 17, should be used.

When delicate electronic equipment is involved in a fire, either Halon or $CO_2$ should be used to limit the damage as much as possible. These agents will cause less damage to computers and sensitive equipment than will water or dry chemicals.

When power distribution lines or transformers are involved in a fire, special care must be taken to ensure the safety of the emergency personnel as well as other facility personnel. No attempt should be made to attack these fires until the power has been disconnected. In some cases, it will be necessary to protect exposures or to extinguish a fire that has spread to other combustible materials. This attack should be attempted only if there is no chance of coming in contact with electrically energized equipment. If a hose stream comes in contact with the energized equipment, the current may flow back through the hose and electrocute brigade members who are in contact with the hose.

*Do not* apply water to a burning transformer. Water can cause the cooling liquid oils in the transformer to spill or splash, contaminating both brigade members and the environment. If the transformer is located on a pole, it should be allowed to burn until plant or power utility professionals arrive and de-energize the transformer. Brigade team members should keep in mind that transformers may retain or hold power for several minutes after the circuit has been de-energized. Once power is disconnected, dry chemical extinguishers should be applied to control the fire. Fires in ground-mounted transformers can also be extinguished using dry chemical agents after the power has been disconnected. Brigade team members should stay out of the smoke and away from any liquids that are discharged from a transformer, and they must wear full PPE and SCBA when attacking the fire.

Some very large transformers contain large quantities of cooling oil and require foam agents to be applied to extinguish fires. The foam can be applied *only* after the power has been disconnected. Until the power is disconnected, brigade team members should concentrate their efforts on protecting exposures, taking precautions to avoid contamination of themselves and the environment.

Underground power lines and transformers are often located in vaults below ground level. Explosive gases can build up within these vaults and be ignited by a spark. When ignited, these trapped gases can burn with explosive force, lifting a manhole cover from a vault and hurling it a considerable distance. Products of combustion can also leak into buildings through the underground conduits and pipe chases. Brigade team members should never enter underground electrical vaults while the equipment is energized. If suppression actions are immediately necessary, brigade team members can discharge Class C agents into the vault and replace the cover. Replacing the lid will help to maintain the concentrations of the agents at levels that will aid in controlling the fire until the vault equipment can be de-energized. Even after the power is disconnected, these vaults should be considered to be confined spaces and should be entered only by team members with confined-space training, full PPE, and SCBA. Vaults will typically contain toxic gases, but could also contain explosive or oxygen-deficient atmospheres.

Industrial complexes often have high-voltage electrical service connections and interior rooms containing transformers and distribution equipment. These areas should be clearly marked with electrical hazard signs and should not be entered by brigade members unless a rescue must be made. Electrical energy can "jump" from the high-voltage equipment to the brigade team member even when no water is being applied or even present. In these circumstances, brigade team members should limit their actions to protecting exposures. They must also wear full PPE and SCBA owing to the inhalation hazard created by the burning of the plastics and cooling liquids that are often used with equipment of this size and type.

Brigade teams should exercise extreme caution when operating around any energized electrical circuits. It is vital that the correct extinguishing agents be selected and properly applied to control the hazard and put out the fire. Team members should ensure the fire is completely extinguished before they leave the area. When departing or backing out of the area, they should never turn their back on the charged equipment or circuits.

## Preservation of Evidence

Facility brigade members have a responsibility to preserve evidence that could indicate the cause or point of origin of a fire. Members who discover something that could be evidence while performing fire suppression actions or other activities at a fire scene should leave the item in place, make sure no one interferes with it or the surrounding area, and notify a brigade team leader or fire investigator immediately.

Responding facility brigade members must be alert and open-minded when they are involved in fire suppression duties. During the course of their fire suppression work, they may witness or come in contact with key evidence that might be used to determine the cause of fire. First-arriving team members have the responsibility to work with fire investigators to ensure that important evidence is neither lost nor destroyed. Fire suppression team members should be trained to recognize and collect important information pertaining to the following issues:

- Behavior of the fire
- Incendiary devices, trailers, and accelerants
- Condition of fire alarm or suppression systems
- Obstacles
- Contents
- Charring and burn patterns

Although most evidence is found during the salvage and overhaul phase of fire operations, the preservation of evidence begins during suppression activities. If evidence could be damaged or destroyed during suppression activities, cover it with a salvage cover or some other type of protection, such as a garbage can. Use barrier tape to keep others from accidentally walking through evidence. Before moving an object to protect it from damage, be sure that witnesses are present, that a location sketch is drawn, and that a photograph is taken.

Evidence should not be contaminated, or altered from its original state, in any way. Fire investigators use special containers to store evidence and prevent contamination from any other products.

### Finding the Point of Origin

The first step in most fire investigations is to determine the point of the fire's origin. The point of origin is the exact physical location where a heat source and a fuel come in contact and a fire begins. It is usually determined by examination of fire damage and fire pattern evidence at the fire scene. Flames, heat, and smoke leave distinct patterns that can often be traced back to identify the area or the specific location where the heat ignited the fuel and the burning first occurred. Determining the point of origin requires the analysis of information from four sources:

- The physical marks, or fire patterns, left by the fire.
- The observations reported by persons who witnessed the fire (including brigade team members) or who were aware of conditions present at the time of the fire.
- The physics and chemistry of fire initiation, development, and growth as they relate to known or hypothesized fire conditions capable of producing those conditions.
- The location where electrical arcing has caused damage as well as the electrical circuit involved. An electrical arc is a luminous discharge of electricity from one object to another, which typically blackens objects in the immediate area.

Burn patterns and smoke residue will often spread outward from the room or area of origin where the damage is the most severe. Because heat rises, the flow of heated gases from a fire will almost always be up and out from the point of origin. This upward, outward flow can usually be recognized even when the fire affected all of the room's contents. Often the point of origin is found directly below the most heavily damaged area on the ceiling, where the heat of the fire was most concentrated.

A charred V-pattern on a wall indicates that the fire spread up and out from something at the base of the "V". An inverted V-pattern can indicate that a flammable liquid was used to start the fire.

The depth of charring can also assist in determining how long a fire burned in a particular location. An area that burned for only a few minutes would have mostly surface damage and a shallow depth of char. Conversely, materials that burned for a longer period would have a deeper depth of char. The area with the deepest char is not always the point of origin, however.

Burn patterns are a better indicator of the point of origin.

# Wrap-Up

## Ready for Review

- Wear appropriate PPE to all fires.
- Understand the fire tetrahedron and how eliminating one part of it will suppress any fire.
- Fire suppression operations can be classified as either offensive or defensive.
- Know and understand your brigade's SOPs on dealing with different types of fires and extinguishing agents.
- Never turn your back on any fire.
- There are three primary fire streams: fog, straight, and solid.
- There are three types of interior attack: direct, indirect, and combination.
- Large handlines are often used in defensive situations to direct a heavy stream of water onto a fire from an exterior position.
- Master stream devices produce high-volume water streams for large fires.
- Exposure protection refers to actions that are taken to prevent the spread of fire to areas not already burning.
- Vehicle fires are common in most communities. Given their explosive potential, proceed with caution.
- Special tactics must be used to extinguish a flammable liquid fire.
- Flammable gas cylinders that are exposed to direct heating are subject to catastrophic failure, commonly called a boiling-liquid, expanding vapor explosion (BLEVE).
- Class C fires involve energized electrical equipment.
- The preservation of evidence begins during suppression activities.

## Hot Terms

**Boiling-liquid, expanding-vapor explosion (BLEVE)** An explosion that can occur when a tank containing a volatile liquid is heated and the tank fails.

**Combination attack** A type of attack employing both the direct and indirect attack methods.

**Deck gun** Apparatus-mounted master stream device intended to flow large amounts of water directly onto a fire or exposed building.

**Direct attack** Firefighting operations involving the application of extinguishing agents directly onto the burning fuel.

**Elevated master stream** Nozzle mounted on the end of an aerial device capable of delivering large amounts of water onto a fire or exposed building from an elevated position.

**Incipient stage** Refers to the severity of a fire where the progression is in the early stage and has not developed beyond that which can be extinguished using portable fire extinguishers or small (up to 1½-inch) water handlines, flowing a maximum of 125 gpm (473 L/min). A fire is considered beyond the incipient stage when the use of thermal protective clothing or self-contained breathing apparatus is required, or a facility fire brigade team member is required to crawl on the ground or floor to stay below smoke and heat.

**Indirect application of water** Using a solid object such as a wall or ceiling to break apart a stream of water, causing it to create more surface area on the water droplets and thereby absorb more heat.

**Indirect attack** Firefighting operations involving the application of extinguishing agents to reduce the buildup of heat released from a fire without applying the agent directly onto the burning fuel.

**Ladder pipe** Nozzle attached to the tip of an aerial ladder truck designed to provide large volumes of water from an elevated position.

**Master stream device** A large capacity nozzle that can be supplied by two or more hose lines or fixed piping. Can flow in excess of 300 gpm. Includes deck guns, portable ground monitors, ladder pipes and other elevated devices.

**Portable monitor** A master stream appliance designed to be set up and operated from the ground. This device is designed to flow large amounts of water onto a fire or exposed building from a ground-level position.

**Solid stream** A stream made by using a smooth bore nozzle to produce a penetrating stream of water.

**Straight stream** A stream made by using an adjustable nozzle that is set to provide a straight stream of water.

**Thermal protective clothing** Protective clothing such as helmets, footwear, gloves, hoods, trousers, and coats that are designed and manufactured to protect the fire brigade member from the adverse effects of fire.

**Water hammer** An event that occurs when flowing water is suddenly stopped; the velocity of the moving water is transferred to everything it is in contact with. These tremendous forces can damage equipment and cause injury.

# Brigade Member in Action

**During a maintenance turnaround period, your fire brigade receives an alert stating that a fire has occurred in a temporary maintenance shop located on the edge of a processing unit. As the first-due engine turns the corner toward the incident location, the brigade leader riding in the front passenger seat informs your crew that the fire not only involves the temporary maintenance shop, which is made of scaffold components and plywood, but has also extended to two vehicles sitting next to the shop. The fire is also impinging on several racks of compressed-gas cylinders and several totes of alcohol-based chemical additives, which are staged adjacent to the process area. The brigade leader informs you that he will assume command and complete a size-up of the fire area. He orders you and the other two crew members on the engine to initiate fire control operations.**

1. Which of the following hazards would concern you related to the vehicles that are on fire?
   A. Shock absorbers
   B. Toxic smoke from burning plastic
   C. Alternative fuels
   D. All of the above

2. Which of the following operations would allow the greatest flow of water with the least amount of manpower?
   A. Deck gun
   B. Large handline
   C. Portable monitor
   D. Multiple attack lines

3. After knocking down the bulk of the fire, a group of brigade members are assigned to extinguish the fire in the engine compartment in one of the vehicles. Which approach should they use?
   A. Attack the fire from directly in front of the vehicle
   B. Attack the fire from a 45° angle from either of the front corners of the vehicle
   C. Attack the fire through the firewall in the vehicle's passenger compartment
   D. None of the above

4. What would be the greatest potential hazard with the fire involving the compressed-gas cylinders?
   A. BLEVE of the cylinders
   B. Activation of a cylinder pressure relief valve
   C. Fire from materials being released from the cylinder valve
   D. The cylinders falling over

# Preincident Planning

# Chapter 22

## NFPA 1081 Standard

### Incipient Facility Fire Brigade Member

**4.3.5** Perform a fire safety survey in a facility, given an assignment, survey forms, and procedures, so that fire and life safety hazards are identified, recommendations for their correction are made, and unresolved issues are referred to the proper authority.

**(A)** *Requisite Knowledge.* Organizational policy and procedures, common causes of fire and their prevention, the importance of fire safety, and referral procedures.

**(B)** *Requisite Skills.* The ability to complete forms, recognize hazards, match findings to preapproved recommendations, and effectively communicate findings to the proper authority.

### Advanced Exterior Facility Fire Brigade Member

**5.1.2.1** Utilize a preincident plan, given preincident plans and an assignment, so that the Facility fire brigade member implements the responses detailed by the plan.

**(A)** *Requisite Knowledge.* The sources of water supply for fire protection or other fire-extinguishing agents, site-specific hazards, the fundamentals of fire suppression and detection systems including specialized agents, and common symbols used in diagramming construction features, utilities, hazards, and fire protection systems.

**(B)** *Requisite Skills.* The ability to identify the components of the prefire plan such as fire suppression and detection systems, structural features, site-specific hazards, and response considerations.

**5.3.1** Perform a fire safety survey in a facility, given an assignment, survey forms, and procedures, so that fire and life safety hazards are identified, recommendations for their correction are made, and unresolved issues are referred to the proper authority.

**(A)** *Requisite Knowledge.* Organizational policy and procedures, common causes of fire and their prevention, and the importance of fire safety and referral procedures.

**(B)** *Requisite Skills.* The ability to complete forms, recognize hazards, match findings to preapproved recommendations, and effectively communicate findings to the proper authority.

**5.3.2 (A)** *Requisite Knowledge.* Site drawing reading, access procedures, forcible entry tools and procedures, and site-specific hazards, such as access to areas restricted by railcar movement, fences, and walls. Procedures associated with special hazard areas such as electrical substation, radiation hazard areas, and other areas specific to the site if needed.

**(B)** *Requisite Skills.* The ability to read site drawings, identify areas of low overhead clearance, identify areas on roadways having load restrictions, identify access routes to water supplies, identify hazardous materials locations, identify electrical equipment locations (overhead and below-grade equipment), ability to open gates by manual and/or automatic means, ability to forcibly gain access to areas, and the ability to identify site hazards.

### Interior Structural Facility Fire Brigade Member

**6.1.2.3** Utilize a preincident plan, given preincident plans and an assignment, so that the Facility fire brigade member implements the preincident plan.

**(A)** *Requisite Knowledge.* The sources of water supply for fire protection or other fire-extinguishing agents, site-specific hazards, the fundamentals of fire suppression and detection systems including specialized agents, and common symbols used in diagramming construction features, utilities, hazards, and fire protection systems.

**(B)** *Requisite Skills.* The ability to identify the components of the preincident plan such as fire suppression and detection systems, structural features, site-specific hazards, and response considerations.

**6.3.10** Perform a fire safety survey in a facility, given an assignment, survey forms, and procedures, so that fire and life safety hazards are identified, recommendations for their correction are made, and unresolved issues are referred to the proper authority.

**(A)** *Requisite Knowledge.* Organizational policy and procedures, common causes of fire and their prevention, and the importance of fire safety and referral procedures.

**(B)** *Requisite Skills.* The ability to complete forms, recognize hazards, match findings to preapproved recommendations, and effectively communicate findings to the proper authority.

## Knowledge Objectives

After completing this chapter, you will be able to:

- Recognize site-specific hazards.
- Identify the components of a preincident plan.
- Describe the advantages of preincident planning and safety.
- Recognize potential scene access restrictions.

## Skills Objectives

After completing this chapter, you will be able to perform the following skills:

- Conduct a preincident survey.
- Prepare an accurate sketch or diagram.
- Obtain the required occupancy information.
- Note any items of concern regarding occupancy.

# You Are the Brigade Member

You've been a probationary brigade member for three months. Today, your brigade leader announces that it is time to review the preincident plans prepared by another crew. You yawn and wish your brigade had something more exciting to do.

The brigade leader begins by saying, "Let me tell you a story. Two months before you joined us, we were dispatched to a fire at the equipment maintenance shop long after the day shift had left for the day. When we arrived, heavy smoke was showing. My first reaction was to begin an aggressive interior attack. However, the preincident plan contained two red flags: It noted that the building was constructed with lightweight trusses and that the air-conditioning units were mounted on the roof. Based on this information, I ordered an exterior attack. Two minutes later, the roof collapsed."

"If there had not been a preincident plan for this building, we could have lost three brigade members that day. I strongly believe that preincident planning is one of the most important things we can do. There is no doubt in my mind that preplanning saves lives." Suddenly you realize that you are no longer tired!

1. *Why is preincident planning so important to a brigade member's survival?*
2. *What types of buildings require preincident planning?*

## Introduction

Preincident planning gives you the tools and knowledge to become an effective brigade member. Without a preincident plan, you are going into an emergency situation "blind." You may not be familiar with the structure, the location of hydrants, or the potential hazards. With a preincident plan, you not only know where the hydrants and exits are but also what hazards to anticipate. A preincident plan puts all that information at your fingertips, either on paper or in a computer file, when you respond to a fire.

Preincident planning helps your brigade to make better command decisions because important information is assembled before the emergency occurs. At the emergency scene, the incident commander (IC) can use the preincident information to direct the emergency operations much more effectively (◄ Figure 22-1). Because a preincident plan identifies potentially hazardous situations before the emergency occurs, brigade members can be made aware of hidden dangers and prepare for them.

Figure 22-1 The preincident information is supplied to the IC at the emergency scene.

## Preincident Plan

Preincident planning is the process of obtaining information about a building or a process and storing the information in a system so that it can be retrieved quickly for future reference. Preincident planning is usually performed under the direction of a brigade leader. The completed preincident plan should be available to all units that would respond to an incident at that location.

The preincident plan is intended to help the IC make informed decisions when an emergency incident occurs at the location (► Figure 22-2). Some of the information

## Tactical Priorities

| | | |
|---|---|---|
| Address: 1500, 1510, 1520 | | |
| Occupancy Name: | | |
| Preplan #: 02-N-01 | Number Drawings: 1 | Revised Date: 12/2002 |
| District: E275A | Subzone: 60208 | By: ACEVEDO |
| Rescue Considerations: Yes ( ) No ( X ) | | |
| Occupancy Load Day: | Occupancy Load Night: | |
| Building Size: | Best Access: | |
| Knox Box: NONE | Knox Switch: NONE | Opticom: NONE |
| Roof Type: X | Attic Space: Yes ( ) No ( ) | Attic Height: X |
| Ventilation Horizontal: | Ventilation Vertical: | |

| | | | | |
|---|---|---|---|---|
| Sprinklers: | Yes ( X ) | No ( ) | Full ( X ) | Partial ( ) |
| Standpipes: | Yes ( X ) | No ( ) | Wet ( ) | Dry ( ) |
| Gas: | Yes ( ) | No ( ) | Lpg ( ) | |

Hazardous Materials: Yes ( X ) No ( )
DIESEL GENERATORS

1,000 GALLON TANKS

BATTERY ROOM

Brigade Member Safety Considerations:
ELEVATOR PIT

Property Conservation and Special Considerations:
VENTILATION: AUTOMATIC SMOKE REMOVAL SYSTEM

3 OFFICE BLDGS; 2 PARKING STRUCTURES

6 FLRS - 1230 W. WASHINGTON ST.
4 FLRS - 1500 N. PRIEST DR.
4 FLRS - PARKING GARAGE

**Figure 22-2** An example of a preincident plan.

would be useful to any crew or unit responding to an incident at that location. A preincident plan can also be used in training activities to help brigade members become familiar with properties within their facility.

The objective of a preincident plan is to make valuable information immediately available during an emergency incident that otherwise would not be readily evident or easily determined. The amount and the nature of the information provided for different areas will depend on the size and complexity of the property, the types of risks that are present, and the particular hazards or challenges that are likely to be encountered.

The use of computers has greatly increased the ability of fire brigades to capture, store, organize, update, and quickly retrieve preincident planning information. Before this technology became widely available, preincident information was often limited by the need to carry hard copy information in fire apparatus and command vehicles. With computers and mobile data terminals, information such as drawings, maps, aerial photographs, descriptive text, lists of hazardous materials, and material safety data sheets is easily available.

It is particularly important to inform brigade members of potential safety hazards. The most critical information, like hydrant locations and life hazards, should be instantly available, with additional data accessible as needed. All of the information must be presented in an understandable format.

A preincident plan usually includes one or more diagrams to show details such as the building location and arrangement, access routes, entry points, exposures, and hydrant locations

## Brigade Member Tips

Preincident planning is different from fire prevention. The goal of fire prevention is to identify hazards and minimize or correct them so that fires do not occur or have limited consequences. Preincident planning assumes that a fire will occur and compiles information that responding brigade members would need. In many cases, a preincident survey identifies the need for a preincident plan update or for a fire prevention inspection.

**Table 22-1 Typical Target Hazard Properties**

- Fuel storage and power houses
- Multistory structures
- Laboratories
- Cafeterias
- Assembly/meeting areas
- Maintenance shops
- Production areas
- Tank farms
- Electrical substations
- Mechanical rooms
- Critical process areas
- Storage structures for hazardous materials and waste
- Warehouses

or alternative water supplies. The location and nature of any special hazards should be particularly noted on the diagrams. Information about the actual building should include the height and overall dimensions, type of construction, nature of the occupancy, and the types of contents in different areas. Additional information should include interior floor plans, stairway and elevator locations, utility shutoff locations, and information about built-in fire protection systems. For more information, see NFPA 1620, *Recommended Practice for Preincident Planning*.

## Target Hazards

Fire brigades should be able to create a preincident plan for every individual facility on the property. They should pay extra attention to properties that are particularly large and/or where unusual risks are present. These properties are identified as target hazards (▲ Table 22-1). Target hazard properties pose an increased risk to brigade members.

A preincident plan should be prepared for every property that involves a high life safety hazard to the occupants or presents safety risks for responding brigade members. Preincident plans should also be prepared for properties that have the potential to create a large fire or conflagration (a large fire involving multiple structures).

Properties that have an increased life safety hazard include:

- Power houses
- Laboratories
- Warehouses
- Flammable liquid production facilities
- Hazardous material storage and waste facilities

## Developing a Preincident Plan

The information that goes into a preincident plan is gathered during a preincident survey. The survey is usually performed by one of the crews that would respond to an emergency incident at the location. This enables the crew to visit the property and become familiar with the location as they collect the preincident information.

The survey data are compiled in a standard preincident plan format and filed by property location in an information management system. The actual preincident plan is usually prepared at the station, where the brigade leader takes the time to organize the information properly and to create drawings that can be used effectively during an emergency.

The information can be stored on paper or entered into a computer system. If the preincident plan is stored on paper, copies should be made available to brigade members and brigade leaders who would respond to the location on an initial alarm. The plans should be kept in binders or in a filing system on each vehicle (▼ Figure 22-3).

The communications center should also have a copy of the completed plan so that the dispatcher has access to the information. In some large operations, the communications center can fax a copy to a portable fax machine in a command unit at the scene of an incident.

**Figure 22-3** Copies of preincident plans can be kept in three-ring binders.

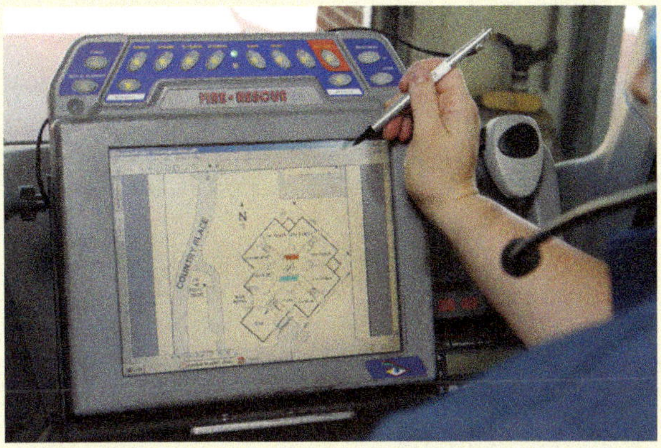


Figure 22-4 Computers allow access to preincident plans from any location.

Using a computer system to develop and store preincident plans has many advantages. Plan data can be accessed directly, through computer terminals at fire stations, and through mobile data terminals in vehicles ▲Figure 22-4. Updating preincident plans is easier when they are stored on computers. As soon as the master copy is updated, everyone has access to the new information. There is no need to make and distribute new copies.

## Conducting a Preincident Survey

A preincident survey should be conducted with the knowledge and cooperation of the area supervisor. The supervisor should be contacted before the preincident survey is conducted. This enables the brigade to schedule an acceptable time, to explain the purpose, the importance of the preincident survey, and to clarify that the information is needed to prepare brigade members in the event that an emergency occurs at the location.

The brigade members who conduct the survey should conduct themselves in a manner appropriate to the brigade's mission. A representative of the facility or area should accompany the survey team to answer questions and provide access to different areas. Every effort should be made to obtain accurate, useful information.

The preincident survey is conducted in a systematic fashion, following a uniform format. Begin with the outside of the building. Gather all of the information about the building's geographic location, external features, and access points. Then survey the inside to collect information about every interior area. A good, systematic approach starts at the roof and works down through the building, covering every level of the structure. If the property is large and complicated, it may be necessary to make more than one visit to ensure that all the required information is obtained and accurately recorded.

**Table 22-2 Information Gathered During the Preincident Survey**

- Building location
- Apparatus access to exterior of the building
- Access points to the interior of the building
- Hydrant locations and/or alternative water supply
- Size of the building (height, number of stories, length, width)
- Exposures to the fire building and separation distances
- Type of building construction
- Building use
- Type of occupancy (assembly, institutional, residential, commercial, industrial)
- Floor plan
- Life hazards
- Building exit plan and exit locations
- Stairway locations (note if enclosed or unenclosed)
- Elevator locations and emergency controls
- Built-in fire protection systems (sprinklers, sprinkler control valves, standpipes, standpipe connections)
- Fire alarm systems and **fire alarm annunciator panel** (part of the fire alarm system that indicates the location of an alarm within the building) location
- Utility shutoff locations
- Ventilation locations
- Presence of hazardous materials
- Presence of unusual contents or hazards
- Type of incident expected
- Sources of potential damage
- Special resources required
- General firefighting concerns
- Roof access

The same set of basic information must be collected for each property that is surveyed ▲Table 22-2. Additional information should be gathered for areas that are unusually large, complicated, or where particularly hazardous situations are likely to occur. Most fire brigades use standard forms to record the survey information. After the brigade returns to the station, they can use the information to develop the preincident plan.

### Site Drawings

The brigade members conducting the survey should prepare site drawings, or sketches to show the process or building layout and the location of important features such as exits. Many brigades also use digital cameras to record information.

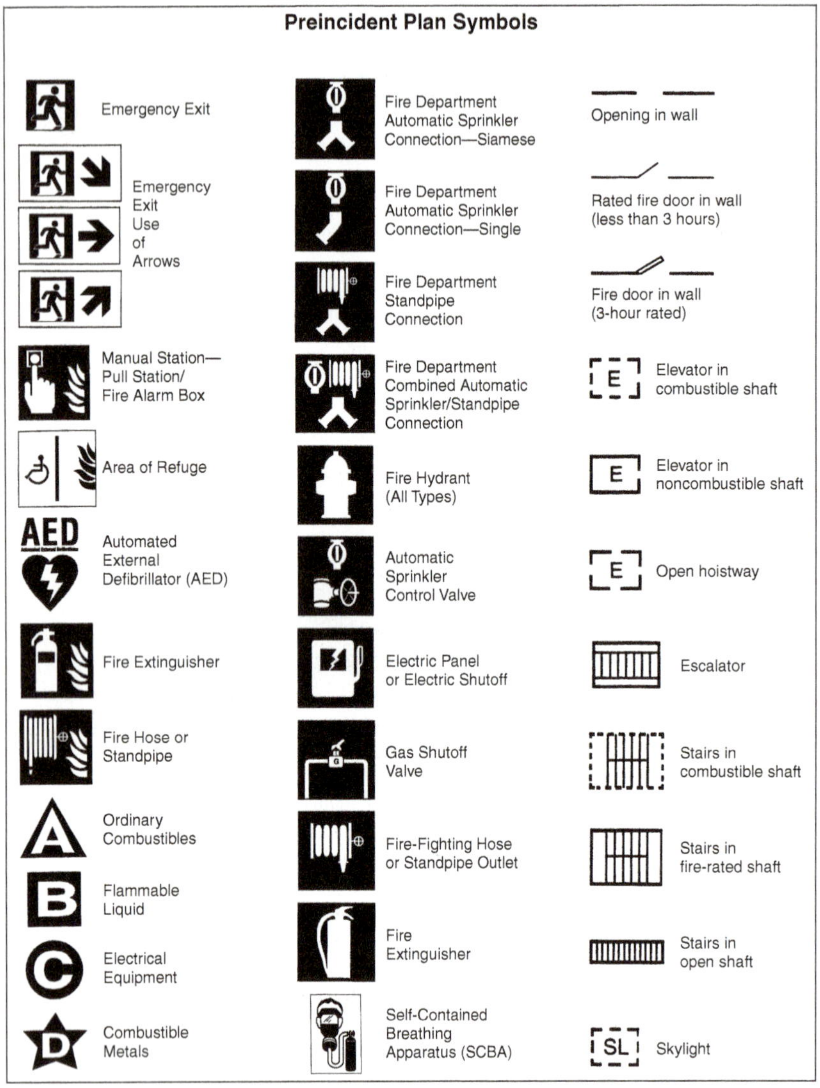


**Figure 22-5** Preincident plan symbols.

It takes practice and experience to learn how to sketch the required information while doing the survey, and then to convert the information to a final drawing. In some cases, the area supervisor or plant engineer can provide the survey team with a copy of a plot plan or a floor plan.

The completed site drawing should use standard, easily understood map symbols (▲ **Figure 22-5** and ▶ **Figure 22-6**). *NFPA 170 Standard for Fire Safety and Emergency Symbols* is an excellent resource for standardized symbols used by the fire service, engineering drawings, and preincident planning sketches. Many fire brigades use computer-assisted drawing software to create and store these diagrams.

Brigade members should also have the ability to read site drawings. These drawings can provide critical safety information. Drawings must be updated whenever remodeling, equipment relocations, and construction activities occur, or at least on an annual basis, so that they will be current. Fire fighters should also be familiar with any other type of form, map, or warning placards that are used in the facility.

The following items should be identified on the facility site drawings:

- Areas of low overhead clearance such as pedestrian bridges, pipe racks, and utility lines
- Barriers such as gates and fences
- Points of access into the building, particularly if there are multiple entrances
- Areas on roadways having load restrictions, such as timber bridges and buried culverts
- Access routes to water supplies that are usable year-round
- Hazardous materials locations such as storage tanks and gas cylinders, satellite hazardous waste stations, and storage rooms
- Electrical equipment locations (overhead and below-grade equipment) such as generator rooms, large outdoor transformer equipment, and equipment located below-grade in a vault that is accessible by a manhole cover or grated cover

Brigade members should have the following abilities:

- Open gates by employing manual and/or automatic means using the proper switch and card keys, or through bypassing the system manually by pulling a pin in the arm assembly
- Gain access to areas using forcible entry procedures as per the facility SOP/SOG
- Identify site hazards using placards and other warning signs as per the facility's site-specific hazard communications program

## Preincident Planning for Response and Access

Building layout and access information is particularly important during the response phase of an emergency incident. A preincident plan should provide information that would be valuable to units en route to an incident. For instance, the plan could identify the most efficient route to the fire building or process area and also note an alternate route if the time of the day and local traffic patterns would affect the primary route. Alternate routes should also be established if primary routes require crossing railroad tracks or other potentially blocked routes.

When conducting a preincident survey, brigade members should ensure that the building identification is easily visible

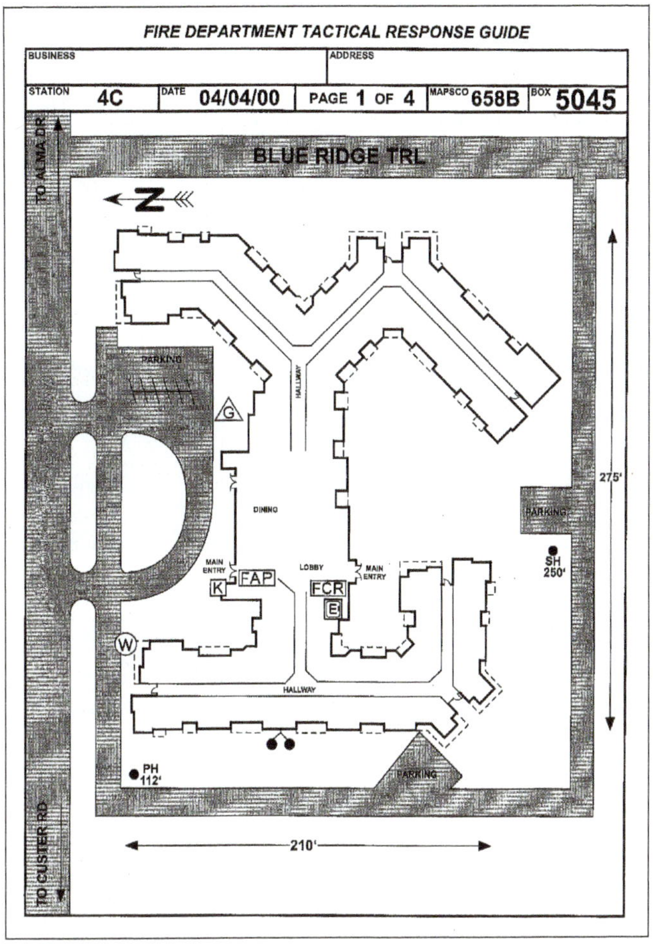


Figure 22-6 The completed drawing should use standard, easily understood map symbols.

to save time in locating an emergency incident. If the building is part of a complex, the best route to each individual building or section should be clearly indicated on a site drawing. It is important to consider the following points before accessing a building:

- Security officers who have not been trained in the use of PPE/SCBA cannot enter hot zones to unlock secure areas. As a consequence, procedures to access locked/secured areas of the structure and facility grounds must be coordinated between the brigade and the security team.
- Some sites have special access challenges owing to fences, walls, railcar movement, and truck parking. Procedures to open, move, or circumvent these hazards can include going over them, through them, and around them.
- Whenever brigade members work near truck or train traffic, they must assure that the vehicle will not be moved. Wheel/rail chocks should be used, and communication must be maintained with the driver of the machinery.
- Special hazard areas, such as electrical substations, radiation hazard areas, chemical storage areas, and other site-specific hazards, must be the subject of preincident planning. Written procedures for handling an emergency in these areas should be developed.
- Lock out/tag out procedures, PPE requirements, and area entry permits should be part of preincident planning for these types of hazardous areas.

## Access to the Exterior of the Building

The preincident survey plan should address access to the exterior of the area. For example, you should consider the following:

- Are there several roads that lead to the building, or just a few?
- Where are hydrants located?
- Where are the fire department connections for automatic sprinkler and standpipe systems located?
- Are there security barriers that will limit access to the site?
- Are there fire lanes to provide access to specific areas?
- Are there barricades, gates, or other obstructions that are not wide enough or high enough to allow passage of apparatus?
- Are there bridges or underground structures that will not support the weight of the apparatus?
- Are there gates that require keys or a code to gain entry?
- Will it be necessary to cut fences?
- Are there natural barriers such as streams, lakes, or rivers that limit access?
- Does the topography limit access to any parts of the building?
- Does the landscaping or snow prevent access to certain parts of the building?
- Are there adequate turnarounds within the complex for fire vehicles?

## Access to the Interior of the Building

The preincident survey should also consider access to the interior of the building. The following questions are helpful:

- Is there a lockbox containing keys to the building? Do the keys work?
- Are key codes needed to gain access to the building? Who has them?
- Does the area have security guards? Is a guard always on duty? Does the guard have access to all areas?
- Where is the fire alarm annunciator panel located?
- Is the fire alarm annunciator panel properly programmed so you can quickly determine the exact location of an alarm?
- Will the fire brigade gain access or wait for maintenance employees with keys?

**Table 22-3 Types of Building Construction**

| | | |
|---|---|---|
| Type I: **Fire Resistive** | Buildings where the structural members are of noncombustible materials that have a specified fire resistance (▶ Figure 22-7). Materials include concrete, steel beams, masonry block walls, etc. |  Figure 22-7 A Type I building. |
| Type II: **Noncombustible** | Buildings where the structure members are of noncombustible materials, but may not have fire resistance protection (▶ Figure 22-8). Includes unprotected steel beams, etc. |  Figure 22-8 A Type II building. |
| Type III: **Ordinary** | Buildings where the exterior walls are noncombustible or limited-combustible, but the interior floors and walls are made of combustible materials (▶ Figure 22-9). |  Figure 22-9 A Type III building. |
| Type IV: **Heavy Timber** | Buildings where the exterior walls are noncombustible or limited-combustible, but the interior walls and floors are made of combustible materials (▶ Figure 22-10). The dimensions of the interior materials are greater than that of ordinary construction (typically with minimum dimensions of 8 × 8 feet). |  Figure 22-10 A Type IV building. |
| Type V: **Wood Frame** | Buildings where the exterior walls, interior walls, floors, and roof are made of combustible wood material (▶ Figure 22-11). |  Figure 22-11 A Type V building. |

## Preincident Planning for Scene Size-Up

The preincident survey must also obtain essential information about the area that is important for size-up (the ongoing observation and evaluation of factors that are used to develop objectives, strategies, and tactics for fire suppression). This information should include the construction, height, area, use, and occupancy, as well as the presence of hazardous materials or other risk factors. The location of other processes or structures that may be jeopardized by a fire in the building should also be recorded.

Preplanning for process areas should include dike containment volumes, grades that indicate liquid runoff directions, operating pressures, vessel or tank volumes and contents, foam flow requirements, exposure protection priorities, and collapse potentials.

Fire protection system information is also important for size-up. The preincident survey should identify areas that are protected by automatic sprinklers, deluge systems, foam systems, clean agent systems or other types of fire extinguishing systems, the locations of standpipes, and the locations of firewalls and other features designed to limit the spread of a fire. The survey should also note any areas where fire protection is lacking, such as an area without sprinklers in a building that has a sprinkler system.

Features inside a building that would allow a fire to spread but are not readily visible from the outside should also be noted. Unprotected openings between floors, or buildings connected by conveyor systems are examples of these features.

## Construction

Building construction is an important factor to identify in the preincident survey. Two similar-looking buildings may be constructed differently and thus will behave differently during a fire. ◄ Table 22-3 lists five different types of construction, as defined in NFPA 220, *Standard on Types of Building Construction*. Building construction is discussed in detail in Chapter 6, Building Construction. Preincident planning enables you to take the time that is needed to determine the exact type of building construction and identify any problem areas. Then, if an emergency occurs, the IC can check the preincident plan for information on construction, instead of hazarding a guess.

### Lightweight Construction

It is important to identify whether or not the building contains lightweight construction. Lightweight construction uses assemblies of small components, such as trusses or fabricated beams, as structural support materials. This type of construction can be found in newer buildings as well as in older buildings with extensive remodeling. In many cases, the lightweight components are located in void spaces or concealed above ceilings and are not readily visible.

A wood truss, constructed from 2 × 4 feet or 2 × 6 feet pieces of wood, can be used to span a wide area and support a floor or roof. A truss uses the principle of a triangle to build a structure that can support a great deal of weight with much less supporting material than conventional construction methods ▼ Figure 22-12 and ► Figure 22-13. Lightweight construction is used in storage buildings, temporary office buildings, and many other types of buildings.

### Remodeled Buildings

Buildings that have been remodeled or renovated present special situations. Remodeling can remove some of the original, built-in fire protection and create new hazards. For example, a remodeled building may contain multiple ceilings with void spaces between them. The new construction may have put a concrete topping over a wooden floor assembly. There may be new openings between floors or through walls that originally were fire resistant.

**Brigade Member Safety Tips**

Structures that are being remodeled, modified or demolished are at higher risk for major fires. You should consider doing a special preincident survey for structures undergoing substantial changes during the construction phase. This will enable you to gather more thorough information for a preincident plan of the completed structure.

**Brigade Member Safety Tips**

Lightweight construction is a relative term. A structure may be "lightweight" in comparison to other methods of construction, but its components are still heavy enough to cause serious injury or death if they collapse on a brigade member.

**Fire Marks**

Because lightweight construction uses truss supports, a floor or roof may appear sturdier than it actually is. Two municipal fire fighters in Houston, Texas died while battling an early-morning blaze in a fast food restaurant. The fire burned through the lightweight wood truss roof supports, and the roof-mounted air conditioner dropped into the building, trapping and killing the fire fighters. The fire was later determined to be arson.

Figure 22-12 A lightweight wood truss roof assembly.

Figure 22-13 A lightweight wood truss floor assembly.

If you can conduct the preincident survey during construction or remodeling, you will be able to see the construction from the inside out, before everything is covered over. It is also important to realize that buildings under construction, whether they are being remodeled or demolished, are especially vulnerable to fire. Unfinished construction is open, without many of the fire resistant features and fire detection/suppression systems that will be part of the finished structure.

## Building Use

The second major consideration during size-up is the building's use and occupancy, so it is important to identify these during a preincident survey. Buildings are used for many different purposes such as offices, warehouses, plant process areas, maintenance shops, and many other purposes. For each use, different types of problems, concerns, and hazards may be present.

For example, building use can help to determine the number of occupants and their ability to escape if a fire occurs. Building use is also a key factor in determining the probable contents of the building. These factors can have a major impact on the problems and hazards encountered by brigade members responding to an emergency incident at the location.

A building is usually classified by major use. This identifies the basic characteristics of the building. Within the major use classification, there are occupancy subclassifications that provide a more specific description of possible uses and their associated characteristics.

Many large building complexes contain multiple occupancy subcategories under one roof. For example, a large industrial building may have training rooms, mechanical shops, and warehouse space. The building also could be part of a larger complex that includes offices, power houses, and a production building. Each area may have very different characteristics and risk factors.

### Occupancy Changes

Building use may change over time. An outdated factory may be transformed into a office, or an unused warehouse may be converted into a packaging building. A warehouse that once stored concrete blocks may now be filled with foam-plastic insulation or plastic packaging materials. This is an important reason why preincident plans should be checked and updated on a regular schedule. A building's current occupancy information must always be determined during a preincident survey.

## Exposures

An exposure is any other building or item that may be in danger if an incident occurs in another building or area. An exposure could be an attached building, separated by a common wall, or it could be a building across an alley or street. Exposures can include other buildings, vehicles, outside storage, or anything else that could be damaged by or involved in a fire. For example, heat from a burning warehouse could ignite adjacent buildings and spread the fire.

A preincident survey should identify any potential exposures to the property that is being evaluated. This should take into account the size, construction, and fire load (the amount of combustible material and the rate of heat release) of the property being evaluated, the distance to the exposure, and the ease of ignition by radiation, convection, or conduction of heat.

## Built-In Fire Protection Systems

The preincident survey should identify built-in fire protection and fire suppression systems on the property. These systems include automatic sprinklers, standpipes, stationary fire pumps, fire alarms, and fire detection systems, as well as systems designed to control or extinguish particular types of fires. Some buildings have automatic smoke control or exhaust systems. Each is covered in detail in Chapter 25, Fire Detection, Protection, and Suppression Systems.

### Automatic Sprinkler Systems

A properly designed and maintained automatic sprinkler system can help control or extinguish a fire before the arrival of the fire brigade. When properly designed and maintained, these systems are extremely effective and can play a major role in reducing the loss of life and property at an incident. They also create a safer situation for brigade members.

The preincident survey should determine if the building has a sprinkler system and what parts of the building the system covers. Note the location of valves that control water

## Brigade Member Tips

Remember that a structure may have multiple uses during its lifetime or can incorporate more than one use within the same structure.

flow to different sections of the system. These valves should always be open.

In addition to the control valves, sprinkler systems should have a fire department connection outside the building. The fire department connection is used to supplement the water supply to the sprinklers by pumping water from a hydrant or other water supply through fire hoses into this connection. The location of the fire department connection for the sprinkler system must be marked on the preincident plan.

### Standpipe Systems

Standpipe systems are installed to deliver water to fire hose outlets on each floor of a building. This eliminates the need to extend hoselines from an engine at the street level up to fire level. Standpipes may also be used in low-rise buildings, such as large warehouses. Brigade members can bring attack hoselines inside the building and connect them to a standpipe outlet close to the fire, while the engine delivers water to a fire department connection outside the building.

The location of the fire department connection, as well as the locations of the outlets on each floor, should be marked on the preincident survey. The locations of nearby hydrants that will be used to supply water to the fire department connection must also be recorded. There may be multiple fire department connections to deliver water to different parts of a large building. The area or floor levels served by each connection should be carefully noted on the preincident plan and labeled at the connection.

### Fire Alarm and Fire Detection Systems

The primary role of a fire alarm system is to alert the occupants of a building so that they can evacuate or take action when an incident occurs. Some fire alarm systems are connected directly to the fire station; others are monitored by a dispatch center that calls the fire brigade when the system is activated.

In some cases, the fire alarm system must be manually activated. In other cases, the alarm system is activated automatically. A smoke or heat detection system can trigger the alarm or it may respond to a device that indicates when water is being discharged from the sprinkler system.

The annunciator panel, which is usually located close to a building entrance, indicates the location and type of device that activated the alarm system. In most cases, brigade members who respond to the alarm must check the annunciator panel to determine the actual source within the building or complex. The preincident survey should identify what type of system is installed, and where the annunciator panel is located.

### Special Fire Extinguishing Systems

There are several different types of fixed fire extinguishing systems that can be installed to protect areas where automatic sprinklers are not suitable. Most computer rooms are protected by clean agent fire suppression systems (▼ Figure 22-14). Special extinguishing systems are also found in many industrial facilities. The type of system and the area that is protected should be identified in the preincident survey.

Areas where flammable liquids are stored or used may have sophisticated foam or dry chemical fire suppression systems. The location of these systems should be noted on the preincident plan. Details about the method of operation, location of equipment, and foam supplies should also be recorded during the preincident survey. More information about built-in systems is presented in Chapter 25, Fire Detection, Protection, and Suppression Systems.

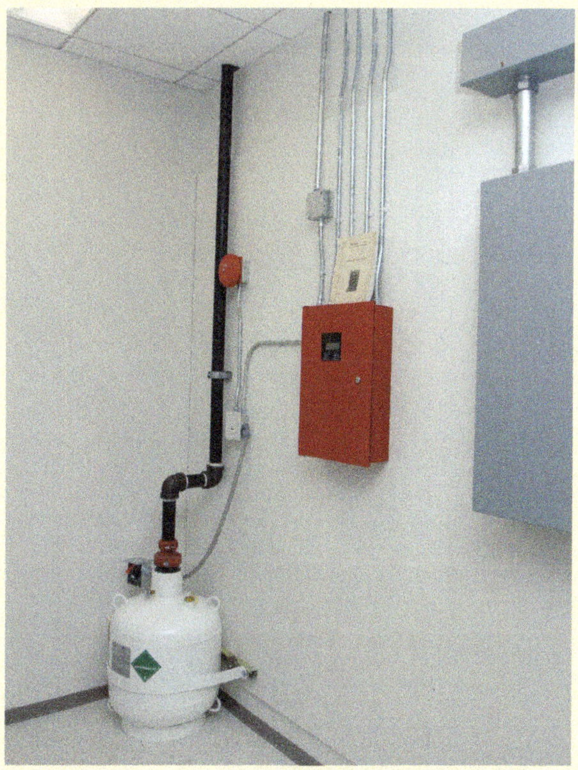

Figure 22-14 Many computer rooms have clean agent fire suppression systems.

## Tactical Information

There are many different categories of tactical information that can be obtained during a preincident survey and documented in a preincident plan. This is information that would be of particular value to an IC in directing operations during an emergency incident or to brigade members who are assigned to perform specific functions. When conducting a preincident survey, all of the factors that could be significant when conducting emergency operations at that location should be considered.

> **Fire Marks**
>
> A private water system may not be able to supply both sprinklers and hoses effectively. A commercial property in Bluffton, Indiana was destroyed when the responding fire department hooked its pumpers to private fire hydrants. The reduction in water flow that resulted rendered the building's sprinklers ineffective.

### Considerations for Water Supply

During the preincident survey, the amount of water needed to fight a fire in a structure or process area should be determined. The water supply source should be identified as well. The required flow rate, measured in gallons per minute or liters per minute, can be calculated, based on size, construction, contents, and exposures.

In most industrial plants, the water supply will come from plant or municipal hydrants. It is important to locate the hydrants closest to the fire area. For large flow rates, it will be necessary to locate enough hydrants to supply the volume of water required to control a fire. The ability of the municipal water system to provide the required flow must also be determined. It may be necessary to use hydrants that are supplied by different water mains of sufficient size to achieve the needed water flow (◄ Figure 22-15).

In areas without municipal water systems, the water may have to be obtained from a static water sources, such as a lake or stream, or delivered by fire brigade tankers. When static water sources are used, the preincident plan must identify drafting sites or the locations of the nearest dry hydrants. It is also important to measure the distance from the water source to the fire area to determine whether a large-diameter hose can be used and whether additional engines will be needed. The preincident plan should also outline the operation that would be required to deliver water to the fire.

If a tender shuttle (a system of tenders transporting water from a water supply to the fire scene) will be used to deliver water, several additional details must be included in the preincident plan. In particular, sites for filling the tenders and for discharging their loads must be identified (► Figure 22-16). The preincident plan should also identify how many tenders are needed based on the distance they must travel, the quantity of water each vehicle can transport, and the time it takes to empty and refill each vehicle.

A large industrial complex may have its own water supply system that provides water for automatic sprinkler systems, standpipes, and private hydrants. These systems usually include storage tanks or reservoirs as well as fixed fire pumps to deliver the water under pressure. The details and arrangement of the in plant water supply system must be determined during the preincident survey. This information will be available from the site engineer or maintenance staff.

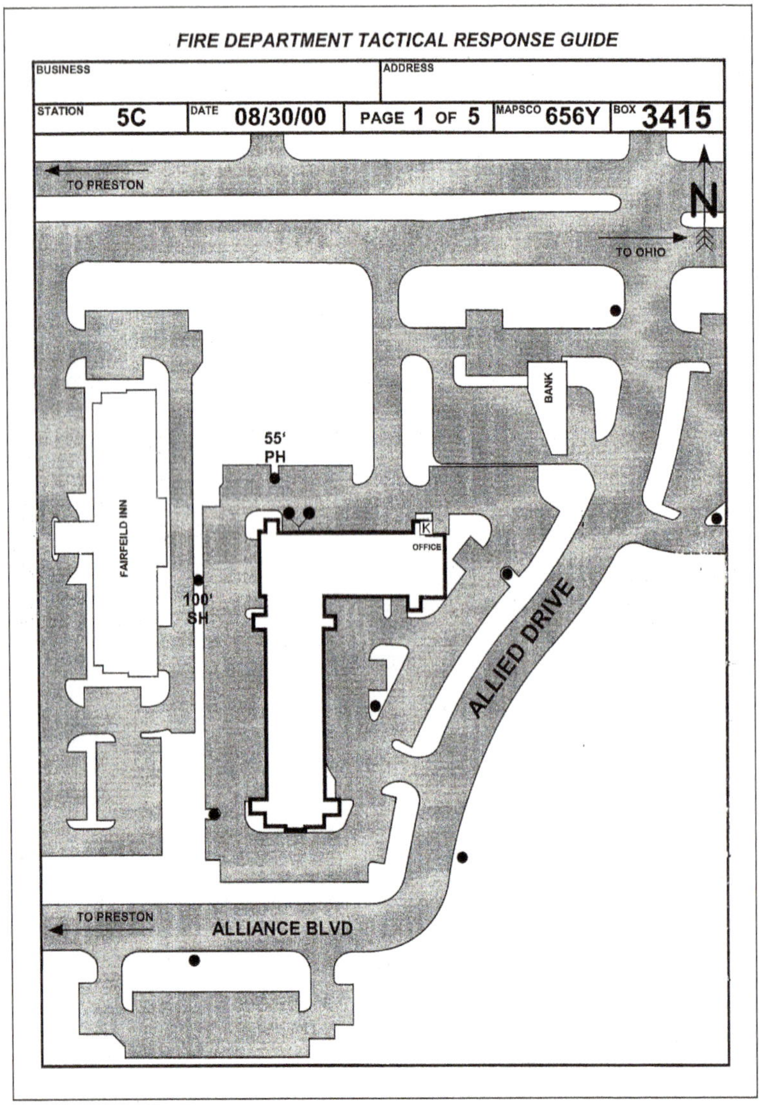


Figure 22-15 A diagram showing hydrant locations around a building.

Figure 22-16 The points where tenders can be filled and where they can discharge their loads must be identified.

## Brigade Member Safety Tips

Only qualified personnel should shutdown utilities.

## Brigade Member Safety Tips

The shutoff of utilities should be coordinated with operations or maintenance personnel to prevent an "upset condition" that could increase the problem.

The preincident survey should indicate how much water is stored on the property and where the tanks are located. If there is a fixed fire pump, its location, its capacity in gallons per minute, and its power source (electricity or a diesel engine) should be noted. Operational instructions should also be noted. It is also important to make sure that these in plant water systems are maintained in good operating condition. One or more brigade members may be assigned to verify proper operation of fire pumps and valves at an incident.

In many cases, the same in plant water main provides water for both the sprinkler system and the in plant hydrants. If the fire brigade uses the in plant hydrants, the water supply to the sprinklers may be compromised. The preincident plan for these sites should note if public, off-site hydrants should be used instead of the in plant hydrants.

## Utilities

During an emergency incident, it may be necessary to turn off utilities such as electricity or natural gas as a safety measure. The preincident survey should note the locations of shutoffs for electricity, natural gas, propane gas, fuel oil, and other energy sources. These sites, as well as contact information for the appropriate operations group or utility company, should be included in the preincident plan. In some cases, special knowledge or equipment is required to disconnect utilities. This must be noted in the preincident plan, along with the procedure for contacting the appropriate individual or organization.

## Brigade Member Tips

The location of utility shutoffs is as varied as the layout of structures. While some structures may have their utility shutoffs outside on a utility pole, other structures may have their utility shutoffs inside in a mechanical room. It is important to note the location of utility shutoffs on the preincident plan in order to prevent lengthy delays in shutting utilities down during an emergency. It is also important to note your plant procedure for shutting down power or utilities.

Electrical wires pose particular problems. Electricity may be supplied by overhead wires or through underground utilities. Overhead wires can be deadly if a ground ladder or aerial ladder comes in contact with them. The preincident plan should show the locations of high-voltage electrical lines and equipment that could be dangerous. If electricity is supplied by underground cables, the shutoff may be located inside the building or in an underground vault. These locations should also be noted on the preincident plan.

If propane gas or fuel oil is stored on the property, the preincident plan should show the locations and note the capacity of each tank. The presence of an emergency generator should also be noted, along with the fuel source and a list of equipment powered by the generator. Shutting off utilities could create process complications. Additionally, backup battery and/or generator systems may be in place to backfeed electrical systems. Prior to disconnecting utilities, operations or maintenance should be consulted.

## Preincident Planning for Search and Rescue

Brigade members conducting search-and-rescue operations will need to know where the occupants of a building are located as well as the location of exits. The preincident survey should identify all entrances and exits to the building, including fire escapes and roof exits. Muster or evacuation areas should be included as well. A large number of people may be moving to these areas, and their routes of travel may take them into the paths of responding apparatus.

In addition to conducting a search for occupants unable to escape on their own, search-and-rescue teams may need to assist occupants who are trying to use the exits. Brigade members may need to assist occupants descending fire

## Brigade Member Tips

Identifying the available water supply for a particular area is important. Consider the following:

- How much water will be needed to fight a fire?
- Where are the nearest hydrants located?
- How much water is available and at what pressure?
- Where are hydrants that are on different water mains located?
- How much hose would be needed to deliver the water to the fire?

If there are no hydrants, considerations should include:

- What is the nearest water supply?
- Is the water readily available year-round? Does it freeze during the winter or dry up during the summer?
- How many tenders would be needed to deliver the water?
- How long will it take to establish a water shuttle?

## Brigade Member Tips

The preincident plan should include information on how to contact utility companies or plant personnel to shutdown gas and electrical service.

escapes or to place ladders near windows or other locations for occupants to use.

An interior floor plan should be obtained during a preincident survey (▶ Figure 22-17). Knowing the floor plan of a building can be a lifesaver when the building is filled with smoke. It is much easier to understand a floor plan when you can tour a building under nonemergency conditions. In large buildings, it may be necessary to plan for the use of ropes during search and rescue to prevent disorientation in conditions of limited visibility.

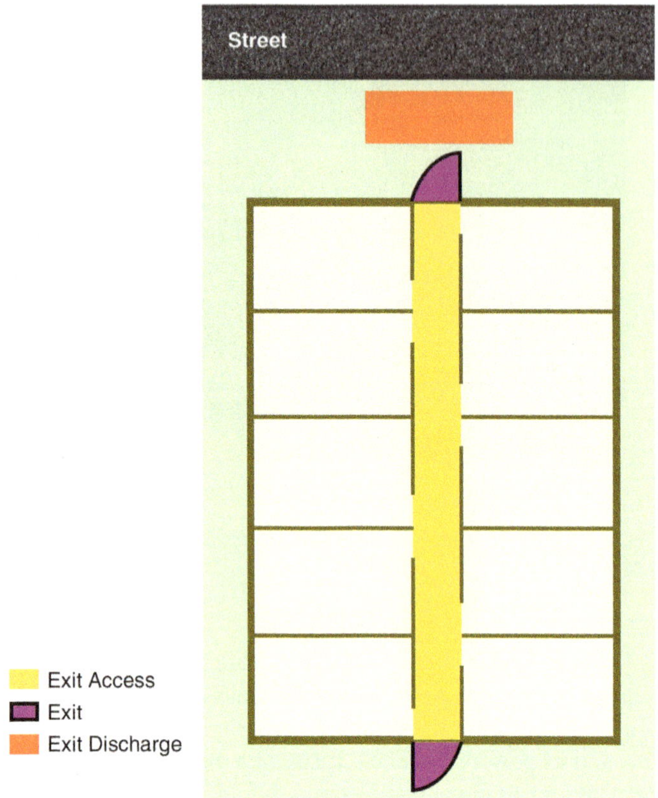


Figure 22-17 An interior floor plan showing the exits should be obtained during a preincident survey.

## Preincident Planning for Forcible Entry

As previously noted, the preincident survey should consider both exterior and interior access problems. Locations where forcible entry may be required should be identified and marked on the site diagrams and building floor plan. Making a note of what tools would be needed to gain entry can save time during the actual emergency. The location of a lock box and instructions on obtaining keys also should be noted.

## Preincident Planning for Ladder Placement

The preincident survey is an excellent time to identify the best locations for placing ground ladders or aerial apparatus (▶ Figure 22-18). The length of ladder needed to reach a roof or entry point should be noted. When planning ladder placement, pay careful attention to electrical wires and other

Figure 22-18 Considerations for use of ladders should include identifying the best locations to place ground ladders or aerial apparatus.

Figure 22-19 Note overhead obstructions, particularly electrical wires that might not be visible at night.

obstructions that might not be visible at night or in a smoky atmosphere (▲ Figure 22-19).

### Preincident Planning for Ventilation

While performing a preincident survey, brigade members should consider what information would be valuable to the members of a ventilation team during a fire. For example, what would be the best means to provide ventilation? How useful are the existing openings for ventilation? Are there windows and doors that would be suitable for horizontal ventilation? Where could fans be placed? Can the roof be opened to provide vertical ventilation? What is the best way to reach the roof? Is the roof safe? Are there ventilators or skylights that can be easily removed or bulkhead doors that can be opened easily? Will saws and axes be needed to cut through the roof? Are there multiple ceilings that will have to be punctured to allow smoke and heat to escape?

It is also important to know if the HVAC (heating, ventilation, and air-conditioning) system can be used to remove smoke without circulating it throughout the building. Many buildings with sealed windows have controls that enable the fire brigade to set the HVAC system to deliver outside air to some areas and exhaust smoke from other areas. The instructions for controlling the HVAC system should be included in the preincident plan. Many brigades dispatch a member to place the HVAC system on the 100% outside air mode on each alarm, generally reducing smoke damage.

Roof construction must also be evaluated to determine whether it would be safe to work on the roof when there is a fire below. If the roof is constructed with lightweight trusses, the risk of collapse may be too great to send brigade members above the fire. The existence of an attic that will allow a fire to spread quickly under the roof should also be noted.

**Brigade Member Tips**

Determine where and how ventilation can be accomplished for different parts of the building.

## Occupancy Considerations

Each type of occupancy involves particular considerations that should be taken into account when preparing a preincident plan. Brigade members should keep these factors in mind when conducting a preincident survey.

### Assembly Occupancies

Assembly venues, such as auditoriums and cafeterias, present the possibility that large numbers of people could become involved in an emergency incident. Gaining access to the location of the fire or emergency situation may be difficult when all of the occupants are trying to evacuate at the same time.

## Locations Requiring Special Considerations

Preincident planning should extend beyond planning for fires and other types of emergency situations that could occur in a building. Preincident planning should also anticipate the types of incidents that could occur at other locations, such as storage tanks, process piping, electric transformers and fuel loading docks, as well as incidents along highways or railroad lines, or at new construction sites (▼ Figure 22-20) in the facility.

Figure 22-20 Preincident planning should also anticipate the types of incidents that are likely to occur at other locations, such as plant construction sites.

# Voices of Experience

**"When we arrived on the scene, 50% of the building was heavily involved in fire, but we were unsure which chemicals were involved in the fire and what impact they would have on the surrounding community."**

My company responded on an automatic aid first alarm call one night following a report from a passing police officer of smoke coming from a 30,000-square-foot warehouse containing pool chemicals. The building had previously been a manufacturing facility with a large paint oven, and we were somewhat familiar with it because we had responded to the manufacturing facility a number of times for paint oven fires. The manufacturing company had moved on, however, and the building sat vacant for a number of years, resulting in the sprinkler system falling into disrepair.

When the pool chemical company moved in, the sprinkler system was not restored, and no detailed preincident planning was completed for it. We were, however, aware of the company's presence. Because our first-due area included both this building and a number of pools, we obtained a booklet developed by pool chemical manufacturers detailing information on handling spills and fires involving pool chemicals. We had also studied information about a similar warehouse that had burned in Florida about 20 years prior and that had resulted in numerous lasting health effects for the fire fighters involved in fighting the fire.

The warehouse was several hundred feet off of the main road and did not have a fire alarm system. When we arrived on the scene, 50% of the building was heavily involved in fire, but we were unsure which chemicals were involved in the fire and what impact they would have on the surrounding community. Owing to the presence of oxidizing chemicals, the fire was rapidly progressing down the remainder of the building. When we asked the owner of the building which chemicals were inside, he couldn't tell us, but we remembered to pull out the pool chemical book. The book indicated that there were four common pool chemicals, and the owner was able to verify that the warehouse contained "thousands of pounds" of three of the four chemicals listed in the book. Based on this information, we were able to develop a plan of attack, a safety and decontamination plan for the brigade members on scene, and a plan for conducting air monitoring in the surrounding community.

The building burned to the ground as the strategy generally became to protect exposures and minimize runoff to a nearby creek. All brigade members were required to be decontaminated before the canteen service was permitted to serve them. More than 1000 residents were advised to shelter in place, and there were no serious injuries.

Following the fire, the warehouse was rebuilt, this time with a full sprinkler system, and the county developed a detailed preplan for hazardous materials emergencies involving this location. The current preplan lists all communities, including schools, nursing homes, hospitals, and other targeted occupancies in the area, that could potentially be affected by a fire or release at this warehouse today. While we had done a limited amount of preplanning for this situation before it happened, having such a detailed plan would have resulted in a much more effective action plan than our ad hoc plan developed under duress that night.

Greg Jakubowski
Pennsylvania State Fire Instructor
Furlong, Pennsylvania

Figure 22-21 Electrical transmission lines.

Figure 22-22 Ships and waterways.

Figure 22-23 Fuel loading/unloading areas.

Figure 22-24 Railroads.

Similar planning should be done for bridges, tunnels, and many other locations where complicated situations could occur.

Other special locations that would require preincident planning include:

- Gas or liquid fuel transmission pipelines
- Electrical transmission lines ▲ Figure 22-21
- Ships and waterways ▲ Figure 22-22
- Fuel Loading/Unloading Areas ► Figure 22-23
- Railroads ► Figure 22-24

## Special Hazards

One of the most important reasons for developing a preincident plan is to identify any special hazards and to provide information that would be valuable during an emergency incident. This includes chemicals or hazardous materials that are stored or used on the premises, structural conditions that could result in a building collapse, industrial processes that pose hazards, high-voltage electrical equipment, and confined spaces. Preincident plans warn brigade members of potentially dangerous situations and could include detailed instructions about what to do. This information could save the lives of brigade members.

Brigade members who are conducting a preincident survey should always look for special hazards and obtain as much information as possible at that time ► Figure 22-25. If the information is not readily available, it may be necessary to contact specialists for advice or conduct further research before completing the preincident plan.

**Figure 22-25** During the preincident survey, look for special hazards and obtain as much information as possible.

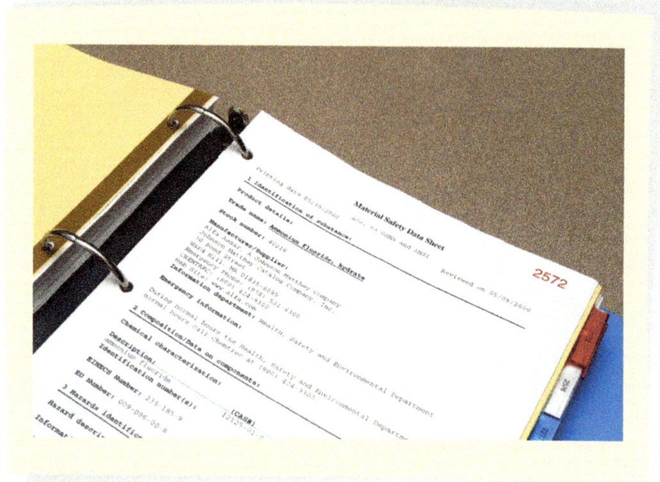


**Figure 22-26** MSDS document.

> **Fire Marks**
>
> The presence of hazardous materials presents special challenges to brigade members. Improper storage of chemicals in a warehouse in Phoenix, AZ resulted in a fire that overwhelmed the sprinkler system. Brigade members were not able to apply enough water to control the fire before it destroyed the warehouse. An NFPA Fire Investigation Report is available on this fire.

### Hazardous Materials

A preincident survey should obtain a complete description of all hazardous materials that are stored, used, or produced at the property. This includes an inventory of the types and quantities of hazardous materials that are on the premises, as well as information about where they are located, how they are used, and how they are stored. The appropriate actions and precautions for brigade members in the event of a spill, leak, fire, or other emergency incident should also be listed.

Brigade members responding to a hazardous materials event must be trained to the level of expected response actions. Most codes or standards require training at the First Responder Awareness Level. This awareness level limits response actions to identifying the hazard and activating response actions, but prohibits the responders from taking action to mitigate the incident.

By contrast, brigade member training at the First Responder Operations Level limits response actions to defensive activities, such establishing control zones, site security, and controlling hazardous material releases from a remote location. Training at the Technician Level allows responders to take offensive actions to control and stop the release, and to take other actions in the hot zone.

Detailed requirements for competencies at each level can be found in NFPA 472, *Standard for Professional Competence of Responders to Hazardous Materials Incidents*. Response plan requirements for hazardous substance release are specified in OSHA 29 CFR 1910.120(q).

Many jurisdictions require a company that stores or uses hazardous materials to obtain a permit. They may also require that hazardous materials specialists conduct special inspections. If the quantity of hazardous materials on hand exceeds a specified limit, federal and state regulations also require a company to provide the local fire department with current inventories and Material Safety Data Sheet (MSDS) documents (▲ **Figure 22-26**). Fire brigade leaders should ensure that the required information has been provided and is up-to-date.

Brigade members should expect to encounter hazardous materials at certain types of facilities, such as chemical plants, waste areas, production facilities, power plants, and maintenance areas. Brigade members should always be alert for hazardous materials.

Some facilities put placards on the outside of any building that contains hazardous materials. The placards should use the marking system specified in NFPA 704, *Standard System for the Identification of the Hazards of Materials for Emergency Response*.

Some hazardous materials require special suppression techniques or extinguishing agents. This information should be noted in the preincident plan. The plan should also contain contact information for individuals or organizations that can provide advice if an incident occurs.

# Fire Prevention Techniques

## Fire Safety Surveys

Fire safety surveys are used as a fire prevention tool. A fire safety survey differs from a preincident survey in that it focuses specifically on identifying hazards and code violations. The goal is to identify and eliminate hazards and violations *before* they become a safety or code violation issue.

By using a site-specific survey form or checklist, the brigade member can conduct a fire safety survey of the facility searching for hazards such as blocked or locked exits and exit corridors, blocked or missing fire protection equipment, improper storage of combustibles and flammables, electrical hazards, and improper housekeeping.

### Conducting a Fire Safety Survey

A team of brigade members who are properly equipped and trained in the site-specific process should conduct fire safety surveys. Team members may schedule the survey during normal production hours or, in some cases, during off hours.

The survey team should prepare in the following ways:

- Assemble and review any pertinent documentation, such as maps, floor plans, previous incident reports and survey reports, preincident plans, and blank survey checklists.
- Gather the equipment needed, including any PPE needed to enter all areas of the facility, survey forms (hard-copy or electronic versions), flashlights, radios, cameras, and measuring tapes.
- Develop a schedule or plan of action that the survey team will follow, thereby allowing team members to save time and be more productive.
- Notify the appropriate facility supervisors of the date and time when the survey will be conducted in their department.

The fire safety survey should begin with the exterior of the facility. Using the survey checklist, the team should assess all parking areas and yards, outbuildings, fuel/chemical storage areas, piping and tanks, fire protection devices such as fire hydrants and fixed monitors, electrical stations such as transformers and substations, and any other facility-specific outdoor items listed in the survey checklist.

The survey should then move to the roof of the facility. All roofs should be checked for the general condition of the roof, chimneys, pipe racks, tanks, fixed fire protection equipment, roof vents, access by ladder or hatches, and other relevant issues.

Moving indoors, the team should survey the facility's basements, mezzanines, attics, and storage rooms. This assessment should also include the areas between the ceiling and the roof and above all production-floor-level offices.

The interior survey checklist should focus on the following issues:

- Housekeeping practices
- Chemical storage areas
- HVAC systems
- Fire protection equipment and systems
- Electrical transmission and distribution systems
- Structural components of the facility
- Maintenance of process and/or production equipment
- Other site-specific areas identified by the facility

At the completion of the fire safety survey, the form is returned to the brigade leader for review and filing. Any hazards that are found are reported immediately. The supervisors who have the responsibility and authority to mitigate each of these hazards should be notified as per the site-specific procedure. Depending on the size and organizational structure of the facility, the individual responsible for maintenance may receive the report about the hazard and then have staff members correct the hazard. In other facilities, the report may go to the engineering or safety department. The facility manager may or may not receive the report in cases where the hazard was mitigated quickly and at little or no cost.

## Fire Safety Inspections

Fire safety inspections are based on the NFPA fire codes for the type of occupancy (such as commercial or manufacturing facilities) and for the type of hazard (such as flammable liquids or combustible dust).

A site-specific checklist or a fire safety survey should be provided to the inspector so that all known hazards in the facility will be inspected and so that each inspection can be documented. This checklist should be designed so that it guides the inspector through the facility in an orderly manner, covering all indoor and outdoor areas. It should also move the inspector through a production process from beginning to end.

Common fire hazards found in industrial facilities include, but are not limited to, the following problems:

- Improper storage and housekeeping of ordinary combustibles
- Improper storage of flammable and combustible liquids
- Improper housekeeping associated with dusts, metal, and plastics
- Incorrect storage and use of chemicals
- Improper use of electrical and heating equipment
- Lack of maintenance and testing of fire protection systems

Preventing these hazards can best be accomplished by instituting strict inspection guidelines that include sending a written report to the brigade leader, the safety or environmental department manager, or the supervisor responsible for the facility where the hazards were found.

Hazards that are deemed to be dangerous to life safety must be corrected immediately by the inspector when possible or by others through the notification of maintenance or housekeeping staff. A hazard correction notice should be used to document the findings. A written plan of correction is then produced and returned to the appropriate office as soon as the hazard is corrected.

Fire safety inspections are more detailed than fire safety surveys in that they may include prefire plans for a particular facility or process. Such inspections may also be used as a compliance tool to help the facility meet the requirements of NFPA 101, *Life Safety Code,* as it applies to the type of occupancy the facility falls under. Other NFPA codes, such as NFPA 54, *National Fuel Gas Code,* and NFPA 79, *Electrical Standard for Industrial Machinery*, can be used to measure how well the facility meets these standards.

Facilities that do not have trained inspectors on-site should work with their local fire services, insurance company representatives, and others to develop a site-specific fire inspection process and forms or checklists.

Also, at this level, members should be thoroughly familiar with the facility's programs for Hazard Communications (OSHA 1910.1200), Hazardous Waste Operations and Emergency Response (OSHA 1910.120), and Process Safety Management of Highly Hazardous Chemicals (OSHA 1910.119), as well as all other OSHA programs with which the facility is required to comply.

Fire safety inspections are also important to facilities' insurance ratings and costs. Facilities that do not conduct the required testing and maintenance of their sprinkler systems, for example, may not be covered by insurance if they suffer a loss. Likewise, a facility may not have its insurance policies renewed if it fails to meet NFPA codes.

# Wrap-Up

## Ready for Review

- Preincident planning enables a fire brigade to evaluate layout and location information as well as the conditions of target hazards before an emergency.
- Preincident plans must be systematically gathered, recorded, updated, and it must be made available to members of the brigade who might respond to that location.
- The brigade leader takes the information collected from the preincident survey and fire safety survey to create a preincident plan.
- The brigade member conducting the preincident survey should prepare sketches or drawings to note the layout and location information.
- The brigade member conducting the fire safety survey will focus specifically on identifying hazards and code violations.
- The preincident plan provides tactical information to the IC during an emergency.
- Brigade members should take into account the specific considerations of each type of occupancy when creating a preincident survey.
- Preincident plans should also prepare for incidents that could occur in locations like airports and subways.

## Hot Terms

**Conflagration** A large fire, often involving multiple structures.

**Drafting sites** Location where an engine can draft water directly from a static source.

**Dry hydrant** An arrangement of pipe that is permanently connected to allow a fire brigade engine to draft water from a static source.

**Exposure** Any person or property that may be endangered by flames, smoke, gases, heat, or runoff from a fire.

**Fire alarm annunciator panel** Part of the fire alarm system that indicates the source of an alarm within a building.

**Fire load** The weight of combustibles in a fire area or on a floor in buildings and structures including either contents or building parts, or both.

**Fire resistive construction** Buildings where the structural members are of noncombustible materials that have a specified fire resistance. Also known as Type I building construction.

**Fire safety inspection** An inspection based on the NFPA fire code for the type of occupancy (such as commercial or manufacturing facilities) and for the type of specific hazard (such as flammable liquids or combustible dust).

**Fire safety survey** The process of identifying hazards before an incident takes place. For example, accessibility of doors, storage of combustible or flammable materials, and electrical equipment are all considered during this process.

**Heavy timber construction** Buildings constructed with noncombustible or limited-combustible exterior walls and interior walls and floors made of large dimension combustible materials. Also known as Type IV building construction.

**Horizontal ventilation** The process of making openings so that the smoke, heat, and gases can escape horizontally from a building through openings such as doors, windows.

**HVAC system** Heating, ventilation, and air conditioning system in large buildings.

**Lightweight construction** The use of small dimension members such as 2 × 4 foot or 2 × 6 foot wood assemblies as structural supports in buildings.

**Material Safety Data Sheet (MSDS)** A form, provided by manufacturers and compounders (blenders) of chemicals, containing information about chemical composition, physical and chemical properties, health and safety hazards, emergency response, and waste disposal of the material.

**Noncombustible construction** Buildings where the structural members are of noncombustible materials without fire resistance. Also known as Type II building construction.

**Ordinary construction** Buildings where the exterior walls are noncombustible or limited-combustible, but the interior floors and walls are made of combustible materials. Also known as Type III building construction.

**Preincident plan** A written document resulting from the gathering of general and detailed information to be used by responders for determining the response to reasonable anticipated emergency incidents at a specific facility.

**Preincident survey** The process used to gather information to develop a preincident plan.

**Size-up** The ongoing observation and evaluation of factors that are used to develop objectives, strategy, and tactics for fire suppression.

**Sprinkler system** An automatic fire protection system designed to turn on sprinklers if a fire occurs.

**Standpipe system** An arrangement of piping, valves, and hose connections installed in a structure to deliver water for fire hoses.

**Static water sources** A water supply that is not under pressure, such as a pond, lake, or stream.

**Target hazard** Any occupancy type or facility that presents a high potential for loss of life or serious impact to the company resulting from fire, explosion, or chemical release.

**Tender shuttle** A method of transporting water from a source to a fire scene using a number of mobile water supply apparatus.

**Truss** A collection of lightweight structural components joined in a triangular configuration that can be used to support either floors or roofs.

**Vertical ventilation** The process of making openings so that the smoke, heat, and gases can escape vertically from a structure.

**Wood frame construction** Buildings where the exterior walls, interior walls, floors, and roof structure are made of wood. Also known as Type V building construction.

# Brigade Member in Action

**Your brigade leader has directed your brigade team to conduct a preincident survey on a new building that was built in your plant area. The building is a new cafeteria. It has a rated occupancy capacity of 500 people. Your leader has contacted the building manager and has scheduled a time to meet with a representative and walk through. Up to this point, you have studied preincident surveys, and this is your first opportunity to participate in the development process.**

1. Which of the following pieces of information should be gathered during the preincident survey?
   A. Access points to the exterior and interior of the building
   B. Hydrant locations
   C. Floor plans
   D. Building location
   E. All of the above

2. Which of the following is not information obtained about the fire protection system for preincident planning for scene size-up?
   A. Areas that are protected by automatic sprinklers
   B. Location of standpipes
   C. Location of firewalls
   D. Location of pull stations

**During your survey, you note that the structural members are made of steel and are not protected by a fire-resistive coating. While speaking with the building manager, you find that the building will also be used for large meetings and presentations.**

3. What type of construction is this?
   A. Fire resistive
   B. Noncombustible
   C. Ordinary
   D. Heavy timber

4. What would be the classification of this building?
   A. Institutional
   B. Commercial
   C. Public assembly
   D. Industrial

# Assisting Special Rescue Teams

# Chapter 23

## NFPA 1081 Standard

**Incipient Facility Fire Brigade Member**

NFPA 1081 contains no Incipient Facility Job Performance Requirements for this chapter.

**Advanced Exterior Facility Fire Brigade Member**

NFPA 1081 contains no Advanced Exterior Facility Job Performance Requirements for this chapter.

**Interior Structural Facility Fire Brigade Member**

NFPA 1081 contains no Interior Structural Facility Job Performance Requirements for this chapter.

## Knowledge Objectives

After completing this chapter, you will be able to:

- Define the types of special rescues encountered by brigade members.
- Describe the steps of a special rescue.
- Describe the general procedures at a special rescue scene.
- Describe how to safely approach and assist at a vehicle or machinery rescue incident.
- Describe how to safely approach and assist at a confined space rescue incident.
- Describe how to safely approach and assist at a rope rescue incident.
- Describe how to safely approach and assist at a trench and excavation rescue incident.
- Describe how to safely approach and assist at a structural collapse rescue incident.
- Describe how to safely approach and assist at a hazardous materials rescue incident.
- Describe how to safely respond to an elevator or escalator rescue.

## Skills Objectives

There are no skill objectives for this chapter.

# You Are the Brigade Member

**You're a member of a large Facility fire brigade, based in a petrochemical facility. The fire brigade has been paged out for a man down on top of one of the ethylene glycol storage tanks. Your team arrives on scene; according to bystanders, the worker collapsed due to exhaustion. Further inquiry reveals that the worker has a "weak heart." You cannot see the worker, but assume that he is still atop the tank. Access to the tank is via a staircase.**

**Your brigade leader calls for the technical rescue team, advising them that they will need to set up for a high angle rescue.**

**You realize how dangerous a technical rescue from such a height can be. A technical rescue incident (TRI) is not an everyday incident and requires specialized knowledge. As you are helping to unload the equipment from the technical rescue team's apparatus, you understand the complexity of technical rescues.**

1. ***What concepts about high angle rescue does this scenario illustrate?***
2. ***What is different about rope rescue incidents compared to virtually any other rescue?***
3. ***Why is redundancy such a critical issue when it comes to high angle rescue?***

## Introduction

In many industrial settings emergency response teams (ERTs) are composed of facility fire brigades which primarily respond to fires and specialized rescue teams that are capable of performing confined space rescue, high angle rope rescue, and to varying degrees trench and excavation rescue. Some fire brigades will also have a separate Hazmat team. On the other hand, numerous industrial organizations have facility fire brigades that are cross-trained in the various roles of technical rescue and hazardous materials response, which enables the facility fire brigade to respond to most incidents. Although the NFPA 1081 standard does not specifically address responding to or working with specialized rescue and/or Hazmat teams, it is beneficial for the facility fire brigade member to be knowledgeable in how to assist either in-house specially trained rescue teams or outside mutual aid agencies tasked with technical rescue or hazardous materials response. Each brigade member must be able to recognize and identify these specialized situations and subsequently call for appropriate assistance.

A technical rescue incident (TRI) is a complex rescue incident involving vehicles or machinery, water or ice, rope techniques, trench or excavation collapse, confined spaces, structural collapse, or hazardous materials that requires specially trained personnel and special equipment.

This chapter shows you how to assist specially trained rescue personnel in carrying out the tasks for which they have been trained. The chapter will not make you an expert in the skills that require specialized training to handle the various types of rescue situations discussed. The more training you receive in these areas, the better you will understand how to conduct yourself at these scenes.

Training in technical rescue areas is conducted at three levels: awareness, operations, and technician.

Most of the training you receive as a beginning brigade member is intended to provide general background knowledge so that you can identify the hazards and secure the scene to prevent additional people from becoming victims.

## Types of Rescues Encountered by Brigade Members

Many facility fire brigades respond to a variety of special rescue situations (▶ Figure 23-1). These include the following:

- Machinery and vehicle rescue
- Confined space rescue
- Rope rescue
- Trench and excavation rescue
- Structural collapse rescue
- Hazardous materials incidents

In order to become proficient in handling these situations, you must attend specific training to gain specialized knowledge and skills. NFPA 1670, *Standard on Operations and Training for Technical Rescue Incidents,* covers these knowledge and skills areas as does NFPA 1006, *Standard for Rescue Technician Professional Qualifications.*

It is important for "nonrescue" responders to have an understanding of these special types of rescues. The initial actions taken by the brigade may determine the safety of the

Figure 23-1 Many fire brigades respond to a variety of special rescue situations.

victims and the safety of the rescuers. Initial actions will also determine how efficiently the rescue is completed.

It is also important for you to understand and recognize many of the tools needed for rescue situations. You may not have to know how to use all of them, but you may be called upon to assist with a tool at any time.

## Guidelines for Operations

When assisting rescue team members, keep in mind the five guidelines that you follow during other firefighting operations:

- Be safe.
- Follow orders.
- Work as a team.
- Think.

### Be Safe

Rescue situations have many hidden hazards. These can include hazards such as oxygen-deficient and toxic atmospheres, weakened floors, and open excavations. Knowledge and training are required to recognize the signs that indicate a hazardous rescue situation exists. Once the hazards are recognized, determine what actions are necessary to ensure your own safety, as well as the safety of your team members, the victims of the incident, and bystanders. It requires experience and skill to determine that a rescue scene is not safe to enter and that determination could save lives.

### Follow Orders

As you begin as a brigade member, you will have limited training and experience. Your brigade leaders and the rescue teams with whom you will work on special rescue incidents have received extensive specialized training. They have been chosen for their duties because they have experience and skills in a particular area of rescue. It is critical to follow the orders of those who understand exactly what needs to be done to ensure safety and to mitigate the dangers involved in the rescue situation. Orders should be followed exactly as given. If you do not understand what is expected of you, *ask*. Have the orders clarified so you will be able to complete your assigned task safely.

To do well as a brigade member, you must have a strong appreciation for the basic philosophy of many firefighting organizations. Most fire brigades are run like a paramilitary organization. Grasping the paramilitary structure will enable you to understand the command and control concept of fire brigades.

- The brigade leader's knowledge base and experience is greater than yours.
- Orders come from superiors. Legitimate orders are only those given by a brigade leader or other designated person.
- Follow rules and procedures. A brigade member is required to follow rules, procedures, and guidelines regardless of personal opinions.
- You must do your own job.
- Get the job done. In emergency situations, time is critical. But you must not act beyond your own skill and training level, and you must not violate any rules, procedures, standard operating guidelines, or orders of a superior in order to get the job done quickly.

You will find that acting with these ideas clear in your head will enable you to deal with the orders of your commanding brigade leaders quickly and confidently.

### Work as a Team

Rescue efforts often require many people to complete a wide variety of tasks. There are personnel that are trained in specific tasks, such as rope rescue or swift water rescue ▶ Figure 23-2. However, they cannot do their jobs without the support and assistance of other brigade members.

Firefighting requires members of the team to work together to complete the goal of fire suppression. Rescue is a different goal, but it requires the same kind of team effort. Your role in the rescue effort is an essential part of this team effort.

### Think

As you are working on a rescue situation, you must constantly assess and reassess the scene. You may see something that your brigade leader does not see. If you think your assigned task may be unsafe, bring this to the attention of your superior leader. Do not try to reorganize the total rescue effort that is being directed by people who are highly trained and experienced, but do not ignore what is going on around you either.

Observations that a brigade member should bring to the attention of a superior brigade leader include changing

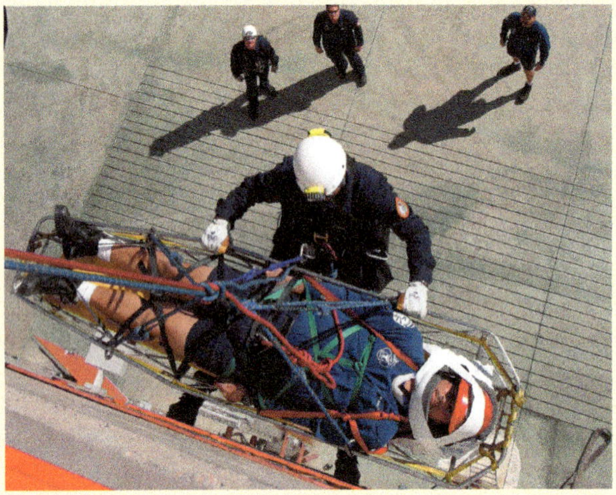

Figure 23-2 There are personnel who are trained in specific tasks, such as rope rescue.

## Brigade Member Tips

### F-A-I-L-U-R-E

In any rescue event, it is important to know why rescuers fail. The reasons for rescue failures are most commonly referred to by the acronym "FAILURE":

**F**—Failure to understand the environment, or underestimating it

**A**—Additional medical problems not considered

**I**—Inadequate rescue skills

**L**—Lack of teamwork or experience

**U**—Underestimating the logistics of the incident

**R**—Rescue versus recovery mode not considered

**E**—Equipment not mastered

If you as a brigade member and your team as a whole can learn to avoid these traps, you will have a greater chance of succeeding.

weather conditions that might affect the rescue scene operations, suspicious packages or items on the scene, and broken equipment. In a hazardous materials rescue incident, wind direction and wind speed are important to know. An increase in wind speed would cause hazardous vapors to be spread more quickly. This knowledge would affect the response plan.

## Steps of Special Rescue

Though special rescue situations may take many different forms, there are basic steps that all rescuers take in order to perform special rescues in a safe, effective, and efficient manner. The ten steps of the special rescue sequence are as follows:

- Preparation
- Response
- Arrival and size-up
- Stabilization
- Access
- Disentanglement
- Removal
- Transport
- Security of the scene and preparation for the next call
- Postincident analysis

### Preparation

You can prepare for responses to emergency rescue incidents by training with other fire brigades and mutual aid municipal fire departments in your area. Doing this will better enable you to respond to a mutual aid call. It will also inform you of the type of rescue equipment other organizations have access to, as well as the training levels of their personnel.

Know the terminology used in the field. This will make communicating with other rescuers easier and more effective. Know also the different types of rescue situations that you could encounter. Prior to a technical rescue call, your organization must consider these issues:

- Does the organization have the personnel and equipment to handle a rescue from start to finish?
- Does the organization meet NFPA and the Occupational Safety and Health Administration (OSHA) standards for technical rescue calls?
- What will the organization send on a technical rescue call?
- Do members of the organization know the hazard areas of the organization response area; have they visited hazard areas with local representatives?

### Response

A rescue should have a dispatch protocol. If your organization has its own technical rescue team, it may respond with a medic unit, an engine company, and a brigade leader. This initial response will satisfy most of the basic requirements of any rescue.

Fire brigades that do not have their own team may respond with an EMS unit and apparatus for support. The rescue squad will come from an outside agency. Often, it is necessary to notify power and utility companies during a rescue for possible assistance. Some technical rescues involve electricity, sewer pipes, or factors that may otherwise create the need for additional heavy equipment, to which utility companies have ready access.

## Arrival and Size-Up

Immediately upon arrival, the first brigade leader will assume command. A rapid and accurate size-up is needed to avoid placing rescuers in danger and to determine what additional resources may be needed. What is the extent of injuries and how many victims are involved? This will help to determine how many medic units and other resources are needed.

When responding within the industrial facility, the brigade leader should make contact with the foreman or supervisor, also known as the responsible party. These individuals can often provide valuable information about the worksite and the emergency situation. The most important part of any rescue is the identification of any hazards and the decision of recovery versus rescue. With this information, a decision is made to call for any additional resources, and actions are taken to stabilize the incident.

Do NOT rush into the incident scene until an assessment can be made of the situation. A brigade member approaching a trench collapse may cause further collapse. A brigade member climbing down into a vessel or tank to evaluate an unconscious victim may be overcome by an oxygen-deficient atmosphere. Stop to think about the dangers that may be present. Do not make yourself part of the problem.

## Stabilization

Once the resources are on the way and the scene is safe to enter, it is time to stabilize the incident. An outer perimeter is established as a safe area for support agencies involved in the operations to establish command, and a smaller perimeter is established directly around the rescue. A rescue area is an area that surrounds the incident site (collapsed structure, or collapsed trench, or hazardous spill area, etc.) and whose size is proportional to the hazards that exist. The areas should be established by identifying and evaluating the hazards that are discovered at the scene, observing the geographical area, noting the routes of access and exit, observing weather and wind conditions, and considering evacuation problems and transport distances. Three controlled zones should be established:

- Hot zone: This area is for entry teams and rescue teams only. The hot zone immediately surrounds the dangers of the site (e.g., hazardous materials releases) to protect personnel outside the zone.
- Warm zone: This area is for properly trained and equipped personnel only. The warm zone is where personnel and equipment decontamination and hot zone support take place.
- Cold zone: This area is for staging vehicles and equipment. The cold zone contains the command post. The public and the media should be kept clear of the cold zone at all times.

To figure out the scope of each of these zones, fire fighters should identify and evaluate the hazards that are discovered at the scene, observe the geographical area, note the routes of access and exit, observe weather and wind conditions, and consider evacuation problems and transport distances.

Figure 23-3 Lockout and tagout systems are methods of ensuring that the electricity has been shutdown.

A common method of establishing the hazardous working areas and safe areas for an emergency incident site is the use of police or fire line tape. Police or fire line tape is available in a variety of colors.

Once the operating areas have been established by the brigade, responders should ensure that these areas are enforced. The scene control activities are sometimes managed by facility security personnel or in some cases, by law enforcement when the event extends "off property."

Lockout and tagout systems should be used at this time to secure a safe environment ▲ Figure 23-3. Lockout and tagout systems are methods of ensuring that the electricity has been shutdown and that electrical switches are "locked" so that they cannot be switched on.

In an emergency situation, all power must be turned off and locked, and switches or valves must be tagged with labels to protect personnel and emergency response workers from accidental machine start-up. Lockout and tagout procedures are used to warn personnel and ensure that the electrical power is disconnected. Only qualified, authorized, and trained personnel can disconnect the source of power, lock it out, and tag it. Locks and tags are used for everyone's protection against electrical dangers. For your safety and that of others, *never* remove or ignore a lock or tag. The OSHA requirements for lock-out tag out can be found in *1910.147, The control of hazardous energy.* Facilities have procedures established to comply with the standard. It is essential to comply with your facility's lock-out/tag-out procedure.

Prior to entry into a confined space, or during a hazardous materials incident, atmospheric monitoring must also be started to identify any **immediate danger to life and health (IDLH)** environments for rescuers and victims. The next steps involve looking at the type of incident and planning how to safely rescue victims. In a trench rescue, this would include setting up ventilation fans for airflow, setting up lights for visibility, or protecting a trench from further collapse.

## Access

Once the scene is stabilized, access to the victim must be gained. How is the victim trapped? In a trench situation, it may be a dirt pile. In a rope rescue, it may be scaffolding that has collapsed. In a confined space, a hazardous atmosphere may have caused someone to collapse. In order to reach a victim who is buried or trapped beneath debris, it is sometimes necessary to dig a tunnel as a means of rescue and escape. Identify the method of rescue, and work toward freeing the victim safely.

Communicate with the victims at all times during the rescue to make sure they are not injured further by the rescue operation. Even if they are not injured, they need to be reassured that the team is working as quickly as possible to free them.

Emergency medical care should be initiated as soon as access is made to the victim. Medical personnel with some rescue knowledge are a vital resource during rescues; not only can these responders start IVs and treat medical conditions, but they also know how to deal with the equipment and procedures that are going on around them. It is important that EMS personnel are effectively coordinated into ongoing operations during rescue incidents. Their main functions are to treat victims and to stand by in case a rescue team member needs medical assistance. As soon as a rescue area or scene is secured and stabilized, the EMS personnel must be allowed access to the victims for medical assessment and stabilization. Some fire brigades have rescuers who are fully trained medical personnel. They can enter hazardous areas and provide direct assistance to the victim. Throughout the course of the rescue operation, which may span many hours, EMS personnel must continually monitor and ensure the stability of the victim and, therefore, must be allowed access to the victim.

Gaining access to the victim depends on the type of incident. For example, in an incident involving a scaffold collapse, the location position and damage to the scaffold, and the position of the victim are important considerations. The means of gaining access to the victim must take into account the victim's injuries and their severity. The chosen means of access may have to be changed during the course of the rescue as the nature or severity of the victim's injuries becomes apparent.

## Disentanglement

Once precautions have been taken and the reason for entrapment has been identified, the victim needs to be freed as safely as possible. A team member should remain with the victim to direct the rescuers who are performing the disentanglement. In a trench incident, this would include digging either with a shovel or by hand to free the victim.

In a vehicle accident, the most important point to remember is that the vehicle is to be removed from around the victim rather than trying to remove the victim through the wreckage. Various parts of the vehicle may trap the occupants, such as the steering wheel, seats, pedals, and dashboard. Disentanglement is the cutting of a vehicle (and or machinery) away from trapped or injured victims. In a vehicle accident, this is accomplished by using extrication and rescue tools with various extrication methods.

## Removal

Once the victim has been disentangled, efforts will be redirected to removing the victim (▼ Figure 23-4). In some instances this may simply amount to having someone assist the victim up a ladder. Usually it will require removal with spinal immobilization due to possible injuries. There are

**Figure 23-4** Victim removal.

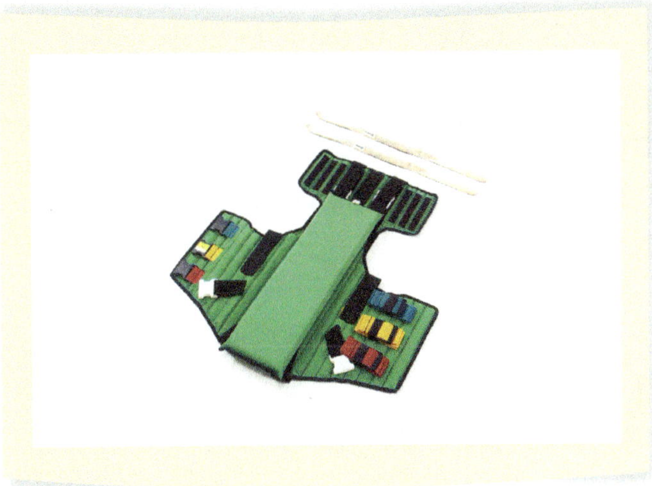

Figure 23-5 The Kendrick's extrication device.

Figure 23-6 The Oregon spine splint.

a wide variety of choices such as Stokes baskets, backboards, stretchers, and other immobilization devices that are commonly used to remove an injured victim from a trench, confined space, or an elevated point.

Preparing a victim with non-life-threatening injuries for removal, or in situations where the victim's life is not in immediate danger, should include the dressing of wounds and the immobilizing of suspected fractures and head and spinal injuries. The use of standard splints in confined areas is difficult and frequently impossible, but stabilization of the victim's trunk and of the legs to each other will often be adequate until the victim is positioned on a long spine board, which may serve as a splint for the whole body. The Kendrick's extrication device or similar devices such as the Oregon spine splint are frequently used for stabilization of the sitting victim (◄ Figure 23-5 and Figure 23-6).

Sometimes a victim may have to be removed quickly (rapid extrication) because the victim's general condition is deteriorating, and time will not permit meticulous splinting and dressing procedures. Quick removal also occurs if there are hazards present, such as spilled fuel or hazardous materials that could endanger the victim or the rescue personnel. The only time the victim should be moved prior to completion of initial care, assessment, stabilization, and treatment is when the victim's life or the emergency responder's life is in immediate danger.

**Packaging** is preparing the victim for movement and is often accomplished by means of a long spine board or similar device. The boards are essential for moving victims with potential or actual spine injuries.

The overriding objective for each rescue, transfer, and removal is to complete the process as safely and efficiently as possible. It is important that the rescuer utilize good body mechanics, victim-packaging, removal, and transportation skills.

### Transport

Once the victim has been removed from the hazard area, transport to an appropriate medical facility will usually be done by EMS. Depending upon the severity of the injuries and the distance to the medical facility, the type of transport will vary. For example, if a victim is critically injured or if the rescue is taking place some distance from the hospital, air transportation may be more appropriate than a ground ambulance.

Transporting victims who have been injured in a hazardous materials incident may pose additional challenges. Prior to transport, it is extremely important that proper decontamination takes place. It is dangerous to the victim and the rescue personnel to place a poorly decontaminated victim inside an ambulance or helicopter and then close the doors. Any toxic substance given off by the victim or by the victim's clothing can contaminate the inside of the transport vehicle, causing injury to personnel. Bags of contaminated clothing or other personal effects must be properly sealed if they are going with the victim. Receiving hospitals should be notified of the impending arrival of victims who have been involved in a hazardous materials exposure incident. This will alert the hospital triage teams so that the appropriate precautions can be taken once the victim arrives.

## Postincident Duties

### Security of the Scene and Preparation for the Next Call

Once the rescue is complete, the scene must be stabilized by the rescue crew to ensure that no one else becomes injured.

# VOICES OF EXPERIENCE

**"There was no reverse on the press; we would have to separate the rollers to remove the patient."**

We were dispatched to a local manufacturing plant for a man trapped in a machine incident. We responded with one engine and a Paramedic rescue company. Upon arrival we met a member of the plant ERT who quickly directed us to the door closest to the incident. Due to the size of the plant, this saved us a few minutes of time and shortened the distance that we would have had to carry our equipment.

At the door we met another member of the ERT who led us several hundred feet, past tanks, piping, and large presses to the machine where our patient was entrapped. Our captain met with the plant ERT Leader who gave us his size up. The size up report stated that we had a man with his arm trapped between a set of heavy stainless steel rollers. His forearm was trapped just past his wrist. There was no reverse on the press; we would have to separate the rollers to remove the patient.

A continued size up and the hazard assessment revealed that this was a heat-treated press machine that used organic peroxides in the process. The unit was 40 feet by 10 feet long. We had high heat; lots of noise and another press operating 12 feet away.

While the captain and team leader were developing their incident action plan, we began to set up medical and extrication equipment while the ERT members began the lockout process on the press, ran tape to isolate the area from onlookers, pulled a fire hose and tended to the comfort of the patient.

The patient remained calm due to the assistance of the first aid members. He was in some pain, but was not bleeding and showed no signs of broken bones. Our paramedic monitored the patient's status continuously.

Plan A was agreed upon and we began to set up our hydraulic spreader system. As soon as we began trying to spread the rollers, we discovered several problems. First, we had to hold the heavy spreader unit over four feet from the floor. Second, we could not get a good bite in between the six inch rollers. When we did get a good bite, the rollers rolled forward and pulled the patients arm further into the machine.

Plan B was quickly developed. We decided to try to disassemble the roller assembly. The ERT leader consulted with the plant maintenance engineer about this plan. The ERT maintenance members quickly gathered tools, retrieved the blue prints of the machine and completed the lockout process.

In a very short time, they had disassembled one side of the roller assembly enough so that we could get our pry bars between the rollers and separate them enough so that we could pull the patient away from the machine.

This incident demonstrates how important it is that ERT members are trained to assist the local responders at a TRI. We used members to identify and control hazards, control the perimeter, retrieve and assist with rescue tools, stand-by with a hose line, comfort the patient and finally to disassemble the equipment.

These efforts greatly reduced the rescue time, which increased the patients Golden Hour time. Because we were able to use disentanglement versus a more destructive extrication technique, they saved the cost of expensive repairs and the loss of production time from that machine.

This incident motivated the ERT to develop and deliver an Industrial Extrication Course for all of their members. They also completed disentanglement preplans for all of their machinery.

Raymond Lussier
Auburn Fire & Rescue
Auburn, Maine

In a trench incident, this includes filling the trench with dirt or roping off the surrounding area.

In a hazardous materials incident, the cleanup of equipment and personnel takes place after the hazardous materials incident has been completely controlled and all victims have been treated and transported. Trained disposal crews should be utilized to clean the site. All equipment, protective gear, and clothing, as well as the rescue personnel, must be decontaminated.

In an industrial setting, the supervisor of the facility is usually responsible for securing the scene, but the technical rescue team still must follow up with the supervisor or the safety coordinator to ensure that further problems are prevented.

Once you have secured the scene and packed up your equipment, it is important to return to the station to inventory, clean and service all of the equipment to prepare it for the next call. Some items will need repair, but most will need simple maintenance before being placed back on the apparatus and considered in service.

Back at the station, complete paperwork and document the rescue incident. Record-keeping serves several important purposes. Adequate reporting and maintaining accurate records ensure the continuity of quality care, guarantee proper transfer of responsibility, and fulfill the administrative needs of the fire brigade. In addition, the reports can be used to evaluate response times, equipment usage, and other areas of administrative interest.

### Postincident Analysis

As with any type of call, the best way to prepare for the next rescue call is to review the last one and identify any strengths and weaknesses. What could have been done better? What equipment would have made the rescue safer or easier? If a death or serious injury occurred during the call, a critical incident stress management (CISM) session may occur to assist brigade members. Reviewing a rescue with everyone involved will allow everyone to learn from the call and make the next call even more successful.

## General Rescue Scene Procedures

At any scene you respond to—whether it is a fire, EMS, or technical rescue call—the safety of you, your brigade, and facility personnel is paramount. At a rescue there are many things that need to be considered. While the temptation may be to approach the victim or the accident area, it is critically important to slow down and properly evaluate the situation.

In confined space rescue incidents, the potential hazards can be deep or isolated spaces, multiple complicating hazards (such as water or chemicals), failure of essential equipment or service, and environmental conditions (such as snow or rain). The primary concern in a confined space is lack of oxygen. In all rescue incidents, brigade members should consider the potential general hazards and risks of utilities, hazardous materials, confined spaces, and environmental conditions, as well as hazards that are IDLH.

### Approaching the Scene

As you approach the scene of a rescue you will not always know what kind of scene you are going into. Is it a construction scene? Are there piles of dirt that would indicate a trench? Has a structural collapse occurred in a building? What actions are the employees taking? Are they attempting to rescue trapped people, possibly placing themselves at great danger?

From the initial dispatch of the rescue call, the brigade member should be compiling facts and factors about the call. Size-up begins with the information gained from the person reporting the incident and then from the bystanders at the scene upon arrival.

The information received when an emergency call is received is important to the success of the rescue operation. The information should include the following:

- Location of the incident
- Nature of the incident (kinds and number of vehicles)
- Condition and position of victims
- Condition and position of equipment
- Number of people trapped or injured, and types of injuries
- Any specific or special hazard information
- Name of person calling and a number where the person can be reached

Once on the scene, life-threatening hazards can be identified and corrective measures can be taken. If additional resources are needed, they should be ordered by the Incident Commander (IC).

A size-up should include the initial and continuous evaluation of the following:

- Scope and magnitude of the incident
- Risk and benefit analysis
- Number of known and potential victims
- Hazards
- Access to the scene
- Environmental factors
- Available and necessary resources
- Establishment of control perimeter

Careful size-up and coordination of rescue efforts are a must in order to avoid further injuries to the victims and to provide for the safety of the brigade members.

### Utility Hazards

Are there any downed electrical wires that are near the scene ▶ Figure 23-7? Is the equipment or machinery present electrically charged so as to present a hazard to the victim or the rescuers? The IC should ensure that the proper

Figure 23-7 Downed electrical wires present a hazard.

Figure 23-8 Technical rescue technicians need to be able to move in their gear.

procedures have been taken to shut off the utilities in the area where the rescuers will be working.

Utility hazards require the assistance of trained personnel. For electrical hazards, such as downed lines, park at least one utility pole span away. Watch for falling utility poles; a damaged pole may bring other poles down with it. Do not touch any wires, power lines, or other electrical sources until they have been deactivated by a power company representative. It is not just the wires that are hazardous; any metal that they touch is also energized. Metal fences that become energized are energized for their entire unbroken length. Be careful around running or standing water, since it is an excellent conductor of electricity.

Both natural gas and liquefied petroleum gas are nontoxic but are classified as asphyxiants because they displace breathing air. In addition, both types of gas are explosive. If a call involves leaking gas, call plant operations or the gas company immediately. If a victim has been overcome by leaking gas, wear positive-pressure self-contained breathing apparatus (SCBA). Remove the victim from the hazardous atmosphere before beginning treatment.

## Scene Security

Has the area been secured to prevent people from entering the area? Often coworkers and sometimes other rescuers will enter an unsafe scene and become additional victims. The IC should coordinate with plant security to help secure the scene and control access. A strict accountability system should be used by the fire brigade to control access to the rescue scene.

## Protective Equipment

Firefighting gear is designed to protect the body from the temperatures of fire. It does, however, restrict movement. Technical rescues generally require the ability to move around freely. Firefighting gear does not work well in a rescue. Most specialized teams also carry items such as harnesses; smaller, lighter helmets; and jumpsuits that are easier to move in than turnout gear (▲Figure 23-8).

A handheld strobe light or reflectors may help brigade members keep track of each other in poorly lit areas. When working along highways, brigade members can hook these lights or reflectors onto their belts or attach them to their upper arm to provide additional visibility to oncoming vehicles. Strobe lights or reflectors are lightweight, quite durable, and readily visible at night.

Other equipment items that are easily carried by brigade members are binoculars, chalk or spray paint for marking searched areas, first aid kits, and cyalume-type light sticks. The IC and the technical rescue team will help to determine what protective equipment you will need to wear while assisting. Also see NFPA 1951, *Standard on Protective Ensembles for Technical Rescue Incidents.*

## Incident Management System (IMS)

The first arriving brigade leader immediately assumes command and implements the incident management system or IMS. This is critically important because many rescues will eventually become very complex and require a large number of assisting units. Without the IMS in place, it will be difficult, if not impossible, to ensure the safety of the rescuers.

## Accountability

Accountability should be practiced at all emergencies, no matter how small. The **accountability system** is the single most important process that any rescuer needs in order to ensure safety. An accountability system tracks the personnel on the scene, including their identities, assignments, and locations. This system ensures that only rescuers who have been given specific assignments are operating within the area where the rescue is taking place. By using an accountability system and working within IMS, an IC can track the resources at the scene, task out assignments, and ensure that every person at the scene operates safely.

### Making Victim Contact

At any rescue scene, try to communicate with the victim if at all possible. Technical rescue situations often last for hours, with the victim left alone for long periods of time. Sometimes it is just not possible, but try to communicate via a radio, cell phone, or by using your voice effectively. Reassure the victim that everything is being done to ensure his or her safety.

It is important for the brigade member to stay in communication with the victim. If possible, it is advisable to have someone assigned to talk to the victim, while others are involved with the rescue. Realize that the victim could be sick or injured and is probably frightened. If the brigade member is calm, this will help calm the victim. To help keep a victim calm do the following:

- Make and keep eye contact with the victim.
- Tell the truth. Lying destroys trust and confidence. You may not always tell the victim everything, but if the victim asks a specific question, answer truthfully.
- Communicate at a level that the victim can understand.
- Be aware of your own body language.
- Always speak slowly, clearly, and distinctly.
- Use the victim's name.
- If a victim is hard of hearing, speak clearly and directly at the person.
- Allow time for the victim to answer or respond to your questions.
- Try to make the victim comfortable and relaxed whenever possible.

Many of the victims at rescues require medical care, but this care should be given only if it can be done so *safely*. Do not become a victim during a rescue attempt. Would-be rescuers have been killed entering a hazardous rescue environment trying to help a dead victim.

## Assisting Rescue Crews

Every rescue will be different. If you have the role of assisting a technical rescue team, training with the team is probably the most important thing you can do. By training with the team you will respond with, you will get a feel for how they operate and they will get an idea of what they can trust you with. The more knowledge you have, the more you will be able to do.

At any rescue, follow the orders of the brigade leader who receives direction from the IC. Many of the tasks will involve moving equipment and objects from one place to another. Other tasks will involve protecting the team and victims. Do not take these tasks lightly; they may not seem important but they are essential to the effort. Without scene security, traffic control, protective hose lines, and people to maintain equipment, the rescue will not be effective.

As you read the information that follows about assisting at specific types of rescue scenes, bear in mind the three factors regarding safely approaching the scene:

- Approach the scene cautiously.
- Position apparatus properly.
- Assist specialized team members as needed.

### Vehicles and Machinery

There are a wide variety of rescue situations involving vehicles and machinery. This section includes rescues involving vehicles and machinery. These include trucks, trains, watercraft, construction machinery, and industrial machinery.

#### Safe Approach

Vehicle and machinery rescues present a wide variety of problems to rescuers and victims. There are hazards such as flammable liquids, electrical hazards, and unstable machines or vehicles. Electricity is an invisible hazard. Any machine that is encountered should be considered electrically charged until proven otherwise. It is important to lockout and tagout any electrical source before approaching. The operator of the machine or a maintenance person who is familiar with the machine can be a valuable resource. For vehicle calls it is extremely important to make sure that traffic has been controlled. Placement of the apparatus is key to the protection of you and those around you.

Stabilization of the rescue scene includes making sure that the machine cannot move. Access to some victims will be easy. In other situations access to the victim can occur only after extensive stabilization of the scene.

Unusual and special rescue situations include extrication or disentanglement operations at incidents involving vehicles on their tops or sides, trucks and large commercial vehicles. To ensure responder safety, beware of fuel spilled at these scenes and protect the scene from ignition.

#### How You Can Assist

At a vehicle or machinery rescue call, you may be called upon to assist in the extrication and treatment of the victim. Protection of the victim (C-Spine immobilization, holding of blankets, etc.) will allow the technical rescue team to operate hydraulic tools and cut the vehicle or machine apart without further hurting the victim. Many of the hydraulic tools are heavy and may require you to help support them while technicians operate them.

You may also be called upon to do the following:

- Assist with controlling site security and the perimeter of the rescue incident
- Obtain information from witnesses

## Brigade Member Safety Tips

Remember, the greatest hazards in a confined space are the lack of oxygen and the presence of poisonous gases. If rescuers fail to identify a confined space and to ensure that it contains a safe atmosphere, injury and death can occur to rescuers as well as to the original victim. There have been cases where well-meaning rescuers have gone into a confined space and become incapacitated because they did not properly evaluate the situation.

- Retrieve more rescue tools and equipment from the fire apparatus or rescue trucks
- Keep bystanders out of the way
- Assist in moving items that are in the way of the rescue team
- Service equipment
- Set up power tools
- Stand by with a hose line
- Extinguish a fire

### Tools Used

There are many tools required for a successful vehicle or machinery rescue situation, and some are inexpensive items. Simple tools such as a spring-loaded punch can safely break out a vehicle window to access a victim. Tools and items that will be required for a vehicle or machinery rescue include the following:

- Personal protective equipment (PPE)
- Hydraulic tools (spreaders, cutters, rams)
- Halligan tool
- Cutting torch
- Air chisels
- Cribbing
- Saber saw
- Windshield cutter
- Spring-loaded punch
- Chains
- Airbags (high and low pressure)
- Basic hand tools
- Come-along
- Portable generator
- Seat belt cutters
- Hand lights
- Hose lines (protection)
- Blanket

## Confined Space

A confined space is defined as a space large enough and configured so an employee can enter, has limited or restricted mean for entry or exit and is not designed for continuous employee occupancy. Confined spaces may have limited ventilation to provide air circulation and exchange. Confined spaces are common in industrial settings. Industrial pits, tanks, vessels and belowground structures are all examples of confined spaces.

A permit required confined space (PRCS) also has the potential for hazardous atmospheres, entrapment or entanglement, or other recognized hazards. PRCS are identified at every facility and require special approval procedures for entry. OSHA 1910.146 *Permit Required Confined Space* provides additional information.

Confined spaces present a special hazard, because they may be oxygen deficient or contain toxic gases. Entering a confined space without testing the atmosphere for safety and without the proper breathing apparatus can result in death. A confined space call is sometimes dispatched as a heart attack or medical illness call, because the caller assumes the person who entered a confined space and became unresponsive suffered a heart attack or medical illness.

A minimum of one rescue trained attendant is required to be at the scene of any permit required confined space when the entry involves an IDLH atmosphere. Other confined space entry conditions may require additional on-site rescue resources.

### Safe Approach

As you approach a rescue scene, look for a bystander who might have witnessed the emergency. Information gathered prior to the technical rescue team's arrival will save valuable time during the actual rescue. Do not believe a person in a pit has simply suffered a heart attack; always assume that there is an IDLH atmosphere at any confined space call. An IDLH atmosphere can immediately incapacitate anyone who enters the confined space without breathing protection. There can be toxic gases present, or there may not be enough oxygen to support life. When a rescue involves a confined space, remember that it will take some time for qualified rescuers to arrive on the scene and prepare for a safe entry into the confined space. The victim of the original incident may have died before your arrival. Do not put your life in danger for a dead victim.

### How You Can Assist

Your main role in confined space rescues is to secure the scene, preventing other people from entering the confined space until additional rescue resources arrive. As additional trained personnel arrive, your crew may provide help by giving the rescuers a situation report.

The first responding crew must share whatever information is discovered at the rescue scene with the arriving crew. Anything that may be important to the response should be noted by the first arriving unit. Observed conditions should be compared to reported conditions and a determination should be made as to the relative change over the time period. Whether an incident appears to be stable or has changed greatly since the first report will affect the operation strategy for the rescue. A size-up should be quickly completed immediately upon arrival, and this information should be relayed to the special rescue team members upon their arrival at the scene. Other items of importance that should be included in a situation report are a description of any rescue attempts that have been made, exposures, hazards, extinguishment of fires, the facts and probabilities of the scene, the situation and resources of the fire brigade, the identity of any hazardous materials present, and a progress evaluation so far.

Each PRCS must have been tested for atmospheric conditions as a condition of entry. These conditions will be listed on the permit on site.

Many facility fire brigades carry atmospheric or gas detection devices; if you have this capability, obtain secondary readings to determine if there is a hazard. If you do encounter an oxygen-deficient atmosphere, set up a ventilation fan to help remove toxic gases and improve airflow to the victim. Sometimes you can help a victim without entering the confined space by passing down an SCBA, oxygen, first aid supplies, or even a ladder to climb out.

Confined space rescues can be complex and can take a long time to complete. You may be asked to assist by bringing rescue equipment to the scene, maintaining a charged hose line, or assisting with crowd control. By understanding the hazards of confined spaces, you will be better prepared to assist a specialized team that is dealing with an emergency involving a confined space.

### Tools Used

One of the key components of any confined space rescue is a supplied air respirator (SAR) ▶ Figure 23-9. These are similar to SCBA, except that instead of carrying the air supply in a cylinder on your back, you are connected by a hose line to an air supply located outside of the confined space. This provides the rescuer with a continuous supply of air that is not limited by the capacity of a back-mounted air cylinder.

There are a number of tools and pieces of equipment that can be used during a confined space rescue operation:

- SAR
- Tripod for raising and lowering rescuers and victims
- Pry bar or manhole cover remover
- Rescue rope
- Personnel harnesses
- Gas monitors
- Ventilation fans
- Explosion-proof lights
- Radios
- Protective gear
- Lockout, tagout, and blankout kits
- Accountability boards
- Victim removal devices such as Stokes baskets, backboards, and other commercially made devices

Figure 23-9 Supplied air Respirator.

- Medical equipment
- Carabiners and locking "D" rings
- Descenders with ears
- Safety goggles/glasses
- Hand light (battery-operated)
- Hearing protection

## Rope Rescue

Rope rescue skills are used in a variety of different rescues. It may be necessary to lift out an injured victim from inside of a tank using a hoisting system. Rescuers may need to lower themselves down to the area where victims are trapped. Rope rescue skills are the most versatile and widely used technical rescue skills.

Rope rescue incidents are divided into low-angle and high-angle operations. Low-angle operations are situations where the slope of the ground over which the rescuers are working is less than 45°, rescuers are dependent on the ground for their primary support, and the rope system becomes the secondary means of support ▶ Figure 23-10. An example of a low-angle system is a rope stretched from the top of an embankment and used for support by rescuers who are carrying a victim up an incline. The rope is the secondary means of support for the rescuers. The primary support is the rescuer's contact with the ground.

Low-angle operations are used when the scene requires ropes to be used only as assistance to pull or haul up a victim or rescuer. This is usually necessary when adequate footing is not present in areas such as a dirt or rock embankment. In such an incident, a rope will be tied to the rescuer's harness and the rescuer will climb the embankment on his or her own, using the rope only to make sure he or she does not fall.

Figure 23-10 Low-angle operations.

Ropes can also be used to assist in carrying a Stokes basket. This is done to aid the rescuers and to free them from having to carry all of the weight over rough terrain. Rescuers at the top of the embankment can help to pull up or lower the basket using a rope system.

High-angle operations are situations where the slope of the ground is greater than 45°, and rescuers or victims are dependent on life safety rope and not a fixed surface of support such as the ground. High-angle rescue techniques are used to raise or lower a person when other means of raising or lowering are not readily available. Sometimes a rope rescue is performed to remove a person from a position of peril. At other times rope rescues are needed to remove ill or injured persons.

In Chapter 9, Ropes and Knots, you learned how to tie some basic knots and use them to lift and lower selected tools. Rope rescue training courses build on the skills you have learned. There is a lot more to learn before you are ready to perform high-angle rescues. You need to be able to perform complex tasks quickly and safely. Rope rescue operations require a cohesive team effort.

### Safe Approach

If you respond to an incident that may require a rope rescue operation, consider your safety and the safety of those around you. Rope rescues are among the most time-consuming calls that you will encounter. There is extensive setup and equipment that needs to be assembled prior to initiating any rescue. Protect your safety by remaining away from the area, under the victim and away from any loose materials that may fall. Work to control the scene so that the bystanders and friends of the victim move to an area where they will not be injured. You can do a lot to stabilize the scene and prevent further injuries by remaining calm and putting your skills to work. A rope can be used to secure a rescuer when making a rescue attempt, haul a stretcher or litter up an embankment, or lower a rescuer into a trench or vessel.

### How You Can Assist

Rope rescues are labor-intensive operations. By remembering the introductory rope skills you have learned, you may be able to assist with many phases of the rescue operation. You may be assigned to a technical rescue team member to tie knots and get anchors ready. Do not be offended if the team members go back and check your work. It is protocol to check a system two or three times to ensure safety. Remember to avoid stepping on ropes. Any damage or friction to the outer layers of a rope can decrease its tensile strength and possibly cause a catastrophic failure. Keep everyone clear of areas where a falling object could cause injuries. Keep in mind the importance of following the IMS. Work to complete your assigned tasks.

### Tools Used

Rope rescues rely almost completely on the equipment used. When rappelling down to rescue a victim 300 feet off of the ground, rescuers need equipment that has been properly designed and maintained to prevent injury. Some of the tools and equipment that you will need to recognize and use are as follows:

- PPE
- Personnel harnesses
- Stokes basket
- Harness for Stokes basket
- Rescue ropes
- Carabiners
- Webbing
- Miscellaneous hardware (racks, pulleys)
- Electrical cords (with plug and adapters)
- Electrical lights and hand lights (battery-operated)
- Ladders
- Tools (hacksaw, rubber mallet, pipe wrench, bolt cutter, crow bar, pry bar)

## Trench and Excavation Collapse

Trench and excavation rescues occur when the earth has been removed for a utility line or for other construction and the sides of the excavation collapse, trapping a worker ► Figure 23-11. Entrapments can occur when people are working around a pile of sand or earth that collapses. Many entrapments occur because the required safety precautions were not taken.

Any time there has been collapse, you need to understand that the collapsed product is unstable and prone to further collapse. Earth and sand are heavy, and a person partly entrapped cannot be pulled out. They must be carefully dug out. This can be done only after shoring has stabilized the sides of the excavation.

**Figure 23-11** Trench rescue.

Vibration or additional weight on top of displaced earth will increase the probability of a **secondary collapse**. A secondary collapse is one that occurs after the initial collapse. It can be caused by equipment vibration, personnel standing at the edge of the trench, or water eroding away the soil. Safe removal of trapped persons requires a special rescue team that is trained and equipped to erect shoring to protect the rescuers and the entrapped person from secondary collapse.

## Safe Approach

Safety is of paramount importance when approaching a trench or excavation collapse. Walking close to the edge of a collapse can trigger a secondary collapse. Stay away from the edge of the collapse and keep all workers and bystanders away. Vibration from equipment and machinery can cause secondary collapses, so shut off all unnecessary heavy equipment.

Soil that has been removed from the excavation and placed in a pile is called the **spoil pile**. This material is unstable and may collapse if placed too close to the excavation. Avoid disturbing the spoil pile.

Make verbal contact with the trapped person if possible, but do NOT place yourself in danger while doing this. If you are going to approach the trench at all, approach it from the narrow end where the soil will be more stable. However, it is best *not* to approach the trench unless absolutely necessary. Stay out of a trench unless properly trained.

Provide reassurance by letting the trapped person know that a trained rescue team is on the way. By removing people from the edges of the excavation, shutting down machinery, and establishing contact with the victim, you start the rescue process.

You can also size up the scene by looking for evidence that would indicate where the trapped victims may be located. Hand tools are an indicator of where the victims may have been working. Hard hats are another indicator. By questioning the bystanders, you can also determine where the victims were last seen.

## How You Can Assist

As the rescue team starts to work, your fire brigade will be assigned certain tasks. These may range from unloading lumber for shoring to assisting with cutting timbers a safe distance away from the entrapment. This type of rescue can take a long time. If it is hot or cold, a rehabilitation sector may need to be set up. Early implementation of the IMS will help make this type of rescue go smoothly. If someone in your brigade has a specialty such as carpentry, this should be made known, since this skill is valuable at this type of rescue. Cutting and measuring timber and shores for a trench operation is an important aspect of the rescue. Just as in a confined space, trenches may have an IDLH atmosphere due to gases like methane or sewer gases. This should be tested with environmental monitoring equipment. Setting up ventilation fans can often make a difference in victim and rescuer survivability.

It may also be necessary to pump water out of a trench. If the collapse occurred because of a ruptured pipe or during a rainstorm, rising water in the trench can endanger the trapped victim and cause additional soil to collapse into the trench. By removing this water, the situation can be stabilized. Always test the atmosphere for CO if a gasoline pump is used.

When extricating victims from the trench, it may be necessary to lift them out using a hoisting system, similar to a high-angle rope rescue operation. This will require that the rescuers have rescue ropes, harnesses, pulleys, carabiners, and other associated equipment. Personnel must be trained in the use and application of the systems in this rescue scenario.

## Tools Used

Trench and excavation rescue shares equipment, rescue techniques, and skills with both confined space rescue and structural collapse rescue. Tools and equipment used in trench and excavation rescue include the following:

- PPE: helmet, gloves, personal protective clothing, harness, flashlight, work boots, knee pads, elbow pads, eye protection, SCBA, SAR
- Hydraulic, pneumatic, and wood shores
- Lumber and plywood for shoring
- Cribbing

- Power cutting tools, saws
- Carpenter hand tools
- Shovels
- Buckets for moving soil
- Rescue rope, harnesses, webbing, associated hardware
- Utility rope
- Ventilation fans
- Pumps
- Lighting
- Ladders
- Shovel
- Extrication equipment such as Stokes baskets or backboards
- Harness set for Stokes baskets
- Medical equipment

## Structural Collapse

Structural collapse is the sudden and unplanned fall of part or all of a building (▼ Figure 23-12). Collapses occur because of fires, removal of supports during construction or renovation, vehicle crashes, explosions, rain, wind or snowstorms, earthquakes, and tornados. Consider the type of building construction when determining the potential for collapse. When any part of a building is compromised, the dynamics of the building change; brigade members should always be alert for signs of a possible building collapse. A partial building collapse may be hazardous to rescuers because of the potential for secondary or further collapse.

### Safe Approach

Because of the variety of factors that can cause building collapses, you must approach the scene carefully. The cause of some collapses, such as a vehicle crashing through the wall of a building, will be evident. Others, like a natural gas explosion, may require extensive investigation before the cause is determined. As you approach a building collapse, consider the need to shut off utilities. Entering a structure with escaping natural gas or propane is extremely hazardous. Electricity from damaged wiring can also present a deadly hazard.

A prime safety consideration is the stability of the building. Even a well-trained engineer cannot always determine the stability of a building by looking only at its exterior. Therefore, you must operate as though the building may experience a secondary collapse at any time. The IC must make the decision regarding whether the building is safe to enter. This is a difficult decision, especially since the bystanders cannot always tell you whether there were people in the building before the collapse. Be aware of the potential for secondary collapse.

Figure 23-12 Structural collapse is the sudden and unplanned fall of part or all of a building.

### How You Can Assist

Rescue operations at a structure collapse vary depending on the size of the building and the amount of damage to the building. In cases of large building collapses, the rescue operation will be sizeable. Urban search and rescue teams or structural collapse teams have received special training in dealing with these types of situations. They are trained in shoring and specialized techniques for gaining access and extricating victims that enable them to systematically search the affected building. These types of rescue situations take a lot of time and require a lot of personnel.

Personnel without special training will usually be assigned to support operations. You may be assigned to be part of a bucket brigade that works to remove debris from the building. There may be a substantial amount of manual digging and searching that is physically taxing. You may be able to assist with this if the work is not in an unstable location.

In order for a rescue effort to be successful, there must be teamwork and a well-organized and well-implemented incident command system. Without a solid command structure, most large-scale rescue efforts are doomed to fail.

### Tools Used

Brigade members should know the tools and equipment designed for structural collapse emergency rescue incidents:

- PPE: helmet, gloves, work boots, harness, elbow pads, knee pads, eye protection, SCBA
- Shoring equipment
- Lumber for shoring and cribbing
- Power tools
- Hand tools
- Lighting
- Rescue ropes, harnesses, webbing, and associated hardware
- Utility ropes
- Buckets
- Shovels

## Brigade Member Tips

Fire brigade members must learn when to offer helpful suggestions and when to remain quiet. There may be times when you see something that the brigade leader does not see, or have an idea that the brigade leader has not considered. However, the need to provide helpful information must be balanced against the risk of overwhelming a brigade leader with "helpful" suggestions. There can be no hard and fast rule about this, but stay mindful of the fact that remaining quiet is at times the most helpful thing you can do.

## Hazardous Materials Incidents

Hazardous materials are defined as any materials or substances that pose a significant risk to the health and safety of persons or to the environment if they are not properly handled during manufacture, processing, packaging, transportation, storage, use, or disposal (▼ Figure 23-13).

While hazardous materials incidents often involve a petroleum product, there are many other chemicals that have toxic effects when not handled properly.

With the current threat of domestic terrorism, many of the agents that could be used as weapons also fall into this category. Many fire brigades are trained and equipped to recognize these incidents, contain the hazards, and evacuate people if necessary. All fire brigade members must be trained to the awareness level, to recognize and identify hazardous materials incidents and take initial steps to call for assistance and contain the incident.

### Safe Approach

Hazardous materials incidents are not always dispatched as hazardous materials incidents. You must be able to recognize the signs that indicate that there may be hazardous materials present as you approach the scene. You may see an escaping chemical or smell a suspicious odor. Warning placards are required for most hazardous materials either for storage or in transit.

Once you have recognized the presence of a hazardous material, you must protect yourself by staying out of the area exposed to the hazardous material. If you have happened upon a hazardous materials incident, it is important to have the hazardous materials team dispatched as soon as possible. In addition to standard action and precautions, a special hazardous materials rescue team may be required to implement site control and scene management, and assist specialized personnel after their arrival according to your training level.

Figure 23-13 Hazardous materials incident.

### How You Can Assist

In order to assist at a hazardous materials incident, you must have formal training. Training at the first responder operations level will provide you with the knowledge and skills you need to be able to recognize the presence of a hazardous material, protect yourself, call for appropriate assistance, and evacuate or secure the affected area. As part of your training, you will learn how to assist other hazardous materials responders.

The four major objectives of training at the operational level are to analyze the magnitude of the hazardous materials incident, plan an initial response, implement the planned response, and evaluate the progress of the actions taken to mitigate the incident.

### Tools Used

Tools used in a hazardous materials incident include the following:

- PPE appropriate to the level of the hazard
- Two-way radios
- Lighting
- Gas monitors
- Various testing and monitoring tools to identify the materials involved
- Research material
- Decontamination equipment
- Hand tools (hammers, screwdrivers, axe, wrenches, and pliers)
- Devices for sealing breached containers
- Leak-control devices
- Binoculars
- Fire barrier line tape

## Elevator Rescue

Elevator and escalator rescues are technical responses that require technical knowledge. Rescue crews should review the American Society of Mechanical Engineers' (ASME) A17.4 Guide for Emergency Personnel, and OSHA's lockout and tagout procedures. All elevator and escalator responses begin with knowledge of the machinery located within your brigade's jurisdiction. This can be accomplished only through a preincident analysis of these structures. Specialized training classes are also an essential component of safe and successful operations.

At times facility fire brigades may be called to assist with an elevator emergency (▶ Figure 23-14). Although this is a relatively rare occurrence in the industrial setting, fire brigade members must be able to contend with such an emergency if it arises.

Brigade members should never attempt to move or relocate an elevator under any circumstances. Only professional elevator technicians who are thoroughly trained and authorized to do so should consider this. When responding to elevator incidents, consider these recommendations:

1. **Always cut power to a malfunctioning elevator,** and secure the power supply through the lockout and tagout procedures. Remember, elevator technicians and building maintenance personnel have been summoned to the same elevator incident. They may try to turn on the elevator equipment. To eliminate any possibility of this happening, use the lockout and tagout procedures at every incident. The elevator machine room is where the power supply should be turned off. Identify the location of the stalled elevator within the hoist-way or elevator shaft. Are there victims in the stalled elevator? If so, communicate with them, and reassure them that the situation is under control, they are not in danger, and the situation will soon be corrected.
2. **Perform incident risk management and risk assessment.** Can the victims in this disabled elevator be removed without increasing the risk of injury or death to brigade members as well as the victims? Elevator cabs are passenger compartments that are securely locked to ensure that the victims are contained within the cab for their own safety. Brigade members must (a) carefully evaluate the situation and the inherent risks involved in a rescue and extrication operation, and (b) understand that the victims are in a much safer environment inside than outside. The best option may be to reassure the victims and wait for the elevator technician to arrive and correct the situation.
3. **Are there enough trained personnel and the proper equipment on scene** to perform the rescue and extrication operation safely? Elevator hall door release keys, safety harnesses, rope bags, ladders, forcible entry tools, portable lighting, and an assortment of hand tools will be necessary. Always have additional trained personnel on hand, and anticipate obstacles and problems.
4. **Once the incident has been resolved, leave the power supply off.** Always leave the building with the elevator power supply turned off.

Figure 23-14 Facility fire brigades may be called to assist with an elevator emergency.

# Wrap-Up

## Ready for Review

- The ten steps of the special rescue sequence are:
  - Preparation
  - Response
  - Arrival and size-up
  - Stabilization
  - Access
  - Disentanglement
  - Removal
  - Transport
  - Security on scene
  - Postincident analysis
- Brigade members encounter many different kinds of rescues. The basic principles of rescue apply to all types.
- Vehicle and machinery rescues occur in many settings. These situations require responders to stabilize the machinery and ensure that the electricity is off.
- It is important to recognize a confined space. Rescuers must ensure there is safe air supply before entering confined spaces.
- Rope rescue skills are the most versatile and widely used technical rescue skills.
- Trench and excavation rescues are hazardous and require responders to minimize the chance for secondary collapses.
- A damaged building is prone to structural collapse. Any time a building has been damaged, assume it may collapse.
- Hazardous materials training at the awareness level is suggested for all levels of facility fire brigade members in order to meet the competencies outlined in NFPA 472 *Standard for Professional Competence of Responders to Hazardous Materials Incidents,* Chapter 4. This includes recognizing hazardous materials, protecting responders, calling for appropriate assistance, and evacuating and securing the affected area. Training at the operations level and above is encouraged for all levels of facility fire brigade members.
- Elevator and escalator rescues are technical responses that require technical knowledge.

## Hot Terms

**Accountability system** A method of accounting for all personnel at an emergency incident and ensuring that only personnel with specific assignments are permitted to work within the various zones.

**Cold zone** This area is for staging vehicles and equipment. The cold zone contains the command post. The public and the media should be kept clear of the cold zone at all times.

**Confined space** A space large enough and configured so an employee can enter, has limited or restricted mean for entry or exit and is not designed for continuous employee occupancy.

**Entrapment** A condition in which a victim is trapped by debris, soil, or other material and is unable to extricate himself or herself.

**Excavation** Any man-made cut, cavity, trench, or depression in an earth surface, formed by earth removal.

**Hazardous materials** Any materials or substances that pose an unreasonable risk of damage or injury to persons, property, or the environment if not properly controlled during handling, storage, manufacture, processing, packaging, use and disposal, or transportation.

**High-angle operations** A rope rescue operation where the angle of the slope is greater than 45°; rescuers depend on life safety rope rather than a fixed support surface such as the ground.

**Hot zone** This area is for entry teams and rescue teams only. The hot zone immediately surrounds the dangers of the site (e.g., hazardous materials releases) to protect personnel outside the zone.

**Immediate danger to life and health (IDLH)** An atmospheric concentration of any toxic, corrosive, or asphyxiant substance that poses an immediate threat to life or could cause irreversible or delayed adverse health effects. There are three general IDLH atmospheres: toxic, flammable, and oxygen-deficient.

**Lockout and tagout systems** Lockout and tagout are methods of ensuring that electricity and other utilities have been shutdown and switches are "locked" so that they cannot be switched on, in order to prevent flow of power or gases into the area where rescue is being conducted.

**Low-angle operations** A rope rescue operation on a mildly sloping surface (less than 45°) or flat land where brigade members are dependent on the ground for their primary support, and the rope system is a secondary means of support.

**Packaging** Preparing the victim for movement, often accomplished with a long spine board or similar device.

**Permit required confined space (PRCS)** Space that is identified at every facility and requires special approval procedures for entry. Has the potential for hazardous atmospheres, entrapment or entanglement, or other recognized hazards.

**Placards** Signage required to be placed on sides of buildings, highway transport vehicles, railroad tank cars, and other forms of hazardous materials transportation that identifies the hazardous contents of the vehicle, using a standardized system.

**Secondary collapse** A collapse that occurs following the primary collapse. This can occur in trench, excavation, and structural collapses.

**Shoring** A method of supporting a trench wall or building components such as walls, floors, or ceilings using either hydraulic, pneumatic, or wood shoring systems; it is used to prevent collapse.

**Spoil pile** The pile of dirt that has been removed from an excavation, which may be unstable and prone to collapse.

**Supplied air respirator (SAR)** Emergency breathing systems similar to SCBA, but which utilize an airline running from the rescuers to a fixed air supply located outside of the confined space.

**Technical rescue incident (TRI)** A complex rescue incident involving vehicles or machinery, water or ice, rope techniques, a trench or excavation collapse, confined spaces, a structural collapse, an SAR, or hazardous materials, and which requires specially trained personnel and special equipment.

**Technical rescue team** A group of rescuers specially trained in the various disciplines of technical rescue.

**Trench** A narrow excavation (in relation to its length) made below the surface of the ground. In general, the depth is greater than the width.

**Warm zone** This area is for properly trained and equipped personnel only. The warm zone is where personnel and equipment decontamination and hot zone support take place.

# Brigade Member in Action

**Your facility fire brigade is called out to assist with the rescue of a contractor who has fallen inside a 100-foot vessel during a recent plant shutdown. According to bystanders, the contractor removed his safety equipment while he was inside cleaning the vessel because it was too hot and cumbersome. The contractor was overcome by residual Xylene and fell approximately 30 ft. to the bottom of the vessel.**

1. What are your priorities in rescuing this victim?
   - **A.** Time is of the essence. You should put on SCBA and have a fellow team member lower you down using a tripod.
   - **B.** Perform a size-up and communicate this information to the technical rescue team.
   - **C.** Without entering the vessel, if available obtain a reading using atmospheric or gas detection devices.
   - **D.** Both B and C.

2. Which of the following is considered a key component of any confined space rescue?
   - **A.** Self-Contained Underwater Breathing Apparatus (SCUBA)
   - **B.** Supplied Air Respiratory System (SAR)
   - **C.** Handline
   - **D.** Come-along

**The technical rescue team arrives and begins to access, remove, and transport the victim.**

3. Confined space rescues can be complex and can take a long time to complete. What are some ways in which you can assist the rescue team?
   A. Bring rescue equipment to the scene.
   B. Maintain a charged hose line.
   C. Assist with crowd control.
   D. All of the above

4. You are asked by the technical rescue team leader to carry out atmospheric testing. The atmospheric test reveals an oxygen-deficient atmosphere. What can you do to help the victim?
   A. Set up a ventilation fan.
   B. Pass down SCBA to the victim.
   C. Pass down oxygen to the victim.
   D. Maintain a charged hose line.

# Terrorism Awareness

# Chapter 24

## NFPA 1081 Standard

### Incipient Facility Fire Brigade Member

NFPA 1081 contains no Incipient Facility Job Performance Requirements for this chapter.

### Advanced Exterior Facility Fire Brigade Member

NFPA 1081 contains no Advanced Exterior Facility Job Performance Requirements for this chapter.

### Interior Structural Facility Fire Brigade Member

NFPA 1081 contains no Interior Structural Facility Job Performance Requirements for this chapter.

## Knowledge Objectives

After completing this chapter, you will be able to:

- Describe the threat posed by terrorism.
- Identify potential terrorist targets in your jurisdiction.
- Describe the dangers posed by explosive devices.
- Describe the difference between chemical and biological agents.
- Describe the dangers posed by radiological incidents.
- Describe the need for decontamination of exposed victims and response personnel.

## Skills Objectives

There are no skills objectives for this chapter.

# You Are the Brigade Member

**During routine switching of the incoming railcars to be filled with raw products, members of your rail shipping department note a cardboard box sitting near the dome of one of the railcars and a compressed gas cylinder located underneath the same car. The facility security department is immediately contacted, and it immediately requests that the facility emergency response team report to the incident site. As response team members arrive on the scene, you note numerous personnel standing in very close proximity to the railcar.**

1. ***What should your first actions be as you arrive on the scene?***
2. ***Which characteristics of this scene should concern you?***
3. ***What information would you immediately request from the personnel involved in the incident?***
4. ***Who would you immediately contact regarding this incident?***

## Introduction

Domestic terrorism, as defined by U.S. Code, Title 18, Part I, Chapter 113B, is activities that "involve acts dangerous to human life that are a violation of the criminal laws of the United States or of any State. These would include activities that are, or appear to be, intended to intimidate or coerce a civilian population, the government, or any segment thereof, in furtherance of political or social objectives."

The Federal Bureau of Investigation (FBI) classifies terrorism as either domestic or international. Domestic terrorism refers to acts committed within the United States, without any influence of foreign interests. International terrorism includes any acts that transcend international boundaries.

Terrorism is a worldwide threat. Throughout the world, there were 1,106 incidents of international terrorism, resulting in the deaths of 2,494 innocent civilians in 2000. Although there were only 864 terrorist incidents recorded in 2001, the attack on the World Trade Center accounted for the majority of the 4,739 worldwide deaths (► **Figure 24-1**).

**Figure 24-1** The September 11th attack on the World Trade Center accounted for the majority of the deaths caused by terrorists in 2001.

Within the United States, there were 24 terrorist incidents between 1994 and 1999 and three additional crimes that were suspected to be terrorist in nature. Most of these incidents were classified as domestic terrorism. During the same period, law enforcement agencies prevented 47 terrorist incidents.

### Fire Service Response to Terrorist Incidents

The fire service has a major role in protecting communities from terrorism. The fire service role includes emergency medical services (EMS), hazardous materials mitigation, technical rescue, and fire suppression. All of these functions may be needed when a terrorist incident occurs. Terrorism presents new challenges for the fire service. It also presents an unparalleled threat to the lives of brigade members and emergency responders.

The terrorist threat requires brigade members to work closely with local, state, and federal law enforcement agencies; emergency management agencies; allied health agencies; and the military. It is critical that all of these agencies work together in a coordinated and cooperative manner. All emergency responders and law enforcement agencies must be prepared to face a wide range of potential situations.

The greatest threat posed by terrorists is the use of weapons of mass destruction (WMD), devices that are designed to cause maximum damage to property or people. These weapons include chemical, biological, and radiological agents, as well as conventional weapons and explosives. Thousands of casualties could result from a WMD attack in an urban area. An incident of this magnitude could quickly overwhelm not only the largest and best-trained emergency response agencies but also the local healthcare system.

The fire service must adapt and be prepared for the threat of terrorist attacks by exploring new approaches and technologies to manage WMD incidents. One of the highest priorities is to improve the ability of first responders to identify and mitigate releases of chemical, biological, and radiological agents.

## Potential Targets and Tactics

Terrorists are usually motivated by a cause and choose targets they believe will help them achieve their goals and objectives. Many terrorist incidents aim to instill fear and panic among the general population and to disrupt daily ways of life. In other cases, they choose a symbolic target, such as a place of worship, a foreign embassy, a monument, or a prominent government building. Sometimes the objective is sabotage, to destroy or disable a facility that is significant to the terrorist cause. The ultimate goal could be to cause economic turmoil by interfering with transportation, trade, or commerce.

Terrorists choose a method of attack they think will make the desired statement or achieve the maximum results. They may vary their methods or change them over time. Explosive devices have been used in thousands of terrorist attacks; in recent years, however, there has been a significant increase in the number of suicide bombings.

Many terrorist incidents in the 1980s involved the taking of hostages on hijacked aircraft or cruise ships. Sometimes, only a few people were held; other times, hundreds of people were taken hostage. Diplomats, journalists, and athletes were targeted in several incidents, and the terrorists often offered to release their hostages in exchange for the release of imprisoned individuals allied with the terrorist cause. More recent terrorist actions have endangered thousands of lives with no opportunity to bargain.

Terrorism may occur in many different ways, including events that may look like an accident or a naturally occurring event. Chemical, biological, radiological, nuclear, and explosive (CBRNE) events may be masked to take many forms and be performed by lone actors or "state-sponsored" groups. Many events have targeted such places as family planning clinics, food production laboratories, and civic buildings. There are many causes with supporters who range from peaceful, nonviolent organizations to fanatical fringe groups.

It is often possible to anticipate likely targets and potential attacks. Law enforcement agencies routinely gather intelligence about terrorist groups, threats, and potential targets. In many cases this information is shared with fire brigades so that preincident plans can be developed for possible targets and scenarios. Even if no specific threats have been made, certain types of occupancies are known to be potential targets, and preincident planning for those locations should include the possibility of a terrorist attack as well as an accidental fire.

News accounts of terrorist incidents abroad can help keep brigade members current with trends in terrorist tactics. They can provide useful information about situations that could occur in your jurisdiction in the future. There are a number of "high-tech" methods available to keep abreast of emerging issues, such as through numerous news groups, list serves, and federal websites (e.g., www.fema.gov, www.llis.gov, and www.dhs.gov).

### Ecoterrorism Targets

Ecoterrorism refers to illegal acts committed by groups supporting environmental or related causes. Examples include spiking trees to sabotage logging operations, vandalizing a university research laboratory that is conducting experiments on animals, or firebombing a store that sells fur coats. Several incidents of domestic ecoterrorism have been attributed to special interest groups such as the Earth Liberation Front and the Animal Liberation Front.

### Infrastructure Targets

Terrorists might strike bridges, tunnels, or subways in an attempt to disrupt transportation and inflict a large number of casualties (► Figure 24-2). They could also attack the public water supply or try to disable the electrical power distribution system, telephones, or the Internet. Disruption of a community's 9-1-1 system or public safety radio network would have a very direct impact on emergency response agencies. Homeland Security Presidential Directive (HSPD) 7 identifies critical infrastructures as agriculture and food, defense industrial base, energy, public health and healthcare, national monuments and icons, banking and finance, drinking water, and water treatment systems. The "Sector-Specific Federal Agencies" responsible for each critical infrastructure are identified in the directive.

Figure 24-2 Subways, airports, bridges, and hospitals are all vulnerable to attack by terrorists who seek to interrupt a nation's infrastructure.

## Symbolic Targets

Monuments such as the Lincoln Memorial, Washington Monument, or Mount Rushmore may be targeted by groups who want to attack symbols of national pride and accomplishment (► Figure 24-3). Foreign embassies and institutions might be attacked by groups promoting revolution within those countries or protesting their international policies. Religious institutions or other visible icons are potential

### Fire Marks

The Earth Liberation Front claimed credit for arson fires that destroyed several mountaintop buildings at a ski resort in Vail, Colorado in 1998. The group was trying to prevent the development of additional ski trails on the mountain.

Figure 24-3 Terrorists might attempt to destroy visible national icons.

## Brigade Member Tips

### Potential Terrorist Targets

**Ecoterrorism Targets**

Controversial development projects
Environmentally sensitive areas
Research facilities

**Infrastructure Targets**

Brigades and tunnels
Emergency facilities
Hospitals
Oil refineries
Pipelines
Power plants
Railroads
Telecommunications systems
Water reservoirs and treatment plants

**Symbolic Targets**

Embassies
Government buildings
Military bases
National monuments
Places of worship

**Civilian Targets**

Arenas and stadiums
Airports and railroad stations
Mass-transit systems
Schools and universities
Shopping malls
Theme parks

**Cyberterrorism Targets**

Banking and finance computer systems
Business computer systems
Court computer systems
Government computer systems
Law-enforcement computer systems
Military computer systems

**Agroterrorism Targets**

Crops
Feed storage
Grain elevators
Livestock and poultry

## Fire Marks

### Major Terrorist Incidents in the United States

**September 1984 The Dalles, Oregon**

To influence local elections, a religious sect spread salmonella on salad bars in four restaurants, resulting in 750 cases of salmonella poisoning.

**February 1993 New York, New York**

A large explosive device was detonated in a van parked in the underground garage of the World Trade Center. Six workers were killed and more than a thousand people were injured.

**April 1995 Oklahoma City, Oklahoma**

The Alfred E. Murrah Federal Building was demolished by a truck bomb that also killed 167 people. (An NFPA Fire Investigations report is available on this incident.)

**1978 to 1995 United States**

Over a period of 17 years, the Unabomber mailed at least 16 packages containing explosives to university professors, corporate executives, and other targeted individuals. These attacks killed 3 individuals and injured 23 others.

**July 1996 Atlanta, Georgia**

A pipe bomb exploded in the Centennial Olympic Park, killing one person and injuring 111 people.

**January 1997 Atlanta, Georgia**

Following the bombing of an abortion clinic in suburban Atlanta, a secondary device exploded, wounding several emergency responders. A month later, another secondary device was found and disarmed at the scene of a bombing at a gay nightclub in Atlanta.

**January 1998 Birmingham, Alabama**

A bomb killed a police officer who was providing security at an abortion clinic.

**October 1998 Vail, Colorado**

Arson destroyed eight buildings at a ski resort. An extremist environmental group opposed to expansion of the resort claimed responsibility.

**September 11, 2001 New York, New York/Washington, D.C./Pennsylvania**

Terrorists hijacked four commercial jets. Two were flown into the World Trade Center, one struck the Pentagon, and the fourth crashed into a field in rural Pennsylvania. More than 3,000 people died in the various incidents.

**Fall 2001 United States**

Five people died after letters containing anthrax virus were sent to various locations in the eastern United States. The Brentwood Post Office in Washington, D.C., finally reopened more than 2 years after anthrax passed through its doors.

targets of hate groups. By targeting these symbols, terrorist groups seek to make people aware of their demands and create a sense of fear in the public.

### Civilian Targets

Terrorists who attack civilian targets such as shopping malls, schools, or stadiums indiscriminately kill or injure the maximum number of potential victims ▼ Figure 24-4. Their goal is to create fear in every member of society and to make citizens feel vulnerable in their daily lives. Letter bombs or letters that contain a biological agent have a similar effect.

### Cyberterrorism Targets

Groups might engage in cyberterrorism by electronically attacking government or private computer systems. Several attempts have been made to disrupt the Internet or to attack government computer systems and other critical networks.

Figure 24-4 By attacking a civilian target like a crowded stadium, terrorists might make citizens feel vulnerable in their everyday lives.

Figure 24-5 Agroterrorism might affect our food supply.

### Agroterrorism Targets

Agroterrorism is the use of a biological, chemical, radiological, or other agent against either the preharvest or postharvest stages of food and fiber production to inspire fear or cause economic damage, public health impact, or other adverse impact against the United States ◄ Figure 24-5. Biological agents can impact animal or plant health or lead to human disease if they are zoonotic (i.e., diseases that are naturally transmitted between vertebrate animals and man with or without an arthropod intermediary).

## Agents and Devices

Terrorists can use several different kinds of weapons in an attack. Bombings are the most frequent terrorist acts, but brigade members also must be aware of other potential weapons. By shooting into a crowd at a shopping mall or train station, a single terrorist with an automatic weapon could cause devastating carnage. A biological agent that is visible only through a microscope could cause thousands to become ill and die. A computer virus that attacks the banking industry could cause tremendous economic losses. Planning should consider the full range of possibilities.

### Explosives and Incendiary Devices

According to the FBI's Bomb Data Center, there has been a substantial increase in the number of bombings in the United States over recent years. Between 1987 and 1997, the FBI recorded 23,613 bombing incidents, resulting in 448 deaths and 4,170 injuries. Incendiary devices account for 20–25% of all bombing incidents in the United States.

Groups or individuals have used explosives to further a cause; to intimidate a coworker or former spouse; to take revenge; or simply to experiment with a recipe found in a book or on the Internet.

Each year thousands of pounds of explosives are stolen from construction sites, mines, military facilities, and other locations. How much of this material makes its way to criminals and terrorists is not known ► Figure 24-6. Terrorists can also use commonly available materials, such as a mixture of ammonium nitrate fertilizer and fuel oil (ANFO), to create their own blasting agents.

An improvised explosive device (IED) is any explosive device that is fabricated in an improvised manner. An IED could be contained in almost any type of package from a letter bomb to a truckload of explosives. The Unabomber constructed at least 16 bombs that were delivered in small packages through the postal service. The bombings of the World Trade Center in 1993 and the Alfred E. Murrah Federal Building in 1995 both involved delivery vehicles

Figure 24-6 Every year, thousands of pounds of explosives are stolen.

Figure 24-8 Pipe bombs can come in many shapes and sizes.

Figure 24-7 The Alfred E. Murrah Federal Building in Oklahoma City was destroyed by a truck bomb.

loaded with ANFO and detonated by a simple timer ▲ Figure 24-7. As of late 2003, 40–60% of all attacks in Iraq began with an IED.

## Pipe Bombs

The most common IED is the pipe bomb. A pipe bomb is simply a length of pipe filled with an explosive substance and rigged with some type of detonator ▶ Figure 24-8. Most pipe bombs are simple devices, made with black powder or smokeless powder and ignited by a hobby fuse. More sophisticated pipe bombs may use a variety of chemicals and incorporate electronic timers, mercury switches, vibration switches, photocells, or remote control detonators as triggers.

### Brigade Member Safety Tips

A terrorist event may be designed to target emergency responders as well as civilians. Explosives do not discriminate. Anyone nearby can be severely injured or killed.

Pipe bombs are sometimes packed with nails or other objects so they will inflict as much injury as possible on anyone in the vicinity. A chemical or biological agent or radiological material could be added to a pipe bomb to create a much more complicated and dangerous incident. Experts can only speculate at the number of casualties that could result from such a weapon.

## Secondary Devices

Emergency responders must realize that terrorists may have placed a secondary device in the area where an initial event has occurred. These devices are intended to explode some time after the initial device explodes. Secondary devices are designed to kill or injure emergency responders, law enforcement personnel, spectators, or news reporters. Terrorists may use this tactic to attack the best-trained and most experienced investigators and emergency responders, or simply to increase the levels of fear and chaos following an attack.

The use of secondary devices is a common tactic in incidents abroad and has occurred at incidents in North America. In 1998, a bomb exploded outside a Georgia abortion clinic. About an hour later, a second explosion injured seven people, including two emergency responders. A similar secondary device was discovered approximately one month later at the scene of a bombing in a nearby community. Responders were able to disable this device before it detonated.

### Working with Other Agencies

Joint training with local, state, and federal agencies charged with handling incidents involving explosive devices should occur on a routine basis. Among these agencies are local and state police; the FBI; the Bureau of Alcohol, Tobacco, and Firearms; and military explosive ordnance disposal (EOD) units. In some jurisdictions, the municipal fire brigade is responsible for handling explosive devices. Designated brigade members should be given the necessary training and equipment.

### Potentially Explosive Devices

Brigade members should consult their organization's Standard Operating Procedures (SOP) for specific policies on bombing incidents. Fire brigade units responding to an incident that involves a potentially explosive device—one that has not yet exploded—should move all employees from the area and establish a perimeter at a safe distance. At no time should brigade members handle a potential explosive device unless they have received special training. Trained EOD personnel should assess the device and render it inoperative.

While waiting for the properly trained personnel to arrive, the fire brigade should establish an initial command post and a staging area (▼ Figure 24-9). Because there may be secondary devices, the staging area should be at least 3,000 feet from the incident site. In this type of incident, a joint command structure, commonly referred to as a unified command, should be established to coordinate the actions of all agencies.

If the bomb disposal team decides to disarm the device, an emergency action plan must be developed in case there is an accidental detonation. The incident commander (IC) will work in concert with law enforcement and EOD personnel to determine a safe perimeter where brigade members and emergency medical personnel should be staged.

Figure 24-9 If the fire brigade is first to arrive at the scene of an explosion, it should establish an initial command post.

## Brigade Member Safety Tips

Radio waves can trigger electric blasting caps, initiating an explosion. Radio transmitters should not be used near a suspicious device.

In some cases, a **forward staging area** will be established with a rapid intervention team standing by to provide immediate assistance to the bomb disposal team if something goes wrong. Other brigade members and emergency medical personnel would remain in a remote staging area.

### Actions Following an Explosion

Unless the cause of an explosion is known to be accidental, brigade members at the scene should always consider the possibility that an explosive device was detonated. The first priority should be to ensure the safety of the scene. Brigade members should also consider the possibility that a secondary device may be in the vicinity. Responders should quickly survey the area for any suspicious bags, packages, or other items.

It is also possible that chemical, biological, or radiological agents may be involved in a terrorist bombing. Qualified personnel with monitoring instruments should be assigned to check the area for potential contaminants. These precautions should be implemented immediately.

The initial size-up should also include an assessment of hazards and dangerous situations. The stability of any building involved in the explosion must be evaluated before anyone is permitted to enter. Entering an unstable area without proper training and equipment may complicate rescue and recovery efforts.

## Chemical Agents

Chemical agents have the potential to kill or injure great numbers of people. Several chemical weapons, including phosgene, chlorine, and mustard agents, were used in World War I, resulting in thousands of battlefield deaths and permanent injuries. Chemical weapons were also used during the Iran–Iraq War (1980 to 1988). Although international agreements prohibit the use of chemical and biological agents on the battlefield, there is great concern that chemical weapons could be used by terrorists or combatants.

In 1995, the religious cult Aum Shinrikyo released a nerve agent, **sarin**, into the Tokyo subway system. This attack resulted in 12 deaths and over 1,000 injuries. Many experts believe that this attack was ineffective, because there were relatively few casualties even though thousands of subway riders were potentially exposed. This attack instilled fear in a large population. A previous sarin attack in Matsumoto, Japan killed seven people and injured approximately 300 others.

The basic instructions for making chemical weapons are readily available through a variety of sources, including the Internet, books, and other publications. Additionally, the required chemicals can be obtained fairly easily. Many of the chemicals classified as WMD are routinely used in legitimate industrial processes. For instance, chlorine gas is used in swimming pools, at water treatment facilities, and in industry ▼ Figure 24-10. However, chlorine is classified as a choking agent, because inhalation causes severe pulmonary damage.

Cyanide compounds such as hydrogen cyanide (HCN) are used in the production of paper and synthetic textiles as well as in photography and printing. These chemicals are stored and transported in various sized containers, ranging from small containers and pressurized cylinders to railroad tank cars.

Chemical weapons can be disseminated in several ways. Simply releasing chlorine gas from a storage tank in an unguarded rail yard could cause thousands of injuries and deaths. A chemical agent can be added to an explosive device. Crop-dusting aircraft, truck-mounted spraying units, or hand-operated pump tanks could be used to disperse an agent over a wide area ► Figure 24-11.

### Protection from Chemical Agents

Dispersion by aerosolization is a key method used in chemical attacks. Agents released outdoors will follow wind currents; agents released inside will be dispersed through a building's air circulation system. Some chemical agents have distinctive odors; if there is any unusual odor at an emergency scene, brigade members must use full personal protective equipment (PPE) including self-contained breathing apparatus (SCBA) ▲ Figure 24-12. In addition, the area should be monitored with the appropriate detection devices. Remember that it is never sufficient to rely strictly on odor to determine the presence of a chemical agent.

Figure 24-10 Although chlorine is regularly used in swimming pools and water treatment facilities, it is also classified as a choking agent that can be used in a terrorist attack.

Figure 24-11 Crop-dusting may be used to distribute chemical or biological agents.

Figure 24-12 If an unusual odor is reported at the scene, a brigade member must use full PPE gear and an SCBA device.

Figure 24-13 In their normal states, nerve agents are liquids. They must be dispersed in aerosol form to be inhaled or absorbed by the skin.

## Nerve Agents

Nerve agents are toxic substances that attack the central nervous system. They were first developed in Germany before World War II. Today, several countries maintain stockpiles of these agents. Nerve agents are similar to some pesticides, but they are extremely toxic. Nerve agents can be 100 to 1,000 times more toxic than similar pesticides. Exposure to these agents can result in injury or death within minutes.

In their normal states, common nerve agents are liquids ▲ Figure 24-13. As liquids, these agents are not likely to contaminate large numbers of people because direct contact with the agent is required. To be an effective weapon, the liquid must be dispersed in aerosol form, or broken down into fine droplets so that it can be inhaled or absorbed by the skin.

Pouring a liquid nerve agent onto the floor of a crowded building would probably not immediately contaminate large numbers of people; however, it would produce fear and panic. The effectiveness of the agent would depend on how long it remains in the liquid state and how widely it is dispersed throughout the building. Sarin, the most volatile nerve agent, evaporates at the same rate as water and is not considered persistent. The most stable nerve agent, V-agent or VX, is considered persistent because it takes several days or weeks to evaporate. Common nerve agents, their methods of contamination, and specific characteristics are shown in ▲ Table 24-1.

When a person is exposed to a nerve agent, symptoms will become evident within minutes. The symptoms include: pinpoint pupils, runny nose, drooling, difficulty breathing, tearing, twitching, diarrhea, convulsions or seizures, and loss of consciousness. Several mnemonics are used to help you remember these symptoms. The mnemonic used most often by the fire service is SLUDGE ▲ Table 24-2.

**Table 24-1 Common Nerve Agents**

| Nerve Agent | Method of Contamination | Characteristics |
|---|---|---|
| Tabun (GA) | Skin contact<br>Inhalation | Disables the chemical connections between nerves and target organs |
| Soman (GD) | Skin contact<br>Inhalation | Odor of camphor |
| Sarin (GB) | Inhalation | Evaporates quickly |
| V-agent (VX) | Skin contact | Oily liquid that can persist for weeks |

**Table 24-2 Symptoms of Nerve Agent Exposure**

**SLUDGE**
- Salivation
- Lacrimation
- Urination
- Diaphoresis
- Gastrointestinal distress (diarrhea, vomiting)
- Emesis

**Triple Bs (BBB) and Miosis (M)**
- Bradycardia
- Bronchorrhea
- Bronchospasm
- Miosis

## SLUDGE and the Triple Bs

SLUDGE is an acronym that stands for salivation, lacrimation (excessive tears), urination, diaphoresis (excessive sweating), gastrointestinal distress/diarrhea, and emesis (vomiting). The triple B's are bradycardia (slow heart rate), bronchorrhea (excessive secretion in the lung membranes), and bronchospasm (muscle spasm in the bronchial tubes). An M is added for miosis (excessive constriction of the pupils).

The U.S. military has developed a nerve agent antidote kit, which contains two medications: atropine and pralidoxime chloride (2-PAM). These antidotes can be quickly injected into a person who has been contaminated by a nerve agent. The Mark 1 Nerve Agent Antidote Kit has been issued

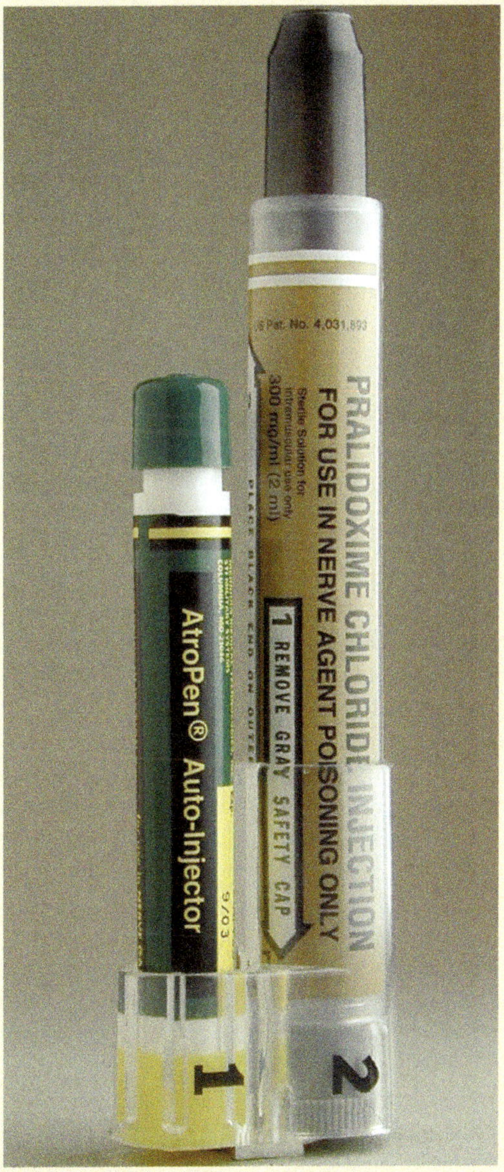


Figure 24-14 Mark 1 Nerve Agent Antidote Kit.

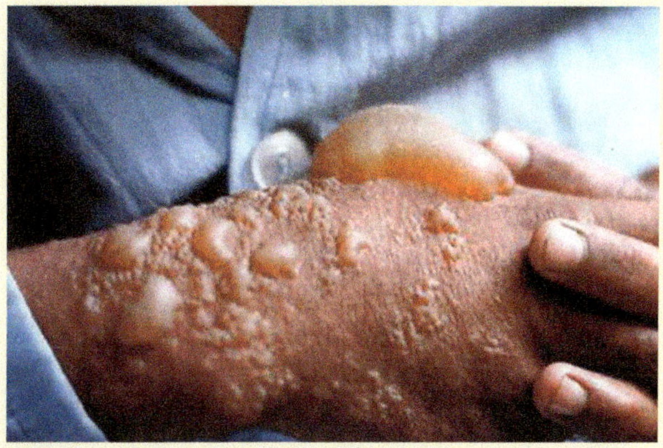

Figure 24-15 Typical effects of a blistering agent.

to many municipal fire departments' hazardous materials teams ◄ Figure 24-14.

## Blister Agents

Two chemical agents are generally used as blistering agents. Contact with either of these chemicals will cause the skin to blister.

Sulfur mustard (H) is a clear, yellow, or amber, oily liquid with a faint sweet odor of mustard or garlic. It vaporizes slowly at temperate climates and may be dispersed in an aerosol form.

Lewisite (L) is an oily, colorless-to-dark-brown liquid with an odor of geraniums.

Blister agents produce painful burns and blisters with even minimal exposure ▲ Figure 24-15. The major difference between these two blister agents is that lewisite causes pain immediately upon contact with the skin, while the signs and symptoms from exposure to mustard gas may not appear for several hours. The patient will complain of burning at the site of the exposure, the skin will redden, and blisters will appear. The eyes may also itch, burn, and turn red. Inhalation of mustard gas produces significant respiratory damage.

## Choking Agents

The two chemicals that are likely to be used as choking agents are phosgene (CG) and chlorine (CL). Both of these agents were used extensively as weapons in World War I, and both have several industrial uses. They cause severe pulmonary damage and asphyxia.

These choking agents are heavier than air, so they tend to settle in low areas. Subways, basements, and sewers are prime areas for these agents to accumulate. Brigade members wearing appropriate PPE should quickly evacuate people

Figure 24-16 Someone who has been exposed to phosgene or chlorine should be moved from the contaminated area and the skin flushed with water.

from such areas when the possibility of phosgene or chlorine exposure exists.

Exposure to high concentrations of phosgene or chlorine will immediately irritate the eyes, nose, and upper airway. Within hours, the exposed individual will begin to develop pulmonary edema (fluid in the lungs). Those exposed to lower concentrations may not have any initial symptoms but can still experience respiratory damage.

Although neither agent is absorbed through the skin, they can both cause skin burns on contact. Decontamination consists of removing exposed individuals from the area and flushing the skin with water ▲ Figure 24-16.

### Blood Agents

Blood agents are highly toxic poisons that can cause death within minutes of exposure. The two most common blood agents are both cyanide compounds: hydrogen cyanide (AC) and cyanogen chloride (CK). Both of these chemicals are commonly used in industry.

Cyanide can be inhaled or ingested. Cyanide gas is typically associated with the death penalty and the gas chamber. A liquid cyanide mixed with valium and a grape-flavored drink was used in the mass suicide of over 900 members of a religious cult in Guyana in 1978.

The few symptoms associated with cyanide exposure occur quickly. Those exposed to the gas will begin gasping for air, and if enough agent is inhaled, the skin may begin to appear red. Seizures are also possible. A complete and rapid evacuation of the area is the most appropriate course of action.

## Biological Agents

Biological agents are organisms that cause diseases and attack the body. Biological agents include bacteria, viruses, and toxins. Some of these organisms, like anthrax, can live in the ground for years, while others are rendered harmless after being exposed to sunlight for only a short period of time. The highest potential for infection is through inhalation, although some biological agents can be absorbed, injected through the skin, or ingested. The effects of a biological agent depend on the specific organism or toxin, the dose, and the route of entry.

Some of these diseases, such as smallpox and pneumonic plague, are contagious and can be passed from person to person. Doctors are very concerned about the use of contagious diseases as weapons, because the resulting epidemic could overwhelm the healthcare system. There is growing concern that a terrorist group will use one of these agents against a civilian population.

### Anthrax

Anthrax is an infectious disease caused by the bacteria *Bacillus anthracis*, which is typically found around farms and infects livestock. For use as a weapon, the bacteria is cultured to develop anthrax spores. The spores, in powdered form, can then be dispersed in a variety of ways ► Figure 24-17. Approximately 8,000 to 10,000 spores are typically required to cause an anthrax infection. Spores infecting the skin cause cutaneous anthrax; ingested spores cause gastrointestinal anthrax; and inhaled spores cause inhalational anthrax.

In 2001, four letters containing anthrax were mailed to locations in New York City; Boca Raton, Florida; and Washington, D.C. Five people died after being exposed to the contents of these letters, including two postal workers who were exposed as the letters passed through postal sorting centers. Several major government buildings had to be shut down for months to be decontaminated.

Because these incidents followed shortly after the events of September 11, 2001, they caused tremendous public concern. Emergency personnel had to respond to thousands of incidents involving suspicious packages and citizens who believed that they might have been exposed to anthrax. Each of these situations had to be evaluated.

A 1970 publication from the World Health Organization is often used to show the dangers of anthrax. According to this publication, 110 pounds of anthrax sprayed over an

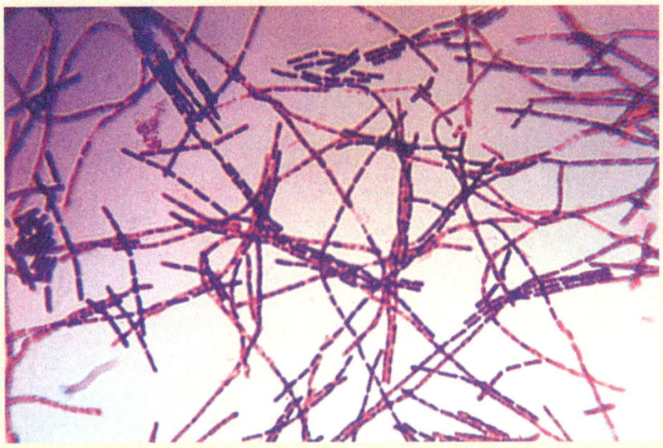

Figure 24-17 Anthrax spores can be dispersed in a variety of ways.

urban center with 500,000 residents could cause 220,000 deaths.

### Smallpox

Smallpox is a highly infectious disease caused by the virus variola. The disease kills approximately 30% of those infected. Although smallpox was once routinely encountered throughout the world, by 1980 it was successfully eradicated as a public health threat through the use of an extremely effective vaccine. Some nations maintained cultures of the disease for research purposes. International terrorist groups may have acquired the virus.

The smallpox virus could be dispersed over a wide area in an aerosol form. Infecting a small number of people could lead to a rapid spread of the disease throughout a targeted population because smallpox is highly contagious. The disease is easily spread by direct contact, droplet, and airborne transmission. Patients are considered highly infectious and should be quarantined until the last scab has fallen off ▶ Figure 24-18.

By 1980, when smallpox was eradicated, the worldwide vaccination program was discontinued. Today millions of people have never been vaccinated and millions more have reduced immunity because decades have passed since their last immunization.

### Plague

Plague is caused by the bacterium *Yersinia pestis,* which is commonly found on rodents. There are three forms of plague: bubonic, septicemic, and pneumonic. The bacterium is most often transmitted to humans by fleas that feed on infected animals and then bite humans. Those who are bitten generally develop bubonic plague, which attacks the lymph nodes ▶ Figure 24-19. Pneumonic plague can be contracted by inhaling the bacterium.

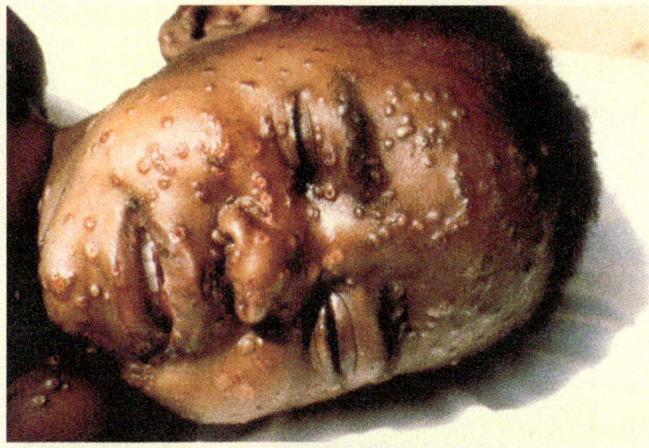

Figure 24-18 Smallpox is a highly contagious disease with a mortality rate of approximately 30%. *(Courtesy of Centers for Disease Control and Prevention.)*

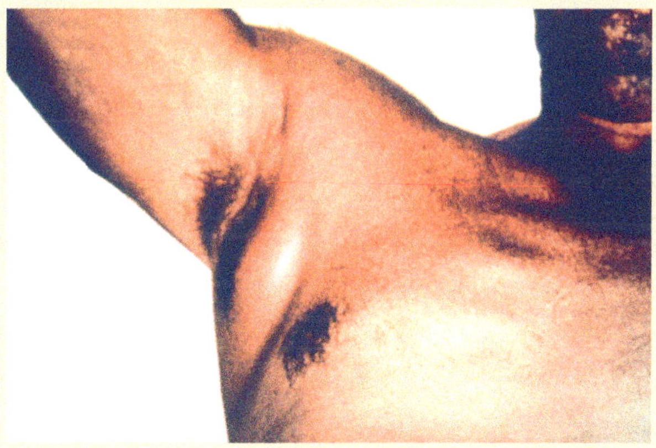

Figure 24-19 A bubo, one of the symptoms of plague, is a swollen, painful lymph node.

The plague bacteria can survive for weeks in water, moist soil, or grains. It has been cultured for distribution as a weapon in aerosol form. Inhalation of the aerosol would put the target population at risk for pneumonic plague.

### Dealing with Biological Agents

It is unlikely that brigade members or emergency responders would even recognize that a biological weapon had been released in their communities. The time period between the actual infection and the appearance of symptoms is known as the incubation period. The symptoms of a biological agent attack would typically manifest over a period of days after the exposure. If left untreated, many of these diseases will cause death in a large proportion of those infected.

# Voices of Experience

**"I looked around, found my bearings, and walked through the ankle-deep dust and debris."**

As I arrived on the scene of New York City World Trade Center (WTC) attacks in September 2001, I saw that the city was quiet—eerily quiet. I walked to the command post to report for duty, and the scene was surreal. I looked north, south, east, and west, and it was like a scene from a bad Hollywood movie. I looked around, found my bearings, and walked through the ankle-deep dust and debris. It covered the streets like snow or ash.

I was sent to the WTC attacks as a member of the then-U.S. Department of Justice, Office of State and Local Domestic Preparedness Support (OSLDPS), to provide support to the New York City Fire Department in any way possible. I reported to the Chief of Logistics, Chief Charlie Blaich, and worked side by side with him and fire fighter Lee Morris until the day before Christmas 2001.

What did I see there, and what did I learn?

I had to use every skill that I had ever learned as a fire fighter and EMS officer. I provided safety support as I had done while a safety officer at the Federal Emergency Management Agency (FEMA), where I had been a Disaster Safety Officer for the previous eight years.

One thing is for certain: When it comes to terrorism, we don't know everything, and we must be constantly alert. During an event involving terrorism, nothing is as it seems—nothing. Revert to your training and practice, practice, practice your skills. Remember that during such an event, other factors may be in play that do not necessarily equate to why you were brought in the first place. There will be many unknowns, and your skills—your professionalism, your analytical qualities, and your demeanor—may well be the *only* thing that makes any sense during the early stages of the incident.

What you read someplace doesn't matter during such an event; the only thing that matters is what you can do to help the situation by being an engaged member of the team. There is no place for lone wolves and no place for grandstanding—it's not the TV show 24. It's real people needing assistance in making informed decisions, and your task may be to help them make the right ones, as part of a team. It's up to you to be well trained, well informed, and well intentioned, and to be a part of the solution.

Be safe, be proactive, and stay alive.

Michael J. Fagel
University of Chicago—Master of Threat Risk Program
Chicago, Illinois

Area hospitals and the local public health departments would typically be the first to recognize the situation, after identifying significant numbers of people arriving at their facilities with similar symptoms. Multiple medical calls with patients exhibiting similar symptoms could provide a clue, especially if the location is considered to be a potential target. Once it has been determined that a situation is quickly unfolding, the Centers for Disease Control and Prevention (CDC) would be notified.

Most experts believe that a biological weapon would probably be spread by a device similar to a garden sprayer or crop-dusting plane. After the release of a biological weapon, residents would begin experiencing flu-like symptoms, although the symptoms may not appear for 2–10 days.

Brigade members or emergency responders treating people who may have been exposed to a biological agent should follow the recommendations of their brigades regarding universal precautions (protective measures for use in dealing with objects, blood, body fluids, or other potential exposure risks of communicable diseases). These measures include wearing appropriate gloves, masks, eye protection, and surgical gowns when treating patients.

Emergency responders who exhibit flu-like symptoms after a potential terrorist incident should seek medical care immediately. It is important that brigade members report any possible exposure to brigade officials. Postincident actions may include medical screening, testing, or vaccinations.

## Radiological Agents

Although unlikely, the possibility that terrorists could detonate a nuclear weapon is a threat that cannot be ignored. Several countries possess nuclear weapons and a number of others are attempting to develop or acquire them. The only limiting factor is the ability to acquire weapons-grade fuel, which is needed to cause a nuclear explosion. Some experts fear that terrorists could obtain and detonate a suitcase nuke, a nuclear weapon that is small enough to fit into a small package.

A more probable scenario is the use of radiological agents by terrorists. Terrorists could employ a variety of alternative approaches to disperse radiological agents and contaminate an area with radioactive materials. This threat is very different from a nuclear detonation and much more likely to occur.

### The Dirty Bomb

The radiation dispersal device (RDD) or "dirty bomb" is an area of serious concern. An RDD is described as "any device that causes the purposeful dissemination of radioactive material across an area without a nuclear detonation." Packing radioactive material around a conventional explosive device could contaminate a wide area. The size of the affected area would depend on the amount of radioactive material and the power of the explosive device.

**Figure 24-20** The materials needed to build a radiation dispersal device can be obtained from a number of places, including a nuclear power plant.

To limit this threat, radioactive materials, even in small amounts, must be kept secure and protected. Radioactive materials are widely used in industry and health care, particularly with x-ray machines. The necessary materials could be obtained from medical clinics, nuclear power plants, university research facilities, and some industrial complexes (▲ Figure 24-20). These materials are also routinely transported in a variety of large and small containers. Terrorists could construct a dirty bomb with a small quantity of stolen radioactive material.

Security experts are also concerned that a large explosive device, such as a truck bomb, could be detonated near a nuclear power plant. Such an explosion could damage the reactor containment vessel sufficiently to release radioactive material. A large aircraft also could be used to damage a nuclear reactor much as aircraft were used to destroy the World Trade Center.

### Radiation

Radioactive materials release energy in the form of electromagnetic waves or energy particles. Radiation cannot be detected by the normal senses of smell and taste. Several instruments can detect the presence of radiation and measure dose rates. Brigade members should become familiar with the radiation detectors used by their brigades. There are three types of radiation: alpha particles, beta particles, and gamma rays.

Alpha particles quickly lose their energy and can travel only 1 to 2 inches from their source. Clothing or a sheet of paper can stop this type of energy. However, if ingested or inhaled, alpha particles can damage a number of internal organs.

Beta particles are more powerful, capable of traveling 10–15 feet. Heavier materials such as metal, plastic, and glass

## Brigade Member Safety Tips

### Casualty Management After Detonation of a Nuclear Weapon

- Do not enter an area with possible contamination until radiation levels have been measured and found to be safe.
- Enter the area only to save lives; radiation levels will be very high.
- Wear PPE and SCBA.
- All personnel and equipment must be decontaminated after leaving the contaminated area. Because radioactive dust collects on clothing, all clothing should be removed and discarded after leaving the area. Failure to do so will result in continued radiation exposure to yourself and others.
- Wash thoroughly with lukewarm water as soon as possible after leaving the scene, even if you have been through a decontamination process.
- Do not eat, drink, or smoke while exposed to potential radioactive dust or smoke.
- Use radiation monitoring devices to map the location of areas with high radioactivity. Monitoring devices should be wrapped in plastic bags to prevent contamination.
- Vehicles should be washed before they leave the scene. The only exception is emergency units transporting critical victims.
- Life-threatening injuries are more serious than radioactive contamination. Treat such injuries before decontaminating the individual.
- Alert local hospitals that they may receive radioactive-contaminated victims.

can stop this type of energy. Beta radiation can be harmful to both the skin and eyes, and ingestion or inhalation will damage internal organs.

Gamma rays can travel significant distances, penetrate most materials, and pass through the body. Gamma radiation is the most destructive to the human body. The only materials with sufficient mass to stop gamma radiation are concrete, earth, and dense metals such as lead.

### Effects of Radiation

Symptoms of low-level radioactive exposure might include nausea and vomiting. Exposure to high levels of radiation can cause bone marrow destruction, nerve and digestive system damage, and radioactive skin burns. An extreme exposure may cause death rapidly; however, it often takes a considerable amount of time before the signs and symptoms of radiation poisoning become obvious. As the effects progress over the years, a prolonged death due to leukemia or carcinoma is likely.

A contaminated person will have radioactive materials on the exposed skin and clothing. In severe cases, the contamination could penetrate to the internal organs of the body. Decontamination procedures are required to remove the radioactive materials. Decontamination must be complete and thorough; the situation could escalate rapidly if decontamination procedures are inadequate. The contamination could quickly spread to ambulances, hospitals, and other locations where contaminated individuals or items are transported.

Radioactive materials expose rescuers to radiation injury as well as continue to affect the exposed person. Anyone who comes in contact with a contaminated person must be protected by appropriate PPE and shielding, depending on the particular contaminant. Rescuers and medical personnel will have to be decontaminated after handling a contaminated person.

A person can be exposed to radiation without coming into direct contact with a radioactive material through inhalation, skin absorption, or ingestion. A person can also be injured by exposure to radiation, without being contaminated by the radioactive material itself. Someone who has been exposed to radiation, but has not been contaminated by radioactive materials, requires no special handling.

There are three ways to limit exposure to radioactivity. Keep the time of the exposure as short as possible. Stay as far away from the source of the radiation as possible, and use shielding to limit the amount of radiation absorbed by the body. Firefighting PPE and breathing apparatus will generally provide adequate shielding against alpha particles; breathing apparatus is needed to provide protection against beta radiation. Heavy shielding is needed to provide protection from sources emitting gamma radiation.

If radioactive contamination is suspected, everyone who enters the area should be equipped with a personal dosi-meter to measure the amount of radioactive exposure (► Figure 24-21). Rescue attempts should be made as quickly as possible to reduce exposure time.

## Operations

Responding to a terrorist incident puts brigade members and other emergency personnel at risk. Responders must ensure their own safety at every incident; however, a terrorist incident has an extra dimension of risk. Because the terrorist's objective is to cause as much harm as possible, emergency responders are just as likely to be targets as ordinary civilians.

In many cases, the first units to respond will not be dispatched for a known WMD or terrorist incident. The initial dispatch might be for an explosion, for a possible hazardous materials incident, for a single person with difficulty breathing, or for multiple sick patients with similar symptoms. Emergency responders will usually not know that a terrorist incident has occurred until personnel on the scene begin to piece together information gained by their own observations and from witnesses.

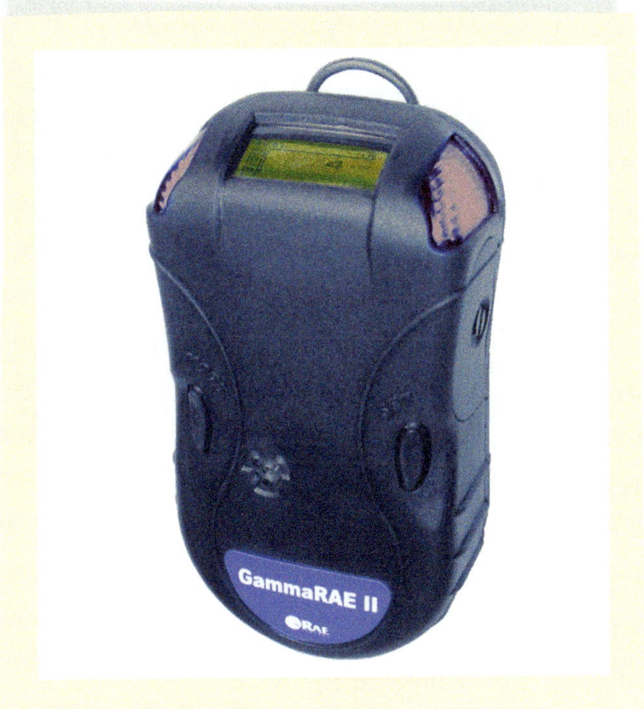


Figure 24-21 A personal dosimeter measures the amount of radioactive exposure received by an individual.

## Fire Marks

Three ways to limit exposure to radioactivity.

T-Decrease time exposed to material
D- Increase distances between personnel and radioactive material
S - Increase shielding.

If appropriate precautions are not taken, the initial responders may find themselves in the middle of a dangerous situation before they realize what has happened. Initial responders should be aware of any factors that suggest the possibility of a terrorist incident and immediately implement appropriate procedures. The possibility of a terrorist incident should be considered when responding to any location that has been identified as a potential terrorist target. It could be difficult to determine the true nature of the situation until a scene size-up is conducted.

## Initial Actions

Responders should approach a known or potential terrorist incident just as they would a hazardous materials incident. The possibility that chemical, biological, or radiological agents are involved cannot be ruled out until a hazardous materials team checks the area with detection and monitoring instruments. Apparatus and personnel should, if possible, approach the scene from a position that is uphill and upwind. Emergency responders should don PPE, including SCBA. Additional arriving units should be staged an appropriate distance away from the incident.

The first units to arrive should establish an outer perimeter to control access to and from the scene. They should deny access to all but emergency responders, and they should prevent potentially contaminated individuals from leaving the area before they have been decontaminated. The perimeter must completely surround the affected area, to keep people who were not initially involved from becoming additional victims.

Incident Command should be established in a safe location, which could be as far as 3,000 feet from the actual incident scene. The command post must be outside the area of possible contamination and beyond the distance where a secondary device may be planted. The initial priority should be to determine the nature of the situation, the types of hazards that could be encountered, and the magnitude of the problems that will have to be faced.

An initial reconnaissance or "recon" team should be sent out to quickly examine the involved area and to determine how many people are involved. Proper use of PPE, including SCBA, is essential for the recon team, and the initial survey must be conducted very cautiously, although as rapidly as possible. The possibility that chemical, biological, or radiological agents are involved cannot be ruled out until qualified personnel with instruments and detection devices have surveyed the area. Responders should take special care not to touch any liquids or solids or to walk through pools of liquid. Emergency responders who become contaminated must not leave the area until they have been decontaminated.

A process of elimination may be required to determine the nature of the situation. Occupants and witnesses should be asked if they observed any unusual packages or detected any strange odors, mists, or sprays. A large number of casualties, who have no outward signs of trauma, could indicate a possible chemical agent exposure. Possible symptoms include trouble breathing, skin irritations, or seizures.

A visible vapor cloud would also be a strong indicator of a chemical release, which could be an accident or a terrorist attack. Responders should look out for dead or dying animals, insects, or plant life as indications of a chemical agent release.

When approaching the scene of an explosion, responders should consider the possibility of a terrorist bombing incident

## Teamwork Tips

A hazardous materials team should respond to any potential or suspected terrorist incident. The hazardous materials personnel should test for chemical, biological, or radiological contaminants using appropriate test instruments.

and recall the potential for secondary explosive devices. Responders should note any suspicious packages and notify the IC. Brigade members who have not been specially trained should never approach a suspicious object. Explosive ordnance disposal personnel should examine any suspicious articles and disable them.

## Interagency Coordination

If a terrorist incident is suspected, the IC should consult immediately with local law enforcement officials. If a mass casualty situation is evident, the IC should notify area hospitals and activate the local medical response system. Local technical rescue teams should be requested to evaluate structural damage and initiate rescue operations.

State emergency management officials should be notified as soon as possible. This will help ensure a quick response by both state and federal resources to a major incident. Large-scale search-and-rescue incidents could require the response of Urban Search and Rescue (USAR) task forces through the Federal Emergency Management Agency (FEMA). Medical response teams, such as Disaster Medical Assistance Teams (DMAT), may be needed for incidents involving large numbers of people.

## Brigade Member Tips

### Additional Resources for Terrorist Incidents

| Resources | Action |
|---|---|
| Air Evacuation | Air transportation of injured |
| Air Supply Unit | Refill SCBA cylinders |
| Bomb Squad | Render explosive devices safe |
| Emergency Medical Services | Emergency medical care |
| Hazardous Materials Unit | Neutralize hazardous substances, including chemical agents |
| Health Department | Technical advice on chemical and biological agents |
| Local Law Enforcement | Law enforcement, perimeter control, and evidence recovery |
| Structural Collapse Teams | Search and stabilize collapsed structures |

Figure 24-22 An EOC is set up in a predetermined location for large-scale incidents.

An Emergency Operations Center (EOC) can coordinate the actions of all involved agencies in a large-scale incident, particularly if terrorism is involved. The EOC is usually set up in a predetermined remote location and is staffed by experienced command and staff personnel (▲ Figure 24-22). The IC, who is at the scene of the incident, should provide detailed situation reports to the EOC and request additional resources as needed.

Brigade members must remember that a terrorist incident is also a crime scene. To avoid destroying important evidence that could lead to a conviction of those responsible, responders should not disturb the scene any more than is necessary. Where possible, law enforcement personnel should be consulted prior to overhaul and before any material is removed from the scene. Brigade members should also realize that one or more terrorists could be among the injured. Be alert for threatening behavior, and be aware of anyone who seems determined to leave the scene.

## Decontamination

Everyone who is exposed to chemical, biological, or radiological agents must be decontaminated to remove or neutralize any chemicals or substances on bodies or clothing. Decontamination should occur as soon as possible to prevent

## Brigade Member Tips

Emergency responders at a terrorist event may later be called to testify in court. Make mental notes if you must disturb or move any evidence to rescue someone in immediate danger. Document this information as soon as possible.

Figure 24-23 Mass decontamination procedures may be required to handle a large group of contaminated victims.

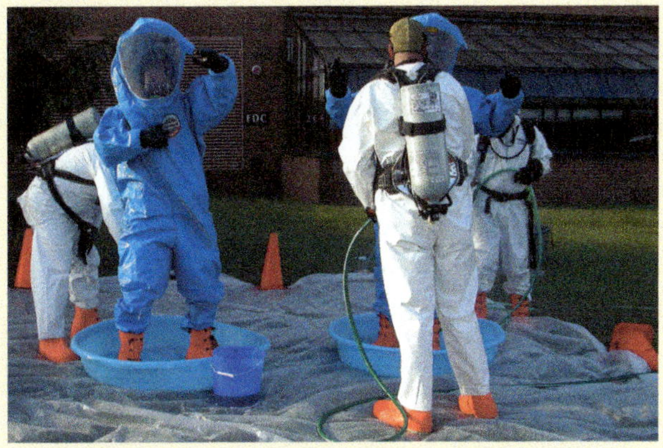

Figure 24-24 All emergency responders and victims who have been exposed to chemical, radiological, or biological agents must be decontaminated.

further absorption of a contaminant and to reduce the possibility of spreading the contamination. Equipment will also have to be decontaminated before it leaves the scene.

The perimeter must fully surround the area of known or suspected contamination. Every effort must be made to avoid contaminating any additional areas, particularly hospitals and medical facilities. Qualified personnel must monitor the perimeter with instruments or detection devices to ensure that contaminants are not spread.

Standard decontamination procedures usually involve a series of stations. At each station, clothing and protective equipment are removed and the individual is cleaned. Some contaminants require only soap and water, but chemical and biological agents may require special neutralizing solutions.

A terrorist incident that results in a large number of casualties may require a different decontamination process. Special procedures for **mass decontamination** have been devised for incidents involving large numbers of people. These procedures use master stream devices from engine companies and aerial apparatus to create high-volume, low-pressure showers. This allows large numbers of people to be decontaminated rapidly (▲ Figure 24-23).

Emergency responders who enter a contaminated area must be decontaminated before they can leave the area (► Figure 24-24). Contaminated clothing and equipment must remain inside the perimeter. Depending on the involved agent, it may be necessary to maintain this perimeter for an extended period, until the entire incident scene can be decontaminated. It could take weeks or months to fully decontaminate some incident scenes.

## Mass Casualties

A terrorist or WMD incident may result in a large number of casualties. Some scenarios could involve hundreds or thousands of injuries. Special mass casualty plans are essential to manage this type of situation, which would quickly overwhelm the normal capabilities of most emergency response systems. Mass casualty plans usually involve resources from multiple agencies to handle large numbers of patients efficiently.

A terrorist incident could involve several additional complications and considerations, including the possibility of contamination by chemical, biological, or radiological agents. The mass casualty plan must be expanded to address these problems.

If contamination is suspected, the plan must ensure that it does not spread beyond a defined perimeter. In some cases, patients may be decontaminated as they are moved to the **triage** and treatment areas (▼ Figure 24-25). This will keep these areas free of contaminants. In other cases, the

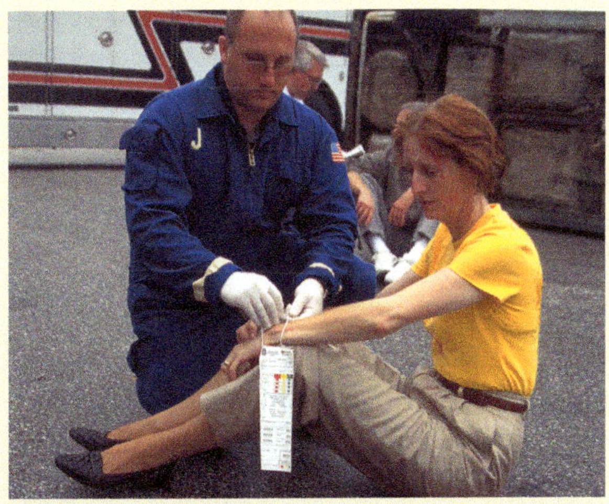

Figure 24-25 Patients should be decontaminated before they are placed in triage and treatment areas.

triage and treatment areas may be considered contaminated zones. Patients would be decontaminated before they are transported from the scene.

During the early stages of an incident, it may be difficult to determine what agent was used so that appropriate treatment and decontamination procedures can be instituted. It may be necessary to quarantine exposed individuals within an area that is assumed to be contaminated until laboratory test results are evaluated. The strategic plan for this situation must assume that any personnel who provide assistance or treatment to patients during this period will also become contaminated, as will any equipment or vehicles used in the contaminated area.

## Additional Resources

A terrorist incident will likely result in a massive response by local, state, and federal government agencies. Following the terrorist attacks of September 11, 2001, HSPDs have made both crisis and consequence management a shared responsibility.

Funding for programs to train and equip local emergency responders has increased significantly since September 11, 2001. A Cabinet-level Department of Homeland Security has been established in the United States to provide a unifying core for the large network of organizations and institutions that have a responsibility to help keep our nation secure.

The Homeland Security Advisory System is designed to provide quick, comprehensive information concerning the potential threat of terrorist attacks or threat levels. Threat conditions can apply nationally, regionally, by industry, or by specific target. You should be aware of the current Homeland Security Threat Level at all times when on duty. Implement listed protective measures as directed in your brigade and recommended listed actions to the public.

### Low Condition (Green)

This condition is declared when there is a low risk of terrorist attacks. Agencies should consider the following general measures in addition to any agency-specific protective measures they develop and implement:

1. Refine and exercise preplanned protective measures.
2. Ensure that personnel receive proper training on the Homeland Security Advisory System and on specific preplanned departmental or agency protective measures.
3. Institute a process to ensure that all facilities and regulated sectors are regularly assessed for vulnerability to terrorist attacks, and that all reasonable measures are taken to mitigate these vulnerabilities.

Advise the public to:

- Develop a family emergency plan, including family meeting locations in the event of an attack, and a contact point, such as the home of a relative in a distant state.
- Keep recommended immunizations up to date.
- Know how to turn off power, gas, and water service to homes or businesses. Keep tools available to do so.
- Know what hazardous materials are stored on the property and how to properly dispose of unneeded chemicals.
- Support the efforts of local emergency responders (brigade members, law enforcement officers, and EMS personnel).
- Know the natural hazards that are prevalent in your area and what measures can be taken to mitigate these hazards.
- Maintain a first-aid kit.

### Guarded Condition (Blue)

This condition is declared when there is a general risk of terrorist attacks. In addition to taking the protective measures listed under the previous Threat Condition, agencies should consider the following general measures as well as the agency-specific protective measures:

1. Check communications with designated emergency response or command locations.
2. Review and update emergency response procedures.
3. Provide the public with any information that would strengthen its ability to act appropriately.

Advise the public to:

- Continue normal activities, but be watchful for suspicious activities. Report criminal activity to local law enforcement.
- Review family emergency plans, including family meeting locations and distant points of contact in the event of an attack.
- Be familiar with the local natural and technological (man-made) hazards in your community and know what measures can be taken to mitigate these hazards.
- Maintain a first-aid kit.
- Increase individual or family emergency preparedness through training, maintaining good physical fitness and health, and storing a three-day supply of food, water, and emergency supplies, such as medications.
- Monitor local and national news for terrorist alerts.

### Elevated Condition (Yellow)

An Elevated Condition Threat Alert is declared when there is a significant risk of terrorist attacks. In addition to the protective measures listed under the previous threat conditions, agencies should consider the following general measures as well as the protective measures that they will develop and implement:

1. Increase surveillance of critical locations.
2. Coordinate emergency plans as appropriate with nearby jurisdictions.
3. Address whether the precise characteristics of the threat require the further refinement of preplanned protective measures.

4. Implement, as appropriate, contingency and emergency response plans.

Advise the public to:

- Continue normal activities, but report suspicious activities to local law enforcement agencies.
- Become trained/certified in first aid or emergency response protocol. Advise the public of local training programs.
- Maintain a first-aid kit.
- Become active in your local Neighborhood Crime Watch program. Report suspicious activities to local law enforcement.
- Network with your family, neighbors, and community for mutual support during a disaster or terrorist attack. Update previously established family emergency plans, including a distant point of contact. Update preparations to shelter in place, such as preparing of materials to seal openings to your home and shut down HVAC systems.
- Know the critical facilities and specific hazards located in your community, and report suspicious activities at or near these sites.
- Increase individual or family emergency preparedness through training, maintaining good physical fitness and health, and storing a three-day supply of food, water, and emergency supplies, such as medications.
- Monitor media reports concerning terrorism situations.

## High Condition (Orange)

A High Condition Threat Alert is declared when there is a high risk of terrorist attacks. In addition to the protective measures listed under the previous Threat Conditions, agencies should consider the following general measures as well as agency-specific protective measures:

1. Coordinate necessary security efforts with federal, state, and local law enforcement agencies or any National Guard or other appropriate armed forces organizations.
2. Take additional precautions at public events and consider alternative venues or even cancellation.
3. Prepare to execute contingency procedures, such as moving to an alternate site or dispersing a work force.
4. Restrict threatened facility access to essential personnel only.

Advise the public to:

- Expect some delays, baggage searches, and restrictions as a result of heightened security at public buildings and facilities.
- Continue to monitor world and local events as well as local government threat advisories.
- Report suspicious activities at or near critical facilities to local law enforcement agencies by calling 9-1-1.
- Avoid leaving unattended packages or briefcases in public areas.
- Inventory and organize emergency supply kits (a three-day food and water supply) and discuss emergency plans with family members. Re-evaluate the meeting location based on any specific threat, and ensure that a mechanism to get to the location exists. Identify a distant point of contact for your family. Ensure that supplies for sheltering in place are maintained. Maintain a first-aid kit.
- Consider taking reasonable personal security precautions. Be alert to your surroundings, avoid placing yourself in a vulnerable situation, and monitor the activities of your children.
- Maintain close contact with your family and neighbors to ensure their safety and emotional welfare.

## Severe Condition (Red)

Declaring a Severe Condition reflects a severe risk of terrorist attacks. Under most circumstances, the protective measures for a Severe Condition are not intended to be sustained for extended periods of time. In addition to the protective measures listed under the previous Threat Conditions, agencies should consider the following general measures as well as the agency-specific protective measures:

1. Increase or redirect personnel to address critical emergency needs.
2. Assign emergency response personnel and preposition and mobilize specially trained teams or resources.
3. Monitor, redirect, or constrain transportation systems.
4. Close public and government facilities.

Advise the public to:

- Report suspicious activities and call 9-1-1 for immediate response.
- Expect delays, searches of purses and bags, and restricted access to public buildings.
- Expect traffic delays and restrictions.
- Take personal security precautions (such as avoiding target facilities) to avoid becoming a victim of a crime or a terrorist attack.
- Avoid crowded public areas and gatherings.
- Do not travel into areas affected by an attack.
- Keep emergency supplies accessible and your automobile's fuel tank full.
- Maintain a first-aid kit.
- Be prepared to evacuate your home or to shelter in place on the order of local authorities. Have supplies on hand for this purpose, including a three-day supply of water, food, and medication, and an identified distant point of contact.
- Be suspicious of persons taking photographs of critical facilities, asking detailed questions about physical security, or who are dressed inappropriately for weather conditions (suicide bombers). Report these incidents immediately to law enforcement.

- Closely monitor news reports and Emergency Alert System radio/TV stations.
- Assist neighbors who may need help.
- Avoid passing unsubstantiated information.

Source: Homeland Security Presidential Directive #3, Washington, DC; March 12, 2002; and Homeland Security Advisory System Recommendations for Individuals, Families, Neighborhoods, Schools, and Businesses, The American Red Cross, www.redcross.org, March 2003.

Several Internet resources provide additional information. The U.S. Army Medical Research Institute of Chemical Defense (http://chemdef.apgea.army.mil/), the U.S. Army Medical Research Institute of Infectious Diseases (http://www.usamriid.army.mil/), and the National Homeland Security Knowledgebase (http://www.twotigersonline.com/resources.html) provide useful information and training materials.

## Homeland Security Presidential Directives

HSPDs are issued by the president on matters pertaining to homeland security.

- **HSPD-1: Organization and Operation of the Homeland Security Council:** Ensures coordination of all homeland security–related activities among executive brigades and agencies and promotes the effective development and implementation of all homeland security policies.
- **HSPD-2: Combating Terrorism Through Immigration Policies in the United States:** Provides for the creation of a task force that will work aggressively to prevent aliens who engage in or support terrorist activity from entering the United States and to detain, prosecute, or deport any such aliens who are within the United States.
- **HSPD-3: Homeland Security Advisory System:** Establishes a comprehensive and effective means to disseminate information regarding the risk of terrorist acts to federal, state, and local authorities and to the American people.
- **HSPD-4: National Strategy to Combat Weapons of Mass Destruction:** Applies new technologies, increases emphasis on intelligence collection and analysis, strengthens alliance relationships, and establishes new partnerships with former adversaries to counter this threat in all of its dimensions.
- **HSPD-5: Management of Domestic Incidents:** Enhances the ability of the United States to manage domestic incidents by establishing a single, comprehensive national incident management system.
- **HSPD-6: Integration and Use of Screening Information:** Provides for the establishment of the Terrorist Threat Integration Center.
- **HSPD-7: Critical Infrastructure Identification, Prioritization, and Protection:** Establishes a national policy for federal departments and agencies to identify and prioritize U.S. critical infrastructure and key resources and to protect them from terrorist attacks.
- **HSPD-8: National Preparedness:** Identifies steps for improved coordination in response to incidents. This directive describes the way federal departments and

agencies will prepare for such a response, including prevention activities during the early stages of a terrorism incident. This directive is a companion to HSPD-5.

- **HSPD-9: Defense of United States Agriculture and Food:** Establishes a national policy to defend the agriculture and food system against terrorist attacks, major disasters, and other emergencies.
- **HSPD-10: Biodefense for the 21st Century:** Provides a comprehensive framework for our nation's biodefense.
- **HSPD-11: Comprehensive Terrorist-Related Screening Procedures:** Implements a coordinated and comprehensive approach to terrorist-related screening that supports homeland security, at home and abroad. This directive builds upon HSPD-6.
- **HSPD-12: Policy for a Common Identification Standard for Federal Employees and Contractors:** Establishes a mandatory, government-wide standard for secure and reliable forms of identification issued by the federal government to its employees and contractors (including contractor employees).
- **HSPD-13: Maritime Security Policy:** Establishes policy guidelines to enhance national and homeland security by protecting U.S. maritime interests.

# Wrap-Up

## Ready for Review

- Potential terrorism targets include
  - Ecoterrorism
  - Infrastructure
  - Symbolic
  - Civilian
  - Cyberterrorism
  - Agroterrorism
- Terrorists use several different kinds of weapons including explosives, chemical agents, biological agents, and radiological agents.
- The possibility of secondary devices must be considered at all terrorist incidents.
- Brigade members should become familiar with potential terrorist targets in their response area.
- Most chemical agents produce symptoms almost immediately upon exposure.
- Biological agents usually produce flu-like symptoms following an incubation period of 2–10 days.
- When the presence of radioactive material is suspected, brigade members should limit their exposure time.
- Brigade members should establish a staging area a safe distance from the scene of a terrorist incident and follow the direction of the IC.
- PPE will provide the brigade member with limited protection from most agents used by terrorists.
- Be aware of the homeland security threat level at all times.
- Homeland Security Presidential Directives are issued by the president on matters pertaining to homeland security.

## Hot Terms

**Agroterrorism** The use of a biological, chemical, radiological, or other agent against either the preharvest or postharvest stages of food and fiber production to inspire fear or cause economic damage, public health impact, or other adverse impact against the United States. Biological agents can impact animal or plant health or lead to human disease if they are zoonotic (i.e., diseases that are naturally transmitted between vertebrate animals and man with or without an arthropod intermediary).

**Alpha particles** A type of radiation that quickly loses energy and can travel only 1–2 inches from its source. Clothing or a sheet of paper can stop this type of energy.

**ANFO** An explosive material containing ammonium nitrate fertilizer and fuel oil.

**Anthrax** An infectious disease spread by the bacteria *Bacillus anthracis*; typically found around farms, infecting livestock.

**Beta particles** A type of radiation that is capable of traveling 10–15 feet. Heavier materials such as metal, plastic, and glass can stop this type of energy.

**Biological agents** Disease-causing bacteria, viruses, and other agents that attack the human body.

**Blistering agents** Chemicals that cause the skin to blister.

**Chlorine** A yellowish gas that is about 2.5 times heavier than air and slightly water-soluble. It has many industrial uses but also damages the lungs when inhaled (a choking agent).

**Choking agent** A chemical designed to inhibit breathing and typically intended to incapacitate rather than kill.

**Cyanide** A highly toxic chemical agent that attacks the circulatory system.

**Cyberterrorism** The intentional act of electronically attacking government or private computer systems.

**Decontamination** The physical or chemical process of removing any form of contaminant from a person, an object, or the environment.

**Ecoterrorism** Terrorism directed against causes that radical environmentalists think would damage the earth or its creatures.

**Forward staging area** A strategically placed area, close to the incident site, where personnel and equipment can be held in readiness for rapid response to an emergency event.

**Gamma rays** A type of radiation that can travel significant distances, penetrating most materials and passing through the body. Gamma radiation is the most destructive to the human body.

**Improvised explosive device (IED)** An explosive or incendiary device that is fabricated in an improvised manner.

**Incubation period** Time period between the initial infection by an organism and the development of symptoms.

**Lewisite** Blister-forming agent that is an oily, colorless-to-dark brown liquid with an odor of geraniums.

**Mark 1 Nerve Agent Antidote Kit** A military kit containing antidotes that can be administered to victims of a nerve agent attack.

**Mass decontamination** Special procedures for incidents that involve large numbers of people. Master stream devices from engine companies and aerial apparatus provide high-volume, low-pressure showers so that large numbers of people can be decontaminated rapidly.

**Nerve agents** Toxic substances that attack the central nervous system in humans.

**Personal dosimeters** Devices that measure the amount of radioactive exposure to an individual.

**Phosgene** A chemical agent that causes severe pulmonary damage. A by-product of incomplete combustion.

**Pipe bomb** A device created by filling a section of pipe with an explosive material.

**Plague** An infectious disease caused by the bacterium *Yersinia pestis*; commonly found on rodents.

**Radiation dispersal device (RDD)** Any device that causes the purposeful dissemination of radioactive material without a nuclear detonation; a dirty bomb.

**Radiological agents** Materials that emit radioactivity.

**Sarin** A nerve agent that is primarily a vapor hazard.

**Secondary device** An explosive device designed to injure emergency responders who have responded to an initial event.

**Smallpox** A highly infectious disease caused by the virus variola.

**Soman** A nerve gas that is both a contact and a vapor hazard that has the odor of camphor.

**Suitcase nuke** A nuclear explosive device that is small enough to fit in a suitcase.

**Sulfur mustard** A clear, yellow, or amber oily liquid with a faint sweet odor of mustard or garlic that may be dispersed in an aerosol form. It causes blistering of exposed skin.

**Tabun** A nerve gas that is both a contact and a vapor hazard that operates by disabling the chemical connection between nerve and target organs.

**Triage** The process of sorting victims based on the severity of injury and medical needs to establish treatment and transportation priorities.

**Universal precautions** Procedures for infection control that treat blood and certain bodily fluids as capable of transmitting bloodborne diseases.

**V-agent** A nerve agent, principally a contact hazard; an oily liquid that can persist for several weeks.

**Weapons of mass destruction (WMD)** A weapon intended to cause mass casualties, damage, and chaos.

# Brigade Member in Action

**Your chemical plant security department has noticed a suspicious vehicle driving around the perimeter of the plant. Before members of the department can contact local law enforcement to report the situation, the vehicle accelerates suddenly and crashes through the fence line near an incoming pipeline manifold. It comes to rest against a natural gas pipeline. The driver jumps from the vehicle and runs into the plant. Your fire brigade is immediately dispatched to the scene of the incident.**

1. What should be the first action of the arriving brigade members?
   - A. Send a reconnaissance team in to evaluate the situation.
   - B. Establish a security perimeter and keep all unneeded personnel clear of the perimeter.
   - C. Immediately start flowing water on the manifold.
   - D. Await local law enforcement.

2. Several brigade members working in close proximity to the vehicle begin to complain of a burning sensation on exposed skin areas. What should be the first action taken as the personnel leave the secured perimeter area?
   - A. Immediately decontaminate all exposed personnel.
   - B. Have local law enforcement interview the personnel.
   - C. Send the personnel to the plant medical facility.
   - D. Have the personnel brief the next entry team.

3. What resource provides technical advice on chemical and biological agents?
   A. Emergency medical services
   B. Bomb squad
   C. Health department
   D. Local law enforcement

4. What resource provides neutralization of hazardous substances, including chemical agents?
   A. Emergency medical services
   B. Hazardous materials unit
   C. Bomb squad
   D. Structural collapse teams

# Fire Detection, Protection, and Suppression Systems

Chapter 25

# NFPA 1081 Standard

### Incipient Facility Fire Brigade Member

**4.3.2*** Activate a fixed fire protection system, given a fixed fire protection system, a procedure, and an assignment, so that the steps are followed and the system operates.

**(A)** *Requisite Knowledge.* Types of extinguishing agents, hazards associated with system operation, how the system operates, sequence of operation, system overrides and manual intervention procedures, and shutdown procedures to prevent damage to the operated system or to those systems associated with the operated system.

**(B)** *Requisite Skills.* The ability to operate fixed fire protection systems via electrical or mechanical means.

### Advanced Exterior Facility Fire Brigade Member

**5.3.7*** Interpret alarm conditions, given an alarm signaling system, a procedure, and an assignment, so that the alarm condition is correctly interpreted and a response is initiated.

**(A)** *Requisite Knowledge.* The different alarm detection systems within the facility; difference between alarm, trouble, and supervisory alarms; hazards protected by the detection systems; hazards associated with each type of alarm condition; knowledge of the emergency response plan; and communication procedures.

**(B)** *Requisite Skills.* The ability to understand the different types of alarms, to implement the response, and to provide information through communications.

**5.3.8*** Activate a fixed fire suppression system, given PPE, a fixed fire protection system, a procedure, and an assignment, so that the correct steps are followed and the system operates.

**(A)** *Requisite Knowledge.* Different types of extinguishing agents, hazards associated with system operation, how the system operates, sequence of operation, system overrides and manual intervention procedures, and shutdown procedures to prevent damage to the operated system or to those systems associated with the operated system.

**(B)** *Requisite Skills.* The ability to operate fixed fire suppression systems via electrical or mechanical means and shutdown procedures for fixed fire suppression systems.

### Interior Structural Facility Fire Brigade Member

**6.3.1*** Interpret alarm conditions, given an alarm signaling system, a procedure, and an assignment, so that the alarm condition is correctly interpreted and a response is initiated.

**(A)** *Requisite Knowledge.* The different alarm detection systems within the facility; difference between alarm, trouble, and supervisory alarms; hazards protected by the detection systems; hazards associated with each type of alarm condition; the emergency response plan; and communication procedures.

**(B)** *Requisite Skills.* The ability to understand the different types of alarms, to implement the response, and to provide information through communications.

**6.3.2*** Activate a fixed fire protection system, given required PPE, a fixed fire protection system, a procedure, and an assignment, so that the procedures are followed and the system operates.

**(A)** *Requisite Knowledge.* Different types of extinguishing agents on site, manual fire suppression activities within areas covered by fixed fire suppression systems, hazards associated with system operation, how the system operates, sequence of operation, system overrides and manual intervention procedures, and shutdown procedures to prevent damage to the operated system or to those systems associated with the operated system.

**(B)** *Requisite Skills.* The ability to operate fixed fire suppression systems via electrical or mechanical means and to shut down fixed fire suppression systems.

# Knowledge Objectives

After completing this chapter, you will be able to:

- Explain why facility brigade members need to have knowledge of fire protection systems.
- Describe the basic functions and components of a fire alarm system.
- Describe the basic types of fire alarms used in an industrial setting, related initiating devices, and where each is most suitable for use.
- Describe the fire brigade's role in resetting alarm systems and putting them back in service after an alarm.
- Explain how alarms are transmitted to the in-site proprietary central station and retransmitted to the municipal fire authority.
- Identify the four different styles of sprinkler heads.
- Identify the different types of sprinkler control valves.
- Describe the operation and application of the following automatic sprinkler systems:
  - Wet-pipe
  - Dry-pipe
  - Preaction
  - Deluge
- Describe when and how water is shut off to a sprinkler system and how to stop or control water coming from a sprinkler head.
- Describe the three types of standpipe systems.
- Describe how clean agents function and their applications.
- Describe how smoke, heat, and flame detectors operate and the best applications for each.
- Describe why fire department connections are important for a sprinkler system.
- Describe the operation and application of wet and dry chemical systems.

# Skills Objectives

After completing this chapter, you will be able to perform the following skills:

- Stop the water from an activated sprinkler head.
- Operate firewater control valves.
- Connect a fire hose to a fire department connection.
- Reset a fire alarm panel.
- Connect a fire hose to a standpipe outlet.
- Show competency in the field by identifying the different styles of sprinkler heads.
- Show competency in the field by identifying the different types of sprinkler systems.
- Show competency in the field by identifying the different types of sprinkler water control valves.
- Show competency in the field by identifying the different types of detectors (e.g., smoke, heat, flame).

# You Are the Brigade Member

**The on-shift security supervisor contacts the fire brigade leader to inform him that the "nuisance" fire alarm at the engineering annex is sounding again. Per standard operating procedures, the brigade leader requests that an engine company be dispatched to investigate the alarm activation. When you arrive on the scene at the annex building, you note that the strobe lights of the detection system have activated. As you walk around the corner of the building, you hear the water flow alarm for the building sprinkler system sounding. The brigade leader immediately requests that a line be connected to the sprinkler system connection.**

1. ***Do you believe that the brigade leader is overreacting to this situation?***
2. ***Which clues indicate that this situation may actually be a real emergency?***
3. ***After the line is connected to the sprinkler system, to which operation do you think that the brigade leader will assign you?***

## Introduction

Today, fire prevention and building codes require that most new structures have some sort of fire protection system installed. This makes it more important than ever for the line brigade member to have a working knowledge of these systems, which include fire alarm systems and automatic fire detection and suppression systems. Understanding how these systems operate is important for the brigade member's safety and necessary to provide effective customer service to the companies we protect.

From a corporate standpoint, brigade members need to understand the operations and limitations of fire detection and suppression systems. A building with a fire protection system will have very different working conditions during a fire than an unprotected building. Brigade members need to know how to fight a fire in a building with a working fire suppression system, and how to shutdown the system after the fire is extinguished.

From a customer service standpoint, brigade members who understand how fire protection systems work can help dispel misconceptions about these systems and advise management and occupants after an alarm is sounded. Most people have no idea of how the fire protection or detection systems in their building work, and there are often more false alarms in buildings with fire protection systems than actual fires. Brigade members can help the company determine what activated the system, how they can prevent future false alarms, and what needs to be done to restore a system to service. This chapter discusses the basic design and operation of fire alarm, fire detection, and fire suppression systems. Fire protection systems have fairly standardized design requirements across North America; most areas follow the applicable NFPA standards. Unfortunately, local fire prevention and building codes may require different types of systems for different buildings, and these requirements still vary considerably.

## Fire Alarm and Detection Systems

Practically all new construction includes some sort of fire alarm and detection system. In most cases, both fire detection and fire alarm components are integrated in a single system. A fire detection system recognizes when a fire is occurring and activates the fire alarm system, which alerts building occupants and, in some cases, the fire brigade. Some fire detection systems also automatically activate fire suppression systems to control the fire.

Fire alarm and detection systems range from simple, single-station smoke alarms, to complex fire detection and control systems for office areas, control rooms, and large production or process areas. Many fire alarm and detection systems in large buildings also control other systems to help protect occupants and control the spread of fire and smoke. Although these systems can be complex, they generally have the same basic components.

### Fire Alarm System Components

A fire alarm system has three basic components—an alarm initiation device, an alarm notification device, and a control panel. The alarm initiation device is either an automatic or

Figure 25-1 Most older fire alarm control panels indicate the zone where an alarm was initiated. The new "addressable" systems give an exact location, often on an LCD read-out screen.

manually operated device that, when activated, causes the system to indicate an alarm. The alarm notification device is generally an audible device, often accompanied by a visual device, that alerts the building occupants when the system is activated. But the most important part of any fire alarm system is the control panel, which links the initiation device to the notification device and performs other essential functions.

## Fire Alarm System Control Panels

Most installed fire alarm systems in buildings have several alarm initiation devices in different areas and use both audible and visible devices to notify the occupants of an alarm. The fire alarm control panel serves as the "brain" of the system, linking the activation devices to the notification devices.

The control panel manages and monitors the proper operation of the system. It can indicate the source of an alarm, so that responding fire personnel will know what activated the alarm and where the initial activation occurred. The control panel also manages the primary power supply and provides a backup power supply for the system. It may perform additional functions, such as notifying the fire brigade, municipal fire department or central station monitoring company when the alarm system is activated, and may interface with other systems within the facility.

Control panels vary greatly depending on the age of the system and the manufacturer. For example, an older system may simply indicate that an alarm has been activated, while a newer system may indicate a specific location within the building (◄ Figure 25-1). The most modern panels (addressable) actually specify the exact location of the activated initiation device.

Fire alarm control panels are used to silence the alarm and reset the system. These panels should always be locked; many newer systems require the use of a password before the alarm can be silenced or reset. Alarms should not be silenced or reset until the activation source has been found and checked by brigade members to ensure that the situation is under control. In some systems, the individual alarm activation devices must be reset individually before the entire system can be reset; other systems reset themselves after the problem is resolved.

Many buildings have an additional display panel in a separate location, usually near the front door of the building. This is called a remote annunciator (▼ Figure 25-2). The remote annunciator enables brigade members to ascertain the type and location of the activated alarm device as they enter the building.

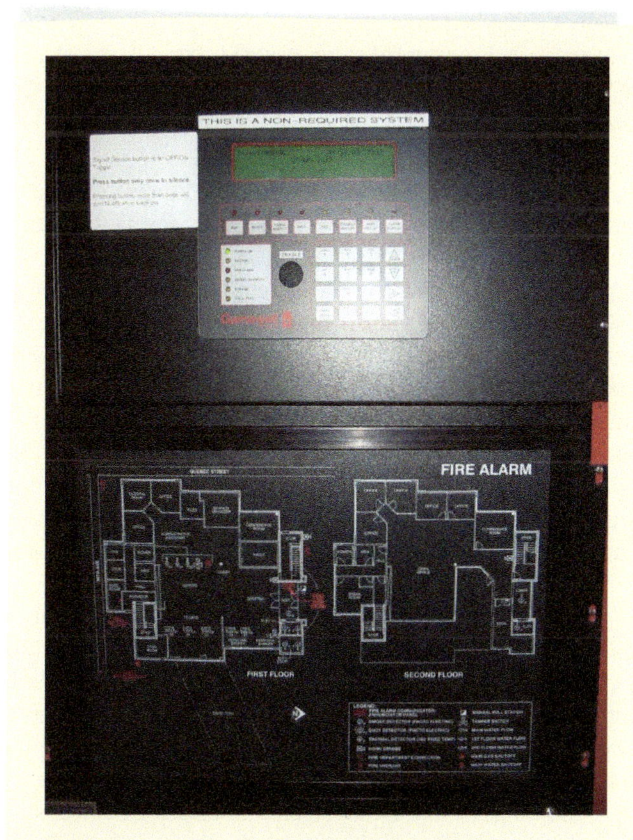


Figure 25-2 A remote annunciator allows brigade members to quickly determine the type and location of the activated alarm device.

Figure 25-3 Fire prevention codes require that alarm systems have a backup power supply, which is automatically activated when the normal electrical power is interrupted.

The fire alarm control panel should also monitor the condition of the entire alarm system to detect any faults. Faults within the system are indicated by a trouble alarm, which shows that a component of the system is not operating properly and requires attention. Trouble alarms do not activate the building's fire alarm, but should make an audible sound and illuminate a special light at the alarm control panel. They may also transmit a notification to a remote service location or an on-site proprietary central station.

A fire alarm system is usually powered by a 110-volt line, even though the system's appliances may use a lower voltage. There are, however, some older alarm systems that require 110 volts for all components.

In addition to the normal power supply, fire prevention codes call for a backup power supply for all alarm systems (▲Figure 25-3). In some systems, a battery in the fire alarm control panel will automatically activate when the external power is interrupted. Fire codes will specify how long the system must be able to function on the battery backup. In large buildings or industrial facilities, the backup power supply could be an emergency generator or a secondary power source. If either the main power supply or the backup power source fails, the trouble alarm should sound.

Depending on the building's size and floor plan, an activated alarm may sound throughout the building or only in particular areas. In a large manufacturing complex, fire alarm systems are often programmed to alert only the occupants in the same area as the activated alarm as well as those in the areas immediately above and below. Alarms on the remaining floors can be manually activated from the system control panel. Some systems have a public address feature, so that a brigade leader can provide specific instructions or information for occupants in different areas.

The control panel in a large facility may be programmed to perform several additional functions. For example, it may automatically shut down or change the operation of air handling systems, recall elevators to the ground floor, and unlock stairwell doors so that a person in an exit stairway can reenter an occupied floor.

## Alarm Initiating Devices

Alarm initiation devices are the components that activate a fire alarm system. Manual initiation devices require human activation; automatic devices function without human intervention. Manual fire alarm boxes are the most common type of alarm initiating devices that require human activation. Automatic initiation devices include various types of heat, flame, and smoke detectors and other devices that automatically recognize the evidence of a fire.

### Manual Initiation Devices

Manual initiation devices are designed so that building occupants can activate the fire alarm system if they discover a fire in the building. Many older alarm systems could only be activated manually. The primary manual initiation device is the manual fire alarm box, or manual pull-station (▼Figure 25-4). These stations have a switch that either opens or closes an electrical circuit to activate the alarm. Pull-stations come in various sizes and designs, depending on the manufacturer. They can be either single-action or double-action devices.

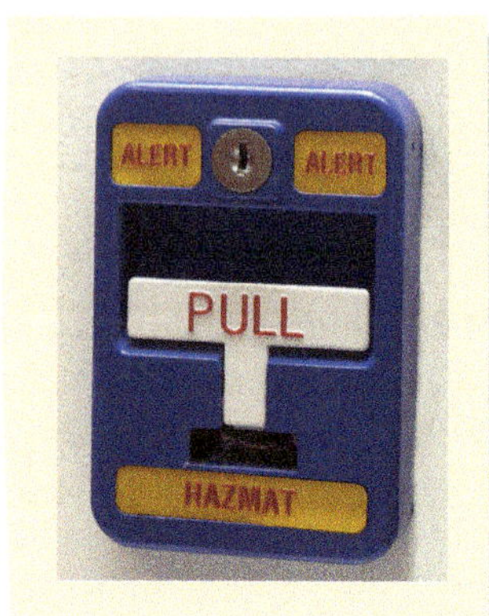

Figure 25-4 There are several different types of manual fire alarm boxes, also known as manual pull-stations.

Figure 25-5 A variation on the double-action pull-station, designed to prevent malicious false alarms, has a clear plastic cover and a separate tamper alarm.

Single-action pull-stations require a person to pull down a lever, toggle, or handle to activate the alarm. The alarm sounds as soon as the pull-station is activated. Double-action pull-stations require a person to perform two steps before the alarm will activate. They are designed to reduce the number of false alarms caused by accidental or intentional pulling of the alarm. The person must move a flap, lift a cover, or break a piece of glass to reach the alarm activation device. Designs that use glass are no longer in favor, because the glass must be replaced each time the alarm is activated and because the broken glass poses the risk of injury.

Once activated, a manual pull-station should stay in the "activated" position until it is reset. This enables responding brigade members to determine which pull-station initiated the alarm. Resetting the pull-station requires a special key, screwdriver, or Allen wrench. In many systems, the pull-station must be reset before the building alarm can be reset at the alarm control panel.

A variation on the double-action pull-station, designed to prevent malicious false alarms, is covered with a piece of clear plastic ▲ Figure 25-5. These covers are often used in areas where malicious false alarms frequently occur. The plastic cover must be opened before the pull-station can be activated. Lifting the cover triggers a loud tamper alarm at that specific location but does not activate the fire alarm system. Snapping the cover back into place resets the tamper alarm. The intent is that a person planning to initiate a false alarm will drop the cover and run when the tamper alarm sounds. In most cases, the pull-station tamper alarm is not connected to the fire alarm system in any way.

Even automatic fire alarm systems can be manually activated by pressing a button or flipping a switch on the alarm system control panel. In large facilities, the main control panel has separate manual activation switches for different areas of the plant or facility.

## Automatic Initiating Devices

Automatic initiating devices are designed to function without human intervention and will activate the alarm system when they detect evidence of a fire. These systems can be programmed to transmit the alarm to the fire brigade or municipal fire department, even if the building is unoccupied, and to perform other functions when a detector is activated.

Automatic initiating devices can use several different types of detectors. Some detectors are activated by smoke or by the invisible products of combustion, and others react to heat, the light produced by an open flame, or specific gases.

### Smoke Detectors

A smoke detector is designed to sense the presence of smoke. Among brigade members, the term "smoke detector" generally refers to a sensing device that is part of a fire alarm system ▼ Figure 25-6.

Smoke detectors come in different designs and styles for different applications. The most common smoke detectors are ionization smoke detectors and photoelectric smoke

Figure 25-6 Commercial ionization smoke detector.

Figure 25-7 Beam detectors are used in large open spaces.

detectors; however, smoke detectors used in commercial fire alarm systems are usually more sophisticated and more expensive than residential smoke alarms.

Each detection device is rated to protect a certain floor area, so in large areas the detectors are often placed in a grid pattern. Newer smoke detectors also have a visual indicator, such as a steady or flashing light, that indicates when the device has been activated.

A beam detector is a type of photoelectric smoke detector used to protect large open areas such as warehouses, airport terminals, and indoor sports arenas. In these facilities, it would be difficult or costly to install large numbers of individual smoke detectors, but a single beam detector could be used (▲ Figure 25-7).

A beam detector has two components: a sending unit that projects a narrow beam of light across the open area and a receiving unit that measures the intensity of the light when the beam strikes the receiver. When smoke interrupts the light beam, the receiver detects a drop in the light intensity and activates the fire alarm system.

Most photoelectric beam detectors are set to respond to a certain obscuration rate, meaning percentage of the light that is blocked. If the light is completely blocked, such as when a solid object is moved across the beam, the trouble alarm will sound, but the fire alarm will not be activated.

Smoke detectors are usually powered by a low-voltage circuit and send a signal to the fire alarm control panel when they are activated. Both ionization and photoelectric smoke detectors are self-restoring. After the smoke condition clears, the alarm system can be reset at the control panel.

## Heat Detectors

Heat detectors are also common automatic alarm initiation devices. Heat detectors can provide property protection but cannot provide reliable life safety protection because they do not react quickly enough to incipient fires. They are generally used in situations where smoke alarms cannot be used, such as dusty environments and areas that experience extreme cold or heat. Heat detectors are often installed in unheated areas, manufacturing areas such as where dust and fumes would cause smoke detectors to alarm falsely.

Heat detectors are generally very reliable and less prone to false alarms than smoke alarms. You may come across heat detectors that were installed 30 or more years ago and are still in service. However, older units have no visual trigger that tells which device was activated, so tracking down the cause of an alarm may be very difficult. Newer models have an indicator light that shows which device was activated.

There are several types of heat detectors, each designed for specific situations and applications. Single-station heat alarms are sometimes installed in unoccupied areas of buildings that do not have fire alarm systems, such as attics or storage rooms. Spot detectors are individual units that can be spaced throughout an occupancy; each detector covers a specific floor area. Spot detectors are usually installed in light commercial settings; the units may be in individual rooms or spaced at intervals along the ceiling in larger areas.

Line detectors use wire or tubing strung along the ceiling of large open areas to detect an increase in heat. An increase in temperature anywhere along the line will activate the detector. Line detectors are found in many warehouses and industrial or manufacturing applications.

Heat detectors can be designed to operate at a fixed temperature or to react to a rapid increase in temperature. Either fixed temperature or rate-of-rise devices can be configured as spot or line detectors.

### *Fixed-Temperature Heat Detectors*

Fixed-temperature heat detectors, as the name implies, are designed to operate at a preset temperature (► Figure 25-8). A typical temperature for a light-hazard occupancy, such as an office building, would be 135°F (57°C). Fixed-temperature detectors usually use a metal alloy that will melt at the preset temperature. The melting alloy releases a lever-and-spring mechanism, to open or close a switch. Most fixed-temperature heat detectors must be replaced after they have been activated, even if the activation was accidental.

### *Rate-of-Rise Heat Detectors*

Rate-of-rise heat detectors will activate if the temperature of the surrounding air rises more than a set amount in a given period of time. A typical rating might be "greater than 12°F in one minute." If the temperature increase is less than this rate, the rate-of-rise heat detector will not activate. An increase greater than this rate will activate the detector and set off the fire alarm. Rate-of-rise heat detectors should not be located in areas that normally experience rapid changes in temperature, such as near

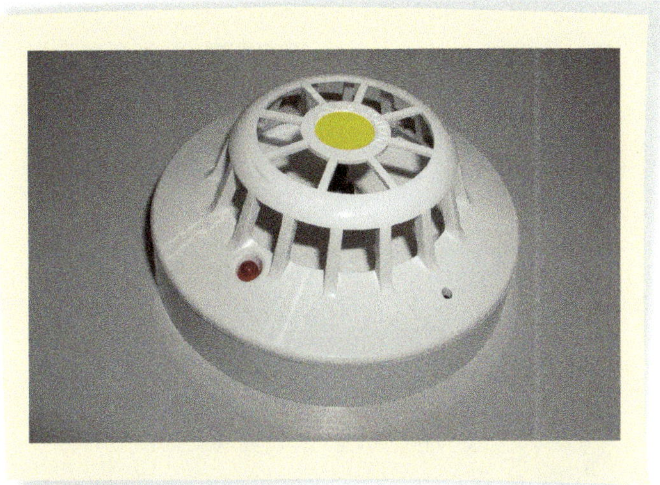

**Figure 25-8** A fixed-temperature heat detector initiates an alarm at a preset temperature.

**Figure 25-9** Flame detectors are specialized devices that detect the electromagnetic light waves produced by a flame.

garage doors or refrigeration units subject to open doors in warm weather.

Some rate-of-rise heat detectors have a **bimetallic strip** made of two metals that respond differently to heat. A rapid increase in temperature causes the strip to bend unevenly, which opens or closes a switch. Another rate-of-rise heat detector uses an air chamber and diaphragm mechanism. As air in the chamber heats up, the pressure increases. Gradual increases in pressure are released through a small hole, but a rapid increase in pressure will press upon the diaphragm and activate the alarm. Most rate-of-rise heat detectors are self-restoring, and do not need to be replaced after an activation unless they were directly exposed to a fire.

Rate-of-rise heat detectors generally respond faster to most fires than fixed-temperature heat detectors. However, a slow burning fire, such as a smoldering couch, may not activate a rate-of-rise heat detector until the fire is well-established.

Combination rate-of-rise and fixed-temperature heat detectors are available. These devices balance the faster response of the rate-of-rise detector with the reliability of the fixed-temperature heat detector.

### *Line Heat Detectors*

Line heat detectors use wires or a sealed tube to sense heat. One wire-type line detector has two wires inside, separated by an insulating material. When heat melts the insulation, the wires short out and activate the alarm. The damaged section of insulation must be replaced with a new piece after activation.

Another wire-type detector measures changes in the electrical resistance of a single wire as it heats up. This device is self-restoring and does not need to be replaced after activation unless it is directly exposed to a fire.

The tube-type line heat detector has a sealed metal tube filled with air or a nonflammable gas. When the tube is heated, the internal pressure increases and activates the alarm. Like the single-wire line heat detector, this device is self-restoring and does not need to be replaced after activation, unless it is directly exposed to a fire.

### Flame Detectors

**Flame detectors** are specialized devices that detect the electromagnetic light waves produced by a flame (▲ **Figure 25-9**). These devices can quickly recognize even a very small fire.

Typically flame detectors are found in places such as aircraft hangars or specialized industrial settings using flammable liquids in which early detection and rapid reaction to a fire are critical. Flame detectors are also used in explosion suppression systems to detect and suppress an explosion as it is occurring.

Flame detectors are complicated and expensive; in addition, other infrared or ultraviolet sources such as the sun or a welding operation can set off an unwanted alarm. Flame detectors that combine infrared and ultraviolet sensors are sometimes used to lessen the chances of a false alarm.

### Gas Detectors

**Gas detectors** are calibrated to detect the presence of a specific gas that is created by combustion or that is used in the facility. Depending on the system, a gas detector may be programmed to activate either the building's fire alarm system or a separate alarm. Gas detectors are specialized instruments that need regular calibration.

### Air Sampling Detectors

**Air sampling detectors** continuously capture air samples and measure the concentrations of specific gases or products

Figure 25-10 Air sampling detector.

Figure 25-11 Automatic sprinkler systems use an electric flow switch to activate the building's fire alarm system and transmit an alarm to the brigade or central station.

of combustion. These devices draw in the air samples, which are then analyzed by exposing each sample to a high-intensity, broad-spectrum light source such as a laser beam ▲ Figure 25-10. Air sampling detectors are often installed in the return air ducts of large structures. They will sound an alarm and shut down the air handling system.

More complex systems are sometimes installed in special hazard areas to draw air samples from rooms, enclosed spaces, or equipment cabinets. The samples pass through gas analyzers that can identify smoke particles, products of combustion, and concentrations of other gases associated with a dangerous condition such as a release of chlorine in a water treatment plant. Air sampling detectors are most often used in areas that contain valuable contents or sensitive equipment where it is important to detect problems early.

### Alarm Initiation by Fire Suppression Systems

Other fire protection systems in a building may be used to activate the fire alarm system. Automatic sprinkler systems are usually connected to the fire alarm system and will activate the alarm if there is a water flow ► Figure 25-11. This system not only alerts building occupants and the fire brigade to a possible fire, but it also ensures that someone is aware water is flowing, in case of an accidental discharge. Any other fire extinguishing systems in a building should also be tied into the building's or facility's fire alarm.

## False, Unwanted, and Nuisance Alarms

As the number of fire detection and alarm systems increases, so does the number of false, unwanted, or nuisance alarms. Knowing how to handle these alarms is just as important as knowing how to deal with an actual fire. Fire brigade personnel who are informed about fire alarm and detection systems can advise management when these systems experience problems.

The term "false alarm" is generally used to describe all fire alarm activations that are not associated with a true emergency. In reality, there are three distinct types of false alarms: malicious false alarms, unwanted alarms, and nuisance alarms. It is important to distinguish among the three types in determining the root cause of the fire alarm activation. The problem must be recognized before it can be rectified.

Regardless of the cause, all three types of false alarms have the same results. They waste fire brigade resources and may delay legitimate responses. Frequent false alarms at the same site can desensitize building occupants to the alarm system so that they may not respond appropriately to a real emergency.

### Malicious False Alarms

Malicious false alarms are caused by individuals who deliberately activate a fire alarm when there is no fire, causing a disturbance. Manual fire alarm boxes are popular targets for pranksters. A malicious false alarm is an illegal act if the municipal fire department is summoned.

### Unwanted Alarms

An unwanted alarm occurs when an alarm system is activated by a condition that is not really an emergency. For example, a smoke alarm placed too close to a steam trap may be triggered by normal activities. An unwanted alarm could also occur if a person who is smoking unknowingly stands under a smoke detector. The smoke detector does its job—recognizing smoke and activating the alarm system—but there is no real emergency.

### Nuisance Alarms

Nuisance alarms are caused by improper functioning of an alarm system or one of its components. Alarm systems must be properly maintained on a continual basis. A mechanical failure or a lack of maintenance that causes an alarm system to activate when there is no emergency could also fail to activate the system when there is a real fire.

### Preventing Unwanted and Nuisance Alarms

Systems that experience unwanted and nuisance alarms should be examined to determine the cause and correct the problem. Many unwanted alarms could be avoided by relocating a detector or using a different type of detector in a particular location. Proper design, installation, and system maintenance are essential to prevent nuisance alarms.

If an alarm system is activated whenever it rains, water is probably leaking into the wiring. Detectors can respond to a buildup of dust, dirt, or other debris by becoming more sensitive and needlessly going off. That is why smoke alarms and smoke detectors must be cleaned periodically.

Several different methods can be used to reduce unwanted and nuisance alarms caused by smoke detection systems. In a cross-zoned system, the activation of a single smoke detector will not sound the fire alarm, although it will usually set off a trouble alarm. A second smoke detector must be activated before the actual fire alarm will sound.

In a verification system, there is a delay of 30–60 seconds between activation and notification. During this time, the system may show a trouble or "prealarm" condition at the system control panel. After the preset interval, the system rechecks the detector. If the condition has cleared, the system returns to normal. If the detector is still sensing smoke, then the fire alarm is activated. Both cross-zoned and verification systems are designed to prevent brief exposures to common occurrences such as steam from activating the alarm system.

## Alarm Notification Appliances

Audible and visual alarm notification devices such as bells, horns, and electronic speakers produce an audible signal when the fire alarm is activated. Some systems also incorporate visual alerting devices. These audible and visual alarms alert occupants of a building to a fire.

Older systems used various sounds as notification devices. However, this often caused confusion over whether the sound was actually an alarm. More recent fire prevention codes have adopted a standardized audio pattern, called the temporal-3 pattern, that must be produced by any audio device used as a fire alarm. This enables people to recognize a fire alarm immediately.

Some systems also have the capability to play a recorded evacuation announcement in conjunction with the temporal-3 pattern. The recorded message is played through the fire alarm speakers and provides safe evacuation instructions (▲ Figure 25-12). In facilities such as airport terminals, this announcement is recorded in multiple languages. This type of system may include a public address feature that fire brigade or site security personnel can use. This feature may be used to provide specific instructions, information about the situation, or notice when the alarm condition is terminated.

Figure 25-12 A combination speaker/strobe device.

Many new fire alarm systems incorporate visual notification devices such as high-intensity strobe lights or other types of flashing lights as well as audio devices (► Figure 25-13). Visual devices alert hearing-impaired occupants to a fire alarm and are very useful in environments where an audible alarm might not be heard, particularly in manufacturing areas with constant background noise.

## Other Fire Alarm Functions

In addition to alerting occupants and summoning the fire brigade, fire alarm systems may also control other building functions, such as air handling systems, fire doors, and elevators. To control smoke movement through the building, the system may shut down or start up air handling systems. Fire doors that are normally held open by magnets may be released to compartmentalize the building and confine the fire to a specific area. Doors allowing re-entry from exit stairways into occupied areas may be unlocked. Elevators will be summoned

Figure 25-13 This alarm notification device has both a loud horn and a high-intensity strobe light, similar to Figure 25-12.

to a predetermined floor, usually the main lobby, so they can be used by fire crews.

Responding fire personnel must understand which building functions are being controlled by the fire alarm, for both safety and fire suppression reasons. This information should be gathered during preincident planning surveys and should also be available in printed form or on a graphic display at the control panel location.

## Fire Alarm Annunciation Systems

Some fire alarm systems give little information at the alarm control panel; others will specify exactly which initiation device activated the fire alarm. The systems can be further subdivided based on whether they are zoned or coded systems.

In a zoned system, the alarm control panel will indicate where in the building the alarm was activated. Almost all alarm systems are now zoned to some extent. Only the most rudimentary alarm systems give no information at the alarm control panel about where the alarm was initiated. In a coded system, the zone is identified not only at the alarm control panel but also through the audio notification device. ▲ Table 25-1 shows how, using these two variables, systems can be broken down into four categories: noncoded alarm, zoned noncoded alarm, zoned coded alarm, and master-coded alarm.

**Table 25-1 Categories of Alarm Annunciation Systems**

| Category | Description |
|---|---|
| Noncoded alarm | No information is given on what device was activated or where it is located. |
| Zoned noncoded alarm | Alarm system control panel indicates the zone in the building that was the source of the alarm. It may also indicate the specific device that was activated. |
| Zoned coded alarm | The system indicates over the audible warning device which zone has been activated. This type of system is often used in hospitals, where it is not feasible to evacuate the entire facility. |
| Master-coded alarm | The system is zoned and coded. The audible warning devices are also used for other emergency-related functions. |

### Noncoded Alarm System

In a noncoded alarm system, the control panel has no information indicating where in the building the fire alarm was activated. The alarm typically sounds a bell or horn. Fire brigade personnel must search the entire building to find which initiation device was activated. This type of system is generally found only in older facilities.

### Zoned Noncoded Alarm System

This is the most common type of system, particularly in newer buildings. The building or plant is divided into multiple zones. The alarm control panel indicates in which zone the activated device is located. It may also indicate the type of device that was activated. Responding personnel can go directly to that part of the building to search for the problem and check the activated device.

Many zoned noncoded alarm systems have an individual indicator light for each zone. When a device in that area is activated, the indicator light goes on. Computerized alarm systems also may use "addressable devices." In these systems, each individual initiation device—whether it is a smoke detector, heat detector, or pull-station—has its own unique identifier. When the device is activated, the identifier is indicated on a display or print-out at the control panel. Responding personnel know exactly which device or devices have been activated.

### Zoned Coded Alarm

In addition to having all the features of a zoned alarm system, a zoned coded alarm system will also indicate which zone has been activated over the announcement system. Hospitals often use this type of system, because it is not possible to evacuate all staff and patients for every fire alarm. The audible notification devices give a numbered code that indicates which zone was activated. A code list tells building personnel which zone is in an alarm condition and which areas must be evacuated.

More modern systems of this type use speakers as the alarm notification devices. This enables a voice message indicating the nature and location of the alarm to accompany the audible alarm signal.

### Master-Coded Alarm

In a master-coded alarm system, the audible notification devices for fire alarms are also used for other purposes. For example, a school may use the same bell to announce a change in classes, to signal a fire alarm, to summon the janitor, or to make other notifications. Most of these systems have been replaced by modern speaker systems that use the temporal-3 pattern fire alarm signal and have public address capabilities. This type of system is not often installed in new buildings or facilities.

## Fire Brigade Notification

The fire brigade should always be notified when a fire alarm system is activated. In some cases, a person must make a telephone call to the fire brigade. Or, the fire alarm system can be connected directly to the fire brigade or to a proprietary central station where someone on duty calls the fire brigade. As shown in ▼Table 25-2, fire alarm systems can be classified in four categories, based on how the fire brigade is notified of an alarm.

Figure 25-14 Buildings with a local alarm system should post notices requesting occupants to call the plant emergency number and report the alarm after they exit. Shown here is a sprinkler water motor gong, a flow alarm that signifies there is water flowing in the system.

### Local Alarm System

A local alarm system does not notify the fire brigade. The alarm sounds only in the building to notify the occupants. Buildings with this type of system should have notices posted requesting occupants to call the fire brigade and plant emergency number and report the alarm after they exit (►Figure 25-14).

### Remote Station System

A remote station system sends a signal directly to the fire brigade or to another monitoring location via a telephone line or

**Table 25-2 Fire Brigade Notification Systems**

| Type of System | Description |
|---|---|
| Local Alarm | The fire alarm system sounds an alarm only in the building where it was activated. No signal is sent out of the building. Someone must call the fire brigade to respond. |
| Remote Station | The fire alarm system sounds an alarm in the building and transmits a signal to a remote location. The signal may go directly to the fire brigade or to another location where someone is responsible for calling the fire brigade, such as the security office. |
| Proprietary System | The fire alarm system sounds an alarm someplace in the facility and transmits a signal to a monitoring location owned and operated by the facility's owner. Depending upon the nature of the alarm and arrangements with the local fire brigade, facility personnel may respond and investigate, or the alarm may be immediately retransmitted to the municipal fire department. These facilities are monitored 24 hours a day. |
| Central Station | The fire alarm system sounds an alarm in the building and transmits a signal to an off-premises alarm monitoring facility. The off-premises monitoring facility is then responsible for notifying the fire brigade and/or outside fire department to respond. |

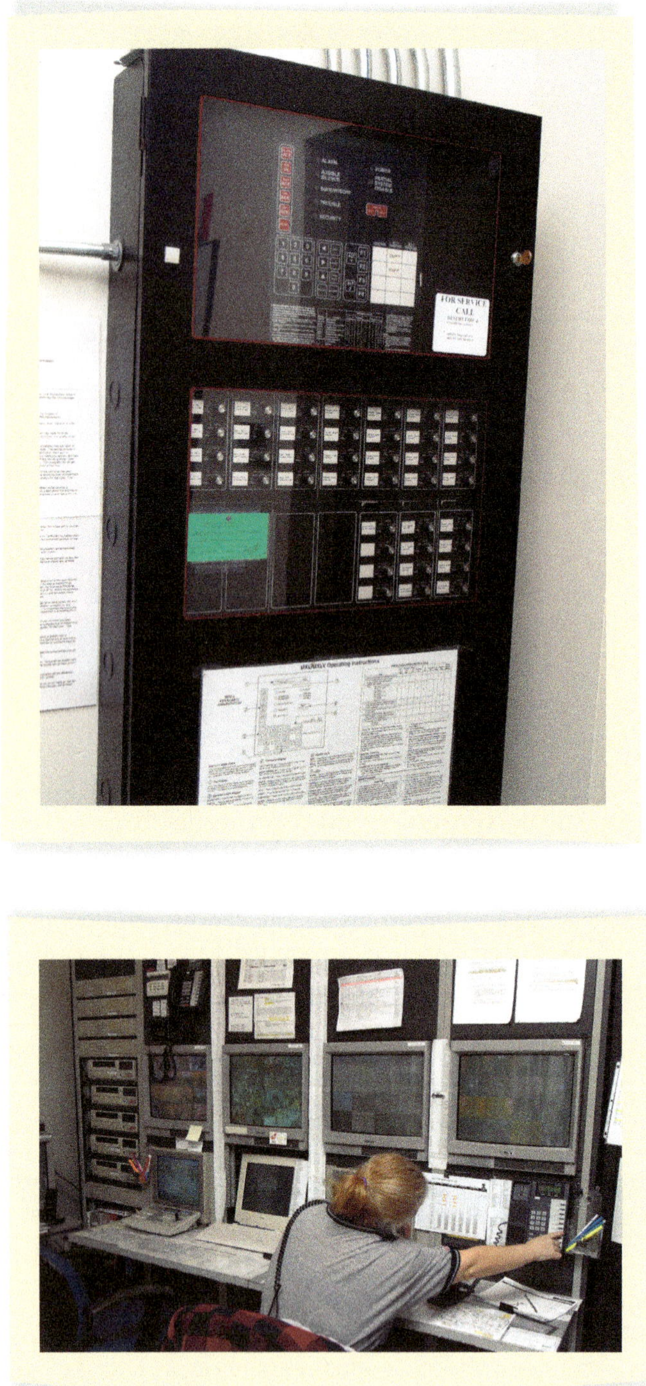

Figure 25-15 In a proprietary system, fire alarms from several buildings are connected to a single monitoring site owned and operated by the buildings' owner.

a radio signal. This type of direct notification can be installed only in facilities where the fire brigade is equipped to handle direct alarms. If the signal goes to a monitoring location, that site must be continually staffed by someone who will notify the fire brigade.

Figure 25-16 A central station monitors alarm systems at many locations.

## Proprietary System

In a proprietary system, the building's fire alarms within the plant are connected directly to a monitoring site that is owned and operated by the building owner. Proprietary systems are often installed at facilities with multiple buildings belonging to the same owner, such as universities or industrial complexes. Each building is connected to a monitoring site on the premises, usually the security center, which is staffed at all times (◀ Figure 25-15). When an alarm sounds, the staff at the monitoring site report the alarm to the fire brigade and may call the municipal fire department by telephone or a direct line.

## Central Station

A central station is a third-party, off-site monitoring facility that monitors multiple alarm systems. Facility owners contract and pay the central station to monitor their facilities (▲ Figure 25-16). An activated alarm transmits a signal to the central station by telephone or radio. Personnel at the central station then notify the appropriate fire brigade or municipal fire department of the fire alarm. The central station facility may be located in the same city as the facility or in a different part of the country.

Usually, building alarms are connected to the central station through leased or standard telephone lines. However, the use of either cellular telephone frequencies or radio frequencies is becoming more common. Cellular or radio connections may be used to back up regular telephone lines in case they fail; in remote areas without telephone lines, they may be the primary transmission method.

Brigades need to develop SOPs that identify the procedures for investigating alarms. The response to all alarms—whether real or false, unwanted, and nuisance—needs to be consistent. In some cases, facility first responders such as security guards have been injured or even killed because they investigated activated fire alarms and were not prepared for the actual fires that they encountered.

Recurring and nuisance alarms may be system maintenance issues, which must be addressed because they consume brigade members' time and other resources.

## Fire Suppression Systems

Fire suppression systems include automatic sprinkler systems, standpipe systems, and specialized extinguishing systems such as dry chemical systems. Construction for most new industrial facilities incorporates at least one of these systems.

Understanding how these systems work is important because they can affect fire behavior. In addition, brigade members should know how to interface with the system and how to shut down a system to prevent unnecessary damage.

### Automatic Sprinkler Systems

The most common type of fire suppression system is the automatic sprinkler system. Automatic sprinklers are reliable and effective, with a history of more than 100 years of successfully controlling fires. When properly installed and maintained, automatic sprinkler systems can help control fires and protect lives and property.

In most automatic sprinkler systems, the sprinkler heads open one at a time as they are heated to their operating temperature. Usually, only one or two sprinkler heads open before the fire is controlled.

The basic operating principles of an automatic sprinkler system are simple. A system of water pipes is installed throughout a building to deliver water to every area where a fire might occur. Depending on the design and occupancy of the building, the pipes may be above or below the ceiling.

Automatic sprinkler heads are located along the system of pipes. Each sprinkler head covers a particular floor area. A fire in that area will activate the sprinkler head, which discharges water on the fire. It is like having a brigade member in every room with a charged hose line, just waiting for a fire.

One of the major advantages of a sprinkler system is that it can function as both a fire detection system and a fire suppression system. An activated sprinkler head not only discharges water on the fire, but it also triggers a water-motor gong, a flow alarm that signifies there is water flowing in the system. In addition, the system prompts electric flow switches to activate the building's fire alarm system, notifying the fire brigade and the occupants. The system is so effective that by the time brigade members arrive, the sprinklers have often already extinguished the fire, or at least controlled it.

#### Automatic Sprinkler System Components

The overall design of automatic sprinkler systems can be complex, especially in large industrial facilities. However, even the largest systems have just four major components: the automatic sprinkler heads, piping, control valves, and a water supply, which may or may not include a fire pump (▲ Figure 25-17).

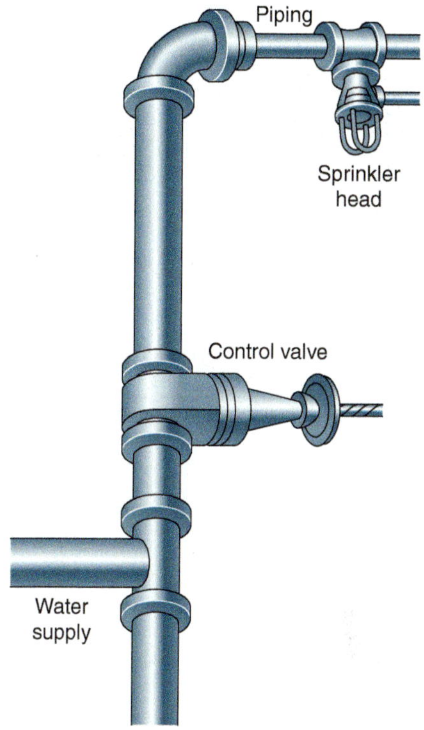


Figure 25-17 The basic components of an automatic sprinkler system include sprinkler heads, piping, control valves, and a water supply.

#### Automatic Sprinkler Heads

Automatic sprinkler heads, commonly referred to as sprinkler heads or just heads, are the working ends of a sprinkler system. In most systems, the heads serve two functions: they activate the sprinkler system and they apply water to the fire. Sprinkler heads are composed of a body, which includes the orifice (opening); a release mechanism that holds a cap in place over the orifice; and a deflector that directs the water in a spray pattern (► Figure 25-18). Standard sprinkler heads have a $\frac{1}{2}$-inch orifice, but several other sizes are available for special applications.

Although sprinkler heads come in several styles, they are all categorized according to the type of release mechanism and the intended mounting position—upright, pendant, or horizontal. They are also rated according to their release temperature. The release mechanisms hold the cap in place until the release temperature is reached. At that point, the mechanism is released, and the water pushes the cap out of the way as it discharges onto the fire.

Fusible link sprinkler heads use a metal alloy, such as solder, that melts at a specific temperature (► Figure 25-19).

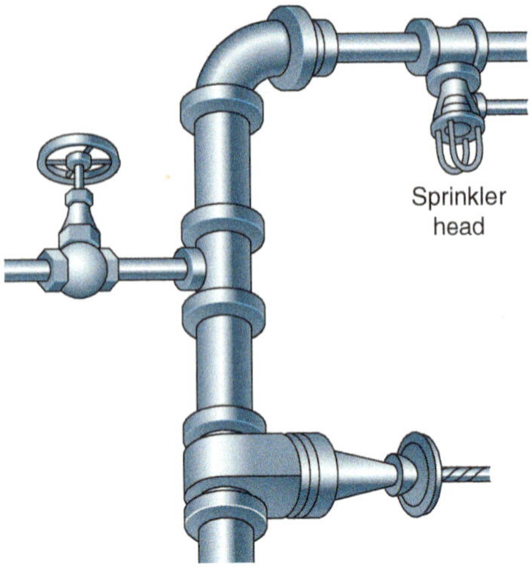


Figure 25-18 Automatic sprinkler heads have a body with an opening, a release mechanism, and a water deflector.

Figure 25-19 Fusible link sprinkler heads use two pieces of metal linked together by an alloy such as solder.

The alloy links two pieces of metal that keep the cap in place. When the designated operating temperature is reached, the solder melts and the link breaks, releasing the cap. Fusible link sprinkler heads come in a wide range of styles and temperature ratings.

Frangible bulb sprinkler heads use a glass bulb filled with glycerin or alcohol to hold the cap in place ▶ Figure 25-20. The bulb also contains a small air bubble. As the bulb is heated, the liquid absorbs the air bubble and expands until it breaks the glass, releasing the cap. The volume and composition of the liquid and the size of the air bubble determine the temperature at which the head activates, as well as how quickly it responds.

A frangible-pellet sprinkler head has a rod between the orifice cap and the sprinkler frame ▶ Figure 25-21. The rod is held in place by a pellet of solder under compression. When the solder melts, the rod moves out of the way of the orifice cap, which is then pushed off by the water pressure.

Figure 25-20 Frangible bulb sprinkler heads activate when the liquid in the bulb expands and breaks the glass.

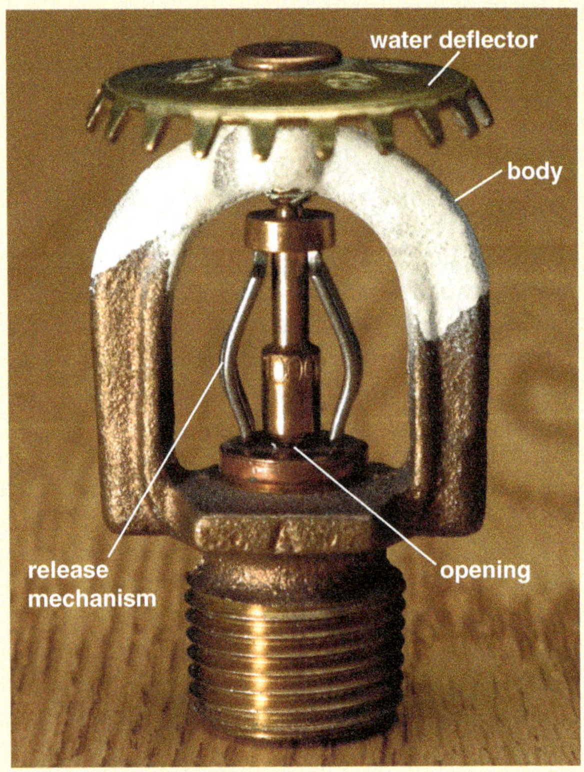


Figure 25-21 Frangible-pellet sprinkler head.

### Special Sprinkler Heads

Sprinkler heads can also be designed for special applications such as covering large areas or discharging the water in extra-large droplets or as a fine mist. Some sprinkler heads have protective coatings to help prevent corrosion. Builders and

Figure 25-22 An early suppression fast response (ESFR) sprinkler head.

Figure 25-23 A deluge sprinkler head has no release mechanism.

installers should consider these characteristics when designing the system and selecting appropriate heads. It is important to ensure that the proper heads are installed and any replacement heads are of the same type.

Early suppression fast response (ESFR) sprinkler heads have improved heat collectors to speed up the response and ensure rapid release ▲ Figure 25-22. They are used in large warehouses and distribution facilities where early fire suppression is important. These heads often have large orifices to discharge water onto a fire and often times require a minimum of 2-inch piping.

### Deluge Heads

Deluge heads are easily identifiable, because they have no cap or release mechanism. The orifice is always open ▶ Figure 25-23. Deluge heads are used only in deluge sprinkler systems, which are covered later in this chapter.

### Temperature Ratings

Sprinkler heads are rated according to their release temperature. A typical rating for sprinkler heads in a light hazard occupancy, such as an office building, would be 165°F (74°C). Sprinkler heads that are used in areas with warmer ambient air temperatures would have higher ratings. The rating should be stamped on the body of the sprinkler head. Frangible bulb sprinkler heads use a color-coding system to identify the temperature rating ▼ Table 25-3. Some fusible link and chemical-pellet sprinklers also use this system.

The temperature rating on a sprinkler head must match the anticipated ambient air temperatures. If the rating is too

**Table 25-3 Temperature Rating Determined by Color of Sprinkler Head**

| Maximum Ceiling Temperature (°F) | Temperature Rating (°F) | Color Code | Glass Bulb Colors |
|---|---|---|---|
| 100 | 135–170 | Uncolored or black | Orange or red |
| 150 | 175–225 | White | Yellow or green |
| 225 | 250–300 | Blue | Blue |
| 300 | 325–375 | Red | Purple |
| 375 | 400–475 | Green | Black |
| 475 | 500–575 | Orange | Black |
| 625 | 650 | Orange | Black |

Figure 25-24 Spare sprinkler heads should be kept in a special box near the main sprinkler system valve with a sprinkler head wrench.

low for the ambient air temperature, accidental alarms may occur. Conversely, if the rating is too high, the system will be slow to react to a fire, and the fire will be able to establish itself and grow before the system activates.

Spare heads that match those used in the system should always be available on site. Usually the spare heads are kept in a clearly marked box near the main control valve ▲ Figure 25-24. Having spare heads handy enables sprinkler systems to be returned to full service quickly, whether they were set off by a fire or by accident.

### Mounting Position

Sprinkler heads with different mounting positions are not interchangeable. Each mounting position has deflectors specifically designed to produce an effective water stream down or out toward the fire. Each automatic sprinkler head is designed to be mounted in one of three positions ▼ Figure 25-25:

- Pendant sprinkler heads are designed to be mounted on the underside of the sprinkler piping, hanging down toward the room. Pendant heads are commonly marked SSP, which stands for Standard Spray Pendant.
- Upright sprinkler heads are designed to be mounted on top of the supply piping as the name suggests. Upright heads are usually marked SSU, for Standard Spray Upright.
- Sidewall sprinkler heads are designed for horizontal mounting, projecting out from a wall and are usually marked "Sidewall."

### Old Style Versus New Style Sprinkler Heads

Up until the early 1950s, deflectors in both pendant and upright sprinkler heads directed part of the water stream up toward the ceiling. It was believed that this helped cool the area up and around the ceiling as well as extinguish or control the fire. Sprinkler heads with this design are called old style sprinklers. There are still many old style heads in service today.

Automatic sprinklers manufactured after the mid-1950s deflect the entire water stream down to the fire. These types of heads are referred to as new style heads or standard spray heads. New style heads can replace old style heads but the reverse is not true. Due to different coverage patterns, old style heads should not be used to replace any new style heads.

## Sprinkler Piping

Sprinkler piping, the network of pipes that delivers water to the sprinkler heads, includes the main water supply lines, risers, feeder lines, and branch lines. Sprinkler pipes are usually

A B C

Figure 25-25 Sprinkler head mounting positions. A. Upright B. Pendant C. Sidewall.

Figure 25-26 Steel pipe is commonly used in industrial sprinkler systems.

made of steel, but other metals can be used (▲ Figure 25-26). Plastic pipe is sometimes used in residential sprinkler systems.

Sprinkler system designers use piping schedules or hydraulic calculations to determine the size of pipe and the layout of the "grid." Most new systems are designed using computer software. Near the main control valve, pipes have a large diameter; as the pipes approach the sprinkler heads, the diameter generally decreases.

## Valves

A sprinkler system includes several different valves such as the main water supply control valve, the alarm valve, and other, smaller valves used for testing and service. Many large systems have zone valves so the water supply to different areas can be shut down without turning off the entire system. All of the valves play a critical role in the design and function of the system.

### Water Supply Control Valves

Every sprinkler system must have at least one main control valve that allows water to enter the system. This water supply control valve must be of the "indicating" type, meaning that the position of the valve itself indicates whether it is open or closed. Four primary types of control valves are available:

- The outside stem and yoke valve (OS&Y) has a stem that moves in and out as the valve is opened or closed (◄ Figure 25-27). If the stem is out, the valve is open. OS&Y valves are often found in a mechanical room in the building, where water to supply the sprinkler system enters the building. In warmer climates, OS&Y valves may be found outside.
- The postindicator valve (PIV) has an indicator that reads either open or shut depending on its position (▼ Figure 25-28). A PIV is usually located in an open area outside the building and controls an underground

Figure 25-27 An outside stem and yoke (OS&Y) valve is often used to control the flow of water into a sprinkler system.

Figure 25-28 A postindicator valve (PIV) is used to open or close an underground valve.

Figure 25-29 A wall postindicator valve controls the flow of water from an underground pipe into a sprinkler system.

Figure 25-30 A tamper switch activates an alarm if someone attempts to close a valve that should remain open.

valve. Opening or closing a PIV requires a wrench, which is usually attached to the side of the valve and locked in the open position.

- The wall postindicator valve (WPIV) is similar to a PIV but is designed to be mounted on the outside wall of a building (▲ Figure 25-29).
- The indicating butterfly valve (IBV) is equipped with a directional arrow to indicate the position of the valve. A gear operator prevents the valve from being closed in less than 5 seconds, preventing a water hammer.

The main control valve must be supervised in the open position. This procedure ensures that the water supply to the sprinkler system is never shut off unless the proper notification is made that the system is out of service. It is critically important that the sprinkler system is always charged with water and ready to operate if needed.

Methods commonly used to secure the control valve include the use of a lock, a lock and chain, or tamper seals. When tamper seals are used, the system must be inspected and documented weekly.

An alternative to locking the valves open is equipping them with tamper switches (► Figure 25-30). These devices monitor the position of the valves. If someone closes the valve, the tamper switch sends a signal to the fire alarm control panel indicating a change in valve position. If this change has not been authorized, the cause of the signal can be investigated and the problem corrected.

### Main Sprinkler System Valves

The type of main sprinkler system valve used depends on the type of sprinkler system installed. Options include an alarm valve, a dry-pipe valve, or a deluge valve. These valves are usually installed on the main riser, above the water supply control valve.

The primary functions of an alarm valve are to signal an alarm when a sprinkler head is activated and to prevent nuisance alarms caused by pressure variations and surges in the water supply to the system. The alarm valve has a clapper mechanism that remains in the closed position until a sprinkler head opens. The closed clapper prevents water from flowing out of the system and back into the public water mains when water pressure drops.

When a sprinkler head is activated, the clapper opens fully and allows water to flow freely through the system. The open clapper also allows water to flow to the water-motor gong, sounding an alarm. Electrical flow or pressure switches activate connections to external alarm systems.

Without a properly functioning alarm valve, sprinkler system flow alarms would occur frequently. The normal pressure changes and surges in a public water system will not open the clapper. This prevents water from flowing to the water-motor gong or tripping the electrical water flow switches.

In dry-pipe and deluge systems, the main valve functions both as an alarm valve and as a dam, holding back the water until the sprinkler system is activated. When the system is activated, the valve opens fully so water can enter the sprinkler piping. Both dry-pipe and deluge systems are described later in this chapter.

### Additional Valves

Sprinkler systems are equipped with various other control valves. Several smaller valves are usually located near the main control valve, with others located elsewhere in the building. These smaller valves include drain valves, test valves, and connections to alarm devices. All of these valves should be properly labeled.

In larger facilities, the sprinkler system may be divided into zones. Each zone has a valve that controls the flow of water to that particular zone. This design makes maintenance easy and also is valuable when a fire occurs. After the fire is extinguished, water flow to the affected area can be shut off so that the activated heads can be replaced. Fire protection in the rest of the building is unaffected by the shutdown.

## Water Supplies

The water used in an automatic sprinkler system may come from a municipal water system, from on-site storage tanks, or from static water sources such as storage ponds or rivers. Whatever the source, the water supply must be able to handle the demand of the sprinkler system, as well as the needs of the fire brigade in the event of a fire.

The preferred water source for a sprinkler system is a municipal water supply, if one is available. If the municipal supply cannot meet the water pressure and volume requirements of the sprinkler system, alternative supplies must be provided.

Fire pumps are often used on large facilities when the water comes from a static source. They may also be used to boost the pressure in some sprinkler systems, particularly for tall buildings (▶ Figure 25-31). Fire pumps will usually turn on automatically when the sprinkler system activates or when the pressure drops to a preset level.

Figure 25-31 In large industrial sites, a fire pump may be needed to maintain appropriate pressure in the sprinkler system.

A large industrial complex could have more than one water source, such as a municipal system and a backup storage tank (▶ Figure 25-32). Multiple fire pumps can provide water to the sprinkler and standpipe systems in different areas through underground pipes. Private hydrants may also be connected to the same underground system.

Figure 25-32 An elevated storage tank ensures that there will be sufficient water and adequate pressure to fight a fire.

Each sprinkler system should also have a fire department connection (FDC). This connection allows the brigade's engine to pump water into the sprinkler system (▶ Figure 25-33). The FDC is used as either a supplement or the main source of water to the sprinkler system if the regular supply is interrupted or a fire pump fails.

The FDC usually has two or more 2½-inch female couplings or one large-diameter hose coupling mounted on an outside wall or placed near the building. It ties directly into the sprinkler system after the main control valve or alarm valve. Each fire brigade should have a standard procedure for first-arriving companies. The procedure should specify how to connect to the FDC and when to charge the system.

In large facilities, a single FDC may be used to deliver water to all fire protection systems in the complex. The water from the FDC flows into the private underground water mains instead of into each system. Water pumped into this type of FDC should come from a source that does not service the complex, such as a public hydrant on a different grid.

Figure 25-34 A water-motor gong sounds when water is flowing in a sprinkler system.

Figure 25-33 A fire department connection delivers additional water and boosts the pressure in a sprinkler system.

Consider pumping "finished foam" into your FDCs to feed sprinkler and standby systems.

### Water Flow Alarms

All sprinkler systems should be equipped with a method for sounding an alarm whenever there is water flowing in the pipes. This is important both in cases of an actual fire, and in cases of accidental activation. Without these alarms, the occupants or the fire brigade might not be aware of the sprinkler activation. If a building is unoccupied, the sprinkler system could continue to discharge water long after a fire is extinguished, resulting in excessive water damage.

Most systems incorporate a mechanical flow alarm called a water-motor gong (▲ Figure 25-34). When the sprinkler system is activated and the main alarm valve opens, some water is fed through a pipe to a water-powered gong located on the outside of the building. This alerts people outside the building that there is water flowing. This type of alarm will function even if there is no electricity.

Accidental soundings of water-motor gongs are rare. If a water-motor gong is sounding, water is probably flowing from the sprinkler system somewhere in the building. Fire companies who arrive and hear the distinctive sound of a water-motor gong know that there is a fire or that something else is causing the sprinkler system to flow water.

Most modern sprinkler systems also are connected to the building's fire alarm system by either an electric **flow switch** or a pressure switch. This connection will trigger the alarm to alert the building's occupants. A monitored system also will notify the fire brigade. Unlike water-motor gongs, flow and pressure switches can be accidentally triggered by water pressure surges in the system. To reduce the risk of accidental activations, a **retard chamber** may be installed. The retard chamber is a valve accessory that collects excess water associated by pressure surges. In systems subject to pressure surges, the retarding chamber collects excess water from the alarm valve and feeds the water back into the main drain before activating the alarm.

## Types of Automatic Sprinkler Systems

Automatic sprinkler systems are divided into four categories, depending on what type of sprinkler head is used, and how the system is designed to activate. The four categories of sprinkler systems are:

- Wet sprinkler systems
- Dry sprinkler systems
- Preaction sprinkler systems
- Deluge sprinkler systems

Many buildings may use the same type of system to protect the entire facility; it is not uncommon, however, to see two or three systems combined in one building. Some facilities use a wet sprinkler system to protect most of the structure, but will have a dry sprinkler or preaction system in a specific area. In many cases, a dry sprinkler or preaction system will branch off from the wet sprinkler system.

### Wet Sprinkler Systems

A wet sprinkler system is the most common and the least expensive type of automatic sprinkler system. As the name implies, the piping in a wet system is always filled with water. As a sprinkler head activates, water is immediately discharged onto the fire. The major drawback to a wet sprinkler system is that it cannot be used in areas where temperatures drop below freezing. They will also flow water if a sprinkler head is accidentally opened or a leak occurs in the piping.

If only a small, unheated area needs to be protected, two options are available. A dry-pendant sprinkler head can be used in very small areas, such as walk-in freezers (▶ Figure 25-35). The bottom part of a dry-pendant head, which resembles a standard sprinkler head, is mounted inside the freezer. The head has an elongated neck, usually 6- to 18-inch long, that extends up and connects to the wet sprinkler piping in the heated area above the freezer. The vertical neck section is filled with air and capped at each end. The top cap prevents water from entering the lower section where it would freeze. The bottom cap acts just like the cap on a standard sprinkler head. When the head is activated and the lower cap drops out, a device inside the neck releases the upper cap, so water can flow down. The entire dry-pendant head assembly must be replaced after it has been activated.

Larger unheated areas, such as a loading dock, can be protected with an antifreeze loop. An antifreeze loop is a small section of the wet sprinkler system that is filled with glycol or glycerin instead of water. A check valve separates the antifreeze loop from the rest of the sprinkler system. When a sprinkler head in the unheated area is activated, the antifreeze sprays out first, followed by water.

### Dry Sprinkler Systems

A dry sprinkler system operates much like a wet sprinkler system, except that the pipes are filled with pressurized air instead of water. A dry-pipe valve keeps water from entering the pipes until the air pressure is released. Dry systems are used in large facilities that may experience below-freezing temperatures, such as unheated warehouses or garages.

Figure 25-35 A dry-pendant sprinkler head can be used to protect a freezer box.

The air pressure is set high enough to hold a clapper inside the dry-pipe valve in the closed position (▶ Figure 25-36). When a sprinkler head opens, the air escapes. As the air pressure drops, the water pressure on the other side of the clapper forces it open and water flows into the pipes. When the water reaches the open sprinkler head, it is discharged onto the fire.

Dry sprinkler systems do not eliminate the risk of water damage from accidental activation. If a sprinkler head breaks, the air pressure will drop and water will flow, just as in a wet sprinkler system.

The clapper assembly inside most dry-pipe valves works on a pressure differential. The system or air side of the clapper has a larger surface area than the supply or wet side. In this way, a lower air pressure can hold back a higher water pressure. A small compressor is used to maintain the air pressure in the system.

Dry sprinkler systems should have an air pressure alarm to alert building personnel if the air pressure drops. This could mean that the air source is not working, or that there is an air leak in the system. If the air pressure in the system is too low, the clapper will open, and the system will fill with water. At that point, the system would essentially be a wet sprinkler system, which could freeze in low temperatures. The system would have to be drained and reset to prevent the pipes from freezing.

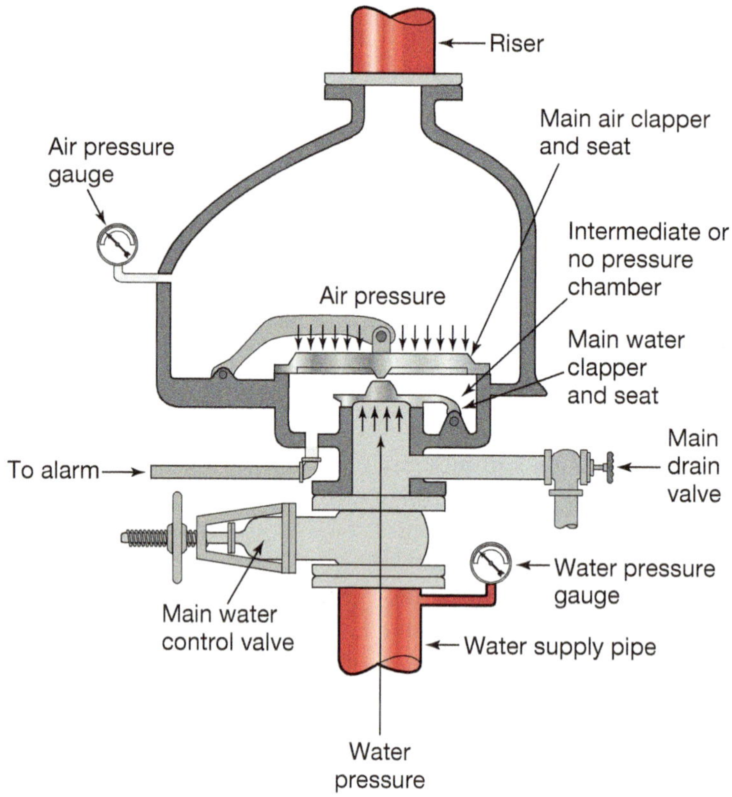


**Figure 25-36** Water pressure on one side of the dry-pipe valve is balanced by air pressure on the other side.

Dry sprinkler systems must be drained after every activation so the dry-pipe valve can be reset. The clapper also must be reset, and the air pressure must be restored before the water is turned back on.

### *Accelerators and Exhausters*

One problem encountered in dry sprinkler systems is the delay between the activation of a sprinkler head and the actual flow of water out of the head. The pressurized air that fills the system must escape through the open head before the water can flow. For personal safety and property protection reasons, any delay longer than 90 seconds is unacceptable. Large systems, however, can take several minutes to empty of air and refill with water. To compensate for this problem, two additional devices are used: accelerators and exhausters.

An accelerator is installed at the dry-pipe valve. The rapid drop in air pressure caused by an open sprinkler head triggers the accelerator, which allows air pressure to flow to the supply side of the clapper valve. This quickly eliminates the pressure differential, opening the dry-pipe valve and allowing the water pressure to force the remaining air out of the piping.

An exhauster is installed on the system side of the dry-pipe valve, often at a remote location in the building. Like an accelerator, the exhauster monitors the air pressure in the piping. If it detects a drop in pressure, it opens a large-diameter portal, so the air in the pipes can escape. The exhauster closes when it detects water, diverting the flow to the open sprinkler heads. Large systems may have multiple exhausters located in different sections of the piping.

### Preaction Sprinkler Systems

A preaction sprinkler system is similar to a dry sprinkler system with one key difference. In a preaction sprinkler system, a secondary device, such as a smoke detector or a manual-pull alarm, must be activated before water is released into the sprinkler piping. When the system is filled with water, it functions as a wet sprinkler system.

A preaction system uses a deluge valve instead of a dry-pipe valve. The deluge valve will not open until it receives a

**Brigade Member Tips**

Some dry-pipe valves will not open if there is water above the **clapper valve**, a mechanical device that allows the water to flow in only one direction. Unless the main valve is designed to operate as either a wet- or dry-pipe valve, a dry system that has been filled with water should be immediately drained and reset for dry-pipe operation.

signal that the secondary device has been activated. Because a detection system usually will activate more quickly than a sprinkler system, water in a preaction system will generally reach the sprinklers before a head is activated.

The primary advantage of a preaction sprinkler system is in preventing accidental water discharges. If a sprinkler head is accidentally broken or the pipe is damaged, the deluge valve will prevent water from entering the system. This makes preaction sprinkler systems well-suited for locations where water damage is a major concern, such as computer rooms, libraries, and museums.

### Deluge Sprinkler Systems

A **deluge sprinkler system** is a type of dry sprinkler system in which water flows from all of the sprinkler heads as soon as the system is activated (► Figure 25-37). A deluge system does not have closed heads that open individually at the activation temperature; all of the heads in a deluge system are always open.

Deluge systems can be activated in three ways. A detection system can release the deluge valve when a detector is activated. The deluge system can also be connected to a separate pilot system of air-filled pipes with closed sprinkler heads. When a head on the pilot line is activated, the air pressure drops, opening the deluge valve. Most deluge valves can be released manually as well.

Deluge systems are found in industry in areas such as aircraft hangars or industrial processes, where rapid fire suppression is critical. In some cases, foam concentrate added to the water will discharge a foam blanket over the hazard. Deluge systems are also used for special hazard applications, such as liquid propane gas loading stations. In these situations, a heavy deluge of water is needed to protect exposures from a large fire that occurs very rapidly.

## Shutting Down Sprinkler Systems

Responding fire brigades should know where and how to shut down any automatic sprinkler system if needed. If the system is accidentally activated, it should be shut down as soon as possible to avoid excessive water damage.

In an actual fire, the order to shut down the sprinkler system should come only from the Incident Commander. Generally, the sprinkler system should not be shut down until the fire is completely extinguished. However, if the system is damaged and is wastefully discharging water, shutting down the sprinkler system could make more water available to fight the fire manually.

When the system is shut down, a brigade member with a portable radio should stand by at the control valve, in case it has to be reopened quickly.

In most cases, the entire sprinkler system can be shut down by closing the main control valve. In a zoned system, the affected zone can be shut down by closing the appropriate valve. The rest of the system can then remain operational while the activated heads are being replaced.

Placing a wooden wedge or a commercial sprinkler stopper into the sprinkler head can quickly stop the flow of water (► Figure 25-38), although this will not work with all types of heads. This shuts off the flow of water until the control valve can be located and shutdown. After the main valve is closed, opening the drain valve allows the system to drain so that the activated head can be replaced.

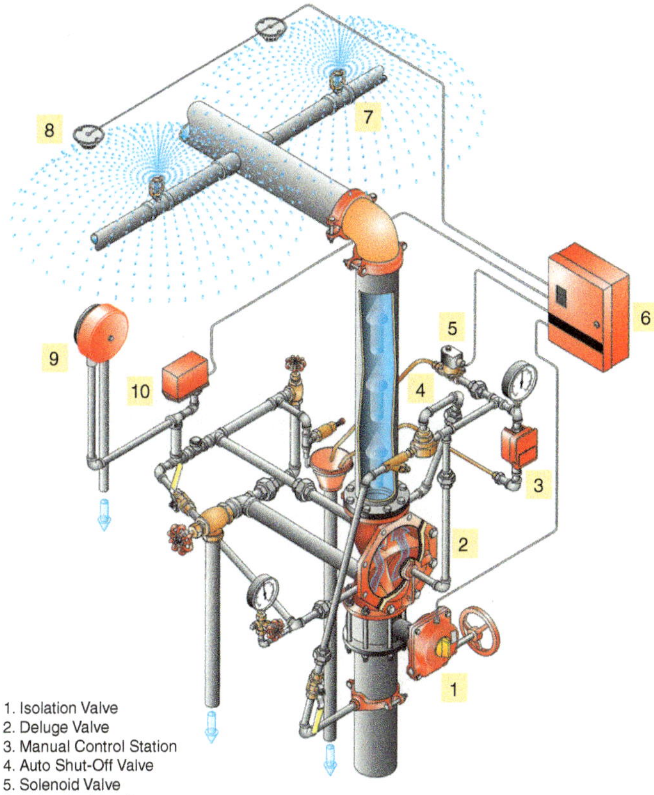


1. Isolation Valve
2. Deluge Valve
3. Manual Control Station
4. Auto Shut-Off Valve
5. Solenoid Valve
6. Releasing Panel
7. Open Nozzle
8. Smoke/Heat Detector
9. Water-Motor Alarm Gong
10. Pressure Alarm Switch

Figure 25-37 Water flows from all of the heads in a deluge system as soon as the system is activated.

**Brigade Member Safety Tips**

Do not shut down the water to a sprinkler system without orders from the Incident Commander. The fire must be completely out and hose lines must be available should the fire reignite. A brigade member should be stationed at the valve with a portable radio, ready to reopen the valve if necessary.

## Standpipe Systems

A **standpipe system** is a network of pipes and outlets for fire hoses built into a structure to provide water for firefighting purposes. Standpipe systems are usually used in high-rise buildings, although they are found in many large industrial facilities as well. At set intervals throughout the building,

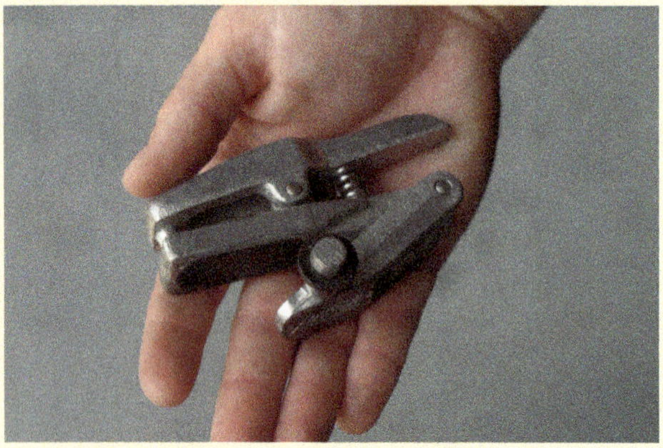

Figure 25-38 A sprinkler stopper can be used to stop the flow of water from a sprinkler head that has been activated.

there will be a valve where brigade members can connect a hose to the standpipe ▶ Figure 25-39.

Standpipes are found in buildings with and without sprinkler systems. In many newer buildings, sprinklers and standpipes are combined in one system. Older buildings more commonly have separate sprinkler and standpipe systems.

The three categories of standpipes—Class I, Class II, and Class III—are defined by their intended use.

Figure 25-39 Standpipe outlets allow fire hoses to be connected inside a building.

## Class I Standpipe

A Class I standpipe is designed for use by fire brigade personnel only. Each outlet has a 2½-inch male coupling and a valve to open the water supply after the hose is connected ▶ Figure 25-40. Often, the connection is located inside a cabinet, which may or may not be locked. Responding fire personnel carry the hose into the building with them, usually in some sort of roll, bag, or backpack. A Class I standpipe system must be able to supply an adequate volume of water with sufficient pressure to operate fire brigade attack lines.

## Class II Standpipes

A Class II standpipe is designed for use by fire departments during initial response and trained personnel only. The outlets are generally equipped with a length of 1½-inch single-jacket hose preconnected to the system ▶ Figure 25-41. Class II standpipe outlets are frequently connected to the domestic water piping system in the building rather than an outside main or a separate system. Instead of using equipment that may not be reliable or adequate, municipal brigade members usually use their own issued equipment.

## Class III Standpipes

A Class III standpipe has the features of both Class I and Class II standpipes in a single system. They have 2½-inch outlets for trained personnel use as well as smaller outlets with attached hoses for occupant use. The occupant hoses have been removed, either intentionally or by vandalism, in many facilities, so the system basically becomes a Class I system.

## Water Flow in Standpipe Systems

Standpipes are designed to deliver a minimum amount of water at a particular pressure to each floor. The design requirements depend on the code requirements in effect when the building was constructed. The actual flow also depends on the water supply, as well as on the condition of the piping system and fire pumps.

Flow-restriction devices ▶ Figure 25-42 or pressure-reducing valves ▶ Figure 25-43 are often installed at the outlets to limit the pressure and flow. A vertical column of water, such as the water in a standpipe riser, exerts a backpressure (also called head pressure). In a tall building, this backpressure can be hundreds of pounds per square inch (psi) at lower floor levels. If a hose line is connected to an outlet without a flow restrictor or a pressure-reducing valve, the water pressure could rupture the hose and excessive nozzle pressure could make the line difficult or dangerous to handle. Building and fire codes limit the height of a single riser and may also require the installation of pressure-reducing valves on lower floors.

If they are not properly installed and maintained, these devices can cause problems for brigade members. An

Figure 25-40 A Class I standpipe provides water for structural fire brigade hose lines.

Figure 25-41 A Class II standpipe is intended to be used by trained personnel to attack incipient-stage fires.

Figure 25-42 A flow-restriction device on a standpipe outlet can cause problems for brigade members if not properly installed.

Figure 25-43 A pressure-reducing valve on a standpipe outlet may be necessary on lower floors to avoid problems caused by backpressure.

improperly adjusted pressure-reducing valve could severely restrict the flow to a hose line. A flow restriction device could also limit the flow of water to fight a fire.

The flow and pressure capabilities of a standpipe system should be determined during preincident planning. Many standpipe systems deliver water at a pressure of only 65 psi at the top of the building. The combination fog-and-straight-stream nozzles used by many fire brigades are designed to operate at 100 psi. As a result, many fire brigades use low-pressure combination nozzles for fighting fires in multistory buildings or require the use of only smooth-bore nozzles when operating from a standpipe system. Current standards require 100 psi at the top outlets.

Preincident plans should include an evaluation of the building's standpipe system and a determination of the anticipated flows and pressures. This information should be used to make decisions about the appropriate nozzles and tactics for those buildings.

Fire brigades that respond to buildings equipped with standpipes should carry a kit that includes the appropriate hose and nozzle, a spanner wrench, and any required

# Voices of Experience

**"Every one of those tough meetings with management about installing these fixed systems was suddenly well worth the time and effort of doing the research and sitting around the table using every negotiation skill I had ever learned."**

On an ice cold day last December about midday, an alarm was received at our proprietary central station from one of our large process buildings on the plant. This building houses large scale-up labs that use large amounts of flammable liquids (H2 occupancy). The labs have sprinklers, and the walk-in fume hoods have high-pressure water mist sprinkler heads and automatic total flooding dry chemical systems. The page went out, "Attention all fire brigade members and plant EMS. We are receiving an evacuation pull station from building 500, first floor, east wing." This building is an 80 × 740, four-story, fireproof building. As we pulled out of the firehouse, plant dispatch notified us that in addition to the evacuation alarm, they just received an alarm for a "dry chemical system discharge." No doubt in our minds we "had a job." As the responding incident commander, I told dispatch to notify the municipal fire department.

Upon arrival, we found the building being evacuated. The first due engine ran a 5-inch foam line to the fire department connection, our SOP for any large process building on site with large amounts of flammable liquids. Command was established and we met with the building emergency coordinator, an employee who volunteered to help with the evacuation and accountability in this building. He confirmed a working fire in the lab and that all employees were accounted for and that there were no injuries. The first interior crew started to stretch a line off of the standpipe (which now had foam in it) from the safe side of a set of fire doors. The brigade leader made his way to the door of the lab and reported that the fire was knocked down and appeared to be out. They donned their SCBA and made entry with a charged hose line in order to protect themselves and mop up. The fire was in fact out, extinguished by the dry chemical system. The hood sprinklers never discharged because the actuation temperature was set lower for the dry chemical system than for the sprinkler heads.

As I reported to upper management at the scene that the fire was out, I also told them that the dry chemical system had discharged and worked as designed, and that the cost of initial installation had just paid for itself as well as the inconvenience they experienced semi-annually for maintenance of said systems. They were ecstatic that there were no injuries, that the fire was contained, and that a one-day clean-up would have the lab back in service.

Every one of those tough meetings with management about installing these fixed systems was suddenly well worth the time and effort of doing the research and sitting around the table using every negotiation skill I had ever learned. My 18 years in industrial fire protection and 13 years as chief of emergency services had paid off again, as it had in the past at similar incidents.

Ronald E. Kanterman
Merck & Co., Inc.
Rahway, NJ

adapters. This kit should also include tools to adjust the settings of pressure-reducing valves or to remove restrictors that are obstructing flows.

### Water Supplies

Both standpipe systems and sprinkler systems are supplied with water in essentially the same way. Many wet standpipe systems in facilities are connected to the plant water supply with an electric or diesel fire pump to provide additional pressure. Many of these systems also have water storage tanks. In these systems the FDC on the outside of a building or process area can be used to increase the flow, boost the pressure, or obtain water from an alternative source.

Dry standpipe systems are found in many older buildings. If freezing weather is a problem, such as in open parking structures, bridges, and tunnels, dry standpipe systems are still acceptable. Most dry standpipe systems do not have a permanent connection to a water supply, so the FDC must be used to pump water into the system. If there is a fire in a building with dry standpipes, connecting the hose lines to the FDC and charging the system with water is a high priority.

Some dry standpipe systems are connected to a water supply through a dry-pipe or deluge valve, similar to a sprinkler system. Opening an outlet valve or tripping a switch next to the outlet releases water into the standpipes in these systems.

Multistory buildings often have complex systems of risers, storage tanks, and fire pumps to deliver the needed flows to upper floors. The details of these systems should be obtained during preincident planning surveys. Brigade procedures should dictate how responding units will supply the standpipes with water as well as how crews should use the standpipes inside.

## Specialized Extinguishing Systems

Automatic sprinkler systems are used to protect whole buildings, or at least major sections of buildings. But in certain situations, more specialized extinguishing systems are needed. Specialized extinguishing systems are often used in areas where water would not be an acceptable extinguishing agent (▶ Figure 25-44). For example, water is not the agent of choice for areas containing sensitive electronic equipment or contents such as computers, valuable books, or documents. Water is also incompatible with materials such as flammable liquids or water-reactive chemicals. Areas where these materials are stored or used may have a separate extinguishing system.

**Brigade Member Safety Tips**

When using a Class III standpipe, always connect your hose line to the $2^1/_2$-inch outlet to ensure that you get as much water as possible. The smaller $1^1/_2$-inch outlet may be equipped with a pressure-reducing device.

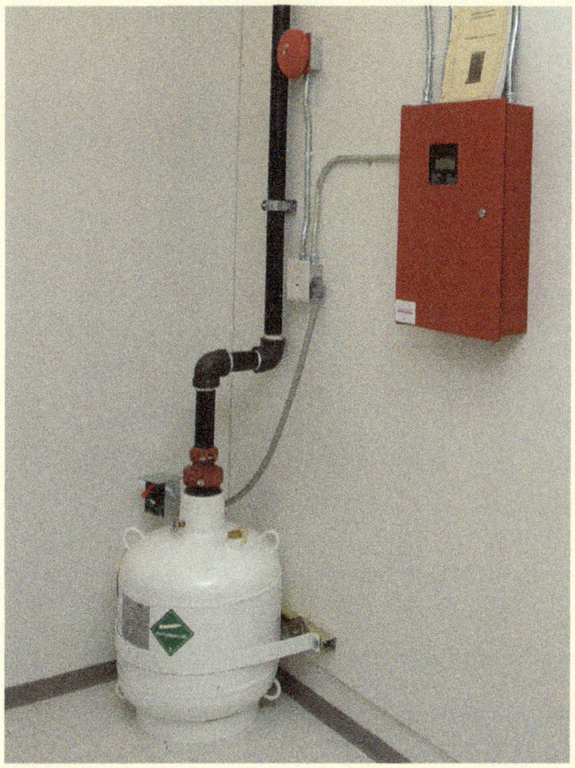

**Figure 25-44** Special extinguishing systems are used in areas where water would not be effective or desirable.

### Dry Chemical and Wet Chemical Extinguishing Systems

Dry chemical and wet chemical extinguishing systems are the most common specialized agent systems. In commercial kitchens, they are used to protect the cooking areas and exhaust systems. Many self-service gas stations have dry chemical systems that protect the dispensing areas. These systems are often installed inside buildings to protect areas where flammable liquids are stored or used. Both dry chemical and wet chemical extinguishing systems are similar in basic design and arrangement.

Dry chemical extinguishing systems use the same types of finely powdered agents as dry chemical fire extinguishers (▶ Figure 25-45). The agent is kept in self-pressurized tanks or in tanks with an external cartridge of carbon dioxide or nitrogen that provides pressure when the system is activated.

Wet chemical extinguishing systems are used in most new commercial kitchens (▶ Figure 25-46). These systems use a proprietary liquid extinguishing agent, which is much more effective on vegetable oils than the dry chemicals used in older kitchen systems. Wet chemical systems are also easier to clean

(Figure 25-45) Dry chemical extinguishing systems are installed at many self-service gasoline filling stations.

(Figure 25-46) Wet chemical extinguishing systems are used in most new commercial kitchens.

up after a discharge, so the kitchen can resume operations more quickly after the system has discharged.

Wet chemical extinguishing agents are not compatible with normal all-purpose dry chemical extinguishing agents. Only wet agents or B:C-rated dry chemical extinguishers should be used where these systems are installed.

Fusible links or other automatic initiation devices are placed above the target hazard to activate both dry chemical and wet extinguishing agent systems (► Figure 25-47). A manual discharge button is also provided so that workers can activate the system if they discover a fire (► Figure 25-48). Open nozzles are located over the target areas to discharge the agent directly onto a fire. When the system is activated, the extinguishing agent flows out of all the nozzles.

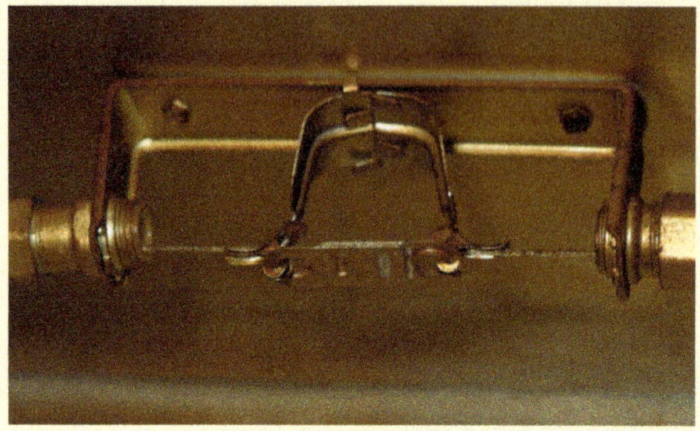

(Figure 25-47) Fusible links can be used to activate a special agent extinguishing system.

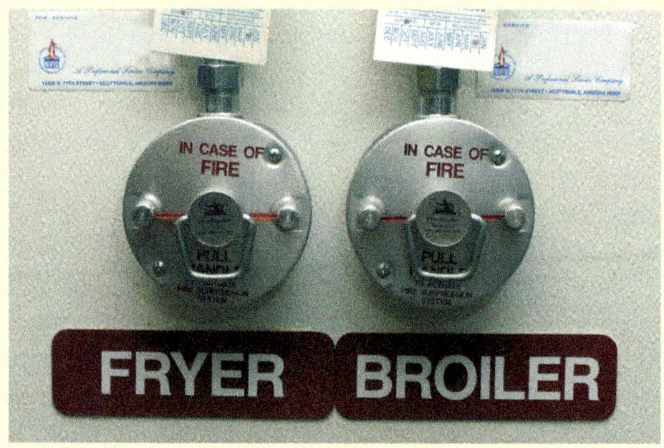


(Figure 25-48) Most special extinguishing systems can also be manually activated.

Many kitchen systems discharge agent into the ductwork above the exhaust hood, as well as onto the cooking surface. This helps prevent a fire from igniting any grease buildup inside the ductwork and spreading throughout the system. Although the ductwork should be cleaned regularly, it is not unusual for a kitchen fire to extend into the exhaust system.

Most dry and wet chemical extinguishing systems are tied into the building's fire alarm system. Kitchen extinguishing systems should also shut down the gas or electricity to the cooking appliances and exhaust fans.

## Clean Agent Extinguishing Systems

Clean agent extinguishing systems are often installed in areas where computers or sensitive electronic equipment are used, or where valuable documents are stored. These agents are nonconductive and leave no residue. Halogenated

## Brigade Member Safety Tips

Halon and clean agents are considered nontoxic. However, you must use self-contained breathing apparatus (SCBA) at all times when working in rooms where these agents have discharged, as they may displace oxygen.

## Brigade Member Safety Tips

Carbon dioxide displaces the oxygen in a room and creates a dangerous situation. Wear SCBA at all times when entering these rooms!

Figure 25-49 Carbon dioxide extinguishes a fire by displacing the oxygen in the room and smothering the fire.

agents or carbon dioxide are generally used because they will extinguish a fire without causing significant damage to the contents.

Clean agent systems operate by discharging a gaseous agent into the atmosphere at a concentration that will extinguish a fire. Smoke detectors or heat detectors installed in these areas activate the system. A manual discharge button is also provided with most installations. Discharge is usually delayed 30–60 seconds after the detector is activated to allow workers in the area to evacuate.

During this delay (the prealarm period), an abort switch can be used to stop the discharge. In some systems, the abort button must be pressed until the detection system is reset; releasing the abort button too soon could cause the system to discharge.

If there is a fire, the clean agent system should be completely discharged before brigade members arrive. Whether the fire was successfully extinguished or not, brigade members entering the area should use SCBA until the area has been properly ventilated. Although these agents are not considered immediately dangerous to life and health, it is better to avoid any unnecessary exposure to them. Toxic products or by-products of combustion could be present in the atmosphere, or the oxygen level could be reduced.

Clean agent systems should be tied to the building's fire alarm system and indicated as a zone on the control panel. This alerts brigade members that they are responding to a situation where a clean agent has discharged. If the system has a preprogrammed delay, the prealarm should activate the building's fire alarm system.

Until the 1990s, Halon 1301 was the agent of choice for protecting areas such as enclosed petrochemical processing facilities, computer rooms, telecommunications rooms, and other sensitive areas. Halon 1301 is a nontoxic, odorless, colorless gas that leaves no residue. It is very effective at extinguishing fires because it interrupts the chemical reaction of combustion. However, this agent has been classified as damaging to the environment and its production has been terminated. Alternative agents, such as halocarbons and inert gas mixes, continue to be developed for use in new systems; they are also used as replacements to Halon 1301 in many existing systems.

### Carbon Dioxide Extinguishing Systems

Carbon dioxide extinguishing systems are similar in design to clean agent systems. The primary difference is that carbon dioxide extinguishes a fire by displacing the oxygen in the room and smothering the fire. Large quantities of carbon dioxide are required, because the area must be totally flooded to extinguish a fire (▲ Figure 25-49).

Carbon dioxide systems are designed to protect a single room or a series of rooms. They usually have the same series of prealarms and abort buttons as Halon systems. Because discharge creates an oxygen-deficient atmosphere in the room, it is immediately dangerous to life. Any occupant who is still in the room when the agent is discharged is likely to be rendered unconscious and asphyxiated. Brigade members responding to a carbon dioxide extinguishing system discharge must use SCBA protection until the area is fully vented.

Carbon dioxide extinguishing systems should be connected to the building's fire alarm system. Responding brigade members should see that a carbon dioxide system discharge has been activated. Using this knowledge, they can deal with the situation safely.

# Wrap-Up

## Ready for Review

- All brigade members, not just brigade leaders and inspectors, need to have a general understanding of how fire alarm, detection, and suppression systems function.
- Brigade members will respond to more false alarms than actual fires; knowing how to handle false alarms is critical for providing customer service and preventing future false alarms.
- Although it is not the brigade member's job to maintain fire protection systems, brigade members should be able to assist the plant to restore systems to service.
- Regardless of how sophisticated a fire protection system is, a serious fire can still occur.
- It is every brigade member's job to try to limit water damage from the activation of a sprinkler system.
- Brigade members must understand the potential shortcomings of using a standpipe system to prevent injuries and fatalities.
- Specialized extinguishing systems are often used in areas where water would not be an acceptable extinguishing agent.

## Hot Terms

**Accelerator** A device that accelerates the removal of the air from a dry-pipe or preaction sprinkler system.

**Air sampling detector** A system that captures a sample of air from a room or enclosed space and passes it through a smoke detection or gas analysis device.

**Alarm initiation device** An automatic or manually operated device in a fire alarm system that, when activated, causes the system to indicate an alarm condition.

**Alarm notification device** An audible and/or visual device in a fire alarm system that makes occupants or other persons aware of an alarm condition.

**Alarm valve** This valve signals an alarm when a sprinkler head is activated and prevents nuisance alarms caused by pressure variations.

**Automatic sprinkler heads** The working ends of a sprinkler system. They serve to activate the system and to apply water to the fire.

**Automatic sprinkler system** A system of pipes filled with water under pressure that discharges water immediately when a sprinkler head opens.

**Beam detector** A smoke detection device that projects a narrow beam of light across a large open area from a sending unit to a receiving unit. When the beam is interrupted by smoke, the receiver detects a reduction in light transmission and activates the fire alarm.

**Bimetallic strip** A device with components made from two distinct metals that respond differently to heat. When heated, the metals will bend or change shape.

**Carbon dioxide extinguishing system** A system designed to protect a single room or series of rooms by flooding the area with carbon dioxide.

**Central station** An off-premises facility that monitors alarm systems and is responsible for notifying the fire brigade of an alarm. These facilities may be geographically located some distance from the protected building(s).

**Clapper valve** A mechanical device installed within a piping system that allows water to flow in only one direction.

**Class I standpipe** A standpipe system designed for use by fire brigade personnel only. Each outlet should have a valve to control the flow of water and a $2^1/_2$-inch male coupling for fire hose.

**Class II standpipe** A standpipe system designed for use by trained personnel only. Each outlet is generally equipped with a length of $1^1/_2$-inch single-jacket hose and a nozzle, preconnected to the system.

**Class III standpipe** A combination system that has features of both Class I and Class II standpipes.

**Clean agent** An electrically nonconducting, rapidly evaporating, or gaseous fire extinguishant that does not leave a residue.

**Coded system** A fire alarm system design that divides a building or facility into zones and has audible notification devices that can be used to identify the area where an alarm originated.

**Cross-zoned system** A fire alarm system that requires activation of two separate detection devices before initiating an alarm condition. If a single detection device is activated, the alarm control panel will usually show a problem or trouble condition.

**Deluge head** A sprinkler head that has no release mechanism; the orifice is always open.

**Deluge sprinkler system** A sprinkler system in which all sprinkler heads are open. When an initiating device, such as a smoke detector or heat detector, is activated, the deluge valve opens and water discharges from all of the open sprinkler heads simultaneously.

**Deluge valve** A valve assembly designed to release water into a sprinkler system when an external initiation device is activated.

**Double-action pull-station** A manual fire alarm activation device that takes two steps to activate the alarm. The person must push in a flap, lift a cover, or break a piece of glass before activating the alarm.

**Dry chemical extinguishing system** An automatic fire extinguishing system that discharges a dry chemical agent.

**Dry-pipe valve** The valve assembly on a dry sprinkler system that prevents water from entering the system until the air pressure is released.

**Dry sprinkler system** A sprinkler system in which the pipes are normally filled with compressed air. When a sprinkler head is activated, it releases the air from the system, which opens a valve so the pipes can fill with water.

**Early suppression fast response (ESFR) sprinkler head** A sprinkler head designed to react quickly and suppress a fire in its early stages.

**Exhauster** A device that accelerates the removal of the air from a dry-pipe or preaction sprinkler system.

**False alarm** The activation of a fire alarm system when there is no fire or emergency condition.

**Fire alarm control panel** That component in a fire alarm system that controls the functions of the entire system.

**Fire department connection (FDC)** A fire hose connection through which the fire brigade can pump water into a sprinkler system or standpipe system.

**Fixed-temperature heat detector** A sensing device that responds when its operating element is heated to a predetermined temperature.

**Flame detector** A sensing device that detects the radiant energy emitted by a flame.

**Flow switch** An electrical switch that is activated by water moving through a pipe on a sprinkler system.

**Frangible bulb sprinkler head** A sprinkler head that uses a glass bulb filled with glycerin or alcohol to hold the cap in place. The bulb also contains an air bubble. As the bulb is heated, liquid absorbs the air bubble and expands until it breaks the glass, releasing the cap.

**Frangible-pellet sprinkler head** A sprinkler head that is activated when the solder pellet melts at a preset temperature.

**Fusible link sprinkler head** A sprinkler head with an activation mechanism that incorporates two pieces of metal held together by low-melting-point solder. When the solder melts, it releases the link and water begins to flow.

**Gas detector** A device that detects and/or measures the concentration of dangerous gases.

**Halocarbon** A compound, such as a fluorocarbon, that is made up of carbon combined with one or more halogens (e.g., fluorine, chlorine, bromine, or iodine).

**Heat detector** A fire alarm device that detects either abnormally high temperatures or rate-of-rise in temperature, or both.

**Indicating butterfly valve (IBV)** A sprinkler control valve that is equipped with a directional arrow to indicate the position of the valve and a gear operator to prevent the valve from being closed in less than 5 seconds, preventing a water hammer.

**Ionization smoke detector** A device containing a small amount of radioactive material that ionizes the air between two charged electrodes to sense the presence of smoke particles.

**Line detector** Wire or tubing that can be strung along the ceiling of large open areas to detect an increase in heat.

**Local alarm system** A fire alarm system that sounds an alarm only in the building where it was activated. No signal is sent out of the building.

**Malicious false alarm** A fire alarm signal when there is no fire, usually initiated by individuals who wish to cause a disturbance.

**Manual pull-station** A device with a switch that either opens or closes a circuit, activating the fire alarm.

**Master-coded alarm** An alarm system in which audible notification devices can be used for multiple purposes, not just for the fire alarm.

**Noncoded alarm** An alarm system that provides no information at the alarm control panel indicating where the activated alarm is located.

**Nuisance alarm** A fire alarm signal caused by malfunction or improper operation of a fire alarm system or component.

**Obscuration rate** A measure of the percentage of light transmission that is blocked between a sender and a receiver unit.

**Outside stem and yoke (OS&Y) valve** A sprinkler control valve with a valve stem that moves in and out as the valve is opened or closed.

**Pendant sprinkler head** A sprinkler head designed to be mounted on the underside of sprinkler piping so the water stream is directed down.

**Photoelectric smoke detector** A device to detect visible products of combustion using a light source and a photosensitive sensor.

**Post indicator valve (PIV)** A sprinkler control valve with an indicator that reads either open or shut depending on its position.

**Preaction sprinkler system** A dry sprinkler system that uses a deluge valve instead of a dry-pipe valve and requires activation of a secondary device before the pipes fill with water.

# Wrap-Up

**Proprietary system** A fire alarm system that transmits a signal to a monitoring location owned and operated by the facility's owner.

**Rate-of-rise heat detector** A device that responds when the temperature rises at a rate that exceeds a predetermined value.

**Remote annunciator** A secondary fire alarm control panel in a different location than the main alarm panel, usually near the front door of a building.

**Remote station system** A fire alarm system that sounds an alarm in the building and transmits a signal to the fire brigade or an off-premise monitoring location.

**Retard chamber** A valve accessory that is used to prevent a sprinkler flow alarm caused by system water pressure surges.

**Sidewall sprinkler head** A sprinkler that is mounted on a wall and discharges water horizontally into a room.

**Single-action pull-station** A manual fire alarm activation device that takes a single step, such as moving a lever, toggle, or handle, to activate the alarm.

**Smoke detector** A device that detects smoke and sends a signal to a fire alarm control panel.

**Spot detector** Single heat-detector devices, spaced throughout an area.

**Sprinkler piping** The network of piping in a sprinkler system that delivers water to the sprinkler heads.

**Standpipe system** A system of pipes and hose outlet valves used to deliver water to various parts of a building for fighting fires.

**Tamper switch** A switch on a sprinkler valve that transmits a signal to the fire alarm control panel if the normal position of the valve is changed.

**Temporal-3 pattern** A standard fire alarm audible signal for alerting occupants of a building.

**Unwanted alarm** A fire alarm signal caused by a device reacting properly to a condition that is not a true fire emergency.

**Upright sprinkler head** A sprinkler head designed to be installed on top of the supply piping and usually marked SSU (Standard Spray Upright).

**Verification system** A fire alarm system that does not immediately initiate an alarm condition when a smoke detector activates. The system will wait a preset interval, generally 30 to 60 seconds, before checking the detector again. If the condition is clear, the system returns to normal status. If the detector is still sensing smoke, the system activates the fire alarm.

**Wall post indicator valve (WPIV)** A sprinkler control valve that is mounted on the outside wall of a building. The position of the indicator tells whether the valve is open or shut.

**Water-motor gong** An audible alarm notification device that is powered by water moving through the sprinkler system.

**Wet chemical extinguishing systems** An extinguishing system that discharges a proprietary liquid extinguishing agent.

**Wet sprinkler system** A sprinkler system in which the pipes are normally filled with water.

**Zoned coded alarm** A fire alarm system that indicates which zone was activated both on the alarm control panel and through a coded audio signal.

**Zoned noncoded alarm** A fire alarm system that indicates the activated zone on the alarm control panel.

**Zoned system** A fire alarm system design that divides a building or facility into zones so the area where an alarm originated can be identified.

# Brigade Member in Action

**As part of your brigade duties, you are asked to assist the facility fire prevention inspector in conducting her reviews of key fire detection and suppression systems. The day you are paired with the inspector, you assist in the inspection of several sprinkler systems, standpipe systems, and fire detection systems.**

**1.** During the inspection of a dry sprinkler system, you are asked to look at the air pressure and water pressure gauges on the dry pipe valve. Which of the following would you expect to find?
- **A.** The air pressure gauge reads higher than the water pressure gauge.
- **B.** The water pressure gauge reads higher than the air pressure gauge.
- **C.** The air pressure gauge and the water gauge both read the same positive pressure.
- **D.** The air pressure gauge and the water gauge both read zero.

**2.** The standpipe system in the multistory office building has $1\frac{1}{2}$" discharge valves and single-jacket fire hose. Which kind of standpipe system is this?
- **A.** Class I standpipe.
- **B.** Class II standpipe.
- **C.** Class II standpipe.
- **D.** Hose line standpipe.

**3.** All of the fire detection systems within the facility sound at a central alarm panel located in the on-site security office. Which type of fire detection system is this?
- **A.** Local alarm system.
- **B.** Remote station.
- **C.** Proprietary system.
- **D.** Central alarm station.

**4.** Which of the following are considered alarm notification devices?
- **A.** Bells.
- **B.** Horns.
- **C.** Strobe lights.
- **D.** All of the above.

**5.** What is the most common method of identifying the temperature rating of a sprinkler head?
- **A.** A paper sticker attached to the sprinkler hat.
- **B.** A stamp on the supply piping.
- **C.** An inspection tag attached to each head.
- **D.** Color coding of the heads.

# Fire Cause Determination

# Chapter 26

## NFPA 1081 Standard

### Incipient Facility Fire Brigade Member

**4.2.1*** Extinguish incipient fires, given an incipient fire and a selection of portable fire extinguishers, so that the correct extinguisher is chosen, the fire is completely extinguished, proper extinguisher-handling techniques are followed, and the area of origin and fire cause evidence are preserved.

**4.3.1*** Attack an incipient stage fire, given a handline flowing up to 473 L/min (125 gpm), appropriate equipment, and a fire situation, so that the fire is approached safely, exposures are protected, the spread of fire is stopped, agent application is effective, the fire is extinguished, and the area of origin and fire cause evidence are preserved.

**(B)** *Requisite Skills.* The ability to recognize inherent hazards related to the material's configuration; operate handlines; prevent water hammers when shutting down nozzles; open, close, and adjust nozzle flow; advance charged and uncharged hose; extend handlines; operate handlines; evaluate and modify water application for maximum penetration; assess patterns for origin determination; and evaluate for complete extinguishment.

### Advanced Exterior Facility Fire Brigade Member

**5.2.6** Overhaul a fire scene, given personal protective equipment, a handline, hand tools, scene lighting, and an assignment, so that structural integrity is not compromised, all hidden fires are discovered, fire cause evidence is preserved, and the fire is extinguished.

**(A)** *Requisite Knowledge.* Types of fire handlines and water application devices most effective for overhaul, water application methods for extinguishment that limit water damage, types of tools and methods used to expose hidden fire, dangers associated with overhaul, obvious signs of area of origin or signs of arson, and reasons for protection of fire scene.

**(B)** *Requisite Skills.* The ability to deploy and operate a handline, expose void spaces without compromising structural integrity, apply water for maximum effectiveness, expose and extinguish hidden fires, recognize and preserve obvious signs of area of origin and fire cause, and evaluate for complete extinguishment.

### Interior Structural Facility Fire Brigade Member

**6.2.4*** Overhaul a fire scene, given personal protective equipment, attack line, hand tools, a flashlight, and an assignment, so that structural integrity is not compromised, all hidden fires are discovered, fire cause evidence is preserved, and the fire is extinguished.

**(A)** *Requisite Knowledge.* Types of fire handlines and application devices most effective for overhaul, application methods for extinguishing agents that limit damage, types of tools and methods used to expose hidden fire, dangers associated with overhaul, obvious signs of area of origin and signs of arson, and reasons for protection of fire scene.

**(B)** *Requisite Skills.* The ability to deploy and operate handlines, expose void spaces without compromising structural integrity, apply extinguishing agents for maximum effectiveness, expose and extinguish hidden fires, recognize and preserve obvious signs of area of origin and fire cause, and evaluate for complete extinguishment.

## Knowledge Objectives

After completing this chapter, you will be able to:

- Describe the role and relationship of the brigade member to criminal investigators and insurance investigators.
- Differentiate accidental fires from incendiary fires.
- Describe the point of origin.
- Define the chain of custody.
- Describe demonstrative, direct, and circumstantial evidence.
- Describe techniques for preserving fire cause evidence.
- Describe the observations brigade members should make during fireground operations.
- Describe the steps needed to secure a property.
- Explain the importance of protecting a fire scene.

## Skills Objectives

There are no skills objectives for this chapter.

# You Are the Brigade Member

**Your industrial plant has just undergone a major reconstruction effort to modernize some of the older equipment and part of the building that was built in the late 1960s. The area has a lot of electrical and hydraulic equipment that had been replaced with new up-to-date equipment. You are working an afternoon shift and about 7:00 pm you hear the fire alarm going off and the PA stating there is a fire in the prefab area.**

**You respond and find that the fire is around one of the new pieces of equipment and appears to be very intense. You grab a fire extinguisher and attempt to control the fire. At that point the fire protection system is activated and the sprinkler system begins to control the fire.**

**The brigade leader asks you to make an initial assessment of the possible cause of the fire. It appears that the main electrical service line has a section missing and there is heavy damage in the area where it connects to the machine. The hydraulic lines in this area are also damaged and could have added fuel to the fire.**

1. ***What will the plant safety officer ask you about your observations?***
2. ***Did you see anything that points to the cause of the fire?***
3. ***What evidence at the scene would need to be protected?***

## Introduction

Brigade members usually arrive at a fire scene before a trained fire investigator arrives. This means that brigade members are able to observe important signs and patterns that the investigators can use in determining how and where the fire started. By identifying and preserving possible evidence, as well as recalling and reporting objective findings, brigade members provide essential assistance to fire investigators. In some cases, the observations and actions of brigade members could be significant in apprehending and convicting arsonists.

Fire brigades determine the causes of fires so they can take steps to prevent future fires. For example, a fire brigade might develop an education program to reduce accidental fires. A series of fires could point to a product defect, such as a design error in a chimney flue or the improper installation of chimney flues. Identifying fires that were intentionally set could lead to the arrest of the person responsible.

Brigade members must understand the basic principles of fire investigation and participate in this important departmental responsibility. Fire cause determination is often difficult, because important evidence can be consumed by the fire or destroyed during fire suppression operations or during salvage and overhaul. Investigators must rely on brigade members to observe, capture, and retain information, as well as to preserve evidence until it can be examined. Evidence that is lost can never be replaced.

## Who Conducts Fire Investigations?

In most jurisdictions, a city or state official (fire marshal) has a legal responsibility to determine the causes of fires. Some fire brigades may ask an investigator whether a person from the brigade can be assigned to assist in or observe the investigation process. Where the cause of the fire is determined not to be arson, many industrial facilities will conduct a root-cause analysis to determine the cause.

Typically, fire brigade leaders initiate fire cause investigations and then turn the materials and information identified through this investigation over to insurance representatives or plant safety officers. If criminal activity is suspected, the appropriate official responsible for investigations will be contacted. If the cause of the fire is obvious and accidental, the fire brigade leader would usually be responsible for gathering the information and filing the necessary reports. If the cause cannot be determined or appears to be suspicious, the fire brigade leader should summon an investigator.

### Investigation Authority

An arson investigation must determine not only the cause and origin of the fire but also who was responsible for starting it and what sequence of events led up to it. Determination of cause and origin is a small but important step in the overall investigation. To prove that a crime took

## Brigade Member Tips

Fire investigations involve much more than simply determining what caused the fire, where it started, and whether it was accidental or incendiary (intentionally set). Many fire investigations also examine many other aspects of a fire incident.

If the fire results in injuries or fatalities, the investigation often will examine every factor that could have contributed to or prevented those losses. In a fire that causes millions of dollars in damage, a whole range of factors could contribute to the large loss. Investigators could examine the construction and contents of the building to determine whether fire and building codes were followed, if the built-in fire protection systems functioned properly, or what lessons could be applied to amending codes or inspection procedures. A fire investigation can teach many lessons, even if the specific cause and origin are never determined.

The NFPA investigates major incidents around the world to identify factors that contributed to an unusual number of deaths or widespread property damage. The NFPA uses this information to amend its codes and standards to help avoid future catastrophes.

place, investigators must establish a link from the fire to the cause and origin and eliminate any other possible cause.

Whether the fire investigators have police powers and can conduct a criminal investigation depends on state and local laws. In some jurisdictions, fire investigators may have police powers during fire investigations. In other areas, police officers are trained as fire investigators. Sometimes fire investigators determine the cause and origin, and turn the investigation over to a law enforcement agency if the fire is determined to be intentionally set. In other jurisdictions, fire brigade and police department personnel work together throughout the investigation.

### Investigation Assistance

Most fire organizations, fire brigades, and government agencies have limited time, personnel, and resources for fire investigations. A state fire marshal or similar authority may have an investigations unit that concentrates on major incidents and supports local investigators on smaller fires. Because these investigators cannot always reach the fire scene quickly, the local fire brigade must be prepared to conduct a thorough preliminary investigation and to protect the scene and preserve evidence.

Federal resources are also available for major investigations. The U.S. Bureau of Alcohol, Tobacco and Firearms (ATF) has individual agents who can assist local jurisdictions with fire investigations. The ATF also has response teams across the country that respond to large-scale incidents. Each of these teams has about 15 agents and equipment that ranges from simple tools for digging out a fire scene to laser-surveying devices for fully documenting the scene.

Insurance companies often investigate fires to determine the validity of a claim or to identify factors that might help prevent future fires. The cost of an investigation is more than offset by the savings the company would realize by identifying a fraudulent claim. Some insurance companies have their own investigators, while others retain independent investigators. These outside investigators often have valuable experience and can provide technical support to determine the cause of a fire.

## Causes of Fires

Every fire has a cause, which the fire investigator tries to uncover. A cause and origin investigation determines where, why, and how the fire originated. Some fires have simple causes that are easily identified and understood; others result from a complex set of circumstances that must be examined carefully to determine what actually happened. In some cases, the cause of a fire will never be determined with absolute certainty.

Basically, every fire has a starting point where ignition occurs and fuel begins to burn. This location is the **point of origin** of the fire. At the point of origin, an ignition source comes into contact with a fuel supply, such as a lighted match touching a piece of paper. The cause of the fire is the particular set of circumstances that brought the ignition source into contact with the fuel.

Fires result when a **competent ignition source** and a fuel come together long enough to ignite. A competent ignition source must have enough heat energy to ignite the fuel and must be in contact with the fuel until the fuel reaches its ignition temperature, which could be a fraction of a second to hours, days, or weeks.

A fire can be caused by an act or by an omission. Igniting a piece of paper with a match is an act. Working in the area and not observing the proper protective measures to ensure the heat is not transmitted to combustible parts and allowing them to heat and ignite combustibles is considered an omission. Leaving a weed burner or propane torch unattended would be considered an omission. The cause of a fire can also be classified as either incendiary or accidental. **Arson**, which is the malicious burning of one's own or another's property with a criminal intent, is an **incendiary fire**. **Accidental fires** do not involve a criminal or malicious intent, even if they are caused by human error or carelessness. Falling asleep with a lit cigarette would be considered an accidental cause. Emptying an ashtray into a wastepaper basket would be considered an accidental cause.

Investigators should always consider a fire to have an **undetermined** cause until the specific cause is established. The evidence from both incendiary and accidental fires can be very similar. Untrained individuals should not attempt to categorize fires as either incendiary or accidental. Brigade members are responsible for helping to identify and preserve possible evidence for the fire investigator to examine.

## Fire Cause Statistics

Most structure fires occur in residential occupancies; home fires represent 73% of all structure fires. The 10 leading causes of structure fires in industrial properties, averaged over the years 1999 to 2002, are shown in ▶ Table 26-1.

## Accidental Fire Causes

Accidental fires have hundreds of possible causes and involve multiple factors and circumstances. The most important reason for investigating and determining the causes of accidental fires is to prevent future fires. To reduce the number of fires, efforts must concentrate on the most frequent causes and those involving the greatest risks of death, injury, and property damage.

Most fires, fire deaths, and injuries occur in residential occupancies. The most commonly reported accidental causes of fire in these occupancies involve smoking, cooking, heating equipment, and electrical equipment. These statistics provide a foundation for fire prevention and public education efforts. Additional analysis can provide more specific information. For instance, fires caused by electrical equipment can be divided into four groups: those caused by worn-out or defective equipment, by improper use of approved equipment, by defective installations, and by other accidents. A proper, thorough investigation of an electrical fire would identify and classify the specific cause within one of these groups. Industrial settings have many varied possible causes. Each may need to be considered.

For example, worn-out or defective electrical equipment would include deteriorating 50-year-old wiring circuits or a computer circuit board with an internal defect. Other worn-out equipment that might still be in use includes electrical motors, switches, appliances, and extension cords. Properly used and maintained equipment that has been tested and listed by a recognized laboratory rarely causes a fire, but should be replaced when it wears out.

Placing a portable heater too close to flammable materials or using a toaster oven to heat a container of flammable glue are examples of the improper use of electrical equipment. Defective installations are those not acceptable under electrical codes or printed instructions, such as using a light-duty extension cord to connect a heavy-duty appliance to a wall outlet.

Some electrical fires result from an accidental misuse or oversight, such as unintentionally leaving a cooking appliance turned on. The improper maintenance of the machinery might be the cause of the fire. Natural events can also cause accidental electrical fires, such as when a tree falls on a wire. A fire caused by an electrical overload or a short circuit can start wherever there is electricity, such as in the electrical panel, fuses, fuse boxes, circuit breakers, wiring, and appliances.

Sometimes it may be difficult or impossible to identify the specific source of ignition because the fire destroyed every trace of evidence. Investigators will classify these fires as having an undetermined cause, rather than use a "best guess." For example, low temperature ignition can occur when wood is subjected to low heat, such as that generated by steam pipes or incandescent light bulbs, for a long period of time. Gradually, the wood can deteriorate and eventually ignite. If the fire destroys the building, the point of origin would be difficult to identify. In an industrial setting, a machine part that is not lubricated or that is wearing out might cause heavy damage and ignite other combustible material. The evidence might be consumed, making a cause determination difficult.

### Incendiary Fire Causes

People may set fires for several reasons and in many different ways. The same type of cause and origin investigation is needed to identify where and how the fire started. With incendiary fires, it is particularly important to rule out possible accidental causes to prove beyond doubt that the fire was deliberately set.

A fire caused by arson requires a second phase of investigation to identify the person responsible. All of the evidence relating to the cause of the fire must be handled in a way that ensures it would be admissible as evidence in a criminal trial. A trained, qualified fire investigator should always be called to determine the cause of any fire that may have been deliberately set.

Arson and the factors that could indicate an incendiary fire cause are discussed in more detail later in this chapter. A fire might be set at an industrial facility for several reasons. A motive is often needed to prove what happened. This will be discussed later. A fire may be set in order to get off shift early or to make a point to the supervisors that things are unsafe.

# Determining the Cause and Origin of a Fire

A systematic analysis is needed to determine the cause and origin of a fire. The investigator must determine where the fire started and how it was ignited. The investigator must look at the situation objectively to be sure that the evidence is convincing and fully explains the situation. If there is more than one possible explanation for the observations, each possibility must be considered. The cause cannot be determined with absolute certainty until all alternative explanations have been ruled out.

## Identifying the Point of Origin

One of the first steps in a fire investigation is identifying the point of origin. At this location, the investigator can look for clues indicating the specific cause of the fire.

The investigation process usually begins with an examination of the building's exterior. The investigator should look for indications that the fire originated outside the

## Table 26-1 Leading Causes of Structure Fires in Industrial and Manufacturing Properties 2000–2004 Annual Averages

| Cause | Fires | | Civilian Deaths | | Civilian Injuries | | Direct Property Damage (in Millions) | |
|---|---|---|---|---|---|---|---|---|
| Shop tools and industrial equipment excluding torches, burners, or soldering irons | 1,790 | (15%) | 2 | (12%) | 145 | (40%) | $83.4 | (11%) |
| Heating equipment fires | 1,430 | (12%) | 3 | (15%) | 40 | (11%) | $48.5 | (6%) |
| *Identified heating equipment* | *580* | *(5%)* | *2* | *(12%)* | *22* | *(6%)* | *$44.6* | *(6%)* |
| *Confined heating equipment* | *850* | *(7%)* | *1* | *(3%)* | *18* | *(5%)* | *$3.9* | *(1%)* |
| Electrical distribution and lighting equipment | 640 | (5%) | 3 | (15%) | 14 | (4%) | $42.8 | (6%) |
| Cooking equipment fires | 400 | (3%) | 0 | (0%) | 2 | (0%) | $16.2 | (2%) |
| *Identified cooking equipment* | *50* | *(0%)* | *0* | *(0%)* | *0* | *(0%)* | *$10.5* | *(1%)* |
| *Confined cooking fire* | *340* | *(3%)* | *0* | *(0%)* | *2* | *(0%)* | *$5.8* | *(1%)* |
| Exposure to other fire | 380 | (3%) | 0 | (0%) | 5 | (1%) | $11.6 | (2%) |
| Intentional | 380 | (3%) | 0 | (0%) | 3 | (1%) | $16.9 | (2%) |
| Torch, burner or soldering iron | 320 | (3%) | 1 | (8%) | 12 | (3%) | $10.4 | (1%) |
| Confined commercial compactor fire | 220 | (2%) | 0 | (0%) | 3 | (1%) | $0.7 | (0%) |
| Spontaneous combustion or chemical reaction | 210 | (2%) | 1 | (9%) | 2 | (1%) | $14.0 | (2%) |
| Confined fire involving incinerator overload or malfunction | 200 | (2%) | 0 | (0%) | 5 | (1%) | $3.1 | (0%) |
| Smoking materials (i.e., lighted tobacco products) | 180 | (2%) | 0 | (0%) | 1 | (0%) | $3.9 | (1%) |
| Contained trash or rubbish fire | 1,060 | (9%) | 0 | (0%) | 4 | (1%) | $0.5 | (0%) |

Note: These are the leading causes, obtained from the following list: intentional (from the NFIRS field "cause"); playing with fire (from factor contributing to ignition); confined heating (including confined chimney and confined fuel burner or boiler fires), confined cooking, confined commercial compactor, confined fire involving incinerator overload or malfunction, and contained trash or rubbish from incident type; identified heating, identified cooking, clothes dryer or washer, torch (including burner and soldering iron), electrical distribution and lighting equipment, medical equipment, and electronic, office or entertainment equipment (from equipment involved in ignition); smoking materials, candles, lightning, and spontaneous combustion or chemical reaction (from heat source). The statistics on smoking materials and candles include a proportional share of fires in which the heat source was heat from an unclassified open flame or smoking material. Exposure fires include fires with an exposure number greater than zero, as well as fires identified by heat source or factor contributing to ignition when no equipment was involved in ignition and the fires were not intentionally set. Because contained trash or rubbish fires are a scenario without causal information on heat source, equipment involved, or factor contributing to ignition, they are shown at the bottom of the table if they account for at least 2% of the fires. Casual information is not routinely collected for these incidents. The same fire can be listed under multiple causes, based on multiple data elements. Details on handling of unknowns, partial unknowns, and other underspecified codes may be found in the Appendix.

These are national estimates of fires reported to U.S. municipal fire departments and so exclude fires reported only to Federal or state agencies or Facility fire brigades. These national estimates are projections based on the detailed information collected in Version 5.0 of NFIRS. Casualty and loss projections can be heavily influenced by the inclusion or exclusion of one unusually serious fire. Fires are rounded to the nearest ten, civilian deaths and injuries are rounded to the nearest one, and direct property damage is rounded to the nearest hundred thousand dollars. Property damage has not been adjusted for inflation.

Source: NFIRS and NFPA survey.

Figure 26-1 Depth of char.

Figure 26-2 Often the point of a V-pattern is near or at the point of origin.

building before looking inside, and should size-up the building to identify important information. The overall size, construction, layout, and occupancy of the building should be noted, as well as the extent of damage that is visible from the exterior. The investigator should look for any openings that might have created drafts that influenced the fire spread and should examine the condition of outside utilities, such as the electrical power connection and gas meter.

The search for indications to the point of origin continues inside the building, beginning with the area of lightest damage to the area of heaviest damage. This is probably the area that was burning for the longest time; areas with less damage were probably not as involved in the fire or were involved for shorter periods.

Where process or manufacturing equipment is involved, the investigation may focus on process control systems. A failure in level control alarms, by-pass piping, relief valves, and other control devices may be either a contributing cause or a direct cause of a fire or explosion.

The depth of char can be used to help determine how long a fire burned in a particular location (▲ Figure 26-1). The depth of char is related to how long the surface of a material was exposed to the fire, from the time of ignition to the time of extinguishment. An area that burned for only a few minutes would have mostly surface damage and a shallow depth of char. Materials that burned for longer periods will show evidence of charring deeper into their cores. Charring is usually deepest at the point of origin; however, flammable liquids and combustible materials in other locations can also leave heavy charring. The area with the deepest char is not necessarily the point of origin.

Burn patterns and smoke residue can be helpful in identifying the area of origin, but again are not conclusive. Burn patterns and damage will often spread outward from the room or area where the damage is most severe. Because heat rises, the flow of heated gases from a fire will almost always be up and out from the point of origin. This upward, outward flow can usually be recognized, even when all of the building's contents were involved in the fire. Often the point of origin is found directly below the most damaged area on the ceiling, where the heat of the fire was most concentrated.

A charred V-pattern on a wall indicates that fire spread up and out from something at the base of the V. The best place to start looking for a specific fire cause in a room with a V-pattern on the wall, a pile of charred debris at the base of the V, and minimal damage to the other contents is in the pile of debris (▲ Figure 26-2).

An experienced fire investigator knows that many factors can influence burn patterns, including ventilation, fire suppression efforts, and the burning materials themselves. A V-pattern on a wall could indicate that there was an easily ignited and intensely burning fuel source at that location. It might also indicate that something fell from a higher level and burned on the floor.

An inverted V-pattern on a wall could indicate that a flammable liquid was used along the base of the wall to set the fire intentionally. A fire burning across a wide area at the floor level can funnel into a thermal column as it rises, creating the inverted V-pattern.

Once the investigator identifies the exact or approximate location of the point of origin, the search for indications of a specific cause can begin. The investigator must determine what happened at that location to cause the fire. This involves identifying both the source of ignition and the fuels that were involved. The investigator must ultimately determine how the source of ignition and fuel came together, either accidentally or intentionally.

Figure 26-3 Fire investigators systematically dig through the debris of a fire to search for evidence.

## Digging Out

After the investigator identifies the area of origin, brigade members could be asked to assist in digging out the fire scene. "Digging out" is a term used to describe the process of carefully looking for evidence within the debris. Sometimes the entire fire scene must be closely examined to determine the cause of the fire and gather evidence.

The fire investigator will take extensive photographs of the fire scene as it first appears. Then, he or she will begin to remove and inspect the debris, layer by layer, from the top of the pile down to the bottom. The type of evidence uncovered and its location in the layers of debris can provide important indications of how the fire originated and progressed.

Removing and inspecting the layers of debris enables the investigator to determine the sequence in which items burned, whether an item burned from the top down or from the bottom up, and how long it burned. Did the fire start at a low point and burn up or did burning items fall down from above and ignite combustible materials below? Did the fire spread along the ceiling or along the floor? Is there a residue of rags containing a flammable liquid under the furniture? Why are papers that should have been in a metal filing cabinet stacked on the floor and partially burned?

Systematically digging through the debris often can uncover the exact point of origin and cause of both accidental and deliberate fires ▲ Figure 26-3. If circumstances or eyewitness accounts indicate a deliberate fire, the investigator may request that brigade members help examine the entire area. Common search methods include the "grid" search (also known as the "double strip" search) and the "strip" search. The investigator will explain generally what to look for, how to search, and what to do with any potential evidence.

Where process equipment is involved, "digging out" may involve the systematic removal of piping, valves, support structures, and associated control systems to determine the cause or point of origin.

When you find possible evidence, stop so the investigator can examine it in place. It is the investigator's job to document, photograph, and remove any potential evidence, whether or not it supports the suspected cause.

To determine the cause and origin of a fire, the investigator must evaluate all potential causes for the fire. The process of eliminating possible alternative causes and documenting the reasons for rejecting them is as important as properly documenting the ultimate cause of the fire.

For example, if the fire started in a cafeteria, potential causes could include:

- The stove
- All of the electrical appliances
- Light fixtures
- Smoking materials
- Cleaning supplies

## Evidence

Evidence refers to all of the information gathered and used by an investigator in determining the cause of a fire. Evidence can be used in a legal process to establish a fact or prove a point. To be admissible in court, evidence must be gathered and processed under strict procedures.

Physical evidence consists of items that can be observed, photographed, measured, collected, examined in a laboratory, and presented in court to prove or demonstrate a point ▼ Figure 26-4. Fire investigators can gather physical evidence, such as a burn pattern on a wall or an empty gasoline can, at the fire scene to explain how the fire started or to document how it burned.

Trace or transfer evidence is a minute quantity of physical evidence that is conveyed from one place to another. For example, a suspect's clothing may contain the residue of the same flammable liquid found at the scene of a fire.

Figure 26-4 Physical evidence at a fire scene.

Demonstrative evidence is anything that can be used to validate a theory or to show how something could have occurred. To demonstrate how a fire could spread, for example, an investigator may use a computer model of the burned building.

Evidence used in court can be considered either direct or circumstantial. Direct evidence includes facts that can be observed or reported first hand. Testimony from an eyewitness who saw a person actually ignite a fire or a videotape from a security camera showing the person starting the fire are examples of direct evidence. Direct evidence is rare in arson cases, but eyewitnesses often can describe the circumstances of an accidental fire.

Circumstantial evidence is information that can be used to prove a theory, based on facts that were observed first hand. For example, an investigation might show that gasoline from a container found at the scene was used to start a fire. Two different witnesses testify that the suspect purchased a container of gasoline before the fire and walked away from the fire scene without the container a few minutes before the fire brigade arrived. Such circumstantial evidence clearly places the suspect at the fire site with an ignition source at the time the fire started. Fire investigators must often work with circumstantial evidence to connect an arson suspect to a fire.

## Preservation of Evidence

Brigade members have a responsibility to preserve evidence that could indicate the cause or point of origin of a fire. Brigade members who discover something that could be evidence while digging out a fire scene or performing other activities should leave it in place, make sure that no one interferes with it or the surrounding area, and notify a brigade leader or fire investigator immediately.

Evidence is most often found during the salvage and overhaul phases of a fire. Salvage and overhaul should always be performed carefully and can often be delayed until an investigator has examined the scene. Do not move debris any more than is absolutely necessary and never discard debris until the investigator gives approval. The fire investigator must decide whether the evidence is relevant, not the brigade member.

Brigade members at the scene are not in the position to decide whether the evidence they find will be admissible in court and therefore worthy of preservation; that is the fire investigator's decision. Your responsibility is to make sure that potential evidence is not destroyed or lost. Too much evidence is better than too little, so no piece of potential evidence should be considered insignificant.

If evidence could be damaged or destroyed during fire suppression activities, cover it with a salvage cover or some other type of protection, such as a garbage can. Use barrier tape to keep others from accidentally walking through evidence. Before moving an object to protect it from damage, be sure that witnesses are present, that a location sketch is drawn, and that a photograph is taken.

Evidence should not be contaminated, or altered from its original state, in any way. Fire investigators use special containers to store evidence and prevent contamination from any other products.

## Chain of Custody

To be admissible in a court of law, physical evidence must be handled according to certain prescribed standards. Because the cause of the fire may not be known when evidence is first collected, all evidence should be handled according to the same procedure. In an incendiary or arson fire, evidence relating to the cause of the fire will probably have to be presented in court. The same is true for accidental fires, because lawsuits might be filed to claim or recover damages.

Chain of custody (also known as chain of evidence or chain of possession) is a legal term that describes the process of maintaining continuous possession and control of the evidence from the time it is discovered until it is presented in court. Every step in the capture, movement, storage, and examination of the evidence must be properly documented. For example, if a gasoline can is found in the debris of a suspected arson fire, documentation must record the person who found it, where and when it was found and under what circumstances. Photographs should be taken to show where it was found and what condition it was in. In court, the investigator must be able to show that the gas can presented is the specific can that was found.

### Brigade Member Tips

To avoid contaminating evidence, fire investigators always wash their tools between taking samples. This ensures that material from one piece of evidence will not be unintentionally transferred to another piece. Investigators also change their gloves each time they take a sample of evidence and place evidence only in absolutely clean containers. Before entering the fire scene, investigators often wash their boots to keep from transporting any contamination into the fire scene.

### Brigade Member Tips

Gasoline-powered tools and equipment, including saws and generators, are often used during firefighting operations. Gasoline from the equipment could contaminate the area and lead to an erroneous assumption that accelerants (materials used to initiate or increase the spread of fire) were used in the fire. Refuel such equipment outside the investigation area to prevent spilled fuel contamination.

Figure 26-5 Evidence should remain where you find it until you can turn it over to your brigade leader or the fire investigator.

The person who takes initial possession of the evidence must keep it under his or her personal control until it is turned over to another official. Each successive transfer of possession must be recorded. If evidence is stored, documentation must indicate where and when it was placed in storage, whether the storage location was secure, and when it was removed. Often, evidence is maintained in a secured evidence locker to ensure that only authorized personnel have access to it.

Everyone who had possession of the evidence must be able to attest that it has not been contaminated, damaged, or changed. If evidence is examined in a laboratory, the laboratory tests must be documented. The documentation for chain of custody must establish that the evidence was never out of the control of the responsible agency and that no one could have tampered with it.

Brigade members are frequently the first link in the chain of custody. The brigade member's responsibility in protecting the integrity of the chain is relatively simple: report everything to a fire brigade leader and disturb nothing needlessly. The individual who finds the evidence should remain with it until it is turned over to a brigade leader or to the fire investigator (▲ Figure 26-5). Remember that you could be called as a witness to state that you are the brigade member who discovered this particular piece of evidence at the specific location pictured.

The fire investigator's standard operating procedures for collecting and processing evidence generally include the following steps:

- Take photographs of each piece of evidence as it is found and collected. If possible, photograph the item as it was found, before it is moved or disturbed.
- On the fire scene, sketch, mark, and label the location of the evidence. Sketch the scene as near to scale as possible.
- Place evidence in appropriate containers to ensure safety and prevent contamination. Unused paint cans with lids that automatically seal when closed are the best containers for transporting evidence. Glass mason jars sealed with a sturdy sealing tape are appropriate for transporting smaller quantities of materials. Plastic containers and plastic bags should not be used for evidence containing petroleum products since it may deteriorate the plastic. Paper bags can be used for dry clothing or metal articles, matches, or papers. Soak up small quantities of liquids with either a cellulose sponge or cotton batting. Protect partially burned paper and ash by placing them between layers of glass (assuming that small sheets or panes of glass are available at the scene).
- Tag all evidence at the fire scene. Evidence being transported to the laboratory should include a label with the date, time, location, discoverer's name, and witnesses' names.
- Record the time the evidence was found, where it was found, and the name of the person who found it. Keep a record of each person who handled the evidence.
- Keep a constant watch on the evidence until it can be stored in a secure location. Evidence that must be moved temporarily should be put in a secure place accessible only to authorized personnel.
- Preserve the chain of custody in handling all the evidence. A broken chain of custody may result in a court ruling that the evidence is inadmissible.

Only one person should be responsible for collecting and taking custody of all evidence at a fire scene, no matter who discovers it. If someone other than the assigned evidence collector must seize the evidence, that person must photograph, mark, and contain the evidence properly and turn it over to the evidence collector as soon as possible. The evidence collector must also document all evidence that is collected, including the date, time, and location of discovery, the name of the finder, known or suspected nature of the evidence, and the evidence number from the evidence tag or label. A log of all photographs taken should also be recorded at the scene.

## Witnesses

Although brigade members do not interview witnesses, you can identify potential witnesses to the investigator. People who were on the scene when brigade members arrived could have invaluable information about the fire. If a brigade member learns something that might be related to the cause of the fire, this information should be passed on to a brigade leader or to the fire investigator.

Interviews with witnesses should be conducted by the fire investigator or by a police officer. If the fire investigator is not on the scene or does not have the opportunity to interview the witness, obtain the person's name, address, and telephone

# VOICES OF EXPERIENCE

**“The workplace fire may cause the company to lose several hours or days of money-making work, but the cause needs to be found to make the workplace safe and deter a possible arsonist.”**

It was night shift at a manufacturing plant when the announcement was made over the PA for the fire brigade members to respond to a fire in Area B. It was strange because over the last four shifts there had been small fires or evidence of fires around the plant that no one could figure out. This was the first one this week and it was in an area that had lots of fuel.

The week before there had been a small fire in the dumpster where the plant's trash was normally dumped. The brigade leader thought it was just an accident, maybe something hot placed in the trash before it had time to cool. According to a fire brigade member, this didn't make sense because the trash was from an area where nothing was hot.

The fire brigade members advised that when they made their way through the plant to Area B they saw workers leaving that area, and a haze was starting to filter into Area A. At the doorway they saw that there was active fire in the area of the boxing machine and some of the workers in Area B were starting to fight the fire. The fire brigade members took a handline toward the area on fire and started to attack the fire. A member observed an area that was burning to the left that was not part of the larger fire. The member did not think it was connected and was out of the way. This stuck in his mind, and it was reported to the brigade leader when everything was under control.

With several members of the brigade on hand, the fire was brought under control. The local fire department arrived and worked with the brigade to overhaul the scene. A member of the fire brigade advised the leader of the separate fire and stated that it seemed strange, considering all of the fires that had taken place lately at the plant. The leader requested that the members make a list of fires they knew had taken place in the past few months.

The plant safety/risk manager arrived at the plant and started an investigation. The brigade leader relayed the information about the other fires. A team member said that he knew of at least three other fires in the plant over the last few months. Each one seemed to have happened in an area where there was no real reason for it to have started.

The safety/risk manager and the FD investigator asked what had been seen. A member told them about what they had seen as they approached that evening's fire. They said they had seen a fire in the area of the doorway to the outside. It was small, but it was strange because the main fire was not spreading toward that area. The investigator asked the fire brigade member to show him this fire area. Viewing the area made it clear to the investigator that the fire had been started separately because it was well protected from anything else starting the fire. This proved to be a critical piece of evidence in the investigation, showing that a person was setting several fires at the same time to confuse the investigators and the fire brigade members.

During the overhaul of the main fire in Area B, investigation of the scene found that there was a lot of what appeared to be trash stuffed into the pallets on the floor. This trash should not have been there. The supervisor who worked the area was questioned about anyone who would have been upset enough to stuff papers into that area.

The supervisor stated that he had problems with an employee who had been transferred to his area the week before. This person did not produce like the other employees did and was always having to be checked to make sure he was working. When he was confronted about his work, he would say the supervisor was picking on him. When the fire was reported, one of the brigade members who responded from outside said they saw this employee on the loading dock, sitting there uninterested in what was going on.

A check of the other fires that had happened over the past few weeks showed they were all in areas where this same person had worked, and from which he had been moved or reprimanded for poor work production. Each fire had been set using just the ordinary combustibles that could have been found in the area or plant and an open flame. Every area had nothing wrong with any of the equipment or other accidental sources of heat.

When confronted, he admitted that he had started the fires because he thought if the fire did enough damage he would be able to leave. He also confessed to several fires away from the plant.

The observations made during the response, attack, and overhaul are always important to the investigation. The separate fire was critical because there was no reason for it. The papers stuffed into the pallets were another suspicious clue. The fire a week before was just as out of place because there was nothing from the area that could have started it. Protecting the scene and providing the investigators with accurate observations are valuable to a thorough and complete investigation. The fire brigade members are a valuable asset to the investigators because brigade members are the eyes of the investigators until they are on the scene working to find the cause of the fire. Without their observations, the fires would have continued and could have eventually caused the loss of the plant and jobs.

Mike Dalton
Knox County Fire Investigation Unit
Knoxville, Tennessee

**Brigade Member Tips**

Do not speculate on the cause of the fire with or in front of bystanders. Speculation that is reported or printed tends to be treated as fact.

**Brigade Member Tips**

Jesting remarks and jokes should never be made at the scene, because your comments could be overheard by the media.

number and give it to the investigator. A witness who leaves the scene without providing this information could be difficult or impossible to locate later.

Brigade members have a primary responsibility to save lives and property. Until the fire is under control, brigade members must concentrate on fighting the fire, not investigating the cause. However, brigade members should pay attention to the situation and make mental notes about any observations. Brigade members must tell the investigator about any odd or unusual happenings. Information, suspicions, or theories about the fire should be shared only with the fire investigator and only in private.

Do not make statements of accusation, personal opinion, or probable cause to anyone other than the investigator. Comments that are overheard by a bystander, a news reporter, or others can impede the efforts of the fire investigator to obtain complete and accurate information. A witness trying to be helpful might report an overheard comment as a personal observation. In this way, inaccurate information can generate a rumor that becomes a theory and turns into a reported "fact" as it passes from person to person.

Never make jesting remarks or jokes at the scene. Careless, unauthorized, or premature remarks could embarrass the company or the fire brigade. Statements to news reporters about the fire's cause should be made only by an official spokesperson after the fire investigator and ranking brigade leader have agreed to their accuracy and validity. Until then, "The fire is under investigation" is a sufficient reply to any questions concerning the cause of the fire.

# Observations During Fireground Operations

Although a brigade member's primary concern is saving lives and property, you will make observations and gather information as you perform your duties. What you observe could be significant in the subsequent investigation of the incident. The following sections will help you identify specific signs, patterns, and evidence during various fireground operations, from dispatch and response to fire attack and overhaul.

## Dispatch and Response

During dispatch and response, form a mental image of the scene you expect to encounter. Note the time of day, weather conditions, and route obstructions. As the incident progresses, pay attention to things that do not match your expectations because they could help fire investigators determine the origin and cause of the fire.

### Time of Day

Time of day and type of occupancy can indicate the number and type of people at an incident. Offices, warehouses, and maintenance shops are filled with people during the day and mostly empty at night. These same areas will most likely be empty or have a smaller number of employees during night shifts.

As victims evacuate the area, brigade members may be able to note any that stand out, because their behavior and demeanor are quite different.

### Weather Conditions

Note whether the day is hot, cold, cloudy, or clear, and whether conditions in the burning structure match the weather. On a cold day, windows should be closed; on a hot day, the furnace should be off.

Lightning, heavy snow, ice, flooding, fog, or other hazardous conditions can help cover an arsonist's activities because they delay the fire brigade's arrival and make a brigade member's job more difficult. Because wind direction and velocity help determine the natural path of fire spread, being aware of these conditions will help you recognize when a fire behaves in an unnatural way.

### Route Obstructions

Unusual traffic patterns or barriers blocking the route to the scene may be early indications of a suspicious fire. Be sure to note these and any other obstructions, such as barricades, downed cables, or trash containers (dumpsters) that cause delays.

## Arrival and Size-Up

Size-up operations can provide valuable information for fire investigators. Pay attention to the fire conditions, building characteristics, and vehicles and people at or leaving the scene.

### Description of the Fire

Compare the dispatcher's description with the actual fire conditions. If the fire has intensified dramatically in a short time, an accelerant could have been used. Note whether flames are visible or only smoke. Also observe the quantity, color, and

## Brigade Member Safety Tips

Safety is always the first and most important priority on the fireground.

source of the smoke. Does the fire appear to be burning in one place, or in multiple locations?

### Vehicles and People on the Scene

The appearance and behavior of people and vehicles at the scene of a fire can provide valuable clues. Do you recognize any bystanders from other fire scenes? Note the attitude and dress of personnel and/or occupants of the building, as well as other individuals at the scene of a fire.

Anyone who seems out of place at a fire or someone who has been observed at several fires in various locations should be reported to the investigator. Some arsonists are emotionally disturbed individuals who receive personal satisfaction in watching a "working" fire. Sometimes investigators will photograph the crowd at a fire scene, particularly if there is a series of similar fires, and look for the same faces at different incidents.

Most people at a fire scene are serious and intent on watching the drama unfold (▼ Figure 26-6). Someone who is talking loudly, laughing, or making light of the situation should be considered suspicious. Fire investigators may also want to know about someone who eagerly volunteered multiple theories or too much information.

### Unusual Items or Conditions

Always note any unusual items or conditions about the property, such as whether windows and doors are open or closed.

Figure 26-6 Most persons at a fire scene are intent on watching brigade members at work.

## Brigade Member Tips

Photographs—whether taken by film or digital cameras—can document conditions before the arrival of fire investigators. However, specific procedures must be followed for the photographs to be admissible in court as evidence. Follow your brigade's standard operating procedure regarding picture taking.

Arsonists commonly draw the shades or cover windows and doors to delay the discovery of a fire. Gasoline cans, forcible entry tools, and a damaged hydrant or sprinkler connection might all suggest an intentionally set fire.

## Entry

As you prepare to enter the fire area, look for evidence of any prior entry, such as shoeprints leading into or out of the structure or tracks from vehicle tires. Note whether the windows and doors are intact, whether they are locked or unlocked, and whether there are any unusual barriers limiting access to the structure. Also note any signs of forced entry by others. If you see evidence of forced entry, ask yourself if the forced entry likely occurred before the fire, or if it could have been caused by brigade members gaining access to the structure.

Forced entry could leave impressions of tools on the windows and doors. Cut or torn edges of wood, metal, or glass also may indicate forced entry. Investigators might be able to determine whether the glass had been broken by heat or by mechanical means. Look for signs of a burglary; the fire may have been set to destroy evidence of another crime.

## Search and Rescue

As you enter the building to perform search and rescue or interior fire suppression activities, consider the location and extent of the fire. Separate or seemingly unconnected fires may indicate multiple points of origin and are often found at arson scenes. The location and condition of the building's contents may also provide clues. Unusual building contents or conditions—such as barriers in doors and windows, the absence of inventory from a warehouse.

The team responsible for shutting off electric power to the building should note whether the circuit breakers were on or off when they arrived. The location of any people found in the building should also be noted.

## Ventilation

The ventilation crew should note whether the windows and doors were open or closed, locked or unlocked. They should also note the color and quantity of the smoke, as well as the presence of any unusual odors.

### Color of Smoke

The color of smoke often indicates what is burning. Unusual smoke might indicate that additional fuel was added to the mix.

### Unusual Odors

Self-contained breathing apparatus protects brigade members from hazardous fumes and toxic odors. However, sometimes an odor is so strong that it can be detected en route or linger after the fire has been extinguished. Fires involving rubber drive belt and overheated light ballasts produce distinctive, identifiable odors familiar to experienced brigade members. Other common odors familiar to brigade members include liquid hydrocarbons, solvents, and natural gas.

Odors often linger in the soil under a building without a basement, particularly if an accelerant has been used and if the ground is wet. Concrete, brick, and plaster will all retain vapors after a fire has been extinguished.

### Effects of Ventilation on Burn Patterns

Fire brigade ventilation operations can dramatically influence the behavior of a fire and alter burn patterns. For example, if the ventilation crew opens a window or makes an opening in the roof, the heat, smoke, and fire are likely to move toward this opening, creating new burn patterns on the walls, floors, and ceilings. Relay such information to the fire investigator so that he or she can correctly interpret these patterns.

## Suppression

Fire behavior, the presence of incendiary devices (materials used to start a fire or cause an explosion), obstacles encountered during fire suppression operations, and charring and burn patterns are among the factors that might help to determine the origin and cause of the blaze.

### Behavior of Fire

During the fire attack, observe the behavior of the fire and how it reacts when an extinguishing agent is applied. Look for unusual flame colors, sounds, or reactions. For example, most flammable liquids will float, continue to burn, and spread the fire when water is applied. Rekindles in the same area or a flare-up when water is applied could indicate the presence of an accelerant.

### Incendiary Devices, Trailers, and Accelerants

While fighting the fire, be aware of streamers or trailers (combustible materials placed to spread the fire). Look for combustible materials like wood, paper, or rags in unusual locations. Note containers of flammable or combustible liquids normally not found in the type of occupancy. Very intense heat or rapid fire spread might indicate the use of an accelerant to increase the fire spread.

#### Brigade Member Safety Tips

If you encounter an incendiary device that has not ignited, notify your supervisor and others in the area immediately! Do not make any attempt to move it or disable it.

#### Brigade Member Safety Tips

Sometimes, incendiary fires contain traps for fire fighter or brigade members, such as steps that have been removed or holes in the floor, deliberately covered. Be alert for this potential hazard.

Incendiary devices can include unusual items in unlikely places, such as a packet of matches tied to a bundle of combustible fibers or attached to a mechanical device. Often, incendiary devices fail to ignite or burn out without igniting other materials. Try to extinguish the fire without unduly disturbing any suspicious contents.

### Condition of Fire Alarm or Suppression Systems

If the building is equipped with a fire alarm or fire suppression system, brigade members should examine it to see whether the system operated properly or was disabled. If the fire suppression system failed to work, the problem could be poor maintenance or deliberate tampering. Notify the fire inspector of any findings.

### Obstacles

An arsonist may place obstacles to hinder the efforts of brigade members. Note whether furniture or equipment was moved to block entry. Arsonists also may prop open fire doors, pull down plaster to expose the wood structure, or punch holes in walls and ceilings to increase the rate of fire spread.

### Contents

Brigade members involved in interior fire suppression, like the members of the search and rescue team, should make note of anything unusual about the contents of the building. The absence of personal items in an office may indicate that they were removed and the fire was intentionally set. Empty boxes in a warehouse may belong there, or they may indicate that the valuable contents were removed. If only one item or a particular stock of products burns, and there is no reasonable explanation for the fire, arson could be suspected. The fire investigator will consider these factors with other evidence before reaching any conclusion.

### Charring and Burn Patterns

Charring in unusual places—like open floor space away from any likely accidental ignition source—could indicate that the fire was deliberately set. Char on the underside of doors or on the underside of a low horizontal surface, such as a tabletop, could indicate that there was a pool of flammable liquid.

## Overhaul

During overhaul, the smoke and steam should begin to dissipate, enabling both brigade members and fire investigators to get a better look at the surroundings. Brigade members should continue to look for indications of the signs, patterns, and evidence previously discussed while conducting overhaul in a way that allows evidence to be identified and preserved. The overhaul process, if not done carefully, can quickly destroy valuable evidence.

If possible, the investigator should take a good look at an area before overhaul begins. The investigator can often quickly identify potential evidence and can help direct or guide the overhaul operation so that it is properly preserved. Evidence located during overhaul should be left where it is found, untouched and undisturbed, until the investigator examines it. Evidence that must be removed from the scene should be properly identified, documented, photographed, packaged, and placed in a secure location.

Fire suppression personnel and investigators must work as a team to ensure that the fire is completely extinguished and properly overhauled, while continually searching for and preserving signs, patterns, and evidence. Taking photographs during this phase of the fire operation is a good idea.

Be careful not to destroy evidence during overhaul ▼ Figure 26-7. Avoid throwing materials into a pile. Use low-velocity hose streams to avoid breaking up and scattering debris needlessly. Thermal imaging devices can be used to find hot spots without tearing apart the interior structure.

Figure 26-7 Be aware of the need to preserve evidence during overhaul.

Watch for evidence that was shielded from the fire and is lying beneath burned debris. For example, a wall clock may have fallen during the fire and been covered with debris. If the clock stopped approximately when the fire broke out, it could be an important piece of evidence.

A fire investigator will always try to determine whether the building contents were changed or removed prior to a fire. If all of a business' computers are missing, the fire might have been set to conceal a theft.

A vehicle fire can start from an accidental electrical short or from a combination of gasoline and matches placed in the passenger compartment. Signs of spilled fuel on the ground around a vehicle may be evidence of an intentional fire. Do not move the vehicle without documenting this spilled fuel; the evidence will be lost.

## Injuries and Fatalities

Any fire that results in an injury or fatality must be thoroughly investigated and the fire scene documented. A search and rescue operation should never be compromised, but it is important to document the location and position of any victims, especially in relation to the fire and the exits.

Clothing removed from any victim should be preserved as evidence. It may contain traces of flammable liquids, and burn patterns can indicate the fire flow. If the clothing is removed in the ambulance or at the hospital, these personnel should be instructed to collect it and keep it as intact as possible.

Document what may be lying under the victim's body after it is removed. This often is a protected area and may reveal important evidence.

# Securing and Transferring the Property

A fire brigade member doesn't need to collect or move any item that might be evidence. They might control the scene but do not need to go beyond that point. Maintaining site integrity is critical to the fire investigation. The building and premises must be properly secured and guarded until the fire investigator has finished gathering evidence and documenting the fire scene. Otherwise, any efforts to determine the cause of a malicious or incendiary fire, no matter how efficient or complete, will be wasted.

If a fire investigator is not immediately available, the premises should be guarded and maintained under the control of the fire brigade or security until the investigation takes place and all evidence is collected. In the interim, take the following steps:

- Suspend salvage and overhaul, and secure the scene. Keep nonessential personnel out of the area. Deny entry to all unauthorized and unnecessary persons.

## Arsonists

Arsonists fall into several categories, with various explanations for their behavior. The fire service has identified two groups who are responsible for a large number of fires: pyromaniacs and juvenile fire-setters. Many other arsonists start fires for a wide range of motives.

### Pyromaniacs

A pyromaniac is a pathological fire-setter. Most are adult males, often loners. They are usually introverted, polite but timid, and have difficulty relating to other people. The fires set by pyromaniacs have the following characteristics:

- Fires are set in easily accessible locations, such as immediately inside entrances, on basement stairs, in trash bins, or on porches.
- Fires are set in structures such as occupied residences of all types, barns, and vacant buildings.
- Accelerants are rarely used. The pyromaniac is impulsive, so materials readily at hand are used.
- Each pyromaniac usually has a pattern, setting fires at the same time of day or night, using the same method, and in similar locations.

### Juvenile Fire-Setters

Juvenile fire setters usually set fires in residential or commercial structures, but they need to be considered in the industrial setting as well. They may set fires along fences or the rear of buildings in remote areas. They may set them simply from curiosity, but the culprits may be juveniles who have a grudge against a security guard and set the fire out of revenge. Listed below are the indicators that the fire brigade member or leader should be aware of if they suspect a juvenile has started a fire in an area of a plant. Juvenile fire-setters are usually divided into three groups according to age: 8 years old and under, 9–12 years old (preadolescent), and 13–17 years old (adolescent). Preadolescent and adolescent fire-setters often exhibit the same personality traits as adult pyromaniacs: introverted, difficulty with interpersonal relationships, and extreme politeness when questioned.

Children under 8 years old are seldom criminally motivated when they set fires; they usually are just curious and experimenting. Children of this age do not really understand the danger of fire. They usually set fires in or near their homes or in nearby fields or vacant lots. They start fires with matches or by sticking combustible material into equipment such as electric heaters that provide an ignition source. The remains of matches, matchboxes, or matchbooks are often found at the point of origin.

Preadolescent fire-starters do not venture far from home, but this group does set most of the fires that involve schools and churches. Preadolescents have motivations other than idle curiosity. The motivations of boys range from spite to revenge and disruptive behavior. Girls usually set less aggressive fires and are motivated by a need for attention or in response to a particular stress, such as a test they don't want to take.

The preadolescent usually does not use elaborate trailers or incendiary devices. They will use common, available accelerants such as gasoline, kerosene, or lighter fluid. The preadolescent often uses whatever materials are at hand and on site, including trash container contents, loose papers, and rags. The fire will show a lack of planning, preparation, and sophistication, particularly if it was a group effort.

The fires set by adolescents are similar to those set by adults. Adolescents have better access to transportation and can travel farther, so they have access to a wider variety of buildings. They also have many of the same motivations of adult fire-setters, such as revenge or attempts to hide larceny, and they often use accelerants. Two-thirds of fires set in vacant buildings are set by adolescents, but no class of property is exempt from harm. Often a great amount of vandalism at the scene will be a clue that the fire was set by an adolescent.

### Arsonist Motives

An arsonist is someone who deliberately sets a fire with criminal intent. Arsonists set all types of fires and have a range of motives. Male arsonists are usually motivated by profit, revenge, or vanity, and often use accelerants. In the past, female fire-setters were seldom motivated by profit and seldom used accelerants, but this seems to be changing.

There are six common motives listed in NFPA 921, *Guide to Fire and Explosion Investigations*. Each of the motives listed applies to an industrial setting and should be considered if the fire appears to be incendiary. Even though the listed items may apply to a residence, individuals set fires for the same reasons in an industrial setting.

1. Vandalism
2. Excitement
3. Revenge
4. Crime concealment
5. Profit
6. Extremism

Arsonists who are motivated by excitement are relatively easy to apprehend because they usually have

## Brigade Member Tips

### Arson Facts

- Arson is often called a "young man's crime;" 54% of those arrested for arson are under the age of 18.
- Only about 2% of all incendiary or suspected incendiary fires lead to conviction.
- Arson is the leading cause of property damage in the United States, resulting in $1.3 billion of property damage annually.
- One out of eight fire fatalities is due to a fire started by arson.

SOURCE: U.S. Arson Trends and Patterns, NFPA

## Brigade Member Tips

### Serial Arson

There are three types of serial (repetitive) arson identified by NFPA 921, *Guide to Fire and Explosion Investigations:*

1. Serial arson involves an offender who sets three or more fires, with a cooling-off period between fires.
2. Spree arson involves an arsonist who sets three or more fires at separate locations with no emotional cooling-off period between fires.
3. Mass arson involves an offender who sets three or more fires at the same site or location during a limited period of time.

SOURCE: NFPA 921, *Guide to Fire and Explosion Investigations,* 2001 Edition, Section 19.4.8.2

some identifiable connection with their targets, such as a disgruntled employee to a company. One category of excitement fires includes fires started by a would-be hero. The would-be hero can be either a discoverer or an assister. The discoverer sets the fire, "discovers" it, and turns in the alarm. The assister sets the fire, sometimes discovers it and reports it, and is always on hand to help or initiate fire extinguishment.

An unfortunate category of excitement arsonists is closely related to the would-be hero. These are fire fighters and would-be fire fighters (persons who have tried to become fire fighters and failed) who intentionally start fires.

Revenge arsonists are often very careful, and make detailed plans for setting the fire and escaping detection afterwards. Arsonists motivated by revenge seldom consider the extent of damage fire can inflict. Their intent is to harm a particular person or group to get revenge for a perceived injustice.

Arson for profit occurs because the arsonist will benefit from the fire either directly or indirectly. Fires that directly benefit the arsonist often involve businesses. The business owner might set the fire or arrange to have it set, to gain an insurance settlement that can be used to cover other financial needs. An increasing problem connected with this arson for profit is fraud. For example, a person may buy and insure a property, only to burn it for the sole purpose of defrauding an insurance company.

The other type of profit arsonist sets a fire to benefit from another person's loss. Included in this motive category are:

- Insurance agents who want to sell more insurance to the victim's neighbors
- Contractors who want to secure a contract for rebuilding or wrecking or to get a salvage
- Competitors who want to drive the victim out of business
- Owners of adjoining property who would like to buy the property and expand their own holdings
- Owners of nearby property who want to prevent an occupancy change, such as the conversion of a building to a drug rehabilitation center

- Photograph the fire scene extensively. Start from the area of least damage and work toward the area of possible origin. Take several pictures of the point of origin from various angles. Photograph any incendiary devices on the premises exactly where they were found.
- If weather, traffic, or other factors could destroy the evidence, take steps to preserve it in the best way possible. Protect tire tracks or footprints by placing boxes over them to prevent dust accumulation. Use barricades to block off the area to further traffic. Rope off areas surrounding plants, trailers, and devices, and post a guard.

The property should be secured by cordoning off the area with fire- or police-line tape. A member of the brigade

Figure 26-8 Some fire brigades have contracts with local companies that provide 24-hour board-up service.

security or law enforcement agency should remain at the scene to ensure that no unauthorized persons cross the line.

Before leaving the scene, make sure that the building is properly secured, and no hazards to public safety exist. Shut off all utilities and seal any openings in the roof to prevent additional water damage. Board up and secure windows and doors to prevent unauthorized entry.

Fire brigades can secure and protect the premises in several ways. Lock and guard gates if necessary, rope off dangerous areas and mark them with signs. Some brigades' companies have contracts with local companies that provide 24-hour board-up services and sometimes hire private security guards to secure a property after a fire ◄ Figure 26-8.

Eventually, when brigade operations are over, the property will be returned to operations as normal. This should not be done, however, until the investigation is complete and all evidence is collected.

## Incendiary Fires

The term incendiary fires refers to all fires that were deliberately started for malicious or criminal intent. In many jurisdictions, arson has a narrower, specific legal definition. This chapter uses arson to mean the malicious burning of property with criminal intent.

Brigade members must be aware of factors that could indicate an intentionally set fire, to report any observations to a brigade leader or fire investigator, and to help protect evidence.

## Indications of Arson

Arson fires have several distinct, recognizable patterns or indications. For example, a deliberate fire might have multiple points of origin or multiple simultaneous fires. An arsonist will ignite fires in different areas to create a large fire as quickly as possible and involve the entire building. Arsonists also use trailers made from combustible materials, such as paper, rags, clothing, curtains, kerosene-soaked rope or other fuels, to spread a fire. Trailers often leave distinctive char and/or burn patterns that investigators can use to trace the spread of a fire.

An incendiary device is a device or mechanism, such as candles, timers, or electrical heaters that is used to start a fire or explosion. Many incendiary devices leave evidence, such as metal parts, electrical components, or mechanical devices, at the point of origin. Often an arsonist will use more than one incendiary device; sometimes, a failed incendiary device is found at the fire scene. If trailers were used, investigators may be able to trace the burn patterns back to an incendiary device.

Evidence of a flammable liquid often indicates an incendiary fire but does not necessarily establish arson as the cause because there could be an accidental explanation for a flammable liquid spill. To prove arson, the fire investigator must determine that a flammable liquid was used and that there is no explanation other than arson for the presence of the flammable liquid.

Extensive burn damage on a floor's surface could indicate that a flammable liquid was poured and ignited. For example, a wood floor might have a pool-shaped burn pattern, or floor tiles could be discolored or blistered with an irregular burn pattern. On concrete and masonry floors, a flammable liquid pattern is usually irregular with various shades of gray to black. Cracked and pitted concrete may also suggest flammable liquid use. Liquids flow to the lowest level possible, so the evidence of pooled liquids could be visible in corners and along the base of walls, where there might also be low levels of charring.

Sometimes the first indications of a possible arson fire are entirely circumstantial. A particular fire could fit into a pattern, such as a series of fires in the same area, at about the same time on the same days of the week. There might be a series of fires in the same location or on the same shift. A brigade member who notices a pattern or a set of similar circumstances should immediately report the observation to a brigade leader or a fire investigator.

# Wrap-Up

## Ready for Review

- Preserving evidence assists brigade members with the primary goal of preventing loss of lives and property loss.
- Fires are caused by either incendiary or accidental causes.
- Basic fire investigation includes locating the point of origin, determining the fuel used, and identifying the ignition source.
- Fire investigation should be performed by one of the following: trained fire brigade investigators, the fire marshal's office, insurance company investigators, or a law enforcement agency.
- Physical evidence must be preserved by maintaining an unbroken chain of custody.
- The brigade member's role in fire investigation is to identify and preserve possible evidence until it can be turned over to a trained fire investigator.
- The brigade member's role in identifying and preserving evidence continues throughout the fire suppression sequence and includes the following factors:
  - Time of day, weather
  - People leaving the scene
  - Extent of fire, number of locations of fire
  - Security of the building
  - Signs of property break-in
  - Vehicles or people in the area
  - Indications of unusual fire situations
  - Unusual color of smoke
  - Position of windows and roof
  - The reaction of the fire during initial attack
  - Abnormal behavior of fire
  - Condition of the building contents
  - Need to coordinate overhaul and evidence preservation activities
  - Need to transfer responsibility of the property from fire suppression personnel to fire investigators
  - Need to secure the property

## Hot Terms

**Accelerants** Materials, usually flammable liquids, used to initiate or increase the spread of fire.

**Accidental fires** Fire cause classification that includes fires with a proven cause that does not involve a deliberate human act.

**Arson** The malicious burning of one's own or another's property with a criminal intent.

**Arsonist** A person who deliberately sets a fire to destroy property with criminal intent.

**Chain of custody** A legal term used to describe the paperwork or documentation describing the movement, storage, and custody of evidence, such as the record of possession of a gas can from a fire scene.

**Circumstantial evidence** The means by which alleged facts are proven by deduction or inference from other facts that were observed first hand.

**Competent ignition source** A competent ignition source is one that can ignite a fuel under the existing conditions at the time of the fire. It must have sufficient heat and be in proximity to the fuel for sufficient time to ignite the fuel.

**Contaminated** A term used to describe evidence that may have been altered from its original state.

**Demonstrative evidence** Term used to describe materials used to demonstrate a theory or explain an event.

**Depth of char** The thickness of the layer of a material that has been consumed by a fire. The depth of char on wood can be used to help determine the duration of a fire.

**Direct evidence** Evidence that is reported first hand, such as statements from an eyewitness who saw or heard something.

**Incendiary device** A device or mechanism used to start a fire or explosion.

**Incendiary fires** Intentionally set fires.

**Mass arson** Involves an offender who sets three or more fires at the same site or location during a limited period of time.

**Physical evidence** Items that can be observed, photographed, measured, collected, examined in a laboratory, and presented in court to prove or demonstrate a point.

**Point of origin** The exact location where a heat source and a fuel come in contact with each other and a fire begins.

**Pyromaniac** A pathological fire-setter.

**Serial arson** A series of fires set by the same offender, with a cooling-off period between fires.

**Spree arson** A series of fires started by an arsonist who sets three or more fires at separate locations with no emotional cooling-off period between fires.

**Trace (transfer) evidence** Evidence of a minute quantity that is conveyed from one place to another.

**Trailers** Combustible material, such as rolled rags, blankets, and newspapers or flammable liquid, used to spread fire from one point or area to other points or areas, often used in conjunction with an incendiary device.

**Undetermined** Cause classification that includes fires for which the cause has not or cannot be proven.

# Brigade Member in Action

**On a third shift you hear the alert that a fire has been discovered in the fabrication area of the plant. As you respond you notice smoke is spreading from that area and you hear the water flow alarm for the sprinkler system. You go to the standpipe, connect a hoseline, and advance it toward the equipment in the fire area.**

**You notice that the fire is still burning under the equipment. Flames are starting to melt hydraulic lines on the machine and the fire is growing. As you advance your hose line and extinguish the fire, you observe that the fire area was all underneath the equipment. You know from working at the plant that this machine does not produce a lot of heat. You observe that there are tools and parts lying nearby. You know from experience that a fire investigator will want to know what you have seen. You make notes of your observations.**

1. Based on your observations of the fire area, what is the most likely reason the sprinkler system didn't extinguish the fire?
   A. The sprinkler system malfunctioned.
   B. The fire was too hot for the sprinkler system water flow to extinguish.
   C. The machinery acted as a shield and prevented water from reaching the fire.
   D. The fire involved a combustible metal.

2. Why does the fire investigator remove and inspect each layer of debris?
   A. To determine how long it burned.
   B. To determine the sequence in which items burned.
   C. To take fingerprints of the material that was burned.
   D. Both A and B.

3. Once the fire scene is secured and the area is safe, the maintenance technicians ask if they can remove their tools and parts. When can the tools and parts be removed?
   A. After you make a list of the tools, parts, and their location.
   B. When the fabrication shop foreman says it is okay to do so.
   C. After you have taken pictures of the area.
   D. After the fire investigator releases the scene.

4. As you consider the circumstances of this fire, you wonder if the cause was intentional. Who should you share your observations with?
   A. The fabrication shop supervisor.
   B. Other brigade members.
   C. The fire investigator.
   D. The plant manager.

# Fire Brigade Leader

# Chapter 27

## NFPA 1081 Standard

**Facility Fire Brigade Leader**

**7.1** General.

**7.1.1** This duty shall involve establishing command, using emergency response procedures, and overseeing the emergency response and other administrative duties as outlined in Chapter 4 of NFPA 600, *Standard on Facility Fire Brigades,* depending on the site organizational statement.

**7.1.2** *Qualification or Certification.* For qualification or certification as an facility fire brigade leader, the member shall meet the JPRs of the level of the facility fire brigade in which they are leading in accordance with the requirements of Chapters 5, 6, or 7 and the JPRs as defined in Sections 8.1 and 8.2.

**7.1.3** *General Requisite Knowledge.* The organizational structure of the facility fire brigade; operating procedures for administration, emergency operations, and safety; information management and record keeping; incident management system; methods used by leaders to obtain cooperation within a group of subordinates; and policies and procedures regarding the operation of the facility fire brigade.

**7.1.4** *General Prerequisite Skills.* The ability to operate at all levels in the incident management system as defined by the National Incident Management System (NIMS) and NFPA 1561, *Standard on Emergency Services Incident Management System.*

**7.2** Supervisory Functions.

**7.2.1** Assign tasks or responsibilities to members, given an assignment at an emergency situation, so that the instructions are complete, clear, and concise; safety considerations are addressed; and the desired outcomes are conveyed.

**(A)** *Requisite Knowledge.* Verbal communications during emergency situations, techniques used to make assignments under stressful situations, and methods of confirming understanding of assigned tasks.

**(B)** *Requisite Skills.* The ability to condense instructions for frequently assigned unit tasks based upon training and SOPs.

**7.2.2** Develop an initial action plan, given size-up information for an incident and assigned emergency response resources, so that resources are deployed to control the emergency.

**(A)*** *Requisite Knowledge.* Elements of a size-up, SOPs for emergency operations, and fire behavior.

**(B)** *Requisite Skills.* The ability to analyze emergency scene conditions, to allocate resources, and to communicate verbally.

**7.2.3*** Implement an action plan at an emergency situation, given assigned resources, type of incident, preliminary plan, and facility fire brigade safety policies and procedures, so that resources are deployed to mitigate the situation and team safety is maintained.

**(A)** *Requisite Knowledge.* SOPs, resources available, basic fire control and emergency operation procedures, an incident management system, rapid intervention crew (RIC) procedures, personnel accountability system, common causes of personal injury during facility fire brigade activities, safety policies and procedures, and basic facility fire brigade member safety.

**(B)*** *Requisite Skills.* The ability to implement an incident management system, to communicate verbally, to supervise and account for assigned personnel under emergency conditions, and to identify safety hazards.

**7.2.4*** Coordinate multiple resources, such as in-house and mutual aid, during emergency situations, given an incident requiring multiple resources and a site incident management system, so that the site incident management system is implemented and the required resources, their assignments, and safety considerations for successful control of the incident are identified.

**(A)** *Requisite Knowledge.* SOPs and local resources available for the handling of the incident under emergency situations, basic fire control and emergency operation procedures, an incident management system, and a personnel accountability system.

**(B)** *Requisite Skills.* The ability to implement the site incident management system, to communicate verbally, and to supervise and account for assigned personnel under emergency conditions.

**7.2.5** Implement support operations at an incident, given an assignment and available resources, so that scene lighting is adequate for the tasks to be undertaken, personnel rehabilitation is facilitated, and the support operations facilitate the incident objectives.

**(A)** *Requisite Knowledge.* Resource management protocols, principles for establishing lighting, and rescuer rehabilitation practices and procedures.

**(B)** *Requisite Skills.* The ability to manage resources, provide power, set up lights, use lighting, select rehab areas, and personnel rotations.

**7.2.6** Direct members during a training evolution, given a training evolution and training policies and procedures, so that the evolution is performed in accordance with safety plans, and the stated objectives or learning outcomes are achieved as directed.

**(A)** *Requisite Knowledge.* Oral communication techniques to facilitate learning.

**(B)** *Requisite Skills.* The ability to distribute issue-guided directions to members during training evolutions.

## Knowledge Objectives

After completing this chapter, you will be able to:

- Describe the responsibilities of the fire brigade leader.
- Describe specific challenges the fire brigade leader faces.
- Describe mechanisms the fire brigade leader can use to face these challenges.

## Skills Objectives

There are no skills objectives for this chapter.

# You Are the Brigade Member

**You are the fire brigade leader at a chemical plant with various processing areas that utilize flammable liquids. You have a six-member advanced exterior brigade on each shift, and a 1000 gallon-per-minute foam pumper. A response is initiated for an explosion and fire involving a tanker truck unloading area for ethanol. There is a running fuel fire, and an adjacent office trailer facility is exposed to the fire.**

1. ***What resources should you access to plan your response?***
2. ***What assignments do you give your brigade members as they respond?***
3. ***What assistance will you need to handle this incident?***

## Introduction

Fire brigade leaders have many responsibilities, both during emergency operations and with the day-to-day operations of the fire brigade (▶ **Figure 27-1**). During an emergency, the fire brigade leader will need to establish command at incidents, oversee the emergency response, and implement the brigade's emergency response procedures. On a daily basis, the fire brigade leader may be responsible for a wide range of administrative duties, including:

- Establishing programs to accomplish the items identified in the facility fire brigade organizational statement
- Establishing the size and organization of the facility fire brigade
- Coordinating and scheduling meetings necessary to accomplish brigade business
- Selecting brigade equipment and establishing and maintaining fire protection equipment inspection programs
- Coordinating the maintenance and review of necessary reports and records
- Maintaining liaison with local fire authorities
- Providing information on hazardous materials and processes to which the brigade can be exposed
- Establishing job-related physical performance requirements for brigade members
- Establishing a chain of command within the brigade to act in the absence of the fire brigade leader
- Assisting in the selection process of brigade members
- Establishing and maintaining a brigade roster
- Selecting assistant brigade leaders as appropriate to the size of the brigade and keeping them informed of all operations of the brigade
- Developing preemergency plans for site-specific hazards
- Issuing written reports on the status of the facility fire brigade to management
- Assisting in fire investigations.

Most importantly, fire brigade leaders must meet the job performance requirements of the level of the facility fire brigade that they are leading.

**Figure 27-1** The company-level fire brigade leader is responsible for the supervision, performance, and safety of a crew of fire brigade members.

Fire brigade leaders must always keep in mind several fundamental concepts. The fire brigade organizational statement defines the brigade and its capabilities/limitations as set forth by plant management. In emergency situations where the risk to facility fire brigade members is unacceptable, the response activities should be limited to defensive operations. Regardless of the risk, actions shall not exceed the scope of the organizational statement and standard operating procedures (SOPs).

## Supervisory Functions

A key role of the fire brigade leader is to supervise the fire brigade during emergency operations. This includes assigning tasks or responsibilities to members so that the instructions are complete, clear, and concise; safety considerations are addressed; and the desired outcomes are conveyed. Fire brigade leaders must be able to verbally communicate effectively during emergencies, make assignments under stressful situations, and confirm understanding of assigned tasks under pressure. For unit tasks that are frequently assigned, fire brigade leaders must be capable of condensing instructions based upon brigade training and SOPs. Brigade leaders must develop a chain of command in advance of incidents and provide for enough subordinate leaders to maintain a command presence during the leader's absence and to account for vacations and shift changes. There is no way to eliminate stress during an emergency response. The brigade leader can reduce stress by supervising response operations using consistent command and control procedures. Brigade leaders must be able to approach response incidents in clinical terms, not emotional terms (▶ Figure 27-2). If brigade leaders are emotionally involved in the incident, they will not be able to make sound decisions. Experience and lessons learned will strengthen a brigade leader's ability to remain focused on the incident and the tasks to be completed. Don't make it personal; you didn't make the world combustible.

The brigade leader should ensure that SOPs for site-specific conditions and hazards are developed in writing, reviewed, and maintained (▶ Figure 27-3). These procedures should address the site-specific functions identified in the fire brigade organizational statement and should include information regarding site-specific hazards to which fire brigade members can be exposed during a fire or other emergency. To ensure safety of personnel, SOPs should address the site-specific limitations of emergency response organizations and be accessible to all fire brigade members. A key item in the procedures should be that the shift facility fire brigade leader is notified of all major fire protection systems and equipment that are out of service. Brigade SOPs are a foundation upon which brigade training, drills, and responses are built.

Figure 27-2 Brigade leaders must be able to approach response incidents in clinical terms, not emotional terms.

Figure 27-3 Site-specific SOPs must be written, reviewed, and maintained.

The fire brigade leader will be required to develop an initial action plan for incidents that should be based upon size-up information for the incident and assigned emergency response resources. Size-up includes the many variables that must be collected from the time of dispatch, during response, and upon arrival and can include structural type and occupancy, fire involvement, number of occupants, materials spilled or involved in fire, wind direction, topography, and other observations relevant to the incident. As this information is being gathered, the response resources must be effectively deployed to control the emergency. As the resources are deployed, team safety must be maintained. To do all of this effectively, the fire brigade leader must be knowledgeable of fire brigade SOPs, the resources that are available for emergency response, basic fire control and emergency operation procedures, incident management systems, rapid intervention crew (RIC) procedures, personnel accountability systems, common causes of personal injury during facility fire brigade activities, safety policies and procedures, and basic facility fire brigade member safety. The fire brigade leader must be able to implement an incident management system, to communicate verbally, to supervise and account for assigned personnel under emergency conditions, and to identify safety hazards to respond to the wide variety of emergency situations a facility fire brigade may encounter (► Figure 27-4).

Along with operational functions, the fire brigade leader needs to ensure that support operations are put into place at an emergency incident. Support functions such as scene lighting adequate for the tasks to be undertaken, personnel rehabilitation, and support operations necessary to facilitate the incident objectives must be put into place. The fire brigade leader must know how these support functions can be activated, and how and where to best locate them at an incident scene.

Fire brigade leaders are also likely to be responsible for planning and scheduling drills and training for the brigade. Fire brigades should not be expected to perform duties that they are not trained to perform. The prevention of accidents, injury, death, and illness during performance of any facility fire brigade function shall be an established goal of the brigade's training and education. The quality and frequency of brigade training and education should ensure that facility fire brigade members are capable of performing their assigned response duties in a manner that does not present a hazard to themselves or endanger other personnel. Drills and training should be based on realistic scenarios for credible site-specific emergencies.

Where mutual aid or other outside agencies play an important role in the emergency response procedures of the site, drills and training should be conducted with these agencies. Drills may be either announced or unannounced. Announced drills have the advantage of brigades being able to train and prepare in advance, which can improve overall performance while still providing a level of performance evaluation. Announced drills are likely to be better received by plant management because they provide for the release of personnel to participate in the drill and allow the use of an actual operations location to perform the drill. Unannounced drills may provide a more effective gauge of the readiness of the brigade, but can cause discontent among both brigade membership and operations management along with difficulties in being able to get full participation of mutual aid or other outside agencies. Unannounced drills can also create safety challenges, because responders may take additional risks during these exercises, such as speed of response, if they are unaware that it is a drill. Brigade leaders must be able to manage the implications of whichever type of drill is conducted. In all cases, drill objectives must be clearly developed in advance. Objectives should include evaluating the firefighting readiness of the fire brigade, brigade leader, and/or fire protection systems and equipment.

## Brigade Member Tips

### Practice Like You Play

One of the keys to a consistent performance at emergency incidents is to "practice like you play." Every training activity and every response should be approached with the same degree of professionalism and strict adherence to SOPs. This approach is essential to ensuring that the brigade performs as expected when the situation demands maximum performance. The brigade leader should ensure that fire brigade members properly use personal protective gear and follow SOPs at every incident.

Most fire brigade responses involve relatively minor situations, such as activated fire alarms and fires that involve less than a room and contents. To prepare the fire brigade for larger fires and critical situations, each response should require the same consistent approach. The following are some examples:

- All fire brigade crew members don their protective clothing, sit down, and fasten their seat belts on the apparatus before it moves, and remain seated and belted as long as the apparatus is moving.
- The first arriving engine company lays a supply line.
- The first arriving ladder company positions and prepares to deploy the aerial.
- Crews entering the building wear full protective gear, including self-contained breathing apparatus; carry their tools; and perform their assignments for a working fire.

If there is any doubt about whether there is a fire in the structure, all operations should be performed with the assumption that there is. Without creating a customer service problem, the crew should search the entire building, including attics and basements. This allows each fire brigade member to develop awareness of the building environment. More importantly, when the crew encounters a working fire, the standard fireground tasks have been practiced and are familiar.

To make drills and training most effective, they should be critiqued for performance of the organization. It is also useful to critique the actual development and execution of the drill or training scenario. Lessons learned from critiques should be incorporated into the training and education program to improve any performance that is below established standards. It should be noted that although training benefits are achieved through drills exercising the knowledge and skills of the fire brigade, drills should not take the place of regular training sessions. It is vital that fire brigade members be trained on equipment and techniques *before* they are expected to use them in a drill or actual emergency. Otherwise it will be difficult for brigade members to demonstrate competence in performing these tasks in a drill. If training is to be conducted off-site, the brigade leader will have to establish contingency plans to ensure the plant is properly protected while personnel are away from the facility.

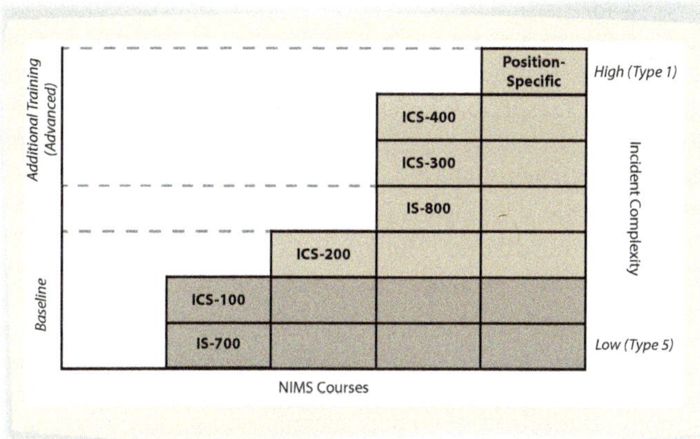


**Figure 27-4** Courses critical for facility fire brigade leaders to operate at all levels of the incident management system.

FEMA http://training.fema.gov/EMIWeb/is/ICSResource/TrainingMaterials.htm

NFPA 1081, *Standard for Facility Fire Brigade Member Professional Qualifications* indicates that responses to actual emergencies can reduce the necessity to conduct drills, providing the actual responses occur with sufficient frequency and as long as the fire brigade performance during these responses is evaluated in accordance with established performance objectives and is properly documented. However, strong prevention programs reduce the frequency of response, and holding training and drills is critical to smooth and safe brigade operations even if the brigade handles emergency responses on a regular basis. Training and drills can also focus on tools and techniques, particularly specialized ones, which are rarely utilized. Drills can also be a valuable tool to determine the frequency of refresher training necessary to maintain fire brigade skills. In all cases, facility fire brigade members designated as leaders should receive training and education commensurate with their response duties, which should be more comprehensive than that provided to the other facility fire brigade members. Fire brigade leaders must be competent not only in technical firefighting skills but also in the management and training skills necessary to lead the brigade. To help gain these skills, fire brigade leaders should take advantage of management leadership and methods of instruction courses.

## Personnel Issues

Brigade leaders must be familiar with personnel issues as they affect brigade operations. Brigade members may be assigned to the brigade on a full-time basis; however, typically they are part-time personnel who hold other jobs in the facility. Full-time fire brigade members can be easier to manage when scheduling training, drills, or other events because they would generally be available to participate as long as they are working on the date and time that the event is being held. Part-time fire brigade members with other jobs present a number of different challenges. First, if the event is being held during the members' normal working hours, the brigade leader must collaborate with the members' supervision to ensure the members can participate in the event. If the event is being held during a time when the members are not scheduled to work, collaboration with the members' supervision is still necessary, and the members must also be consulted to ensure that they are available to participate on their time off. In all cases, the fire brigade leader must be familiar with any personnel contracts or agreements that are in place that deal with issues such as overtime and callback, to ensure that compliance with these contracts or agreements is maintained. Brigade leaders must also be familiar with grievance and similar procedures, and agreements should be determined in advance as to how these are to be handled for both employees who are directly supervised by the fire brigade leader and those who are supervised by the fire brigade leader either indirectly or on a part-time basis.

Frequent use of part-time personnel during their normal work hours may conflict with the needs of their normal operating departments, particularly during busy shifts or seasons for the plant. It is important for the brigade leader to be aware of this, and attempt, where possible, to schedule events so that part-time personnel are not utilized during these busy times. Another option is to use part-time personnel during overtime hours. However, some part-time personnel may not even be able to be utilized during overtime hours because their normal department may be busy enough that they are needed to work overtime in their normally assigned area. Besides training, drills, and other events, it is likely that brigade leaders will need to schedule meetings of all or part of the brigade to accomplish specific tasks such as brigade officers' meetings, committees to specify new equipment, or recognition of various achievements. Leaders will need to be sensitive to include personnel on all shifts and schedule these activities in a way that makes it as convenient as possible for members on the various shifts to attend.

Technologies such as teleconferences, videoconferences, and webcasts can be utilized to include personnel who are stationed at different sites or who work different shifts and may find it difficult to come to a central meeting location, or come back to the plant to attend a meeting that may only run for a relatively short period of time.

If part-time brigade members are released during their normal work hours for training, drills, meetings, and emergency assignments, brigade leaders should keep in mind the importance of maximizing the use of the part-time members' time, and attempt to demobilize during the incident to release part-time members back to their normal job functions as soon as it can be safely done. At the same time, it is important that brigades don't take undue advantage of mutual aid assistance just to allow part-time brigade members to return to their jobs, unless this is agreed to in advance by both the brigade leadership and the mutual aid organization leadership. It is likely that members of the mutual aid organization will have their own contractual, overtime, and other personnel issues that they have to deal with. Being sensitive to these issues will maintain good relationships with plant and mutual aid organization management.

Fire brigade leaders leading voluntary fire brigades have to rely on their leadership skills even more than those leading full-time brigades. Pride and personal commitment are the main forces that compel a volunteer to remain active and loyal to the organization. The volunteer fire brigade leader must pay attention to the satisfaction level of every member and be alert for issues that create conflict or frustration. Some of the unique issues that a fire brigade leader of volunteer members must consider are as follows:

- Work and personal issues often affect volunteer availability. Busy shifts or departmental emergencies may conflict with training or meeting availability. Family situations, such as child or elder care, can profoundly affect the time a volunteer is able to devote to the fire brigade.
- Extensive training requirements can have an impact on volunteer availability. It is easy to assign a full-time brigade member to attend a 40-hour training class that occurs during the regular workweek. It is difficult to accomplish the same training when the opportunities are restricted to times when the volunteer member is available.
- Interpersonal conflicts can exist between members. Conflicts can drive away members and erode fire brigade preparedness. The volunteer brigade leader must act quickly when such problems are identified. The rights and reasonable expectations of individual members must be considered, without compromising on the mission or good of the organization.

## Mutual Aid

In most cases, fire brigades will not be able to handle all scenarios at a facility by themselves. It simply is not cost-effective to maintain all of the tools, equipment, and staffing necessary on a round-the-clock basis, except perhaps in some of the most remote locations, to deal with all possible incidents. It is important for the brigade to conduct an evaluation of the types and nature of mutual aid that they would need to call upon in an emergency, and the fire brigade leader needs to ensure that this evaluation is completed.

Once the evaluation is completed, brigade management must develop a close working relationship with all of the emergency response organizations they have identified that could reasonably be expected to respond to the facility during an emergency. Written mutual aid agreements should be developed and approved with each of the organizations that the brigade anticipates requesting assistance from. A joint incident management system should be put into place. This management system should clearly identify the roles and responsibilities of each organization at an incident. Walkthroughs of key hazard areas, together with drills that exercise the management system and operational relationships of each organization, should be held regularly but at the very least on an annual basis. During the walkthroughs, compatibility of emergency equipment among the various response organizations should be reviewed, including fire hose threads and adapters, foam or other firefighting chemicals, protective equipment, or other equipment organizations would be expected to share or utilize at responses at facilities to which mutual aid is provided.

Documented preplans should be made available, showing access, information on hazards to brigade members, and information regarding the use of fixed systems to protect the hazards. Mechanisms for communication between the facility fire brigade and mutual aid organizations must be determined in advance. Are common radio frequencies available? In some cases, brigades are permitted to access municipal radio frequencies, although normal plant operations may occur on plant operational frequencies. In other cases, brigades on mutual aid frequencies should consider the exchange of respective portable radios, although this will mean greater communication coordination efforts will be needed by both the fire brigade leadership and mutual aid leadership at incidents. Finally, it is important for both the facility brigade and mutual aid organizations to be cognizant of not only what resources and equipment are available from each other, but also how the equipment works. Drills and training exercises should encourage interaction with both organizations so that all personnel are familiar with operating each other's equipment.

## Safety

The incident management system a brigade leader puts into place at most emergency scenes must include mechanisms to provide for the brigade's safety and should include a safety officer as a command staff member. In the United States, Occupational Safety and Health Administration and Environmental Protection Agency regulations require an incident safety officer at hazardous materials incidents. If an incident safety officer is not appointed, the Incident Commander will also need to function as the incident safety officer. A system to manage accountability of personnel on the scene must be in place and should be incorporated into the site incident management system or SOPs. This system should include the interface between the site personnel and the outside mutual aid personnel, recognizing that the personnel accountability system for the site can be different from that of the outside mutual aid organizations. The incident management system should also include a mechanism to provide data on the hazards of chemicals involved in an incident as part of the incident demobilization process.

A rapid intervention crew (RIC) must be implemented for structural fires, confined space rescues, hazardous materials incidents, or other incidents that place personnel in environments that are immediately dangerous to life and health (IDLH). A RIC must be a minimum of two personnel who are provided with personal protective equipment that is at least equal to, if not greater than, the level of protective equipment being worn by personnel operating within the IDLH atmosphere. Depending upon the size and complexity of the incident scene, the facility fire brigade leader may need to assign additional RICs.

## Preplanning

The fire brigade leader must be the facility's champion for preemergency planning (▼ Figure 27-5). The brigade leader ensures that the preincident plan is done effectively and kept current, and that preplan manuals are accessible to brigade members for responses. NFPA 1620, *Recommended Practice for Preincident Planning* provides an excellent guideline for conducting the preplanning process. The hazards of chemicals and other materials present in the plant must be reviewed with personnel as part of the preplanning and training process. The brigade leader should also ensure that preemergency planning is conducted in conjunction with mutual aid or other outside agencies where these agencies play an important role in the emergency response procedures of the site. Finally, the brigade leader should ensure that preplans are routinely reviewed and used as a basis for training evaluations that the brigade conducts.

Figure 27-5 Fire brigade leaders often conduct preincident planning during the course of their workday.

## Administrative Issues

Brigade leaders may be expected to perform a number of administrative tasks. The brigade leader should routinely evaluate the size and organization of the brigade. Are full-time brigade members needed to adequately perform the tasks assigned to the brigade? A brigade may be able to assume additional duties related to maintenance, inspection, and hazard standby (such as confined space entries) that can justify the use of full-time members. It may be appropriate to utilize full-time members on some shifts, while covering other shifts with part-time members. Are part-time brigade members volunteers, or are they required by their particular job position to be brigade members? Using volunteers has proven in many cases to provide a greater level of dedication to the brigade's goals and objectives. Brigade members who are required by their particular job position to be brigade members provide a more stable number of responders per shift. The job position requirement brigade may limit brigade participation for employees who may be well-qualified and would like to participate, but are unable because their particular job is not assigned to be part of the brigade membership. Should management–employee friction develop, there is the potential that voluntary brigade members may be unwilling to volunteer for brigade activities; the fire brigade leader needs to be aware of this potential.

Selection of brigade members is a challenging task. In the case of a job position requirements brigade member, it can be rather simple. If someone performs a certain job, they are on the brigade. It then becomes the responsibility of the fire brigade leader to properly prepare and train that member to be a part of the brigade. It will be useful for the brigade leader to work with plant management to include some form of prerequisites related to brigade functions in the job requirements for those jobs designated to be brigade members. For voluntary brigade members, the fire brigade leader will have a bit more control of who becomes a fire brigade member. Physical requirements for the job are important. Brigade members should be physically

# Voices of Experience

**"I discovered shortly after releasing my grip on the rail of the platform that the spring-loaded clip on my truck-style belt had apparently disengaged as I climbed over the flooring."**

Brigade management is not just limited to organizing selected routine and emergency scene functions; it also requires leadership. As a leader, you need to accumulate life lessons and incorporate them into the organization.

Years ago, I was a young fireman in a volunteer community fire department. The department operated an articulating elevated platform equipped with a portable controlled-descent self-rescue device. On numerous occasions, several of my peers and I would raise the (80-foot) platform and descend out of the basket using this rescue device. On my last occasion to do so, I discovered shortly after releasing my grip on the rail of the platform that the spring-loaded clip on my truck-style belt had apparently disengaged as I climbed over the flooring. Fortunately, the clip had fastened again and just caught the edge of the lowering strap. In essence, my life was literally hanging within a quarter-inch away from disengagement, dropping me to the pavement below.

In reflection and upon later examination of that very moment at which I came close to becoming a training death statistic, it's easy to see several areas ripe for improvement. I had not been properly trained in the equipment, nor had any of my peers. The "training" was not supervised by any qualified individual. The truck-style belt was an inappropriate piece of equipment for this operation. And as it turned out, the manufacturer recommended that this rescue device be for one-time use only.

I took these lessons learned to heart. As the chief, I insist that my team receive pertinent training for their expected responsibilities that is conducted by qualified and knowledgeable instructors. Equipment is routinely inspected for operability, and everyone must understand the use and limitations of personal protective equipment and other tools. In our emergency response business, training needs to start on day 1 and only conclude the day you retire.

Leading an organization is much like emergency scene size-up. Size-up never ends and really requires satisfying three potential questions:

1. What is going on?
2. What can we do to improve the situation?
3. Does this require more resources?

William C. Kelly
Northeast Refining, Sunoco Inc.
Marcus Hook, Pennsylvania

qualified to perform brigade duties, and this should be included in brigade selection criteria. If a brigade leader has multiple applicants to choose from, prerequisites such as previous emergency training can be a part of the selection criteria.

The brigade's equipment must be maintained and tested. Will this be done by brigade members, plant maintenance personnel, or contracted out? Maintenance and testing must be completed by trained or certified personnel. It is challenging, except in larger or remote facilities, to have in-plant trained or certified personnel for testing and maintenance on complicated equipment such as foam systems, apparatus, aerial ladders, and breathing apparatus. In this case, the best option may be to contract this work to outside qualified vendors. Annual testing, such as hose pressure testing, can easily be scheduled well in advance by either contracting this service out, conducting it during normal work hours, or scheduling it on an overtime shift. Although it is tempting to replace normally scheduled drills or training with equipment maintenance and testing, brigade leaders should be careful not to compromise the completion of necessary training activities for this purpose. Testing and maintenance of equipment facilitate familiarity of the equipment for brigade members but in most cases do not replace actual training with the equipment.

Selection of brigade equipment is also an important administrative task. There are several considerations when selecting equipment, including practicality, budget, and availability of similar equipment from mutual aid organizations.

Where possible, determine what support equipment already exists at the facility and determine if it can be made available in an emergency. Many plant maintenance staff have a variety of tools such as pumps and cutting and prying tools that have utility in emergency response. A thermal imaging camera is a useful tool for electricians to evaluate transformers and other electrical equipment for overheating from a maintenance perspective. It is vital that the fire brigade leader reviews with the equipment "owners" the ability to access this equipment in an emergency. If new equipment will be purchased, the brigade leader will find strong value in getting input from various brigade members on what is practical for your facility. For larger equipment, such as apparatus, it may be best to form a committee to evaluate options and develop the specifications for the equipment.

Brigade leaders must maintain reports and records for the brigade. First and foremost, reports and records required by their employer must be maintained. This may include time sheets, injury reports, maintenance and inspection records, and purchasing records. The plant's normal systems should be used for these records wherever possible, and the brigade leader must coordinate with the supervisors of part-time brigade members to ensure who is responsible for completing reports such as time sheets and injury reports related to these members. Reports of responses should be maintained, wherever possible in accordance with NFPA standards. Response reports should indicate which brigade members responded to the incident for both time reporting and future trending and analysis of response capabilities.

## Investigations

Brigade leaders are likely to have some responsibility for conducting investigations. This includes investigations into the cause of incidents the brigade responds to, as well as near-miss or injury investigations related to brigade personnel and equipment. It is vital that brigade leaders do not attempt to assign blame for an incident while the incident is occurring. Deal with the emergency situation at hand, and conduct the detailed investigation later. It is likely that an investigation team will need to be established for serious incidents (▼ Figure 27-6). Work with plant management and staff groups to provide proper representation for this team.

Scenes must be preserved so that effective incident investigations can be conducted (► Figure 27-7). There may be legal requirements from law enforcement or other outside agencies that make them responsible for the investigation, and the fire brigade leader should provide full cooperation to them. The brigade leader can assist with expertise related to plant procedures, systems, and documentation of brigade activities at the emergency scene to assist in incident reconstruction. Be prepared to gather information such as dispatch logs, radio and telephone record data, and similar information that will help the team fully understand what occurred during the incident.

Figure 27-6 The investigator is interested in determining fire cause and origin to help prevent future fires and prosecute criminal actions.

Figure 27-7 The fire scene evidence must be protected from the public and from excessive overhaul and salvage.

## Brigade Member Tips

### Building Trust in the New Fire Brigade Leader

It is vital for a new supervisor to work to build trust with the workers. Failure to establish trust from the beginning of the relationship can quickly undermine a supervisor's career. This is particularly true for a new fire brigade leader. Fire brigade members make or break a new fire brigade leader, and they will make every effort to protect themselves from a brigade leader who appears to be unsafe, unstable, or unprepared for the job.

How can the new fire brigade leader build trust? Here are five suggestions:

1. Know the fire brigade leader job, both administrative and tactical.
2. Be consistent. Strive to provide a measured response to any problem, emergency, or challenge.
3. Walk your talk. Actions speak much louder than words.
4. Support your fire brigade members. Make sure that you help meet their physiological, safety, security, and order needs.
5. Make fire brigade members feel strong. Help them to become competent and confident in their emergency service skills. Show how they can control their destiny.

Although brigades should strive to achieve zero incidents, injuries and near-miss incidents involving brigade members and equipment should be investigated and reported in accordance with plant procedures. Root causes should be determined, and corrective action plans developed and implemented as appropriate. Many plants have safety departments with experts in investigating injury and near-miss incidents; these individuals can be used as resources to assist the brigade leader in conducting such investigations.

## Summary

Leadership requires the fire brigade leader to provide purpose, direction, and motivation to fire brigade members. Success depends on the brigade members not only understanding and accepting the vision and mission but also having them become a part of their own personal and group culture. The fire brigade leader applies personal attributes to the leadership process. This is a lifelong endeavor as the brigade leader develops and refines the beliefs, values, ethics, character, knowledge, and skills that make up the sum of a person.

## Additional Resources

Angle, James S. Gala, Michael. Harlow, David. Lombardo, William. Maciuba, Craig. (2015). *Firefighting Strategies and Tactics*, Third Edition. Burlington, MA: Jones & Bartlett Learning.

FEMA Emergency Management Institute ICS Training Courses:

- ICS-100 Introduction to the Incident Command System
- ICS-200 ICS for Single Resources and Initial Action Incidents
- ICS-300 Intermediate ICS for Expanding Incidents
- ICS-400 Advanced ICS
- IS-700 NIMS, an Introduction
- IS-800 National Response Framework (NRF), an Introduction

Hildebrand, Michael S. Noll, Gregory G. (2014). *Hazardous Materials: Managing the Incident, Fourth Edition*. Burlington, MA: Jones & Bartlett Learning.

International Association of Fire Chiefs, National Fire Protection Association. (2015). *Fire Officer: Principles and Practice, Third Edition*. Burlington, MA: Jones & Bartlett Learning.

International Association of Fire Chiefs, National Fire Protection Association. (2012). *Chief Officer: Principles and Practice*. Burlington, MA: Jones & Bartlett Learning.

International Association of Fire Chiefs, International Association of Arson Investigators, National Fire Protection Association. (2012). *Fire Investigator: Principles and Practice to NFPA 921 and 1033, Third Edition*. Burlington, MA: Jones & Bartlett Learning.

International Association of Fire Chiefs, National Fire Protection Association. (2012). *Fire Inspector: Principles and Practice*. Burlington, MA: Jones & Bartlett Learning.

Lesage, Paul. Dyer, Jeff. Evans, Bruce. (2011). *Crew Resource Management: Principles and Practice*. Burlington, MA: Jones & Bartlett Learning.

Noll, Gregory. Hildebrand, Michael. Rudner, Glen. Schnepp, Rob. (2014). *Hazardous Materials: Managing the Incident, Fourth Edition*. Burlington, MA: Jones & Bartlett Learning.

Schnepp, Rob. (2010). *Hazardous Materials Awareness and Operations*. Burlington, MA: Jones & Bartlett Learning.

Walsh, Donald. (2012). *National Incident Management System: Principles and Practice, Second Edition*. Burlington, MA: Jones & Bartlett Learning.

Zimmerman, Don. (2015). *Fire Fighter Safety and Survival, Second Edition*. Burlington, MA: Jones & Bartlett Learning.

# Wrap-Up

## Ready for Review

- Leading a fire brigade requires skills in a number of different areas. First and foremost, the brigade leader must meet the performance requirements for fire brigade members at the level of brigade response being led.
- Plant management will set the organizational structure for the fire brigade and provide the resources to staff, equip, and train the brigade.
- The brigade leader must be the champion for preplanning the facility and must effectively interface with mutual aid organizations that are expected to support the brigade's operations during incidents.
- Voluntary brigade members may have job conflicts that can reduce the amount of time they are able to participate in brigade activities.
- Plant management needs to be kept apprised of brigade activities and status of roster, training, and equipment.
- The brigade leader is responsible for conducting investigations into the cause of incidents as well as near-miss or injury investigations related to personnel or equipment.

## Hot Terms

**Budget** An itemized summary of estimated or intended expenditures for a given period along with proposals for financing them.

**Chain of command** The superior-subordinate authority relationship that starts at the top of the organization hierarchy and extends to the lowest levels.

**Full-time fire brigade member** A fire brigade member who is assigned to the fire brigade full-time; normally supervised by the fire brigade leader.

**Job position requirements brigade member** An employee who is a member of a fire brigade because his or her job position requires them to be a fire brigade member.

**Part-time fire brigade member** A fire brigade member who holds another job with the employer and participates in the fire brigade on a part-time basis.

**Preincident plan** A written document resulting from the gathering of general and detailed data to be used by responding personnel for determining the resources and actions necessary to mitigate anticipated emergencies at a specific facility.

**Rapid intervention crew (RIC)** A minimum of two fully equipped personnel on site, in a ready state, for immediate rescue of disoriented, injured, lost, or trapped rescue personnel.

**Standard operating procedures (SOPs)** A written organizational directive that establishes or prescribes specific operational or administrative methods to be followed routinely for the performance of designated operations or actions.

**Voluntary brigade member** An employee who volunteers to be a member of the fire brigade.

# Brigade Member in Action

**Due to cost-cutting measures, your fire brigade, which has previously been a full-time, interior structural brigade, is being downgraded by plant management to a voluntary, part-time advanced exterior brigade.**

1. Which of these items does not need to be changed as a result of this fire brigade downgrade?
   - A. Facility preemergency plans
   - B. Fire brigade organizational statement
   - C. Fire brigade SOPs
   - D. Mutual aid agreements

2. Previously, the brigade trained daily as a part of their regular shift. What is the best way to complete training under the new organizational structure?
   - A. No training is necessary for advanced exterior brigades.
   - B. Continue to hold training daily as a part of their regular shift.
   - C. Monthly scheduled training classes.
   - D. Weekly, unannounced training drills to test the brigade's competency.

3. As an advanced exterior brigade, the brigade leader can allow the brigade members to do the following at an incident:
   - A. Enter an office building that is on fire to search for a missing employee.
   - B. Set up a master stream to protect exposures from a chemical tank fire.
   - C. Act as the RIC for municipal fire fighters members called in to assist the plant at a structure fire in the manufacturing area.
   - D. Perform vertical ventilation at a plant warehouse building fire.

4. Several voluntary members of the newly reorganized brigade appear to be only volunteering for the brigade due to extra overtime afforded to brigade members to attend training. They are not actively participating in incidents as valued responders. How can the brigade leader realistically handle this?
   - A. Cancel training on overtime.
   - B. Meet with the voluntary brigade members on a one-on-one basis to determine their value to and participation with the brigade.
   - C. Request plant management return the brigade to full-time status.
   - D. Meet with the voluntary brigade members' supervisors and request that they discipline the brigade members for their lack of participation.

# NFPA 1081: *Standard for Facility Fire Brigade Member Professional Qualifications*, 2018 Edition

## Chapter 4: Incipient Facility Fire Brigade Member

**4.1 General.** This duty shall involve initiating communications, using facility communications equipment to effectively relay oral or written information, responding to alarms, returning equipment to service, and completing incident reports, according to the JPRs in 4.1.1 through 4.2.3.

**4.1.1 Qualification or Certification.** For qualification or certification at the incipient facility fire brigade level, the facility fire brigade member shall meet the entrance requirements in Chapter 1 and Sections 4.1 and 4.2, the site-specific requirements in Section 4.3 as defined by the management of the facility fire brigade, and the requirements defined in Chapter 4 of NFPA 1072.

**4.1.2 Basic Incipient Facility Fire Brigade Member JPRs.** All basic incipient facility fire brigade members shall have a general knowledge of basic fire behavior, operation within an incident management system, operation within the emergency response operations plan for the site, the standard operating and safety procedures for the site, and site-specific hazards.

**4.1.2.1** Initiate a response to a reported emergency, given the report of an emergency, facility standard operating procedures, and communications equipment, so that all necessary information is obtained and communications equipment is operated properly.

**(A)** *Requisite Knowledge.* Procedures for reporting an emergency.

**(B)** *Requisite Skills.* The ability to operate facility communications equipment, relay information, and record information.

**4.1.2.2*** Transmit and receive messages via the facility communications system, given facility communications equipment and operating procedures, so that the information is promptly relayed and is accurate, complete, and clear.

**(A)** *Requisite Knowledge.* Facility communications procedures and etiquette for routine traffic, emergency traffic, and emergency evacuation signals.

**(B)** *Requisite Skills.* The ability to operate facility communications equipment and discriminate between routine and emergency communications.

**4.1.2.3** Respond to a facility emergency, given the necessary equipment and facility response procedures, so that the team member arrives in a safe manner.

**(A)** *Requisite Knowledge.* Facility layout, special hazards, and emergency response procedures.

**(B)** *Requisite Skills.* The ability to recognize response hazards and to safely use each piece of response equipment provided.

**4.1.2.4*** Return equipment to service, given an assignment, policies, and procedures, so that the equipment is inspected, damage is noted, the equipment is cleaned, and the equipment is placed in a ready state for service or is reported otherwise.

**(A)** *Requisite Knowledge.* Types of cleaning methods for various equipment, correct use of cleaning materials, and manufacturer's or facility guidelines for returning equipment to service.

**(B)** *Requisite Skills.* The ability to clean, inspect, and maintain equipment and to complete recording and reporting procedures.

# Appendix A

**4.1.2.5*** Complete a basic incident report, given the report forms, guidelines, and incident information, so that all pertinent information is recorded, the information is accurate, and the report is complete.

**(A)** *Requisite Knowledge.* Content requirements for basic incident reports, the purpose and usefulness of accurate reports, consequences of inaccurate reports, and how to obtain necessary information.

**(B)** *Requisite Skills.* The ability to collect necessary information, proof reports, and operate facility equipment necessary to complete reports.

**4.1.3 Activating an Emergency Call.** Activate an emergency call for assistance, given vision-obscured conditions, personal protective equipment (PPE), and department standard operating procedures (SOPs), so that the fire fighter can be located and rescued.

**(A)** *Requisite Knowledge.* Personnel accountability systems, emergency communication procedures, and emergency evacuation methods.

**(B)** *Requisite Skills.* The ability to initiate an emergency call for assistance in accordance with the AHJ's procedures, the ability to use other methods of emergency calls for assistance. [**1001**:4.2.4(B)]

**4.2 Manual Fire Suppression.** This duty shall involve tasks related to the manual control of fires and property conservation activities by the incipient facility fire brigade member.

**4.2.1*** Extinguish incipient fires, given an incipient fire and a selection of portable fire extinguishers, so that the correct extinguisher is chosen, the fire is completely extinguished, proper extinguisher-handling techniques are followed, and the area of origin and fire cause evidence are preserved.

**(A)** *Requisite Knowledge.* The classifications of fire; risks associated with each class of fire; and the types, rating systems, operating methods, and limitations of portable fire extinguishers.

**(B)** *Requisite Skills.* The ability to select, carry, and operate portable fire extinguishers, using the appropriate extinguisher based on the size and type of fire.

**4.2.2*** Conserve property, given special tools and equipment and an assignment within the facility, so that the facility and its contents are protected from further damage.

**(A)** *Requisite Knowledge.* The purpose of property conservation and its value to the organization, methods used to protect property, methods to reduce damage to property, types of and uses for salvage covers, and operations at properties protected with automatic sprinklers or special protection systems.

**(B)** *Requisite Skills.* The ability to deploy covering materials, control extinguishing agents, and cover building openings, including doors, windows, floor openings, and roof openings.

**4.2.3** Exit hazardous area, given that the fire has progressed beyond the incipient stage, so that a safe haven is found and the team members' safety is maintained.

**(A)** *Requisite Knowledge.* Communication procedures, emergency evacuation methods, what constitutes a safe haven, and elements that create or indicate a hazard.

**(B)** *Requisite Skills.* The ability to follow facility evacuation routes, evaluate areas for hazards, and identify a safe haven.

## Chapter 4: Incipient Facility Fire Brigade Member (continued)

**4.3* Site-Specific Requirements.** The management of the facility fire brigade shall determine the site-specific requirements that are applicable to the incipient facility fire brigade members operating on their site. The process used to determine the site-specific requirements shall be documented, and these additional JPRs added to those identified in Sections 4.1 and 4.2.

**4.3.1*** Attack an incipient stage fire, given a handline flowing up to 473 L/min (125 gpm), appropriate equipment, and a fire situation, so that the fire is approached safely, exposures are protected, the spread of fire is stopped, agent application is effective, the fire is extinguished, and the area of origin and fire cause evidence are preserved.

**(A)** *Requisite Knowledge.* Types of handlines used for attacking incipient fires, precautions to be followed when advancing handlines to a fire, observable results that a fire stream has been properly applied, dangerous building conditions created by fire, principles of exposure protection, and dangers such as exposure to products of combustion resulting from fire condition.

**(B)** *Requisite Skills.* The ability to recognize inherent hazards related to the material's configuration; operate handlines; prevent water hammers when shutting down nozzles; open, close, and adjust nozzle flow; advance charged and uncharged hose; extend handlines; operate handlines; evaluate and modify water application for maximum penetration; assess patterns for origin determination; and evaluate for complete extinguishment.

**4.3.2*** Activate a fixed fire protection system, given a fixed fire protection system, a procedure, and an assignment, so that the steps are followed and the system operates.

**(A)** *Requisite Knowledge.* Types of extinguishing agents, hazards associated with system operation, how the system operates, sequence of operation, system overrides and manual intervention procedures, and shutdown procedures to prevent damage to the operated system or to those systems associated with the operated system.

**(B)** *Requisite Skills.* The ability to operate fixed fire protection systems via electrical or mechanical means.

**4.3.3*** Utilize master stream appliances, given an assignment, an extinguishing agent, and a master stream device, so that the agent is applied to the fire as assigned.

**(A)** *Requisite Knowledge.* Safe operation of master stream appliances, uses for master stream appliances, tactics using fixed master stream appliances, and property conservation.

**(B)** *Requisite Skills.* The ability to put into service a fixed master stream appliance, and to evaluate and forecast a fire's growth and development.

**4.3.4*** Establish a water supply for fire-fighting operations, given an assignment, a water source, and tools, so that a water supply is established and maintained.

**(A)** *Requisite Knowledge.* Water sources, operation of site water supply components, hydraulic principles, and the effect of mechanical damage and temperatures on the operability of the water supply source.

**(B)** *Requisite Skills.* The ability to operate the site water supply components and to identify damage or impairment.

**4.3.5** Perform a fire safety survey in a facility, given an assignment, survey forms, and procedures, so that fire and life safety hazards are identified, recommendations for their correction are made, and unresolved issues are referred to the proper authority.

**(A)** *Requisite Knowledge.* Organizational policy and procedures, common causes of fire and their prevention, the importance of fire safety, and referral procedures.

**(B)** *Requisite Skills.* The ability to complete forms, recognize hazards, match findings to preapproved recommendations, and effectively communicate findings to the proper authority.

## Chapter 5: Advanced Exterior Facility Fire Brigade Member

### 5.1* General.

**5.1.1 Qualification or Certification.** For qualification or certification at the advanced exterior facility fire brigade member level, the facility fire brigade member shall meet the entrance requirements in Chapter 1 and Sections 4.1, 4.2, 5.1, and 5.2, the site-specific requirements in Sections 4.3 and 5.3 as defined by the management of the facility fire brigade, and the requirements defined in Chapter 5 of NFPA 1072.

**5.1.2 Basic Advanced Exterior Facility Fire Brigade Member JPRs.**

**5.1.2.1*** Utilize a pre-incident plan, given pre-incident plans and an assignment, so that the facility fire brigade member implements the responses detailed by the plan.

**(A)** *Requisite Knowledge.* The sources of water supply for fire protection or other fire-extinguishing agents, site-specific hazards, the fundamentals of fire suppression and detection systems including specialized agents, and common symbols used in diagramming construction features, utilities, hazards, and fire protection systems.

**(B)** *Requisite Skills.* The ability to identify the components of the pre-incident plan such as fire suppression and detection systems, structural features, site-specific hazards, and response considerations.

**5.1.2.2*** Interface with outside mutual aid organizations, given SOPs for mutual aid response and communication protocols, so that a unified command is established and maintained.

**(A)** *Requisite Knowledge.* Mutual aid procedures and the structure of the mutual aid organization, site SOPs, and incident management systems.

**(B)** *Requisite Skills.* The ability to communicate with mutual aid organizations and to integrate operational personnel into teams under a unified command.

### 5.2 Manual Fire Suppression.

**5.2.1** Use thermal protective clothing during exterior firefighting operations, given thermal protective clothing, so that the clothing is correctly donned within 2 minutes (120 seconds), worn, and doffed.

**(A)** *Requisite Knowledge.* Conditions that require personal protection, uses and limitations of thermal protective clothing, components of thermal protective clothing ensemble, and donning and doffing procedures.

**(B)** *Requisite Skills.* The ability to correctly don and doff thermal protective clothing and to perform assignments while wearing thermal protective clothing.

**5.2.2*** Use self-contained breathing apparatus (SCBA) during emergency operations, given SCBA and other personal protective equipment (PPE), so that the SCBA is correctly donned, the SCBA is correctly worn, controlled breathing techniques are used, emergency procedures are enacted if the SCBA fails, all low-air warnings are recognized, respiratory protection is not intentionally compromised, and hazardous areas are exited prior to air depletion.

**(A)** *Requisite Knowledge.* Conditions that require respiratory protection, uses and limitations of SCBA, components of SCBA, donning procedures, breathing techniques, indications for and emergency procedures used with SCBA, and physical requirements of the SCBA wearer.

**(B)** *Requisite Skills.* The ability to control breathing, replace SCBA air cylinders, use SCBA to exit through restricted passages, initiate and complete emergency procedures in the event of SCBA failure or air depletion, and complete donning procedures.

**5.2.3*** Attack an exterior fire operating as a member of a team, given a water source, a handline, PPE, tools, and an assignment, so that team integrity is maintained, the attack line is correctly deployed for advancement, access is gained into

## Chapter 5: Advanced Exterior Facility Fire Brigade Member (continued)

the fire area, appropriate application practices are used, the fire is approached in a safe manner, attack techniques facilitate suppression given the level of the fire, hidden fires are located and controlled, the correct body posture is maintained, hazards are avoided or managed, and the fire is brought under control.

**(A)** *Requisite Knowledge.* Principles of fire streams; types, design, operation, nozzle pressure effects, and flow capabilities of nozzles; precautions to be followed when advancing handlines to a fire; observable results that a fire stream has been correctly applied; dangerous conditions created by fire; principles of exposure protection; potential long-term consequences of exposure to products of combustion; physical states of matter in which fuels are found; the application of each size and type of attack line; the role of the backup team in fire attack situations; attack and control techniques; and exposing hidden fires.

**(B)** *Requisite Skills.* The ability to prevent water hammers when shutting down nozzles; open, close, and adjust nozzle flow and patterns; apply water using direct, indirect, and combination attacks; advance charged and uncharged 38 mm (1½ in.) diameter or larger handlines; extend handlines; replace burst hose sections; operate charged handlines of 38 mm (1½ in.) diameter or larger; couple and uncouple various handline connections; carry hose; attack fires; and locate and suppress hidden fires.

**5.2.4** Conduct search and rescue operations as a member of a team, given an assignment, obscured vision conditions, PPE, scene lighting, forcible entry tools, handlines, and ladders when necessary, so that all equipment is correctly used, all assigned areas are searched, all victims are located and removed, team integrity is maintained, and team members' safety, including respiratory protection, is not compromised.

**(A)** *Requisite Knowledge.* Use of appropriate tools and equipment, psychological effects of operating in obscured conditions and ways to manage them, methods to determine if an area is tenable, primary and secondary search techniques, team members' roles and goals, methods to use and indicators of finding victims, victim removal methods, and considerations related to respiratory protection.

**(B)** *Requisite Skills.* The ability to use SCBA to exit through restricted passages, use tools and equipment for various types of rescue operations, rescue a facility fire brigade member with functioning respiratory protection, rescue a facility fire brigade member whose respiratory protection is not functioning, rescue a person who has no respiratory protection, and assess areas to determine tenability.

**5.2.5*** Conserve property operating as a member of a team, given special tools and equipment and an assignment within the facility, so that exposed property and the environment are protected from further damage.

**(A)** *Requisite Knowledge.* The purpose of property conservation and its value to the organization, methods used to protect property, methods to reduce damage to property, operations at properties protected with automatic sprinklers or special protection systems, understanding the impact of using master streams and multiple hose streams on property conservation, particularly as it can relate to the impact on outside facilities.

**(B)** *Requisite Skills.* The ability to deploy covering materials, control extinguishing agents, and cover openings and equipment such as doors, windows, floor openings, and roof openings related to the impact of outside facilities.

**5.2.6** Overhaul a fire scene, given PPE, a handline, hand tools, scene lighting, and an assignment, so that structural integrity is not compromised, all hidden fires are discovered, fire cause evidence is preserved, and the fire is extinguished.

**(A)** *Requisite Knowledge.* Types of fire handlines and water application devices most effective for overhaul, water application methods for extinguishment that limit water damage, types of tools and methods used to expose hidden fire, dangers associated with overhaul, obvious signs of area of origin or signs of arson, and reasons for protection of a fire scene.

**(B)** *Requisite Skills.* The ability to deploy and operate a handline, expose void spaces without compromising structural integrity, apply water for maximum effectiveness, expose and extinguish hidden fires, recognize and preserve obvious signs of area of origin and fire cause, and evaluate for complete extinguishment.

**5.2.7*** Establish a water supply for fire-fighting operations, given a water source and tools, so that a water supply is established and maintained.

**(A)** *Requisite Knowledge.* Water sources, correct operation of site water supply components, hydraulic principles, and the effect of mechanical damage and temperatures on the operability of the water supply source.

**(B)** *Requisite Skills.* The ability to operate the site water supply components and identify damage or impairment.

**5.2.8*** Exit a hazardous area as a team, given vision-obscured conditions, so that a safe haven is found before exhausting the air supply, others are not endangered, and the team integrity is maintained.

**(A)** *Requisite Knowledge.* Personnel accountability systems, communication procedures, emergency evacuation methods, what constitutes a safe haven, elements that create or indicate a hazard, and emergency procedures for loss of air supply.

**(B)** *Requisite Skills.* The ability to operate as a team member in vision-obscured conditions, locate and follow a guideline, conserve air supply, evaluate areas for hazards, and identify a safe haven.

**5.2.9*** Operate as a member of a rapid intervention crew, given size-up information, basic rapid intervention tools and equipment, and an assignment, so that strategies to effectively rescue the facility brigade member(s) are identified and implemented; hazard warning systems are established and understood by all participating personnel; incident-specific PPE is identified, provided, and utilized; physical hazards are identified; and confinement, containment, and avoidance measures are discussed.

**(A)** *Requisite Knowledge.* Identification and care of PPE; specific hazards associated with the facility; strategic planning for rescue incidents; communications and safety protocols; atmospheric monitoring equipment needs; identification, characteristics, expected behavior, type, causes, and associated effects of personnel becoming incapacitated or trapped; and recognition of, potential for, and signs of impending building collapse.

**(B)** *Requisite Skills.* The ability to use PPE, determine resource needs, select and operate basic and specialized tools and equipment, implement communications and safety protocols, and mitigate specific hazards associated with rescue of trapped or incapacitated personnel.

**5.3* Site-Specific Requirements.** The JPRs in 5.3.1 through 5.3.11 shall be considered as site-specific functions of the advanced exterior facility fire brigade member. The management of the facility fire brigade shall determine the site-specific requirements that are applicable to the advanced exterior facility fire brigade member operating on their site. The process used to determine the site-specific requirements shall be documented, and these additional JPRs added to those identified in Sections 5.1 and 5.2. Based on the assessment of the site-specific hazards of the facility and the duties that facility fire brigade members are expected to perform, the management of the facility fire brigade shall determine the specific requirements of Chapters 5 or 6 of NFPA 472 or the corresponding requirements in OSHA 29 CFR 1910.120(q) that apply.

**5.3.1** Perform a fire safety survey in a facility, given an assignment, survey forms, and procedures, so that fire and life safety hazards are identified, recommendations for their correction are made, and unresolved issues are referred to the proper authority.

**(A)** *Requisite Knowledge.* Organizational policy and procedures, common causes of fire and their prevention, and the importance of fire safety and referral procedures.

## Chapter 5: Advanced Exterior Facility Fire Brigade Member (continued)

**(B)** *Requisite Skills.* The ability to complete forms, recognize hazards, match findings to pre-approved recommendations, and effectively communicate findings to the proper authority.

**5.3.2*** Gain access to facility locations, given keys, forcible entry tools (e.g., bolt cutters, small hand tools, and ladders), and an assignment, so that areas are accessed and remain accessible during advanced exterior facility fire brigade operations.

**(A)** *Requisite Knowledge.* Site drawing reading, access procedures, forcible entry tools and procedures, and site-specific hazards, such as access to areas restricted by railcar movement, fences, and walls. Procedures associated with special hazard areas such as electrical substations, radiation hazard areas, and other areas specific to the site, if needed.

**(B)** *Requisite Skills.* The ability to read site drawings, identify areas of low overhead clearance, identify areas on roadways having load restrictions, identify access routes to water supplies, identify hazardous materials locations, identify electrical equipment locations (overhead and belowgrade equipment), ability to open gates by manual and/or automatic means, ability to forcibly gain access to areas, and the ability to identify site hazards.

**5.3.3** Utilize master stream appliances, given an assignment, an extinguishing agent, and a master stream device and supply hose, so that the appliance is set up correctly and the agent is applied as assigned.

**(A)** *Requisite Knowledge.* Correct operation of master stream appliances, uses for master stream appliances, tactics using master stream appliances, selection of the master stream appliance for different fire situations, the effect of master stream appliances on search and rescue, ventilation procedures, and property conservation.

**(B)** *Requisite Skills.* The ability to correctly put in service a master stream appliance and evaluate and forecast a fire's growth and development.

**5.3.4*** Extinguish an ignitible (or simulated ignitible) liquid fire operating as a member of a team, given an assignment, a handline, PPE, a foam proportioning device, a nozzle, foam concentrates, and a water supply, so that the correct type of foam concentrate is selected for the given fuel and conditions, a correctly proportioned foam stream is applied to the surface of the fuel to create and maintain a foam blanket, the fire is extinguished, re-ignition is prevented, and team protection is maintained.

**(A)** *Requisite Knowledge.* Methods by which foam prevents or controls a hazard; principles by which foam is generated; causes for poor foam generation and corrective measures; difference between hydrocarbon and polar solvent fuels and the concentrates that work on each; the characteristics, uses, and limitations of fire-fighting foams; the advantages and disadvantages of using fog nozzles versus foam nozzles for foam application; foam stream application techniques; hazards associated with foam usage; and methods to reduce or avoid hazards.

**(B)** *Requisite Skills.* The ability to prepare a foam concentrate supply for use, assemble foam stream components, master various foam application techniques, and approach and retreat from fires and spills as part of a coordinated team.

**5.3.5*** Control a flammable gas fire operating as a member of a team, given an assignment, a handline, PPE, and tools, so that crew integrity is maintained, contents are identified, the flammable gas source is controlled or isolated, hazardous conditions are recognized and acted upon, and team safety is maintained.

**(A)** *Requisite Knowledge.* Characteristics of flammable gases, components of flammable gas systems, effects of heat and pressure on closed containers, boiling liquid expanding vapor explosion (BLEVE) signs and effects, methods for identifying contents, water stream usage and demands for pressurized gas fires, what to do if the fire is prematurely extinguished, alternative actions related to various hazards, and when to retreat.

**(B)** *Requisite Skills.* The ability to execute effective advances and retreats, apply various techniques for water application, assess gas storage container integrity and changing conditions, operate control valves, and choose effective procedures when conditions change.

**5.3.6*** Extinguish an exterior fire using special extinguishing agents other than foam operating as a member of a team, given an assignment, a handline, PPE, and an extinguishing agent supply, so that fire is extinguished, re-ignition is prevented, and team protection is maintained.

**(A)** *Requisite Knowledge.* Methods by which special agents, such as dry chemical, dry powder, and carbon dioxide, prevent or control a hazard; principles by which special agents are generated; the characteristics, uses, and limitations of firefighting special agents; the advantages and disadvantages of using special agents; special agents application techniques; hazards associated with special agents usage; and methods to reduce or avoid hazards.

**(B)** *Requisite Skills.* The ability to operate a special agent supply for use, master various special agents application techniques, and approach and retreat from hazardous areas as part of a coordinated team.

**5.3.7*** Interpret alarm conditions, given an alarm signaling system, a procedure, and an assignment, so that the alarm condition is correctly interpreted and a response is initiated.

**(A)** *Requisite Knowledge.* The different alarm detection systems within the facility; difference between alarm, trouble, and supervisory alarms; hazards protected by the detection systems; hazards associated with each type of alarm condition; knowledge of the emergency response plan; and communication procedures.

**(B)** *Requisite Skills.* The ability to understand the different types of alarms, to implement the response, and to provide information through communications.

**5.3.8*** Activate a fixed fire suppression system, given PPE, a fixed fire protection system, a procedure, and an assignment, so that the correct steps are followed and the system operates.

**(A)** *Requisite Knowledge.* Different types of extinguishing agents, hazards associated with system operation, how the system operates, sequence of operation, system overrides and manual intervention procedures, and shutdown procedures to prevent damage to the operated system or to those systems associated with the operated system.

**(B)** *Requisite Skills.* The ability to operate fixed fire suppression systems via electrical or mechanical means and shutdown procedures for fixed fire suppression systems.

**5.3.9*** Extinguish a Class C (electrical) or simulated Class C fire as a member of a team, given an assignment, a Class C fire-extinguishing appliance/extinguisher, and PPE, so that the proper type of Class C agent is selected for the condition, the selected agent is correctly applied to the fuel, the fire is extinguished, re-ignition is prevented, team protection is maintained, and the hazard is faced until retreat to safe haven is reached.

**(A)** *Requisite Knowledge.* Methods by which a Class C agent prevents or controls a hazard; methods by which Class C fires are de-energized; causes of injuries from Class C fire fighting on live Class C fires with Class A agents and the Class C agents; the extinguishing agents' characteristics, uses, and limitations; the advantages and disadvantages of de-energizing as using water fog nozzles on a Class A or Class B fire; and methods to reduce or avoid hazards.

**(B)** *Requisite Skills.* The ability to operate Class C fire extinguishers or fixed systems and approach and retreat from Class C fires as part of a coordinated team.

**5.3.10*** Utilize tools and equipment assigned to the facility fire brigade, given an assignment and specific tools, so that tools are selected and correctly used under adverse conditions in accordance with manufacturer's recommendations and the policies and procedures of the facility fire brigade.

**(A)** *Requisite Knowledge.* Available tools and equipment, their storage locations, and their correct

## Chapter 5: Advanced Exterior Facility Fire Brigade Member (continued)

use in accordance with recognized practices, and selection of tools and equipment given different conditions.

**(B)** *Requisite Skills.* The ability to select and use the correct tools and equipment for various tasks, follow guidelines, and restore tools and equipment to service after use.

**5.3.11** Set up and use portable ladders, given an assignment, single and extension ladders, and team members as appropriate, so that hazards are assessed, the ladder is stable, the angle is correct for climbing, extension ladders are extended to the correct height with the fly locked, the top is placed against a reliable structural component, and the assignment is accomplished.

**(A)** *Requisite Knowledge.* Parts of a ladder, hazards associated with setting up ladders, what constitutes a stable foundation for ladder placement, different angles for various tasks, safety limits to the degree of angulation, and what constitutes a reliable structural component for top placement.

**(B)** *Requisite Skills.* The ability to carry ladders, raise ladders, extend ladders and lock flies, determine that a wall and roof will support the ladder, judge extension ladder height requirements, and place the ladder to avoid obvious hazards.

**N 5.3.12*** Respond on apparatus to an emergency scene, given personal protective clothing and other necessary PPE, so that the apparatus is correctly mounted and dismounted, seat belts are used while the vehicle is in motion, and other personal protective equipment is correctly used. [**1001**:4.3.2]

**(A)** *Requisite Knowledge.* Mounting and dismounting procedures for riding fire apparatus, hazards and ways to avoid hazards associated with riding apparatus, prohibited practices, and types of department PPE and the means for usage. [**1001**:4.3.2(A)]

**(B)** *Requisite Skills.* The ability to use each piece of provided safety equipment. [**1001**:4.3.2(B)]

**5.3.13** Attack a vehicle fire operating as a member of a team, given PPE, an attack line, and hand tools, so that hazards are avoided, leaking flammable liquids are identified and controlled, protection from flash fires is maintained, all vehicle compartments are overhauled, and the fire is extinguished.

**(A)** *Requisite Knowledge.* Principles of fire streams as they relate to fighting fires; precautions to be followed when advancing hose lines; observable results that a fire stream has been properly applied; identifying alternative fuels and the hazards associated with them; dangerous conditions created during a fire; common types of accidents or injuries related to fighting fires and how to avoid them; how to access locked passenger, trunk, and engine compartments; and methods for overhauling.

**(B)** *Requisite Skills.* The ability to identify fuel type; assess and control fuel leaks; open, close, and adjust the flow and pattern on nozzles; apply water for maximum effectiveness while maintaining flash fire protection; advance 38 mm (1½in.) or larger diameter attack lines; and expose hidden fires by opening all compartments.

## Chapter 6: Interior Structural Facility Fire Brigade Member

### 6.1 General.

**6.1.1 Qualification or Certification.** For qualification or certification at the interior structural facility fire brigade member level, the member shall meet the entrance requirements in Chapter 1 and Sections 4.1, 4.2, 6.1, and 6.2, the site-specific requirements in Sections 4.3 and 6.3 as defined by the management of the facility fire brigade, and requirements defined by Chapter 5 of NFPA 1072.

**6.1.2 Basic Interior Structural Fire Brigade Member JPRs.**

**6.1.2.1*** Use thermal protective clothing during structural fire-fighting operations, given thermal protective clothing, so that the clothing is correctly donned within 2 minutes (120 seconds), worn, and doffed.

**(A)** *Requisite Knowledge.* Conditions that require personal protection, uses and limitations of thermal protective clothing, components of thermal protective clothing ensemble, and donning and doffing procedures.

**(B)** *Requisite Skills.* The ability to correctly don and doff thermal protective clothing and perform assignments while wearing thermal protective clothing.

**6.1.2.2*** Use self-contained breathing apparatus (SCBA) during emergency operations, given SCBA and other personal protective equipment (PPE), so that the SCBA is correctly donned, the SCBA is correctly worn, controlled breathing techniques are used, emergency procedures are enacted if the SCBA fails, all low-air warnings are recognized, respiratory protection is not intentionally compromised, and hazardous areas are exited prior to air depletion.

**(A)** *Requisite Knowledge.* Conditions that require respiratory protection, uses and limitations of SCBA, components of SCBA, donning procedures, breathing techniques, indications for and emergency procedures used with SCBA, and physical requirements of the SCBA wearer.

**(B)** *Requisite Skills.* The ability to control breathing, replace SCBA air cylinders, use SCBA to exit through restricted passages, initiate and complete emergency procedures in the event of SCBA failure or air depletion, and complete donning and doffing procedures.

**6.1.2.3** Utilize a pre-incident plan, given pre-incident plans and an assignment, so that the facility fire brigade member implements the pre-incident plan.

**(A)** *Requisite Knowledge.* The sources of water supply for fire protection or other fire-extinguishing agents, site-specific hazards, the fundamentals of fire suppression and detection systems including specialized agents, and common symbols used in diagramming construction features, utilities, hazards, and fire protection systems.

**(B)** *Requisite Skills.* The ability to identify the components of the pre-incident plan such as fire suppression and detection systems, structural features, site-specific hazards, and response considerations.

**N 6.1.2.4** Given a structural fire-fighting ensemble including respiratory protection and decontamination equipment on the fireground or at the fire station, the interior structural fire brigade member shall describe and perform field reduction of contaminants of PPE so that the PPE is handled in the proper manner and maintained in a safe working condition.

**N (A)** *Requisite Knowledge.* Working knowledge of field reduction of contaminants, policies and procedures for the reduction of carcinogenic or radiological particulates, etiological and chemical hazards of contaminated PPE on the fireground or at the fire station, manufacturer's specifications, and industry best practices for transporting, cleaning, inspecting, and identifying needed repairs.

**N (B)** *Requisite Skills.* The ability to perform PPE field reduction of contaminants at the fireground or at the fire station.

### 6.2 Manual Fire Suppression.

**6.2.1*** Attack an interior structural fire operating as a member of a team, given a water source, a handline, PPE, tools, and an assignment, so that team integrity is maintained, the handline is deployed for advancement, access is gained into the fire area, correct application practices are used, the fire is approached safely, attack techniques facilitate suppression given the level of the fire, hidden fires are located and controlled, the correct body posture is maintained, hazards are avoided or managed, and the fire is brought under control.

**(A)** *Requisite Knowledge.* Principles of conducting initial fire size-up; principles of fire streams; types, design, operation, nozzle pressure effects, and flow capabilities of nozzles; precautions to be followed when advancing hose lines to a fire; observable results that a fire stream has been correctly applied; dangerous building conditions created by fire; principles of exposure protection; potential long-term consequences of exposure to products of combustion; physical states of matter in which fuels are found; common types of accidents or

## Chapter 6: Interior Structural Facility Fire Brigade Member (continued)

injuries and their causes; and the application of each size and type of handlines, the role of the backup team in fire attack situations, attack and control techniques, and exposing hidden fires.

**(B)** *Requisite Skills.* The ability to prevent water hammers when shutting down nozzles; open, close, and adjust nozzle flow and patterns; apply water using direct, indirect, and combination attacks; advance charged and uncharged 38 mm (1½ in.) diameter or larger handlines; extend handlines; replace burst hose sections; operate charged handlines of 38 mm (1½ in.) diameter or larger; couple and uncouple various handline connections; carry hose; attack fires; and locate and suppress hidden fires.

**6.2.2** Force entry into a structure, given PPE, tools, and an assignment, so that the tools are used, the barrier is removed, and the opening is in a safe condition and ready for entry.

**(A)** *Requisite Knowledge.* Basic construction of typical doors, windows, and walls within the facility; operation of doors, windows, and their associated locking mechanisms; and the dangers associated with forcing entry through doors, windows, and walls.

**(B)** *Requisite Skills.* The ability to transport and operate site-specific tools to force entry through doors, windows, and walls using assorted methods and tools.

**6.2.3*** Perform ventilation on a structure operating as a member of a team, given an assignment, PPE, and tools, so that a sufficient opening is created, all ventilation barriers are removed, structural integrity is not compromised, and products of combustion are released from the structure.

**(A)** *Requisite Knowledge.* The principles, advantages, limitations, and effects of horizontal and vertical ventilation; safety considerations when venting a structure; the methods of heat transfer; the principles of thermal layering within a structure on fire; fire behavior in a structure; the products of combustion found in a structure fire; the signs, causes, effects, and prevention of backdrafts; and the relationship of oxygen concentration to life safety and fire growth.

**(B)** *Requisite Skills.* The ability to transport and operate tools and equipment to create an opening and implement ventilation techniques.

**6.2.4*** Overhaul a fire scene, given PPE, attack line, hand tools, a flashlight, and an assignment, so that structural integrity is not compromised, all hidden fires are discovered, fire cause evidence is preserved, and the fire is extinguished.

**(A)** *Requisite Knowledge.* Types of fire handlines and application devices most effective for overhaul, application methods for extinguishing agents that limit damage, types of tools and methods used to expose hidden fire, dangers associated with overhaul, obvious signs of area of origin and signs of arson, and reasons for protection of fire scene.

**(B)** *Requisite Skills.* The ability to deploy and operate handlines, expose void spaces without compromising structural integrity, apply extinguishing agents for maximum effectiveness, expose and extinguish hidden fires, recognize and preserve obvious signs of area of origin and fire cause, and evaluate for complete extinguishment.

**6.2.5*** Exit a hazardous area as a team, given vision-obscured conditions, so that a safe haven is found before exhausting the air supply, others are not endangered, and the team integrity is maintained.

**(A)** *Requisite Knowledge.* Personnel accountability systems, communication procedures, emergency evacuation methods, what constitutes a safe haven, elements that create or indicate a hazard, and emergency procedures for loss of air supply.

**(B)** *Requisite Skills.* The ability to operate as a team member in vision-obscured conditions, locate and follow a guideline, conserve air supply, and evaluate areas for hazards and identify a safe haven.

**6.2.6*** Establish a water supply for fire-fighting operations, given a water source and tools, so that a water supply is established and maintained.

**(A)** *Requisite Knowledge.* Water sources, correct operation of site water supply components, hydraulic principles, and the effect of mechanical damage and temperatures on the operability of the water supply source.

**(B)** *Requisite Skills.* The ability to operate the site water supply components and take action to address damage or impairment.

**6.2.7** Interface with outside mutual aid organizations, given SOPs for mutual aid response and communication protocols, so that a unified command is established and maintained.

**(A)** *Requisite Knowledge.* Mutual aid procedures and the structure of the mutual aid organization, site SOPs, and incident management systems.

**(B)** *Requisite Skills.* The ability to communicate with mutual aid organizations and to integrate operational personnel into teams under a unified command.

**6.2.8** Conduct search and rescue operations as a member of a team, given an assignment, obscured vision conditions, PPE, a flashlight, forcible entry tools, handlines, and ladders when necessary, so that all equipment is correctly used, all assigned areas are searched, all victims are located and removed, team integrity is maintained, and team members' safety, including respiratory protection, is not compromised.

**(A)** *Requisite Knowledge.* Use of appropriate tools and equipment, psychological effects of operating in obscured conditions and ways to manage them, methods to determine if an area is tenable, primary and secondary search techniques, team members' roles and goals, methods to use and indicators of finding victims, victim removal methods, and considerations related to respiratory protection.

**(B)** *Requisite Skills.* The ability to use SCBA to exit through restricted passages, use tools and equipment for various types of rescue operations, rescue a facility fire brigade member whose respiratory protection is not functioning, rescue a person who has no respiratory protection, and assess areas to determine tenability.

**6.2.9*** Conserve property operating as a member of a team, given special tools and equipment and an assignment within the facility, so that exposed property and the environment are protected from further damage.

**(A)** *Requisite Knowledge.* The purpose of property conservation and its value to the organization, methods used to protect property, methods to reduce damage to property, types of and uses for salvage covers, operations at properties protected with automatic sprinklers or special protection systems, and understanding the impact of using master streams and multiple hose streams on property conservation, particularly as it can relate to the impact on outside facilities.

**(B)** *Requisite Skills.* The ability to deploy covering materials, control extinguishing agents, and cover building openings, including doors, windows, floor openings, and roof openings.

**6.2.10*** Operate as a member of a rapid intervention crew, given size-up information, basic rapid intervention tools and equipment, and an assignment, so that strategies to effectively rescue the brigade member(s) are identified and implemented; hazard warning systems are established and understood by all participating personnel; incident-specific PPE is identified, provided, and utilized; physical hazards are identified; and confinement, containment, and avoidance measures are discussed.

**(A)** *Requisite Knowledge.* Identification and care of PPE; specific hazards associated with the facility; strategic planning for rescue incidents; communications and safety protocols; atmospheric monitoring equipment needs; identification, characteristics, expected behavior, type, causes, and associated effects of personnel becoming incapacitated or trapped; and recognition of, potential for, and signs of impending building collapse.

**(B)** *Requisite Skills.* The ability to use PPE, determine resource needs, select and operate basic and specialized tools and equipment, implement communications and safety protocols, and mitigate specific hazards associated with rescue of trapped or incapacitated personnel.

## Chapter 6: Interior Structural Facility Fire Brigade Member (continued)

**6.3*** *Site-Specific Requirements.* The management of the facility fire brigade shall determine the site-specific requirements that are applicable to the interior structural facility fire brigade member operating on their site. The process used to determine the site-specific requirements shall be documented, and these additional JPRs added to those identified in Sections 6.1 and 6.2. Based on the assessment of the site-specific hazards of the facility and the duties that facility fire brigade members are expected to perform, the management of the facility fire brigade shall determine the specific requirements of Chapters 5 and 6 of NFPA 472 or the corresponding requirements in OSHA 29 CFR 1910.120(q) that apply.

**6.3.1*** Interpret alarm conditions, given an alarm signaling system, a procedure, and an assignment, so that the alarm condition is correctly interpreted and a response is initiated.

**(A)** *Requisite Knowledge.* The different alarm detection systems within the facility; difference between alarm, trouble, and supervisory alarms; hazards protected by the detection systems; hazards associated with each type of alarm condition; the emergency response plan; and communication procedures.

**(B)** *Requisite Skills.* The ability to understand the different types of alarms, to implement the response, and to provide information through communications.

**6.3.2*** Activate a fixed fire protection system, given required PPE, a fixed fire protection system, a procedure, and an assignment, so that the procedures are followed and the system operates.

**(A)** *Requisite Knowledge.* Different types of extinguishing agents on site, manual fire suppression activities within areas covered by fixed fire suppression systems, hazards associated with system operation, how the system operates, sequence of operation, system overrides and manual intervention procedures, and shutdown procedures to prevent damage to the operated system or to those systems associated with the operated system.

**(B)** *Requisite Skills.* The ability to operate fixed fire suppression systems via electrical or mechanical means and to shut down fixed fire suppression systems.

**6.3.3** Utilize master stream appliances, given an assignment, an extinguishing agent, a master stream device, and a supply hose, so that the appliance is set up correctly and the agent is applied as assigned.

**(A)** *Requisite Knowledge.* Correct operation of master stream appliances, uses for master stream appliances, tactics using master stream appliances, selection of the master stream appliances for different fire situations, and the effect of master stream appliances on search and rescue, ventilation procedures, and property conservation.

**(B)** *Requisite Skills.* The ability to correctly put in service a master stream appliance and to evaluate and forecast a fire's growth and development.

**6.3.4*** Extinguish an ignitible liquid fire operating as a member of a team, given an assignment, a handline, PPE, a foam proportioning device, a nozzle, foam concentrates, and a water supply, so that the correct type of foam concentrate is selected for the given fuel and conditions, a correctly proportioned foam stream is applied to the surface of the fuel to create and maintain a foam blanket, fire is extinguished, reignition is prevented, and team protection is maintained.

**(A)** *Requisite Knowledge.* Methods by which foam prevents or controls a hazard; principles by which foam is generated; causes for poor foam generation and corrective measures; difference between hydrocarbon and polar solvent fuels and the concentrates that work on each; the characteristics, uses, and limitations of fire-fighting foams; the advantages and disadvantages of using fog nozzles versus foam nozzles for foam application; foam stream application techniques; hazards associated with foam usage; and methods to reduce or avoid hazards.

**(B)** *Requisite Skills.* The ability to prepare a foam concentrate supply for use, assemble foam stream components, master various foam application techniques, and approach and retreat from fires and spills as part of a coordinated team.

**6.3.5*** Control a flammable gas fire operating as a member of a team, given an assignment, a handline, PPE, and tools, so that team integrity is maintained, contents are identified, the flammable gas source is controlled or isolated, hazardous conditions are recognized and acted upon, and team safety is maintained.

**(A)** *Requisite Knowledge.* Characteristics of flammable gases, components of flammable gas systems, effects of heat and pressure on closed containers, BLEVE signs and effects, methods for identifying contents, water stream usage and demands for pressurized gas fires, what to do if the fire is prematurely extinguished, alternative actions related to various hazards, and when to retreat.

**(B)** *Requisite Skills.* The ability to execute effective advances and retreats, apply various techniques for water application, assess gas storage container integrity and changing conditions, operate control valves, and choose effective procedures when conditions change.

**6.3.6*** Extinguish a fire using special extinguishing agents other than foam operating as a member of a team, given an assignment, a handline, PPE, and an extinguishing agent supply, so that fire is extinguished, re-ignition is prevented, and team protection is maintained.

**(A)** *Requisite Knowledge.* Methods by which special agents, such as dry chemical, dry powder, and carbon dioxide, prevent or control a hazard; principles by which special agents are generated; the characteristics, uses, and limitations of firefighting special agents; the advantages and disadvantages of using special agents; special agent application techniques; hazards associated with special agent usage; and methods to reduce or avoid hazards.

**(B)** *Requisite Skills.* The ability to operate a special agent supply for use, master various special agents application techniques, and approach and retreat from hazardous areas as part of a coordinated team.

**6.3.7*** Utilize tools and equipment assigned to the facility fire brigade, given an assignment and specific tools, so that tools are selected and correctly used under adverse conditions in accordance with manufacturer's recommendations and the policies and procedures of the facility fire brigade.

**(A)** *Requisite Knowledge.* Available tools and equipment, their storage locations, and their correct use in accordance with recognized practices; and selection of tools and equipment given different conditions.

**(B)** *Requisite Skills.* The ability to select and use the correct tools and equipment for various tasks, follow guidelines, and restore tools and equipment to service after use.

**6.3.8** Set up and use portable ladders, given an assignment, single and extension ladders, and team members as appropriate, so that hazards are assessed, the ladder is stable, the angle is correct for climbing, extension ladders are extended to the correct height with the fly locked, the top is placed against a reliable structural component, and the assignment is accomplished.

**(A)** *Requisite Knowledge.* Parts of a ladder, hazards associated with setting up ladders, what constitutes a stable foundation for ladder placement, different angles for various tasks, safety limits to the degree of angulation, and what constitutes a reliable structural component for top placement.

**(B)** *Requisite Skills.* The ability to carry ladders, raise ladders, extend ladders and lock flies, determine that a wall and roof will support the ladder, judge extension ladder height requirements, and place the ladder to avoid obvious hazards.

**6.3.9*** Interface with outside mutual aid organizations, given SOPs for mutual aid response and communication protocols, so that a unified command is established and maintained.

## Chapter 6: Interior Structural Facility Fire Brigade Member (continued)

**(A)** *Requisite Knowledge.* Mutual aid procedures and the structure of the mutual aid organization, site SOPs, and incident management systems.

**(B)** *Requisite Skills.* The ability to communicate with mutual aid organizations and to integrate operational personnel into teams under a unified command.

**6.3.10** Perform a fire safety survey in a facility, given an assignment, survey forms, and procedures, so that fire and life safety hazards are identified, recommendations for their correction are made, and unresolved issues are referred to the proper authority.

**(A)** *Requisite Knowledge.* Organizational policy and procedures, common causes of fire and their prevention, and the importance of fire safety and referral procedures.

**(B)** *Requisite Skills.* The ability to complete forms, recognize hazards, match findings to pre-approved recommendations, and effectively communicate findings to the proper authority.

**6.3.11*** Extinguish a Class C (electrical) fire as a member of a team, given an assignment, a Class C fire-extinguishing appliance/extinguisher, and PPE, so that the type of Class C agent is selected for the condition, a selected agent is correctly applied to the fuel, fire is extinguished, re-ignition is prevented, team protection is maintained, and the hazard is faced until retreat to safe haven is reached.

**(A)** *Requisite Knowledge.* Methods by which a Class C agent prevents or controls a hazard; methods by which Class C fires are de-energized; causes of injuries from Class C fire fighting on live Class C fires with Class A agents and the Class C agents; the extinguishing agents' characteristics, uses, and limitations; the advantages and disadvantages of de-energizing using water fog nozzles on a Class A or Class B fire; and methods to reduce or avoid hazards.

**(B)** *Requisite Skills.* The ability to operate Class C fire extinguishers or fixed systems and approach and retreat from Class C fires as part of a coordinated team.

## Chapter 7: Facility Fire Brigade Leader

### 7.1 General.

**7.1.1** This duty shall involve establishing command, using emergency response procedures, and overseeing the emergency response and other administrative duties as outlined in Chapter 4 of NFPA 600 depending on the site organizational statement.

**7.1.2 Qualification or Certification.** For qualification or certification as a facility fire brigade leader, the member shall meet the JPRs of the level of the facility fire brigade in which they are leading in accordance with the requirements of Chapters 1, 5, or 6 and the JPRs as defined in Sections 7.1 and 7.2.

**7.1.3 General Requisite Knowledge.** The organizational structure of the facility fire brigade; operating procedures for administration, emergency operations, and safety; information management and record keeping; incident management system; methods used by leaders to obtain cooperation within a group of subordinates; and policies and procedures regarding the operation of the facility fire brigade.

**7.1.4 General Prerequisite Skills.** The ability to operate at all levels in the incident management system as defined by the National Incident Management System (NIMS) and NFPA 1561.

### 7.2 Supervisory Functions.

**7.2.1** Assign tasks or responsibilities to members, given an assignment at an emergency situation, so that the instructions are complete, clear, and concise; safety considerations are addressed; and the desired outcomes are conveyed.

**(A)** *Requisite Knowledge.* Verbal communications during emergency situations, techniques used to make assignments under stressful situations, and methods of confirming understanding of assigned tasks.

**(B)** *Requisite Skills.* The ability to condense instructions for frequently assigned unit tasks based upon training and SOPs.

**7.2.2** Develop an initial action plan, given size-up information for an incident and assigned emergency response resources, so that resources are deployed to control the emergency.

**(A)*** *Requisite Knowledge.* Elements of a size-up, SOPs for emergency operations, and fire behavior.

**(B)** *Requisite Skills.* The ability to analyze emergency scene conditions, to allocate resources, and to communicate verbally.

**7.2.3*** Implement an action plan at an emergency situation, given assigned resources, type of incident, preliminary plan, and facility fire brigade safety policies and procedures, so that resources are deployed to mitigate the situation and team safety is maintained.

**(A)** *Requisite Knowledge.* SOPs, resources available, basic fire control and emergency operation procedures, an incident management system, rapid intervention crew (RIC) procedures, personnel accountability system, common causes of personal injury during facility fire brigade activities, safety policies and procedures, and basic facility fire brigade member safety.

**(B)*** *Requisite Skills.* The ability to implement an incident management system, to communicate verbally, to supervise and account for assigned personnel under emergency conditions, and to identify safety hazards.

**7.2.4*** Coordinate multiple resources, such as in-house and mutual aid, during emergency situations, given an incident requiring multiple resources and a site incident management system, so that the site incident management system is implemented and the required resources, their assignments, and safety considerations for successful control of the incident are identified.

**(A)** *Requisite Knowledge.* SOPs and local resources available for the handling of the incident under emergency situations, basic fire control and emergency operation procedures, an incident management system, and a personnel accountability system.

**(B)** *Requisite Skills.* The ability to implement the site incident management system, to communicate verbally, and to supervise and account for assigned personnel under emergency conditions.

**7.2.5** Implement support operations at an incident, given an assignment and available resources, so that scene lighting is adequate for the tasks to be undertaken, personnel rehabilitation is facilitated, and the support operations facilitate the incident objectives.

**(A)** *Requisite Knowledge.* Resource management protocols, principles for establishing lighting, and rescuer rehabilitation practices and procedures.

**(B)** *Requisite Skills.* The ability to manage resources, provide power, set up lights, use lighting, select rehab areas, and personnel rotations.

**7.2.6** Direct members during a training evolution, given a training evolution and training policies and procedures, so that the evolution is performed in accordance with safety plans, and the stated objectives or learning outcomes are achieved as directed.

**(A)** *Requisite Knowledge.* Oral communication techniques to facilitate learning.

**(B)** *Requisite Skills.* The ability to distribute issue-guided directions to members during training evolutions.

# NFPA 1081: *Standard for Facility Fire Brigade Member Professional Qualifications*, 2018 Edition Correlation Guide

| NFPA 1081, *Standard for Facility Fire Brigade Member Professional Qualifications*, 2018 Edition | Corresponding Textbook Chapter |
|---|---|

## Chapter 4: Incipient Facility Fire Brigade Member

| NFPA 1081 | Chapter |
|---|---|
| 4.1 | General |
| 4.1.1 | 1 |
| 4.1.2 | 4, 5 |
| 4.1.2.1 | 3 |
| 4.1.2.1 (A) | 3 |
| 4.1.2.1 (B) | 3 |
| 4.1.2.2 (A) | 3 |
| 4.1.2.2 | 3 |
| 4.1.2.2 (B) | 3 |
| 4.1.2.3 | 10 |
| 4.1.2.3 (A) | 10 |
| 4.1.2.3 (B) | 10 |
| 4.1.2.4 | 8 |
| 4.1.2.4 (A) | 8 |
| 4.1.2.4 (B) | 8 |
| 4.1.2.5 | 3 |
| 4.1.2.5 (A) | 3 |
| 4.1.2.5 (B) | 3 |
| 4.1.3 | 3 |
| 4.2 | 7 |
| 4.2.1 | 7, 26 |
| 4.2.1 (A) | 7 |
| 4.2.1 (B) | 7 |
| 4.2.2 | 19 |
| 4.2.2 (A) | 19 |
| 4.2.2 (B) | 19 |
| 4.2.3 | 18 |
| 4.2.3 (A) | 18 |
| 4.2.3 (B) | 18 |
| 4.3 | 1 |
| 4.3.1 | 21, 26 |
| 4.3.1 (A) | 21 |
| 4.3.1 (B) | 21, 26 |
| 4.3.2 | 25 |
| 4.3.2 (A) | 25 |
| 4.3.2 (B) | 25 |
| 4.3.3 | 21 |
| 4.3.3 (A) | 21 |
| 4.3.3 (B) | 21 |
| 4.3.4 | 15 |
| 4.3.4 (A) | 15 |
| 4.3.4 (B) | 15 |
| 4.3.5 | 22 |
| 4.3.5 (A) | 22 |
| 4.3.5 (B) | 22 |

## Chapter 5: Advanced Exterior Facility Fire Brigade Member

| NFPA 1081 | Chapter |
|---|---|
| 5.1 | General |
| 5.1.1 | 1 |
| 5.1.2 | 1 |
| 5.1.2.1 | 22 |
| 5.1.2.1 (A) | 22 |
| 5.1.2.1 (B) | 22 |
| 5.1.2.2 | 4 |
| 5.1.2.2 (A) | 4 |
| 5.1.2.2 (B) | 4 |
| 5.2 | Manual Fire Suppression |
| 5.2.1 | 1 |
| 5.2.1 (A) | 1 |
| 5.2.1 (B) | 1 |
| 5.2.2 | 1 |
| 5.2.2 (A) | 1 |
| 5.2.2 (B) | 1 |
| 5.2.3 | 21 |
| 5.2.3 (A) | 1, 5, 16, 21 |
| 5.2.3 (B) | 16, 21 |
| 5.2.4 | 13 |

# Appendix B

| NFPA 1081, *Standard for Facility Fire Brigade Member Professional Qualifications,* 2018 Edition | Corresponding Textbook Chapter |
|---|---|
| 5.2.4 (A) | 13 |
| 5.2.4 (B) | 13 |
| 5.2.5 | 19 |
| 5.2.5 (A) | 19 |
| 5.2.5 (B) | 19 |
| 5.2.6 | 19, 26 |
| 5.2.6 (A) | 19, 26 |
| 5.2.6 (B) | 19, 26 |
| 5.2.7 | 15 |
| 5.2.7 (A) | 15 |
| 5.2.7 (B) | 15 |
| 5.2.8 | 18 |
| 5.2.8 (A) | 18 |
| 5.2.8 (B) | 18 |
| 5.2.9 | 18 |
| 5.2.9 (A) | 18 |
| 5.2.9 (B) | 18 |
| 5.3 | 1 |
| 5.3.1 | 22 |
| 5.3.1 (A) | 22 |
| 5.3.1 (B) | 22 |
| 5.3.2 | 11 |
| 5.3.2 (A) | 11, 22 |
| 5.3.2 (B) | 22 |
| 5.3.3 | 21 |
| 5.3.3 (A) | 21 |
| 5.3.3 (B) | 21 |
| 5.3.4 | 17 |
| 5.3.4 (A) | 17 |
| 5.3.4 (B) | 17 |
| 5.3.5 | 21 |
| 5.3.5 (A) | 21 |
| 5.3.5 (B) | 21 |
| 5.3.6 | 7 |
| 5.3.6 (A) | 7 |
| 5.3.6 (B) | 7 |
| 5.3.7 | 25 |
| 5.3.7 (A) | 25 |
| 5.3.7 (B) | 25 |
| 5.3.8 | 25 |
| 5.3.8 (A) | 25 |
| 5.3.8 (B) | 25 |
| 5.3.9 | 21 |
| 5.3.9 (A) | 21 |
| 5.3.9 (B) | 21 |
| 5.3.10 | 8 |
| 5.3.10 (A) | 8 |
| 5.3.10 (B) | 8 |
| 5.3.11 | 12 |
| 5.3.11 (A) | 12 |
| 5.3.11 (B) | 12 |
| 5.3.12 | 12 |
| 5.3.12 (A) | 12 |
| 5.3.12 (B) | 12 |
| 5.3.13 | 21 |
| 5.3.13 (A) | 21 |
| 5.3.13 (B) | 21 |

## Chapter 6: Interior Structural Facility Fire Brigade Member

| | |
|---|---|
| 6.1 | General |
| 6.1.1 | 1 |
| 6.1.2 | 1 |
| 6.1.2.1 | 1 |
| 6.1.2.1 (A) | 1 |
| 6.1.2.1 (B) | 1 |
| 6.1.2.2 | 1 |
| 6.1.2.2 (A) | 1 |
| 6.1.2.2 (B) | 1 |
| 6.1.2.3 | 22 |

| NFPA 1081, *Standard for Facility Fire Brigade Member Professional Qualifications*, 2018 Edition | Corresponding Textbook Chapter |
|---|---|
| 6.1.2.3 (A) | 22 |
| 6.1.2.3 (B) | 22 |
| 6.1.2.4 | 1 |
| 6.1.2.4 (A) | 1 |
| 6.1.2.4 (B) | 1 |
| 6.2 | Manual Fire Suppression |
| 6.2.1 | 21 |
| 6.2.1 (A) | 1, 5, 10, 16, 21 |
| 6.2.1 (B) | 16, 21 |
| 6.2.2 | 11 |
| 6.2.2 (A) | 6, 11 |
| 6.2.2 (B) | 11 |
| 6.2.3 | 14 |
| 6.2.3 (A) | 5, 14 |
| 6.2.3 (B) | 14 |
| 6.2.4 | 19, 26 |
| 6.2.4 (A) | 19, 26 |
| 6.2.4 (B) | 21, 26 |
| 6.2.5 | 18 |
| 6.2.5 (A) | 18 |
| 6.2.5 (B) | 18 |
| 6.2.6 | 15 |
| 6.2.6 (A) | 15 |
| 6.2.6 (B) | 15 |
| 6.2.7 | 4 |
| 6.2.7 (A) | 4 |
| 6.2.7 (B) | 4 |
| 6.2.8 | 13 |
| 6.2.8 (A) | 13 |
| 6.2.8 (B) | 13 |
| 6.2.9 | 19 |
| 6.2.9 (A) | 19 |
| 6.2.9 (B) | 19 |
| 6.2.10 | 18 |
| 6.2.10 (A) | 18 |
| 6.2.10 (B) | 18 |
| 6.3 | 1 |
| 6.3.1 | 25 |
| 6.3.1 (A) | 25 |
| 6.3.1 (B) | 25 |
| 6.3.2 | 25 |
| 6.3.2 (A) | 25 |
| 6.3.2 (B) | 25 |
| 6.3.3 | 21 |
| 6.3.3 (A) | 21 |
| 6.3.3 (B) | 21 |
| 6.3.4 | 17 |
| 6.3.4 (A) | 17 |
| 6.3.4 (B) | 17 |
| 6.3.5 | 21 |
| 6.3.5 (A) | 21 |
| 6.3.5 (B) | 21 |
| 6.3.6 | 21 |
| 6.3.6 (A) | 21 |
| 6.3.6 (B) | 21 |
| 6.3.7 | 8 |
| 6.3.7 (A) | 8 |
| 6.3.7 (B) | 8 |
| 6.3.8 | 12 |
| 6.3.8 (A) | 12 |
| 6.3.8 (B) | 12 |
| 6.3.9 | 4 |

| NFPA 1081, *Standard for Facility Fire Brigade Member Professional Qualifications*, 2018 Edition | Corresponding Textbook Chapter |
|---|---|
| 6.3.9 (A) | 4 |
| 6.3.9 (B) | 4 |
| 6.3.10 | 22 |
| 6.3.10 (A) | 22 |
| 6.3.10 (B) | 22 |
| 6.3.11 | 21 |
| 6.3.11 (A) | 21 |
| 6.3.11 (B) | 21 |

## Chapter 7: Facility Fire Brigade Leader

| NFPA 1081, *Standard for Facility Fire Brigade Member Professional Qualifications*, 2018 Edition | Corresponding Textbook Chapter |
|---|---|
| 7.1 | General |
| 7.1.1 | 27 |
| 7.1.2 | 27 |
| 7.1.3 | 27 |
| 7.1.4 | 27 |
| 7.2 | Supervisory Functions |
| 7.2.1 | 27 |
| 7.2.1 (A) | 27 |
| 7.2.1 (B) | 27 |
| 7.2.2 | 27 |
| 7.2.2 (A) | 27 |
| 7.2.2 (B) | 27 |
| 7.2.3 | 27 |
| 7.2.3 (A) | 27 |
| 7.2.3 (B) | 27 |
| 7.2.4 | 27 |
| 7.2.4 (A) | 27 |
| 7.2.4 (B) | 27 |
| 7.2.5 | 27 |
| 7.2.5 (A) | 27 |
| 7.2.5 (B) | 27 |
| 7.2.6 | 27 |
| 7.2.6 (A) | 27 |
| 7.2.6 (B) | 27 |

# Glossary

A tool A cutting tool with a pry bar built into the cutting part of the tool.

Accelerants Materials, usually flammable liquids, used to initiate or increase the spread of fire.

Accelerator A device that accelerates the removal of the air from a dry-pipe or preaction sprinkler system.

Accidental fires Fire cause classification that includes fires with a proven cause that does not involve a deliberate human act.

Accordion hose load A method of loading hose on a vehicle whose appearance resembles accordion sections. It is achieved by standing the hose on its edge, then placing the next fold on its edge and so on.

Accountability system A method of accounting for all personnel at an emergency incident and ensuring that only personnel with specific assignments are permitted to work within the various zones.

Activity logging system Device that keeps a detailed record of every incident and activity that occurs.

Adaptor A device that joins hose couplings of the same type, such as male to male or female to female.

Adjustable gallonage fog nozzle A nozzle that allows the operator to select a desired flow from several settings.

Adz The prying part of the Halligan tool.

Aerial ladder A power-operated ladder permanently mounted on a piece of apparatus.

Agroterrorism The use of a biological, chemical, radiological, or other agent against either the preharvest or postharvest stages of food and fiber production to inspire fear or cause economic damage, public health impact, or other adverse impact against the United States. Biological agents can impact animal or plant health or lead to human disease if they are zoonotic (i.e., diseases that are naturally transmitted between vertebrate animals and man with or without an arthropod intermediary).

Air cylinder The component of the SCBA that stores the compressed air supply.

Air line The hose through which air flows, either within an SCBA or from an outside source to a supplied air respirator.

Air management The way in which an individual utilizes a limited air supply to be sure it will last long enough to enter a hazardous area, accomplish needed tasks, and return safely.

Air sampling detector A system that captures a sample of air from a room or enclosed space and passes it through a smoke detection or gas analysis device.

Alarm initiation device An automatic or manually operated device in a fire alarm system that, when activated, causes the system to indicate an alarm condition.

Alarm notification device An audible and/or visual device in a fire alarm system that makes occupants or other persons aware of an alarm condition.

Alarm valve This valve signals an alarm when a sprinkler head is activated and prevents nuisance alarms caused by pressure variations.

Alpha particles A type of radiation that quickly loses energy and can travel only 1" to 2" from its source. Clothing or a sheet of paper can stop this type of energy.

Ammonium phosphate An extinguishing agent used in dry chemical fire extinguishers that can be used on Class A, B, and C fires.

ANFO An explosive material containing ammonium nitrate fertilizer and fuel oil.

Annealed The process of forming standard glass.

Anthrax An infectious disease spread by the bacteria *Bacillus anthracis*; typically found around farms, infecting livestock.

Aqueous film-forming foam (AFFF) A water-based extinguishing agent used on Class B fires that forms a foam layer over the liquid and stops the production of flammable vapors.

Arched roof A rounded roof usually associated with a bow-truss.

Arson The malicious burning of one's own or another's property with a criminal intent.

Arsonist A person who deliberately sets a fire to destroy property with criminal intent.

Attack engine The engine from which the attack lines have been pulled.

Attack hose (attack line) The hose that delivers water from a fire pump to the fire. Attack hoses range in size from 1" to 2 $^1/_2$".

Automatic adjusting fog nozzle A nozzle that can deliver a wide range of water stream flows. It operates by an internal spring-loaded piston.

Automatic sprinkler heads The working ends of a sprinkler system. They serve to activate the system and to apply water to the fire.

Automatic sprinkler system A system of pipes filled with water under pressure that discharges water immediately when a sprinkler head opens.

Awning window Windows that have one large or two medium panels operated by a hand crank from the corner of the window.

Backdraft The sudden explosive ignition of fire gases when oxygen is introduced into a superheated space previously deprived of oxygen.

Backpack The harness of the SCBA that supports the components worn by a brigade member.

Ball valves Valves used on nozzles, gated wyes, and engine discharge gates. Made up of a ball with a hole in the middle of the ball.

Balloon-frame construction An older type of wood frame construction in which the wall studs extend vertically from the basement of a structure to the roof without any fire stops.

Bam-bam tool A tool with a case-hardened screw, which is secured in the keyway of a lock, to remove the keyway from the lock.

Bangor ladder A ladder equipped with tormentor poles or staypoles that stabilize the ladder during raising and lowering operations.

Base station Radios are permanently mounted in a building, such as a fire station, communications center, or remote transmitter site.

Batch mixing Pouring foam concentrate directly into the booster tank of an apparatus.

Battering ram A tool made of hardened steel with handles on the sides used to force doors and to breach walls. Larger versions are used by up to four people; smaller versions are made for one or two people.

Beam detector A smoke detection device that projects a narrow beam of light across a large open area from a sending unit to a receiving unit. When the beam is interrupted by smoke, the receiver detects a reduction in light transmission and activates the fire alarm.

Beam One of the two main structural pieces running the entire length of each ladder or ladder section. The beams support the rungs.

Bearing wall A wall that is designed to support the weight of a floor or roof.

Bed section The lowest and widest section of an extension ladder. The fly sections of the ladder extend from the bed section.

Bends Knots used to join two ropes together.

Beta particles A type of radiation that is capable of traveling 10' to 15'. Heavier materials such as metal, plastic, and glass can stop this type of energy.

Bight A U-shape created by bending a rope with the two sides parallel.

Bimetallic strip A device with components made from two distinct metals that respond differently to heat. When heated, the metals will bend or change shape.

Biologic agents Disease-causing bacteria, viruses, and other agents that attack the human body.

Bite A small opening made to enable better tool access in forcible entry.

Blistering agents Chemicals that cause the skin to blister.

Block creel construction Rope constructed without knots or splices in the yarns, ply yarns, strands, braids, or rope.

Boiling-liquid, expanding-vapor explosion (BLEVE) An explosion that can occur when a tank containing a liquid fuel overheats.

Boilover Violent ejection of fuel from a tank when hot heavy hydrocarbons contact water in a tank bottom, causing a steam explosion.

Bolt cutter A cutting tool used to cut through thick metal objects such as bolts, locks, and wire fences.

Booster hose (booster lines) A rigid hose that is 3/4" or 1" in diameter. This hose delivers only 30 to 60 gpm, but can do so at high pressures. It is used for small fires.

Bounce off Foam application utilizing an object to bounce foam off of to gently flow onto the fuel surface.

Bowstring truss Trusses that are curved on the top and straight on the bottom.

Box-end wrench A hand tool used to tighten or loosen bolts. The end is enclosed, as opposed to an open-end wrench. Each wrench is a specific size and most have ratchets for easier use.

Braided rope Rope constructed by intertwining or weaving strands together.

Branch A supervisory level established to manage the span of control above the division or group level; usually applied to operations or logistics functions.

Branch director Officer in charge of all resources operating within a specified branch, responsible to the next higher level in the incident organization (either a Section Chief or the Incident Commander).

Breakaway nozzle A nozzle with a tip that can be separated from the shut-off valve.

Bresnan distributor nozzle A device that can be placed in confined spaces. The nozzle spins, spreading water over a large area.

Brigade member The brigade member may be assigned any task, from placing hose lines to extinguishing fires. Generally, the brigade member is not responsible for any command functions and does not supervise other personnel, except on a temporary basis when promoted to an acting officer.

British thermal unit (BTU) The amount of heat required to raise the temperature of one pound of water by one degree Fahrenheit.

Buddy system A system in which two brigade members always work as a team for safety purposes.

Budget An itemized summary of estimated or intended expenditures for a given period along with proposals for financing them.

Bunker coat (turnout coat) The protective coat worn by a brigade member for interior structural firefighting.

Bunker pants (turnout pants) The protective trousers worn by a brigade member for interior structural firefighting.

Burnback resistance The ability of a foam blanket to resist direct flame impingement.

Butt Often called the heel or base, the butt is the end of the ladder that is placed against the ground when the ladder is raised.

Butt plate (footpad) An alternative to a simple butt spur; a swiveling plate with both a spur and a cleat or pad that is attached to the butt of the ladder.

Butt spurs The metal spikes attached to the butt of a ladder. The spurs help prevent the butt from slipping out of position.

Butterfly valves Valves that are found on the large pump intake valve where the hard or soft sleeve hose connects.

Call box System of telephones connected by phone lines, radio equipment, or cellular technology to communicate with a communications center or fire organization.

Carabiner (snap link) A piece of metal hardware used extensively in rope rescue operations. It is generally an oval-shaped device with a spring-loaded clip that can be used for connecting together pieces of rope, webbing, or other hardware.

Carbon dioxide ($CO_2$) fire extinguisher A fire extinguisher that uses carbon dioxide gas as the extinguishing agent.

Carbon dioxide extinguishing system A system designed to protect a single room or series of rooms by flooding the area with carbon dioxide.

Carbon monoxide (CO) A toxic gas produced through incomplete combustion.

Carpenter's handsaw A saw designed for cutting wood.

Carryall A piece of heavy canvas with handles, which can be used to tote debris, ash, embers, and burning materials out of a structure.

Cartridge-operated extinguisher A fire extinguisher that has the expellant gas in a cartridge separate from the extinguishing agent storage shell. The storage shell is pressurized by a mechanical action that releases the expellant gas.

Cascade system An apparatus consisting of multiple tanks used to store compressed air and fill SCBA cylinders.

Case-hardened steel A process that uses carbon and nitrogen to harden the outer core of a steel component, while the inner core remains soft. Case-hardened steel can only be cut with specialized tools.

Casement window Windows in a steel or wood frame that open away from the building via a crank mechanism.

Ceiling hook A tool with a long wooden or fiberglass pole that has a metal point with a spur at right angles at one end. It can be used to probe ceilings and pull down plaster lath material.

Cellar nozzles Nozzles used to fight fires in cellars and other inaccessible places. These devices work by spreading water in a wide pattern as the nozzle is lowered by a hole into the cellar.

Central station An off-premises facility that monitors alarm systems and is responsible for notifying the fire brigade of an alarm. These facilities may be geographically located some distance from the protected building(s).

Chain of command The superior-subordinate authority relationship that starts at the top of the organization hierarchy and extends to the lowest levels.

Chain of custody A legal term used to describe the paperwork or documentation describing the movement, storage, and custody of evidence, such as the record of possession of a gas can from a fire scene.

Chain saw A power saw that uses the rotating movement of a chain equipped with sharpened cutting edges. Typically used to cut through wood.

Chase Open space within walls for wires and pipes.

Chemical energy Energy created or released by the combination or decomposition of chemical compounds.

Chisel A metal tool with one sharpened end used to break apart material in conjunction with a hammer, mallet, or sledgehammer.

Chlorine A yellowish gas that is about 2.5 times heavier than air and slightly water-soluble. It has many industrial uses but also damages the lungs when inhaled (a choking agent).

Choking agent A chemical designed to inhibit breathing and typically intended to incapacitate rather than kill.

Churning Recirculation of exhausted air that is drawn back into a negative-pressure fan in a circular motion.

Circumstantial evidence The means by which alleged facts are proven by deduction or inference from other facts that were observed first hand.

Clapper valve A mechanical device installed within a piping system that allows water to flow in only one direction.

**Class I standpipe** A standpipe system designed for use by fire brigade personnel only. Each outlet should have a valve to control the flow of water and a 2 $^1/_2$" male coupling for fire hose.

**Class II standpipe** A standpipe system designed for use by trained personnel only. Each outlet is generally equipped with a length of 1 $^1/_2$" single-jacket hose and a nozzle, preconnected to the system.

**Class III standpipe** A combination system that has features of both Class I and Class II standpipes.

**Class A fires** Fires involving ordinary combustible materials such as wood, cloth, paper, rubber, and many plastics.

**Class B fires** Fires involving flammable and combustible liquids, oils, greases, tars, oil-based paints, lacquers, and flammable gases.

**Class C fires** Fires that involve energized electrical equipment where the electrical conductivity of the extinguishing media is of importance.

**Class D fires** Fires involving combustible metals such as magnesium, titanium, zirconium, sodium, and potassium.

**Class K fires** Fires involving combustible cooking media such as vegetable oils, animal oils, and fats.

**Claw** The forked end of a tool.

**Claw bar** A tool with a pointed claw-hook on one end and a forked- or flat-chisel pry on the other end. It is often used for forcible entry.

**Clean agent** Gaseous fire extinguishing agent that does not leave a residue when it evaporates. Also known as halogenated agents.

**Clemens hook** A multipurpose tool that can be used for several forcible entry and ventilation applications because of its unique head design.

**Closed-circuit breathing apparatus** SCBA designed to recycle the user's exhaled air. The system removes carbon dioxide and generates fresh oxygen.

**Closet hook** A type of pike pole intended for use in tight spaces, commonly 2' to 4' in length.

**Cockloft** The concealed space between the top floor ceiling and the roof of a building.

**Coded system** A fire alarm system design that divides a building or facility into zones and has audible notification devices that can be used to identify the area where an alarm originated.

**Cold zone** This area is for staging vehicles and equipment. The cold zone contains the command post. The public and the media should be kept clear of the cold zone at all times.

**Combination attack** A type of attack employing both the direct and indirect attack methods.

**Combination ladder** A ladder that converts from a straight ladder to a step ladder configuration (A-frame) or from an extension ladder to a step ladder configuration.

**Combustibility** Determines whether or not a material will burn.

**Combustion** A chemical process of oxidation that occurs at a rate fast enough to produce heat and usually light in the form of either a glow or flames.

**Come along** A hand-operated tool used for dragging or lifting heavy objects. Sometimes known as lever blocks.

**Command** The first component of the IMS system. This is the only position in the IMS system that must always be staffed.

**Command post** The location at the scene of an emergency where the Incident Commander is located and where command, coordination, control, and communications are centralized.

**Command staff** Staff positions established to assume responsibility for key activities in the incident management system; individuals at this level report directly to the IC. Command staff include the Safety Officer, Public Information Officer, and Liaison Officer.

**Communications center** Facility that receives emergency or non-emergency reports from citizens. Many communications centers are responsible for dispatching fire brigade units as well.

**Company officer** Usually a lieutenant or captain in charge of a team of brigade members, both on scene and at the station. The brigade leader is responsible for firefighting strategy, safety of personnel, and the overall activities of the brigade members on their apparatus.

**Competent ignition source** A competent ignition source is one that can ignite a fuel under the existing conditions at the time of the fire. It must have sufficient heat and be in proximity to the fuel for sufficient time to ignite the fuel.

**Compressor** A mechanical device that increases the pressure and decreases the volume of atmospheric air; used to refill SCBA cylinders.

**Computer-aided dispatch (CAD)** Computer-based, automated systems used by telecommunicators to obtain and assess dispatch information. The system recommends the type of response required.

**Conduction** Heat transfer to another body or within a body by direct contact.

**Confined space** A space large enough and configured so an employee can enter, has limited or restricted mean for entry or exit and is not designed for continuous employee occupancy.

**Conflagration** A large fire, often involving multiple structures.

**Contaminated** A term used to describe evidence that may have been altered from its original state.

**Convection** Heat transfer by circulation within a medium such as a gas or a liquid.

**Coping saw** A saw designed to cut curves in wood.

**Coupling** One set or a pair of connection devices attached to a fire hose that allows the hose to be interconnected to additional lengths of hose.

**Crew** An organized group of brigade members under the leadership of a brigade leader, crew leader, or other designated official.

**Critical incident stress debriefing (CISD)** A confidential group discussion among those who served at a traumatic incident to address emotional, psychological, and stressful issues; usually occurs within 24 to 72 hours of the incident.

**Critical incident stress management (CISM)** A system to help personnel deal with critical incident stress in a positive manner; its aim is to promote long-term mental and emotional health after a critical incident.

**Critique** A process that examines the overall effectiveness of a policy, drill, or emergency response and is used in the development of recommendations for improving an organization's day-to-day and emergency response procedures.

**Cross-zoned system** A fire alarm system that requires activation of two separate detection devices before initiating an alarm condition. If a single detection device is activated, the alarm control panel will usually show a problem or trouble condition.

**Crowbar** A straight bar made of steel or iron with a forked-like chisel on the working end suitable for performing forcible entry.

**Curtain walls** Nonbearing walls used to separate the inside and outside of the building, but not part of the support structure for the building.

**Curved roofs** Roofs that have a curved shape.

**Cutting tool** Tools that are designed to cut into metal or wood.

**Cutting torch** A torch that produces a high temperature flame capable of heating metal to its melting point, thereby cutting through an object. Because of the high temperatures (5,700ºF) that these torches produce, the operator must be specially trained before using this tool.

**Cyanide** A highly toxic chemical agent that attacks the circulatory system.

**Cyberterrorism** The intentional act of electronically attacking government or private computer systems.

**Cylinder** The body of the fire extinguisher where the extinguishing agent is stored.

**Cylindrical lock** The most common fixed lock in use today. The locks and handles are placed into predrilled holes in the doors. One side of the door will usually have a key-in-the-knob lock, and the other will have a keyway, a button, or some other type of locking/unlocking device.

**Deadbolt** A secondary locking device used to secure a door in its frame.

**Dead load** The weight of a building; the dead load consists of the weight of all materials of construction incorporated into a building, including but not limited to walls, floors, roofs, ceilings, stairways, built-in partitions, finishes, cladding, and other similarly incorporated architectural and structural items, as well as fixed service equipment, including the weight of cranes.

**Decay phase** The phase of fire development where the fire has consumed either the available fuel or oxygen and is starting to die down.

**Deck gun** Apparatus-mounted master stream device intended to flow large amounts of water directly onto a fire or exposed building.

**Decontamination** The physical or chemical process of removing any form of contaminant from a person, an object, or the environment.

**Defensive attack** Exterior fire suppression operations directed at protecting exposures.

**Dehydration** A state in which fluid losses are greater than fluid intake into the body, leading to shock and even death if untreated.

**Deluge head** A sprinkler head that has no release mechanism; the orifice is always open.

**Deluge sprinkler system** A sprinkler system in which all sprinkler heads are open. When an initiating device, such as a smoke detector or heat detector, is activated, the deluge valve opens and water discharges from all of the open sprinkler heads simultaneously.

**Deluge valve** A valve assembly designed to release water into a sprinkler system when an external initiation device is activated.

**Demonstrative evidence** Term used to describe materials used to demonstrate a theory or explain an event.

**Depressions** Indentations felt on a kernmantle rope that indicate damage to the interior core, or kern, of the rope.

**Depth of char** The thickness of the layer of a material that has been consumed by a fire. The depth of char on wood can be used to help determine the duration of a fire.

**Designated incident facilities** Assigned locations where specific functions are always performed.

**Direct attack** Firefighting operations involving the application of extinguishing agents directly onto the burning fuel.

**Direct evidence** Evidence that is reported first hand, such as statements from an eyewitness who saw or heard something.

**Direct line** Telephone that connects two predetermined points.

**Dispatch** A summons to fire brigade units to respond to an emergency. Also known as alerting, dispatch is performed by the telecommunicator at the communications center.

**Distributors** Relatively small-diameter underground pipes that deliver water to systems or devices.

**Division** An organizational level within IMS that divides an incident into geographic areas of operational responsibility.

**Division supervisor** The officer in charge of all resources operating within a specified division, responsible to the next higher level in the incident organization, and the point-of-contact for the division within the organization.

**Doff** To take off an item of clothing or equipment.

**Dogs (pawls, ladder locks, and rung locks)** A mechanical locking device used to secure the fly section(s) of a ladder after they have been extended.

**Don** To put on an item of clothing or equipment.

**Door** An entryway; the primary choice for forcing entry into a vehicle or structure.

**Double-action pull-station** A manual fire alarm activation device that takes two steps to activate the alarm. The person must push in a flap, lift a cover, or break a piece of glass before activating the alarm.

**Double-female adaptor** A hose adaptor that is equipped with two female connectors. It allows two hoses with male couplings to be connected together.

**Double-hung window** Windows that have two movable sashes that can go up and down.

**Double jacket hose** A hose constructed with two layers of woven fibers.

**Double-male adaptor** A hose adaptor that is equipped with two male connectors. It allows two hoses with female couplings to be connected together.

**Double-pane glass** Window design that traps air or inert gas between two pieces of glass to improve insulation.

**Drafting sites** Location where an engine can draft water directly from a static source.

**Drainage rate** The rate at which solution drains from the foam blanket. For foam quality test purposes, it is the time it takes for 25% of the solution to drain from the foam blanket.

**Driver/operator** Often called an engineer or technician, the driver is responsible for getting the fire apparatus to the scene safely, setting up, and running the pump or operating the aerial ladder once it arrives on the scene.

**Dry-barrel hydrant** A type of hydrant used in areas subject to freezing weather. The valve that allows water to flow into the hydrant is located underground, and the barrel of the hydrant is normally dry.

**Dry chemical extinguishing system** An automatic fire extinguishing system that discharges a dry chemical agent.

**Dry chemical fire extinguisher** An extinguisher that uses a mixture of finely divided solid particles to extinguish fires. The agent is usually sodium bicarbonate-, potassium bicarbonate-, or ammonium phosphate-based, with additives to provide resistance to packing and moisture absorption and to promote proper flow characteristics.

**Dry hydrant** An arrangement of pipe that is permanently connected to allow a fire brigade engine to draft water from a static source.

**Dry-pipe valve** The valve assembly on a dry sprinkler system that prevents water from entering the system until the air pressure is released.

**Dry powder extinguishing agent** Extinguishing agent used in putting out Class D fires. The common dry powder extinguishing agents include sodium chloride and graphite-based powders.

**Dry powder fire extinguisher** A fire extinguisher that uses an extinguishing agent in powder or granular form, designed to extinguish Class D combustible metal fires by crusting, smothering, or heat-transferring means.

**Dry sprinkler system** A sprinkler system in which the pipes are normally filled with compressed air. When a sprinkler head is activated, it releases the air from the system, which opens a valve so the pipes can fill with water.

**Drywall hook** A specialized version of a pike pole that can remove drywall more effectively because of its hook design.

**Duck-billed lock breaker** A tool with a point that can be inserted in the shackles of a padlock. As the point is driven further into the lock, it gets larger and forces the shackles apart until they break.

**Duplex channel** A radio system that uses two frequencies per channel. One frequency transmits and the other receives messages. The system uses a repeater site to transmit messages over a greater distance than a simplex system.

**Dutchman** A term used for a short fold placed in a hose when loading it into the bed. This fold prevents the coupling from turning in the hose bed.

**Dynamic rope** A rope generally made out of synthetic materials that is designed to be elastic and stretch when loaded. Used often by mountain climbers.

**Early suppression fast response (ESFR) sprinkler head** A sprinkler head designed to react quickly and suppress a fire in its early stages.

**Ecoterrorism** Terrorism directed against causes that radical environmentalists think would damage the earth or its creatures.

**Edge roller** A device used to prevent damage to a rope from jagged edges or friction.

**Eductor** A foam proportioner that operates as a venturi to draw foam concentrate into the water stream.

**Egress** A method of exiting from an area or a building.

**Ejectors** Electrical fans used in negative-pressure ventilation.

**Electrical energy** Heat produced by electricity.

**Electrolytes** Certain salts and other chemicals that are dissolved in body fluids and cells.

**Elevated master stream** Nozzle mounted on the end of an aerial device capable of delivering large amounts of water onto a fire or exposed building from an elevated position.

**Elevated water storage tank** An above-ground water storage tank that is designed to maintain pressure on a water distribution system.

**Elevation pressure** The amount of pressure created by gravity.

**Emergency by-pass mode** Operating mode that allows an SCBA to be used even if part of the regulator fails to function properly.

**Emergency incident rehabilitation** A function on the emergency scene that cares for the well-being of the brigade members. It includes physical assessment, revitalization, medical evaluation and treatment, and regular monitoring of vital signs.

Emergency Medical Services (EMS) personnel EMS personnel trained to administer prehospital care to people who are sick or injured.

Emergency response operations plan Plan designed to identify the levels of response needed for certain locations with emergency situations.

Emergency traffic An urgent message, such as a call for help or evacuation, transmitted over a radio that takes precedence over all normal radio traffic.

Employee assistance program (EAP) Program adopted by many organizations for brigade members to receive confidential help with problems such as substance abuse, stress, depression, or burnout that can affect their work performance.

Endothermic reactions Reactions that absorb heat or require heat to be added.

Entrapment A condition in which a victim is trapped by debris, soil, or other material and is unable to extricate himself or herself.

Evacuation signal Warn all personnel to pull back to a safe location.

Excavation Any man-made cut, cavity, trench, or depression in an earth surface, formed by earth removal.

Exhauster A device that accelerates the removal of the air from a dry-pipe or preaction sprinkler system.

Exothermic reactions Reactions that result in the release of energy in the form of heat.

Expanded incident report narrative A report in which all company members submit a narrative describing what they observed and which activities they performed during an incident.

Exposure Any person or property that may be endangered by flames, smoke, gases, heat, or runoff from a fire.

Extension Fire that moves into areas not originally involved, including walls, ceilings, and attic spaces; also the movement of fire into uninvolved areas of a structure.

Extension ladder An adjustable-length, multiple-section ladder.

Exterior wall A wall—often made of wood, brick, metal, or masonry—that makes up the outer perimeter of a building. Exterior walls are often load-bearing.

Extinguishing agent Material used to stop the combustion process; extinguishing agents may include liquids, gases, dry chemical compounds, and dry powder compounds.

Extra hazard locations Occupancies where the total amounts of Class A combustibles and Class B flammables are greater than expected in occupancies classed as ordinary (moderate) hazard.

Face piece Component of SCBA that fits over the face.

False alarm The activation of a fire alarm system when there is no fire or emergency condition.

Federal Communications Commission (FCC) United States federal regulatory authority that oversees radio communications.

Film-forming fluoroprotein (FFFP) A water-based extinguishing agent used on Class B fires that forms a foam layer over the liquid and stops the production of flammable vapors.

Finance/administration section The command-level section of IMS responsible for all costs and financial aspects of the incident, as well as any legal issues that arise.

Finished foam The homogeneous blanket obtained by mixing water, foam concentrate, and air.

Fire A rapid, persistent chemical reaction that releases both heat and light.

Fire alarm annunciator panel Part of the fire alarm system that indicates the source of an alarm within a building.

Fire alarm control panel That component in a fire alarm system that controls the functions of the entire system.

Fire department connection (FDC) A fire hose connection through which the fire brigade can pump water into a sprinkler system or standpipe system.

Fire enclosures Fire-rated assemblies used to enclose vertical openings such as stairwells, elevator shafts, and chases for building utilities.

Fireground command (FGC) An incident management system developed in the 1970s for day-to-day fire brigade incidents (generally handled with fewer than 25 units or companies).

Fire helmet Protective head covering worn by brigade members to protect the head from falling objects, blunt trauma, and heat.

Fire hydraulics The physical science of how water flows through a pipe or hose.

Fire load The weight of combustibles in a fire area or on a floor in buildings and structures including either contents or building parts, or both.

Fire partitions Interior walls extending from the floor to the underside of the floor above.

Fire point (flame point) The lowest temperature at which a liquid releases enough vapors to ignite and sustain combustion.

Fire protection engineer The fire protection engineer usually has an engineering degree, reviews plans, and works with building owners to ensure that their fire suppression and detection systems will meet the relevant codes and function as needed.

Fire-rated glass Special glass formulated to achieve a fire rating.

Fire resistive construction (Type 1) Buildings where the structural members are of noncombustible materials that have a specified fire resistance.

Fire safety inspection An inspection based on the NFPA fire code for the type of occupancy (such as commercial or manufacturing facilities) and for the type of specific hazard (such as flammable liquids or combustible dust).

Fire safety survey The process of identifying hazards before an incident takes place. For example, accessability of doors, storage of combustible or flammable materials and electrical equipment are all considered during this process.

FIRESCOPE (FIre RESources of California Organized for Potential Emergencies) An organization of agencies established in the early 1970s to develop a standardized system for managing fire resources at large-scale incidents such as wildland fires.

Fire tetrahedron A geometric shape used to depict the four components required for a fire to occur: fuel, oxygen, heat, and chemical chain reactions.

Fire wall A wall with a fire-resistive rating and structural stability that separates buildings or subdivides a building to prevent the spread of fire.

Fire window A window or glass block assembly with a fire-resistive rating.

Fixed gallonage fog nozzle A nozzle delivers a set number of gallons per minute that the nozzle was designed for, no matter what pressure is applied to the nozzle.

Fixed-temperature heat detector A sensing device that responds when its operating element is heated to a predetermined temperature.

Flame detector A sensing device that detects the radiant energy emitted by a flame.

Flammability limits (explosive limits) The upper and lower concentration limits of a flammable gas or vapor in air that can be ignited, expressed as a percentage of fuel by volume.

Flashover The condition where all combustibles in a room or confined space have been heated to the point at which they release vapors that will support combustion, causing all combustibles to ignite simultaneously.

Flash point The minimum temperature at which a liquid releases sufficient vapor to form an ignitable mixture with the air.

Flat bar A specialized type of prying tool made of flat steel with prying ends suitable for performing forcible entry.

Flat-head axe A tool that has an axe head on one side and a flat head on the opposite side.

Flat hose load A method of putting a hose on apparatus in which the hose is laid flat and stacked on top of the previous section.

Flat roofs Horizontal roofs often found on commercial or industrial occupancies.

Floodlight A light that can illuminate a broad area.

Floor runner A piece of canvas or plastic material, usually 3' to 4' wide and in various lengths, used to protect flooring from dropped debris and/or dirt from shoes and boots.

Flow pressure The amount of pressure created by moving water.

Flow switch An electrical switch that is activated by water moving through a pipe on a sprinkler system.

Fluorochemical surfactant A chemical compound containing fluorine that is used to reduce surface tension when dissolved in a solution.

Fly section A section of an extension ladder that is raised or extended from the base section or from another fly section. Some extension ladders have more than one fly section.

**Foam concentrate** The foaming agent that is mixed with the appropriate amounts of water and air to produce mechanical foam.

**Foam proportioner** The device that mixes foam concentrate into a water stream in the correct percentage.

**Foam solution** A homogeneous mixture of water and foam concentrate in the correct proportions.

**Fog stream nozzle** Device placed at the end of a fire hose that separates water into fine droplets to aid in heat absorption.

**Folding ladder** A ladder that collapses by bringing the two beams together for portability. Unfolded, the folding ladder is narrow and used for access to attic scuttle holes and confined areas.

**Forcible entry** Techniques used by brigade members to gain entry into buildings, vehicles, aircraft, or other areas when normal means of entry are locked or blocked.

**Forward lay** A method of laying a supply line where the line starts at the water source and is laid toward the fire.

**Forward staging area** A strategically placed area, close to the incident site, where personnel and equipment can be held in readiness for rapid response to an emergency event.

**Four-way hydrant valve** A specialized type of valve that can be placed on a hydrant that allows another engine to increase the supply pressure without interrupting flow.

**Frangible bulb sprinkler head** A sprinkler head that uses a glass bulb filled with glycerin or alcohol to hold the cap in place. The bulb also contains an air bubble. As the bulb is heated, liquid absorbs the air bubble and expands until it breaks the glass, releasing the cap.

**Frangible-pellet sprinkler head** A sprinkler head that is activated when the solder pellet melts at a preset temperature.

**Freelancing** Dangerous practice of acting independently of command instructions.

**Fresno ladder (attic ladder)** A narrow, two-section extension ladder that has no halyard. Because of its limited length, it can be extended manually.

**Friction loss** The reduction in pressure due to the water being in contact with the side of the hose. This contact requires force to overcome the drag the wall of the hose reates.

**Frostbite** Damage to tissues as the result of exposure to cold; frozen or partially frozen body parts.

**Frothover** A frothing of burning crude oil caused when water contacts superheated fuel and flashes to steam.

**Fuel** All combustible materials. The actual material that is being consumed by a fire, allowing the fire to take place.

**Fully developed phase** The phase of fire development where the fire is free-burning and consuming much of the fuel.

**Fully encapsulated suits** A protective suit that completely covers the brigade member, including the breathing apparatus, and does not let any vapor or fluids enter the suit.

**Full-time fire brigade member** A fire brigade member who is assigned to the fire brigade full-time; normally supervised by the fire brigade leader.

**Fusible link sprinkler head** A sprinkler head with an activation mechanism that incorporates two pieces of metal held together by low-melting-point solder. When the solder melts, it releases the link and water begins to flow.

**Gamma rays** A type of radiation that can travel significant distances, penetrating most materials and passing through the body. Gamma radiation is the most destructive to the human body.

**Gas** One of the three phases of matter. A substance that will expand indefinitely and assume the shape of the container that holds it.

**Gas detector** A device that detects and/or measures the concentration of dangerous gases.

**Gated wye** A valved device that splits a single hose into two separate hoses, allowing each hose to be turned on and off independently.

**Gate valves** Valves found on hydrants and sprinkler systems.

**Generator** An engine-powered device that provides electricity.

**Glass blocks** Thick pieces of glass similar to bricks or tiles.

**Glazed** Transparent glass.

**Glucose** The source of energy for the body. One of the basic sugars, it is the primary fuel, along with oxygen.

**Grade** The level at which the ground intersects the foundation of a structure.

**Gravity feed system** A water distribution system that depends on gravity to provide the required pressure. The system storage is usually located at a higher elevation than the end users.

**Gripping pliers** A hand tool with a pincer-like working end that can be used to bend wire or hold smaller objects.

**Group** An organization level within IMS that divides an incident according to functional areas of operation.

**Group supervisor** The brigade leader in charge of all resources operating within a specified group, responsible to the next higher level in the incident organization, and the point-of-contact for the group within the organization.

**Growth phase** The phase of fire development where the fire is spreading beyond the point of origin and beginning to involve other fuels in the immediate area.

**Guideline** A rope used for orientation when inside a structure when there is low or no visibility; it is attached to a fixed object outside the hazardous area.

**Guides** Strips of metal or wood that serve to guide a fly section during extension. Channels or slots in the bed or fly section may also serve as guides.

**Gusset plates** The connecting plate made of wood or lightweight metal used in trusses.

**Gypsum** A naturally occurring material composed of calcium sulfate and water molecules.

**Gypsum board** The generic name for a family of sheet products consisting of a noncombustible core primarily of gypsum with paper surfacing.

**Hacksaw** A cutting tool designed for use on metal. Different blades can be used for cutting different types of metal.

**Halligan tool** A prying tool that incorporates a pick and a claw, designed for use in the fire service. Sometimes known as a Hooligan tool.

**Halocarbon** A compound, such as a fluorocarbon, that is made up of carbon combined with one or more halogens (e.g., fluorine, chlorine, bromine, or iodine).

**Halogenated-agent extinguisher** An extinguisher that uses halogenated extinguishing agents.

**Halogenated extinguishing agent** A liquefied gas extinguishing agent that puts out fires by interrupting the chemical chain reaction.

**Halon 1211** Bromochlorodifluoromethane ($CBrClF_2$), a halogenated agent that is effective on Class A, B, and C fires.

**Halyard** The rope or cable used to extend or hoist the fly section(s) of an extension ladder.

**Hammer** A striking tool.

**Handle** The grip used for holding and carrying a portable fire extinguisher.

**Hand light** Small, portable light carried by brigade members to improve visibility at emergency scenes, often powered by rechargeable batteries.

**Handline nozzles** Used on hoses ranging from 1 $^1/_2$" to 2 $^1/_2$" hose lines, usually flow between 90 and 350 gallons per minute.

**Handsaw** A manually powered saw designed to cut different types of materials. Examples include hacksaws, carpenter's handsaws, keyhole saws, and coping saws.

**Hard suction hose** A hose designed to prevent collapse under vacuum conditions so that it can be used for drafting water from below the pump (lakes, rivers, wells, or sea water, etc.).

**Hardware** The parts of a door or window that enable it to be locked or opened.

**Harness** A piece of equipment worn by a rescuer that can be attached to a life safety rope.

**Hazardous materials** Any materials or substances that pose an unreasonable risk of damage or injury to persons, property, or the environment if not properly controlled during handling, storage, manufacture, processing, packaging, use and disposal, or transportation.

**Hazardous materials technician** "Hazmat" technicians have training and certification in chemical identification, leak control, decontamination, and clean-up procedures.

Heat detector A fire alarm device that detects either abnormally high temperatures or rate-of-rise in temperature, or both.

Heat sensor label A piece of heat-sensitive material on each section of a ladder that identifies when the ladder has been exposed to high heat conditions.

Heat wave A build-up of heavy hydrocarbons that collect as crude oil burns.

Heavy timber construction (Type IV) Buildings constructed with noncombustible or limited-combustible exterior walls and interior walls and floors made of large dimension combustible materials.

Higbee indicators An indicator on the male and female threaded couplings that indicates where the threads start. These indicators should be aligned before starting to thread the couplings together.

High-angle operations A rope rescue operation where the angle of the slope is greater than 45°; rescuers depend on life safety rope rather than a fixed support surface such as the ground.

Hitches Knots that attach to or wrap around an object.

Hollow-core A door made of panels that are honeycombed inside, creating an inexpensive and lightweight design.

Horizontal-sliding window Windows that slide open horizontally.

Horizontal ventilation The process of making openings so that smoke, heat, and gases can escape horizontally from a building through openings such as doors and windows.

Horn The tapered discharge nozzle of a carbon dioxide-type fire extinguisher.

Horseshoe hose load A method of loading hose where the hose is laid into the bed along the three walls of the bed, resembling a horseshoe.

Hose appliance Any device used to connect to a fire hose for the purpose of delivering water.

Hose clamp A device used to compress a fire hose to stop water flow.

Hose jacket A device used to stop a leak in a fire hose or to join hoses that have damaged couplings.

Hose liner (hose inner jacket) The inside portion of a hose that is in contact with the flowing water.

Hose roller A device that is placed on the edge of a roof and is used to protect hose as it is hoisted up and over the roof edge.

Hot zone This area is for entry teams and rescue teams only. The hot zone immediately surrounds the dangers of the site (e.g., hazardous materials releases) to protect personnel outside the zone.

Hux bar A multipurpose tool that can be used for several forcible entry and ventilation applications because of its unique design. Also can be used as a hydrant wrench.

HVAC system Heating, ventilation, and air conditioning system in large buildings.

Hydrant wrench A hand tool used to operate the valves on a hydrant; may also be used as a spanner wrench. Some are plain wrenches, and others have a ratchet feature.

Hydraulic shears A lightweight hand-operated tool that can produce up to 10,000 pounds of cutting force.

Hydraulic spreader A lightweight hand-operated tool that can produce up to 10,000 pounds of prying and spreading force.

Hydraulic ventilation Ventilation that relies upon the movement of air caused by a fog stream.

Hydrocarbons Organic chemical compounds composed of carbon and hydrogen atoms.

Hydrogen cyanide Toxic gas produced by combustion of plastics and synthetics.

Hydrolyzed Decomposition of a chemical compound by reaction with water.

Hydrostatic testing Periodic certification test performed on pressure vessels, including SCBA cylinders.

Hypothermia A condition in which the internal body temperature falls below 95°F, usually a result of prolonged exposure to cold or freezing temperatures.

Hypoxia A state of inadequate oxygenation of the blood and tissue.

I-beam A ladder beam constructed of one continuous piece of I-shaped metal or fiberglass to which the rungs are attached.

Ignition point The minimum temperature at which a substance will burn.

Ignition temperature The minimum temperature at which a fuel, when heated, will spontaneously ignite and continue to burn.

Immediately dangerous to life and health (IDLH) Any condition that poses an immediate or delayed threat to life, might cause irreversible adverse health effects, or might interfere with an individual's ability to escape unaided from a hazardous environment.

Improvised explosive device (IED) An explosive or incendiary device that is fabricated in an improvised manner.

IMS general staff The chiefs of each of the four major sections of IMS: Operations, Planning, Logistics, and Finance/Administration.

Incendiary device A device or mechanism used to start a fire or explosion.

Incendiary fires Intentionally set fires.

Incident action plan (IAP) The objectives for the overall incident strategy, tactics, risk management, and member safety that are developed by the IC. Incident Action Plans are updated throughout the incident.

Incident commander (IC) The person in charge of the incident site who is responsible for all decisions relating to the management of the incident.

Incident command system (ICS) The first standard system for organizing large, multi-jurisdictional and multi-agency incidents involving more than 25 resources or operating units; eventually developed into the Incident Management System (IMS).

Incident management system (IMS) The combination of facilities, equipment, personnel, procedures, and communications under a standard organizational structure to manage assigned resources effectively to accomplish stated objectives for an incident. Also known as Incident Command System (ICS).

Incipient The initial stage of a fire.

Incipient phase The phase of fire development where the fire is limited to the immediate point of origin.

Incipient stage Refers to the severity of a fire where the progression is in the early stage and has not developed beyond that which can be extinguished using portable fire extinguishers or small (up to 1 $^1/_2$") water handlines, flowing a maximum of 125 gpm (473 L/min).

Incomplete combustion A burning process in which the fuel is not completely consumed, usually due to a limited supply of oxygen.

Incubation period Time period between the initial infection by an organism and the development of symptoms.

Indicating butterfly valve (IBV) A sprinkler control valve that is equipped with a directional arrow to indicate the position of the valve and a gear operator to prevent the valve from being closed in less than 5 seconds, preventing a water hammer.

Indirect application of water Using a solid object such as a wall or ceiling to break apart a stream of water, causing it to create more surface area on the water droplets and thereby absorb more heat.

Indirect attack Firefighting operations involving the application of extinguishing agents to reduce the buildup of heat released from a fire without applying the agent directly onto the burning fuel.

Industrial water system A water distribution system that is designed to deliver water to either fire protection pumps or both process and fire systems.

Institutional memory A situation in which behaviors, policies, and procedures have been structured and formalized in an organization's culture and practices.

Integrated communications The ability of all appropriate personnel at the emergency scene to communicate with their supervisor and their subordinates.

Interior attack The assignment of a team of brigade members to enter a structure and attempt fire suppression.

Interior finish Any coating or veneer applied as a finish to a bulkhead, structural insulation, or overhead, including the visible finish, all intermediate materials, and all application materials and adhesives.

Interior wall (partition or nonbearing wall) A wall inside a building that divides a large space into smaller areas.

Inverter A device that converts the direct current from an apparatus electrical system into alternating current.

Ionization smoke detector A device containing a small amount of radioactive material that ionizes the air between two charged electrodes to sense the presence of smoke particles.

Irons A combination tool, normally the Halligan tool and the flat-head axe.

J tool A tool that is designed to fit between double doors equipped with panic bars.

Jalousie window Windows made of small slats of tempered glass that overlap each other when closed. Jalousie windows are held by a metal frame and operated by a small hand wheel or crank found in the corner of the window.

Jamb The part of a doorway that secures the door to the studs in a building.

Job position requirements brigade member An employee who is a member of a fire brigade because his or her job position requires them to be a fire brigade member.

Junction box A device that attaches to an electrical cord to provide additional outlets.

K tool Used to remove lock cylinders from structural doors so the locking mechanism can be unlocked.

Kelly tool A steel bar with two main features: a large pick and a large chisel or claw.

Kerf cut A cut that is only the width and depth of the saw blade. It is used to make inspection holes.

Kernmantle rope Rope made of two parts—the kern (interior core) and the mantle (the outside sheath).

Kevlar® Strong, synthetic material used in the construction of protective clothing and equipment.

Keyhole saw A saw designed to cut keyholes in wood.

Knots A fastening made by tying together lengths of rope or webbing in a prescribed way, used for a variety of purposes.

Ladder belt A belt specifically designed to secure a brigade member to a ladder or elevated surface.

Ladder gin An A-shaped structure formed with two ladder sections. It can be used as a makeshift lift when raising a trapped person. One form of the device is called an A-frame hoist.

Ladder halyards Rope used on extension ladders to raise and lower a fly section.

Ladder pipe Nozzle attached to the tip of an aerial ladder truck designed to provide large volumes of water from an elevated position.

Laminated glass Also known as safety glass; the lamination process places a thin layer of plastic between two layers of glass, so that the glass does not shatter and fall apart when broken.

Laminated wood Pieces of wood that are glued together.

Large-diameter hose (LDH) Hose in the 4" to 12" range.

Latch The part of the door lock that secures into the jamb.

Laths Thin strips of wood used to make the supporting structure for roof tiles.

Leap-frogging A fire spread from one floor to the other through exterior windows (auto-exposure).

Lewisite Blister-forming agent that is an oily, colorless-to-dark brown liquid with an odor of geraniums.

Liaison officer The position within IMS that establishes a point of contact with outside agency representatives.

Lifeline A rope secured to a brigade member that enables him or her to retrace his or her steps out of a structure.

Life safety rope Rope used solely for the purpose of supporting people during firefighting, rescue, other emergency operations, or during training exercises.

Light-emitting diode (LED) An electronic semiconductor that emits a single-color light when activated.

Light hazard locations Occupancies where the total amount of combustible materials is less than expected in an ordinary hazard location.

Lightweight construction The use of small dimension members such as 2' x 4' or 2'x 6' wood assemblies as structural supports in buildings.

Line detector Wire or tubing that can be strung along the ceiling of large open areas to detect an increase in heat.

Liquid One of the three phases of matter. A nongaseous substance that is composed of molecules that move and flow freely and assumes the shape of the container that holds it.

Litigious Prone to engage in lawsuits.

Live load The weight of the building contents.

Load-bearing wall A wall that supports structural members or upper floors of a building.

Loaded-stream extinguisher A water-based fire extinguisher that uses an alkali metal salt as a freezing point depressant.

Lob A method of discharging extinguishing agent in an arc to avoid splashing or spreading the burning fuel.

Local alarm system A fire alarm system that sounds an alarm only in the building where it was activated. No signal is sent out of the building.

Lock body The part of a padlock that holds the main locking mechanisms and secures the shackles.

Locking mechanism A device that locks an extinguisher's trigger to prevent accidental discharge.

Lockout and tagout systems Lockout and tagout are methods of ensuring that electricity and other utilities have been shut down and switches are "locked" so that they cannot be switched on, in order to prevent flow of power or gases into the area where rescue is being conducted.

Logistics section The section within IMS responsible for providing facilities, services, and materials for the incident.

Logistics section chief The general staff position responsible for directing the logistics function; generally assigned on complex, resource-intensive, or long-duration incidents.

Loop A piece of rope formed into a circle.

Louver cut A cut that is made using power saws and axes to cut along and between roof supports so that the sections created can be tilted into the opening.

Low volume nozzles Nozzles that flow 40 gpm or less.

Low-angle operations A rope rescue operation on a mildly sloping surface (less than 45°) or flat land where brigade members are dependent on the ground for their primary support, and the rope system is a secondary means of support.

Lower explosive limit (LEL) The minimum amount of fuel vapor mixed in air that will ignite or explode.

Malicious false alarm A fire alarm signal when there is no fire, usually initiated by individuals who wish to cause a disturbance.

Mallet A short-handled hammer.

Manual pull-station A device with a switch that either opens or closes a circuit, activating the fire alarm.

Manufactured trailer structures A factory-assembled structure or structures transportable in one or more sections that is built on a permanent chassis and designed to be used as a dwelling without a permanent foundation when connected to the required utilities, including the plumbing, heating, air-conditioning, and electric systems contained therein.

Mark 1 Nerve Agent Antidote Kit A military kit containing antidotes that can be administered to victims of a nerve agent attack.

Masonry Built-up unit of construction or combination of materials such as brick, clay tiles, or stone set in mortar.

Mass arson Involves an offender who sets three or more fires at the same site or location during a limited period of time.

Mass decontamination Special procedures for incidents that involve large numbers of people. Master stream devices from engine companies and aerial apparatus provide high-volume, low-pressure showers so that large numbers of people can be decontaminated rapidly.

Master-coded alarm An alarm system in which audible notification devices can be used for multiple purposes, not just for the fire alarm.

Master stream device A large capacity nozzle that can be supplied by two or more hose lines or fixed piping. Can flow in excess of 300 gpm. Includes deck guns, portable ground monitors, ladder pipes and other elevated devices.

Material Safety Data Sheet (MSDS) A form, provided by manufacturers and compounders (blenders) of chemicals, containing information about chemical composition, physical and chemical properties, health and safety hazards, emergency response, and waste disposal of the material.

Maul A specialized striking tool, weighing six pounds or more, with an axe head on one side and a sledgehammer head on the other side.

Mechanical energy Heat energy created by friction.

Mechanical saw Usually powered by electric motors or gasoline engines. The three primary types are chain saws, rotary saws, and reciprocating saws.

Mechanical ventilation Ventilation that uses mechanical devices to move air.

Medium diameter hose (MDH) Hose of 2 $^1/_2$" or 3" size.

Mildew A condition that can occur on hose if it is stored wet. Mildew can damage the jacket of a hose.

Miscible Readily mixes with water.

Mobile data terminals (MDTs) Technology that allows brigade members to receive information in the apparatus or at the station.

Mobile radio Two-way radio that is permanently mounted in a fire apparatus.

Mortise lock Door locks with both a latch and a bolt built into the same mechanism; the two locking devices operate independently of each other. This is a common lock found in places such as hotel rooms.

Multipurpose dry chemical extinguishers Extinguishers rated to fight Class A, B, and C fires.

Multipurpose hook A long pole with a wooden or fiberglass handle and a metal hook on one end used for pulling.

Mushrooming The process that occurs when rising smoke, heat, and gases encounter a horizontal barrier such as a ceiling and begin to move out and back down.

National Fire Incident Reporting System (NFIRS) A nationwide database held at the National Fire Data Center under the U.S. Fire Administration that collects fire-related data so as to provide information on the national fire problem.

National Institute for Occupational Safety and Health (NIOSH) A U.S. Federal agency responsible for research and development of occupational safety and health issues.

Natural ventilation Ventilation that relies upon the natural movement of heated smoke and wind currents.

Negative-pressure ventilation Ventilation that relies upon electric fans to pull or draw the air from a structure or area.

Nerve agents Toxic substances that attack the central nervous system in humans.

Nomex® A fire-resistant synthetic material used in the construction of personal protective equipment for fire fighting.

Nonbearing wall (partition or interior wall) A wall that does not support a ceiling or structural member, but simply divides a space.

Non-coded alarm An alarm system that provides no information at the alarm control panel indicating where the activated alarm is located.

Noncombustible construction (Type II) Buildings where the structural members are of noncombustible materials without fire resistance.

Normal operating pressure The observed static pressure in a water distribution system during a period of normal demand.

Nose cups An insert inside the face piece of an SCBA that fits over the user's mouth and nose.

Nozzle The discharge orifice of a portable fire extinguisher; attachment to the discharge end of an attack hose to give fire stream shape and direction.

Nozzle shut off Device that enables the person at the nozzle to start or stop the flow of water.

Nuisance alarm A fire alarm signal caused by malfunction or improper operation of a fire alarm system or component.

Obscuration rate A measure of the percentage of light transmission that is blocked between a sender and a receiver unit.

Occupancy The purpose for which a building or other structure, or part thereof, is used or intended to be used.

Occupational Safety and Health Administration (OSHA) The federal agency that regulates worker safety and, in some cases, responder safety. OSHA is part of the U.S. Department of Labor.

Offensive attack An advance into the fire building by brigade members with hose lines or other extinguishing agents to overpower the fire.

Oleophobic Oil hating; having the ability to shed hydrocarbon liquids.

One-person rope A rope rated to carry the weight of a single person (300 lbs).

Open-circuit breathing apparatus SCBA in which the exhaled air is released into the atmosphere and is not reused.

Open-end wrench A hand tool used to tighten or loosen bolts. The end is open, as opposed to a box-end wrench. Each wrench is a specific size.

Operating lever The handle of a door that turns the latch to open it.

Operations section The section within IMS responsible for all tactical operations at the incident.

Operations section chief The general staff position responsible for managing all operations activities; usually assigned when complex incidents involve more than 20 single resources or when the IC cannot be involved in the details of tactical operations.

Ordinary construction (Type III) Buildings where the exterior walls are noncombustible or limited-combustible, but the interior floors and walls are made of combustible materials.

Ordinary hazard locations Occupancies that contain more Class A and Class B materials than are found in light hazard locations.

Outside stem and yoke (OS&Y) valve A sprinkler control valve with a valve stem that moves in and out as the valve is opened or closed.

Overhaul Examination of all areas of the building and contents involved in a fire to ensure that the fire is completely extinguished.

Oxidation A chemical reaction initiated by combining an element with oxygen, resulting in the form of the element or one of its compounds.

Oxidizing agent A substance that will release oxygen or act in the same manner as oxygen in a chemical reaction.

Oxygen deficiency Any atmosphere where the oxygen level is below 19.5%. Low oxygen levels can have serious effects on people, including adverse reactions such as poor judgment and lack of muscle control.

Oxygenated To treat, combine, or infuse with oxygen.

Packaging Preparing the victim for movement, often accomplished with a long spine board or similar device.

Padlock The most common lock on the market today, built to provide regular-duty or heavy-duty service. Several types of locking mechanism are available, including a keyway, combination wheels, or combination dials.

Parallel chord truss A truss in which the top and bottom chords are parallel.

Parapet walls Walls on a flat roof that extend above the roof line.

Partition (interior wall or nonbearing wall) A wall that does not support a ceiling or structural member, but simply divides a space.

Part-time fire brigade member A fire brigade member who holds another job and participates in the fire brigade on a part-time basis.

Party walls Walls constructed on the line between two properties.

P-A-S-S Acronym used for operating a portable fire extinguisher: Pull pin, Aim nozzle, Squeeze trigger, Sweep the nozzle across burning fuel.

Passing command Option that can be used by the first-arriving brigade leader to direct the next arriving unit to assume command.

PBI® A fire-retardant synthetic material used in the construction of personal protective equipment.

Pendant sprinkler head A sprinkler head designed to be mounted on the underside of sprinkler piping so the water stream is directed down.

Permit required confined space (PRCS) Space that is identified at every facility and requires special approval procedures for entry. Has the potential for hazardous atmospheres, entrapment or entanglement, or other recognized hazards.

Personal alert safety system (PASS) Device worn by a brigade member that sounds an alarm if the brigade member is motionless for a period of time.

Personal dosimeters Devices that measure the amount of radioactive exposure to an individual.

Personal escape rope An emergency use rope designed to carry the weight of only one person and to be used only once.

Personal protective equipment (PPE) Gear worn by brigade members that includes helmet, gloves, hood, coat, pants, SCBA, and boots. The personal protective equipment provides a thermal barrier for brigade members against intense heat.

Personnel Accountability Report (PAR) A roll call report confirming that all members of a company are present.

Personnel accountability system A method of tracking the identity, assignment, and location of brigade members operating at an incident scene.

**Personnel accountability tag (PAT)** Identification card used to track the location of a brigade member on an emergency incident.

**Phosgene** A chemical agent that causes severe pulmonary damage. A by-product of incomplete combustion.

**Photoelectric smoke detector** A device to detect visible products of combustion using a light source and a photosensitive sensor.

**Physical evidence** Items that can be observed, photographed, measured, collected, examined in a laboratory, and presented in court to prove or demonstrate a point.

**Pick** The pointed end of a pick axe that can be used to make a hole or bite in a door, floor, or wall.

**Pick-head axe** A tool that has an axe head on one side and a pointed "pick" on the opposite side.

**Piercing nozzle** A nozzle that can be driven through sheet metal or other material to deliver a water stream to that area.

**Pike pole** A pole with a sharp point, or pike, on one end coupled with a hook. Used to make openings in ceilings, walls, etc.

**Pipe bomb** A device created by filling a section of pipe with an explosive material.

**Pipe wrench** A wrench having one fixed grip and one movable grip that can be adjusted to fit securely around pipes and other tubular objects.

**Pitched chord truss** Type of truss typically used to support a sloping roof.

**Pitched roof** A roof with sloping or inclined surfaces.

**Pitot gauge** A type of gauge that is used to measure the velocity pressure of water that is being discharged from an opening. Used to determine the flow of water from a hydrant.

**Placards** Signage required to be placed on sides of buildings, highway transport vehicles, railroad tank cars, and other forms of hazardous materials transportation that identifies the hazardous contents of the vehicle, using a standardized system.

**Plague** An infectious disease caused by the bacterium *Yersinia pestis*, commonly found on rodents.

**Planning section** The section within IMS responsible for the collection, evaluation, and dissemination of tactical information related to the incident and for preparation and documentation of incident management plans.

**Planning section chief** The general staff position responsible for planning functions; assigned when the IC needs assistance in managing information.

**Plaster hook** A long pole with a pointed head and two retractable cutting blades on the side.

**Plate glass** A type of glass with additional strength so it can be formed in larger sheets, but will still shatter upon impact.

**Platform-frame construction** Construction technique for building the frame of the structure one floor at a time. Each floor has a top and bottom plate that acts as a fire stop.

**Platform-frame techniques** Subflooring is laid on the joists, and the frame for the first floor walls is erected on the first floor.

**Plume** The column of hot gases, flames, and smoke that rises above a fire, also called a convection column, thermal updraft, or thermal column.

**Point of origin** The exact location where a heat source and a fuel come in contact with each other and a fire begins.

**Polar solvent** A water-soluble flammable liquid such as alcohol, acetone, ester, and ketone.

**Polymer** A naturally occurring or synthetic compound consisting of large molecules made up of a linked series of repeated simple monomers.

**Pompier ladder (scaling ladder)** A lightweight, single beam ladder.

**Portable ladder** Ladder carried on fire apparatus, but designed to be removed from the apparatus and deployed by brigade members where needed.

**Portable monitor** A master stream appliance designed to be set up and operated from the ground. This device is designed to flow large amounts of water onto a fire or exposed building from a ground-level position.

**Portable radio** A battery-operated, hand-held transceiver.

**Positive-pressure ventilation** Ventilation that relies upon fans to push or force clean air into a structure.

**Post indicator valve (PIV)** A sprinkler control valve with an indicator that reads either open or shut depending on its position.

**Pounds per square inch (psi)** Standard unit used in measuring pressure.

**Preaction sprinkler system** A dry sprinkler system that uses a deluge valve instead of a dry-pipe valve and requires activation of a secondary device before the pipes fill with water.

**Preincident plan** A written document resulting from the gathering of general and detailed information to be used by responders for determining the response to reasonable anticipated emergency incidents at a specific facility.

**Preincident survey** The process used to gather information to develop a preincident plan.

**Premixed foam** Mixed foam and water used in portable extinguishers and dual agent systems.

**Pressure gauge** A device that measures and displays pressure readings. In an SCBA, the pressure gauges indicate the quantity of breathing air that is available at any time.

**Pressure indicator** A gauge on a pressurized portable fire extinguisher that indicates the internal pressure of the expellant.

**Primary cut** The main ventilation opening made in a roof to allow smoke, heat, and gases to escape.

**Primary feeder** The largest diameter pipes in a water distribution system, carrying the greatest amounts of water.

**Primary search** An initial search conducted to determine if there are victims who must be rescued.

**Private water system** A privately owned water system operated separately from the municipal water system.

**Products of combustion** Heat, smoke, and toxic gases.

**Projected windows (factory windows)** Usually found in older warehouse or commercial buildings. They project inward or outward on an upper hinge.

**Proprietary system** A fire alarm system that transmits a signal to a monitoring location owned and operated by the facility's owner.

**Protection plates** Reinforcing material placed on a ladder at chaffing and contact points to prevent damage from friction and contact with other surfaces.

**Protective hood** A part of a brigade member's PPE designed to be worn over the head and under the helmet to provide thermal protection for the neck and ears.

**Pry axe (multipurpose axe)** A specially designed hand axe that serves multiple purposes. Similar to a Halligan bar, it can be used to pry, cut, and force doors, windows, and many other types of objects.

**Pry bar** A specialized prying tool made of a hardened steel rod with a tapered end that can be inserted into a small area.

**Public information officer** The position within IMS responsible for gathering and releasing incident information to the media and other appropriate agencies.

**Public safety answering point (PSAP)** The community's emergency response communications center.

**Pulley** A small, grooved wheel through which the halyard runs. The pulley is used to change the direction of the halyard pull, so that a downward pull on the halyard creates an upward force on the fly section(s).

**Pump tank extinguishers** Nonpressurized, manually operated water extinguishers, usually have a nozzle at the end of a short hose.

**Pump tank water-type extinguisher** A nonpressurized portable water fire extinguisher. Discharge pressure is provided by a hand-operated double-acting piston pump.

**Pyrolysis** The chemical decomposition of a compound into one or more substances by heat alone; pyrolysis often precedes combustion.

**Pyromaniac** A pathological fire-setter.

**Rabbit** A type of door frame that has the stop for the door cut into the frame.

**Rabbit tool** Hydraulic spreading tool designed to pry open doors that swing inward.

**Radiation dispersal device (RDD)** Any device that causes the purposeful dissemination of radioactive material without a nuclear detonation; a dirty bomb.

**Radiation** The combined process of emission, transmission, and absorption of energy traveling by electromagnetic wave propagation between a region of higher temperature and a region of lower temperature.

Radio repeater system A radio system that automatically retransmits a radio signal on a different frequency.

Radiological agents Materials that emit radioactivity.

Rafters Joists that are mounted in an inclined position to support a roof.

Rail The top or bottom piece of a trussed-beam assembly used in the construction of a trussed ladder. The term rail is also sometimes used to describe the top and bottom surfaces of an I-beam ladder. Each beam will have two rails.

Rain-down Foam application method to apply a raised foam stream to allow the foam to gently fall onto the fuel surface.

Rapid intervention company/crew (RIC) A minimum of two fully equipped personnel on site, in a ready state, for immediate rescue of injured or trapped brigade members. In some organizations, this is also known as Rapid Intervention Team.

Rapid oxidation Chemical process that occurs when a fuel is combined with oxygen, resulting in the formation of ash or other waste products and the release of energy as heat and light.

Rate-of-rise heat detector A device that responds when the temperature rises at a rate that exceeds a predetermined value.

Reciprocating saw Powered by electric or battery motors, a reciprocating saw's blade moves back and forth.

Reconnaissance report The inspection and exploration of a specific area in order to gather information for the incident commander.

Reducer A device that can join two hoses of different sizes.

Rehabilitate To restore to a condition of health or to a state of useful and constructive activity.

Rehabilitation A systematic process to provide periods of rest and recovery for emergency workers during an incident; usually conducted in a designated area away from the hazardous area.

Rekindle A situation where a fire, which was thought to be completely extinguished, reignites.

Remote annunciator A secondary fire alarm control panel in a different location than the main alarm panel, usually near the front door of a building.

Remote station system A fire alarm system that sounds an alarm in the building and transmits a signal to the fire brigade or an off-premise monitoring location.

Rescue Those activities directed at locating endangered persons at an emergency incident, removing those persons from danger, treating the injured, and providing for transport to an appropriate health care facility.

Reservoir A water storage facility.

Residual pressure The pressure remaining in a water distribution system while water is flowing. The residual pressure indicates how much more water is potentially available.

Resource management A standard system of assigning and keeping track of the resources involved in the incident.

Respirator A protective device used to provide safe breathing air to a user in a hostile or dangerous atmosphere.

Response Activities that occur in preparation for an emergency and continue until the arrival of emergency apparatus at the scene.

Retard chamber A valve accessory that is used to prevent a sprinkler flow alarm caused by system water pressure surges.

Reverse lay A method of laying a supply line where the line starts at the fire and ends at the water source.

Rim lock Surface or interior mounted locks on or in a door with a bolt that provide additional security.

Risk-benefit analysis The process of weighing predicted risks against potential benefits and making decisions based on the outcome of that analysis.

Rocker lug (pin lug) Fittings on threaded couplings that aid in coupling the hoses.

Roll on Method of applying foam at the front edge of the fuel and allowing the foam to flow across the fuel surface.

Rollover (flameover) The condition where unburned products of combustion from a fire have accumulated in the ceiling layer of gas to a sufficient concentration (i.e., at or above the lower flammable limit) that they ignite momentarily.

Roof covering The material or assembly that makes up the weather-resistant surface of a roof.

Roof decking The rigid component of a roof covering.

Roof hooks The spring-loaded, retractable, curved metal pieces that allow the tip of a roof ladder to be secured to the peak of a pitched roof. The hooks fold outward from each beam at the top of a roof ladder.

Roof ladder (hook ladder) A straight ladder equipped with retractable hooks so that the ladder can be secured to the peak of a pitched roof. Once secured, the ladder lies flat against the surface of the roof, providing secure footing.

Roofman's hook A long pole with a solid metal hook used for pulling.

Rope bag A bag used to protect and store rope so that the rope can be easily and rapidly deployed without kinking.

Rope record A record for each piece of rope that includes a history of when the rope was placed in service, when it was inspected, when and how it was used, and what types of loads were placed on it.

Rotary saw Powered by electric motors or gasoline engines, a rotary saw uses a large rotating blade to cut through material. The blades can be changed depending upon the material that is being cut.

Round turn A piece of rope looped to form a complete circle with the two ends parallel.

Rubber-covered hose (rubber-jacket hose) Hose whose outside covering is made of rubber, said to be more resistant to damage.

Run cards Cards used to determine a predesignated response to an emergency.

Rung A ladder crosspiece that provides a climbing step for the user. The rung transfers the weight of the user out to the beams of the ladder or back to a center beam on an I-beam ladder.

Running end The part of a rope used for lifting or hoisting.

Safe haven A temporary place of refuge to await rescue.

Safety knot (overhand knot or keep knot) A knot used to secure the leftover working end of the rope.

Safety officer The position within IMS responsible for identifying and evaluating hazardous or unsafe conditions at the scene of the incident. Safety officers have the authority to stop any activity deemed unsafe.

Salvage Removing or protecting property that could be damaged during firefighting or overhaul operations.

Salvage cover Large square or rectangular sheets made of heavy canvas or plastic material spread over furniture and other items to protect them from water run-off and falling debris.

San Francisco hook A multipurpose tool that can be used for several forcible entry and ventilation applications because of its unique design, which includes a built-in gas shut-off and directional slot.

Saponification The process of converting the fatty acids in cooking oils or fats to soap or foam.

Sarin A nerve agent that is primarily a vapor hazard.

SCBA harness The straps and fasteners used to attach the SCBA to the brigade member.

SCBA regulators Part of the SCBA that reduces the high pressure in the cylinder to a usable lower pressure and controls the flow of air to the user.

Screwdriver A tool used for turning screws.

Scupper An opening through which water can be removed from a building.

Search The process of looking for victims who are in danger.

Search and rescue The process of searching a building for a victim and extricating the victim from the building.

Search rope A guide rope used by brigade members that allows them to maintain contact with a fixed point.

Seat of the fire The main area of the fire origin.

Seatbelt cutter A specialized cutting device that cuts through seatbelts.

Secondary collapse A collapse that occurs following the primary collapse. This can occur in trench, excavation, and structural collapses.

Secondary cut An additional ventilation opening to create a larger opening, or to limit fire spread.

**Secondary device** An explosive device designed to injure emergency responders who have responded to an initial event.

**Secondary feeder** Smaller diameter pipes that connect the primary feeders to the distributors.

**Secondary loss** Property damage that occurs due to smoke, water, or other measures taken to extinguish the fire.

**Secondary search** A more thorough search undertaken after the fire is under control. This search is done to ensure that there are no victims still trapped inside the building.

**Self-contained breathing apparatus (SCBA)** Respirator with independent air supply used by brigade members to enter toxic and otherwise dangerous atmospheres.

**Self-contained underwater breathing apparatus (SCUBA)** Respirator with independent air supply used by underwater divers.

**Self-expelling** A fire extinguisher in which the agents have sufficient vapor pressure at normal operating temperatures to expel themselves.

**Self-rescue** The activity of a brigade member using techniques and tools to remove himself or herself from a hazardous situation.

**Serial arson** A series of fires set by the same offender, with a cooling-off period between fires.

**Shackles** The U-shaped part of a padlock that runs through a hasp and secures back into the lock body.

**Shock load** An instantaneous load that places a rope under extreme tension, such as when a falling load is suddenly stopped when the rope becomes taut.

**Shoring** A method of supporting a trench wall or building components such as walls, floors, or ceilings using either hydraulic, pneumatic, or wood shoring systems; it is used to prevent collapse.

**Shut-off valve** Any valve that can be used to shut down water flow to a water user or system.

**Siamese connection** A device that allows two hoses to be connected together and flow into a single hose.

**Sidewall sprinkler head** A sprinkler that is mounted on a wall and discharges water horizontally into a room.

**Simplex channel** Radio system that uses one frequency to transmit and receive all messages.

**Single resource** An individual vehicle and the personnel that arrive on that unit.

**Single-action pull-station** A manual fire alarm activation device that takes a single step, such as moving a lever, toggle, or handle, to activate the alarm.

**Size-up** The ongoing observation and evaluation of factors that are used to develop objectives, strategy, and tactics for fire suppression.

**Sledgehammer** A long, heavy hammer that requires the use of both hands.

**Slopover** Burps of crude oil caused by steam explosions when the heat wave contacts small areas of water in the fuel column below the surface.

**Small diameter hose (SDH)** Hose in the 1" to 2" range.

**Smallpox** A highly infectious disease caused by the virus variola.

**Smoke** An airborne particulate product of incomplete combustion suspended in gases, vapors, or solid and liquid aerosols.

**Smoke detector** A device that detects smoke and sends a signal to a fire alarm control panel.

**Smoke inversion** Smoke hanging low to the ground due to the cold air.

**Smoke particles** Airborne solid material consisting of ash and unburned or partially burned fuel released by a fire.

**Smooth bore nozzle** Nozzles that produce a solid stream of water.

**Smooth bore tip** A nozzle device that is a smooth tube, used to deliver a solid stream of water.

**Socket wrench** A wrench that fits over a nut or bolt and uses the ratchet action of an attached handle to tighten or loosen the nut or bolt.

**Soft sleeve hose** A large diameter hose that is designed to be connected to the large port on a hydrant (steamer connection) and into the engine.

**Solid** One of the three phases of matter. A substance that has three dimensions and is firm in substance.

**Solid beam** A ladder beam constructed of a solid rectangular piece of material, typically wood, to which the ladder rungs are attached.

**Solid-core** A door design that consists of wood filler pieces inside the door. This creates a stronger door that may be fire-rated.

**Solid stream** A stream made by using a smooth bore nozzle to produce a penetrating stream of water.

**Soman** A nerve gas that is both a contact and a vapor hazard that has the odor of camphor.

**Sounding** The process of striking a roof with a tool to determine if the roof is solid enough to support the weight of a brigade member.

**Spalling** Chipping or pitting of concrete or masonry surfaces.

**Spanner wrench** A type of tool used in coupling or uncoupling hoses by turning the rocker lugs on the connections.

**Span of control** The number of people that a single person supervises. The maximum number of people that one person can effectively supervise is about five.

**Split hose bed** A hose bed that is divided into two or more sections.

**Split hose lay** A scenario where the attack engine will lay a supply line from a point away from the fire, and the supply engine will lay a supply line from the hose left by the attack engine to the water source.

**Spoil pile** The pile of dirt that has been removed from an excavation, which may be unstable and prone to collapse.

**Spot detector** Single heat-detector devices, spaced throughout an area.

**Spotlight** A light designed to project a narrow, concentrated beam of light.

**Spree arson** A series of fires started by an arsonist who sets three or more fires at separate locations with no emotional cooling-off period between fires.

**Spring-loaded center punch** A spring-loaded punch used to break automobile glass.

**Sprinkler piping** The network of piping in a sprinkler system that delivers water to the sprinkler heads.

**Sprinkler stop** A mechanical device inserted between the deflector and the orifice of a sprinkler head to stop the flow of water.

**Sprinkler system** An automatic fire protection system designed to turn on sprinklers if a fire occurs.

**Sprinkler wedge** A piece of wedge-shaped wood placed between the deflector and the orifice of a sprinkler head to stop the flow of water.

**Squelch** An electric circuit designed to cut off weak radio transmissions that are only capable of generating noise.

**Staging area** A prearranged, strategically placed area where support personnel, vehicles, and other equipment can be held in an organized state of readiness for use during an emergency.

**Standard operating procedures (SOPs)** Written rules, policies, regulations, and procedures enforced to structure the normal operations of most fire brigades.

**Standing part** The part of a rope between the working end and the running end.

**Standpipe system** A system of pipes and hose outlet valves used to deliver water to various parts of a building for fighting fires.

**Static pressure** The pressure in a water pipe when there is no water flowing.

**Static rope** A rope generally made out of synthetic material that stretches very little under load.

**Static water sources** A water source that is not under pressure, such as a pond, lake, or stream.

**Staypole (tormentor)** A long piece of metal attached to the top of the bed section of an extension ladder and used to help stabilize the ladder during raising and lowering. The pole attaches to a swivel point and has a spur on the other end. One pole is attached to each beam of long (40' or longer) extension ladders.

**Steamer port** The large diameter port on a hydrant.

**Stop** A piece of material that prevents the fly section(s) of a ladder from overextending and collapsing the ladder.

**Stored-pressure extinguisher** A fire extinguisher in which both the extinguishing agent and the expellant gas are kept in a single container; generally equipped with a pressure indicator or gauge.

**Stored-pressure water-type extinguisher** A fire extinguisher in which water or a water-based extinguishing agent is stored under pressure.

**Storz-type coupling** A hose coupling that has the property of being both the male and female coupling. It is connected by engaging the lugs and turning the coupling one-third of a turn.

**Straight stream** A stream made by using an adjustable nozzle that is set to provide a straight stream of water.

**Strike team** Five units of the same resource category, such as engines or ambulances, with a leader.

**Strike team leader** The person in charge of a strike team, responsible to the next higher level in the incident organization, and the point-of-contact for the strike team within the organization.

**Striking tool** Tools designed to strike other tools or objects such as walls, doors, or floors.

**Suitcase nuke** A nuclear explosive device that is small enough to fit in a suitcase.

**Sulfur mustard** A clear, yellow, or amber oily liquid with a faint sweet odor of mustard or garlic that may be dispersed in an aerosol form. It causes blistering of exposed skin.

**Supplied-air respirator (SAR)** A respirator that gets its air through a hose from a remote source, such as a compressor or compressed air cylinder.

**Supply hose (supply line)** The hose used to deliver water from a source to a fire pump.

**Surface tension** The attractive force exerted upon the surface molecules of a liquid by the molecules beneath.

**Surface to mass ratio (STMR)** The surface area of a material in proportion to the mass of that material.

**Surface to volume ratio** The surface area of a liquid in proportion to the mass.

**Tabun** A nerve gas that is both a contact and a vapor hazard that operates by disabling the chemical connection between nerve and target organs.

**Talk around channel** A simplex channel used for on-site communications.

**Tamper seal** A retaining device that breaks when the locking mechanism is released.

**Tamper switch** A switch on a sprinkler valve that transmits a signal to the fire alarm control panel if the normal position of the valve is changed.

**Target hazard** Any occupancy type or facility that presents a high potential for loss of life or serious impact to the company resulting from fire, explosion, or chemical release.

**Task force** Any combination of single resources assembled for a particular tactical need; has common communications and a leader.

**Task force leader** The person in charge of a task force, responsible to the next higher level in the incident organization, and the point-of-contact for the task force within the organization.

**Technical rescue incident (TRI)** A complex rescue incident involving vehicles or machinery, water or ice, rope techniques, a trench or excavation collapse, confined spaces, a structural collapse, an SAR, or hazardous materials, and which requires specially trained personnel and special equipment.

**Technical rescue team** A group of rescuers specially trained in the various disciplines of technical rescue.

**Technical rescue technician** A "tech rescue" technician is trained in special rescue techniques for incidents involving structural collapse, trench rescue, vehicle/machinery rescue, confined-space rescue, high-angle rescue, and other unusual situations.

**Telecommunicator** A trained individual responsible for answering requests for emergency and nonemergency assistance from citizens. This individual assesses the need for response and alerts responders to the incident.

**Telephone interrogation** Phase in a 9-1-1 call when a telecommunicator asks questions to obtain vital information such as the location of the emergency.

**Tempered glass** A type of safety glass that is heat-treated so that it will break into small pieces that are not as dangerous.

**Temporal-3 pattern** A standard fire alarm audible signal for alerting occupants of a building.

**Ten-codes** System of predetermined, coded messages, such as "What is your 10-20?" used by responders over the radio.

**Tender shuttle** A method of transporting water from a source to a fire scene using a number of mobile water supply apparatus.

**Thermal column** A cylindrical area above a fire in which heated air and gases rise and travel upward.

**Thermal conductivity** Describes how quickly a material will conduct heat.

**Thermal imaging devices** Electronic devices that detect differences in temperature based on infrared energy and then generate images based on that data. Commonly used in obscured environments to locate victims and hot spots.

**Thermal layering** The stratification, or heat layers, that occur in a room as a result of a fire.

**Thermal protective clothing** Protective clothing such as helmets, footwear, gloves, hoods, trousers, and coats that are designed and manufactured to protect the fire brigade member from the adverse effects of fire.

**Thermoplastic material** Plastic material capable of being repeatedly softened by heating and hardened by cooling and, that in the softened state, can be repeatedly shaped by molding or forming.

**Thermoset material** Plastic material that, after having been cured by heat or other means, is substantially infusible and cannot be softened and formed.

**Threaded hose couplings** A type of coupling that requires a male and female fitting to be screwed together.

**Tie rod** A metal rod that runs from one beam of the ladder to the other to keep the beams from separating. Tie rods are typically found in wood ladders.

**Time marks** Status updates provided to the communications center every 10 to 20 minutes. This update should include the type of operation, the progress of the incident, the anticipated actions, and the need for additional resources.

**Tip** The very top of the ladder.

**Trace (transfer) evidence** Evidence of a minute quantity that is conveyed from one place to another.

**Trailers** Combustible material, such as rolled rags, blankets, and newspapers or flammable liquid, used to spread fire from one point or area to other points or areas, often used in conjunction with an incendiary device.

**Training officer** Responsible for updating the training of current brigade members and for training new members.

**Transfer of command** Reassignment of command authority and responsibility from one individual to another.

**Trench** A narrow excavation (in relation to its length) made below the surface of the ground. In general, the depth is greater than the width.

**Trench cut** A cut that is made from bearing wall to bearing wall to prevent horizontal fire spread in a building.

**Triage** The process of sorting victims based on the severity of injury and medical needs to establish treatment and transportation priorities.

**Triangular cut** A triangle-shaped ventilation cut in the roof decking that is made using saws or axes.

**Trigger** The button or lever used to discharge the agent from a portable fire extinguisher.

**Triple layer load** A hose loading method that utilizes folding the hose back onto itself to reduce the overall length to one-third before loading in the bed. This load method reduces deployment distances.

**Trunking system** A radio system that uses a shared bank of frequencies to make the most efficient use of radio resources.

**Truss** A collection of lightweight structural components joined in a triangular configuration that can be used to support either floors or roofs.

**Truss block** A piece of wood or metal that ties the two rails of a trussed beam ladder together and serves as the attachment point for the rungs.

**Trussed beam** A ladder beam constructed of top and bottom rails joined by truss blocks that tie the rails together and support the rungs.

**Turnout coat (bunker coat)** Protective coat that is part of a protective clothing ensemble for structural firefighting.

**Turnout pants (bunker pants)** Protective trousers that are part of a protective clothing ensemble for structural firefighting.

**Twisted rope** Rope constructed of fibers twisted into strands, which are then twisted together.

**Two-in/two-out rule** A safety procedure that requires a minimum of two personnel to enter a hazardous area and a minimum of two back-up personnel to remain outside the hazardous area during the initial stages of an incident.

**Two-person rope** A rope rated to carry the weight of two people (600 lbs).

**Two-way radio** A portable communication device. Every firefighting team should carry at least one radio to communicate distress, progress, changes in fire conditions, and other pertinent information.

**Type I construction (fire resistive)** Buildings with structural members made of noncombustible materials that have a specified fire resistance.

**Type II construction (noncombustible)** Buildings with structural members made of noncombustible materials without fire resistance.

**Type III construction (ordinary construction)** Buildings with the exterior walls made of noncombustible or limited-combustible materials, but interior floors and walls made of combustible materials.

**Type IV construction (heavy timber)** Buildings constructed with noncombustible or limited-combustible exterior walls, and interior walls and floors made of large dimension combustible materials.

**Type V construction (wood frame)** Buildings with exterior walls, interior walls, floors, and roof structures made of wood.

**Underwriters Laboratories Inc. (UL)** The U.S. organization that tests and certifies that fire extinguishers (among many other products) meet established standards.

**Undetermined** Cause classification that includes fires for which the cause has not or cannot be proven.

**Unified command** IMS option that allows representatives from multiple jurisdictions and/or agencies to share command authority and responsibility, working together as a "joint" incident command team.

**Unity of command** A characteristic of the IMS structure that has each individual reporting to a single supervisor and everyone reporting to the IC directly or through the chain of command.

**Universal precautions** Procedures for infection control that treat blood and certain bodily fluids as capable of transmitting bloodborne diseases.

**Unlocking device** A key way, combination wheel, or combination dial.

**Unwanted alarm** A fire alarm signal caused by a device reacting properly to a condition that is not a true fire emergency.

**Upper explosive limit (UEL)** The maximum amount of fuel vapor mixed in air that will ignite or explode.

**Upright sprinkler head** A sprinkler head designed to be installed on top of the supply piping and usually marked SSU (Standard Spray Upright).

**Utility rope** Rope used for securing objects, for hoisting equipment, or for securing a scene to prevent bystanders from being injured. It is never to be used in life safety operations.

**V-agent** A nerve agent, principally a contact hazard; an oily liquid that can persist for several weeks.

**Vapor density** The weight of an airborne concentration (vapor or gas) as compared to an equal volume of dry air.

**Vapor pressure** The pressure exerted by a vapor.

**Ventilation** The process of removing smoke, heat, and toxic gases from a burning structure and replacing them with clean air.

**Venturi** A tube with a constricted throat that causes an increase in the velocity of water, creating a low-pressure area.

**Verification system** A fire alarm system that does not immediately initiate an alarm condition when a smoke detector activates. The system will wait a preset interval, generally 30 to 60 seconds, before checking the detector again. If the condition is clear, the system returns to normal status. If the detector is still sensing smoke, the system activates the fire alarm.

**Vertical ventilation** The process of making openings so that the smoke, heat, and gases can escape vertically from a structure.

**Viscosity** The degree to which a fluid resists flow under an applied force. The lower the viscosity, the easier a fluid will flow.

**Voice recording system** Recording devices or computer equipment connected to telephone lines and radio equipment in a communications center to record telephone calls and radio traffic.

**Volatility** The ready ability of a substance to produce combustible vapors.

**Voluntary brigade member** An employee who volunteers to be a member of the fire brigade.

**Wall post indicator valve (WPIV)** A sprinkler control valve that is mounted on the outside wall of a building. The position of the indicator tells whether the valve is open or shut.

**Warm zone** This area is for properly trained and equipped personnel only. The warm zone is where personnel and equipment decontamination and hot zone support take place.

**Water catch-all** A salvage cover folded to form a container to hold water until it can be removed.

**Water chute** A salvage cover folded to direct water flow out of a building or away from sensitive items or areas.

**Water curtain nozzles** Nozzles used to deliver a flat screen of water to form a protective sheet of water.

**Water hammer** An event that occurs when flowing water is suddenly stopped; the velocity force of the moving water is transferred to everything it is in contact with. These can be tremendous forces that can damage equipment and cause injury.

**Water main** The generic term for any above ground or underground water supply pipe.

**Water supply** A source of water.

**Water thief** A device with an inlet and an outlet of the same size and several additional outlets of smaller size.

**Water vacuum** A device similar to a household vacuum cleaner, but with the ability to pick up liquids. Used to remove water from buildings.

**Water-motor gong** An audible alarm notification device that is powered by water moving through the sprinkler system.

**Weapons of mass destruction (WMD)** A weapon intended to cause mass casualties, damage, and chaos.

**Wet chemical extinguisher** A fire extinguisher for use on Class K fires that contains wet chemical extinguishing agents.

**Wet chemical extinguishing agent** An extinguishing agent for Class K fires; commonly uses solutions of water and potassium acetate, potassium carbonate, potassium citrate, or any combination thereof.

**Wet chemical extinguishing systems** An extinguishing system that discharges a proprietary liquid extinguishing agent.

**Wet-barrel hydrant** A hydrant used in areas that are not susceptible to freezing. The barrel of the hydrant is normally filled with water.

**Wet sprinkler system** A sprinkler system in which the pipes are normally filled with water.

**Wetting-agent water-type extinguisher** An extinguisher that expels water combined with a chemical or chemicals to reduce its surface tension.

**Wheeled fire extinguisher** A portable fire extinguisher equipped with a carriage and wheels intended to be transported to the fire by one person.

**Wooden beams** Load-bearing members assembled from individual wood components.

**Wood-frame construction (Type V)** Buildings with exterior walls, interior walls, floors, and roof made of combustible wood material.

**Wired glass** Glass made by molding glass around a special wire mesh.

**Wood panels** Thin sheets of wood glued together.

**Wood trusses** Assemblies of small pieces of wood or wood and metal.

**Working end** The part of the rope used for forming the knot.

**Wye** A device used to split a single hose into two separate lines.

**Zoned coded alarm** A fire alarm system that indicates which zone was activated both on the alarm control panel and through a coded audio signal.

**Zoned non-coded alarm** A fire alarm system that indicates the activated zone on the alarm control panel.

**Zoned system** A fire alarm system design that divides a building or facility into zones so the area where an alarm originated can be identified.

# Index

Note: pages followed by *f* and *t* refers figures and tables respectively.

## A



## B

## C



## D

## G



## H



## I

## J



## K



## L



## M

# N



# O



# P

## Q



## R

## S

# T



# U



# V

## W



## Z

## Credits

*Voices of Experience* © Michael Guo/ShutterStock, Inc; *You are the Brigade Member* and *Brigade Member in Action* © Jack Dagley Photography/ShutterStock, Inc.

**Chapter 1**
Opener © Glen E. Ellman; 1-2 Courtesy of NFPA; 1-3 © Peter Willott, *St. Augustine Record*/AP Photos; 1-24 Courtesy of Draeger Safety, Inc; 1-30 Air-Pak® NxG7™ SCBA courtesy of Scott Health & Safety. Used with permission; 1-33 Courtesy of Arleigh Movitz, Santa Clara County Fire Department

**Chapter 3**
Opener Courtesy of Captain David Jackson, Saginaw Township Fire Department; 3-5 © Chase Swift/Corbis

**Chapter 4**
Opener Courtesy of Captain David Jackson, Saginaw Township Fire Department; 4-1 Courtesy of the USDA Forest Service; 4-2 © Keith D. Cullom; 4-3 © Dan Myers; 4-7 © Chitose Suzuki/AP Photos; 4-13 © Glen E. Ellman; 4-16 © Keith D. Cullom

**Chapter 5**
Opener, 5-14 © Glen E. Ellman

**Chapter 6**
Opener Courtesy of Dr. Edwin P. Ewing, Jr/CDC; 6-1 © AbleStock; 6-2 Courtesy of Captain David Jackson, Saginaw Township Fire Department; 6-4 © Adam Helweh/ShutterStock, Inc; 6-5 © John Foxx/Alamy Images; 6-6 Courtesy of Blaze-Tech; 6-8 © Ken Hammon/USDA; 6-9 Courtesy of NFPA; 6-12 © Shi Yali/ShutterStock, Inc; 6-13 © Calvin College. Used with permission. Photograph by Luke Robinson; 6-14A © Thaddeus Robertson/ShutterStock, Inc; 6-14B Courtesy of Pacific Northwest National Laboratory; 6-14C Courtesy of Doyal Dum/US Army Corps of Engineers; 6-14D © AbleStock; 6-15A © Paul Springett/Up the Resolution/Alamy Images; 6-15B © Ron Chapple/Thinkstock/Alamy Images; 6-15C © TH Photo/Alamy Images; 6-16 © Bo Vilmos Widerberg/ShutterStock, Inc.

**Chapter 7**
7-1A&B Courtesy of Amerex Corporation; 7-7 © Andrew Lambert Photography/Photo Researchers, Inc; 7-8 Courtesy of Vent-A-Hood Company; 7-13 © Image Ideas/Jupiterimages; 7-14 © PhotoCreate/ShutterStock, Inc; 7-15 © Shanta Giddens/ShutterStock, Inc; 7-18, 7-19 Courtesy of Amerex Corporation; 7-20 Courtesy of Ansul Incorporated; 7-SD3 © 2003 Berta A. Daniels; 7-SD7 Courtesy of Ansul Incorporated

**Chapter 8**
8-12, 8-17, 8-30 © 2003 Berta A. Daniels; 8-33 Courtesy of ISG Thermal Systems USA, Inc

**Chapter 9**
9-9A, 9-9B Courtesy of Yale Cordage; 9-10A Courtesy of Michael Nevins/US Army Corps of Engineers; 9-10B Anzac Bridge, Sydney, Australia, courtesy of John Symonds; 9-16 Courtesy of Scott Dornan, ConocoPhillips Alaska; 9-18 © Jeff Cooper, *Salina Journal*/AP Photos; 9-19 Courtesy of Captain David Jackson, Saginaw Township Fire Department; 9-22 © AbleStock; 9-23 Courtesy of Scott Dornan, ConocoPhillips Alaska

**Chapter 10**
Opener © Thinkstock/age fotostock; 10-5 © Robert Pernell/ShutterStock, Inc; 10-7 © 2003 Berta A. Daniels; 10-8, 10-10 © Glen E. Ellman

**Chapter 11**
11-8 © Carsten Medom Madsen/ShutterStock, Inc; 11-13 Courtesy of Horton Automatics; 11-14A Courtesy of St. Cloud Overhead Door Company; 11-14B © Photodisc/Creatas; 11-16, 11-19, 11-20 Used with permission of Andersen Corporation; 11-21 Courtesy of Potomac Communications Group, Inc; 11-23, 11-26, 11-27 Courtesy of IR Safety and Security-Americas; 11-29 © Arthur Eugene Preston/ShutterStock, Inc.

**Chapter 12**
12-2 Courtesy of NASA Ames Disaster Assistance and Rescue Team; 12-6 © Dennis Wetherhold, Jr; 12-7 © Christa Knijff/Alamy Images; 12-8, 12-10, 12-13, 12-14 Courtesy of Duo-Safety Ladder Corporation; 12-20 Courtesy of Underwriters Laboratories Inc. © 2003 Underwriters Laboratories Inc®; 12-26 © Dennis Wetherhold, Jr

**Chapter 13**
13-1 © Steven Townsend/Code 3 Images; 13-2 © Adam Alberti, NJFirePictures.com; 13-3A © Clive Watkins/ShutterStock, Inc; 13-3B © Disco Volante/Alamy Images; 13-11 Courtesy of Scott Dornan, ConocoPhillips Alaska; 13-13 © Rick Feld/AP Photos; 13-14 Courtesy of Scott Dornan, ConocoPhillips Alaska; 13-SD13 Courtesy of AAOS

**Chapter 14**
Opener © Glen E. Ellman; 14-4 Courtesy of Cpl. K. A. Thompson/U.S. Marines; 14-8 © Glen E. Ellman; 14-9 © Ian Shaw/Alamy Images; 14-10 © Patricia Marks/ShutterStock, Inc; 14-13 Courtesy of Super Vacuum Mfg. Co., Inc; 14-18 © Joseph McCullar/ShutterStock, Inc; 14-19 © Glen E. Ellman; 14-21 Courtesy of Captain David Jackson, Saginaw Township Fire Department; 14-22 © Clynt Garnham/Alamy Images; 14-24 Courtesy of Four Seasons Solar Products LLC; 14-27 Courtesy of Captain David Jackson, Saginaw Township Fire Department

**Chapter 15**
15-2 Courtesy of Scott Dornan, ConocoPhillips Alaska; 15-3 Courtesy of District Chief Chris E. Mickal/New Orleans Fire Department, Photo Unit; 15-5 Courtesy of Tom Fields, Douglas Forest Protective Association; 15-09, 15-10A Courtesy of American AVK Company; 15-11, 15-12 Courtesy of Captain David Jackson, Saginaw Township Fire Department; 15-16, 15-17 Jones & Bartlett Learning. Photographed by Glen E. Ellman

**Chapter 16**
Opener Courtesy of the University of Nevada, Reno Fire Science Academy; 16-11, 16-12 Courtesy of Akron Brass Company; 16-14, 16-15, 16-16 © 2003 Berta A. Daniels; 16-24, 16-25, 16-26 Courtesy of Scott Dornan, ConocoPhillips Alaska; 16-27 Courtesy of Thomas Products, Inc. Used with permission; 16-32 Courtesy of Scott Dornan, ConocoPhillips Alaska; 16-33 Courtesy of Flamefighter Corporation; 16-34B Courtesy of Akron Brass Company; 16-35 Courtesy of POK of North America, Inc.

**Chapter 17**
Opener © Cyrus J. Cornell/ShutterStock, Inc; 17-1, 17-2, 17-3, 17-4 Modified from *The Foam Fire Fighting Guide* and used with permission of Kidde National Foam, Inc; 17-6A Courtesy of Elkhart Brass; 17-6B Courtesy of Scott Dornan, ConocoPhillips Alaska; 17-7 Courtesy of Roy Robichaux, ConocoPhillips Alliance Refinery; 17-8 Courtesy of Elkhart Brass; 17-9, 17-10 Courtesy of Ansul Incorporated; 17-12, 17-13, 17-14 Modified from *The Foam Fire Fighting Guide* and used with permission of Kidde National Foam, Inc.

**Chapter 18**
Opener © Glen E. Ellman; 18-1 © Mark Sykes/Alamy Images; 18-2 Courtesy of NFPA

**Chapter 19**
Opener © Glen E. Ellman; 19-1 Courtesy of Akron Brass Company; 19-3 © 2003 Berta A. Daniels; 19-5 Courtesy of Tyco Fire and Building Products; 19-8 Courtesy of Cascade Fire Equipment Company; 19-SD2 Courtesy of Scott Dornan, ConocoPhillips Alaska

**Chapter 20**
20-2, 20-7 Courtesy of University of Nevada, Reno-Fire Science Academy; 20-8 © Glenn Ogilvie, *The Observer*/AP Photos; 20-10 © Michael Newman/PhotoEdit, Inc; 20-12 © Tom Carter/PhotoEdit, Inc; 20-14 © Photodisc/Creatas; 20-15 © Mark Adams/SuperStock, Inc; 20-17 Courtesy of Neil Malcomm Winkelmann

**Chapter 21**
Opener Courtesy of District Chief Chris E. Mickal/New Orleans Fire Department, Photo Unit; 21-1, 21-2, 21-7 Courtesy of the University of Nevada, Reno-Fire Science Academy; 21-10 © Kurt Hegre, *The Fresno Bee*/AP Photos

**Chapter 22**
Opener © Dennis Wetherhold, Jr; 22-5 Courtesy of NFPA; 22-7 © John Foxx/Alamy Images; 22-8 © Douglas Litchfield/ShutterStock, Inc; 22-9 © Ken Hammon/USDA; 22-10 Courtesy of APA—The Engineered Wood Association; 22-11 © R/ShutterStock, Inc; 22-12 Courtesy of Underwriters Laboratories Inc. © 2003 Underwriters Laboratories Inc®; 22-13 © NREL/DOE; 22-16 © Jim Smalley; 22-19 © Anson Hung/ShutterStock, Inc; 22-20 Courtesy of Doyal Dunn/U.S. Army Corps of Engineers; 22-22 © Donald Joski/ShutterStock, Inc; 22-23 © Brand X Pictures/Alamy Images; 22-24 © Greg Biggs/ShutterStock, Inc.

**Chapter 23**
Opener © Glen E. Ellman; 23-4 © Patrick Schneider, *Charlotte Observer*/AP Photos; 23-5 Courtesy of Ferno Canada; 23-6 Courtesy of Skedco, Inc; 23-7 © Ron Niebrugge/Alamy Images; 23-10 Courtesy of Scott Dornan, ConocoPhillips Alaska; 23-11 Courtesy of Captain David Jackson, Saginaw Township Fire Department; 23-12 Courtesy of Mark Wolfe/FEMA; 23-13 © Glen E. Ellman; 23-14 © Dennis Wetherhold, Jr.

**Chapter 24**
Opener Courtesy of Andrea Booher/FEMA; 24-1 © Todd Hollis/AP Photos; 24-2A © Arthur S. Aubry/Photodisc/Getty Images; 24-2B © AbleStock; 24-2C © Steve Allen/Brand X Pictures/Alamy Images; 24-2D © Galina Barskaya/ShutterStock, Inc; 24-3 © James P. Blair/Photodisc/Getty Images; 24-4 © Photodisc/Getty Images; 24-5 © Larry Rana/USDA; 24-6 From *When Violence Erupts: A Survival Guide for Emergency Responders*, courtesy of Dennis R. Krebs; 24-7 Courtesy of FEMA; 24-8 Courtesy of Captain David Jackson, Saginaw Township Fire Department; 24-9, 24-10 Courtesy of Scott Dornan, ConocoPhillips Alaska; 24-11 © Tim McCabe/USDA; 24-13 © AbleStock; 24-14 Mark I Nerve Agent Antidote Kit courtesy of Meridian Medical Technologies, Inc, a subsidiary of King Pharmaceuticals, Inc; 24-15 Courtesy of Dr. Saeed Keshavarz/RCCI (Research Center of Chemical Injuries)/Iran; 24-16 © John Mokrzycki/Reuters/Landov; 24-17, 24-18, 24-19 Courtesy of CDC; 24-20 © AbleStock; 24-21 Courtesy of RAE Systems, Inc. Used with permission; 24-23 © Jim Mone/AP Photos; 24-24 Courtesy of Captain David Jackson, Saginaw Township Fire Department

**Chapter 25**
Opener Courtesy of Tyco Fire and Building Products; 25-1 Courtesy of Honeywell/Gamewell; 25-2 Courtesy of Gamewell-FCI by Honeywell; 25-3 Courtesy of Honeywell/Gamewell; 25-4A Courtesy of Honeywell/Fire-Lite Alarms; 25-5 Courtesy of STI-USA; 25-7, 25-8 Courtesy of Firetronics Pte Ltd; 25-12 Courtesy of Honeywell/Fire-Lite Alarms; 25-16 Courtesy of www.acimonitoring.com, Doug Beaulieu; 25-23 Courtesy of Tyco Fire Products; 25-27 Courtesy of Ralph G. Johnson (nyail.com/fsd); 25-37 Courtesy of Tyco Fire and Building Products; 25-40 Courtesy of NFPA; 25-41 Courtesy of Ralph G. Johnson (nyail.com/fsd); 25-42, 25-43 Courtesy of Dixon Valve and Coupling; 25-49 Courtesy of Chemetron Fire Systems

**Chapter 26**
Opener © Doug McSchooler, *The Indianapolis Star*/AP Photos; 26-1, 26-2 Courtesy of Charles B. Hughes/Unified Investigations & Sciences, Inc; 26-3 Courtesy of Captain David Jackson, Saginaw Township Fire Department; 26-4 Courtesy of Charles B. Hughes/Unified Investigations & Sciences, Inc; 26-5 Courtesy of NFPA; 26-6 © Kristian Gonyea, *The Press of Atlantic City*/AP Photos; 26-7 © Joe Giblin/AP Photos; 26-8 © Ron Hilton/ShutterStock, Inc.

**Chapter 27**
Opener Courtesy of Scott Dornan, ConocoPhillips Alaska; 27-1 Courtesy of Captain David Jackson, Saginaw Township Fire Department; 27-2, 27-3 Courtesy of Scott Dornan, ConocoPhillips Alaska; 27-5 Courtesy of Dennis Wetherhold, Jr; 27-6 Courtesy of Captain David Jackson, Saginaw Township Fire Department; 27-7 © Ron Hilton/ShutterStock, Inc.

Unless otherwise indicated, photographs are under copyright of Jones and Bartlett Publishers, courtesy of the Maryland Institute of Emergency Medical Services Systems.